Mathematical Methods and Algorithms for Signal Processing

Mathematical Methods and Algorithms for Signal Processing

Todd K. Moon
Utah State University

Wynn C. Stirling
Brigham Young University

PRENTICE HALL
Upper Saddle River, NJ 07458

Library of Congress Cataloging-in-Publication Data

Moon, Todd K.
    Mathematical methods and algorithms for signal processing /
Todd K. Moon, Wynn C. Stirling.
      p.   cm.
    Includes bibliographical references and index.
    ISBN 0-201-36186-8
    1. Signal processing—Mathematics.   2. Algorithms.   I. Stirling,
Wynn C.   II. Title.
    TK5102.9 .M63 1999
    621.382′2′0151—dc21
                                                                                                                                     99-31038
                                                                                                                                     CIP

Editor-in-Chief: *Marcia Horton*
Production editor: *Brittney Corrigan-McElroy*
Managing editor: *Vince O'Brien*
Assistant managing editor: *Eileen Clark*
Art director: *Kevin Berry*
Cover design: *Karl Miyajima*
Manufacturing manager: *Trudy Pisciotti*
Assistant vice president of production and manufacturing: *David W. Riccardi*

© 2000 by Prentice Hall
Prentice-Hall, Inc.
Upper Saddle River, New Jersey 07458

All rights reserved. No part of this book may be reproduced in any form or by any means, without permission in writing from the publisher.

The author and publisher of this book have used their best efforts in preparing this book. These efforts include the development, research, and testing of the theories and programs to determine their effectiveness. The author and publisher make no warranty of any kind, expressed or implied, with regard to these programs or the documentation contained in this book. The author and publisher shall not be liable in any event for incidental or consequential damages in connection, or arising out of, the furnishing, performance, or use of these programs.

Printed in the United States of America

10 9 8 7 6 5 4 3 2 1

ISBN 0-201-36186-8

PRENTICE-HALL INTERNATIONAL (UK) LIMITED, *London*
PRENTICE-HALL OF AUSTRALIA PTY. LIMITED, *Sydney*
PRENTICE-HALL CANADA INC., *Toronto*
PRENTICE-HALL HISPANOAMERICANA, S.A., *Mexico*
PRENTICE-HALL OF INDIA PRIVATE LIMITED, *New Delhi*
PRENTICE-HALL OF JAPAN, INC., *Tokyo*
PRENTICE-HALL (SINGAPORE) PTE. LTD., *Singapore*
EDITORA PRENTICE-HALL DO BRASIL, LTDA., *Rio de Janeiro*

Contents

I Introduction and Foundations   1

1 Introduction and Foundations   3
- 1.1   What is signal processing? 3
- 1.2   Mathematical topics embraced by signal processing 5
- 1.3   Mathematical models 6
- 1.4   Models for linear systems and signals 7
  - 1.4.1   Linear discrete-time models 7
  - 1.4.2   Stochastic MA and AR models 12
  - 1.4.3   Continuous-time notation 20
  - 1.4.4   Issues and applications 21
  - 1.4.5   Identification of the modes 26
  - 1.4.6   Control of the modes 28
- 1.5   Adaptive filtering 28
  - 1.5.1   System identification 29
  - 1.5.2   Inverse system identification 29
  - 1.5.3   Adaptive predictors 29
  - 1.5.4   Interference cancellation 30
- 1.6   Gaussian random variables and random processes 31
  - 1.6.1   Conditional Gaussian densities 36
- 1.7   Markov and Hidden Markov Models 37
  - 1.7.1   Markov models 37
  - 1.7.2   Hidden Markov models 39
- 1.8   Some aspects of proofs 41
  - 1.8.1   Proof by computation: direct proof 43
  - 1.8.2   Proof by contradiction 45
  - 1.8.3   Proof by induction 46
- 1.9   An application: LFSRs and Massey's algorithm 48
  - 1.9.1   Issues and applications of LFSRs 50
  - 1.9.2   Massey's algorithm 52
  - 1.9.3   Characterization of LFSR length in Massey's algorithm 53
- 1.10   Exercises 58
- 1.11   References 67

II Vector Spaces and Linear Algebra   69

2 Signal Spaces   71
- 2.1   Metric spaces 72
  - 2.1.1   Some topological terms 76
  - 2.1.2   Sequences, Cauchy sequences, and completeness 78

		2.1.3	Technicalities associated with the L_p and L_∞ spaces	82
	2.2	Vector spaces		84
		2.2.1	Linear combinations of vectors	87
		2.2.2	Linear independence	88
		2.2.3	Basis and dimension	90
		2.2.4	Finite-dimensional vector spaces and matrix notation	93
	2.3	Norms and normed vector spaces		93
		2.3.1	Finite-dimensional normed linear spaces	97
	2.4	Inner products and inner-product spaces		97
		2.4.1	Weak convergence	99
	2.5	Induced norms		99
	2.6	The Cauchy–Schwarz inequality		100
	2.7	Direction of vectors: Orthogonality		101
	2.8	Weighted inner products		103
		2.8.1	Expectation as an inner product	105
	2.9	Hilbert and Banach spaces		106
	2.10	Orthogonal subspaces		107
	2.11	Linear transformations: Range and nullspace		108
	2.12	Inner-sum and direct-sum spaces		110
	2.13	Projections and orthogonal projections		113
		2.13.1	Projection matrices	115
	2.14	The projection theorem		116
	2.15	Orthogonalization of vectors		118
	2.16	Some final technicalities for infinite dimensional spaces		121
	2.17	Exercises		121
	2.18	References		129

3 Representation and Approximation in Vector Spaces — 130

	3.1	The Approximation problem in Hilbert space		130
		3.1.1	The Grammian matrix	133
	3.2	The Orthogonality principle		135
		3.2.1	Representations in infinite-dimensional space	136
	3.3	Error minimization via gradients		137
	3.4	Matrix Representations of least-squares problems		138
		3.4.1	Weighted least-squares	140
		3.4.2	Statistical properties of the least-squares estimate	140
	3.5	Minimum error in Hilbert-space approximations		141

Applications of the orthogonality theorem

	3.6	Approximation by continuous polynomials	143
	3.7	Approximation by discrete polynomials	145
	3.8	Linear regression	147
	3.9	Least-squares filtering	149
		3.9.1 Least-squares prediction and AR spectrum estimation	154
	3.10	Minimum mean-square estimation	156
	3.11	Minimum mean-squared error (MMSE) filtering	157
	3.12	Comparison of least squares and minimum mean squares	161
	3.13	Frequency-domain optimal filtering	162
		3.13.1 Brief review of stochastic processes and Laplace transforms	162

		3.13.2	Two-sided Laplace transforms and their decompositions................................	165
		3.13.3	The Wiener–Hopf equation	169
		3.13.4	Solution to the Wiener–Hopf equation	171
		3.13.5	Examples of Wiener filtering	174
		3.13.6	Mean-square error	176
		3.13.7	Discrete-time Wiener filters...........................	176
	3.14	A dual approximation problem		179
	3.15	Minimum-norm solution of underdetermined equations		182
	3.16	Iterative Reweighted LS (IRLS) for L_p optimization		183
	3.17	Signal transformation and generalized Fourier series................		186
	3.18	Sets of complete orthogonal functions...........................		190
		3.18.1	Trigonometric functions	190
		3.18.2	Orthogonal polynomials...............................	190
		3.18.3	Sinc functions	193
		3.18.4	Orthogonal wavelets	194
	3.19	Signals as points: Digital communications		208
		3.19.1	The detection problem	210
		3.19.2	Examples of basis functions used in digital communications....................................	212
		3.19.3	Detection in nonwhite noise	213
	3.20	Exercises..		215
	3.21	References		228
4	**Linear Operators and Matrix Inverses**			**229**
	4.1	Linear operators		230
		4.1.1	Linear functionals	231
	4.2	Operator norms		232
		4.2.1	Bounded operators...................................	233
		4.2.2	The Neumann expansion	235
		4.2.3	Matrix norms	235
	4.3	Adjoint operators and transposes		237
		4.3.1	A dual optimization problem..........................	239
	4.4	Geometry of linear equations..................................		239
	4.5	Four fundamental subspaces of a linear operator....................		242
		4.5.1	The four fundamental subspaces with non-closed range	246
	4.6	Some properties of matrix inverses..............................		247
		4.6.1	Tests for invertibility of matrices	248
	4.7	Some results on matrix rank...................................		249
		4.7.1	Numeric rank	250
	4.8	Another look at least squares		251
	4.9	Pseudoinverses		251
	4.10	Matrix condition number		253
	4.11	Inverse of a small-rank adjustment		258
		4.11.1	An application: the RLS filter	259
		4.11.2	Two RLS applications................................	261
	4.12	Inverse of a block (partitioned) matrix		264
		4.12.1	Application: Linear models	267
	4.13	Exercises..		268
	4.14	References		274

5	**Some Important Matrix Factorizations**		**275**
	5.1	The LU factorization	275
		5.1.1 Computing the determinant using the LU factorization	277
		5.1.2 Computing the LU factorization	278
	5.2	The Cholesky factorization	283
		5.2.1 Algorithms for computing the Cholesky factorization	284
	5.3	Unitary matrices and the QR factorization	285
		5.3.1 Unitary matrices	285
		5.3.2 The QR factorization	286
		5.3.3 QR factorization and least-squares filters	286
		5.3.4 Computing the QR factorization	287
		5.3.5 Householder transformations	287
		5.3.6 Algorithms for Householder transformations	291
		5.3.7 QR factorization using Givens rotations	293
		5.3.8 Algorithms for QR factorization using Givens rotations	295
		5.3.9 Solving least-squares problems using Givens rotations	296
		5.3.10 Givens rotations via CORDIC rotations	297
		5.3.11 Recursive updates to the QR factorization	299
	5.4	Exercises	300
	5.5	References	304
6	**Eigenvalues and Eigenvectors**		**305**
	6.1	Eigenvalues and linear systems	305
	6.2	Linear dependence of eigenvectors	308
	6.3	Diagonalization of a matrix	309
		6.3.1 The Jordan form	311
		6.3.2 Diagonalization of self-adjoint matrices	312
	6.4	Geometry of invariant subspaces	316
	6.5	Geometry of quadratic forms and the minimax principle	318
	6.6	Extremal quadratic forms subject to linear constraints	324
	6.7	The Gershgorin circle theorem	324

Application of Eigendecomposition methods

	6.8	Karhunen–Loève low-rank approximations and principal methods	327
		6.8.1 Principal component methods	329
	6.9	Eigenfilters	330
		6.9.1 Eigenfilters for random signals	330
		6.9.2 Eigenfilter for designed spectral response	332
		6.9.3 Constrained eigenfilters	334
	6.10	Signal subspace techniques	336
		6.10.1 The signal model	336
		6.10.2 The noise model	337
		6.10.3 Pisarenko harmonic decomposition	338
		6.10.4 MUSIC	339
	6.11	Generalized eigenvalues	340
		6.11.1 An application: ESPRIT	341
	6.12	Characteristic and minimal polynomials	342
		6.12.1 Matrix polynomials	342
		6.12.2 Minimal polynomials	344
	6.13	Moving the eigenvalues around: Introduction to linear control	344
	6.14	Noiseless constrained channel capacity	347

	6.15	Computation of eigenvalues and eigenvectors	350
		6.15.1 Computing the largest and smallest eigenvalues	350
		6.15.2 Computing the eigenvalues of a symmetric matrix	351
		6.15.3 The QR iteration	352
	6.16	Exercises	355
	6.17	References	368
7	**The Singular Value Decomposition**		**369**
	7.1	Theory of the SVD	369
	7.2	Matrix structure from the SVD	372
	7.3	Pseudoinverses and the SVD	373
	7.4	Numerically sensitive problems	375
	7.5	Rank-reducing approximations: Effective rank	377

Applications of the SVD

	7.6	System identification using the SVD	378
	7.7	Total least-squares problems	381
		7.7.1 Geometric interpretation of the TLS solution	385
	7.8	Partial total least squares	386
	7.9	Rotation of subspaces	389
	7.10	Computation of the SVD	390
	7.11	Exercises	392
	7.12	References	395
8	**Some Special Matrices and Their Applications**		**396**
	8.1	Modal matrices and parameter estimation	396
	8.2	Permutation matrices	399
	8.3	Toeplitz matrices and some applications	400
		8.3.1 Durbin's algorithm	402
		8.3.2 Predictors and lattice filters	403
		8.3.3 Optimal predictors and Toeplitz inverses	407
		8.3.4 Toeplitz equations with a general right-hand side	408
	8.4	Vandermonde matrices	409
	8.5	Circulant matrices	410
		8.5.1 Relations among Vandermonde, circulant, and companion matrices	412
		8.5.2 Asymptotic equivalence of the eigenvalues of Toeplitz and circulant matrices	413
	8.6	Triangular matrices	416
	8.7	Properties preserved in matrix products	417
	8.8	Exercises	418
	8.9	References	421
9	**Kronecker Products and the Vec Operator**		**422**
	9.1	The Kronecker product and Kronecker sum	422
	9.2	Some applications of Kronecker products	425
		9.2.1 Fast Hadamard transforms	425
		9.2.2 DFT computation using Kronecker products	426
	9.3	The vec operator	428
	9.4	Exercises	431
	9.5	References	433

III Detection, Estimation, and Optimal Filtering — 435

10 Introduction to Detection and Estimation, and Mathematical Notation — 437
- 10.1 Detection and estimation theory — 437
  - 10.1.1 Game theory and decision theory — 438
  - 10.1.2 Randomization — 440
  - 10.1.3 Special cases — 441
- 10.2 Some notational conventions — 442
  - 10.2.1 Populations and statistics — 443
- 10.3 Conditional expectation — 444
- 10.4 Transformations of random variables — 445
- 10.5 Sufficient statistics — 446
  - 10.5.1 Examples of sufficient statistics — 450
  - 10.5.2 Complete sufficient statistics — 451
- 10.6 Exponential families — 453
- 10.7 Exercises — 456
- 10.8 References — 459

11 Detection Theory — 460
- 11.1 Introduction to hypothesis testing — 460
- 11.2 Neyman–Pearson theory — 462
  - 11.2.1 Simple binary hypothesis testing — 462
  - 11.2.2 The Neyman–Pearson lemma — 463
  - 11.2.3 Application of the Neyman–Pearson lemma — 466
  - 11.2.4 The likelihood ratio and the receiver operating characteristic (ROC) — 467
  - 11.2.5 A Poisson example — 468
  - 11.2.6 Some Gaussian examples — 469
  - 11.2.7 Properties of the ROC — 480
- 11.3 Neyman–Pearson testing with composite binary hypotheses — 483
- 11.4 Bayes decision theory — 485
  - 11.4.1 The Bayes principle — 486
  - 11.4.2 The risk function — 487
  - 11.4.3 Bayes risk — 489
  - 11.4.4 Bayes tests of simple binary hypotheses — 490
  - 11.4.5 Posterior distributions — 494
  - 11.4.6 Detection and sufficiency — 498
  - 11.4.7 Summary of binary decision problems — 498
- 11.5 Some M-ary problems — 499
- 11.6 Maximum-likelihood detection — 503
- 11.7 Approximations to detection performance: The union bound — 503
- 11.8 Invariant Tests — 504
  - 11.8.1 Detection with random (nuisance) parameters — 507
- 11.9 Detection in continuous time — 512
  - 11.9.1 Some extensions and precautions — 516
- 11.10 Minimax Bayes decisions — 520
  - 11.10.1 Bayes envelope function — 520
  - 11.10.2 Minimax rules — 523
  - 11.10.3 Minimax Bayes in multiple-decision problems — 524

		11.10.4	Determining the least favorable prior	528
		11.10.5	A minimax example and the minimax theorem	529
	11.11	Exercises		532
	11.12	References		541

12 Estimation Theory — 542

- 12.1 The maximum-likelihood principle ... 542
- 12.2 ML estimates and sufficiency ... 547
- 12.3 Estimation quality ... 548
  - 12.3.1 The score function ... 548
  - 12.3.2 The Cramér–Rao lower bound ... 550
  - 12.3.3 Efficiency ... 552
  - 12.3.4 Asymptotic properties of maximum-likelihood estimators ... 553
  - 12.3.5 The multivariate normal case ... 556
  - 12.3.6 Minimum-variance unbiased estimators ... 559
  - 12.3.7 The linear statistical model ... 561
- 12.4 Applications of ML estimation ... 561
  - 12.4.1 ARMA parameter estimation ... 561
  - 12.4.2 Signal subspace identification ... 565
  - 12.4.3 Phase estimation ... 566
- 12.5 Bayes estimation theory ... 568
- 12.6 Bayes risk ... 569
  - 12.6.1 MAP estimates ... 573
  - 12.6.2 Summary ... 574
  - 12.6.3 Conjugate prior distributions ... 574
  - 12.6.4 Connections with minimum mean-squared estimation ... 577
  - 12.6.5 Bayes estimation with the Gaussian distribution ... 578
- 12.7 Recursive estimation ... 580
  - 12.7.1 An example of non-Gaussian recursive Bayes ... 582
- 12.8 Exercises ... 584
- 12.9 References ... 590

13 The Kalman Filter — 591

- 13.1 The state-space signal model ... 591
- 13.2 Kalman filter I: The Bayes approach ... 592
- 13.3 Kalman filter II: The innovations approach ... 595
  - 13.3.1 Innovations for processes with linear observation models ... 596
  - 13.3.2 Estimation using the innovations process ... 597
  - 13.3.3 Innovations for processes with state-space models ... 598
  - 13.3.4 A recursion for $P_{t|t-1}$ ... 599
  - 13.3.5 The discrete-time Kalman filter ... 601
  - 13.3.6 Perspective ... 602
  - 13.3.7 Comparison with the RLS adaptive filter algorithm ... 603
- 13.4 Numerical considerations: Square-root filters ... 604
- 13.5 Application in continuous-time systems ... 606
  - 13.5.1 Conversion from continuous time to discrete time ... 606
  - 13.5.2 A simple kinematic example ... 606
- 13.6 Extensions of Kalman filtering to nonlinear systems ... 607

	13.7	Smoothing	613
		13.7.1 The Rauch–Tung–Streibel fixed-interval smoother	613
	13.8	Another approach: H_∞ smoothing	616
	13.9	Exercises	617
	13.10	References	620

IV Iterative and Recursive Methods in Signal Processing 621

14 Basic Concepts and Methods of Iterative Algorithms 623

	14.1	Definitions and qualitative properties of iterated functions	624
		14.1.1 Basic theorems of iterated functions	626
		14.1.2 Illustration of the basic theorems	627
	14.2	Contraction mappings	629
	14.3	Rates of convergence for iterative algorithms	631
	14.4	Newton's method	632
	14.5	Steepest descent	637
		14.5.1 Comparison and discussion: Other techniques	642

Some Applications of Basic Iterative Methods

	14.6	LMS adaptive Filtering	643
		14.6.1 An example LMS application	645
		14.6.2 Convergence of the LMS algorithm	646
	14.7	Neural networks	648
		14.7.1 The backpropagation training algorithm	650
		14.7.2 The nonlinearity function	653
		14.7.3 The forward–backward training algorithm	654
		14.7.4 Adding a momentum term	654
		14.7.5 Neural network code	655
		14.7.6 How many neurons?	658
		14.7.7 Pattern recognition: ML or NN?	659
	14.8	Blind source separation	660
		14.8.1 A bit of information theory	660
		14.8.2 Applications to source separation	662
		14.8.3 Implementation aspects	664
	14.9	Exercises	665
	14.10	References	668

15 Iteration by Composition of Mappings 670

	15.1	Introduction	670
	15.2	Alternating projections	671
		15.2.1 An applications: bandlimited reconstruction	675
	15.3	Composite mappings	676
	15.4	Closed mappings and the global convergence theorem	677
	15.5	The composite mapping algorithm	680
		15.5.1 Bandlimited reconstruction, revisited	681
		15.5.2 An example: Positive sequence determination	681
		15.5.3 Matrix property mappings	683
	15.6	Projection on convex sets	689
	15.7	Exercises	693
	15.8	References	694

Contents

16 Other Iterative Algorithms — 695
- 16.1 Clustering — 695
  - 16.1.1 An example application: Vector quantization — 695
  - 16.1.2 An example application: Pattern recognition — 697
  - 16.1.3 k-means Clustering — 698
  - 16.1.4 Clustering using fuzzy k-means — 700
- 16.2 Iterative methods for computing inverses of matrices — 701
  - 16.2.1 The Jacobi method — 702
  - 16.2.2 Gauss–Seidel iteration — 703
  - 16.2.3 Successive over-relaxation (SOR) — 705
- 16.3 Algebraic reconstruction techniques (ART) — 706
- 16.4 Conjugate-direction methods — 708
- 16.5 Conjugate-gradient method — 710
- 16.6 Nonquadratic problems — 713
- 16.7 Exercises — 713
- 16.8 References — 715

17 The EM Algorithm in Signal Processing — 717
- 17.1 An introductory example — 718
- 17.2 General statement of the EM algorithm — 721
- 17.3 Convergence of the EM algorithm — 723
  - 17.3.1 Convergence rate: Some generalizations — 724

Example applications of the EM algorithm

- 17.4 Introductory example, revisited — 725
- 17.5 Emission computed tomography (ECT) image reconstruction — 725
- 17.6 Active noise cancellation (ANC) — 729
- 17.7 Hidden Markov models — 732
  - 17.7.1 The E- and M-steps — 734
  - 17.7.2 The forward and backward probabilities — 735
  - 17.7.3 Discrete output densities — 736
  - 17.7.4 Gaussian output densities — 736
  - 17.7.5 Normalization — 737
  - 17.7.6 Algorithms for HMMs — 738
- 17.8 Spread-spectrum, multiuser communication — 740
- 17.9 Summary — 743
- 17.10 Exercises — 744
- 17.11 References — 747

V Methods of Optimization — 749

18 Theory of Constrained Optimization — 751
- 18.1 Basic definitions — 751
- 18.2 Generalization of the chain rule to composite functions — 755
- 18.3 Definitions for constrained optimization — 757
- 18.4 Equality constraints: Lagrange multipliers — 758
  - 18.4.1 Examples of equality-constrained optimization — 764
- 18.5 Second-order conditions — 767
- 18.6 Interpretation of the Lagrange multipliers — 770
- 18.7 Complex constraints — 773
- 18.8 Duality in optimization — 773

	18.9	Inequality constraints: Kuhn–Tucker conditions	777
		18.9.1 Second-order conditions for inequality constraints	783
		18.9.2 An extension: Fritz John conditions	783
	18.10	Exercises	784
	18.11	References	786
19	**Shortest-Path Algorithms and Dynamic Programming**		**787**
	19.1	Definitions for graphs	787
	19.2	Dynamic programming	789
	19.3	The Viterbi algorithm	791
	19.4	Code for the Viterbi algorithm	795
		19.4.1 Related algorithms: Dijkstra's and Warshall's	798
		19.4.2 Complexity comparisons of Viterbi and Dijkstra	799
		Applications of path search algorithms	
	19.5	Maximum-likelihood sequence estimation	800
		19.5.1 The intersymbol interference (ISI) channel	800
		19.5.2 Code-division multiple access	804
		19.5.3 Convolutional decoding	806
	19.6	HMM likelihood analysis and HMM training	808
		19.6.1 Dynamic warping	811
	19.7	Alternatives to shortest-path algorithms	813
	19.8	Exercises	815
	19.9	References	817
20	**Linear Programming**		**818**
	20.1	Introduction to linear programming	818
	20.2	Putting a problem into standard form	819
		20.2.1 Inequality constraints and slack variables	819
		20.2.2 Free variables	820
		20.2.3 Variable-bound constraints	822
		20.2.4 Absolute value in the objective	823
	20.3	Simple examples of linear programming	823
	20.4	Computation of the linear programming solution	824
		20.4.1 Basic variables	824
		20.4.2 Pivoting	826
		20.4.3 Selecting variables on which to pivot	828
		20.4.4 The effect of pivoting on the value of the problem	829
		20.4.5 Summary of the simplex algorithm	830
		20.4.6 Finding the initial basic feasible solution	831
		20.4.7 MATLAB® code for linear programming	834
		20.4.8 Matrix notation for the simplex algorithm	835
	20.5	Dual problems	836
	20.6	Karmarker's algorithm for LP	838
		20.6.1 Conversion to Karmarker standard form	842
		20.6.2 Convergence of the algorithm	844
		20.6.3 Summary and extensions	846
		Examples and applications of linear programming	
	20.7	Linear-phase FIR filter design	846
		20.7.1 Least-absolute-error approximation	847
	20.8	Linear optimal control	849

	20.9	Exercises	850
	20.10	References	853

A Basic Concepts and Definitions — 855

A.1	Set theory and notation	855
A.2	Mappings and functions	859
A.3	Convex functions	860
A.4	O and o Notation	861
A.5	Continuity	862
A.6	Differentiation	864
	A.6.1 Differentiation with a single real variable	864
	A.6.2 Partial derivatives and gradients on $\mathbb{R}^m$	865
	A.6.3 Linear approximation using the gradient	867
	A.6.4 Taylor series	868
A.7	Basic constrained optimization	869
A.8	The Hölder and Minkowski inequalities	870
A.9	Exercises	871
A.10	References	876

B Completing the Square — 877

B.1	The scalar case	877
B.2	The matrix case	879
B.3	Exercises	879

C Basic Matrix Concepts — 880

C.1	Notational conventions	880
C.2	Matrix Identity and Inverse	882
C.3	Transpose and trace	883
C.4	Block (partitioned) matrices	885
C.5	Determinants	885
	C.5.1 Basic properties of determinants	885
	C.5.2 Formulas for the determinant	887
	C.5.3 Determinants and matrix inverses	889
C.6	Exercises	889
C.7	References	890

D Random Processes — 891

D.1	Definitions of means and correlations	891
D.2	Stationarity	892
D.3	Power spectral-density functions	893
D.4	Linear systems with stochastic inputs	894
	D.4.1 Continuous-time signals and systems	894
	D.4.2 Discrete-time signals and systems	895
D.5	References	895

E Derivatives and Gradients — 896

E.1	Derivatives of vectors and scalars with respect to a real vector	896
	E.1.1 Some important gradients	897
E.2	Derivatives of real-valued functions of real matrices	899
E.3	Derivatives of matrices with respect to scalars, and vice versa	901
E.4	The transformation principle	903
E.5	Derivatives of products of matrices	903

	E.6	Derivatives of powers of a matrix	904
	E.7	Derivatives involving the trace	906
	E.8	Modifications for derivatives of complex vectors and matrices	908
	E.9	Exercises	910
	E.10	References	912

F	**Conditional Expectations of Multinomial and Poisson r.v.s**		**913**
	F.1	Multinomial distributions	913
	F.2	Poisson random variables	914
	F.3	Exercises	914

Bibliography   915

Index   929

List of Figures

1.1	Input/output relation for a transfer function	10
1.2	Realization of the AR part of a transfer function	15
1.3	Realization of a transfer function	15
1.4	Realization of a transfer function with state-variable labels	16
1.5	Prediction error	23
1.6	Linear predictor as an inverse system	24
1.7	PSD input and output	26
1.8	Representation of an adaptive filter	28
1.9	Identification of an unknown plant	29
1.10	Adapting to the inverse of an unknown plant	29
1.11	An adaptive predictor	30
1.12	Configuration for interference cancellation	30
1.13	The Gaussian density	31
1.14	Demonstration of the central limit theorem	33
1.15	Plot of two-dimensional Gaussian distribution	35
1.16	A simple Markov model	38
1.17	A hidden Markov model	39
1.18	An HMM with four states	40
1.19	Binary symmetric channel model	47
1.20	LFSR realization	48
1.21	Alternative LFSR realization	48
1.22	A binary LFSR and its output	51
1.23	Simple feedback configuration	62
2.1	Illustration of the triangle inequality	72
2.2	Quantization of the vector $\mathbf{x}$	73
2.3	Comparison of d_∞ and d_2 metrics	75
2.4	x_0 is interior, x_2 is exterior, and x_1 is neither interior nor exterior	77
2.5	Illustration of open and closed sets	78
2.6	The function $f_n(t)$	80
2.7	Illustration of Gibbs phenomenon	83
2.8	A subspace of $\mathbb{R}^3$	89
2.9	A triangle inequality interpretation	94
2.10	Unit spheres in $\mathbb{R}^2$ under various l_p norm	95
2.11	Chebyshev polynomials $T_0(t)$ through $T_5(t)$ for $t \in [-1, 1]$	104
2.12	A space and its orthogonal complement	107
2.13	Disjoint lines in $\mathbb{R}^2$	111
2.14	Decomposition of $\mathbf{x}$ into disjoint components	113
2.15	Orthogonal projection finds the closest point in V to $\mathbf{x}$	115
2.16	Orthogonal projection onto the space spanned by several vectors	115

2.17	The projection theorem	117
2.18	The first steps of the Gram–Schmidt process	119
2.19	Third step of the Gram–Schmidt process	119
2.20	The parallelogram law	125
2.21	Functions to orthogonalize	128
3.1	The approximation problem	130
3.2	Approximation with one and two vectors	131
3.3	An error surface for two variables	138
3.4	Projection solution	139
3.5	Statistician's Pythagorean theorem	142
3.6	Comparison of LS, WLS, and Taylor series approximations to e^t	144
3.7	A discrete function and the error in its approximation	147
3.8	Data for regression	147
3.9	Illustration of least-squares and weighted least-squares lines	149
3.10	Least-squares equalizer example	154
3.11	An equalizer problem	159
3.12	Contour plot of an error surface	161
3.13	Pole–zero plot of rational $S_y(s)$ ($\times$ = poles, $\circ$ = zeros)	164
3.14	y_t as the output of a linear system driven by white noise	167
3.15	ν_t as the output of a linear system driven by y_t	167
3.16	The optimal filter as the cascade of a whitening filter and a Wiener filter with white-noise inputs	173
3.17	Minimum norm to a linear variety	179
3.18	Magnitude response for filters designed using IRLS	186
3.19	Legendre polynomials $p_0(t)$ through $p_5(t)$ for $t \in [-1, 1]$	192
3.20	A function $f(t)$ and its projection onto V_0 and V_{-1}	196
3.21	The simplest scaling and wavelet functions	198
3.22	Illustration of scaling and wavelet functions	199
3.23	Illustration of a wavelet transform	201
3.24	Multirate interpretation of wavelet transform	203
3.25	Illustration of the inverse wavelet transform	204
3.26	Filtering interpretation of an inverse wavelet transform	205
3.27	Perfect reconstruction filter bank	206
3.28	Two basis functions, and some functions represented by using them	209
3.29	Implementations of digital receiver processing	211
3.30	Digital receiver processing	211
3.31	Implementation of a matched filter receiver	211
3.32	PSK signal constellation and detection example	212
3.33	Illustration of concepts of various signal constellations	213
3.34	Block diagram for detection processing	213
4.1	Geometry of the operator norm	232
4.2	Intersections of lines form solutions of systems of linear equations	240
4.3	Intersecting planes: (a) no solution (b) infinite number of solutions	241
4.4	The four fundamental subspaces of a matrix operator	246
4.5	Operation of the pseudoinverse	252
4.6	Demonstration of an ill-conditioned linear system	253
4.7	Condition of the Hilbert matrix	256
4.8	Condition number for a bad idea	257
4.9	RLS adaptive equalizer	262

List of Figures

4.10	Illustration of RLS equalizer performance	263
4.11	System identification using the RLS adaptive filter	264
4.12	Illustration of system identification using the RLS filter	265
5.1	The Householder transformation of a vector	288
5.2	Zeroing elements of a vector by a Householder transformation	289
5.3	Two-dimensional rotation	293
6.1	The direction of eigenvectors is not modified by A	306
6.2	The geometry of quadratic forms	319
6.3	Level curves for a Gaussian distribution	320
6.4	The maximum principle	322
6.5	Illustration of Gershgorin disks	325
6.6	Scatter data for principal component analysis	330
6.7	Noisy signal to be filtered using an eigenfilter h	331
6.8	Magnitude response specifications for a lowpass filter	332
6.9	Eigenfilter response	334
6.10	Response of a constrained eigenfilter	335
6.11	The MUSIC spectrum for example 6.10.2	340
6.12	Plant with reference input and feedback control	345
6.13	State diagram for a constrained channel	348
6.14	Direct and indirect transmission through a noisy channel	362
6.15	Expansion and interpolation using multirate processing	363
6.16	Transformation from a general matrix to first companion form	365
7.1	Illustration of the sensitive direction	376
7.2	Comparison of least-squares and total least-squares fit	382
7.3	PTLS linear parameter identification	389
7.4	A data set rotated relative to another data set	389
8.1	The first two stages of a lattice prediction filter	405
8.2	The kth stage of a lattice filter	406
8.3	Comparison of $S(\omega)$ and the eigenvalues of R_n for $n = 30$ and $n = 100$	417
9.1	4-point fast Hadamard transform	426
9.2	6-point DFT using Kronecker decomposition	428
10.1	Loss function (or matrix) for "odd or even" game	438
10.2	Elements of the statistical decision game	440
10.3	A simple binary communications channel	440
10.4	A typical payoff matrix for the Prisoner's Dilemma game	456
11.1	Illustration of threshold for Neyman–Pearson test	465
11.2	Scalar Gaussian detection of the mean	471
11.3	Error probabilities for Gaussian variables with different means and equal variances	473
11.4	ROC for Gaussian detection	473
11.5	Test for vector Gaussian random variables with different means	475
11.6	Probability of error for BPSK signaling	476

11.7	An orthogonal and antipodal binary signal constellation	477
11.8	ROC: normal variables with equal means and unequal variances	481
11.9	Demonstration of the concave property of the ROC	481
11.10	Illustration of even–odd observations	488
11.11	Risk function for statistical odd or even game	488
11.12	A binary channel	489
11.13	Risk function for binary channel	489
11.14	Loss function	496
11.15	Bayes risk for a decision	497
11.16	Geometry of the decision space for multivariate Gaussian detection	502
11.17	Decision boundaries for a quaternary decision problem	502
11.18	Venn diagram for the union of two sets	504
11.19	Bound on the probability of error for PSK signaling	504
11.20	A test biased by $\gamma \mathbf{c}$	505
11.21	Channel gain and rotation	506
11.22	Incoherent binary detector	512
11.23	Probability of error for BPSK	516
11.24	A projection approach to signal detection	518
11.25	Bayes envelope function	521
11.26	Bayes envelope function: normal variables with unequal means and equal variances	522
11.27	Bayes envelope function for example 11.4.5	523
11.28	Bayes envelope for binary channel	523
11.29	Geometrical interpretation of the risk set	526
11.30	Geometrical interpretation of the minimax rule	527
11.31	The risk set and its relation to the Neyman–Pearson test	528
11.32	Risk function for statistical odd or even game	529
11.33	Risk set for odd or even game	529
11.34	Risk set for the binary channel	531
11.35	Regions for bounding the Q function	535
11.36	Channel with Laplacian noise and decision region	535
11.37	Some signal constellations	537
11.38	Signal constellation with three points	538
12.1	Empiric distribution function	544
12.2	Explicitly computing the estimate of the phase	567
12.3	A phase-locked loop	567
12.4	Illustration of the update and propagate steps in sequential estimation	582
12.5	Acoustic level framework	587
12.6	Equivalent representations for the Gaussian estimation problem	589
13.1	Illustration of Kalman filter	595
14.1	Illustration of an orbit of a function with an attractive fixed point	625
14.2	Illustration of an orbit of a function with a repelling fixed point	625
14.3	Examples of dynamical behavior on the quadratic logistic map	627
14.4	Illustration of $g(x) = f(f(x))$ when $\lambda = 3.2$	628
14.5	Iterations of an affine transformation, acting on a square	630
14.6	Illustration of Newton's method	634

14.7	Contour plots of Rosenbrock's function and Newton's method	637
14.8	A function with local and global minima	638
14.9	Convergence of steepest descent on a quadratic function	639
14.10	Error components in principal coordinates for steepest descent	641
14.11	Error in the LMS algorithm for $\mu = 0.075$ and $\mu = 0.0075$, compared with the RLS algorithm, for an adaptive equalizer problem	645
14.12	Optimal equalizer coefficients and adaptive equalizer coefficients	646
14.13	Representation of the layers of an artificial neural network	648
14.14	An artificial neuron	649
14.15	Notation for a multilayer neural network	650
14.16	The sigmoidal nonlinearity	654
14.17	Pattern-recognition problem for a neural network	656
14.18	Desired output (solid line) and neural network output (dashed line)	657
14.19	Effect of convergence rate on μ and α	658
14.20	The blind source-separation problem	660
14.21	The binary entropy function $H(p)$	661
15.1	Illustration of a projection on a set	672
15.2	Projection on convex sets in two dimensions	673
15.3	Results of the bandlimited reconstruction algorithm	676
15.4	Property sets in X and their intersection $\mathcal{P}$	677
15.5	Illustration of the composition of point-to-set mappings	679
15.6	Projection onto a non-convex	681
15.7	Producing a positive sequence from the Hamming window	683
15.8	Results from the application of a composite mapping algorithm to sinusoidal data	689
15.9	Geometric properties of convex sets	691
15.10	Projection onto two convex sets	692
16.1	Demonstration of clustering	697
16.2	Clusters for a pattern recognition problem	698
16.3	Illustration of iterative inverse computation	704
16.4	Residual error in the ART algorithm as a function of iteration	708
16.5	Convergence of conjugate gradient on a quadratic function	711
17.1	An overview of the EM algorithm	718
17.2	Illustration of a many-to-one mapping from $\mathcal{X}$ to $\mathcal{Y}$	721
17.3	Representation of emission tomography	726
17.4	Detector arrangement for tomographic reconstruction example	728
17.5	Example emission tomography reconstruction	729
17.6	Single-microphone ANC system	730
17.7	Processor block diagram of the ANC system	730
17.8	$\log P(\mathbf{y}_1^T \mid \theta^{[k]})$ for an HMM	740
17.9	Representation of signals in an SSMA system	741
17.10	Multiple-access receiver matched-filter bank	741
18.1	Examples of minimizing points	752
18.2	Contours of $f(x_1, x_2)$, showing minimum and constrained minimum	754

18.3	Relationships between variables in composite functions	756
18.4	Illustration of functional dependencies	756
18.5	Surface and contour plots of $f(x_1, x_2)$	759
18.6	Tangent plane to a surface	759
18.7	Curves on a surface	760
18.8	Minimizing the distance to an ellipse	766
18.9	The projection of $\mathbf{Ly}$ into P to form L_P	770
18.10	Duality: the nearest point to K is the maximum distance to a separating hyperplane	774
18.11	The dual function $g(\lambda)$	775
18.12	Saddle surface for minimax optimization	776
18.13	Illustration of the Kuhn–Tucker condition in a single dimension	778
18.14	Illustration of "waterfilling" solution	783
19.1	Graph examples	788
19.2	A multistage graph	789
19.3	A trellis diagram	792
19.4	State machine corresponding to a trellis	793
19.5	State-machine output observed after passing through a noisy channel	793
19.6	Steps in the Viterbi algorithm	794
19.7	A trellis with irregular branches	797
19.8	A trellis with multiple outputs	798
19.9	MLSE detection in ISI	801
19.10	Trellis diagram and detector structure for ISI detection	803
19.11	CDMA signal model	805
19.12	CDMA detection	806
19.13	Convolutional coding	807
19.14	Comparing HMM training algorithms	811
19.15	Illustration of the warping alignment process	812
19.16	Probability of failure of network links	816
20.1	A linear programming problem	819
20.2	Illustration of Karmarker's algorithm	840
20.3	Filter design constraints	847
20.4	Frequency and impulse response of a filter designed using linear programming ($n = 45$ coefficients)	848
A.1	Illustration of convex and nonconvex sets	858
A.2	Indicator functions for some simple sets	859
A.3	Illustration of a convex function	860
A.4	Illustration of the definition of continuity	863
A.5	A constrained optimization problem	869
A.6	The indicator function for a fuzzy number "near 10"	872
A.7	The set sum	872

List of Algorithms

1.1	Massey's algorithm (pseudocode)	56
1.2	Massey's algorithm	56
2.1	Gram–Schmidt algorithm (QR factorization)	120
3.1	Least-squares filter computation	153
3.2	Forward–backward linear predictor estimate	156
3.3	Two-tap channel equalizer	161
3.4	Iterative reweighted least-squares	185
3.5	Filter design using IRLS	185
3.6	Some wavelet coefficients	199
3.7	Demonstration of wavelet decomposition	204
3.8	Demonstration of wavelet decomposition (alternative indexing)	205
3.9	Nonperiodic wavelet transform	207
3.10	Nonperiodic inverse wavelet transform	207
3.11	Periodic wavelet transform	207
3.12	Inverse periodic wavelet transform	207
4.1	The RLS algorithm	261
4.2	The RLS algorithm (MATLAB® implementation)	261
5.1	LU factorization	282
5.2	Cholesky factorization	285
5.3	Householder transformation functions	291
5.4	QR factorization via Householder transformations	292
5.5	Computation of $Q^H \mathbf{b}$	292
5.6	Computation of Q from V	293
5.7	Finding $\cos\theta$ and $\sin\theta$ for a Givens rotation	295
5.8	QR factorization using Givens rotations	295
5.9	Computation of $Q^H \mathbf{b}$ for the Givens rotation factorization	296
5.10	Computation of Q from θ	296
6.1	Eigenfilter design	334
6.2	Constrained eigenfilter design	335
6.3	Pisarenko harmonic decomposition	339
6.4	Computation of the MUSIC spectrum	339
6.5	Computation of the frequency spectrum of a signal using ESPRIT	342
6.6	Computation of the largest eigenvalue using the power method	350
6.7	Computation of the smallest eigenvalue using the power method	351
6.8	Tridiagonalization of a real symmetric matrix	352
6.9	Implicit QR shift	354
6.10	Complete eigenvalue/eigenvector function	355
7.1	System identification using SVD	380
7.2	Total least squares	384

7.3	Partial total least squares, part 1	387
7.4	Partial total least squares, part 2	388
7.5	Computing the SVD	392
8.1	Durbin's algorithm	403
8.2	Conversion of lattice FIR to direct-form	407
8.3	Conversion of direct-form FIR to lattice	407
8.4	Levinson's algorithm	409
11.1	Example Bayes minimax calculations	532
12.1	Maximum-likelihood ARMA estimation	565
13.1	Kalman filter I	594
13.2	Kalman filter example	595
14.1	Logistic function orbit	628
14.2	LMS adaptive filter	645
14.3	Neural network forward-propagation algorithm	655
14.4	Neural network backpropagation training algorithm	655
14.5	Neural network test example	656
14.6	Blind source separation test	665
15.1	Bandlimited reconstruction using alternating projections	675
15.2	Mapping to a positive sequence	682
15.3	Mapping to the nearest stochastic matrix	686
15.4	Mapping to a Hankel matrix of given rank	687
15.5	Mapping to a Toeplitz/Hankel matrix stack of given rank	688
16.1	k-means clustering (LGB algorithm)	699
16.2	Jacobi iteration	703
16.3	Gauss–Seidel iteration	703
16.4	Successive over-relaxation	706
16.5	Algebraic reconstruction technique	707
16.6	Conjugate-gradient solution of a symmetric linear equation	711
16.7	Conjugate-gradient solution for unconstrained minimization	713
17.1	EM algorithm example computations	720
17.2	Simulation and reconstruction of emission tomography	729
17.3	Overview of HMM data structures and functions	738
17.4	HMM likelihood computation functions	739
17.5	HMM model update functions	739
17.6	HMM generation functions	739
18.1	A constrained optimization of a racing problem	782
19.1	Forward dynamic programming	791
19.2	The Viterbi algorithm	796
19.3	Initializing the Viterbi algorithm	796
19.4	Flushing the shortest path in the VA	796
19.5	Dijkstra's shortest-path algorithm	798
19.6	Warshall's transitive closure algorithm	799
19.7	Norm and initialization for Viterbi HMM computations	809
19.8	Best-path likelihood for the HMM	810
19.9	HMM training using Viterbi methods	810
19.10	Use of the Viterbi methods with HMMs	811
19.11	Warping code	813

List of Algorithms

20.1	The simplex algorithm for linear programming	834
20.2	Tableau pivoting for the simplex algorithm	834
20.3	Elimination and backsubstitution of free variables for linear programming	834
20.4	Karmarker's algorithm for linear programming	842
20.5	Conversion of standard form to Karmarker standard form	844
20.6	Optimal filter design using linear programming	847

List of Boxes

Box 1.1	Notation for complex quantities	7
Box 1.2	Notation for vectors	9
Box 1.3	Notation for random variables and vectors	31
Box 1.4	Groups, rings, and fields	49
Box 1.5	$GF(2)$	50
Box 2.1	Sup and inf	74
Box 2.2	The measure of a set	82
Box 2.3	David Hilbert (1862–1943)	107
Box 2.4	Isomorphism	112
Box 3.1	Positive-definite matrices	134
Box 4.1	James H. Wilkinson (1919–1986)	254
Box 5.1	Carl Friedrich Gauss (1777–1855)	278
Box 6.1	Arg max and arg min	326
Box 7.1	Commutative diagrams	375
Box 11.1	The Q function	472
Box 11.2	The Γ function	478
Box 11.3	The t distribution	507
Box 11.4	The function $I_0(x)$	510
Box 12.1	The β distribution	575
Box 12.2	The Γ distribution	576
Box 14.1	Isaac Newton (1642–1727)	633

Preface

Rationale

The purpose of this book is to bridge the gap between introductory signal processing classes and the mathematics prevalent in contemporary signal processing research and practice, by providing a unified *applied* treatment of fundamental mathematics, seasoned with demonstrations using MATLAB®. This book is intended not only for current students of signal processing, but also for practicing engineers who must be able to access the signal processing research literature, and for researchers looking for a particular result to apply. It is thus intended both as a textbook *and* as a reference.

Both the theory and the practice of signal processing contribute to and draw from a variety of disciplines: controls, communications, system identification, information theory, artificial intelligence, spectroscopy, pattern recognition, tomography, image analysis, and data acquisition, among others. To fulfill its role in these diverse areas, signal processing employs a variety of mathematical tools, including transform theory, probability, optimization, detection theory, estimation theory, numerical analysis, linear algebra, functional analysis, and many others. The practitioner of signal processing—the "signal processor"—may use several of these tools in the solution of a problem; for example, setting up a signal reconstruction algorithm, and then optimizing the parameters of the algorithm for optimum performance. Practicing signal processors must have knowledge of both the *theory* and the *implementation* of the mathematics: how and why it works, and how to make the computer do it. The breadth of mathematics employed in signal processing, coupled with the opportunity to apply that math to problems of engineering interest, makes the field both interesting and rewarding.

The mathematical aspects of signal processing also introduce some of its major challenges: how is a student or engineering practitioner to become versed in such a variety of mathematical techniques while still keeping an eye toward applications? Introductory texts on signal processing tend to focus heavily on transform techniques and filter-based applications. While this is an essential part of the training of a signal processor, it is only the tip of the iceberg of material required by a practicing engineer. On the other hand, more advanced texts typically develop mathematical tools that are specific to a narrow aspect of signal processing, while perhaps missing connections between these ideas and related areas of research. Neither of these approaches provides sufficient background to read and understand broadly in the signal processing research literature, nor do they equip the student with many signal processing tools.

The signal processing literature has moved steadily toward increasing sophistication: applications of the singular value decomposition (SVD) and wavelet transforms abound; everyone knows something about these by now, or should! Part of this move toward sophistication is fueled by computer capabilities, since computations that formerly required considerable effort and understanding are now embodied in convenient mathematical packages. A naive view might held that this automation threatens the expertise of the engineer: Why hire a specialist to do what anyone can do in ten minutes with a MATLAB toolbox? Viewed more positively, the power of the computer provides a variety of new opportunities, as engineers are freed from computational drudgery to pursue new applications. Computer software provides platforms upon which innovative ideas may be developed with ever greater ease. Taking advantage of this new freedom to develop useful concepts will require a solid understanding of mathematics, both to appreciate what is in the toolboxes and to extend beyond their limits. This book is intended to provide a foundation in the requisite mathematics.

We assume that students using this text have had a course in traditional transform-based digital signal processing at the senior or first-year graduate level, and a traditional course in stochastic processes. Though basic concepts in these areas are reviewed, this book does not supplant the more focused coverage that these courses provide.

Features

- Vector-space geometry, which puts least-squares and minimum mean-squares in the same framework, and the concept of signals as vectors in an appropriate vector space, are both emphasized. This vector-space approach provides a natural framework for topics such as wavelet transforms and digital communications, as well as the traditional topics of optimum prediction, filtering, and estimation. In this context, the more general notion of metric spaces is introduced, with a discussion of signal norms.
- The linear algebra used in signal processing is thoroughly described, both in concept and in numerical implementation. While software libraries are commonly available to perform linear algebra computations, we feel that the numerical techniques presented in this book exercise student intuition regarding the geometry of vector spaces, and build understanding of the issues that must be addressed in practical problems.

    The presentation includes a thorough discussion of eigen-based methods of computation, including eigenfilters, MUSIC, and ESPRIT; there is also a chapter devoted to the properties and applications of the SVD. Toeplitz matrices, which appear throughout the signal processing literature, are treated both from a numerical point of view—as an example of recursive algorithms—and in conjunction with the lattice-filtering interpretation.

    The matrices in linear algebra are viewed as operators; thus, the important concept of an operator is introduced. Associated notions, such as the range, nullspace, and norm of an operator are also presented. While a full coverage of operator theory is not provided, there is a strong foundation that can serve to build insight into other operators.
- In addition to linear algebraic concepts, there is a discussion of *computation*. Algorithms are presented for computing the common factorizations, eigenvalues, eigenvectors, SVDs, and many other problems, with some numerical consideration for implementation. Not all of this material is necessarily intended for classroom use in a conventional signal processing course—there will not be sufficient time in most cases. Nonetheless, it provides an important

perspective to prospective practitioners, and a starting point for implementations on other platforms. Instructors may choose to emphasize certain numeric concepts because they highlight particular topics, such as the geometry of vector spaces.

- The Cauchy–Schwartz inequality is used in a variety of places as an optimizing principle.
- Recursive least square and least mean square adaptive filters are presented as natural outgrowths of more fundamental concepts: matrix inverse updates and steepest descent. Neural networks and blind source separation are also presented as applications of steepest descent.
- Several chapters are devoted to iterative and recursive methods. Though iterative methods are of great theoretical and practical significance, no other signal processing textbook provides a similar breadth of coverage. Methods presented include projection on convex sets, composite mapping, the EM algorithm, conjugate gradient, and methods of matrix inverse computation using iterative methods.
- Detection and estimation are presented with several applications, including spectrum estimation, phase estimation, and multidimensional digital communications.
- Optimization is a key concept in signal processing, and examples of optimization, both unconstrained and constrained, appear throughout the text. Both a theoretical justification for Lagrange multiplier methods and a physical interpretation are explicitly spelled out in a chapter on optimization. A separate chapter discusses linear programming and its applications. Optimizations on graphs (shortest-path problems) are also examined, with a variety of applications in communications and signal processing.
- The EM algorithm as presented here is the only treatment in a signal processing textbook that we are aware of. This powerful algorithm is used for many otherwise intractable estimation and learning problems.

In general, the presentation is at a more formal level than in many recent digital signal processing texts, following a "theorem/proof" format throughout. At the same time, it is less formal than many math texts covering the same material. In this, we have attempted to help the student become comfortable with rigorous thinking, without overwhelming them with technicalities. (A brief review of methods of proofs is also provided to help students develop a sense of how to approach the proofs.) Ultimately, the aim of this book is to teach its reader how to think about problems. To this end, some material is covered more than once, from different perspectives (e.g., with more than one proof for certain results), to demonstrate that there is usually more than one way to approach a problem.

Throughout the text, the intent has been to explain the "what" and the "why" of the mathematics, but not become overwrought with some of the more technical mathematical preoccupations. In this regard, the book does not always thoroughly treat questions of "how well." (For example, in our coverage of linear numerical analysis, the perturbation analysis that characterizes much of the research literature has been largely ignored. Nor do issues of computational complexity form a major consideration.) To visualize this approach, consider an automotive analogy: Our intent is to "get under the hood" to a sufficient degree that it is clear why the engine runs and what it can do, but not to provide a molecular-level description of the metallurgical structure of the piston rings. Such fine-grained investigations might be a necessary part of research into fine-tuning the performance of the engine—or the algorithm—but are not appropriate for a student learning the basic mechanics.

Throughout the chapters and in the appendices, there is a great deal of material that will be of reference value to practicing engineers. For example, there are facts regarding matrix rank, the invertibility of matrices, properties of Hermitian matrices, properties of structured matrices preserved under multiplication, and an extensive table of gradients. Not all of this material is necessarily intended for classroom use, but is provided to enhance the value of the book as a reference. Nevertheless, where such reference material is provided, it is usually accompanied by an explanation of its derivation, so that related facts may often be derived by the reader.

Though this book does not provide the final word in any research area, for many research paths it will at least provide a good first step. The contents of the book have been selected according to a variety of criteria. The primary criterion was whether material has been of use or interest to us in our research; questions from students and the need to find clear explanations, exceptional writings found in other textbooks and papers, have also been determining factors. Some of the material has been included for its practicality, and some for its outstanding beauty.

In the ongoing debate regarding the teaching of mathematics to engineers, recent proposals suggest using "just in time" mathematics: provide the mathematical concept only when the need for it arises in the solution of an engineering problem. This approach has arisen as a response to the charge that mathematical pedagogy has been motivated by a "just in case" approach: we'll teach you all this stuff just in case you ever happen to need it. In reality, these approaches are neither fully desirable nor achievable, potentially lacking rigor and depth on the one hand, and motivation and insight on the other. As an alternative, we hope that the presentation in this book is "justified," so that the level of mathematics is suited to its application, and the applications are seen in conjunction with the concepts.

Programs

The algorithms found throughout the text, written in MATLAB, allow the reader to see how the concepts developed in the text might be implemented, allow easy exploration of the concepts (and, sometimes, of the limitations of the theory), and provide a useful library of core functionality for a variety of signal processing research applications. With thorough theoretical and applied discussion surrounding each algorithm, this is not simply a book of recipes; raw ingredients are provided to stir up some interesting stews!

In most cases, the algorithms themselves have not been presented in the text. Instead, an icon (as shown below)

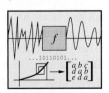

is used to indicate that the text an algorithm is to be found on the included CD-ROM (in some instances the algorithm consists of several related files).

In the interest of brevity, type-checking of arguments has not been incorporated into the functions. Otherwise, we believe that all of the code provided works, at least to produce the examples described in the book. Of course, information regarding program bugs, fixes, and improvements is always welcome. Nevertheless, we are required to make the standard disclaimer of warranty which can be found in its entirety on pg. 946.

Readers are free to use the programs or any derivatives of them for any scientific purpose, with appropriate citation of this book. Updated versions of the programs, and other information, can be found at the following website: www.prenhall.com/moon

Exercises

The exercises found at the end of each chapter are loosely divided into sections, but it may be necessary to draw from material in other sections (or even other chapters) in order to solve some of the problems.

There are relatively few merely numerical exercises. With the computer performing automated computations in many cases, simply running numbers doesn't provide an informative exercise. Readers are encouraged, of course, to play around with the algorithms to get a sense of how they work. Insight frequently can be gained on some difficult problems by trying several related numerical approaches.

The intent of the exercises is to engage the reader in the development of the theory in the book. Many of the exercises require derivations of results presented in the chapters, or proofs of some of the lemmas and theorems; other exercises require programming an extension or modification of a MATLAB algorithm presented in the chapter; and still others lead the student through a step-by-step process leading to some significant result (for example, a derivation of Gaussian quadrature or linear prediction theory, extension of inverses of Toeplitz matrices, or another derivation of the Kalman filter). As students work through these exercises, they should develop skill in organizing their thinking (which can help them to approach other problems) as well as acquire background in a variety of important topics.

Most of the exercises require a fair degree of insight and effort to solve—students should plan on being challenged. Wherever possible, students are encouraged to interact with the computer for computational assistance, insight, and feedback.

A solutions manual is available to instructors who have adopted the book for classroom use. Not only are solutions provided but, in many cases, MATLAB and MATHEMATICA™ code is also provided, indicating how a problem might be approached using the computer. Solutions to selected exercises can also be found on the CD-ROM.

Courses of study

There is clearly more information in this book than can be covered in a single semester, or even a full year. Several different courses of study could be devised based on this material, giving instructors the opportunity to choose the material suitable for the needs and development of their students. For example, depending on the focus of the class, instructors might choose to skip completely the numerical aspects of algorithms or, conversely, make them a focus of the course.

Several possible course options are described in the following list.

1. In a straightforward overview, the material in the first two parts is regarded as the foundation upon which the major concepts of signal processing are built. The first part provides a review of signal models and representations (e.g., difference equations, transfer functions, state-space form), and introduces several important signal processing problems, such as spectrum estimation and system identification. The second part provides a thorough foundation in linear algebra, working from an undergraduate level up through several applications. Selections from these first two parts, with possible additions from the first appendix on mathematical fundamentals, would make a solid single-semester course in "mathematical methods for signals and systems." A possible course sequence might be as follows:

   - Move fairly quickly through Chapter 1 (with sections 1.8 and 1.10 optional, depending on interest (1–2 weeks)).
   - In Chapter 2, move quickly to the vector-space concepts, then focus on the concept of orthogonality. It may be useful to skip the more technical sections associated with infinite-dimensional vector spaces (for example, sections 2.1.2, 2.1.3, and 2.16). (2 weeks)
   - Spend time in Chapter 3 on least-squares and minimum mean-square filtering and estimation concepts, and the dual approximation problem (sections 3.1–3.14). (2–3 weeks) Depending on interest, examine either wavelet transforms or digital communications from this geometric viewpoint. (1 week)
   - In Chapter 4, focus on sections 4.1–4.5 to get the geometry of the operators, 4.9 for a return to the least-squares idea, and 4.10 for practical computation issues. Introduce the recursive least square filter in section 4.11, and visit partitioned matrix inverses in section 4.12. (2–3 weeks)
   - In Chapter 5, focus on sections 5.2 and 5.3. The QR factorization, in particular, is a foundation for many signal processing algorithms. If a numeric implementation viewpoint is not of interest, then material after section 5.3.5 may be omitted. (2–3 weeks)
   - Sections 6.1–6.5 constitute the principal theory of chapter 6. After these sections have been covered, applications may be drawn from sections 6.7–6.12, with those in 6.8 and 6.9 probably of the most interest. If a numeric focus is desired, section 6.14 may be covered. (2–3 weeks)
   - The theory of the SVD in sections 7.1–7.5 should be covered, followed by a subset of applications from sections 7.6–7.9. (2–3 weeks)
   - Topics related to special matrices (with special emphasis on Toeplitz matrices) can fill any remaining time.

2. Chapters 10–14 would fit well into a first course on detection and estimation, especially when supplemented by some of the material on linear algebra (such as eigendecompositions and the SVD).

3. This book can be the basis for a one-semester tools course that selects topics from parts I, II, and III. Assuming prior familiarity with continuous-time and discrete-time systems, topics in such a course could include the following.

   (a) The multivariate Gaussian density (section 1.7). (< 1 week)

(b) Essential vector-space notions (sections 2.1–2.6, 2.10, 2.13, 2.14–2.15). (2 weeks)

(c) Applications of vector-space concepts; for example, least-squares and minimum mean-squares filtering (sections 3.1, 3.2, 3.4, 3.8–3.12). (3 weeks)

(d) Matrix factorizations (sections 5.2 and 5.3, no numeric discussion). (<1 week)

(e) Singular value decompositions (sections 7.1–7.3, 7.5), with some applications (such as section 7.6). (2 weeks)

(f) Introduction to detection and estimation (sections 10.1–10.3, 10.5–10.6). (1 week)

(g) Detection theory (sections 11.1–11.6). (3 weeks)

(h) Estimation theory (sections 12.1–12.2, 12.4–12.6). (2 weeks)

(i) Kalman filtering (sections 13.1, 13.2, or 13.3). (1 week)

4. A course in "iterative methods for signal processing" could focus on chapters in part IV. The course material could well be accompanied by a student research project.

5. A course in "methods of optimization for signal processing" could focus on chapters in part V.

6. Yet another alternative is a wrap-up course for students in the signals and systems area, who are familiar with their topic areas but wish to sharpen their analytical skills. This course could be similar to the first one outlined, with less time spent in Chapter 1 and more time spent examining numerical implementations. Topics from the last parts of the book could also be selected.

Acknowledgments

> If I have seen further it is by standing on ye shoulders of Giants.

> I do not know what I may appear to the world; but to myself I seem to have been only like a boy, playing on the sea-shore, and diverting myself in now and then finding a smoother pebble or a prettier shell than ordinary, whilst the great ocean of truth lay undiscovered before me.
>
> — *Isaac Newton*

For providing a challenging and stimulating environment in which the development of this book could occur, I offer my appreciation to the late Dr. Richard Harris, Chairman of the Electrical and Computer Engineering Department at Utah State University. The suggestions, comments, and much-needed criticism of the many reviewers has strengthened the presentation considerably and for that I am grateful. Paul Becker and his erstwhile group at Addison Wesley Longman provided friendly encouragement, and it has been a pleasure working with them. The production staff at Interactive Composition Corporation have been monumentally productive, and I thank them for making this all come together.

For stimulating and baffling conversations and questions, I thank my students. I am grateful for comments, suggestions, encouragements, and advice from friends and colleagues who have read portions of the text in progress.

The material in part on detection and estimation theory comes from Wynn Stirling, and I am grateful and honored that his notes can be incorporated into this book, and for the opportunity to collaborate with him.

Despite the assistance, review, oversight, and editing of so many people, I have no doubt that errors still lurk undetected. These are mine alone, and it is my parent hope that the reader of this book will discover them and bring them to my attention, so that they all may be eradicated.

To those who have played on the shores of knowledge and found so many brilliant shells, I extend enthusiastic appreciation. I also thank those who, by their writing and interpretations, by their teaching and dedication, have extended my views and helped me climb up toward the shoulders of the giants. My parents instilled in me the curiosity and wonder about the world around me: to my mother I give thanks for an insatiable curiosity about life; to my father, thanks for providing the pattern.

My most heartfelt thanks go to Barbara, who more than anyone has shouldered with me the burden of seeing this through and has shared me with this book. She also appreciates the need to know. Thanks also to our children—Leslie, Kyra, Kaylie, Jennie, Kiana, and Spencer—who provide more than sufficient reason for joy in my life.

— T.K.M.

Mathematical Methods and Algorithms
for
Signal Processing

Part I

Introduction and Foundations

In this first part, we set the stage for what follows by presenting some commonly-used signal processing models for applications developed throughout the book. We also provide some background on proofs.

Chapter 1

Introduction and Foundations

> There is full-time employment for all simply in exploring the world without destroying it, and by the time we begin to understand something of its marvelous richness and complexity, we'll also begin to see that it does have uses we never suspected ...
>
> — *Hugh Nibley*
>
> At this point I am reminded of a paper described in Littlewood's *Mathematician's Miscellany*. The paper began "The aim of this paper is to prove ..." and it transpired only much later that this aim was not achieved (the author hadn't claimed that it was). What I have outlined above is the content of a book the realization of whose plan and the incorporation of whose details would perhaps be impossible; what I have written is a second or third draft of a preliminary version of this book.
>
> — *Michael Spivak*
> A Comprehensive Introduction to Differential Geometry

1.1 What is signal processing?

The scope of signal processing far exceeds the capability of any single book to contain it. Though the subject has grown so broad as to obviate a perfect and precise definition of what is entailed in it, certain concepts must be considered indispensable for rudimentary understanding. Certainly, signal processing includes the material taught in traditional, DSP courses (see, e.g., [262, 244]), such as transforms of many varieties (Z, Laplace, Fourier, etc.) and the concepts of frequency response, impulse response, and convolution, for both deterministic and random signals. It also includes the basic concepts of filtering and filter design. These concepts are assumed as a background to this text and are used, as necessary, throughout the text. Traditional areas in signal processing include (as taken from the IEEE *Transactions on Signal Processing* classifications): filter design, fast filtering algorithms, time-frequency analysis, multi-rate filters, signal reconstruction, adaptive filters, nonlinear signals and systems, spectral analysis, and extensions of these concepts to multidimensional systems. These topics are employed in a variety of application areas. Implementation, in hardware or software, is also an important facet of signal processing. Providing a thorough coverage of these topics alone requires multiple volumes.

But, in the view of this book, signal processing has an even greater reach, because of its influence on related disciplines. Signal processing overlaps with the study traditionally known as *controls*, since control ultimately involves producing a signal based upon measured output of a plant by means of some processing upon that signal. Before a system can be controlled, the particular parameters of that system usually must be determined, so *system identification* is an aspect of signal processing. This in turn relates to *spectrum estimation* and all of its applications. Signal processing has strong ties to *communications theory* and, recently, especially to digital communication, since the capabilities of modern communication systems are the result of the signal processing performed within them. Related to digital communication are questions of *detection* and *estimation* theory: how to get the best information out of signals measured in the presence of random noise. Detection and estimation theory in turn relate to *pattern recognition*. Digital communication also spills over into the areas of *information theory* and *coding theory*. System identification and estimation theory treat questions of solving overdetermined systems of equations that, in turn, have application in *tomography*. These, in turn, have some bearing on questions of approximation and smoothing of signals. If a treatment of fundamental signal processing topics requires several volumes, then inclusion of these latter topics requires a library.

Signal processing covers a large territory. However, there is a common thread among all the areas mentioned: they all involve a fair degree of mathematical sophistication, and in both theory and practice assume an analytical and a computational component. Most of these areas share a large overlap in conceptual content. We propose the following as a tentative definition of signal processing, at least for the purposes of this book.

Definition 1.1 Signal processing is that area of applied mathematics that deals with operations on or analysis of signals, in either discrete or continuous time, to perform useful operations on those signals. □

With its focus on "applied mathematics," this book neglects several important aspects of signal processing, including hardware design and implementation on signal processing chips. "Useful operation" is deliberately left ambiguous. Depending upon the application, a useful operation could be control, data compression, data transmission, denoising, prediction, filtering, smoothing, deblurring, tomographic reconstruction, identification, classification, or a variety of other operations.

The primary intent of this book is **to present a treatment of relevant mathematics such that students and practitioners of signal processing and related fields are able to read, apply, and ultimately contribute to the literature in a variety of areas of signal processing research and practice**. The intent is not to explore pure mathematics, however, but rather to provide a mathematical modicum sufficient to explain and explore the more important mathematical paradigms used in signal processing *algorithms*. A student with a background from this book should be able to move expeditiously to a particular area of interest and begin making effective progress in the specialized literature of that area. We have endeavored to maintain a precarious balance: purists in mathematics will find some of the analytical methods deficient, while pragmatists will argue that there are far too many equations. To use a garage analogy, we have provided enough information to get under the hood of the car, taking apart for examination many of the engine components, but without getting into detail at the level of metallurgical phenomena. Such minute investigations are best conducted after the student understands how the car operates.

In addition to the primary goal of this book, there are two others. First, to develop within the student a degree of "mathematical maturity." The student with this maturity will (it is hoped) be able to organize effective approaches of his/her own to a variety of problems. This maturity will be developed by working problems, following and doing proofs, and writing and running programs. Second, the book is intended as a useful reference, with reference material gathered on several areas in signal processing, such as derivatives, linear algebra, optimization, inequalities, etc.

This statement of intent should make clear what this book is not. There are several very good books available on application areas in signal processing, such as spectrum estimation, adaptive filtering, array processing, and so on. This book does not choose any of those particular areas as its focus. Thus, while many different techniques of spectrum estimation will be presented as applications of the techniques discovered, issues central to the study of spectrum estimation (such as comparisons of the different techniques in terms of spectral resolution, bias, etc.) are not presented here. Similarly, the major paradigms of adaptive filtering are presented as applications of other important concepts (e.g., least-squares and minimum mean-squares, and recursive computation of matrix inverses), but a thorough treatment of the convergence of the filters is avoided. Rather than focusing on one particular area of research interest, this book presents the tools that are used in these research areas, enabling the interested student to move into a variety of different areas.

1.2 Mathematical topics embraced by signal processing

So what does a signal processor—that is, an individual who wants to design signal processing algorithms, not the specialized microprocessor that might be used to implement the algorithms—need to know, to be effective? Depending on the problem, several mathematical tools can be employed.

Linear signals and systems, and transform theory These topics, core to many undergraduate and introductory graduate courses, are assumed as background to this book. Familiarity with both continuous- and discrete-time systems is assumed (although a review of some topics is provided in section 1.4).

Probability and stochastic processes This is a critically important area that is also assumed as background. Students should be acquainted with probability, and have had a course in stochastic processes as a prerequisite to this book. Probability is an important tool, and students are advised to continue sharpening their skills with it. A brief review of important topics in stochastic processes is provided in appendix D.

Programming A signal processor must know how to program in at least one high-level language. In most cases, signal processing ultimately boils down to a software or hardware implementation on some kind of computing platform. This requires deployment of the concept, simulation, and testing, all usually software-related activities. An understanding of basic programming concepts such as variables, program flow, recursion, data structures, and program complexity, is assumed.

Calculus and analysis These foundation concepts occur repeatedly in the signal processing literature. A broad and shallow coverage of analysis appears in appendix A.

Vector spaces and linear algebra While every undergraduate engineer has some exposure to linear algebra, these topics are so important to signal processing that additional exposure is critical. Many of the basic concepts are reviewed in this book, with an eye toward applications in signal processing. Because of its importance, chapters 2 through 9 are devoted largely to linear algebra and its applications.

Numerical methods With the increasing penetration of computers into engineering culture there is, paradoxically, a decrease in many students' exposure to numerical methods. And yet, a significant portion of signal processing consists of nothing more than numerical methods applied to a particular set of problems involving signals. Many of the techniques described in this book are borrowed from the numerical methods literature.

Functional analysis In signal processing, a signal is a function. The tools from functional analysis provide a framework from which to view the signal, leading the way to powerful signal transforms and signal spaces in digital communications. In this book we present concepts from functional analysis in the context of vector spaces, particularly in chapters 2 and 3.

Optimization A common theme running through many signal processing applications is optimization: whatever is being computed, we wish to do it in the best possible way. Or, if we cannot get to the optimal operation point in one step, we will progress toward it as we continue to process data (that is, we will adapt). Because of its ubiquity in application, in Part IV we present fundamental concepts in optimization, including constrained optimization, linear programming, and path search algorithms.

Statistical decision theory Statistical decision theory can be described as the science of making decisions in the face of random uncertainty. Such decision-making also describes what is done in many signal processing applications. The application of statistics to signal processing can be divided into two major overlapping areas, **detection theory** and **estimation theory**. Detection theory is a framework for making decisions in the presence of noise. Estimation theory provides a means of determining the value of a quantity in the presence of noise. Detection and estimation are covered in chapters 10 through 13.

Iterative methods Many signal processing methods converge to their solution after several iterations—for example, adaptive filters and neural networks. We present some basic concepts and examples of iterative methods in chapters 14 through 17.

These topics cover a very large territory. In each of these topic areas, numerous volumes have been written. Our intent is to not to provide an exhaustive treatment in each area, but to present enough information to provide a useful set of tools with broad application. Our approach is different from many other books on signal processing, in that we do not exhaustively examine a particular discipline of signal processing—for example, spectrum estimation—bringing in mathematical tools as necessary to treat issues that arise. Instead, we present the mathematical perspective first, introducing new signal processing problems and enhancing understanding of already-introduced problems as the material permits. By this means, parallels may be drawn between areas that share mathematical tools, but that are not commonly presented together.

1.3 Mathematical models

Throughout most of the remainder of this chapter, we present examples of several different models that are commonly used in signal processing. The models are roughly categorized as follows:

1. Linear signal models for discrete and continuous time, including transfer function and state space representations. Also, applications of these models to signal processing problems such as prediction, spectrum estimation, and so on.
2. Adaptive filtering models, and applications to prediction, system identification, and so forth.
3. The Gaussian random variable, including the important idea of conditioning upon an observation.
4. Hidden Markov models.

These examples illustrate some of the notation used throughout this book, and provide a starting point for several of the signal processing applications that are examined. The material here is presented partly by way of review, and partly as a partial survey and motivator of concepts to be developed throughout this book.

1.4 Models for linear systems and signals

After this introductory material, we present a discussion of proofs. The chapter ends with the development of a fast algorithm—finally, an algorithm!—for fast solution of a system of Toeplitz equations. This algorithm—more commonly discussed in the error control literature than the signal processing literature—ties together several themes of the chapter: linear systems notation, autoregressive models, algorithms, and proofs.

1.4 Models for linear systems and signals

Most of the systems treated in signal processing are assumed to be linear, a concept that should be familiar from introductory signal processing courses. We will focus principally on systems that are also time invariant; such systems are said to be linear time-invariant (LTI). Systems are divided according to whether they operate in continuous time or discrete time. In discrete time, the data associated with time t are indicated by either square brackets, such as $x[t]$, or by subscripts, such as x_t, where t is an integer. We will also employ other variables as a discrete-time index, such as n or k. For continuous-time signals, the notation $x(t)$ or x_t is commonly employed, where t is a real number. The material in this section is intended primarily as a review.

1.4.1 Linear discrete-time models

Difference equations

Let $f[t]$ denote the (scalar) input to a discrete-time linear system, and let $y[t]$ denote the (scalar) output. It is common to assume an input/output relation of the form of the difference equation

$$y[t] = -\bar{a}_1 y[t-1] - \bar{a}_2 y[t-2] - \cdots - \bar{a}_p y[t-p] + \bar{b}_0 f[t] \\ + \bar{b}_1 f[t-1] + \cdots + \bar{b}_q f[t-q]. \tag{1.1}$$

The equation is shown under general assumption of complex signals, and the bar over the coefficients denotes *complex conjugation*. (See box 1.1.) By redefining each coefficient $\bar{a}_i$ and $\bar{b}_i$ in terms of its conjugate, (1.1) could also be written without the conjugates as

$$y[t] = -a_1 y[t-1] - a_2 y[t-2] - \cdots - a_p y[t-p] + b_0 f[t] \\ + b_1 f[t-1] + \cdots + b_q f[t-q].$$

With consistent and careful use of the notation, the question of whether the coefficients are conjugated in the definition of the linear model is of no ultimate significance—the answers obtained are invariably the same. However, the bulk of signal processing literature seems to favor the conjugated representation in (1.1), and we follow that convention. Of course,

---

Box 1.1: Notation for complex quantities

We use the engineer's notation $j = \sqrt{-1}$, rather than the mathematician's i. However, in some places j will be used as an index of summation; context should make clear what is intended.

A bar over a quantity denotes *complex conjugation*. Other authors commonly indicate complex conjugation using a superscript asterisk, as a^*. However, the $\bar{a}$ notation is used in this book to indicate conjugation, since a^* is also commonly used to denote a particular value of a, such as a minimizing value, or to indicate the adjoint of a linear operator.

when the signals and coefficients are strictly real, the conjugation is superfluous and the system can also be written in the form

$$y[t] = -a_1 y[t-1] - a_2 y[t-2] - \cdots - a_p y[t-p] + b_0 f[t]$$
$$+ b_1 f[t-1] + \cdots + b_q f[t-q]$$

without the conjugates on the coefficients.

In the case of a system that is not time invariant, the coefficients may be a function of the time index t. We will assume, for the most part, constant coefficients. The relation (1.1) can be written as

$$\sum_{k=0}^{p} \overline{a}_k y[t-k] = \sum_{k=0}^{q} \overline{b}_k f[t-k], \tag{1.2}$$

with $a_0 = 1$.

In (1.2), when $p = 0$,

$$y[t] = \sum_{k=0}^{q} \overline{b}_k f[t-k]; \tag{1.3}$$

the signal $y[t]$ is called in the statistical literature a *moving average* (MA) signal, since it is formed by simply adding up (scaled versions of) the input signal over a window of $q+1$ values. The number q is the *order* of the MA signal. The signal is denoted either as *MA* or *MA(q)*. We can also write (1.3) using a convenient vector notation. Let

$$\mathbf{f}[t] = \begin{bmatrix} f[t] \\ f[t-1] \\ \vdots \\ f[t-q] \end{bmatrix} \quad \text{and} \quad \mathbf{b} = \begin{bmatrix} b_0 \\ b_1 \\ \vdots \\ b_q \end{bmatrix}.$$

Then

$$y[t] = \mathbf{b}^H \mathbf{f}[t] = \overline{(\mathbf{f}^T[t]\mathbf{b})}.$$

The vector notation used in this example is summarized in box 1.2. In equation 1.2, when $q = 0$, so that

$$y[t] = \overline{b}_0 u[n] - \sum_{k=1}^{p} \overline{a}_k y[t-k],$$

the signal y is said to (AR) an *autoregressive* be signal of order p. *Auto* because it expresses the signal in terms of itself; *regressive* in the sense that a functional relationship exists between two or more variables. An autoregressive model is denoted as *AR* or *AR(p)*. Writing

$$\mathbf{y}[t] = \begin{bmatrix} y[t-1] \\ y[t-2] \\ \vdots \\ y[t-p] \end{bmatrix} \quad \text{and} \quad \mathbf{a} = \begin{bmatrix} a_1 \\ a_2 \\ \vdots \\ a_p \end{bmatrix},$$

we can write the AR signal as

$$y[t] = \overline{b}_0 u[t] - \mathbf{a}^H \mathbf{y}[t].$$

The general form in (1.2), combining both the autoregressive and the moving average components, is called an *autoregressive moving average*, or *ARMA*, or *ARMA(p, q)*. Where all the signals are deterministic, the term DARMA (deterministic ARMA) is sometimes employed.

1.4 Models for Linear Systems and Signals

> **Box 1.2: Notation for vectors**
>
> 1. Vectors in a finite-dimensional vector space are denoted in bold font, such as **b**.
> 2. All vectors in this book are assumed to be column vectors. In some cases a vector will be typeset in horizontal format, with ^{T} (transpose) to indicate that it should be transposed. Thus we could have equivalently written
> $$\mathbf{b} = [b_0, b_1, \ldots, b_q]^T \quad \text{or} \quad \mathbf{b}^T = [b_0, b_1, \ldots, b_q].$$
> 3. In general, the ith component of a vector **b** will be designated as b_i. Whether the index i starts with 0 or 1 (or some other value) depends on the needs of the particular problem.
> 4. The notation $\mathbf{b}^H$ denotes the *Hermitian* transpose, in which **b** is transposed and its elements are conjugated:
> $$\mathbf{b}^H = [\bar{b}_0, \bar{b}_1, \ldots, \bar{b}_q].$$
>
> These rules notwithstanding, for notational convenience we will sometimes denote the vector with n elements as an n-tuple, so that
> $$\mathbf{x} = [x_1 \ \ x_2 \ \ \ldots \ \ x_n]^T \quad \text{and} \quad \mathbf{x} = (x_1, x_2, \ldots, x_n)$$
> are occasionally used synonymously. This n-tuple notation is used particularly when **x** is regarded as a point in $\mathbb{R}^n$. Furthermore, since we will generalize the concept of vectors to include functions, the math italic notation x will be used in the most general case to represent vectors, either in $\mathbb{R}^n$ or as functions.
>
> Matrices are represented with capital letters, as in A or X. The matrix I is an identity matrix. The notation **0** is used to indicate a vector or matrix of zeros, with the size determined by context. Similarly, the notation **1** is used to indicate a vector or matrix of ones, with the size determined by context.

System function and impulse response

In the interest of getting a system function that does not depend upon initial conditions, we assume that the initial conditions are zero, and take the Z-transform to obtain

$$Y(z) \sum_{k=0}^{p} \bar{a}_k z^{-k} = F(z) \sum_{k=0}^{q} \bar{b}_k z^{-k},$$

which we write as

$$Y(z) A(z) = F(z) B(z).$$

We will occasionally write the transform relationship as

$$y[t] \leftrightarrow Y(z),$$

where the particular transform intended is determined by context. We will also denote Z-transforms by

$$Y(z) = \mathcal{Z}[y[t]].$$

The *system function* is

$$H(z) = \frac{Y(z)}{F(z)} = \frac{\sum_{k=0}^{q} \overline{b}_k z^{-k}}{\sum_{k=0}^{p} \overline{a}_k z^{-k}} = \frac{\sum_{k=0}^{q} \overline{b}_k z^{-k}}{1 + \sum_{k=1}^{p} \overline{a}_k z^{-k}} = \frac{B(z)}{A(z)}. \quad (1.4)$$

This is also called (usually interchangeably) the *transfer function* of the system. We write

$$Y(z) = H(z)F(z), \quad (1.5)$$

and represent this as shown in figure 1.1. If the system is AR, then

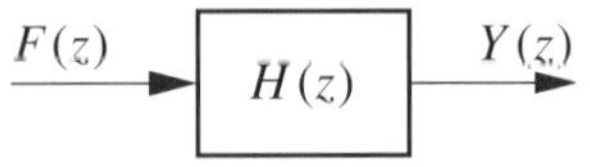

Figure 1.1: Input/output relation for a transfer function

$$H(z) = \frac{1}{1 + \sum_{k=1}^{p} \overline{a}_k z^{-k}} = \frac{1}{A(z)},$$

and $H(z)$ is said to be an *all-pole* system. If the system is MA, then

$$H(z) = \sum_{k=0}^{q} \overline{b}_k z^{-k} = B(z),$$

which is called an *all-zero* system. The corresponding difference equation (1.3) has only a finite number of nonzero outputs when the input is a delta function $f[t] = \delta[t]$, where

$$\delta[t] = \begin{cases} 1 & t = 0 \\ 0 & t \neq 0. \end{cases}$$

We will also write the delta function as δ_t. Occasionally the function $\delta[t - \tau]$ will be written as $\delta_{t,\tau}$.

A system that has only a finite number of nonzero outputs in response to a delta function is referred to as a finite impulse response (FIR) system. A system which is not FIR is infinite impulse response (IIR).

We can view signal $Y(z)$ as the output of a system with system function $H(z)$ driven by an input $F(z)$. Taking the inverse Z-transform of (1.5), and recalling the convolution property (multiplication in the transform domain corresponds to convolution in the time domain) we obtain

$$y[t] = \sum_{k=-\infty}^{\infty} u[k]h[t-k],$$

where $h[t]$, the impulse response, is the inverse transform of $H(z)$.

To compute the inverse transform of $H(z)$, we first factor $H(z)$ into monomial factors using the roots of the numerator and denominator polynomials,

$$H(z) = \frac{\overline{b}_0 \prod_{k=1}^{q}(1 - z_i z^{-1})}{\prod_{k=1}^{p}(1 - p_i z^{-1})} = \frac{B(z)}{A(z)},$$

where the z_i are the nonzero roots of $B(z)$ (called the *zeros* of the system function) and the p_i are the nonzero roots of $A(z)$ (called the *poles* of the system function). In this form, we observe that if a pole is equal to a zero, the factors can be canceled out of both the numerator and denominator to obtain an equivalent transfer function. A word of caution: even though terms may cancel from the numerator and denominator as seen from the transfer function, the physical components that these terms model may still exist and could introduce difficulty.

1.4 Models for Linear Systems and Signals

A system with the smallest degree numerator and denominator is said to be a *minimal system*.

Example 1.4.1 The system function

$$H(z) = \frac{1 - .7z^{-1} + .12z^{-2}}{1 - .5z^{-1} + .06z^{-2}}$$

can be factored as

$$H(z) = \frac{(1 - .3z^{-1})(1 - .4z^{-1})}{(1 - .2z^{-1})(1 - .3z^{-1})} = \frac{1 - .4z^{-1}}{1 - .2z^{-1}}.$$

Thus the $H(z)$ is not a minimal realization. □

Partial Fraction Expansion (PFE)

Assuming for the moment that the poles are unique (no repeated poles) and that $q < p$, then, by partial fraction expansion (PFE), the system function can be expressed as

$$H(z) = \sum_{k=1}^{p} \frac{N_k}{1 - p_k z^{-1}}, \qquad (1.6)$$

where

$$N_k = H(z)(1 - p_k z^{-1})|_{z=p_k}.$$

Taking the causal inverse Z-transform of (1.6), we obtain

$$h[t] = \sum_{k=1}^{p} N_k (p_k)^t \qquad t \geq 0.$$

The functions p_k^n are the natural modes of the system $H(z)$. Clearly, for the causal modes to be bounded in time, we must have $|p_k| \leq 1$. In general, the output of a linear time-invariant system is the sum of the natural modes of the system plus the input modes of the system.

Example 1.4.2 Let

$$H(z) = \frac{1 - .3z^{-1}}{1 - 1.1z^{-1} + .3z^{-2}} = \frac{1 - .3z^{-1}}{(1 - .5z^{-1})(1 - .6z^{-1})}.$$

Then, a partial fraction expansion is

$$H(z) = \frac{-2}{1 - .5z^{-1}} + \frac{3}{1 - .6z^{-1}}.$$

The impulse response is

$$h[t] = [(-2)(.5)^t + 3(.6)^t] u[t],$$

where $u[t]$ is the unit-step function,

$$u[t] = \begin{cases} 1 & t \geq 1 \\ 0 & t < 0. \end{cases}$$

□

To compute the PFE when $q \geq p$, the ratio of polynomials is first divided out. When there are repeated poles, somewhat more care is required. For example, a root repeated r

times, as in
$$H(z) = \frac{B(z)}{(1-pz^{-1})^r},$$
gives rise to the partial fraction expansion
$$H(z) = \frac{k_0}{(1-pz^{-1})^r} + \frac{k_1}{(1-pz^{-1})^{r-1}} + \cdots + \frac{k_{r-1}}{(1-pz^{-1})}, \quad (1.7)$$
where[1]
$$k_j = \frac{1}{p^j j!}(-1)^j \frac{d^j}{d(z^{-1})^j}(1-pz^{-1})^r H(z). \quad (1.8)$$

The inverse Z-transform corresponding to (1.7) is of the form
$$h[t] = [c_0 p^t + c_1 t p^t + \cdots + c_{r-1} t^r p^t]u[t],$$
where the coefficients $\{c_i\}$ are linearly related to the PFE coefficients $\{k_i\}$.

Using computer software, such as the `residue` or `residuez` command in MATLAB, is recommended to compute partial fraction expansions.

Example 1.4.3 Let
$$H(z) = \frac{3 + 2.4z^{-1} + 6z^{-2}}{1 - .7z^{-1} + .1z^{-2}}.$$

We desire to find the impulse response $h[t]$. Since the degree of the numerator is the same as the degree of the denominator, we divide, then find the partial fraction expansion.
$$H(z) = 60 + \frac{44.4z^{-1} - 57}{(1 - .2z^{-1})(1 - .5z^{-1})}$$
$$= 60 + \frac{-110}{1 - .2z^{-1}} + \frac{53}{1 - .5z^{-1}},$$
then,
$$h[t] = 60\delta[t] - 110(.2)^t + 53(.5)^t \qquad t \geq 0. \qquad \square$$

1.4.2 Stochastic MA and AR models

In stochastic MA and AR models, the input $f[t]$ is assumed to be a white discrete-time random process that is usually zero mean. (The reader is encouraged to review the concepts of random processes summarized in appendix D.) The input coefficient b_0 is set to 1, with the input power determined by the variance of the signal. Thus,
$$E[f[t]] = 0 \qquad \text{for all } t$$
and
$$E[f[t]\overline{f}[s]] = \begin{cases} \sigma_f^2 & t = s \\ 0 & \text{otherwise}. \end{cases}$$

---

[1] The symbol j here does not represent $\sqrt{-1}$. In instances where confusion is unlikely, we may use j as an index value.

1.4 Models for Linear Systems and Signals

Autocorrelation function

Signal processing often involves comparing two signals; one means of comparison is by means of correlation. When a signal is compared with itself, the correlation is called autocorrelation. For stochastic signals, we define the autocorrelation of a zero-mean (wide-sense) stationary signal $y[t]$ as

$$r_{yy}[l-k] = E[y[t-k]\bar{y}[t-l]], \qquad (1.9)$$

or, equivalently, $r_{yy}[k] = E[y[t]\bar{y}[t-k]]$. The autocorrelation function has the property that

$$r_{yy}[k] = \bar{r}_{yy}[-k]. \qquad (1.10)$$

For even random processes, $r_{yy}[k] = r_{yy}[-k]$; a function that has this property is said to be *even*.

For the MA process

$$y[t] = f[t] + \bar{b}_1 f[t-1] + \cdots \bar{b}_q f[t-q],$$

it is straightforward to show that the autocorrelation function is

$$r_{yy}[k] = \sigma_f^2 \sum_l b_{l-k} \bar{b}_l. \qquad (1.11)$$

For the AR model

$$y[t] + \bar{a}_1 y[t-1] + \cdots + \bar{a}_p y[t-p] = f[t], \qquad (1.12)$$

multiply both sides by $\bar{y}[t-l]$ and take expectations, to obtain

$$E\left[\sum_{k=0}^{p} \bar{a}_k y[t-k]\bar{y}[t-l]\right] = E[f[t]\bar{y}[t-l]]. \qquad (1.13)$$

We recognize that $E[y[t-k]\bar{y}[t-l]] = r_{yy}[l-k]$, and that the RHS

$$E[f[t]\bar{y}[t-l]] = 0$$

for $l > 0$, since $f[t]$ is a white-noise process. Then, using the fact that $a_0 = 1$, we can write

$$r_{yy}[l] = -\bar{a}_1 r_{yy}[l-1] - \bar{a}_2 r_{yy}[l-2] - \cdots - \bar{a}_p r_{yy}[l-p] \qquad \text{for } l > 0. \qquad (1.14)$$

This difference equation for the autocorrelation is similar to the equation for the original difference equation in (1.12). Stacking (1.14) for $l = 1, 2, \ldots, p$, we obtain

$$\begin{bmatrix} r_{yy}[0] & r_{yy}[-1] & \cdots & r_{yy}[-(p-1)] \\ r_{yy}[1] & r_{yy}[0] & \cdots & r_{yy}[-(p-2)] \\ \vdots & & & \\ r_{yy}[p-1] & r_{yy}[p-2] & \cdots & r_{yy}[0] \end{bmatrix} \begin{bmatrix} -\bar{a}_1 \\ -\bar{a}_2 \\ \vdots \\ -\bar{a}_p \end{bmatrix} = \begin{bmatrix} r_{yy}[1] \\ r_{yy}[2] \\ \vdots \\ r_{yy}[p] \end{bmatrix}. \qquad (1.15)$$

Conjugating both sides using (1.10), we obtain

$$\begin{bmatrix} r_{yy}[0] & r_{yy}[1] & \cdots & r_{yy}[(p-1)] \\ \bar{r}_{yy}[1] & r_{yy}[0] & \cdots & r_{yy}[(p-2)] \\ \vdots & & & \\ \bar{r}_{yy}[p-1] & \bar{r}_{yy}[p-2] & \cdots & r_{yy}[0] \end{bmatrix} \begin{bmatrix} -a_1 \\ -a_2 \\ \vdots \\ -a_p \end{bmatrix} = \begin{bmatrix} \bar{r}_{yy}[1] \\ \bar{r}_{yy}[2] \\ \vdots \\ \bar{r}_{yy}[p] \end{bmatrix}. \qquad (1.16)$$

These equations are known as the *Yule–Walker equations*. We commonly write (1.16) as

$$R\mathbf{w} = \mathbf{r},$$

where

$$\mathbf{w} = [-a_1 \quad -a_2 \quad \cdots \quad -a_p]^T \qquad \mathbf{r} = [\bar{r}_{yy}[1] \quad \bar{r}_{yy}[2] \quad \cdots \quad \bar{r}_{yy}[p]].$$

The matrix R is said to be the *autocorrelation matrix* of y. We will have considerable to say about the properties of R and algorithms that operate on it. For now, we make the following observations.

1. R is *Hermitian symmetric*, which means that

$$R = R^H.$$

We will see that this means that the eigenvalues of R are real and the eigenvectors corresponding to distinct eigenvalues are orthogonal. If R is real, then R is *symmetric*: $R^T = R$.

2. R is a *Toeplitz matrix*, which means that R is constant along the diagonals. If r_{ij} denotes the i, jth element of R, then

$$r_{jj} = r_{i-j};$$

which is to say, the element of R depends only on the difference between the index values. We shall see that the Toeplitz structure of R leads to efficient algorithms for solving equations similar to the Yule–Walker equations.

Realizations

A block diagram, or *realization*, of (1.2) can be easily derived. The realization presented here is known in the control literature as the *controller canonical form*. Write the system function as

$$H(z) = \frac{Y(z)}{W(z)} \frac{W(z)}{F(z)} = \left(\sum_{k=0}^{q} \bar{b}_k z^{-k} \right) \left(\frac{1}{1 + \sum_{k=1}^{p} \bar{a}_k z^{-k}} \right) = H_1(z) H_2(z), \qquad (1.17)$$

where the signal $W(z)$ has been artificially introduced. From the transfer function $H_2(z)$ we get the relationship

$$W(z) \left(1 + \sum_{k=1}^{p} \bar{a}_k z^{-k} \right) = F(z), \qquad (1.18)$$

corresponding to the difference equation

$$w[t] + \bar{a}_1 w[t-1] + \cdots + \bar{a}_p w[t-p] = f[t],$$

or,

$$w[t] = f[t] - \bar{a}_1 w[t-1] - \bar{a}_2 w[t-2] - \cdots - \bar{a}_p w[t-p].$$

A block diagram of a realization of (1.18) is shown in figure 1.2. From $H_1(z)$ in (1.17), we have

$$Y(z) = W(z) B(z),$$

with the corresponding difference equation

$$y[t] = \bar{b}_0 w[t] + \bar{b}_1 w[t-1] + \cdots + \bar{b}_q w[t-q].$$

1.4 Models for Linear Systems and Signals

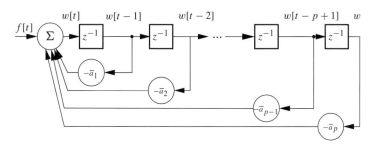

Figure 1.2: Realization of the AR part of a transfer function

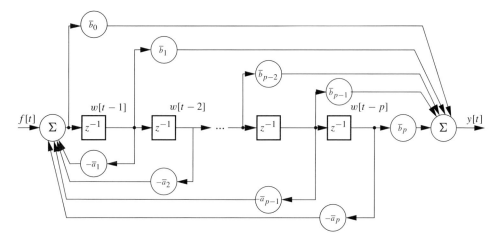

Figure 1.3: Realization of a transfer function

This realization (drawn assuming that $p = q$) is shown in figure 1.3. We explore other possible realizations in the exercises.

State-space form

Consider the block diagram in figure 1.4, in which the outputs of the delay blocks are labeled $x_1, x_2, \ldots, x_p$, from right to left. From this block diagram, we obtain the following equations:

$$x_1[t+1] = x_2[t]$$
$$x_2[t+1] = x_3[t]$$
$$\vdots$$
$$x_{p-1}[t+1] = x_p[t] \qquad (1.19)$$
$$x_p[t+1] = f[t] - \overline{a}_1 x_p[t] - \overline{a}_2 x_{p-1}[t] - \cdots - \overline{a}_{p-1} x_2[t] - \overline{a}_p x_1[t]$$
$$y[t] = \overline{b}_p x_1[t] + \overline{b}_{p-1} x_2[t] + \cdots + \overline{b}_2 x_{p-1}[t] + \overline{b}_1 x_p[t]$$
$$\qquad + \overline{b}_0 (f[t] - \overline{a}_1 x_p[t] - \overline{a}_2 x_{p-1}[t] - \cdots - \overline{a}_p x_1[t]).$$

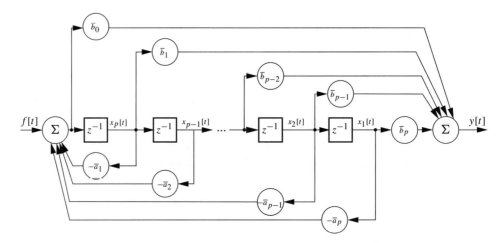

Figure 1.4: Realization of a transfer function with state-variable labels

Observe that the direct connection from input f to output y is via b_0. The variables $x_1, x_2, \ldots, x_p$ are the **state variables**. Let $\mathbf{x}[t]$ be the **state vector**

$$\mathbf{x}[t] = \begin{bmatrix} x_1[t] \\ x_2[t] \\ \vdots \\ x_p[t] \end{bmatrix}.$$

We also introduce the vectors

$$\mathbf{b} = \underbrace{[0, 0, \ldots, 0, 1]^T}_{p \text{ elements}}$$

$$\mathbf{c} = \begin{bmatrix} \bar{b}_p - \bar{b}_0 \bar{a}_p \\ \bar{b}_{p-1} - \bar{b}_0 \bar{a}_{p-1} \\ \vdots \\ \bar{b}_1 - \bar{b}_0 \bar{a}_1 \end{bmatrix} \quad \text{and} \quad d = \bar{b}_0$$

and the matrix

$$A = \begin{bmatrix} 0 & 1 & 0 & 0 & \cdots & 0 & 0 \\ 0 & 0 & 1 & 0 & \cdots & 0 & 0 \\ \vdots & & & & & & \\ 0 & 0 & 0 & 0 & \cdots & 0 & 1 \\ -\bar{a}_p & -\bar{a}_{p-1} & -\bar{a}_{p-2} & -\bar{a}_{p-3} & \cdots & -\bar{a}_2 & -\bar{a}_1 \end{bmatrix}. \quad (1.20)$$

If $b_0 = 0$, then $\mathbf{c}$ is

$$\mathbf{c}^T = [\bar{b}_p, \bar{b}_{p-1}, \ldots, \bar{b}_1],$$

which explicitly displays the numerator coefficients of $H(z)$. The equations in (1.19) can be written using these definitions as

$$\begin{aligned} \mathbf{x}[t+1] &= A\mathbf{x}[t] + \mathbf{b} f[t] \\ y[t] &= \mathbf{c}^T \mathbf{x}[t] + d f[t]. \end{aligned} \quad (1.21)$$

1.4 Models for Linear Systems and Signals

An equation of the form (1.21) is in *state-space* form. The system is denoted as $(A, \mathbf{b}, \mathbf{c}^T, d)$ or, when $d = 0$, as $(A, \mathbf{b}, \mathbf{c}^T)$. The particular form of the state-space system in (1.21) is called the controller form. The form of the matrix A, with ones above the diagonal and coefficients on the last row, is called a first *companion* matrix.

System transformations; similar matrices

The state-variable representation is not unique. In fact, an infinite number of possible realizations exist which are mathematically equivalent, although not necessarily identical in physical operation. We can create a new state-variable representation by letting $\mathbf{x} = T\mathbf{z}$ for any invertible $p \times p$ matrix T. Then (1.21) becomes

$$T\mathbf{z}[t+1] = AT\mathbf{z}[t] + \mathbf{b}f[t]$$
$$y[t] = \mathbf{c}^T T\mathbf{z} + df[t],$$

which can be written as

$$\begin{aligned}\mathbf{z}[t+1] &= \overline{A}\mathbf{z}[t] + \overline{\mathbf{b}}f[t] \\ y[t] &= \overline{\mathbf{c}}^T \mathbf{z}[t] + \overline{d} f[t],\end{aligned} \quad (1.22)$$

where

$$\overline{A} = T^{-1}AT \qquad \overline{\mathbf{b}} = T^{-1}\mathbf{b} \qquad \overline{\mathbf{c}} = T^T \mathbf{c} \qquad \overline{d} = d.$$

(The bar does not indicate conjugation in this instance.) Matrices A and $\overline{A}$ that are related as $\overline{A} = T^{-1}AT$ are said to be *similar*. It is straightforward to show that the system $(\overline{A}, \overline{\mathbf{b}}, \overline{\mathbf{c}}^T, \overline{d})$ has the same input/output relationships (dynamics and transfer function) as does the system $(A, \mathbf{b}, \mathbf{c}^t, d)$—which means, as we shall see, that A and $\overline{A}$ have the same eigenvalues.

Time-varying state-space model

When the system is time-varying, the state-space representation is

$$\begin{aligned}\mathbf{x}[t+1] &= A[t]\mathbf{x}[t] + \mathbf{b}[t]f[t] \\ y[t] &= \mathbf{c}^T[t]\mathbf{x}[t] + d[t]f[t],\end{aligned} \quad (1.23)$$

in which the explicit dependence of $(A[t], \mathbf{b}[t], \mathbf{c}^T[t], d[t])$ on the time index t is shown.

Transformed state-space model

The time-invariant state-space form can be represented using a system function. We can take the Z-transform of (1.21). The Z-transform of a vector is simply the transform of each component. We obtain the equations

$$z\mathbf{X}(z) = A\mathbf{X}(z) + \mathbf{b}F(z) \quad (1.24)$$
$$Y(z) = \mathbf{c}^T \mathbf{X}(z) + dF(z). \quad (1.25)$$

From (1.24), we obtain

$$(zI - A)\mathbf{X}(z) = \mathbf{b}F(z).$$

The matrix I is the identity matrix. Then

$$\mathbf{X}(z) = (zI - A)^{-1}\mathbf{b}F(z),$$

where $(zI - A)^{-1}$ is the matrix inverse of $zI - A$. (Matrix inverses are discussed in chapter 4.) Substituting $\mathbf{X}(z)$ into (1.25), we obtain

$$Y(z) = (\mathbf{c}^T (zI - A)^{-1}\mathbf{b} + d)F(z).$$

Since $Y(z)$ and $F(z)$ are scalar signals, we can form their ratio to obtain the system function

$$H(z) = \frac{Y(z)}{F(z)} = (\mathbf{c}^T(zI - A)^{-1}\mathbf{b} + d). \quad (1.26)$$

Example 1.4.4 We will go from a system function to state-space form, and back. Let

$$H(z) = \frac{3 + 2z^{-1} + 4z^{-2}}{1 + 3z^{-1} + 5z^{-2}}.$$

In some literature, it is common to eliminate negative powers of z in the system functions. This can be done by multiplying by z^2/z^2:

$$H(z) = \frac{3z^2 + 2z + 4}{z^2 + 3z + 5}.$$

Placing the system in controller form, we have

$$\mathbf{b} = \begin{bmatrix} 0 \\ 1 \end{bmatrix} \quad \mathbf{c} = \begin{bmatrix} 4 - (3)(5) \\ 2 - (3)(3) \end{bmatrix} = \begin{bmatrix} -11 \\ -7 \end{bmatrix}$$

$$A = \begin{bmatrix} 0 & 1 \\ -5 & -3 \end{bmatrix} \quad d = 3.$$

To return to a transfer function, we first compute

$$(zI - A) = \begin{bmatrix} z & -1 \\ 5 & z+3 \end{bmatrix}$$

and

$$(zI - A)^{-1} = \frac{1}{z(z+3) + 5} \begin{bmatrix} z+3 & 1 \\ -5 & z \end{bmatrix}.$$

The inverse of a 2×2 matrix is

$$\boxed{\begin{bmatrix} a & b \\ c & d \end{bmatrix} = \frac{1}{ad - bc} \begin{bmatrix} d & -b \\ -c & a \end{bmatrix}}$$

Then, using (1.26), we obtain

$$H(z) = \frac{1}{z^2 + 3z + 5}[-11, -7]\begin{bmatrix} z+3 & 1 \\ -5 & z \end{bmatrix}\begin{bmatrix} 0 \\ 1 \end{bmatrix} + d = \frac{3z^2 + 2z + 4}{z^2 + 3z + 5},$$

as expected.

To emphasize that the state-space representation is not unique, let

$$\tilde{A} = \begin{bmatrix} .5 & 4.5 \\ -1.5 & -3.5 \end{bmatrix} \quad \tilde{\mathbf{b}} = \begin{bmatrix} -1 \\ 1 \end{bmatrix} \quad \tilde{\mathbf{c}} = \begin{bmatrix} -2 \\ -9 \end{bmatrix} \quad \tilde{d} = 3.$$

This system is not in controller form. We may verify that

$$\tilde{H}(z) = (\tilde{\mathbf{c}}^T(zI - \tilde{A})^{-1}\tilde{\mathbf{b}} + \tilde{d}) = H(z). \qquad \square$$

1.4 Models for Linear Systems and Signals

Solution of the state-space difference equation

It is also possible to determine an explicit expression for the state of a system in state-variable form. It can be shown (see exercise 1.4-19) that, starting from an initial state $\mathbf{x}[0]$,

$$\mathbf{x}[t] = A^t \mathbf{x}[0] + \sum_{k=0}^{t-1} A^k \mathbf{b} f[t-1-k]. \tag{1.27}$$

The sum is simply the convolution of $A^t \mathbf{b}$ with $f[t-1]$. The output is

$$y[t] = \mathbf{c}^T A^t \mathbf{x}[0] + \sum_{k=0}^{t-1} \mathbf{c}^T A^k \mathbf{b} f[t-1-k] + df[t].$$

The quantities $\mathbf{c}^T A^k \mathbf{b}$ are called the *Markov parameters* of the system; they correspond to the impulse response of the system $(A, \mathbf{b}, \mathbf{c}^T)$.

Multiple inputs and outputs

State-space representation can be used to represent signals with multiple inputs and outputs. For example, a system might be described by

$$\mathbf{x}[t+1] = \begin{bmatrix} x_1[t+1] \\ x_2[t+1] \\ x_3[t+1] \end{bmatrix} = \begin{bmatrix} 3 & 2 & -1 \\ 1 & 2 & -5 \\ 2 & 1 & 1 \end{bmatrix} \mathbf{x}[t] + \begin{bmatrix} 2 & 1 \\ -1 & -5 \\ -1 & 1 \end{bmatrix} \begin{bmatrix} f_1[t] \\ f_2[t] \end{bmatrix}$$

$$\mathbf{y}[t] = \begin{bmatrix} y_1[t] \\ y_2[t] \end{bmatrix} = \begin{bmatrix} 2 & 4 & 6 \\ 1 & 2 & 0 \end{bmatrix} \mathbf{x}[t].$$

This system has three state variables, two inputs, and two outputs. In general, a multi-input, multioutput system is of the form

$$\begin{aligned} \mathbf{x}[t+1] &= A\mathbf{x}[t] + B\mathbf{u}[t] \\ \mathbf{y}[t] &= C\mathbf{x}[t] + D\mathbf{u}[t]. \end{aligned} \tag{1.28}$$

If there are p state variables and l inputs and m outputs, then

$$\begin{aligned} A &\text{ is } p \times p \\ B &\text{ is } p \times l \\ C &\text{ is } m \times p \\ D &\text{ is } m \times l. \end{aligned}$$

State-space systems in noise

A signal model that arises frequently in practice is

$$\begin{aligned} \mathbf{x}[t+1] &= A\mathbf{x}[t] + B\mathbf{u}[t] + \mathbf{w}[t] \\ \mathbf{y}[t] &= C\mathbf{x}[t] + D\mathbf{u}[t] + \mathbf{v}[t]. \end{aligned} \tag{1.29}$$

The signals $\mathbf{w}[t]$ and $\mathbf{v}[t]$ represent noise present in the system. The vector $\mathbf{w}[t]$ is an input to the system that represents unknown random components. For example, in modeling airplane dynamics, $\mathbf{w}[t]$ might represent random gusts of wind. The vector $\mathbf{v}[t]$ represents measurement noise. Measurement noise is a fact of life in most practical circumstances. Getting useful results out of noisy measurements is an important aspect of signal processing. It has been said that noise is the signal processor's bread and butter: without the noise, many problems would be too trivial to be of significant interest.

This book will touch on some aspects of systems in state-space form, but a thorough study of linear systems, including state-space concepts, is beyond the scope of this book. (For supplementary treatments, see the reference section at the end of this chapter.)

1.4.3 Continuous-time notation

For continuous-time signals and systems, the concepts for input/output relations, transfer functions, and state-space representations translate directly, with z^{-1} (unit delay) replaced by $1/s$ (integration). The reader is encouraged to review the discrete-time notations presented above and reformulate the expressions given, in terms of continuous-time signals. The principal difference between discrete time and continuous time arises in the explicit solution of the differential equation

$$\dot{\mathbf{x}}(t) = A(t)\mathbf{x}(t) + B(t)\mathbf{f}(t)$$
$$\mathbf{y}(t) = C(t)\mathbf{x} + D(t)\mathbf{f}(t). \tag{1.30}$$

For the time-invariant system (when (A, B, C, D) is constant), the solution is

$$\mathbf{x}(t) = e^{At}\mathbf{x}(0) + \int_0^t e^{A(t-\lambda)} B\mathbf{f}(\lambda)\,d\lambda, \tag{1.31}$$

where e^{At} is the *matrix exponential*, defined in terms of its Taylor series,

$$e^{At} = I + At + A^2\frac{t^2}{2} + A^3\frac{t^3}{3!} + \cdots, \tag{1.32}$$

where I is the *identity matrix*. (See section A.6.5 for a review of Taylor series, and section 6.2 for more on the matrix exponential.) The matrix exponential can also be expressed in terms of Laplace transforms,

$$e^{At} = \mathcal{L}^{-1}(sI - A)^{-1},$$

where $sI - A$ is known as the *characteristic matrix* of A and $\mathcal{L}[\cdot]$ denotes the Laplace transform operator,

$$\mathcal{L}[f(t)] = \int_{0^-}^{\infty} f(t)e^{-st}\,dt.$$

An interesting and fruitful connection is the following. Recall the geometric expansion

$$\frac{1}{1-x} = 1 + x + x^2 + x^3 + \cdots, \tag{1.33}$$

which converges for $|x| < 1$. This also applies to general operators (including matrices) to

$$(I - F)^{-1} = I + F + F^2 + F^3 + \cdots; \tag{1.34}$$

when $\|F\| < 1$. The notation $\|F\|$ signifies the operator norm; it is discussed in section 4.2. The expansion (1.34) is known as the Neumann expansion (see section 4.2.2). Using (1.34), the expression $(sI - A)^{-1}$ is

$$\frac{1}{s}(I + A/s + A^2/s^2 + \cdots)$$

from which the Taylor series formula (1.32) follows immediately, from the inverse Laplace transform.

For the time-invariant single-input, single-output system

$$\dot{\mathbf{x}}(t) = A\mathbf{x}(t) + \mathbf{b}f(t)$$
$$y(t) = \mathbf{c}^T\mathbf{x}(t)$$

1.4 Models for Linear Systems and Signals

the transfer function is

$$H(s) = \mathbf{c}(sI - A)^{-1}\mathbf{b}.$$

Using (1.34), we write

$$H(s) = \sum_{i=1}^{\infty} h_i s^i,$$

where $h_i = \mathbf{c}^T A^{i-1} \mathbf{b}$ are the Markov parameters of the continuous-time system.

The first term of (1.31) is the solution of the homogeneous differential equation

$$\dot{\mathbf{x}}(t) = A\mathbf{x}(t),$$

while the second term of (1.31) is the particular solution of

$$\dot{\mathbf{x}} = A(t)\mathbf{x}(t) + B(t)\mathbf{f}(t).$$

It is straightforward to show (see exercise 1.4-22) that, starting from a state $\mathbf{x}(\tau)$, the state at time t can be determined as

$$\mathbf{x}(t) = e^{A(t-\tau)}\mathbf{x}(\tau) + \int_{\tau}^{t} e^{A(t-\tau)} B \mathbf{u}(\lambda) \, d\lambda. \tag{1.35}$$

Since $e^{A(t-\tau)}$ provides the mechanism for moving from state $\mathbf{x}(\tau)$ to state $\mathbf{x}(t)$, it is called the *state-transition matrix*.

For the time-varying system (1.30), the solution can be written as

$$\mathbf{x}(t) = \Phi(t, 0)\mathbf{x}(0) + \int_{0}^{t} \Phi(t, \lambda) B(\lambda) \mathbf{u}(\lambda) \, d\lambda, \tag{1.36}$$

where $\Phi(t, \tau)$ is the state-transition matrix—not determined by the matrix exponential in the time-varying case. The function $\Phi(t, \tau)$ has the following properties:

1. $\Phi(t, t) = I$,
2. $\dfrac{\partial \Phi(t, \tau)}{\partial t} = A(t)\Phi(t, \tau)$,
3. $\Phi(t, \tau) = [\Phi(\tau, t)]^{-1}$ (the matrix inverse).

1.4.4 Issues and applications

The notation introduced in the previous sections allows us now to discuss a variety of issues of both practical and theoretical importance. Here are a few examples:

- Given a desired frequency response specification—either

$$H(e^{j\omega})$$

for discrete-time systems, or

$$H(j\omega)$$

for continuous-time systems—determine the coefficients $\{a_i\}$ and $\{b_i\}$ to meet, or closely approximate, the response specification. This is the *filter design* problem.
- Given a sequence of output data from a system, how can the parameters of the system be determined if the input signal is known? If the input signal is not known?
- Determine a "minimal" representation of a system.
- Given a signal output from a system, determine a predictor for the signal.

- Determine a means of efficiently coding (representing) a signal modeled as the output of an LTI system.
- Determine the spectrum of the output of an LTI system.
- Determine the modes of the same system.
- For algorithms of the sort just prescribed, develop computationally efficient algorithms.
- Suppose the modes of a signal are not what we want them to be; develop a means of using feedback to bend them to suit our purposes.

Examination of many of these issues is taken up at appropriate places throughout this book, with varying degrees of completeness.

Estimation of parameters; linear prediction

It may occur that a signal can be modeled as the output of a discrete-time system with system function $H(z)$, for which the parameters $\{p, q, b_0, \ldots, b_q, a_1, \ldots, a_p\}$ are not known. Given a sequence of observations $y[0], y[1], \ldots$, we want to determine, if possible, the parameters of the system. This basic problem has two major variations:

- The input $f[t]$ is deterministic and known.
- The input $f[t]$ is random.

Other complications may also be modeled in practice. For example, it may be that the output $y[t]$ is corrupted by noise, so that the data available is

$$z[t] = y[t] + e[t],$$

where $w[t]$ is a noise (or error) signal. This is a "signal plus noise" model that we will employ frequently.

In the case where the input is known and there is negligible or no measurement noise, it is straightforward to set up a system of linear equations to determine the system parameters. For the ARMA(p, q) system of (1.2), if the order (p, q) is known, a system of equations to find the unknown parameters can be set up as

$$A\mathbf{x} = \mathbf{b}, \qquad (1.37)$$

in which

$$A = \begin{bmatrix} y[p-1] & y[p-2] & \cdots & y[0] & f[p] & f[p-1] & \cdots & f[p-q-1] \\ y[p] & y[p-1] & \cdots & y[1] & f[p+1] & f[p-1] & \cdots & f[p-q] \\ \vdots & & & & & & & \\ y[N-1] & y[N-2] & \cdots & y[N-p] & f[N] & f[N-1] & \cdots & f[N-q-1] \end{bmatrix},$$

$$\mathbf{x} = \begin{bmatrix} a_1 \\ a_2 \\ \vdots \\ a_p \\ b_0 \\ b_1 \\ \vdots \\ b_q \end{bmatrix} \quad \text{and} \quad \mathbf{b} = \begin{bmatrix} y[p] \\ y[p+1] \\ \vdots \\ y[N] \end{bmatrix},$$

1.4 Models for Linear Systems and Signals

where N is large enough that there are as many equations as unknowns. When there is measurement noise in the system, N can be increased so that there are more equations than unknowns, and a least-squares solution can be computed, as discussed in chapters 3 and 5.

An important special case in this parameter estimation problem in which the input is assumed to be noise, is when $H(z)$ is known to be, or assumed to be, an $AR(p)$ system, with p known:

$$H(z) = \frac{1}{1 + \sum_{k=1}^{p} a_k z^{-k}}.$$

Such a model is commonly assumed in speech processing, where a speech signal is modeled as the output of an all-pole system driven by either a zero-mean uncorrelated signal in the case of unvoiced speech (such as the letter "s"), or by a periodic pulse sequence in the case of voiced speech (such as the letter "a"). We assume that the signal is generated according to

$$y[t] = -\mathbf{a}^T \mathbf{y}[t-1] + f[t]$$

(Further assuming here the model uses real data). Our estimated model has output $\hat{y}[t]$, where

$$\hat{y}[t] = -\hat{\mathbf{a}}^T \mathbf{y}[t],$$

and

$$\hat{\mathbf{a}} = \begin{bmatrix} \hat{a}_1 \\ \hat{a}_2 \\ \vdots \\ \hat{a}_p \end{bmatrix}.$$

The mark "^" over a quantity indicates an estimated or approximate value. We can interpret the estimated AR system as a *linear predictor*: the value $\hat{y}[t]$ is the prediction of $y[t]$ given the past data $y[t-1], y[t-2], \ldots, y[t-p]$. The prediction problem can be stated as follows: determine the parameters $\hat{a}_1, \ldots, \hat{a}_p$ to get the "best" prediction. There is an error between what is actually produced by the system and the predicted value:

$$e[t] = y[t] - \hat{y}[t].$$

This is illustrated in figure 1.5. A "good" predictor will make the error as "small" in some sense as possible. The solution to the prediction problem is discussed in chapter 3.

One application of linear prediction is in data compression. We desire to represent a sequence of data using the smallest number of bits possible. If the sequence were completely deterministic, so that $y[t]$ is a deterministic function of prior outputs, we would not need to send any bits to determine $y[t]$ if the prior outputs were known: we could simply use a perfect predictor to reproduce the sequence. If $y[t]$ is not deterministic, we predict $y[t]$,

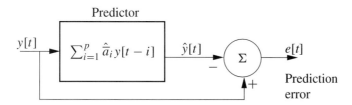

Figure 1.5: Prediction error

then code (quantize) only the prediction error. If the prediction error is small, then only a few bits are required to accurately represent it. Coding in this way is called differential pulse code modulation. When particular focus is given to the process of determining the parameters $\hat{\mathbf{a}}$, it may be called linear predictive coding (LPC). To be successful, it must be possible to determine the coefficients inside the predictor.

Linear prediction also has applications to pattern recognition. Suppose there are several classes of signals to be distinguished (for example, several speech sounds to be recognized). Each signal will have its own set of prediction coefficients: signal 1 has $\mathbf{a}_1$, signal 2 has $\mathbf{a}_2$, and so forth. An unknown input signal can be reduced (by estimating the prediction coefficients that represent it) to a vector $\mathbf{a}$. Then $\mathbf{a}$ can be compared with $\mathbf{a}_1$, $\mathbf{a}_2$, and so forth, using some comparison function, to determine which signal the unknown input is most similar to.

We can examine the linear prediction problem from another perspective. If

$$Y(z) = H(z)F(z),$$

then

$$F(z) = Y(z)\frac{1}{H(z)}.$$

That is,

$$f[t] = y[t] + \mathbf{a}^T \mathbf{y}[t-1].$$

If we regard $y[t]$ as the input, then $f[t]$ is the output of an inverse system. If we have an estimated system

$$\hat{H}(z) = \frac{1}{1 + \sum_{k=1}^{p} \hat{a}_k z^{-k}},$$

then the output

$$\hat{f}[t] = y[t] + \hat{\mathbf{a}}^T \mathbf{y}[t-1]$$

should be close (in some sense) to $f[t]$. A block diagram is shown in figure 1.6. In this case, we would want to choose the parameters $\hat{\mathbf{a}}$ to minimize (in some sense) the error $f[t] - \hat{f}[t]$. That is, we want to determine a good inverse filter for $H(z)$.

Interestingly, using either the point of view of finding a good predictor or of finding a good inverse filter produces the same estimate. It is also interesting that computationally efficient algorithms exist for solving the equations that arise in the linear prediction problem; these are discussed in chapter 8.

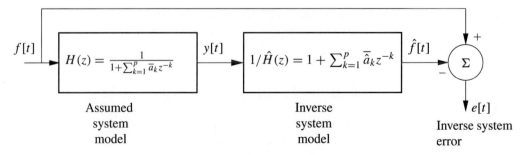

Figure 1.6: Linear predictor as an inverse system

1.4 Models for Linear Systems and Signals

Estimation of parameters: spectrum analysis

It is common in signal analysis to consider that a general signal is composed of sinusoidal signals added together. Determining these frequency components based upon measured signals is called *spectrum estimation* or *spectral analysis*. There are two general approaches to spectral analysis. The first approach is by means of Fourier transforms, in particular, the discrete Fourier transform. This approach is called nonparametric spectrum estimation. The second approach is a parametric approach, in which a model for the signal is proposed (such as the one in (1.2)), and then the parameters are estimated from the measured data. Once these are known, the spectrum of the signal can be determined. Provided that the modeling assumptions are accurate, it is possible to obtain better spectral resolution with fewer parameters using parametric methods.

Discussion of spectrum analysis requires some familiarity with the concepts of energy and power spectral densities. For a discrete-time deterministic signal $y[t]$, the discrete-time Fourier transform (DTFT) is

$$\boxed{Y(\omega) = \sum_{t=-\infty}^{\infty} y[t] e^{-j\omega t}}$$

where $j = \sqrt{-1}$. The *energy spectral density* (ESD) is a measure of how much energy there is at each frequency. An *energy signal* $y[t]$ has finite energy,

$$\sum_{t=-\infty}^{\infty} |y[t]|^2 < \infty.$$

For a deterministic energy signal, the ESD is defined by

$$G_{yy}(\omega) = |Y(\omega)|^2,$$

where the subscript in S_y indicates the signal whose ESD is represented. The *autocorrelation function* of a deterministic sequence is

$$\rho_{yy}[k] = \sum_{t=-\infty}^{\infty} y[t] \overline{y}[t-k].$$

Then (see exercise 1.4-27),

$$G_{yy}(\omega) = \sum_{k=-\infty}^{\infty} \rho_{yy}[k] e^{-j\omega k}; \tag{1.38}$$

that is, the energy spectral density is the DTFT of the autocorrelation function.

The power spectral density (PSD) is employed for spectral analysis of stochastic signals. It provides an indication of how much (average) power there is in the signal, as a function of frequency. We assume that the signal is zero mean, $E[y[t]] = 0$. For the signal $y[t]$ with autocorrelation function $r_{yy}[k]$, we also assume that

$$\lim_{N \to \infty} \frac{1}{N} \sum_{k=-N}^{N} |k| \, |r_{yy}[k]| = 0. \tag{1.39}$$

The PSD is defined as

$$S_{yy}(\omega) = \sum_{k=-\infty}^{\infty} r_{yy}[k] e^{-j\omega k}.$$

That is, the PSD is the DTFT of the autocorrelation sequence. One of the important properties of the PSD is that

$$S_{yy}(\omega) \geq 0 \qquad \text{for all } \omega.$$

This corresponds to the physical fact that real power cannot be negative.

A signal $f[t]$ with PSD $S_f(\omega)$, input to a system with system function $H(z)$, produces the signal $y[t]$, as shown in figure 1.7. Let us define

$$H(\omega) = H(e^{j\omega}) = H(z)|_{z=e^{j\omega}}.$$

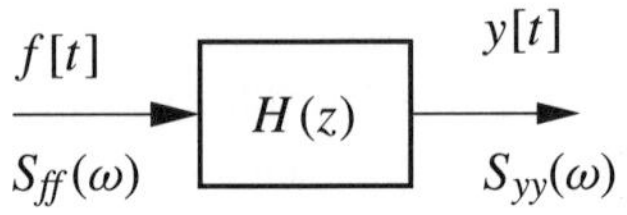

Figure 1.7: PSD input and output

The first equality is "by definition," and is actually an abuse of notation. However, it affords some notational simplicity and is very common. Then (see appendix D), the PSD of the output is

$$S_{yy}(\omega) = |H(\omega)|^2 S_{ff}(\omega).$$

The spectrum estimation problem is as follows: given a set of observations from a random signal, $y[0], y[1], \ldots, y[N]$, determine (estimate) the PSD. In the parametric approach to spectrum estimation, we regard $y[t]$ as the output of a system $H(z)$. It is common to assume that the input signal is a zero-mean white signal, so that

$$S_{ff}(\omega) = \text{constant} = \sigma_f^2.$$

The parameters of $H(z)$ and the input power provide the information necessary to estimate the output spectrum $S_{yy}(\omega)$.

1.4.5 Identification of the modes

Related to spectrum estimation is the identification of the modes in a system. We present the fundamental concept using a second-order system without the complication of noise in the signal. Assume that a signal $y[t]$ is the output of a second-order homogeneous system

$$y[t+2] + a_1 y[t+1] + a_2 y[t] = 0, \qquad (1.40)$$

subject to certain initial conditions. The characteristic equation of this system is

$$z^2 + a_1 z + a_2 = 0. \qquad (1.41)$$

The modes of the system are determined by the roots of the characteristic equation. Writing

$$z^2 + a_1 z + a_2 = (z - p_1)(z - p_2),$$

and assuming that $p_1 \neq p_2$, then

$$y[t] = c_1 (p_1)^t + c_2 (p_2)^t \qquad t \geq 0,$$

where the mode strengths (amplitudes) c_1 and c_2 are determined by the initial conditions.

1.4 Models for Linear Systems and Signals

Based upon the (noise-free) equation (1.40), we can write a set of equations to determine the system parameters (a_1, a_2),

$$\begin{bmatrix} -y[1] & -y[0] \\ -y[2] & -y[1] \\ \vdots & \vdots \end{bmatrix} \begin{bmatrix} a_1 \\ a_2 \end{bmatrix} = \begin{bmatrix} y[2] \\ y[3] \\ \vdots \end{bmatrix}.$$

Provided that the matrix in this equation has full rank, the parameters a_1 and a_2 can be found by solving this set of equations, from which the modes can be identified by finding the roots of (1.41). Using this method, two modes can be identified using as few as four measurements. Two real sinusoids (with two complex exponential modes in each) can be identified with as few as eight measurements, and they can (in principle, and in the absence of noise) be distinguished no matter how close in frequency they are.

Example 1.4.5 Suppose that $y[t]$ is known to consist of two real sinusoidal signals,

$$y[t] = A\cos(\omega_1 t + \theta_1) + B\cos(\omega_2 t + \theta_2).$$

Each cosine function contributes two modes,

$$\cos(\omega_1 t) = \frac{e^{j\omega_1 t} + e^{-j\omega_1 t}}{2}$$

so we will assume that $y[t]$ is governed by the fourth-order difference equation

$$y[t] + a_1 y[t-1] + a_2 y[t-2] + a_3 y[t-3] = 0.$$

Then, assuming that clean, noise-free measurements are available, we can solve for the coefficients of the difference equation by

$$\begin{bmatrix} -y[3] & -y[2] & -y[1] & -y[0] \\ -y[4] & -y[3] & -y[2] & -y[1] \\ -y[5] & -y[4] & -y[3] & -y[2] \\ -y[6] & -y[5] & -y[4] & -y[3] \end{bmatrix} \begin{bmatrix} a_1 \\ a_2 \\ a_3 \\ a_4 \end{bmatrix} = \begin{bmatrix} y[4] \\ y[5] \\ y[6] \\ y[7] \end{bmatrix}. \quad (1.42)$$

If the measured output data set is

$$\mathbf{y} = \{y[0], y[1], \ldots, y[7]\}$$
$$= \{2.55433, 1.91774, 1.15137, 0.33427, -0.451325, -1.1354, -1.67244, -2.0477\},$$

substitution in (1.42) yields

$$(a_1, a_2, a_3, a_4) = (-3.7153, 5.4404, -3.7153, 1)$$
$$z^4 - 3.7153z^3 + 5.4404z^2 - 3.7153z + 1,$$

which has roots at

$$e^{\pm j0.5} \quad \text{and} \quad e^{\pm j0.2}.$$

So, the frequencies of the modes are $\omega_1 = 0.5$ and $\omega_2 = 0.2$. Once the frequencies are known, the amplitudes and phases can also be determined. □

Generalization of these concepts to a system of any order is discussed in section 8.1. Treatment of the measurement noise is discussed in sections 6.9. and 6.10.1.

1.4.6 Control of the modes

Suppose we have a system described by the dynamics

$$\begin{bmatrix} x_1[t+1] \\ x_2[t+1] \end{bmatrix} = \begin{bmatrix} 0.5 & 0 \\ 0 & 3 \end{bmatrix} \begin{bmatrix} x_1[t] \\ x_2[t] \end{bmatrix} + \begin{bmatrix} 1 \\ 1 \end{bmatrix} f[t].$$

Because the A matrix is a diagonal matrix, the state variable equations are said to be uncoupled:

$$x_1[t+1] = 0.5x_1[t] + f[t]$$

does not depend on x_2, and

$$x_2[t+1] = 3x_2[t] + f[t]$$

does not depend upon x_1. (The question of how to put a general system into diagonal form is addressed in section 6.2.) The homogeneous responses (zero-input) of the modes separately are

$$x_1[t] = (0.5)^n x_1[0] \qquad x_2[t] = (3)^n x_2[0].$$

The state variable $x_1[t]$ decays to zero as $n \to \infty$, while the state variable $x_2[t]$ blows up. If this represented the state of a mechanical system, such exponential growth would probably be undesirable. A natural question arises: Is it possible to determine an input sequence $f[t]$ (in conjunction with feedback) that controls the system so that both state variables remain stable? The means of accomplishing this falls very naturally into place using some techniques from linear algebra; see section 6.12.

1.5 Adaptive filtering

An adaptive filter is a filter, usually with an FIR impulse response, in which the coefficients are obtained by attempting to force the output of the filter $y[t]$ to match some desired input signal $d[t]$. (Several examples of desired input signals are given below.) Schematically, the filter is shown in figure 1.8. The error signal

$$e[t] = d[t] - y[t]$$

is used in specialized algorithms (the adaptation rule) to adjust the coefficients of the adaptive filter. A variety of adaptation rules are employed; in particular, we will study the recursive least-squares (RLS) algorithm presented in section 4.11.1 and the least mean squares (LMS) algorithm presented in section 15.6. Adaptive filters are employed in a variety of configurations, some of which are highlighted in this section.

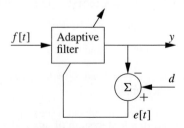

Figure 1.8: Representation of an adaptive filter

1.5 Adaptive Filtering

1.5.1 System identification

An adaptive filter can estimate the the transfer function of an unknown plant, using the configuration shown in figure 1.9. The adaptive filter and the plant are both driven by the same input signal, and the desired signal $d[t]$ is the plant output. The adaptive filter will converge to a "best" representation of the unknown system. If the system is an IIR system and the adaptive filter is an FIR system, or if the order of the adaptive filter is less than the order of the system, then the adaptive filter can be at best an approximation of the true system response.

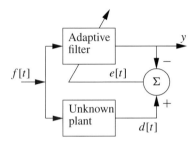

Figure 1.9: Identification of an unknown plant

1.5.2 Inverse system identification

When the adaptive filter is configured as shown in figure 1.10, then it will converge when the output of the adaptive filter matches the delayed input of the inverse system as closely as possible. Ideally, the adaptive filter will converge to the inverse of the plant, so that the cascade of the plant and the adaptive filter is simply a delay. This configuration is employed in some modems to reduce the effect of the channel on the transmitted signal. The signal representing a sequence of input bits ($f[t]$) passes through a channel with an unknown transfer function $H(z)$. At the receiver, the signal is processed by an adapted inverse system before detecting the bits.

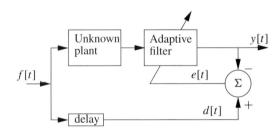

Figure 1.10: Adapting to the inverse of an unknown plant

1.5.3 Adaptive predictors

In the configuration shown in figure 1.11, the input to the adaptive filter is a delayed version of the desired signal. In this case, the adaptive filter converges in such a way as to provide a predictor of the input signal (if prediction is possible). In this mode it can be used for all the applications mentioned previously for linear predictors, including data compression, pattern recognition, or spectrum estimation.

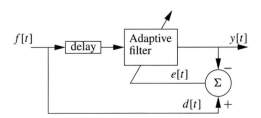

Figure 1.11: An adaptive predictor

1.5.4 Interference cancellation

In the context of interference cancellation, the signal $d[t]$ is commonly referred to as the "primary signal," while the filter input is referred to as the "secondary signal." The primary $d[t]$ is modeled as the sum of a signal of interest, $x[t]$, plus noise:

$$d[t] = x[t] + w[t].$$

The secondary input consists of a noise signal,

$$f[t] = n[t]$$

(see figure 1.12). As an example, suppose that a background acoustic noise source (say the hum of a fan), $w[t]$ is superimposed on a desired audio signal, $x[t]$, which is recorded using a microphone to form the primary input. A second microphone placed far from the desired signal records the noise, $n[t]$, but not the desired signal. There is a different acoustic transfer function for each of the two microphones, hence $n[t]$ is not the same as $w[t]$. The adaptive filter is driven to minimize the error, which adapts to accommodate this difference in transfer function from the noise source. Thus, the resulting difference signal, $e[t]$, will have (insofar as possible) the noise from the reference signal subtracted from the noise from the primary signal.

The interference cancellation configuration has been used in several applications, such as noise cancellation, echo cancellation, and adaptive beamforming in array processing.

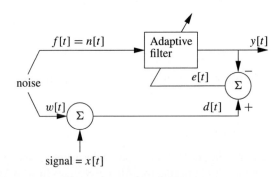

Figure 1.12: Configuration for interference cancellation

1.6 Gaussian Random Variables and Random Processes

> **Box 1.3: Notation for random variables and vectors**
>
> Scalar random variables are represented using capital letters, while a particular outcome value for a random variable is indicated in lower case, usually the same letter. Thus X is a random variable, and x may be an outcome of the random variable. Random vectors are usually presented as bold capital letters. Where the notation of the literature commonly employs lower case, we follow suit.
>
> A probability density function (pdf) or probability mass function (pmf) for a random variable X is written as $f_X(x)$. However, it will be common throughout the text to suppress the subscript notation, letting the argument of the function provide the indication of the random variable. Thus we will frequently write $f(x)$ to mean $f_X(x)$.

1.6 Gaussian random variables and random processes

We begin by reviewing the basic properties of single Gaussian random variables. (See box 1.3 for typographical notation.) Let W be a Gaussian random variable with mean μ and variance σ^2. Notationally, we write

$$W \sim \mathcal{N}(\mu, \sigma^2).$$

The scalar Gaussian probability density function (pdf) should be familiar,

$$f_W(w) = \frac{1}{\sigma\sqrt{2\pi}} e^{-(w-\mu)^2/2\sigma^2}$$

where μ is the mean and σ^2 is the variance of the distribution. That is,

$$\mu = E[W] = \int_{-\infty}^{\infty} w f_W(w)\, dw = \frac{1}{\sqrt{2\pi}\sigma} \int_{-\infty}^{\infty} w e^{-(w-\mu)^2/2\sigma^2}\, dw$$

and

$$\sigma^2 = E[(w-\mu)^2] = E[w^2] - \mu^2 = \frac{1}{\sqrt{2\pi}\sigma} \int_{-\infty}^{\infty} w^2 e^{-(w-\mu)^2/2\sigma^2}\, dw - \mu^2.$$

Figure 1.13 illustrates a Gaussian pdf with $\mu = 0$ and $\sigma^2 = 1$.

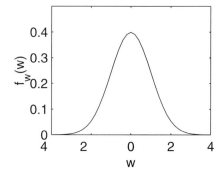

Figure 1.13: The Gaussian density

Associated with the Gaussian pdf are the following useful integrals, true for all values of μ and $\sigma \neq 0$:

$$\boxed{\frac{1}{\sigma\sqrt{2\pi}} \int_{-\infty}^{\infty} e^{-(x-\mu)^2/2\sigma^2} \, dx = 1} \tag{1.43}$$

$$\boxed{\frac{1}{\sigma\sqrt{2\pi}} \int_{-\infty}^{\infty} x e^{-(x-\mu)^2/2\sigma^2} \, dx = \mu} \tag{1.44}$$

$$\boxed{\frac{1}{\sigma\sqrt{2\pi}} \int_{-\infty}^{\infty} x^2 e^{-(x-\mu)^2/2\sigma^2} \, dx = \sigma^2 + \mu^2} \tag{1.45}$$

Measured signals are commonly corrupted by noise. If $\mathbf{Y}[t]$ represents a vector system output, the measured value is often modeled as

$$\mathbf{Z}[t] = \mathbf{Y}[t] + \mathbf{W}[t],$$

where $\mathbf{W}[t]$ is a vector of noise samples,

$$\mathbf{W}[t] = \begin{bmatrix} W_1[t] \\ W_2[t] \\ \vdots \\ W_k[t] \end{bmatrix}.$$

This is the "signal plus noise" model.

In the absence of specific reasons to the contrary, it is common to assume that additive noise signals are distributed with a *Gaussian* (or normal) distribution. (Quantization noise is an exception to this assumption; it is usually modeled as a uniform random variable.) There are reasons for assuming that random variables and random processes are Gaussian. First, Gaussian noise occurs physically. For example, the thermal noise at the front end of a radio receiver is often Gaussian. Second, Gaussian noise signals have a variety of useful properties which simplify several theoretical developments. Some of these properties are described in the following list.

1. By the central limit theorem, the distribution of the sums of several random variables tends toward a Gaussian distribution. That is, if $X_1, X_2, \ldots, X_N$ are independent random variables, with means $\mu_1, \mu_2, \ldots, \mu_N$ and variances $\sigma_1^2, \sigma_2^2, \ldots, \sigma_N^2$, respectively, then

$$Y = \sum_{i=1}^{N} \frac{X_i - \mu_i}{\sigma_i}$$

is distributed almost like a Gaussian with mean 0 and variance 1, if N is large enough. In the limit, as $N \to \infty$ then $Y \sim \mathcal{N}(0, 1)$. The central limit theorem accounts, in large measure, for the occurrence of Gaussian noise in practice; the measured noise is actually the sum of many small independent effects.

Example 1.6.1 An appreciation of the central limit theorem can be gained by looking at the sum of only three variables. Let X_1, X_2, and X_3 be independent random variables uniformly distributed from $-1/2$ to $1/2$. Notationally, we write $X_i \sim \mathcal{U}(-1/2, 1/2)$. The pdf for this uniform random variable is shown in figure 1.14(a). Let $Z = X_1 + X_2$. (Keep in mind that the pdf of the sum of independent random variables is the convolution of the pdfs.) The pdf of Z is thus the "hat" shaped function shown in figure 1.14(b), the convolution of two flat pulses.

1.6 Gaussian Random Variables and Random Processes

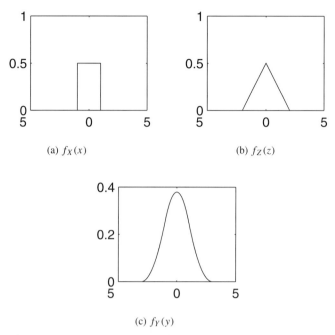

Figure 1.14: Demonstration of the central limit theorem

Let $Y = Z + X_3 = X_1 + X_2 + X_3$. The pdf of Y, obtained again by convolution, is shown in figure 1.14(c). This is a piecewise quadratic function, but observe how it is already beginning to look like the Gaussian density in figure 1.13. □

2. A Gaussian random variable W is entirely determined by its mean and its variance. A Gaussian random process $w(t)$ is determined by its mean

$$m_w(t) = E[w(t)]$$

and autocorrelation

$$r_w(t, s) = E[w(t)\overline{w}(s)]. \qquad (1.46)$$

A Gaussian random process with constant mean and $r_w(t, s) = r_w(s - t)$ (that is, with the autocorrelation dependent upon the time difference in sample points) is stationary.

3. Linear operations on Gaussian random variables produce Gaussian random variables. That is, if X and Y are jointly Gaussian, then

$$Z = aX + bY$$

is also Gaussian for any constants a and b. In particular, the sum of Gaussians is Gaussian. (This follows since the convolution of Gaussians is Gaussian.)

Furthermore, if a Gaussian random process is input to a linear system, then the output is also a Gaussian random process. All that must be determined is the mean and autocorrelation of the output signal, and it is fully characterized.

4. Maximum likelihood detection or estimation involving Gaussian random variables corresponds to a Euclidean distance metric. This is generally geometrically palatable and analytically tractable.

5. Wide-sense stationary (WSS) Gaussian random processes are also strict-sense stationary (SSS). (See appendix D.)

6. Uncorrelated Gaussian random variables are also independent.
7. A Gaussian conditioned upon a Gaussian is Gaussian.

Justifications for some of these properties are provided later.

For a Gaussian random vector $\mathbf{W}$ of dimension k with mean μ and covariance matrix R, we write $\mathbf{W} \sim \mathcal{N}(\mu, R)$. The pdf is

$$f_{\mathbf{w}}(\mathbf{w}) = \frac{1}{(2\pi)^{k/2}|R|^{1/2}} \exp\left[-\frac{1}{2}(\mathbf{w} - \mu)^T R^{-1}(\mathbf{w} - \mu)\right] \quad (1.47)$$

where μ is the mean,

$$\mu = E[\mathbf{W}] = \begin{bmatrix} E[w_1] \\ E[w_2] \\ \vdots \\ E[w_k] \end{bmatrix},$$

and R is the $k \times k$ covariance matrix,

$$R = E[(\mathbf{W} - \mu)(\mathbf{W} - \mu)^T] = E[\mathbf{W}\mathbf{W}^T] - \mu\mu^T.$$

The notation $|R|$ in (1.46) indicates the absolute value of the determinant of the matrix R (see section C.5). (In other contexts, the notation $|R|$ will indicate the determinant, but the absolute value is needed in this case since a density function is always nonnegative.)

Many of the significant concepts associated with Gaussian random vectors can be obtained by examination of two-dimensional vectors. When $\mathbf{W} = [w_1, w_2]^T$,

$$R = \begin{bmatrix} \sigma_1^2 & \sigma_{12} \\ \sigma_{12} & \sigma_2^2 \end{bmatrix}, \quad (1.48)$$

where

$$\sigma_1^2 = E[w_1^2] - \mu_1^2 \qquad \sigma_2^2 = E[w_2^2] - \mu_2^2$$

and

$$\sigma_{12} = E[w_1 w_2] - \mu_1 \mu_2.$$

The *correlation coefficient* is defined as

$$\rho = \frac{E[w_1 w_2] - \mu_1 \mu_2}{\sigma_1 \sigma_2}. \quad (1.49)$$

Using the Cauchy-Schwarz inequality introduced in section 2.6, it can be shown that

$$-1 \leq \rho \leq 1.$$

The correlation coefficient provides information about how w_1 varies with w_2. If $\rho = 1$, then $w_1 = w_2$, and w_1 tells everything there is to know about w_2 (and vice versa). If $\rho = -1$, then $w_1 = -w_2$. If $\rho = 0$, then the variables are said to be *uncorrelated*: w_1 does not provide any information about w_2. More generally, for a k-dimensional random vector $\mathbf{w}$, if the correlation matrix R is diagonal, the components of $\mathbf{w}$ are uncorrelated.

We can write the inverse of the covariance matrix (1.48) in terms of the correlation coefficient and variances, as

$$R^{-1} = \frac{1}{1-\rho^2} \begin{bmatrix} \frac{1}{\sigma_1^2} & \frac{-\rho}{\sigma_1 \sigma_2} \\ \frac{-\rho}{\sigma_1 \sigma_2} & \frac{1}{\sigma_2^2} \end{bmatrix}. \quad (1.50)$$

1.6 Gaussian Random Variables and Random Processes

The joint pdf of w_1 and w_2 can now be written as

$$f(w_1, w_2) = \frac{1}{2\pi \sigma_1 \sigma_2 \sqrt{1-\rho^2}} \exp\left[-\frac{1}{2(1-\rho^2)}\left\{\frac{(w_1-\mu_1)^2}{\sigma_1^2} + \frac{(w_2-\mu_2)^2}{\sigma_2^2}\right.\right.$$
$$\left.\left. - \frac{2\rho(w_1-\mu_1)(w_2-\mu_2)}{\sigma_1 \sigma_2}\right\}\right]. \tag{1.51}$$

A surface-curve plot of this function is shown in figure 1.15 for $\mu_x = \mu_y = 0, \sigma_x^2 = \sigma_y^2 = 1$, for two values of ρ.

In (1.51), if $\rho = 0$, then

$$f(w_1, w_2) = \frac{1}{2\pi \sigma_1 \sigma_2} \exp\left[-\frac{1}{2}\left\{\frac{(w_1-\mu_1)^2}{\sigma_1^2} + \frac{(w_2-\mu_2)^2}{\sigma_2^2}\right\}\right] = f(w_1)f(w_2),$$

substantiating the claim made previously that uncorrelated Gaussian random variables are independent.

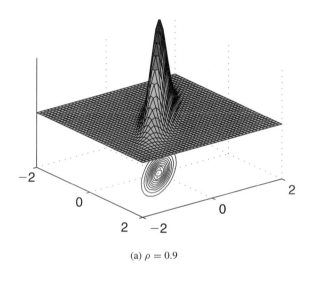

(a) $\rho = 0.9$

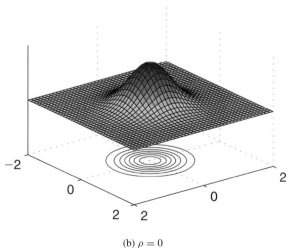

(b) $\rho = 0$

Figure 1.15: Plot of two-dimensional Gaussian distribution

1.6.1 Conditional Gaussian densities

Conditional probabilities constitute the core of many detection and estimation algorithms. In this section, we present a simple example of conditioning as forerunner to the more complete development of statistical decision making in part II.

Suppose that X and Y are jointly Gaussian random variables, $X \sim \mathcal{N}(\mu_x, \sigma_x^2)$, $Y \sim \mathcal{N}(\mu_y, \sigma_y^2)$, with correlation coefficient ρ. We want to *estimate* a value for X, which we will denote as $\hat{x}$. In the absence of any measurements, a reasonable value for $\hat{x}$ is simply the mean of X, so

$$\hat{x} = \mu_x.$$

Such an estimate—obtainable without the benefit of any measurements—is a *prior* or *a priori* estimate, and the density $f_X(x)$ is known as the *a priori* density for X. When a measurement of Y is available, say $Y = y$, then this can be used to modify our prior estimate of X, since X and Y are correlated. One approach to this is to form the conditional pdf $f_{X|Y}(x|y)$, the density of X given that $Y = y$ is known, and determine our estimate $\hat{x}$ by the mean of this new density. The conditional density is defined as

$$f_{X|y}(x|y) = f(x|y) = \frac{f(x, y)}{f(y)}.$$

From (1.51), with $x = w_1$ and $y = w_2$, we obtain

$$f(x|y) = \frac{\frac{1}{2\pi \sigma_x \sigma_y \sqrt{1-\rho^2}} \exp\left[-\frac{1}{2(1-\rho^2)}\left(\frac{(x-\mu_x)^2}{\sigma_x^2} + \frac{(y-\mu_y)^2}{\sigma_y^2} - \frac{2\rho}{\sigma_x \sigma_y}(x-\mu_x)(y-\mu_y)\right)\right]}{\frac{1}{\sqrt{2\pi}\sigma_y} \exp\left[-\frac{1}{2\sigma_y^2}(y-\mu_y)^2\right]}$$

$$= \frac{1}{\sqrt{(2\pi)1-\rho^2}\sigma_x} \exp\left[-\frac{1}{2\sigma_x^2 \sqrt{1-\rho^2}}\left(x - \left(\mu_x + \frac{\sigma_x}{\sigma_y}\rho(y-\mu_y)\right)\right)^2\right]. \quad (1.52)$$

The algebra here requires completing the square, as described in appendix B. From the form of the pdf we recognize that $f(x|y)$ is Gaussian, with mean

$$E[X|y] = \mu_x + \frac{\sigma_x}{\sigma_y}\rho(y-\mu_y) \quad (1.53)$$

and variance

$$\text{var}(X|y) = \sigma_x^2 \sqrt{1-\rho^2}. \quad (1.54)$$

If x and y are correlated (that is, $\rho \neq 0$), then knowing y should tell us something about x. Based on the information available about y, a reasonable estimate of X is the conditional mean,

$$\hat{x} = \mu_x + \frac{\sigma_x}{\sigma_y}\rho(y-\mu_y). \quad (1.55)$$

The variance of this estimate is the conditional variance of (1.54). We can make a meaningful interpretation of the estimate (1.55). If there is no correlation, the conditional mean is the same as the prior mean. If ρ is small, we make only a small modification to the prior mean. If σ_y is large, then the correction to the prior mean is small, as it should be if we have large uncertainty about the outcome y. We also observe that incorporating information about y reduces the variance in x:

$$\sigma_x^2 \sqrt{1-\rho^2} \leq \sigma_x^2,$$

since $|\rho| \leq 1$.

This conditional density with only two variables is extended in section 4.12 to Gaussian vectors conditioned on Gaussian vectors.

This example introduces an important part of estimation theory. An observed (or measured) variable such as y in the foregoing can be used to modify our understanding of variables that we have not measured (or cannot measure). A powerful extension of this simple example is the Kalman filter, in which the state of a system in random noise, such as in (1.29), is estimated based upon observations that are also in noise. In the Kalman filter, the density of the state variable, $f(\mathbf{x}[t])$, is modified by the observation $\mathbf{y}[t]$, taking into account the dynamics of the system and the mechanism for observation. The Kalman filter is discussed in chapter 13.

Several other extensions and issues now arise, among them:

- Given a sequence of data from some source, which is assumed to be drawn according to a Gaussian distribution, how can the parameters of the Gaussian distribution be estimated? How can the quality of the estimates be assessed? These questions are answered in part by *estimation theory*. (An early answer is explored in exercise 1.6-37.)
- If a signal is chosen at random from among a discrete set of signals, and then observed in additive noise, how can the chosen signal be discriminated? This is the *detection* problem which lies at the heart of digital communication.
- Given correlated random vectors $\mathbf{x}$ and $\mathbf{y}$, how can the conditional density $f(\mathbf{x}|\mathbf{y})$ be computed? How may this be applied?
- How can Gaussian random variables of given parameters be generated, and used in simulation, for testing of signal processing algorithms? (An answer for scalar Gaussian r.v.s is found in exercise 1.6-36.)

1.7 Markov and hidden Markov models

A hidden Markov model (HMM) is a stochastic model that is used to model time-varying random phenomena. It is based upon a Markov model, and can be understood in terms of the state-space models already derived. We now present the basic concepts, providing resolution to the issues raised here in chapters 17 and 19. Placement here serves several purposes: it provides a demonstration of the utility of the state-space formulation to yet another system; it smoothes the development of HMM algorithms in later chapters; and it provides introduction and motivation for two important algorithms, the EM algorithm and the Viterbi algorithm.

1.7.1 Markov models

The Markov model is used to model the evolution of random phenomena that can be in discrete states as a function of time, where the transition from one state to the next is random. Suppose that a system can be in one of S distinct states, and that at each step of discrete time it can move to another state at random, with the probability of the transition at time t dependent only upon the state of the system at time t. It is convenient to represent this concept using a probabilistic state diagram, as shown in figure 1.16. In this figure, the Markov model has three states. From state 1, transitions to each of the states are possible; from state 1 to state 1 with probability 0.5, and so forth. Let $S[t]$ denote the state at time t, where $S[t]$ takes on one of the values $1, 2, \ldots, S$. The initial state is selected according to a probability π_i,

$$\pi_i = P(S[1] = i) \qquad i = 1, 2, \ldots, S.$$

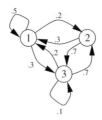

Figure 1.16: A simple Markov model

By the foregoing description, the probability of transition depends only upon the current state:

$$P(S[t+1] = j | S[t] = i, S[t-1] = k, S[t-2] = l, \ldots) = P(S[t+1] = j | S[t] = i).$$

This structure on the probabilities is called the *Markov* property, and the random sequence of state values $S[0], S[1], S[2], \ldots$, is called a *Markov sequence* or a *Markov chain*. This sequence is the output of the Markov model.

We can determine the probability of arriving in the next state by adding up all the probabilities of the ways of arriving there.

$$\begin{aligned} P(S[t+1] = j) &= P(S[t+1] = j | S[t] = 1) P(S[t] = 1) \\ &+ P(S[t+1] = j | S[t] = 2) P(S[t] = 2) + \cdots \\ &+ P(S[t+1] = j | S[t] = S) P(S[t] = S). \end{aligned} \quad (1.56)$$

The computation in (1.56) can be made conveniently in matrix notation. Let

$$\mathbf{p}[t] = \begin{bmatrix} P(S[t] = 1) \\ P(S[t] = 2) \\ \vdots \\ P(S[t] = S) \end{bmatrix}$$

be the vector of probabilities for each state, and let the matrix A contain the transition probabilities

$$A = \begin{bmatrix} P(1|1) & P(1|2) & \cdots & P(1|S) \\ P(2|1) & P(2|2) & \cdots & P(2|S) \\ \vdots & & & \\ P(S|1) & P(S|2) & \cdots & P(S|S) \end{bmatrix}, \quad (1.57)$$

where $P(i|j)$ is an abbreviation for $P(S[t+1] = i | S[t] = j)$, or $a_{ij} = P(S[t+1] = i | S[t] = j)$. For example, for the Markov model of figure 1.16

$$A = \begin{bmatrix} .5 & .3 & .2 \\ .2 & 0 & .7 \\ .3 & .7 & .1 \end{bmatrix}. \quad (1.58)$$

A *steady-state probability* assignment is one that does not change from one time step to the next, so the probability must satisfy the equation $A\mathbf{p} = \mathbf{p}$. This is a particular eigenequation, with an eigenvalue of 1. (More will be said about eigenvalue problems in chapter 6.)

By the law of total probability, each column of A must sum to 1.

1.7 Markov and Hidden Markov Models

Definition 1.2 An $m \times m$ matrix P, such that $\sum_{j=1}^{m} p_{ij} = 1$ (each row sums to 1) and each element of P is nonnegative, is called a **stochastic matrix**. If the rows and columns each sum to 1, then P is **doubly stochastic**. □

The matrix A of (1.57) is the transpose of a stochastic matrix. The vector $\boldsymbol{\pi}$ contains the initial probabilities. Thus, we can write the probabilistic update equation as

$$\mathbf{p}[t+1] = A\mathbf{p}[t] \quad \text{with} \quad \mathbf{p}[0] = \boldsymbol{\pi}.$$

Or, to put it another way,

$$\mathbf{p}[t+1] = A\mathbf{p}[t] + \boldsymbol{\pi}\,\delta_t, \tag{1.59}$$

with $\mathbf{p}[t] = \mathbf{0}$ for $t \leq 0$. The similarity of (1.59) to the first equation of (1.21) should be apparent. In comparing these two, it should be noted that the "state" represented by (1.59) is actually the vector of probabilities $\mathbf{p}[t]$, not the state of the Markov sequence $S[t]$.

1.7.2 Hidden Markov models

The idea behind the HMM can be illustrated using the urn problems of elementary probability, as shown in figure 1.17. Suppose we have S different urns, each of which contains its own set of colored balls. At each instant of time, an urn is selected at random according to the state it was in at the previous instant of time. (That is, according to a Markov model.) Then, a ball is drawn at random from the urn selected at time t. The ball is what we observe as the output, and the actual state is hidden.

The distinction between Markov models and hidden Markov models can be further clarified by continuing the analogy with the state-space equations in (1.21). Equation (1.59) provides for the state update of the Markov system. In most linear systems, however, the state vector is not directly observable; instead, it is observed only through the observation matrix C (assuming for the moment that D is zero),

$$\mathbf{y}[t] = C\mathbf{x}[t],$$

so the state is hidden from direct observation. Similarly, in the HMM we do not observe the state directly. Instead, each state has a probability distribution associated with it. When the HMM moves into state $s[t]$ at time t, the observed output $y[t]$ is an outcome of a random variable $Y[t]$ that is selected according to distribution $f(y[t]|S[t] = s)$, which we will

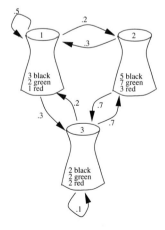

Figure 1.17: The concept of a hidden Markov model

represent using the notation

$$f(y|S[t] = s) = f_s(y).$$

(This idea is illustrated in figure 1.18.) In the urn example of the preceeding paragraph, the output probabilities depend on the contents of the urns. A sequence of outputs from an HMM is $y[0], y[1], y[2], \ldots$. The underlying state information is not seen directly; it is hidden. The probability distribution in each state can be of any type and, in general, each state could have its own type of distribution. Most often in practice, however, each state has the same type of distribution, but with different parameters.

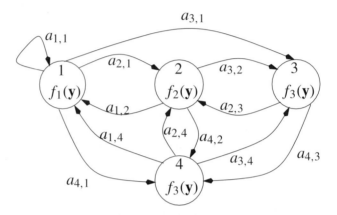

Figure 1.18: An HMM with four states

Let M denote the number of possible outcomes from all of the states, and let $Y[t]$ be the random variable output at time t, with outcome $y[t]$. We can determine the probability of each possible output by adding up all the probabilities,

$$\begin{aligned}P(Y[t] = j) &= P(Y[t] = j|S[t] = 1)P(S[t] = 1) \\ &+ P(Y[t] = j|S[t] = 2)P(S[t] = 2)) + \cdots \\ &+ P(Y[t] = j|S[t] = S)P(S[t] = S).\end{aligned}$$

Let

$$\mathbf{q}[t] = \begin{bmatrix} P(Y[t] = 1) \\ P(Y[t] = 2) \\ \vdots \\ P(Y[t] = M) \end{bmatrix}$$

and

$$C = \begin{bmatrix} P(Y[t] = 1)|S[t] = 1) & \cdots & P(Y[t] = 1|S[t] = S) \\ P(Y[t] = 2)|S[t] = 1) & \cdots & P(Y[t] = 2|S[t] = S) \\ \vdots & & \\ P(Y[t] = M)|S[t] = 1) & \cdots & P(Y[t] = M|S[t] = S) \end{bmatrix};$$

so, $c_{i,j} = P(Y[t] = i|S[t] = j)$. For the urns shown in figure 1.17, with the ball colors black, green, and red corresponding to values 1, 2, and 3, respectively,

$$C = \begin{bmatrix} 1/2 & 1/3 & 1/3 \\ 1/3 & 7/15 & 1/3 \\ 1/6 & 1/5 & 1/3 \end{bmatrix}.$$

Each of the columns must sum to one. Therefore, the output probabilities can be computed by
$$\mathbf{q}[t] = C\mathbf{p}[t].$$
The similarity with (1.21) should be clear. Based on this discussion, the HMM parameters are described by the triple (A, π, C), much like our state-space models.

The HMM can be applied to pattern recognition, where the patterns occur as events occurring sequentially in time. The most successful application is to speech processing. Each word or sound (phoneme) to be recognized is represented by an HMM, where the output is some feature vector that is derived from the speech data. The random variability in the feature vector and the amount of time each feature is produced is modeled by the HMM. The variability in the duration of the word is modeled by the Markov model. The variability in the outputs is modeled by the random selection from within each state. For example, in a small vocabulary system with N words, there are N HMMs, (A_i, π_i, C_i), each being trained (or adapted) to represent the parameters for that word. This is the training phase of the pattern recognition problem.

To perform recognition of an unknown word, its sequence of feature vectors is computed, and the likelihood (probability) that this sequence of feature vectors was produced by the HMM (A_i, π_i, C_i) is computed for each i. That HMM which produces the highest probability selects the recognized word.

The HMM has also been applied to handwriting recognition, speaker identification, and other areas.

Based on this simple discussion, there are several questions that can be posed in conjunction with HMMs.

1. How can the parameters (A, π, C) be estimated based upon observations of the data? (Or, more generally, how can the parameters of other output distributions be computed?) In other words, how can we train the parameters of the models in the pattern recognition problem?

2. Suppose we have an HMM and we observe a sequence of data. How can we determine how well the data fits the model? In other words, can we (efficiently) determine the likelihood of the data?

3. Related somewhat to the previous, suppose we have an HMM and we observe some data supposedly generated from it. How can we determine the sequence of states of the underlying Markov model? (That is, we want to uncover the hidden states.)

These issues are explored in chapters 17 and 19, where the EM algorithm and the Viterbi algorithm are introduced and applied to this problem.

1.8 Some aspects of proofs

> Mathematics is simply sustained logical thinking.
> — *H.P.P. Ferguson*

> There is no royal road to geometry.
> — *Plato*

> Some people believe that a theorem is proved when a logically correct proof is given; but some people believe it is proved only when the students sees why it is inevitably true.
> — *Richard W. Hamming*
> Coding and Information Theory, p. 164

In engineering classes that require proofs, it almost inevitably arises that a student will complain that he or she does "not know how to do proofs." The way it is usually stated, of "doing proofs," seems to suggest that the student perhaps believes there is some universally applicable method of doing proofs that will prove all problems. On the one hand, there is no one that knows how to "do proofs" of everything. A proof requires insight, understanding, background, and creativity, and some plausible conjectures have thus far eluded proof (and will continue to do so: that itself is a theorem). Some proofs have the subtlety and beauty of a well-crafted sonnet. On the other hand, most proofs consist of clarifications of patterns that have been previously observed, or are precise statements of some fact. Every engineering student should be able to "do proofs" to some extent.

Signal processing, employing mathematical concepts to accomplish engineering purposes, often presents a difficult challenge to engineering students who want to know how to use the material, but resist the mathematical formalities—in particular, theorems and proofs. Nevertheless, throughout this book, many of the concepts are presented in a theorem–proof format as a means of organization, and opportunities for proving some concepts are provided in the exercises. The following justifications are provided for requiring proofs of engineering students:

1. Because an engineer puts things together, with an eye to design and utility, the ability to move from a requirement specification to a finished design is an important skill. In its restricted domain, proving a theorem is nothing more than design; taking specifications and using available components to produce a result. The specifications are the hypotheses of the theorem, and the available components are whatever knowledge can be brought to bear on the problem. Like most design problems, there may be many correct solutions, and many incorrect approaches. (It is perhaps the flexibility of choice exercised against inflexible logic that makes proofs challenging.) Like design, a proof may require trying many different avenues before a fruitful approach is encountered.

2. A proof provides an opportunity to review and deepen understanding of concepts and definitions that have been presented. Tools that don't get used or are not understood correctly will never become useful tools.

3. As new algorithms are developed, they must be evaluated. Often this is done empirically, by means of computer simulation or by testing of prototypes. However, it is better to have a sense of the correctness of a design before too many resources are expended in its prototyping. The skills developed in learning to do proofs of theorems may assist in evaluating and improving signal processing algorithms.

4. There is no escaping the fact that the signal processing literature is very mathematical. A broad mathematical vocabulary and the ability to read mathematics are necessary to draw meaningful information from the literature. Should the occasion arise when student wish to publish their own results in signal processing literature, they will need to speak the language.

5. Doing a proof is a good chance to stretch some intellectual muscles.

The intent of this section is to provide some suggestions on methods of proof that appear in the literature. This is by no means an exhaustive list; new and important concepts can arise as new ways of answering questions are created. As an example, consider Shannon's channel-coding theorem, which states (basically) that there is a code which can be used to transmit data over a channel with arbitrarily low probability of error, provided that the rate of transmission is less than the capacity of the channel. In proving the theorem, Shannon took an unprecedented step. Instead of looking for a particular code to answer the question, he instead averaged over all possible codes. This particular trick made the analysis fall right into place. Such "tricks," or creative insights, cannot be taught. There are, however, some logical approaches which can be taught and exercised.

1.8 Some Aspects of Proofs

A theorem may be stated something like: if P, then Q. In this, P is called the *hypothesis* and Q is called the *conclusion*. We say that P implies Q, and may write $P \Rightarrow Q$. The statement "if P, then Q" is not logically equivalent to saying that, because Q occurs, P must also occur. For example, consider the following syllogism:

If a book falls on Frank's head, his head will hurt.
Frank's head hurts.

We cannot conclude that a book has fallen on Frank's head; he may simply have a headache. In the implication $P \Rightarrow Q$, we say that P is sufficient for Q: knowledge that P occurs is sufficient to establish the presence of Q. However, P is not necessary for Q: Q could (perhaps) have happened another way.

Note that if $P \Rightarrow Q$ and if Q is not true, then P cannot be true. Based on the syllogism above, if Frank's head does not hurt, we *can* conclude that a book did not fall on his head.

Equivalent ways of expressing this implication are:

P implies Q
if P, then Q
$P \Rightarrow Q$
Q, if P
P only if Q
P is a sufficient (but not necessary) condition for Q
not Q implies not P (this is the *contrapositive*)
Q is a necessary condition for P

For the statement $P \Rightarrow Q$, the statement obtained by reversing the roles of P and Q

$$Q \Rightarrow P$$

is called the *converse*. That fact that $P \Rightarrow Q$ and its converse $Q \Rightarrow P$ are both true can be stated in a variety of equivalent ways:

P implies Q and Q implies P
P implies Q, and conversely
P if and only if Q
P is a necessary and sufficient condition for Q
$P \Leftrightarrow Q$

The statement "P if and only if Q" is often abbreviated P iff Q.

We now present some comments about proofs in a general framework. These suggestions do not provide an exhaustive bag of tricks, but are merely intended to suggest some approaches that might work.

1.8.1 Proof "by computation": direct proof

Proofs of some statements may be mostly computational, may involve such techniques as integration (often using change of variables), properties of integration, linear algebra, Taylor series, etc. As a simple example, to prove that convolution commutes, that is, that

$$\int_{-\infty}^{\infty} f(t-\tau) h(\tau) \, d\tau = \int_{-\infty}^{\infty} f(\tau) h(t-\tau) \, d\tau,$$

it suffices to make a change of variable $x = t - \tau$ in the first integral. If you were approaching the problem without knowing the "trick," the best thing to do would be to simply try several approaches. If what you are trying to prove is true, sooner or later you may stumble across the correct approach. While this may lack polish, it mirrors the way things are discovered in the real world: rarely does a useful concept or product spring forth full-blown, as if from the head of Zeus. Discovery requires exploration, thought, and trial-and-error. Of course, experience in an area can shorten the time between concept and execution. To experienced mathematicians, some things become transparently obvious because they have solved so many related problems. A student starting out in an area may not have the benefit of that insight. What is often required is the determination to try things out, possibly without being able to foresee at the outset what will result. Experience will lengthen the number of steps you can see ahead.

Example 1.8.1 Here is an example of a direct proof.

Let $X = \mathbf{x}_1, \mathbf{x}_2, \ldots, \mathbf{x}_m$ be a set of discrete points in $\mathbb{R}^n$. The sets defined by

$$V_i = \{\mathbf{x} \in \mathbb{R}^n : \mathbf{x} \text{ is closer to } \mathbf{x}_i \text{ than to any other } \mathbf{x}_j, i \neq j\},$$

that is,

$$V_i = \{\mathbf{x} \in \mathbb{R}^n : d(\mathbf{x}, \mathbf{x}_i) < d(\mathbf{x}, \mathbf{x}_j), i \neq j\},$$

are called the *Voronoi regions* of X. The vector $\mathbf{x}_i$ in V_i is called the cell representative. Voronoi regions arise in vector quantization and data compression (see section 16.1). We will prove that Voronoi regions are convex sets. Pick a Voronoi cell; without loss of generality we will call the cell V_1, with its cell representative $\mathbf{x}_1$.

Let $\mathbf{p}$ and $\mathbf{q}$ be arbitrary points in V_1, and let us designate $\mathbf{p}$ as the point which is further from $\mathbf{x}_1$. If every point on the line between $\mathbf{p}$ and $\mathbf{q}$ is in V_1, then the set is convex. Let $\mathbf{x}$ be a point on the line between $\mathbf{p}$ and $\mathbf{q}$,

$$\mathbf{x} = \lambda \mathbf{p} + (1 - \lambda)\mathbf{q}, \qquad 0 \leq \lambda \leq 1.$$

Then,

$$\begin{aligned}
\|\mathbf{x}_1 - \mathbf{x}\| &= \|\mathbf{x}_1 - (\lambda \mathbf{p} + (1-\lambda)\mathbf{q})\| \\
&= \|\lambda(\mathbf{x}_1 - \mathbf{p}) + (1-\lambda)(\mathbf{x}_1 - \mathbf{q})\| \\
&\leq \lambda \|\mathbf{x}_1 - \mathbf{p}\| + (1-\lambda)\|\mathbf{x}_1 - \mathbf{q}\| \\
&\leq \lambda \|\mathbf{x}_1 - \mathbf{p}\| \leq \|\mathbf{x}_1 - \mathbf{p}\|,
\end{aligned}$$

where the first inequality follows from the triangle inequality. Thus $\mathbf{x}$ is closer to $\mathbf{x}_1$ than is $\mathbf{p}$, which is in the Voronoi cell. By the definition of the Voronoi cell, if $\mathbf{p}$ is in the Voronoi cell, then $\mathbf{x}$ must also be. □

Of course, the trial-and-error aspect of finding the correct computation in this example is not shown, only the finished product.

Some standard "tricks" that are employed in proofs are worth mentioning:

1. Counting and lists. Make an exhaustive list of all the elements, and consider what you are trying to do applied to all of them.
2. To show that A and B are the same, it may work to show that $A \subset B$ and $B \subset A$. Similarly, to show that $x = y$, show that $x \geq y$ and $y \geq x$. (See, for example, the proof to theorem 2.2.)
3. In analytical work, the Taylor series and the mean value theorem are excellent tools.
4. Exhaustive checking. For example, to verify that a set satisfies certain properties, simply validate that the properties hold individually.

1.8.2 Proof by contradiction

> Contradictions do not exist. Whenever you think that you are facing a contradiction, check your premises. You will find that one of them is wrong.
> — *Ayn Rand*
> Atlas Shrugged

A powerful proof technique is proof by contradiction. In order to show that $P \Rightarrow Q$, we take as true the hypothesis P and *assume* that Q is not true. The proof follows by showing that this assumption leads to a logical contradiction.

Example 1.8.2 We will prove a millennia-old theorem known to the Pythagoreans of Greece. Recall that a rational number is a number that can be expressed as a ratio of integers. Thus 3/7 is a rational number. □

Theorem: $\sqrt{2}$ *is irrational*.

Prior to establishing this theorem, the Pythagoreans held the viewpoint that the harmonies of the cosmos could be expressed as ratios of integers. This theorem lead to considerable religious upheaval in its day.

Proof We will assume a result contrary to the statement of the theorem, and show that this leads to a contradiction. We assume that $\sqrt{2}$ *is* rational, that is, that

$$\sqrt{2} = m/n \qquad (1.60)$$

for some integers m and n. Now we show that this leads to a contradiction. Squaring (1.60), we obtain

$$2 = \frac{m^2}{n^2}, \qquad (1.61)$$

so

$$2n^2 = m^2.$$

From this we see that m^2 must be an even number, and hence that m must be even (show this!). Let us write $m = 2k$ for some integer k. Substituting this into (1.61), we obtain

$$2 = \frac{4k^2}{n^2},$$

or,

$$2 = \frac{n^2}{k^2}.$$

This is equivalent to

$$\sqrt{2} = \frac{n}{k}.$$

Now we have returned an expression having the same form as (1.60), but with $k < n$. Being now in a position to repeat the operation, we have reached the precipice leading to a contradiction, because the numbers in the ratio will be reduced by iteration of these same steps, down to absurdly small values. By this contradiction, we must conclude that the original assumption (1.60) is false. □

One of the issues over which mathematicians sometimes fret is the uniqueness of a solution to a given problem. Proving uniqueness is very commonly done using contradiction. Two distinct solutions to the problem are proposed, and it is shown that these solutions are equal, a contradiction which points out that only one solution is possible. This method is exemplified in the proof of theorem 2.1.

1.8.3 Proof by induction

> The essential characteristic of reasoning by recurrence is that it contains, condensed so to speak, in a single formula, an infinite number of syllogisms.
>
> — *Henri Poincarè*
> Science and Hypothesis

Proof by induction allows one to establish general conclusions from a limited set of test cases. Suppose you have some statement that depends upon an integer n. We will denote this statement by $S(n)$—statement S is a function of n. You begin by showing that $S(n)$ is true for $n = 1$ (sometimes another small value of n is the starting point). Then you show that assuming $S(n)$ is true leads to an implication that $S(n+1)$ is also true. What is amazing and powerful is that you get to *assume* the truth of $S(n)$, and use this to show the truth of $S(n+1)$. The assumed hypothesis $S(n)$ is called the *inductive hypothesis*.

Example 1.8.3 The first example should be familiar. We want to show that the sum of the first n integers is

$$\sum_{k=0}^{n} k = \frac{n(n+1)}{2}.$$

Clearly this is true for $n = 0$, and also clearly it is true for $n = 1$. Let us assume its truth for n. That is, we now *assume* that

$$\sum_{k=0}^{n} k = \frac{n(n+1)}{2},$$

and show that this implies the truth for $n+1$. That is, we need to show that

$$\sum_{k=0}^{n+1} k = \frac{(n+1)(n+2)}{2}.$$

We have

$$\sum_{k=0}^{n+1} k = \left(\sum_{k=0}^{n} k\right) + (n+1)$$
$$= \frac{n(n+1)}{2} + (n+1)$$
$$= \frac{n^2 + 3n + 2}{2} = \frac{(n+1)(n+2)}{2}$$

where the second equality comes by assumption of the inductive hypothesis. □

We do another inductive proof of mathematical flavor to illustrate another point.

Example 1.8.4 We will show that,

$$\text{if } n \geq 5, \text{ then } 2^n > n^2.$$

What makes this example fundamentally different from the previous is that the starting point is not $n = 0$, but $n = 5$.

1.8 Some Aspects of Proofs

The statement is clearly true when $n = 5$. Let us assume that it holds for n; that is, our inductive hypothesis is

$$2^n > n^2$$

and show that it must be true for $n + 1$, that is,

$$2^{n+1} > (n + 1)^2.$$

We have

$$\begin{aligned}
2^{n+1} &= 2 \cdot 2^n \\
&> 2n^2 \quad \text{(by the inductive hypothesis)} \\
&= n^2 + n^2 \geq n^2 + 5n \quad \text{(because } n \geq 5\text{)} \\
&= n^2 + 2n + 3n > n^2 + 2n + 1 \\
&> (n + 1)^2.
\end{aligned}$$
□

We now offer an example with a little more of an engineering flavor.

Example 1.8.5 Suppose there is a communication link in which errors can be made with probability p. (This link is diagrammed in figure 1.19(a).) When a 0 is sent, it is received as a 0 with probability $1 - p$, and as a 1 with probability p. This communication-link model is called a binary symmetric channel (BSC). Now, suppose that n BSCs are placed end to end, as in figure 1.19(b). Denote the probability of error after n channels by $P_n(e)$. We wish to show that the end-to-end probability of error is

$$P_n(e) = \frac{1}{2}[1 - (1 - 2p)^n]. \tag{1.62}$$

When $n = 1$, we compute $P_1(e) = p$, as expected. Let us now assume that $P_n(e)$ as given in (1.62) is true for n, and show that this provides a true formula for $P_{n+1}(e)$.

In $n + 1$ stages, we can make an error if there are no errors in the first n stages and an error occurs in the last stage, or if an error has occurred over the first n stages and no error occurs in the last stage. Thus,

$$\begin{aligned}
P_{n+1}(e) &= (1 - p)P_n(e) + p(1 - P_n(e)) \\
&= (1 - p)\frac{1}{2}[1 - (1 - 2p)^n] + p\left(1 - \frac{1}{2}(1 - (1 - 2p)^n\right) \\
&\quad \text{(by the inductive hypothesis)} \\
&= \frac{1}{2}[1 - (1 - 2p)^{n+1}].
\end{aligned}$$
□

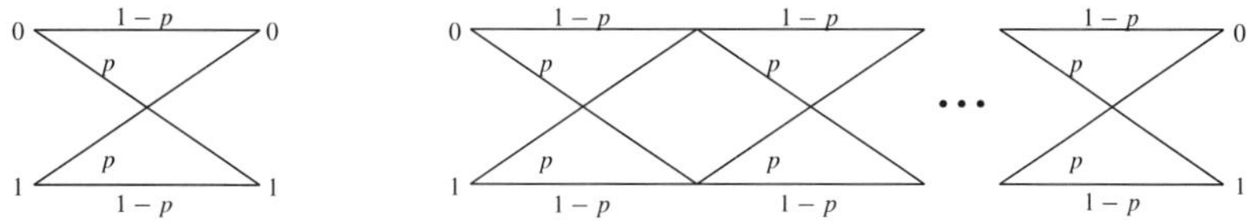

(a) A single channel   (b) n channels end-to-end

Figure 1.19: Binary symmetric channel model

Proof by induction is very powerful and works in a remarkable number of cases. It requires that you be able to state the theorem: you must start with the inductive hypothesis, which is usually the difficult part. In practice, statement of the theorem must come by some initial grind, some insight, and a lot of work. Then induction is used to prove that the result is correct. Some simple opportunities for stating an inductive hypothesis and then proving it are provided in the exercises.

1.9 An application: LFSRs and Massey's algorithm

In this section we introduce the linear feedback shift register (LFSR), which is nothing more than a deterministic autoregressive system. The concepts presented here will illustrate some of the linear systems theory presented in this chapter, provide a demonstration of some methods of proof, and introduce our first algorithm.

An LFSR is simply an autoregressive filter over a field F (see box 1.4) that has no input signal. An LFSR is shown in figure 1.20. An alternative realization, preferred in high-speed implementations because the addition operations are not cascaded, is shown in figure 1.21. If the contents are binary, it is helpful to view the storage elements as D flip-flops, so that the memory of the LFSR is simply a shift register and the LFSR is a digital state machine. For a binary LFSR, the connections are either 1 or 0 (connection or no connection), and all operations are carried out in $GF(2)$; that is, modulo 2 (see box 1.5). Massey's algorithm applies over any field, but most commonly it is used in connection with the binary field.

The output of the LFSR is

$$y_j = -\sum_{i=1}^{p} c_i y_{j-i} \qquad j = p, p+1, p+2, \ldots . \tag{1.63}$$

The number of feedback coefficients p is called the *length* of the LFSR.

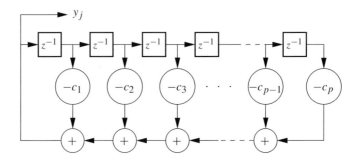

Figure 1.20: LFSR realization

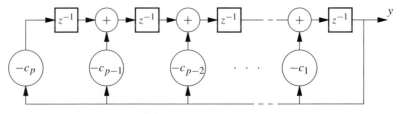

Figure 1.21: Alternative LFSR realization

1.9 An Application: LFSRs and Massey's Algorithm

Box 1.4: Groups, rings, and fields

At various places throughout the text, we will have occasion to refer to algebraic systems different from the familiar real or complex numbers. In these different systems, the operations are organized into particular sets of arithmetic rules. We define here three important sets of operations.

Groups. A set S equipped with a binary operation $*$ is a group if it satisfies the following:

1. There is an identity element $e \in S$, such that for any $a \in S$,
$$a * e = e * a = a.$$
That is, the identity element leaves every element unchanged under the operation $*$.

2. For every element $a \in S$ there is an element $b \in S$ called its *inverse*, such that
$$a * b = e \qquad b * a = e.$$

3. The binary operation is associative. For every $a, b, c \in S$,
$$(a * b) * c = a * (b * c).$$

We denote the group by $\langle S, * \rangle$.

If it is true that $a * b = b * a$ for every $a, b \in S$, then the group is said to be a **commutative** or **Abelian** group.

Rings. A set R equipped with two operations, which we will denote as $+$ and $*$, is a ring if it satisfies the following:

1. $\langle R, + \rangle$ is an Abelian group.
2. The operation $*$ is associative.
3. Left and right distributive laws hold. For all $a, b, c \in R$,
$$a(b + c) = ab + ac \qquad (a + b)c = ac + bc.$$

The operator $*$ is not necessarily associative; nor is an identity or inverse required for the operation $*$. We denote the ring by $\langle R, +, * \rangle$.

Fields. incorporate the algebraic operations we are familiar with from working with real and complex numbers. A set F equipped with two operations $+$ and $*$ is a field if it satisfies the following:

1. $\langle F, + \rangle$ is an Abelian group.
2. The set F excluding 0 (the additive identity) is a commutative group under $*$.
3. The operations $+$ and $*$ distribute.

Example 1.9.1 The LFSR over $GF(2)$ shown in figure 1.22(a) satisfies
$$y_j = y_{j-1} + y_{j-3}.$$

With initial register contents $y_{-3} = 1$, $y_{-2} = 0$ $y_{-1} = 0$, the LFSR output sequence is shown in figure 1.22(b), where the notation $D = z^{-1}$ is employed. The alternative realization is shown in figure 1.21.

> **Box 1.5:** $GF(2)$
>
> An important class of fields are those that have a finite number of elements. These are known as Galois fields. All Galois fields have a number of elements equal to p^m, where p is prime and m is an integer. Of these, arguably the most important is $GF(2)$, the field of binary arithmetic done without carry. The addition and multiplication tables for $GF(2)$ are shown here:
>
> | a | b | a+b | | a | b | a·b |
> |---|---|-----|-----|---|---|-----|
> | 0 | 0 | 0 | | 0 | 0 | 0 |
> | 0 | 1 | 1 | | 0 | 1 | 0 |
> | 1 | 0 | 1 | | 1 | 0 | 0 |
> | 1 | 1 | 0 | | 1 | 1 | 1 |
>
> The addition operation is the `exclusive-or` operation familiar from digital logic, and the multiplication operation is the `and` operation. The reader can verify that these operations satisfy the requirements of a field.

After $j = 6$ the sequence repeats, so that seven distinct states occur in this digital state machine. Note that for this LFSR, the register contents assume all possible nonzero sequences of three digits. □

Taking the Z-transform of (1.63), we obtain

$$Y(z)(1 + c_1 z^{-1} + c_2 z^{-2} + \cdots + c_p z^{-p}) = 0. \tag{1.64}$$

It will be convenient to represent the LFSR, using the polynomial in (1.64), in the form

$$C(D) = 1 + c_1 D + c_2 D^2 + \cdots + c_p D^p,$$

where $D = z^{-1}$ is a delay operator. We note that the output sequence produced by the LFSR depends upon both the feedback coefficients and the initial contents of the storage registers.

1.9.1 Issues and applications of LFSRs

With a correctly designed feedback polynomial $C(D)$, the output sequence of a binary LFSR is a "maximal-length" sequence, producing $2^p - 1$ outputs before the sequence repeats. This sequence, although not truly random, exhibits many of the characteristics of noise, such as producing runs of zeros and ones of different lengths, having a correlation function that approximates a delta function, and so forth. The sequence produced is sometimes called a pseudonoise sequence. Pseudonoise sequences are applied in spread-spectrum communications, error detection, ranging, and so on. The global position system based on an array of satellites in geosynchronous orbit, employs pseudonoise sequences to carry timing information used for navigational purposes.

In some of these applications, the following problem arises: given a sequence $\{y_0, y_1, \ldots, y_{N-1}\}$ deemed to be the output of an LFSR, determine the feedback connection polynomial $C(D)$ and the initial register contents of the shortest LFSR that could produce the sequence. Solving this problem is the focus of the remainder of this section. The algorithm we develop is known as Massey's algorithm. Not only does it solve the particular problem stated here but, as we shall see, it provides an efficient algorithm for solving a particular set of Toeplitz equations.

An LFSR that produces the sequence

$$\{y_0, y_1, \ldots, y_{N-1}\}$$

1.9 An Application: LFSRs and Massey's Algorithm

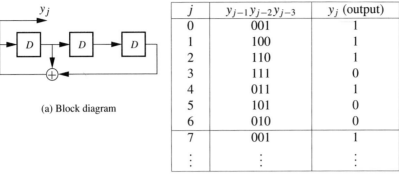

(a) Block diagram

j	$y_{j-1}y_{j-2}y_{j-3}$	y_j (output)
0	001	1
1	100	1
2	110	1
3	111	0
4	011	1
5	101	0
6	010	0
7	001	1
⋮	⋮	⋮

(b) Output sequence

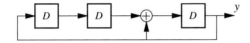

(c) Alternate block diagram

j	$y_{j-1}y_{j-2}y_{j-3}$	y_j (output)
0	001	1
1	101	1
2	111	1
3	110	0
4	011	1
5	100	0
6	010	0
7	001	1
⋮	⋮	⋮

(d) Output sequence for alternative realization

Figure 1.22: A binary LFSR and its output

could clearly be obtained from an LFSR of length N, each storage element containing one of the values. However, this may not be the shortest possible LFSR. Another approach to the system synthesis is to set up a system of equations of the following form (assuming that the length of the LFSR is $p = 3$):

$$\begin{bmatrix} y_2 & y_1 & y_0 \\ y_3 & y_2 & y_1 \\ y_4 & y_3 & y_2 \end{bmatrix} \begin{bmatrix} c_1 \\ c_2 \\ c_3 \end{bmatrix} = \begin{bmatrix} -y_3 \\ -y_4 \\ -y_5 \end{bmatrix}.$$

These equations are in the same form as the Yule–Walker equations in (1.16); in particular, the matrix on the left is a Toeplitz matrix. Whereas the Yule–Walker equations were originally developed in this book in the context of a stochastic signal model, we observe that there is a direct parallel with deterministic autoregressive signal models.

Knowing the value of p, the Yule–Walker equations could be solved by any means available to solve p equations in p unknowns. However, directly solving this set of equations is inefficient in at least two ways:

1. A general solution of a $p \times p$ set of equations requires $O(p^3)$ operations. We are interested in developing an algorithm that requires fewer operations. The algorithm we develop requires $O(p^2)$ operations.

2. The order p is not known in advance. The value of p could be determined by starting with a small value of p, and increasing the size of the matrix until an LFSR is obtained that produces the entire sequence. This could be done without taking into account the result for smaller values of p. More desirable would be an algorithm that builds recursively on previously-obtained solutions to obtain a new solution. This is, in fact, how we proceed.

Since we buildup the LFSR using information from prior computations, we need a notation to represent the polynomial used at different stages of the algorithm. Let

$$C^{[n]}(D) = 1 + c_1^{[n]} D + \cdots + c_{L_n}^{[n]} D^{L_n}$$

denote the feedback connection polynomial for the LFSR capable of producing the output sequence $\{y_0, y_1, \ldots, y_{n-1}\}$, where L_n is the degree of the feedback connection polynomial.

The algorithm we obtain provides an efficient way of solving the Yule–Walker equations when p is not known. In chapter 8 we encounter an algorithm for solving Toeplitz matrix equations with fixed p, the Levinson–Durbin algorithm. A third general approach, based on the Euclidean algorithm, is also known (see, e.g., [36]). Each of these algorithms has $O(p^2)$ complexity, but they have tended to be used in different application areas, the Levinson–Durbin algorithm being used most commonly with linear prediction and speech processing, and the Massey or Sugiyama algorithm being used in finite-field applications, such as error-correction coding.

1.9.2 Massey's algorithm

We build the LFSR that produces the entire sequence by successively modifying an existing LFSR, if necessary, to produce increasingly longer sequences. We start with an LFSR that could produce y_0. We determine if that LFSR could also produce the sequence $\{y_0, y_1\}$; if it can, then no modifications are necessary. If the sequence cannot be produced using the current LFSR configuration, we determine a new LFSR that can produce the entire sequence. We proceed this way inductively, eventually constructing an LFSR configuration that can produce the entire sequence $\{y_0, y_1, \ldots, y_{N-1}\}$. By this process, we obtain a sequence of polynomials and their degrees,

$$(C^{[1]}(D), L_1)$$
$$(C^{[2]}(D), L_2)$$
$$\vdots$$
$$(C^{[N]}(D), L_N),$$

where the last LFSR produces $\{y_0, \ldots, y_{N-1}\}$.

At some intermediate step, suppose we have an LFSR $C^{[n]}(D)$ that produces $\{y_0, y_1, \ldots, y_{n-1}\}$ for some $n < N$. We check if this LFSR will also compute y_n by computing the output

$$\hat{y}_n = -\sum_{i=1}^{L_n} c_i^{[n]} y_{n-i}.$$

If $\hat{y}_n$ is equal to y_n, then there is no need to update the LFSR, and $C^{[n+1]}(D) = C^{[n]}(D)$.

1.9 An Application: LFSRs and Massey's Algorithm

Otherwise, there is some nonzero *discrepancy*,

$$d_n = y_n - \hat{y}_n = y_n + \sum_{i=1}^{L_n} c_i^{[n]} y_{n-i} = \sum_{i=0}^{L_n} c_i^{[n]} y_{n-i}.$$

In this case, we will update our LFSR using the formula

$$C^{[n+1]}(D) = C^{[n]}(D) + AD^l C^{[m]}(D), \tag{1.65}$$

where A is some element in the field, l is an integer, and $C^{[m]}(D)$ is one of the prior LFSRs produced by our process that also had a nonzero discrepancy d_m. Using this new LFSR, we compute the new discrepancy, denoted by d'_n, as

$$\begin{aligned}
d'_n &= \sum_{i=0}^{L_{n+1}} c_i^{[n+1]} y_{n-i} \\
&= \sum_{i=0}^{L_n} c_i^{[n]} y_{n-i} + A \sum_{i=0}^{L_m} c_i^{[m]} y_{n-i-l}.
\end{aligned} \tag{1.66}$$

Now, let $l = n - m$. Then the second summation gives

$$A \sum_{i=0}^{L_m} c_i^{[m]} y_{m-i} = A d_m.$$

Thus, if we choose $A = -d_m^{-1} d_n$, then the summation in (1.66) gives

$$d'_n = d_n - d_m^{-1} d_n d_m = 0.$$

So the new LFSR produces the sequence $\{y_0, y_1, \ldots, y_n\}$.

1.9.3 Characterization of LFSR length in Massey's algorithm

The update in (1.65) is, in fact, the heart of Massey's algorithm. From an operational point of view, no further analysis is necessary. However, the problem was to find the shortest LFSR producing a given sequence. We have produced a means of finding an LFSR, but have no indication yet that it is the shortest. Establishing this will require some additional effort in the form of two theorems. The proofs are challenging, but it is worth the effort to think them through.

(In general, considerable signal processing research follows this general pattern. An algorithm may be established that can be shown to work empirically for some problem, but characterizing its performance limits often requires significant additional effort.)

Theorem 1.1 *Suppose that an LFSR of length L_n produces the sequence $\{y_0, y_1, \ldots, y_{n-1}\}$, but not the sequence $\{y_0, y_1, \ldots, y_n\}$. Then any LFSR that produces the latter sequence must have a length L_{n+1} satisfying*

$$L_{n+1} \geq n + 1 - L_n.$$

Proof The theorem is only of practical interest if $L_n < n$ (otherwise it is trivial to produce the sequence). Let us take, then, $L_n < n$. Let

$$C^{[n]}(D) = 1 + c_1^{[n]} D + \cdots + c_{L_n}^{[n]} D^{L_n}$$

represent the connections for the LFSR which produces $\{y_0, y_1, \ldots, y_{n-1}\}$, and let

$$C^{[n+1]}(D) = 1 + c_1^{[n+1]} D + \cdots + c_{L_{n+1}}^{[n+1]} D^{L_{n+1}}$$

denote the connections for the LFSR which produces $\{y_0, y_1, \ldots, y_n\}$. Now we do a proof by contradiction:

Assume (contrary to the theorem) that

$$L_{n+1} \leq n - L_n. \tag{1.67}$$

From the definitions of the connection polynomials, we observe that

$$-\sum_{i=1}^{L_n} c_i^{[n]} y_{j-i} \quad \begin{cases} = y_j & j = L_n, L_n+1, \ldots, n-1 \\ \neq y_n & j = n \end{cases} \tag{1.68}$$

and

$$-\sum_{i=1}^{L_{n+1}} c_i^{[n+1]} y_{j-i} = y_j \qquad j = L_{n+1}, L_{n+1}+1, \ldots, n. \tag{1.69}$$

From (1.69), we have

$$y_n = -\sum_{i=1}^{L_{n+1}} c_i^{[n+1]} y_{n-i}.$$

The indices in this summation range from $n-1$ to $n-L_{n+1}$ which, because of the (contrary) assumption made in (1.67), is a subset of the range $L_n, L_n+1, \ldots, n-1$. Thus, the equality in (1.68) applies, and we can write

$$y_n = -\sum_{i=1}^{L_{n+1}} c_i^{[n+1]} y_{n-i} = \sum_{i=1}^{L_{n+1}} c_i^{[n+1]} \sum_{k=1}^{L_n} c_k^{[n]} y_{n-i-k}.$$

Interchanging the order of summation we have

$$y_n = -\sum_{k=1}^{L_n} c_k^{[n]} \sum_{i=1}^{L_{n+1}} c_i^{[n+1]} y_{n-i-k}. \tag{1.70}$$

Setting $j = n$ in (1.68), we obtain

$$y_n \neq -\sum_{k=1}^{L_n} c_k^{[n]} y_{n-k}.$$

In this summation the indices range from $n-1$ to $n-L_n$ which, because of (1.67), is a subset of the range $L_{n+1}, L_{n+1}+1, \ldots, n$ of (1.69). Thus, we can write

$$y_n \neq \sum_{k=1}^{L_n} c_k^{[n]} \sum_{i=1}^{L_{n+1}} c_i^{[n+1]} y_{n-k-i}. \tag{1.71}$$

Comparing (1.70) with (1.71), we observe a contradiction. Hence, the assumption on the length of the LFSRs must have been incorrect. By this contradiction, we must have

$$L_{n+1} \geq n + 1 - L_n. \qquad \square$$

Since the shortest LFSR that produces the sequence $\{y_0, y_1, \ldots, y_n\}$ must also produce the first part of that sequence, we must have $L_{n+1} \geq L_n$. Combining this with the result of the theorem, we obtain

$$L_{n+1} \geq \max(L_n, n+1-L_n). \tag{1.72}$$

In other words, the shift register cannot become shorter as more outputs are produced.

1.9 An Application: LFSRs and Massey's Algorithm

We have seen how to update the LFSR to produce a longer sequence using (1.65), and also have seen that there is a lower bound on the length of the LFSR. We now show that this lower bound can be achieved *with equality*, thus providing the *shortest* LFSR that produces the desired sequence.

Theorem 1.2 *Let $\{(L_i, C^{[i]}(D)), i = 0, 2, \ldots, n\}$ be a sequence of minimum-length LFSRs that produce the sequence $\{y_0, y_1, \ldots, y_{i-1}\}$. If $C^{[n+1]}(D) \neq C^{[n]}(d)$, then a new LFSR can be found that satisfies*

$$L_{n+1} = \max(L_n, n + 1 - L_n).$$

Proof We will do a proof by induction, taking as the inductive hypothesis that

$$L_{k+1} = \max(L_k, k + 1 - L_k) \qquad (1.73)$$

for $k = 0, 1, \ldots, n$. This clearly holds when $k = 0$, since $L_0 = 0$.

Let $C^{[m]}$, $m < n$, denote the *last* connection polynomial before $C^{[n]}(D)$ that can produce the sequence $\{y_0, y_1, \ldots, y_{m-1}\}$ but not the sequence $\{y_0, y_1, \ldots, y_m\}$, such that

$$L_m < L_n.$$

Then

$$L_{m+1} = L_n;$$

hence, in light of (1.73),

$$L_{m+1} = L_n = m + 1 - L_m. \qquad (1.74)$$

If $C^{[n+1]}(D)$ is updated from $C^{[n]}(D)$ according to (1.65), with $l = n - m$, we have already observed that it is capable of producing the sequence $\{y_0, y_1, \ldots, y_n\}$. By the update formula (1.65), we note that

$$L_{n+1} = \max(L_n, n - m + L_m).$$

Using (1.74) we find that

$$L_{n+1} = \max(L_n, n + 1 - L_n). \qquad \square$$

In the update step, we observe that if

$$2L_n > n$$

then, using (1.73), $c^{[n+1]}$ has length $L_{n+1} = L_n$, that is, the polynomial is updated, but there is no change in length.

The shift-register synthesis algorithm, known as Massey's algorithm, is presented first in pseudocode as Algorithm 1.1, where we use the notations

$$c(D) = C^{[n]}(D) \qquad p(D) = C^{[m]}(D).$$

Algorithm 1.1 Massey's algorithm (pseudocode)

Input: $y_0, y_1, \ldots, y_{N-1}$
Initialize:
$L = 0$
$c(D) = 1$ (the current connection polynomial)
$p(D) = 1$ (the connection polynomial before last length change)
$s = 1$ (s is $n - m$, the amount of shift in update)
$d_m = 1$ (previous discrepancy)
for $n = 0$ to $N - 1$
    $d = y_n + \sum_{i=1}^{L} c_i y_{n-i}$
    if ($d = 0$)
        $s = s + 1$
    else
        if ($2L > n$) then (no length change in update)
            $c(D) = c(D) - d d_m^{-1} D^s p(D)$
            $s = s + 1$
        else (update c with length change)
            $t(D) = c(D)$ (temporary store)
            $c(D) = c(D) - d d_m^{-1} D^s p(D)$
            $L = n + 1 - L$
            $p(D) = t(D)$
            $d_m = d$
            $s = 1$
        end
end

A MATLAB implementation of Massey's algorithm with computations over $GF(2)$ is shown in Algorithm 1.2. The vectorized structure of MATLAB allows the pseudocode implementation to be expressed almost directly in executable code. The statement `c = mod([c zeros(1,Lm + s - Ln)] + [zeros(1,s)p],2);` simply aligns the polynomials represented in `c` and `p` by appending and prepending the appropriate number of zeros, after which they can be added directly (addition is mod 2 since operations are in $GF(2)$).

Algorithm 1.2 Massey's algorithm

```
function [c] = massey(y)
% function [c] = massey(y)
% This function runs Massey's algorithm (in GF(2)), returning
% the shortest-length LFSR
%
% y = input sequence
% c = LFSR connections, c = 1 + c(2)D + c(3)D^2 + ... c(L+1)D^L
%     (Note: opposite from usual Matlab order)

N = length(y);
% Initialize the variables
```

1.9 An Application: LFSRs and Massey's Algorithm

```
Ln = 0;        % current length of LFSR
Lm = 0;        % length before last change
c = 1;         % feedback connections
p = 1;         % c before last change
s = 1;         % amount of shift

for n=1:N      % N = current matching output sequence length
  d = mod(c*y(n:-1:n-Ln)',2);   % compute the discrepancy (binary arith.)
  if(d == 0)                    % no discrepancy
    s = s+1;
  else
    if(2*Ln > n-1)              % no length change in update
      c = mod(c + [zeros(1,s) p zeros(1,Ln-(Lm+s))],2);
      s = s+1;
    else                        % update with new length
      t = c;
      c = mod([c zeros(1,Lm+s-Ln)] + [zeros(1,s) p],2);
      Lm = Ln;  Ln = n - Ln;    p = t;   s = 1;
    end
  end
end
```

Because the MATLAB code so closely follows the pseudocode, only a few of the algorithms throughout the book will be shown using pseudocode, with preference given to MATLAB code to illustrate and define the algorithms.

To conserve page space, subsequent algorithms are not explicitly displayed. Instead, the icon

is used to indicate that the algorithm is to be found on the CD-ROM.

Example 1.9.2 For the sequence of example 1.9.1,

$$y = \{1, 1, 1, 0, 1, 0, 0\},$$

the feedback connection polynomial obtained by a call to `massey` is

$$c = \{1, 1, 0, 1\},$$

which corresponds to the polynomial

$$C(D) = 1 + D + D^3.$$

Thus,

$$Y(z)(1 + z^{-1} + z^{-3}) = 0,$$

or

$$y_j = y_{j-1} + y_{j-3},$$

as expected. □

1.10 Exercises

1.4-1 (Complex arithmetic) This exercise gives a brief refresher on complex multiplication, as well as matrix multiplication. Let $z_1 = a + jb$ and $z_2 = c + jd$ be two complex numbers. Let $z_3 = z_1 z_2 = e + jf$.

(a) Show that the product can be written as

$$\begin{bmatrix} e \\ f \end{bmatrix} = \begin{bmatrix} c & -d \\ d & c \end{bmatrix} \begin{bmatrix} a \\ b \end{bmatrix}.$$

In this form, four real multiplies and two real adds are required.

(b) Show that the complex product can also be written as

$$e = (a-b)d + a(c-d) \qquad f = (a-b)d + b(c+d).$$

In this form, only three real multiplies and five real adds are required. (If addition is significantly easier than multiplication in hardware, then this saves computations.)

(c) Show that this modified scheme can be expressed in matrix notation as

$$\begin{bmatrix} e \\ f \end{bmatrix} = \begin{bmatrix} 1 & 0 & 1 \\ 0 & 1 & 1 \end{bmatrix} \begin{bmatrix} (c-d) & 0 & 0 \\ 0 & (c+d) & 0 \\ 0 & 0 & d \end{bmatrix} \begin{bmatrix} 1 & 0 \\ 0 & 1 \\ 1 & -1 \end{bmatrix} \begin{bmatrix} a \\ b \end{bmatrix}.$$

1.4-2 Show that (1.8) for the partial fraction expansion of a Z-transform with repeated roots is correct.

1.4-3 Determine the partial fraction expansion of the following.

(a) $H(z) = \dfrac{1 - 3z^{-1}}{1 - 1.5z^{-1} + .56z^{-2}}$
(b) $H(z) = \dfrac{1 - 5z^{-1} - 6z^{-2}}{1 - 1.5z^{-1} + .56z^{-2}}$

(c) $H(z) = \dfrac{2 - 3z^{-1}}{(1 - .3z^{-1})^2}$
(d) $H(z) = \dfrac{5 - 6z^{-1}}{(1 - .3z^{-1})^2(1 - .4z^{-1})}$

Check your results using residuez in MATLAB.

1.4-4 (Inverses of higher-order modes)

(a) Prove the following property for Z transforms: If

$$x[t] \leftrightarrow X(z),$$

then

$$tx[t] \leftrightarrow -z \frac{dX(z)}{dz}.$$

(b) Using the fact that $p^t u[t] \leftrightarrow 1/(1 - pz^{-1})$, show that

$$tp^t u[t] \leftrightarrow \frac{pz^{-1}}{(1 - pz^{-1})^2}.$$

(c) Determine the Z-transform of $t^2 p^t u[t]$.

(d) By extrapolation, determine the order of the pole of a mode of the form $t^k p^t u[t]$.

1.4-5 Show that the autocorrelation function defined in (1.9) has the property that

$$r_{yy}[k] = \bar{r}_{yy}[-k].$$

1.4-6 Show that (1.11) is correct.

1.4-7 For the MA process

$$y[t] = f[t] + 2f[t-1] + 3f[t-2],$$

1.10 Exercises

where $f[t]$ is a zero-mean white random process with $\sigma_f^2 = .1$, determine the 3 × 3 autocorrelation matrix R.

1.4-8 For the first-order real AR process

$$y[t+1] + a_1 y[t] = f[t+1],$$

with $|a_1| < 1$ and $E[f[t]] = 0$, show that

$$\sigma_y^2 = E[y^2[t]] = \frac{\sigma_f^2}{1 - a_1^2}. \tag{1.75}$$

1.4-9 For an AR process (1.12) driven by a white-noise sequence $f[t]$ with variance σ_f^2, show that

$$\sigma_f^2 = \sum_{i=0}^{p} a_i r_{yy}[i]. \tag{1.76}$$

1.4-10 (Second-order AR processes) Consider the second-order real AR process

$$y[t+2] + a_1 y[t+1] + a_2 y[t] = f[t+2], \tag{1.77}$$

where $f[t]$ is a zero-mean white-noise sequence. The difference equation in (1.14) has a characteristic equation with roots

$$p_1, p_2 = \frac{1}{2}\left(-a_1 \pm \sqrt{a_1^2 - 4a_2}\right).$$

(a) Using the Yule–Walker equations, show that if the autocorrelation values

$$r_{yy}[l-k] = E[y[t-k]\overline{y}[t-l]]$$

are known, then the model parameters may be determined from

$$\begin{aligned} a_1 &= -\frac{r_{yy}[1](r_{yy}[0] - r_{yy}[2])}{r_{yy}^2[0] - r_{yy}^2[1]} \\ a_2 &= -\frac{r_{yy}[0]r[2] - r_{yy}^2[1]}{r_{yy}^2[0] - r_{yy}^2[1]}. \end{aligned} \tag{1.78}$$

(b) On the other hand, if $\sigma_y^2 = r_{yy}[0]$ and a_1 and a_2 are known, show that the autocorrelation values can be expressed as

$$\begin{aligned} r_{yy}[1] &= -\frac{a_1}{1 + a_2}\sigma_y^2 \\ r_{yy}[2] &= \sigma_y^2 \left(\frac{a_1^2}{1 + a_2} - a_2\right). \end{aligned} \tag{1.79}$$

(c) Using (1.76) and the results of this problem, show that

$$r_{yy}[0] = \sigma_y^2 = \left(\frac{1 + a_2}{1 - a_2}\right) \frac{\sigma_f^2}{[(1 + a_2)^2 - a_1^2]}. \tag{1.80}$$

(d) Using $r_{yy}[0] = \sigma_y^2$ and $r_{yy}[1] = -a_1 \sigma_y^2/(1 + a_2)$ as initial conditions, find an explicit solution to the Yule–Walker difference equation

$$r_{yy}[k] + a_1 r_{yy}[k-1] + a_2 r_{yy}[k-2] = 0$$

in terms of p_1, p_2, and σ_y^2.

1.4-11 For the second-order difference equation

$$y[t+2] - .7y[t+1] + .12y[t] = f[t+2],$$

where $f[t]$ is a zero-mean white sequence with $\sigma_f^2 = .1$, determine $\sigma_y^2 = r_{yy}[0]$, $r_{yy}[1]$ and $r_{yy}[2]$.

1.4-12 A random process $y[t]$, having zero-mean and $m \times m$ autocorrelation matrix R, is applied to an FIR filter with impulse response vector $\mathbf{h} = [h_0, h_1, h_2, \ldots, h_{m-1}]^T$. Determine the average power of the filter output $x[t]$.

1.4-13 Place the following into state variable form (controller canonical form), and draw a realization.

(a) $H(z) = \dfrac{1 - 3z^{-1}}{1 - 1.5z^{-1} + .56z^{-2}}$

(b) $H(z) = \dfrac{1 - 5z^{-1} - 6z^{-2}}{1 - 1.5z^{-1} + .56z^{-2}}$

1.4-14 In addition to the block diagram shown in figure 1.3, there are many other forms. This problem introduces one of them, the *observer canonical form*.

(a) Show that the Z-transform relation implied by (1.2) can be written as
$$Y(z) = \bar{b}_0 F(z) + [\bar{b}_1 F(z) - \bar{a}_1 Y(z)]z^{-1} + [\bar{b}_2 F(z) - \bar{a}_2 Y(z)]z^{-2} + \cdots$$
$$+ [\bar{b}_p F(z) - \bar{a}_p Y(z)]z^{-1}. \tag{1.81}$$

(b) Draw a block diagram representing (1.81), containing p delay elements.

(c) Label the outputs of the delay elements from right to left as $x_1, x_2, \ldots, x_p$. Show that the system can be put into state space form with

$$A = \begin{bmatrix} -\bar{a}_1 & 1 & 0 & \cdots & 0 \\ -\bar{a}_2 & 0 & 1 & \cdots & 0 \\ \vdots & & & & \\ -\bar{a}_{p-1} & 0 & 0 & \cdots & 0 \\ -\bar{a}_p & 0 & 0 & \cdots & 1 \end{bmatrix} \quad \mathbf{b} = \begin{bmatrix} \bar{b}_1 - \bar{a}_1 \bar{b}_0 \\ \bar{b}_2 - \bar{a}_2 \bar{b}_0 \\ \cdots \\ \bar{b}_{p-1} - \bar{a}_{p-1} \bar{b}_0 \\ \bar{b}_p - \bar{a}_p \bar{b}_0 \end{bmatrix} \quad \mathbf{c} = \begin{bmatrix} 1 \\ 0 \\ 0 \\ \vdots \\ 0 \end{bmatrix} \quad d = \bar{b}_0.$$

A matrix A of this form is said to be in *second companion form*.

(d) Draw the block diagram in observer canonical form for
$$H(z) = \dfrac{2 + 3z^{-1} + 4z^{-2}}{1 + z^{-1} - 6z^{-2} - 7z^{-3}},$$
and determine the system matrices $(A, \mathbf{b}, \mathbf{c}^T, d)$.

1.4-15 Another block diagram representation is based upon the partial fraction expansion. Assume initially that there are no repeated roots, so that
$$H(z) = \sum_{k=1}^{p} \dfrac{N_k}{1 - p_k z^{-1}}.$$

(a) Draw a block diagram representing the partial fraction expansion, by using the fact that
$$\dfrac{Y(z)}{F(z)} = \dfrac{1}{1 - pz^{-1}}$$
has the block diagram

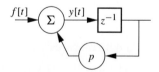

(b) Let $x_i, i = 1, 2, \ldots, p$ denote the outputs of the delay elements. Show that the system can be put into state-space form, with

$$A = \begin{bmatrix} p_1 & 0 & 0 & \cdots & 0 \\ 0 & p_2 & 0 & \cdots & 0 \\ \vdots & & & & \\ 0 & 0 & 0 & \cdots & p_p \end{bmatrix} \quad \mathbf{b} = \begin{bmatrix} 1 \\ 1 \\ \vdots \\ 1 \end{bmatrix} \quad \mathbf{c} = \begin{bmatrix} N_1 \\ N_2 \\ \vdots \\ N_p \end{bmatrix} \quad d = \bar{b}_0.$$

A matrix A in this form is said to be a *diagonal* matrix.

(c) Determine the partial fraction expansion of

$$H(z) = \frac{1 - 2z^{-1}}{1 + .5z^{-1} + .06z^{-2}}$$

and draw the block diagram based upon it. Determine $(A, \mathbf{b}, \mathbf{c}, d)$.

(d) When there are repeated roots, things are slightly more complicated. Consider, for simplicity, a root appearing only twice. Determine the partial fraction expansion of

$$H(z) = \frac{1 + z^{-1}}{(1 - .2z^{-1})(1 - .5z^{-1})^2}.$$

Be careful about the repeated root!

(e) Draw the block diagram corresponding to $H(z)$ in partial fraction form using only three delay elements.

(f) Show that the state variables can be chosen so that

$$A = \begin{bmatrix} .5 & 0 & 0 \\ 1 & .5 & 0 \\ 0 & 0 & .2 \end{bmatrix}.$$

A matrix in this form (blocks along the diagonal, each block being either diagonal or diagonal with ones in it as shown) is in *Jordan* form.

1.4-16 Show that the system in (1.22) has the same transfer function and solution as does the system in (1.21).

1.4-17 For a system in state-space representation,

(a) Show by induction that (1.27) is correct.

(b) For a time-varying system, as in (1.23), determine a representation similar to (1.27).

1.4-18 (Interconnection of systems)[164] Let $(A_1, \mathbf{b}_1, \mathbf{c}_1^T)$ and $(A_2, \mathbf{b}_2, \mathbf{c}_2^T)$ be two systems. Determine the system $(A, \mathbf{b}, \mathbf{c}^T)$ obtained by connecting the two systems:

(a) In series.

(b) In parallel.

(c) In a feedback configuration with $(A_1, \mathbf{b}_1, \mathbf{c}_1^T)$ in the forward loop and $(A_2, \mathbf{b}_2, \mathbf{c}_2^T)$ in the feedback loop.

1.4-19 Show that

$$\begin{bmatrix} A & A_1 \\ 0 & A_2 \end{bmatrix} \quad \begin{bmatrix} \mathbf{b} \\ \mathbf{0} \end{bmatrix} \quad [\mathbf{c}^T \mathbf{q}^T]$$

and

$$\begin{bmatrix} A & 0 \\ A_1 & A_2 \end{bmatrix} \quad \begin{bmatrix} \mathbf{b} \\ \mathbf{q} \end{bmatrix} \quad [\mathbf{c}^T \mathbf{0}]$$

and $(A, \mathbf{b}, \mathbf{c}^T)$ all have the same transfer function, for all values of A_1, A_2, and $\mathbf{q}$ that lead to valid matrix operations. Conclude that realizations can have different numbers of states.

1.4-20 Consider the system function

$$H(z) = \frac{z^3 + 3z^2 + 2z}{z^3 + 10z^2 + 31z + 30}.$$

(a) Draw the controller canonical block diagram.

(b) Draw the block diagram in Jordan form (diagonal form).

(c) How many modes are really present in the system? The problem here is that a *minimal* realization of A is not obtained directly from the $H(z)$ as given.

1.4-21 [164] If $(A, \mathbf{b}, \mathbf{c}^T, d)$ with $d \neq 0$ describes a system $H(s)$ in state-space form, show that
$$(A - \mathbf{b}\mathbf{c}^T/d, \mathbf{b}/d, \mathbf{c}^T/d, 1/d)$$
describes a system with system function $1/H(s)$.

1.4-22 (State-space solutions)
  (a) Show that (1.35) is a solution to the differential equation in (1.30), for constant (A, B, C, D).
  (b) Show that (1.36) is a solution to the differential equation in (1.30), for non-constant (A, B, C, D), provided that Φ satisfies the properties given.

1.4-23 Find a solution to the differential equation described by the state-space equations
$$\dot{\mathbf{x}}(t) = \begin{bmatrix} 0 & 1 \\ -1 & 0 \end{bmatrix} \mathbf{x}(t)$$
$$y(t) = [1 \quad 0]\mathbf{x}(t),$$
with $\mathbf{x}(0) = \mathbf{x}_0$. These equations describe simple harmonic motion.

1.4-24 Consider the system described by
$$\dot{\mathbf{x}}(t) = \begin{bmatrix} -2 & 0 \\ 1 & 1 \end{bmatrix} \mathbf{x}(t) + \begin{bmatrix} 2 \\ -1 \end{bmatrix} f(t)$$
$$y(t) = [0 \quad 2]\mathbf{x}(t).$$
  (a) Determine the transfer function $H(s)$.
  (b) Find the partial fraction expansion of $H(s)$.
  (c) Verify that the modes of $H(s)$ are the same as the eigenvalues of A.

1.4-25 Verify (1.33) by long division.

1.4-26 (System identification) In this exercise you will develop a technique for identification of the parameters of a continuous-time second-order system, based upon frequency response measurements (Bode plots). Assume that the system to be identified has an open-loop transfer function
$$H_o(s) = \frac{b}{s(s+a)}.$$
  (a) Show that with the system in a feedback configuration as shown in figure 1.23, the transfer function can be written as
$$H_c(s) = \frac{Y(s)}{F(s)} = \frac{1}{1 + (a/b)s + (1/b)s^2}.$$

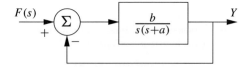

Figure 1.23: Simple feedback configuration

  (b) Show that
$$\frac{1}{H_c(j\omega)} = A(j\omega) \angle \phi(j\omega),$$
where
$$A(j\omega) = \frac{1}{b}\sqrt{(b - \omega^2)^2 + (a\omega)^2} \quad \text{and} \quad \tan \phi(j\omega) = \frac{a\omega}{b - \omega^2}.$$

1.10 Exercises

The quantities $A(j\omega)$ and $\phi(j\omega)$ correspond to the reciprocal amplitude and the phase difference between input and output.

(c) Show that if amplitude/phase measurements are made at n different frequencies $\omega_1, \omega_2, \ldots, \omega_n$, then the unknown parameters a and b can be estimated by solving the overdetermined set of equations

$$\begin{bmatrix} A(j\omega_1) & -\omega_1\sqrt{1+1/\tan^2\phi(j\omega_1)} \\ \tan\phi(j\omega_1) & -\omega_1 \\ A(j\omega_2) & -\omega_2\sqrt{1+1/\tan^2\phi(j\omega_2)} \\ \tan\phi(j\omega_2) & -\omega_2 \\ \vdots & \\ A(j\omega_n) & -\omega_n\sqrt{1+1/\tan^2\phi(j\omega_n)} \\ \tan\phi(j\omega_n) & -\omega_n \end{bmatrix} \begin{bmatrix} b \\ a \end{bmatrix} = \begin{bmatrix} 0 \\ \omega_1^2 \tan\phi(j\omega_1) \\ 0 \\ \omega_2^2 \tan\phi(j\omega_2) \\ \vdots \\ 0 \\ \omega_n^2 \tan\phi(j\omega_n) \end{bmatrix}.$$

1.4-27 Verify (1.38).

1.4-28 Show that

$$\sum_{n=-\infty}^{\infty} |y[t]|^2 = \frac{1}{2\pi} \int_{-\pi}^{\pi} G_{yy}(\omega)\, d\omega.$$

Hint: Recall the inverse Fourier transform

$$y[t] = \frac{1}{2\pi} \int_{-\pi}^{\pi} Y(\omega) e^{j\omega t}\, d\omega.$$

1.4-29 Show that under the condition that (1.39) is true, the PSD satisfies

$$S_{yy}(\omega) = \lim_{N\to\infty} E\left[\frac{1}{N}\left|\sum_{n=1}^{N} y[n] e^{-j\omega n}\right|^2\right].$$

Hint: Show and use the fact that

$$\sum_{n=1}^{N}\sum_{m=1}^{N} f(n-m) = \sum_{l=-N+1}^{N-1} (N-|l|) f(l).$$

1.4-30 (Modal analysis) The following data is measured from a third-order system:
$$y = \{0.3200, 0.2500, 0.1000, -0.0222, 0.0006, -0.0012, 0.0005, -0.0001\}.$$
Assume that the first time index is 0, so that $y[0] = 0.32$.

(a) Determine the modes in the system, and plot them in the complex plane.

(b) The data can be written as
$$y[t] = c_1(p_1)^t + c_2(p_2)^t + c_3(p_3)^t \qquad t \geq 0.$$
Determine the constants c_1, c_2, and c_3.

(c) To explore the effect of noise on the system, add random Gaussian noise to each data point with variance $\sigma^2 = 0.01$, then find the modes of the noisy data. Repeat several times (with different noise), and comment on how the modal estimates move.

1.4-31 (Modal analysis) If $y[t]$ has two real sinusoids,
$$y[t] = A\cos(\omega_1 t + \theta_1) + B\cos(\omega_2 t + \theta_2),$$
and the frequencies are known, determine a means of computing the amplitudes and phases from measurements at time instants $t_1, t_2, \ldots, t_N$..

1.6-32 Show that R^{-1} from (1.50) is correct.

1.6-33 Show that (1.51) follows from (1.47) and (1.50).

1.6-34 Suppose that $X \sim \mathcal{N}(\mu_x, \sigma_x^2)$ and $N \sim \mathcal{N}(0, \sigma_n^2)$ are independently distributed Gaussian r.v.s. Let
$$Y = X + N.$$
(a) Determine the parameters of the distribution of Y.

(b) If $Y = y$ is measured, we can estimate X by computing the conditional density $f(X|y)$. Determine the mean and variance of this conditional density. Interpret these results in terms of getting information about X if (i) $\sigma_n^2 \gg \sigma_x^2$, and (ii) $\sigma_n^2 \ll \sigma_x^2$.

1.6-35 Suppose that $X \sim \mathcal{N}(\mu_x, \sigma_x^2)$ and $Y \sim \mathcal{N}(\mu_y, \sigma_y^2)$ are jointly distributed Gaussian r.v.s with correlation ρ. Determine the parameters of the distribution of $Z = aX + bY$.

1.6-36 If $X \sim \mathcal{N}(0, 1)$, show that
$$Y = \sigma X + \mu$$
is distributed as $Y \mathcal{N}(\sigma^2, \mu)$.

1.6-37 Let $x_1, x_2, \ldots, x_n$ be n independent observations of a Gaussian random variable X with unknown mean and variance. We desire to estimate the mean and variance of X. The joint density of n independent Gaussian r.v.s, conditioned on knowing the mean μ and the variance σ^2, is
$$f(x_1, x_2, \ldots, x_n | \mu, \sigma^2) = \frac{1}{(2\pi)^{n/2} \sigma^n} \exp\left[-\frac{1}{2\sigma^2} \sum_{i=1}^{n} (x_i - \mu)^2\right].$$

(a) Determine a *maximum likelihood* estimate of μ by maximizing this joint density with respect to μ (i.e., take the derivative with respect to μ). Call the estimate of the mean thus obtained $\hat{\mu}$.

(b) Since $\hat{\mu}$ is a function of random variables, it is itself a random variable. Determine the mean (expected value) of $\hat{\mu}$. An estimate whose expected value is equal to the value it is estimated is said to be *unbiased*.

(c) Determine the variance of $\hat{\mu}$.

(d) Determine an estimate for σ^2.

It is natural to ask if there is a better estimator for the mean than the "obvious" one just obtained. However, as will be shown in section 12.3.2, this estimator is dependably the best, in that it has the lowest possible variance for any unbiased estimate.

1.7-38 A Markov random process $X(t)$ has the property that
$$P(X(t_3) = x_2 | X(t_2) = x_2, X(t_1) = x_1) = P(X(t_3) = x_3 | X(t_2) = x_2)$$
when $t_3 > t_2 > t_1$; that is, the probability depends only upon the most recent conditioning event. We will abbreviate this using the notation
$$f(x_3 | x_2, x_1) = f(x_3 | x_2).$$
(a) For a Markov process, show that
$$f(x_3, x_1 | x_2) = f(x_3 | x_2) f(x_2 | x_1).$$
This is the property of conditional independence (x_3 is independent of x_1, provided that they are each conditioned on an intermediate observation x_2).

(b) Now suppose $X(t)$ is a Gaussian random process, and assume (for convenience only) that it is zero-mean. Let
$$r_x(t, s) = E[X(t)X(s)].$$
If $X(t)$ is also Markov, show that
$$r_x(t_3, t_1) = \frac{r_x)t_3, t_2) r_x(t_2, t_1)}{r_x(t_2, t_2)}.$$
Hint: Use the fact that $E[E[X(t_3)X(t_1)|X(t_2)]] = E[X(t_3)X(t_1)]$, and use the formula for conditional expectation derived in (1.52).

1.10 Exercises

1.7-39 For the state-transition probability matrix A given in (1.58), find a probability vector $\mathbf{p}$ such that

$$A\mathbf{p} = \mathbf{p}.$$

Such a probability vector is called the *steady-state* probability of the Markov model.

1.8-40 Show that $\sqrt{3}$ is irrational.

1.8-41 Show that there are an infinite number of primes. Hint: Use a proof by contradiction, assuming that there are only a finite number of primes. Then build a number $2 \cdot 3 \cdot 5 \cdots p + 1$, where p is the assumed last prime, and show that this is not divisible by any of the listed primes.

1.8-42 Show that if m^2 is even, then m must be even.

1.8-43 By trial and error, determine a plausible formula for

$$\sum_{i=0}^{n} 2^i.$$

Then prove by induction that your formula is correct.

1.8-44 Determine (by experiment) a plausible formula for the sum of the first n odd integers

$$1 + 3 + 5 + \cdots + (2n - 1).$$

Then prove by induction that your formula is correct.

1.8-45 Determine (by experiment) a plausible formula for

$$\sum_{i=1}^{n} \frac{1}{i^2 + i}.$$

Then prove by induction that your formula is correct.

1.8-46 Show by induction, for every positive integer n, that $n^3 - n$ is divisible by 3.

1.8-47 The quantity

$$\boxed{\binom{n}{k} = \frac{n!}{k!(n-k)!}}$$

is the number of ways of choosing k objects out of n objects, where $n \geq k$. The quantity $\binom{n}{k}$ is also known as the *binomial coefficient*. We read the notation $\binom{n}{k}$ as "n choose k."
Show by induction that, for $1 \leq k \leq n$,

$$\binom{n+1}{k} = \binom{n}{k} + \binom{n}{k-1} \tag{1.82}$$

1.8-48 Show by induction that, for $n \geq 0$,

$$\sum_{i=0}^{n} \binom{n}{k} = 2^n.$$

1.8-49 Show by induction that

$$\boxed{(x+y)^n = \sum_{k=0}^{n} \binom{n}{k} x^k y^{n-k}} \tag{1.83}$$

This important formula is known as the *binomial theorem*.

1.8-50 Prove the following by induction:
$$\sum_{k=1}^{n} k^2 = \frac{n(n+1)(2n+1)}{6}.$$

1.8-51 Prove the following by induction:
$$\boxed{\sum_{k=1}^{n} r^n = \frac{r^n - 1}{r - 1} \qquad r \neq 1}$$

1.8-52 Prove by induction that
$$\frac{1}{\sqrt{4n+1}} < \frac{1}{2} \cdot \frac{3}{4} \cdots \cdots \frac{2n-3}{2n-2} \cdot \frac{2n-1}{2n} < \frac{1}{\sqrt{3n+1}},$$
for integers $n \geq 1$.

1.8-53 Prove by induction that, for $x, y, n \in \mathbb{Z}$, $(x - y)$ divides $x^n - y^n$. This is written as
$$(x - y)|(x^n - y^n).$$

1.9-54 Prepare a table showing the storage contents and outputs for the LFSR shown in the accompanying illustration, with initial conditions shown in the delay elements. Also, determine the connection polynomial $C(D)$.

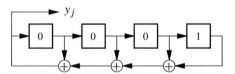

1.9-55 Consider the LFSR described by the polynomial
$$C(D) = 1 + D + D^2 + D^3.$$
(a) Draw the LFSR block diagram using both the realization shown in figure 1.20 and the realization shown in 1.21.
(b) For the initial condition $\{0, 0, 1\}$ trace the operation of both realizations of the LFSR and verify that the output sequence of each is the same. How many district states are there?

1.9-56 Consider the LFSR described by the the polynomial
$$1 + D^2 + D^3.$$
(a) Draw the LFSR block diagram using both the realization shown in figure 1.20 and the realization shown in 1.21.
(b) For the initial condition $\{0, 0, 1\}$ trace the operation of both realizations of the LFSR and verify that the output sequence of each is the same. How many district states are there?

1.9-57 Given the sequence $\{0, 0, 0, 1, 0, 1, 0\}$,
(a) Determine the shortest-length LFSR that could produce this sequence, performing the computations by hand.
(b) Check your work using Algorithm 1.2 in MATLAB.

1.9-58 Show that for $j = 0, 1, \ldots, n$, the output of the LFSR with connection polynomial $C^{[n+1]}(D)$ as in (1.65) with $A = -d_m^{-1} d_n$ and $l = n - m$ satisfies $d_j = 0$ (no discrepancy).

1.9-59 Write the output sequence as a polynomial
$$Y(D) = y_0 + y_1 D + y_2 D^2 + \cdots.$$

(a) Using (1.63), show that the jth coefficient in $Y(D)C(D)$ vanishes for $j = p, p+1, \ldots$, where $\deg(C(D)) = p$. Hence, we can write
$$C(D)Y(D) = Z(D),$$
where
$$Z(D) = z_0 + z_1 D + \cdots + z_{p-1} D^{p-1}.$$
Thus, knowing $Z(D)$, we can find the output by polynomial long division:
$$Y(D) = \frac{Z(D)}{C(D)}. \tag{1.84}$$

(b) Show that the coefficients of $Z(D)$ can be related to the initial conditions of the LFSR by
$$\begin{bmatrix} 1 & 0 & \cdots & 0 \\ c_1 & 1 & \cdots & 0 \\ c_2 & c_1 & \cdots & 0 \\ \vdots & & & \\ c_{p-1} & c_{p-2} & \cdots & c_1 & 1 \end{bmatrix} \begin{bmatrix} y_0 \\ y_1 \\ y_2 \\ \vdots \\ y_{p-1} \end{bmatrix} = \begin{bmatrix} z_0 \\ z_1 \\ z_2 \\ \vdots \\ z_{p-1} \end{bmatrix}.$$

1.9-60 Let $C(D) = 1 + D^2 + D^3$, with initial contents $\{y_0, y_1, y_2\} = \{1, 0, 0\}$. Determine the first six outputs using polynomial long division (1.84). Compare the results to those obtained directly from the LFSR.

1.9-61 Determine the sequence $\{y_i\}$ of length seven generated by $C(D) = 1 + D + D^3$, and call its length N. Then compute the cyclic autocorrelation function
$$\rho(k) = \frac{1}{N} \sum_{i=0}^{N-1} y_i y_{((i-k))},$$
where $y_{((i-k))}$ means that the subscript is computed modulo N. Plot this autocorrelation function.

1.11 References

The linear systems theory presented here in broad strokes is painted in considerably finer detail in [284] and [164]. Our brief introduction to linear prediction is more extensively presented in [68, 132], while considerably more on spectrum analysis appears in [174, 220]. The applications of adaptive filtering highlighted here are discussed in depth in [132] and [368]. The hidden Markov model is presented in [266, 68] and [265]. For an enjoyable and readable introduction to proofs, with a variety of suggestions and examples and some good mathematical background, [352] is recommended. A thought-provoking book on mathematical thinking is [256].

Massey's algorithm is presented in [221]. An excellent presentation of the algorithm is in [32]. The book [109] provides an introduction to LFSRs, and the paper [288] an interesting discussion of decimated maximal-length sequences. Applications of LFSRs to spread-spectrum communications are discussed in [387].

Part II

Vector Spaces and Linear Algebra

The first important theme of this part is that **signals are vectors**, allowing us to apply the powerful tools of vector analysis and linear algebra to signal analysis. This identification leads to a variety of applications, including optimal filtering, approximation, interpolation, data compression, and transforms.

The second important theme is the existence and nature of the solution of linear equations that arise in signal processing. Results are discussed for matrix linear operators and other linear operators.

The third important theme is how solutions to linear problems are *computed* in a reliable and efficient manner. Examination of this issue leads to useful matrix factorizations, including LU, Cholesky, QR, and SVD, and specialized techniques for matrices which arise in signal processing.

The concept of invariance under linear transformation—the eigenspace of an operator—forms a fourth theme. A variety of applications of eigenvalue and eigenvector concepts are presented, including modal estimation, controls, and filter design.

Before embarking on the material in this part, the reader is encouraged to review basic matrix notation and concepts in appendix C.

Chapter 2

Signal Spaces

> Language makes a mighty loose net with which to go fishing for simple facts, when facts are infinite.
> — *Edward Abbey*
> Desert Solitaire

> Beginners are not prepared for real mathematical rigor; they would see in it nothing but empty, tedious subtleties. It would be a waste of time to try to make them more exacting; they have to pass rapidly and without stopping over the road which was trodden slowly by the founders of the science.
> — *Henri Poincarè*
> Science and Hypothesis

This chapter is mostly about two kinds of mathematical objects: metric spaces and linear vector spaces. The idea behind a metric space is simply that we provide a way of measuring the distance between mathematical objects, such as sets, points, functions, or sequences. With this notion of distance we will be able to generalize some of the familiar concepts of calculus, such as continuity or convergence, beyond operations on a single dimension to operations in higher dimensions.

The concept of a vector space is also simple: it is a set of objects that can be combined together using linear combinations. But the theory of vector spaces has far-reaching ramifications, covering a significant portion of the theory of signal processing. A key insight in vector space theory is that, in a geometrically useful sense, **functions (i.e., signals) can be regarded as vectors**. This geometric understanding provides a powerful tool for signal analysis. In this chapter, the basic theory and notation of vector spaces is developed. In chapter 3 we put this notion to work in a variety of applications, including optimal filtering (both least squares and minimum mean squares), transforms, data compression, sampling, and interpolation.

In our study of metric spaces and vector spaces, the intent is to provide a framework for the general discussion of signals. Before embarking on this chapter, the reader is encouraged to review the basic definitions of functions and sets appearing in appendix A. In this study, matrix notation is heavily employed sections so review of the basic matrix notations presented in appendix C is also recommended.

In the development of this chapter, we build successively from **metric spaces**, to **vector spaces**, to **normed vector spaces**, to **normed inner-product** spaces. This will lead us to the important idea of projections and orthogonal projections. Orthogonal projection will be a tool of tremendous importance to us in the next chapter, where it will be used as the geometrical basis for both least-squares and minimum mean-squares filtering and prediction.

2.1 Metric spaces

We may consider that the signals (functions) of interest to us in a particular problem are members of some set X. In studying and applying these signals, we may be interested in understanding how a signal compares with other signals in this set. One way to do this is to measure a "distance" between the signals using a measure of distance that is both mathematically practically and physically meaningful. The mathematical aspects of a useful measuring function are expressed in the following definition.

Definition 2.1 A **metric** $d\colon X \times X \to \mathbb{R}$ is a function that is used to measure distance between elements in a set X. In order to be a metric, it must satisfy the following properties, for all $x, y \in X$:

M1 $d(x, y) = d(y, x)$.

M2 $d(x, y) \geq 0$.

M3 $d(x, y) = 0$ if and only if $x = y$.

M4 For all points $x, y, z \in X$,

$$d(x, z) \leq d(x, y) + d(y, z). \tag{2.1}$$

□

Example 2.1.1 For $x, y \in \mathbb{R}$ we can define a metric using the absolute value function by

$$d(x, y) = |x - y|.$$

The required properties of a metric are all satisfied. The last property follows from the **triangle inequality**, so called because of the relationship it imposes on the sides of a planar triangle. Let x, y, and z denote the corners of a triangle, as shown in figure 2.1. Then $d(x, z)$ is the length of one side,

Figure 2.1: Illustration of the triangle inequality

$d(y, z)$ is the length of the second side, and $d(x, z)$ is the length of the third side. The length of the third side cannot be longer than the lengths of the first two sides. □

There are a variety of metrics used; the following example demonstrates a few of them.

2.1 Metric Spaces

Example 2.1.2 Let X be the set of numbers in $\mathbb{R}^n$. Let $\mathbf{x} \in \mathbb{R}^n$ and $\mathbf{y} \in \mathbb{R}^n$.

1. The metric $d_1 \colon \mathbb{R}^n \times \mathbb{R}^n \to \mathbb{R}$ defined by
$$d_1(\mathbf{x}, \mathbf{y}) = \sum_{i=1}^{n} |x_i - y_i|$$
is called the l_1 metric, also known as the Manhattan metric, since distance measured in a city laid out on a Cartesian grid must follow straight along the streets. Satisfaction of property (2.1) for this metric follows from the triangle inequality applied to each term.

2. The metric $d_2 \colon \mathbb{R}^n \times \mathbb{R}^n \to \mathbb{R}$ defined by
$$d_2(\mathbf{x}, \mathbf{y}) = \left(\sum_{i=1}^{n} (x_i - y_i)^2 \right)^{1/2}$$
is called the l_2 metric. It represents the Euclidean distance between the points. The fact that this metric satisfies property (2.1) is proved in section 2.6.

3. Generalizing the first two metrics, we have
$$d_p(\mathbf{x}, \mathbf{y}) = \left(\sum_{i=1}^{n} (x_i - y_i)^p \right)^{1/p}.$$
This is the l_p metric. The fact that this metric satisfies (2.1) follows from the Minkowski inequality, which is proved in appendix A.

4. As $p \to \infty$, the l_p metric becomes the l_∞ metric,
$$d_\infty(\mathbf{x}, \mathbf{y}) = \max_{i=1,2,\ldots,n} |x_i - y_i|.$$
□

Example 2.1.3 Consider a vector $\mathbf{x} \in \mathbb{R}^n$ which is to be approximated (quantized) by a vector as $\hat{\mathbf{x}}$ illustrated in figure 2.2. To have a good representation of the data, we desire that $\hat{\mathbf{x}}$ "look like" $\mathbf{x}$, according to some criterion, and the quantizer should be designed with this in mind. While many different metrics have been examined, frequently the metrics employed in quantizer design turn out to be one of these mentioned above, such as $d_1(\mathbf{x}, \hat{\mathbf{x}})$ or $d_2(\mathbf{x}, \hat{\mathbf{x}})$. □

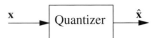

Figure 2.2: Quantization of the vector $\mathbf{x}$

Example 2.1.4 Let $\mathbf{x}$ be a binary sequence, $\mathbf{x} = \{x_0, x_1, \ldots, x_{n-1}\}$, where x_i is either 0 or 1. This sequence is transmitted through a channel where it may be corrupted by some noise. The received sequence is $\mathbf{y} = \{y_0, y_1, \ldots, y_{n-1}\}$. In receiving such sequences, the goal for good reception is that the bits in $\mathbf{y}$ should match the bits in $\mathbf{x}$. An appropriate metric for this criterion is the *Hamming distance* between the sequences, which is the number of places that x_i and y_i are different,
$$d_H(x, y) = \sum_{i=0}^{n-1} h(x_i - y_i)$$
where
$$h(x - y) = \begin{cases} 1 & \text{if } x - y \neq 0 \\ 0 & \text{if } x - y = 0. \end{cases}$$
When $\mathbf{x}$ and $\mathbf{y}$ are binary sequences, then the Hamming distance between them can be written as
$$d_H(x, y) = \sum_{i=0}^{n-1} x_i \oplus y_i,$$
in which $\oplus$ denotes addition modulo 2. □

Definition 2.2 A **metric space** (X, d) is a set X together with a metric d. □

There are many possible metric spaces. We begin with metric spaces defined for sequences.

Example 2.1.5

1. The set $\mathbb{R}^n$ equipped with the metric $d_2(\mathbf{x}, \mathbf{y})$ is a metric space.

2. Let $l_p = l_p(0, \infty)$ be the set consisting of all infinite sequences of real or complex numbers $\{x_0, x_1, x_2, \ldots\}$ such that $\sum_{i=0}^{\infty} |x_i|^p < \infty$. We will take $1 \leq p < \infty$. The function

$$d_p(x, y) = \left[\sum_{i=0}^{\infty} |x_i - y_i|^p\right]^{1/p}$$

defines a metric on l_p, which we will call the l_p metric. We refer to this metric space as the $l_p(0, \infty)$ space, or simply the l_p space. This is an infinite-dimensional space known as a *sequence space*.

The set of two-sided sequences $\{\ldots, x_{-1}, x_0, x_1, \ldots\}$ with metric d_p gives the metric space $l_p(-\infty, \infty)$.

In discrete-time signal-processing applications, we deal most frequently with l_1 space or with l_2 space, the former because absolute values are easy to compute, and the latter because the quadratic metric function is easily differentiable.

3. The space $l_\infty(0, \infty)$ consists of all sequences of numbers $\{x_0, x_1, x_2, \ldots\}$ such that $|x_n| \leq M$ for some finite bound M, equipped with the metric

$$d_\infty(x, y) = \sup_n |x_n - y_n|. \tag{2.2}$$

See box 2.1. The corresponding space of two-sided sequences is denoted as $l_\infty(-\infty, \infty)$.

□

There are also many useful metric spaces defined over functions. These infinite-dimensional spaces are called *function spaces*.

The metric space $(C[a, b], d_p)$. Let $X = C[a, b]$ be the set of real-valued (or complex-valued) functions defined on the interval $[a, b]$, with $b > a$. We can define a metric on

---

> **Box 2.1: Sup and inf**
>
> For a set $S \subset \mathbb{R}$, the least upper bound (LUB) is the smallest number z such that $z \geq x$ for every $x \in S$. The LUB of a set S is called the **sup** (supremum) of the set. If there is no number that is greater than all the elements of S, then $\sup(S) = \infty$. Similarly, the greatest lower bound (GLB) of a set is the largest number w such that $w \leq x$ for every $x \in S$. The GLB is called the **inf** (infimum) of S. If there is no number less than all the elements of S, then $\inf(S) = -\infty$.
>
> The inf and sup are generalizations of min and max, respectively. Generally the inf and sup are used when there is a continuum of values over which to find the max or min, or where the extrema may be infinite.
>
> **Example 2.1.6** Let $S = (2, 5) \subset \mathbb{R}$. (This is an open set, and does not contain the endpoints.) Then,
>
> $$\sup(S) = 5 \quad \text{and} \quad \inf(S) = 2.$$
>
> Let $T = [4, 7)$. Then $\inf(T) = 4$ and $\sup(T) = 7$. Let $U = (1, \infty)$. Then $\inf(U) = 1$ and $\sup(U) = \infty$ □

2.1 Metric Spaces

functions x and y in X by

$$d_p(x, y) = \left[\int_a^b |x(t) - y(t)|^p \, dt \right]^{1/p}, \tag{2.3}$$

where $1 \leq p < \infty$. This gives the metric space $(C[a, b], d_p)$. The metric d_p between functions is referred to as the L_p metric. It must be established that (2.3) is, in fact, a metric. For $p = 2$ this is established using the Cauchy–Schwarz inequality (see section 2.6). For other values of p, the Minkowski inequality proved in appendix A is used.

The metric space $(C[a, b], d_\infty)$. Letting $p \to \infty$ in the definition of the last metric, we obtain (see box 2.1),

$$d_\infty(x, y) = \sup\{|x(t) - y(t)| : a \leq t \leq b\}. \tag{2.4}$$

In other words, the distance between the functions is obtained at the point where the functions are farthest apart. This metric space is denoted as $(C[a, b], d_\infty)$ or, more simply, as $C[a, b]$ (the metric being understood by convention).

The difference between the metric spaces $(C[a, b], d_\infty)$ and $(C[a, b], d_p)$ can be appreciated by considering the functions illustrated in figure 2.3. Let $X = C[0, T]$, and let x_0 be a point in X (a function). Figure 2.3(a) shows the region within which all functions x that satisfy

$$d_\infty(x_0, x) < \epsilon$$

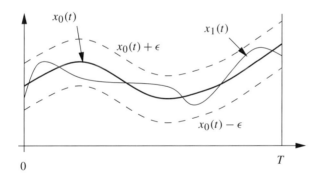

(a) d_∞ approximation

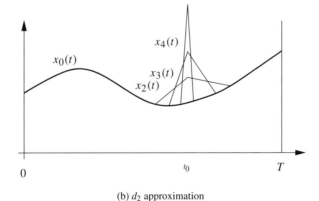

(b) d_2 approximation

Figure 2.3: Comparison of d_∞ and d_2 metrics

must fall. For example, the function $x_1(t)$, as shown, falls in the region. While there may be some wiggling around within the region, the function is never allowed to escape.

By contrast, figure 2.3(b) illustrates some functions that satisfy

$$d_2(x_0, x) < \epsilon.$$

That is, these are functions for which

$$\int_0^T (x_0(t) - x(t))^2 \, dt < \epsilon.$$

At any given point t_0, there may be significant deviation from $x_0(t)$, as long as the region over which the deviation occurs is not too long. The narrower the region of deviation, the bigger the deviation might be. If $x(t)$ is an approximation to $x_0(t)$, using the d_∞ metric in expressing the approximation criterion provides an upper bound to the approximation error $x(t) - x_0(t)$ that cannot be obtained when using the d_p metric for $1 \leq p < \infty$.

The metric space $L_p[a, b]$.   Let $L_p[a, b]$ denote the set of real- or complex-valued functions $x(t)$ defined on the interval $t \in [a, b]$ such that

$$\int_a^b |x(t)|^p \, dt < \infty,$$

where $1 \leq p < \infty$. This set, equipped with the metric d_p of (2.3), forms the metric space $(L_p[a, b], d_p)$ or, more simply, $L_p[a, b]$. When the interval is understood, this is often written simply as L_p. The metric (2.3) is often referred to as the L_p metric.

Several technicalities associated with the L_p space are discussed in section 2.1.3. For many problems of engineering interest, these technicalities do not present a difficulty, but they do bear some consideration.

The metric space $L_\infty[a, b]$.   Let $L_\infty[a, b]$ denote the set of real- or complex-valued functions $x(t)$ defined on the interval $[a, b]$ such that

$$\sup_{t \in [a,b]} |x(t)| < \infty.$$

This set, equipped with the metric d_∞ of (2.4), is a metric space.

2.1.1 Some topological terms

With the notion of a metric established, we can introduce some elementary concepts from point-set topology.

In a metric space X, the **ball** or **sphere** centered at x_0 of radius δ is the set of points which are within a distance δ of x_0:

$$B(x_0, \delta) = \{x \in X : d(x_0, x) < \delta\}. \tag{2.5}$$

Such a ball is also said to be a **neighborhood** of x_0: it is the set of points that live close to x_0.

Definition 2.3   A point $x_0 \in X$ is **interior** to a set $S \subset X$ if all points sufficiently near to x_0 are in X. That is, there is some $\delta > 0$ such that $B(x_0, \delta) \subset S$.

The **interior** of a set S is the set of all points in x that are interior to the set. A point $x_0 \notin S$ is **exterior** if there is neighborhood of x_0 that is outside (does not intersect) S.   □

Figure 2.4 illustrates an interior point, an exterior point, and a point which is neither interior nor exterior.

2.1 Metric Spaces

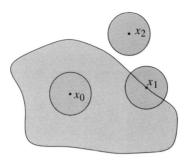

Figure 2.4: x_0 is interior, x_2 is exterior, and x_1 is neither interior nor exterior

Definition 2.4 A set X is **open** if every point in X is interior. □

Example 2.1.7 The set $X = (0, 1) \subset \mathbb{R}$ is an open set. We will show that every point is interior. Let $x_0 \in X$. Then the neighborhood of x_0

$$B(x_0, x_0/2) = \{x \in X : |x - x_0| < x_0/2\}$$

is a subset of S for any x_0.

The set $Y = [0, 1] \subset \mathbb{R}$ (including the endpoints) is not open. The point 0 has no neighborhood surrounding it that lies entirely in Y. □

It is straightforward to show that finite unions and intersections of open sets are open.

Definition 2.5 A set $X \subset X$ is said to be **closed** if the complement of X is an open set. □

Example 2.1.8 Let $X = (0, 4)$, and let $S = [1, 2] \subset X$. Then $\overline{S} = (0, 1) \cup (2, 4)$. This is the union of two open sets, and hence is open. Thus S must be closed. □

For many purposes in this book, we will use open sets because they cannot contain only a single point. For example, in some results on optimization, we might state something like: "$f(t)$ is continuous in an open neighborhood around t_0." What this means is that we can look at the points around t_0—in at least some neighborhood—and use continuity there.

Definition 2.6 A **boundary point** of a set X is a point x_0 such that every neighborhood of x_0 contains elements both in S and not in S. A boundary point is not necessarily an element of X.

The **boundary** of a set X is the collection of all the boundary points of X. The boundary of a set X is sometimes denoted as $\mathrm{bdy}(X)$. □

Example 2.1.9 For the set $X = [0, 1) \subset \mathbb{R}$, the point 0 is a boundary point, since every neighborhood of 0 has points in X and points not in X. The point 1 is also a boundary point (which is not an element of X). The boundary of X is $\mathrm{bdy}(X) = \{0, 1\}$.

This set is neither open nor closed in $\mathbb{R}$. □

Definition 2.7 The **closure** of a set X is the union of the set X with its boundary. The closure of X is denoted as $\mathrm{closure}(X)$. (Other texts use $\overline{X}$ to indicate closure.)

$$\mathrm{closure}(X) = X \cup \mathrm{bdy}(X).$$

The closure of a set is always closed. □

Example 2.1.10 For the set $X = [0, 1)$, the closure is

$$\mathrm{closure}(X) = [0, 1].$$

Figure 2.5 illustrates open and closed sets.

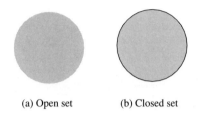

(a) Open set   (b) Closed set

Figure 2.5: Illustration of open and closed sets

Example 2.1.11 Some examples of open and closed sets:

1. The set of (x, y) such that $x^2 - 2yx = 0$ is closed in $\mathbb{R}^2$. (Every point is a boundary point.)
2. The set of (x, y) such that $x^2 y > 3(x - y)$ is open in $\mathbb{R}^2$.
3. The set $\mathbb{Z}$ is closed in $\mathbb{R}$. (Every point is isolated from every other point; every point is a boundary point.)

In addition to the simple sets of points in $\mathbb{R}^n$, it is interesting to examine open and closed sets over more complicated metric spaces.

Example 2.1.12 Let $X = C[0, T]$. The set of functions $x \in X$ such that

$$d_\infty(x_0, x) < \epsilon,$$

which is portrayed in figure 2.3(a), is an open set. This is the open neighborhood of functions around $x_0(t)$.

Definition 2.8 A point $x \in X$ is said to be a **cluster point** in X if every neighborhood around x contains infinitely many points of X.

Definition 2.9 The **support** of a function $f : A \to B$ is the closure of the set of elements $a \in A$ where $f(a) \neq 0$.

In concluding this section of definitions, we summarize some of the basic topological properties of sets, as follows.

1. The union of any number (even an infinite number) of open sets is open. The intersection of any number (even an infinite number) of closed sets is closed.
2. The intersection of an infinite number of open sets need not be open. To see this, let $A_k = (0, 1 + 1/k)$. Then, $A_1 \supset A_2 \supset A_3 \supset \cdots$. The intersection of all these intervals, $B = \cap_{k=1}^\infty A_k$ is the interval $(0, 1]$, which is not an open set.
3. The union of an infinite number of closed sets need not be closed.

2.1.2 Sequences, Cauchy sequences, and completeness

Sequences of numbers or functions arise frequently in signal processing theory and practice. As an example, an iterative algorithm such as an adaptive filter produces a sequence of vectors (filter weights).

2.1 Metric Spaces

Many sequences are generated as follows: Starting from some initial point x_0 in a metric space X, a sequence is obtained by updating the last point, possibly incorporating some new data. The update for an iterative algorithm can be written abstractly as

$$x_{n+1} = f(x_n, u_n),$$

where f is an update function and u_n is the input data at the nth iteration. Repeated iteration gives the sequence $x_0, x_1, \ldots$.

If x_n ultimately gets close to some value for large enough n, we can say that the sequence $\{x_n\}$ converges. This is stated more precisely in the following definition.

Definition 2.10 If for every $\delta > 0$, there is an n_0 such that $d(x_n, x^*) < \delta$ for every $n > n_0$ for some fixed value x^*, then the sequence $\{x_n\}$ is said to converge to x^*. In this case we write

$$x_n \to x^*.$$

We say in this case that x^* is the **limit** of x_n. □

Another way of stating this is as follows: The sequence $\{x_n\}$ converges to x^* if and only if every neighborhood around x^* contains all the terms x_n for $n > n_0$. For every neighborhood N around x^*, there is an n_0 such that $x_n \in N$ when $n \geq n_0$.

Example 2.1.13 Convergence can be appreciated by considering sequences that do not converge. The sequences

$$a_n = n^2,$$
$$b_n = 1 + (-1)^n,$$

do not converge: the first sequence is not bounded, and the second sequence oscillates between 0 and 2. □

The following facts about convergent sequences are important:

1. Let (X, d) be a metric space. The closure of a set $A \subset X$ is the set of all limits of converging sequences of points from A.
2. A set $A \subset X$ is closed if and only if it contains the limit of every converging sequence $\{x_n\}$ whose points lie in A.

Example 2.1.14 Consider the following sequence of numbers:

$$\{1, 1.41, 1.414, 1.4142, 1.41421, \ldots\}.$$

Each number in this sequence is a rational number, an element of $\mathbb{Q}$. This sequence is converging to $\sqrt{2}$, which is an irrational number. Since the limit of the sequence is not in the set $\mathbb{Q}$, we conclude that $\mathbb{Q}$ is not closed. However, the set of real numbers $\mathbb{R}$ is closed: every convergent sequence in $\mathbb{R}$ has its limit in $\mathbb{R}$. □

Similar to a limit is a **limit point**: if the sequence x_n returns infinitely often to a neighborhood of a point x^*, then x^* is a limit point. In the sequence

$$b_n = 1 + (-1)^n,$$

the points 0 and 2 are both limit points (but not limits) of the sequence. If there are limit points of a sequence, however, we can take a **subsequence** which converges to a limit.

The largest limit point of a sequence $\{x_n\}$ is called the limit superior, or **limsup**. It is often written as

$$\limsup_{n \to \infty} x_n.$$

The smallest limit point of a sequence is called the limit inferior, or **liminf**. It is often written as
$$\liminf_{n \to \infty} x_n.$$
Obviously, if $\limsup x_n = \liminf x_n$ then the sequence is convergent.

Example 2.1.15 Consider the sequence
$$c_n = 1 + \frac{1}{n} + (-1)^n.$$
There are two limit points: 2 and 0. The subsequence $\{c_0, c_2, c_4, \ldots\}$ has the limit 2, and the subsequence $\{c_1, c_3, c_5, \ldots,\}$ has the limit 0. For the sequence $\{c_n\}$,
$$\limsup_{n \to \infty} c_n = 2,$$
$$\liminf_{n \to \infty} c_n = 0.$$
□

Definition 2.11 A sequence $\{x_n\}$ in $\mathbb{R}$ is **monotonic** if
$$x_1 \leq x_2 \leq x_3 \leq \cdots$$
or
$$x_1 \geq x_2 \geq x_3 \geq \cdots.$$
□

For sequences over the real numbers, the following fact is clear: every bounded monotonic sequence is convergent. Since the sequence is bounded, the monotonic sequence "runs out of room," and hence must have a limit point, which (because the sequence is monotonic) must be unique.

Definition 2.12 A sequence $\{x_n\}$ in a metric space (X, d) is said to be a **Cauchy sequence** if, for any $\epsilon > 0$, there is an $N > 0$ (which may depend upon ϵ) such that $d(x_n, x_m) < \epsilon$ for every $m, n > N$.
□

It can be shown (see the exercises) that if a sequence converges, it is a Cauchy sequence. On the other hand, it is possible for a sequence to be a Cauchy sequence and not be convergent in X.

Example 2.1.16 Let $C[a, b]$ be the set of continuous functions defined on the interval $[a, b]$. Let $X = C[-1, 1]$, and consider the sequence of functions $f_n(t)$ defined by
$$f_n(t) = \begin{cases} 0 & t < -1/n, \\ nt/2 + 1/2 & -1/n \leq t \leq 1/n, \\ 1 & t > 1/n. \end{cases} \quad (2.6)$$
A typical function is shown in figure 2.6. In the metric space (X, d_2), where d_2 is the metric defined by
$$d_2(f, g) = \int_{-1}^{1} (f(t) - g(t))^2 \, dt,$$

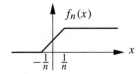

Figure 2.6: The function $f_n(t)$

2.1 Metric Spaces

we find that
$$d_2(f_n, f_m) = \frac{1}{6m^3n}(m^3 + 4m^2n + mn^2 + 2n^3) \qquad m > n,$$
which $\to 0$ for m and n large. Thus, the sequence is a Cauchy sequence, but the limit function
$$f(t) = \begin{cases} 0 & t < 0, \\ 1/2 & t = 0, \\ 1 & t > 0. \end{cases}$$
is a discontinuous function and hence is not in X. Therefore, we cannot say that $f_n(t)$ is a convergent sequence in X. □

The failure of a Cauchy sequence to converge is a deficiency—a "hole"—in the underlying metric space.

Definition 2.13 A metric space (X, d) is **complete** if every Cauchy sequence in X is convergent in X. □

By this definition, the metric space $(C[a, b], d_2)$ of example 2.1.16 is not complete: There exist Cauchy sequences in it where limit is not in the metric space.

Example 2.1.17 Whether a metric space is complete or not depends on the metric employed. Consider the metric space (X, d_∞), and d_∞ is the metric
$$d_\infty(f, g) = \sup_{t \in [-1,1]} |f(t) - g(t)|.$$
It can be shown that the sequence $f_n(t)$ is not a Cauchy sequence in this metric space, so we cannot use this sequence to test the completeness of $(C[a, b], d_\infty)$. But we can still argue for the completeness of the space. Let $x_n(t)$ be a Cauchy sequence in $(C[a, b], d_\infty)$; then for any $\epsilon > 0$,
$$\sup_{t \in [-1,1]} |x_m(t) - x_n(t)| < \epsilon$$
for m and n sufficiently large, so that $|x_m(t) - x_n(t)| < \epsilon$ for every $t \in [-1, 1]$. Hence, for every fixed $t_0 \in [-1, 1]$, $x_n(t_0)$ is convergent to some number $x(t_0)$. Collectively, these define a function $x(t)$. To show completeness, we must show that $x(t) \in C[a, b]$; in other words, that it is continuous. Let n be sufficiently large that $|x_n(t) - x(t)| < \epsilon < 3$. Let δ be determined so that $|x_n(t) - x_n(t_0)| < \epsilon/3$ when $|t - t_0| < \delta$. (Since $x_n(t)$ is continuous, such a δ exists.) Then
$$|x(t) - x(t_0)| = |(x(t) - x_n(t)) + (x_n(t) - x_n(t_0)) + (x_n(t_0) - x(t_0))|$$
$$\leq |x(t) - x_n(t)| + |x_n(t) - x_n(t_0)| + |x_n(t_0) - x(t_0)| < \epsilon,$$
where the first inequality follows from the triangle inequality. Thus we see that for $|t - t_0| < \delta$ we have $|x(t) - x(t_0)| < \epsilon$, so $x(t)$ must be continuous. □

In examining the convergence of sequences (such as the result of an iterative algorithm), it is usually easier to show that a sequence is a Cauchy sequence than to show that it is a convergent sequence. To determine if a sequence is Cauchy requires only that we examine the sequence, and establish that points become sufficiently close. On the other hand, establishing convergence requires information apart from the sequence; namely, the limiting value of the sequence. However, if the underlying space is complete, then establishing that a sequence is a Cauchy sequence is sufficient to establish convergence. For this reason, we shall usually assume that the work on function spaces is carried out in a complete metric space.

Example 2.1.18 An example of an incomplete space is the metric space $(\mathbb{Q}, d_1)$, the set of rational numbers. In this space, the sequence $\{1, 1.4, 1.41, 1.414, 1.4142, \ldots\}$, the sequence approaching $\sqrt{2}$, is a Cauchy sequence, but it is not convergent in $\mathbb{Q}$, since $\sqrt{2}$ is not rational. □

> **Box 2.2: The measure of a set**
>
> Given a real interval $S = [a, b]$, the measure of S is simply the length of the interval, $\mu(S) = b - a$. For a set that is the union of disjoint intervals, $S = S_1 \cup S_2 \cup \cdots$, where $S_i \cap S_j = \emptyset$, the measure is the sum of the individual measures,
>
> $$\mu(S) = \mu(S_1) + \mu(S_2) + \cdots.$$
>
> A set of real numbers is said to have measure zero if, for every $\epsilon > 0$, the set can be covered by a collection of open intervals whose total length is $< \epsilon$. A single point has measure 0; so does a finite collection of isolated points. Any countable set of points has measure zero, since around the nth point an open interval of length $\epsilon/2^n$ can be placed. The total length of the countable set is thus less than or equal to
>
> $$\epsilon \left(\frac{1}{2} + \frac{1}{4} + \frac{1}{8} + \cdots \right) = \epsilon.$$
>
> The measure of sets in $\mathbb{R}^n$ is defined by finding areas, volumes, and so forth, of sets in $\mathbb{R}^n$.

Generalizing the results of example 2.1.16 it can be shown that, $(C[a, b], d_p)$ is not a complete metric space for $p < \infty$. However, the space $(L_p[a, b], d_p)$ *is* a complete metric space.

2.1.3 Technicalities associated with the L_p and L_∞ spaces*

There are several technicalities associated with the L_p space that bear at least brief consideration.

1. Consider the functions defined by

$$f_1(t) = \begin{cases} \sin(t) & 0 \leq t \leq 4, \\ 0 & \text{otherwise}; \end{cases} \qquad f_2(t) = \begin{cases} \sin(t) & 0 \leq t \leq 4, t \neq 2, \\ 5 & t = 2, \\ 0 & \text{otherwise}. \end{cases}$$

   These functions are clearly not equal at every point. However, for any p in the range $1 \leq p < \infty$, $d_p(f_1, f_2) = 0$. Thus we have functions which are not equal but for which the metric is zero, in violation of requirement M3 for metrics, as stated in definition 2.1. The functions $f_1(t)$ and $f_2(t)$ are said to differ on a *set of measure zero*, or to be equal *almost everywhere*, abbreviated as a.e. (See box 2.2.)

   For our purposes, functions f and g for which $d_p(f, g) = 0$ are said to be equal, even though they may not be equal at every point. Thus, when we talk of a function, we are actually referring to a whole *class* of functions which differ on a set of measure zero. So "equality" does not necessarily mean equality at every point!

   Example 2.1.19 It is understood from elementary signals theory that a periodic function can be represented using a Fourier series. The periodic square-wave function defined by

$$f(t) = \begin{cases} 0 & -\frac{1}{2} \leq t < -\frac{1}{4}, \\ 1 & -\frac{1}{4} \leq t \leq \frac{1}{4}, \\ 0 & \frac{1}{4} < t < \frac{1}{2}, \end{cases}$$

---

*Note: This section can be skipped on a first reading.

2.1 Metric Spaces

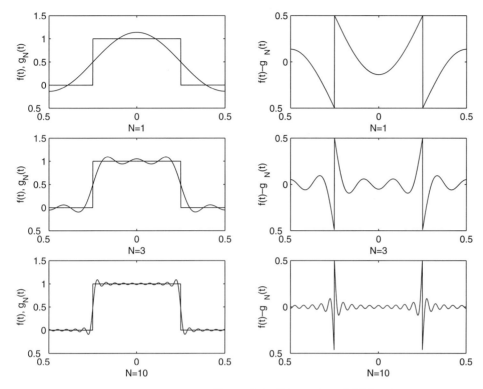

Figure 2.7: Illustration of Gibbs phenomenon

has the Fourier series

$$f(t) = \frac{1}{2} + \frac{1}{\pi}\left[2\cos 2\pi t - \frac{2}{3}\cos 6\pi t + \frac{2}{5}\cos 10\pi t + \cdots\right] = g(t). \qquad (2.7)$$

Then, by the convergence of the Fourier series we have

$$d_2(f(t), g(t)) = \left[\int_{-1/2}^{1/2}(f(t) - g(t))^2 \, dt\right]^{1/2} = 0.$$

However, it is also known that for discontinuous functions, the Gibbs phenomenon occurs: at a point of discontinuity there is an overshoot or an undershoot, no matter how many terms are taken in the summation. Figure 2.7 illustrates the nature of the convergence by showing the sum in (2.7) truncated to N terms, for $N = 1, 3,$ and 10 terms in the summation, with the plots on the left showing the function and its N-term Fourier representation $g_N(t)$, and the plots on the right showing the error $f(t) - g_N(t)$. The point-by-point error is converging to zero everywhere except at the points of discontinuity, where it *never converges to zero*. However, since the width of the error location becomes narrower and narrower, the *integral* of the square of the error approaches 0 as $N \to \infty$. □

2. The space $(L_p[a,b], d_p)$ is larger than the space $(C[a,b], d_p)$, in the sense that the former contains the latter. This is true because there are functions in L_p that are not continuous, whereas all functions in $C[a,b]$ are also functions in L_p (why?). L_p is a "completion" of $C[a,b]$: sequences in $C[a,b]$ that do not have a limit in $C[a,b]$ do have a limit in L_p.

3. In fact, L_p is a large enough space of functions that the concept of integration that we learn in basic calculus, the Riemann integral, does not consistently apply to every function in L_p. Recall that the Riemann integral is defined as the limit of a sum

$\sum f(x_i) \Delta x_i$, where the x_i are chosen as points inside Δx_i. There are functions in L_p that cannot be integrated by this limiting process.

Example 2.1.20 The usual pathological example of such a non-integrable function is the function defined on the interval $[0, 1]$ by

$$f(t) = \begin{cases} 1 & \text{if } t \text{ is rational,} \\ 0 & \text{if } t \text{ is irrational.} \end{cases} \tag{2.8}$$

In the Riemann integral, if the points x_i are chosen to be rational, then $\int_0^1 f(t)\,dt = 1$. If the points x_i are chosen to be irrational, then $\int_0^1 f(t)\,dt = 0$. By careful selection of the points x_i, the integral can take *any* value between 0 and 1. □

The integral appropriate for use in L_p spaces is the *Lebesgue integral*. For our purposes, we will not need to worry (beyond this brief mention) about the distinctions. Letting $\int_R$ denote Riemann integration and $\int_L$ denote Lebesgue integration, the following rules apply:

(a) If $\int_R f(t)\,dt$ exists, then $\int_L f(t)\,dt$ exists, and $\int_R f(t)\,dt = \int_L f(t)\,dt$.
(b) The Lebesgue integral is linear: For a scalar α,

$$\int_L \alpha f = \alpha \int_L f \qquad \int_L (f+g) = \int_L f + \int_L g.$$

(c) If $\int_L |f(t)|^2\,dt$ and $\int_L |g(t)|^2\,dt$ exist (are finite), then so are $\int_L f(t)g(t)\,dt$ and $\int_L (f+g)^2\,dt$.
(d) If f and g are equal *except on a set of measure zero*, then

$$\int_L (f-g) = 0 \qquad \int_L (f-g)^2 = 0.$$

This last rule suffices to cover many of the pathological functions for which the Riemann integral has no value. For example, the function $f(t)$ defined in (2.8) is equal to the function $g(t) = 0$, except on a set of measure zero (since the rational numbers form a countable set). Thus, using the Lebesgue integral there is no ambiguity and $\int_0^1 f(t)\,dt = 0$.

4. When dealing with the L_∞ norm, yet another issue arises. Consider the function

$$x(t) = \begin{cases} 1 & t = 0, \\ 0 & t \neq 0. \end{cases} \tag{2.9}$$

For this function $\sup x(t) = 1$. However, $x(t)$ differs from the all-zero function only on a set of measure zero. As for the case of the L_p norms, it is convenient to define the L_∞ norm so that functions that are equal almost everywhere have the same norm. We accordingly define the L_∞ norm by finding the function $y(t)$ that is equal to $x(t)$ almost everywhere and which has the smallest supremum,

$$\|x\|_\infty = \inf_{y(t)=x(t) \text{ a.e.}} \sup |y(t)|.$$

For the function $x(t)$ in (2.9), we find that $y(t) = 0$ satisfies this; hence,

$$\|x(t)\|_\infty = 0.$$

The quantity $\inf_{y(t)=x(t)} \sup |y(t)|$ is called the *essential supremum* of $x(t)$.

2.2 Vector spaces

A finite-dimensional vector **x** may be written as

$$\mathbf{x} = \begin{bmatrix} x_1 \\ x_2 \\ \vdots \\ x_n \end{bmatrix}.$$

The elements of the vector are x_i, $i = 1, 2, \ldots, n$. Each of the elements of the vector lies in some set, such as the set of real numbers $x_i \in \mathbb{R}$, or the set of integers $x_i \in \mathbb{Z}$. This set of numbers is called the set of scalars of the vector space.

The finite-dimensional vector representation is widely used, especially for discrete-time signals, in which the discrete-time signal components form elements in a vector. However, for representing and analyzing continuous-time signals, a more encompassing understanding of vector concepts is useful. It is possible to regard the function $x(t)$ as a vector and to apply many of the same tools to the analysis of $x(t)$ that might be applied to the analysis of a more conventional vector **x**. We will therefore use the symbol x (or $x(t)$) also to represent vectors as well as the symbol **x**, preferring the symbol **x** for the case of finite-dimensional vectors. Also, in introducing new vector space concepts, vectors are indicated in bold font to distinguish the vectors from the scalars. Note: in handwritten notation (such as on a blackboard), the bold font is usually denoted in the signal processing community by an underscore, as in $\underline{x}$, or, for brevity, by no additional notation. Denoting handwritten vectors with a superscripted arrow $\vec{x}$ is more common in the physics community.

Definition 2.14 A **linear vector space** S over a set of scalars R is a collection of objects known as vectors, together with an additive operation $+$ and a scalar multiplication operation $\cdot$, that satisfy the following properties:

VS1 S forms a group under addition. That is, the following properties are satisfied.

  (a) For any **x** and **y** $\in S$, $\mathbf{x} + \mathbf{y} \in S$. (The addition operation is closed.)[1]

  (b) There is an identity element in S, which we will denote as **0**, such that for any $\mathbf{x} \in S$,

$$\mathbf{x} + \mathbf{0} = \mathbf{0} + \mathbf{x} = \mathbf{x}.$$

  (c) For every element $\mathbf{x} \in S$ there is another element $\mathbf{y} \in S$ such that

$$\mathbf{x} + \mathbf{y} = \mathbf{0}.$$

  The element **y** is the additive inverse of **x**, and is usually denoted as $-\mathbf{x}$.

  (d) The addition operation is associative; for any **x**, **y**, and $\mathbf{z} \in S$,

$$(\mathbf{x} + \mathbf{y}) + \mathbf{z} = \mathbf{x} + (\mathbf{y} + \mathbf{z}).$$

VS2 For any $a, b \in R$ and any **x** and **y** in S,

$$a\mathbf{x} \in S,$$
$$a(b\mathbf{x}) = (ab)\mathbf{x},$$
$$(a + b)\mathbf{x} = a\mathbf{x} + b\mathbf{x},$$
$$a(\mathbf{x} + \mathbf{y}) = a\mathbf{x} + a\mathbf{y}.$$

---

[1] A closed operation is a distinct concept from a closed set.

VS3 There is a multiplicative identity element $1 \in R$ such that $1\mathbf{x} = \mathbf{x}$. There is an element $0 \in R$ such that $0\mathbf{x} = 0$.

The set R is the set of scalars of the vector space. □

The set of scalars is most frequently taken to be the set of real numbers or complex numbers. However, in some applications, other sets of scalars are used, such as polynomials or numbers modulo 256. The only requirement on the set of scalars is that the operations of addition and multiplication can be used as usual (although no multiplicative inverse is needed), and that there is a number 1 that is a multiplicative identity. In this chapter, when we talk about issues such as closed subspaces, complete subspaces, and so on, it is assumed that the set of scalars is either the real numbers $\mathbb{R}$ or the complex numbers $\mathbb{C}$, since these are complete.

We will refer interchangeably to *linear vector space* or *vector space*.

Example 2.2.1 The most familiar vector space is $\mathbb{R}^n$, the set of n-tuples. For example, if $\mathbf{x}_1, \mathbf{x}_2 \in \mathbb{R}^4$, and

$$\mathbf{x}_1 = \begin{bmatrix} 1 \\ 5 \\ 4 \\ 2 \end{bmatrix} \qquad \mathbf{x}_2 = \begin{bmatrix} 5 \\ 2 \\ 0 \\ -2 \end{bmatrix},$$

then

$$\mathbf{x}_1 + \mathbf{x}_2 = \begin{bmatrix} 6 \\ 7 \\ 4 \\ 0 \end{bmatrix} \qquad 3\mathbf{x}_1 + 2\mathbf{x}_2 = \begin{bmatrix} 13 \\ 19 \\ 12 \\ 2 \end{bmatrix}.$$

□

Several other finite-dimensional vector spaces exist, of which we mention a few.

Example 2.2.2

1. The set of $m \times n$ matrices with real elements.
2. The set of polynomials of degree up to n with real coefficients.
3. The set of polynomials with real coefficients, with the usual addition and multiplication modulo the polynomial $p(t) = 1 + t^8$, forms a linear vector space. We denote this vector space as $\mathbb{R}[t]/(t^8 + 1)$. □

In addition to these examples (which will be shown subsequently to have finite dimensionality), there are many important vector spaces that are infinite-dimensional (in a manner to be made precise in the following).

Example 2.2.3

1. Sequence spaces: The set of all infinitely-long sequences $\{x_n\}$ forms an infinite-dimensional vector space.
2. Continuous functions: The set of continuous functions defined over the interval $[a, b]$ forms a vector space. We denote this vector space as $C[a, b]$.
3. $L_p[a, b]$: The functions in L_p form the elements of an infinite-dimensional vector space. □

Definition 2.15 Let S be a vector space. If $V \subset S$ is a subset such that V is itself a vector space, then V is said to be a **subspace** of S. □

2.2 Vector Spaces

Example 2.2.4

1. Let S be the set of all polynomials, and let V be the set of polynomials of degree less than 6. The V is a subspace of S.

2. Let S consist of the set of 5-tuples
$$S = \{(0,0,0,0,0), (0,1,0,0,1), (1,0,0,0,1), (1,1,0,0,0)\}$$
and let V be the set
$$V = \{(0,0,0,0,0), (0,1,0,0,1)\},$$
where the addition is done modulo 2. Then S is a vector space (check this!) and V is a subspace. □

Throughout this chapter and the remainder of the book, we will use interchangeably the words "vector" and "signal." For a discrete-time signal, we may think of the vector composed of the samples of the function as a vector in $\mathbb{R}^n$ or $\mathbb{C}^n$. For a continuous-time signal $s(t)$, the vector is the signal itself, an element of a space such as $L_p[a, b]$. Thus

> the study of vector spaces is the study of signals.

2.2.1 Linear combinations of vectors

Let S be a vector space over R, and let $\mathbf{p}_1, \mathbf{p}_2, \ldots, \mathbf{p}_m$ be vectors in S. Then for $c_i \in R$, the linear combination
$$\mathbf{x} = c_1\mathbf{p}_1 + c_2\mathbf{p}_2 + \cdots c_m\mathbf{p}_m$$
is in S. The set of vectors $\{\mathbf{p}_i\}$ can be regarded as *building blocks* or ingredients for other signals, and the linear combination synthesizes $\mathbf{x}$ from these components. If the set of ingredients is sufficiently rich, than a wide variety of signals (vectors) can be constructed. If the ingredient vectors are known, then the vector $\mathbf{x}$ is entirely characterized by the representation $(c_1, c_2, \ldots, c_m)$, since knowing these tells how to synthesize $\mathbf{x}$.

Definition 2.16 Let S be a vector space over R, and let $T \subset S$ (perhaps with infinitely many elements). A point $\mathbf{x} \in S$ is said to be a **linear combination** of points in T if there is a *finite* set of points $\mathbf{p}_1, \mathbf{p}_2, \ldots, \mathbf{p}_m$ in T and a finite set of scalars $c_1, c_2, \ldots, c_m$ in R such that
$$\mathbf{x} = c_1\mathbf{p}_1 + c_2\mathbf{p}_2 + \cdots + c_m\mathbf{p}_m.$$
□

It is significant that the linear combination entails only a finite sum.

Example 2.2.5 Let $S = C(\mathbb{R})$, the set of continuous functions defined on the real numbers. Let $p_1(t) = 1$, $p_2(t) = t$, and $p_3(t) = t^2$. Then a linear combination of these functions is
$$x(t) = c_1 + c_2 t + c_3 t^2.$$
These functions can be used as building blocks to create any second-degree polynomial. (As will be seen in the following, there are other functions better suited to the task of building polynomials.) If the function $p_4(t) = t^2 - 1$ is added to the set of functions then, a function of the form
$$x(t) = c_1 + c_2 t + c_3 t^2 + c_4(t^2 - 1) = (c_1 - c_4) + a_2 t + (c_3 + c_4)t^2$$
can be constructed, which is still just a quadratic polynomial. That is, the new function does not expand the set of functions that can be constructed, so $p_4(t)$ is, in some sense, redundant. This means that there is more than one way to represent a polynomial. For example, the polynomial
$$x(t) = 6 + 5t + t^2$$

can be represented as
$$x(t) = 8p_1(t) + 5p_2(t) - p_3(t) + 2p_4(t)$$
or as
$$x(t) = 9p_1(t) + 5p_2(t) - 2p_3(t) + 3p_4(t).$$
□

Example 2.2.6 Let $\mathbf{p}_1, \mathbf{p}_2 \in \mathbb{R}^3$, with $\mathbf{p}_1 = [1, 0, 1]^T$, $\mathbf{p}_2 = [1, 1, 0]^T$. Then
$$\mathbf{x} = c_1 \mathbf{x}_1 + c_2 \mathbf{x}_2 = \begin{bmatrix} c_1 + c_2 \\ c_2 \\ 1 \end{bmatrix}.$$

The set of vectors that can be constructed with $\{\mathbf{p}_1, \mathbf{p}_2\}$ does not cover the set of all vectors in $\mathbb{R}^3$. For example, the vector
$$\mathbf{x} = \begin{bmatrix} 5 \\ 2 \\ 6 \end{bmatrix}$$
cannot be formed as a linear combination of $\mathbf{p}_1$ and $\mathbf{p}_2$. □

Several questions related to linear combinations are addressed in this and succeeding sections, among them:

- Is the representation of a vector as a linear combination of other vectors unique?
- What is the smallest set of vectors that can be used to synthesize any vector in S?
- Given the set of vectors $\mathbf{p}_1, \mathbf{p}_2, \ldots, \mathbf{p}_m$, how are the coefficients $c_1, c_2, \ldots, c_m$ found to represent the vector $\mathbf{x}$ (if in fact it can be represented)?
- What are the requirements on the vectors $\mathbf{p}_i$ in order to be able to synthesize any vector $x \in S$?
- Suppose that $\mathbf{x}$ cannot be represented exactly using the set of vectors $\{\mathbf{p}_i\}$. What is the best approximation that can be made with a given set of vectors?

In this chapter we examine the first two questions, leaving the rest of the questions to the applications of the next chapter.

2.2.2 Linear independence

We will first examine the question of the uniqueness of the representation as a linear combination.

Definition 2.17 Let S be a vector space, and let T be a subset of S. The set T is **linearly independent** if for each finite nonempty subset of T (say $\{\mathbf{p}_1, \mathbf{p}_2, \ldots, \mathbf{p}_m\}$) the only set of scalars satisfying the equation
$$c_1 \mathbf{p}_1 + c_2 \mathbf{p}_2 + \cdots + c_m \mathbf{p}_m = 0$$
is the trivial solution $c_1 = c_2 = \cdots = c_m = 0$.

The set of vectors $\mathbf{p}_1, \mathbf{p}_2, \ldots, \mathbf{p}_m$ is said to be **linearly dependent** if there exists a set of scalar coefficients $c_1, c_2, \ldots, c_m$ which are not all zero, such that
$$c_1 \mathbf{p}_1 + c_2 \mathbf{p}_2 + \cdots + c_m \mathbf{p}_m = 0.$$
□

Example 2.2.7

1. The functions $p_1(t), p_2(t), p_3(t), p_4(t) \in S$ of example 2.2.5 are linearly dependent, because
$$p_4(t) + p_1(t) - p_3(t) = 0;$$
that is, there is a nonzero linear combination of the functions which is equal to zero.

2.2 Vector Spaces

2. The vectors $\mathbf{p}_1 = [2, -3, 4]^T$, $\mathbf{p}_2 = [-1, 6, -2]$, and $\mathbf{p}_3 = [1, 6, 2]^T$ are linearly dependent since
$$4\mathbf{p}_1 + 5\mathbf{p}_2 + 3\mathbf{p}_3 = 0.$$

3. The functions $p_1(t) = t$ and $p_2(t) = 1 + t$ are linearly independent. □

Definition 2.18 Let T be a set of vectors in a vector space S over a set of scalars R (the number of vectors in T could be infinite). The set of vectors V that can be reached by all possible (finite) linear combinations of vectors in T is the **span** of the vectors. This is denoted by
$$V = \text{span}\{T\}.$$
That is, for any $\mathbf{x} \in V$, there is some set of coefficients $\{c_i\}$ in R such that
$$\mathbf{x} = \sum_{i=1}^{m} c_i \mathbf{p}_i,$$
where each $\mathbf{p}_i \in T$. □

It may be observed that V is a subspace of S. We also observe that $V = \text{span}(T)$ is the smallest subspace of S containing T, in the sense that, for every subspace $M \subset S$ such that $T \subset M$, then $V \subset M$.

The span of a set of vectors can be thought of as a line (if it occupies one dimension), or as a plane (if it occupies two dimensions), or as a hyperplane (if it occupies more than two dimensions). In this book we will speak of the *plane* spanned by a set, regardless of its dimensionality.

Example 2.2.8

1. Let $\mathbf{p}_1 = [1, 1, 0]^T$ and $\mathbf{p}_2 = [0, 1, 0]^T$ be in $\mathbb{R}^3$. Linear combinations of these vectors are
$$\mathbf{x} = \begin{bmatrix} c_1 + c_2 \\ c_2 \\ 0 \end{bmatrix},$$
for $c_1, c_2 \in \mathbb{R}$. The space $V = \text{span}\{\mathbf{p}_1, \mathbf{p}_2\}$ is a subset of the space $\mathbb{R}^3$: it is the plane in which the vectors $[1, 1, 0]^T$ and $[0, 1, 0]^T$ lie, which is the xy plane in the usual coordinate system, as shown in figure 2.8.

2. Let $p_1(t) = 1 + t$ and $p_2(t) = t$. Then $V = \text{span}\{p_1, p_2\}$ is the set of all polynomials up to degree 1. The set V could be envisioned abstractly as a "plane" lying in the space of all polynomials. □

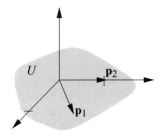

Figure 2.8: A subspace of $\mathbb{R}^3$

Definition 2.19 Let T be a set of vectors in a vector space S and let $V \subset S$ be a subspace. If every vector $\mathbf{x} \in V$ can be written as a linear combination of vectors in T, then T is a **spanning set** of V. □

Example 2.2.9

1. The vectors $\mathbf{p}_1 = [1, 6, 5]^T, \mathbf{p}_2 = [-2, 4, 2]^T, \mathbf{p}_3 = [1, 1, 0]^T, \mathbf{p}_4 = [7, 5, 2]^T$ form a spanning set of $\mathbb{R}^3$.
2. The functions $p_1(t) = 1 + t$, $p_2(t) = 1 + t^2$, $p_3(t) = t^2$, and $p_4(t) = 2$ form a spanning set of the set of polynomials up to degree 2. □

Linear independence provides us with what we need for a unique representation as a linear combination, as the following theorem shows.

Theorem 2.1 *Let S be a vector space, and let T be a nonempty subset of S. The set T is linearly independent if and only if for each nonzero $\mathbf{x} \in \mathrm{span}(T)$, there is exactly one finite subset of T, which we will denote as $\{\mathbf{p}_1, \mathbf{p}_2, \ldots, \mathbf{p}_m\}$, and a unique set of scalars $\{c_1, c_2, \ldots, c_m\}$ such that*

$$\mathbf{x} = c_1 \mathbf{p}_1 + c_2 \mathbf{p}_2 + \cdots + c_m \mathbf{p}_m.$$

Proof We will show that "T linearly independent" implies a unique representation. Suppose that there are two sets of vectors in T,

$$\{\mathbf{p}_1, \mathbf{p}_2, \ldots, \mathbf{p}_m\} \quad \text{and} \quad \{\mathbf{q}_1, \mathbf{q}_2, \ldots, \mathbf{q}_n\}$$

and corresponding nonzero coefficients such that

$$\mathbf{x} = c_1 \mathbf{p}_1 + c_2 \mathbf{p}_2 + \cdots + c_m \mathbf{p}_m \quad \text{and} \quad \mathbf{x} = d_1 \mathbf{q}_1 + d_2 \mathbf{q}_2 + \cdots + d_n \mathbf{q}_n.$$

We need to show that $n = m$ and $\mathbf{p}_i = \mathbf{q}_i$ for $i = 1, 2, \ldots, m$, and that $c_i = d_i$.

We note that

$$c_1 \mathbf{p}_1 + c_2 \mathbf{p}_2 + \cdots + c_m \mathbf{p}_m - d_1 \mathbf{q}_1 - d_2 \mathbf{q}_2 - \cdots - d_n \mathbf{q}_n = 0.$$

Since $c_1 \neq 0$, by the definition of linear independence the vector $\mathbf{p}_1$ must be an element of the set $\{\mathbf{q}_1, \mathbf{q}_2, \ldots, \mathbf{q}_n\}$ and the corresponding coefficients must be equal; say, $\mathbf{p}_1 = \mathbf{q}_1$ and $c_1 = d_1$. Similarly, since $c_2 \neq 0$ we can say that $\mathbf{p}_2 = \mathbf{q}_2$ and $c_2 = d_2$. Proceeding similarly, we must have $\mathbf{p}_i = \mathbf{q}_i$ for $i = 1, 2, \ldots, m$, and $c_i = d_i$.

Conversely, suppose that for each $\mathbf{x} \in \mathrm{span}(T)$ the representation $\mathbf{x} = c_1 \mathbf{p}_1 + \cdots c_m \mathbf{p}_m$ is unique. Assume to the contrary that T is linearly dependent, so that there are vectors $\mathbf{p}_1, \mathbf{p}_2, \ldots, \mathbf{p}_m$ such that

$$\mathbf{p}_1 = -a_2 \mathbf{p}_2 - a_3 \mathbf{p}_3 - \cdots - a_m \mathbf{p}_m. \tag{2.10}$$

But this gives two representations of the vector $\mathbf{p}_1$: itself, and the linear combination (2.10). Since this contradicts the unique representation, T must be linearly independent. □

2.2.3 Basis and dimension

Up to this point we have used the term "dimension" freely and without a formal definition. We have not clarified what is meant by "finite-dimensional" and "infinite-dimensional" vector spaces. In this section we amend this omission by defining the Hamel basis of a vector space.

Definition 2.20 Let S be a vector space, and let T be a set of vectors from S such that $\mathrm{span}(T) = S$. If T is linearly independent, then T is said to be a **Hamel basis** for S. □

2.2 Vector Spaces

Example 2.2.10

1. The set of vectors in the last example is not linearly independent, since
$$-4\mathbf{p}_1 + 5\mathbf{p}_2 - 21\mathbf{p}_3 + 5\mathbf{p}_4 = 0.$$
However, the set $T = \{\mathbf{p}_1, \mathbf{p}_2, \mathbf{p}_3\}$ is linearly independent and spans the space $\mathbb{R}^3$. Hence T is a (Hamel) basis for $\mathbb{R}^3$.

2. The vectors
$$\mathbf{e}_1 = \begin{bmatrix} 1 \\ 0 \\ 0 \end{bmatrix} \quad \mathbf{e}_2 = \begin{bmatrix} 0 \\ 1 \\ 0 \end{bmatrix} \quad \mathbf{e}_3 = \begin{bmatrix} 0 \\ 0 \\ 1 \end{bmatrix}$$
form another (Hamel) basis for $\mathbb{R}^3$. This basis is often called the **natural basis**.

3. The vectors $p_1(t) = 1$, $p_2(t) = t$, $p_3(t) = t^2$ form a (Hamel) basis for the set
$$S = \{\text{all polynomials of degree} \leq 2\}.$$
Another (Hamel) basis for S is the set of polynomials $\{q_1(t) = 2, q_2(t) = t + t^2, q_3(t) = t\}$. □

As this example shows, there is not necessarily a unique (Hamel) basis for a vector space. However, the following theorem shows that every basis for a vector space have a common attribute: the cardinality, or number of elements in the basis.

Theorem 2.2 *If T_1 and T_2 are Hamel bases for a vector space S, then T_1 and T_2 have the same cardinality.*

The proof of this theorem is split into two pieces: the finite-dimensional case, and the infinite-dimensional case. The latter may be omitted on a first reading.

Proof (Finite-dimensional case) Suppose
$$T_1 = \{\mathbf{p}_1, \mathbf{p}_2, \ldots, \mathbf{p}_m\} \quad \text{and} \quad T_2 = \{\mathbf{q}_1, \mathbf{q}_2, \ldots, \mathbf{q}_n\}$$
are two Hamel bases of S. Express the point $\mathbf{q}_1 \in T_2$ as
$$\mathbf{q}_1 = c_1 \mathbf{p}_1 + c_2 \mathbf{p}_2 + \cdots + c_m \mathbf{p}_m.$$
At least one of the coefficients c_i must be nonzero; let us take this as c_1. We can then write
$$\mathbf{p}_1 = \frac{1}{c_1}(\mathbf{q}_1 - c_2 \mathbf{p}_2 - \cdots - c_m \mathbf{p}_m).$$
By this means we can eliminate $\mathbf{p}_1$ as a basis vector in T_1 and use instead the set $\{\mathbf{q}_1, \mathbf{p}_2, \ldots, \mathbf{p}_m\}$ as a basis. Similarly, we write
$$\mathbf{q}_2 = d_1 \mathbf{q}_1 + d_2 \mathbf{p}_2 + \cdots + c_m \mathbf{p}_m$$
and as before eliminate $\mathbf{p}_2$, so that $\{\mathbf{q}_1, \mathbf{q}_2, \mathbf{p}_3, \ldots, \mathbf{p}_m\}$ forms a basis. Continuing in this way, we can eliminate each $\mathbf{p}_i$, showing that $\{\mathbf{q}_1, \ldots, \mathbf{q}_m\}$ spans the same space as $\{\mathbf{p}_1, \ldots, \mathbf{p}_m\}$. We can conclude that $m \geq n$. Suppose, to the contrary, that $n > m$. Then a vector such as $\mathbf{q}_{m+1}$, which does not fall in the basis set $\{\mathbf{q}_1, \ldots, \mathbf{q}_m\}$, would have to be linearly dependent with that set, which violates the fact that T_2 is itself a basis.

Reversing the argument, we find that $n \geq m$. In combination, then, we conclude that $m = n$.

(Infinite-dimensional case) Let T_1 and T_2 be bases. For an $\mathbf{x} \in T_1$, let $T_2(\mathbf{x})$ denote the unique finite set of points in T_2 needed to express $\mathbf{x}$.

Claim: If $\mathbf{y} \in T_2$, then $\mathbf{y} \in T_2(\mathbf{x})$ for some $\mathbf{x} \in T_1$. Proof: Since a point $\mathbf{y}$ is in S, then $\mathbf{y}$ must be a finite linear combination of vectors in T_1; say,

$$\mathbf{y} = c_1 \mathbf{x}_1 + c_2 \mathbf{x}_2 + \cdots + c_m \mathbf{x}_m$$

for some set of vectors $\mathbf{x}_i \in T_1$. Then, for example,

$$\mathbf{x}_1 = \frac{1}{c_1}(\mathbf{y} - c_2 \mathbf{x}_2 - \cdots - c_m \mathbf{x}_m),$$

so that, by the uniqueness of the representation, $\mathbf{y} \in B_2(\mathbf{x})$.

Since for every $\mathbf{y} \in T_2$ there is some $\mathbf{x} \in T_1$ such that $\mathbf{y} \in T_2(\mathbf{x})$, it follows that

$$T_2 = \bigcup_{\mathbf{x} \in T_1} T_2(\mathbf{x}).$$

Noting that there are $|T_1|$ sets in this union[2], each of which contributes at least one element to T_2, we conclude that $|T_2| \geq |T_1|$.

Now turning the argument around, we conclude that $|T_1| \geq |T_2|$. By these two inequalities we conclude that $|T_1| = |T_2|$. □

On the strength of this theorem, we can state a consistent definition for the dimension of a vector space.

Definition 2.21 Let T be a Hamel basis for a vector space S. The cardinality of T is the **dimension** of S. This is denoted as dim(S). It is the *number of linearly independent vectors required to span the space.* □

Since the dimension of a vector space is unique, we can conclude that a basis T for a subspace S is a *smallest set* of vectors whose linear combinations can form every vector in a vector space S, in the sense that a basis of $|T|$ vectors is contained in every other spanning set for S.

The last remaining fact, which we will not prove, shows the importance of the Hamel basis: *Every vector space has a Hamel basis.* So, for many purposes, whatever we want to do with a vector space can be done to the Hamel basis.

Example 2.2.11 Let S be the set of all polynomials. Then a polynomial $x(t) \in S$ can be written as a linear combination of the functions $\{1, t, t^2, \dots\}$. It can be shown (see exercise 2.2-32) that this set of functions is linearly independent. Hence the dimension of S is infinite. □

Example 2.2.12 [Bernard Friedman, *Principles and Techniques of Applied Mathematics*, Dover, 1990.] To illustrate that infinite dimensional vector spaces can be difficult to work with, and particular care is required, we demonstrate that for an infinite-dimensional vector space S, an infinite set of linearly independent vectors which span S need not form a basis for S.

Let X be the infinite-sequence space, with elements of the form $(x_1, x_2, x_3, \dots)$, where each $x_i \in \mathbb{R}$. The set of vectors

$$\mathbf{p}_j = (1, 0, 0, \dots, 0, 1, 0, \dots), \qquad j = 2, 3, \dots$$

where the second 1 is in the jth positions forms a set of linearly independent vectors.

We first show the set $\{\mathbf{p}_j, j = 2, 3, \dots\}$ spans X. Let $x = (x_1, x_2, x_3, \dots)$ be an arbitrary element of X. Let

$$\sigma_n = x_1 - x_2 - \cdots - x_n,$$

and let τ_n be an integer larger than $n|\sigma_n|^2$. Now consider the sequence of vectors

$$\mathbf{y}_n = x_2 \mathbf{p}_2 + x_3 \mathbf{p}_3 + \cdots + x_n \mathbf{p}_n + \frac{\sigma_n}{\tau_n}(\mathbf{p}_{n+1} + \cdots + \mathbf{p}_p),$$

---

[2] Recall that the notation $|S|$ indicates the cardinality of the set S; see section A.1.

2.3 Norms and Normed Vector Spaces

where $p = n + \tau_n$. For example,

$$\mathbf{y}_3 = \mathbf{x}_2\mathbf{p}_2 + \mathbf{x}_3\mathbf{p}_3 + \frac{x_1 - x_2 - x_3}{\tau_n}(\mathbf{p}_4 + \mathbf{p}_5 + \cdots + \mathbf{p}_{4+\tau_n})$$

$$= \mathbf{x}_2\mathbf{p}_2 + \mathbf{x}_3\mathbf{p}_3 + (x_1 - x_2 - x_3)\left(1, \frac{1}{\tau_n}, \frac{1}{\tau_n}, \ldots, \frac{1}{\tau_n}\right).$$

In the limit as $n \to \infty$, the residual term becomes

$$(x_1 - x_2 - \cdots)(1, 0, 0, \ldots)$$

and $\mathbf{y}_n \to \mathbf{x}$. So there is a representation for $\mathbf{x}$ using this infinite set of basis functions.

However—this is the subtle but important point—the representation exists as a result of a limiting process. There is no finite set of fixed scalars $c_2, c_3, \ldots, c_N$ such that the sequence $\mathbf{x} = (1, 0, 0, \ldots)$ can be written in terms of the basis functions as

$$\mathbf{x} = (1, 0, 0, \ldots) = c_2\mathbf{p}_2 + c_3\mathbf{p}_3 + \cdots + c_N\mathbf{p}_N.$$

When we introduced the concept of linear combinations in definition 2.15, only *finite* sums were allowed. Since representing $\mathbf{x}$ would require an infinite sum, the set of functions $\mathbf{p}_2, \mathbf{p}_3 \ldots$ *does not form a basis*.

It may be objected that it would be straightforward to simply express an infinite sum $\sum_{j=2}^{\infty} c_2\mathbf{p}_2$, and have done with the matter. But dealing with infinite series always requires more care than does finite series, so we consider this as a different case. □

2.2.4 Finite-dimensional vector spaces and matrix notation

The major focus of our interest in vector spaces will be on finite-dimensional vector spaces. Even when dealing with infinite-dimensional vector spaces, we shall frequently be interested in finite-dimensional representations. In the case of finite-dimensional vector spaces, we shall refer to the Haumel basis simply as the basis.

One particularly useful aspect of finite-dimensional vector spaces is that matrix notation can be used for convenient representation of linear combinations. Let the matrix A be formed by stacking the vectors $\mathbf{p}_1, \mathbf{p}_2, \ldots, \mathbf{p}_m$ side by side,

$$A = [\mathbf{p}_1 \quad \mathbf{p}_2 \quad \cdots \quad \mathbf{p}_m].$$

For a vector

$$\mathbf{c} = \begin{bmatrix} c_1 \\ c_2 \\ \vdots \\ c_m \end{bmatrix},$$

the product $\mathbf{x} = A\mathbf{c}$ computes the linear combination

$$\mathbf{x} = c_1\mathbf{p}_1 + c_2\mathbf{p}_2 + \cdots + c_m\mathbf{p}_m.$$

The question of the linear dependence of the vectors $\{\mathbf{p}_i\}$ can be examined by looking at the rank of the matrix A, as discussed in section 4.7.

2.3 Norms and normed vector spaces

When dealing with vector spaces it is common to talk about the length and direction of the vector, and there is an intuitive geometric concept as to what the length and direction are. There are a variety of ways of defining the length of a vector. The mathematical concept associated with the length of a vector is the **norm**, which is discussed in this section. In section 2.4 we introduce the concept of an inner product, which is used to provide an interpretation of angle between vectors, and hence direction.

Definition 2.22 Let S be a vector space with elements $\mathbf{x}$. A real-valued function $\|\mathbf{x}\|$ is said to be a **norm** if $\|\mathbf{x}\|$ satisfies the following properties.

N1 $\|\mathbf{x}\| \geq 0$ for any $\mathbf{x} \in S$.
N2 $\|\mathbf{x}\| = 0$ if and only if $\mathbf{x} = \mathbf{0}$.
N3 $\|\alpha \mathbf{x}\| = |\alpha| \, \|\mathbf{x}\|$, where α is an arbitrary scalar.
N4 $\|\mathbf{x} + \mathbf{y}\| \leq \|\mathbf{x}\| + \|\mathbf{y}\|$ (triangle inequality).

The real number $\|\mathbf{x}\|$ is said to be the norm of $\mathbf{x}$, or the length of $\mathbf{x}$. □

The triangle inequality N4 can be interpreted geometrically using figure 2.9, where $\mathbf{x}$, $\mathbf{y}$, and $\mathbf{z}$ are the sides of a triangle.

Figure 2.9: A triangle inequality interpretation

A norm "feels" a lot like a metric, but actually requires more structure than a metric. For example, the definition of a norm requires that addition $\mathbf{x} + \mathbf{y}$ and scalar multiplication $\alpha \mathbf{x}$ are defined, which was not the case for a metric.

Nevertheless, because of their similar properties, norms and metrics can be defined in terms of each other. For example, if $\|\mathbf{x}\|$ is a norm, then

$$d(\mathbf{x}, \mathbf{y}) = \|\mathbf{x} - \mathbf{y}\|$$

is a metric. The triangle inequality for metrics is established by noting that

$$\|\mathbf{x} - \mathbf{y}\| = \|\mathbf{x} - \mathbf{z} + \mathbf{z} - \mathbf{x}\| \leq \|\mathbf{x} - \mathbf{z}\| + \|\mathbf{y} - \mathbf{z}\|.$$

(This trick of adding and subtracting the quantity to make the answer come out right is often used in analysis.) Alternatively, given a metric d defined on a vector space, a norm can be written as

$$\|\mathbf{x}\| = d(\mathbf{x}, \mathbf{0}),$$

the distance that $\mathbf{x}$ is from the origin of the vector space.

Example 2.3.1 Based upon the metrics we have already seen, we can readily define some useful norms for n-dimensional vectors.

1. The l_1 norm: $\|\mathbf{x}\|_1 = \sum_{i=1}^{n} |x_i|$.
2. The l_2 norm: $\mathbf{x}\|_p = \left(\sum_{i=1}^{n} |x_i|^p \right)^{1/p}$.
3. The l_∞ norm: $\|\mathbf{x}\|_\infty = \max_{i=1,2,\ldots,n} |x_i|$.

Each of these norms introduces its own geometry. Consider, for example, the unit "sphere" defined by

$$S_p = \{\mathbf{x} \in \mathbb{R}^2 : \|\mathbf{x}\|_p \leq 1\}.$$

Figure 2.10 illustrates the shape of such spheres for various values of p. □

2.3 Norms and Normed Vector Spaces

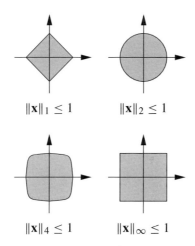

Figure 2.10: Unit spheres in $\mathbb{R}^2$ under various l_p norms

Example 2.3.2 We can also define norms of functions defined over the interval $[a, b]$.

1. The L_1 norm: $\|x(t)\|_1 = \int_a^b |x(t)|\, dt$.
2. The L_p norm: $\|x(t)\|_2 = \left(\int_a^b |x(t)|^p\, dt \right)^{1/p}$ for $1 \leq p < \infty$.
3. The L_∞ norm: $\|x(t)\|_\infty = \sup_{t \in [a,b]} |x(t)|$. $\square$

The l_∞ and L_∞ norms are referred to as the *uniform* norms.

Definition 2.23 A **normed linear space** is a pair $(S, \|\cdot\|)$, where S is a vector space and $\|\cdot\|$ is a norm defined on S. A normed linear space is often denoted simply by S. $\square$

When discussing the metrical properties of a normed linear space, the metric is defined in terms of the norm, $d(\mathbf{x}, \mathbf{y}) = \|\mathbf{x} - \mathbf{y}\|$.

Definition 2.24 A vector $\mathbf{x}$ is said to be **normalized** if $\|\mathbf{x}\| = 1$. It is possible to normalize any vector except the zero vector: $\mathbf{y} = \mathbf{x}/\|\mathbf{x}\|$ has $\|\mathbf{y}\| = 1$. A normalized vector is also referred to as a **unit vector**. $\square$

With a variety of norms to choose from, it is natural to address the issue of which norm should be used in a particular case. Often the l_2 or L_2 norm is used, for reasons which become clear subsequently. However, occasions may arise in which other norms or norm-like functions are used. For example, in a high-speed signal-processing algorithm, it may be necessary to use the l_1 norm, since it may be easier in the available hardware to compute an absolute value than to compute a square. Or, in a problem of data representation of audio information (quantization), it may be appropriate to use a norm for which a representation is chosen that is best as perceived by human listeners. Ideally, a norm that measured exactly the distortion perceived by the human ear would be desired in such an application. (This is only approximately achievable, since it depends upon so many psychoacoustic effects, of which only a few are understood.) Similar comments could be made regarding norms for video coding. In short, the norm should be chosen that is best suited to the particular application.

The exact norm values computed for a vector $\mathbf{x}$ change depending on the particular norm used, but a vector that is small with respect to one norm is also small with respect to another norm. Norms are thus equivalent in the sense described in the following theorem.

Theorem 2.3 *(Norm equivalence theorem) If $\|\cdot\|$ and $\|\cdot\|'$ are two norms on $\mathbb{R}^n$ (or $\mathbb{C}^n$), then*

$$\|\mathbf{x}_k\| \to 0 \quad \text{as } k \to \infty \quad \text{if and only if} \quad \|\mathbf{x}_k\|' \to 0 \quad \text{as } k \to \infty.$$

The proof of this theorem makes use of the Cauchy–Schwarz inequality, which is introduced in section 2.6. You may want to come back to this proof after reading that section.

Proof It suffices to show that there are constants $c_1, c_2 > 0$ such that

$$c_1 \|\mathbf{x}\| \le \|\mathbf{x}\|' \le c_2 \|\mathbf{x}\|. \tag{2.11}$$

To prove (2.11), it suffices to assume that $\|\cdot\|'$ is the l_2 norm. To see this, observe that if

$$d_1 \|\mathbf{x}\| \le \|\mathbf{x}\|_2 \le d_2 \|\mathbf{x}\| \quad \text{and} \quad d_1' \|\mathbf{x}\|' \le \|\mathbf{x}\|_2 \le d_2' \|\mathbf{x}\|'$$

then

$$d_1 \|\mathbf{x}\| \le d_2' \|\mathbf{x}\|'$$

and

$$d_1' \|\mathbf{x}\|' \le d_2 \|\mathbf{x}\|,$$

so (2.11) holds with $c_1 = d_1/d_2'$ and $c_2 = d_2/d_1'$. Let $\mathbf{x}$ be expressed as a linear combination of basis vectors

$$\mathbf{x} = \sum_{i=1}^{n} x_i \mathbf{e}_i.$$

Then, by the properties of the norm,

$$\|\mathbf{x}\| = \left\| \sum_{i=1}^{n} x_i \mathbf{e}_i \right\| \le \sum_{i=1}^{n} |x_i| \, \|\mathbf{e}_i\|.$$

The sum on the right is simply the inner product of the vector composed of the magnitudes of the x_i's with the vector composed of the magnitudes of the basis vectors. Being an inner product, the Cauchy–Schwarz inequality applies, and

$$\|\mathbf{x}\| \le \|\mathbf{x}\|_2 \left(\sum_{i=1}^{n} \|\mathbf{e}_i\|^2 \right)^{1/2}.$$

Let

$$\beta = \left(\sum_{i=1}^{n} \|\mathbf{e}_i\|^2 \right)^{1/2}.$$

Then the left inequality of (2.11) applies with $c_1 = 1/\beta$.

For points $\mathbf{x}$ on the unit sphere $S = \{\mathbf{x}: \|\mathbf{x}\|_2 = 1\}$, the norm $\|\cdot\|$ must be greater than 0 (by the properties of norms) and, hence, $\|\mathbf{x}\| \ge \alpha$ for some $\alpha > 0$ for $\mathbf{x} \in S$. Then

$$\|\mathbf{x}\| = \left\| \frac{\mathbf{x}}{\|\mathbf{x}\|_2} \right\| \, \|\mathbf{x}\|_2 \ge \alpha \|\mathbf{x}\|_2,$$

so the right-hand inequality holds with $c_2 = 1/\alpha$. □

For example,

$$\|\mathbf{x}\|_2 \le \|\mathbf{x}\|_1 \le \sqrt{n} \|\mathbf{x}\|_2,$$
$$\|\mathbf{x}\|_\infty \le \|\mathbf{x}\|_2 \le \sqrt{n} \|\mathbf{x}\|_\infty, \tag{2.12}$$
$$\|\mathbf{x}\|_\infty \le \|\mathbf{x}\|_1 \le n \|\mathbf{x}\|_\infty.$$

Definition 2.25 For a sequence $\{x_n\}$, if there exists a number $M < \infty$ such that

$$\|x_n\| < M \qquad \forall n$$

then the sequence is said to be **bounded**. □

2.3.1 Finite-dimensional normed linear spaces

The notion of a closed set and a complete set were introduced in section 2.1.2. As pointed out, having complete sets is advantageous because all Cauchy sequences converge, so that convergence of a sequence can be established simply by determining whether a sequence is Cauchy.

Finite-dimensional normed linear spaces have several very useful properties:

1. Every finite-dimensional subspace of a vector space is closed.
2. Every finite-dimensional subspace of a vector space is complete.
3. If $L: X \to Y$ is a linear operator and X is a finite dimensional normed vector space, then L is continuous. (This is true even if Y is not finite dimensional.) As we shall see in chapter 4, this means that the operator is also bounded.
4. As observed above, different norms are equivalent on $\mathbb{R}^n$ or $\mathbb{C}^n$. In fact, in any finite dimensional space, any two norms are equivalent.

A lot of the issues over which a mathematician would fret entirely disappear in finite-dimensional spaces. This is particularly useful, since many of the problems of interest in signal processing are finite dimensional.

We will not prove these useful facts here. Interested readers should consult, for example, [238, section 5.10].

2.4 Inner products and inner-product spaces

An inner product is an operation on two vectors that returns a scalar value. Inner products can be used to provide the geometric interpretation of the "direction" of a vector in an arbitrary vector space. They can also be used to define a norm known as the induced norm.

We will define the inner product in the general case, in which the vector space S has elements that are complex.

Definition 2.26 Let S be a vector space defined over a scalar field R. An **inner product** is a function $\langle \cdot, \cdot \rangle : S \times S \to R$ with the following properties:

IP1 $\langle \mathbf{x}, \mathbf{y} \rangle = \overline{\langle \mathbf{y}, \mathbf{x} \rangle}$, where the overbar indicates complex conjugation. For real vectors this simplifies to $\langle \mathbf{x}, \mathbf{y} \rangle = \langle \mathbf{y}, \mathbf{x} \rangle$.

IP2 $\langle \alpha \mathbf{x}, \mathbf{y} \rangle = \alpha \langle \mathbf{x}, \mathbf{y} \rangle$.

IP3 $\langle \mathbf{x} + \mathbf{y}, \mathbf{z} \rangle = \langle \mathbf{x}, \mathbf{z} \rangle + \langle \mathbf{y}, \mathbf{z} \rangle$.

IP4 $\langle \mathbf{x}, \mathbf{x} \rangle > 0$ if $\mathbf{x} \neq 0$, and $\langle \mathbf{x}, \mathbf{x} \rangle = 0$ if and only if $\mathbf{x} = 0$. □

Definition 2.27 A vector space equipped with an inner product is called an **inner-product space**. □

Inner-product spaces are sometimes called pre-Hilbert spaces. We encounter in section 2.9 what a Hilbert space is.

There are a variety of ways that an inner product can be defined. Notational advantage and algorithmic expediency can be obtained by suitable selection of an inner product. We begin with the most straightforward examples of inner products.

Example 2.4.1 For finite-dimensional vectors $\mathbf{x}, \mathbf{y} \in \mathbb{R}^n$, the conventional inner product between the vectors

$$\mathbf{x} = \begin{bmatrix} x_1 \\ x_2 \\ \vdots \\ x_n \end{bmatrix} \quad \text{and} \quad \mathbf{y} = \begin{bmatrix} y_1 \\ y_2 \\ \vdots \\ y_n \end{bmatrix}$$

is

$$\langle \mathbf{x}, \mathbf{y} \rangle = x_1 y_1 + x_2 y_1 + \cdots + x_n y_n$$
$$= \sum_{i=1}^{n} x_i y_i$$
$$= \mathbf{y}^T \mathbf{x} = \mathbf{x}^T \mathbf{y}.$$

This inner product is the **Euclidean inner product**. This is also the **dot product** used in vector calculus, and is sometimes written

$$\langle \mathbf{x}, \mathbf{y} \rangle = \mathbf{x} \cdot \mathbf{y}.$$

If the vectors are in $\mathbb{C}^n$ (with complex elements), then the Euclidean inner product is

$$\langle \mathbf{x}, \mathbf{y} \rangle = \sum_{k=1}^{n} x_k \overline{y}_k = \mathbf{y}^H \mathbf{x}.$$
□

Example 2.4.2 Extending the "sum of products" idea to functions, the following is an inner product for the space of functions defined on [0, 1]:

$$\langle x(t), y(t) \rangle = \int_0^1 x(t) \overline{y}(t)\, dt.$$

For functions defined over $\mathbb{R}$, an inner product is

$$\langle x(t), y(t) \rangle = \int_{-\infty}^{\infty} x(t) y(t)\, dt.$$
□

Example 2.4.3 Consider a causal signal $x(t)$ which is passed through a causal filter with impulse response $h(t)$. The output at a time T is

$$y(T) = x(t) * h(t)|_{t=T} = \int_0^T x(\tau) h(T - \tau)\, d\tau.$$

Let $g(\tau) = h(T - \tau)$. Then

$$y(T) = \int_0^T x(\tau) g(\tau)\, d\tau = \langle x, g \rangle.$$

where the inner product is

$$\langle f, g \rangle = \int_0^T f(t) g(t)\, dt.$$

So the operation of filtering (and taking the output at a fixed time) is equivalent to computing an inner product.
□

An inner product can also be defined on matrices. Let S be the vector space of $m \times n$ matrices. Then we can define an inner product on this vector space by

$$\langle A, B \rangle = \operatorname{tr}(A^H B).$$

2.4.1 Weak convergence*

When we have a sequence of vectors $\{\mathbf{x}_n\}$, as we saw in section 2.1.2, we can talk about convergence of the sequence to some value, say $\mathbf{x}_n \to \mathbf{x}$, which means that

$$\|\mathbf{x}_n - \mathbf{x}\| \to 0$$

for some norm $\|\cdot\|$. It is interesting to examine the question of convergence in the context of inner products.

Lemma 2.1 *The inner product is continuous. That is, if $\mathbf{x}_n \to \mathbf{x}$ in some inner product space, S then $\langle \mathbf{x}_n, \mathbf{y} \rangle \to \langle \mathbf{x}, \mathbf{y} \rangle$ for any $\mathbf{y} \in S$.*

Proof Since $\mathbf{x}_n$ is convergent, it must be bounded, so that $\|\mathbf{x}_n\| \le M < \infty$. Then

$$\|\langle \mathbf{x}_n, \mathbf{y} \rangle - \langle \mathbf{x}, \mathbf{y} \rangle\| = |\langle \mathbf{x}_n - \mathbf{x}, \mathbf{y} \rangle|$$
$$\le \|\mathbf{x}_n - \mathbf{x}\| \|\mathbf{y}\|.$$

Since $\|\mathbf{x}_n - \mathbf{x}\| \to 0$, the convergence of $\langle \mathbf{x}_n, \mathbf{y} \rangle$ is established. □

From this we note that convergence $\mathbf{x}_n \to \mathbf{x}$ (called *strong* convergence) implies $\langle \mathbf{x}_n, \mathbf{y} \rangle \to \langle \mathbf{x}, \mathbf{y} \rangle$ (which is called *weak* convergence). On the other hand, it does not follow necessarily that if a sequence converges weakly, so that

$$\langle \mathbf{x}_n, \mathbf{y} \rangle \to \langle \mathbf{x}, \mathbf{y} \rangle,$$

that it also converges strongly.

Example 2.4.4 Let $\mathbf{x}_n = (0, 0, 0, \ldots, 1, 0, 0, \ldots)$ be the sequence that is all 0 except for a 1 at position n, and let $\mathbf{y} = (1, 1/2, 1/4, 1/8, \ldots)$. Then

$$\langle \mathbf{x}_n, \mathbf{y} \rangle \to 0,$$

but the sequence $\{\mathbf{x}_n\}$ has no limit. The sequence thus converges weakly but not strongly. □

2.5 Induced norms

We have seen that the Euclidean norm of a vector $\mathbf{x} \in \mathbb{R}^n$ is defined as

$$\|\mathbf{x}\|_2^2 = x_1^2 + x_2^2 + \cdots + x_n^2.$$

We observe that the inner product of $\mathbf{x}$ with itself is

$$\langle \mathbf{x}, \mathbf{x} \rangle = x_1^2 + x_2^2 + \cdots + x_n^2.$$

Hence, we can use the inner product to produce a special norm, called the **induced norm**. More generally, given an inner product $\langle \cdot, \cdot \rangle$ in a vector space S, we have the induced norm defined by

$$\boxed{\|\mathbf{x}\| = \langle \mathbf{x}, \mathbf{x} \rangle^{1/2}}$$

for every $x \in S$. It should be pointed out that not every norm is an induced norm. For example, the l_p and L_p norms are only induced norms when $p = 2$.

---

*The concepts in this section are used briefly in section 2.10 and mostly in chapter 15; it is recommended that this section be skipped on a first reading.

Example 2.5.1 Another example of an induced norm is for functions in $L_2[a, b]$,

$$\|x(t)\|_2 = \langle x(t), x(t) \rangle^{1/2} = \left(\int_a^b |x(t)|^2 \, dt \right)^{1/2}.$$

□

For an induced norm, we have the following useful fact (for an inner product over a complex vector space):

$$\|\mathbf{x} - \mathbf{y}\|^2 = \langle \mathbf{x} - \mathbf{y}, \mathbf{x} - \mathbf{y} \rangle = \langle \mathbf{x}, \mathbf{x} \rangle - \langle \mathbf{x}, \mathbf{y} \rangle - \langle \mathbf{y}, \mathbf{x} \rangle + \langle \mathbf{y}, \mathbf{y} \rangle$$
$$= \|\mathbf{x}\|^2 - 2 \operatorname{Re} \langle \mathbf{x}, \mathbf{y} \rangle + \|\mathbf{y}\|^2.$$

For a vector over a real vector space, this simplifies to

$$\|\mathbf{x} - \mathbf{y}\|^2 = \|\mathbf{x}\|^2 - 2\langle \mathbf{x}, \mathbf{y} \rangle + \|\mathbf{y}\|^2.$$

2.6 The Cauchy–Schwarz inequality

In the definition of a norm, one of the key requirements of the function $\|\cdot\|$ is that

$$\|\mathbf{x} + \mathbf{y}\| \leq \|\mathbf{x}\| + \|\mathbf{y}\|.$$

Up to this point, we have assumed that the metrics mentioned do satisfy this property. We are now ready to prove this result for the important special case of the l_2 or L_2 norm, or more generally for a norm induced from any inner product. In the interest of generality we shall express this result in terms of inner products first.

The key inequality in our proof is the *Cauchy–Schwarz inequality*. This inequality will prove to be one of the cornerstones of signal processing analysis. It will provide the basis for the important projection theorem, and be the key step in the derivation of the matched filter. It can be used to prove the important geometrical fact that the gradient of a function points in the direction of steepest increase, which is the key fact used in the development of gradient descent optimization techniques. Not only is it specifically useful, but the analysis and optimization performed using the Cauchy–Schwarz inequality provides a powerful archetype for many other optimization problems: optimizing values can often be obtained by establishing an inequality, then satisfying the conditions for which the inequality achieves equality. If the Cauchy–Schwarz inequality does not serve the purpose, other inequalities often will, such as the Cauchy–Schwarz's big brothers, the Hölder and Minkowski inequalities which are presented in Appendix A.

Theorem 2.4 *(Cauchy–Schwarz inequality) In an inner product space S with induced norm $\|\cdot\|$,*

$$|\langle \mathbf{x}, \mathbf{y} \rangle| \leq \|\mathbf{x}\| \, \|\mathbf{y}\| \quad (2.13)$$

for any $\mathbf{x}, \mathbf{y} \in S$, with equality if, and only if, $\mathbf{y} = \alpha \mathbf{x}$ for some α.

Proof By expressing our proof in terms of inner products, we cover both the case of finite- and infinite-dimensional vectors. For generality, we assume complex vectors.

First, note that if $\mathbf{x} = 0$ or $\mathbf{y} = 0$, the theorem is trivial, so we exclude these cases. Form the quantity

$$\|\mathbf{x} - \alpha \mathbf{y}\|^2 = \|\mathbf{x}\| - 2 \operatorname{Re} \langle \mathbf{x}, \alpha \mathbf{y} \rangle + |\alpha|^2 \|\mathbf{y}\|^2. \quad (2.14)$$

This is always positive. We want to choose α to make this as small as possible. For real vectors, this can be done simply by taking the derivative with respect to α, and equating the derivative to zero. We demonstrate another technique by completing the square (see appendix B). We can write

$$0 \leq \|\mathbf{x} - \alpha \mathbf{y}\|^2 = \|\mathbf{y}\|^2 \left[\left(\alpha - \frac{\langle \mathbf{x}, \mathbf{y} \rangle}{\|\mathbf{y}\|^2} \right) \left(\overline{\alpha} - \frac{\overline{\langle \mathbf{x}, \mathbf{y} \rangle}}{\|\mathbf{y}\|^2} \right) \right] - \frac{|\langle \mathbf{x}, \mathbf{y} \rangle|^2}{\|\mathbf{y}\|^2} + \|\mathbf{x}\|^2.$$

Then the minimum value of $\|\mathbf{x} - \alpha \mathbf{y}\|^2$ is obtained when

$$\alpha = \frac{\langle \mathbf{x}, \mathbf{y} \rangle}{\|\mathbf{y}\|^2},$$

in which case the completion of the square leaves

$$-\frac{|\langle \mathbf{x}, \mathbf{y} \rangle|^2}{\|\mathbf{y}\|^2} + \|\mathbf{x}\|^2 \geq 0,$$

from which the desired inequality follows.

Now examine the condition for equality. If $\mathbf{y} = \alpha \mathbf{x}$, then equality in (2.13) is immediate. On the other hand, suppose that the equality in (2.13) is satisfied. Then working backward through (2.14) indicates that $\|\mathbf{x} - \alpha \mathbf{y}\| = 0$. But by the properties of a norm, this means that $\mathbf{x} = \alpha \mathbf{y}$ for some α. □

This theorem applies to *any* normed linear vector space with an induced norm. For the vector space $\mathbb{R}^n$ with the Euclidean inner product, the Cauchy–Schwarz inequality is

$$\boxed{(\mathbf{x}^T \mathbf{y})^2 \leq (\mathbf{x}^T \mathbf{x})(\mathbf{y}^T \mathbf{y})}$$

For the vector space $\mathbb{C}^n$ with the Euclidean inner product, the Cauchy–Schwarz inequality is $(\mathbf{x}^H \mathbf{y})^2 \leq (\mathbf{x}^H \mathbf{x})(\mathbf{y}^H \mathbf{y})$. For the vector space of real functions defined over $[a, b]$, the Cauchy–Schwarz inequality is

$$\boxed{\left(\int_a^b f(t) g(t) \, dt \right)^2 \leq \int_a^b f^2(t) \, dt \int_a^b g^2(t) \, dt}$$

Using the Cauchy–Schwarz inequality, we can now show that the induced norm satisfies the required triangle inequality property. For vectors $\mathbf{x}$ and $\mathbf{y}$ (which we assume for convenience to be real), we have

$$\|\mathbf{x} + \mathbf{y}\|^2 = \langle \mathbf{x} + \mathbf{y}, \mathbf{x} + \mathbf{y} \rangle = \langle \mathbf{x}, \mathbf{x} \rangle + 2\langle \mathbf{x}, \mathbf{y} \rangle + \langle \mathbf{y}, \mathbf{y} \rangle$$
$$\leq \langle \mathbf{x}, \mathbf{x} \rangle + 2\|\mathbf{x}\| \|\mathbf{y}\| + \langle \mathbf{y}, \mathbf{y} \rangle = (\|\mathbf{x}\| + \|\mathbf{y}\|)^2.$$

2.7 Direction of vectors; orthogonality

The inner product can be used to define a direction of angular separation between vectors, and hence a concept of direction.

For vectors $\mathbf{x}$ and $\mathbf{y}$ in $\mathbb{R}^3$ or $\mathbb{R}^2$, it is well known that the cosine of the angle between the vectors is

$$\cos \theta = \frac{\langle \mathbf{x}, \mathbf{y} \rangle}{\|\mathbf{x}\|_2 \|\mathbf{y}\|_2}.$$

Note that the 2-norm—which is the induced norm—is used in defining the length. Using the Cauchy–Schwarz inequality, it can be shown that

$$-1 \leq \frac{\langle \mathbf{x}, \mathbf{y} \rangle}{\|\mathbf{x}\|_2 \|\mathbf{y}\|_2} \leq 1, \qquad (2.15)$$

so the angle θ is real. This same expression, with the appropriate inner product, defines direction in any inner product space.

Example 2.7.1 Consider the vectors
$$\mathbf{x} = [1 \quad 2 \quad 3 \quad 4]^T \qquad \mathbf{y} = [4 \quad 2 \quad 4 \quad 5]^T.$$
Then the angle θ between the vectors is determined by
$$\cos\theta = \frac{\langle \mathbf{x}, \mathbf{y}\rangle}{\|\mathbf{x}\|\|\mathbf{y}\|} = 0.935.$$
□

Example 2.7.2 For functions defined on $[0, 1]$, find the angle between the functions
$$x_1(t) = 1 + t^2 \qquad \text{and} \qquad x_2(t) = t^2 - 2t.$$
First compute
$$\|x_1\| = \left(\int_0^1 (x_1(t))^2\, dt\right)^{1/2} = \sqrt{28/15}$$
and
$$\|x_2\| = \left(\int_0^1 (x_2(t))^2\, dt\right)^{1/2} = \sqrt{8/15}.$$
Then
$$\cos\theta = \frac{\int_0^1 x_1(t)x_2(t)\, dt}{\|x_1\|\|x_2\|} = -\frac{29}{8\sqrt{14}}.$$
□

Definition 2.28 If $\mathbf{x}$ and $\mathbf{y}$ are nonzero vectors and $\mathbf{x} = \alpha\mathbf{y}$ for some scalar α, then $\mathbf{x}$ and $\mathbf{y}$ are said to be **colinear**. In an inner-product space, this means that the angle between $\mathbf{x}$ and $\mathbf{y}$ satisfies $\cos\theta = \pm 1$. □

A geometric concept which will be of considerable importance to us is the idea of orthogonal vectors.

Definition 2.29 Vectors x and y in an inner product space are said to be **orthogonal** if
$$\langle x, y\rangle = 0.$$
Notationally, this is denoted as $x \perp y$. The words "perpendicular" and "normal" are synonymous with "orthogonal." □

The zero vector is orthogonal to every other vector.

Definition 2.30 A set of vectors $\{\mathbf{p}_1, \mathbf{p}_2, \ldots, \mathbf{p}_m\}$ is said to be **orthonormal** if they are mutually (pairwise) orthogonal and each have unit length,
$$\langle \mathbf{p}_i, \mathbf{p}_j\rangle = \delta_{i,j},$$
where $\delta_{i,j}$ is the **Kronecker delta** function, defined by
$$\delta_{i,j} = \begin{cases} 1 & i = j, \\ 0 & \text{otherwise.} \end{cases}$$
□

For orthogonal vectors, regardless of the inner product, the familiar Pythagorean theorem holds:

Lemma 2.2 *(The Pythagorean theorem) If $\mathbf{x} \perp \mathbf{y}$ and $\|\cdot\|$ is an induced norm, then for the norm $\|\cdot\|$ induced from the inner product,*
$$\|\mathbf{x} + \mathbf{y}\|^2 = \|\mathbf{x}\|^2 + \|\mathbf{y}\|^2. \tag{2.16}$$

The proof is straightforward.

Example 2.7.3 Consider the set of polynomials

$$P_0(t) = 1 \qquad P_1(t) = t \qquad P_2(t) = \frac{1}{2}(3t^2 - 1) \qquad P_3(t) = \frac{1}{2}(5t^3 - 3t)$$

$$P_4(t) = \frac{1}{8}(35t^4 - 30t^2 + 3).$$

Then it may be verified by direct computation, when the inner product is defined as

$$\langle f, g \rangle = \int_{-1}^{1} f(t)g(t)\,dt,$$

these polynomials are orthogonal,

$$\langle P_m, P_n \rangle = \begin{cases} 0 & m \neq n, \\ \frac{2}{2n+1} & m = n. \end{cases}$$

These polynomials are the first few *Legendre* polynomials, all of which are orthogonal over $[-1, 1]$. □

2.8 Weighted inner products

For a finite-dimensional vector space, a **weighted inner product** can be obtained by inserting a Hermitian weighting matrix W between the elements:

$$\langle \mathbf{x}, \mathbf{y} \rangle_W = \mathbf{x}^H W \mathbf{y} = \overline{\mathbf{y}^H W \mathbf{x}}.$$

The concept of orthogonality is defined with respect to the particular inner product used: changing the inner product may change the orthogonality relationship between vectors.

Example 2.8.1 Consider the vectors

$$\mathbf{x}_1 = \begin{bmatrix} 1 \\ 1 \end{bmatrix} \qquad \mathbf{x}_2 = \begin{bmatrix} 2 \\ 1 \end{bmatrix}.$$

It is easily verified that these vectors are not orthogonal with respect to the usual inner product $\mathbf{x}_1^T \mathbf{x}_2$. However, for the weighted inner product

$$\langle \mathbf{x}, \mathbf{y} \rangle_W = \mathbf{x}^T \begin{bmatrix} 2 & -2 \\ -2 & 2 \end{bmatrix} \mathbf{y},$$

the vectors $\mathbf{x}_1$ and $\mathbf{x}_2$ are orthogonal. □

In order for the weighted inner product to be used to define a norm, as in

$$\|\mathbf{x}\|_W^2 = \langle \mathbf{x}, \mathbf{x} \rangle_W = \mathbf{x}^H W \mathbf{x},$$

it is necessary that $\mathbf{x}^H W \mathbf{x} > 0$ for all $\mathbf{x} \neq 0$. A matrix W with this property is said to be **positive definite**.

Example 2.8.2 The weighted inner product of the previous example cannot be used as a norm because, for any vector of the form

$$\mathbf{x} = \begin{bmatrix} \alpha \\ \alpha \end{bmatrix},$$

the product $\mathbf{x}^T W \mathbf{x} = 0$, which violates the conditions for a norm. □

Weighting can also be applied to integral inner products. If there is some function $w(t) \geq 0$ over $[a, b]$, then an inner product can be defined as

$$\langle f, g \rangle_w = \int_a^b w(t) f(t) g(t)\,dt.$$

The weighting can be used to place more emphasis on certain parts of the function. (More precisely, we must have $w(t) \geq 0$, with $w(t) = 0$ only on a set of measure zero.)

Example 2.8.3 Let us define a set of polynomials by
$$T_n(t) = \cos(n \cos^{-1}(t))$$
for $t \in [-1, 1]$. The first few of these (obtained by application of trigonometric identities) are
$$T_0(t) = 1 \qquad T_1(t) = t \qquad T_2(t) = 2t^2 - 1 \qquad T_3(t) = 4t^3 - 3t.$$
A plot of the first few of these is shown in figure 2.11. These polynomials are the *Chebyshev polynomials*. They have the interesting property that over the interval $[-1, 1]$, all the extrema of the functions have the values -1 or 1. This property makes them very useful for approximation of functions. Furthermore, the Chebyshev polynomials are orthogonal with weight function
$$w(t) = \frac{1}{\sqrt{1-t^2}}$$
over the interval $[-1, 1]$. The orthogonality relationship between the Chebyshev polynomials is
$$\int_{-1}^{1} \frac{1}{\sqrt{1-t^2}} T_n(t) T_m(t) \, dt = \pi \delta_{n-m}. \qquad \square$$

We can define a weighted inner product on the vector space of $m \times n$ matrices by
$$\langle A, B \rangle = \text{tr}(A^H W B),$$
where W is a symmetric positive-definite $m \times m$ matrix.

Using a norm induced from a weighted inner product, we can define a weighted distance between two vectors:
$$d_W(\mathbf{x}, \mathbf{y})^2 = \|\mathbf{x} - \mathbf{y}\|_W^2 = (\mathbf{x} - \mathbf{y})^H W (\mathbf{x} - \mathbf{y}). \tag{2.17}$$

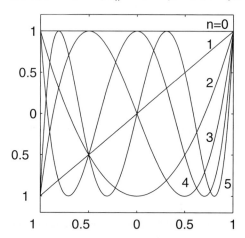

Figure 2.11: Chebyshev polynomials $T_0(t)$ through $T_5(t)$ for $t \in [-1, 1]$

Example 2.8.4 A weighted distance arises naturally in many certain signal detection, estimation, and pattern recognition problems in non-white Gaussian noise. In this example, a detection problem is considered. Detection problems are discussed more fully in chapter 11.

Let $\mathbf{S} \in \mathbb{R}^n$ be a signal which takes on one of two different values, either $\mathbf{S} = \mathbf{s}_0$ or $\mathbf{S} = \mathbf{s}_1$. One of these signals is chosen at random with equal probability—either by a binary data transmitter or by nature. The signal $\mathbf{S}$ is observed in the presence of additive Gaussian noise $\mathbf{N}$ which has mean $\mathbf{0}$ and covariance matrix R. The observation $\mathbf{Y}$ can be modeled as
$$\mathbf{Y} = \mathbf{S} + \mathbf{N}.$$

2.8 Weighted Inner Products

From the observation of $\mathbf{Y} = \mathbf{y}$, we desire to determine which value of $\mathbf{S}$ actually occurred. This is the **detection** problem.

Conditioned upon a value of $\mathbf{S} = \mathbf{s}$, the observation is Gaussian with mean $\mathbf{S}$ and the same covariance:

$$f(\mathbf{y}|\mathbf{S}=\mathbf{s}) = \frac{1}{(2\pi)^{n/2} \det(R)^{1/2}} \exp\left[-\frac{1}{2}(\mathbf{y}-\mathbf{s})^T R^{-1}(\mathbf{y}-\mathbf{s})\right],$$

where either $\mathbf{s} = \mathbf{s}_0$ or $\mathbf{s} = \mathbf{s}_1$. From the observation $\mathbf{y}$, we can compute the *likelihood* that the signal was produced by $\mathbf{s}$ for each of the possible values of $\mathbf{s}$, then select the one with the highest likelihood. That is, we compare

$$f(\mathbf{y}|\mathbf{S}=\mathbf{s}_0) \quad \text{with} \quad f(\mathbf{y}|\mathbf{S}=\mathbf{s}_1), \qquad (2.18)$$

and determine our decision about $\mathbf{S}$ on the basis of which likelihood function is largest. (This is the maximum likelihood decision rule.) Canceling common factors in the comparison, this is equivalent to comparing

$$(\mathbf{y}-\mathbf{s}_0)^T R^{-1}(\mathbf{y}-\mathbf{s}_0) \quad \text{with} \quad (\mathbf{y}-\mathbf{s}_1)^T R^{-1}(\mathbf{y}-\mathbf{s}_1) \qquad (2.19)$$

and choosing either $\mathbf{s}_0$ or $\mathbf{s}_1$, depending upon which quantity is smaller. These quantities can be observed to be weighted distances of the form (2.17). Let $W = R^{-1}$, and define the weighted inner product in $\mathbb{R}^n$ by

$$\langle \mathbf{x}, \mathbf{y} \rangle_W = \mathbf{x}^T W \mathbf{y}.$$

This induces a weighted norm

$$\|\mathbf{x}\|_W^2 = \mathbf{x}^T W \mathbf{x}.$$

The comparison in (2.19) corresponds to computing

$$\|\mathbf{y}-\mathbf{s}_0\|_W \quad \text{and} \quad \|\mathbf{y}-\mathbf{s}_1\|_W,$$

with the maximum likelihood choice being that which has the minimum weight distance. This weighed distance measure arises commonly in pattern recognition problems and is known as the *Mahalonobis distance*.

Further simplifications are often possible in this comparison.

$$\|\mathbf{y}-\mathbf{s}_0\|_W = \mathbf{y}^T W \mathbf{y} - \mathbf{y}^T W \mathbf{s}_0 - \mathbf{s}_0^T W \mathbf{y} + \mathbf{s}_0^T W \mathbf{s}_0$$
$$= \mathbf{y}^T W \mathbf{y} - 2\mathbf{y}^T W \mathbf{s}_0 + \mathbf{s}_0^T W \mathbf{s}_0,$$

and similarly for $\|\mathbf{y}-\mathbf{s}_1\|_W$. If $\mathbf{s}_0$ and $\mathbf{s}_1$ have the same inner product norm so $\mathbf{s}_0^T W \mathbf{s}_0 = \mathbf{s}_1^T W \mathbf{s}_1$, then, when comparing $\|\mathbf{y}-\mathbf{s}_0\|_W$ with $\|\mathbf{y}-\mathbf{s}_1\|_W$, these terms cancel, as well as the $\mathbf{y}^T W \mathbf{y}$ term. The choice is made depending on whether

$$\mathbf{y}^T W \mathbf{s}_0 \quad \text{or} \quad \mathbf{y}^T W \mathbf{s}_1$$

is larger, that is, depending on which weighted inner product is largest. The inner product is thus seen to be a similarity measure: the signal $\mathbf{s}$ is chosen that is most similar to the received signal vector, where the similarity is determined by the weighted inner product. □

2.8.1 Expectation as an inner product

The examples of weighted inner products up until now have been of deterministic functions. An important generalization develops when a joint density is used as a weighting function in the inner product. Let X and Y be random variables with joint density $f_{X,Y}(x, y)$. We define an inner product between them as

$$\langle X, Y \rangle = \int xy f_{X,Y}(x, y)\, dx\, dy.$$

This inner product is, of course, an expectation, and introduction of this inner product allows the conceptual power of vector spaces to be applied to mean-square estimation theory. Thus

$$\langle X, Y \rangle = E[XY]$$

(E is the expectation operator). Orthogonality is defined for random variables as it is for deterministic quantities; the random variables X and Y are orthogonal if $E[XY] = 0$. The inner product induces a norm,

$$\langle X, X \rangle = EX^2.$$

If X is a zero-mean r.v., then $\langle X, X \rangle = \text{var}(X)$ is an induced norm.[3]

We can also define an inner product between random *vectors*. Let $\mathbf{X} = [X_1, X_2, \ldots, X_n]^T$ and $\mathbf{Y} = [Y_1, Y_2, \ldots, Y_n]^T$ be n-dimensional random vectors. Then we can define an inner product between these vectors as

$$\langle \mathbf{X}, \mathbf{Y} \rangle = E \sum_{i=1}^{n} X_i \overline{Y}_i.$$

Note that we can write this inner product as

$$\langle \mathbf{Y}, \mathbf{Y} \rangle = E[\mathbf{Y}^H \mathbf{Y}].$$

Another notation that is sometimes convenient is to write

$$\langle \mathbf{Y}, \mathbf{Y} \rangle = \text{tr}(E[\mathbf{Y}\mathbf{Y}^H]),$$

where the $\text{tr}(X)$ is the trace operator, the sum of the elements on the diagonal of the square matrix X. (See section C.3.)

When the vector-space viewpoint is applied to problems of minimization, as discussed subsequently, there are two major approaches to the problem. In the first case, an inner product is used that is not based on an expectation; minimization of this sort is referred to as *least-squares* (LS) in the signal processing literature. When an inner product is used that is defined as an expectation, then the approximation obtained is referred to as a *minimum mean-squares* (MMS) approximation. In fact, both approximation techniques rely on precisely the same theory, but simply employ inner products suited to the needs of the particular problem.

2.9 Hilbert and Banach spaces

With the definitions of metric spaces and inner-product spaces behind us, we are now ready to introduce the spaces in which most of the work in signal processing is performed.

Definition 2.31 A complete normed vector space is called a **Banach space**. A complete normed vector space with an inner product (in which the norm is the induced norm) is called a **Hilbert space**. □

See box 2.3 for an introduction to the man Hilbert.
Some examples of Banach and Hilbert spaces:

1. The space of continuous functions ($C[a, b], d_\infty$) forms a Banach space. (Recall that in example 2.1.17 ($C[-1, 1], d_\infty$) was shown to be complete.)

2. However, the space of functions $C[a, b]$ with the L_p norm, $p < \infty$, does *not* form a Banach space, since it is not complete. (We saw in example 2.1.16 a sequence of continuous functions that does not have a limit in $C[-1, 1]$.)

---

[3]As with other function spaces, there are some technical problems associated with vector spaces over probability spaces, since there may be random variables X and Y such that $\|X - Y\| = 0$ but $X \neq Y$ always. However, it can be shown that if $\|X - Y\| = 0$, then $X = Y$ a.s. (almost surely, that is, except on a set of probability measure 0).

> **Box 2.3: David Hilbert (1862–1943)**
>
> David Hilbert has been called the "greatest mathematician of recent times." Born and educated at Königsberg, he received a professorship in Göttingen in 1895.
>
> Throughout his life he worked in a variety of areas, including algebraic invariants, algebraic numbers, calculus of variations, spectral theory and Hilbert space, and axiomatics. He is well known for proposing, in 1900, 23 significant mathematical problems. Work on these problems since that time has tremendously enriched mathematics.
>
> He spent considerable effort working on the foundations of mathematics, attempting to prove that mathematics provides an internally consistent system, so that it is not possible, for example, to prove that "F and not-F" is true. His efforts were doomed, however; Kurt Gödel demonstrated, in 1931, that it is impossible for any sufficiently rich formal deductive system to prove consistency of the system by the system itself. There are, Gödel showed, formally undecidable propositions, which cannot be proven to be either true or false, and the consistency of the system is one of these propositions.

3. The sequence space $l_p(0, \infty)$ is a Banach space. When $p = 2$, it is a Hilbert space.
4. The space $L_p[a, b]$ is a Banach space. When $p = 2$ it is a Hilbert space. The Hilbert space of functions with domain over the whole real line is denoted $L_p(\mathbb{R})$.

Because of the utility of having the norm induced from an inner product, the emphasis in this and succeeding chapters is on Hilbert spaces.

It can be shown that if a normed vector space is finite-dimensional, then it is complete [238, p. 267]. Hence, every normed finite-dimensional space is a Banach space; if the norm is induced from an inner product then it is also a Hilbert space. Furthermore, every finite-dimensional subspace of a space is complete.

2.10 Orthogonal subspaces

Definition 2.32 Let S be a vector space, and let V and W be subspaces of S. V and W are **orthogonal** if every vector $\mathbf{v} \in V$ is orthogonal to every vector $\mathbf{w} \in W$: $\langle \mathbf{v}, \mathbf{w} \rangle = 0$. ☐

Definition 2.33 For a subset V of an inner product space S, the space of all vectors orthogonal to V is called the **orthogonal complement** of V. This is denoted as $V^\perp$. ☐

Example 2.10.1 Let V be the plane shown in figure 2.12. Then the orthogonal space $W = V^\perp$ is spanned by the vector $\mathbf{w}$. ☐

The orthogonal complement of a subspace is itself a subspace (see exercise 2.10-52). The orthogonal complement has the following properties:

Theorem 2.5 Let V and W be subsets of an inner product space S (not necessarily complete). Then:

1. $V^\perp$ is a closed subspace of S.
2. $V \subset V^{\perp\perp}$.
3. If $V \subset W$, then $W^\perp \subset V^\perp$.
4. $V^{\perp\perp\perp} = V^\perp$.

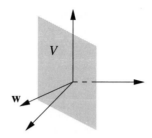

Figure 2.12: A space and its orthogonal complement

5. If $x \in V \cap V^\perp$, then $x = 0$.
6. $\{0\}^\perp = S$ and $S^\perp = \{0\}$.

Proof We will prove part 1. The rest of the properties are to be proved as an exercise (see exercise 2.10-53). To show closure of $V^\perp$, let $\{\mathbf{x}_n\}$ be a convergent sequence in $V^\perp$, so that $\mathbf{x}_n \to \mathbf{x}$. Then by the continuity of the inner product shown in lemma 2.1, we have, for any $v \in V$,
$$0 = \langle \mathbf{x}_n, \mathbf{v} \rangle \to \langle \mathbf{x}, \mathbf{v} \rangle,$$
so that $\mathbf{x} \in V^\perp$. □

What is perhaps a little surprising at first about this theorem is the fact that it may *not* be the case that $V^{\perp\perp} = V$. What is lacking is the completeness: $V^{\perp\perp}$ may have Cauchy sequences in it that V does not.

2.11 Linear transformations: range and nullspace

We pause in our development of vector spaces to reintroduce a concept that should be familiar.

Definition 2.34 A transformation $L: X \to Y$ from a vector space X to a vector space Y (where X and Y have the same scalar field R) is a **linear** transformation if for all vectors $x, x_1, x_2 \in X$:

1. $L(\alpha x) = \alpha L(x)$ for all $\mathbf{x} \in X$ and all scalars $\alpha \in R$, and
2. $L(x_1 + x_2) = L(x_1) + L(x_2)$. □

We will think of linear transformations as *operators*.

Example 2.11.1 We will begin with several examples from vector spaces of functions.

1. Let X be the set of continuous real-valued functions, and define $L: X \to X$ by
$$Lx(t) = \int_0^t h(\tau) x(t - \tau) \, d\tau$$
for all $x(t) \in X$. Then L is a linear transformation which convolves the signal x with the signal h.

2. Let X be the set of continuous real-valued functions defined on $[0, 1]$. Then $L: X \to \mathbb{R}$ defined by
$$Lx(t) = \int_0^1 h(\tau) x(\tau) \, d\tau$$
is a linear transformation (an inner product).

2.11 Linear Transformations: Range and Nullspace

3. Let X be the set of continuous real-valued functions, and let $T_{T_0}: X \to X$ be defined by

$$T_{T_0} x(t) = \begin{cases} x(t) & |t| < T_0, \\ 0 & \text{otherwise,} \end{cases}$$

where T_0 is a parameter of the transformation. Then T_{T_0} is a linear transformation. This transformation truncates a signal in time.

4. Let X be the set of all Fourier transformable functions, and let Y be the set of Fourier transforms of elements in X. Define $F: X \to Y$ by

$$Fx(t) = \int_{-\infty}^{\infty} x(t) e^{-j\omega t} \, dt.$$

The operator F is a linear operator.

5. Let $B: X \to X$ be defined by

$$B_{B_0} x(t) = \mathcal{F}^{-1} T_{B_0} X(\omega),$$

where $X(\omega)$ is the Fourier transform of $x(t)$, $\mathcal{F}^{-1}$ is the inverse Fourier transform operator, and $T_{B_0} X(\omega)$ truncates the Fourier transform. Thus $B_{B_0} x(t)$ is a bandlimited signal. □

Example 2.11.2 Perhaps more commonly, we see linear transformations between vector spaces of finite dimension. In general a linear transformation L from the vector space R^n to R^m can be expressed using the notation of an $m \times n$ matrix L. That is, the matrix becomes the linear transformation.

1. Let $L: \mathbb{R}^3 \to \mathbb{R}^2$ be defined by

$$L(x_1, x_2, x_3) = (x_1 + 2x_2, 3x_2 + 4x_3).$$

This linear transformation can be placed in matrix notation. By writing an element in $\mathbb{R}^3$ in vector form as $[x_1, x_2, x_3]^T \in \mathbb{R}^3$, we can write

$$L = \begin{bmatrix} 1 & 2 & 0 \\ 0 & 3 & 4 \end{bmatrix}.$$

Then,

$$L\mathbf{x} = \begin{bmatrix} x_1 + 2x_2 \\ 3x_2 + 4x_3 \end{bmatrix}.$$

2. Let $L: \mathbb{R}^3 \to \mathbb{R}^3$ be defined by the matrix

$$L = \begin{bmatrix} 0 & 0 & 1 \\ 0 & 1 & 0 \\ 1 & 0 & 0 \end{bmatrix}.$$

Then L is the linear transformation that reverses the coordinates of a vector $\mathbf{x} \in \mathbb{R}^3$. □

Considerably more is said about linear transformations between finite-dimensional vectors spaces in chapter 4.

Associated with any operator (linear or otherwise) are two important spaces. These spaces are the range and the nullspace. (Two more spaces associated with linear operators are presented in section 4.5.)

Definition 2.35 Let $L: X \to Y$ be an operator (linear or otherwise). The **range space** $\mathcal{R}(L)$ is

$$\mathcal{R}(L) = \{\mathbf{y} = L\mathbf{x} : \mathbf{x} \in X\};$$

that is, it is the set of values in Y that are reached from X by application of L. The **nullspace** $\mathcal{N}(L)$ is

$$\mathcal{N}(L) = \{\mathbf{x} \in X : L\mathbf{x} = \mathbf{0}\};$$

that is, it is the set of values in $\mathbf{x}$ that are transformed to $\mathbf{0}$ in Y by L. The nullspace of an operator is also called the **kernel** of the operator. □

Let A be an $n \times m$ matrix,
$$A = [\mathbf{p}_1, \mathbf{p}_2, \ldots, \mathbf{p}_m],$$
which we regard as a linear operator. Then a point $\mathbf{x} \in \mathbb{R}^m$ is transformed as
$$A\mathbf{x} = x_1\mathbf{p}_1 + x_2\mathbf{p}_2 + \cdots + x_m\mathbf{p}_m,$$
which is a linear combination of the columns of A. Thus, the range may be expressed as
$$\mathcal{R}(A) = \text{span}(\{\mathbf{p}_1, \mathbf{p}_2, \ldots, \mathbf{p}_m\}).$$
The range of a matrix is also referred to as the *column space* of A. The nullspace is that set of vectors such that $A\mathbf{x} = \mathbf{0}$.

Example 2.11.3 Let
$$A = \begin{bmatrix} 1 & 0 & 0 \\ 0 & 0 & 0 \\ 1 & 0 & 1 \end{bmatrix}.$$
Then the range of A is
$$\text{span}([1, 0, 1]^T, [0, 0, 1]^T).$$
The nullspace of A is
$$\mathcal{N}(A) = \text{span}([0, 1, 0]^T). \qquad \Box$$

Example 2.11.4

1. Let $Lx(t) = \int_0^t x(\tau)h(t - \tau)\,d\tau$. Then the nullspace of L is the set of all functions $x(t)$ that result in zero when convolved with $h(t)$. From systems theory, we realize that we can transform the convolution operation and multiply in the frequency domain. From this perspective, we perceive that the nullspace of L is the set of functions whose Fourier transforms do not share any support with the support of the Fourier transform of h.

2. Let $Lx(t) = \int_0^t x(\tau)h(\tau)\,d\tau$, where X is the set of continuous functions. Then $\mathcal{R}(\mathcal{L})$ is the set of real numbers, unless $h(t) \equiv 0$.

3. The range of the operator
$$A = \begin{bmatrix} 1 & 0 \\ 0 & 0 \end{bmatrix}$$
is the set of all vectors of the form $[c, 0]^T$. The nullspace of this operator is $\text{span}([0, 1])$. $\qquad \Box$

2.12 Inner-sum and direct-sum spaces

Definition 2.36 If V and W are linear subspaces, the space $V + W$ is the **(inner) sum** space, consisting of all points $\mathbf{x} = \mathbf{v} + \mathbf{w}$, where $\mathbf{v} \in V$ and $\mathbf{w} \in W$. $\qquad \Box$

Example 2.12.1 Consider $S = (GF(2))^3$, that is, the set of all 3-tuples of elements of $GF(2)$ (see box 1.5). Then, for example,
$$\mathbf{x} = (1, 0, 1) \in S \quad \text{and} \quad \mathbf{y} = (0, 0, 1) \in S,$$
and $\mathbf{x} + \mathbf{y} = (1, 0, 0)$.

Let $W = \text{span}[(0, 1, 0)]$ and $V = \text{span}[(1, 0, 0)]$ be two subspaces in S. Then
$$W = \{(0, 0, 0), (0, 1, 0)\}$$
and
$$V = \{(0, 0, 0), (1, 0, 0)\}.$$
These two subspaces are orthogonal.

2.12 Inner-Sum and Direct-Sum Spaces

The orthogonal complement to V is

$$V^\perp = \{(0,0,0), (0,1,0), (0,0,1), (0,1,1)\}.$$

Thus, $W \subset V^\perp$.

The inner sum space of V and W is

$$V + W = \{(0,0,0), (0,1,0), (1,0,0), (1,1,0)\}.$$

Definition 2.37 Two linear subspaces V and W of the same dimensionality are **disjoint** if $V \cap W = \{0\}$. That is, the only vector they have in common is the zero vector. (This definition is slightly different from disjoint sets, since they must have the zero vector in common.)

Example 2.12.2 In figure 2.13, the plane S is a vector space in two dimensions, and V and W are two one-dimensional subspaces, indicated by the lines in the figure. The only point they have in common is the origin, so they are disjoint. (Note that they are not necessarily orthogonal.)

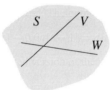

Figure 2.13: Disjoint lines in $\mathbb{R}^2$

When $S = V + W$ and V and W are disjoint, W is said to be the *algebraic complement* of V. The last example illustrates an algebraic complement: the inner sum of the two lines gives the entire vector space S. On the other hand, the sets V and W in example 2.12.1 are not algebraic complements, since $V + W$ is not the same as S. An algebraic complement to the set V of that example would be the set

$$Z = \text{span}(\{(0,1,0), (0,0,1)\}) = \{(0,0,0), (0,1,0), (0,0,1), (0,1,1)\}.$$

It is straightforward to show that in any vector space S every linear subspace has an algebraic complement. Let B be a (Hamel) basis for S, and let $B_1 \subset B$ be a (Hamel) basis for V. Then let $B_2 = B - B_1$ (the set difference), so that $B_1 \cap B_2 = \emptyset$. Then

$$W = \text{span}(B_2)$$

is a (Hamel) basis for the algebraic complement of V.

The direct sum of disjoint spaces can be used to provide a unique representation of a vector.

Lemma 2.3 *([238]) Let V and W be subspaces of a vector space S. Then for each $\mathbf{x} \in V + W$, there is a unique $\mathbf{v} \in V$ and a unique $\mathbf{w} \in W$ such that $\mathbf{x} = \mathbf{v} + \mathbf{w}$ if and only if V and W are disjoint.*

Another way of combining vector spaces is by the direct sum.

> **Box 2.4: Isomorphism**
>
> > What's in a name? that which we call a rose,
> > By any other name would smell as sweet.
> > — *William Shakespeare*
>
> Isomorphism denotes the fact that two objects may have the same operational behavior, even if they have different names.
>
> As an example, consider the following two operations for two groups called $\langle G_1, + \rangle$ and $\langle G_2, * \rangle$.
>
> | + | (0,0) | (0,1) | (1,0) | (1,1) |
> |-------|-------|-------|-------|-------|
> | (0,0) | (0,0) | (0,1) | (1,0) | (1,1) |
> | (0,1) | (0,1) | (0,0) | (1,1) | (1,0) |
> | (1,0) | (1,0) | (1,1) | (0,0) | (0,1) |
> | (1,1) | (1,1) | (1,0) | (0,1) | (0,0) |
>
> | * | a | b | c | d |
> |---|---|---|---|---|
> | a | a | b | c | d |
> | b | b | a | d | c |
> | c | c | d | a | b |
> | d | d | c | b | a |
>
> Careful comparison of these addition tables reveals that the same operation occurs in both tables, but the names of the elements and the operator have been changed.
>
> More generally, we describe an isomorphism as follows. Let G_1 and G_2 be two algebraic objects (e.g., groups, fields, vector spaces, etc.). Let $*$ be a binary operation on G_1 and let $\circ$ be the corresponding operation on G_2. Let $\phi: G_1 \to G_2$ be a one-to-one and onto function. For any $x, y \in G_1$, let
>
> $$s = \phi(x) \quad \text{and} \quad t = \phi(y),$$
>
> where $s \in G_2$ and $t \in G_2$. Then ϕ is an isomorphism if
>
> $$\phi(x * y) = \phi(x) \circ \phi(y).$$
>
> Note that the operation on the left takes place in G_1 while the operation on the right takes place in G_2.

Definition 2.38 The **direct sum** of linear spaces V and W, denoted $V \oplus W$, is defined on the Cartesian product $V \times W$, so a point in $V \oplus W$ is an ordered pair (v, w) with $v \in V$ and $w \in W$. Addition is defined component-wise: $(v_1, w_1) + (v_2, w_2) = (v_1 + v_2, w_1 + w_2)$. Scalar multiplication is defined as $\alpha(v, w) = (\alpha v, \alpha w)$. □

The sum $V + W$ and the direct sum $V \oplus W$ are different linear spaces. However, if V and W are *disjoint*, then $V + W$ and $V \oplus W$ have exactly the same structure mathematically; they are simply different representations of the same thing. When different mathematical objects behave the same, only varying in the name, the objects are said to be *isomorphic* (see box 2.4).

Example 2.12.3 Using the vector space of example 2.12.1, we find

$$V \oplus W = \{(0,0,0,0,0,0), (1,0,0,0,0,0), (0,0,0,0,1,0), (1,0,0,0,1,0)\}.$$

Under the mapping $\phi[(\mathbf{v}, \mathbf{w})] = \mathbf{v} + \mathbf{w}$, we find

$$\phi(V \oplus W) = \{(0,0,0), (1,0,0), (0,1,0), (1,1,0)\},$$

which is the same as found in $V + W$ in the previous example. Vector space operations (addition, multiplication by a scalar, etc.) on $V \oplus W$ have exactly analogous results on $\phi(V \oplus W)$, so $V \oplus W$ and $V + W$ are isomorphic. □

The direct sum $V \oplus W$ is commonly employed between orthogonal vector spaces in the "isomorphic" form, that is, as the sum of the elements. This is justified because orthogonal spaces are disjoint (see exercise 2.12-64).

The following theorem indicates when $V + W$ and $V \oplus W$ are isomorphic:

Theorem 2.6 *[238, page 199] Let V and W be linear subspaces of a linear space S. Then $V + W$ and $V \oplus W$ are isomorphic if and only if V and W are disjoint.*

Because of this theorem, it is common to write $V + W$ in place of $V \oplus W$, and vice versa. Care should be taken, however, to understand what space is actually intended.

2.13 Projections and orthogonal projections

As pointed out in lemma 2.3, if V and W are disjoint subspaces of a linear space S, then any vector $\mathbf{x} \in S$ can be uniquely written as

$$\mathbf{x} = \mathbf{v} + \mathbf{w},$$

where $\mathbf{v} \in V$ and $\mathbf{w} \in W$. This representation is illustrated in figure 2.14.

Let us introduce projection operator $P: S \to V$ with the following operation: for any $\mathbf{x} \in S$ with the decomposition

$$\mathbf{x} = \mathbf{v} + \mathbf{w},$$

let

$$P\mathbf{x} = \mathbf{v}.$$

That is, the projection operator returns that component of $\mathbf{x}$ which lies in V. If $\mathbf{x} \in V$ to begin with, then operation by P does not change the value of $\mathbf{x}$. Thus since $P\mathbf{x} \in V$, we see that $P(P\mathbf{x}) = P\mathbf{x}$. This motivates the following definition.

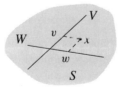

Figure 2.14: Decomposition of **x** into disjoint components

Definition 2.39 *A linear transformation P of a linear space into itself is a **projection** if $P^2 = P$.* □

An operator P such that $P^2 = P$ is said to be *idempotent*.

If V is a linear subspace and P is an operator that projects onto V, the projection of a vector $\mathbf{x}$ onto V is sometimes denoted as $x_{\text{proj } V}$.

The range and nullspace of a projection operator provide a disjoint decomposition of a vector space, as the following theorem shows.

Example 2.13.1 Let $x(t)$ be a signal with Fourier transform $X(\omega)$. Then the transformation P_{ω_0}, $\omega_0 \geq 0$ defined by

$$P[X(\omega)] = \begin{cases} X(\omega) & \text{for } -\omega_0 \leq \omega \leq \omega_0, \\ 0 & \text{otherwise,} \end{cases}$$

which corresponds to filtering the signal with a "brick-wall" lowpass filter, is a projection operation. □

Example 2.13.2 Let $P_T, T \geq 0$ be the transformation on the function $x(t)$ defined by

$$(P_T x)(t) = \begin{cases} x(t) & \text{for } -T \leq t \leq T, \\ 0 & \text{otherwise.} \end{cases}$$

This is a time-truncation operation and is a projection. □

Example 2.13.3 A matrix A is said to be a *smoothing* matrix if there is a space of "smooth" vectors V such that, for a vector $\mathbf{x} \in V$,

$$A\mathbf{x} = \mathbf{x};$$

that is, a smooth vector unaffected by a smoothing operation. Also, the limit

$$A^\infty = \lim_{p \to \infty} A^p$$

exists. As an arbitrary vector that is not already smooth is repeatedly smoothed, it becomes increasingly smooth. By the requirement that $A\mathbf{x} = \mathbf{x}$ for $\mathbf{x} \in V$, it is clear that the set of smooth vectors is in fact $\mathcal{R}(A)$, and A is a projection matrix. (Smoothing matrices are discussed further in [120, 121].) □

Theorem 2.7 *Let P be a projection operator defined on a linear space S. Then the range and nullspace of P are disjoint linear subspaces of S, and $S = \mathcal{R}(P) + \mathcal{N}(P)$. That is, $\mathcal{R}(P)$ and $\mathcal{N}(P)$ are algebraic complements.*

Let P be a projection onto a closed subspace V of S. Then $I - P$ is also a projection (see exercise 2.13-73). Then we can write

$$\mathbf{x} = P\mathbf{x} + (I - P)\mathbf{x}.$$

This decomposes $\mathbf{x}$ into the two parts,

$$P\mathbf{x} \in V$$

and

$$(I - P)\mathbf{x} \in W.$$

As figure 2.14 suggests, the subspaces V and W involved in the projection are not necessarily orthogonal. However, in most applications, orthogonal subspaces are needed. This leads to the following definition.

Definition 2.40 Let P be a projection operator on an inner product space S. P is said to be an *orthogonal projection* if its range and nullspace are orthogonal, $\mathcal{R}(P) \perp \mathcal{N}(P)$. □

The need for an orthogonal projection matrix is provided by the following problem: Given a point $\mathbf{x}$ in a vector space S and a subspace $V \subset S$, what is the nearest point in V to $\mathbf{x}$? Consider the various representations of $\mathbf{x}$ shown in figure 2.15. As suggested by the figure, decomposition of $\mathbf{x}$ as

$$\mathbf{x} = \mathbf{v}_0 + \mathbf{w}_0$$

provides the point $\mathbf{v}_0 \in V$ that is closest to $\mathbf{x}$. The vector $\mathbf{w}_0$ is orthogonal to V, with respect to the inner product appropriate to the problem. Of the various $\mathbf{w}$ vectors that might be

2.13 Projections and Orthogonal Projections

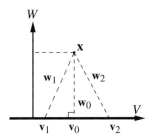

Figure 2.15: Orthogonal projection finds the closest point in V to $\mathbf{x}$

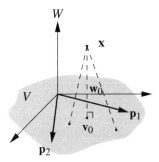

Figure 2.16: Orthogonal projection onto the space spanned by several vectors

used in the representation (the vectors $\mathbf{w}_0$, $\mathbf{w}_1$, or $\mathbf{w}_2$ in the figure), the vector $\mathbf{w}_0$ is the vector of the shortest length, as determined by the norm induced by the inner product. Proof of this geometrically appealing and intuitive notion is presented in the next section as the projection theorem. It is difficult to overstate the importance of the notion of projection. Projection is the key concept of most stochastic filtering and prediction theory in signal processing. Chapter 3 is entirely concerned with applications of this important concept.

Another viewpoint of the projection theorem is represented in figure 2.16. Suppose that V is the span of the basis vectors $\{\mathbf{p}_1, \mathbf{p}_2\}$, as shown. Then the nearest point to $\mathbf{x}$ in V is the point $\mathbf{v}_0$, and the vector $\mathbf{w}_0$ is the difference. If $\mathbf{w}_0$ is orthogonal to $\mathbf{v}_0$, then it must be orthogonal to $\mathbf{p}_1$ and $\mathbf{p}_2$.

If we regard $\mathbf{v}_0$ as an approximation to $\mathbf{x}$ that must lie in the span of $\mathbf{p}_1$ and $\mathbf{p}_2$, then

$$\mathbf{w}_0 = \mathbf{x} - \mathbf{v}_0$$

is the approximation error. Consider the vectors $\mathbf{p}_1$ and $\mathbf{p}_2$ as the data from which the approximation is to be formed. Then *the length of the approximation error vector $\mathbf{w}_0$ is minimized when the error is orthogonal to the data.*

2.13.1 Projection matrices

Let us restrict our attention for the moment to finite-dimensional vector spaces. Let A be an $m \times n$ matrix written as

$$A = [\mathbf{p}_1, \mathbf{p}_2, \ldots, \mathbf{p}_n],$$

and let the subspace V be the column space of A,

$$V = \text{span}(\mathbf{p}_1, \mathbf{p}_2, \ldots, \mathbf{p}_n) = \mathcal{R}(A).$$

Assume that we are using the usual inner product, $\langle \mathbf{x}, \mathbf{y} \rangle = \mathbf{x}^H \mathbf{y}$. Then, as we see in the next chapter, the projection matrix P_A that projects orthogonally onto the column space of A is

$$P_A = A(A^H A)^{-1} A^H. \tag{2.20}$$

We can characterize projection matrices as follows.

Theorem 2.8 *Any symmetric matrix with $P^2 = P$ is an orthogonal projection matrix.*

Proof The operation $P\mathbf{x}$ is a linear combination of the columns of P. To show that P is an orthogonal projection we must show that $\mathbf{x} - P\mathbf{x}$ is orthogonal to the column space of P. For any vector $P\mathbf{c}$ in the column space of P,

$$(\mathbf{x} - P\mathbf{x})^T P\mathbf{c} = \mathbf{x}^T (P - P^2)\mathbf{c} = 0,$$

so $\mathbf{x} - P\mathbf{x}$ is orthogonal to the column space of P. $\square$

It will occasionally be useful to do the projection using a weighted inner product. Let the inner product be

$$\langle \mathbf{x}, \mathbf{y} \rangle_W = \mathbf{x}^H W \mathbf{y}, \tag{2.21}$$

where W is a positive definition Hermitian symmetric matrix. The induced norm is

$$\|\mathbf{x}\|_W^2 = \langle \mathbf{x}, \mathbf{x} \rangle_W = \mathbf{x}^H W \mathbf{x}.$$

Let A be an $m \times n$ matrix, as before. Then the projection matrix which projects orthogonally onto the column space of A, where the orthogonality is established using the inner product (2.21), is the matrix

$$P_{A,W} = A(A^H W A)^{-1} A^H W. \tag{2.22}$$

2.14 The projection theorem

> Important attributes of many fully evolved major theorems:
> 1. It is trivial.
> 2. It is trivial because the terms appearing in it have been properly defined.
> 3. It has significant consequences.
>
> — *Michael Spivak*
> Calculus on Manifolds

The main purpose of this section is to prove the geometrically intuitive notion introduced in the previous section: the point $\mathbf{v}_0 \in V$ that is closest to a point $\mathbf{x}$ is the orthogonal projection of $\mathbf{x}$ onto V.

Theorem 2.9 *(The projection theorem) ([209]) Let S be a Hilbert space and let V be a closed subspace of S. For any vector $\mathbf{x} \in S$, there exists a* unique *vector $\mathbf{v}_0 \in V$ closest to $\mathbf{x}$; that is, $\|\mathbf{x} - \mathbf{v}_0\| \leq \|\mathbf{x} - \mathbf{v}\|$ for all $\mathbf{v} \in V$. Furthermore, the point $\mathbf{v}_0$ is the minimizer of $\|\mathbf{x} - \mathbf{v}_0\|$ if and only if $\mathbf{x} - \mathbf{v}_0$ is orthogonal to V.*

The idea behind the theorem is shown in figure 2.17.

2.14 The Projection Theorem

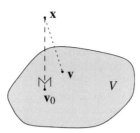

Figure 2.17: The projection theorem

Proof There are several aspects of this theorem.

1. The first (and most technical) aspect is the *existence* of the minimizing point $\mathbf{v}_0$. Assume $\mathbf{x} \notin V$, and let $\delta = \inf_{\mathbf{v} \in V} \|\mathbf{x} - \mathbf{v}\|$. We need to show that there is a $\mathbf{v}_0 \in V$ with $\|\mathbf{x} - \mathbf{v}_0\| = \delta$. Let $\{\mathbf{v}_i\}$ be a sequence of vectors in V such that $\|x - v_i\| \to \delta$. We will show that $\{v_i\}$ is a Cauchy sequence, hence has a limit in S. By (2.27),

$$\|(\mathbf{v}_j - \mathbf{x}) + (\mathbf{x} - \mathbf{v}_i)\|^2 + \|(\mathbf{v}_j - \mathbf{x}) - (\mathbf{x} - \mathbf{v}_i)\|^2 = 2\|\mathbf{v}_j - \mathbf{x}\|^2 + 2\|\mathbf{x} - \mathbf{v}_i\|^2.$$

The latter can be rearranged as

$$\|\mathbf{v}_j - \mathbf{v}_i\|^2 = 2\|\mathbf{v}_j - \mathbf{x}\|^2 + 2\|\mathbf{x} - \mathbf{v}_i\|^2 - 4\|\mathbf{x} - (\mathbf{v}_i + \mathbf{v}_j)/2\|^2.$$

Since S is a vector space, $(\mathbf{v}_i + \mathbf{v}_j)/2 \in S$. Also, by the definition of δ,

$$\|\mathbf{x} - (\mathbf{v}_i + \mathbf{v}_j)/2\| \geq \delta,$$

so that

$$\|\mathbf{v}_i - \mathbf{v}_j\|^2 \leq 2\|\mathbf{v}_j - \mathbf{x}\| + 2\|\mathbf{x} - \mathbf{v}_i\| - 4\delta^2.$$

Then, since $\{\mathbf{v}_i\}$ is defined so that $\|\mathbf{v}_j - \mathbf{x}\| \to \delta^2$, we conclude that

$$\|\mathbf{v}_i - \mathbf{v}_j\|^2 \to 0,$$

so $\{\mathbf{v}_i\}$ is a Cauchy sequence. Since V is a Hilbert space (a subspace of S), the limit exists, and $\mathbf{v}_0 \in V$.

2. Let us now show that if $\mathbf{v}_0$ minimizes $\|\mathbf{x} - \mathbf{v}_0\|$, then $(\mathbf{x} - \mathbf{v}_0) \perp V$. Let $\mathbf{v}_0$ be the nearest vector to $\mathbf{x}$ in V. Let $\mathbf{v}$ be a unit-norm vector in V such that (contrary to the statement of the theorem)

$$\langle \mathbf{x} - \mathbf{v}_0, \mathbf{v} \rangle = \delta \neq 0.$$

Let $\mathbf{z} = \mathbf{v}_0 + \delta \mathbf{v} \in V$ for some number δ. Then

$$\|\mathbf{x} - \mathbf{z}\|^2 = \|\mathbf{x} - \mathbf{v}_0\|^2 - 2 \operatorname{Re} \langle \mathbf{x} - \mathbf{v}_0, \delta \mathbf{v} \rangle + \|\delta \mathbf{v}\|^2$$
$$= \|\mathbf{x} - \mathbf{v}_0\|^2 - |\delta|^2 < \|\mathbf{x} - \mathbf{v}_0\|^2.$$

This is a contradiction, hence $\delta = 0$.

3. Conversely, suppose that $(\mathbf{x} - \mathbf{v}_0) \perp V$. Then for any $\mathbf{v} \in V$ with $\mathbf{v} \neq \mathbf{v}_0$,

$$\|\mathbf{x} - \mathbf{v}\|^2 = \|\mathbf{x} - \mathbf{v}_0 + \mathbf{v}_0 - \mathbf{v}\|^2$$
$$= \|\mathbf{x} - \mathbf{v}_0\|^2 + \|\mathbf{v}_0 - \mathbf{v}\|^2, \tag{2.23}$$
$$\geq \|\mathbf{x} - \mathbf{v}_0\|^2, \tag{2.24}$$

where orthogonality is used to obtain (2.23).

4. Uniqueness of the nearest point in V to $\mathbf{x}$ may be shown as follows. Suppose that $\mathbf{x} = \mathbf{v}_1 + \mathbf{w}_1 = \mathbf{v}_2 + \mathbf{w}_2$, where $\mathbf{w}_1 = \mathbf{x} - \mathbf{v}_1 \perp V$ and $\mathbf{w}_2 = \mathbf{x} - \mathbf{v}_2 \perp V$ for some $\mathbf{v}_1, \mathbf{v}_2 \in V$. Then $0 = \mathbf{v}_1 - \mathbf{v}_2 + \mathbf{w}_1 - \mathbf{w}_2$, or

$$\mathbf{v}_2 - \mathbf{v}_1 = \mathbf{w}_1 - \mathbf{w}_2.$$

But, since $\mathbf{v}_2 - \mathbf{v}_1 \in V$, it follows that $\mathbf{w}_1 - \mathbf{w}_2 \in V$, so $\mathbf{w}_1 = w_2$, hence $\mathbf{v}_1 = \mathbf{v}_2$. □

Based on the projection theorem, every vector in a Hilbert space S can be expressed uniquely as that part which lies in a subspace V, and that part which is orthogonal to V.

Theorem 2.10 *([209]) Let V be a closed linear subspace of a Hilbert space S. Then*

$$S = V \oplus V^\perp$$

and

$$V = V^{\perp\perp}.$$

(The isomorphic interpretation of the direct sum is implied in this notation.)

Proof Let $\mathbf{x} \in S$. Then by the projection theorem, there is a unique $\mathbf{v}_0 \in V$ such that $\|\mathbf{x} - \mathbf{v}_0\| \leq \|\mathbf{x} - \mathbf{v}\|$ for all $\mathbf{v} \in V$, and $\mathbf{w}_0 = \mathbf{x} - \mathbf{v}_0 \in V^\perp$. We can thus decompose any vector in S into

$$\mathbf{x} = \mathbf{v}_0 + \mathbf{w}_0 \quad \text{with} \quad \mathbf{v}_0 \in V, \, \mathbf{w}_0 \in V^\perp.$$

To show that $V = V^{\perp\perp}$, we need to show only that $V^{\perp\perp} \subset V$, since we already know by theorem 2.10 that $V \subset V^{\perp\perp}$. Let $\mathbf{x} \in V^{\perp\perp}$. We will show that it is also true that $\mathbf{x} \in V$. By the first part we can write $\mathbf{x} = \mathbf{v} + \mathbf{w}$, where $\mathbf{v} \in V$ and $\mathbf{w} \in V^\perp$. But, since $V \subset V^{\perp\perp}$ we have $\mathbf{v} \in V^{\perp\perp}$, so that

$$\mathbf{w} = \mathbf{x} - \mathbf{v} \in V^{\perp\perp}.$$

Since $\mathbf{w} \in V^\perp$ and $\mathbf{w} \in V^{\perp\perp}$, we must have $\mathbf{w} \perp w$, or $w = \mathbf{0}$. Thus $\mathbf{x} = \mathbf{v} \in V$. □

This theorem applies to Hilbert spaces, where both completeness and an inner product (defining orthogonality) are available.

2.15 Orthogonalization of vectors

In many applications, computations involving basis vectors are easier if the vectors are orthogonal. Since vector space computations are more conveniently done with orthonormal vectors, it is useful to be able to take a set of vectors T and produce an orthogonal set of vectors T' with the same span as T. This is what the Gram–Schmidt orthogonalization procedure does. Gram–Schmidt can also be used to determine the dimension of the space spanned by a set of vectors, since a vector linearly dependent on other vectors examined prior in the procedure yields a zero vector.

Given a set of vectors $T = \{\mathbf{p}_1, \mathbf{p}_2, \ldots, \mathbf{p}_n\}$, we want to find a set of vectors $T' = \{\mathbf{q}_1, \mathbf{q}_2, \ldots, \mathbf{q}_{n'}\}$ with $n' \leq n$ so that

$$\text{span}\{\mathbf{q}_1, \mathbf{q}_2, \ldots, \mathbf{q}_{n'}\} = \text{span}\{\mathbf{p}_1, \mathbf{p}_2, \ldots, \mathbf{p}_n\}$$

and

$$\langle \mathbf{q}_i, \mathbf{q}_j \rangle = \delta_{i,j}.$$

Assume that none of the $\mathbf{p}_i$ vectors are zero vectors.

2.15 Orthogonalization of Vectors

The process will be developed stepwise. The norm $\|\cdot\|$ in this section is the induced norm.

1. Normalize the first vector:
$$\mathbf{q}_1 = \frac{\mathbf{p}_1}{\|\mathbf{p}_1\|}.$$

2. Compute the difference between the projection of $\mathbf{p}_2$ onto $\mathbf{q}_1$ and $\mathbf{p}_2$. By the orthogonality theorem, this is orthogonal to p_1:
$$\mathbf{e}_2 = \mathbf{p}_2 - \frac{\langle \mathbf{p}_2, \mathbf{q}_1 \rangle}{\|\mathbf{q}_1\|^2} \mathbf{q}_1 = \mathbf{p}_2 - \langle \mathbf{p}_2, \mathbf{q}_1 \rangle \mathbf{q}_1.$$

If $\mathbf{e}_2 = 0$, then $\mathbf{q}_2 \in \text{span}(\mathbf{q}_1)$ and can be discarded; we will assume that such discards are done as necessary in what follows. If $\mathbf{e}_2 \neq 0$, then normalize
$$\mathbf{q}_2 = \frac{\mathbf{e}_2}{\|\mathbf{e}_2\|}.$$

These steps are shown in figure 2.18.

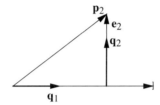

Figure 2.18: The first steps of the Gram–Schmidt process

3. At the next stage, a vector orthogonal to $\mathbf{q}_1$ and $\mathbf{q}_2$ obtained from the error between $\mathbf{p}_3$ and its projection onto $\text{span}(\mathbf{q}_1, \mathbf{q}_2)$:
$$\mathbf{e}_3 = \mathbf{p}_3 - \langle \mathbf{p}_3, \mathbf{q}_1 \rangle \mathbf{q}_1 - \langle \mathbf{p}_3, \mathbf{q}_2 \rangle \mathbf{q}_2.$$

This is normalized to produce $\mathbf{q}_3$:
$$\mathbf{q}_3 = \frac{\mathbf{q}_3}{\|\mathbf{q}_3\|}.$$

See figure 2.19

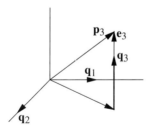

Figure 2.19: Third step of the Gram-Schmidt process

4. Now proceed inductively: To form the next orthogonal vector using $\mathbf{p}_k$, determine the component orthogonal to all previously found vectors.
$$\mathbf{e}_k = \mathbf{p}_k - \sum_{i=1}^{k-1} \langle \mathbf{p}_k, \mathbf{q}_i \rangle \mathbf{q}_i \qquad (2.25)$$

and normalize
$$\mathbf{q}_k = \frac{\mathbf{e}_k}{\|\mathbf{e}_k\|_2}. \qquad (2.26)$$

Example 2.15.1 The set of functions $\{1, t, t^2, \ldots, t^m\}$ defined over $[-1, 1]$ forms a linearly independent set. Let the inner product be

$$\langle f, g \rangle = \int_{-1}^{1} f(t)g(t)\, dt.$$

By the Gram–Schmidt procedure, we find
$$y_0(t) = 1$$
$$y_1(t) = t$$
$$y_2(t) = t^2 - 1/3$$
$$y_3(t) = t^3 - 3t/5$$
$$\vdots$$

The functions so obtained are known as the *Legendre polynomials*.

If we change the inner product to include a weighting function,

$$\langle f, g \rangle = \int_{-1}^{1} \frac{1}{\sqrt{1 - t^2}} f(t)g(t)\, dt,$$

then the orthogonal polynomials obtained by applying the Gram–Schmidt process to the polynomials $\{1, t, \ldots, t^n\}$ are the Chebyshev polynomials, described in example 2.8.3. □

A matrix-based implementation

For finite-dimensional vectors, the Gram–Schmidt process can be represented in a matrix form. Let $A = [\mathbf{p}_1, \mathbf{p}_2, \ldots, \mathbf{p}_n]$. The orthogonal vectors obtained by the Gram–Schmidt process are stacked in a matrix $Q = [\mathbf{q}_1, \mathbf{q}_2, \ldots, \mathbf{q}_{n'}]$, to be determined. We let the upper triangular matrix R hold the inner products and norms from (2.25) and (2.26):

$$R = \begin{bmatrix} \|\mathbf{p}_1\| & \langle \mathbf{p}_2, \mathbf{q}_1 \rangle & \langle \mathbf{p}_3, \mathbf{q}_1 \rangle & \cdots & \langle \mathbf{p}_n, \mathbf{q}_1 \rangle \\ & \|\mathbf{e}_2\| & \langle \mathbf{p}_3, \mathbf{q}_2 \rangle & \cdots & \langle \mathbf{p}_n, \mathbf{q}_2 \rangle \\ & & \|\mathbf{e}_3\| & \cdots & \langle \mathbf{p}_n, \mathbf{q}_3 \rangle \\ & & & & \vdots \\ & & & \cdots & \|\mathbf{e}_n\| \end{bmatrix}.$$

The inner products in the summation $\sum_{i=1}^{k-1} \langle \mathbf{p}_k, \mathbf{q}_i \rangle \mathbf{q}_i$ are represented by $R(1{:}k-1, k) = Q(:, 1{:}k-1)^H A(:, k)$, and the sum is then $Q(:, 1{:}k-1)R(1{:}k-1, k)$. Algorithm 2.1 illustrates a MATLAB implementation of this Gram–Schmidt process.

Algorithm 2.1 Gram–Schmidt algorithm (QR factorization)
File: `gramschmidt1.m`

With the observation from (2.25) that

$$\mathbf{p}_k = \mathbf{q}_k r_{kk} + \sum_{i=1}^{k-1} r_{ik} \mathbf{q}_i,$$

we note that we can write A in a factored form as $A = QR$, and that Q satisfies $Q^H Q = I$. The matrix Q provides an orthogonal basis to the column space of A.

For finite-dimensional vectors, the computations of the Gram–Schmidt process may be numerically unstable for poorly conditioned matrices. Exercise 2.15-80 discusses a modified Gram–Schmidt, while other more numerically stable methods of orthogonalization are explored in chapter 5.

2.16 Some final technicalities for infinite-dimensional spaces

The concept of basis that was introduced in section 2.2.3 was based upon the stipulation that linear combinations are *finite* sums. With the additional concepts of orthogonality and normality, we can introduce a slightly modified notion of a basis. A set $T = \{\mathbf{p}_1, \mathbf{p}_2, \ldots\}$ is said to be orthonormal if $\langle \mathbf{x}_i, \mathbf{x}_j \rangle = \delta_{i-j}$. For an orthonormal set T, it can be shown that the infinite sum

$$\sum_{i=1}^{\infty} c_i \mathbf{p}_i$$

converges if and only if the series $\sum_{i=1}^{n} |c_i|^2$ converges.

An orthonormal set of basis functions $\{\mathbf{p}_1, \mathbf{p}_2, \ldots\}$ is said to be a **complete** set for a Hilbert space S if every $\mathbf{x} \in S$ can be represented as

$$\mathbf{x} = \sum_{i=1}^{\infty} c_i \mathbf{p}_i$$

for some set of coefficients c_i. Several sets of complete basis functions are presented in chapter 3, after a means has been presented for finding the coefficients $\{c_i\}$. A complete set of functions will be called a **basis** (more strictly, an orthonormal basis). The basis and the Hamel basis are not identical for infinite-dimensional spaces. In practice, it is the basis, not the Hamel basis, which is of most use. It can be shown that any orthonormal basis is a subset of a Hamel basis.

In finite dimensions, none of these issues have any bearing. An orthonormal Hamel basis *is* an orthonormal basis. Only the notion of "basis" needs to be retained for finite-dimensional spaces. In the future, we will drop the adjective "Hamel" and refer only to a "basis" for a finite-dimensional vector space.

2.17 Exercises

2.1-1 We will examine the l_∞ metric to get a sense as to why it selects the maximum value. Given the vector $\mathbf{x} = \{1, 2, 3, 4, 5, 6\}$, compute the l_p metric $d_p(\mathbf{x}, \mathbf{0})$ for $p = 1, 2, 4, 10, 100, \infty$. Comment on why $d_p(\mathbf{x}, \mathbf{0}) \to \max(x_i)$ as $p \to \infty$.

2.1-2 Let X be an arbitrary set. Show that the function defined by

$$d(x, y) = \begin{cases} 1 & x = y \\ 0 & x \neq 0 \end{cases}$$

is a metric.

2.1-3 Verify that the Hamming distance $d_H(\mathbf{x}, \mathbf{y})$, introduced in example 2.1.4, is a metric.

2.1-4 Proof of the triangle inequality:

(a) For $x, y \in \mathbb{R}$, prove the triangle inequality in the form

$$|x + y| \leq |x| + |y|.$$

What is the condition for equality?

(b) For $\mathbf{x}, \mathbf{y} \in \mathbb{R}^n$, prove the triangle inequality

$$\|\mathbf{x} + \mathbf{y}\| \leq \|\mathbf{x}\| + \|\mathbf{y}\|.$$

where $\|\cdot\|$ is the usual Euclidean norm. Hint: Use the fact that $\sum_{i=1}^{n} x_i y_i \leq \|\mathbf{x}\| \|\mathbf{y}\|$. (i.e., the Cauchy–Schwarz inequality).

2.1-5 Let (X, d) be a metric space. Show that
$$d_b(x, y) = \frac{d(x, y)}{1 + d(x, y)}$$
is a metric on X. What significant feature does this metric possess?

2.1-6 Let (X, d) be a metric space. Show that
$$d_m(x, y) = \min(1, d(x, y))$$
is a metric on X. What significant feature does this metric possess?

2.1-7 In defining the metric of the sequence space $l_\infty(0, \infty)$ in (2.2), "sup" was used instead of "max." To see the necessity of this definition, define the sequences **x** and **y** by
$$x_n = \frac{1}{n+1} \qquad y_n = \frac{n}{n+1}$$
Show that $d_\infty(x, y) > |x_n - y_n|$ for all $n \geq 1$.

2.1-8 For the metric space $(\mathbb{R}^n, d_p)$, show that $d_p(\mathbf{x}, \mathbf{y})$ is decreasing with p. That is, $d_p(\mathbf{x}, \mathbf{y}) \geq d_q(\mathbf{x}, \mathbf{y})$ if $p \leq q$. Hint: Take the derivative with respect to p, and show that it is ≤ 0. Use the *log sum inequality* [56], which states that for non-negative sequences $a_1, a_2, \ldots, a_n$ and $b_1, b_2, \ldots, b_n$,
$$\sum_{i=1}^n a_i \log\left(\frac{a_i}{b_i}\right) \geq \left(\sum_{i=1}^n a_i\right) \log \frac{\sum_{i=1}^n a_i}{\sum_{i=1}^n b_i}.$$
Use $b_i = 1$ and $a_i = |x_i - y_i|^p$. Also use the fact that for nonnegative sequence $\{\alpha_i\}$ such that $\sum_{i=1}^n \alpha_i = 1$, the maximum value of
$$\sum_{i=1}^n \alpha_i \log \alpha_i$$
is 0.

2.1-9 If requirement M3 in the definition of a metric is relaxed to the requirement
$$d(x, y) = 0, \qquad \text{if } x = y,$$
allowing the possibility that $d(x, y) = 0$ even when $x \neq y$, then a *pseudometric* is obtained. Let $f: X \to \mathbb{R}$ be an arbitrary function defined on a set X. Show that $d(x, y) = |f(x) - f(y)|$ is a pseudometric.

2.1-10 Show that if A and B are open sets:
(a) $A \cup B$ is open.
(b) $A \cap B$ is open.

2.1-11 Devise an example to show that the union of an infinite number of closed sets need not be closed.

2.1-12 Let
$$B = \{\text{all points } p \in \mathbb{R}^2 \text{ with } 0 < |p| \leq 2\} \cup \{\text{the point } (0, 4)\}.$$
(a) Draw the set B.
(b) Determine the boundary of B.
(c) Determine the interior of B.

2.1-13 Explain why the set of real numbers is both open and closed.

2.1-14 Determine inf and sup for the following sets of real numbers:
$$A = (0, 4) \qquad B = (0, \infty) \qquad C = (-\infty, 5].$$

2.17 Exercises

2.1-15 Show that the boundary of a set S is a closed set.

2.1-16 Show that the boundary of a set S is the intersection of the closure of S and the closure of the complement of S.

2.1-17 Show that $S \subset \mathbb{R}^n$ is closed if and only if every cluster point of S belongs to S.

2.1-18 Find $\limsup_{n \to \infty} a_n$ and $\liminf_{n \to \infty} a_n$ for
  (a) $a_n = \cos\left(\frac{2\pi}{3}n\right)$.
  (b) $a_n = \cos(\sqrt{2}n)$.
  (c) $a_n = 2 + (-1)^n(3 - 2/n)$.
  (d) $a_n = n^2(-1)^n$.

2.1-19 If $\limsup_{n \to \infty} a_n = A$ and $\limsup_{n \to \infty} b_n = B$, then is it necessarily true that
$$\limsup_{n \to \infty}(a_n + b_n) = A + B?$$

2.1-20 Show that if $\{x_n\}$ is a sequence such that
$$d(x_{n+1}, x_n) < Cr^n$$
for $0 \le r < 1$, then $\{x_n\}$ is a Cauchy sequence.

2.1-21 Let $\mathbf{p}_n = (x_n, y_n, z_n) \in \mathbb{R}^3$. Show that if $\{\mathbf{p}_n\}$ is a Cauchy sequence using the metric
$$d(\mathbf{p}_j, \mathbf{p}_k) = \sqrt{(x_j - x_k)^2 + (y_j - y_k)^2 + (z_j - z_k)^2},$$
then so are the sequences $\{x_n\}, \{y_n\}$ and $\{z_n\}$ using the metric $d(x_j, x_k) = |x_j - xk|$.

2.1-22 Show that if a sequence $\{x_n\}$ is convergent, then it is a Cauchy sequence.

2.1-23 Show that the sequence $x_n = \int_1^n \frac{\cos t}{t^2} dt$ is convergent using the metric $d(x, y) = |x - y|$. Hint: show that x_n is a Cauchy sequence. Use the fact that
$$\int \frac{|\cos t|}{t^2} dt \le \int \frac{1}{t^2} dt.$$
(Note: this is an example of knowing that a sequence converges, without knowing what it converges to.)

2.1-24 The fact that a sequence is Cauchy depends upon the metric employed. Let $f_n(t)$ be the sequence of functions defined in (2.6) in the metric space $(C[a, b], d_\infty)$, where
$$d_\infty(f, g) = \sup_t |f(t) - g(t)|.$$
Show that
$$d_\infty(f_n, f_m) = \frac{1}{2} - \frac{n}{2m} \qquad m > n.$$
Hence, conclude that in this metric space, f_n is not a Cauchy sequence.

2.1-25 In this problem we will show that the set of continuous functions is complete with respect to the uniform (sup) norm. Let $\{f_n(t)\}$ be a Cauchy sequence of continuous functions. Let $f(t)$ be the pointwise limit of $\{f_n(t)\}$. For any $\epsilon > 0$ let N be chosen so that $\max |f_n(t) - f_m(t)| \le \epsilon/3$. Since $f_k(t)$ is continuous, there is a $D > 0$ such that $|f(t + \delta) - f(t)| < \epsilon/3$ whenever $\delta \le D$. From this, conclude that
$$|f(t + \delta) - f(t)| < \epsilon,$$
and hence that $f(t)$ is continuous.

2.1-26 Find the essential supremum of the function $x(t)$ defined by
$$x(t) = \begin{cases} \sin(\pi t) & t \in [-1, 1], t \ne 0, \\ 3 & t = 0. \end{cases}$$

2.2-27 An equivalent definition for linear independence follows: A set T is linearly independent if, for each vector $\mathbf{x} \in T$, $\mathbf{x}$ is not a linear combination of the points in the set $T - \{\mathbf{x}\}$; that is, the set T with the vector $\mathbf{x}$ removed. Show that this definition is equivalent to that of definition 2.16.

2.2-28 Let S be a finite-dimensional vector space with $\dim(S) = m$. Show that every set containing $m + 1$ points is linearly dependent. Hint: use induction.

2.2-29 Show that if T is a subset of a vector space S with $\text{span}(T) = S$, then T contains a Hamel basis of S.

2.2-30 Let S denote the set of all solutions of the following differential equation defined on $C^3[0, \infty)$ (see definition A.8.);
$$\frac{d^3x}{dt^3} + b\frac{d^2x}{dt^2} + c\frac{dx}{dt} + dx = 0.$$
Show that S is a linear subspace of $C^3[0, \infty)$..

2.2-31 Let S be $L_2[0, 2\pi]$, and let T be the set of all functions $x_n(t) = e^{jnt}$ for $n = 0, 1, \ldots$. Show that T is linearly independent. Conclude that $L_2[0, 2\pi]$ is an infinite-dimensional space. Hint: assume that $c_1 e^{jn_1 t} + c_2 e^{jn_2 t} + \cdots + c_m e^{jn_m t} = 0$ for $n_i \neq n_j$ when $i \neq j$. Differentiate $(m - 1)$ times, and use the properties of Vandermonde matrices (section 8.4).

2.2-32 Show that the set $1, t, t^2, \ldots, t^m$ is a linearly independent set. (Hint: the fundamental theorem of algebra states that a polynomial $f(x)$ of degree m has exactly m roots, counting multiplicity.)

2.3-33 Show that in a normed linear space,
$$\boxed{|\,\|x\| - \|y\|\,| \leq \|x - y\|}$$

2.3-34 Show that a norm is a convex function. (See section A.3.)

2.3-35 Show that every Cauchy sequence $\{x_n\}$ in a normed linear space is bounded.

2.3-36 Let X be the vector space of *finitely* nonzero sequences $\mathbf{x} = \{x_1, x_2, x_3, \ldots, x_n, 0, 0, \ldots\}$. Define the norm on X as $\|\mathbf{x}\| = \max |x_i|$. Let $\mathbf{x}_n$ be a point in X (a sequence) defined by
$$\mathbf{x}_n = \left\{1, \frac{1}{2}, \frac{1}{3}, \ldots, \frac{1}{n-1}, 0, 0, \ldots, \right\}.$$
(a) Show that the sequence $\mathbf{x}_n$ is a Cauchy sequence.
(b) Show that X is not complete.

2.3-37 Let p be in the range $0 < p < 1$, and consider the space $L_p[0, 1]$ of all functions with
$$\|x\| = \int_0^1 |x(t)|^p \, dt < \infty.$$
Show that $\|x\|$ is not a norm on $L_p[0, 1]$. Hint: for a real number α such that $0 \leq \alpha \leq 1$, note that $\alpha \leq \alpha^p \leq 1$.

2.3-38 Let S be a normed linear space. Show that the norm function $\|\cdot\|: S \to \mathbb{R}$ is continuous. Hint: See exercise 2.3-33.

2.3-39 For each of the inequality relationships between norms in (2.12), determine a vector $\mathbf{x} \in \mathbb{R}^n$ for which each inequality is achieved with equality.

2.4-40 Compute the inner products $\langle f, g \rangle$ for the following, using the definition
$$\langle f, g \rangle = \int_0^1 f(t)g(t)\,dt.$$

(a) $f(t) = t^2 + 2t$, $g(t) = t + 1$.

(b) $f(t) = e^{-t}$, $g(t) = t + 1$.

(c) $f(t) = \cos(2\pi t)$, $g(t) = \sin(2\pi t)$.

2.4-41 Compute the inner products $\mathbf{x}^T\mathbf{y}$ of the following, using the Euclidean inner product.

(a) $\mathbf{x} = [1, 2, -3, 4]^T$, $\mathbf{y} = [2, 3, 4, 1]^T$.

(b) $\mathbf{x} = [2, 3]$, $\mathbf{y} = [1, -2]^T$.

2.4-42 Determine which of the following determines an inner product over the space of real continuous functions with continuous first derivatives.

(i) $\langle f, g \rangle = \int_0^1 f'(t)g'(t)\,dt + f(0)g(0)$ (ii) $\langle f, g \rangle = \int_0^1 f'(t)g'(t)\,dt$.

2.5-43 Show that for an induced norm $\|\cdot\|$ over a real vector space:

(a) The *parallelogram law* is true:
$$\|x + y\|^2 + \|x - y\|^2 = 2\|x\|^2 + 2\|y\|^2. \tag{2.27}$$

In two-dimensional geometry, as shown in figure 2.20, the result says that the sum of squares of the lengths of the diagonals is equal to twice the sum of the squares of the adjacent sides, a sort of two-fold Pythagorean theorem.

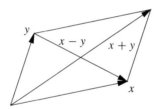

Figure 2.20: The parallelogram law

(b) Show that
$$\langle x, y \rangle = \frac{\|x + y\|^2 - \|x - y\|^2}{4}.$$

This is known as the polarization identity.

2.6-44 For the inner produce $\langle f, g \rangle = \int_0^1 f(t)g(t)\,dt$, verify the Cauchy–Schwarz inequality if:

(a) $f(t) = e^{-t}$, $g(t) = t + 1$.

(b) $f(t) = e^{-t}$, $g(t) = -5e^{-t}$.

2.6-45 Show that the inequality (2.15) is true.

2.7-46 Prove lemma 2.2.

2.7-47 Let $x_1(t) = 3t^2 - 1$, $x_2(t) = 5t^3 = 3t$ and $x_3(t) = 2t^2 - t$, and define the inner product as $\langle f, g \rangle = \int_{-1}^1 f(t)g(t)\,dt$. Compute the angles of each pairwise combination of these functions, and identify functions that are orthogonal.

2.7-48 Let
$$\mathbf{x}_1 = [1, 2, 4, -2]^T,$$
$$\mathbf{x}_2 = [5, -2, -3, 1]^T,$$
$$\mathbf{x}_3 = [1, 2, 1, 2]^T,$$

and compute the angles between these vectors, using the Euclidean inner product, and identify which vectors are orthogonal.

2.7-49 Show that a set of nonzero vectors $\{p_1, p_2, \ldots, p_m\}$ that are mutually orthogonal, so that
$$\langle p_i, p_j \rangle = 0 \quad \text{if } i \neq j,$$
is linearly independent. (Orthogonality implies linear independence.)

2.8-50 Perform the simplifications to go from the comparison in (2.18) to the comparison in (2.19).

2.8-51 Show by integration that
$$\int_{-1}^{1} \frac{1}{\sqrt{1-t^2}} T_n(t) T_m(t) = \begin{cases} \pi & n = m = 0, \\ \pi/2 & n = m, n \neq 0, \\ 0 & n \neq m. \end{cases}$$
Hint: use $t = \cos x$ in the integral.

2.10-52 Show that the orthogonal complement of a subspace is a subspace.

2.10-53 Prove items 2 through 6 of theorem 2.5. Hint: for item 5, use theorem 2.10.

2.11-54 Determine the range and nullspace of the following linear operators (matrices):
$$A = \begin{bmatrix} 1 & 0 \\ 5 & 4 \\ 2 & 4 \end{bmatrix} \quad B = \begin{bmatrix} 1 & 0 & 1 \\ 5 & 4 & 9 \\ 2 & 4 & 6 \end{bmatrix}.$$

2.11-55 Let X and Y be vector spaces over the same set of scalars. Let $LT[X, Y]$ denote the set of all linear transformations from X to Y. Let L and M be operators from $LT[X, Y]$. Define an addition operator between L and M as
$$(L + M)(x) = L(x) + M(x)$$
for all $x \in X$. Also define scalar multiplication by
$$(aL)(x) = a(L(x)).$$
Show that $LT[X, Y]$ is a linear vector space.

2.12-56 Prove lemma 2.3.

2.12-57 Show that if V and W are subspaces of a vector space S then, their intersection $V \cap W$ is a subspace.

2.12-58 Show that if V and W are subspaces of a vector space S then, their sum $V + W$ is a subspace.

2.12-59 [238, p. 200] Let $X = L_2[-\pi, \pi]$, and let
$$S_1 = \text{span}(1, \cos t, \cos 2t, \ldots) \quad S_2 = \text{span}(\sin t, \sin 2t, \ldots).$$
(a) Show that $S_1 \oplus S_2$ and $S_1 + S_2$ are isomorphic.
(b) Show that $\dim(S_1 \oplus S_2) = \dim(S_1) + \dim(S_2)$.

2.12-60 Show that:
(a) If V and W are orthogonal subspaces, then they are disjoint.
(b) If V and W are disjoint, they are not necessarily orthogonal.

2.12-61 Let S be a linear space and assume that $S = S_1 + S_2 + \cdots + S_n$, where the S_i are mutually disjoint linear subspaces of S. Let B_i be a Hamel basis of S_i. Show that $B = B_1 \cup B_2 \cup \cdots \cup B_n$ is a Hamel basis for S.

2.12-62 Prove theorem 2.6.

2.17 Exercises

2.12-63 Let V and W be linear subspace of a finite-dimensional linear space S. Show that
$$\dim(V + W) = \dim(V) + \dim(W) - \dim(V \cap W).$$
Then conclude that $\dim(V \oplus W) = \dim(V) + \dim(W)$.

2.13-64 If $\mathbf{v}$ is a vector, show that the matrix which projects onto span($\mathbf{v}$) is
$$P_v = \frac{\mathbf{v}\mathbf{v}^H}{\mathbf{v}^H\mathbf{v}}.$$

2.13-65 Show that the matrix P_A in (2.20) is a projection matrix.

2.13-66 For the projection matrix $P_{A,W}$ in (2.22):
  (a) Show that $P_{A,W}^2 = P_{A,W}$.
  (b) Show that $P_{A,W}^\perp = I - P_{A,W}$ is orthogonal to $P_{A,W}$, using the weighted inner product (that is, $P_{A,W}^H W P_{A,W}^\perp = 0$).

2.13-67 Let
$$\mathbf{p}_1 = \begin{bmatrix} 1 \\ 2 \\ 3 \\ 4 \end{bmatrix} \quad \mathbf{p}_2 = \begin{bmatrix} 4 \\ -2 \\ -6 \\ -7 \end{bmatrix} \quad \mathbf{p}_3 = \begin{bmatrix} 3 \\ 4 \\ -2 \\ 1 \end{bmatrix}$$
and
$$\mathbf{x} = \begin{bmatrix} 1 \\ 2 \\ 3 \\ 7 \end{bmatrix}.$$
Determine the nearest vector $\hat{\mathbf{x}}$ in span$[\mathbf{p}_1, \mathbf{p}_2, \mathbf{p}_3]$. Also determine the orthogonal complement of $\mathbf{x}$ in span$[\mathbf{p}_1, \mathbf{p}_2, \mathbf{p}_3]$.

2.13-68 Let A be a matrix which can be factored as
$$A = U\Sigma V^H, \tag{2.28}$$
such that
$$U^H U = I \quad V^H V = I$$
and Σ_A is a diagonal matrix with real values. The factorization in (2.28) is the singular value decomposition (see chapter 2.7). Show that $P_A = P_{U_A}$.

2.13-69 Two orthogonal projection operators P_A and P_B are said to be orthogonal if $P_A P_B = 0$. This is denoted as $P_A \perp P_B$. Show that:
  (a) P_A and P_B are orthogonal if and only if their ranges are orthogonal.
  (b) $(P_A + P_B)$ is a projection operator if and only if P_A and P_B are orthogonal.

2.13-70 Prove theorem 2.7.

2.13-71 Let $P_1, P_2, \ldots, P_m$ be a set of orthogonal projections with $P_i P_j = 0$ for $i \neq j$. Show that $Q = P_1 + P_2 + \cdots + P_m$ is an orthogonal projection.

2.13-72 If P is a projection operator, show that $I - P$ is a projection operator. Determine the range and nullspace of $I - P$.

2.13-73 Let S be a vector space, and let $V_1, V_2, \ldots, V_n$ be linear subspaces such that V_i is orthogonal from $\sum_{j \neq i} V_j$, for each i, and where
$$S = V_1 + V_2 + \cdots + V_n.$$
Let P_j be the projection on S for which $\mathcal{R}(P_j) = V_j$ and $\mathcal{N}(P_j) = \sum_{j \neq k} V_k$. Define an operator
$$P = \lambda_1 P_1 + \lambda_2 P_2 + \cdots + \lambda_n P_n.$$
  (a) Show that if $\mathbf{x} \in V_j$, then $P\mathbf{x} = \lambda_j \mathbf{x}$.
  (b) Show that P is a projection if and only if λ_j is either 0 or 1.

2.13-74 Let A and B be matrices such that $A^H B = 0$. Then $V = \mathcal{R}(A)$ and $W = \mathcal{R}(B)$ are orthogonal. Show that $P_A = I - P_B$.

2.15-75 Using a symbolic manipulation package, write a function which performs the Gram–Schmidt orthogonalization of a set of functions.

2.15-76 Determine the first four polynomials orthogonal over $[0, 1]$. A symbolic manipulation package is recommended.

2.15-77 For the functions shown in figure 2.21, determine a set of orthogonal functions spanning the same space, using the functions in the order shown.

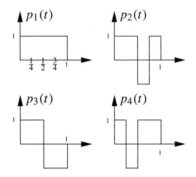

Figure 2.21: Functions to orthogonalize

2.15-78 Modify algorithm 2.1 so that it only retains columns of Q that are nonzero, making corresponding adjustments to R. Comment on the product QR in this case.

2.15-79 Modify algorithm 2.1 to compute a set of orthogonal vectors with respect to the weighted inner product $\langle \mathbf{x}, \mathbf{y} \rangle = \mathbf{x}^T W \mathbf{y}$ for a positive definite symmetric matrix W.

2.15-80 (Modified Gram–Schmidt) The computations of the Gram–Schmidt algorithm can be reorganized to be more stable numerically. In these modified computations, a column of Q and a *row* of R is produced at each iteration. (The regular Gram–Schmidt process produces a column of Q and a column of R at each iteration.) Let the kth column of Q be denoted as $\mathbf{q}_k$, and let the kth row of R be denoted as $\mathbf{r}_k^T$.

(a) Show that for an $m \times n$ matrix A,
$$A - \sum_{i=1}^{k-1} \mathbf{q}_i \mathbf{r}_i^T = \sum_{i=k}^{n} \mathbf{q}_i \mathbf{r}_i^T = \begin{bmatrix} \mathbf{0} & A^{[k]} \end{bmatrix},$$
where $A^{[k]}$ is $m \times (n - k + 1)$.

(b) Let $A^{[k]} = [\mathbf{z}_k \ B]$, where B is $m \times n - k$, and explain why the kth column of Q and the kth row of R are given by
$$r_{kk} = \|\mathbf{e}_k\| \qquad \mathbf{q}_k = \mathbf{e}_k / r_{kk} \qquad (r_{k,k+1}, \ldots, r_{k,n}) = \mathbf{q}_k^T B.$$

(c) Then show that the next iteration can be started by computing
$$A - \sum_{i=1}^{k} \mathbf{q}_i \mathbf{r}_i^T = \begin{bmatrix} \mathbf{0} & A^{[k+1]} \end{bmatrix},$$
where $A^{[k+1]} = B - \mathbf{q}_k(r_{k,k+1}, \ldots, r_{k,n})$.

(d) Code the modified Gram–Schmidt algorithm in MATLAB.

2.18 References

Much of the material on metric spaces, Hilbert spaces, and Banach spaces presented here is significantly compressed from [238]. In their expanded treatment they provide proofs of several points that we have merely mentioned. An excellent historical source on vector spaces and their applications to signal processing and engineering is [209]. Function spaces with an emphasis on series representations are discussed in [177]. A similar treatment of metric and vector spaces is found in [92].

Extensive properties of the orthogonal polynomials introduced here are discussed and tabulated in [2]; see also [358].

An extension of the concept of a basis is that of a *frame*, which provides an overdetermined set of representational functions. A tutorial introduction to frames, with applications in signal processing, appears in [253].

Chapter 3

Representation and Approximation in Vector Spaces

> Any good mathematical commodity is worth generalizing.
> — *Michael Spivak*
> *Calculus on Manifolds*

3.1 The approximation problem in Hilbert space

Let $(S, \|\cdot\|)$ be a normed linear vector space for some norm $\|\cdot\|$. Let $T = \{\mathbf{p}_1, \mathbf{p}_2, \ldots, \mathbf{p}_m\} \subset S$ be a set of linearly independent vectors in a vector space S and let $V = \text{span}(T)$. The analysis problem is this: given a vector $\mathbf{x} \in S$, find the coefficients $c_1, c_2, \ldots, c_m$ so that

$$\hat{x} = c_1 \mathbf{p}_1 + c_2 \mathbf{p}_2 + \cdots + c_m \mathbf{p}_m \tag{3.1}$$

approximates $\mathbf{x}$ as closely as possible. The ˆ (caret) indicates that this is (or may be) an approximation. That is, we wish to write

$$\mathbf{x} = \hat{\mathbf{x}} + \mathbf{e}$$
$$= c_1 \mathbf{p}_1 + c_2 \mathbf{p}_2 + \cdots + c_m \mathbf{p}_m + \mathbf{e},$$

where $\mathbf{e}$ is the approximation error, so that

$$\|\mathbf{x} - \hat{\mathbf{x}}\| = \|\mathbf{e}\|$$

is as small as possible. The problem is diagrammed in figure 3.1 for $m = 2$. Of course,

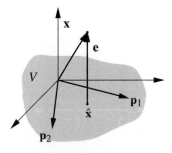

Figure 3.1: The approximation problem

if $\mathbf{x} \in V$ then it is possible to find coefficients so that $\|\mathbf{x} - \hat{\mathbf{x}}\| = 0$. The particular norm chosen in performing the minimization affects the analytic approach to the problem and the

3.1 The Approximation Problem in Hilbert Space

final answer. If the l_1 (or L_1) norm is chosen, then the analysis involves absolute values, which makes an analytical solution involving derivatives difficult. If the l_∞ (or L_∞) norm is chosen, the analysis may involve derivatives of the max function, which is also difficult. If the l_2 (or L_2) norm is chosen, many of the analytical difficulties disappear. The norm is the induced norm, and the properties of the projection theorem can be used to formulate the solution. Alternatively, the solution can be obtained using calculus techniques. (Actually, for problems posed using the l_p norms, a generalization of the projection theorem can be used, optimizing in Banach space rather than Hilbert space, but this lies beyond the scope of this book.) Choosing the l_2 norm allows familiar Euclidean geometry to be used to develop insight. The approximation problem when the induced norm is used (for example, either an l_2 or L_2 norm) is known as the Hilbert space approximation problem.

To develop geometric insight into the approximation problem, the analysis formulas are presented by starting with the approximation problem with one element in T, aided by a key observation: the error is orthogonal to the data. The analysis is then extended to two dimensions, then to arbitrary dimensions. We will begin first with geometric plausibility and calculus, then prove the result using the Cauchy–Schwarz inequality.

To begin, let $T \in \mathbb{R}^2$ consist of only one vector, $T = \{\mathbf{p}_1\}$. For a vector $\mathbf{x} \in \mathbb{R}^2$, we wish to represent $\mathbf{x}$ as a linear combination of T,

$$\mathbf{x} = c_1 \mathbf{p}_1 + \mathbf{e},$$

in such a way as to minimize the norm of the approximation error $\|\mathbf{e}\|$. In this simplest case, there is only the parameter c_1 to identify. The situation is illustrated in figure 3.2(a). If the l_2 or L_2 norm is used, it may be observed geometrically that **the error is minimized when the error is orthogonal to** V; that is, when the error is orthogonal to the data that forms our estimate. Written mathematically, the norm of the error $\|\mathbf{e}\|$ is minimized when

$$\mathbf{e} \perp \mathbf{p}_1,$$

or

$$\langle \mathbf{x} - c_1 \mathbf{p}_1, \mathbf{p}_1 \rangle = 0.$$

Using the properties of inner products,

$$c_1 = \frac{\langle \mathbf{x}, \mathbf{p}_1 \rangle}{\|\mathbf{p}_1\|_2^2}. \tag{3.2}$$

Geometrically, the quantity

$$\frac{\langle \mathbf{x}, \mathbf{p}_1 \rangle}{\|\mathbf{p}_1\|_2^2}$$

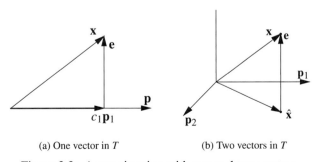

(a) One vector in T (b) Two vectors in T

Figure 3.2: Approximation with one and two vectors

is the **projection** of the vector $\mathbf{x}$ in the direction of $\mathbf{p}_1$; it is the length of the shadow that $\mathbf{x}$ casts onto $\mathbf{p}_1$ (expressed as a proportion of the length of $\mathbf{p}_1$).

The same approximation formula may also be obtained by calculus. We find c_1 to minimize

$$\|\mathbf{x} - c_1\mathbf{p}_1\|_2^2 = \langle \mathbf{x} - c_1\mathbf{p}_1, \mathbf{p} - c_1\mathbf{x}_1 \rangle$$

by taking the derivative with respect to c_1 and equating the result to zero. This gives the same answer as (3.2).

Continuing our development, when T contains two vectors we can write the approximation as

$$\mathbf{x} = c_1\mathbf{p}_1 + c_2\mathbf{p}_2 + \mathbf{e}.$$

Figure 3.2(b) illustrates the concept for vectors in $\mathbb{R}^3$. It is clear from this figure that if Euclidean distance is used, the error is orthogonal to the data $\mathbf{p}_1$ and $\mathbf{p}_2$. This gives the following orthogonality conditions:

$$\langle \mathbf{x} - (c_1\mathbf{p}_1 + c_2\mathbf{p}_2), \mathbf{p}_1 \rangle = 0,$$
$$\langle \mathbf{x} - (c_1\mathbf{p}_1 + c_2\mathbf{p}_2), \mathbf{p}_2 \rangle = 0.$$

Expanding these using the properties of inner products gives

$$\langle \mathbf{x}, \mathbf{p}_1 \rangle = c_1 \langle \mathbf{p}_1, \mathbf{p}_1 \rangle + c_2 \langle \mathbf{p}_2, \mathbf{p}_1 \rangle,$$
$$\langle \mathbf{x}, \mathbf{p}_2 \rangle = c_1 \langle \mathbf{p}_1, \mathbf{p}_2 \rangle + c_2 \langle \mathbf{p}_2, \mathbf{p}_2 \rangle,$$

which can be written more concisely in matrix form as

$$\begin{bmatrix} \langle \mathbf{p}_1, \mathbf{p}_1 \rangle & \langle \mathbf{p}_2, \mathbf{p}_1 \rangle \\ \langle \mathbf{p}_1, \mathbf{p}_2 \rangle & \langle \mathbf{p}_2, \mathbf{p}_2 \rangle \end{bmatrix} \begin{bmatrix} c_1 \\ c_2 \end{bmatrix} = \begin{bmatrix} \langle \mathbf{x}, \mathbf{p}_1 \rangle \\ \langle \mathbf{x}, \mathbf{p}_2 \rangle \end{bmatrix}. \qquad (3.3)$$

Solution of this matrix equation provides the desired coefficients.

Example 3.1.1 Suppose $\mathbf{x} = [1, 2, 3]^T$, $\mathbf{p}_1 = [1, 1, 0]^T$, and $\mathbf{p}_2 = [2, 1, 0]^T$. It is clear that

$$\hat{\mathbf{x}} = c_1\mathbf{p}_1 + c_2\mathbf{p}_2$$

cannot be an exact representation of $\mathbf{x}$ since there is no way to match the third element. Using (3.3), we obtain

$$\begin{bmatrix} 2 & 3 \\ 3 & 5 \end{bmatrix} \begin{bmatrix} c_1 \\ c_2 \end{bmatrix} = \begin{bmatrix} 3 \\ 4 \end{bmatrix}.$$

This can be solved to give

$$c_1 = 3 \qquad c_2 = -1.$$

Then the approximation vector is

$$\hat{\mathbf{x}} = c_1\mathbf{p}_1 + c_2\mathbf{p}_2 = 3[1, 1, 0]^T - [2, 1, 0]^T = [1, 2, 0]^T.$$

Note that the approximation $\hat{\mathbf{f}}$ is the same as $\mathbf{f}$ in the first two coefficients. The vector has been **projected** onto the plane formed by the vectors $\mathbf{p}_1$ and $\mathbf{p}_2$. The error in this case has length 3. □

Jumping now to higher numbers of vectors, what we can do for two vectors in T, we can do for m ingredient vectors. We approximate $\mathbf{x}$ as

$$\mathbf{x} = \sum_{i=1}^{m} c_i \mathbf{p}_i + \mathbf{e} = \hat{\mathbf{x}} + \mathbf{e}$$

3.1 The Approximation Problem in Hilbert Space

to minimize $\|\mathbf{e}\| = \|\mathbf{x} - \hat{\mathbf{x}}\|$. If the norm used is the l_2 or L_2 norm, this is the *linear least-squares* problem. Whenever the norm measuring the approximation error $\|\mathbf{e}\|$ is induced from an inner product, we can express the minimization in terms of an orthogonality condition: the minimum-norm error must be orthogonal to each vector p_j:

$$\left\langle \mathbf{x} - \sum_{i=1}^{m} c_i \mathbf{p}_i, p_j \right\rangle = 0, \qquad j = 1, 2, \ldots, m.$$

This gives us m equations in the m unknowns, which may be written as

$$\begin{bmatrix} \langle \mathbf{p}_1, \mathbf{p}_1 \rangle & \langle \mathbf{p}_2, \mathbf{p}_1 \rangle & \cdots & \langle \mathbf{p}_m, \mathbf{p}_1 \rangle \\ \langle \mathbf{p}_1, \mathbf{p}_2 \rangle & \langle \mathbf{p}_2, \mathbf{p}_2 \rangle & \cdots & \langle \mathbf{p}_m, \mathbf{p}_2 \rangle \\ \vdots & & & \vdots \\ \langle \mathbf{p}_1, \mathbf{p}_m \rangle & \langle \mathbf{p}_2, \mathbf{p}_m \rangle & \cdots & \langle \mathbf{p}_m, \mathbf{p}_m \rangle \end{bmatrix} \begin{bmatrix} c_1 \\ c_2 \\ \vdots \\ c_m \end{bmatrix} = \begin{bmatrix} \langle \mathbf{x}, \mathbf{p}_1 \rangle \\ \langle \mathbf{x}, \mathbf{p}_2 \rangle \\ \vdots \\ \langle \mathbf{x}, \mathbf{p}_m \rangle \end{bmatrix}. \qquad (3.4)$$

We define the vector

$$\mathbf{p} = \begin{bmatrix} \langle \mathbf{x}, \mathbf{p}_1 \rangle \\ \langle \mathbf{x}, \mathbf{p}_2 \rangle \\ \vdots \\ \langle \mathbf{x}, \mathbf{p}_m \rangle \end{bmatrix} \qquad (3.5)$$

as the *cross-correlation vector*, and

$$\mathbf{c} = \begin{bmatrix} c_1 \\ c_2 \\ \vdots \\ c_m \end{bmatrix} \qquad (3.6)$$

as the vector of coefficients. Then (3.4) can be written as

$$R\mathbf{c} = \mathbf{p},$$

where R is the matrix of inner products in (3.4). Equations of this form are known as the **normal equations**. Since the solution minimizes the square of the error, it is known as a *least-square* or *minimum mean-square* solution, depending on the particular inner product used.

3.1.1 The Grammian matrix

The $m \times m$ matrix

$$R = \begin{bmatrix} \langle \mathbf{p}_1, \mathbf{p}_1 \rangle & \langle \mathbf{p}_2, \mathbf{p}_1 \rangle & \cdots & \langle \mathbf{p}_m, \mathbf{p}_1 \rangle \\ \langle \mathbf{p}_1, \mathbf{p}_2 \rangle & \langle \mathbf{p}_2, \mathbf{p}_2 \rangle & \cdots & \langle \mathbf{p}_m, \mathbf{p}_2 \rangle \\ \vdots & & & \vdots \\ \langle \mathbf{p}_1, \mathbf{p}_m \rangle & \langle \mathbf{p}_2, \mathbf{p}_m \rangle & \cdots & \langle \mathbf{p}_m, \mathbf{p}_m \rangle \end{bmatrix} \qquad (3.7)$$

in the left hand side of (3.4) is said to be the **Grammian** of the set T. Since the (i, j)th element of the matrix is

$$R_{ij} = \langle \mathbf{p}_j, \mathbf{p}_i \rangle,$$

it follows that the Grammian is a Hermitian symmetric matrix; that is,

$$R^H = R$$

(where ^{H} indicates conjugate-transpose). Some implications of the Hermitian structure are examined in section 6.2. Solution of (3.4) requires that R be invertible. The following theorem determines conditions under which R is invertible. Recall that a matrix R for which

$$\mathbf{x}^H R \mathbf{x} > 0$$

> **Box 3.1: Positive-definite matrices**
>
> We will encounter several times in the course of this book the notion of positive-definite matrices. We collect together here several important facts related to positive definite matrices.
>
> **Definition 3.1** A matrix A is said to be **positive definite** (PD) if $\mathbf{x}^H A \mathbf{x} > 0$ for all $\mathbf{x} \neq 0$. This is sometimes denoted as $A > 0$. (Caution: the notation $A > 0$ is also sometimes used to indicate that all the elements of A are greater than zero, which is not the same as being PD.) If $\mathbf{x}^H A \mathbf{x} \geq 0$ for all $\mathbf{x}$, then A is **positive semidefinite** (PSD). If $>$ is replaced by $<$, the matrix is said to be **negative definite** (ND), and if $\geq$ is replaced by $\leq$, the matrix is **negative semidefinite** (NSD). □
>
> Here are some properties of positive-definite (or semidefinite) matrices.
>
> 1. All diagonal elements of a PD (PSD) matrix are positive (nonnegative). (Caution: this does not mean that positive diagonal elements imply that a matrix is PD).
> 2. A Hermitian matrix A is PD (PSD) if and only if all of the eigenvalues are positive (nonnegative). Hence, a PD matrix has a positive determinant. Hence, a PD matrix is invertible.
> 3. A Hermitian matrix P is PD if and only if all principal minors are positive.
> 4. If A is PD, then the pivots obtained in the LU factorization are positive.
> 5. If $A > 0$ and $B > 0$, then $A + B > 0$. If A is PD and B is PSD, then $A + B$ is PD.
> 6. A Hermitian PD matrix A can be factored as $A = B^H B$ (using the Cholesky factorization, for instance), where B is full rank. This is a matrix square root.

for any nonzero vector $\mathbf{x}$ is said to be positive-definite (see box 3.1). An important aspect of positive-definite matrices is that they are always invertible. If R is such that

$$\mathbf{x}^H R \mathbf{x} \geq 0$$

for any nonzero vector $\mathbf{x}$, then R is said to be *positive-semidefinite*.

Theorem 3.1 *A Grammian matrix R is always positive-semidefinite (that is, $\mathbf{x}^H R \mathbf{x} \geq 0$ for any $\mathbf{x} \in \mathbb{C}^m$). It is positive-definite if and only if the vectors $\mathbf{p}_1, \mathbf{p}_2, \ldots, \mathbf{p}_m$ are linearly independent.*

Proof Let $\mathbf{y} = [y_1, y_2, \ldots, y_m]^T$ be an arbitrary vector. Then

$$\mathbf{y}^H R \mathbf{y} = \sum_{i=1}^m \sum_{j=1}^m \overline{y}_i y_j \langle \mathbf{p}_j, \mathbf{p}_i \rangle = \sum_{i=1}^m \sum_{j=1}^m \langle y_j \mathbf{p}_j, y_i \mathbf{p}_i \rangle$$
$$= \left\langle \sum_{j=1}^m y_j \mathbf{p}_j, \sum_{i=1}^m y_i \mathbf{p}_i \right\rangle = \left\| \sum_{j=1}^m y_j \mathbf{p}_j \right\|^2 \geq 0. \qquad (3.8)$$

Hence R is positive-semidefinite.

3.2 The Orthogonality Principle

If R is not positive-definite, then there is a nonzero vector $\mathbf{y}$ such that
$$\mathbf{y}^H R \mathbf{y} = 0,$$
so that (by (3.8))
$$\sum_{i=1}^{m} y_i \mathbf{p}_i = 0;$$
thus, the $\{\mathbf{p}_i\}$ are linearly dependent.

Conversely, if R is positive-definite, then
$$\mathbf{y}^H R \mathbf{y} > 0$$
for all nonzero $\mathbf{y}$ and by (3.8)
$$\sum_{i=1}^{m} y_i \mathbf{p}_i \neq 0.$$
This means that the $\{\mathbf{p}_i\}$ are linearly independent. $\square$

As a corollary to this theorem, we get another proof of the Cauchy–Schwarz inequality. The 2×2 Grammian
$$R = \begin{bmatrix} \langle x, x \rangle & \langle x, y \rangle \\ \langle y, x \rangle & \langle y, y \rangle \end{bmatrix}$$
is positive-semidefinite, which means that its determinant is nonnegative:
$$\langle x, x \rangle \langle y, y \rangle - \langle x, y \rangle \langle y, x \rangle \geq 0,$$
which is equivalent to (2.13).

The concept of using orthogonality for the Euclidean inner product to find the minimum norm solution generalizes to *any induced norm* and its associated inner product.

If the set of vectors $\{\mathbf{p}_1, \mathbf{p}_2, \ldots, \mathbf{p}_m\}$ are orthogonal, then the Grammian in (3.7) is diagonal, significantly reducing the amount of computation required to find the coefficients of the vector representation. In this case, the coefficients are obtained simply by
$$c_j = \frac{\langle \mathbf{x}, \mathbf{p}_j \rangle}{\langle \mathbf{p}_j, \mathbf{p}_j \rangle}. \tag{3.9}$$
Each coefficient uses the same projection formula that was used in (3.3) for a single dimension. The coefficients can also be readily interpreted: for orthogonal vectors, the coefficient of each vector indicates the strength of the vector component in the signal representation.

3.2 The orthogonality principle

The **orthogonality principle** for least-squares (LS) optimization introduced in section 3.1 is now formalized.

Theorem 3.2 *(The orthogonality principle) Let $\mathbf{p}_1, \mathbf{p}_2, \ldots, \mathbf{p}_m$ be data vectors in a vector space S. Let $\mathbf{x}$ be any vector in S. In the representation*
$$\mathbf{x} = \sum_{i=1}^{m} c_i \mathbf{p}_i + \mathbf{e} = \hat{\mathbf{x}} + \mathbf{e},$$
the induced norm of the error vector $\|\mathbf{e}\|$ is minimized when the error $\mathbf{e} = \mathbf{x} - \hat{\mathbf{x}}$ is orthogonal to each of the data vectors,
$$\left\langle \mathbf{x} - \sum_{i=1}^{m} c_i \mathbf{p}_i, \mathbf{p}_j \right\rangle = 0 \qquad j = 1, 2, \ldots, m.$$

Proof One proof relies on the projection theorem, theorem 2.8, with the observation that $V = \text{span}(\mathbf{p}_1, \mathbf{p}_2, \ldots, \mathbf{p}_m)$ is a subspace of S. We present a more direct proof using the Cauchy–Schwarz inequality.

In the case that $\mathbf{x} \in \text{span}(\mathbf{p}_1, \mathbf{p}_2, \ldots, \mathbf{p}_m)$, the error is zero and hence is orthogonal to the data vectors. This case is therefore trivial and is excluded from what follows.

If $\mathbf{x} \notin \text{span}(\mathbf{p}_1, \mathbf{p}_2, \ldots, \mathbf{p}_m)$, let $\mathbf{y}$ be a fixed vector that is orthogonal to all of the data vectors,

$$\langle \mathbf{y}, \mathbf{p}_i \rangle = 0 \qquad i = 1, 2, \ldots, m,$$

such that

$$\mathbf{x} = \sum_{i=1}^{m} a_i \mathbf{p}_i + \mathbf{y}$$

for some set of coefficients $\{a_1, a_2, \ldots, a_m\}$. Let $\mathbf{e}$ be a vector satisfying

$$x = \sum_{i=1}^{m} c_i \mathbf{p}_i + \mathbf{e} \qquad (3.10)$$

for some set of coefficients $\{c_1, c_2, \ldots, c_m\}$. Then by the Cauchy–Schwarz inequality,

$$\|\mathbf{e}\|^2 \|\mathbf{y}\|^2 \geq |\langle \mathbf{e}, \mathbf{y} \rangle|^2 \qquad \text{(Cauchy–Schwarz)}$$

$$= \left| \langle \mathbf{x}, \mathbf{y} \rangle - \left\langle \sum_{i=1}^{m} c_i \mathbf{p}_i, \mathbf{y} \right\rangle \right|^2$$

$$= |\langle \mathbf{x}, \mathbf{y} \rangle|^2 \qquad \text{(orthogonality of } \mathbf{y}\text{).} \qquad (3.11)$$

The lower bound is independent of the coefficients $\{c_i\}$, and hence no set of coefficients can make the bound smaller. By the equality condition for the Cauchy–Schwarz inequality, the lower bound is achieved—implying the minimum $\|\mathbf{e}\|$—when

$$\mathbf{e} = \alpha \mathbf{y}$$

for some scalar α. Since $\mathbf{e}$ must satisfy (3.10), it must be the case that $\mathbf{e} = \mathbf{y}$, hence the error is orthogonal to the data. □

When $\mathbf{c}$ is obtained via the principle of orthogonality, the optimal estimate

$$\hat{\mathbf{x}} = \sum_{i=1}^{m} c_i \mathbf{p}_i$$

is also orthogonal to the error $\mathbf{e} = \mathbf{x} - \hat{\mathbf{x}}$, since it is a linear combination of the data vectors $\{\mathbf{p}_i\}$. Thus,

$$\langle \hat{\mathbf{x}}, \mathbf{e} \rangle = 0. \qquad (3.12)$$

3.2.1 Representations in infinite-dimensional space

If there are an infinite number of vectors in $T = \{\mathbf{p}_1, \mathbf{p}_2, \ldots, \}$, then the representation

$$\hat{\mathbf{x}} = \sum_{i=1}^{\infty} c_i \mathbf{p}_i$$

is suspect, because a linear combination is defined, technically, only in terms of a finite

3.3 Error minimization via gradients

While the orthogonality theorem is used principally throughout this chapter as the geometrical basis for finding a minimum error approximation under an induced norm, it is pedagogically worthwhile to consider another approach based on gradients, which reaffirms what we already know but demonstrates the use of some new tools.

Minimizing $\|\mathbf{e}\|^2$ for the induced norm in

$$\mathbf{x} = \sum_{i=1}^{m} c_i \mathbf{p}_i + \mathbf{e}$$

requires minimizing

$$J(\mathbf{c}) = \left\langle \mathbf{x} - \sum_{j=1}^{m} c_j \mathbf{p}_j, \mathbf{x} - \sum_{i=1}^{m} c_i \mathbf{p}_i \right\rangle$$

$$= \langle \mathbf{x}, \mathbf{x} \rangle - 2\operatorname{Re}\left(\sum_{i=1}^{m} \overline{c}_i \langle \mathbf{x}, \mathbf{p}_i \rangle\right) + \sum_{i=1}^{m}\sum_{j=1}^{m} c_j \overline{c}_i \langle \mathbf{p}_j, \mathbf{p}_i \rangle. \quad (3.13)$$

Using the vector notations defined in (3.5), (3.6), and (3.7), we can write (3.13) as

$$J(\mathbf{c}) = \|\mathbf{x}\|^2 - 2\operatorname{Re}(\mathbf{c}^H \mathbf{p}) + \mathbf{c}^H R^T \mathbf{c}. \quad (3.14)$$

Gradient formulas appropriate for this optimization are presented in section E.1.1 of appendix E. In particular, the following gradient formulas are derived:

$$\frac{\partial}{\partial \overline{\mathbf{c}}} \mathbf{d}^H \mathbf{c} = \mathbf{0} \qquad \frac{\partial}{\partial \overline{\mathbf{c}}} \mathbf{c}^H \mathbf{d} = \mathbf{d} \qquad \frac{\partial}{\partial \overline{\mathbf{c}}} \operatorname{Re}(\mathbf{c}^H \mathbf{d}) = \frac{1}{2}\mathbf{d} \qquad \frac{\partial}{\partial \overline{\mathbf{c}}} \mathbf{c}^H R \mathbf{c} = R\mathbf{c}.$$

Taking the gradient of (3.14) using the last two of these, we obtain

$$\frac{\partial}{\partial \overline{\mathbf{c}}}(\|x\|^2 - 2\operatorname{Re}(\mathbf{c}^H \mathbf{p}) + \mathbf{c}^H R \mathbf{c}) = -\mathbf{p} + R\mathbf{c}. \quad (3.15)$$

Equating this result to zero we obtain

$$R\mathbf{c} = \mathbf{p},$$

giving us again the normal equations.

To determine whether the extremum we have obtained by the gradient is in fact a minimum, we compute the gradient a second time. We have the Hessian matrix

$$\frac{\partial}{\partial \overline{\mathbf{c}}} R\mathbf{c} = R,$$

which is a positive-semidefinite matrix, so the extremum must be a minimum.

Restricting attention for the moment to real variables, consider the plot of the norm of the error $J(\mathbf{c})$ as a function of the variables $c_1, c_2, \ldots, c_m$. Such a plot is called an *error surface*. Because $J(\mathbf{c})$ is quadratic in $\mathbf{c}$ and R is positive semidefinite, the error surface is a parabolic bowl. Figure 3.3 illustrates such an error surface for two variables c_1 and c_2. Because of its parabolic shape, any extremum must be a minimum, and is in fact a global minimum.

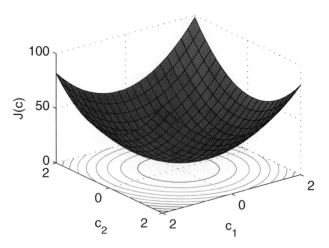

Figure 3.3: An error surface for two variables

3.4 Matrix representations of least-squares problems

While vector space methods apply to both infinite- and finite-dimensional vectors (signals), the notational power of matrices can be applied when the basis vectors are finite dimensional. The linear combination of the finite set of vectors $\{\mathbf{p}_1, \mathbf{p}_2, \ldots, \mathbf{p}_m\}$ can be written as

$$\hat{\mathbf{x}} = \sum_{i=1}^{m} c_i \mathbf{p}_i = [\mathbf{p}_1 \quad \mathbf{p}_2 \quad \cdots \quad \mathbf{p}_m] \begin{bmatrix} c_1 \\ c_2 \\ \vdots \\ c_m \end{bmatrix}.$$

This is the linear combination of the columns of the matrix A defined by

$$A = [\mathbf{p}_1 \quad \mathbf{p}_2 \quad \cdots \quad \mathbf{p}_m],$$

which we compute by

$$\hat{\mathbf{x}} = A\mathbf{c}.$$

The approximation problem can be stated as follows:

> Determine $\mathbf{c}$ to minimize $\|\mathbf{e}\|_2^2$ in the problem $\qquad \mathbf{x} = A\mathbf{c} + \mathbf{e} = \hat{\mathbf{x}} + \mathbf{e}.$ (3.16)

The minimum $\|\mathbf{e}\|_2^2 = \|\mathbf{x} - A\mathbf{c}\|^2$ occurs when $\mathbf{e}$ is orthogonal to each of the vectors

$$\langle \mathbf{x} - A\mathbf{c}, \mathbf{p}_j \rangle = 0, \qquad j = 1, 2, \ldots, m.$$

Stacking these orthogonality conditions, we obtain

$$\begin{bmatrix} \mathbf{p}_1^H \\ \mathbf{p}_2^H \\ \vdots \\ \mathbf{p}_m^H \end{bmatrix} (\mathbf{x} - A\mathbf{c}) = \mathbf{0}.$$

Recognizing that the stack of vectors is simply A^H, we obtain

$$A^H A \mathbf{c} = A^H \mathbf{x}. \tag{3.17}$$

3.4 Matrix Representations of Least-Squares Problems

The matrix $A^H A$ is the Grammian R, and the vector $A^H \mathbf{x}$ is the cross-correlation $\mathbf{p}$. We can write (3.17) as

$$R\mathbf{c} = A^H \mathbf{x} = \mathbf{p}. \tag{3.18}$$

These equations are the normal equations. Then the optimal (least-squares) coefficients are

$$\boxed{\mathbf{c} = (A^H A)^{-1} A^H \mathbf{x} = R^{-1} \mathbf{p}} \tag{3.19}$$

By theorem 3.1, $A^H A$ is positive definite if the $\mathbf{p}_1, \ldots, \mathbf{p}_m$ are linearly independent. The matrix $(A^H A)^{-1} A^H$ is called a *pseudoinverse* of A, and is often denoted $A^\dagger$. More is said about pseudoinverses in section 4.9. While (3.19) provides an analytical prescription for the optimal coefficients, it should rarely be computed explicitly as shown, since many problems are numerically unstable (subject to amplification of roundoff errors). Numerical stability is discussed in section 4.10. Stable methods for computing pseudoinverses are discussed in sections 5.3 and 7.3. In MATLAB, the pseudoinverse may be computed using the command `pinv`.

Using (3.19), the approximation is

$$\boxed{\hat{\mathbf{x}} = A\mathbf{c} = A(A^H A)^{-1} A^H \mathbf{x}} \tag{3.20}$$

The matrix $P = A(A^H A)^{-1} A^H$ is a *projection matrix*, which we encountered in section 2.13. The matrix P projects onto the range of A. Consider geometrically what is taking place: we wish to solve the equation $A\mathbf{c} = \mathbf{x}$, but there is no exact solution, since $\mathbf{x}$ is not in the range of A. So we project $\mathbf{x}$ orthogonally down onto the range of A, and find the best solution in that range space. The idea is shown in figure 3.4

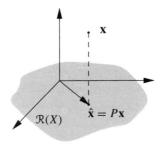

Figure 3.4: Projection solution

A useful representation of the Grammian $R = A^H A$ can be obtained by considering A as a stack of rows,

$$A = \begin{bmatrix} \mathbf{q}_1^H \\ \mathbf{q}_2^H \\ \vdots \\ \mathbf{q}_n^H \end{bmatrix}, \tag{3.21}$$

so that $A^H = [\mathbf{q}_1, \mathbf{q}_2, \ldots, \mathbf{q}_n]$ and

$$A^H A = [\mathbf{q}_1 \quad \mathbf{q}_2 \quad \cdots \quad \mathbf{q}_n] \begin{bmatrix} \mathbf{q}_1^H \\ \mathbf{q}_2^H \\ \vdots \\ \mathbf{q}_n^H \end{bmatrix} = \sum_{i=1}^{n} \mathbf{q}_i \mathbf{q}_i^H. \tag{3.22}$$

3.4.1 Weighted least-squares

A weight can also be applied to the data points, reflecting the confidence in the data, as illustrated by the next example. This is naturally incorporated into the inner product. Define a weighted inner product as

$$\langle \mathbf{x}, \mathbf{y} \rangle_W = \mathbf{x}^H W \mathbf{y}.$$

Then minimizing $\|\mathbf{e}\|_W^2 = \|A\mathbf{c} - \mathbf{x}\|_W^2$ leads to the weighted normal equations

$$A^H W A \mathbf{c} = A^H W \mathbf{x}, \qquad (3.23)$$

so the coefficients which minimize the weighted squared error are

$$\mathbf{c} = (A^H W A)^{-1} A^H W \mathbf{y}. \qquad (3.24)$$

Another approach to (3.24) is to presume that we have a factorization of the weight $W = S^H S$ (see section 5.2). Then we weight the equation

$$S A \mathbf{c} \approx S \mathbf{y}.$$

Multiplying through by $(SA)^H$ and solving for $\mathbf{c}$, we obtain

$$\mathbf{c} = ((SA)^H SA)^{-1} (SA)^H S \mathbf{y},$$

which is equivalent to (3.24).

3.4.2 Statistical properties of the least-squares estimate

The matrix-least squares solution (3.20) has some useful statistical properties. Suppose that the signal $\mathbf{x}$ has the true model according to the equation

$$\mathbf{x} = A \mathbf{c}_0 + \mathbf{e}, \qquad (3.25)$$

for some "true" model parameter vector $\mathbf{c}_0$; and that we assume a statistical model for the model error $\mathbf{e}$: assume that each component of $\mathbf{e}$ is a zero-mean, i.i.d. random variable with variance σ_e^2. The estimated parameter vector is

$$\mathbf{c} = (A^H A)^{-1} A^H \mathbf{x}. \qquad (3.26)$$

This least-squares estimate, being a function of the random vector $\mathbf{x}$, is itself a random vector. We will determine the mean and covariance matrix for this random vector.

Mean of $\mathbf{c}$. Substituting the "true" model of (3.25) into (3.26), we obtain

$$\mathbf{c} = (A^H A)^{-1} A^H A \mathbf{c}_0 + (A^H A)^{-1} A^H \mathbf{e}$$
$$= \mathbf{c}_0 + (A^H A)^{-1} A^H \mathbf{e}.$$

If we now take the expected value of our estimated parameter vector, we obtain

$$E[\mathbf{c}] = E[\mathbf{c}_0 + (A^H A)^{-1} A^H \mathbf{e}] = \mathbf{c}_0,$$

since each component of $\mathbf{e}$ has zero mean. Thus, the expected value of the estimate is equal to the true value. Such an estimate is said to be **unbiased**.

Covariance of $\mathbf{c}$. The covariance can be written as

$$\mathrm{Cov}[\mathbf{c}] = E[(\mathbf{c} - \mathbf{c}_0)(\mathbf{c} - \mathbf{c}_0)^H]$$
$$= (A^H A)^{-1} A^H E[\mathbf{e}\mathbf{e}^H] A (A^H A)^{-1}.$$

Since the components of $\mathbf{e}$ are i.i.d., it follows that $E[\mathbf{e}\mathbf{e}^H] = \sigma_e^2 I$, so that

$$\mathrm{Cov}[\mathbf{c}] = \sigma_e^2 (A^H A)^{-1} = \sigma_e^2 R^{-1}.$$

3.5 Minimum Error in Vector-Space Approximations

Smallest covariance. Another interesting fact: of all possible unbiased linear estimates, the estimator (3.19) has the "smallest" covariance. Suppose we have another unbiased linear estimator $\tilde{\mathbf{c}}$ given by

$$\tilde{\mathbf{c}} = L\mathbf{x},$$

where $E[\tilde{\mathbf{c}}] = \mathbf{c}_0$. Using our statistical model (3.25), we obtain

$$\tilde{\mathbf{c}} = LA\mathbf{c}_0 + L\mathbf{e}.$$

In order for the estimate $\tilde{\mathbf{c}}$ to be unbiased, we must have $E[\tilde{\mathbf{c}}] = \mathbf{c}_0$, so

$$LA = I.$$

We therefore obtain $\tilde{\mathbf{c}} = \mathbf{c}_0 + L\mathbf{e}$. The covariance of $\tilde{\mathbf{c}}$ is

$$\operatorname{Cov}[\tilde{\mathbf{c}}] = E[(\tilde{\mathbf{c}} - \mathbf{c}_0)(\tilde{\mathbf{c}} - \mathbf{c}_0)^H] = \sigma_e^2 LL^H.$$

We will show that $LL^H > R^{-1}$, in the sense that the matrix $LL^H - R^{-1}$ is positive semidefinite. Let

$$Z = L - R^{-1}A^H.$$

Then for any $\mathbf{z}$,

$$0 \leq \|Z^H\mathbf{z}\|^2 = \langle Z^H\mathbf{z}, Z^H\mathbf{z}\rangle = \mathbf{z}^H ZZ^H \mathbf{z}.$$

But

$$ZZ^H = LL^H - R^{-1},$$

where we have used the fact that $LA = I$. Thus, for any $\mathbf{z}$,

$$\mathbf{z}^H(LL^H - R^{-1})\mathbf{z} \geq 0,$$

so $LL^H - R^{-1}$ is positive semidefinite, or R^{-1} is a smaller covariance matrix. The estimator $\mathbf{c}$ is said to be a best linear unbiased estimator (BLUE).

3.5 Minimum error in vector-space approximations

In this section we examine how much error is left when an optimal (minimal-norm) solution is obtained. Under the model that

$$\mathbf{x} = \sum_{i=1}^{m} c_i \mathbf{p}_i + \mathbf{e},$$

when the coefficients are found so that the estimation error is orthogonal to the data, we have

$$\mathbf{x} = \hat{\mathbf{x}} + \mathbf{e}_{\min},$$

where $\mathbf{e}_{\min}$ denotes the minimum achievable error. Taking the squared norm of both sides, we obtain

$$\|\mathbf{x}\|^2 = \|\hat{\mathbf{x}}\|^2 + \|\mathbf{e}_{\min}\|^2. \qquad (3.27)$$

This result, sometimes called the statistician's Pythagorean theorem, follows because $\hat{\mathbf{x}}$ is orthogonal to the minimum-norm error,

$$\langle \hat{x}, \mathbf{e}_{\min} \rangle = 0.$$

The statistician's Pythagorean theorem is illustrated in figure 3.5. (See also lemma 2.2.)

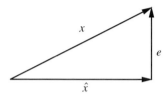

Figure 3.5: Statistician's Pythagorean theorem

The squared norm of the minimum error is

$$\|\mathbf{e}_{\min}\|^2 = \|\mathbf{x}\|^2 - \|\hat{\mathbf{x}}\|^2.$$

When we use the matrix formulation, we can obtain a more explicit representation for the minimum error. Then $\hat{\mathbf{x}} = A\mathbf{c}$, so

$$\|\hat{\mathbf{x}}\|^2 = \mathbf{c}^H A^H A \mathbf{c} = \mathbf{c}^H R \mathbf{c} = \mathbf{c}^H \mathbf{p}, \qquad (3.28)$$

where $\mathbf{p}$ from (3.18) has been employed. This gives

$$\|\mathbf{e}_{\min}\|^2 = \mathbf{x}^H \mathbf{x} - \mathbf{c}^H \mathbf{p}.$$

Another form for $\|\hat{\mathbf{x}}\|^2$ is obtained from (3.20),

$$\|\hat{\mathbf{x}}\|^2 = (A\mathbf{c})^H (A\mathbf{c}) = \mathbf{x}^H A(A^H A)^{-1} A^H \mathbf{x}. \qquad (3.29)$$

Then

$$\|\mathbf{e}_{\min}\|^2 = \mathbf{x}^H \mathbf{x} - \mathbf{x}^H A(A^H A)^{-1} A^H \mathbf{x}$$
$$= \mathbf{x}^H (I - A(A^H A)^{-1} A^H) \mathbf{x}.$$

It can be shown (see exercise 3.5-2) that

$$(I - A(A^H A)^{-1} A^H) \qquad (3.30)$$

is a positive-semidefinite matrix, from which we can conclude that $\|\mathbf{e}_{\min}\|^2$ is smaller than $\|\mathbf{x}\|^2$.

Applications of the orthogonality theorem

Because a number of vector spaces and inner products can be formulated, the orthogonality principle is used in a variety of applications. The orthogonality theorem provides the foundation for a good part of signal processing theory, since it provides a prescription for an optimum estimator: **in the optimum (least-squares) estimator, the error is orthogonal to the data.** The theorem is applied by defining an inner product, and hence the induced norm, to match the needs of the problem. Under various inner-product definitions, much of approximation theory, estimation theory, and prediction theory can be accommodated. Examples are given in the next several sections.

3.6 Approximation by continuous polynomials

Suppose we want to find the best polynomial approximation of a real continuous function $f(t)$ over an interval $t \in [a, b]$, in the sense that

$$\int_a^b (f(t) - p(t))^2 \, dt$$

is minimized for a polynomial $p(t)$ of degree $m-1$. The vector space underlying the problem is $S = C[a, b]$. We will (naively) take as basis vectors the functions $\{1, t, t^2, \ldots, t^{m-1}\}$, so that

$$p(t) = c_0 + c_1 t + c_2 t^2 + \cdots + c_{m-1} t^{m-1}.$$

The optimal coefficients can be determined (for example) directly by calculus, but the orthogonality theorem applies, using the inner product

$$\langle f, g \rangle = \int_a^b f(t) g(t) \, dt.$$

Then, using (3.4) we obtain

$$\begin{bmatrix} \langle 1, 1 \rangle & \langle 1, t \rangle & \cdots & \langle 1, t^{m-1} \rangle \\ \langle t, 1 \rangle & \langle t, t \rangle & \cdots & \langle t, t^{m-1} \rangle \\ \vdots & & & \\ \langle t^{m-1}, 1 \rangle & \langle t^{m-1}, t \rangle & \cdots & \langle t^{m-1}, t^{m-1} \rangle \end{bmatrix} \begin{bmatrix} c_0 \\ c_1 \\ \vdots \\ c_m \end{bmatrix} = \begin{bmatrix} \langle f, 1 \rangle \\ \langle f, t \rangle \\ \vdots \\ \langle f, t^{m-1} \rangle \end{bmatrix}. \quad (3.31)$$

If we take the specific case that the function is to be approximated over the interval $[0, 1]$, then the Grammian matrix in (3.31) can be computed explicitly as

$$\langle t^i, t^j \rangle = \int_0^1 t^{i+j} \, dt = \frac{1}{i+j+1}, \quad i, j = 0, 1, \ldots, m-1,$$

so that

$$R = \begin{bmatrix} 1 & \frac{1}{2} & \frac{1}{3} & \cdots & \frac{1}{m} \\ \frac{1}{2} & \frac{1}{3} & \frac{1}{4} & \cdots & \frac{1}{m+1} \\ \vdots & & & & \\ \frac{1}{m} & \frac{1}{m+1} & \frac{1}{m+2} & \cdots & \frac{1}{2m} \end{bmatrix}. \quad (3.32)$$

A matrix of this particular form is known as a **Hilbert matrix**. The Hilbert matrix is famous as a classic example of a matrix that is ill conditioned: as m increases, the matrix becomes ill conditioned exponentially fast, which means (as discussed in section 4.10) that it will suffer from severe numerical problems if m is even moderately large, no matter how it is inverted. Because of this, the particular set of basis functions chosen is not recommended. The use of the Legendre polynomials described in example 2.15.1, or other orthogonal polynomials, is preferred for polynomial approximation.

Example 3.6.1 Let $f(t) = e^t$ and $m = 3$. (For only three parameters, the Hilbert matrix (3.32) is still well conditioned.) The vector on the right hand of (3.31) is

$$\mathbf{b} = \begin{bmatrix} e - 1 \\ 1 \\ e - 2 \end{bmatrix},$$

and the coefficients in (3.31) are computed as

$$\begin{bmatrix} c_0 \\ c_1 \\ c_2 \end{bmatrix} = R^{-1}\mathbf{b} = \begin{bmatrix} 1.0130 \\ 0.8511 \\ 0.8392 \end{bmatrix}.$$

The approximating polynomial is

$$e^t \approx 1.0130 + .8511t + .8392t^2.$$

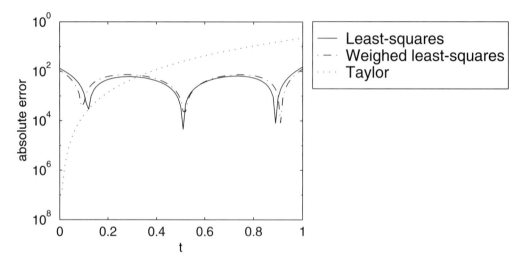

Figure 3.6: Comparison of LS, WLS, and Taylor series approximations to e^t

Figure 3.6 shows the absolute error $|e^t - p(t)|$ for this polynomial for $t \in [0, 1]$. For comparison, the error we would get by approximating e^t by the first three terms of the Taylor series expansion,

$$e^t \approx 1 + t + t^2/2,$$

is also shown, as is the weighed least-squares (WLS) approximation discussed subsequently. The error in the Taylor series starts small, but increases to a larger value than does the least-squares approximation. (How would the Taylor series have compared if the series had been expanded about the midpoint of the region, at $t_0 = \frac{1}{2}$?) □

The basis functions of the previous example give rise to the Hilbert matrix as the Grammian. However, a set of *orthogonal* polynomials can be used that has a diagonal (and hence well-conditioned) Grammian.

Now suppose that for some reason it is more important to get the approximation more correct on the extremes of the interval of approximation. We will denote the approximating polynomial in this case by $p_w(t)$. To attempt to make the approximation more exact on the

extremes of the interval of approximation, we use a weighted norm
$$\int_a^b w(t)(f(t) - p_w(t))^2 dt,$$
which is induced from the inner product
$$\langle f, g \rangle = \int_a^b \sqrt{w(t)} f(t) g(t) \, dt.$$

Example 3.6.2 Continuing the example above with $f(t) = e^t$ over $[0, 1]$, take the weighting function as
$$w(t) = 10(t - 0.5)^2.$$
Then the Grammian matrix is
$$R = \begin{pmatrix} \frac{1}{2}\sqrt{5/2} & \frac{1}{4}\sqrt{5/2} & \frac{3}{16}\sqrt{5/2} \\ \frac{1}{4}\sqrt{5/2} & \frac{3}{16}\sqrt{5/2} & \frac{5}{32}\sqrt{5/2} \\ \frac{3}{16}\sqrt{5/2} & \frac{5}{32}\sqrt{5/2} & \frac{13}{96}\sqrt{5/2} \end{pmatrix}$$
and the Right Hand vector (computed numerically) is
$$\mathbf{b} = [1.38603 \quad 0.860513 \quad 0.690724].$$
The approximating polynomial is now
$$p_w(t) = 1.0109 + .8535t + .8415t^2.$$
Figure 3.6 shows the error $e^t - p_w(t)$ and $e^t - p(t)$. As expected, the error is smaller (though only slightly) for $p_w(t)$ near the endpoints, but larger in between. □

As various weightings are imposed, the error at some values of t is reduced, while error for other values of t may increase. This raises the following interesting (and important) question: Is there some way to design the approximation so that the maximum error is minimized? This is what L_∞ approximation is all about:
$$\min \|f(t) - p(t)\|_\infty.$$
The approximation is chosen so that the maximum error is minimized.

3.7 Approximation by discrete polynomials

We can approximate discrete (sampled) data using polynomials in a manner similar to the continuous polynomial approximations of section 3.6 using a set of discrete-time basis functions $\{1, k, \ldots, k^{m-1}\}$. We desire to fit an $(m-1)$st order polynomial through the data points $x_1, x_2, \ldots, x_n$, so that
$$x_k \approx p(k), \quad k = 1, 2, \ldots, n,$$
where
$$p(k) = c_0 + c_1 k + c_2 k^2 + \cdots + c_{m-1} k^{m-1}.$$
The polynomial $p(k)$ can be written as
$$p(k) = \begin{bmatrix} 1 & k & k^2 & \cdots & k^{m-1} \end{bmatrix} \begin{bmatrix} c_0 \\ c_1 \\ c_2 \\ \vdots \\ c_{m-1} \end{bmatrix}.$$

If $m = n$ and the x_k are distinct, then there exists a polynomial, the *interpolating polynomial*, passing exactly through all n points. If $m < n$, then there is probably not a polynomial that will pass through all n points, in which case we desire to find the polynomial to minimize the squared error,

$$\sum_{k=1}^{n} |x_k - p(k)|^2.$$

This can be expressed as a vector norm

$$\|\mathbf{x} - \mathbf{p}\|_2,$$

which is induced from the Euclidean inner product $\langle \mathbf{x}, \mathbf{y} \rangle = \mathbf{x}^H \mathbf{y}$, where

$$\mathbf{x} = \begin{bmatrix} x_1 \\ x_2 \\ \vdots \\ x_n \end{bmatrix} \quad \text{and} \quad \mathbf{p} = \begin{bmatrix} p(1) \\ p(2) \\ \vdots \\ p(n) \end{bmatrix}.$$

We can write $\mathbf{p}$ in terms of the coefficients of the polynomial as

$$\mathbf{p} = \begin{bmatrix} 1 & 1 & 1 & \cdots & 1 \\ 1 & 2 & 4 & \cdots & 2^{m-1} \\ 1 & 3 & 9 & \cdots & 3^{m-1} \\ \vdots & & & & \vdots \\ 1 & n & n^2 & \cdots & n^{m-1} \end{bmatrix} \begin{bmatrix} c_0 \\ c_1 \\ c_2 \\ \vdots \\ c_{m-1} \end{bmatrix} = [\mathbf{p}_1 \ \mathbf{p}_2 \ \mathbf{p}_3 \ \cdots \ \mathbf{p}_m] \begin{bmatrix} c_0 \\ c_1 \\ c_2 \\ \vdots \\ c_{m-1} \end{bmatrix} = P\mathbf{a}.$$

The vectors $\mathbf{p}_i, i = 1, 2, \ldots, m$ represent the data in this approximation problem. If P is square, it is called a *Vandermonde* matrix, about which more is presented in section 8.4. As with the continuous-time polynomial approximation, there may be better basis functions for this problem from a numerical point of view.

Using this notation, the approximation problem becomes

$$\mathbf{x} = P\mathbf{c} + \mathbf{e},$$

which is a problem in the same form as (3.2), from which observe that the $\mathbf{c}$ which minimizes $\|\mathbf{e}\|^2$ is

$$\mathbf{c} = (P^T P)^{-1} P^T \mathbf{x}.$$

The approximated vector $\mathbf{p}$ is thus

$$\mathbf{p} = P\mathbf{c} = P(P^T P)^{-1} P^T \mathbf{x}.$$

Example 3.7.1 We desire to approximate the function

$$x[k] = \sin(k\pi/7)$$

using a quadratic polynomial ($m = 3$) to obtain the best match for $k = 1:7$. The P matrix is

$$\begin{bmatrix} 1 & 1 & 1 \\ 1 & 2 & 4 \\ 1 & 3 & 9 \\ 1 & 4 & 16 \\ 1 & 5 & 25 \\ 1 & 6 & 36 \\ 1 & 7 & 49 \end{bmatrix}$$

3.8 Linear Regression

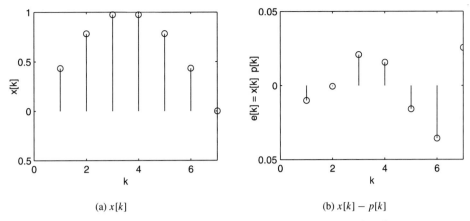

(a) $x[k]$

(b) $x[k] - p[k]$

Figure 3.7: A discrete function and the error in its approximation

and the coefficients are computed as $\mathbf{c}^T = [-0.0612, 0.5885, -0.0833]$. Figure 3.7(a) shows $x[k]$ and figure 3.7(b) shows the error $p[k] - x[k]$. □

3.8 Linear regression

From the data in figure 3.8(a), where there are n points $\mathbf{x}_i$, $i = 1, 2, \ldots, n$ with each $\mathbf{x}_i = [x_i, y_i]^T$, it would appear that we can approximately fit a line of the form

$$y_i \approx ax_i + b, \qquad i = 1, 2, \ldots, n \tag{3.33}$$

for suitably chosen slope a and intercept b. As stated, this is a *linear regression* problem; that is, a problem of determining a functional relation between the measured variables x_i and y_i. Nonlinear regressions are also used, such as the quadratic regression,

$$y_i \approx a_0 + a_1 x_i + a_2 x_i^2. \tag{3.34}$$

Or we may have data vectors $\mathbf{x}_i \in \mathbb{R}^3$, with $\mathbf{x}_i = [x_i, y_i, z_i]^T$, and we may regress among the points as

$$z_i \approx ax_i + by_i + c. \tag{3.35}$$

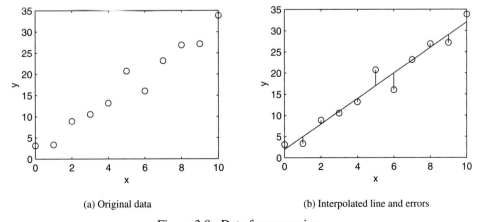

(a) Original data

(b) Interpolated line and errors

Figure 3.8: Data for regression

In all such regression problems, we desire to choose the regression parameters so that the Right Hand Side of the regression equations provides a good representation of the Left Hand Side.

We will consider in detail the linear regression problem (3.33). We can stack the equations to obtain

$$\begin{bmatrix} y_1 \\ y_2 \\ \vdots \\ y_n \end{bmatrix} = \begin{bmatrix} ax_1 + b \\ ax_2 + b \\ \vdots \\ ax_n + b \end{bmatrix} + \begin{bmatrix} e_1 \\ e_2 \\ \vdots \\ e_n \end{bmatrix} \quad (3.36)$$

for some error terms e_i. Let

$$\mathbf{y} = [y_1, y_2, \ldots, y_n]^T \qquad \mathbf{e} = [e_1, e_2, \ldots, e_n]^T \qquad \mathbf{c} = \begin{bmatrix} a \\ b \end{bmatrix}$$

and

$$A = \begin{bmatrix} x_1 & 1 \\ x_2 & 1 \\ \vdots & \\ x_n & 1 \end{bmatrix}.$$

Then (3.36) is of the form

$$\mathbf{y} = A\mathbf{c} + \mathbf{e}, \quad (3.37)$$

which again is in the form (3.16), so the best (in the least-squares sense) estimate of $\mathbf{c}$ is

$$\mathbf{c} = (A^H A)^{-1} A^H \mathbf{y}. \quad (3.38)$$

The line found by (3.38) minimizes the sums of the squares of the *vertical* distances between the data abscissas and the line, as shown in figure 3.8(b). To minimize *shortest* distances of the data to the interpolating line, the method of *total least squares* discussed in section 7.7 must be used.

Since $A^H A$ in (3.38) is a 2×2 matrix, explicit closed-form expressions for m and b in $\mathbf{c}$ can be found. The slope and intercept (for real data) are

$$a = \frac{n \sum_{i=1}^{n} x_i y_i - \left(\sum_{i=1}^{n} x_i\right)\left(\sum_{j=1}^{n} y_i\right)}{n \sum_{i=1}^{n} x_i^2 - \left(\sum_{i=1}^{n} x_i\right)^2},$$

$$b = \frac{\left(\sum_{i=1}^{n} x_i^2\right)\left(\sum_{j=1}^{n} y_i\right) - \left(\sum_{i=1}^{n} x_i\right)\left(\sum_{i=1}^{n} x_i y_i\right)}{n \sum_{i=1}^{n} x_i^2 - \left(\sum_{i=1}^{n} x_i\right)^2}. \quad (3.39)$$

Example 3.8.1 (Weighted least-squares) Five measurements (x_i, y_i), $i = 1, 2, \ldots, 5$, are made in a system, of which the first three are believed to be fairly accurate, and two are known to be somewhat corrupted by measurement noise. The measurements are

$$(1, 2.5) \quad (3, 3.5) \quad (6, 5) \quad (5, 3) \quad (3, 4).$$

3.9 Least-Squares Filtering

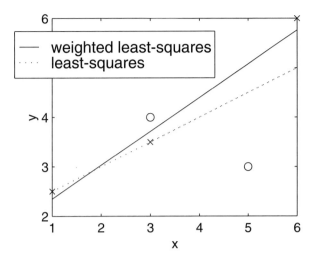

Figure 3.9: Illustration of least-squares and weighted least-squares

From these five measurements, the data are to be fitted to a line according to the model $y = ax + b$. The measurements stack up in the model equation as

$$\begin{bmatrix} 1 & 1 \\ 3 & 1 \\ 6 & 1 \\ 5 & 1 \\ 3 & 1 \end{bmatrix} \begin{bmatrix} a \\ b \end{bmatrix} = \begin{bmatrix} 2.5 \\ 3.5 \\ 5 \\ 3 \\ 4 \end{bmatrix} + \mathbf{e},$$

or

$$A\mathbf{c} = \mathbf{y} + \mathbf{e}.$$

In finding the best (minimum squared-error) solution to this problem, it is appropriate to weight most heavily those equations which are believed to be the most accurate. Let

$$W = \text{diag}\{10, 10, 10, 1, 1\}.$$

Then using (3.24), we can determine the optimal (under the weighted inner product) set of coefficients. Figure 3.9 illustrates the data and the least-squares lines fitted to them. The accurate data are plotted with ×, and the inaccurate data are plotted with ○. The weighted least-squares line fits more closely (on average) to the more accurate data, while the unweighted least-squares line is pulled off significantly by the inaccurate data at $x = 5$. □

3.9 Least-squares filtering

In the least-squares filter problem, we desire to filter a sequence of input data $\{f[t]\}$, using a filter with impulse response $h[t]$ of length m to produce an output that matches a desired sequence $\{d[t]\}$ as closely as possible. (Examples in which such a circumstance arises are given in section 1.5, in the context of adaptive filtering.) If we call the output of the filter $y[t]$, we have the filter expression

$$y[t] = \sum_{i=0}^{m-1} h[i] f[t-i].$$

We can write $d[t] = y[t] + e[t]$, where $e[t]$ is the error between the filter output and the desired filter output,

$$d[t] = \sum_{i=0}^{m-1} h[i]f[t-i] + e[t].$$

We want to choose the filter coefficients $\{h[i]\}$ in such a way that the error between the filter output and the desired signal should be as small as possible; that is, we want to make

$$e[t] = d[t] - y[t]$$

small for each t.

When doing *least-squares filtering*, the criterion of minimal error is that the sum of the squared errors is as small as possible:

$$\min \sum_{i=i_1}^{i_2} |e[i]|^2, \tag{3.40}$$

where i_1 is the starting index and i_2 the ending index over which we desire to minimize. The squared norm in (3.40) is induced from the inner product defined by

$$\langle \mathbf{x}, \mathbf{y} \rangle = \sum_{i=i_1}^{i_2} x_i \bar{y}_i \tag{3.41}$$

Letting

$$\mathbf{y} = \begin{bmatrix} y[i_1] \\ y[i_1+1] \\ \vdots \\ y[i_2] \end{bmatrix} \qquad \mathbf{h} = \begin{bmatrix} h[0] \\ h[1] \\ \vdots \\ h[m-1] \end{bmatrix} \qquad \mathbf{x} = \begin{bmatrix} x[i_1] \\ x[i_1+1] \\ \vdots \\ x[i_2] \end{bmatrix},$$

the inner product (3.41) can be written as

$$\langle \mathbf{x}, \mathbf{y} \rangle = \mathbf{y}^H \mathbf{x},$$

and the filtered outputs can be written as

$$\mathbf{y} = A\mathbf{h},$$

where A is a matrix of the input data, $f[t]$. The matrix A takes various forms, depending on the assumptions made on the data, as described in the following. Let

$$\mathbf{d} = \begin{bmatrix} d[i_1] \\ d[i_1+1] \\ \vdots \\ d[i_2] \end{bmatrix}$$

be a vector of desired outputs. Then we want

$$\mathbf{d} \approx \mathbf{y} = A\mathbf{h}.$$

We can represent our approximation problem as

$$\mathbf{d} = A\mathbf{h} + \mathbf{e},$$

where $\mathbf{e}$ is the difference between the output $\mathbf{y}$ and the desired output $\mathbf{d}$. We desire to find the filter coefficients $\mathbf{h}$ to minimize $\|\mathbf{e}\|$. By comparison with (3.16), observe that the solution is

$$\mathbf{h} = (A^H A)^{-1} A^H \mathbf{y}. \tag{3.42}$$

We now examine the form of the A matrix under various assumptions about the inputs. Assume that we have available to us, for the purpose of finding the coefficients, the data $f[1], f[2], \ldots, f[N]$, with a total of N data points.

3.9 Least-Squares Filtering

The "covariance" method. In this method, we use only data that is explicitly available, not making any assumptions about data outside this segment of observed data. The data matrix A in this case is the $(N - m + 1) \times m$ matrix

$$A = \begin{bmatrix} f[m] & f[m-1] & f[m-2] & \cdots & f[1] \\ f[m+1] & f[m] & f[m-1] & \cdots & f[2] \\ \vdots & & & & \\ f[N] & f[N-1] & f[N-2] & \cdots & f[N-m+1] \end{bmatrix}.$$

Let $\mathbf{q}[i]$ be the $m \times 1$ data vector corresponding to a (conjugated) row of A, as in (3.21); then

$$\mathbf{q}[i] = \begin{bmatrix} \overline{f}[i] \\ \overline{f}[i-1] \\ \cdots \\ \overline{f}[i-m+1] \end{bmatrix}, \tag{3.43}$$

with the notation that $f[i] = 0$ where i is outside the range 1 to N, and we can represent the data matrix as

$$A = \begin{bmatrix} \mathbf{q}[m]^H \\ \mathbf{q}[m+1]^H \\ \vdots \\ \mathbf{q}[N]^H \end{bmatrix}.$$

The Grammian can be written as

$$R = A^H A = \sum_{i=m}^{N} \mathbf{q}[i]\mathbf{q}^H[i]. \tag{3.44}$$

The Grammian R is a Hermitian matrix.

The "autocorrelation" method. In this case, we assume that data prior to $f[1]$ and after $f[N]$ are all zero, and fill up the data matrix A with these assumed values. The output is taken from $i_1 = 1$ up through $i_2 = N + m - 1$. The data matrix is the $(N + m - 1) \times m$ matrix

$$A = \begin{bmatrix} f[1] & 0 & 0 & \cdots & 0 \\ f[2] & f[1] & 0 & \cdots & 0 \\ f[3] & f[2] & f[1] & \cdots & 0 \\ \vdots & & & & \\ f[m] & f[m-1] & f[m-2] & \cdots & f[1] \\ f[m+1] & f[m] & f[m-2] & \cdots & f[2] \\ \vdots & & & & \\ f[N] & f[N-1] & f[N-2] & \cdots & f[N-m+1] \\ 0 & f[N] & F[N-1] & \cdots & f[N-m+2] \\ \vdots & & & & \\ 0 & 0 & 0 & \cdots & f[N] \end{bmatrix}.$$

The terms "covariance method" and "autocorrelation method" do not produce, respectively, a covariance matrix and an autocorrelation matrix in the usual sense. Rather, these are the terms for these methods commonly employed in the speech processing literature (see, e.g., [215]). Using the notation of (3.43), we can write the data matrix as

$$A = \begin{bmatrix} \mathbf{q}^H[1] \\ \mathbf{q}^T[2] \\ \vdots \\ \mathbf{q}^T[N+m-1] \end{bmatrix}.$$

In a manner similar to (3.44), we can write

$$R = A^H A = \sum_{i=1}^{N+m-1} \mathbf{q}[i]\mathbf{q}^H[i].$$

This is a Toeplitz matrix.

Pre-windowing method. In this method we assume that $f[t] = 0$ for $t < 1$, and use data up to $f[N]$, so that $i_1 = 1$ and $i_2 = N$. Then the data matrix is the $N \times q$ matrix

$$A = \begin{bmatrix} f[1] & 0 & 0 & \cdots & 0 \\ f[2] & f[1] & 0 & \cdots & 0 \\ f[3] & f[2] & f[1] & \cdots & 0 \\ \vdots & & & & \\ f[m] & f[m-1] & f[m-2] & \cdots & f[1] \\ f[m+1] & f[m] & f[m-2] & \cdots & f[2] \\ \vdots & & & & \\ f[N] & f[N-1] & f[N-2] & \cdots & f[N-m+1] \end{bmatrix} = \begin{bmatrix} \mathbf{q}^H[1] \\ \mathbf{q}^H[2] \\ \vdots \\ \mathbf{q}^H[N] \end{bmatrix}, \quad (3.45)$$

and

$$R = \sum_{i=1}^{N} \mathbf{q}[i]\mathbf{q}^H[i].$$

Post-windowing method. We begin with $i_1 = m$, and assume that data after N are equal to zero. Then A is the $N \times m$ matrix

$$A = \begin{bmatrix} f[m] & f[m-1] & f[m-2] & \cdots & f[1] \\ f[m+1] & f[m] & f[m-2] & \cdots & f[2] \\ \vdots & & & & \\ f[N] & f[N-1] & f[N-2] & \cdots & f[N-m+1] \\ 0 & f[N] & F[N-1] & \cdots & f[N-m+2] \\ \vdots & & & & \\ 0 & 0 & 0 & \cdots & f[N] \end{bmatrix},$$

and

$$R = \sum_{i=m}^{m+N} \mathbf{q}[i]\mathbf{q}^H[i].$$

Example 3.9.1 Suppose we observe the data sequence

$$\{f[1], \ldots, f[5]\} = \{1, -2, 3, -4, 5\}$$

and want to filter these data with a filter of length $m = 3$. The data matrices corresponding to each interpretation, labeled respectively A_{cov}, A_{ac}, A_{pre}, and A_{post}, with their corresponding Grammians,

3.9 Least-Squares Filtering

are shown here:

$$A_{\text{cov}} = \begin{bmatrix} 3 & -2 & 1 \\ -4 & 3 & -2 \\ 5 & -4 & 3 \end{bmatrix} \qquad A_{\text{ac}} = \begin{bmatrix} 1 & 0 & 0 \\ -2 & 1 & 0 \\ 3 & -2 & 1 \\ -4 & 3 & -2 \\ 5 & -4 & 3 \\ 0 & 5 & -4 \\ 0 & 0 & 5 \end{bmatrix}$$

$$A_{\text{pre}} = \begin{bmatrix} 1 & 0 & 0 \\ -2 & 1 & 0 \\ 3 & -2 & 1 \\ -4 & 3 & -2 \\ 5 & -4 & 3 \end{bmatrix} \qquad A_{\text{post}} = \begin{bmatrix} 3 & -2 & 1 \\ -4 & 3 & -2 \\ 5 & -4 & 3 \\ 0 & 5 & -4 \\ 0 & 0 & 5 \end{bmatrix}$$

$$R_{\text{cov}} = \begin{bmatrix} 50 & -38 & 26 \\ -38 & 29 & -20 \\ 26 & -20 & 14 \end{bmatrix} \qquad R_{\text{ac}} = \begin{bmatrix} 55 & -40 & 26 \\ -40 & 55 & -40 \\ 26 & -40 & 55 \end{bmatrix}$$

$$R_{\text{pre}} = \begin{bmatrix} 55 & -40 & 26 \\ -40 & 30 & -20 \\ 26 & -20 & 14 \end{bmatrix} \qquad R_{\text{post}} = \begin{bmatrix} 50 & -38 & 26 \\ -38 & 54 & -40 \\ 26 & -40 & 55 \end{bmatrix}.$$

Observe that while all of the data matrices are Toeplitz (constant along the diagonals), the only Grammian which is Toeplitz is the one which arises from the autocovariance form of the data matrix.

MATLAB code to compute the least-squares filter coefficients is given in algorithm 3.1.

Algorithm 3.1 Least-squares filter computation
File: `lsfilt.m`

Example 3.9.2 For the input data of the previous example, the following desired data are known:

$$\mathbf{d} = [2, -5, 11, -17, 23, -17, 15]^T.$$

We want to find a filter of length $m = 3$ that produces this data. Using the four different data sets in the example, with selections of $\mathbf{d}$ corresponding to the data used, we obtain from the MATLAB commands

```
hcv = lsfilt(f,d(3:5),3,1)
hac = lsfilt(f,d,3,2)
hpre = lsfilt(f,d(1:5),3,3)
hpost = lstilf(f,d(3:7),3,4)
```

the filter coefficients

$$\mathbf{h}_{\text{cov}} = [1.5 \quad -2 \quad 2.5]^T \qquad \mathbf{h}_{\text{auto}} = [2 \quad -1 \quad 3]^T$$
$$\mathbf{h}_{\text{pre}} = [2 \quad -1 \quad 3]^T \qquad \mathbf{h}_{\text{post}} = [2 \quad -1 \quad 3]^T,$$

respectively.

Example 3.9.3 An application of least-squares filtering is illustrated in figure 3.10 in a channel equalizer application. A sequence of bits $\{b[t]\}$ is passed through a discrete-time channel with unknown

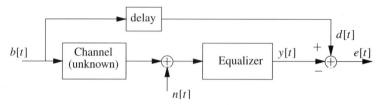

Figure 3.10: Least-squares equalizer example

impulse response, the output of which is corrupted by noise. To counteract the effect of the channel, the signal is passed through an equalizer, which in this case is an FIR filter whose coefficients have been determined using a least-squares criterion. In order to determine what the coefficients are, some set of known data—a *training sequence*—is used at the beginning of the transmission. This sequence is delayed and used as the desired signal $d[t]$. Using this training sequence, the filter coefficients $h[k]$ are computed by using (3.42), after which the coefficients are loaded into the equalizer filter.

This example is more a demonstration of a concept than a practical reality. While equalizers are common on modern modem technology, they are more commonly implemented using adaptive filters. Adaptive equalizers are examined in section 4.11.2 (RLS adaptive equalizer) and section 14.6 (LMS adaptive equalizer). □

3.9.1 Least-squares prediction and AR spectrum estimation

Consider now the estimation problem in which we desire to predict $x[t]$ using a linear predictor based upon $x[t-1], x[t-2], \ldots, x[t-m]$. We then have

$$x[t] = -\sum_{i=1}^{m} a_i x[t-i] + f[t], \tag{3.46}$$

using $a_i = -h_i$ as the coefficients, where $f[t]$ is now used to denote the (forward) predictor error. The predictor of (3.46) is called a *forward predictor*. This is essentially the problem solved in the last section, in which the desired signal is the sample $d[t] = x[t]$, and the data used are the *previous* data samples. We can model the signal $x[t]$ as being the output of a signal with input $f[t]$, where the system function is

$$H(z) = \frac{X(z)}{F(z)} = \frac{1}{1 + \sum_{i=1}^{n} a_i z^{-i}} = \frac{1}{A(z)}.$$

If $f[t]$ is a random signal with power spectral density (PSD) $S_f(z)$, then the PSD of $x[t]$ is

$$S_x(z) = \frac{1}{\left(1 + \sum_{i=1}^{m} a_i z^{-i}\right)\left(1 + \sum_{i=1}^{m} \overline{a}_i z^{i}\right)} P_f(z) = \frac{1}{A(z)\overline{A}(1/z)}. \tag{3.47}$$

If $f[t]$ is assumed to be a white-noise sequence with variance σ_f^2, then the random process $x[t]$ has the PSD

$$S_x(z) = \frac{\sigma_f^2}{A(z)\overline{A}(1/z)}.$$

Evaluating this on the unit circle $z = e^{j\omega}$, we obtain

$$S_x(\omega) \triangleq S_x(z)\big|_{z=e^{j\omega}} = \frac{\sigma_f^2}{\left|1 + \sum_{i=1}^{m} a_i e^{-j\omega i}\right|^2} = \frac{\sigma_f^2}{|A(\omega)|^2}. \tag{3.48}$$

Thus, by finding the coefficients of the linear predictor, we can determine an estimate of the spectrum, under the assumption that the signal is produced by the AR model (3.46).

We can obtain more data to put in our data matrix (and usually decrease the variance of the estimate) by using a *backward predictor* in addition to a forward predictor. In the

3.9 Least-Squares Filtering

backward predictor, the m data points $x[t], x[t-1], \ldots, x[t-m+1]$ are used to estimate $x[t-m]$, by

$$x[t-m] = -\sum_{i=1}^{m} a_i x[t-m+i] + b[t],$$

where $b[t]$ is the backward prediction error. As before, if we view $x[t-m]$ as the output of a system driven by an input $b[t]$, we obtain a system function

$$H_b(z) = \frac{x(z)}{b(z)} = \frac{1}{z^{-m}\left(1 + \sum_{i=1}^{m} \overline{a}_i z^i\right)} = \frac{1}{z^{-m}\overline{A}(1/z)}.$$

If $b[t]$ is a white-noise sequence with variance $\sigma_b^2 = \sigma_f^2$, then the PSD of the signal $x[t-m]$ is

$$S_x(z) = \sigma_b^2 \frac{1}{\overline{A}(1/z) A(z)}, \tag{3.49}$$

the same as in (3.47). Since both the forward predictor and the backward predictor use the same predictor coefficients (just conjugated and in a different order), we can use the backward predictor information to improve our estimate of the coefficients. If we have measured data $x[1], x[2], \ldots, x[N]$, we write our prediction equations as follows (using the covariance method employing only measured data):

$$\begin{bmatrix} x[m] & x[m-1] & \cdots & x[1] \\ x[m+1] & x[m] & \cdots & x[2] \\ \vdots & & & \\ x[N-1] & x[N-2] & \cdots & x[N-m] \\ \overline{x}[2] & \overline{x}[3] & \cdots & \overline{x}[m+1] \\ \overline{x}[3] & \overline{x}[4] & \cdots & \overline{x}[m+2] \\ \vdots & & & \\ \overline{x}[N-m+1] & \overline{x}[N-m+2] & \cdots & \overline{x}[N] \end{bmatrix} \begin{bmatrix} -a_1 \\ -a_2 \\ \vdots \\ -a_m \end{bmatrix}$$

$$= \begin{bmatrix} x[m+1] \\ x[m+2] \\ \vdots \\ x[N] \\ \overline{x}[1] \\ \overline{x}[2] \\ \vdots \\ \overline{x}[N-m] \end{bmatrix} + \begin{bmatrix} f[m+1] \\ f[m+2] \\ \vdots \\ f[N] \\ \overline{b}[N-m+1] \\ \overline{b}[N-m+2] \\ \vdots \\ \overline{b}[N-m] \end{bmatrix}.$$

Let us write this as

$$\mathbf{x} = A\mathbf{h} + \mathbf{e},$$

where $\mathbf{x}$ and $\mathbf{e}$ now are $2(N-m) \times 1$ and A is $2(N-m) \times n$. In the data matrix, the first $N-m$ rows correspond to the forward predictor and the second $N-m$ rows correspond to the backward predictor. Our optimization criterion is to minimize

$$\sum_{i=n+1}^{N} |f[i]|^2 + |b[i]|^2.$$

As before, a least-squares solution is straightforward. This technique of spectrum estimation is known as the forward–backward linear prediction (FBLP) technique, or the modified covariance technique. An estimate of the variance is

$$\hat{\sigma}_f^2 = \hat{\sigma}_b^2 = \|\mathbf{e}_{\min}\|^2/2.$$

A MATLAB function that computes the AR parameters using the modified covariance technique is shown in algorithm 3.2.

Algorithm 3.2 Forward–backward linear predictor estimate
File: `fblp.m`

3.10 Minimum mean-square estimation

In the least-squares estimation of the preceding sections, we have not employed, nor assumed the existence of, any probabilistic model. The optimization criterion has been to minimize the sum of squared error. In this section, we change our viewpoint somewhat by introducing a probabilistic model for the data.

Let $P_1, P_2, \ldots, P_m$ be random variables. We desire to find coefficients $\{c_i\}$ to estimate the random variable X, using

$$X = c_1 P_1 + c_2 P_2 + \cdots + c_m P_m + e$$

in such a way that the norm of the squared error is minimized. Using the inner product

$$\langle X, Y \rangle = E[X\overline{Y}], \tag{3.50}$$

the minimum mean-square estimate of $\mathbf{c}$ is given by

$$R\mathbf{c} = \mathbf{p},$$

where

$$R = \begin{bmatrix} E[P_1\overline{P}_1] & E[P_2\overline{P}_1] & \cdots & E[P_m\overline{P}_1] \\ E[P_1\overline{P}_2] & E[P_2\overline{P}_2] & \cdots & E[P_m\overline{P}_2] \\ \vdots & & & \\ E[P_1\overline{P}_m] & E[P_2\overline{P}_m] & \cdots & E[P_m\overline{P}_m] \end{bmatrix} \quad \text{and} \quad \mathbf{p} = \begin{bmatrix} E[X\overline{P}_1] \\ E[X\overline{P}_2] \\ \vdots \\ E[X\overline{P}_m] \end{bmatrix}. \tag{3.51}$$

The minimum mean-squared error in this case is given using (3.29) as

$$\begin{aligned} \|e\|_{\min} &= \sigma_x^2 - \mathbf{p}^H R^{-1} \mathbf{p} \\ &= \sigma_x^2 - \mathbf{p}^H \mathbf{c}. \end{aligned} \tag{3.52}$$

Example 3.10.1 Suppose that

$$\mathbf{Z} = [X_1, X_2, X_3]^T$$

is a real Gaussian random vector with mean zero and covariance

$$R_{zz} = \text{cov}(\mathbf{Z}) = E[\mathbf{Z}\mathbf{Z}^T] = \begin{bmatrix} 1 & .2 & .1 \\ .2 & 2 & .3 \\ .1 & .3 & 4 \end{bmatrix}.$$

Given measurements of X_1 and X_2, we wish to estimate X_3 using a linear estimator,

$$\hat{X}_3 = c_1 X_1 + c_2 X_2.$$

The necessary correlation values in (3.51) can be obtained from the covariance R_{zz},

$$R = \begin{bmatrix} E[X_1 X_1] & E[X_1 X_2] \\ E[X_2 X_1] & E[X_2 X_2] \end{bmatrix} = \begin{bmatrix} 1 & .2 \\ .2 & 2 \end{bmatrix} \quad \text{and} \quad \mathbf{p} = \begin{bmatrix} E[X_3 X_1] \\ E[X_3 X_2] \end{bmatrix} = \begin{bmatrix} .1 \\ .3 \end{bmatrix},$$

from which the optimal coefficients are

$$\mathbf{c} = \begin{bmatrix} 0.0714 \\ 0.1429 \end{bmatrix}.$$

The minimum mean-squared error is

$$\|e\|_{\min} = 4 - \mathbf{p}^T R^{-1} \mathbf{p} = 3.95.$$

□

3.11 Minimum mean-squared error (MMSE) filtering

A minimum mean-square (MMS) filter is called a *Wiener filter*. It is mathematically similar to a least-squares filter, except that the expectation operator is used as the inner product. Given a sequence of data $\{f[t]\}$, we desire to design a filter in such a way that we get as close as possible to some desired sequence $d[t]$. In the interest of generality, we assume the possibility of an IIR filter,

$$y[t] = \sum_{l=0}^{\infty} h[l] f[t-l]. \tag{3.53}$$

In adopting a statistical model, we assume that the signals involved are wide-sense stationary so that, for example,

$$E[x[t]] = E[x[t-l]] \quad \text{for all } l$$

and

$$E[x[t]\overline{x}[t-l]]$$

depends only upon the time difference l and not upon the sample instant t.

Using

$$e[t] = d[t] - y[t] \tag{3.54}$$

as the estimator error, by the orthogonality principle, the squared norm of error, which in this case is termed the *mean-squared error*,

$$\|e[t]\|^2 = E[e[t]\overline{e}[t]],$$

is minimized when the error is orthogonal to the data. That is, the optimal estimator satisfies

$$\left\langle d[t] - \sum_{l=0}^{\infty} h[l] f[t-l], f[t-i] \right\rangle = 0$$

for $i = 0, 1, 2, \ldots$; or,

$$\langle d[t], f[t-i] \rangle = \sum_{l=0}^{\infty} h[l] \langle f[t-l], f[t-i] \rangle = 0. \tag{3.55}$$

Using the inner product (3.50), we obtain

$$\sum_{l=0}^{\infty} h[l] E[f[t-l]\overline{f}[t-i]] = E[\overline{f}[t-i]d[t]]. \tag{3.56}$$

Equation (3.56) is an infinite set of normal equations. For this case in which the inner product is defined using the expectation, the normal equations are referred to as the *Wiener–Hopf* equations. We can place the normal equations into a more standard form by expressing the Grammian in the form of an autocorrelation matrix. Define

$$r(i - l) = E[f[t - l]\overline{f}[t - i]] = \langle f[t - l], f[t - i]\rangle$$

and

$$p(i) = E[\overline{f}[t - i]d[t]] = \langle d[t], f[t - i]\rangle,$$

and observe that $r(-k) = \overline{r}(k)$. Then (3.56) can be written as

$$\sum_{l=0}^{\infty} h[l]r(i - l) = p(i), \qquad i = 0, 1, \ldots. \tag{3.57}$$

Solution of this problem for an IIR filter is reexamined in section 3.13.

For now, we focus on the solution when $\{h\}$ is an FIR filter with m coefficients. Then the filter output can be written as

$$y[t] = \mathbf{f}[t]^H \mathbf{h},$$

where

$$\mathbf{f}[t] = [\overline{f}[t] \quad \overline{f}[t - 1] \quad \ldots \quad \overline{f}[t - m + 1]]^T \tag{3.58}$$

(note the conjugates in this definition) and

$$\mathbf{h} = [h[0] \quad h[1] \quad \ldots \quad h[m - 1]]^T.$$

Under the assumption of an FIR filter, (3.57) can be written as

$$\sum_{l=0}^{m-1} h[l]r(i - l) = p(i), \qquad i = 0, 1, \ldots, \tag{3.59}$$

which we can express in matrix form with $R_{il} = r(i - l)$ as

$$R\mathbf{h} = \mathbf{p}, \tag{3.60}$$

where

$$R = \begin{bmatrix} r(0) & \overline{r}(1) & \overline{r}(2) & \cdots & \overline{r}(m - 1) \\ r(1) & r(0) & \overline{r}(1) & \cdots & \overline{r}(m - 2) \\ r(2) & r(1) & r(0) & \cdots & \overline{r}(m - 3) \\ \vdots & & & & \\ r(m - 1) & r(m - 2) & r(m - 3) & \cdots & r(0) \end{bmatrix}$$

$$= E[\mathbf{f}[t]\mathbf{f}^H[t]] \tag{3.61}$$

and

$$\mathbf{p} = \begin{bmatrix} p(0) \\ p(1) \\ p(2) \\ \ldots \\ p(m - 1) \end{bmatrix}$$

$$= E[\mathbf{f}[t]d[t]]. \tag{3.62}$$

The optimal weights from (3.60) are $\mathbf{h} = R^{-1}\mathbf{p}$.

3.11 Minimum Mean-Squared Error (MMSE) Filtering

The matrix R is the Grammian matrix and has the special form of a Toeplitz matrix: the diagonals are equal to each other. Because of this special form, fast algorithms exist for inverting the matrix and solving for the optimum filter coefficients. Toeplitz matrices are discussed further in section 8.3. (We have already seen one example of the solution of Toeplitz equations with a special right-hand side, in Massey's algorithm in section 1.9.)

The minimum mean-squared error can be determined using (3.52) to be

$$\|e\|^2_{\min} = E[e^2]_{\min} = \|d\|^2 - \|y\|^2.$$

Using the notation $\|e\|^2 = \sigma_e^2$ and $\|d\|^2 = \sigma_d^2$, and noting that

$$\|y[t]\|^2 = E[y[t]\overline{y}[t]] = E[\mathbf{h}^H[t]\mathbf{x}[t]\mathbf{x}^H[t]]\mathbf{h} = \mathbf{h}^H R \mathbf{h} = \mathbf{p}^H \mathbf{h},$$

we obtain

$$\left(\sigma_e^2\right)_{\min} = \sigma_d^2 - \mathbf{p}^H \mathbf{h}. \tag{3.63}$$

Example 3.11.1 In this example we explore a simple equalizer. Suppose we have a channel with transfer function

$$H_c(z) = \frac{1}{1 + .6z^{-1}}.$$

Passing into the channel is a desired signal $d[t]$. The output of the channel is $u[t]$, so that we have

$$u[t] - 0.6u[t-1] = d[t]. \tag{3.64}$$

However, we observe only a noise-corrupted version of the channel output,

$$f[t] = u[t] + n[t],$$

where $n[t]$ is a zero-mean white-noise sequence with variance $\sigma_n^2 = 0.16$, which is uncorrelated with $v[t]$. Suppose, furthermore, that we have a statistical model for the desired signal, in which we know that $d[t]$ is a first-order AR signal generated by

$$d[t] = -.5d[t-1] + v[t],$$

where $v[t]$ is a zero-mean white-noise signal with variance $\sigma_v^2 = 0.1$. Based on this information, we desire to find an optimal Wiener filter to estimate $d[t]$, using the observed sequence $f[t]$. The diagram is shown in figure 3.11. The cascade of the AR process and the channel gives the combined transfer

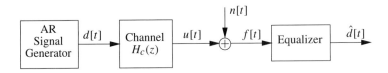

Equivalent Model

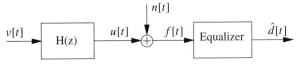

Figure 3.11: An equalizer problem

function from $v[t]$ to $u[t]$ as

$$H(z) = \frac{1}{(1 + .5z^{-1})(1 - .6z^{-1})} = \frac{1}{1 - .1z^{-1} - .3z^{-2}} = \frac{1}{1 + a_1 z^{-1} + a_2 z^{-2}}$$

so that
$$u[t] - .1u[t-1] - .3u[t-2] = v[t].$$

In this example, since the channel output is an AR(2) process, the equalizer used is a two-tap FIR filter.

We need the matrix R, containing autocorrelations of the signal $f[t]$, and the cross-correlation vector $\mathbf{p}$. Since $f[t] = u[t] + n[t]$, and since $v[t]$ and $n[t]$ are uncorrelated, we have

$$R = R_{ff} = R_{uu} + R_{nn},$$

where R_{uu} is the autocorrelation matrix for the signal $u[t]$ and R_{nn} is the autocorrelation matrix for the signal $n[t]$. Since $n[t]$ is a white-noise sequence, $R_{nn} = \sigma_n^2 I$, where I is the 2×2 identity matrix. To find

$$R_{uu} = \begin{bmatrix} r_u(0) & r_u(1) \\ r_u(1) & r_u(0) \end{bmatrix},$$

we use the results from section 1.4.2. Specifically, from (1.79) and (1.78) we find

$$\sigma_u^2 = r_u(0) = \left(\frac{1+a_2}{1-a_2}\right) \frac{\sigma_v^2}{(1+a_2)^2 - a_1^2} = 0.1122$$

$$r_u(1) = \frac{-a_1}{1+a_2}\sigma_u^2 = 0.0160.$$

Thus,

$$R = \begin{bmatrix} .16 & 0 \\ 0 & .16 \end{bmatrix} + \begin{bmatrix} .1122 & .0160 \\ .0160 & .1122 \end{bmatrix} = \begin{bmatrix} .2722 & .0160 \\ .0160 & .2722 \end{bmatrix}.$$

For the cross-correlation vector,

$$\mathbf{p} = E\begin{bmatrix} \overline{f}[t]d[t] \\ \overline{f}[t-1]d[t] \end{bmatrix} = E\begin{bmatrix} (\overline{u}[t] + \overline{n}[t])d[t] \\ (\overline{u}[t-1] + \overline{n}[t-1])d[t] \end{bmatrix}$$

$$= E\begin{bmatrix} \overline{u}[t]d[t] \\ \overline{u}[t-1]d[t] \end{bmatrix},$$

since $d[t]$ is uncorrelated with $n[t-n]$. Multiplying (3.64) through by $\overline{u}[t-k]$ and taking expectations, we obtain

$$p(k) = E[\overline{u}[t-k]d[t]] = r_u(k) - 0.6r_u(k-1),$$

from which we can determine

$$\mathbf{p} = \begin{bmatrix} 0.1206 \\ -0.0513 \end{bmatrix}.$$

The optimal filter coefficients are

$$\mathbf{h} = R^{-1}\mathbf{p} = \begin{bmatrix} 0.3893 \\ -0.2113 \end{bmatrix}.$$

To compute the minimum mean-squared error from (3.63) we need σ_d^2. This is found using (1.75) as

$$\sigma_d^2 = \frac{\sigma_v^2}{1 - .5^2}.$$

Then

$$\sigma_e^2 = 0.0826.$$

The error surface is obtained by plotting (see (3.14))

$$J(\mathbf{h}) = \sigma_d^2 - 2\mathbf{p}^T \begin{bmatrix} h[0] \\ h[1] \end{bmatrix} + [h[0], h[1]]R \begin{bmatrix} h[0] \\ h[1] \end{bmatrix}$$

3.12 Comparison of Least-Squares and Minimum Mean-Squares

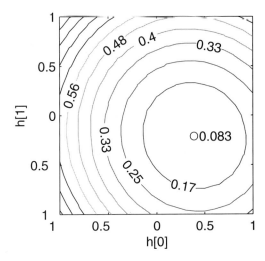

Figure 3.12: Contour plot of an error surface

as a function of $\{h[0], h[1]\}$. Figure 3.12 shows a contour plot of the error surface. Algorithm 3.3 is MATLAB code demonstrating these computations.

Algorithm 3.3 Two-tap channel equalizer
File: `wftest.m`

Another example of MMSE filter design is given in conjunction with the RLS filter in 4.11.2.

3.12 Comparison of least-squares and minimum mean-squares

It is interesting to contrast the method of least-squares and the method of minimum mean squares, both of which are widely used in signal processing. For the method of least-squares, we make the following observations:

1. Only the sequence of data observed at the time of the estimate is used in forming the estimate.
2. Depending upon assumptions made about the data before and after the observation interval, the Grammian matrix may not be Toeplitz.
3. No statistical model is necessarily assumed.

For the method of minimum mean-squares, we make the following observations:

1. A statistical model for the correlations and cross-correlations is necessary. This must be obtained either from explicit knowledge of the channel and signal (as was seen in example 3.11.1), or on the basis of the multivariable distribution of the data (as was seen in example 3.10.1). In the absence of such knowledge, it is common to estimate the necessary autocorrelation and cross-correlation values. An example of an estimate of the autocorrelation $r(n) = E[x(k)\bar{x}(k-n)]$ using the data $\{x(1), x(2), \ldots, x(N)\}$

is

$$\hat{r}(n) = \frac{1}{N} \sum_{k=1+n}^{N} x(k)\bar{x}(k-n). \tag{3.65}$$

This is actually a biased estimate of $r(n)$ (see exercise 3.12-25), but it has been found (see, e.g., [38]) to produce a lower variance when the lag n is close to N.

In order for (3.65) to be a reasonable estimate of $r(n)$, the random process $x(k)$ must be *ergodic*, so that the time average asymptotically approaches the ensemble average. This assumption of ergodicity is usually made tacitly, but it is vital.

When the data sequence used to compute the estimate of the correlations' parameters is the same as the data sequence for which the filter coefficients are computed, the minimum mean-squared error technique is essentially the same as the least-squares technique.

2. Commonly, the coefficients of the MMS technique are computed using a separate set of data whose statistics are *assumed* to be the same as those of the real data set of interest. This set of data is used as a *training set* to find the autocorrelation functions and the filter coefficients. Provided that the training data does have the same (or very similar) statistics as the data set of interest, this works well. However, if the training data is significantly different from the data set of interest, finding the optimum filter coefficients can actually lead to poor performance, because one has found the best solution to the wrong problem.

3. We also note that the (true) Grammian matrix R used in prediction and optimal FIR filtering problems is always a Toeplitz matrix, and hence fast algorithms apply to finding the coefficients.

In section 4.11.1 we examine how the coefficients of the LS filter can be updated adaptively, so that the coefficients are modified as new data arrives. In section 14.6, we develop an algorithm so that the coefficients of the MMS filter can be updated adaptively by approximating the expectation. These two concepts form the heart of adaptive filtering theory.

3.13 Frequency-domain optimal filtering

We have seen several examples of FIR minimum mean-squared filters, in which the equations obtained involve a finite number of unknowns. In this section, we take a different viewpoint, and develop optimal filtering techniques for scalar signals in the frequency domain. This allows us to extend the minimum mean-squared error filters of section 3.11 to IIR filters. Following a brief review of stochastic processes and their processing by linear systems, we present the notion of two-sided Laplace transforms, and some decompositions of these that are critical to the solution of the Wiener filter equations. This is followed by the development of the continuous-time Wiener filter. Finally, we present analogous results for discrete-time Wiener filters.

3.13.1 Brief review of stochastic processes and Laplace transforms

To expedite our development of frequency-domain filtering, it will be helpful to review briefly some fundamental results from stochastic processes associated with linear systems (see also appendix D).

3.13 Frequency-Domain Optimal Filtering

Power spectral density functions and filtering stochastic processes

Let $\{x_t, -\infty < t < \infty\}$ and $\{y_t, -\infty < t < \infty\}$ be two wide-sense stationary, zero-mean, scalar stochastic processes. Throughout this development, we will assume that all processes are real. The *auto- and cross-correlation* functions are

$$R_x(t) = E x_{\alpha+t} x_\alpha \qquad R_y(t) = E y_{\alpha+t} y_\alpha$$

$$R_{xy}(t) = E x_{\alpha+t} y_\alpha \qquad R_{yx}(t) = E y_{\alpha+t} x_\alpha.$$

The bilateral Laplace transforms of these functions are denoted by

$$S_x(s) = \int_{-\infty}^{\infty} R_x(t) e^{-st} dt \qquad S_y(s) = \int_{-\infty}^{\infty} R_y(t) e^{-st} dt$$

$$S_{xy}(s) = \int_{-\infty}^{\infty} R_{xy}(t) e^{-st} dt \qquad S_{yx}(s) = \int_{-\infty}^{\infty} R_{yx}(t) e^{-st} dt,$$

where $s = \sigma + j\omega$ is a complex variable. These bilateral Laplace transforms exist whenever s is in the region of convergence. For all of our applications, the region of convergence will include the imaginary axis, and we may obtain the Fourier transform of these functions by restricting s to the imaginary axis, that is, setting $s = j\omega$. The resulting function, $S_x(j\omega)$, etc., is the usual *power spectral density* function. By an abuse of notation, we will usually drop the explicit inclusion of the imaginary unit in the argument, and simply refer to the power spectral density as $S_x(\omega)$, and so on.

We observe that, since the autocovariance is real and even, its bilateral Laplace transform is even; that is,

$$S_x(s) = S_x(-s).$$

Furthermore, when $s = j\omega$, the power spectral density has the property

$$S_x(-\omega) = S_x^*(\omega).$$

Filtering of stochastic processes

Let h_t be the impulse response function of a time-invariant linear system Laplace transform $H(s)$. We will be concerned (as usual) mainly with causal systems, in which $h(t) = 0$ for $t < 0$.

Let y_t be the output of a system with impulse response driven by the wide-sense stationary stochastic process $\{x_t, -\infty < t < \infty\}$. The output of this system, denoted $\{y_t, -\infty < t < \infty\}$, is also a wide-sense stationary stochastic processes. The correlation functions $R_{xy}(\tau)$, $R_{yx}(\tau)$ and $R_y(\tau)$ are given by

$$R_{xy}(\tau) = R_x(\tau) * h(-\tau) \qquad R_{yx}(\tau) = h(\tau) * R_x(\tau)$$

$$R_y(\tau) = h(\tau) * R_x(\tau) * h(-\tau).$$

The equivalent relationships in the spectral domain are

$$S_{xy}(s) = S_x(s) H(-s) \qquad S_{yx}(s) = H(s) S_x(s)$$

$$S_y(s) = H(s) S_x(s) H(-s).$$

Lumped systems and processes

A linear system is said to be *lumped* if it has a *rational* transfer function; that is, its transfer function is a ratio of polynomials in s. Thus, if $G(s)$ is a rational transfer function, then it is of the form

$$G(s) = \frac{\prod_{i=1}^{m}(s - z_i)}{\prod_{i=1}^{n}(s - p_i)},$$

where z_i and p_i are the roots of the numerator (the zeros) and the denominator (the poles), respectively. We require that $n \geq m$.

A stochastic process is said to be *lumped* if its power spectral density is a rational function. Let $\{y_t\}$ be a lumped stochastic process. Its spectral density function, $S_y(\omega)$, is even and nonnegative. We will sometimes refer to $S_y(s)$ as the power spectral density, although the nonnegativeness only holds for $s = j\omega$. The evenness and nonnegativeness of $S_y(\omega)$, however, means that the poles and zeros of $S_y(s)$ have a particular quadrantal symmetry:

- The poles and zeros are symmetric about the real axis of the complex plane, because $S_y(\omega)$ is real.
- The poles and zeros are symmetric about the imaginary axis of the complex plane, because $S_y(\omega)$ is even.
- The imaginary axis of the complex plane has zeros of even multiplicity, because $S_y(\omega)$ is nonnegative.
- There are no poles on the imaginary axis, because the inverse Fourier transform cannot be a covariance function.

Figure 3.13 illustrates the pole–zero structure of a rational power spectral density function.

The region of convergence for a stable inverse of $S_y(s)$ is a strip in the complex plane containing the $j\omega$ axis. The inverse Laplace transform of $S_y(s)$ is of the form

$$R_y(t) = \sum_{i=1}^{n} c_i e^{-d_i |t|},$$

which is a sum of damped exponentials for positive as well as for negative t. If any coefficient d_i is complex, then its complex conjugate, d_i^*, must also be one of the coefficients. Purely imaginary d_i are excluded since R_y must be a correlation function. The coefficients c_i must be real.

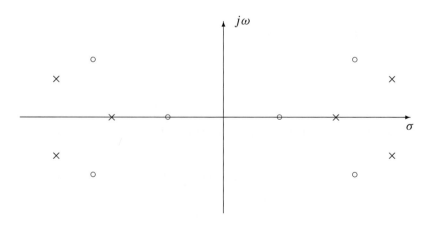

Figure 3.13: Pole–zero plot of rational $S_y(s)$ ($\times$ = poles, $\circ$ = zeros).

3.13.2 Two-sided Laplace transforms and their decompositions

The one-sided Laplace transform should be familiar to students of signal processing. Less familiar, but applicable to our current study, is the two-sided, or bilateral, Laplace transform, defined as

$$F(s) = \mathcal{L}_{\text{two-sided}}(f(t)) = \int_{-\infty}^{\infty} f(t)e^{-st}\, dt.$$

Of course, for a causal function $f(t)$, the bilateral transform is equivalent to the one-sided transform.

Like the two-sided Z-transform (which should be somewhat more familiar), different inverses of a given function $F(s)$ can be obtained depending upon the region of convergence that is selected. We make the following summarizing observations, where $f(t)$ and $F(s)$ are Laplace transform pairs.

1. If the region of convergence includes the $j\omega$ axis, then the inverse transform $f(t)$ is *stable*.
2. If the region of convergence is to the *right* of all poles of $F(s)$, then the inverse $f(t)$ is *causal*. That is, the region of convergence is a region of the form $\mathrm{Re}(s) > \mathrm{Re}(p)$, for all poles p of $F(s)$. Conversely, if $f(t)$ is a causal, stable function, then there are no poles in the RHP.
3. If the region of convergence is to the *left* of all poles of $F(s)$, then the inverse $f(t)$ is *anticausal*. Conversely, if $f(t)$ is an anticausal, stable function, then there are no poles in the RHP.
4. If the region of convergence is neither to the right nor to the left of all of the poles, the inverse transform is two-sided.

Some simple examples will demonstrate these concepts.

Example 3.13.1 1. The transform $F(s) = 1/(s+\alpha)$, $\alpha > 0$, has its poles in the LHP, and the region of convergence to the right of the poles contains the $j\omega$ axis, indicating that $f(t)$ is stable. In fact, the inverse (one-sided) Laplace transform is $f(t) = e^{-\alpha t}u(t)$, a stable, causal function.

2. Let $f(t) = -e^{\alpha t}u(-t)$. Then the two-sided transform of $F(s)$ is

$$F(s) = \frac{1}{s - \alpha},$$

with region of convergence $\mathrm{Re}(s) < \mathrm{Re}(\alpha)$. If $\mathrm{Re}(\alpha) > 0$, then $f(t)$ is stable, and $F(s)$ has no poles in the LHP.

3. Let

$$F(s) = \frac{2\alpha}{\alpha^2 - s^2} = \frac{1}{s + \alpha} - \frac{1}{s - \alpha}.$$

The region of convergence of $F(s)$ containing the $j\omega$ axis has poles both to the right and to the left, hence the inverse using this region of convergence is stable, but not causal. In fact, it can be verified that the inverse corresponding to this region of convergence is $f(t) = e^{-\alpha|t|}$.

4. Suppose

$$F(s) = \frac{1}{\alpha + s}e^{s\lambda}, \qquad \alpha > 0,\ \lambda > 0.$$

This is *not* the transfer function of a lumped system. Let the region of convergence be $\mathrm{Re}(s) > -\alpha$, which includes the $j\omega$ axis and hence is stable. The stable inverse transform

of $F(s)$ is
$$f(t) = e^{-\alpha(t+\lambda)} u(t+\lambda) = e^{-\alpha\lambda} e^{\alpha t} u(t+\lambda),$$
which is not causal. The causal portion of this function is
$$f(t)u(t) = e^{-\alpha\lambda} e^{\alpha t} u(t).$$
This causal function has Laplace transform
$$f(t)u(t) \leftrightarrow \frac{e^{-\alpha\lambda}}{s+\alpha}.$$
□

Canonical factorizations

Let $\{z_1, \ldots, z_m\}$ be the LHP zeros of a lumped system $F(s)$, and let $\{p_1, \ldots, p_m\}$ be the LHP poles of $F(s)$ for some Laplace transform function $F(s)$. We may then express $F(s)$ as
$$F(s) = F^+(s) F^-(s), \tag{3.66}$$
where
$$F^+(s) = \frac{\prod_{i=1}^m (s - z_i)}{\prod_{i=1}^n (s - p_i)}.$$

Then, since the $j\omega$ axis is to the right of all the poles of $F^+(s)$, the stable inverse Laplace transform of $F^+(s)$ is causal.

For a power spectral density $S_y(s)$ with LHP zeros and poles $\{z_1, \ldots, z_m\}$ and $\{p_1, \ldots, p_m\}$, respectively, the zeros and poles occur in mirror images, so that $S_y(s)$ has the canonical factorization
$$S_y(s) = S_y^+(s) S_y^-(s),$$
where
$$S_y^+(s) = \frac{\prod_{i=1}^m (s - z_i)}{\prod_{i=1}^n (s - p_i)} \qquad S_y^-(s) = S_y^+(-s).$$

$S_y^+(s)$ is often called the *canonical spectral factor* of $S_y(s)$. Since $S_y^+(s)$ has all of its poles and zeros in the LHP, its reciprocal, $W(s) = \frac{1}{S_y^+(s)}$, also has its poles and zeros in the LHP. Functions that have both poles and zeros in the left-half plane are said to be of *minimum phase*, and such functions may be viewed as transfer functions of causal systems that possess the property that their inverse is also causal. Thus, we may view $\{y_t\}$ as the output of a linear system with transfer function $S_y^+(s)$, driven by a white noise, $\{v_t\}$, as illustrated in figure 3.14. We take the spectral density of the white noise process $\{v_t\}$ to be unity ($S_v(s) = 1$), so the spectral density of $\{y_t\}$ is
$$S_y(s) = S_y^+(s) S_v(s) S_y^+(-s) = S_y^+(s) S_y^-(s),$$
which agrees with the canonical factorization (3.66).

Since $S_y^+(s)$ is causally invertible, we may also view $\{v_t\}$ as the output of a causal and causally invertible linear system with transfer function $\frac{1}{S_y^+(s)}$, driven by $\{y_t\}$, as illustrated in figure 3.15.

The relationship between $\{y_t\}$ and $\{v_t\}$ is very important. Since the transfer function $S_y^+(s)$ is causal, we can obtain v_t from $\{y_\alpha, \alpha < t\}$; and since the transfer function $\frac{1}{S_y^+(s)}$ is causal, we can obtain y_t from $\{v_\alpha, \alpha < t\}$. Thus, $\{y_\alpha, \alpha < t\}$ and $\{v_\alpha, \alpha < t\}$ contain exactly the same information—nothing is lost or destroyed as a result of the filtering operations.

3.13 Frequency-Domain Optimal Filtering

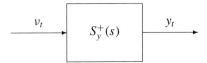

Figure 3.14: y_t as the output of a linear system driven by white noise

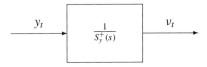

Figure 3.15: v_t as the output of a linear system driven by y_t.

We will say that two signals that enjoy this relationship are *informationally equivalent*.[1] The main difference between the two processes is that, while y_t may be dependent on $\{y_\alpha, \alpha < t\}$, v_t is *not* dependent on $\{v_\alpha, \alpha < t\}$, v_t. In other words, $\{y_t\}$ is a correlated process, and $\{v_t\}$ is an uncorrelated process. The action of filtering by $\frac{1}{S_y^+(s)}$ is to decorrelate $\{y_t\}$ by, essentially, removing all redundant information (that is, the part y_t that can be obtained as a function of y_α for $\alpha < t$) from y_t at each time t. The process $\{v_t\}$ is called the *innovations* process, and contains only new information about y_t that cannot be predicted from past values. Since $v(t)$ is a white-noise signal, we say that the filter $W(s) = \frac{1}{S_y^+(s)}$ is a whitening filter. The process $\{v_t\}$ is a very special white-noise process, since it represents exactly the same information as is contained in the original signal.

Additive decompositions

Let $f(t)$ be any function whose bilateral Laplace transform, $\mathcal{L}\{f(t)\}$, exists in a region containing the $j\omega$ axis. The auto- and cross-correlation functions associated with lumped processes and transfer functions of lumped systems all satisfy this constraint. We may decompose $f(t)$ into its left- and right-hand components

$$f(t) = f(t)u(t) + f(t)u(-t) - f(0),$$

where

$$u(t) = \begin{cases} 1 & t \geq 0 \\ 0 & t < 0 \end{cases}$$

is the unit step function. The bilateral Laplace transform of $f(t)$, denoted

$$F(s) = \mathcal{L}\{f(t)\} = \int_{-\infty}^{\infty} f(t)e^{-st}dt,$$

may be decomposed into

$$F(s) = \{F(s)\}_+ + \{F(s)\}_-,$$

where

$$\{F(s)\}_+ = \mathcal{L}\{f(t)u(t)\} = \int_{0-}^{\infty} f(t)e^{-st}dt$$

---

[1] This notion of information is not the same as either Shannon information or Fisher information.

is the Laplace transform of the causal part of $f(t)$, and

$$\{F(s)\}_- = \mathcal{L}\{f(t)u(-t)\} = \int_{-\infty}^{0-} f(t)e^{-st}dt$$

is the Laplace transform of the anticausal part of $f(t)$. Here $0-$ signifies taking a left-hand limit (this is necessary to account for impulsive autocorrelation functions).

Since $f(t)u(t)$ is a right-sided function, the region of convergence for its Laplace transform includes the RHP; that is, the transform $\{F(s)\}_+$ has no poles in the RHP. Similarly, since $f(t)u(-t)$ is an anticausal (left-sided) function, the region of convergence for its Laplace transform includes the LHP. Thus, for rational $F(s)$,

$$\{F(s)\}_+ = \mathcal{L}\{\text{impulsive functions}\} + \sum_{\text{LHP poles}} \{\text{partial fraction expansion of } F(s)\},$$

$$\{F(s)\}_- = \sum_{\text{RHP poles}} \{\text{partial fraction expansion of } F(s)\}.$$

Despite the confusing notation, the canonical factorization and the additive decomposition should not be confused: the canonical factorization is a *multiplicative* decomposition. For a function $F(s)$, we have the canonical factorization

$$F(s) = F^+(s)F^-(s),$$

and the additive decomposition

$$F(s) = \{F(s)\}_+ + \{F(s)\}_-,$$

and it is *not* the case that $F^+(s) = \{F(s)\}_+$. What is true is that they both have poles only in the RHP, and causal inverse transforms. (Note that the canonical factorization places the $+$ and $-$ in the exponent, while the additive decomposition placed the $+$ and $-$ in the subscript.)

Example 3.13.2 Let

$$f(t) = e^{-\alpha|t|},$$

where $\alpha > 0$. Then

$$F(s) = \frac{2\alpha}{\alpha^2 - s^2} = \frac{1}{s + \alpha} - \frac{1}{s - \alpha}.$$

This has the canonical factorization

$$F(s) = \frac{\sqrt{2\alpha}}{s + \alpha} \frac{\sqrt{2\alpha}}{\alpha - s} = F^+(s)F^-(s).$$

The causal part of $f(t)$ is $f(t)u(t)$, which has Laplace transform

$$\{F(s)\}_+ = \frac{1}{s + \alpha}, \qquad \text{Re}(s) > -\alpha,$$

leading to the additive decomposition

$$F(s) = \frac{1}{s + \alpha} - \frac{1}{s - \alpha} = \{F(s)\}_+ + \{F(s)\}_-. \qquad \square$$

Example 3.13.3 Let $S_y^-(s)$ be the canonical factor with its poles and zeros in the RHP, of the form

$$S_y^-(s) = \frac{s^2 - 3s + 2}{s^2 - 7s + 12} = 1 + \frac{4s - 10}{s^2 - 7s + 2}.$$

3.13 Frequency-Domain Optimal Filtering

We desire to find $\{S_y^-(s)\}_+$, the transform due to the causal part of the inverse Laplace transform of $S_y^-(s)$. We first find

$$\mathcal{L}^{-1}(S_y^-(s)) = \delta(t) - 6e^{4t}u(-t) + 2e^{3t}u(-t),$$

so that the "causal" part is simply $\delta(t)$. Taking the transform of the causal part, we thus have

$$\{S_y^-(s)\}_+ = 1. \qquad \square$$

As this last example shows, $\{S_y^-(s)\}_+$ may have a nonzero part. In fact, if $S_y^-(s)$ is rational,

$$S_y^-(s) = \frac{\beta(s)}{\alpha(s)}$$

for polynomials $\beta(s)$ and $\alpha(s)$ *of equal degree*, then with a little thought, we realize that

$$\{S_y^-(s)\}_+ = 1. \tag{3.67}$$

3.13.3 The Wiener–Hopf equation

Let x_t and y_t be zero-mean, stationary stochastic processes, and let $\mathcal{Y}_t = \{y_\alpha, \alpha \leq t\}$ be observed. Suppose we wish to estimate $x_{t+\lambda}$, given $\mathcal{Y}_t$. If $\lambda > 0$, we wish to predict future values of x_t given past and present values of y_t. This is called the *prediction* problem. If $\lambda = 0$, we wish to estimate x_t in real time; this is called the *filtering* problem. If $\lambda < 0$, we wish to estimate the signal λ time units in the past; this is called the *smoothing* problem. The prediction and filtering problems are *causal*, and can be implemented in real time, while the smoothing problem is *noncausal*, and cannot be implemented in real time.

We first formulate the integral

$$\hat{x}_{t+\lambda} = \int_{-\infty}^{t} h(t,s) y_s \, ds, \tag{3.68}$$

where $h(t,s)$ is to be chosen such that

$$E(x_{t+\lambda} - \hat{x}_{t+\lambda})^2 \tag{3.69}$$

is minimized and $h(t,s)$ is causal (that is, $h(t,s) = 0$ for $t < s$). The integral in (3.68) is to be taken in the mean-square sense. We address this problem by appealing to the **orthogonality principle**, as we have done so many times, whereby we require

$$E[(x_{t+\lambda} - \hat{x}_{t+\lambda}) y_\sigma] = 0, \qquad \forall \sigma \leq t; \tag{3.70}$$

that is, the *estimation error must be perpendicular to all data used to generate the estimate*. This condition implies that

$$E x_{t+\lambda} y_\sigma = E \hat{x}_{t+\lambda} y_\sigma = \int_{-\infty}^{t} h(t,\tau) E y_\tau y_\sigma \, d\tau, \qquad \forall \sigma \leq t,$$

or

$$R_{xy}(t + \lambda - \sigma) = \int_{-\infty}^{t} h(t,\tau) R_y(\tau - \sigma) \, d\tau, \qquad \forall \sigma \leq t.$$

We can render this expression more simply by making some changes of variable. First, let $\alpha = t - \tau$; then

$$R_{xy}(t + \lambda - \sigma) = \int_{0}^{\infty} h(t, t - \alpha) R_y(t - \alpha - \sigma) \, d\alpha, \qquad \forall \sigma \leq t.$$

Next, let $\xi = t - \sigma$, to obtain

$$R_{xy}(\xi + \lambda) = \int_{0-}^{\infty} h(\xi + \sigma, \xi + \sigma - \alpha) R_y(\xi - \alpha) \, d\alpha, \qquad \forall \xi \geq 0.$$

Since the left-hand side of this expression is independent of σ, the right-hand side must also be independent of σ, which in turn implies that h is not a function of σ. The only way this can happen is if h is a function of the difference of its first and second arguments; that is, if h is a function of α only. So, we introduce the (abuse of) notation $h(z_1, z_2) = h(z_1 - z_2)$ and, reverting back to t as the independent variable, we obtain

$$R_{xy}(t + \lambda) = \int_{0-}^{\infty} h(\tau) R_y(t - \tau) d\tau, \qquad \forall t \geq 0, \tag{3.71}$$

the celebrated *Wiener–Hopf* equation.

Equation (3.71), describing the solution of an optimal filter in continuous time, should be contrasted with (3.56) of chapter 3. In chapter 3, a set of matrix equations is obtained, whereas in the present case an integral equation is obtained. However, the structure in both cases is equivalent: the optimal filter coefficients are operated on by the autocorrelation of the input function to obtain the cross-correlation between the input and output.

Once the Wiener–Hopf equation is solved for h, then

$$\hat{x}_{t+\lambda} = \int_{-\infty}^{t} h(t - \tau) y_\tau \, d\tau \tag{3.72}$$

represents the minimum mean-square estimate of $x_{t+\lambda}$. Solving (3.71), however, involves more than simply taking Fourier or even bilateral Laplace transforms. To see why this is so, take the Laplace transform of both sides of (3.71):

$$\int_{0-}^{\infty} R_{xy}(t + \lambda) e^{-st} dt = \int_{0-}^{\infty} \int_{0-}^{\infty} h(\tau) R_y(t - \tau) e^{-s(t-\tau)} e^{-s\tau} d\tau \, dt$$

$$= \int_{0-}^{\infty} e^{-s\tau} h(\tau) \int_{0-}^{\infty} R_y(t - \tau) e^{-s(t-\tau)} dt \, d\tau$$

$$= \int_{0-}^{\infty} e^{-s\tau} h(\tau) \int_{-\tau}^{\infty} R_y(\sigma) e^{-s\sigma} d\sigma \, d\tau, \tag{3.73}$$

where we make the change of variable $\sigma = t - \tau$ for the last integral. We observe that the right-hand side of (3.73) is *not* equal to the product of the Laplace transforms of h and R_y, since the limits of the inner integral depend on τ. This condition arises from the requirement that $t > 0$ in (3.71). If we did not worry about physical realizability (that is, causality), we could relax the condition that $h(t) = 0$ for $t < 0$. In *this* case only, we may obtain, via Fourier analysis, the result that the optimal filter transfer function is given by

$$H(\omega) = \frac{S_{xy}(\omega)}{S_y(\omega)}; \tag{3.74}$$

the resulting impulse response function is noncausal unless x_t and y_t are white. For applications where causality is not a constraint, this result is perfectly valid. For example, let x_t be an image (here, t represents spatial coordinates), and suppose we observe

$$y_t = x_t + v_t,$$

where $\{v_t, -\infty < t < \infty\}$ is a white-noise process with $R_v(t) = \sigma^2 \delta(t)$. It is easy to see that $R_y(\tau) = R_x(\tau) + \sigma^2 \delta(\tau)$ and $R_{xy}(\tau) = R_x(\tau)$, so

$$H(\omega) = \frac{S_x(\omega)}{S_x(\omega) + \sigma^2}.$$

This result admits a very intuitive interpretation. Over frequencies where the signal energy is high compared to the noise, the filter acts as an identity filter and passes the signal without change. Over frequencies where the noise power dominates, the signal filter attenuates the observation.

3.13 Frequency-Domain Optimal Filtering

In some contexts, (3.74) is called a Wiener filter, but that is not quite accurate. More precisely, the Wiener filter is the solution to (3.71), and more sophistication is needed to obtain that solution. The solution comes via the celebrated Wiener–Hopf technique.

As we examine (3.71), we observe that we could solve this equation with transform techniques if $R_y(\sigma) = 0$ for $\sigma < 0$. Unfortunately, since R_y is a correlation function, this situation generally will not occur. One notable and important situation in which this does occur, however, is when $\{y_t\}$ is a white-noise process, for then $R_y(t) = \delta(t)$. In this case, the solution to (3.71) is trivial:

$$R_{xy}(t+\lambda) = \int_0^\infty h(\tau)\delta(t-\tau)\,d\tau = h(t), \qquad \forall t \geq 0, \tag{3.75}$$

so

$$h(t) = \begin{cases} R_{xy}(t+\lambda) & t \geq 0, \\ 0 & t < 0. \end{cases} \tag{3.76}$$

3.13.4 Solution to the Wiener–Hopf equation

We will present two approaches to the solution of the Wiener-Hopf equation. The first is based upon careful consideration of the locations of poles. The second is based upon the innovations representation of a process. The second is easier, pointing out again the utility of placing signals in the proper coordinate frame (i.e., a set of orthogonal functions).

Theorem 3.3 *The solution to the Wiener–Hopf equation,*

$$R_{xy}(t+\lambda) = \int_{0-}^\infty h(\tau)R_y(t-\tau)\,d\tau, \qquad \forall t \geq 0, \tag{3.77}$$

where

$$h(t) = 0, \qquad t < 0,$$

is

$$H(s) = \frac{1}{S_y^+(s)} \left\{ \frac{S_{xy}(s)e^{s\lambda}}{S_y^-(s)} \right\}_+. \tag{3.78}$$

Proof We first observe that since $h(t)$ is to be stable and causal, its bilateral Laplace transform will have no poles in the RHP. The transform of $R_{xy}(t)$ is

$$S_{xy}(s) = \int_{-\infty}^\infty R_{xy}(t)e^{-st}\,dt.$$

Consequently,

$$\begin{aligned} S_{xy}(s)e^{s\lambda} &= e^{s\lambda} \int_{-\infty}^\infty R_{xy}(\tau)e^{-s\tau}\,d\tau \\ &= \int_{-\infty}^\infty R_{xy}(\tau)e^{-s(\tau-\lambda)}\,d\tau \\ &= \int_{-\infty}^\infty R_{xy}(t+\lambda)e^{-st}\,dt, \end{aligned} \tag{3.79}$$

where we have made the change of variable $t = \tau - \lambda$ in the last integral.

Let
$$g(t) = R_{xy}(t+\lambda) - \int_{0-}^{\infty} h(\tau) R_y(t-\tau)\, d\tau.$$

From (3.77), the right-hand side of this equation is zero for $t \geq 0$, so
$$g(t) = \begin{cases} 0 & t \geq 0, \\ \text{unknown} & t < 0. \end{cases}$$

We will establish our result by examining the bilateral Laplace transform of $g(t)$. Since $g(t)$ is an anticausal (left-sided) function, its region of convergence is the LHP; consequently $G(s)$ has no poles in the LHP. Taking the bilateral Laplace transform of $g(t)$ and using (3.79),
$$G(s) = S_{xy}(s) e^{s\lambda} - H(s) S_y(s).$$

Now, observing that $S_y(s) = S_y^+(s) S_y^-(s)$ (canonical factorization) and dividing both sides of this equation by $S_y^-(s)$, we obtain
$$\frac{G(s)}{S_y^-(s)} = \frac{S_{xy}(s) e^{s\lambda}}{S_y^-(s)} - H(s) S_y^+(s). \qquad (3.80)$$

Since $G(s)$ has no LHP poles and $S_y^-(s)$ has no LHP zeros, $\frac{G(s)}{S_y^-(s)}$ has no LHP poles. Furthermore, $H(s)$ has no RHP poles, and neither does $S_y^+(s)$, so the product $H(s) S_y^+(s)$ has no RHP poles. The quantity $\frac{S_{xy}(s) e^{s\lambda}}{S_y^-(s)}$, however, may have poles in both the RHP and the LHP. The only way equality can obtain is for the LHP poles of $\frac{S_{xy}(s) e^{s\lambda}}{S_y^-(s)}$ to be equal to the poles of $H(s) S_y^+(s)$. Let $\phi(t)$ be the inverse Laplace transform of $\frac{S_{xy}(s) e^{s\lambda}}{S_y^-(s)}$; that is,
$$\phi(t) = \frac{1}{2\pi j} \int_{-j\infty}^{j\infty} \frac{S_{xy}(s) e^{s\lambda}}{S_y^-(s)} e^{st}\, ds.$$

$\phi(t)$ will, in general, be a two-sided function. The LHP poles of the bilateral Laplace transform of $\phi(t)$, however, may be obtained by taking the Laplace transform of $\phi(t) u(t)$. In other words, applying the $\{\cdot\}_+$ operation to both sides of (3.80) yields
$$0 = \left\{ \frac{S_{xy}(s) e^{s\lambda}}{S_y^-(s)} \right\}_+ - H(s) S_y^+(s),$$

and, consequently,
$$H(s) = \frac{1}{S_y^+(s)} \left\{ \frac{S_{xy}(s) e^{s\lambda}}{S_y^-(s)} \right\}_+. \qquad \square$$

It may be useful to compare the causal Wiener filter,
$$H(s) = \frac{1}{S_y^+(s)} \left\{ \frac{S_{xy}(s) e^{s\lambda}}{S_y^-(s)} \right\}_+,$$

with the noncausal "Wiener filter,"
$$H(s) = \frac{S_{xy}(s)}{S_y(s)}.$$

We note that, except for the $\{\cdot\}_+$ operation, they are the same, so the structure is not so foreign as it might seem upon first exposure.

3.13 Frequency-Domain Optimal Filtering

Let v_t be a white-noise process, and let us use this process to form the estimate $\hat{x}_t + \lambda$. The second derivation of the Wiener filter is based upon two observations:

- The Wiener–Hopf equation is trivial to solve if the observed process is a white noise, since (repeating (3.75) and (3.76)),

$$R_{xy}(t+\lambda) = \int_0^\infty h(\tau)\delta(t-\tau)\,d\tau = h(t), \quad \forall t \geq 0,$$

so

$$h(t) = \begin{cases} R_{xy}(t+\lambda) & t \geq 0 \\ 0 & t < 0 \end{cases}.$$

- A stationary lumped process may be transformed into a white noise without loss of information by means of a causal and causally invertible transform (see figure 3.15).

These two observations permit us to adopt a two-step procedure: (a) first, we "prewhiten" the observed signal, $\{y_t\}$, to create an innovations process, $\{v_t\}$; (b) we then apply the trivial Wiener filter to the pre-whitened signal. In other words, the Wiener filter can be obtained by cascading the pre-whitening filter and the Wiener filter for white-noise observations, as illustrated in figure 3.16.

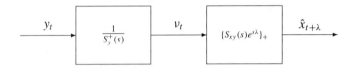

Figure 3.16: The optimal filter as the cascade of a pre-whitening filter and a Wiener filter with white-noise inputs

From our earlier development, the pre-whitening filter is simply $W(s) = \frac{1}{S_y^+(s)}$, the canonical spectral factor of $S_y(s)$, and the optimal filter based on white-noise observations is $\{S_{xv}(s)e^{s\lambda}\}_+$, so

$$H(s) = \frac{1}{S_y^+(s)} \left\{ \frac{S_{xv}(s)e^{s\lambda}}{S_v^-(s)} \right\}_+$$

$$= \frac{1}{S_y^+(s)} \left\{ S_{xv}(s)e^{s\lambda} \right\}_+,$$

since $S_v^-(s) = 1$.

The only thing left to compute is $\{S_{xv}(s)e^{s\lambda}\}_+$. But

$$R_{xv}(t) = Ex_{\alpha+t}v_\alpha = E\left[x_{\alpha+t}\int_{-\infty}^\infty w(\beta)y_{\alpha-\beta}\,d\beta\right]$$

$$= \int_{-\infty}^\infty w(\beta)E[x_{\alpha+t}y_{\alpha-\beta}]\,d\beta = \int_{-\infty}^\infty w(\beta)R_{xy}(t+\beta)\,d\beta = R_{xy}(t) * w(-t),$$

where $w(t)$ is the inverse transform of $W(s)$. Consequently,

$$S_{xv}(s) = S_{xy}(s)W(-s) = S_{xy}(s)\frac{1}{S_y^+(-s)} = S_{xy}(s)\frac{1}{S_y^-(s)},$$

and therefore

$$H(s) = \frac{1}{S_y^+(s)} \left\{ \frac{S_{xy}(s)e^{s\lambda}}{S_y^-(s)} \right\}_+,$$

which is the same formula we obtained with our original derivation.

3.13.5 Examples of Wiener filtering

Example 3.13.4 Suppose $x_t \equiv y_t$ and $\lambda > 0$ in (3.68). This is a problem of pure prediction: we wish to obtain an expression for $x_{t+\lambda}$, given $\{y_\alpha, \alpha \leq t\}$. Since $S_y(s) = S_x(s) = S_{xy}(s)$, (3.78) becomes

$$H(s) = \frac{1}{S_y^+(s)} \left\{ S_y^+(s) e^{s\lambda} \right\}_+.$$

Now let y_t be an *Ornstein–Uhlenbeck* process, which is a process $\{y_t, -\infty < t < \infty\}$ with zero-mean and with correlation function

$$R_y(t) = e^{-\alpha |t|},$$

with $\alpha > 0$. Then

$$S_y(s) = \frac{\sqrt{2\alpha}}{\alpha^2 - s^2} \qquad S_y^+(s) = \frac{\sqrt{2\alpha}}{\alpha + s}.$$

Our task is to compute $\{\frac{\sqrt{2\alpha}}{\alpha+s} e^{s\lambda}\}_+$. Since this transform is not a rational function of s, we cannot use partial fractions directly, and must appeal to the definition by finding the inverse Laplace transform, then taking the causal part:

$$F(s) = \frac{\sqrt{2\alpha} e^{s\lambda}}{\alpha + s},$$

and specifying the region of convergence as $\operatorname{Re} s > -\alpha$. Taking the (stable) inverse Laplace transform, we find

$$f(t) = \sqrt{2\alpha} e^{-\alpha \lambda} e^{-\alpha t} u(t + \lambda),$$

which is not causal. We find that the causal part of $f(t)$ is

$$f(t) u(t) = \sqrt{2\alpha} e^{-\alpha \lambda} e^{-\alpha t} u(t),$$

and so, taking transforms,

$$\{F(s)\}_+ = \frac{\sqrt{2\alpha} e^{-\alpha \lambda}}{s + \alpha}, \qquad \operatorname{Re} s > -\alpha.$$

Therefore,

$$H(s) = \frac{1}{S_y^+(s)} \left\{ S_y^+ e^{s\lambda} \right\}_+$$

$$= \frac{\alpha + s}{\sqrt{2\alpha}} \frac{\sqrt{2\alpha} e^{-\alpha \lambda}}{\alpha + s}$$

$$= e^{-\alpha \lambda},$$

so the impulse response function of the optimal filter is

$$h(t) = e^{-\alpha \lambda} \delta(t),$$

and the optimal predictor is

$$\hat{x}_{t+\lambda} = \int_{-\infty}^{\infty} e^{-\alpha \lambda} \delta(t - \tau) x_\tau d\tau$$

$$= e^{-\alpha \lambda} x_t.$$

Thus, the predicted value of x_t decays from its last observed value exponentially at a rate governed by the correlation time-constant. □

Example 3.13.5 **Filtering in White Noise**. Let

$$y_t = x_t + v_t,$$

3.13 Frequency-Domain Optimal Filtering

$$S_v(s) = 1,$$
$$S_x(s) = \frac{b(s^2)}{a(s^2)},$$

where $b(s^2)$ and $a(s^2)$ are polynomials in s^2 with the degree of $b(s^2)$ strictly **lower** than the degree of $a(s^2)$. Furthermore, assume $R_{xv}(t) \equiv 0$. Direct calculation yields

$$S_{xy}(s) = S_x(s),$$
$$S_y(s) = S_x(s) + 1,$$

so

$$H(s) = \frac{1}{S_y^+(s)} \left\{ \frac{S_{xy}(s)}{S_y^-(s)} \right\}_+$$
$$= \frac{1}{S_y^+(s)} \left\{ \frac{S_y(s) - 1}{S_y^-(s)} \right\}_+$$
$$= \frac{1}{S_y^+(s)} \left\{ S_y^+(s) - \frac{1}{S_y^-(s)} \right\}_+$$
$$= \frac{1}{S_y^+(s)} \left[\{S_y^+(s)\}_+ - \left\{ \frac{1}{S_y^-(s)} \right\}_+ \right].$$

Now, observe that

$$S_y(s) = \frac{b(s^2)}{a(s^2)} + 1 = \frac{a(s^2) + b(s^2)}{a(s^2)}.$$

Since the degree of $b(s^2)$ is lower than the degree of $a(s^2)$, the degrees of the numerator and denominator of $S_y(s)$ are the same, say of degree $2n$. The canonical factors of $S_y(s)$ will therefore be rational functions with numerators and denominators of degree n. Thus, $S_y^-(s)$ is of the form

$$S_y^-(s) = \frac{\beta(s)}{\alpha(s)} = 1 + \frac{\gamma(s)}{\alpha(s)},$$

since the leading coefficients of both $\beta(s)$ and $\alpha(s)$ are the same. Since the rational function $\frac{\gamma(s)}{\alpha(s)}$ has all of its poles in the RHP, we immediately obtain (see (3.67))

$$\left\{ \frac{1}{S_y^-(s)} \right\}_+ = 1. \tag{3.81}$$

Thus,

$$H(s) = \frac{1}{S_y^+(s)} \left[\{S_y^+(s)\}_+ - \left\{ \frac{1}{S_y^-(s)} \right\}_+ \right]$$
$$= 1 - \frac{1}{S_y^+(s)}. \tag{3.82}$$

$\square$

Example 3.13.6 As an application of the results of the previous problem, consider the case when

$$S_x(s) = \frac{s^2 - 9}{s^4 - 5s^2 + 4} = \frac{(s+3)(s-3)}{(s^2 + 3s + 2)(s^2 - 3s + 2)},$$

so that

$$S_y(s) = \frac{s^4 - 4s^2 - 5}{s^4 - 5s^2 + 4} = \frac{s^2 + (\sqrt{5}+1)s + \sqrt{5}}{s^2 + 3s + 2} \frac{s^2 - (\sqrt{5}+1)s + \sqrt{5}}{s^2 - 3s + 2} = S_y^+(s) S_y^-(s)$$

and, thus,
$$\frac{1}{S_y^+(s)} = \frac{s^2 - 3s + 2}{s^2 - (\sqrt{5}+1)s + \sqrt{5}} = 1 + \frac{\sqrt{5}-2}{s^2 - (\sqrt{5}+1)s + \sqrt{5}}.$$

The inverse Laplace transform of $1/S_y^-(s)$ is anticausal, except for the constant term 1, so that, as in (3.81), the portion due to the causal part is

$$\left\{\frac{1}{S_y^-(s)}\right\}_+ = 1 + \left\{\frac{\gamma(s)}{\alpha(s)}\right\}_+ = 1.$$

Then
$$H(s) = 1 - \frac{1}{S_y^+(s)} = \frac{(\sqrt{5}-2)s + \sqrt{5} - 2}{s^2 + (\sqrt{5}+1)s + \sqrt{5}}. \qquad \square$$

3.13.6 Mean-square error

The error associated with the Wiener filtering problem is given by
$$\tilde{x}_{t+\lambda} = x_{t+\lambda} - \hat{x}_{t+\lambda},$$
with
$$\hat{x}_{t+\lambda} = \int_{-\infty}^{t} h(t-\tau) y_\tau \, d\tau,$$
where
$$h(t) = \mathcal{L}^{-1}\{H(s)\},$$
which is the inverse Laplace transform of the optimal transfer function given by (3.78). The covariance of the estimation error is

$$E\tilde{x}_{t+\lambda}^2 = E[x_{t+\lambda} - \hat{x}_{t+\lambda}]^2 = E\left[x_{t+\lambda} - \int_{-\infty}^{t} h(t-\tau) y_\tau \, d\tau\right]^2. \qquad (3.83)$$

Since $\hat{x}_{t+\lambda}$ is a function of $\{y_\alpha, \alpha \leq t\}$, the orthogonality condition (3.70) requires that the estimation error be orthogonal to the estimate, that is,

$$E\left\{\left[x_{t+\lambda} - \int_{-\infty}^{t} h(t-\tau) y_\tau \, d\tau\right] \hat{x}_{t+\lambda}\right\} = 0,$$

so (3.83) becomes

$$E\tilde{x}_{t+\lambda}^2 = E\left\{\left[x_{t+\lambda} - \int_{-\infty}^{t} h(t-\tau) y_\tau \, d\tau\right] x_{t+\lambda}\right\}$$
$$= R_x(0) - \int_{-\infty}^{t} h(t-\tau) R_{yx}(\tau - t - \lambda) \, d\tau$$
$$= R_x(0) - \int_{-\infty}^{t} h(t-\tau) R_{xy}(t - \tau + \lambda) \, d\tau$$
$$= R_x(0) - \int_{0}^{\infty} h(\alpha) R_{xy}(\alpha + \lambda) \, d\alpha,$$

where the last equality holds by making the change of variable $\alpha = t - \tau$.

3.13.7 Discrete-time Wiener filters

The Wiener filter theory also applies in discrete time. We have already seen, the Wiener filter results for FIR filters. We now apply the notion of spectral factorization to the Wiener–Hopf equations with causal IIR filters. We summarize the results for this development.

3.13 Frequency-Domain Optimal Filtering

Canonical factorization

Let $S_y(z)$ be the power spectral density of a discrete-time random process. Then $S_y(z)$ has poles inside and outside the unit circle. The canonical factorization is

$$S_y(z) = S_y^+(z) S_y^-(z),$$

where $S_y^+(z)$ has all of its poles and zeros inside the unit circle.

Additive decomposition

Let $f[t], t = \ldots, -1, 0, 1, \ldots$, be a discrete-time function. Then

$$f[t] = f[t]u[t] + f[t]u[-k] - f[0],$$

where $u[t]$ is the discrete-time unit-step function. The Z-transform of $f[t]$ is

$$F(z) = \sum_{n=-\infty}^{\infty} f[n] z^{-n}$$

$$= \underbrace{\sum_{n=-\infty}^{-1} f[n] z^{-n}}_{\{F(z)\}_-} + \underbrace{\sum_{n=0}^{\infty} f[n] z^{-n}}_{\{F(z)\}_+}.$$

Wiener–Hopf equation

Let x_t and y_t be zero-mean, jointly stationary discrete-time stochastic processes. We wish to estimate $x_{t+\lambda}$, given $\{y_j, j \leq t\}$, with an estimator of the form

$$\hat{x}_{t+\lambda} = \sum_{i=-\infty}^{t} h[t-i] y_i, \qquad \lambda \text{ an integer } \geq 0,$$

where $h(i)$ is the solution (from orthogonality) to

$$R_{xy}[t+p] = \sum_{i=0}^{\infty} h[i] R_y[t-i], \qquad t \geq 0.$$

To solve this equation for h we follow the Wiener–Hopf technique of defining the function

$$g[t] = R_{xy}[t+\lambda] = \sum_{i=0}^{\infty} h[i] R_y[t-i], \qquad \text{all } t,$$

$$= \begin{cases} 0 & t \geq 0, \\ \text{unknown} & t < 0. \end{cases}$$

Since $g[t]$ is an anticausal function, its region of convergence is the interior of the unit circle—it has no poles within the unit circle.

Taking the bilateral Z-transform of $g[t]$,

$$G(z) = \sum_{n=-\infty}^{\infty} R_{xy}[n+\lambda]z^{-n} + H(z)S_y(z)$$

$$= z^\lambda \sum_{j=-\infty}^{\infty} R_{xy}(j)z^{-j} + H(z)S_y(z)$$

$$= z^\lambda S_{xy}(z) + H(z)S_y(z). \qquad (3.84)$$

The canonical spectral factorization of $S_y(z)$ is of the form

$$S_y(z) = S_y^+(z)S_y^-(z),$$

where $S_y^-(z)$ has poles and zeros outside the unit circle and $S_{+y}(z)$ has poles and zeros inside the unit circle. Dividing both sides of (3.84) by $S_y^-(z)$ and applying the $\{\cdot\}_+$ operation to both sides yields

$$H(z) = \frac{1}{S_y^+(z)} \left\{ \frac{z^\lambda S_{xy}(z)}{S_y^-(z)} \right\}_+,$$

the discrete-time Wiener filter.

Example 3.13.7 Let $\{y_t, -\infty < t < \infty\}$ be a discrete-time, zero-mean, wide-sense stationary process with correlation function

$$R_y[t] = \frac{4}{3}\left(\frac{1}{2}\right)^{|t|}.$$

Let $x_t = y_t$, and predict $x_{t+\lambda}$ for $\lambda \geq 0$.

We seek a predictor of the form

$$\hat{x}_{t+\lambda} = \sum_{i=-\infty}^{\infty} h[t-i]y_i, \qquad \lambda \geq 0.$$

We have $R_x[t] = R_{xy}[t] = R_y[t]$, and

$$S_y(z) = \sum_{n=-\infty}^{\infty} R_y[n]z^{-n} = \sum_{n=-\infty}^{-1} \frac{4}{3}\left(\frac{1}{2}\right)^{-n} z^{-n} + \sum_{n=0}^{\infty} \frac{4}{3}\left(\frac{1}{2}\right)^n z^{-n}$$

$$= \sum_{j=0}^{\infty} \frac{4}{3}\left(\frac{z}{2}\right)^j - \frac{4}{3} + \sum_{n=0}^{\infty} \frac{4}{3}\left(\frac{1}{2}\right)^n z^{-n} = \frac{4}{3}\left[\frac{1}{1-\frac{z}{2}} - 1 + \frac{1}{1-\frac{1}{2z}}\right]$$

$$= \underbrace{\frac{1}{1-\frac{z}{2}}}_{S_y^-(z)} \underbrace{\frac{1}{1-\frac{1}{2z}}}_{S_y^+(z)}.$$

Next, we calculate

$$\frac{z^\lambda S_{xy}(z)}{S_y^-(z)} = \frac{z^\lambda}{\left(1-\frac{z}{2}\right)\left(1-\frac{1}{2z}\right)} = \frac{2z^{\lambda+1}}{2z-1}.$$

By long division, we obtain

$$\frac{2z^{\lambda+1}}{2z-1} = z^\lambda + \frac{1}{2}z^{\lambda-1} + \frac{1}{4}z^{\lambda-2} + \cdots + \frac{1}{2^\lambda}z^0 + \frac{1}{2^{\lambda+1}}z^{-1} + \cdots$$

$$= \frac{1}{2^p}\sum_{n=-p}^{\infty} \frac{1}{2^n}z^{-n}.$$

3.14 A Dual Approximation Problem

We may obtain the inverse Z-transform as the coefficients of z^{-n}. The operation $\{\cdot\}_+$ is effected by discarding all samples before $n = 0$ and returning to the transform domain:

$$\left\{\frac{z^\lambda S_{xy}(z)}{S_y^-(z)}\right\}_+ = \frac{1}{2^\lambda}\sum_{n=0}^{\infty}\left(\frac{1}{2}\right)^n z^{-n} = \frac{1}{2^\lambda}\frac{1}{1-\frac{1}{2z}},$$

and

$$H(z) = \frac{1}{S_y^+(z)}\left\{\frac{z^\lambda S_{xy}(z)}{S_y^-(z)}\right\}_+ = \frac{1}{2^\lambda},$$

so

$$h[t] = \frac{1}{2^\lambda}\delta_t$$

and

$$\hat{x}_{t+\lambda} = \left(\frac{1}{2}\right)^\lambda y_t.$$

$\square$

3.14 A dual approximation problem

The approximation problems we have seen up till now have selected a point from a finite-dimensional subspace of the Hilbert space of the problem. In each case, because the solution was in a finite-dimensional subspace, solving an $m \times m$ system of equations was sufficient. In some approximation problems, the subspace in which the solution lies is not finite dimensional, so a simple finite set of equations cannot be solved to obtain the solution. There are some problems, however, in which a finite set of constraints provides us with sufficient information to solve the problem from a finite set of equations.

We begin with a definition.

Definition 3.2 Let M be a subspace of a linear space S, and let $x_0 \in S$. The set $V = x_0 + M$ is said to be a *translation* of M by x_0. This translation is called a **linear variety**. $\square$

A linear variety is not in general a subspace.

Example 3.14.1 Let $M = \{(0,0,0),(0,1,0)\}$ in the vector space $(GF(2))^3$ introduced in example 2.12.1, and let $\mathbf{x}_0 = (1,1,1) \in S$. Then

$$\mathbf{x} + M = \{(1,1,1),(1,0,1)\}$$

is a linear variety. $\square$

A version of the orthogonality theorem appropriate for linear varieties is illustrated in figure 3.17. Let $V = \mathbf{x}_0 + M$ be a closed linear variety in a Hilbert space H. Then there is a *unique* vector $\mathbf{v}_0 \in V$ of minimum norm. The minimizing vector $\mathbf{v}_0$ is orthogonal to M. This result is an immediate consequence of the projection theorem for Hilbert spaces (simply translate the variety and the origin by $-\mathbf{x}_0$).

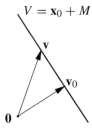

Figure 3.17: Minimum norm to a linear variety

Let S be a Hilbert space. Given a set of linearly independent vectors $\mathbf{y}_1, \mathbf{y}_2, \ldots, \mathbf{y}_m \in S$, let $M = \text{span}(\mathbf{y}_1, \mathbf{y}_2, \ldots, \mathbf{y}_m)$. The set of $\mathbf{x} \in S$ such that

$$\langle \mathbf{x}, \mathbf{y}_1 \rangle = 0$$
$$\langle \mathbf{x}, \mathbf{y}_2 \rangle = 0$$
$$\vdots$$
$$\langle \mathbf{x}, \mathbf{y}_m \rangle = 0$$

is a subspace, which (because of these inner-product constraints) must be $M^\perp$. Suppose now we have problem in which there are inner-product constraints of the form

$$\begin{aligned}
\langle \mathbf{x}, \mathbf{y}_1 \rangle &= a_1 \\
\langle \mathbf{x}, \mathbf{y}_2 \rangle &= a_2 \\
&\vdots \\
\langle \mathbf{x}, \mathbf{y}_m \rangle &= a_m.
\end{aligned} \qquad (3.85)$$

If we can find any point $\mathbf{x} = \mathbf{x}_0$ that satisfies the constraints in (3.14), then for any $\mathbf{v} \in M^\perp$, $\mathbf{x}_0 + \mathbf{v}$ also satisfies the constraints. Hence the space of solutions of (3.14) is the linear variety $V = \mathbf{x}_0 + M^\perp$. A linear variety V satisfying the m constraints in (3.14) is said to have *codimension m*, since the orthogonal complement of the subspace $M^\perp$ producing it has dimension m.

Example 3.14.2 In $\mathbb{R}^3$, let $\mathbf{y}_1 = (1, 0, 0)$ and $\mathbf{y}_2 = (0, 1, 0)$, and let $M = \text{span}(\mathbf{y}_1, \mathbf{y}_2)$. The set of points such that

$$\langle \mathbf{x}, \mathbf{y}_1 \rangle = 0 \qquad \langle \mathbf{x}, \mathbf{y}_2 \rangle = 0$$

is

$$\text{span}(0, 0, 1) = M^\perp.$$

Now, for the constraints

$$\langle \mathbf{x}, \mathbf{y}_1 \rangle = 3, \qquad \langle \mathbf{x}, \mathbf{y}_2 \rangle = 4.$$

observe that if $\mathbf{x} = (3, 4, s)$ for any $s \in \mathbb{R}$ then the constraints are satisfied. The set $V = (3, 4, 0) + M^\perp$ is a linear variety of codimension 2. $\square$

We are now in a position to state the minimization problem.

Theorem 3.4 *(Dual approximation) Let $\{\mathbf{y}_1, \mathbf{y}_2, \ldots, \mathbf{y}_m\}$ be linearly independent in a Hilbert space S, and let $M = \text{span}(\mathbf{y}_1, \ldots, \mathbf{y}_m)$. The element $\mathbf{x} \in S$ satisfying*

$$\begin{aligned}
\langle \mathbf{x}, \mathbf{y}_1 \rangle &= a_1 \\
\langle \mathbf{x}, \mathbf{y}_2 \rangle &= a_2 \\
&\vdots \\
\langle \mathbf{x}, \mathbf{y}_m \rangle &= a_m
\end{aligned} \qquad (3.86)$$

with minimum norm lies in M; specifically,

$$\mathbf{x} = \sum_{i=1}^m c_i \mathbf{y}_i,$$

3.14 A Dual Approximation Problem

where the coefficients in this linear combination satisfy

$$\begin{bmatrix} \langle y_1, y_1 \rangle & \langle y_2, y_1 \rangle & \cdots & \langle y_m, y_1 \rangle \\ \langle y_1, y_2 \rangle & \langle y_2, y_2 \rangle & \cdots & \langle y_m, y_2 \rangle \\ \vdots & & & \\ \langle y_1, y_m \rangle & \langle y_2, y_m \rangle & \cdots & \langle y_m, y_m \rangle \end{bmatrix} \begin{bmatrix} c_1 \\ c_2 \\ \vdots \\ c_m \end{bmatrix} = \begin{bmatrix} a_1 \\ a_2 \\ \vdots \\ a_m \end{bmatrix}. \tag{3.87}$$

Proof By the discussion above, the solution lies in the linear variety $V = \mathbf{x}_0 + M^\perp$ for some $\mathbf{x}_0$. Furthermore, the optimal solution $\mathbf{x}_0$ is orthogonal to $M^\perp$, so that $\mathbf{x}_0 \in M^{\perp\perp} = M$. Thus, $\mathbf{x}_0$ is of the form

$$\mathbf{x}_0 = \sum_{i=1}^m c_i \mathbf{y}_i.$$

Taking inner products of this equation with $\mathbf{y}_1, \mathbf{y}_2, \ldots, \mathbf{y}_m$, and recognizing that, for the solution, $\langle \mathbf{x}_0, \mathbf{y}_i \rangle = a_i$, we obtain the set of equations in (3.87). $\square$

Example 3.14.3 For the linear variety of the previous problem, let us find the solution of minimum norm. Using (3.87), we find $x = (3, 4, 0)$ to be the minimum norm solution satisfying the constraints. $\square$

Example 3.14.4 We examine here a problem in which the solution space is infinite dimensional. Suppose we have an LTI system with causal impulse response $h(t) = e^{-2t} + 3e^{-4t}$, in which the initial conditions are $y(0) = 0$ and $\dot{y}(0) = 0$. We desire to determine an input signal $x(t)$ so that the output $y(t) = x(t) * h(t)$ satisfies the constraints

$$y(1) = 1 \qquad \int_0^1 y(t)\,dt = 0$$

in such a way that the input energy $\int_0^1 |x(t)|^2\,dt$ is minimized. Writing the convolution integral for the first output, the first constraint can be written

$$\int_0^1 (e^{-2(1-\tau)} + 3e^{-4(1-\tau)}) x(\tau)\,d\tau = 1.$$

Using the inner product

$$\langle f, g \rangle = \int_0^1 f(\tau) g(\tau)\,dt,$$

the first constraint can be written as

$$\langle x, y_1 \rangle = 1,$$

where

$$y_1(\tau) = e^{-2(1-\tau)} + 3e^{-4(1-\tau)}.$$

The second constraint can be written using the integral of the impulse response (see exercise 14.27),

$$k(t) = \int_0^t h(\tau)\,dt = \frac{5}{4} - \frac{3}{4}e^{-4t} - \frac{1}{2}e^{-2t}.$$

Then the second constraint is

$$\langle x, y_2 \rangle = 0,$$

where
$$y_2(\tau) = \frac{5}{4} - \frac{3}{4}e^{-4(1-\tau)} - \frac{1}{2}e^{-2(1-\tau)}.$$

The solution $x_0(t)$ must lie in the space spanned by y_1 and y_2,
$$x_0 = c_1 y_1(t) + c_2 y_2(t).$$

Then the equation (3.87) becomes
$$\begin{bmatrix} \langle y_1, y_1 \rangle & \langle y_1, y_2 \rangle \\ \langle y_1, y_2 \rangle & \langle y_2, y_2 \rangle \end{bmatrix} \begin{bmatrix} a_1 \\ a_2 \end{bmatrix} = \begin{bmatrix} 2.36756 & 0.682808 \\ 0.682808 & 0.818254 \end{bmatrix} \begin{bmatrix} c_1 \\ c_2 \end{bmatrix} = \begin{bmatrix} 0 \\ 1 \end{bmatrix},$$

which has solution
$$[a_1 \; a_2]^T = [-0.464166 \; 1.60945]. \qquad \square$$

3.15 Minimum-norm solution of underdetermined equations

The solution to the dual approximation problem provides a method of finding a least-squares solution to an underdetermined set of equations.

Example 3.15.1 Suppose that we are to solve the set of equations
$$\begin{bmatrix} 1 & 2 & -3 \\ -5 & 4 & 1 \end{bmatrix} \begin{bmatrix} x_1 \\ x_2 \\ x_3 \end{bmatrix} = \begin{bmatrix} -4 \\ 6 \end{bmatrix}, \qquad (3.88)$$

One solution is
$$\mathbf{x} = \begin{bmatrix} 1 \\ 2 \\ 3 \end{bmatrix}.$$

However, observe that the vector $\mathbf{v} = [1, 1, 1]^T$ is in the nullspace of A, so that $A\mathbf{v} = 0$; any vector of the form
$$\begin{bmatrix} 1 \\ 2 \\ 3 \end{bmatrix} + t \begin{bmatrix} 1 \\ 1 \\ 1 \end{bmatrix}$$

for $t \in \mathbb{R}$ is also a solution to (3.88). $\qquad \square$

When solving m equations with n unknowns with $m < n$, unless the equations are inconsistent, as in the example
$$\begin{bmatrix} 1 & 2 & 3 \\ 2 & 4 & 6 \end{bmatrix} \begin{bmatrix} x_1 \\ x_2 \\ x_3 \end{bmatrix} = \begin{bmatrix} 4 \\ 7 \end{bmatrix},$$

there will be an infinite number of solutions.

Let $\mathbf{x}$ be a solution of $A\mathbf{x} = \mathbf{b}$, where A is an $m \times n$ matrix with $m < n$, and let $N = \mathcal{N}(A)$. Then, if $\mathbf{x}_0$ is a solution to $A\mathbf{x} = \mathbf{b}$, so is any vector of the form $\mathbf{x}_0 + \mathbf{n}$, where $\mathbf{n} \in N$. If the nullspace is not trivial, a variety of solutions are possible. In order to have a well-determined algorithm for uniquely solving the problem, some criterion must be established regarding which solution is desired. A reasonable criterion is to find the solution $\mathbf{x}$ of smallest norm. That is, we want to

$$\text{minimize} \quad \|\mathbf{x}\|$$
$$\text{subject to} \quad A\mathbf{x} = \mathbf{b}.$$

3.16 Iterative Reweighted LS (IRLS) for L_p Optimization

The minimum norm solution is appealing from a numeric standpoint, because representations of small numbers are usually easier than representations of large numbers. It also leads to a unique solution that can be computed using the formulation of the dual problem of the previous section.

Let us write A in terms of its rows as

$$A = \begin{bmatrix} \mathbf{y}_1^H \\ \mathbf{y}_2^H \\ \vdots \\ \mathbf{y}_m^H \end{bmatrix}.$$

Then we observe that the equation $A\mathbf{x} = \mathbf{b}$ is equivalent to

$$\mathbf{y}_1^H \mathbf{x} = b_1$$
$$\mathbf{y}_2^H \mathbf{x} = b_2$$
$$\vdots$$
$$\mathbf{y}_m^H \mathbf{x} = b_m.$$

Our constraint equation therefore corresponds to m inner-product constraints of the sort shown in (3.14). By theorem 3.4, the minimum-norm solution must be of the form

$$\mathbf{x} = \sum_{i=1}^{m} c_i \mathbf{y}_i, \tag{3.89}$$

where the c_i are the solution to (3.87). We can write (3.89) as

$$\mathbf{x} = A^H \mathbf{c}, \tag{3.90}$$

where

$$A^H = \begin{bmatrix} \mathbf{y}_1 & \mathbf{y}_2 & \cdots & \mathbf{y}_m \end{bmatrix}.$$

Furthermore, in matrix notation we can write (3.87) in the form

$$(AA^H)\mathbf{c} = \mathbf{b}.$$

Provided that the rows are linearly independent, the matrix AA^H is invertible and we can solve for $\mathbf{c}$ as

$$\mathbf{c} = (AA^H)^{-1}\mathbf{b}.$$

Substituting this into (3.90), we obtain the minimum-norm solution

$$\mathbf{x} = A^H(AA^H)^{-1}\mathbf{b}. \tag{3.91}$$

Example 3.15.2 The minimum norm solution to (3.88) found using (3.91) is

$$\mathbf{x} = \begin{bmatrix} -1 \\ 0 \\ 1 \end{bmatrix}.$$

□

The matrix $A^H(AA^H)^{-1}$ is a pseudoinverse of the matrix A.

3.16 Iterative reweighted LS (IRLS) for L_p optimization

This chapter has focused largely on L_2 optimization, because the power of the orthogonality theorem allows analytical expressions to be determined in this case. In this section, we examine an algorithm for determining solutions to L_p optimization problems for $p \neq 2$. The method relies upon weighted least-squares techniques, but using a different weighting for each iteration.

We begin by examining a weighted least-squares problem. Suppose, as in section 3.1, we wish to determine a coefficient vector $\mathbf{c} \in \mathbb{R}^m$ to minimize the weighted norm of the error $\mathbf{e}$ in

$$\mathbf{x} = A\mathbf{c} + \mathbf{e}.$$

Let $W = S^T S$ be a weighting matrix. Then, to find

$$\min_{\mathbf{c}} \mathbf{e}^T W \mathbf{e} = \min_{\mathbf{c}} \mathbf{e}^T S^T S \mathbf{e},$$

we use (3.24) to obtain

$$\mathbf{c} = (A^T S^T S A)^{-1} A^T S^T S \mathbf{x}. \tag{3.92}$$

Now consider the L_p optimization problem

$$\min_{\mathbf{c}} \|\mathbf{x} - A\mathbf{c}\|_p^p = \min_{\mathbf{c}} \sum_{i=1}^{m} |x_i - (A\mathbf{c})_i|^p. \tag{3.93}$$

Let $\mathbf{c}^*$ be the solution to this optimization problem. The problem (3.93) can be written using a weighting as

$$\sum_{i=1}^{m} w_i |x_i - (A\mathbf{c})_i|^2$$

where $w_i = |x_i - (A\mathbf{c}_i^*)|^{p-2}$, producing a weighted least-squares problem which has a tractable solution. However, the solution cannot be found in one step, because $\mathbf{c}^*$ is needed to compute the appropriate weight. In iterative reweighted least-squares, the current solution is used to compute a weight which is used for the next iteration.

To this end, let $S^{[k]}$ be the weight matrix for the kth iteration, and let $c^{[k]}$ be the corresponding weighted least-squares solution obtained via (3.92). The error at the kth iteration is

$$\mathbf{e}^{[k]} = \mathbf{x} - A\mathbf{c}^{[k]}.$$

Then a new weight matrix $S^{[k+1]}$ is created according to

$$S^{[k+1]} = \text{diag}\bigl[|e_1^{[k]}|^{(p-2)/2} \quad |e_2^{[k]}|^{(p-2)/2} \quad \cdots \quad |e_m^{[k]}|^{(p-2)/2}\bigr].$$

Using this weight, the weighted error measure at the $(k+1)$st iteration is

$$\mathbf{e}^{[k+1]}(S^{[k+1]})^T S^{[k+1]} \mathbf{e}^{[k+1]} = \sum_{i=1}^{m} |x_i - (A\mathbf{c}^{[k+1]})_i|^p.$$

If this algorithm converges, then the weighted least-squares solution provides a solution to the L_p approximation problem.

However, it is known that the algorithm as described has slow convergence [45]. A more stable approach has been found; let

$$\hat{\mathbf{c}}^{[k+1]} = (A^T (S^{[k+1]})^T S^{[k+1]} A)^{-1} A^T (S^{[k+1]})^T S^{[k+1]} \mathbf{x}$$

and

$$\mathbf{c}^{[k+1]} = \lambda \hat{\mathbf{c}}^{[k+1]} + (1 - \lambda) \mathbf{c}^{[k]},$$

for some $\lambda \in (0, 1]$. It has been found [89, 162] that choosing

$$\lambda = \frac{1}{p-1}$$

3.16 Iterative Reweighted LS (IRLS) for L_p Optimization

leads to convergence properties of the algorithm similar to Newton's method (see section 14.4).

One final enhancement has been suggested [43]. A time-varying value of p is chosen, such that near the beginning of the iterative process, p is chosen to be small, then gradually increased until the desired p is obtained. Thus

$$p^{[k]} = \min(p, \gamma p^{[k-1]})$$

is used for some small $\gamma > 1$ (a typical value is $\gamma = 1.5$). Algorithm 3.4 incorporates these ideas.

Algorithm 3.4 Iterative reweighted least-squares
File: `irwls.m`

Example 3.16.1 L_p optimization methods have been used for filter design [43]. In this example we consider an odd tap-length filter

$$H(z) = \sum_{n=0}^{N} h[n] z^{-n},$$

with N even. The filter frequency response can be written (see section 6.8.2) as

$$H(e^{j\omega}) = e^{-jN\omega/2} H_R(\omega),$$

where

$$H_R(\omega) = \sum_{n=0}^{N/2} b_n \cos(\omega n) = \mathbf{b}^T \mathbf{c}(\omega).$$

Let $|H_d(\omega)|$ be the magnitude response of the desired filter. We desire to minimize

$$\int_0^\pi ||H_r(\omega)| - |H_d(\omega)||^p \, d\omega.$$

This can be closely approximated by sampling the frequency range at L_f frequencies $\omega_0, \omega_1, \ldots, \omega_{L_f-1}$, and minimizing

$$\sum_{k=0}^{L_f-1} ||H_r(\omega)| - |H_d(\omega)||^p.$$

This is now expressed as a finite-dimensional L_p optimization problem, and the methods of this section apply. Sample code that sets up the matrices, finds the solution, then plots the solution is shown in algorithm 3.5. Results of this for $p = 4$ and $p = 10$ are shown in figures 3.18(a) and (b), respectively. The $p = 10$ result shown closely approximates L_∞ (equiripple) design. □

Algorithm 3.5 Filter design using IRLS
File: `testirwls.m`

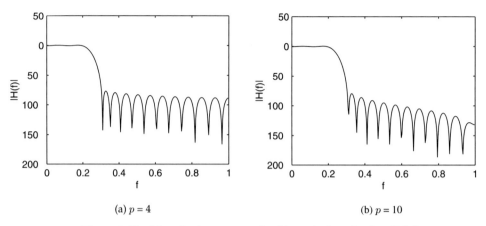

(a) $p = 4$ (b) $p = 10$

Figure 3.18: Magnitude response for filters designed using IRLS

3.17 Signal transformation and generalized Fourier series

Much of the transform theory employed in signal processing is encompassed by representations in an appropriate linear vector space. The set of basis functions for the transformation is chosen so that the coefficients convey desired information about the signal. By determining the basis functions appropriately, different information can be extracted from a signal by finding a representation of the signal in the basis.

In this section, we are largely (but not entirely) interested in approximating continuous-time functions. The metric space is L_2, and we deal with an infinite number of basis functions, so somewhat more care is needed than in the previous sections of this chapter.

Finding the best representation (in an L_2 norm sense) of a function $x(t)$ as

$$x(t) \approx \sum_{i=0}^{m} c_i p_i(t),$$

where $p_i(t)$ is a set of basis functions, is the approximation problem we have seen already many times. If the basis functions are orthonormal, the coefficients which minimize $\|x - \sum_{i=0}^{m} c_i p_i\|_2$ can be found as $c_i = \langle x, p_i \rangle$. The set of coefficients $\{c_i, i = 1, 2, \ldots, m\}$ provides the best representation (in the least-squares sense) of x. The minimum squared error of the series representation is

$$\left\| x - \sum_{i=1}^{m} c_i p_i \right\|^2 = \|x\|^2 - \sum_{i=1}^{m} \langle x, p_i \rangle^2.$$

Since the error is never negative, it follows that

$$\sum_{i=1}^{m} |c_i|^2 = \sum_{i=1}^{m} \langle x, p_i \rangle^2 \leq \|x\|^2. \qquad (3.94)$$

This inequality is known as *Bessel's inequality*.

The function $\sum_{i=1}^{m} c_i p_i$, obtained as a best L_2 approximation of $x(t)$, is said to be the *projection* of $x(t)$ onto the space spanned by $\{p_1, p_2, \ldots, p_m\}$. This may be written as

$$x_{\text{proj}(p_1, p_2, \ldots, p_m)}(t).$$

Assume that x and $\{p_i\}$ are in some Hilbert space H. If the set of basis functions $\{p_i\}$ is infinite, we can take the limit in (3.94) as $m \to \infty$. The representation of this limit is the

3.17 Signal Transformation and Generalized Fourier Series

infinite series
$$y(t) = \sum_{i=1}^{\infty} c_i p_i(t).$$

Since
$$y_m(t) = \sum_{i=1}^{m} c_i p_i(t)$$

is a Cauchy sequence and the Hilbert space is complete, we conclude that $y(t)$ is in the Hilbert space. For any orthonormal set $\{p_i\}$, the best approximation of x (in the L_2 sense) is the function y. We now want to address the question of when $x = y$ for an arbitrary $x \in H$. We must first point out that by the "equality" $x = y$, what we mean is that

$$\|x - y\| = 0,$$

where the norm is the L_2 norm (since we are dealing with a Hilbert space). Functions that differ on a set of measure zero are "equal" in the sense of the L_2 norm. Thus "equal" does not necessarily mean "point-for-point equal," as discussed in section 2.1.3.

We now define a condition under which it is possible to represent every x using the basis set $\{p_i\}$.

Definition 3.3 An orthonormal set $\{p_i, i = 1, 2, \ldots, \infty\}$ in a Hilbert space S is **complete**[2] if

$$x = \sum_{i=1}^{\infty} \langle x, p_i \rangle p_i$$

for every $x \in S$. □

Example 3.17.1 It is straightforward to show (by means of a simple counterexample) that simply having an infinite set of orthonormal functions is not sufficient to establish completeness. In $L_2[0, 2\pi]$, consider the function $x(t) = \cos t$. An infinite set of orthogonal functions is $T = \{p_n(t) = \sin(nt), n = 1, 2, \ldots, \}$. In the generalized Fourier series representation

$$\hat{x}(t) = \sum_{i=1}^{\infty} c_i p_i(t),$$

we find that the coefficients are proportional to

$$\langle \cos t, \sin nt \rangle = \int_0^{2\pi} \cos(t) \sin(nt) \, dt = 0.$$

Hence $\hat{x}(t) = 0$, which is not a good representation. We conclude that the set is not complete. □

Some results regarding completeness are expressed in the following theorem, which we state without proof.

Theorem 3.5 *[177] A set of orthonormal functions $\{p_i, i = 1, 2, \ldots\}$ is complete in an inner product space S with induced norm if any of the following equivalent statements holds:*

1. *For any $x \in S$,*
$$x = \sum_{i=1}^{\infty} \langle x, p_i \rangle p_i.$$

---

[2] This concerns completeness of the set of functions, which refers to the representational ability of the functions, not the completeness of the space, which is used to describe the fact that all Cauchy sequences converge. Some authors use "total" in place of complete here.

2. For any $\epsilon > 0$, there is an $N < \infty$ such that for all $n \geq N$,

$$\left\| x - \sum_{i=1}^{N} \langle x, p_i \rangle p_i \right\| < \epsilon.$$

   (In other words, we can approximate arbitrarily closely.)
3. Parseval's equality holds: $\|x\|^2 = \sum_{i=1}^{\infty} \langle x, p_i \rangle^2$ *for all* $x \in S$.
4. *If* $\langle x, p_i \rangle = 0$ *for all* i, *then* $x = 0$. *(This was shown to fail in the last example.)*
5. *There is no nonzero function* $f \in S$ *for which the set* $\{p_i, i = 1, 2, \ldots\} \cup f$ *forms an orthogonal set.*

For a finite-dimensional space S of dimension m, to have m linearly independent functions p_k, $k = 1, 2, \ldots, m$, is sufficient for completeness.

When $\{p_i\}$ is a complete basis set, then the sequence $\{c_1, c_2, \ldots, \}$ completely describes x; there is a one-to-one relationship between x and $\{c_1, c_2, \ldots\}$. (Except that x is only unique "up to" a set of measure zero.) We sometimes say that the sequence $\{c_1, c_2, \ldots, \}$ is the **transform** or the **generalized Fourier series** of x. Writing

$$\mathbf{c} = \{c_1, c_2, \ldots, \}$$

we can represent the transform relationship as

$$x \leftrightarrow \mathbf{c}.$$

We can define *different* transformations depending upon the set of orthonormal basis functions we choose. Since each coefficient in the transform is a projection of x onto the basis function, the transform coefficient p_i determines how much of p_i is in x. If we want to look for particular features of a signal, one way is to design a set of orthogonal basis functions that have those features and compute a transform using those signals.

If $\{p_i, i = 1, 2, \ldots\}$ is a complete set, there is no error in the representation, so Bessel's inequality (3.94) becomes an equality,

$$\|x\|^2 = \sum_{i=1}^{\infty} |c_i|^2. \qquad (3.95)$$

This relationship is known as *Parseval's equality*; it should be familiar in various special cases to signal processors. We can write this as

$$\|x\| = \|\mathbf{c}\|,$$

where the norm on the left is the L_2 norm (if x is a function) and the norm on the right is the l_2 norm.

For transformations using orthonormal basis sets, the angles are also preserved:

Lemma 3.1 *If x and y have a generalized Fourier series representation using some orthonormal basis set $\{p_i, i = 1, 2, \ldots\}$ in a Hilbert space S, with*

$$x \leftrightarrow \mathbf{c} \quad \text{and} \quad y \leftrightarrow \mathbf{b}$$

then

$$\langle x, y \rangle = \langle \mathbf{c}, \mathbf{b} \rangle. \qquad (3.96)$$

Proof We can write

$$x = \sum_{i=1}^{\infty} c_i p_i \quad \text{and} \quad y = \sum_{i=1}^{\infty} b_i p_i.$$

3.17 Signal Transformation and Generalized Fourier Series

Then

$$\langle x, y \rangle = \left\langle \sum_{i=1}^{\infty} c_i p_i, \sum_{j=1}^{\infty} b_j p_j \right\rangle$$

$$= \sum_{i=1}^{\infty} c_i b_i = \langle \mathbf{c}, \mathbf{b} \rangle, \qquad (3.97)$$

where the cross products in the inner product in (3.97) are zero because of orthogonality. □

Example 3.17.2 (Fourier series.) The set of functions which are periodic on $[0, 2\pi)$ can be represented using the series

$$f(t) = \sum_{n=-\infty}^{\infty} c_n \frac{1}{\sqrt{2\pi}} e^{jnt}.$$

The basis functions $p_n(t) = e^{jnt}/\sqrt{2\pi}$ are orthonormal, since

$$\int_0^{2\pi} e^{jnt} e^{-jmt} \, dt = \begin{cases} 0 & n \neq m, \\ 2\pi & n = m. \end{cases}$$

Then from (3.9),

$$c_n = \frac{1}{\sqrt{2\pi}} \int_0^{2\pi} f(t) e^{-jnt} \, dt.$$

By Parseval's relationship, we have

$$\int_0^{2\pi} |f(t)|^2 \, dt = \sum_n |c_n|^2.$$

More commonly, we use the nonnormalized basis functions $y_n(t) = e^{jnt}$, so the series is

$$f(t) = \sum_n b_n e^{jnt},$$

absorbing the normalizing constant into the coefficient as

$$b_n = \frac{1}{2\pi} \int_0^{2\pi} f(t) e^{-jnt} \, dt.$$

In this case, Parseval's relationship must be normalized as

$$\int_0^{2\pi} |f(t)|^2 \, dt = \frac{1}{2\pi} \sum_i |b_i|^2.$$

More generally, for a function periodic with period T_0, we have the familiar formulas

$$f(t) = \sum_n b_n e^{-jn\omega_0 t},$$

where $\omega_0 = 2\pi/T_0$, and

$$b_n = \frac{1}{T_0} \int_0^{T_0} f(t) e^{-jn\omega_0 t} \, dt.$$

□

Example 3.17.3 (Discrete Fourier transform (DFT)) A discrete-time sequence $x[t]$, $t = 0, 1, \ldots, N-1$, is to be represented as a linear combination of the functions $p_k[t] = (1/\sqrt{N}) e^{j2\pi tk/N}$, by

$$x[t] = \frac{1}{\sqrt{N}} \sum_{k=0}^{N-1} c_k e^{j2\pi tk/N}.$$

The inner product in this case is

$$\langle x[t], y[t]\rangle = \sum_{k=0}^{N-1} x[t]\overline{y}[t].$$

It can be shown (see exercise 3.17-36) that the set of basis functions $\{p_k[t]\}$ are orthogonal, with

$$\langle p_k[t], p_l[t]\rangle = \begin{cases} 1 & k \bmod l \pmod{N}, \\ 0 & \text{otherwise.} \end{cases}$$

The coefficients are therefore computed by

$$c_k = \frac{1}{\sqrt{N}} \sum_{t=0}^{N-1} x[t] e^{-j2\pi tk/N}.$$

More commonly we use the *nonnormalized* basis functions $e^{j2\pi tk/N}$, and shift all of the normalization into the reconstruction formula. Then we have

$$x[t] = \frac{1}{N} \sum_{k=0}^{N-1} d_k e^{j2\pi nk/N}$$

and

$$d_k = \sum_{t=0}^{N-1} x[t] e^{-j2\pi tk/N},$$

which is the usual Fourier transform pair. Parseval's relationship (under this normalization) is

$$\sum_{t=0}^{N-1} |x[t]|^2 = \frac{1}{N} \sum_{k=0}^{N-1} |d_k|^2. \qquad \square$$

3.18 Sets of complete orthogonal functions

There are several sets of complete orthogonal functions that are used in common applications. We will examine a few of the more commonly-used sets, mostly stating results without proofs.

3.18.1 Trigonometric functions

As seen in example 3.17.2, the familiar trigonometric functions employed in Fourier series are orthogonal. They form a complete set of orthogonal functions.

3.18.2 Orthogonal polynomials

As we have seen, one way to obtain orthogonal functions is by means of polynomials. Different sets of orthogonal polynomials are obtained by using different weighting functions, and the inner product is taken over some given interval. Some kinds of orthogonal polynomials arise commonly enough that they have been given names.

Let $f(t)$ and $g(t)$ be polynomials, and let I be a domain of interest, $I = [a, b]$. The polynomials $f(t)$ and $g(t)$ are orthogonal with respect to the weighting function $w(t)$ if

$$\langle f, g\rangle_w = 0,$$

where

$$\langle f, g\rangle_w = \int_a^b w(t) f(t) g(t)\, dt.$$

Using the Gram–Schmidt procedure it is possible to orthogonalize any set of polynomials with respect to any inner product; in particular, the set of polynomials $1, t, t^2, \ldots, t^n$ can

3.18 Sets of Complete Orthogonal Functions

be orthogonalized with respect to the weighted inner product. We will denote a set of orthogonal polynomials by $p_0(t), p_1(t), p_2(t)$, and so forth, where the subscript denotes the degree of the polynomial.

It can be shown that the orthogonal polynomials (properly normalized) form a complete orthonormal basis for $L_2[a, b]$. The proof of this (which we do not present here) relies on the Weierstrass theorem, which states that any continuous function on an interval $[a, b]$ can be approximated arbitrarily closely by a polynomial. By this theorem we can establish a basis for $C[a, b]$. Extending this to $L_2[a, b]$ (which contains non-continuous functions) makes use of the fact that every discontinuous function over the interval $[a, b]$ is arbitrarily close to a continuous function.

An interesting fact about orthogonal polynomials is the following.

Lemma 3.2 *Orthogonal polynomials satisfy the recursion*

$$t p_n(t) = a_n p_{n+1}(t) + b_n p_n(t) + c_n p_{n-1}(t) \tag{3.98}$$

for $n = 1, 2, \ldots$.

Proof Choose a_n so that $t p_n(t) - a_n p_{n+1}(t)$ is of degree n,

$$t p_n(t) - a_n p_{n+1}(t) = g_n(t).$$

Then $g_n(t)$ can be written as a linear combination of $p_0, p_1, \ldots, p_n$:

$$g_n(t) = \sum_{i=0}^{n} d_i p_i(t),$$

where the coefficients are obtained by

$$d_i = \langle g_n(t), p_i(t) \rangle.$$

But for $i < n - 1$, the coefficients are zero, since

$$\langle t p_n, p_i \rangle = \langle p_n, t p_i \rangle,$$

and that p_n is orthogonal to all polynomials of lesser degree, including $t p_i$. When $i = n - 1$ and $i = n$, the coefficients are not zero,

$$b_n = \langle g_n, p_n \rangle \qquad c_n = \langle g_n, p_{n-1} \rangle. \qquad \Box$$

Families of orthogonal polynomials

A variety of types of orthogonal polynomials have been explored over the years. One of the general motivations for this is that orthogonal polynomials can be used to provide solutions to particular differential equations. Since these orthogonal polynomials form a complete orthogonal basis, they can be used to form series solutions for any boundary conditions and input function. Details of this kind of analysis are not discussed here, but may be found in applied mathematics or partial differential equations books, such as [259, 177]. However, the differential equations and several common orthogonal polynomials are presented in the exercises. Another important use of orthogonal polynomials is for Gaussian quadrature, which is an efficient method of numerical integration. This is also derived in the exercises; more details can be found in [265]. In this section we examine only two families of orthogonal polynomials, the Legendre and the Chebyshev polynomials.

Legendre polynomials

The Legendre polynomials are not the most commonly used orthogonal polynomials in signal processing, but occasional uses do arise. The Legendre polynomials use a weighting

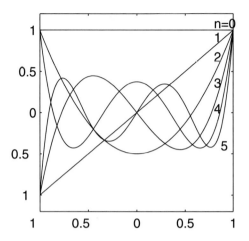

Figure 3.19: Legendre polynomials $p_0(t)$ through $p_5(t)$ for $t \in [-1, 1]$

function $w(t) = 1$ over the interval $[a, b] = [-1, 1]$. The first three are

$$p_0(t) = 1 \qquad p_1(t) = 1 \qquad p_2(t) = x^2 - \frac{1}{3} \qquad p_3(t) = x^3 - \frac{3}{5}x.$$

Additional values can be obtained using the recurrence (3.98), which specializes for Legendre polynomials to

$$tp_n(t) = \frac{n+1}{2n+1} p_{n+1}(t) + \frac{n}{2n+1} p_{n-1}(t).$$

Figure 3.19 shows p_0, p_1, p_2, p_3, p_4, and p_5. Observe that not all polynomials have the same amount of "ripple." This is to be contrasted with the Chebyshev polynomials, discussed next.

Chebyshev polynomials

Chebyshev polynomials are orthogonal with respect to the weighting function $w(t) = \frac{1}{\sqrt{1-t^2}}$ over the interval $I = [-1, 1]$. In particular, if $T_r(t)$ and $T_s(t)$ are Chebyshev polynomials, then

$$\int_{-1}^{1} \frac{1}{\sqrt{1-t^2}} T_r(t) T_s(t)\, dt = \begin{cases} 0 & r \neq s, \\ \pi & r = s = 0, \\ \frac{\pi}{2} & r = s \neq 0. \end{cases} \qquad (3.99)$$

(See exercise 3.18-38) The recurrence relation for Chebyshev polynomials is

$$T_{n+1}(t) = 2t T_n(t) - T_{n-1}(t), \qquad T_0(t) = 1, T_1(t) = t. \qquad (3.100)$$

The Chebyshev polynomials can be expressed as

$$T_n(t) = \cos(n \cos^{-1} t). \qquad (3.101)$$

Using either (3.100) or (3.101), the next few Chebyshev polynomials can be found:

$$T_2(t) = 2t^2 - 1 \qquad T_3(t) = 4t^3 - 3t \qquad T_4(t) = 8t^4 - 8t^2 + 1.$$

The leading coefficient of the Chebyshev polynomial $T_n(t)$ is 2^{n-1}, so that $\frac{1}{2^{n-1}} T_n(t)$ is a monic polynomial. From (3.101), is clear that the zeros of $T_n(t)$ are at

$$t = \cos \frac{2k+1\pi}{2n}, \qquad k = 0, 1, \ldots, n-1.$$

Over $[-1, 1]$ there are $n + 1$ extrema (counting the end points) of magnitude 1 at

$$t = \cos\frac{k\pi}{n}, \quad k = 0, 1, n.$$

Figure 2.11 illustrates the first six Chebyshev polynomials for $t \in [-1, 1]$. What is remarkable about these polynomials is that over this interval, each of the local extrema (maximum or minimum) takes on the value ± 1. This is an important feature in the Chebyshev polynomials, and accounts for most of their applications. This is called the minimum maximum amplitude property: the maximum amplitude (deviation from zero) is minimized.

Theorem 3.6 *Of all monic polynomials of degree n, only the polynomial $Q_n(t) = \frac{1}{2^{n-1}} T_n(t)$ (the scaled Chebyshev polynomial) oscillates with the minimum maximum amplitude on the interval $[-1, 1]$.*

Proof The proof is by contradiction. Suppose there exists a monic polynomial $q_n(t)$ of degree n with smaller minimum maximum amplitude on $[-1, 1]$. Let

$$p_{n-1}(t) = Q_n(t) - q_n(t).$$

Since both q_n and Q_n are monic, p_{n-1} must have degree not exceeding $n-1$. The polynomial Q_n has $n + 1$ extrema, each of magnitude $1/2^{n-1}$. By assumption, $q_n(t)$ has a smaller magnitude at each of these extrema, so that $p_{n-1}(t)$ has the *same sign* as $Q_n(t)$ at each of these extrema. Note that that $n + 1$ extrema of $T_n(t)$, and hence $Q_n(t)$, alternate in sign. Thus $p_{n-1}(t)$ alternates in sign from one extremum of $Q_n(t)$ to the next. Since there are $n + 1$ extrema, there must be n zeros of $p_{n-1}(t)$ in $[-1, 1]$. But $p_{n-1}(t)$ is a polynomial of degree $n - 1$, which has only $n - 1$ zeros, which is a contradiction.

Now suppose that $q_n(t)$ is another polynomial having the same minimum maximum amplitude as $Q_n(t)$. If $|q_n(t)| < |Q_n(t)|$ at an extremum, then we again arrive, as before, at a contradiction. On the other hand, if $q_n(t_0) = Q_n(t_0)$ at an extremum t_0, then $p_{n-1}(t_0) = 0$ and $p'_{n-1}(t_0) = 0$. Then $p_{n-1}(t)$ has (at least) a double zero at t_0. Counting the zeros of $p_{n-1}(t)$ again leads to a contradiction. □

One application of Chebyshev polynomials is as basis functions in a series expansion, such as

$$f(t) = \sum_{j=0}^{\infty} c_j T_j(t).$$

This series converges uniformly whenever $f(t)$ is continuous and of bounded variation in $[-1, 1]$. Because of the minimum maximum property of Chebyshev polynomials, the approximate representation up to mth degree polynomials,

$$f(t) \approx \sum_{j=0}^{m} c_j T_j(t),$$

usually has less error than a corresponding representation using either the basis $1, t, \ldots, t^m$ or the Legendre polynomials.

3.18.3 Sinc functions

The function commonly known as a sinc function,

$$\text{sinc}(t) = \frac{sin(\pi t)}{(\pi t)},$$

can be used to form a set of orthogonal functions

$$p_k(t) = \text{sinc}(2B(t - k/2B)). \tag{3.102}$$

It can be shown (see exercise 3.18-46) for the inner product

$$\langle f, g \rangle = \int_{-\infty}^{\infty} f(t)\overline{g}(t)\,dt$$

that $\langle p_k(t), p_l(t) \rangle = \frac{1}{2B}\delta_{k,l}$. If $f(t)$ is a bandlimited function such that its Fourier transform satisfies

$$F(\omega) = 0 \quad \text{for } \omega \notin (-2\pi B, 2\pi B),$$

then, in the series representation

$$f(t) = \sum_k c_k p_k(t),$$

the coefficients are found to be (see exercise 3.18-47)

$$c_k = \frac{\langle f, p_k \rangle}{\langle p_k, p_k \rangle} = f(k/2B). \tag{3.103}$$

This gives rise to the familiar sampling theorem representation of a bandlimited function,

$$f(t) = \sum_k f(k/2B) \frac{\sin(2\pi B(t - k/(2B)))}{2\pi B(t - k/(2B))}.$$

3.18.4 Orthogonal wavelets

Recently, a set of functions known as *wavelets* has sparked considerable interest. Like the Fourier transform, the wavelet transform can provide information about the spectral content of a signal. However, unlike a sinusoidal signal with infinite support, wavelets are pulses which are well localized in the time domain so that they can provide different spectral information at different time locations of a signal. In doing this, they sacrifice some of their spectral resolution: by the uncertainty principle, we cannot localize perfectly well in both the time domain and the frequency domain. Wavelets have another property that make wavelets them practically useful. When used to analyze lower-frequency components, a wide wavelet signal is used; to analyze higher-frequency components, a narrow wavelet signal is used. Thus wavelets can (in principle) identify short bursts of high-frequency signals imposed on top of ongoing low-frequency signals. One of the major principles of wavelet analysis is that it takes place on several scales, using basis functions of different widths.

There are, in fact, several families of wavelets, each with its own properties and associated transforms. Not all families of wavelets form orthogonal waveforms. A particular family of wavelets that has perhaps attracted the most attention is known as the Daubechies wavelets. These wavelets, which form a complete set, have some very some very nice orthogonality properties that lead to fast computational algorithms. The Daubechies wavelets can be understood best in the context of a Hilbert space, using what is known as a multiresolution analysis. This involves projecting a function onto a whole series of spaces with different resolutions. We now present a brief introduction to the construction of these wavelets. Considerably more information is provided in the literature cited in the references, including generalization in a variety of useful ways of the concepts outlined here.

Characterization of wavelets

Throughout this section we will assume real functions for convenience. Most of these concepts can be generalized to functions of complex numbers. Suppose we have a set of closed subspaces of the Hilbert space $L_2(\mathbb{R})$, denoted by $\ldots, V_{-1}, V_0, V_1, \ldots$, with the

3.18 Sets of Complete Orthogonal Functions

following properties:

1. Nesting:
$$\cdots V_2 \subset V_1 \subset V_0 \subset V_{-1} \subset V_{-2} \cdots.$$

2. Closure:
$$\text{closure}\left(\bigcup_{j \in \mathbb{Z}} V_j\right) = L_2(\mathbb{R});$$

   that is, the closure of the set of spaces covers all of $L_2(\mathbb{R})$, so that every function in L_2 has a representation using elements in one of these nested spaces.

3. Shrinking:
$$\bigcap_{j \in \mathbb{Z}} V_j = \{0\}.$$

4. The "multiresolution" property is obtained by the requirement that if $f(t) \in V_j$, then $f(2^j t) \in V_0$.

5. If $f(t) \in V_0$, then $f(t - n) \in V_0$ for all $n \in \mathbb{Z}$.

6. Finally, there is some $\phi \in V_0$ such that the integer shifts of ϕ form an orthonormal basis for V_0:
$$V_0 = \text{span}\{\phi(t - n), n \in \mathbb{Z}\}.$$

The function $\phi(t)$ is said to be a **scaling function**. The property that $\phi(t) \perp \phi(t - n)$ for $n \in \mathbb{Z}$ is called the *shift orthogonality* property.

We will use the notation $P_j f(t)$ to denote the projection of the function $f(t)$ onto V_j.

Example 3.18.1 Let
$$\phi(t) = u(t) - u(t - 1) \tag{3.104}$$

(a unit pulse), and form
$$V_0 = \text{span}\{\phi(t - n), n \in \mathbb{Z}\}.$$

The set of functions $\{\phi(t-n), n \in \mathbb{Z}\}$ forms an orthonormal set. Then functions in V_0 are functions that are *piecewise constant* on the integers. Figure 3.20 shows a function $f(t)$, the projection $P_0 f(t)$—the nearest function to $f(t)$ that is piecewise constant in the integers—and $P_{-1} f(t)$—which is piecewise constant on the half-integers. □

As j decreases, the projection $P_j f(t)$ represents $f(t)$ with increasing fidelity.

Let us define the scaled and shifted version of the function ϕ by
$$\phi_{j,k}(t) = 2^{-j/2} \phi(2^{-j/2} t - k).$$

The index j controls the *scale* and the index k controls the location of the function $\phi_{j,k}$. If $\phi(t)$ is normalized so that $\|\phi(t)\| = 1$, then so is $\phi_{j,k}(t)$ for any j and k. Since $\phi(t) \in V_0 \subset V_{-1}$ and $\phi_{-1,k}(t)$ form an orthonormal basis for V_{-1}, it must be possible to express $\phi(t)$ as a linear combination of $\phi_{-1,k}(t)$:
$$\phi(t) = \sum_k h_k \phi_{-1,k}(t) = \sqrt{2} \sum_k h_k \phi(2t - k). \tag{3.105}$$

The set of coefficients in (3.105) determines the particular properties of the scaling function and the entire wavelet decomposition. Let N denote the total number of coefficients h_n in

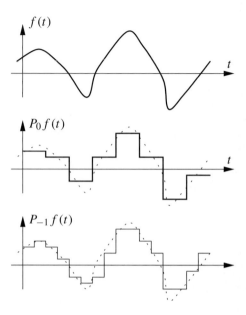

Figure 3.20: A function $f(t)$ and its projection onto V_0 and V_{-1}

(3.105). In general, N could be infinite, but in practice it is always a finite number. We also generally assume that the coefficients h_k are indexed so that $h_k = 0$ for $k < 0$. Let us define $c_k = \sqrt{2} h_k$. Then we can write

$$\phi(t) = \sum_k c_k \phi(2t - k); \tag{3.106}$$

or, given our assumptions, we can write this more precisely as

$$\phi(t) = \sum_{k=0}^{N-1} c_k \phi(2t - k). \tag{3.107}$$

An equation of the form (3.107) is known as a *two-scale equation*.

Example 3.18.2 In (3.107) let us have two coefficients, $c_0 = 1$ and $c_1 = 1$. Then the two-scale equation becomes

$$\phi(t) = \phi(2t) + \phi(2t - 1).$$

It is straightforward to verify that the pulse in (3.104) satisfies this equation. □

Lemma 3.3 *If $\phi(t)$ satisfies a two-scale equation (3.106) and $\phi(t) \perp \phi(t - n)$ for all $n \in \mathbb{Z}$ with $n \neq 0$, then*

$$\sum_k c_k c_{k-2p} = 2\delta_{0p}. \tag{3.108}$$

Proof Using (3.106), we have

$$\int \phi(t) \phi(t-n) \, dt = \int \sum_k c_k \phi(2t - k) \sum_j c_j \phi(2(t-n) - j) \, dt$$

$$= \frac{1}{2} \sum_j \left[\sum_k c_k c_{k+j-2n} \right] \int \phi(t) \phi(t - j) \, dt.$$

3.18 Sets of Complete Orthogonal Functions

In order for this to be zero (because of the orthogonality), the bracketed term must be zero when $j = 0$ and $2n \neq 0$. Then $\sum_k c_k c_{k-2n} = 2\delta_{0n}$. □

In going from a projection $P_{j-1}f(t)$ to a lower-resolution projection $P_j f(t)$, there is some detail information that is lost in the orthogonal complement of V_j. We can represent this detail by saying that

$$V_{j-1} = V_j \oplus W_j, \qquad (3.109)$$

where $W_j = V_j^\perp$ in V_{j-1}. (The direct sum is interpreted in the isomorphic sense.) Thus, W_j contains the detail lost in going from V_{j-1} to V_j. Also (as we shall see), the W_j spaces are orthogonal, so $W_j \perp W_{j'}$ if $j \neq j'$.

Now we introduce the set of functions $\psi_{j,k}(t) = 2^{-j/2}\psi(2^{-j}t - k)$ as an orthonormal basis set for W_j, with $\psi(t) \in W_0$. The function $\psi(t)$ is known as a **wavelet** function, or sometimes as the **mother wavelet**, since the functions $\psi_{j,k}(t)$ are derived from it. Since $V_{-1} = V_0 \oplus W_0$ and $\psi(t) \in V_{-1}$, we have

$$\psi(t) = \sum_k g_k \phi_{-1,k}(t) = \sqrt{2} \sum_k g_k \phi(2t - k). \qquad (3.110)$$

We desire to choose the g_n coefficients to enforce the orthogonality of the spaces. It will be convenient to write

$$d_k = \sqrt{2} g_k.$$

Theorem 3.7 *If $\{\phi(t - n), n \in \mathbb{Z}\}$ forms an orthogonal set and*

$$d_k = (-1)^k c_{2M+1-k}$$

for any $M \in \mathbb{Z}$, then $\{\psi_{j,k}(t)\}$ forms an orthogonal set for all $j, k \in \mathbb{Z}$. Furthermore, $\psi_{j,k}(t) \perp \phi_{l,m}(t)$ for $l \leq j$.

Proof We begin by showing that $\{\psi_{j,k}(t)\}$ forms an orthogonal set for fixed j.

$$\int 2^{-j} \psi(2^{-j}t) \psi(2^{-j}t - k)\, dt = \int \sum_l d_l \phi(2u - l) \sum_m d_m \phi(2(u-k) - m)\, du$$

(where $u = 2^{-j}t$)

$$= \int \left(\frac{1}{2} \sum_l (-1)^l c_{2M+1-l} \phi(x) \right)$$

$$\times \left(\sum_m (-1)^m c_{2M+1-m} \phi(x + l - 2k - m) \right) dx$$

(where $x = 2u - l$)

$$= \frac{1}{2} \sum_j c_j c_{j+2k} \int \phi^2(x)\, dx$$

(by orthogonality, with $j = 2M + 1 - l$)

$$= \delta_{0k}. \quad \text{(using (3.108))}$$

Now we show that $\phi_{j,k}(t) \perp \psi_{j,m}(t)$ for all $k, m \in \mathbb{Z}$, for fixed j. We have

$$\int \psi_{j,k}(t) \phi_{j,m}(t)\, dt = \int 2^{-j} \psi(2^{-j}t - k) \phi(2^{-j}t - m)\, dt$$

$$= \int \psi(u - k) \phi(u - m)\, dt \quad \text{(where } u = 2^{-j}t\text{)}$$

$$= \int \sum_l (-1)^l c_{2M+1-l} \phi(2(u-k)-l) \sum_j c_j \phi(2(u-m)-j)\,du$$

$$= \frac{1}{2} \int \sum_l (-1)^l c_{2M+1-l} \phi(x) \sum_j c_j \phi(x+l+2k-2m-j)\,dx$$

(where $x = 2u - l - 2k$)

$$= \frac{1}{2} \sum_l (-1)^l c_{2M+1-l} c_{l+2k-2m} \int \phi^2(t)\,dt \qquad \text{(by orthogonality)}$$

(3.111)

In the summation in (3.111), let $p = m - k$, so the summation is

$$S = \sum_l (-1)^l c_{2M+1-l} c_{l-2p}.$$

Now, letting $j = 2M + 1 - l + 2p$, we can write

$$S = \sum_j (-1)^{1-j} c_{j-2p} c_{2M+1-j} = -\sum_j (-1)^j c_{2M+1-j} c_{j-2p} = -S.$$

Since $S = -S$, we must have

$$0 = S = \sum_l (-1)^l c_{2M+1-l} c_{l-2p}, \qquad (3.112)$$

establishing the desired orthogonality.

Finally, we show that $\psi_{j,k} \perp \psi_{l,m}$ for all $j, k, l, m \in \mathbb{Z}$ if $j \neq l$ and $k \neq m$. We have already established this for $j = l$. By the multiscale relationship, $\psi_{j,k}(t) \in W_j$. Let $j' > j$, so that $W_j \subset V_{j'}$. But $V_{j'} \perp W_{j'}$, so that $\psi_{j'k}(t)$, which is in $W_{j'}$, must be orthogonal to $\psi_{jk}(t)$. □

Example 3.18.3 We have seen that a scaling function $\phi(t)$ can be formed when $c_0 = c_1 = 1$. The wavelet $\psi(t)$ corresponding to this scaling function is

$$\psi(t) = \phi(2t) - \phi(2t - 1).$$

A plot of $\phi(t)$ and $\psi(t)$ is shown in figure 3.21. The function $\psi(t)$ is also known as the *Haar* basis function. □

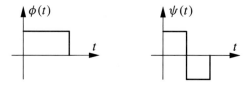

Figure 3.21: The simplest scaling and wavelet functions

There are several families of orthonormal compactly supported wavelets. Algorithm 3.6 provides coefficients for several Daubechies wavelets (there exist wavelets in this family with coefficients of every positive even length). The transform for these coefficients is called the D_N, where there are N coefficients. Plots of some of the corresponding scaling and wavelet functions are shown in figure 3.22. We observe that the functions become smoother as the number of coefficients increases.

3.18 Sets of Complete Orthogonal Functions

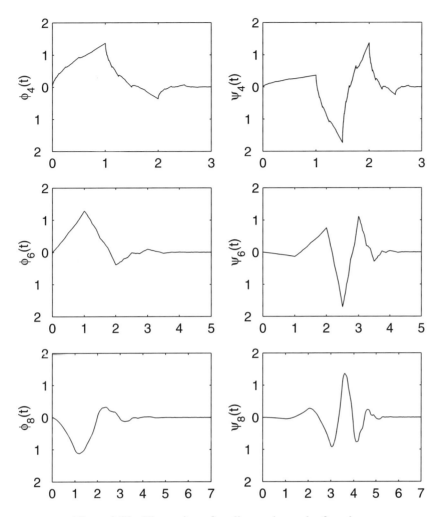

Figure 3.22: Illustration of scaling and wavelet functions

Algorithm 3.6 Some wavelet coefficients [63, page 195]
File: `wavecoeff.m`

Wavelet transforms

In the wavelet transform, a function $f(t)$ is expressed as a linear combination of scaling and wavelet functions. Both the scaling functions and the wavelet functions are complete sets. However, it is common to employ both wavelet and scaling functions in the transform representation.

Suppose that we have a projection of $f(t)$ onto some space V_j of sufficient resolution that it provides an adequate representation of the data. Then we have

$$f(t) \approx P_j f(t) = \sum_k \langle f(t), \phi_{jk}(t) \rangle \phi_{j,k}(t).$$

Commonly we assume that the data has been scaled so that the initial scale is $j = 0$, so that our starting point is $P_0 f(t)$. Let us call this starting function $f_0(t)$, so that

$$f_0(t) = \sum_n \langle f(t), \phi_{0,n}(t) \rangle \phi_{0,n}(t).$$

For the purposes of the transform, we regard the *coefficients* of this representation as the representation of $f(t)$. In practice, the set of initial coefficients are simply *samples* of $f(t)$ obtained by sampling every T seconds. That is, we assume that $\langle f(t), \phi_{0,n}(t) \rangle \approx f(nT)$ for some sampling interval T. Under this approximation, the wavelet transform deals with discrete-time sequences. (Further discussion of this point is provided in [63, page 166].) For convenience of notation, let us denote the sequence $\{\langle f_0, \phi_{0,n}(t) \rangle\}$ as $\{c_n^0\}$, and let us denote the vector of these values as $\mathbf{c}^0$:

$$\mathbf{c}^0 = \begin{bmatrix} c_0^0 & c_1^0 & c_2^0 & \ldots \end{bmatrix}^T.$$

In the wavelet transform, we express $f_0(t)$ in terms of wavelets on longer scales. For example, using (3.109) we have $V_0 = V_1 \oplus W_1$, so that $f_0(t) \in V_0$ can be represented as

$$f_0(t) = \sum_n \langle f_0(t), \psi_{1,n}(t) \rangle \psi_{1,n}(t) + \sum_n \langle f_0(t), \phi_{1,n}(t) \rangle \phi_{1,n}(t).$$

Let $c_n^1 = \langle f_0(t), \phi_{1,n}(t) \rangle$ and $d_n^1 = \langle f_0(t), \psi_{1,n}(t) \rangle$, and let us denote

$$f_1(t) = \sum_n \langle f_0(t), \phi_{1,n}(t) \rangle \phi_{1,n}(t) = \sum_n c_n^1 \phi_{1,n}$$

and

$$\delta_1(t) = \sum_n \langle f_0(t), \psi_{1,n}(t) \rangle \psi_{1,n}(t) = \sum_n d_n^1 \psi_{1,n},$$

where $f_1 \in V_1$ and $\delta_1 \in W_1$. Then

$$f_0(t) = f_1(t) + \delta_1(t). \tag{3.113}$$

Since $f_1 \in V_1$ and $V_1 = V_2 + W_2$, we can split f_1 into its projection onto V_2 and W_2 as

$$f_1(t) = \sum_n \langle f_1(t), \phi_{2,n}(t) \rangle \phi_{2,n}(t) + \sum_n \langle f_1(t), \psi_{2,n}(t) \rangle \psi_{2,n}(t)$$
$$= \sum_n c_n^2 \phi_{2,n} + \sum_n d_n^2 \psi_{2,n}$$
$$= f_2(t) + \delta_2(t), \tag{3.114}$$

where $f_2(t) \in V_2$, and $\delta_2(t) \in W_2$, and $c_n^2 = \langle f_1(t), \phi_{2,n} \rangle$ and $d_n^2 = \langle f_1(t), \psi_{2,n} \rangle$. Substituting (3.114) into (3.113), we have

$$f_0(t) = \delta_1(t) + \delta_2(t) + f_2(t).$$

We will use the notation $\mathbf{c}^j$ and $\mathbf{d}^j$ to represent the coefficients c_n^j and d_n^j, respectively. We can repeat this decomposition for up to J scales, writing $f_j(t) \in V_j$ on each scale $j = 1, 2, \ldots, J$ as

$$f_j(t) = f_{j+1}(t) + \delta_{j+1}(t), \tag{3.115}$$

so

$$f_0(t) = \sum_{j=1}^{J} \delta_j(t) + f_J(t).$$

The set of coefficients $\{\mathbf{d}^1, \mathbf{d}^2, \ldots, \mathbf{d}^J, \mathbf{c}^J\}$ collectively are the **wavelet transform** of the function $f_0(t)$.

3.18 Sets of Complete Orthogonal Functions

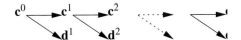

Figure 3.23: Illustration of a wavelet transform

The computations just described are outlined in figure 3.23. Starting from the initial set of coefficients $\mathbf{c}^0$, the algorithm successively produces $\mathbf{c}^{j+1}$ and $\mathbf{d}^{j+1}$ until the Jth level is reached. The set of coefficients $\{\mathbf{d}^1, \mathbf{d}^2, \ldots, \mathbf{d}^J, \mathbf{c}^J\}$ is the wavelet transform of the original data. The coefficients at scale $\mathbf{d}^j$ represent the signal on longer scales (lower-frequency band) than the coefficients at scale $\mathbf{d}^{j-1}$. The coefficients $\mathbf{c}^J$ represents an average of the original data.

While it is conceivable to compute the transform by directly evaluating the indicated inner products, a significantly faster algorithm exists. We note that by (3.107)

$$\psi_{j,k}(t) = 2^{-j/2}\psi(2^{-j}t - k) = 2^{-j/2}\sqrt{2}\sum g_n \phi(2(2^{-j}t - k) - n)$$
$$= \sum_n g_n \phi_{j-1, 2k+n}(t)$$
$$= \sum_n g_{n-2k} \phi_{j-1,n}(t). \tag{3.116}$$

When we compute the wavelet transform coefficient $\langle f_0(t), \psi_{1,k}(t) \rangle$, we get

$$\langle f_0(t), \psi_{1,k}(t) \rangle = \sum_n g_{n-2k} \langle f_0(t), \phi_{0,n}(t) \rangle = \sum_n g_{n-2k} c_n^0. \tag{3.117}$$

To understand this sum better, let us write

$$x_n = g_{-n},$$

and form the vector $\mathbf{x} = [x_0, x_1, \ldots, x_{N-1}]$. Let $\mathbf{y} = \mathbf{x} * \mathbf{c}^0$ (convolution); then

$$y_j = \sum_n x_{j-n} c_n^0 = \sum_n g_{n-j} c_n^0.$$

From this we observe that the summation in (3.117) is the convolution of the sequence $\{g_{-n}\}$ with the sequence $\{c_n^0\}$, in which we retain only the even-numbered outputs.

At a general scale j, we compute the wavelet coefficients as

$$\langle f_0, \psi_{j,k}(t) \rangle = \sum_n g_{n-2k} \langle f_0, \phi_{j-1,n} \rangle, \tag{3.118}$$

which is a convolution of the sequence $\{g_{-n}\}$ with the sequence $\{\langle f_0, \phi_{j-1,n} \rangle\}$, retaining even samples. To compute the coefficients in (3.118), we need to know $\langle f_0, \phi_{j-1,n} \rangle$. However, these can also be obtained efficiently, since

$$\phi_{j,k}(t) = 2^{-j/2}\phi(2^{-j}t - k)$$
$$= \sum_n h_{n-2k} \phi_{j-1,n}(t), \tag{3.119}$$

so that

$$\langle f_0, \phi_{j,k} \rangle = \sum_n h_{n-2k} \langle f_0, \phi_{j-1,n} \rangle,$$

which is again a convolution followed by decimation by 2.

Putting all the pieces together, the wavelet transform is outlined as follows:

1. Let $c_k^0 = \langle f_0, \phi_{0,k} \rangle$ be the given initial data. (Normally a sequence of samples of $f(t)$.)
2. Compute the set of wavelet coefficients on scale 1, $d_k^1 = \langle f_0, \psi_{1,k} \rangle$, using

$$d_k^1 = \sum_n g_{n-2k} c_n^0. \tag{3.120}$$

Also compute the scaling coefficients on this scale, $c_k^1 = \langle f_0, \phi_{1,k} \rangle$, using

$$c_k^1 = \sum_n h_{n-2k} c_n^0. \tag{3.121}$$

3. Now, proceed up through level J similarly,

$$d_k^j = \sum_n g_{n-2k} c_n^{j-1}, \tag{3.122}$$

$$c_k^j = \sum_n h_{n-2k} c_n^{j-1}, \qquad j = 1, 2, \ldots, J. \tag{3.123}$$

The wavelet transform computations can be represented in matrix notation. The operation (3.123) can be represented as a matrix L, where $L_{ij} = h_{j-2i}$ for i and j in some suitable range. The operation (3.122) can be represented as a matrix H, where $H_{ij} = g_{j-2i}$.

Example 3.18.4 We will demonstrate this matrix notation for a wavelet with four coefficients, h_0, h_1, h_2, h_3. We choose M so that $\{g_0, g_1, g_2, g_3\} = \{h_3, -h_2, h_1, h_0\}$. Also, for the sake of a specific representation, we assume that $\{c_n^0\}$ has six elements in it. From (3.121),

$$\mathbf{c}^1 = \begin{bmatrix} c_{-1}^1 \\ c_0^1 \\ c_1^1 \\ c_2^1 \end{bmatrix} = \begin{bmatrix} h_2 & h_3 & & & & \\ h_0 & h_1 & h_2 & h_3 & & \\ & & h_0 & h_1 & h_2 & h_3 \\ & & & & h_0 & h_1 \end{bmatrix} \begin{bmatrix} c_0^0 \\ c_1^0 \\ c_2^0 \\ c_3^0 \\ c_4^0 \\ c_5^0 \end{bmatrix} = L\mathbf{c}^0.$$

(The truncation evident in the first and last rows of the matrix corresponds to an assumption that data outside the samples are equal to zero. As discussed below, there is another assumption that can be made.)

From (3.120),

$$\mathbf{d}^1 = \begin{bmatrix} d_{-1}^1 \\ d_0^1 \\ d_1^1 \\ d_2^1 \end{bmatrix} = \begin{bmatrix} g_2 & g_3 & & & & \\ g_0 & g_1 & g_2 & g_3 & & \\ & & g_0 & g_1 & g_2 & g_3 \\ & & & & g_0 & g_1 \end{bmatrix} \begin{bmatrix} c_0^0 \\ c_1^0 \\ c_2^0 \\ c_3^0 \\ c_4^0 \\ c_5^0 \end{bmatrix}$$

$$= \begin{bmatrix} h_1 & -h_0 & & & & \\ h_3 & -h_2 & h_1 & -h_0 & & \\ & & h_3 & -h_2 & h_1 & -h_0 \\ & & & & h_3 & -h_2 \end{bmatrix} \begin{bmatrix} c_0^0 \\ c_1^0 \\ c_2^0 \\ c_3^0 \\ c_4^0 \\ c_5^0 \end{bmatrix} = H\mathbf{c}^0.$$

3.18 Sets of Complete Orthogonal Functions

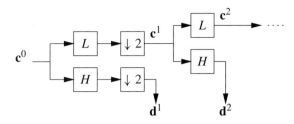

Figure 3.24: Multirate interpretation of wavelet transform

The transform data at the next resolution $\mathbf{d}^2$, and the data $\mathbf{c}^2$, can be obtained (using the same indexing convention as before) as

$$\mathbf{c}^2 = \begin{bmatrix} h_3 & & & \\ h_1 & h_2 & h_3 & \\ & h_0 & h_1 & h_2 \end{bmatrix} \mathbf{c}^1, \quad \mathbf{d}^2 = \begin{bmatrix} g_3 & & & \\ g_1 & g_2 & g_3 & \\ & g_0 & g_1 & g_2 \end{bmatrix} \mathbf{c}^1.$$

It is perhaps worthwhile to point out that the indexing convention on $\mathbf{c}^1$ could be changed (with a corresponding change in (3.123)), so that we interpret $\mathbf{c}^1$ as the vector

$$\mathbf{c}^1 = \begin{bmatrix} c_0^1 & c_1^1 & c_2^1 & c_3^1 \end{bmatrix}.$$

Making this change, the matrix for the second stage transformation would be written as

$$\mathbf{c}^2 = \begin{bmatrix} h_1 & -h_0 & & \\ h_3 & -h_2 & h_1 & -h_0 \\ & & h_3 & -h_2 \end{bmatrix} \mathbf{c}^1,$$

with similar changes for $\mathbf{d}^2$ and its associated transformation matrix. Provided that the same indexing convention is used for the forward transformation as the inverse transformation, the transform is still fully reversible. □

The notation L and H for the matrix operators is deliberately suggestive. The L matrix is a lowpass operator, and the data sequence $\mathbf{c}^1$ is a lowpass sequence. It corresponds to a "blurring" of the original data $\mathbf{c}^0$. The G matrix is a highpass operator, and the data $\mathbf{d}^1$ is highpass (or bandpass) data.

The filtering/subsampling operation represented by these matrices can continue through several stages. The transform coefficients at the end of the process are the collection of data $\mathbf{d}^1, \mathbf{d}^2, \ldots, \mathbf{d}^J$ and $\mathbf{c}^J$, where $\mathbf{c}^J$ is a final course approximation of the original starting data $\mathbf{c}^0$. The wavelet transform computations can also be represented as a filtering/decimation operation, as shown in figure 3.24. The signal $\mathbf{c}^0$ passes through a lowpass and highpass filter, whose outputs are decimated, as indicated by $\boxed{\downarrow 2}$, taking every other sample.

Inverse wavelet transform

The inverse wavelet transform can be obtained by working backwards. Given $\mathbf{d}^j$ and $\mathbf{c}^j$, we wish to find $\mathbf{c}^{j-1}$. We note from (3.115) that

$$\begin{aligned} f^{j-1} &= f^j + \delta^j \\ &= \sum_k c_k^j \phi_{j,k} + \sum_k d_k^j \psi_{j,k}. \end{aligned} \quad (3.124)$$

Then, using the fact that $c_n^{j-1} = \langle f^{j-1}, \phi_{j-1,n} \rangle$ and taking inner-products of both sides of (3.124), we have

$$\begin{aligned} c_n^{j-1} &= \langle f^{j-1}, \phi_{j-1,n} \rangle \\ &= \sum_k c_k^j \langle \phi_{j,k}, \phi_{j-1,k} \rangle + \sum_k d_k^j \langle \psi_{j,k} \phi_{j-1,k} \rangle. \end{aligned} \quad (3.125)$$

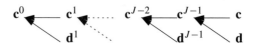

Figure 3.25: Illustration of the inverse wavelet transform

Taking inner products on both sides of (3.119) with $\phi_{j-1,m}$ we observe that

$$\langle \phi_{j,k}, \phi_{j-1,m} \rangle = \sum_n h_{n-2k} \langle \phi_{j-1,n}, \phi_{j-1,m} \rangle = h_{m-2k}$$

by the orthogonality of the ϕ function. Similarly, from (3.116),

$$\langle \psi_{j,k}, \phi_{j-1,m} \rangle = g_{m-2k}.$$

Substituting these into (3.125), we find that

$$c_n^{j-1} = \sum_k c_k^j h_{n-2k} + \sum_k d_k^j g_{n-2k}. \qquad (3.126)$$

This tells us how to go upstream from $\mathbf{c}^j$ and $\mathbf{d}^j$ to $\mathbf{c}^{j-1}$. The process is outlined in figure 3.25. As before, the reconstruction can be expressed in matrix form,

$$\mathbf{c}^{j-1} = L^* \mathbf{c}^j + H^* \mathbf{d}^j,$$

where L^* is the adjoint (conjugate transpose) of L and H^* is the adjoint of H (see section 4.3).

Example 3.18.5 Let us consider a specific numeric example. Using the wavelet with four coefficients, the code in algorithm 3.7 finds the two-scale wavelet transform data $\mathbf{d}^1, \mathbf{d}^2, \mathbf{c}^2$ for the data set $\mathbf{c} = [1, 2, 3, 4, 5, 6]^T$. Also, the inverse transform is found. The pertinent variables of the execution are

$$\mathbf{c}_0 = \begin{bmatrix} 1 \\ 2 \\ 3 \\ 4 \\ 5 \\ 6 \end{bmatrix} \quad \mathbf{d}_1 = \begin{bmatrix} -0.482963 \\ 7.62188e-09 \\ 2.28656e-08 \\ 3.38074 \\ -0.776457 \end{bmatrix} \quad \mathbf{d}_2 = \begin{bmatrix} -0.541266 \\ -0.670753 \\ 0.270032 \\ -0.375 \end{bmatrix} \quad \mathbf{c}_2 = \begin{bmatrix} -0.145032 \\ 0.557132 \\ 8.68838 \\ 1.39952 \end{bmatrix}.$$

Observe that there are six points in the original data, and thirteen points in this transform. The reconstructed signal c0new is equal to the original signal $\mathbf{c}_0$.

Algorithm 3.7 Demonstration of wavelet decomposition
    File: `wavetest.m`

For comparison, algorithm 3.8 shows a decomposition and reconstruction with a different indexing convention. In this case, the transform data is

$$\mathbf{d}_1 = \begin{bmatrix} -0.12941 \\ 0 \\ 0 \\ -1.99191 \end{bmatrix} \quad \mathbf{d}_2 = \begin{bmatrix} -1.14503 \\ 0.195272 \\ -2.33133 \end{bmatrix} \quad \mathbf{c}_2 = \begin{bmatrix} -0.30681 \\ 2.10617 \\ 8.70064 \end{bmatrix}.$$

The reconstructed signal c0new is equal to the original signal. This transform has ten points in it.

3.18 Sets of Complete Orthogonal Functions

Algorithm 3.8 Demonstration of wavelet decomposition (alternative indexing)
File: `wavetesto.m`

The L and H matrices have some interesting properties. In the following theorem, the L and H matrices are assumed to be infinite, so that partial sequences of coefficients do not appear on any rows.

Theorem 3.8 *The L and H operators defined by the operations*

$$L\mathbf{c} = \sum_n h_{n-2k} c_n \qquad H\mathbf{c} = \sum_n g_{n-2k} c_n$$

have the following properties:

1. $HL^H = 0$,
2. $LL^H = I$ and $HH^H = I$, and
3. $L^H L$ and $H^H H$ are mutually orthogonal projections.

Proof Let h_{n-2k} denote the kth column of L^H, and let g_{n-2l} denote the lth row of H. The inner product of these can be written

$$\sum_n h_{n-2k} g_{n-2l} = \sum_n h_{n-2k}(-1)^n h_{2(M+l)+1-n},$$

which is zero by (3.112). Since this is true for any l and k it follows that $HL^H = 0$.

The fact that $LL^H = I$ and $HH^H = I$ is shown by multiplication, using (3.108).

Then we note that $(L^H L L^H L) = L^H(LL^H)L = L^H L$, so $L^H L$ is a projection, and similarly for HH^H. By the fact that $HL^H = 0$ it follows that $L^H L$ and $H^H H$ are orthogonal. Now note that

$$H(L^H L + H^H H) = H(H * H) = H$$

and

$$L(L^H L + H^H H) = L.$$

Thus $L^H L + H^H H$ acts as an identity on the ranges of both H and L, so it is an identity. □

The filtering interpretation for the reconstruction is shown in figure 3.26: the samples are expanded by inserting a zero between every sample, then filtering. When the forward operation and the backward operation are placed together, as shown in figure 3.27, an

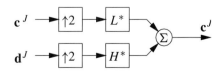

Figure 3.26: Filtering interpretation of an inverse wavelet transform

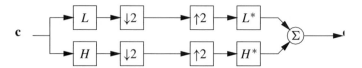

Figure 3.27: Perfect reconstruction filter bank

identity operation from end to end results. One family of such filtering configurations is known as a quadrature mirror filter; it is an example of a perfect reconstruction filter. This multirate configuration is used in data compression, in which the lowpass and highpass signals are quantized using quantizers specialized for the frequency range of the signals.

Periodic wavelet transform

The wavelet transform produces more output coefficients than input coefficients, due to the convolution. If there are n input points and the filters are m points long, then the convolution/decimation operation produces $\lfloor (n+m)/2 \rfloor$ points (or one less, depending on how the indexing is interpreted), so each stage of the transform produces more than half the number of points from the previous stage. Having more transform data than original data is troubling an many circumstances, such as data compression. It is common to assume that the data is periodic and to perform a periodized transform. Suppose that there are L points in $\mathbf{c}^0$,

$$\mathbf{c}^0 = [c_0^0, c_1^0, \ldots, c_{L-1}^0]^T.$$

Then periodized data $\tilde{\mathbf{c}}^0$ is formed (conceptually) by stacking $\mathbf{c}^0$,

$$\tilde{\mathbf{c}}^0 = [\ldots, (\mathbf{c}^0)^T, (\mathbf{c}^0)^T, (\mathbf{c}^0)^T, \ldots]^T.$$

Then an L-point wavelet transform is computed on the periodized data. The effect is that the wavelet transform coefficients appear cyclically shifted around the L and H matrices. For example, with four coefficients and eight data points, the L and H matrices would look like the following:

$$L = \begin{bmatrix} h_0 & h_1 & h_2 & h_3 & & & & \\ & & h_0 & h_1 & h_2 & h_3 & & \\ & & & & h_0 & h_1 & h_2 & h_3 \\ h_2 & h_3 & & & & & h_0 & h_1 \end{bmatrix},$$

$$H = \begin{bmatrix} g_0 & g_1 & g_2 & g_3 & & & & \\ & & g_0 & g_1 & g_2 & g_3 & & \\ & & & & g_0 & g_1 & g_2 & g_3 \\ g_2 & g_3 & & & & & g_0 & g_1 \end{bmatrix}.$$

The same equations used to represent the nonperiodized transforms (3.122) and (3.123), and the inverse transform (3.126), also apply for the periodized transform and its inverse, provided that the indices are taken modulo the appropriate data size.

Wavelet transform implementations

Algorithm 3.9 performs a nonperiodic wavelet transform. The first function, `wavetrans`, sets up some data that is used by the recursively-called function `wave`. Implementation of `wave` is straightforward, with some caution needed to get the indexing started correctly. Since different levels have different lengths of coefficients, an array is also returned indexing the transform coefficients for each level.

3.18 Sets of Complete Orthogonal Functions

Algorithm 3.9 Non-periodic wavelet transform
File: `wavetrans.m`

An inverse nonperiodic wavelet transform is shown in algorithm 3.10.

Algorithm 3.10 Nonperiodic inverse wavelet transform
File: `invwavetrans.m`

Example 3.18.6 The two-level nonperiodic wavelet transform $\mathbf{c} = [1, 2, 3, 4, 5]^T$ using the D_4 coefficients is computed using `[C,ap] = wavetrans(c,d4coeff,2)`, which gives

$C = [-0.1294 \quad 0 \quad 2.8978 \quad -0.647 \quad -1.145 \quad 3.2688 \quad -1.3068 \quad -0.3068 \quad 2.9297 \quad 4.8771]$

$\text{ap} = [5 \quad 1 \quad 5 \quad 8]$,

from which we interpret

$\mathbf{d}_1 = [-0.1294 \quad 0.0000 \quad 2.8978 \quad -0.6470]$,
$\mathbf{d}_2 = [-1.1450 \quad 3.2688 \quad -1.3068]$,
$\mathbf{c}_2 = [-0.3068 \quad 2.9297 \quad 4.8771]$.

The inverse transform computed by `invwave(C,ap,d4coeff)` returns the original data vector. □

Code for the periodized wavelet transform appears in algorithm 3.11, and the periodized inverse wavelet transform is in algorithm 3.12.

Algorithm 3.11 Periodic wavelet transform
File: `invwavetransper.m`

Algorithm 3.12 Inverse periodic wavelet transform
File: `invwavetransper.m`

Applications of wavelets

Wavelets have been used in a variety of applications, of which we mention only a few.

Data compression. One of the most common applications of wavelets is to data compression. A set of data $\mathbf{f}$ is transformed using a wavelet transform. The wavelet transform coefficients smaller than some prescribed threshold are set to zero, and the remaining coefficients are quantized using some uniform quantizer. It is a matter of empirical fact that in most data sets, a large proportion of the coefficients are zeroed out. The truncated/quantized coefficients are then passed through a run-length encoder (and perhaps other lossless encoding techniques), which represents runs of zeros by a single digit indicating how many zeros are in the run.

A more sophisticated version of this algorithm is employed for image compression, in which a two-dimensional wavelet transform is employed. In this case, the hierarchical structure of the wavelet transform is exploited, so that if coefficients on one stage are small, there is a high probability that coefficients underneath are also small. Details of an algorithm of this sort are given in [305].

Time/frequency analysis. Wavelets are naturally employed in the analysis of signals which have a time-varying frequency content, such as speech or geophysical signals.

3.19 Signals as points; digital communications

The vector space viewpoint allows us to view signals, either in discrete or continuous time, as points in a vector space. This signals-as-points interpretation is especially useful in digital communications. In digital communications, a small set of basis functions is chosen—not a complete set—to have certain desired spectral properties. Signals that are transmitted are represented as linear combinations of these points.

As a particular example, let $\phi_1(t)$ and $\phi_2(t)$ be two orthonormal functions as illustrated in figure 3.28(a). (Note: the use of the notation $\phi(t)$ as a basis function in this section is distinct from the notation for $\phi(t)$ as a scaling function in section 3.18.4.) Then a variety of functions, such as those shown in figure 3.28(b), can be formed as linear combinations of $\phi_1(t)$ and $\phi_2(t)$:

$$f_1(t) = 2\sqrt{2}\phi_1(t) + 4\sqrt{2}\phi_2(t) \longrightarrow \mathbf{f}_1 = (2\sqrt{2}, 4\sqrt{2}),$$
$$f_2(t) = 3\sqrt{2}\phi_1(t) - 3\sqrt{2}\phi_2(t) \longrightarrow \mathbf{f}_2 = (3\sqrt{2}, -3\sqrt{2}),$$
$$f_3(t) = 3\sqrt{2}\phi_1(t) + 3\sqrt{2}\phi_2(t) \longrightarrow \mathbf{f}_3 = (3\sqrt{2}, 3\sqrt{2}).$$

Figure 3.28(c) shows the points in $\mathbb{R}^2$ corresponding to the coordinates of the functions.

A function represented by a generalized Fourier series of *m orthonormal* functions

$$f(t) = \sum_{i=1}^{m} c_i \phi_i(t)$$

may be equivalently represented by the set of coordinates

$$f(t) \longleftrightarrow (c_1, c_2, \ldots, c_m) = \mathbf{c},$$

and be conceptualized as a point in $\mathbb{R}^m$. As shown in (3.96), the inner-product relationship between the functions is the same as the inner product between the vectors: if $f_1(t)$ has the coordinate representation $\mathbf{c}_1$ and $f_2(t)$ has the coordinate representation $\mathbf{c}_2$, then

$$\langle f_1, f_2 \rangle = \langle \mathbf{c}_1, \mathbf{c}_2 \rangle,$$

where the inner product on the left is defined for functions and the inner product on the right is defined for vectors. This means that

$$\|f_1\| = \|\mathbf{c}_1\| \tag{3.127}$$

and

$$\|f_1 - f_2\| = \|\mathbf{c}_1 - \mathbf{c}_2\|.$$

3.19 Signals as Points; Digital Communications

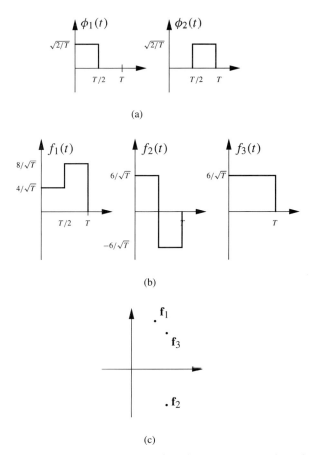

Figure 3.28: Two basis functions, and some functions represented by using them

So the distance can be computed for either the function or the vector. Note that if the basis functions are not normalized, then $\|\mathbf{c}_1\|$ in (3.127) must be normalized according to the norm of the basis functions.

Suppose we have m orthogonal basis functions $\phi_i(t)$, $i = 1, 2, \ldots, m$, and assume that they have support over $[0, T)$. (It is not strictly necessary to deal with orthogonal basis functions, but it makes several of the computations easier. Of course, by the Gram–Schmidt orthogonalization procedure, we can always determine an orthonormal set spanning the same space as a set of nonorthogonal functions, so assuming orthonormality does not represent any loss of generality.)

In the m-dimensional space S spanned by these functions, a set of $M = 2^k$ signal points, known as the *signal constellation*, is selected. Let $\mathbf{s}_1, \mathbf{s}_2, \ldots, \mathbf{s}_M$ denote the signal constellation points, where the points are

$$\mathbf{s}_l = [s_{l1}, s_{l2}, \ldots, s_{lm}]^T.$$

These points in the signal constellation represent the signals that can be sent, $s_l(t)$, $i = 1, 2, \ldots, M$, where

$$s_l(t) = \sum_{l=1}^{m} s_{li} \phi_l(t) \in S.$$

The vector $\mathbf{s}_l$ is sometimes referred to as the *symbol*, while the corresponding $s_l(t)$ is referred to as the *signal*. Normally (though not always), the basis functions $\phi_i(t)$ are designed to last T seconds, and such that $\phi_i(t)$ has support over $[0, T)$. The time T is called the *symbol time*.

Every T seconds, k bits are accepted into the transmitter. These k bits are used to select one of $M = 2^k$ signal points, with its corresponding signal. The transmitted signal $s(t)$ is obtained by concatenating these signals together in time, which we can write as

$$s(t) = \sum_n s_{l_n}(t - nT),$$

where $s_{l_n}(t - nT)$ is the signal that starts at time nT and has support over $t \in [nT, (n+1)T)$, and l_n is the index of the signal selected at the nth symbol time. We will denote the signal that is transmitted at the nth symbol time as $s^n(t)$,

$$s^n(t) = s_{l_n}(t - nT).$$

In a practical system, it is customary to produce the signal $s(t)$ at baseband, then mix it up to some appropriate carrier frequency. In this presentation, we will focus only on the baseband signal $s(t)$. For additional simplicity, we will assume that all signals are real.

3.19.1 The detection problem

In a channel model that is commonly assumed, the signal $s(t)$ is delayed by some delay τ as it passes through the channel, and corrupted by additive noise $v(t)$. The received signal is modeled as

$$r(t) = s(t - \tau) + v(t).$$

Most of the intuitive discussion that follows in this section is accurate only in the case that the noise is Gaussian. We assume that the delay τ is known.

The signal $r(t)$ for $t \in [nT + \tau, (n+1)T + \tau)$ does not, in general, lie in S because of the additive noise. The problem of reliable reception (the detection problem) is to determine the *best* estimate of the transmitted signal $\hat{s}^n(t)$, given $r(t)$. A more formal exploration of this problem is conducted in chapter 11. However, for the purposes of this section we can employ our intuition about how the detection problem should work.

The first step in detection is to project the received signal over one symbol time onto S. The component of the nth received signal in the lth direction (assuming that τ is known) is

$$r_l = \langle r(t), \phi_l(t - \tau - nT) \rangle = \int_{\tau + nT}^{\tau + (n+1)T} r(t)\phi_l(t - \tau - nT)\, dt. \tag{3.128}$$

The processing accomplished by (3.128) is termed a correlator, illustrated in figure 3.29(a). It is also possible to implement the correlator by using a filter with impulse response $h_i(t) = \phi_i(T - t)$. In this case, the filter is termed a *matched filter*. The output of the filter is

$$y(t) = \int r(u) h_i(t-u)\, du = \int r(u) \phi_i(u - t + T)\, dt.$$

Sampling the output at the instant $t = \tau + (n+1)T$ produces the output value r_l (see exercise 3.18-47). The coordinates $\mathbf{r}^n = [r_1, r_2, \ldots, r_m]^T$ represent the projection of the received signal onto S for the nth symbol interval. The detector determines which of the signal points $\mathbf{s}_1, \mathbf{s}_2, \ldots, \mathbf{s}_M$ is closest to $\mathbf{r}$. The closest point corresponds to $\hat{s}^n(t)$, from which it can be determined which bits were sent.

3.19 Signals as Points; Digital Communications

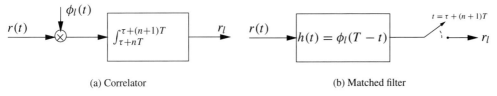

(a) Correlator (b) Matched filter

Figure 3.29: Implementations of digital receiver processing

The projection onto the signal space is illustrated geometrically in figure 3.30 for $m = 2$. The signal $r(t)$ is projected onto the signal-space point **r**. The nearest point in S to **r** is then determined as the estimate of the transmitted signal. The overall processing (using a matched filter implementation) is shown in figure 3.31.

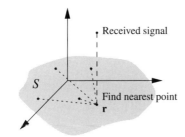

Figure 3.30: Digital receiver processing

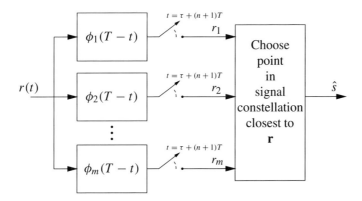

Figure 3.31: Implementation of a matched filter receiver

Example 3.19.1 Let $\phi_1(t) = \sqrt{2/T}\cos(2\pi t)$ and $\phi_2(t) = \sqrt{2/T}\sin(2\pi t)$ for $t \in [0, T)$. These are orthonormal signals. Let the signal constellation be as in figure 3.32. This type of constellation, in which every signal has the same amplitude but different components of phase (due to the combinations of the basis functions), is known as *phase-shift keying* (PSK).

In this signal constellation, suppose that the symbol $\mathbf{s}_0$ is sent, and the projected received signal **r** is as shown. The vector **r** falls in the *decision region* of $\mathbf{s}_0$, shown shaded in the figure. Thus **r** is detected as the signal $\mathbf{s}_0$. □

Of course, it is possible that the noise is severe enough that a received signal is incorrectly detected, and so there is still a nonzero probability of error. Nevertheless, the

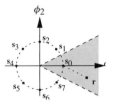

Figure 3.32: PSK signal constellation and detection example

operation of projection and finding nearest neighbor is (for white Gaussian noise) the optimal decision rule. Some examination of the computation of probability of error takes place in chapter 11.

Another way of looking at the detection problem is to find the signal point $s_i(t-\tau-nT)$ that is closest to $r(t)$ for $t \in [nT+\tau, (n+1)T+\tau)$. That is, we wish to find $s_i(t)$ to minimize

$$\int_{\tau+nT}^{\tau+(n+1)T} (r(t) - s_i(t-\tau-nT))^2 \, dt.$$

Expanding this, we want to minimize

$$\int_{\tau+nT}^{\tau+(n+1)T} r^2(t)\,dt - 2\int_{\tau+nT}^{\tau+(n+1)T} r(t)s_i(t-\tau-nT)\,dt + \int_{\tau+nT}^{\tau+(n+1)T} s_i^2(t)\,dt.$$

The first term does not depend upon s_i, and the last term represents signal energy $\|s_i\|^2$, which can be precomputed. The *decision statistic* that we use is

$$z_i = \int_{\tau+nT}^{\tau+(n+1)T} r(t)s_i(t-\tau-nT)\,dt \qquad (3.129)$$

The processing in (3.129) can be done either by a correlator or a matched filter with impulse response $h_i(t) = s_i(T-t)$, sampling the output of the filter at $t = \tau + (n+1)T$. The decision rule, in terms of this nearest signal interpretation, becomes: select the point s_i such that

$$-2z_i + \|s_i\|^2$$

is minimized.

One of the particularly interesting aspects about the vector space viewpoint for digital communications is that it allows different aspects of the problem to be addressed separately. The probability of error for a signal constellation depends ultimately on the geometry of the points in the signal constellation, and the average energy required to send the signals in comparison to the strength of the noise signal. The probability of error is thus completely unaffected by the particular waveforms underlying the signal constellation, provided only that orthonormal waveforms are selected. In contrast, the power spectral density of the transmitted signal depends very strongly on the waveform shapes of the signals transmitted. This separation of probability of error performance from spectral performance leads to better designs.

3.19.2 Examples of basis functions used in digital communications

A variety of waveforms can be used in digital communications. We met in example 3.19.1 the basis functions used for phase-shift keying. Here is a brief survey of some other simple signaling waveforms.

On–off keying, or OOK. When a single basis function $\phi_1(t)$ is used (regardless of its waveshape), with one point in the signal constellation at the origin and the other somewhere along the $\phi_1(t)$ axis, a signaling technique known as on–off keying is produced. (See figure 3.33(a).)

3.19 Signals as Points; Digital Communications

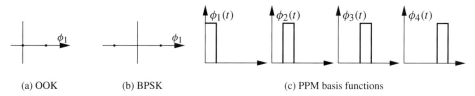

(a) OOK (b) BPSK (c) PPM basis functions

Figure 3.33: Illustration of concepts of various signal constellations

Binary phase-shift keying, or BPSK. When a single basis function is used with two points in the signal constellation $s_1 = -s_0$, the resulting signaling is known as BPSK. (See figure 3.33(b).)

Pulse-position modulation. or PPM is obtained by using a set of N short orthogonal pulses, as shown in figure 3.33(c), which shows a four-dimensional set of basis functions. Usually, a single amplitude is employed along each orthogonal axis.

Frequency-shift keying. or FSK is obtained by using M sinusoidal signals of different frequencies which are spaced so that they are orthogonal over the interval $[0, T)$.

Quadrature-amplitude modulation. or QAM is obtained by using two orthogonal basis functions, as for PSK, but by employing both amplitude and phase modulation. Figure 3.33(d) shows a QAM signal constellation with $M = 16$ points.

3.19.3 Detection in nonwhite noise

In the last section, the channel noise was assumed to be white and the optimal detector was obtained by simply projecting the received signal $r(t)$ onto the signal space S with an orthogonal projection. When the noise is not white, however, the noise may tend to pull the received signal predilectably toward different spectral components. In this case, a more sophisticated filter must be used to obtain a projection onto the signal space to compensate for any bias introduced by the noise. The design of the filter provides yet another application of the Cauchy–Schwarz inequality.

We desire to find a filter with impulse response $h(t)$, so that when $r(t)$ is passed through the filter, the ratio of the signal power to the noise power is maximized at some particular sample time t_0, as shown in figure 3.34. When

$$r(t) = s(t) + v(t),$$

then the output of the filter is

$$r(t) * h(t) = s(t) * h(t) + v(t) * h(t) = s_o(t) + n_o(t).$$

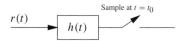

Figure 3.34: Block diagram for detection processing

Signal power

Assuming a causal signal $s(t)$, the portion of the output due to the input signal is

$$s_o(t) = \int_0^t s(\tau) h(t - \tau) \, d\tau.$$

At the time instant t_0, we have

$$s_o(t_0) = \int_0^{t_0} s(\tau) h(t_0 - \tau) \, d\tau.$$

If we assume in addition that $s(t)$ is supported only over $[0, t_0)$ (so that we are using the entire signal s to make our decision), then we can write

$$s_o(t_0) = \int_{-\infty}^{\infty} s(\tau) h(t_0 - \tau) \, d\tau.$$

Let $w(\tau) = \overline{h}(t_0 - \tau)$. Then

$$s_o(t_0) = \int_{-\infty}^{\infty} s(\tau) \overline{w}(\tau) \, dt = \int_{-\infty}^{\infty} S(f) \overline{W}(f) \, dt,$$

where $S(f)$ and $W(f)$ are the Fourier transforms, respectively, of $s_0(t)$ and $w(t)$, and where the equality follows by Parseval's theorem. Using the definition of w, we have

$$s_o(t_0) = \int_{-\infty}^{\infty} S(f) H(f) e^{j 2\pi f t_0} \, dt.$$

The signal power S at some time instant t_0 is $|s_o(t_o)|^2$, or

$$S = \left| \int_0^{\infty} S(f) H(f) e^{j 2\pi f t_0} \, df \right|^2. \tag{3.130}$$

Noise power

Let the PSD of $\nu(t)$ be $S_\nu(f)$. Then the PSD of the noise component at the output of the filter is

$$S_\nu(f) |H(f)|^2,$$

and the total noise power is

$$N = \int_{-\infty}^{\infty} S_\nu(f) |H(f)|^2 \, df. \tag{3.131}$$

The ratio of the signal to noise (SNR) power is, from (3.130) and (3.131),

$$\frac{S}{N} = \frac{\left| \int_0^{\infty} S(f) H(f) e^{j 2\pi f t_0} \, df \right|^2}{\int_{-\infty}^{\infty} S_\nu(f) |H(f)|^2 \, df}. \tag{3.132}$$

The problem can now be stated as: determine a filter with transfer function $H(f)$ that maximizes S/N in (3.132). There is a tradeoff here: the wider the bandwidth of $H(f)$, the more signal gets through, but the more noise also gets through. This is a maximization problem that looks difficult, since the approach to maximization usually involves taking a derivative, and at this stage of development it is difficult to see what it would mean to take a derivative with respect to a transfer function. As we shall see, we will not have to take a derivative at all.

Maximizing S/N

The key to maximizing (3.132) is to use the Cauchy–Schwarz inequality in its integral form,

$$\left|\int x(f)\overline{y}(f)\,df\right|^2 \leq \int |x(f)|^2\,df \int |y(f)|^2\,df. \quad (3.133)$$

We can write

$$\frac{\left|\int_0^\infty S(f)H(f)e^{j2\pi f t_0}\,df\right|^2}{\int_{-\infty}^\infty S_v(f)|H(f)|^2\,df} = \frac{\left|\int_{-\infty}^\infty (H(f)\sqrt{S_v(f)})\left(\frac{S(f)}{\sqrt{S_v(f)}}e^{j2\pi f t_0}\right)df\right|}{\int_{-\infty}^\infty S_v(f)|H(f)|^2\,df}$$

$$\leq \frac{\int_{-\infty}^\infty S_v(f)|H(f)|^2\,df \int_{-\infty}^\infty \frac{|S(f)|^2}{S_v(f)}\,df}{\int_{-\infty}^\infty S_v(f)|H(f)|^2\,df}$$

$$= \int_{-\infty}^\infty \frac{|S(f)|^2}{S_v(f)}\,df, \quad (3.134)$$

where (3.134) comes from the Cauchy–Schwarz inequality, using (by comparison with (3.133))

$$x(f) = \sqrt{S_v(f)}H(f) \quad \text{and} \quad \overline{y}(f) = \frac{S(f)}{\sqrt{S_v(f)}}e^{j2\pi f t_0}.$$

By this inequality, an upper bound on the SNR has been obtained which is independent of any filter and which, therefore, must be the largest possible regardless of the filter employed. The filter that can be used to achieve this upper bound with equality is found by employing the conditions under which the Cauchy–Schwarz inequality is satisfied with equality; in this case, that means that we must have $x(f) = Cy(f)$ for some nonzero complex constant C, or

$$H(f) = C\frac{\overline{S(f)}}{S_v(f)}e^{-j2\pi f t_0}$$

for any nonzero complex constant C.

If $v(t)$ is white, so that $S_v(f) = \frac{1}{2}N_0$, then we have

$$H(f) = C\frac{\overline{S(f)}}{N_0/2}e^{-j2\pi f t_0}.$$

Assume for ease of notation that $C = 2/N_0$. Then, taking the inverse transform, we have

$$h(t) = \overline{s}(t_0 - t).$$

The output of the filter with this impulse response when $t_0 = T$ is

$$\int_0^T r(t)\overline{s}(t)\,dt.$$

3.20 Exercises

3.1-1 There is a connection between Grammians and linear independence, as demonstrated in theorem 3.1. We explore this connection further in this problem.

Let $\{\mathbf{p}_1, \mathbf{p}_2, \ldots, \mathbf{p}_n\}$ be a set of vectors, and let us suppose that the first $k-1$ vectors of this set have passed a test for linear independence. We form

$$\mathbf{e}_k = c_{k-1}^k \mathbf{p}_1 + c_{k-2}^k \mathbf{p}_2 + \cdots c_1^k \mathbf{p}_{k-1} + \mathbf{p}_k$$

and want to know if $\mathbf{e}_k$ is equal to zero for any set of coefficients

$$\mathbf{c}^k = \begin{bmatrix} c_{k-1}^k, c_{k-2}^k, \ldots, c_1^k, 1 \end{bmatrix}^T.$$

If so, then $\mathbf{p}_k$ is linearly dependent. Let
$$A_k = [\mathbf{p}_1, \mathbf{p}_2, \ldots, \mathbf{p}_k]$$
be a data matrix, and let $R_k = A_k^H A_k$ be the corresponding Grammian.

(a) Show that the squared error can be written as
$$\mathbf{e}_k^H \mathbf{e}_k = \sigma_k^2 = \mathbf{c}^H \begin{bmatrix} R_{k-1} & \mathbf{h}_k \\ \mathbf{h}_k^H & r_{kk} \end{bmatrix} \mathbf{c}_k \qquad (3.135)$$
for some $\mathbf{h}_k$ and r_{kk}. Identify $\mathbf{h}_k$ and r_{kk}.

(b) Determine the minimum value of σ_k^2 by minimizing (3.135) with respect to $\mathbf{c}_k$, subject to the constraint that the last element of $\mathbf{c}_k$ is equal to 1. Hint: take the gradient of
$$\mathbf{c}^H \begin{bmatrix} R_{k-1} & \mathbf{h}_k \\ \mathbf{h}_k^H & r_{kk} \end{bmatrix} \mathbf{c}_k - \lambda(\mathbf{c}^H \mathbf{d} - 1),$$
where λ is a Lagrange multiplier and $\mathbf{d} = [0, 0, \ldots, 0, 1]^T$. Show that we can write the corresponding equations as
$$\begin{bmatrix} R_{k-1} & \mathbf{h}_k \\ \mathbf{h}_k^H & r_{kk} \end{bmatrix} \mathbf{c}_k = \sigma_k^2 \mathbf{d}. \qquad (3.136)$$

(c) Show that (3.136) can be manipulated to become
$$\sigma_k^2 = r_{kk} - \mathbf{h}_k^H R_{k-1}^{-1} \mathbf{h}_k.$$

The quantity σ_k^2 is called the *Schur complement* of R_k. If $\sigma_k^2 = 0$, then $\mathbf{p}_k$ is linearly dependent.

3.5-2 Referring to (3.30), show that
$$(I - A(A^H A)^{-1} A^H)$$
is positive semidefinite, and hence that the minimum error $\mathbf{e}_{\min}$ has smaller norm than the original vector $\mathbf{x}$. Hint: consider $0 \leq \|B\mathbf{x}\|^2$, where $B = I - A(A^H A)^{-1} A^H$.

3.8-3 Consider the set of data
$$x = \{2, 2.5, 3, 5, 9\} \qquad y = \{-4.2, -5, 2, 1, 24.3\}.$$

(a) Make a plot of the data.

(b) Determine the best least-squares line that fits this data and plot the line.

(c) Assuming that the first and last points are believed to be the most accurate, formulate a weighting matrix and compute a weighted least-squares line that fits the data. Plot this line.

3.8-4 Formulate the regression problem (3.34) in a linear form as in (3.37).

3.8-5 Formulate the regression problem (3.35) in a linear form as in (3.37).

3.8-6 Formulate the regression $y \approx ce^{ax}$ as a linear regression problem, with regression parameters c and a.

3.8-7 Formulate the regression $y \approx ax^b$ as a linear regression problem.

3.8-8 Perform the computations to verify the slope and intercept of the linear regression in (3.39).

3.8-9 As a measure of fit in a correlation problem, the correlation coefficient, analogous to (1.49), can be obtained as

$$\rho = \frac{\langle \mathbf{x}, \mathbf{y} \rangle - \langle \mathbf{x}, \mathbf{1} \rangle \langle \mathbf{y}, \mathbf{1} \rangle}{(\|\mathbf{x}\| - \langle \mathbf{x}, \mathbf{1} \rangle)(\|\mathbf{y}\| - \langle \mathbf{y}, \mathbf{1} \rangle)}.$$

The correlation coefficient $\rho = \pm 1$ if x and y are exactly functionally related, and $\rho = 0$ if they are independent. For the linear regression in (3.38), determine an explicit expression for ρ.

3.8-10 Define an inner product between matrices X and Y as

$$\langle X, Y \rangle = \text{tr}(XY^H),$$

where $\text{tr}(\cdot)$ is the sum of the diagonal elements (see section C.3). We want to approximate the matrix Y by the scalar linear combination of matrices $X_1, X_2, \ldots, X_m$, as

$$Y = c_1 X_1 + c_2 X_2 + \cdots + c_m X_m + E.$$

Using the orthogonality principle, determine a set of normal equations that can be used to find $c_1, c_2, \ldots, c_m$ that minimize the induced norm of E.

3.8-11 For the ARMA input/output relationship of (1.2), determine a set of linear equations for determining the ARMA model parameters $\{a_1, a_2, \ldots, a_p, b_0, b_1, \ldots, b_q\}$, assuming that the model or (p, q) is known, and that the input is known.

3.8-12 For the data sequence $\{1, 1, 2, 3, 5, 8, 13\}$:

(a) Write down the data matrix A and the Grammian $A^H A$ using (i) the covariance, and (ii) the autocorrelation methods.

(b) We desire to use this sequence to train a simple linear predictor. The "desired signal" $d[t]$ is the value of $x[t]$, and the data used are the two prior samples. That is,

$$x[t] = a_1 x[t-1] + a_2 x[t-2] + e[t],$$

where $e[t]$ is the prediction error. Determine the least-squares coefficients for the predictor using the covariance and autocorrelation methods

(c) Determine the minimum least-squares error for both methods.

3.8-13 Consider a data sequence $\{x[t]\}$, the correlation matrix R is

$$R = \begin{bmatrix} .5 & .3 \\ .3 & .5 \end{bmatrix}$$

and the cross-correlation vector $\mathbf{p}$ with a desired signal is

$$\mathbf{p} = \begin{bmatrix} .2 \\ .5 \end{bmatrix}.$$

Determine the optimal weight vector.

3.8-14 Consider a zero-mean random vector $\mathbf{x} = [x_1, x_2, x_3]$ with covariance

$$\text{cov}(\mathbf{x}) = E[\mathbf{x}\mathbf{x}^T] = \begin{bmatrix} 1 & .7 & .5 \\ .7 & 4 & .2 \\ .5 & .2 & 3 \end{bmatrix}.$$

(a) Determine the optimal coefficients of the predictor of x_1 in terms of x_2 and x_3,

$$\hat{x}_1 = c_1 x_2 + c_2 x_3.$$

(b) Determine the minimum mean-squared error.

(c) How is this estimator modified if the mean of **x** is $E\mathbf{x} = [1, 2, 3]^T$?

3.8-15 [132] A discrete-time radar signal is transmitted as

$$s[t] = A_0 e^{-j\omega_0 t}.$$

The sampled noisy received signals are represented as

$$x[t] = A_1 e^{-j\omega_1 t} + v[t],$$

where ω_1 is the received signal frequency, in general different from ω_0 because of Doppler shift, and $v[t]$ is a white-noise signal with variance σ_n^2. Let

$$\mathbf{x}[t] = [x[0], x[1], \ldots, x[m-1]]^T$$

be a vector of received signal samples.

(a) Show that

$$R = E[\mathbf{x}[t]\mathbf{x}^H[t]] = \sigma_v^2 I + \sigma_1 \mathbf{s}(\omega_1)\mathbf{s}^H(\omega_1),$$

where

$$\mathbf{s}(\omega_1) = [1, e^{-j\omega_1}, e^{-j2\omega_1}, \ldots, e^{-(m-1)j\omega_1}]^T \quad \text{and} \quad \sigma_1^2 = E[|A_1|^2].$$

(b) The time series $x[t]$ is applied to an FIR Wiener filter with m coefficients, in which the cross-correlation between $x[t]$ and the desired signal $d[t]$ is preset to

$$\mathbf{p} = \mathbf{s}(\omega_0).$$

Determine an expression for the tap-weight vector of the Wiener filter.

3.11-16 A channel with transfer function

$$H_2(z) = \frac{1}{1 - .2z^{-1}}$$

and output $u[t]$ is driven by an $AR(1)$ signal $d[t]$ generated by

$$d[t] - .4d[t-1] = v(t),$$

where $v(t)$ is a zero-mean white-noise signal with $\sigma_v^2 = 2$. The channel output is corrupted by noise $n[t]$ with variance $\sigma_n^2 = 1.5$, to produce the signal

$$f[t] = u[t] + n[t].$$

Design a second-order Wiener equalizer to minimize the average squared error between $f[t]$ and $d[t]$.

3.11-17 **Linear prediction** A common application of Wiener filtering is in the context of linear prediction. Let $d[t] = x[t]$ be the desired value, and let

$$\hat{x}[t] = \sum_{i=1}^{m} w_{f,i} x[t-i]$$

be the predicted value of $x[t]$ using an mth order predictor based upon the measurements $\{x[t-1], x[t-2], \ldots, x[t-m]\}$, and let

$$f_m[t] = x[t] - \hat{x}[t]$$

be the *forward prediction error*. Then

$$f_m[t] = \sum_{i=0}^{m} a_{f,i} x[t-i],$$

where $a_{f,0} = 1$ and $a_{f,i} = -w_{f,i}$, $i = 1, 2, \ldots, M$.

Assume that $x[t]$ is a zero-mean random sequence. We desire to determine the optimal set of coefficients $\{w_{f,i}, i = 1, 2, \ldots, M\}$ to minimize $E[|f_M[t]|^2]$.

(a) Using the orthogonality principle, write down the normal equations corresponding to this minimization problem. Use the notation $r[j - l] = E[x[t - l]\overline{x}[t - j]]$ to obtain the Wiener–Hopf equation

$$R\mathbf{w}_f = \mathbf{r},$$

where $R = E[\mathbf{x}[t-1]\mathbf{x}^H[t-1]]$, $\mathbf{r} = E[\mathbf{x}[t-1]x[t]]$, and $\mathbf{x}[t-1] = [\overline{x}[t-1], \overline{x}[t-2], \ldots, \overline{x}[t-m]]^T$.

(b) Determine an expression for the minimum mean-squared error, $P_m = \min E[|f_M[t]|^2]$.

(c) Show that the equations for the optimal weights and the minimum mean-squared error can be combined into *augmented Wiener–Hopf* equations, as

$$\begin{bmatrix} r[0] & \mathbf{r}^H \\ \mathbf{r} & R \end{bmatrix} \begin{bmatrix} 1 \\ -\mathbf{w}_f \end{bmatrix} = \begin{bmatrix} P_m \\ \mathbf{0} \end{bmatrix}$$

(d) Suppose that $x[t]$ happens to be an $AR(m)$ process driven by white noise $v[t]$, such that it is the output of a system with transfer function

$$H(z) = \frac{1}{1 + \sum_{k=1}^{m} a_k z^{-k}},$$

Show that the prediction coefficients are $w_{f,k} = -a_k$, and hence the coefficients of the prediction error filter $f_m[t]$ are

$$a_{f,i} = a_i.$$

(Hint: see section 1.4.2; write down the Yule-Walker equations.) Hence, conclude that in this case the forward prediction error $f_m[t]$ is a *white-noise sequence*. The prediction-error filter can thus be viewed as a *whitening filter* for the signal $x[t]$.

(e) Now let

$$\hat{x}[t - m] = \sum_{i=1}^{m} w_{b,i} x[t - i + 1]$$

be the *backward predictor* of $x[t - m]$ using the data $x[t - m + 1], x[t - m + 2], \ldots, x[t]$, and let

$$b_m[t] = x[t - m] - \hat{x}[t - m]$$

be the backward prediction error. A backward predictor seems strange—after all, why predict what we should have already seen—but the concept will have useful applications in fast algorithms for inverting the autocorrelation matrix. Show that the Wiener–Hopf equations for the optimal backward predictor can be written as

$$R\mathbf{w} = \overline{\mathbf{r}^B}, \tag{3.137}$$

where $\mathbf{r}^B$ is the backward ordering of $\mathbf{r}$ defined above.

(f) From (3.137), show that

$$R^H \overline{\mathbf{w}_b^B} = \mathbf{r},$$

where $\mathbf{w}_b^B$ is the backward ordering of $\mathbf{w}_b$. Hence, conclude that

$$\overline{\mathbf{w}_b^B} = \mathbf{w}_f;$$

that is, the optimal backward prediction coefficients are the reversed conjugated optimal forward prediction coefficients.

3.11-18 Let
$$x[t] = 0.8x[t-1] + v[t],$$
where $v[t]$ is a white-noise zero-mean, unit-variance noise process. We want to determine an optimal predictor.

(a) If the order of the predictor is 2, determine the optimal predictor $\hat{x}[t]$.

(b) If the order of the predictor is 1, determine the optimal predictor $\hat{x}[t]$.

3.11-19 **Random vectors** The mean-squared methods to this point have been for random scalars. Suppose we have the random vector approximation problem
$$\mathbf{y} = c_1 \mathbf{p}_1 + c_2 \mathbf{p}_2 + \cdots + c_m \mathbf{x}_m + \mathbf{e},$$
in which we desire to find an approximation $\mathbf{y}$ in such a way that the norm of $\mathbf{e}$ is minimized. Let us define an inner product between random vectors as
$$\langle \mathbf{x}, \mathbf{y} \rangle = \mathrm{tr}(E[\mathbf{x}\mathbf{y}^H]).$$

(a) Based upon this inner product and its induced norm, determine a set of normal equations for finding $c_1, c_2, \ldots, c_m$.

(b) As an exercise in computing gradients, use the formula for the gradient of the trace (see appendix E) to arrive at the same set of normal equations.

3.11-20 **Multiple gain-scaled vector quantization** Let X and Y be vector spaces of the same dimensionality. Suppose that there are two sets of vectors $\mathcal{X}_1, \mathcal{X}_2 \subset X$. Let $\mathcal{Y}$ be the set of vectors *pooled* from $\mathcal{X}_1$ and $\mathcal{X}_2$ by the invertible matrices T_1 and T_2, respectively. That is, if $\mathbf{x} \in \mathcal{X}_i$, then $\mathbf{y} = T_i \mathbf{x}$ is a vector in $\mathcal{Y}$. Indicate that a vector $\mathbf{y} \in \mathcal{Y}$ came from a vector in $\mathcal{X}_i$ by a superscript i, so $\mathbf{y}^i \in \mathcal{Y}$ means that there is a vector $\mathbf{x} \in \mathcal{X}$ such that $\mathbf{y}^i = T_i \mathbf{x}$. Distances relative to a vector $\mathbf{y}^i \in \mathcal{Y}$ are based upon the l_2 norm of the vectors obtained by mapping back to $\mathcal{X}_i$, so that

$$d(\mathbf{y}^i, \mathbf{y}) = \|\mathbf{y}^i - \mathbf{y}\| \triangleq \|\mathbf{y}^i - \mathbf{y}\|_i = \|T_i^{-1}(\mathbf{y}^i - \mathbf{y})\| = (\mathbf{y}^i - \mathbf{y})^T W_i (\mathbf{y}^i - \mathbf{y}),$$

where $W_i = T_i^{-T} T_i^{-1}$. This is a weighted norm, with the weighting dependent upon the vector in $\mathcal{Y}$. (Note: in this problem $\|\cdot\|_1$ and $\|\cdot\|_2$ refer to the weighted norm for each data set, not the l_1 and l_2 norms, respectively.)

We desire to find a *single* vector $\mathbf{y}_0 \in Y$ that is the best representation of the data pooled from both data sets, in the sense that

$$\sum_{\mathbf{y} \in \mathcal{Y}} \|\mathbf{y} - \mathbf{y}_0\| = \sum_{\mathbf{y}^1 \in \mathcal{Y}} \|\mathbf{y}^1 - \mathbf{y}_0\|_1 + \sum_{\mathbf{y}^2 \in \mathcal{Y}} \|\mathbf{y}^2 - \mathbf{y}_0\|_2$$

is minimized. Show that
$$\mathbf{y}_0 = Z^{-1} \mathbf{r},$$
where
$$Z = \sum_{\mathbf{y}^1 \in \mathcal{Y}} W_1 + \sum_{\mathbf{y}^2 \in \mathcal{Y}} W_2 \quad \text{and} \quad \mathbf{r} = \sum_{\mathbf{y}^1 \in \mathcal{Y}} W_1 \mathbf{y}^1 + \sum_{\mathbf{y}^2 \in \mathcal{Y}} W_2 \mathbf{y}^2.$$

Hint: this is probably easier using gradients than trying to identify the appropriate inner product.

3.20 Exercises

3.12-21 Assume the estimated autocorrelation

$$\hat{r}[n] = \frac{1}{N} \sum_{k=1+n}^{N} x(k)\bar{x}(k-n).$$

(a) Take the expectation $E[\hat{r}[n]]$ and show that it is not equal to $r[n]$, the true value of the autocorrelation.

(b) Determine a scaling factor to make the $\hat{r}[n]$ an unbiased estimate.

(c) Write a MATLAB function that computes $\hat{r}(n)$ from (3.65).

3.13-22 Let x_t, y_t, and v_t be continuous-time random processes, with $y_t = x_t + v_t$, and $S_v(s) = 1$. Determine an optimal causal filter $h(t)$ to determine $x(t)$ when:

(a) The PSD of $x(t)$ is

$$S_x(s) = \frac{s^2 - 16}{s^4 - 53s^2 + 196}.$$

(b) The PSD of $x(t)$ is

$$S_x(s) = \frac{s^4 - 10s^2 + 9}{s^4 - 53s^2 + 196}.$$

3.13-23 **(Spectral factorization; the Fejèr–Riesz theorem)** Because of the importance of the canonical factorization in signal processing, it is of interest to determine when a "square root" of a function exists. In this problem you will prove the following: If $W(z) = \sum_{n=-m}^{m} w[n]z^{-n}$ is real and $W(e^{j\omega}) \geq 0$ for all ω, then there is a function

$$Y(z) = \sum_{n=0}^{} m y_n z^{-n}$$

such that $W(e^{j\omega}) = |Y(e^{j\omega})|^2$.

(a) Show that $w_{-n} = \overline{w}_n$.
(b) Show that $\overline{W}(z) = W(1/\bar{z})$.
(c) Show that if z_i is a root of $W(z)$, then $1/\bar{z}_i$ is a root of $W(z)$.
(d) Argue that if $z_i = e^{j\theta_i}$ is a root on the unit circle, then it must have even multiplicity.
(e) Let $\mathcal{Z} = \{z_i : W(z_i) = 0; |z_i| \leq 1, $ (only half the roots on $|z| = 1$)$\}$ be the set of roots inside, and half those on, the unit circle. Then $\mathcal{Z}$ has m elements and

$$W(z) = Az^{-m} \prod_{i=1}^{m}(z - z_i) \prod_{i=1}^{m}(z\bar{z}_i - 1),$$

From this form, find $Y(z)$.

3.13-24 **Filtering in White Noise** Let x_t, y_t and v_t be discrete-time random processes with

$$y_t = x_t + v_t$$

and

$$S_v(z) = 1,$$
$$S_x(z) = \frac{b(z)}{a(z)},$$

where $b(z)$ and $a(z)$ are polynomials in z with the degree of $b(z)$ strictly **lower** than the degree of $a(z)$. Furthermore, assume $R_{xv}(t) \equiv 0$. Show that (3.82) holds in the discrete-time

case; that is, show that
$$H(z) = 1 - \frac{1}{S_y^+(z)}.$$

3.13-25 Let
$$y_t = x_t + v_t$$
where
$$R_v(t) = \frac{2}{3}\delta_t,$$
$$R_y(t) = \frac{10}{27}\left(\frac{1}{2}\right)^{|t|},$$

with $Ex_t = Ev_t = Ex_tv_t = 0$. Show that

(a) $S_y(z) = \frac{(1-\frac{2}{3}z)(1-\frac{2}{3}\frac{1}{z})}{(1-\frac{1}{2}z)(1-\frac{1}{2}\frac{1}{z})}$. and thus obtain $S_y^-(z)$ and $S_y^+(z)$.

(b) $\left\{\frac{S_{xy}(z)}{S_y^-(z)}\right\}_+ = \frac{\frac{1}{3}}{1-\frac{1}{2z}}$ and, thus, that the Wiener filter is
$$H(z) = \frac{\frac{1}{3}}{1 - \frac{1}{3z}}.$$

3.13-26 Let x_t, y_t, and v_t be discrete-time random processes with $y_t = x_t + v_t$, $S_v(z) = 1$, and
$$S_x(z) = \frac{z^4 - 9.0067z^3 + 28.04z^2 - 9.0067z + 1}{z^4 - 2.0111z^3 + 3.0446z^2 - 2.0111z + 1}.$$

Determine the filter $h[z]$ to optimally predict x_{t+2}.

3.14-27 Let $h(t)$ be the impulse response of a system, and let $y(t) = x(t) * h(t)$. Show that
$$\int_0^T y(t)\,dt = y(t) * k(t),$$
where $k(t)$ is the integral of the impulse response,
$$k(t) = \int_0^t h(t)\,dt.$$

3.14-28 A system is known to have impulse response $h(t) = 3e^{-2t} + 4e^{-5t}$, and is initially relaxed (initial conditions are zero). Determine an input $x(t)$ so that the output satisfies the conditions
$$y(2) = 2 \quad \text{and} \quad \int_0^2 y(t)\,dt = 3,$$
in such a way that the input energy $\|x(t)\|^2$ is minimized.

3.14-29 Let $h[t] = (0.2)^t + 3(0.4)^t$ for $k \geq 0$ be the impulse response of a discrete-time system with zero initial conditions. It is desired to determine a causal input sequence $x[t]$, such that the output $y[t] = h[t] * f[t]$ satisfies the constraints
$$y[10] = 5,$$
$$\sum_{j=0}^{10} y[j] = 2,$$
and such that the input energy $\sum_{k=0}^{10} |x[t]|^2$ is minimized. Formulate this as a dual approximation problem and find the minimizing sequence $x[t]$.

3.20 Exercises

3.15-30 [209] Using the projection theorem, solve the finite dimensional problem

$$\text{minimize } \mathbf{x}^H Q\mathbf{x}$$

$$\text{subject to } A\mathbf{x} = \mathbf{b},$$

where $\mathbf{x} \in \mathbb{C}^n$, Q is a positive-definite symmetric matrix, and A is an $m \times n$ matrix with $m < n$.

3.15-31 [209] Let $\mathbf{x}$ be a vector in a Hilbert space S and let $\{\mathbf{x}_1, \mathbf{x}_2, \ldots, \mathbf{x}_n\}$ and $\{\mathbf{y}_1, \mathbf{y}_2, \ldots, \mathbf{y}_m\}$ be sets of linearly independent vectors in S. We desire to minimize $\|\mathbf{x} - \hat{\mathbf{x}}\|$, while satisfying

$$\mathbf{x} \in M = \text{span}(\{\mathbf{x}_1, \mathbf{x}_2, \ldots, \mathbf{x}_m\})$$

and $\langle \hat{\mathbf{x}}, \mathbf{y}_i \rangle = c_i$, $i = 1, 2, \ldots, m$. Find equations for the solution which are similar to the normal equations.

3.17-32 Show that the functions defined by

$$p_k[t] = \frac{1}{\sqrt{N}} e^{j2\pi kt/N}$$

are orthonormal with respect to the inner product

$$\langle x[t], y[t] \rangle = \sum_{k=0}^{N-1} x[t]\overline{y}[t].$$

3.17-33 Let $g(t) = e^{-t/2}$ for $0 \leq t \leq \pi$, and let $f(t)$ be the π-periodic extension of $g(t)$,

$$f(t) = \sum_k g(t - k\pi).$$

(a) Find the Fourier series coefficients of $f(t)$.

(b) Find the sum of the series

$$\sum_n \left(\frac{2^2}{\pi^2} \frac{1}{1 + 16n^2} \right).$$

Hint: Use Parseval's theorem.

3.18-34 Show that the definition of Chebyshev polynomials (3.101) satisfies the recurrence in (3.100) for $|t| < 1$. Show for $|t| > 1$ that $T_n(x) = \cosh(n \cosh^{-1} x)$ satisfies the recursion (3.100).

3.18-35 **The Christoffel–Darboux formula**

(a) Using (3.98), show that the polynomials $p_k(t)$, orthogonal with respect to the inner product $\langle f, g \rangle_w = \int_a^b f(t)g(t)\,dt$, satisfy

$$\int_a^b p_n(t) p_{n+1}(t) w(t)\,dt = a_n.$$

Also show that

$$c_n = a_{n-1}.$$

(b) Consider the partial sum

$$S_n(t) = \sum_{k=0}^n \langle f, p_k \rangle_w p_k(t).$$

Show that the sum can be written as

$$S_n(t) = \int_a^b f(y) K_n(x, y) w(y)\,dy,$$

where

$$K_n(x, y) = \frac{a_n(p_{n+1}(x)p_n(y) - p_n(x)p_{n+1}(y))}{x - y}$$

and where a_n comes from (3.98). This formula for $K_n(x, y)$ is known as the *Christoffel–Darboux* formula, and is analogous to the Dirichlet kernel of Fourier series. Hint: form $(x - y)K(x, y)$ and use the results from part (a).

3.18-36 Show that the each of the polynomials produced by orthogonalizing $\{1, t, t^2, \ldots\}$ using the Gram–Schmidt procedure over the interval $[a, b]$ has zeros which are real, simple, and located in (a, b).

3.18-37 In this exercise we introduce the idea of *Gaussian quadrature*, a fast and important method of numerical integration. The idea is to approximate the integral as a summation:

$$\int_a^b f(t)\, dt \approx \sum_{i=1}^m a_i f(t_i).$$

Unlike many conventional numerical integration formulas, in Gaussian quadrature the abscissas are not evenly spaced. The problem is to find the $\{t_i\}$ (abscissas) and $\{a_i\}$ (weights) so that the integral is as accurate as possible. In the Gaussian quadrature method of numeric integration, for polynomials up to degree $2m - 1$ the result of the integration is *exact*. For sufficiently smooth nonpolynomial functions the method is often very accurate. The solution makes significant use of orthogonal polynomials. For the purposes of this exercise, we will assume the inner product $\langle f, g \rangle = \int_{-1}^1 f(t)g(t)\, dt$.

(a) As this first part shows, without loss of generality, we may restrict attention to the interval $a = -1, b = 1$. Show that for the integral

$$\int_a^b g(x)\, dx$$

the substitution

$$t = \frac{1}{b - a}(2x - a - b)$$

leads to an integral of the form

$$\int_{-1}^1 f(t)\, dt.$$

(Hence the limits of a and b can be converted to limits of -1 to 1.)

(b) If $\{p_n(t)\}$ is a set of polynomials orthogonal over $[-1, 1]$, where $p_n(t)$ is a polynomial of degree n, show that

$$\langle p(t), p_m(t) \rangle = 0$$

for all polynomials $p(t)$ of degree $\leq m - 1$.

(c) Let $f(t)$ be a polynomial of degree $2m - 1$. Show that $f(t)$ can be written as

$$f(t) = q(t)p_n(t) + r(t),$$

where $q(t)$ and $r(t)$ are of degree $\leq m - 1$. Hint: divide.

(d) Show that there are series expansions

$$q(t) = \sum_{k=0}^{n-1} \alpha_k p_k(t) \quad \text{and} \quad r(t) = \sum_{k=0}^{n-1} \beta_k p_k(t).$$

(e) Show that
$$\int_{-1}^{1} f(t)\,dt = \beta_0 \int_{-1}^{1} p_0^2(t)\,dt. \tag{3.138}$$

(f) Let $t_1, t_2, \ldots, t_m$ be the roots of $p_m(t)$. Show that
$$\sum_{i=1}^{m} a_i f(t_i) = \sum_{k=0}^{m-1} \beta_k \sum_{i=1}^{m} a_i p_k(t_i). \tag{3.139}$$

(g) Show that if the weights a_i are chosen so that
$$\sum_{i=1}^{m} a_i p_k(x_i) = \begin{cases} \int_{-1}^{1} p_0^2(t)\,dt & k = 0, \\ 0 & k = 1, 2, \ldots, n-1, \end{cases}$$
then (3.139) can be written as
$$\sum_{i=1}^{m} a_i f(t_i) = b_0 \int_{-1}^{1} p_0(t)\,dt. \tag{3.140}$$

(h) Write (3.140) as a matrix equation for the weights $\{a_i\}$.

(i) Hence, equating (3.138) and (3.140), write down the formula for Gaussian quadrature.

(j) Generalize this to finding $\int_{-1}^{1} w(t) f(t)\,dt$, where the polynomials $p_k(t)$ are orthogonal with respect to the inner product $\langle f, g \rangle = \int_{-1}^{1} f(t) g(t) w(t)\,dt$.

3.18-38 Prove Parseval's theorem for Fourier transforms: If $y_1(t) \leftrightarrow Y_1(\omega)$ and $y_2(t) \leftrightarrow Y_2(\omega)$, then
$$\int_{-\infty}^{\infty} y_1(t) \overline{y}_2(t)\,dt = \frac{1}{2\pi} \int_{-\infty}^{\infty} Y_1(\omega) \overline{Y}_2(\omega)\,d\omega.$$

3.18-39 Sampling theorem representations.

(a) Show for $p_k(t)$ defined as in (3.102) that $\langle p_k, p_l \rangle = \frac{1}{2B} \delta_{k,l}$. Along the way, show that
$$\int_{-\infty}^{\infty} \frac{\sin t}{t} \frac{\sin(t-z)}{t-z}\,dt = \frac{\pi \sin z}{z}.$$
Hint: use Parseval's theorem and Fourier transforms.

(b) Show that (3.103) is correct for a bandlimited function $f(t)$.

(c) Show that if $f(t)$ is badnlimited to B Hz,
$$f(z) = 2B \int_{-\infty}^{\infty} f(t) p_0(t-z)\,dt,$$
Thus, for bandlimited functions, $p_0(t)$ behaves like a δ function.

3.18-40 Show that if $\phi(t)$ is normalized then $2^{-j/2} \phi(2^{-j} t)$ is normalized.

3.18-41 In (3.106), show that the coefficients c_n must satisfy
$$\sum_{n} c_n = 2.$$

3.18-42 Show that there is no orthogonal scaling function defined by a two-scale equation (3.106) with exactly three nonzero coefficients c_0, c_1, and c_2.

3.18-43 For the multiresolution analysis:

(a) Show that $W_j \perp W_{j'}$.

(b) Show that for $j < J$,
$$V_j = V_J \oplus \bigoplus_{k=0}^{J-j-1} W_{J-k}.$$

3.18-44 Show that if $\phi(t)$ obeys the two-scale relationship in (3.105), and if $\hat{\phi}(\omega)$ represents the Fourier transform of $\phi(t)$, then
$$\hat{\phi}(\omega) = m_0(\omega/2)\hat{\phi}(\omega/2),$$
where
$$m_0(\omega) = \frac{1}{\sqrt{2}} \sum_n h_n e^{-jn\omega} \qquad (3.141)$$
is the scaled discrete-time Fourier transform of the coefficient sequence.

3.18-45 **Decimation** Because of the connection of wavelet transforms to multirate signaling, it is worthwhile to examine the transform of decimated signals. You will show that if $y[n]$ is a decimation of $x[n]$,
$$y[n] = x[nD],$$
then
$$Y(z) = \frac{1}{D} \sum_{k=0}^{D-1} X(e^{-j2\pi k/D} z^{1/D}). \qquad (3.142)$$

(a) Let $p[n]$ be the periodic sampling sequence
$$p[n] = \begin{cases} 1 & n = 0, \pm D, \pm 2D, \ldots, \\ 0 & \text{otherwise.} \end{cases}$$
Show that
$$p[n] = \frac{1}{D} \sum_{k=0}^{D-1} e^{j2\pi kn/D}.$$

(b) Let $z[n] = x[n]p[n]$. Then $y[n] = z[nD]$. Show that
$$Y(z) = \sum_m y[m] z^{-m} = \sum_m z[m].$$

(c) Finally, show that (3.142) is true.

3.18-46 Show that the orthogonality condition (3.108) is equivalent to
$$|m_0(\omega/2)|^2 + |m_0(\omega/2 + \pi)|^2 = 1.$$
Hint: recognize that (3.108) is a decimated convolution, and use the fact that if the Fourier transform of a sequence z_n is $Z(\omega)$, then the Fourier transform of z_{2n} is
$$\frac{1}{2}[Z(\omega/2) + Z(\omega/2 + \pi)].$$

3.18-47 Let $\phi(t)$ be a one-dimensional basis function for digital transmission, of the form
$$\phi(t) = u(t) - u(t-1)$$

3.20 Exercises

(a unit pulse). Assume that $s(t) = \phi(t)$ is transmitted. Let $r(t) = s(t)$ (noise-free reception). Show the output of the correlator

$$y_1(t) = \int_0^t r(s)\phi(s)\,ds$$

and the output of the matched filter with impulse response $h(t) = \phi(T-t)$,

$$y_2(t) = r(t) * h(t).$$

Show that at the sample instant $t = T$, $y_1(t) = y_2(t)$.

3.18-48 Let

$$\phi_m(t) = \begin{cases} \cos[(2\pi f_c + 2\pi m\Delta f)t] & 0 \le t \le T, \\ 0 & \text{otherwise} \end{cases}$$

for $m = 0, 1, \ldots, M-1$ be a set of basis functions. Determine the minimum frequency separation Δf such that

$$\int_0^T \phi_m(t)\phi_k(t)\,dt = 0$$

for $k \ne m$. Assume that $f_c T = n$ for some integer n. (Digital transmission with such signals is called frequency-shift keying.)

3.18-49 **(Spread-spectrum multiple access)** In this exercise, we examine matched filters for a more complicated scenario: spread spectrum multiple access. In this model, K users are transmitting simultaneously, with the kth user transmitting a signal

$$s_k(t) = \sum_n b_k(n)\sqrt{2}w_k\phi_k(t - nT).$$

where $\phi_k(t)$ is the kth user's unique waveform, a signal with support over $[0, T]$. The received signal consists of the sum of each user's delayed signal, appearing in additive noise:

$$r(t) = \sum_{k=1}^K \sum_n b_k(n)w_k\phi_k(t - nT - \tau_k) + z(t).$$

The users' basis functions are *not* orthogonal. Assume that the users are ordered so that $\tau_1 \le \tau_2 \le \cdots \le \tau_K < T$. A matched-filter (or correlator) output is obtained for each user over the nth bit interval, as

$$y_k(n) = \int_{-\infty}^{\infty} r(t)\phi_k(t - nT - \tau_k).$$

Let $\mathbf{y}(n) = [y_1(n), y_2(n), \ldots, y_k(n)]^T$ be the vector of matched filter outputs for all users at interval n.

(a) Show that

$$\mathbf{y}(n) = [H(1)B(n-1) + H(0)B(n) + H(-1)B(n+1)]\mathbf{w} + \mathbf{z}(n)$$

where $H(m)$ is a correlation matrix with elements

$$H_{ij}(m) = \int_{-\infty}^{\infty} \phi_i(t - \tau_i)\phi_j(t - mT - \tau_j)\,dt,$$

B is a diagonal matrix of bits, $B(n) = \text{diag}(b_1(n), b_2(n), \ldots, b_k(n))$, $\mathbf{w} = [w_1, w_2, \ldots, w_k]^T$, and $\mathbf{z}(n) = [z_1(n), z_2(n), \ldots, z_K(bn)]^T$, where

$$z_k(n) = \int z(t)\phi_k(t - nT - \tau_k)\,dt.$$

(b) If $z(t)$ is white with $E[z(t)z(t-s)] = \sigma_z^2 \delta(t-s)$, show that $\mathbf{z}(n)$ satisfies

$$E[\mathbf{z}(n)\mathbf{z}^T(m)] = \begin{cases} \sigma_z^2 H(0) & n = m, \\ \sigma_z^2 H(1) & n = m+1, \\ \sigma_z^2 H(-1) & n = m-1, \\ 0 & \text{otherwise.} \end{cases}$$

3.21 References

The Hilbert approximation theory presented here is summarized from [209] and [177]. Some of the discussion about the Grammian matrix was drawn from [291].

The various windowing methods are described in [132, chapter 11]. A discussion of least-squares and minimum mean-squares filtering is in [132, 263, 291]. Our discussion of Wiener filtering is drawn from [165] and [316]. A thorough discussion of the spectral factorization problem appears in [248].

The Gram–Schmidt is discussed in most books on linear algebra. Specific results on numeric accuracy of the method can be found in [114].

Several variants on least-squares and constrained least-squares, including pseudocode for several useful algorithms, are in [197].

Orthogonal functions are widely discussed in [2], including an extensive table of polynomials orthogonal with respect to many weighting functions, and their properties. In addition to orthogonal polynomials in continuous time, there are also orthogonal polynomials in discrete variables. These are summarized in [2] and examined more thoroughly in [79] and [337]. A recent book describing a variety of orthogonal functions and their smoothness properties is [358].

The use of the function $\sin(x)/x$ (the sinc function) as an orthogonal basis is introduced in [177]. An extensive discussion occurs in [323] and [322].

There has been an explosion of literature on wavelets and wavelet transforms. The definitive reference is probably [62]; see also [63]. Another book with a broad base of coverage is [53]. Among the generalizations discussed in these books are biorthogonal wavelets (in which different filters are used to reconstruct the signal than to analyze it), wavelet packets (choosing different trees of coefficients), and several other families of wavelets. A recent tutorial is [44]. A thorough discussion of implementation of wavelet transforms (and a variety of other useful transforms as well) is provided in [362]. A definitive reference on multirate signal processing is [342]; for a solid introduction to this area see [341].

IRLS is discussed in [43] and references therein, where the number of iterations required to design a filter is closely examined. An alternative viewpoint on estimation, using the L_1 norm for spectral estimation, is investigated in [294, 295, 73, 359]. A more thorough treatment is presented in [37].

The vector space viewpoint, signal constellations, and matched filters are presented in every text on digital communications. See, for example, [373], [261] or [35]. A historical treatment of orthogonal functions used in signaling is given in [127], which also presents some useful orthogonal functions other than those presented here.

There is a tremendous literature on orthogonal polynomials. A recent survey is [358]. A classic reference is [337]. Additional information is found in [2].

Chapter 4

Linear Operators and Matrix Inverses

> Everything that goes on in spacetime has its geometric description...
> — *Misner, Thorne, and Wheeler*
> Gravitation

In this chapter we begin a study of linear operators. The most familiar linear operator is a matrix, appearing in a linear equation of the form

$$A\mathbf{x} = \mathbf{b}, \qquad (4.1)$$

of which several instances have already arisen in this text. However, in the interest of getting the most "bang for the buck," we introduce more general linear operators than matrices. We examine the questions of the existence and uniqueness of the solution of a linear equation of the form (4.1). These results are general, and apply to any linear operator. Finding a solution to (4.1) becomes more complicated, and more interesting, if A is not an invertible operator. In this case there may be no solution, or multiple solutions. We will examine some important theoretical results related to these solutions, and present the important Fredholm alternative theorem.

We then narrow our focus to matrix linear operators, and consider matrix characteristics related to the existence and uniqueness of matrix equations, such as the rank and determinant of a matrix. Even in the apparently simple situation in which A is invertible, there are important issues to address regarding the quality of the result. If the matrix A is "nearly not invertible," then the answers obtained by a numerical algorithm are likely to be unreliable. Some qualitative results regarding how close a matrix is to being noninvertible are obtained by introducing the *condition number* of a matrix.

We conclude this chapter with a discussion of properties of matrix inverses, including inverses of block matrices and small-rank updates to a matrix. These will lead us directly to the RLS and Kalman filters. The trend of this chapter is thus from the more theoretical towards the more applied. We stop short of presenting a numerical algorithm for solving linear matrix equations. This topic is taken up in the next chapter, in association with other matrix factorizations.

The basic goals of this chapter are as follows: to understand when solutions of linear equations exist and are unique, and to develop insight into factors that can affect the quality of the solution.

Another problem that arises frequently in linear systems theory is an equation of the form

$$A\mathbf{x} = \lambda\mathbf{x}.$$

The vector **x** which solves this equation is called an eigenvector of A, and the associated value λ is called an eigenvalue. Examples and applications of eigenvectors and eigenvalues are presented in chapter 6.

4.1 Linear operators

In (4.1), we can regard A as a linear transformation (or operator), which maps a vector **x** into the column space (range) of A. There are a variety of other linear transformations that arise in practice, producing equations of the form

$$Ax = b,$$

with $x \in X$ and $b \in Y$ where X and Y are vector spaces, and A is a linear transformation (operator) from X to Y. For example, the equation

$$\frac{dx}{dt} - ax(t) = b(t)$$

may be written in operator notation as $Ax(t) = b(t)$, where A is the operator

$$A = \frac{d}{dt} - a.$$

Given an input signal $b(t)$, the problem is to find the solution $x(t)$. Solving this linear operator equation amounts to finding the solution to the differential equation. Another example of a linear equation is

$$b(t) = \int_a^b k(t,\tau) x(\tau) \, d\tau. \tag{4.2}$$

The problem again is to find $x(t)$. Again, this may be written as $Ax(t) = b(t)$, where A is now the integral operator

$$Ax(t) = \int_a^b k(t,\tau) x(\tau) \, d\tau.$$

An integral equation of the form (4.2) is called a *Fredholm integral equation of the first kind*. Another example is an equation of the form

$$x(t) = \int_a^b k(t,\tau) x(\tau) \, d\tau + b(t).$$

This is an integral equation known as a *Fredholm integral equation of the second kind*.

Obviously, these different kinds of equations will have different methods of solution. The algorithmic focus of this and succeeding chapters is on linear operators that can be expressed in matrix form, so we will not present details of the solution of these more general linear operator equations. Nevertheless, the principles regarding the existence of the solutions apply to most of these operators. Furthermore, by examining these more general forms of operators, we can gain further understanding of matrices.

Having seen several examples of linear transformations, it is appropriate now to introduce a formal definition.

Definition 4.1 A transformation $A: X \to Y$, where X and Y are vector spaces over a ring R, is said to be **linear** if for every $x_1, x_2 \in X$ and all scalars $\alpha_1, \alpha_2 \in R$,

$$A(\alpha_1 x_1 + \alpha_2 x_2) = \alpha_1 A(x_1) + \alpha_2 A(x_2). \qquad \square$$

The most important example of a linear transformation, in fact the algorithmic focus of this part of the book is the transformation $A: \mathbb{R}^n \to \mathbb{R}^m$. In this case, A is the $m \times n$ matrix with elements from $\mathbb{R}$.

4.1 Linear Operators

Example 4.1.1 Let
$$A = \begin{bmatrix} 1 & 2 \\ 4 & 2 \\ 5 & -2 \end{bmatrix}.$$
Then $A: \mathbb{R}^2 \to \mathbb{R}^3$. □

Several different notations can be used to represent a matrix A with elements from R. We can write $A: R^n \to R^m$. We can say A is an $m \times n$ matrix (where implicitly the elements are usually either real or complex). Or we can say that $A \in R^{m \times n} = R^{mn}$. If we want to emphasize the underlying ring of scalars R, we will also write $A \in M_{m,n}(R)$ to indicate that A is an $m \times n$ matrix with elements from R. We will also use the shorthand $A \in M_m(R)$ if $m = n$, or $A \in M_{m,n}$ if the ring is implicitly understood.

Example 4.1.2 (Another example of a linear operator). Let $X = C[0, 1]$ (the continuous functions defined on $[0, 1]$), and let $Y = \mathbb{R}^n$. Define the operator $A: X \to Y$ by
$$Ax = (x(t_1), x(t_2), \ldots, x(t_n)),$$
where $0 \leq t_1 < t_2 < \cdots < t_n \leq 1$ are fixed. This is a sampling operator, and is linear. □

4.1.1 Linear functionals

Definition 4.2 A **functional** $f: X \to \mathbb{R}$ is a mapping from a vector space X to a real scalar value. (More generally, the range of the mapping could be any set of scalars.)

If f has the property that $f(\alpha \mathbf{x} + \beta \mathbf{y}) = \alpha f(\mathbf{x}) + \beta f(\mathbf{y})$ for all real α and β and all $\mathbf{x}, \mathbf{y} \in X$, then f is said to be a linear functional. □

Example 4.1.3 (Examples of functionals). Let $x(t)$ be a function. Then the following are examples of functionals on $x(t)$:

$$f_1(x) = \frac{1}{T} \int_0^T x(t)\, dt$$

$$f_2(x) = \int_a^b x(t)\phi(t)\, dt$$

$$f_3(x) = \int_a^b e^{-j\omega t}\, dt \quad \text{for fixed } \omega$$

$$f_4(x) = \int_{-\infty}^{\infty} x(t)\delta(t - t_0)\, dt = x(t_0)$$

$$f_5(x) = \|x\|_p$$

All of these examples of functionals are expressible as integrals. □

Example 4.1.4 (More examples of functionals).

1. Let $X = \mathbb{R}^n$. Then for $\mathbf{x} \in \mathbb{R}^n$ and a fixed set of scalars $(\alpha_1, \alpha_2, \ldots, \alpha_n)$, the functional
$$f(\mathbf{x}) = \sum_{i=1}^{n} x_i \alpha_i$$
is a linear functional.

2. Let $X = C[0, 1]$. Then the functional $f: C[0, 1] \to \mathbb{R}$ defined by
$$f(x) = x\left(\frac{1}{3}\right)$$
is a linear functional.

3. Let $X = L_2(\mathbb{R})$, and let $y \in L_2(\mathbb{R})$. Then the functional defined by

$$f(x) = \int_{-\infty}^{\infty} x(t)y(t)\,dt$$

is a linear functional. □

As these examples show, functionals can be used to represent measurements made of functions.

If $\phi \in X$, then we can define a functional f_ϕ by

$$f_\phi(x) = \langle x, \phi \rangle.$$

In other words, inner products are functionals. Conversely (and surprisingly), if X is a Hilbert space (complete), then *any* continuous linear functional can be expressed as an inner product. This is known as the Riesz representation theorem.

If $\|\phi\|$ is bounded, so that $\|\phi\| < K < \infty$ (using the induced norm), then $f_\phi(x)$ is continuous, since by the Cauchy–Schwarz inequality

$$\|f_\phi(x) - f_\phi(x_0)\| \leq \|x - x_0\| \|\phi\|.$$

As shown in theorem 4.1, this means that f must also be bounded.

4.2 Operator norms

An operator norm, like any norm, must satisfy the properties described in section 2.3. There are several different ways of defining the norm of a transformation (operator). One way is to define the norm so that it provides an indication of the maximal amount of change of length of a vector that it operates on. Let X and Y be l_p or L_p, and let A be a linear operator $A: X \to Y$. The p **operator norm**, or p norm, or l_p norm, of A is

$$\|A\|_p = \sup_{x \in X, \neq 0} \frac{\|Ax\|_p}{\|x\|_p} = \sup_{x \in X, \|x\|_p = 1} \|Ax\|_p,$$

where $\|\cdot\|_p$ is the p norm defined in section 2.3. (Note: $Ax \in Y$, so the norm $\|Ax\|_p$ is the norm on Y. We could in general use different norms for $\|x\|$ and $\|Ax\|$, but usually this is not done.) The norm on A so obtained is said to be *subordinate* to the norm on **X**. For a subordinate norm it is straightforward to verify that $\|I\| = 1$, where I is the identity operator. Geometrically, the $\|A\|$ subordinate norm measures the maximum extent that A transforms the unit circle. The concept is shown in figure 4.1. The p norms have the property that

$$\|A\mathbf{x}\|_p \leq \|A\|_p \|\mathbf{x}\|_p.$$

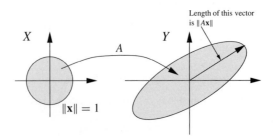

Figure 4.1: Geometry of the operator norm

4.2 Operator Norms

Thus $\|A\|$ "bounds the amplifying power" of the matrix A. Also, the p norms satisfy the **submultiplicative property**,

$$\|AB\|_p \leq \|A\|_p \|B\|_p.$$

This is straightforward to show, since by the definition of the p norm, for all $x \in X$,

$$\|ABx\| \leq \|A\| \|Bx\| \leq \|A\| \|B\| \|x\|.$$

4.2.1 Bounded operators

This section is somewhat technical, and many readers may need only the first definition.

Definition 4.3 If the norm of a transformation is finite, the transformation is said to be *bounded*. □

The following theorem presents a rather remarkable fact about bounded linear operators.

Theorem 4.1 *A linear operator $A: X \to Y$ is bounded if and only if it is continuous.*

Since a linear functional is a linear operator, the same theorem applies to functionals.

Proof Suppose that A is bounded, with M such that $\|Ax\| \leq M\|x\|$ for all $x \in X$. Let $\{x_n\}$ be a sequence approaching zero: $x_n \to 0$. Then $Ax_n \leq M\|x_n\| \to 0$. By the properties of continuity (continuous functions preserve convergence), it follows that A is continuous.

Conversely, assume A is continuous. Then there is a $\delta > 0$ such that $\|Ax\| < 1$ for $\|x\| < \delta$. Then, since the norm of $\delta x/\|x\|$ is equal to δ,

$$\|Ax\| = \|Ax(\delta x/\|x\|)\| \, \|x\|/\delta < \|x\|/\delta.$$

The value $M = 1/\delta$ serves as a bound for A. □

The following theorem is of great utility: by showing that linear operators from finite-dimensional spaces are continuous, we can conclude from the previous theorem that they are also bounded. Since many of the results of this chapter rely on bounded linear operators, this theorem reassures us that matrices—operators on finite-dimensional spaces—will work.

Theorem 4.2 *Let $A: X \to Y$ be a linear operator, where X and Y are normed linear spaces. If X is finite dimensional, then A is continuous.*

Note that this theorem does not assume that Y is finite dimensional.

Proof of theorem 4.2 makes use of the following lemma, which is the most technical part of this section.

Lemma 4.1 *[238, page 265] Let X be a finite-dimensional normed linear space, and let $\{\mathbf{x}_1, \mathbf{x}_2, \ldots, \mathbf{x}_n\}$ be a (Hamel) basis for X. Then for $\mathbf{x} \in X$, each coefficient α_i in the expansion*

$$\mathbf{x} = \alpha_1 \mathbf{x}_1 + \alpha_2 \mathbf{x}_2 + \cdots + \alpha_n \mathbf{x}_n$$

is a continuous linear function of $\mathbf{x}$. Being continuous, it is bounded, so there is a constant M such that $|\alpha_i| \leq M\|\mathbf{x}\|$.

Proof Showing linearity is straightforward, and is omitted.

It will suffice to show that there is an $m > 0$ such that

$$m(|\alpha_1| + |\alpha_2| + \cdots + |\alpha_n|) \leq \|\mathbf{x}\|, \tag{4.3}$$

since it follows that $|\alpha_i| \leq m^{-1}\|\mathbf{x}\|$. We will demonstrate (4.3) first for coefficients $\{\alpha_1, \ldots, \alpha_n\}$ satisfying the condition $|\alpha_1| + \cdots + |\alpha_n| = 1$. Let

$$A = \{(\alpha_1, \ldots, \alpha_n) : |\alpha_1| + \cdots + |\alpha_n| = 1\}.$$

This set is closed and bounded (compact). Now define a function $f: A \to \mathbb{R}$ by

$$f(\alpha_1, \ldots, \alpha_n) = \|\alpha_1 \mathbf{x}_1 + \cdots + \alpha_n \mathbf{x}_n\|. \tag{4.4}$$

It can be shown that f continuous, and it is clear that $f > 0$. Let

$$m = \min_{(\alpha_1, \ldots, \alpha_n) \in A} f(\alpha_1, \ldots, \alpha_n).$$

Since f is continuous on a closed bounded set, this minimum does exist for some point $(\alpha_1^*, \ldots, \alpha_n^*) \in A$. Hence we have found a point m that satisfies (4.3). If $m = 0$, then $\alpha_1^* \mathbf{x}_1 + \cdots + \alpha_n^* \mathbf{x}_n = 0$, contradicting the fact that $\{\mathbf{x}_i\}$ is a basis (linearly independent). Hence $m > 0$.

For general sets of coefficients $\{\alpha_i\}$, set $\beta = |\alpha_1| + \cdots + |\alpha_n|$. If $\beta = 0$, the result is trivial. If $\beta > 0$ then we write

$$\|\alpha_1 \mathbf{x}_1 + \cdots + \alpha_n \mathbf{x}_n\| = \beta \|(\alpha_1/\beta)\mathbf{x}_1 + \cdots + (\alpha_n/\beta)\mathbf{x}_n\|$$
$$= \beta f(\alpha_1/\beta, \ldots, \alpha_n/\beta) \geq m\beta \geq m(|\alpha_1| + \cdots + |\alpha_n|). \qquad \square$$

Proof Now we proceed with the proof of theorem 4.2. Let $\{\mathbf{x}_1, \mathbf{x}_2, \ldots, \mathbf{x}_n\}$ be a (Hamel) basis for X. Let $\mathbf{x} \in X$ be expressed in terms of this basis as

$$\mathbf{x} = \alpha_1 \mathbf{x}_1 + \alpha_2 \mathbf{x}_2 + \cdots + \alpha_n \mathbf{x}_n.$$

Let $D = \max_{1 \leq i \leq n} \|A\mathbf{x}_i\|$. Then

$$\|A\mathbf{x}\| = \|A(\alpha_1 \mathbf{x}_1 + \alpha_2 \mathbf{x}_2 + \cdots + \alpha_n \mathbf{x}_n)\|$$
$$\leq |\alpha_1|\|A\mathbf{x}_1\| + |\alpha_2|\|A\mathbf{x}_2\| + \cdots + |\alpha_n|\|A\mathbf{x}_n\|$$
$$\leq D(|\alpha_1| + |\alpha_2| + \cdots + |\alpha_n|).$$

By the lemma above, there is an M such that $|\alpha_1| + \cdots + |\alpha_n| \leq M\|\mathbf{x}\|$, so that

$$\|A\mathbf{x}\| \leq DM\|\mathbf{x}\|. \qquad \square$$

Before considering the important special case of matrix transformations, we will consider some more generalized transformations.

Example 4.2.1 Let $X = C[0, 1]$, and define $A: X \to X$ by

$$Ax(t) = \int_0^1 k(t, \tau) x(\tau) \, d\tau,$$

where $t \in [0, 1]$, and K is continuous. We will compute the L_∞ norm of this operator.

$$\|Ax\| = \max_{t \in [0,1]} \left| \int_0^1 k(t, \tau) x(\tau) \, d\tau \right|$$
$$\leq \max_{t \in [0,1]} \int_0^1 |k(t, \tau)| \, d\tau \max_{t \in [0,1]} |x(t)|$$
$$= \max_{t \in [0,1]} \int_0^1 |k(t, \tau)| \, d\tau \, \|x\|.$$

It can be shown that equality can be achieved, so that

$$\|A\| = \max_{t \in [0,1]} \int_0^1 |k(t, \tau)| \, d\tau.$$

Since $k(t, \tau)$ is continuous, then A is bounded. □

Example 4.2.2 Let $A: C^1[0, 1] \to C[0, 1]$ be the operator

$$Ax = \frac{d}{dt}x.$$

The function $x(t) = \sin \omega_0 t \in C^1[0, 1]$ has uniform norm 1 for any value of ω_0, but

$$\|Ax\| = \max_{t \in [0,1]} \omega_0 |\cos \omega_0 t|$$

may have norm arbitrarily large by choosing ω_0 to be arbitrarily large. Thus the differential operator is not bounded (and hence not continuous). □

4.2.2 The Neumann expansion

The Neumann expansion provides a useful expansion for the inverse of the linear operator $(I - A)^{-1}$. For a scalar x such that $|x| < 1$, it is straightforward to show using the geometric series that

$$1 + x + x^2 + \cdots = \sum_{i=0}^{\infty} x^i = \frac{1}{1-x} = (1-x)^{-1}.$$

There is a direct extension to linear operators, as follows.

Theorem 4.3 *Suppose $\|\cdot\|$ is a norm satisfying the submultiplicative property and A is an operator with $\|A\| < 1$. Then $(I - A)^{-1}$ exists, and*

$$(I - A)^{-1} = \sum_{i=0}^{\infty} A^i.$$

Proof Let $\|A\| < 1$. If $I - A$ is singular, then there is a vector $\mathbf{x}$ such that $(I - A)\mathbf{x} = 0$. But this means that $\|\mathbf{x}\| = \|A\mathbf{x}\| \le \|\mathbf{x}\|\|A\|$, or $\|A\| \ge 1$. This is a contradiction.

By multiplication, it is clear that

$$(I - A)(I + A + A^2 + \cdots + A^{k-1}) = I - A^k.$$

With $\|A\| < 1$, $\lim_{k \to \infty} A^k = 0$, since

$$\|A^k\| \le \|A\|^k \to 0 \text{ as } k \to \infty.$$

Thus,

$$(I - A) \left(\sum_{i=0}^{\infty} A^i \right) = I.$$

Hence, the quantity

$$\sum_{i=0}^{\infty} A^i$$

must be the inverse of $I - A$. □

4.2.3 Matrix norms

In specializing the foregoing results to matrix operators, we consider the cases $p = 1$, $p = 2$, and $p = \infty$, which are of particular interest.

$$\boxed{\|A\|_\infty = \max_{\|\mathbf{x}\|_\infty = 1} \|A\mathbf{x}\|_\infty = \max_i \sum_j |a_{ij}|} \quad (4.5)$$

that is, it is the largest row sum;

$$\|A\|_1 = \max_{\|\mathbf{x}\|_1=1} \|A\mathbf{x}\|_1 = \max_j \sum_i |a_{ij}| \qquad (4.6)$$

that is, it is the largest column sum.

To deal with the l_2 matrix norm requires an understanding of eigenvalues and constrained optimization. We want to maximize $\|A\mathbf{x}\|_2$, subject to the constraint that $\|\mathbf{x}\|_2 = 1$. This can be written as

$$\text{maximize } \|A\mathbf{x}\|_2^2 = \mathbf{x}^H A^H A \mathbf{x}$$
$$\text{subject to } \mathbf{x}^H \mathbf{x} = 1.$$

The constraint can be incorporated using a Lagrange multiplier to create the functional

$$J = \mathbf{x}^H A^H A \mathbf{x} - \lambda \mathbf{x}^H \mathbf{x}.$$

Taking the gradient with respect to $\mathbf{x}$ and equating the result to zero, we obtain the equation

$$A^H A \mathbf{x} = \lambda \mathbf{x}. \qquad (4.7)$$

The corresponding $\mathbf{x}$ must be an eigenvector of $A^H A$. Multiplying (4.7) by $\mathbf{x}^H$, and recalling the constraint that $\mathbf{x}^H \mathbf{x} = 1$, we obtain

$$\mathbf{x}^H A^H A \mathbf{x} = \lambda \mathbf{x}^H \mathbf{x} = \lambda.$$

Since we are maximizing the quantity on the left, λ must be the largest eigenvalue of $A^H A$. For an $n \times n$ matrix A with eigenvalues $\lambda_1, \lambda_2, \ldots, \lambda_n$, the **spectral radius** $\rho(A)$ is defined as

$$\rho(A) = \max_i |\lambda_i|.$$

The spectral radius is the smallest radius of a circle centered at the origin that contains all the eigenvalues of A. Then the l_2 norm is defined by

$$\|A\|_2 = \sqrt{\rho(A^H A)}$$

Because the l_2 norm requires computation of eigenvalues, it is much more difficult to compute than the l_1 or l_∞ norms. However, it is of significant theoretical value. When A is Hermitian,

$$\|A\|_2 = \rho(A).$$

The l_2 norm is also called the **spectral norm**.

For the subordinate norms, we can also say something about the norm of the inverse A^{-1}, when it exists. For the equation $A\mathbf{x} = \mathbf{b}$, assume that A^{-1} exists, so $\mathbf{x} = A^{-1}\mathbf{b}$. Then

$$\|A^{-1}\| = \max_{\mathbf{b} \neq 0} \frac{\|A^{-1}\mathbf{b}\|}{\|\mathbf{b}\|} = \max_{\|\mathbf{x}\| \neq 0} \frac{\|\mathbf{x}\|}{\|A\mathbf{x}\|}$$
$$= \frac{1}{\min_{x \neq 0} \frac{\|A\mathbf{x}\|}{\|\mathbf{x}\|}}.$$

From this we conclude that

$$\|A^{-1}\|^{-1} = \min_{\|\mathbf{x}\|=1} \|A\mathbf{x}\|. \qquad (4.8)$$

For example, $(\|A^{-1}\|_2)^{-1} = \sqrt{\lambda_{\min}}$, where $\lambda_{\min}$ is the *smallest* eigenvalue of $A^H A$.

A matrix norm which is not a p norm is the **Frobenius norm**,

$$\|A\|_F = \left(\sum_{i=1}^{m} \sum_{j=1}^{n} |a_{ij}|^2 \right)^{1/2}$$

This norm is also called the **Euclidean norm**. It should not be confused with the l_2 norm. The Frobenius norm is often used in matrix analysis, since it is relatively easy to compute. It is a natural norm, for example, to use when comparing how close two matrices A and B are, using $\|A - B\|_F$. For the Frobenius norm, $\|I\| = \sqrt{n}$. The Frobenius norm can also be written using

$$\|A\|_F^2 = \text{tr}(A^H A) \tag{4.9}$$

The following relationships exist between the norms for an $m \times n$ matrix A:

$$\|A\|_2 \leq \|A\|_F \leq \sqrt{n} \|A\|_2, \tag{4.10}$$

$$\max_{i,j} |a_{ij}| \leq \|A\|_2 \leq \sqrt{mn} \max_{i,j} |a_{ij}|, \tag{4.11}$$

$$\frac{1}{\sqrt{n}} \|A\|_\infty \leq \|A\|_2 \leq \sqrt{m} \|A\|_\infty, \tag{4.12}$$

$$\frac{1}{\sqrt{m}} \|A\|_1 \leq \|A\|_2 \leq \sqrt{n} \|A\|_1. \tag{4.13}$$

4.3 Adjoint operators and transposes

Let $A: X \to Y$ be a linear operator (not necessarily merely a matrix), and let $\langle \cdot, \cdot \rangle: Y \times Y \to \mathbb{C}$ be an inner product. Since $Ax \in Y$, the inner product $\langle Ax, y \rangle$ is defined, and it may be viewed as a scalar linear operation (a functional) on $x \in X$. The elements in Y are an example of a *dual space*. For our purposes, we can think of a dual space as the range space of an operator. (For linear operations on Hilbert spaces, this definition is sufficiently precise; for other Banach spaces there are some technical details that are hidden by this interpretation. There is a remarkable body of theory that can be developed related to dual spaces, providing for optimization techniques similar to those developed in chapter 3 for inner-product spaces, for which the norm is not an induced norm, but is another L_p norm. However, an extensive treatment of dual spaces, while powerful, is beyond the scope of this text. Interested readers should consult [209].)

Definition 4.4 Let $A: X \to Y$ be a bounded linear operator from the Hilbert space X to the Hilbert space Y. The **adjoint** of the operator A, denoted A^*, is the operator $A^*: Y \to X$ such that

$$\langle Ax, y \rangle = \langle x, A^* y \rangle$$

for all $x \in X$ and $y \in Y$. An operator A is **self-adjoint** if $A^* = A$. □

Example 4.3.1 Let $A: \mathbb{R}^n \to \mathbb{R}^m$. Then A can be represented as an $m \times n$ matrix. The ith component of $A\mathbf{x}$ is

$$(A\mathbf{x})_i = \sum_{j=1}^{n} a_{ij} x_j, \qquad i = 1, 2, \ldots, m.$$

To find the adjoint of A, we form the inner product using the usual Euclidean inner product,

$$\langle A\mathbf{x}, \mathbf{y} \rangle = \sum_{i=1}^{m} \sum_{j=1}^{n} \overline{y}_i a_{ij} x_j = \sum_{j=1}^{n} x_j \sum_{i=1}^{m} a_{ij} \overline{y}_i = \langle \mathbf{x}, A^H \mathbf{y} \rangle.$$

Thus, the *adjoint of a matrix is the conjugate transpose* of the matrix,

$$A^* = \overline{(A^T)} = A^H.$$ □

For a real matrix, the adjoint is simply the transpose,

$$A^* = A^T.$$

A real matrix which is self-adjoint is said to be *symmetric*. A complex matrix which is self-adjoint is said to be *Hermitian*. Self-adjoint matrices are necessarily square.

A matrix (or operator) A which satisfies $A^*A = AA^*$ is said to be *normal*. Obviously, a (Hermitian) symmetric matrix is normal.

Example 4.3.2 The matrices

$$A = \begin{bmatrix} 2 & 3 & 0 \\ 3 & 6 & 2 \\ 0 & 2 & -3 \end{bmatrix} \quad \text{and} \quad B = \begin{bmatrix} 2 & 2+3j & 5-j \\ 2-3j & 4 & 6+4j \\ 5+j & 6-4j & 9 \end{bmatrix}$$

are self-adjoint (symmetric and Hermitian, respectively). □

Example 4.3.3 Let $X = Y = L_2[0, 1]$, and define

$$Ax = \int_0^1 k(t, \tau) x(\tau) \, d\tau.$$

Using the usual integral inner product, we have

$$\langle Ax, y \rangle = \int_0^1 y(t) \int_0^1 k(t, \tau) x(\tau) \, d\tau \, dt$$

$$= \int_0^1 x(\tau) \int_0^1 k(t, \tau) y(t) \, dt \, d\tau.$$

Since t and τ are dummy variables, we can simply reverse their roles and write

$$\langle Ax, y \rangle = \int_0^1 x(t) \int_0^1 k(\tau, t) y(\tau) \, d\tau \, dt = \langle x, A^*y \rangle,$$

where

$$A^*y = \int_0^1 k(\tau, t) y(\tau) \, d\tau$$

is the adjoint operator. □

In a Hilbert space, we always have the property that $A^{**} = A$. Adjoints have the following properties, which are straightforward to show from the definition. Let A_1 and A_2 be bounded linear operators. Then

1. $(A_1 + A_2)^* = A_1^* + A_2^*$,
2. $(\alpha A)^* = \overline{\alpha} A^*$,
3. $(A_2 A_1)^* = A_1^* A_2^*$ (note the reverse order),
4. If A has an inverse then $(A^{-1})^* = (A^*)^{-1}$.

4.4 Geometry of Linear Equations

For the last property associated with matrices, we commonly write

$$(A^T)^{-1} = A^{-T} \quad \text{or} \quad (A^H)^{-1} = A^{-H},$$

since the order of transposition and inversion does not matter.

The concept of an adjoint provides for the formal extension of least-squares methods to equations for general linear operators. The equation $Ax = b$ has the least-squares solution $x = (A^*A)^{-1}A^*b$ for any bounded linear operator. The adjoint depends upon the inner product, with a weighted inner product leading to a different adjoint.

Example 4.3.4 For the weighted inner product

$$\langle \mathbf{x}, \mathbf{y} \rangle_W = \mathbf{y}^H W \mathbf{x} \tag{4.14}$$

with an invertible Hermitian weight matrix W, the adjoint satisfies

$$A^* = W^{-1} A^H W. \tag{4.15}$$

□

The adjoint notation makes it natural to express the weighted least-squares formulas of section 3.4. The least-squares solution of $A\mathbf{x} = \mathbf{b}$, with respect to any inner product is (by comparison with (3.19))

$$\mathbf{x} = (A^*A)^{-1} A^* \mathbf{b}. \tag{4.16}$$

When the inner product is the weighted inner product, the least-squares solution is (using (4.15))

$$\mathbf{x} = (A^H W A)^{-1} A^H W \mathbf{b}, \tag{4.17}$$

as in (3.24).

4.3.1 A dual optimization problem

The adjoint allows us to express the solution to the problem of finding the minimum-norm solution to the linear operator equation $Ax = y$ when there is more than one solution. This is exactly analogous to the problem addressed in section 3.14.

Theorem 4.4 *[209, p. 161] Let $A: X \to Y$ be a bounded linear operator from the Hilbert space G to the Hilbert space Y. The vector $\mathbf{x}$ of minimum norm satisfying $Ax = y$ is given by*

$$x = A^* z,$$

where z is any solution of $(AA^*)z = y$.

Proof Let $n \in \mathcal{N}(A)$. If x_0 is a solution of $Ax = y$, then so is $x + n$. Since $\mathcal{N}(A)$ is closed (because X is a Hilbert space), there must be a unique vector x of minimum norm satisfying $Ax = y$. This solution must be orthogonal to $\mathcal{N}(A)$. Thus

$$x \in [\mathcal{N}(A)]^\perp = \mathcal{R}(A).$$

Hence $x = A^* z$ for some $z \in Y$. Since $Ax = y$, we must have $AA^* z = y$. □

4.4 Geometry of linear equations

We turn now to the solution of the matrix equation $A\mathbf{x} = \mathbf{b}$, and consider it from two different perspectives to gain insight into the solution of linear equations.

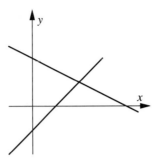

Figure 4.2: Intersections of lines form solutions of systems of linear equations

The "intersecting lines" point of view

Consider the matrix equation

$$\begin{bmatrix} 1 & 2 \\ 1 & -1 \end{bmatrix} \begin{bmatrix} x \\ y \end{bmatrix} = \begin{bmatrix} 8 \\ -1 \end{bmatrix},$$

which represents the pair of equations

$$\begin{aligned} x + 2y &= 4, \\ x - y &= 1. \end{aligned} \tag{4.18}$$

Each equation represents a line on the xy plane, and the solution occurs where the lines intersect, at the point (2,1), as shown in figure 4.2. This it the familiar interpretation of multiple equations in multiple unknowns.

An equation in three unknowns, such as

$$3x + 2y + z = 1,$$

describes a plane in three dimensions. Two such equations describe two planes, whose intersection (if they intersect) is a line. A third equation describes a third plane, and the intersection point (if there is one) of the three planes defines the simultaneous solution to the three equations. It is possible that there is no solution, even if the planes intersect, since the lines of intersection might be parallel. In general, an equation in n unknowns describes a hyperplane in n-dimensional space, and n such planes must intersect at the point of simultaneous solution.

From this interpretation of intersecting lines or planes, three different cases can be considered, both in two dimensions and in more dimensions.

A single solution. The two lines are not parallel; they intersect at one point which gives the solution. In n dimensions, n hyperplanes meet at a single point.

No solution. The two lines are parallel and have no intersection. There is no solution to the set of equations. An example of this is the set of equations

$$\begin{aligned} x + 2y &= 1, \\ 2x + 4y &= 4. \end{aligned}$$

Of course, the dependence is rarely as easily observed as in this simple example. In more dimensions, the hyperplanes have no common point of intersection. Even though two hyperplanes may intersect, other hyperplanes may intersect at another location. Figure 4.3(a) shows three planes, each of which intersects with two other planes but such that there is no common point of intersection for all three planes.

4.4 Geometry of Linear Equations

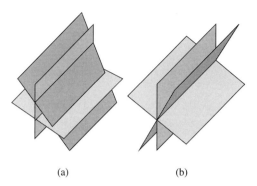

Figure 4.3: Intersecting planes: (a) no solution (b) infinite number of solutions

An infinite number of solutions. The two lines are parallel and right on top of each other. Then there are an infinite number of solutions: any point lying on the first line is also lying on the second line. These equations exhibit the problem:

$$x + 2y = 1,$$
$$2x + 4y = 2.$$

In more dimensions, the planes intersect to form a line, as shown in figure 4.3(b).

The "vector space" point of view

Another way of looking at a set of linear equations is to consider the multiplication as a linear combination of the columns:

$$\begin{bmatrix} 1 & 2 \\ 1 & -1 \end{bmatrix} \begin{bmatrix} x \\ y \end{bmatrix} = \begin{bmatrix} 1 \\ 1 \end{bmatrix} x + \begin{bmatrix} 2 \\ -1 \end{bmatrix} y.$$

The result of the matrix product is a linear combination of the columns of the matrix. In general, *the product of a matrix times a vector is a linear combination of the columns of the matrix.*

Definition 4.5 The space spanned by the columns of a matrix is called the **column space** of the matrix. It is also called the **range** of the matrix, and is denoted by $\mathcal{R}(A)$. If A is written in terms of its columns, $A = [\mathbf{a}_1 \ \mathbf{a}_2 \ \cdots \ \mathbf{a}_n]$, then

$$\mathcal{R}(A) = \text{span}(\mathbf{a}_1, \mathbf{a}_2, \ldots, \mathbf{a}_n).$$

For $A \in M_{m,n}(R)$, the range is a subspace of R^m.

More generally, we can define the range of any linear operator $A: X \to Y$ as

$$\mathcal{R}(A) = \{y \in Y : Ax = y \text{ for } x \in X\}. \qquad \square$$

If $\mathbf{b} \in \mathcal{R}(A)$ for a matrix A, then there must be some linear combination of the columns of A that is equal to $\mathbf{b}$:

> The equation
> $$A\mathbf{x} = \mathbf{b}$$
> has a solution only if $\mathbf{b}$ lies in the column space of A.

If $\mathbf{b} \notin \mathcal{R}(A)$, then no there is no solution to the equation. If $\mathbf{b} \in \mathcal{R}(A)$ but the columns of A are linearly dependent, then there are an infinite number of solutions.

Example 4.4.1 Suppose $A = \begin{bmatrix} 1 & 2 \\ 2 & 4 \end{bmatrix}$. If $\mathbf{b} = \begin{bmatrix} 2 \\ 3 \end{bmatrix}$, then there is no linear combination of the columns of A that can be equal to $\mathbf{b}$, since any linear combination of the columns of A must be of the form $\begin{bmatrix} a \\ 2a \end{bmatrix}$ for some a. On the other hand, if $\mathbf{b} = \begin{bmatrix} 3 \\ 6 \end{bmatrix}$, then

$$\begin{bmatrix} x \\ y \end{bmatrix} = \begin{bmatrix} 1 \\ 1 \end{bmatrix}$$

is a solution, as is

$$\begin{bmatrix} x \\ y \end{bmatrix} = \begin{bmatrix} 3 \\ 0 \end{bmatrix} \quad \text{or} \quad \begin{bmatrix} x \\ y \end{bmatrix} = \begin{bmatrix} 0 \\ 1.5 \end{bmatrix}.$$

□

4.5 Four fundamental subspaces of a linear operator

The first two subspaces we have already met. The first is the range, and the second is the nullspace, which we reintroduce here.

Definition 4.6 The **nullspace** of a linear operator $A: X \to Y$ consists of all vectors $x \in X$ such that $Ax = 0$. It is denoted as $\mathcal{N}(A)$, and is a subspace of X. The dimension of $\mathcal{N}(A)$ is called the **nullity** of A. □

If the nullspace of A is nontrivial (that is, it consists of more than just the zero vector), then the equation $Ax = b$ has an infinite number of solutions, since if x_b is a solution to $Ax = b$, and x_0 is any vector in $\mathcal{N}(A)$, then $x = x_b + cx_0$ is also a solution:

$$Ax = A(x_b + x_0) = Ax_b = b.$$

The next fundamental subspace of A is the **range of the adjoint** A^*, denoted $\mathcal{R}(A^*)$. If A is a matrix, then the adjoint is simply the conjugate transpose, and $\mathcal{R}(A^*)$ is simply the *linear combinations of the conjugates of the rows of A*. For this reason, we may refer to $\mathcal{R}(A^*)$ as the row space of A, but the concept applies to more general linear operators. The row space of $A \in M_{m,n}(R)$ is a subspace of R^n.

The fourth fundamental subspace of A is the **nullspace of** A^*, denoted $\mathcal{N}(A^*)$. For an $m \times n$ matrix A, this space is the set of vectors $\mathbf{y}$ such that $A^H \mathbf{y} = \mathbf{0}$, which can also be written as

$$\mathbf{y}^H A = \mathbf{0}.$$

Because of the multiplication on the left, we will refer to this as the *left nullspace*.

The various subspaces of the operator $A: X \to Y$ are summarized as follows:

$$\begin{aligned} \mathcal{R}(A) &\subset Y, \\ \mathcal{N}(A) &\subset X, \\ \mathcal{R}(A^*) &\subset X, \\ \mathcal{N}(A^*) &\subset Y. \end{aligned} \qquad (4.19)$$

The fundamental subspaces of a linear operator have the following orthogonality properties.

Theorem 4.5 Let $A: X \to Y$ be a bounded linear operator between two Hilbert spaces X and Y, and let $\mathcal{R}(A)$ and $\mathcal{R}(A^*)$ be closed. Then

1. *The range is the orthogonal complement of the left nullspace:*

$$[\mathcal{R}(A)]^\perp = \mathcal{N}(A^*). \qquad (4.20)$$

2. *The orthogonal complement of the row space is the nullspace:*
$$\mathcal{R}(A^*)^\perp = \mathcal{N}(A). \quad (4.21)$$

Complementing these two we have
$$\mathcal{R}(A) = [\mathcal{N}(A^*)]^\perp, \quad (4.22)$$
$$\mathcal{R}(A^*) = [\mathcal{N}(A)]^\perp. \quad (4.23)$$

Proof We show first that $\mathcal{N}(A^*) \subset [\mathcal{R}(A)]^\perp$. Let $n \in \mathcal{N}(A^*)$, and let $y \in \mathcal{R}(A)$, so that $y = Ax$ for some $x \in X$. Then
$$\langle y, n \rangle = \langle Ax, n \rangle = \langle x, A^*n \rangle = \langle x, 0 \rangle = 0.$$
Thus $n \perp y$, so $\mathcal{N}(A^*) \subset [\mathcal{R}(A)]^\perp$. Now let $y \in [\mathcal{R}(A)]^\perp$. Then for every $x \in X$,
$$\langle Ax, y \rangle = 0,$$
and so by the definition of the adjoint $\langle x, A^*y \rangle = 0$. Since this is true for every $x \in X$, we must have $A^*y = 0$, so $y \in \mathcal{N}(A^*)$. We thus have $[\mathcal{R}(A)]^\perp \subset \mathcal{N}(A^*)$. Combining these two inclusions, we conclude that $[\mathcal{R}(A)]^\perp = \mathcal{N}(A^*)$.

To prove part (4.21), replace A by A^*, using the fact that $A^{**} = A$.

To prove (4.22) and (4.23), take the orthogonal complement of both sides of (4.20) and (4.21). □

By theorem 2.10 and (4.19), if $A: X \to Y$, and X and Y are Hilbert spaces with $\mathcal{R}(A)$ and $\mathcal{R}(A^*)$ closed, then
$$\begin{aligned} X &= \mathcal{R}(A^*) \oplus \mathcal{N}(A), \\ Y &= \mathcal{R}(A) \oplus \mathcal{N}(A^*). \end{aligned} \quad (4.24)$$

Theorem 4.5 provide a means of determining whether an equation of the form $Ax = b$ has a solution, and whether that solution is unique. As mentioned previously, in order to have an exact solution b must lie in the column space (range) of A. But if b is in the column space of A, by theorem 4.5 it must be orthogonal to the left nullspace of A. This fact regarding the existence of the solution to the linear equation is fundamental and important. It is known as the Fredholm alternative theorem. We state it first for general linear operators, then specialize it to matrix notation.

Theorem 4.6 *(Fredholm alternative theorem) Let A be a bounded linear operator. The equation $Ax = b$ has a solution if and only if $\langle b, v \rangle = 0$ for every vector $v \in \mathcal{N}(A^*)$. More succinctly,*
$$b \in \mathcal{R}(A) \Leftrightarrow b \perp \mathcal{N}(A^*).$$

In matrix notation, $A\mathbf{x} = \mathbf{b}$ has a solution if and only if $\mathbf{v}^H \mathbf{b} = 0$ for every vector $\mathbf{v}$ such that $A^H \mathbf{v} = 0$.

Proof Assume that $Ax = b$, and let $v \in \mathcal{N}(A^*)$. Then
$$\langle b, v \rangle = \langle Ax, v \rangle = \langle x, A^*v \rangle = \langle x, 0 \rangle = 0.$$
To prove the converse, suppose to the contrary that $\langle b, v \rangle = 0$ when $v \in \mathcal{N}(A^*)$, but $Ax = b$ has no solution. Since $b \notin \mathcal{R}(A)$, let $b = b_r + b_0$, where $b_r \in \mathcal{R}(A)$ and b_0 is the component of b orthogonal to $\mathcal{R}(A)$. Then $\langle Ax, b_0 \rangle = 0$ for all x, from which we conclude that $A^*b_0 = 0$; that is, $b_0 \in \mathcal{N}(A^*)$. By the hypothesis of the theorem, it must be the case, therefore, that $\langle b, b_0 \rangle = 0$, or
$$\begin{aligned} 0 &= \langle b_r + b_0, b_0 \rangle \\ &= \langle b_r, b_0 \rangle + \langle b_0, b_0 \rangle. \end{aligned}$$

Since b_r and b_0 are orthogonal, it must be the case that $b_0 = 0$, so b lies in the range of A. □

The uniqueness of solutions is determined by the following theorem.

Theorem 4.7 *(Uniqueness of solutions) The solution to $Ax = b$ (if it exists) is unique if and only if the only solution of $Ax = 0$ is $x = 0$, that is, if $\mathcal{N}(A) = \{0\}$.*

The proof is given as an exercise.

Example 4.5.1 We demonstrate the concept of the four fundamental subspaces as applied to matrix operators. Consider the matrix

$$A = \begin{bmatrix} 0 & 0 & 1 \\ 0 & 0 & 0 \end{bmatrix}.$$

The column space is $\mathcal{R}(A) = \text{span}\left(\begin{bmatrix} 1 \\ 0 \end{bmatrix}\right)$, which geometrically is the line in two dimensions through the point $(1, 0)$. The null space is

$$\mathcal{N}(A) = \text{span}\left(\begin{bmatrix} 0 \\ 1 \\ 0 \end{bmatrix}, \begin{bmatrix} 1 \\ 0 \\ 0 \end{bmatrix}\right).$$

The left nullspace contains the vector $\begin{bmatrix} 0 \\ 1 \end{bmatrix}$, and so

$$\mathcal{N}(A^T) = \text{span}\left(\begin{bmatrix} 0 \\ 1 \end{bmatrix}\right).$$

The row space is $\mathcal{R}(A^T) = \text{span}([0\ 0\ 1]^T)$, which is the line in three dimensions through $(0, 0, 1)$. □

In this example, the dimension of the column space is the dimension of the row space. This is not an isolated example: *in all matrices, the dimension of the column space is the dimension of the row space.*

Definition 4.7 The dimension of the column space (or the row space) of the matrix A is the **rank** of the matrix. □

The rank of a matrix A is the number of linearly independent columns or rows of A. For an $m \times n$ matrix of rank r, the following size relationships hold:

1. Column space: $\dim(\mathcal{R}(A)) = r$.
2. Row space: $\dim(\mathcal{R}(A^H)) = r$.
3. Nullspace: $\dim(\mathcal{N}(A)) = n - r$.
4. Left nullspace: $\dim(\mathcal{N}(A^H)) = m - r$.

Example 4.5.2 Another example may help to illustrate the relationships among the subspaces. Let

$$A = \begin{bmatrix} 1 & 4 \\ 5 & 20 \end{bmatrix}.$$

Then $m = n = 2$, and $r = 1$.

- The column space is

$$\mathcal{R}(A) = \text{span}\left(\begin{bmatrix} 1 \\ 5 \end{bmatrix}\right)$$

4.5 Four Fundamental Subspaces of a Linear Operator

- The null space is
$$\mathcal{N}(A) = \text{span}\left(\begin{bmatrix} -5 \\ 1 \end{bmatrix}\right)$$

- The row space is
$$\mathcal{R}(A^T) = \text{span}\left(\begin{bmatrix} 1 \\ 4 \end{bmatrix}\right).$$

- The left nullspace is
$$\mathcal{N}(A^T) = \text{span}\left(\begin{bmatrix} -4 \\ 1 \end{bmatrix}\right).$$

□

Example 4.5.3 The matrix
$$A = \begin{bmatrix} 1 & 3 \\ 4 & 12 \end{bmatrix}$$
has a nullspace spanned by the vector
$$\begin{bmatrix} 3 \\ -1 \end{bmatrix},$$
so solutions, if they exist, are not unique. The nullspace of A^T is spanned by $\mathbf{v} = \begin{bmatrix} 4 \\ -1 \end{bmatrix}$. A vector $\mathbf{b}$ is in the column space of A if it is orthogonal to $\mathbf{v}$, so $\mathbf{b}$ must be of the form
$$\mathbf{b} = c \begin{bmatrix} 1 \\ 4 \end{bmatrix}.$$
(In light of the form of the matrix, this is obvious.) The equation
$$A\mathbf{x} = \begin{bmatrix} 1 \\ 4 \end{bmatrix}$$
has solutions
$$\mathbf{x} = \begin{bmatrix} 1 \\ 1 \end{bmatrix} + \beta \begin{bmatrix} 3 \\ -1 \end{bmatrix}$$
for any $\beta \in \mathbb{R}$.

□

Example 4.5.4 The matrix in the equation
$$\begin{bmatrix} 1 & -1 & 0 \\ 0 & 1 & 1 \\ -1 & 0 & -1 \end{bmatrix} \begin{bmatrix} x_1 \\ x_2 \\ x_3 \end{bmatrix} = \begin{bmatrix} b_1 \\ b_2 \\ b_3 \end{bmatrix}$$
has a left nullspace spanned by
$$\mathbf{v} = \begin{bmatrix} 1 \\ 1 \\ 1 \end{bmatrix}.$$
Any right hand side $\mathbf{b}$ must be orthogonal to $\mathbf{v}$,
$$\mathbf{b}^T \mathbf{y} = 0,$$
which means that the sum of the components of $\mathbf{b}$ must be zero.

□

Figure 4.4 summarizes the relationships between the four fundamental subspaces of a matrix $A: \mathbb{R}^n \to \mathbb{R}^m$. (It is straightforward to generalize this to a linear operator $A: X \to Y$.) On the left is the space $\mathbb{R}^n$, which can be decomposed into the row space $\mathcal{R}(A^T)$ and the nullspace

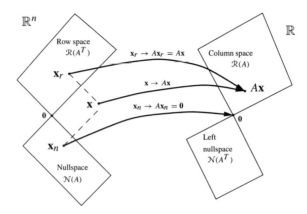

Figure 4.4: The four fundamental subspaces of a matrix operator

$\mathcal{N}(A)$, which are orthogonal. A vector $\mathbf{x} \in \mathbb{R}^n$ can be decomposed into its rowspace component and its nullspace component,

$$\mathbf{x} = \mathbf{x}_r + \mathbf{x}_n,$$

as shown. The action of the matrix A on $\mathbf{x}$ is $A\mathbf{x} = A\mathbf{x}_r + A\mathbf{x}_n$, as follows.

- The nullspace component goes to zero: $A\mathbf{x}_n = 0$.
- The row space component goes to the column space: $A\mathbf{x}_r = A\mathbf{x} \in \mathcal{R}(A)$.

On the right of figure 4.4 is the space $\mathbb{R}^m$, which is decomposed into the orthogonal spaces of column space and the left nullspace. In light of this figure, we observe that a matrix A *maps its row space to its column space*. The mapping from the row space to the column space is invertible:

Theorem 4.8 *Every vector* $\mathbf{b} \in \mathcal{R}(A)$ *comes from one and only one vector* $\mathbf{x}_r$ *in the row space of A.*

The proof is given as an exercise.

The four fundamental spaces of a matrix may be determined numerically by use of the singular value decomposition (SVD), as discussed in section 7.2.

4.5.1 The four fundamental subspaces with non-closed range*

Theorem 4.5 concludes that

$$[\mathcal{R}(A)]^\perp = (A^*).$$

Complementing both sides, we obtain

$$[\mathcal{R}(A)]^{\perp\perp} = [(A^*)]^\perp.$$

If $\mathcal{R}(A)$ is closed, then $[\mathcal{R}(A)]^{\perp\perp} = \mathcal{R}(A)$. However if $\mathcal{R}(A)$ is *not* closed, then

$$[\mathcal{R}(A)]^{\perp\perp} = \overline{\mathcal{R}(A)},$$

---

*This section contains some technical details that correspond to infinite-dimensional vector spaces, and may be skipped on a first reading.

4.6 Some Properties of Matrix Inverses

the closure of $\mathcal{R}(A)$. If $\mathcal{R}(A^*)$ is not closed, we similarly conclude that

$$\overline{\mathcal{R}(A^*)} = [\mathcal{N}(A)]^\perp.$$

Example 4.5.5 As an example of an operator with a range that is not closed, let X and Y be the space l_1, and define the operator $A: X \to Y$ by

$$A(x_1, x_2, x_3, \ldots) = (x_1, x_2/2, x_3/3, \ldots),$$

where $(x_1, x_2, x_3, \ldots) \in X$ and $(x_1, x_2/2, x_3/3, \ldots) \in Y$. Then $\mathcal{R}(A)$ contains all finitely nonzero sequences, so $\overline{\mathcal{R}(A)} = Y = l_1$. However, there are sequences in Y that are not in the range of A. The sequence

$$y = (1, 1/2^2, 1/3^2, \ldots)$$

would be the image of

$$x = (1, 1/2, 1/3, \ldots),$$

but this x is not an element of l_1, since $\sum_{i=1}^{\infty}(1/i)$ does not converge. $\square$

Fortunately, for the problems of interest in this book, our operators are mostly finite dimensional, hence have closed range.

4.6 Some properties of matrix inverses

With the basic theorems regarding the existence and uniqueness of solutions to linear equations in place, we focus now on matrix operators.

Definition 4.8 A matrix A is said to have a **left inverse** if there is a matrix B such that $BA = I$, and a **right inverse** if there is a matrix C such that $AC = I$. $\square$

If A has a left inverse B then the equation $A\mathbf{x} = \mathbf{b}$ has the solution $\mathbf{x} = B\mathbf{b}$, for some matrix B. If A has a right inverse then the equation has the solution $\mathbf{x} = C\mathbf{b}$, for some matrix C. We make the following observations:

1. (Existence) Let $A \in M_{m,n}(F)$ for some field F. In order for the equation

$$A\mathbf{x} = \mathbf{b}$$

    to have **at least one** solution for *any* $\mathbf{b}$, the columns of A must span the space F^m. This means that the rank of the matrix $r = m$. In this case there is an $n \times m$ right inverse C and a solution $\mathbf{x} = C\mathbf{b}$. This right inverse can exist only when $m \leq n$. However, there could exist more than one solution.

2. (Uniqueness) The system $A\mathbf{x} = \mathbf{b}$ has **at most one** solution for any $\mathbf{b}$ if and only if the columns are linearly independent. This means that the rank is $r = n$. In this case there is an $n \times m$ left inverse B. This is possible only if $m \geq n$.

Example 4.6.1 The matrix

$$A = \begin{bmatrix} 2 & 0 & 0 \\ 0 & 7 & 0 \end{bmatrix}$$

has rank $r = 2 = m$; thus, by the theorem, it has a right inverse,

$$AC = \begin{bmatrix} 2 & 0 & 0 \\ 0 & 7 & 0 \end{bmatrix} \begin{bmatrix} \frac{1}{2} & 0 \\ 0 & \frac{1}{7} \\ c_1 & c_1 \end{bmatrix} = \begin{bmatrix} 1 & 0 \\ 0 & 1 \end{bmatrix}.$$

The last row of C is arbitrary, and there are an infinite number of right inverses, and hence an infinite number of solutions $\mathbf{x} = C\mathbf{b}$. □

Example 4.6.2 Consider the matrix transposed from the previous example:
$$A = \begin{bmatrix} 2 & 0 \\ 0 & 7 \\ 0 & 0 \end{bmatrix}.$$

Then
$$BA = \begin{bmatrix} \frac{1}{2} & 0 & b_1 \\ 0 & \frac{1}{7} & b_2 \end{bmatrix} = \begin{bmatrix} 1 & 0 \\ 0 & 1 \end{bmatrix}.$$

The last column of B is arbitrary in this case.

The uniqueness result does not go so far as to say that there *is* a solution. Consider the problem
$$\begin{bmatrix} 2 & 0 \\ 0 & 7 \\ 0 & 0 \end{bmatrix} \begin{bmatrix} x_1 \\ x_2 \\ x_3 \end{bmatrix} = \begin{bmatrix} b_1 \\ b_2 \\ b_3 \end{bmatrix}.$$

If $b_3 \neq 0$, there is no solution. However, in the case that $b_3 = 0$ and there is a solution, since $r = 2 = n$, that solution must be unique. □

One possible formula for the left and right inverses, if they exist, is

$B = (A^H A)^{-1} A^H$ (left inverse),

$C = A^H (AA^H)^{-1}$ (right inverse).

The left inverse corresponds to the solution
$$\mathbf{x} = B\mathbf{b} = (A^H A)^{-1} A^H \mathbf{b},$$

which is the least-squares solution. The guarantee of at least one solution is obtained by projection of $\mathbf{b}$ into the column space of A. Both of these inverses correspond to setting the free variables equal to zero.

4.6.1 Tests for invertibility of matrices

In order for *both* existence and uniqueness of a solution, we must have $r = m \geq n$ and $r = n \geq m$; that, is $m = n$: the matrix must be square. If A has both a left inverse and a right inverse they are the same. In this case A is said to be **invertible**.

Definition 4.9 A matrix $A \in M_{m,n}(F)$ (where F is a field) is said to be **nonsingular** if the only $\mathbf{x} \in F^n$ such that $A\mathbf{x} = \mathbf{0}$ is the zero vector $\mathbf{x} = \mathbf{0}$. □

If A is a nonsingular $n \times n$ matrix, then its nullspace has dimension zero and $\text{rank}(A) = n$. It is therefore invertible. Based on this discussion, as well as some results from later sections, we present a collection of tests for invertibility of an $n \times n$ matrix A:

1. $A\mathbf{x} = 0$ implies $\mathbf{x} = 0$.
2. $\text{rank}(A) = n$.
3. The rows of A are linearly independent; the columns of A are linearly independent.
4. The determinant of A is not zero.
5. There are no zero eigenvalues of A.
6. $A^H A$ is positive definite; that is,
$$\mathbf{x}^H A^H A \mathbf{x} > 0$$
for any nonzero $\mathbf{x} \in R^n$.

4.7 Some results on matrix rank

Definition 4.10 An $m \times n$ matrix is said to be **full rank** if the rank is as large as possible:
$$\text{rank}(A) = \min(m, n).$$
An $m \times n$ matrix is said to be **rank deficient** if it is not full rank. □

The following theorem provides a characterization of the four fundamental spaces of the product of matrices AB.

Theorem 4.9 *For matrices A and B such that AB exists:*

1. $\mathcal{N}(B) \subset \mathcal{N}(AB)$,
2. $\mathcal{R}(AB) \subset \mathcal{R}(A)$,
3. $\mathcal{N}(A^*) \subset \mathcal{N}((AB)^*)$,
4. $\mathcal{R}((AB)^*) \subset \mathcal{R}(B)$.

Proof

1. If $B\mathbf{x} = 0$ then $AB\mathbf{x} = 0$: every $\mathbf{x} \in \mathcal{N}(B)$ is also in $\mathcal{N}(AB)$. Thus
$$\dim \mathcal{N}(AB) \geq \dim \mathcal{N}(B).$$
2. If $\mathbf{x} \in \mathcal{R}(AB)$, then there is some $\mathbf{y}$ so that $\mathbf{x} = (AB)\mathbf{y} = A(B\mathbf{y})$, so $\mathbf{x} \in \mathcal{R}(A)$.
3. If $\mathbf{y}^* A = 0$ then $\mathbf{y}^* AB = 0$.
4. If $\mathbf{x} \in \mathcal{R}((AB)^T)$, then there is some $\mathbf{y}$ so that $\mathbf{x} = (AB)^*\mathbf{y} = B^T(A^*\mathbf{y})$, so $\mathbf{x} \in \mathcal{R}(B^*)$. □

Combining the second and fourth items, we obtain the following fact:

$$\boxed{\text{rank}(AB) \leq \text{rank}(A) \qquad \text{rank}(AB) \leq \text{rank}(B)} \qquad (4.25)$$

Example 4.7.1 An obvious but important example of the fact expressed in (4.25) is that the matrix $B = \mathbf{x}\mathbf{y}^H$, where $\mathbf{x}$ and $\mathbf{y}$ are nonzero vectors, must have rank ≤ 1, since each vector has rank 1. A computation of the form
$$A + \mathbf{x}\mathbf{y}^H$$
is said to be a *rank-one update* to the matrix A. Similarly, if X is a $m \times 2$ matrix and Y is a $2 \times n$ matrix, the update
$$A + XY^H$$
is said to be a rank-two update.

A question explored in section 4.11 is how to compute the inverse of a low-rank update of A, if we already know the inverse of A. □

Example 4.7.2 Let A be a 3×4 matrix of zeros, and let B be a 4×3 matrix of zeros. Then the nullity of B is 3, while the nullity of AB is 4. □

These results provide a "cancellation" theorem:

Theorem 4.10 *[Campbell and Meyer (see page 274)] Suppose A is $m \times n$ and B and C are $n \times p$. Then $A^H AB = A^H AC$ if and only if $AB = AC$.*

Proof The result can be stated equivalently as: $A^*A(B - C) = 0$ if only if $A(B - C)$. This becomes a question of comparing the nullspace of A^*A and A. We need to show that $\mathcal{N}(A^*A) = \mathcal{N}(A)$. Since $\mathcal{N}(A^*) = [\mathcal{R}(A)]^\perp$, if $A^*A\mathbf{x} = 0$, then $A\mathbf{x} = 0$ (and conversely). □

Definition 4.11 A **submatrix** of a matrix A is obtained by removing zero or more columns of A and zero or more rows of A. □

Notationally, when the retained rows and columns of a matrix are adjacent, the ":" notation can be employed, as discussed in section C.1. Since the submatrix cannot be larger than the matrix, we have the following results.

> For an $m \times n$ matrix A of rank r, every submatrix C is of rank $\leq r$.

> For an $m \times n$ matrix A of rank r, there is at least one $r \times r$ matrix of rank exactly r.

Based upon the latter result, we can give an equivalent definition of the rank:

> The rank of a matrix is the size of the largest nonsingular square submatrix. There is a $k \times k$ submatrix with nonzero determinant, but all $(k+1) \times (k+1)$ submatrices of A have determinant 0.

The following facts are also true about rank [142]:

- If $A \in M_{m,k}$ and $B \in M_{k,n}$ then
$$\text{rank}(A) + \text{rank}(B) - k \leq \text{rank}(AB) \leq \min\{\text{rank}(A), \text{rank}(B)\}.$$

- If $A, B \in M_{m,n}$, then
$$\text{rank}(A + B) \leq \text{rank}(A) + \text{rank}(B).$$

- (Result generally attributed to Frobenius) If $A \in M_{m,k}$, $B \in M_{k,p}$, and $C \in M_{p,n}$, then
$$\text{rank}(AB) + \text{rank}(BC) \leq \text{rank}(B) + \text{rank}(ABC).$$

- Rank is unchanged upon either left or right multiplication by a nonsingular matrix. If $A \in M_m$ and $C \in M_n$ are both nonsingular and $B \in M_{m,n}$, then
$$\text{rank}(B) = \text{rank}(AB) = \text{rank}(BC) = \text{rank}(ABC).$$

- If $A, B \in M_{m,n}$, then $\text{rank}(A) = \text{rank}(B)$ if and only if there exist *nonsingular* $X \in M_m$ and $Y \in M_n$ such that $B = XAY$.

- If $A \in M_{m,n}$ has $\text{rank}(A) = k$, then there is a nonsingular $B \in M_k$, $X \in M_{m,k}$, and $Y \in M_{k,n}$, such that
$$A = XBY.$$

- A matrix $A \in M_{m,n}(F)$ of rank 1 can be written as
$$A = \mathbf{x}^T \mathbf{y}$$
for $\mathbf{x} \in F^m$ and $\mathbf{y} \in F^n$.

4.7.1 Numeric rank

Even though the rank of a matrix is well defined mathematically, numerical difficulties may arise when actually trying to compute the rank of a matrix with real-valued elements, due to roundoff.

Example 4.7.3 The matrix
$$A = \begin{bmatrix} 2 & 4 \\ 1 & 2 + \epsilon \end{bmatrix}$$

is full rank for any $\epsilon \neq 0$. However, it is *close* (using some matrix norm) to a matrix that is rank deficient. □

There are a variety of ways of numerically computing the rank of a matrix, including the QR decomposition with column pivoting (see, e.g., [114]). However, one of the best ways is to use the SVD, which can provide information not only on what the rank is (numerically), but also whether the matrix is close to another matrix that is rank deficient.

4.8 Another look at least squares

Let $A: X \to Y$ be a bounded linear operator. If the equation $Ax = b$ has no solution (i.e., $b \notin \mathcal{R}(A)$), then, as we saw in section 3.4, we can find a solution $\hat{x}$ that minimizes

$$\|A\hat{x} - b\|_2^2.$$

The following theorem ties the idea of the projection theorem to the Fredholm alternative theorem, and provides another view of the least-squares pseudoinverse operator.

Theorem 4.11 *The vector $\hat{x} \in X$ minimizes $\|b - Ax\|_2$ if and only if*

$$A^*A\hat{x} = A^*b.$$

Proof Minimizing $\|b - Ax\|$ is equivalent to minimizing $\|b - \hat{b}\|_2$, where $\hat{b} \in \mathcal{R}(A)$. By the projection theorem, we must have

$$b - \hat{b} \in [\mathcal{R}(A)]^\perp.$$

But, by theorem 4.5, this is equivalent to

$$b - \hat{b} \in \mathcal{N}(A^*);$$

that is,

$$A^*(b - \hat{b}) = 0,$$

or

$$A^*\hat{b} = A^*b.$$

Conversely, if $A^*A\hat{x} = A^*b$, then

$$A^*(A\hat{x} - b) = 0,$$

so that $A\hat{x} - b \in \mathcal{N}(A^*)$ and, hence, the error is orthogonal to the subspace $\mathcal{R}(A)$ and has minimal length by the projection theorem. □

If A is a matrix operator such that A^*A is invertible, by theorem 4.11 the least-squares solution is

$$\mathbf{x} = (A^*A)^{-1}A^*\mathbf{b}, \qquad (4.26)$$

which is equivalent to what we obtained in (3.19).

4.9 Pseudoinverses

The matrix $(A^H A)^{-1}A^H$ of (4.26) is an example of a pseudoinverse, sometimes called a Moore–Penrose pseudoinverse. We formally define the pseudoinverse in a manner that captures the smallest error and smallest norm of the solution, as follows.

Definition 4.12 Let $A: X \to Y$ be a bounded linear operator, where X and Y are Hilbert spaces, and let $\mathcal{R}(A)$ be closed. For some $b \in Y$, let $\hat{x}$ be the vector of minimum norm $\|\hat{x}\|_2$ that minimizes $\|Ax - b\|$. The pseudoinverse $A^\dagger$ is the operator mapping b to $\hat{x}$ for each $b \in Y$. □

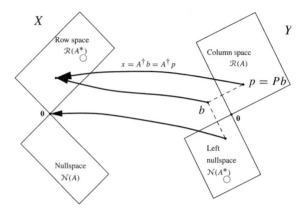

Figure 4.5: Operation of the pseudoinverse

The operation of the pseudoinverse operator is shown in figure 4.5. The pseudoinverse operator takes a point from Y back to a point in $\hat{x} \in \mathcal{R}(A^*)$, in such a way that $\hat{x}$ has minimum norm. The operation of the pseudoinverse operation on a point $b \notin \mathcal{R}(A)$ is to first project b onto $\mathcal{R}(A)$ using the projection P, then to map back to $\mathcal{R}(A^*)$ to a vector $\hat{x}$ of minimum length; by this projection onto $\mathcal{R}(A)$ the error $b - A\hat{x}$ is minimized.

By the definition of the pseudoinverse, with the aid of figure 4.5, the following properties of pseudoinverses can be verified.

1. $A^\dagger$ is linear.
2. $A^\dagger$ is bounded.
3. $(A^\dagger)^\dagger = A$.
4. $(A^\dagger)AA^\dagger = A^\dagger$.
5. $AA^\dagger A = A$.
6. $(A^\dagger A)^* = A^\dagger A$.
7. $A^\dagger = (A^*A)^\dagger A^*$.
8. $A^\dagger = A^*(AA^*)^\dagger$.

The last two properties, as applied to matrices, give an "explicit" formula for computing the pseudoinverse of a matrix A. The formula $A^\dagger = (A^*A)^\dagger A^*$ is appropriate for overdetermined sets of equations (more equations than unknowns); the formula $A^\dagger = A^*(AA^*)^\dagger$ is most appropriate for underdetermined sets of equations. However, as cautioned before, explicitly computing the product $A^H A$ or AA^H is poor practice numerically. In fact, it may happen (if the columns of A are not linearly independent) that $A^H A$ is not full rank, and cannot be inverted. For these reasons, the pseudoinverse should be computed using a factorization approach such as the QR factorization or the SVD.

There are some technical details requiring clarification regarding definition 4.12. (This paragraph may be skipped on a first reading.) For a consistent definition, it must be established that there is a unique x_0 that minimizes $\|Ax - b\|$. Since $\mathcal{R}(A)$ is closed, the minimum $\|Ax - b\|$ is actually achieved, since y is approximated by some vector in $\hat{y} \in \mathcal{R}(A)$. Let x_0 be any vector satisfying $Ax = \hat{y}$. Then any point x on the linear variety $V = x_0 + \mathcal{N}(A)$ also satisfies $Ax = \hat{y}$. Since $\mathcal{N}(A)$ is closed, there is a point x^* in V of minimum norm. Thus $A^\dagger$ is well-defined.

4.10 Matrix condition number

We now focus our attention on matrix linear equations, and consider some numerical aspects. We have seen conditions related to the existence of solution to the equation $A\mathbf{x} = \mathbf{b}$. However, in some cases the conditions may be true in a mathematical sense, but not usefully true in a numerical sense: it may not be possible to compute a reliable solution using finite-precision arithmetic on a useful physical computation device. In the equation $A\mathbf{x} = \mathbf{b}$, the solution can be thought of as the point which is at the intersection of the lines (or planes) determined by each equation. In some equations it may happen that the lines (or planes) are nearly parallel. If this is the case, then a slight change in the coefficients could lead to a substantial change in the location of the solution. This problem is of practical significance because numerical representations in computers are rarely, if ever, exact. The problem may be so severe that the results of solving $A\mathbf{x} = \mathbf{b}$ are completely useless.

Example 4.10.1 Consider the equation

$$\begin{bmatrix} 1+3\epsilon & 1-3\epsilon \\ 3-\epsilon & 3+\epsilon \end{bmatrix} \begin{bmatrix} x_1 \\ x_2 \end{bmatrix} = \begin{bmatrix} b_1 \\ b_2 \end{bmatrix}.$$

For small ϵ the lines represented by these two equations are almost parallel, as shown in figure 4.6 for $b_1 = 1$ and $b_2 = 3$. Small changes in $\mathbf{b}$ will result in significant shifts in the location of the intersection of the two lines. For example, suppose the true value of $\mathbf{b} = [1, 3]^T$. The solution is $\mathbf{x} = [.5, .5]^T$. However, if the right hand side is slightly corrupted (perhaps by a small amount of measurement noise) to $\mathbf{b} = [1.1, 3]^T$ then $\mathbf{x} = [2.005, -.995]^T$: a miniscule change in the input has been amplified to a drastic change in the output. For practical problems, the answer obtained would probably be completely untrustworthy. □

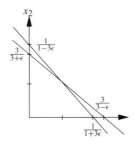

Figure 4.6: Demonstration of an ill-conditioned linear system

Matrices in which small errors in the coefficients are substantially amplified to produce large deviation in the solution are referred to as **ill-conditioned** matrices. Ill-conditioned matrices are those which are "close" to being rank deficient. Associated with each square matrix is a number, the **condition number**, which indicates the degree to which the matrix is ill conditioned. (The condition of a matrix was first explored by J. H. Wilkinson; see box 4.1.) We will get an analytical handle on ill-conditioned matrices by some perturbation analysis. Suppose that A is the matrix for which we wish to solve $A\mathbf{x} = \mathbf{b}$. Instead of the true matrix A, the representation we must deal with (because of numerical representation problems) is a matrix $A + \epsilon E$, where ϵE is "small." The solution that is actually computed, then, is based upon the perturbed equation

$$(A + \epsilon E)\mathbf{x} = \mathbf{b}. \tag{4.27}$$

We will examine bounds on how the solution can change for small perturbations ϵE.

Let $\mathbf{x}_0$ denote the true (unperturbed) solution

$$\mathbf{x}_0 = A^{-1}\mathbf{b}.$$

> **Box 4.1: James H. Wilkinson (1919–1986)**
>
> James H. Wilkinson was a pioneer of numerical analysis and explored the foundations of numerical representation. During the 1940s he worked with Alan Turing on the design of a computing machine, the ACE (Automatic Computing Engine). To test the arithmetic of the ACE, Wilkinson wrote a program to find roots of polynomials. To test the program, he used it to compute the roots of a polynomial whose roots he knew:
>
> $$p(x) = (x-1)(x-2)\cdots(x-20).$$
>
> The computed roots were not even close to the true roots. After eliminating the possibility of a bug in his software, he determined that the *computed roots of the polynomial are incredibly sensitive to the values of the coefficients.* The roots he obtained were in fact the roots of another polynomial that is close to the original one.
>
> Wilkinson later worked on the solution of the equation $A\mathbf{x} = \mathbf{b}$. What he found is that the result $\mathbf{x}$ computed using the available algorithms (which are still used today) provides an exact answer for a nearby problem: the A matrix is perturbed. This means that even though the error residual $\mathbf{b} - A\mathbf{x}$ might be small (a good solution to a nearby problem), the difference between the $\mathbf{x}$ actually obtained and the "true" $\mathbf{x}$ to the original problem could possibly be large, if A is such that a "nearby" matrix has a very different solution. We say that such a matrix A is ill conditioned.

If A is nonsingular and $\|\epsilon E\|$ is small enough that $A + \epsilon E$ is nonsingular (see exercise 4.2-11), then the solution computed from (4.27) is

$$\mathbf{x}_E = (A + \epsilon E)^{-1}\mathbf{b} = [A(I + \epsilon A^{-1}E)]^{-1}\mathbf{b}.$$

The vector $\mathbf{x}_E$ shows how far the perturbed solution deviates from the unperturbed (true) solution as ϵE changes. Using the Neumann formula from section 4.2.2,

$$\mathbf{x}_E = [I - \epsilon A^{-1}E]A^{-1}\mathbf{b} + O(\|\epsilon E\|^2) = \mathbf{x}_0 - \epsilon A^{-1}E\mathbf{x}_0 + O(\|\epsilon E\|^2). \quad (4.28)$$

The relative error between the true solution and the computed solution is thus

$$\frac{\|\mathbf{x}_E - \mathbf{x}_0\|}{\|\mathbf{x}_0\|} \leq \|A^{-1}\|\|\epsilon E\| + O(\|\epsilon E\|^2).$$

Now define the **condition number** $\kappa(A)$ of the matrix A as

$$\kappa(A) = \|A\|\|A^{-1}\|.$$

When A is singular, the convention is to take $\kappa(A) = \infty$. Also, let

$$\rho = \epsilon \frac{\|E\|}{\|A\|}$$

denote the relative error in the perturbation of A. Then the relative error in the solution can be bounded as

$$\frac{\|\mathbf{x}_E - \mathbf{x}_0\|}{\|\mathbf{x}_0\|} \leq \kappa(A)\rho + O(\|\epsilon E\|^2). \quad (4.29)$$

A matrix with a large condition number is said to be ill conditioned. For an ill-conditioned matrix, the relative error in the solution can be large, even when the perturbation ϵE is small.

4.10 Matrix Condition Number

For very small changes in the matrix, there may be very large changes in the answer. It should be pointed out that a large condition number does not *guarantee* a large relative error—(4.29) is an upper bound—but a poorly conditioned matrix very often leads to unacceptable results.

Another interpretation of the condition number is obtained as follows. Let

$$M = \max_{\mathbf{x} \neq 0} \frac{\|A\mathbf{x}\|}{\|\mathbf{x}\|}$$

and

$$m = \min_{\mathbf{x}} \frac{\|A\mathbf{x}\|}{\|\mathbf{x}\|}.$$

Recall (see (4.8)) that

$$\|A^{-1}\| = 1/m.$$

Then the condition number can be written as

$$\kappa(A) = \frac{M}{m}.$$

Thus, for example, using the l_2 norm we can write

$$\kappa(A) = \frac{\sqrt{\lambda_{\max}}}{\sqrt{\lambda_{\min}}},$$

where $\lambda_{\max}$ and $\lambda_{\min}$ are the largest and smallest eigenvalues, respectively, of $A^H A$. If A is self-adjoint, then

$$\kappa(A) = \frac{\lambda_{\max}}{\lambda_{\min}},$$

where the eigenvalues now are of A. In the signal processing literature, reference is made to a matrix with a large eigenvalue spread. This is another way of saying that the matrix has a large condition number.

If we now perturb $A\mathbf{x} = \mathbf{b}$ on the right hand side, replacing $\mathbf{b}$ by $\mathbf{b} + \Delta\mathbf{b}$ for some small $\Delta\mathbf{b}$, then the solution $\mathbf{x}$ is perturbed so we obtain the equation

$$A(\mathbf{x} + \Delta\mathbf{x}) = (\mathbf{b} + \Delta\mathbf{b}),$$

where $A(\Delta\mathbf{x}) = \mathbf{b}$. Then we have

$$\|\mathbf{b}\| \leq M\|\mathbf{x}\| \quad \text{and} \quad \|\Delta\mathbf{b}\| \geq m\|\Delta\mathbf{x}\|.$$

Taking ratios (if $m \neq 0$),

$$\frac{\|\Delta\mathbf{x}\|}{\|\mathbf{x}\|} \leq \kappa(A) \frac{\|\Delta\mathbf{b}\|}{\|\mathbf{b}\|},$$

which shows again how the relative error of the solution is related to the relative perturbation and the condition number: a large condition number has the potential of a large relative error, even for small changes in the right-hand side.

Example 4.10.2 The *machine epsilon* ϵ_{mach} is the smallest number on a computer that can be added to 1.0 and obtain a number larger than 1.0. It is a representation of the relative accuracy of computer arithmetic.

Suppose that $\mathbf{b}$ has a single element that is non-accurate, and all other elements are integers. Then $\|\Delta\mathbf{b}\|/\|\mathbf{b}\|$ may be as large as ϵ_{mach}. Suppose that for some machine $\epsilon_{\text{mach}} = 2^{-22}$, and that $\kappa(A) = 10^5$. The solution could have a relative error of

$$\frac{\|\Delta\mathbf{x}\|}{\|\mathbf{x}\|} \approx \kappa(A) 2^{-22} \approx 2 \times 10^{-2}.$$

Thus simply *storing* the coefficients of the problem on this computer could lead to an error in the second significant digit in the solution, even if all other computations are carried out without any error! □

If both the right hand side and the left hand side of the equation are perturbed (as usually happens when real numbers are represented in a computer), the solution is computed for the equation

$$(A + \epsilon E)\mathbf{x} = \mathbf{b} + \epsilon \mathbf{e}$$

for perturbations ϵE and $\epsilon \mathbf{e}$. It can be shown (see exercise 4.10-42) that the relative error is bounded by

$$\frac{\|\mathbf{x} - \mathbf{x}_0\|}{\|\mathbf{x}_0\|} \leq \kappa(A)(\rho + \rho_b) + O(\epsilon^2), \qquad (4.30)$$

where $\rho = \epsilon \|E\|/\|A\|$ is the relative change in A and $\rho_b = \epsilon \|\mathbf{e}\|/\|\mathbf{b}\|$ is the relative change in $\mathbf{b}$.

As a rough rule of thumb for determining the effect of the condition number on the solution to an equation, the following applies: Let $p = \log_{10} \kappa(A)$. If the solution is computed to n decimal places, then only about $n - p$ places can be considered to be accurate.

Example 4.10.3 Suppose double-precision arithmetic is used to compute the solution to a linear equation. If $n = 18$ places are used in the computation, and the matrix has a condition number of $\kappa(A) = 10^6$, then the computed solution is accurate only to about 12 places. If $\kappa(A) = 10^{20}$, then any solution obtained is probably useless. $\square$

In practice, when solving $A\mathbf{x} = \mathbf{b}$, it is always advisable to determine the condition number of A. Of course, actually computing the inverse to determine the norm $\|A^{-1}\|$ is impractical: there is no guarantee that the inverse obtained is meaningful. However, it is possible to obtain reliable estimates of the condition number of an $n \times n$ matrix using an algorithm that requires $O(n^2)$ operations. This algorithm is described in [161] and in [114]. The condition number using the l_2 norm can also be reliably computed using the SVD; this is implemented in the cond command of MATLAB. An estimate of the reciprocal of the condition is available using the MATLAB command rcond: if rcond(A) is near 1 then A is well conditioned.

Example 4.10.4 The Hilbert matrix that arises in polynomial approximation (as presented in section 3.6),

$$H_m = \begin{bmatrix} 1 & \frac{1}{2} & \frac{1}{3} & \cdots & \frac{1}{m+1} \\ \frac{1}{2} & \frac{1}{3} & \frac{1}{4} & \cdots & \frac{1}{m+2} \\ \vdots & & & & \vdots \\ \frac{1}{m+1} & \frac{1}{m+2} & \frac{1}{m+3} & \cdots & \frac{1}{2m+1} \end{bmatrix},$$

is notoriously ill conditioned. Figure 4.7 shows the condition number of the Hilbert matrix as a function of m: the condition number goes up exponentially with m, and for $m = 12$ any solution

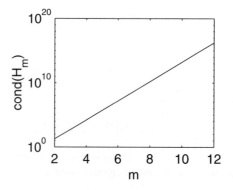

Figure 4.7: Condition of the Hilbert matrix

4.10 Matrix Condition Number

of the equations becomes essentially useless. In fact, the condition is approximately $\kappa(H_n) \approx e^{3.5n}$. Because of the rapid growth in condition number otherwise, orthogonal polynomials such as Legendre polynomials should be used for polynomial approximation. □

The inverse of the Hilbert matrix is discussed in Man-Duen Choi, "Tricks or Treats with the Hilbert Matrix," *SIAM Review*, 952:301–312, 1983.

Example 4.10.5 An intrepid but ill-informed engineer has devised a scheme by which he plans to send virtually unlimited amounts of information over the telephone channel. Every T seconds he sends the sum of n closely spaced sinusoids, each with a different amplitude. By encoding information in the amplitudes, he can send many bits per signal. For example, if 8 amplitudes (representing 3 bits) are allowed using n sinusoidal signals, then $3n$ bits per symbol can be sent. All that the receiver must do is to estimate the original amplitudes, then map them back to bits. By increasing n, the engineer dreams of sending millions of bits per second.

Alas, the scheme is ill advised. In the first place, channel noise places a strict upper bound on the amount of information that can be transferred. The channel capacity theorem for a channel with bandwidth W Hz and signal to noise ratio of S/N says that reliable transmission can only be achieved if the data rate is less than the channel capacity,

$$C = W \log_2(1 + S/N).$$

However, even in a perfectly *noiseless* channel (with $S/N = \infty$) there is little hope of recovering the information in any practical way. Let the signal sent over one symbol interval be

$$s(t) = \sum_{i=0}^{n-1} a_i \cos((\omega_{\max} - i\Delta\omega)t).$$

By sampling this signal n times at $t_1, t_2, \ldots, t_n$ within one symbol time, a set of n equations in n unknowns can be obtained:

$$\begin{bmatrix} \cos(\omega_0 t_1) & \cos((\omega_0 - \Delta\omega)t_1) & \cdots & \cos((\omega_0 - (n-1)\Delta\omega)t_1) \\ \cos(\omega_0 t_2) & \cos((\omega_0 - \Delta\omega)t_2) & \cdots & \cos((\omega_0 - (n-1)\Delta\omega)t_2) \\ \vdots & & & \vdots \\ \cos(\omega_0 t_n) & \cos((\omega_0 - \Delta\omega)t_n) & \cdots & \cos((\omega_0 - (n-1)\Delta\omega)t_n) \end{bmatrix} \begin{bmatrix} a_1 \\ a_2 \\ \vdots \\ a_n \end{bmatrix} = \begin{bmatrix} s(t_1) \\ s(t_2) \\ \vdots \\ s(t_n) \end{bmatrix}. \quad (4.31)$$

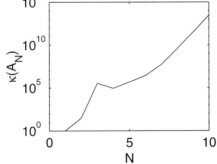

Figure 4.8: Condition number for a bad idea

Alas, as n become even moderately large, the matrix in this equation becomes ill conditioned beyond use for most parameters of the problem. As a particular example, let $f_{\max} = 2000$ Hz and $\Delta f = 50$ Hz, and sample periodically at a rate so that a full period of the highest-frequency signal is sampled. Figure 4.8 shows $\log(\kappa(A_n))$, where A_n is the matrix of (4.31). For $N = 10$, the condition is approximately 3×10^{13}, which is too high for most practical uses. □

As we have mentioned, direct computation of the least-squares solution

$$(A^H A)^{-1} A^H$$

is discouraged. Let us examine now why this is the case. Assume that A is a square matrix, which we assume to be invertible so that the condition number is finite. The condition of number A using the l_2 norm is

$$\kappa(A) = \sqrt{\lambda_{\max}}/\sqrt{\lambda_{\min}},$$

while the condition of $A^H A$ is

$$\kappa(A^H A) = \lambda_{\max}/\lambda_{\min},$$

where the eigenvalues are of the matrix $A^H A$. Thus, the condition of $A^H A$ is the square of the condition of A.

Example 4.10.6 Suppose that $\kappa(A) = 1000$ (a generally acceptable value), and $\epsilon_{\text{mach}} = 10^{-6}$. Then, in direction computation of a least-squares solution, the error in machine representation could be multiplied to become an error as large as $(1000)^2 10^{-6} = 1$. The solution might have no accurate digits! □

We mention two other facts about condition number. It has been shown [160] that relative to the p norm,

$$\frac{1}{\kappa_p(A)} = \min_{A+E \text{ singular}} \frac{\|E\|_p}{\|A\|_p}.$$

In other words, the condition number is a measure of how far the matrix A is away from an ill-conditioned matrix $A + E$, relative to the size of E. Also, although the actual value of the condition number depends upon the particular norm used, a matrix which is poorly conditioned under one norm will also be poorly conditioned under another norm.

4.11 Inverse of a small-rank adjustment

It may happen that for a matrix A the inverse A^{-1} is known, and then A is changed by the addition of some rank 1 matrix to produce a matrix

$$B = A + \mathbf{x}\mathbf{y}^H.$$

In this circumstance, it may be convenient to use what is known about A^{-1}, rather than computing the inverse of B from scratch. This can be accomplished using the following formula:

$$B^{-1} = (A + \mathbf{x}\mathbf{y}^H)^{-1} = A^{-1} - \frac{A^{-1}\mathbf{x}\mathbf{y}^H A^{-1}}{1 + \mathbf{y}^H A^{-1}\mathbf{x}}. \tag{4.32}$$

This is known as the **matrix inversion lemma**, the **Sherman–Morrison formula**, or sometimes as *Woodbury's identity*.

More generally, if $A, B \in M_n$, and

$$B = A + XRY$$

(where X is $n \times r$, Y is $r \times n$, and R is $r \times r$), then the inverse (if it exists) can be computed by

$$B^{-1} = (A + XRY)^{-1} = A^{-1} - A^{-1}X(R^{-1} + YA^{-1}X)^{-1}YA^{-1}. \tag{4.33}$$

If r is small enough compared with n, then computation of the inverses in this formula may be much easier than direct computation of the inverse of B. We shall have occasion to use (4.33) many times in the derivation of the Kalman filter and for estimation and detection problems in Gaussian noise. A useful identity that can be derived from this is

$$(A + XRY)^{-1}XR = A^{-1}X(R^{-1} + YA^{-1}X)^{-1}. \tag{4.34}$$

4.11 Inverse of a Small-Rank Adjustment

4.11.1 An application: the RLS filter

The matrix inversion lemma can be used to provide an update to the inverse of the Grammian matrix in a least-squares problem, to produce a simple version of what is known as the recursive least-squares (RLS) adaptive filter.

Consider the least-squares filtering problem posed in section 3.9, in which a signal $f[t]$ is passed through a filter $h[t]$ to produce an output $y[t]$,

$$y[t] = \sum_{i=0}^{m-1} h[i] f[t-i],$$

where the filter coefficients are to be found to minimize the least-squares error between $y[t]$ and a desired signal $d[t]$. It was found there that the coefficients minimizing the least-squares error (or $e[t] = d[t] - y[t]$) satisfy the normal equation

$$R\mathbf{h} = A^H \mathbf{d}, \qquad (4.35)$$

where A is a data matrix, and

$$R = A^H A = \sum_{i=1}^{N} \mathbf{q}[i] \mathbf{q}^H[i],$$

with

$$\mathbf{q}[i] = \begin{bmatrix} \overline{f}[i] \\ \overline{f}[i-1] \\ \vdots \\ \overline{f}[i-m+1] \end{bmatrix}.$$

Let

$$\mathbf{p} = A^H \mathbf{d} = \sum_{i=1}^{N} \mathbf{q}[i] d[i].$$

Computation of the coefficients is obtained from

$$\mathbf{h} = R^{-1} A^H \mathbf{d} = R^{-1} \mathbf{p}.$$

This least-squares solution assumes the availability of some set of data $\mathbf{q}[1], \mathbf{q}[2], \ldots, \mathbf{q}[N]$ (where the prewindowing method is used), and the computations are done on the entire block of data.

This least-squares technique is turned into an adaptive algorithm by computing a new update to $\mathbf{h}$ as new data become available. Assume that a sequence of data $\mathbf{q}[1], \mathbf{q}[2], \ldots, \mathbf{q}[t]$ up to time t is available. We define the Grammian matrix $R[t]$ as

$$R[t] = \sum_{i=1}^{t} \mathbf{q}[i] \mathbf{q}^H[i].$$

The least-squares filter coefficients that are computed using the data up to time t are indicated by $\mathbf{h}[t]$. The term $A^H \mathbf{y}$ from (4.35), using the data up to time t, we will denote $\mathbf{p}[t]$, where

$$\mathbf{p}[t] = A^H \mathbf{y} = \sum_{i=1}^{t} \mathbf{q}[i] d[i] = \mathbf{p}[t-1] + \mathbf{q}[t] d[i].$$

Then by (4.35) we have

$$\mathbf{h}[t] = R^{-1}[t] \mathbf{p}[t].$$

At each time step t, a new filter update is computed. The algorithm is now adaptive.

As the algorithm now stands, it requires a large number of computations at each time step, since $R^{-1}[t]$ must be computed each time. The complexity may be considerably reduced by using the Sherman–Morrison formula to obtain an adaptive *recursive* algorithm. We can break $R[t]$ up as

$$R[t] = \sum_{i=1}^{t-1} \mathbf{q}[i]\mathbf{q}^H[i] + \mathbf{q}[t]\mathbf{q}^H[t]$$
$$= R[t-1] + \mathbf{q}[t]\mathbf{q}^H[t].$$

Thus $R[t]$ is obtained from $R[t-1]$ by a rank-one update. By the Sherman–Morrison formula,

$$R^{-1}[t] = R^{-1}[t-1] - \frac{R^{-1}[t-1]\mathbf{q}[t]\mathbf{q}^H[t]R^{-1}[t-1]}{1 + \mathbf{q}^H[t]R^{-1}[t-1]\mathbf{q}[t]}. \tag{4.36}$$

For notational convenience, let

$$P[t] = R^{-1}[t]$$

and

$$\mathbf{k}[t] = \frac{R^{-1}[t-1]\mathbf{q}[t]}{1 + \mathbf{q}^H[t]R^{-1}[t-1]\mathbf{q}[t]}. \tag{4.37}$$

Then (4.36) becomes

$$P[t] = P[t-1] - \mathbf{k}[t]\mathbf{q}^H[t]P[t-1]. \tag{4.38}$$

The vector $\mathbf{k}[t]$ is referred to as the *Kalman gain* vector, or sometimes as simply the gain vector.

The coefficients for the filter can be computed as

$$\mathbf{h}[t] = P[t]\mathbf{p}[t] = P[t](\mathbf{p}[t-1] + \mathbf{q}[t]d[t]). \tag{4.39}$$

The term $P[t]\mathbf{p}[t-1]$ in this can be written using (4.38) as

$$P[t]\mathbf{p}[t-1] = P[t-1]\mathbf{p}[t-1] - \mathbf{k}[t]\mathbf{q}^H[t]P[t-1]\mathbf{p}[t-1]$$
$$= \mathbf{h}[t-1] - \mathbf{k}[t]\mathbf{q}^H[t]\mathbf{h}[t-1].$$

Substituting this into (4.39), and using the fact that

$$\mathbf{k}[t] = P[t]\mathbf{q}[t]$$

(see exercise 4.11-50), we obtain

$$\mathbf{h}[t] = \mathbf{h}[t-1] - \mathbf{k}[t]\mathbf{q}^H[t]\mathbf{h}[t-1] + P[t]\mathbf{q}[t]d[t]$$
$$= \mathbf{h}[t-1] - \mathbf{k}[t]\mathbf{q}^H[t]\mathbf{h}[t-1] + \mathbf{k}[t]d[t]$$
$$= \mathbf{h}[t-1] + \mathbf{k}[t](d[t] - \mathbf{q}^H[t]\mathbf{h}[t-1]).$$

The quantity

$$\varepsilon[t] = d[t] - \mathbf{q}^H[t]\mathbf{h}[t-1]$$

represents the filter error when the output of the filter using data at time t is used with the filter coefficients from time $t-1$. Because the filter has not been updated using the data at time t, this error is sometimes termed the *a priori estimation error*, while the error $d[t] - \mathbf{q}^H[t]\mathbf{h}[t]$ is termed the *a posteriori estimation error*.

Getting the RLS algorithm started

In order to get the algorithm started, the inverse $R[0]^{-1}$ is needed. To do this precisely would require the inversion of a matrix. Rather than expend these computations, we make an approximation. Let us slightly perturb the correlation matrix $R[t]$ by writing

$$R[t] = \sum_{i=1}^{t} \mathbf{q}[i]\mathbf{q}^H[i] + \delta I$$

for some small δ. Then

$$R[0] = \delta I,$$

so that the initial inverse is

$$P[0] = \delta^{-1} I.$$

As the algorithm progresses, the error made in approximating $P[0]$ becomes less and less significant. It is also common to assume that the initial coefficients of the filter are all zero, $\mathbf{h}[0] = \mathbf{0}$.

On the basis of these assumptions, the RLS algorithm is summarized in algorithm 4.1.

Algorithm 4.1 The RLS algorithm

Initialization:
    Choose a small positive δ.
    Set $P[0] = \delta^{-1} I$ $\mathbf{h}[0] = \mathbf{0}$.
For $t = 1, 2, 3, \ldots$, compute

$$\mathbf{k}[t] = \frac{P[t]\mathbf{q}[t]}{1 + \mathbf{q}^H[t]P[t-1]\mathbf{q}[t]}$$

$$\varepsilon = d[t] - \mathbf{q}^H[t]\mathbf{h}[t-1]$$

$$\mathbf{h}[t] = \mathbf{h}[t-1] + \mathbf{k}[t]\varepsilon[t]$$

$$P[t] = P[t-1] - \mathbf{k}[t]\mathbf{q}^H[t]P[t-1]$$

MATLAB code to initialize and run the RLS algorithm is shown in algorithm 4.2.

Algorithm 4.2 The RLS algorithm (MATLAB implementation)
    File: `rls.m`
            `rlsinit.m`

A discussion of the convergence of the RLS algorithm is provided in [132].

4.11.2 Two RLS applications

We provide two brief examples of the RLS. The first is in an equalizer (inverse system identification) application, the second in a system identification application.

RLS Equalizer

Consider the system shown in figure 4.9. A sequence of random bits $b[t]$ is passed through a channel with unknown response, and white noise $v[t]$ is added to the signal at the output

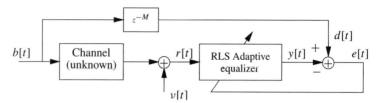

Figure 4.9: RLS adaptive equalizer

to produce the received signal $r[t]$, where $v[t]$ is a zero-mean signal with variance σ_v^2. At the receiver, the signal is passed through an equalizer that attempts to match the inverse of the channel response. It is desired that the output of the equalizer, $y[t]$, match some delayed version of the input data, $d[t] = b[t - M]$.

For the sake of comparing the optimum filter with the results of the RLS algorithm, we will first compute the optimum coefficients of the filter under the assumption that the channel is a known FIR filter with p coefficients $f_0, f_1, \ldots, f_{p-1}$. If a minimum mean-squares FIR filter (Wiener filter) with m coefficients is used as the equalizer, the optimal equalizer coefficients are obtained (see section 3.11) from

$$\mathbf{h}_{\text{opt}} = R^{-1}\mathbf{p}$$

where $R = [r_{ij}]$ is the $m \times m$ correlation of $\mathbf{x}[t]$ with $\mathbf{x}[t]$, where

$$\mathbf{x}[t] = [r[t], r[t-1], \ldots, r[t-m+1]]^T,$$

and $\mathbf{p}$ is the cross-correlation between $\mathbf{x}[t]$ and the desired signal $d[t]$. Assuming that the input data is an independent sequence of real binary bits (± 1) with zero mean, then

$$r_{ij} = \sum_{k=0}^{p-1} f_k f_{|k-(i-j)|} + \sigma_v^2 \delta_{ij}, \qquad (4.40)$$

where $f_{|k-(i-j)|}$ is taken to be zero for indices outside the range $[0, p-1]$. The cross-correlation vector $\mathbf{p}$ is

$$\mathbf{p} = \begin{bmatrix} \mathbf{0} \\ \mathbf{f}^B \\ \mathbf{0} \end{bmatrix},$$

where $\mathbf{f}^B$ is the time-reversed version of the channel filter coefficients, and the zeros correspond to the length of the delay M.

As a particular example, consider the channel in which $\mathbf{f} = \{.2, 1, -.2\}$ (where the first tap is taken at $k = 1$) with an equalizer of $m = 11$ taps. The optimum filter with $n = 11$ coefficients is shown in figure 4.10(a), where a delay of $M = 6$ has been selected and the noise variance is $\sigma_v^2 = 0.001$. The minimum mean-squared error is, from (3.63),

$$J_{\min} = 1 - \mathbf{p}^T \mathbf{h}_{\text{opt}} \approx 1 \times 10^{-3}.$$

The results of running the RLS adaptive filter with 11 unknown coefficients are shown in figure 4.10(b), showing the mean-squared error for the RLS filter as it adapts, along with the minimum mean-squared error for the optimal filter. Shown in the figure is an ensemble average of the square of the *a priori* error $\varepsilon[t]$, averaged over 200 runs of the algorithm with the initial conditions reset each time. The initial covariance was set using $\delta = 0.01$. The plot begins at time $t = 6$, corresponding to when the first nonzero sample was passed to the adaptive filter algorithm. Observe that the algorithm has converged within about $2m$ iterations. Figure 4.10(c) shows the RLS equalizer coefficients after 100 iterations.

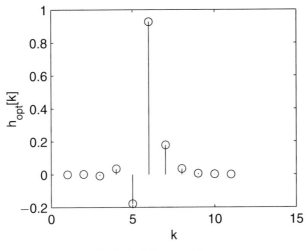
(a) Optimal filter coefficients

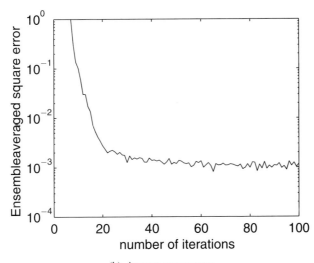
(b) Average square error

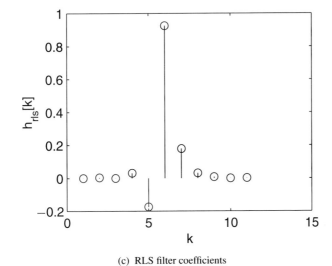
(c) RLS filter coefficients

Figure 4.10: Illustration of RLS equalizer performance

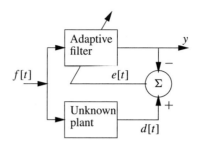

Figure 4.11: System identification using the RLS adaptive filter

Plant identification

The second illustration of the RLS is in a forward system identification application, as shown in figure 4.11. The same input is provided to both the unknown system and the RLS filter, and the RLS filter adapts until its output matches the unknown system. Figure 4.12(a) illustrates an example impulse response, with the ensemble-averaged square error shown in figure 4.12(b) and the RLS estimate of the filter response with an 11-tap filter in figure 4.12(c). Observe that the last few coefficients adapt to zero, since they are not needed.

4.12 Inverse of a block (partitioned) matrix

A discussion of simple operations of block matrices appears in section C.4. In this section we discuss inverses of block matrices.

If A is partitioned as

$$A = \begin{bmatrix} A_{11} & A_{12} \\ A_{21} & A_{22} \end{bmatrix}, \tag{4.41}$$

then

$$A^{-1} = \begin{bmatrix} A_{11}^{-1} + A_{11}^{-1} A_{12} S^{-1} A_{21} A_{11}^{-1} & -A_{11}^{-1} A_{12} S^{-1} \\ -S^{-1} A_{21} A_{11}^{-1} & S^{-1} \end{bmatrix}, \tag{4.42}$$

where $S = A_{22} - A_{21} A_{11}^{-1} A_{12}$. S is known as the *Schur complement* of A (see also exercises 3.3-1 and 4.12-57). Equation (4.42) can be validated by multiplication. When A_{22} is invertible, the (1, 1) block of A^{-1} can also be written as $(A_{11} - A_{12} A_{22}^{-1} A_{21})^{-1}$, using the matrix inversion lemma.

In the particular case in which the A_{21} block is zero,

$$A^{-1} = \begin{bmatrix} A_{11} & A_{12} \\ 0 & A_{22} \end{bmatrix}^{-1} = \begin{bmatrix} A_{11}^{-1} & -A_{11}^{-1} A_{12} A_{22}^{-1} \\ 0 & A_{22}^{-1} \end{bmatrix}. \tag{4.43}$$

Another useful form for the inverse of a partitioned matrix is given by the following.

Lemma 4.2 *For a matrix A partitioned as in (4.41), the inverse is*

$$\begin{bmatrix} A_{11} & A_{12} \\ A_{21} & A_{22} \end{bmatrix}^{-1} = \begin{bmatrix} A_{11}^{-1} & 0 \\ 0 & 0 \end{bmatrix} + \begin{bmatrix} -A_{11}^{-1} A_{12} \\ I \end{bmatrix} S^{-1} \begin{bmatrix} -A_{21} A_{11}^{-1} & I \end{bmatrix}, \tag{4.44}$$

where $S = A_{22} - A_{21} A_{11}^{-1} A_{12}$.

Proof Multiplication and addition as shown in (4.44) leads to (4.42). □

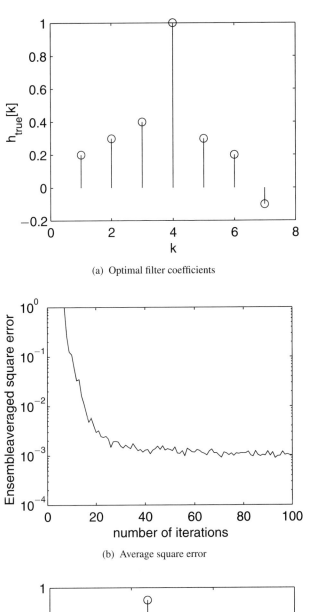

(a) Optimal filter coefficients

(b) Average square error

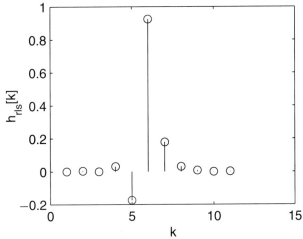

(c) RLS filter coefficients

Figure 4.12: Illustration of system identification using the RLS filter

The formulas for inverses of block matrices will be extensively used in the derivation of the Kalman filter. We will begin some of that work here. Let $\mathbf{X}$ and $\mathbf{Y}$ be random vectors with covariance matrices R_{xx} and R_{yy}, respectively, and let $\mathbf{Z} = [\mathbf{X}^T, \mathbf{Y}^T]$ be a random vector. Then

$$R_{zz} = \operatorname{cov}(\mathbf{Z}) = E[(\mathbf{Z} - \mu_z)(\mathbf{Z} - \mu_z)^T]$$
$$= \begin{bmatrix} R_{xx} & R_{xy} \\ R_{yx} & R_{yy} \end{bmatrix}.$$

Then by (4.42)

$$R_{zz}^{-1} = \begin{bmatrix} R_{xx}^{-1} + R_{xx}^{-1} R_{xy} S^{-1} R_{yx} R_{xx}^{-1} & -R_{xx}^{-1} R_{xy} S^{-1} \\ -S^{-1} R_{yx} R_{xx}^{-1} & S^{-1} \end{bmatrix}, \qquad (4.45)$$

where $S = R_y - R_{yx} R_{xx}^{-1} R_{xy}$ is the Schur complement. Another representation from lemma 4.2 is

$$R_{zz}^{-1} = \begin{bmatrix} R_{xx}^{-1} & 0 \\ 0 & 0 \end{bmatrix} + \begin{bmatrix} -R_{xx}^{-1} R_{xy} \\ I \end{bmatrix} \left[\left(R_{yy}^{-1} - R_{yx} R_{xx}^{-1} R_{xy} \right)^{-1} \right] \begin{bmatrix} -R_{yx} R_{xx}^{-1} & I \end{bmatrix}. \qquad (4.46)$$

We can further use the matrix inversion lemma (4.33) to rewrite some of the components of (4.46). The inverse of the Schur complement can be written

$$S^{-1} = R_{yy}^{-1} + R_{yy}^{-1} R_{yx} \left(R_{xx} + R_{xy} R_{yy}^{-1} R_{yx} \right)^{-1} R_{xy} R_{yy}^{-1}.$$

The inverse in the upper left corner element of R_{zz}^{-1} can be written as

$$\left(R_{xx}^{-1} + R_{xx}^{-1} R_{xy} S^{-1} R_{yx} R_{xx}^{-1} \right)^{-1} = R_{xx} + R_{xx} R_{xx}^{-1} R_{xy} \left(S + R_{yx} R_{xx}^{-1} R_{xx} R_{xx}^{-1} \right) R_{yx} R_{xx}^{-1} R_{xx}$$
$$= R_{xx} + R_{xy} R_{yy}^{-1} R_{yx}.$$

Example 4.12.1 In section 1.6.1 we found a formula for $f(x|Y = y)$ when X and Y are jointly distributed scalar Gaussian random variables. We can use the block inverse formula to generalize this result to Gaussian random vectors. Let $\mathbf{X}$ and $\mathbf{Y}$ be real Gaussian random vectors of dimension m and n, respectively. Then

$$\mathbf{Z} = \begin{bmatrix} \mathbf{X} \\ \mathbf{Y} \end{bmatrix}$$

is a Gaussian random vector with

$$E[\mathbf{Z}] = \mu_z = \begin{bmatrix} \mu_x \\ \mu_y \end{bmatrix}$$

and

$$R_{zz} = \operatorname{cov}(\mathbf{z}) = E[(\mathbf{Z} - \mu_z)(\mathbf{Z} - \mu_z)^T]$$
$$= \begin{bmatrix} R_{xx} & R_{xy} \\ R_{yx} & R_{yy} \end{bmatrix}.$$

The density of $\mathbf{Z}$ (see section 1.6) is

$$f(\mathbf{z}) = f(\mathbf{x}, \mathbf{y}) = \frac{1}{(2\pi)^{p/2} |R_{zz}|^{1/2}} \exp\left[-\frac{1}{2}(\mathbf{z} - \mu_z)^T R_{zz}^{-1} (\mathbf{z} - \mu_z) \right],$$

where $p = m + n$. If $\mathbf{Y} = \mathbf{y}$ is observed, and we wish to obtain an updated distribution on $\mathbf{X}$, then

$$f(\mathbf{x}|\mathbf{y}) = \frac{f(\mathbf{x}, \mathbf{y})}{f(\mathbf{y})}$$

$$= C_1 \frac{\exp\left[-\frac{1}{2}[(\mathbf{x} - \mu_x)^T, (\mathbf{y} - \mu_y)^T] \begin{bmatrix} R_{xx} & R_{xy} \\ R_{yx} & R_{yy} \end{bmatrix}^{-1} \begin{bmatrix} \mathbf{x} - \mu_x \\ \mathbf{y} - \mu_y \end{bmatrix} \right]}{\exp\left[-\frac{1}{2}(\mathbf{y} - \mu_y)^T R_{yy}^{-1} (\mathbf{y} - \mu_y) \right]}.$$

4.12 Inverse of a Block (Partitioned) Matrix

Let us examine the exponent E of this conditional density. Using (4.45), we obtain

$$E = [(\mathbf{x} - \boldsymbol{\mu}_x)^T, (\mathbf{y} - \boldsymbol{\mu}_y)^T] \begin{bmatrix} R_{xx} & R_{xy} \\ R_{yx} & R_{yy} \end{bmatrix}^{-1} \begin{bmatrix} \mathbf{x} - \boldsymbol{\mu}_x \\ \mathbf{y} - \boldsymbol{\mu}_y \end{bmatrix} - (\mathbf{y} - \boldsymbol{\mu}_y)^T R_{yy}^{-1}(\mathbf{y} - \boldsymbol{\mu}_y)$$

$$= (\mathbf{x} - \boldsymbol{\mu}_x)^T \left(R_{xx}^{-1} + R_{xx}^{-1} R_{xy} S^{-1} R_{yz} R_{xx}^{-1} \right)(\mathbf{x} - \boldsymbol{\mu}_x) - 2(\mathbf{x} - \boldsymbol{\mu}_x)^T R_{xx}^{-1} R_{xy} S^{-1}(\mathbf{y} - \boldsymbol{\mu}_y) + \text{stuff}$$

$$\overset{\triangle}{=} (\mathbf{x} - \boldsymbol{\mu}_x)^T A(\mathbf{x} - \boldsymbol{\mu}_x) + (\mathbf{x} - \boldsymbol{\mu}_x)\mathbf{b} + \text{stuff}, \tag{4.47}$$

where "stuff" does not depend upon $\mathbf{x}$. By completing the square (as described in appendix B), we find that

$$E = (\mathbf{x} - \boldsymbol{\mu}_x - \mathbf{z})^T A(\mathbf{x} - \boldsymbol{\mu}_x - \mathbf{z}) + \text{stuff},$$

where, using (4.12),

$$\mathbf{z} = A^{-1}\mathbf{b} = (R_{xx} + R_{xy} R_{yy}^{-1} R_{yz})(R_{xx}^{-1} R_{xy} S^{-1})(\mathbf{y} - \boldsymbol{\mu}_y)$$

$$= -R_{xy} R_{yy}^{-1}(\mathbf{y} - \boldsymbol{\mu}_y).$$

Based on the form of the solution, we conclude that $f(\mathbf{x}|\mathbf{y})$ is Gaussian distributed with mean

$$E[\mathbf{X}|\mathbf{Y}] = \boldsymbol{\mu}_x + R_{xy} R_{yy}^{-1}(\mathbf{y} - \boldsymbol{\mu}_y) \tag{4.48}$$

and covariance

$$\text{cov}(\mathbf{X}|\mathbf{Y}) = A^{-1} = R_{xx} - R_{xy} R_{yy}^{-1} R_{yx}. \tag{4.49}$$

$\square$

It is interesting to compare the results in (4.48) and (4.49) with the analogous results for conditional scalar random variables presented in equations (1.53) and (1.54) in section 1.6.1.

4.12.1 Application: Linear models

Let $\mathbf{X}$ be a real Gaussian random vector with $E[\mathbf{X}] = \boldsymbol{\mu}_x$ and $\text{cov}(\mathbf{X}) = R_x$, and let $\boldsymbol{\nu}$ be a real zero-mean Gaussian random vector with $\text{cov}(\boldsymbol{\nu}) = R_\nu$. Let

$$\mathbf{Y} = H\mathbf{X} + \boldsymbol{\nu} \tag{4.50}$$

for some known matrix H. Then (see exercise 4.12-58),

$$R_y = \text{cov}(\mathbf{y}) = H R_x H^T + R_\nu \quad \text{and} \quad \boldsymbol{\mu}_y = E[\mathbf{y}] = H\boldsymbol{\mu}_x \tag{4.51}$$

and

$$R_{xy} = E[(\mathbf{x} - \boldsymbol{\mu}_x)(\mathbf{y} - \boldsymbol{\mu}_y)^T] = R_x H^T. \tag{4.52}$$

Given a measurement of $\mathbf{Y} = \mathbf{y}$, the random vector $\mathbf{X}|\mathbf{Y} = \mathbf{y}$ is Gaussian with mean (by (4.48))

$$E[\mathbf{X}|\mathbf{Y} = \mathbf{y}] = \boldsymbol{\mu}_x + R_x H^T (H R_x H^T + R_\nu)^{-1}(\mathbf{y} - H\boldsymbol{\mu}_x) \tag{4.53}$$

and covariance

$$P = \text{cov}(\mathbf{X}|\mathbf{Y} = \mathbf{y}) = R_x - R_x H^T (H R_x H^T + R_\nu)^{-1} H R_x. \tag{4.54}$$

We can use $E[\mathbf{X}|\mathbf{Y} = \mathbf{y}]$ as an estimate of $\mathbf{X}$, based upon the measurement $\mathbf{y}$. Thus we denote

$$\hat{\mathbf{x}} = E[\mathbf{X}|\mathbf{Y} = \mathbf{y}].$$

This estimate is known as the Bayes estimate. The error $\mathbf{E} = \mathbf{X} - \hat{\mathbf{x}}$ between the true value of $\mathbf{X}$ and the estimate $\hat{\mathbf{x}}$ satisfies

$$E[\mathbf{E}\hat{\mathbf{x}}^T] = 0 \tag{4.55}$$

Thus, the estimate $\hat{\mathbf{x}}$ (after removing the mean) satisfies the orthogonality theorem, with the error being orthogonal to the estimate.

From this point, arriving at the Kalman filter is essentially just one more step. However, the final development is deferred to chapter 13, where it can be placed in its most useful context.

4.13 Exercises

4.1-1 Write the differential equation
$$\dddot{x}(t) - 2\ddot{x}(t) - \dot{x}(t) = b(t)$$
in operator form.

4.1-2 Show that the solution to the differential equation
$$\frac{dx}{dt} - ax = b,$$
with initial condition $x(0) = x_0$, is
$$x = x_0 e^{at} + \int_0^t e^{a(t-\tau)} b(\tau)\, d\tau.$$
Thus, the inverse of the differential operator $A = \frac{d}{dt} - a$ is an integral operator.

4.1-3 [209] Define the functional f on $L_2[0, 1]$ to be
$$f(x) = \int_0^1 a(t) \int_0^t b(s)x(s)\, ds\, dt,$$
where $a, b \in L_2[0, 1]$. Show that f is a bounded linear functional on $L_2[0, 1]$, and find an element $y \in L_2$ such that
$$f(x) = \langle x, y \rangle.$$

4.2-4 Determine the l_1, l_2, Frobenius, and l_∞ norms of the following matrices:
$$A_1 = \begin{bmatrix} 4 & 3 \\ 3 & 6 \end{bmatrix} \quad A_2 = \begin{bmatrix} 1 & 2 \\ 3 & 0 \end{bmatrix} \quad A_3 = \begin{bmatrix} 1 & 2 \\ 0 & 1 \end{bmatrix}.$$

4.2-5 Show that (4.5) is true; that is, that the l_∞ matrix norm is the largest row sum.

4.2-6 Show that (4.6) is true; that is, that the l_1 matrix norm is the largest column sum.

4.2-7 Show that the function f defined in (4.4) is a continuous function of α_i. Hint: Show that $|f(\alpha_1, \ldots, \alpha_n) - f(\beta_1, \ldots, \beta_n)| \leq M(|\alpha_1 - \beta_1| + \cdots + |\alpha_n - \beta_n|)$ for some M.

4.2-8 [238] Using lemma 4.1, show that:
   (a) If X is a finite-dimensional normed linear space, it is complete (i.e., it is a Banach space). Hint: Let $\{z_k\}$ be a Cauchy sequence in X. Write $\mathbf{z}_k$ as a linear combination of the basis vectors $\{\mathbf{x}_i\}$ using the coefficients $\{\alpha_{kj}\}$, and apply the lemma to show that $\{\alpha_{kj}\}$ is a Cauchy sequence of real numbers, and hence is convergent.
   (b) If X is a normed linear space, show that every finite-dimensional subspace M of X is closed.

4.2-9 Show that not all norms satisfy the submultiplicative property. (Hint: Simply provide a single counterexample.)

4.2-10 Show that for a square matrix F and a norm satisfying the submultiplicative property,
$$\|(I - F)^{-1}\| \leq \frac{1}{1 - \|F\|}.$$
Hint: Use the Neumann expansion.

4.13 Exercises

4.2-11 Show that if $\|\cdot\|$ is a norm satisfying the submultiplicative property and F is a matrix with $\|F\| < 1$, then $I - F$ is nonsingular. Hint: If $I - F$ is singular, there is a vector $\mathbf{x}$ such that $(I - F)\mathbf{x} = 0$.

4.2-12 Show that for a square matrix F with $\|F\| < 1$,
$$\|I - (I - F)^{-1}\| \leq \frac{\|F\|}{1 - \|F\|}.$$
Hint: Show that $I - (I - F)^{-1} = -F(I - F)^{-1}$.

4.2-13 Let A be nonsingular and let E be such that $\|A^{-1}E\| < 1$.
  (a) Show that $A + E$ is nonsingular. Hint: Use exercise 4.2-11.
  (b) Let $F = -A^{-1}E$. Show that F satisfies
  $$(A + E)^{-1} = (I - F)^{-1}A^{-1}.$$
  (c) Show that $(A + E)^{-1} - A^{-1} = -A^{-1}E(A + E^{-1})$.
  (d) Finally, show that
  $$\|(A + E)^{-1} - A^{-1}\| \leq \frac{\|A^{-1}\|^2 \|E\|}{1 - \|F\|}.$$

4.2-14 Provide examples demonstrating that the separate inequalities in (4.11), (4.11), (4.12), and (4.13) can be achieved with equality.

4.2-15 Show that $\|A\|_F = \mathrm{tr}(A^H A)$.

4.2-16 [114] For an $m \times m$ matrix A and a nonzero $m \times 1$ vector $\mathbf{x}$, show that
$$\left\| A \left(I - \frac{\mathbf{x}\mathbf{x}^H}{\mathbf{x}^H \mathbf{x}} \right) \right\|_F^2 = \|A\|_F^2 - \frac{\|A\mathbf{x}\|_2^2}{\mathbf{x}^H \mathbf{x}}.$$

4.2-17 Let B be a submatrix of A. Show that $\|B\|_p \leq \|A\|_p$.

4.2-18 Let P be a projection operator (see section 2.13). Show that $\|P\| = 1$ for a norm satisfying the submultiplicative property.

4.2-19 Show that for an $m \times m$ matrix D,
$$\frac{1}{\sqrt{n}} |\mathrm{tr}(D)| \leq \|D\|_F. \tag{4.56}$$
Hint: Use Cauchy–Schwarz.

4.2-20 [118] Show that for $m \times m$ matrices A and B,
$$\|AB\|_F \leq \frac{1}{m} \|A\|_2 \|B\|_F. \tag{4.57}$$

4.2-21 **Weighted norms** We have seen that we can define a weighted norm by $\|\mathbf{x}\|_W = \|W\mathbf{x}\|$. Show that using the weighted norm $\|\cdot\|_W$, the corresponding subordinate matrix norm is $\|A\|_W = \|WAW^{-1}\|$.

4.2-22 A matrix A such that $\|A\mathbf{x}\| = \|\mathbf{x}\|$ is called *norm-preserving* or *isometric*. Show that a square matrix A is isometric in the spectral norm if and only if it is orthogonal (or unitary if A is complex).
Note: an orthogonal matrix A satisfies $A^T A = I$. A unitary matrix A satisfies $A^H A = I$.

4.2-23 ([114]) Show that if $A \in \mathbb{R}^{m \times n}$ has rank n then $\|A(A^H A)^{-1} A^H\|_2 = 1$.

4.3-24 (Finite dimensional adjoints)
  (a) Show that the adjoint in (4.15) is correct.
  (b) Show that (4.17) follows from (4.16).

4.3-25 Show that $(A^*)^{-1} = (A^{-1})^*$ (when the inverse exists).

4.3-26 Let $X = Y = L_2[0, 1]$, and define the linear operator $A: X \to Y$ by

$$Ax = \int_0^t k(t, \tau) x(\tau) \, d\tau.$$

Show that the adjoint operator using the usual integral inner product is

$$A^* y = \int_t^1 k(\tau, t) y(\tau) \, d\tau.$$

4.3-27 [209] **Least squares** Suppose a linear dynamic system is governed by the differential equation $\dot{\mathbf{x}}(t) = A\mathbf{x}(t) + \mathbf{b} f(t)$. Assume the initial state is $\mathbf{x}(0) = \mathbf{0}$. It is desired to provide an input signal $f(t)$ to move the state at time T to $\mathbf{x}(T) = \mathbf{x}_T$. The explicit solution of the state is

$$x(T) = \int_0^T e^{A(T-t)} \mathbf{b} f(t) \, dt.$$

We define the operator $L: L_2[0, T] \to \mathbb{R}^n$ by

$$Lf = \int_0^T e^{A(T-t)} \mathbf{b} f(t) \, dt.$$

We desire to find a minimum-energy input that moves to the target position. The problem can be expressed as: find f such that $Lf = x_T$, subject to $\|f\|$ being minimum.

(a) Show that the adjoint operator L^* is $L^* = \mathbf{b}^T e^{A^T(T-t)}$. Hint: The appropriate inner product is over $\mathbb{R}^n$, since $Lf \in \mathbb{R}^n$.

(b) Show that $LL^* = \int_0^T e^{A(T-t)} \mathbf{b}\mathbf{b}^T e^{A^T(T-t)} \, dt$.

(c) Of the possible control signals, we seek that which minimizes the signal energy

$$\int_0^T f^2(t) \, dt.$$

Determine an expression for the minimum-energy $f(t)$. Hint: By theorem 4.4 $f = L^* \mathbf{z}$ for some $\mathbf{z}$.

4.3-28 Let $A: H \to H$ be a bounded linear operator on a Hilbert space H. Show that:

(a) The adjoint operator A^* is linear.

(b) The adjoint operator A^* is bounded (using the induced norm).

(c) $\|A\| = \|A^*\|$.

4.5-29 For the matrix

$$A = \begin{bmatrix} 1 & 4 \\ 2 & 8 \\ 3 & 12 \end{bmatrix},$$

determine the four fundamental subspaces.

4.5-30 Show that the nullspace of an operator A is a vector space.

4.5-31 Prove theorem 4.7.

4.5-32 Prove theorem 4.8.

4.6-33 Show that $(AB)^{-1} = B^{-1} A^{-1}$.

4.6-34 Show that if the linear operator $A: X \to Y$ has an inverse, then the inverse is linear.

4.13 Exercises

4.6-35 Show that if A has both a left inverse and a right inverse, they must be the same.

4.6-36 If $AB = 0$ for matrices A and B, show that the $\mathcal{R}(B) \subset \mathcal{N}(A)$.

4.9-37 ([333]) On pseudoinverses:
  (a) Show that, if A has independent columns, its left inverse $(A^T A)^{-1} A^T$ is its pseudoinverse.
  (b) Show that, if A has independent rows, its right inverse $A^T (A A^T)^{-1}$ is its pseudoinverse.

4.9-38 ([333]) Explain why $AA^\dagger$ and $A^\dagger A$ are projection matrices. What fundamental subspaces do they project onto?

4.10-39 Show that $\kappa(AB) \leq \kappa(A)\kappa(B)$, and that $\kappa(\alpha A) = \kappa(A)$ where α is some nonzero scalar. (Hence, that scaling the whole matrix cannot improve the condition number.)

4.10-40 Demonstrate that the determinant of a matrix cannot be used to determine ill conditioning (other than determining if a matrix is singular) by considering two cases:
  (a) Find the determinant and $\kappa_\infty(B_n)$ for the matrix
  $$B_n = \begin{bmatrix} 1 & -1 & -1 & \cdots & -1 \\ 0 & 1 & -1 & \cdots & -1 \\ \vdots & \vdots & \ddots & & \vdots \\ 0 & 0 & 0 & \cdots & 1 \end{bmatrix} \in \mathbb{R}^{n \times n}.$$
  (The notation κ_∞ denotes the condition number computed using the L^∞ norm.)
  (b) Find the determinant and condition number of the matrix
  $$D_n = \mathrm{diag}(10^{-1}, \ldots, 10^{-1}) \in \mathbb{R}^{n \times n}.$$

4.10-41 If U is unitary and the spectral norm is used, show that
$$\kappa(A) = \kappa(UA).$$

4.10-42 Show that (4.30) is true.

4.10-43 This exercise deals with the relative error in solving sets of linear equations subject to perturbations.
  (a) Show that if $\|\epsilon A^{-1} E\| < 1$, then $A^{-1} - (A + \epsilon E)^{-1} = \sum_{k=1}^\infty (-1)^{k+1} (\epsilon A^{-1} E)^k$.
  (b) Show that
  $$\|A^{-1} - A + \epsilon E)^{-1}\| \leq \frac{\kappa(A)\rho \|A^{-1}\|}{1 - \kappa(A)\rho},$$
  where $\rho = \epsilon \|E\|/\|A\|$.
  (c) Hence, show that if $(A + \epsilon E)\mathbf{x} = \mathbf{b}$, then
  $$\frac{\|\mathbf{x} - \mathbf{x}_0\|}{\|\mathbf{x}_0\|} \leq \frac{\kappa(A)\rho}{1 - \kappa(A)\rho},$$
  where $\mathbf{x}_0 = A^{-1}\mathbf{b}$.

4.10-44 Suppose that an approximate solution $\hat{\mathbf{x}}$ is found to the system of equations $A\mathbf{x} = \mathbf{b}$, so that $A\hat{\mathbf{x}}$ does not equal $\mathbf{b}$ exactly. There is a residual vector $\mathbf{r} = \mathbf{b} - A\hat{\mathbf{x}}$. Based on the residual (which can be readily computed), how close is the approximate solution $\hat{\mathbf{x}}$ to the true solution $\mathbf{x}$? Show that the relative error is bounded by
$$\frac{\|\mathbf{x} - \hat{\mathbf{x}}\|}{\|\mathbf{x}\|} \leq \kappa(A) \frac{\|\mathbf{r}\|}{\|\mathbf{b}\|}.$$

If A is well conditioned, the relative error of the solution is not much different than the relative size of the residual. However, if A is ill conditioned, even a solution yielding a small residual could be far from the true solution.

4.10-45 Show that a unitary matrix U is perfectly conditioned for the spectral norm ($\kappa(U) = 1$), with respect to the spectral norm. For the condition with respect to the Frobenius norm, show that $\kappa(U) = n$ for an $n \times n$ unitary matrix.

4.11-46 Show that
$$B^{-1} = A^{-1} - B^{-1}(B - A)A^{-1}.$$

4.11-47 Show that (4.32) is true.

4.11-48 Show that (4.33) is true.

4.11-49 Show that (4.34) is true.

4.11-50 Show that the RLS gain of (4.37) can be written as
$$\mathbf{k}[t] = P[t]\mathbf{q}[t].$$

4.11-51 In many RLS applications, it is desirable to weight the error, so that more recent error terms count for more. The total squared error is computed as
$$E[t] = \sum_{i=0}^{n} \lambda^{t-i} |e[i]|^2,$$
where λ is a constant less than 1—a "forgetting factor." This weighting leads to a Grammian matrix and correlation vector
$$R[t] = \sum_{i=0}^{t} \lambda^{t-i} \mathbf{q}[i]\mathbf{q}^H[i] \qquad \mathbf{p}[t] = \sum_{i=1}^{t} \lambda^{t-i} \mathbf{q}^H[i] d[t].$$
Show that under this weighting, the RLS algorithm can be expressed as:
$$\mathbf{k}[t] = \frac{\lambda^{-1} P[t-1]\mathbf{q}[t]}{1 + \lambda^{-1}\mathbf{q}^H[t] P[t-1]\mathbf{q}[t]},$$
$$P[t] = \lambda^{-1} P[t-1] - \lambda^{-1} \mathbf{k}[t]\mathbf{q}^H[t] P[t-1],$$
$$\epsilon[t] = d[t] - \mathbf{q}^H[t]\mathbf{h}[t-1],$$
$$\mathbf{h}[t] = \mathbf{h}[t-1] + \mathbf{k}[t]\epsilon[t].$$

4.11-52 (Computer exercise) Modify the `rls.m` function so that it can incorporate a weighting factor λ as in exercise 4.11-51.

4.11-53 Consider two sequences of vectors
$$\mathbf{s}_{1,1}, \mathbf{s}_{1,2}, \ldots, \mathbf{s}_{1,N}$$
and
$$\mathbf{s}_{2,1}, \mathbf{s}_{2,2}, \ldots, \mathbf{s}_{2,N}.$$

(a) Determine a transformation T so that the overall squared distance
$$J = \sum_{i=1}^{N} \|T\mathbf{s}_{1,i} - \mathbf{s}_{2,i}\|^2$$
is minimized. (Hint: Use the fact that for a scalar J, $J = \text{tr}[J]$. Use the gradient formulas in appendix E.)

4.13 Exercises

(b) Now take your solution and make it recursive and computationally efficient. Determine initial conditions for the recursive algorithm.

(c) Code and test the algorithm in MATLAB.

4.11-54 (Computer exercise). Based upon the block diagram shown in figure 4.9, write a MATLAB program to identify the inverse response of a system with impulse response $\{.2, .3, .4, -.3, -.2\}$ (starting from coefficient 1). Use binary ± 1 values of input. Compare the results of the program with the optimum filter coefficients. Determine the effect of different energy in the additive noise.

4.11-55 (Computer exercise). Based upon the block diagram in figure 4.11, write a MATLAB program to identify the system with impulse response $\{1, 2, 3, 4, 5\}$. Examine the effect of having the RLS filter length shorter than the length of the desired impulse response.

4.12-56 Show that (4.42) is true.

4.12-57 Let an $n \times n$ covariance matrix be partitioned as

$$R_n = \begin{bmatrix} R_{n-1} & \mathbf{r}_n \\ \mathbf{r}_n^H & r_{nn} \end{bmatrix}$$

for a scalar r_{nn}. Show that R_n^{-1} can be obtained recursively from R_{n-1}^{-1} by

$$R_n^{-1} = \begin{bmatrix} R_{n-1}^{-1} & \mathbf{0} \\ \mathbf{0}^T & 0 \end{bmatrix} + s_n^{-1} \begin{bmatrix} -R_{n-1}^{-1} \mathbf{r}_n \\ 1 \end{bmatrix} \begin{bmatrix} -\mathbf{r}_n^H R_{n-1}^{-1} & 1 \end{bmatrix},$$

where s_n is the Schur complement

$$s_n = r_{nn} - \mathbf{r}_n^H R_{n-1}^{-1} \mathbf{r}_n.$$

4.12-58 Let $\mathbf{X}$ be a real Gaussian random vector with $E[\mathbf{X}] = \mu_x$ and $\text{cov}(\mathbf{X}) = R_x$, and let ν be a real zero-mean Gaussian random vector with $\text{cov}(\nu) = R_\nu$. Let

$$\mathbf{Y} = H\mathbf{X} + \nu$$

for some known matrix H.

(a) Show that

$$R_y = \text{cov}(\mathbf{Y}) = H R_x H^T + R_\nu \quad \text{and} \quad \mu_y = E[\mathbf{Y}] = H\mu_x,$$

and

$$R_{xy} = E[(\mathbf{X} - \mu_x)(\mathbf{Y} - \mu_y)^T] = R_x H^T.$$

4.12-59 Show that (4.55) is true.

4.12-60 Another approach to estimating $\mathbf{x}$ given an observation from the linear model (4.50) is to find a value of $\mathbf{x}$ to minimize the quadratic form

$$J(\mathbf{x}) = (\mathbf{x} - \mu)^T R_x^{-1} (\mathbf{x} - \mu) + (\mathbf{y} - H\mathbf{x})^T R_\nu^{-1} (\mathbf{y} - H\mathbf{x}),$$

which is a combined measure of how close $\mathbf{x}$ is to its mean and how close $\mathbf{y}$ is to its mean.

(a) Using the gradient techniques of section A.6.4, determine $\hat{\mathbf{x}}$ that minimizes $J(\mathbf{x})$. Determine $J(\hat{\mathbf{x}})$, and compare it to P in (4.54).

(b) Determine the minimizing value of $J(\mathbf{x})$ by completing the square, and the corresponding minimum value of $J(\hat{\mathbf{x}})$.

4.14 References

The Riesz representation theorem alluded to in section 4.1.1 is proven, for example, in [209]. The concepts of the four fundamental subspaces of a linear operator are clearly described in [333] and [209]. A good discussion of matrix operator norms is found in [114] and [245]. Theorem 4.2 and its proof are taken from [238]. Our discussion of matrix condition number is drawn mostly from [114]. An early (but still important) discussion of matrix computations is [369].

Pseudoinverses are exhaustively covered in S. L. Campbell and C. D. Meyer, Jr., *Generalized Inverses of Linear Transformations* (New York: Dover, 1979). In the case of least squares when A does not have full column rank or $m < n$, see [114]. Our discussion of rank is drawn from [333] and [142]. Inverses of partitioned matrices are discussed, for example, in [291]. The RLS algorithm is well developed in [132]. The linear model discussed in section 4.12.1 appears also in [41].

Chapter 5

Some Important Matrix Factorizations

Certain matrix factorizations arise commonly enough in matrix analysis in general, and signal processing in particular, that they warrant specific attention. In this chapter, factorizations that form the heart of many signal processing routines are discussed. The factorizations presented in this chapter are as follows:

LU. A square matrix A can be factored as $A = LU$, where L is a lower-triangular matrix with ones on the main diagonal and U is upper triangular. Its main application is in the numerical solution of the problem $A\mathbf{x} = \mathbf{b}$.

Cholesky. A Hermitian (symmetric) positive-definite matrix A can be factored as

$$A = LL^H,$$

where L is lower triangular. The Cholesky factors of a matrix A may be regarded as the "square root" of the matrix. Closely related is the factorization $A = LDL^H$, where D is diagonal, or $A = UDU^H$, where U is upper triangular. The Cholesky factorization is used in simulation (to compute a vector noise of desired covariance) and in some estimation and Kalman filtering routines.

QR. A general matrix A can be factored as

$$A = QR$$

where Q is a unitary matrix, $QQ^H = I$, and R is upper triangular. The QR factorization is used in the solution of least-squares problems.

A factorization important enough to warrant its own chapter is the singular value decomposition (SVD), in which A is factored as

$$A = U\Sigma V^H,$$

where U and V are unitary and Σ is diagonal. The SVD and its applications are presented in chapter 7.

5.1 The LU factorization

We have seen, and will see, several instances where linear equations of the form

$$A\mathbf{x} = \mathbf{b} \tag{5.1}$$

arise in signal processing. In many circumstances, the matrix A has special structure for which special algorithms can be employed to find the solution $\mathbf{x}$, such as being a Toeplitz, Vandermonde, or Hankel matrix. When A lacks such special structure—or is not known to have a special structure—then a general algorithm must be employed. The LU factorization is the preferred numerical method used to solve a general matrix linear equation. For an $m \times m$ matrix A, solving the linear equation requires $O(m^3)$ operations. (Matrices with a particular structure, such as a Toeplitz or Hankel matrix, may require only $O(m^2)$ operations for solution. If A is Hermitian and known to be positive definite, then the Cholesky factorization is preferred over the LU factorization approach.) The solution of (5.1) is obtained *without* explicitly finding A^{-1}, since finding A^{-1} then multiplying $A^{-1}\mathbf{b}$ actually requires more computations than solution via the LU factorization.

The LU factorization of an $m \times m$ matrix A is

$$PA = LU, \tag{5.2}$$

where L is lower triangular with ones on the main diagonal and U is upper triangular,

$$L = \begin{bmatrix} 1 & 0 & 0 & \cdots & 0 \\ l_{21} & 1 & 0 & \cdots & 0 \\ \vdots & & & & \\ l_{m1} & l_{m2} & l_{m3} & \cdots & 1 \end{bmatrix} \quad U = \begin{bmatrix} u_{11} & u_{12} & u_{13} & \cdots & u_{1m} \\ 0 & u_{22} & u_{23} & \cdots & u_{2m} \\ \vdots & & & & \\ 0 & 0 & 0 & \cdots & u_{mm} \end{bmatrix},$$

and P is a permutation matrix which represents the *pivoting* that takes place in the factorization. As discussed below, the purpose of pivoting is to stabilize the numerical computations. Since permutation matrices are orthogonal (see 8.2), this can also be written as

$$A = P^T LU.$$

Using the LU decomposition, (5.1) can be solved as follows. First, write (5.1) using (5.2) as $LU\mathbf{x} = P^T \mathbf{b}$, and let $U\mathbf{x} = \mathbf{y}$. This leads to the system

$$L\mathbf{y} = P^T \mathbf{b} = \mathbf{c}. \tag{5.3}$$

This is a triangular system of equations,

$$\begin{bmatrix} l_{11} & & & & \\ l_{21} & l_{22} & & & \\ l_{31} & l_{32} & l_{33} & & \\ \vdots & & & \vdots & \\ l_{m1} & l_{m2} & l_{m3} & \cdots & l_{mm} \end{bmatrix} \begin{bmatrix} y_1 \\ y_2 \\ \vdots \\ y_m \end{bmatrix} = \mathbf{c},$$

which can be easily solved. From the first row,

$$y_1 = \frac{c_1}{l_{11}} = c_1$$

since L has ones on the main diagonal. Knowing y_1, the second equation

$$l_{21} y_1 + l_{22} y_2 = c_2$$

can be easily solved:

$$y_2 = \frac{c_2 - l_{21} y_1}{l_{22}} = c_2 - l_{21} y_1.$$

The jth equation can be solved as

$$y_j = \left(c_j - \sum_{i=1}^{j-1} l_{ji} y_i \right).$$

5.1 The LU Factorization

This procedure is called *forward substitution*, and it can be computed in approximately $m^2/2$ floating operations.

Once (5.2) is solved for $\mathbf{y}$, then the final solution can be obtained by solving

$$U\mathbf{x} = \mathbf{y}, \tag{5.4}$$

or

$$\begin{bmatrix} u_{11} & u_{12} & u_{13} & \cdots & u_{1m} \\ 0 & u_{22} & u_{23} & \cdots & u_{2m} \\ \vdots & & & & \\ 0 & 0 & 0 & \cdots & u_{mm} \end{bmatrix} \begin{bmatrix} x_1 \\ x_2 \\ \vdots \\ x_m \end{bmatrix} = \mathbf{y}.$$

This is an upper triangular system of equations, and can be readily solved using *back substitution*:

$$x_m = \frac{1}{u_{mm}} y_m,$$

$$x_{m-1} = \frac{1}{u_{(m-1)(m-1)}} (y_{m-1} - x_m u_{m-1,m}),$$

$$\vdots$$

$$x_j = \frac{1}{u_{jj}} \left(y_j - \sum_{k=j+1}^{m} u_{jk} x_k \right).$$

This computation requires approximately $m^2/2$ floating operations. Clearly, solution of (5.3) requires that there be no zeros on the diagonals of U. The diagonal elements of U are called the *pivots*.

If there are equations involving different right-hand sides, such as

$$A\mathbf{x}_1 = \mathbf{b}_1,$$
$$A\mathbf{x}_2 = \mathbf{b}_2,$$

then the solutions can be found using the same LU decomposition for A. Thus $\mathbf{x}_1$ can be obtained in $O(m^2)$ operations by forward and backward substitution, as can $\mathbf{x}_2$, after the LU factorization is obtained. This operation count is essentially the same as would be necessary to compute $A^{-1}\mathbf{b}_1$ if A^{-1} were explicitly known. Not only that, but the solution via the LU factorization has better numerical properties than explicitly computing the inverse and multiplying. For this reason, the following has been suggested as a rule of thumb for a good numerical analyst:

> **Never explicitly invert a matrix numerically.**

5.1.1 Computing the determinant using the LU factorization

The determinant of the matrix A can also be computed using the factorization

$$\det(A) = \det(P^T LU) = \det(P)\det(L)\det(U) = \pm \prod_{i=1}^{m} u_{ii},$$

where the sign is determined by the number of row transpositions in P. For matrices with very large or very small elements, the log of the absolute value of the determinant can be computed as

$$\log|\det(A)| = \sum_{i=1}^{m} \log|u_{ii}|.$$

5.1.2 Computing the LU factorization

The LU factorization is described in virtually every book on linear algebra and numerical analysis. A particularly complete description, including roundoff analysis, is provided in [114]. This section is included to provide insight into how the algorithm works, and as an aid for those who may be developing numerical libraries for new processors. The LU factorization of an $m \times m$ matrix can be computed in approximately $\frac{2}{3}m^3$ floating operations.

The LU decomposition is understood best by means of Gaussian elimination. In Gaussian elimination, a matrix is modified using row operations to produce a matrix that is upper triangular; this produces the U matrix. (The man Gauss is briefly introduced in box 5.1.) By keeping track of the row operations, the L matrix can be revealed. A row operation consists of *replacing a row of a matrix with a linear combination of other rows*. For a matrix A written in terms of its rows as

$$A = \begin{bmatrix} \mathbf{a}_1^T \\ \mathbf{a}_2^T \\ \vdots \\ \mathbf{a}_m^T \end{bmatrix} = \begin{bmatrix} \boxed{\text{row 1}} \\ \boxed{\text{row 2}} \\ \vdots \\ \boxed{\text{row } n} \end{bmatrix},$$

a row operation is of the form

$$\boxed{\text{row } i} \leftarrow \boxed{\text{row } i} - \alpha \boxed{\text{row } j},$$

where α is some scalar. A row operation (being a linear combination of rows) leaves the row space unchanged.

Box 5.1: Carl Friedrich Gauss (1777–1855)

Carl Friedrich Gauss may have been the greatest mathematician of all time. Gaussian elimination, which bears his name, is but a minor point among his many contributions.

A prodigy, at a young age he summed the numbers $1 + 2 + \cdots + 100$ (a task given by a teacher as busy-work) in a matter of moments using the formula $n(n+1)/2$, which he derived for himself on the spot. He independently developed the method of least squares (which was later used to plot the path of the asteroid Ceres), and developed a technique for constructing the 17-sided regular polygon using compass and straightedge, before he was nineteen years old. The latter problem had been unsolved for more than 2000 years.

His doctoral dissertation proved the fundamental theorem of algebra, that every polynomial of degree n with real coefficients has n solutions over the complex numbers. He produced a work on number theory, *Disquisitiones arithmeticae*, in which he introduced the concept of congruences, and presented results on quadratic reciprocity and the fundamental theorem of arithmetic. He studied the distribution of primes, elliptic functions, and made astronomical calculations.

Gauss published much less than he actually created. His seal bore the motto *pauca sed matura*—"few, but ripe." So insightful and creative (and unpublished) was he that for the first half of the nineteenth century when a new development was announced, it was frequently discovered that Gauss had found it earlier but left it unpublished.

5.1 The LU Factorization

In the LU factorization, row operations successively place zeros down the columns in such a way as to leave an upper-triangular matrix U. Keeping track of the row operations employed leads to the matrix L. The LU decomposition is illustrated using a 3×3 matrix. Suppose

$$A = \begin{bmatrix} 2 & 4 & -5 \\ 6 & 8 & 1 \\ 4 & -8 & -3 \end{bmatrix}.$$

To transform this to an upper-triangular matrix, we must zero the elements that currently have the values 6 and 4 in the first column, and -8 in the second column.

1. The first step is to modify $\boxed{\text{row 2}}$ by

$$\boxed{\text{row 2}} \leftarrow \boxed{\text{row 2}} - (3)\boxed{\text{row 1}}.$$

The number 3, which is a_{21}/a_{11}, is called a *multiplier*. (The case when the multiplier cannot be computed because of division by zero is discussed below.) The first step produces

$$\boxed{\text{row 2}} \leftarrow [6 \ 8 \ 1] - 3[2 \ 4 \ -5],$$

which, when inserted back in the matrix, gives

$$\begin{bmatrix} 2 & 4 & -5 \\ 0 & -4 & 16 \\ 4 & -8 & -3 \end{bmatrix}.$$

2. Proceeding down the first column, the next step is to modify $\boxed{\text{row 3}}$ by

$$\boxed{\text{row 3}} \leftarrow \boxed{\text{row 3}} - (2)\boxed{\text{row 1}}.$$

This produces the matrix

$$\begin{bmatrix} 2 & 4 & 5 \\ 0 & -4 & 16 \\ 0 & -16 & 7 \end{bmatrix}.$$

3. The final step is to modify the second column and $\boxed{\text{row 3}}$ by

$$\boxed{\text{row 3}} \leftarrow \boxed{\text{row 3}} - (4)\boxed{\text{row 2}}.$$

The result is that A has been transformed into the upper triangular matrix

$$\begin{bmatrix} 2 & 4 & -5 \\ 0 & -4 & 16 \\ 0 & 0 & -57 \end{bmatrix} = U.$$

The diagonal elements of U are known as the **pivots**.

The U matrix is revealed explicitly by this process. To determine L, the steps of elimination are represented in terms of matrix multiplication. This is for expository purposes only, as a means of presenting the LU factorization. The first modified matrix can be represented as

$$\begin{bmatrix} 2 & 4 & -5 \\ 0 & -4 & 16 \\ 4 & -8 & -3 \end{bmatrix} = \begin{bmatrix} 1 & 0 & 0 \\ -3 & 1 & 0 \\ 0 & 0 & 1 \end{bmatrix} \begin{bmatrix} 2 & 4 & -5 \\ 6 & 8 & 1 \\ 4 & -8 & -3 \end{bmatrix} = E_1 A.$$

The matrix E_1 is called an **elementary matrix**, which is a matrix that is an identity except for a single off-diagonal element. Elementary matrices applied to the left of a matrix represent row operations on that matrix.

The next step can be written as

$$\begin{bmatrix} 2 & 4 & -5 \\ 0 & -4 & 16 \\ 0 & -16 & 7 \end{bmatrix} = \begin{bmatrix} 1 & 0 & 0 \\ 0 & 1 & 0 \\ -2 & 0 & 1 \end{bmatrix} \begin{bmatrix} 2 & 4 & -5 \\ 0 & -4 & 16 \\ 4 & -8 & -3 \end{bmatrix} = E_2 E_1 A.$$

The final step is

$$U = \begin{bmatrix} 2 & 4 & -5 \\ 0 & -4 & 16 \\ 0 & 0 & -57 \end{bmatrix} = \begin{bmatrix} 1 & 0 & 0 \\ 0 & 1 & 0 \\ 0 & -4 & 1 \end{bmatrix} \begin{bmatrix} 2 & 4 & -5 \\ 0 & -4 & 16 \\ 0 & -16 & 7 \end{bmatrix} = E_3 E_2 E_1 A.$$

Solving this for A, we obtain

$$A = E_1^{-1} E_2^{-1} E_3^{-1} U.$$

It is straightforward to verify that the inverses of the elementary matrices are obtained by simply changing the sign of the off-diagonal element,

$$E_1^{-1} = \begin{bmatrix} 1 & 0 & 0 \\ 3 & 1 & 0 \\ 0 & 0 & 1 \end{bmatrix} \quad E_2^{-1} = \begin{bmatrix} 1 & 0 & 0 \\ 0 & 1 & 0 \\ 2 & 0 & 1 \end{bmatrix} \quad E_3^{-1} = \begin{bmatrix} 1 & 0 & 0 \\ 0 & 1 & 0 \\ 0 & 4 & 1 \end{bmatrix},$$

and that

$$E_1^{-1} E_2^{-1} E_3^{-1} = \begin{bmatrix} 1 & 0 & 0 \\ 3 & 1 & 0 \\ 2 & 4 & 1 \end{bmatrix} = L.$$

Observe that L is lower triangular, and that the elements of L in the lower triangle are exactly the multiplier values $l_{ij} = a_{ij}/a_{jj}$ computed in the Gaussian elimination. Thus

$$A = \begin{bmatrix} 1 & 0 & 0 \\ 3 & 1 & 0 \\ 2 & 4 & 1 \end{bmatrix} \begin{bmatrix} 2 & 4 & 5 \\ 0 & -4 & 16 \\ 0 & 0 & -57 \end{bmatrix} = LU.$$

Pivoting

It is possible, even when the matrix is well conditioned, for the procedure just described to be numerically poor. Severe roundoff problems can result if the LU factorization is used as just described on some matrices (see exercise 5.1-4). The problem is that in the computation of the multiplier $l_{ij} = a_{ij}/a_{jj}$, if a pivot (the denominator of the multiplier) is very small in comparison with the numerator of the multiplier, a large multiplier results. This could lead to roundoff problems. The solution to this problem of having a poorly conditioned algorithm, and the solution to division by zero in the multipliers, is found by *permuting the rows of the matrix so that the pivot is the largest (in absolute value) element in the unreduced part of the kth column*. If the largest element in the kth column is zero to machine accuracy, then the matrix is singular.

To repeat the example with pivoting, we begin by interchanging the first two rows of A to put 6 (the largest element in the first column) on the first row. The interchange is done by means of the permutation matrix P_{12} (see section 8.2), where

$$P_{12} A = \begin{bmatrix} 0 & 1 & 0 \\ 1 & 0 & 0 \\ 0 & 0 & 1 \end{bmatrix} A = \begin{bmatrix} 6 & 8 & 1 \\ 2 & 4 & -5 \\ 4 & -8 & -3 \end{bmatrix}.$$

5.1 The LU Factorization

Elimination in the first column is accomplished by the two steps

$$\boxed{\text{row 2}} \leftarrow \boxed{\text{row 2}} - \frac{1}{3}\boxed{\text{row 1}},$$

$$\boxed{\text{row 3}} \leftarrow \boxed{\text{row 3}} - \frac{2}{3}\boxed{\text{row 1}},$$

which in matrix notation is

$$E_2 E_1 P_{12} A = \begin{bmatrix} 1 & 0 & 0 \\ 0 & 1 & 0 \\ -\frac{2}{3} & 0 & 1 \end{bmatrix} \begin{bmatrix} 1 & 0 & 0 \\ -\frac{1}{3} & 1 & 0 \\ 0 & 0 & 1 \end{bmatrix} A = \begin{bmatrix} 6 & 8 & 1 \\ 0 & \frac{4}{3} & -\frac{16}{3} \\ 0 & -\frac{40}{3} & -\frac{11}{3} \end{bmatrix}.$$

Pivoting should now be done to interchange rows 2 and 3, to put the largest (in absolute value) column element in the pivot position. The permutation matrix is

$$P_{23} = \begin{bmatrix} 1 & 0 & 0 \\ 0 & 0 & 1 \\ 0 & 1 & 0 \end{bmatrix}$$

and the permuted matrix is

$$P_{23} E_2 E_1 P_{12} A = \begin{bmatrix} 6 & 8 & 1 \\ 0 & -\frac{40}{3} & -\frac{11}{3} \\ 0 & \frac{4}{3} & -\frac{16}{3} \end{bmatrix}.$$

The second column can be finished by an operation on $\boxed{\text{row 3}}$:

$$\boxed{\text{row 3}} \leftarrow \boxed{\text{row 3}} + \frac{1}{10}\boxed{\text{row 2}}.$$

In matrix form, this is

$$E_3 P_{23} E_2 E_1 P_{12} A = \begin{bmatrix} 1 & 0 & 0 \\ 0 & 1 & 0 \\ 0 & \frac{1}{10} & 1 \end{bmatrix} \begin{bmatrix} 6 & 8 & 1 \\ 0 & -\frac{40}{3} & -\frac{11}{3} \\ 0 & \frac{4}{3} & \frac{14}{3} \end{bmatrix} = \begin{bmatrix} 6 & 8 & 1 \\ 0 & -\frac{40}{3} & -\frac{11}{3} \\ 0 & 0 & -\frac{57}{10} \end{bmatrix} = U.$$

Solving for A, the equations can be written as

$$A = P_{12}^{-1} E_1^{-1} E_2^{-1} P_{23}^{-1} E_3^{-1} \begin{bmatrix} 6 & 8 & 1 \\ 0 & -\frac{40}{3} & -\frac{11}{3} \\ 0 & 0 & \frac{43}{10} \end{bmatrix}$$

$$= \begin{bmatrix} \frac{1}{3} & -\frac{1}{10} & 1 \\ 1 & 0 & 0 \\ \frac{2}{3} & 1 & 0 \end{bmatrix} U$$

$$= VU.$$

Note that the matrix V which is produced is *not* lower triangular. However, the matrix

$$L = (P_{23} P_{12}) V = \begin{bmatrix} 1 & 0 & 0 \\ \frac{2}{3} & 1 & 0 \\ \frac{1}{3} & -\frac{1}{10} & 1 \end{bmatrix}$$

is lower triangular. The LU factorization thus amounts to

$$PA = LU,$$

where

$$P = P_{23}P_{12} = \begin{bmatrix} 0 & 1 & 0 \\ 0 & 0 & 1 \\ 1 & 0 & 0 \end{bmatrix} \quad (5.5)$$

corresponds to the permutation $1 \to 2$, $2 \to 3$, and $3 \to 1$.

A general $m \times m$ matrix A can be factored as

$$E_{m,m-1}P_{m-1} \cdots E_{2,1}P_1 A = U.$$

If we let M_i represent the product of the elementary transformations in the ith column, then we can write the factorization as

$$M_{m-1}P_{m-1}M_{m-2}P_{m-2} \cdots M_1 P_1 A = U.$$

Now let

$$\tilde{M}_{m-1} = M_{m-1} \qquad \tilde{M}_{m-2} = P_{m-1}M_{m-2}P_{m-1} \qquad \tilde{M}_{m-3} = P_{m-1}P_{m-2}M_{m-3}P_{m-2}P_{m-1}$$

and, in general, $\tilde{M}_k = P_{m-1} \cdots P_{k+1} M_k P_{k+1} \cdots P_{k+1}$. Then

$$M_{m-1}P_{m-1}M_{m-2}P_{m-2} \cdots M_1 P_1 = \tilde{M}_{m-1}\tilde{M}_{m-2} \cdots \tilde{M}_1 P,$$

where $P = P_{m-1}P_{m-2} \cdots P_1$, since each permutation matrix is an exchange which satisfies $P_i^2 = I$. Then, letting

$$U = (\tilde{M}_{m-1}\tilde{M}_{m-2} \cdots \tilde{M}_1)^{-1},$$

we can write

$$PA = LU.$$

With a little care, it is possible to code the LU factorization so that the elements of L and U overwrite the elements of A. The permutation is stored in a scalar index array. The MATLAB code in algorithm 5.1 demonstrates the algorithm. (This is for demonstration purposes only, since MATLAB has a built-in `lu` command.) The U matrix is shown explicitly. The lower triangle L matrix is fitted into the lower half of LU.

Algorithm 5.1 LU factorization
File: `newlu.m`

In this algorithm, the multipliers are stored in the lower triangle of A, and the permutations implied by computing $\tilde{M}_i$ are obtained by the exchange

```
dum = A(k,1:n);  A(k,1:n) = A(m,1:n);  A(m,1:n) = dum;
```

Example 5.1.1 For example, calling `newlu` with the A from the preceding example yields

$$lu = \begin{bmatrix} 6 & 8 & 1 \\ \boxed{0.3333} & -13.3333 & -3.6667 \\ \boxed{0.6667} & \boxed{-0.1000} & -5.7000 \end{bmatrix}$$

and `indx = [2 3 1]`, where the boxed elements are the multipliers from L, and the unboxed elements form U. The `indx` can be interpreted in light of the permutation matrix (5.5). □

5.2 The Cholesky factorization

Pivoting may be avoided if A is diagonally dominant. A matrix A is said to be **diagonally dominant** if

$$a_{ii} > \sum_{j \neq i} |a_{ij}| \qquad \text{for } i = 1, 2, \ldots, m.$$

5.2 The Cholesky factorization

The Cholesky factorization is used to compute a "square root" of a positive-definite $m \times m$ Hermitian matrix as

$$B = LL^H,$$

where L is lower triangular. Occasionally, the L matrix is normalized to produce a matrix $\tilde{L}$ that has ones along the main diagonal, and the scaling factor is incorporated in a diagonal matrix factor as

$$L^H = \begin{bmatrix} l_{11} & & & \\ & l_{22} & & \\ & & \ddots & \\ & & & l_{mm} \end{bmatrix} \tilde{L}^H = \sqrt{D} U.$$

Then we can write

$$B = U^H D U,$$

where $D = \text{diag}(l_{11}^2, l_{22}^2, \ldots, l_{mm}^2)$.

Example 5.2.1 For the B shown, we have

$$B = \begin{bmatrix} 4 & 8 & 12 \\ 8 & 20 & 20 \\ 12 & 20 & 41 \end{bmatrix} = \begin{bmatrix} 2 & 0 & 0 \\ 4 & 2 & 0 \\ 6 & -2 & 1 \end{bmatrix} \begin{bmatrix} 2 & 4 & 6 \\ 0 & 2 & -2 \\ 0 & 0 & 1 \end{bmatrix} = LL^T$$

$$= \begin{bmatrix} 1 & 0 & 0 \\ 2 & 1 & 0 \\ 3 & -1 & 1 \end{bmatrix} \begin{bmatrix} 4 & 0 & 0 \\ 0 & 4 & 0 \\ 0 & 0 & 1 \end{bmatrix} \begin{bmatrix} 1 & 2 & 3 \\ 0 & 1 & -1 \\ 0 & 0 & 1 \end{bmatrix} = U^T D U.$$

□

If the Cholesky factorization does not exist (say, as determined by algorithm 5.2) then, to the precision available, the matrix B is not positive definite.

Example 5.2.2 In a simulation of a signal processing algorithm, it is necessary to generate Gaussian random vectors with covariance R. System libraries often provide generators which simulate independent $\mathcal{N}(0, 1)$ random variables. These can be used to generate $\mathcal{N}(0, R)$ random vectors as follows. First, factor R as

$$R = LL^T,$$

where L is lower triangular. For each random vector desired, create a vector $\mathbf{x}$ of $\mathcal{N}(0, 1)$ independent random variables using the Gaussian random number generator, and let

$$\mathbf{z} = L\mathbf{x}.$$

Then, since $E[\mathbf{x}\mathbf{x}^T] = I$,

$$E[\mathbf{z}\mathbf{z}^T] = L E[\mathbf{x}\mathbf{x}^T] L^T = LL^T = R,$$

so $\mathbf{z}$ has the desired covariance.

□

Example 5.2.3 The Cholesky factorization can be used to solve systems of equations. For the equation
$$A\mathbf{x} = \mathbf{b},$$
where A is Hermitian and positive definite, write
$$A = LL^H.$$
Solution then requires solving the two sets of triangular systems
$$L\mathbf{y} = \mathbf{b},$$
$$L^H\mathbf{x} = \mathbf{y},$$
much as was done for the LU decomposition. □

Example 5.2.4 (Application of Cholesky factorization to normal equations) The least-squares solution (3.20),
$$(A^H A)\mathbf{x} = A^H\mathbf{b},$$
can be solved using the Cholesky factorization, where $A^H A = LL^H$. Let $A^H\mathbf{b} = \mathbf{p}$. First solve (by substitution)
$$L\mathbf{y} = \mathbf{p},$$
then solve (by back-substitution)
$$L^H\mathbf{x} = \mathbf{y}.$$
Solving the normal equations using the Cholesky factorization is sometimes called the "normal equation" approach. □

We will see in section 5.3.2 that the QR decomposition can be used to solve least-squares problems. Why, then, would we consider using the Cholesky factorization? In favor of using the QR is the fact that computing $A^H A$ (required for the Cholesky factorization) requires a good dynamic range capability, essentially double the word size for a fixed-point representation, in order not to be hurt by an increase in condition number. On the other hand, for an $m \times n$ matrix A, if $m \gg n$, then $A^H A$ and its factorizations will require less storage and approximately half the computation of the QR representation. In this case, if it can be determined that the system of equations is sufficiently well conditioned, solution using Cholesky factorization may be justified.

The Cholesky factorization is also used in "square root" Kalman filtering applications, which are numerically stable methods of computing Kalman filter updates (see, e.g., [356]).

5.2.1 Algorithms for computing the Cholesky factorization

There are several algorithms that can be used to compute the Cholesky factorization, which are mentioned, for example, in [114]. The algorithm presented here requires $m^3/3$ floating operations, and requires no additional storage. The algorithm is developed recursively. Write
$$B = \begin{bmatrix} \alpha & \mathbf{v}^H \\ \mathbf{v} & B_1 \end{bmatrix}$$
and note that it can be factored as
$$B = \begin{bmatrix} \sqrt{\alpha} & 0 \\ \mathbf{v}/\sqrt{\alpha} & I_{n-1} \end{bmatrix} \begin{bmatrix} 1 & 0 \\ 0 & B_1 - \mathbf{v}\mathbf{v}^H/\alpha \end{bmatrix} \begin{bmatrix} \sqrt{\alpha} & \mathbf{v}^H/\sqrt{\alpha} \\ 0 & I_{n-1} \end{bmatrix}. \quad (5.6)$$
If we could find the Cholesky factorization of $B_1 - \mathbf{v}\mathbf{v}^H/\alpha$ as $G_1 G_1^H$, we would have
$$B = \begin{bmatrix} \sqrt{\alpha} & 0 \\ \mathbf{v}/\sqrt{\alpha} & G_1 \end{bmatrix} \begin{bmatrix} \sqrt{\alpha} & \mathbf{v}^H/\sqrt{\alpha} \\ 0 & G_1^H \end{bmatrix} = GG^H.$$
We therefore proceed recursively, decomposing B into successively smaller blocks. The

MATLAB code is demonstrated in algorithm 5.2 (for purposes of illustration, since MATLAB has a built-in Cholesky factorization via the function `chol`).

Algorithm 5.2 Cholesky factorization
File: `cholesky.m`

5.3 Unitary matrices and the QR factorization

We begin with a description of the "Q" in the QR factorization.

5.3.1 Unitary matrices

Definition 5.1 An $m \times m$ matrix Q with complex elements is said to be **unitary** if
$$Q^H Q = I.$$
If Q has real elements and $Q^T Q = I$, then Q is said to be an **orthogonal** matrix. □

For a unitary (or orthogonal) matrix, we also have $QQ^H = I$.

Lemma 5.1 *If* $\mathbf{y} = Q\mathbf{x}$ *for an* $m \times m$ *matrix* Q, *then:* $\|\mathbf{y}\| = \|\mathbf{x}\|$ *for all* $\mathbf{x} \in \mathbb{R}$ *if and only if* Q *is unitary, where the norm is the usual Euclidean norm.*

A transformation which does not change the length of a vector is said to be **isometric**, or "length preserving." The proof of the lemma is straightforward and is given as an exercise. This lemma allows us to make transformations on variables *without* changing their length. The lemma provides the basis for Parseval's theorem for finite-dimensional vectors.

Lemma 5.2 *If* $Y = QX$ *for an* $m \times m$ *unitary matrix* Q, *then*
$$\|Y\|_F = \|X\|_F$$
where $\|\cdot\|_F$ *is the Frobenius norm.*

There is a useful analogy that can now be introduced.

Hermitian matrices. satisfying $A^H = A$ are analogous to real numbers, numbers whose complex conjugates are equal to themselves.

Unitary matrices. satisfying $U^H U = I$ are analogous to complex numbers z on the unit circle satisfying $|z|^2 = 1$.

Orthogonal matrices. satisfying $Q^T Q = I$ are analogous to the real numbers $z = \pm 1$, such that $z^2 = 1$.

The bilinear transformation
$$z = \frac{1 + jr}{1 - jr} \tag{5.7}$$
takes real numbers r into the unit circle $|z| = 1$, mapping the number $r = \infty$ to $z = -1$. Analogously, by *Cayley's formula*,
$$U = (I + jR)(I - jR)^{-1}, \tag{5.8}$$
a Hermitian matrix R is mapped to a unitary matrix (which does not have an eigenvalue of -1).

5.3.2 The QR factorization

In the QR factorization, an $m \times n$ matrix A is written as

$$A = QR,$$

where Q is an $m \times m$ unitary matrix and R is upper triangular $m \times n$. As discussed in the following, there are several ways in which the QR factorization can be computed. In this section we focus on some of the uses of the factorization.

The most important application of QR is to full-rank least-squares problems. Consider

$$A\mathbf{x} \approx \mathbf{b},$$

where $m > n$ and the columns of A are linearly independent. In this case, the problem is said to be a full-rank least-squares problem. The solution $\hat{\mathbf{x}}$ that minimizes $\|A\hat{\mathbf{x}} - \mathbf{b}\|_2$ is

$$\hat{\mathbf{x}} = (A^H A)^{-1} A^H \mathbf{b}.$$

However, the condition number of $A^H A$ is the square of the condition number of A, so direct computation is not advised. This poor conditioning can be mitigated using the QR decomposition. When $m > n$, the QR decomposition can be written as

$$A = QR = Q \begin{bmatrix} R_1 \\ \mathbf{0} \end{bmatrix},$$

where R_1 is $n \times n$, and the $\mathbf{0}$ denotes an $(m - n) \times n$ block of zeros. Also let

$$Q^H \mathbf{b} = \begin{bmatrix} \mathbf{c} \\ \mathbf{d} \end{bmatrix}, \tag{5.9}$$

where $\mathbf{c}$ is $n \times 1$ and $\mathbf{d}$ is $(m - n) \times 1$. Then

$$\|A\mathbf{x} - \mathbf{b}\|_2^2 = \|QR\mathbf{x} - \mathbf{b}\|_2^2$$

$$= \|Q(R\mathbf{x} - Q^H \mathbf{b})\|_2^2 \tag{5.10}$$

$$= \left\| \begin{bmatrix} R_1 \\ 0 \end{bmatrix} \mathbf{x} - \begin{bmatrix} \mathbf{c} \\ \mathbf{d} \end{bmatrix} \right\|_2^2 \tag{5.11}$$

$$= \|R_1 \mathbf{x} - \mathbf{c}\|_2^2 + \|\mathbf{d}\|_2^2,$$

where (5.10) follows since $\mathbf{b} = QQ^H \mathbf{b}$, and (5.11) follows from lemma 5.1. (Such pulling of orthogonal matrices "out of thin air" to suit an analytical purpose is quite common.) The value $\hat{\mathbf{x}}$ that minimizes (5.11) satisfies

$$R_1 \hat{\mathbf{x}} = \mathbf{c},$$

which can be readily computed since R_1 is a triangular matrix. If A does not have full column rank, or if $m < n$, computing the QR decomposition and solving least-squares problems thereby is more difficult. There are algorithms to compute the QR decomposition in this case, which involve column pivoting. However, in this circumstance it is recommended to use the SVD, and hence those techniques are not discussed here.

5.3.3 QR factorization and least-squares filters

As an example of the use of the QR factorization, consider the least-squares problem

$$\mathbf{d}[k] = A[k]\mathbf{h}[k] + \mathbf{e}[k], \tag{5.12}$$

5.3 Unitary Matrices and the QR Factorization

where we wish to minimize $\|\mathbf{e}[k]\|_2^2$, in which

$$\mathbf{d}[k] = \begin{bmatrix} d[1] & d[2] & \cdots & d[k] \end{bmatrix}^T,$$

$$A = \begin{bmatrix} \mathbf{q}[1]^H \\ \mathbf{q}[2]^H \\ \vdots \\ \mathbf{q}[k]^H \end{bmatrix}.$$

(See section 3.9.) The least squares solution can be obtained by finding the QR factorization of $X[k]$,

$$A[k] = Q[k] \begin{bmatrix} R_1[k] \\ \mathbf{0} \end{bmatrix}.$$

Then

$$R_1[k]\mathbf{h} = (Q[k]\mathbf{d}[k])_{m \times 1}.$$

Thus, we can find $\mathbf{h}$ by back-substitution.

5.3.4 Computing the QR factorization

At least four major ways of computing the QR factorization are widely reported. These are:

1. The Gram–Schmidt algorithm.
2. The modified Gram–Schmidt algorithm. The Gram–Schmidt algorithms are discussed in section 2.15.
3. Householder transformations.
4. Givens rotations.

The Gram–Schmidt methods provide an orthonormal basis spanning the column space of A. The QR factorizations using the Householder transformation and the Givens rotations rely on simple invertible (and orthogonal) geometric transformations. The Householder transformation is simply a reflection operation that is used to zero most of a column of a matrix, while the Givens rotation is a simple two-dimensional rotation that is used to zero a particular single element of a matrix. These operations may be applied in succession to obtain an upper triangular matrix in the QR factorization. They may also be used in other circumstances where zeroing particular elements of a matrix (while preserving the eigenvalues of the matrix) is necessary. Of these four types, the Gram–Schmidt methods are the least complex computationally, but are also the most poorly conditioned.

5.3.5 Householder transformations

Recall from section 2.13.1 that the matrix that projects orthogonally onto span($\mathbf{v}$) (see (2.20)) is

$$P_v = \frac{\mathbf{v}\mathbf{v}^H}{\mathbf{v}^H\mathbf{v}},$$

and the orthogonal projection matrix is

$$P_v^\perp = I - P_v.$$

These are similar to the Householder transformation with respect to a nonzero vector $\mathbf{v}$,

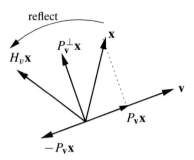

Figure 5.1: The Householder transformation of a vector

which is a transformation of the form

$$H_v = I - 2\frac{vv^H}{v^H v}$$
$$= I - 2P_v. \tag{5.13}$$

It is straightforward to show that H_v is unitary and $H_v^H = H_v$ (see exercise 5.3.4-14). The vector **v** is called a **Householder vector**. Observe that $H_v \mathbf{v} = -\mathbf{v}$, and that if $\mathbf{z} \perp \mathbf{v}$ (with respect to the Euclidean inner product) then $H_v \mathbf{z} = \mathbf{z}$. Write **x** as

$$\mathbf{x} = P_\mathbf{v}\mathbf{x} + P_\mathbf{v}^\perp \mathbf{x};$$

then

$$H_v \mathbf{x} = P_\mathbf{v}^\perp \mathbf{x} - P_\mathbf{v}\mathbf{x},$$

which corresponds to a *reflection* of the vector **x** across the space perpendicular to **v**, as shown in figure 5.1. Reflecting twice returns the original point: $H_v^2 \mathbf{x} = \mathbf{x}$. As an operator, we can write

$$H_v = P_\mathbf{v}^\perp - P_\mathbf{v}.$$

The Householder transformation can be used to zero out all the elements of a vector except for one component. That is, for a vector $\mathbf{x} = [x_1\ x_2\ \ldots\ x_n]^T$, there is a vector **v** in the Householder transformation H_v such that

$$H_v \mathbf{x} = \begin{bmatrix} \alpha \\ 0 \\ \vdots \\ 0 \end{bmatrix}$$

for some scalar α. Since H_v is unitary, $\|\mathbf{x}\|_2 = \|H_v \mathbf{x}\|_2$, hence $\alpha = \pm\|\mathbf{x}\|_2$. One way of viewing the Householder transformation is as a unitary transformation which compresses all of the energy in a vector into a single component, zeroing out the other components of the vector. To find the vector **v** in the transformation H_v, write

$$H_v \mathbf{x} = \alpha \begin{bmatrix} 1 \\ 0 \\ \vdots \\ 0 \end{bmatrix} = \alpha \mathbf{e}_1.$$

5.3 Unitary Matrices and the QR Factorization

Then

$$H_v \mathbf{x} = \left(I - 2\frac{\mathbf{v}\mathbf{v}^H}{\mathbf{v}^H \mathbf{v}}\right) \mathbf{x}$$

$$= \mathbf{x} - 2\frac{\mathbf{v}^H \mathbf{x}}{\mathbf{v}\mathbf{v}^H}\mathbf{v}$$

$$= \alpha \mathbf{e}_1,$$

so that

$$\left(2\frac{\mathbf{v}^H \mathbf{x}}{\mathbf{v}\mathbf{v}^H}\right)\mathbf{v} = \mathbf{x} - \alpha \mathbf{e}_1.$$

This means that $\mathbf{v}$ is a scalar multiple of $\mathbf{x} - \alpha \mathbf{e}_1$. Since we know that $\alpha = \pm \|\mathbf{x}\|_2$, and since scaling $\mathbf{v}$ by a nonzero scalar does not change the Householder transformation, we will take

$$\mathbf{v} = \mathbf{x} \pm \|\mathbf{x}\|_2 \mathbf{e}_1.$$

Although either sign may be taken, numerical considerations suggest a preferred value. For real vectors, if $\mathbf{x}$ is close to a multiple of $\mathbf{e}_1$, then $\mathbf{v} = \mathbf{x} - \text{sign}(x_1)\|\mathbf{x}\|_2 \mathbf{e}_1$ has a small norm, which could lead to a large relative error in the computation of the factor $2/\mathbf{v}^T \mathbf{v}$. This difficulty can be avoided by choosing the sign by

$$\mathbf{v} = \mathbf{x} + \text{sign}(x_1)\|\mathbf{x}\|_2 \mathbf{e}_1.$$

By this selection $\|\mathbf{v}\| \geq \|\mathbf{x}\|$. For complex vectors, choosing according to the sign of the real part is appropriate.

The operation of H_v on $\mathbf{x}$ can be understood geometrically using figure 5.2, where the sign is taken so that $\mathbf{v} = \mathbf{x} + \|\mathbf{x}\|_2 \mathbf{e}_1$. Since $\mathbf{v}$ is the sum of two equal-length vectors, it is the diagonal of an equilateral parallelogram. The other diagonal (orthogonal to the first—see exercise 5.3.4-28) runs from the vector $\mathbf{x}$ to the vector $\|\mathbf{x}\|_2 \mathbf{e}_1$. From the figure, it is clear that $P_\mathbf{v} \mathbf{x} = \mathbf{v}/2$ and $P_\mathbf{v}^\perp \mathbf{x} = \mathbf{x} - \mathbf{v}/2$.

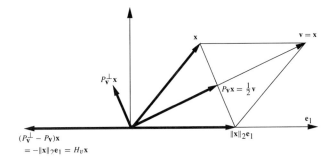

Figure 5.2: Zeroing elements of a vector by a Householder transformation

In the QR factorization, we want to convert A to an upper-triangular form using a sequence of orthogonal transformations. To use the Householder transformation to compute the QR factorization of a matrix, first choose a Householder transformation H_1 to zero out all but the first element of the first column of A, using the vector $\mathbf{v}_1$. For the sake of illustration, let A be 4×3. Then

$$H_1 A = \begin{bmatrix} \alpha_1 & \times & \times \\ 0 & \times & \times \\ 0 & \times & \times \\ 0 & \times & \times \end{bmatrix},$$

where × indicates elements of the matrix which are not zero (in general). Let $Q_1 = H_1$. To continue the process, for the 3×2 matrix on the lower right, choose a Householder transformation matrix H_2 to zero out the last 2 elements, using the vector $\mathbf{v}_2$. Combination with the first transformation is done by

$$\begin{bmatrix} 1 & \mathbf{0} \\ \mathbf{0} & H_2 \end{bmatrix} \begin{bmatrix} \alpha_1 & \times & \times \\ 0 & \times & \times \\ 0 & \times & \times \\ 0 & \times & \times \end{bmatrix} = Q_2 Q_1 A = \begin{bmatrix} \alpha_1 & \times & \times \\ 0 & \alpha_2 & \times \\ 0 & 0 & \times \\ 0 & 0 & \times \end{bmatrix},$$

where

$$Q_2 = \begin{bmatrix} 1 & \mathbf{0} \\ \mathbf{0} & H_2 \end{bmatrix}.$$

For the sake of implementation (described subsequently), note that Q_2 can be formed as a Householder matrix as

$$Q_2 = I - 2 \frac{\tilde{\mathbf{v}}_2 \tilde{\mathbf{v}}_2^H}{\tilde{\mathbf{v}}_2^H \tilde{\mathbf{v}}_2}, \tag{5.14}$$

where

$$\tilde{\mathbf{v}}_2 = \begin{bmatrix} 0 \\ \mathbf{v}_2 \end{bmatrix}.$$

The last two elements in the third column can be reduced with a third Householder transformation H_3. In conjunction with the other elements of the matrix, this can be written as

$$\begin{bmatrix} 1 & 0 & \mathbf{0} \\ 0 & 1 & \mathbf{0} \\ \mathbf{0} & \mathbf{0} & H_3 \end{bmatrix} \begin{bmatrix} \alpha_1 & \times & \times \\ 0 & \alpha_2 & \times \\ 0 & 0 & \times \\ 0 & 0 & \times \end{bmatrix} = Q_3 Q_2 Q_1 A = \begin{bmatrix} \alpha_1 & \times & \times \\ 0 & \alpha_2 & \times \\ 0 & 0 & \alpha_3 \\ 0 & 0 & \times \end{bmatrix},$$

where

$$Q_3 = \begin{bmatrix} 1 & 0 & \mathbf{0} \\ 0 & 1 & \mathbf{0} \\ \mathbf{0} & \mathbf{0} & H_3 \end{bmatrix} = I - 2 \frac{\tilde{\mathbf{v}}_3 \tilde{\mathbf{v}}_3^H}{\tilde{\mathbf{v}}_3^H \tilde{\mathbf{v}}_3}$$

and

$$\tilde{\mathbf{v}}_3 = \begin{bmatrix} 0 \\ 0 \\ \mathbf{v}_3 \end{bmatrix}.$$

Since H_2 and H_3 are orthogonal, so are Q_2 and Q_3 (see exercise 5.3.4-16), and so is $Q^H = Q_3 Q_2 Q_1$. Thus A has been reduced to the product of an orthogonal matrix times an upper-triangular matrix,

$$A = QR = Q_1 Q_2 Q_3 R.$$

For a general $m \times n$ matrix, computation of the QR algorithm involves forming n orthogonal matrices Q_j, $j = 1, 2, \ldots, n$. Then

$$Q = Q_1 Q_2 \cdots Q_n,$$

where

$$Q_j = I - 2 \frac{\tilde{\mathbf{v}}_j \tilde{\mathbf{v}}_j^H}{\tilde{\mathbf{v}}_j^H \tilde{\mathbf{v}}_j}$$

5.3 Unitary Matrices and the QR Factorization

and
$$\tilde{\mathbf{v}}_j = \big[\underbrace{0, 0, \ldots, 0}_{j-1}, \mathbf{v}_j^T\big]^T.$$

5.3.6 Algorithms for Householder transformations

In this section, sample MATLAB code is developed to compute the QR decomposition using Householder transformations. (The code is for demonstration purposes only, since MATLAB has the built-in function qr.)

In the interest of efficiency, the Householder transformation matrix Q is not explicitly formed. Rather than explicitly forming H_v and then multiplying $H_v A$, we note that

$$H_v A = \left(I - 2\frac{\mathbf{v}\mathbf{v}^H}{\mathbf{v}^H \mathbf{v}}\right) A = A + \beta \mathbf{v}\mathbf{w}^H \tag{5.15}$$

where $\beta = -2/\mathbf{v}^H \mathbf{v}$ and $\mathbf{w} = A^H \mathbf{v}$. It is often the case that only the R matrix is explicitly needed, so the Q is represented implicitly by the sequence of $\mathbf{v}_j$ vectors, from which Q can be computed as desired. Algorithm 5.3 illustrates a function which applies a Householder transformation H_v, represented only by the Householder vector $\mathbf{v}$, on the left of A as $H_v A$ and also shows the function makehouse which computes the Householder vector $\mathbf{v}$. Also shown is the function houseright, which applies the Householder transformation on the right to zero out rows of A.

Algorithm 5.3 Householder transformation functions. Compute (1) $\mathbf{v}$, (2) $H_v A$ given $\mathbf{v}$ and (3) $A H_v$ given $\mathbf{v}$
File: makehouse.m
houseleft.m
houseright.m

Example 5.3.1 Let
$$A = \begin{bmatrix} 1 & 2 & 3 \\ 4 & 5 & 6 \\ 6 & 7 & 8 \end{bmatrix}.$$

Then the MATLAB function calls v1 = makehouse(A(:,1)) and vr = makehouse(A(1,:)) return the vectors

$$v_l = [8.28011 \quad 4 \quad 6]^T \qquad v_r = [4.74166 \quad 2 \quad 3]^T.$$

Then $H_v A$ can be computed using houseleft(A,vl) and $A H_v$ can be computed from houseright(A,vr). The results are

$$\text{houseleft}(A,vl) = \begin{bmatrix} -7.28011 & -8.79108 & -10.302 \\ 0 & -0.213011 & -0.426022 \\ 0 & -0.819517 & -1.63903 \end{bmatrix},$$

$$\text{houseright}(A,vr) = \begin{bmatrix} -3.74166 & 0 & 0 \\ -8.55236 & -0.294503 & -1.94175 \\ -11.7595 & -0.490838 & -3.23626 \end{bmatrix}. \qquad \square$$

Algorithm 5.4 computes the QR factorization using the simplifications noted here. The return values are the matrix R and the vector of $\mathbf{v}$ vectors. The complexity of the algorithm is approximately $2n^2(m - n/3)$ floating operations.

Algorithm 5.4 QR factorization via Householder transformations
File: `qrhouse.m`

In order to solve the least-squares equation as previously described, we must be able to compute $Q^H \mathbf{b}$. Since $Q = Q_1 Q_2 \cdots Q_n$ and each Q is Hermitian symmetric,

$$Q^H \mathbf{b} = Q_n^H Q_{n-1}^H \cdots Q_1^H \mathbf{b},$$

which may be accomplished (conceptually) using the following algorithm, which overwrites $\mathbf{b}$ with $Q^H \mathbf{b}$:

for $j = 1 : n$
    $\mathbf{b} = Q_j \mathbf{b}$
end

The multiplication can be accomplished without explicitly forming the Q_j matrices by using the idea shown in (5.15). Computation of $Q^H \mathbf{b}$ is thus accomplished as shown in algorithm 5.5.

Algorithm 5.5 Computation of $Q^H \mathbf{b}$
File: `qrqtb.m`

Example 5.3.2 Suppose it is desired to find the least-squares solution to

$$\begin{bmatrix} 7 & 8 & 8 \\ 8 & 6 & 2 \\ 1 & 7 & 3 \\ 0 & 7 & 3 \\ 6 & 9 & 5 \end{bmatrix} \mathbf{x} = \begin{bmatrix} 47 \\ 26 \\ 24 \\ 23 \\ 39 \end{bmatrix}.$$

Using `[V,R] = qrhouse(A)`, we obtain

$$V = \begin{bmatrix} 19.2474 & 0 & 0 \\ 8 & -12.7989 & 0 \\ 1 & 5.8844 & -6.1096 \\ 0 & 7 & -2.3046 \\ 6 & 2.3065 & -1.9142 \end{bmatrix} \quad R = \begin{bmatrix} -12.2474 & -13.4722 & -8.5732 \\ 0 & 9.8742 & 4.8105 \\ 0 & 0 & 3.7893 \\ 0 & 0 & 0 \\ 0 & 0 & 0 \end{bmatrix}.$$

Using `qrqtb(b,V)`, we obtain

$$Q^H \mathbf{b} = \begin{bmatrix} -64.9115 \\ 34.1800 \\ 11.3680 \\ 0 \\ 0 \end{bmatrix}.$$

The least-squares solution comes from solving the 3×3 upper-triangular system of equations using

5.3 Unitary Matrices and the QR Factorization

back-substitution,

$$\begin{bmatrix} -12.2474 & -13.4722 & -8.5732 \\ 0 & 9.8742 & 4.8105 \\ 0 & 0 & 3.7893 \end{bmatrix} \hat{\mathbf{x}} = \begin{bmatrix} -64.9115 \\ 34.1800 \\ 11.3680 \end{bmatrix},$$

which leads to

$$\hat{\mathbf{x}} = \begin{bmatrix} 1 \\ 2 \\ 3 \end{bmatrix}.$$

□

Where the Q matrix is explicitly desired from V, it can be computed by backward accumulation. To compute $Q = Q_1 Q_2 \cdots Q_r$, we iterate as follows:

$$Q^{[0]} = I$$
$$Q^{[1]} = Q_r Q^{[0]}$$
$$Q^{[2]} = Q_{r-1} Q^{[1]}$$
$$\vdots$$
$$Q = Q^{[r]} = Q_1 Q^{[r-1]}.$$

This is implemented in algorithm 5.6.

Algorithm 5.6 Computation of Q from V
File: `qrmakeq.m`

5.3.7 QR factorization using Givens rotations

Unlike the Householder transformation which zeros out entire columns at a stroke, the Givens technique more selectively zeros one element at a time, using a rotation.

A two-dimensional rotation by an angle θ is illustrated in figure 5.3(a). The figure demonstrates that the point $(1, 0)$ is rotated into the point $(\cos\theta, \sin\theta)$, and the point $(0, 1)$ is rotated into the point $(-\sin\theta, \cos\theta)$. By these points we identity that a matrix G_θ that rotates $[x, y]^T$ is

$$G_\theta = \begin{bmatrix} \cos\theta & \sin\theta \\ -\sin\theta & \cos\theta \end{bmatrix}. \tag{5.16}$$

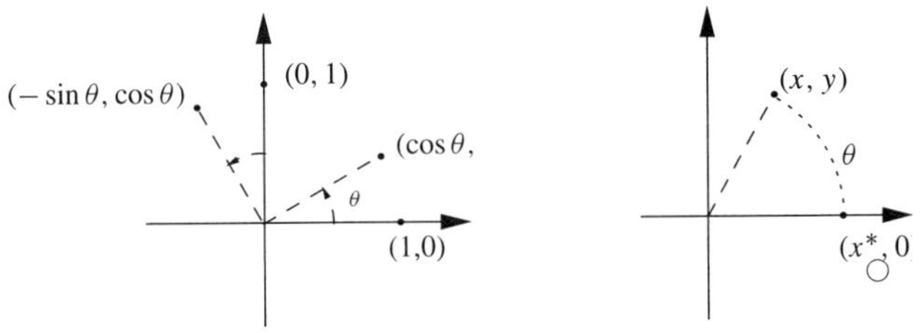

(a) A general rotation

(b) Rotation of the second coordinate to zero

Figure 5.3: Two-dimensional rotation

The rotation matrix is orthogonal: $G_\theta^T G_\theta = I$. It should be clear that any point (x, y) in two dimensions can be rotated by some rotation matrix G so that its second coordinate is zero. This is illustrated in figure 5.3(b). For a vector $\mathbf{x} = [x \ y]^T$, its second coordinate can be zeroed by multiplication by the orthogonal matrix G_θ, where

$$\theta(x, y) = -\tan^{-1} \frac{y}{x}. \tag{5.17}$$

In the Givens rotation approach to the QR factorization, a matrix A is zeroed out one element at a time, starting at the bottom of the first column and working up the columns. To zero a_{ik}, we use $x = a_{jk}$ and $y = a_{ik}$, applying the 2×2 rotation matrix across the jth and ith rows of A. Such a rotation matrix is called a *Givens rotation*. We will denote by $G_\theta(i, k, j)$ the rotation matrix that zeros a_{ik}. For brevity we will also write $G(i, k, j)$. In the QR factorization, a sequence of these rotation matrices are used. A sequence of matrices produced by successive operation of Givens rotations might have the following form, where the convention of taking $j = i - 1$ is used. The rotation is shown above the arrow, and the rows affected by the preceding transformation are shown in boldface.

$$\begin{bmatrix} \times & \times & \times \\ \times & \times & \times \\ \times & \times & \times \\ \times & \times & \times \end{bmatrix} \xrightarrow{G(4,1,3)} \begin{bmatrix} \times & \times & \times \\ \times & \times & \times \\ \times & \times & \times \\ \mathbf{0} & \times & \times \end{bmatrix} \xrightarrow{G(3,1,2)} \begin{bmatrix} \times & \times & \times \\ \times & \times & \times \\ \mathbf{0} & \times & \times \\ 0 & \times & \times \end{bmatrix} \xrightarrow{G(2,1,1)} \begin{bmatrix} \times & \times & \times \\ \mathbf{0} & \times & \times \\ 0 & \times & \times \\ 0 & \times & \times \end{bmatrix} \xrightarrow{G(4,2,3)}$$

$$\begin{bmatrix} \times & \times & \times \\ 0 & \times & \times \\ 0 & \times & \times \\ 0 & \mathbf{0} & \times \end{bmatrix} \xrightarrow{G(3,2,2)} \begin{bmatrix} \times & \times & \times \\ 0 & \times & \times \\ 0 & \mathbf{0} & \times \\ 0 & 0 & \times \end{bmatrix} \xrightarrow{G(4,3,3)} \begin{bmatrix} \times & \times & \times \\ 0 & \times & \times \\ 0 & 0 & \times \\ 0 & 0 & \mathbf{0} \end{bmatrix} \tag{5.18}$$

The two-dimensional rotation (5.16) can be modified to form $G_\theta(i, k, j)$, by defining

$$G_\theta(i, k, j) = \begin{bmatrix} 1 & \cdots & 0 & \cdots & 0 & \cdots & 0 \\ \vdots & \ddots & \vdots & & \vdots & & \vdots \\ 0 & \cdots & c & \cdots & s & \cdots & 0 \\ \vdots & & \vdots & \ddots & \vdots & & \vdots \\ 0 & \cdots & -s & \cdots & c & \cdots & 0 \\ \vdots & & \vdots & & \vdots & \ddots & \vdots \\ 0 & \cdots & 0 & \cdots & 0 & \cdots & 1 \end{bmatrix} \begin{matrix} \\ \\ j \\ \\ i \\ \\ \end{matrix}, \tag{5.19}$$

$$\phantom{G_\theta(i, k, j) = \begin{bmatrix}}j i$$

where $c = \cos \theta$ and $s = \sin \theta$. As is apparent from the form of $G_\theta(i, k, j)$, the operation $G_\theta(i, k, j)A$ sets the (i, k)th element to zero and modifies the ith and jth rows of A, leaving the other rows of A unmodified. The value of θ in $G_\theta(i, k, k)$ is determined from $(x, y) = (A(j, k), A(i, k))$ in (5.17). As is apparent by studying (5.18), taking $j = i - 1$ lets the diagonalization already accomplished in prior columns be unaffected by Givens rotations on later columns. Since this is the most common case, we will henceforth use the abbreviated notation $G(i, k)$ or $G_\theta(i, k)$ for $G_\theta(i, k, j)$.

For the 4×3 example, the factorization is accomplished by

$$G(4, 1)G(3, 1)G(2, 1)G(4, 2)G(3, 2)G(4, 3)A = R.$$

5.3 Unitary Matrices and the QR Factorization

The Q matrix is thus obtained as

$$Q = (G(4,1)G(3,1)G(2,1)G(4,2)G(3,2)G(4,3))^T$$
$$= G(4,3)^T G(3,2)^T G(4,2)^T G(2,1)^T G(3,1)^T G(4,1)^T.$$

Example 5.3.3 Let

$$A = \begin{bmatrix} 1 & 2 \\ 3 & 4 \\ 1 & 3 \end{bmatrix}.$$

A rotation matrix that modifies the last two rows of A to zero the (3, 1) element is

$$G(3,1) = G(3,1,2) = \begin{bmatrix} 1 & 0 & 0 \\ 0 & 0.9487 & 0.3162 \\ 0 & -0.3162 & 0.9487 \end{bmatrix}.$$

Then

$$GA = \begin{bmatrix} 1 & 2 \\ 3.1623 & 4.7434 \\ 0 & 1.5811 \end{bmatrix}.$$
□

5.3.8 Algorithms for QR factorization using Givens rotations

Several aspects of the mathematics outlined for Givens rotations may be streamlined for a numerical implementation. Explicit computation of θ is not necessary; what are needed are $\cos\theta$ and $\sin\theta$, which may be determined from (x, y) without any trigonometric functions,

$$\cos\theta = \cos\tan\left(-\frac{y}{x}\right) = \frac{x}{\sqrt{x^2+y^2}} \qquad \sin\theta = \frac{-y}{\sqrt{x^2+y^2}}.$$

See algorithm 5.7 for a numerically stable method of computing these quantities.

Algorithm 5.7 Finding $\cos\theta$ and $\sin\theta$ for a Givens rotation
File: `qrtheta.m`

In computing the multiplication $G(i,k)A$, it is clearly much more efficient to modify only rows i and k of the product. An explicit Q matrix is never constructed. Instead, the $\cos\theta$ and $\sin\theta$ information is saved. It would also be possible to represent both numbers using a single quantity, and store the Q matrix information in the lower triangle. However, in the interest of speed, this is not done. Algorithm 5.8 computes the QR factorization, and algorithm 5.9 computes $Q^H \mathbf{b}$, for use in solving least-squares problems. Finally, for those instances in which it is needed, algorithm 5.10 computes Q from the θ information by computing

$$Q = G(m,1)^T G(m-1,1)^T \cdots G(2,1)^T G(m,2)^T G(m-1,2)^T \cdots G(m,n)^T,$$

with the multiplication done from left to right.

Algorithm 5.8 QR factorization using Givens rotations
File: `qrgivens.m`

Algorithm 5.9 Computation of $Q^H \mathbf{b}$ for the Givens rotation factorization
File: `qrqtbgiv.m`

Algorithm 5.10 Computation of Q from θ
File: `qrmakeqgiv.m`

5.3.9 Solving least-squares problems using Givens rotations

Givens rotations can be used to solve least-squares problems in a way that is well suited for pipelined implementation in VLSI [263]. Rewrite the equation

$$A\mathbf{x} \approx \mathbf{b}$$

as

$$[A|\mathbf{b}] \begin{bmatrix} \mathbf{x} \\ -1 \end{bmatrix} \approx 0.$$

Let this be written as

$$B\mathbf{h} \approx 0,$$

where $B = [A|\mathbf{b}]$ and $\mathbf{h}^T = [\mathbf{x}^T, -1]$. Then the least-squares solution is the one that minimizes $\|B\mathbf{h}\|_2^2 = \mathbf{h}^H B^H B \mathbf{h}$. Since multiplication by an orthogonal matrix does not change the norm, $\|QB\mathbf{h}\|_2^2 = \|B\mathbf{h}\|_2^2$ for an orthogonal matrix Q. The matrix Q can be selected as a Givens rotation that selectively zeros out elements of the matrix B. By this means, we can transform the problem successively as

$$B\mathbf{h} \approx 0,$$
$$Q_1 B\mathbf{h} \approx 0,$$
$$(Q_2 Q_1)B\mathbf{h} \approx 0,$$
$$(Q_2 Q_1)B\mathbf{h} \approx 0,$$
$$\ldots (Q_p \cdots Q_2 Q_1)B\mathbf{h} \approx 0.$$

With an appropriately chosen sequence of Q_i matrices, the result is that $(Q_p \cdots Q_2 Q_1)B$ is mostly upper triangular, so that we obtain a set of equations of the following form:

$$\begin{bmatrix} \hat{a}_{11} & \hat{a}_{12} & \hat{a}_{13} & \cdots & \hat{a}_{1n} & \hat{b}_1 \\ & \hat{a}_{22} & \hat{a}_{23} & \cdots & \hat{a}_{2n} & \hat{b}_2 \\ & & \hat{a}_{33} & \cdots & \hat{a}_{3n} & \hat{b}_3 \\ \vdots & & & & & \\ & & & & \hat{a}_{nn} & \hat{b}_n \\ \times & \times & \times & \times & \times & \hat{b}_{n+1} \\ \times & \times & \times & \times & \times & \hat{b}_{n+2} \\ \vdots & & & & & \end{bmatrix} \begin{bmatrix} x_1 \\ x_2 \\ x_3 \\ \vdots \\ x_n \\ -1 \end{bmatrix} \approx \begin{bmatrix} 0 \\ 0 \\ 0 \\ \vdots \\ 0 \\ \vdots \end{bmatrix}.$$

5.3 Unitary Matrices and the QR Factorization

Practically speaking, multiplication by the orthogonal matrices can stop when the top n rows are mostly triangularized, as shown. While it would be possible to complete the QR factorization to zero the lower portion of the matrix (the part indicated with ×s), this is not necessary, since the structure allows the solution to be obtained. From this, the least-square solution is

$$x_n = \frac{\hat{b}_n}{\hat{a}_{n,n}},$$

$$x_{n-1} = \frac{\hat{b}_{n-1} - \hat{a}_{n-1,n} x_n}{\hat{a}_{n-1,n-1}},$$

$$\vdots$$

$$x_i = \frac{\hat{b}_i - \sum_{j=i+1}^{n} a_{i,j} x_j}{\hat{a}_{i,i}},$$

$$\vdots$$

$$x_1 = \frac{\hat{b}_1 - \sum_{j=2}^{n} a_{1,j} x_j}{\hat{a}_{1,1}}.$$

5.3.10 Givens rotations via CORDIC rotations

For high-speed real-time applications, it may be necessary to go with pipelined and parallel algorithms for QR decomposition. The method known as CORDIC rotations provides for pipelined implementations of the Givens rotations without the need to compute trigonometric functions or square roots. CORDIC is an acronym for COordinate Rotation DIgital Computation. CORDIC methods have also been applied to a variety of other signal processing problems, including DFTs, FFTs, digital filtering, and array processing. A survey article with a variety of references is [146]. A detailed application of CORDIC techniques to array processing using a VLSI hardware implementation, including some very clever designs for solution of linear equations, appears in [267].

The fundamental step in Givens rotations is the two dimensional rotation

$$\begin{bmatrix} x' \\ y' \end{bmatrix} = \begin{bmatrix} \cos\theta & -\sin\theta \\ \sin\theta & \cos\theta \end{bmatrix} \begin{bmatrix} x \\ y \end{bmatrix}, \quad (5.20)$$

where θ is chosen so that $y' = 0$. This transformation is applied successively to appropriate pairs of rows to obtain the QR factorization. Since it is used repeatedly, it is important to make the computation as efficient as possible. The rotation in (5.20) can be rewritten as

$$\begin{bmatrix} x' \\ y' \end{bmatrix} = \cos\theta \begin{bmatrix} 1 & -\tan\theta \\ \tan\theta & 1 \end{bmatrix} \begin{bmatrix} x \\ y \end{bmatrix}, \quad (5.21)$$

which still requires four multiplications. However, if the angle θ is such that $\tan\theta$ is a power of 2, then the multiplication can be accomplished using only bit-shift operations. A general angle can be constructed as a series of angles whose tangents are powers of 2,

$$\theta = \sum_{i=0}^{\infty} \rho_i \theta_i,$$

where $\rho_i = \pm 1$ and θ_i is constrained so that $\tan\theta_i = 2^{-i}$. In practice, the sum is truncated after a few terms, usually about five or six:

$$\theta \approx \sum_{i=0}^{i_{max}} \rho_i \theta_i.$$

Table 5.1: Power-of-2 angles for CORDIC computations

i	$\tan \theta_i$	θ_i (degrees)	κ_i
0	1	45	0.70711
1	$\frac{1}{2}$	26.5605	0.63245
2	$\frac{1}{4}$	14.0362	0.61357
3	$\frac{1}{8}$	7.12502	0.60883
4	$\frac{1}{16}$	3.57633	0.60764
5	$\frac{1}{32}$	1.78991	0.60728
6	$\frac{1}{64}$	0.89517	0.60726

The power-of-2 angles for CORDIC rotations are shown in table 5.1 up to θ_6. Higher accuracy in the representation can be obtained by taking more terms, although for most practical purposes up to five terms is often adequate.

Example 5.3.4 An angle such as 37° can be represented using the angles in table 5.1 as

$$37 \approx \theta_0 - \theta_1 + \theta_2 + \theta_3 - \theta_4 + \theta_5 - \theta_6 = 36.91832$$

An efficient representation is simply the sequence of signs: $37 \sim \{1, -1, 1, 1, -1, 1, -1\}$. □

The rotation by θ in (5.20) is accomplished stagewise, by a series of *microrotations*. What makes this more efficient is the fact that the factors $\cos \theta_i$ from each microrotation can be combined into a precomputed constant,

$$\kappa_{i_{\max}} = \prod_{i=0}^{i_{\max}} \cos \theta_i.$$

Table 5.1 shows the values of κ for the first few values of $i_{\max}$. The microrotations result in a series of intermediate results. In a CORDIC implemented in $i_{\max}$ stages, the following results are obtained by successive application of (5.21):

$$\begin{bmatrix} x^{[0]} \\ y^{[0]} \end{bmatrix} = \kappa \begin{bmatrix} x \\ y \end{bmatrix},$$

$$\begin{bmatrix} x^{[1]} \\ y^{[1]} \end{bmatrix} = \begin{bmatrix} x^{[0]} \\ y^{[0]} \end{bmatrix} + \rho_0 2^0 \begin{bmatrix} -y^{[0]} \\ x^{[0]} \end{bmatrix},$$

$$\begin{bmatrix} x^{[2]} \\ y^{[2]} \end{bmatrix} = \begin{bmatrix} x^{[1]} \\ y^{[1]} \end{bmatrix} + \rho_1 2^{-1} \begin{bmatrix} -y^{[1]} \\ x^{[1]} \end{bmatrix},$$

$$\begin{bmatrix} x^{[3]} \\ y^{[3]} \end{bmatrix} = \begin{bmatrix} x^{[2]} \\ y^{[2]} \end{bmatrix} + \rho_2 2^{-2} \begin{bmatrix} -y^{[2]} \\ x^{[2]} \end{bmatrix},$$

$$\vdots$$

$$\begin{bmatrix} x' \\ y' \end{bmatrix} = \begin{bmatrix} x^{[i_{\max}]} \\ y^{[i_{\max}]} \end{bmatrix} + \rho_{i_{\max}} 2^{-i_{\max}} \begin{bmatrix} -y^{[i_{\max}]} \\ x^{[i_{\max}]} \end{bmatrix}.$$

The effect of multiplication by κ is to normalize the vector so that the final vector $[x', y']^T$ has the same length as $[x, y]^T$. In circumstances where the angle of the vector is important but its length is not, the first step may be eliminated.

5.3 Unitary Matrices and the QR Factorization

When doing rotation for the QR algorithm, the angle θ through which to rotate is determined by the first element of each of the two rows being rotated. These elements are referred to as the *leaders* of the pair of rows. The rest of the elements on the row are rotated at an angle determined by the leader. For the regular Givens rotation, it is necessary to compute the angle, which at a minimum requires computation of a square root. However, for the CORDIC implementation, it is possible to determine the angle to rotate through implicitly, using the microrotations, simply by examining the signs of the components of the leader. The goal is to rotate a vector $\mathbf{x}^T = [x, y]$ to a vector $[x', 0]$. If $\mathbf{x}$ is in quadrant I or quadrant III, then the rotation is negative. If $\mathbf{x}$ is in quadrant II or quadrant IV, then the rotation is positive. The sign of the microrotation is determined by

$$\rho_i = -\text{sign}(x_{i-1})\text{sign}(y_{i-1}). \tag{5.22}$$

In a pipelined implementation of the CORDIC architecture, a sequence of 2-vectors from a pair of rows of the matrix A are passed through a sequential computational structure. As the first vector from each row—the leader—is passed, the microrotation angle is computed according to (5.22). This information is latched and used for each succeeding vector in the row. Because buffering is used between each stage, the computations may be done in a pipelined manner. As a vector passes through a stage, another vector may immediately be passed into the stage; there is no need to wait for a single vector to pass all the way through. It is the pipelined nature of the architecture that leads to its efficiency.

When using CORDIC for the QR algorithm, several rows must be modified in succession. This may be accomplished by cascading several pipelined CORDIC structures in such a way that a modified row from one stage is passed on to the next stage. This allows for more parallelism in the computation. Additional details are provided in [263] and [267].

5.3.11 Recursive updates to the QR factorization

Consider again the least-squares filtering problem of (5.12), only now consider the problem of updating the estimate. That is, suppose that data $\mathbf{q}[1], \mathbf{q}[2], \ldots, \mathbf{q}[k]$ are used to form an estimate $\mathbf{h}[k]$ by the QR method,

$$R_1[k]\mathbf{h}[k] = Q[k]\mathbf{d}[k].$$

A new data point becomes available, and we desire to compute $\mathbf{h}[k+1]$, using as much of the previous work as possible.

In this case, it is most convenient to reorder the data from last to first, so we will let

$$A[k] = \begin{bmatrix} \mathbf{q}[k]^H \\ \mathbf{q}[k-1]^H \\ \vdots \\ \mathbf{q}[1] \end{bmatrix},$$

$$\mathbf{y}[k] = [y[k] \quad y[k-1] \quad \cdots \quad y[1]]^T,$$

and $\mathbf{d}[k]$ similarly. As before, let

$$X[k] = Q[k]R[k] = Q[k]\begin{bmatrix} R_1[k] \\ 0 \end{bmatrix}.$$

When the new data comes, the A matrix is updated as

$$A[k+1] = \begin{bmatrix} \mathbf{q}[k+1]^H \\ A[k] \end{bmatrix}.$$

Observe that

$$\begin{bmatrix} 1 & \\ & Q^H[k] \end{bmatrix} A[k+1] = \begin{bmatrix} \mathbf{q}[k+1]^H \\ R[k] \end{bmatrix} = \begin{bmatrix} \mathbf{q}[k+1]^H \\ R_1[k] \\ 0 \end{bmatrix} = H.$$

The matrix H has the property that $h_{ij} = 0$ for $i > j + 1$. Such a matrix is known as an upper *Hessenburg* matrix. By forcing a zero down the subdiagonal elements of H, it can be converted to an upper-triangular matrix. This can be accomplished using a series of Givens rotations, one for each subdiagonal element. Let the Givens rotations be indicated as $J_1 J_2, \ldots, J_m$. We thus obtain

$$J_1 J_2 \cdots J_m \begin{bmatrix} 1 & \\ & Q^H[k] \end{bmatrix} A = R[k+1] = \begin{bmatrix} R_1[k+1] \\ 0 \end{bmatrix},$$

from which $Q[k+1]$ can also be identified.

5.4 Exercises

5.1-1 For the matrix

$$A = \begin{bmatrix} 2 & 5 & 9 \\ 1 & 4 & 7 \\ 3 & 2 & 1 \end{bmatrix},$$

determine the LU factorization both with and without pivoting.

5.1-2 Write a MATLAB routine to solve the system of equations $A\mathbf{x} = \mathbf{b}$, assuming that the LU factorization is obtained using `newlu`.

5.1-3 Verify the following facts about triangular matrices:

(a) The inverse of an upper-triangular matrix is upper triangular. The inverse of a lower-triangular matrix is lower triangular.

(b) The product of two upper-triangular matrices is upper triangular.

5.1-4 This exercise illustrates the potential difficulty of LU factorization without pivoting. Suppose it is desired to solve the system of equations

$$\begin{bmatrix} 2 & 4 & -5 \\ 6 & 12.001 & 1 \\ 4 & -8 & -3 \end{bmatrix} \mathbf{x} = \begin{bmatrix} -5 \\ -33.002 \\ -21 \end{bmatrix}.$$

The true solution to this system of equations is $\mathbf{x} = [1\ 2\ 3]^T$, and the matrix A is very well conditioned. Compute the solution to this problem using the LU decomposition without pivoting, using arithmetic rounded to three significant places. Then compute using pivoting, and compare the answers with the exact result.

5.2-5 Compute the Cholesky factorization of

$$A = \begin{bmatrix} 4 & 6 & 4 \\ 6 & 25 & 18 \\ 4 & 18 & 22 \end{bmatrix}$$

as $A = LL^T$. Then write this as $A = U^T DU$, where U is an upper-triangular matrix with ones along the diagonal.

5.2-6 Show that (5.6) is true.

5.2-7 Given a zero-mean discrete-time input signal $f[t]$, which we form into a vector

$$\mathbf{q}[t] = \begin{bmatrix} \overline{f}[t] & \overline{f}[t-1] & \cdots & \overline{f}[t-m] \end{bmatrix}^T,$$

we desire to form a set of outputs

$$\mathbf{b}[t] = [b_0[t] \quad b_1[t] \quad \cdots \quad b_m[t]]^T$$

by $\mathbf{b}[t] = H\mathbf{q}[t]$ that are uncorrelated; that is, so that

$$E\left[b_i[t]\overline{b}_j[t]\right] = 0 \qquad \text{if } i \neq j.$$

Let $R = E[\mathbf{q}[t]\mathbf{q}^H[t]]$ be the correlation matrix of the input data. Determine the matrix H that decorrelates the input data.

5.2-8 Let $X = [\mathbf{x}_1, \mathbf{x}_2, \ldots, \mathbf{x}_n]$ be a set of real-valued zero-mean data, with correlation matrix

$$R_{xx} = \frac{1}{n} X X^T.$$

Determine a transformation on X that produces a data set Y,

$$Y = HX,$$

such that

$$R_{yy} = \frac{1}{n} Y Y^T$$

is equal to an identity.

5.2-9 Write MATLAB routines x forsub(L,b) and backsub(U,b) to solve $L\mathbf{x} = \mathbf{b}$ for a lower-triangular matrix L and $U\mathbf{x} = \mathbf{b}$ for an upper-triangular matrix U.

5.2-10 Develop a means of computing the solution to the weighted least-squares problem $\mathbf{x} = (A^H W A)^{-1} A^H W \mathbf{b}$ using the Cholesky factorization.

5.3.4-11 Show that for a unitary matrix Q,

$$\det(Q) = 1$$

5.3.4-12 (Column-space projectors) Let X be a rank-r matrix. Show that the matrix that projects orthogonally onto the column space of X is

$$P_X = Q(:, 1:r)[Q(:, 1:r)]^H,$$

where $X = QR$ is the QR factorization of X, and $Q(:, 1:r)$ is the MATLAB notation for the first r columns of Q.

5.3.4-13 Regarding the Cayley formula:
   (a) Show that z in (5.7) has $|z|^2 = 1$.
   (b) Show that U in (5.8) satisfies $UU^H = I$.
   (c) Solve (5.8) for R, thus finding a mapping from unitary matrices to Hermitian matrices.
   (d) A matrix S is *skew symmetric* if $S^T = -S$. Show that if S is skew symmetric then $Q = (I + S)(I - S)^{-1}$ is orthogonal.

5.3.4-14 For the Householder transformation (reflection) matrix $H = I - 2\mathbf{v}\mathbf{v}^H/(\mathbf{v}^H\mathbf{v})$, verify the following properties, and provide a geometric interpretation:
   (a) $H\mathbf{v} = -\mathbf{v}$.
   (b) If $\mathbf{z} \perp \mathbf{v}$ then $H\mathbf{z} = \mathbf{z}$.
   (c) $H^H = H$.
   (d) $H^H H = H H^H = I$.

(e) For vectors **x** and **y**,
$$\langle \mathbf{x}, \mathbf{y} \rangle = \langle H\mathbf{x}, H\mathbf{y} \rangle.$$
Thus
$$\|\mathbf{x}\|_2 = \|H\mathbf{x}\|_2.$$

5.3.4-15 Determine a rotation θ in $c = \cos\theta$ and $s = \sin\theta$ such that
$$\begin{bmatrix} c & -s \\ s & c \end{bmatrix} \begin{bmatrix} 3 \\ 4 \end{bmatrix} = \begin{bmatrix} 5 \\ 0 \end{bmatrix}.$$

5.3.4-16 Show that if Q is an orthogonal matrix, then
$$\begin{bmatrix} 1 & \mathbf{0} \\ \mathbf{0} & Q \end{bmatrix}$$
is also orthogonal.

5.3.4-17 [114, page 74] Show that if $Q = Q_1 + jQ_2$ is unitary, where $Q_i \in \mathbb{R}^{m \times m}$, then the $2m \times 2m$ matrix
$$Z = \begin{bmatrix} Q_1 & -Q_2 \\ Q_2 & Q_1 \end{bmatrix}$$
is orthogonal.

5.3.4-18 The Householder matrix defined in (5.13) uses a reflection with respect to an orthogonal projection. In this problem we will explore the Householder matrix using a weighted projection and its associated inner product. Let W be a Hermitian matrix, and define (see (2.22))
$$H_{v,W} = I - 2P_{v,W} = I - 2\frac{\mathbf{v}\mathbf{v}^H W}{\mathbf{v}^H W \mathbf{v}}.$$

(a) Show that $H_{v,W}^H W H_{v,W} = W$ and that $\|H_{v,W}\mathbf{x}\|_W = \|\mathbf{x}\|_W$, where $\|\mathbf{x}\|_W = \mathbf{x}^H W \mathbf{x}$.
(b) Show that $H_{v,W}\mathbf{v} = -\mathbf{v}$.
(c) Show that $H_{v,W}H_{v,W} = I$, so $H_{v,W}$ is a reflection.
(d) Determine a means of choosing **v** so that
$$H_{v,W}\mathbf{x} = \begin{bmatrix} 1 \\ 0 \\ \vdots \\ 0 \end{bmatrix} \alpha$$
for some α.

5.3.4-19 Consider the problem $\mathbf{y}^T = \mathbf{x}^T A$.

(a) Determine **x** such that the first component of **y** is maximized, subject to the constraint that $\|\mathbf{x}\|_2 = 1$. What is the maximum value of $y(1)$ in this case?
(b) Let H be a Householder matrix operating on the first column of A. Comment on the nonzero value of the first column HA, compared with $y(1)$ obtained in the previous part.

5.3.4-20 Let **x** and **y** be nonzero vectors in $\mathbb{R}^n$. Determine a Householder matrix P such that $P\mathbf{x}$ is a multiple of **y**. Give a geometric interpretation of your answer.

5.3.4-21 The computation in (5.15) applies the Householder matrix to the *left* of a matrix, as $H_v A$. Develop an efficient means (such as in (5.15)) for computing AH_v, with the Householder matrix on the right.

5.3.4-22 In this problem, you will demonstrate that direct computation of the pseudoinverse is numerically inferior to computation using a matrix factorization such as the QR or Cholesky

5.4 Exercises

factorization. Suppose it is desired find the least-squares solution to

$$A = \begin{bmatrix} 10000 & 10001 \\ 10001 & 10002 \\ 10002 & 10003 \\ 10003 & 10004 \\ 10004 & 10005 \end{bmatrix} \mathbf{x} = \begin{bmatrix} 20001 \\ 20003 \\ 20005 \\ 20007 \\ 20009 \end{bmatrix}.$$

The exact solution is $\mathbf{x} = [1, 1]^T$.

(a) Determine the condition numbers of A and $A^T A$ (if possible).

(b) Compute the least-squares solution using the formula $\hat{\mathbf{x}} = (A^T A)^{-1} A^T \mathbf{b}$ explicitly.

(c) Compute the solution using the QR decomposition, where the QR decomposition is computed using Householder transformations.

(d) Compute the solution using the Cholesky factorization.

(e) Compare the answers, and comment.

5.3.4-23 (Rotation matrices) Verify the stated properties about rotation matrices of the following form, and provide a geometric interpretation:

$$G_\theta = \begin{bmatrix} \cos\theta & -\sin\theta \\ \sin\theta & \cos\theta \end{bmatrix}.$$

(a) $G_\theta G_{-\theta} = I$.

(b) $G_\theta G_\phi = G_{\theta+\phi}$.

Note that $\{G_\theta, \theta \in \mathbb{R}\}$ forms a group.

5.3.4-24 [267] The Grammian matrix for a least-squares problem is

$$R_k = A^H A,$$

where

$$A = \begin{bmatrix} \mathbf{q}[1]^H \\ \mathbf{q}[2]^H \\ \vdots \\ \mathbf{q}[k]^H \end{bmatrix}.$$

Let $Z = A^H$, so we can write $R_k = ZZ^H$.

(a) Show that if $Z_1 = ZQ_1$, where Q_1 is a unitary matrix, then we can write $R_k = Z_1 Z_1^H$.

(b) Describe how to convert Z to a lower-triangular matrix L by a series of orthogonal transformations. Thus, we can write

$$R_k = LL^H.$$

(c) Describe how to solve the equation $R_k \mathbf{x} = \mathbf{y}$ for $\mathbf{y}$ based upon this representation of R_k.

Note that since we never have to compute R explicitly, the numerical problems of computing $A^H A$ never arise. For fixed-point arithmetic, the wordlength requirements are approximately half that of computing the Grammian and then factoring it [268].

5.3.4-25 Determine a representation of $\theta = 23°$ using the angles in the CORDIC representation.

5.3.4-26 We have seen that in the LU factorization, it is possible to overwrite the original matrix A with information about the L and U factors (with possibly some permutation information stored separately). In this exercise we determine that the same overwriting representation of A also works for Householder and Givens approaches to the QR factorization.

(a) Determine a means by which the Q and R factors computed using Householder transformations can be overwritten in the original A matrix. Hint: let $\mathbf{v}(1) = 1$.

(b) Determine how the Q and R factors computed using Givens transformations can be overwritten in the original A matrix.

5.3.4-27 (Fast Givens transformations) Let D be a diagonal matrix, let M be a matrix such that $M^T M = D$, and let $Q = MD^{-1/2}$.

(a) Show that Q is orthogonal.

(b) For a 2×2 matrix M_1 of the form

$$M_1 = \begin{bmatrix} \beta & 1 \\ 1 & \alpha \end{bmatrix},$$

show how to choose α and β so that for a 2-vector $\mathbf{x}$,

$$M_1 \mathbf{x} = \begin{bmatrix} \times \\ 0 \end{bmatrix}$$

(that is, M sets the second component of $\mathbf{x}$ to zero), and

$$M_1 D M_1^H = D_1$$

is diagonal. Thus $M_1 \mathbf{x}$ acts like a Givens rotation, but without the need to compute a square root.

(c) Describe how to apply the 2×2 matrix to perform a "fast QR" decomposition of a matrix A.

Further information on fast Givens, including some important issues of stabilizing the numerical computations, are given in [114].

5.3.4-28 In relation to figure 5.2, it was stated that the diagonals of an equilateral parallelogram are orthogonal. Prove that this is true.

5.3.4-29 (Matrix spaces from the QR factorization.) If $A \in M_{m,n}$, where $m > n$ and A has full column rank, the QR factorization can be written as

$$A = [Q_1 Q_2] \begin{bmatrix} R_1 \\ 0 \end{bmatrix},$$

where $Q_1 \in M_{m,n}$, $Q_2 \in M_{m,m-n}$, and $R_1 \in M_{nn}$. Show that:

(a) $A = Q_1 R_1$. (This is known as the "skinny" QR factorization.) Observe that the columns of Q_1 are orthogonal.

(b) $\mathcal{R}(A) = \mathcal{R}(Q_1)$.

(c) $\mathcal{R}(A)^\perp = \mathcal{R}(Q_2)$.

5.5 References

Computation of matrix factorizations is widely discussed in a variety of numerical analysis texts. The connection of the LU with Gaussian elimination is described well in [333]. Most of the material on the QR factorization has been drawn from [114]. In addition to factorizations, this source also provides perturbation analyses of the algorithms, and comparisons of variants of the algorithms. A "fast" Givens rotation algorithm that does not require square roots is also presented there, and variants on the Cholesky algorithm presented here are presented in [114] as well. Update algorithms for the QR factorization, in addition to the one for update by adding a row, and including updates for a rank-one modification, and column modifications, are also presented in [114].

The Householder transformation appeared in [144]. Application of Householder transformations with weighted projections is discussed in [268]. Application of QR factorizations to least-squares filtering is extensively discussed in [263] and [132]. [263] also demonstrates application of Gram–Schmidt and modified Gram–Schmidt to least squares and recursive updates of least squares. A discussion of applications of Householder transforms to signal processing appears in [321].

Chapter 6

Eigenvalues and Eigenvectors

> ... neither Heisenberg nor Born knew what to make of the appearance of matrices in the context of the atom. (David Hilbert is reported to have told them to go look for a differential equation with the same eigenvalues, if that would make them happier. —They did not follow Hilbert's well-meant advice and thereby may have missed discovering the Schrödinger wave equation.)
> — *Manfred Schroeder*
> *Number Theory in Science and Communication*

6.1 Eigenvalues and linear systems

The word "eigen" is a German word that can be translated as "characteristic." The eigenvalues of a linear operator are those values which characterize the modes of the operator. Being characteristic, the eigenvalues and associated eigenvectors of a system indicate something that is intrinsic and invariant in the system.

Example 6.1.1 To motivate this description somewhat, consider the following coupled difference equations:

$$y_1[t+1] = -y_1[t] - 1.5y_2[t], \tag{6.1}$$

$$y_2[t+1] = 0.5y_1[t] + y_2[t], \tag{6.2}$$

which can be written in matrix form as

$$\mathbf{y}[t+1] = \begin{bmatrix} -1 & -1.5 \\ 0.5 & 1 \end{bmatrix} \mathbf{y}[t] = A\mathbf{y}[t],$$

where $\mathbf{y}[t] = [y_1[t], y_2[t]]^T$. It is desired to find a solution to these equations. The form of the equations suggests that a good candidate solution is

$$y_1[t] = \lambda^t x_1 \qquad y_2[t] = \lambda^t x_2$$

for some λ, x_1, and x_2 to be determined. Substitution of these candidate solutions into the equation gives

$$\lambda^{t+1} x_1 = -\lambda^t x_1 - 1.5\lambda^t x_2, \tag{6.3}$$

$$\lambda^{t+1} x_2 = 0.5\lambda^t x_1 + \lambda^t x_2, \tag{6.4}$$

which can be written more conveniently as

$$A \begin{bmatrix} x_1 \\ x_2 \end{bmatrix} = \lambda \begin{bmatrix} x_1 \\ x_2 \end{bmatrix}$$

or
$$Ax = \lambda x \qquad (6.5)$$

Equation (6.5) is the equation of interest in eigenvalue problems. The difference equation has been reduced to an algebraic equation, where we wish to solve for λ and x. The scalar quantity λ is called the **eigenvalue** of the equation, and the vector x is called the **eigenvector** of the equation.

Equation (6.5) may be regarded as an operator equation. The eigenvectors of A are those vectors that are not changed in direction by the operation of A: they are simply scaled by the amount λ. This is illustrated in figure 6.1. A vector is an eigenvector if it is not modified in direction, only in magnitude, when operated on by A. The vectors thus form an *invariant* of the operator A. This is fully analogous with the concept from the theory of linear time-invariant systems, either in continuous time or discrete time. The steady-state output of an LTI system to a sinusoidal input is a sinusoidal signal at the *same frequency*, but with possibly different amplitude and phase. The system preserves the frequency of the signal (analogous to preserving the direction of a vector) while modifying its amplitude. Sinusoidal signals are therefore sometimes referred to as the *eigenfunctions* of an LTI system. In the study of linear operators, searching for their eigenfunctions is an important first step to understanding what the operators do.

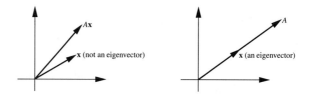

Figure 6.1: The direction of eigenvectors is not modified by A

Definition 6.1 A *nonzero* vector x is called a **right eigenvector** for the eigenvalue λ if $Ax = \lambda x$, and a **left eigenvector** if $x^H A = \lambda x^H$. Unless otherwise stated, "eigenvector" means "right eigenvector." □

Equation (6.5) can be written in the form
$$(A - \lambda I)x = 0. \qquad (6.6)$$

One solution of (6.6) is the solution $x = 0$. This is known as the trivial solution and is not of much interest. The other way that a solution may be obtained is to make sure that x is in the nullspace of $A - \lambda I$, which means that we must make sure that $A - \lambda I$ actually has a nontrivial nullspace. The particular values of λ that cause $A - \lambda I$ to have a nontrivial nullspace are the eigenvalues of A, and the corresponding vectors in the null space are the eigenvectors. In order to have a nontrivial null space, the matrix $A - \lambda I$ must be singular. The values of λ that cause $A - \lambda I$ to be singular are precisely the eigenvalues of A. As discussed in section 4.6.1, we can determine if a matrix is singular by examining its determinant.

Definition 6.2 The polynomial $\chi_A(\lambda) = \det(\lambda I - I)$ is called the **characteristic polynomial** of A. The equation $\det(\lambda I - A) = 0$ is called the characteristic equation of A. The eigenvalues of A are the roots of the characteristic equation. The set of roots of the characteristic polynomial is called the **spectrum** of A, and is denoted $\lambda(A)$. □

6.1 Eigenvalues and Linear Systems

Example 6.1.2 For the matrix A of example 6.1.1, the eigenvalues can be found from

$$\det(A - \lambda I) = \det \begin{bmatrix} -1 - \lambda & -1.5 \\ 0.5 & 1 - \lambda \end{bmatrix} = (-1 - \lambda)(1 - \lambda) + (0.5)(1.5) = 0.$$

Expanding the determinant we obtain

$$\lambda^2 - .25 = 0,$$

which has roots $\lambda = 0.5$ or $\lambda = -0.5$. □

In the study of LTI systems, the characteristic polynomial appears in the denominators of transfer functions. The dynamics of the system are therefore governed by the roots of the characteristic polynomial—the eigenvalues. This is one reason why the eigenvalues are of interest in signal processing.

Example 6.1.3 The LTI system described by the difference equation

$$\mathbf{x}[t + 1] = A\mathbf{x}[t] + B\mathbf{u}[t], \tag{6.7}$$

$$\mathbf{y}[t] = C\mathbf{x}[t], \tag{6.8}$$

has the Z-transform (see section 1.4)

$$H[z] = C(zI - A)^{-1}B.$$

The matrix inverse can be written as

$$H[z] = C \,\text{adj}(zI - A) B \frac{1}{\det(zI - A)}.$$

The notation $\text{adj}(zI - A)$ indicates the *adjugate* of the matrix $zI - A$ (not to be confused with the adjoint). The adjugate is introduced in section C.5.3. The denominator is the characteristic equation of A, and the poles of the system are the eigenvalues of the matrix A. □

Often, eigenvalues are found using an iterative numerical procedure. Once the eigenvalues are found, the eigenvectors are determined by finding vectors in the nullspace of $A - \lambda I$.

Example 6.1.4 For the system of example 6.1.1, we have found the eigenvalues to be $\lambda = \pm 0.5$. To find the eigenvectors, substitute the eigenvalues individually into (6.6) and find the vectors in the nullspace. When $\lambda = 0.5$, we get

$$\begin{bmatrix} -1.5 & -1.5 \\ 0.5 & 0.5 \end{bmatrix} \mathbf{x}_1 = 0.$$

It is clear that $\mathbf{x}_1 = [1, -1]^T$ will satisfy this equation, as will any multiple of this. *The eigenvectors are only determined up to a nonzero scalar constant.* The eigenvectors can be scaled to different magnitudes. Often it is convenient to scale the vectors so they have unit norm. This would lead to the vector $\mathbf{x}_1 = [1/\sqrt{2}, 1/\sqrt{2}]^T$.

For the other eigenvector, substitute $\lambda = -0.5$ into (6.6),

$$\begin{bmatrix} -0.5 & -1.5 \\ 0.5 & 1.5 \end{bmatrix} \mathbf{x}_2 = 0.$$

A solution is $\mathbf{x}_2 = [-3, 1]^T$. Scaling to have unit norm provides the solution $\mathbf{x}_2 = [-3/\sqrt{10}, 1/\sqrt{10}]^T$.

We have determined the eigenvalues and eigenvectors of the system defined in (6.2) and have actually come up with two solutions, one for each eigenvalue. When $\lambda = 0.5$ a solution is

$$\begin{bmatrix} y_1[t] \\ y_2[t] \end{bmatrix} = (0.5)^t \begin{bmatrix} 1 \\ -1 \end{bmatrix} \frac{1}{\sqrt{2}} = (0.5)^t \mathbf{x}_1,$$

and when $\lambda = -0.5$ a solution is

$$\begin{bmatrix} y_1[t] \\ y_2[t] \end{bmatrix} = (-0.5)^t \begin{bmatrix} -3 \\ 1 \end{bmatrix} \frac{1}{\sqrt{10}} = (-0.5)^t \mathbf{x}_2.$$

What do we do with this wealth of solutions? Since the system is linear, the response due to the sum of several inputs is the sum of the responses, so we can take linear combinations of these solutions for a total solution:

$$\mathbf{y}[t] = c_1 (0.5)^t \mathbf{x}_1 + c_2 (-0.5)^t \mathbf{x}_2.$$

The constants c_1 and c_2 can be found to match auxiliary conditions on the system, such as initial conditions. Note that in this solution, the behavior of the system is governed by the eigenvalues: there is one "mode" that goes as $(0.5)^t$ and another mode that goes as $(-0.5)^t$. □

6.2 Linear dependence of eigenvectors

The eigenvectors of a matrix are often used as a set of basis vectors for some space. In order to be able to say something about the dimensionality of the space spanned by the eigenvectors, it is important to tell when the eigenvectors are linearly independent. The first lemma provides part of the story.

Lemma 6.1 *If the eigenvalues of an $m \times m$ matrix A are all distinct, then the eigenvectors of A are linearly independent.*

Proof Start with $m = 2$ and assume, to the contrary, that the eigenvectors are linearly dependent. Then there exist constants c_1 and c_2 such that

$$c_1 \mathbf{x}_1 + c_2 \mathbf{x}_2 = 0. \tag{6.9}$$

Multiply by A to obtain

$$c_1 A \mathbf{x}_1 + c_2 A \mathbf{x}_2 = c_1 \lambda_1 \mathbf{x}_1 + c_2 \lambda_2 \mathbf{x}_2 = 0.$$

Now take λ_2 times equation (6.9), and subtract it from the last equation to obtain

$$c_1 (\lambda_1 - \lambda_2) \mathbf{x}_1 = 0.$$

Since $\lambda_1 \neq \lambda_2$ and $\mathbf{x}_1 \neq 0$, this means that $c_1 = 0$. Similarly it can be shown that $c_2 = 0$. The two vectors must be linearly independent.

Generalization to the case for $m > 2$ proceeds similarly. □

If the eigenvalues are not distinct, then the eigenvectors may or may not be linearly independent. The matrix $A = I$ has m repeated eigenvalues $\lambda = 1$, and n linearly independent eigenvectors can be chosen. On the other hand, the matrix

$$A = \begin{bmatrix} 4 & 2 \\ 0 & 4 \end{bmatrix}$$

has repeated eigenvalues of 4, 4 and both eigenvectors proportional to $\mathbf{x} = [1, 0]^T$: they are linearly dependent.

6.3 Diagonalization of a matrix

In this section we introduce a factorization of a matrix A as

$$A = S\Lambda S^{-1},$$

where S is a diagonal or mostly diagonal matrix. We will begin by assuming that the $m \times m$ matrix A has m linearly independent eigenvectors. Let the eigenvectors of A be $\mathbf{x}_1, \mathbf{x}_2, \ldots, \mathbf{x}_m$, so that

$$A\mathbf{x}_i = \lambda_i \mathbf{x}_i \qquad i = 1, 2, \ldots, m.$$

These equations can be stacked side-by-side to obtain

$$[A\mathbf{x}_1 \quad A\mathbf{x}_2 \quad \cdots \quad A\mathbf{x}_m] = [\lambda_1 \mathbf{x}_1 \quad \lambda_2 \mathbf{x}_2 \quad \cdots \quad \lambda_m \mathbf{x}_m].$$

The stacked matrix on the left can be written as

$$[A\mathbf{x}_1 \quad A\mathbf{x}_2 \quad \cdots \quad A\mathbf{x}_m] = A[\mathbf{x}_1 \quad \mathbf{x}_2 \quad \cdots \quad \mathbf{x}_m] \tag{6.10}$$

and the stacked matrix on the right can be written as

$$[\lambda_1 \mathbf{x}_1 \quad \lambda_2 \mathbf{x}_2 \quad \cdots \quad \lambda_m \mathbf{x}_m] = [\mathbf{x}_1 \quad \mathbf{x}_2 \quad \cdots \quad \mathbf{x}_m] \begin{bmatrix} \lambda_1 & & & \\ & \lambda_2 & & \\ & & \ddots & \\ & & & \lambda_m \end{bmatrix}.$$

Let S be the side-by-side stacked matrix of eigenvectors, and let Λ be the diagonal matrix formed from the eigenvalues:

$$S = [\mathbf{x}_1 \quad \mathbf{x}_2 \quad \cdots \quad \mathbf{x}_n] \qquad \Lambda = \text{diag}(\lambda_1, \lambda_2, \ldots, \lambda_n).$$

Then (6.10) can be written as

$$\boxed{AS = S\Lambda}$$

This equation is true whether or not the eigenvectors are linearly independent. However, if the eigenvectors *are* linearly independent, then S is full rank and invertible, and we can write

$$\boxed{A = S\Lambda S^{-1}} \tag{6.11}$$

or

$$\boxed{\Lambda = S^{-1}AS} \tag{6.12}$$

This is said to be a **diagonalization** of A, and a matrix which has a diagonalization is said to be **diagonalizable**.

The particular form of the transformation from A to Λ arises in a variety of contexts. More generally, if there are matrices A and B with an invertible matrix T such that

$$\boxed{A = TBT^{-1}} \tag{6.13}$$

then A and B are said to be **similar**. It can be shown that

$$\boxed{\text{If } A \text{ and } B \text{ are similar, then they have the same eigenvalues.}}$$

The diagonalization (6.11) shows that A and Λ are similar and, hence, have the same eigenvalues. (This is clear in this case, since the eigenvalues of A appear on the diagonal of Λ.) Other transformations can be used to find matrices similar to A, but the similar matrices will not be diagonal unless they are formed from the eigenvectors of A.

There are a variety of uses for the factorization $A = S\Lambda S^{-1}$. One simple example is that powers of A are easy to compute. For example,

$$A^2 = (S\Lambda S^{-1})(S\Lambda S^{-1}) = S\Lambda^2 S^{-1},$$

and, more generally,

$$\boxed{A^n = S\Lambda^n S^{-1}} \tag{6.14}$$

This allows a means for defining functions operating on matrices. For a function $f(x)$ with the power series representation

$$f(t) = \sum_i f_i t^i,$$

the function operating on a diagonalizable matrix can be defined as

$$f(A) = \sum_i f_i A^i = S\left(\sum_i f_i \Lambda^i\right) S^{-1}.$$

Since Λ is diagonal, Λ^i is obtained simply by computing the elements on the diagonal. An important example of this is

$$e^A = \sum_{i=0}^\infty \frac{A^i}{i!} = S\left(\sum_{i=0}^\infty \frac{\Lambda^i}{i!}\right) S^{-1} = Se^\Lambda S^{-1},$$

where

$$e^\Lambda = \begin{bmatrix} e^{\lambda_1} & & & \\ & e^{\lambda_2} & & \\ & & \ddots & \\ & & & e^{\lambda_m} \end{bmatrix}.$$

Example 6.3.1 Let

$$A = \begin{bmatrix} 23 & -6 \\ -18 & 26 \end{bmatrix}.$$

Then A has the eigendecomposition

$$\lambda_1 = 14 \quad \mathbf{x}_1 = \begin{bmatrix} -.5547 \\ -.8321 \end{bmatrix},$$

$$\lambda_2 = 35 \quad \mathbf{x}_2 = \begin{bmatrix} .4472 \\ -.8944 \end{bmatrix}.$$

Then

$$e^A = \begin{bmatrix} -0.5547 & 0.447214 \\ -0.83205 & -0.894427 \end{bmatrix} \begin{bmatrix} e^{14} & \\ & e^{35} \end{bmatrix} \begin{bmatrix} -0.5547 & 0.447214 \\ -0.83205 & -0.894427 \end{bmatrix}.$$

The MATLAB function expm computes the matrix exponential. This computation arises frequently enough in practice that considerable effort has been dedicated to effective numerical solutions. A treatment of these can be found in [230], of which the method presented here is but one example. □

The homogeneous vector differential equation $\dot{\mathbf{x}}(t) = A\mathbf{x}(t)$ has the solution $\mathbf{x}(t) = e^{At}\mathbf{x}_0$, and the homogeneous vector difference equation $\mathbf{x}[t+1] = A\mathbf{x}[t]$ has the solution $\mathbf{x}[t] = A^t \mathbf{x}_0$. In light of the diagonalization discussed, the differential equation is stable if the eigenvalues of A are in the left-half plane, and the difference equation is stable if the eigenvalues of A are inside the unit circle.

6.3 Diagonalization of a Matrix

6.3.1 The Jordan form

If A has repeated eigenvalues then it is not always possible to diagonalize A. If the eigenvectors are linearly independent, however, then even with repeated eigenvalues, A can be diagonalized. If some of the eigenvectors are linearly dependent, then A cannot be exactly diagonalized. Instead, a matrix which is nearly diagonal is found to which A is similar. This matrix is known as the *Jordan form* of A.

Theorem 6.1 *(Jordan form) An $m \times m$ matrix A with $k \leq m$ linearly independent eigenvectors can be written as*

$$A = TJT^{-1},$$

where J is a block-diagonal matrix,

$$J = \begin{bmatrix} J_1 & & & \\ & J_2 & & \\ & & \ddots & \\ & & & J_k \end{bmatrix}.$$

The blocks J_i are known as Jordan blocks. *Each Jordan block is of the form*

$$J_i = \begin{bmatrix} \lambda_i & 1 & & & \\ & \lambda_i & 1 & & \\ & & \ddots & \ddots & \\ & & & \lambda_i & 1 \\ & & & & \lambda_i \end{bmatrix}.$$

If J_i is $l \times l$, then the eigenvalue λ_i is repeated l times along the diagonal, and 1 appears $l - 1$ times above the diagonal. Two matrices are similar if they have the same Jordan form.

An inductive proof of this theorem appears in [333, appendix B]. Rather than reproduce the proof here, we consider some examples and applications.

Example 6.3.2

1. The matrix

$$A = \begin{bmatrix} 4 & 1 & 3 \\ 0 & 4 & 1 \\ 0 & 0 & 4 \end{bmatrix}$$

   has a single eigenvalue $\lambda = 4$, and all three eigenvectors are the same, $\mathbf{x}_1 = \mathbf{x}_2 = \mathbf{x}_3 = [1, 0, 0]^T$. There is thus a single Jordan block, and A is similar to

$$J = \begin{bmatrix} 4 & 1 & 0 \\ 0 & 4 & 1 \\ 0 & 0 & 4 \end{bmatrix}.$$

2. The matrix

$$B = \begin{bmatrix} 3 & 0 & 1 \\ 0 & 3 & 0 \\ 0 & 0 & 3 \end{bmatrix}$$

   has a single eigenvalue $\lambda = 3$ and two eigenvectors,

$$\mathbf{x}_1 = [1, 0, 0]^T \quad \text{and} \quad \mathbf{x}_2 = [0, 1, 0]^T.$$

The Jordan form has two Jordan blocks,

$$J_1 = \begin{bmatrix} 3 & 1 \\ 0 & 3 \end{bmatrix} \quad \text{and} \quad J_2 = 3,$$

so

$$J = \begin{bmatrix} 3 & 1 & 0 \\ 0 & 3 & 0 \\ 0 & 0 & 3 \end{bmatrix}.$$

□

If A has the Jordan form representation

$$A = SJS^{-1},$$

then

$$A^n = SJ^n S^{-1}$$

and

$$e^{At} = Se^{Jt} S^{-1},$$

but computing J^n is somewhat more complicated if J is not strictly diagonal. As an example, for a 3×3 Jordan block,

$$\begin{bmatrix} \lambda & 1 & 0 \\ 0 & \lambda & 1 \\ 0 & 0 & \lambda \end{bmatrix}^n = \begin{bmatrix} \lambda^n & n\lambda^{n-1} & \frac{1}{2}n(n-1)\lambda^{n-2} \\ 0 & \lambda^n & n\lambda^{n-1} \\ 0 & 0 & \lambda^n \end{bmatrix}. \quad (6.15)$$

The presence of terms which grow as a polynomial function of n can be understood by comparison with repeated roots in a differential or difference equation: the repeated roots give rise to solutions of the form $te^{\lambda t}$ for the differential equation, and $t\lambda^t$ for the difference equation.

Example 6.3.3 A signal has transfer function

$$Y(z) = \frac{3z^2 + .3z}{(z - .3)^2},$$

with time function

$$y[t] = 3(.3)^t u[t] + 4t(.3)^t u[t].$$

Placing the system into state-variable form (as in (1.20)), we find

$$A = \begin{bmatrix} 0 & 1 \\ -0.09 & .6 \end{bmatrix},$$

which has repeated eigenvalues $\lambda = .3$, and only a single eigenvector. The presence of the linearly growing term $4t(.3)^t$ is equivalent to the fact that the Jordan form for A is not strictly diagonal. □

6.3.2 Diagonalization of self-adjoint matrices

Hermitian symmetric matrices arise in a variety of contexts as a mathematical representation of symmetric interactions: if a affects b as much as b affects a, then a matrix describing their interactions will be symmetric. Throughout this section we employ inner-product notation interspersed with more traditional matrix notation, to reinforce its use and to avoid, as much as possible, having to say "symmetric or Hermitian." As discussed in section 4.3, self-adjoint matrices are matrices for which

$$\langle A\mathbf{x}, \mathbf{x} \rangle = \langle \mathbf{x}, A^*\mathbf{x} \rangle.$$

6.3 Diagonalization of a Matrix

Self-adjoint matrices are symmetric if the elements are real: $A^T = A$; and are Hermitian if the elements are complex: $A^H = A$. The first useful result concerning self-adjoint matrices is that their eigenvalues are real.

Lemma 6.2

> *The eigenvalues of a self-adjoint matrix are real.*

Proof Let λ and $\mathbf{x}$ be an eigenvalue and eigenvector of a self-adjoint matrix A. Then

$$\langle A\mathbf{x}, \mathbf{x}\rangle = \lambda \langle \mathbf{x}, \mathbf{x}\rangle \tag{6.16}$$

and

$$\langle \mathbf{x}, A^*\mathbf{x}\rangle = \overline{\lambda} \langle \mathbf{x}, \mathbf{x}\rangle. \tag{6.17}$$

Since $\langle A\mathbf{x}, \mathbf{x}\rangle = \langle \mathbf{x}, A^*\mathbf{x}\rangle$, we must have $\lambda = \overline{\lambda}$, so λ is real. $\square$

Lemma 6.3

> *For a self-adjoint matrix, the eigenvectors corresponding to distinct eigenvalues are orthogonal.*

Proof Let λ_1 and λ_2 be distinct eigenvalues of a self-adjoint matrix A, with corresponding eigenvectors $\mathbf{x}_1$ and $\mathbf{x}_2$. Then

$$\langle A\mathbf{x}_1, \mathbf{x}_2\rangle = \langle \mathbf{x}_1, A^*\mathbf{x}_2\rangle = \langle \mathbf{x}_1, \lambda_2\mathbf{x}_2\rangle = \lambda_2 \langle \mathbf{x}_1, \mathbf{x}_2\rangle.$$

We also have

$$\langle A\mathbf{x}_1, \mathbf{x}_2\rangle = \lambda_1 \langle \mathbf{x}_1, \mathbf{x}_2\rangle,$$

so that

$$(\lambda_1 - \lambda_2)\langle \mathbf{x}_1, \mathbf{x}_2\rangle = 0.$$

Since $\lambda_1 \neq \lambda_2$, we must have $\mathbf{x}_1 \perp \mathbf{x}_2$. $\square$

We have already observed for Hermitian matrices with distinct eigenvalues that diagonalization is possible, and the unitary diagonalizing matrix U is simply formed from the eigenvectors of A. However, this theorem is true *even for matrices with repeated eigenvalues*. This theorem is known as the *spectral theorem*, and the set of eigenvalues of a Hermitian matrix is known as its *spectrum*.

Theorem 6.2 *Every Hermitian $m \times m$ matrix A can be diagonalized by a unitary matrix:*

$$U^H A U = \Lambda, \tag{6.18}$$

where U is unitary and Λ is diagonal.

It follows that every real symmetric matrix A can be diagonalized by an orthogonal matrix:

$$Q^T A Q = \Lambda. \tag{6.19}$$

When A has distinct eigenvalues, theorem (6.18) is immediate, in light of the discussion in the previous section. However, the result is true even when A has repeated eigenvalues. We can write (6.18) as

$$A = U \Lambda U^H = \sum_{i=1}^{m} \lambda_i \mathbf{u}_i \mathbf{u}_i^H. \tag{6.20}$$

The proof of theorem 6.2 is left as an exercise (6.3-24). The proof follows in two simple steps from the following key lemma, which is interesting in its own right. It should be observed that this lemma applies not only to Hermitian matrices, but to *any square matrix*.

Lemma 6.4 *(Schur's lemma) For any square matrix A, there is a unitary matrix U such that*

$$U^H A U = T,$$

where T is upper triangular. Every matrix is similar to an upper-triangular matrix.

Observe that since the eigenvalues of a diagonal matrix appear on the diagonal, this lemma provides one method of computing the eigenvalues of any matrix.

Proof The proof is constructive. For typographical convenience the lemma will be demonstrated using a 3×3 matrix; extension to an arbitrary square matrix is straightforward. Let A be a 3×3 matrix. It must have at least one eigenvalue λ_1 (which may be repeated, but this does not matter) and a corresponding eigenvector $\mathbf{u}_1$, which we assume to be normalized to a unit vector. By the Gram–Schmidt process it is possible to find two unit vectors $\mathbf{x}_{1,2}$, $\mathbf{x}_{1,3}$ that are orthogonal to $\mathbf{u}_1$ and form a unitary matrix U_1 with $\mathbf{u}_1$ in the first column. Then

$$AU_1 = A[\mathbf{u}_1 \quad \mathbf{x}_{1,2} \quad \mathbf{x}_{1,3}] = U_1 \begin{bmatrix} \lambda_1 & \times & \times \\ 0 & \times & \times \\ 0 & \times & \times \end{bmatrix} = U_1 \begin{bmatrix} \lambda_1 & \times & \times \\ 0 & & \\ 0 & & A_2 \end{bmatrix},$$

where $\times$ denotes an element which takes on an arbitrary value. Now consider the 2×2 matrix A_2 in the lower right of the matrix on the right. It also has at least one eigenvalue λ_2 and a corresponding eigenvector $\mathbf{u}_{2,2}$. Again using Gram–Schmidt, a 2×2 unitary matrix M_2 can be constructed,

$$M_2 = [\mathbf{u}_{2,2}, \mathbf{x}_{2,3}],$$

so that

$$A_2 M_2 = \begin{bmatrix} \lambda_2 & \times \\ 0 & \times \end{bmatrix}.$$

Then a 3×3 unitary matrix can be constructed by

$$U_2 = \begin{bmatrix} 1 & 0 & 0 \\ 0 & & \\ 0 & & M_2 \end{bmatrix}.$$

Then

$$AU_1 U_2 = U_2 U_1 \begin{bmatrix} \lambda_1 & \times & \times \\ 0 & \lambda_2 & \times \\ 0 & 0 & \times \end{bmatrix},$$

which is upper triangular. The matrix $U = U_1 U_2$ is unitary, so the theorem is proved (for the 3×3 case). $\square$

Lemma 6.5 *Let A be an $m \times m$ matrix of rank $r < m$. Then at least $m - r$ of the eigenvalues of A are equal to zero.*

The proof is required in exercise 6.3-25.

Example 6.3.4 Let

$$A = \begin{bmatrix} 1 & 0 & 0 \\ 0 & 0 & 1 \\ 0 & 1 & 0 \end{bmatrix},$$

6.3 Diagonalization of a Matrix

which has eigenvalues $\lambda_1 = \lambda_2 = 1$ and $\lambda_3 = -1$. Following the steps in the proof of lemma 6.4, we first find an eigenvector of A corresponding to $\lambda_1 = 1$,

$$\mathbf{u}_1 = \begin{bmatrix} 1 \\ 0 \\ 0 \end{bmatrix}.$$

Two vectors that are orthogonal to this are $\mathbf{x}_{1,2} = \mathbf{e}_2$ and $\mathbf{x}_{1,3} = \mathbf{e}_3$, giving $U_1 = I$. Then the 2×2 matrix A_2 in the lower-right corner of AU_1 is

$$A_2 = \begin{bmatrix} 0 & 1 \\ 1 & 0 \end{bmatrix},$$

which has an eigenvalue of $\lambda_2 = 1$ with a corresponding eigenvector $\mathbf{u}_{2,2} = \frac{1}{\sqrt{2}}[1, 1]^T$. Then

$$M_2 = \frac{1}{\sqrt{2}} \begin{bmatrix} 1 & 1 \\ 1 & -1 \end{bmatrix}$$

and

$$U = U_1 U_2 = \begin{bmatrix} 1 & 0 & 0 \\ 0 & \frac{1}{\sqrt{2}} & \frac{1}{\sqrt{2}} \\ 0 & \frac{1}{\sqrt{2}} & -\frac{1}{\sqrt{2}} \end{bmatrix} = [\mathbf{u}_1, \mathbf{u}_2, \mathbf{u}_3].$$

Thus, A has the representation

$$A = U \Lambda U^T = \sum_{i=1}^{3} \lambda_i \mathbf{u}_i \mathbf{u}_i^T$$

$$= \lambda_1 \begin{bmatrix} 1 & 0 & 0 \\ 0 & 0 & 0 \\ 0 & 0 & 0 \end{bmatrix} + \lambda_2 \begin{bmatrix} 0 & 0 & 0 \\ 0 & \frac{1}{2} & \frac{1}{2} \\ 0 & \frac{1}{2} & \frac{1}{2} \end{bmatrix} + \lambda_3 \begin{bmatrix} 0 & 0 & 0 \\ 0 & \frac{1}{2} & -\frac{1}{2} \\ 0 & -\frac{1}{2} & \frac{1}{2} \end{bmatrix}$$

$$= \lambda_1 \begin{bmatrix} 1 & 0 & 0 \\ 0 & \frac{1}{2} & \frac{1}{2} \\ 0 & \frac{1}{2} & \frac{1}{2} \end{bmatrix} + \lambda_2 \begin{bmatrix} 0 & 0 & 0 \\ 0 & \frac{1}{2} & -\frac{1}{2} \\ 0 & -\frac{1}{2} & \frac{1}{2} \end{bmatrix}$$

$$= \lambda_1 P_1 + \lambda_3 P_2, \tag{6.21}$$

since $\lambda_1 = \lambda_2$.

We see subsequently that the matrices P_1 and P_2 that appear in (6.21) are in fact projection matrices and that P_1 and P_2 are orthogonal: $P_1^T P_2 = 0$. P_1 projects onto the space spanned by the vectors $\{[1, 0, 0]^T, [0, 1, 1]^T\}$, the eigenvectors corresponding to the eigenvalue $\lambda = 1$; and P_2 projects onto the space spanned by $[0, 1, -1]$, the eigenvector corresponding to the eigenvalue $\lambda = -1$. □

The diagonalization $A = U \Lambda U^H$ illustrates an important principle, that of finding an appropriate coordinate system in which to solve a problem. Many problems in mathematics can be simplified by expressing them in an appropriate orthogonal coordinate system, where the global problem can be addressed as a series of scalar problems. This is one reason why efforts are made to find sets orthogonal basis functions, as described in chapter 3. The convolution theorem, which states that the transform of a convolution is the product of the transforms, is another example of the application of this concept. Rather than convolving two signals, which involves more-or-less global interaction of the signals, the signals are represented in a transform domain (a new coordinate system) where the convolution can be represented as multiplication. The importance of this in real signal processing is profound, as this leads to fast convolution using the FFT. Exercise 6.3-28 examines this topic in more detail.

Sylvester's Law of Inertia

Definition 6.3 Let A be a Hermitian matrix, with $\lambda_+(A)$ positive eigenvalues, $\lambda_-(A)$ negative eigenvalues, and $\lambda_0(A)$ zero eigenvalues. The **inertia** of A is the triple

$$(\lambda_+(A), \lambda_-(A), \lambda_0(A))$$

(the number of positive, negative, and zero eigenvalues). The **signature** of A is $\lambda_+(A) - \lambda_-(A)$. □

Theorem 6.3 *(Sylvester's law of inertia) Let A and B be $m \times m$ Hermitian matrices. Then there is a nonsingular matrix S such that $A = SBS^H$ if and only if A and B have the same inertia.*

Proof [142] The converse is presented as an exercise.

Suppose that $A = SBS^H$ for some nonsingular matrix S. Then

$$\text{rank}(A) = \text{rank}(SBS^H) = \text{rank}(B),$$

so $\lambda_0(A) = \lambda_0(B)$. It remains to show that $\lambda_+(A) = \lambda_+(B)$. Let $\mathbf{u}_1, \mathbf{u}_2, \ldots, \mathbf{u}_{\lambda_+(A)}$ be the orthonormal eigenvectors of A corresponding to the positive eigenvalues of A, which we denote as $\lambda_1(A), \lambda_2(A), \ldots, \lambda_{\lambda_+(A)}(A)$. Let

$$S_+(A) = \text{span}(\mathbf{u}_1, \mathbf{u}_2, \ldots, \mathbf{u}_{\lambda_+(A)}).$$

Then $\dim(S_+(A)) = \lambda_+(A)$. Let $\mathbf{x} = \alpha_1 \mathbf{u}_1 + \cdots + \alpha_{\lambda_+(A)} \mathbf{u}_{\lambda_+(A)} \neq 0$. Then

$$\mathbf{x}^H A \mathbf{x} = \lambda_1(A)|\alpha_1|^2 + \cdots + \lambda_{\lambda_+(A)}|\alpha_{\lambda_+(A)}|^2 > 0.$$

We also have

$$\mathbf{x}^H SBS^H \mathbf{x} = (S^H \mathbf{x})^H B (S^H \mathbf{x}) > 0,$$

so $\mathbf{y}^H B \mathbf{y} > 0$ for all nonzero vectors $\mathbf{y} \in \text{span}(S^H \mathbf{v}_1, \ldots, S^H \mathbf{v}_{\lambda_+(A)}\}$, which has dimension $\lambda_+(A)$. Then (see exercise 6.5-34), B must have at least $\lambda_+(A)$ eigenvalues, $\lambda_+(B) \geq \lambda_+(A)$. Reversing the roles of A and B in this argument, we see that $\lambda_+(A) = \lambda_+(B)$. □

6.4 Geometry of invariant subspaces

Definition 6.4 Let A be a matrix. If $S \subset \mathcal{R}(A)$ is such that $\mathbf{x} \in S$ means that $A\mathbf{x} \in S$, then S is said to be an **invariant subspace** for A. □

Subspaces formed by sets of eigenvectors form the invariant subspaces of a matrix. For an $m \times m$ matrix A with $k \leq m$ distinct eigenvalues, let $\mathbf{x}_1, \mathbf{x}_2, \ldots, \mathbf{x}_m$ denote the normalized eigenvectors, and let $X_i, i = 1, 2, \ldots, k$ denote the set of eigenvectors associated with the eigenvalue λ_i. We can denote the ith invariant subspace of A by

$$R_i = \text{span}(X_i).$$

The matrix

$$P_i = \sum_{\mathbf{x}_j \in X_i} \mathbf{x}_j \mathbf{x}_j^H$$

6.4 Geometry of Invariant Subspaces

is the projection matrix which projects onto R_i. By means of the projectors onto invariant subspaces, we can decompose an operator A into simple pieces, so that the operation of A can be expressed as the sum of simple projection operations. This is what the following theorem does for us.

Theorem 6.4 *Let A be an $m \times m$ self-adjoint matrix with $k \leq m$ distinct eigenvalues. Then*

1. *Spectral decomposition:*

$$A = \sum_{i=1}^{k} \lambda_i P_i. \qquad (6.22)$$

2. *Resolution of identity:*

$$I = \sum_{i=1}^{k} P_i. \qquad (6.23)$$

The proof of this theorem is left as an exercise.

By theorem 6.4, the action of A on the vector $\mathbf{x}$ can be written as

$$A\mathbf{x} = \sum_{i=1}^{k} \lambda_i P_i \mathbf{x}.$$

This can be interpreted as follows:

1. Find the components of $\mathbf{x}$ in each of the invariant subspaces $R_1, R_2, \ldots, R_k$, by projecting $\mathbf{x}$ into each of these spaces,

$$\mathbf{x} = P_1 \mathbf{x} + P_2 \mathbf{x} + \cdots + P_k \mathbf{x},$$

where $P_i \mathbf{x} \in R_i$.

2. Stretch these components by $\lambda_1, \lambda_2, \ldots, \lambda_k$, respectively.
3. Add all the pieces together.

Theorem 6.4 also provides a means of constructing a self-adjoint matrix with a given eigenstructure.

Example 6.4.1 We want to construct a 2×2 self-adjoint matrix with eigenvalues $\lambda_1 = 5$ and $\lambda_2 = 10$, with eigenvectors pointing in the directions

$$\mathbf{x}_1 = \begin{bmatrix} 3 \\ 4 \end{bmatrix} \qquad \mathbf{x}_2 = \begin{bmatrix} -4 \\ 3 \end{bmatrix}.$$

Note that the eigenvectors point in orthogonal directions, as they must. Since the vectors are not normalized, we must normalize them. Then

$$P_1 = \frac{1}{25} \begin{bmatrix} 3 \\ 4 \end{bmatrix} \begin{bmatrix} 3 & 4 \end{bmatrix} = \frac{1}{25} \begin{bmatrix} 9 & 12 \\ 12 & 16 \end{bmatrix},$$

$$P_2 = \frac{1}{25} \begin{bmatrix} -4 \\ 3 \end{bmatrix} \begin{bmatrix} -4 & 3 \end{bmatrix} = \frac{1}{25} \begin{bmatrix} 16 & -12 \\ -12 & 9 \end{bmatrix}.$$

Then

$$A = 5P_1 + 10P_2$$

has the desired eigenvalues and eigenvectors. □

6.5 Geometry of quadratic forms and the minimax principle

Definition 6.5 A *quadratic form* of a self-adjoint matrix A is a scalar of the form $\langle A\mathbf{y}, \mathbf{y} \rangle = \mathbf{y}^H A \mathbf{y}$. This will also be written as

$$Q_A(\mathbf{y}) = \mathbf{y}^H A \mathbf{y}.$$
□

Quadratic forms arise in a variety of signal processing applications where squared-error terms or Gaussian densities are employed. An understanding of the geometry induced by quadratic forms can also aid in understanding some iterative optimization and filtering operations.

Example 6.5.1 Consider the least-squares functional from (3.14), where we assume for convenience that all variables are real,

$$J(\mathbf{c}) = \|x\|^2 - 2\mathbf{c}^T \mathbf{p} + \mathbf{c}^T R \mathbf{c}. \tag{6.24}$$

We can write this as a quadratic form with a scalar offset by completing the square. From section B, we see that we can write

$$J(\mathbf{c}) = (\mathbf{c} - \mathbf{c}_0)^T R (\mathbf{c} - \mathbf{c}_0) + d,$$

where $\mathbf{c}_0 = R^{-1}\mathbf{p}$ and $d = \|x\| - \mathbf{p}^T R^{-1} \mathbf{p}$. We now make a translation of the coordinate system by $\mathbf{y} = \mathbf{c} - \mathbf{c}_0$, and write (with some abuse of notation)

$$J(\mathbf{y}) = \mathbf{y}^T R \mathbf{y} + d.$$

This is an offset quadratic form. □

Example 6.5.2 We desire to make a plot of the contours of constant probability for the two-dimensional Gaussian vector $\mathbf{X} \sim \mathcal{N}(\boldsymbol{\mu}, R)$. That is, we want to plot

$$f_\mathbf{x}(\mathbf{x}) = \frac{1}{2(\pi)^{m/2}|R|^{1/2}} \exp\left[-\frac{1}{2}(\mathbf{x} - \boldsymbol{\mu})^T R^{-1}(\mathbf{x} - \boldsymbol{\mu})\right] = C$$

for different values of the constant C. After some algebraic reduction, this reduces to

$$(\mathbf{x} - \boldsymbol{\mu})^T R^{-1}(\mathbf{x} - \boldsymbol{\mu}) = C',$$

where $C' = -2 \log C((2\pi)^{m/2}|R|^{1/2})$. Letting $\mathbf{y} = \mathbf{x} - \boldsymbol{\mu}$, we obtain

$$\mathbf{y}^T R^{-1} \mathbf{y} = C'. \tag{6.25}$$

This is an equation of the form $Q_{R^{-1}}(\mathbf{y}) = C'$. □

By diagonalizing the matrix A in the quadratic form $Q_A(\mathbf{y})$, we transform to a new coordinate system in which the geometry becomes more apparent. For convenience, we will assume that real vectors are used. Using the decomposition $A = Q \Lambda Q^T$, where

$$Q = [\mathbf{q}_1 \quad \mathbf{q}_2 \quad \cdots \quad \mathbf{q}_m],$$

we can observe that

$$Q_A(\mathbf{y}) = \mathbf{y}^T (Q \Lambda Q^T) \mathbf{y} = \mathbf{z}^T \Lambda \mathbf{z} = \sum_{i=1}^{m} \lambda_i z_i^2, \tag{6.26}$$

where

$$\mathbf{z} = Q^T \mathbf{y}.$$

6.5 Geometry of Quadratic Forms

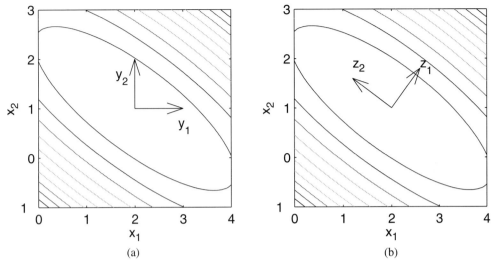

Figure 6.2: The geometry of quadratic forms. (a) The original and translated coordinates. (b) The rotated coordinates

The new variable $\mathbf{z}$ is in a coordinate system in which the interaction between the components of the vector are eliminated.

The variable $\mathbf{z}$ can be interpreted geometrically in two dimensions by observing that when $\mathbf{z} = \begin{bmatrix} 1 \\ 0 \end{bmatrix}$ then $\mathbf{y} = \mathbf{q}_1$, the first eigenvector of A, and when $\mathbf{z} = \begin{bmatrix} 0 \\ 1 \end{bmatrix}$ then $\mathbf{y} = \mathbf{q}_2$. The *orthogonal eigenvectors of A thus provide the orthogonal bases of a new coordinate system.* Figure 6.2 illustrates this concept. In figure 6.2(a), level curves of the quadratic form $(\mathbf{x} - \mathbf{x}_0)^T A (\mathbf{x} - \mathbf{x}_0)$ are shown, where

$$A = \begin{bmatrix} 3.88 & 3.84 \\ 3.84 & 6.12 \end{bmatrix} \tag{6.27}$$

has the eigendecomposition

$$\lambda_1 = 9 \quad \mathbf{x}_1 = \frac{1}{5}[3, 4]^T,$$

$$\lambda_2 = 1 \quad \mathbf{x}_2 = \frac{1}{5}[-4, 3]^T,$$

and $\mathbf{x}_0 = [2, 1]^T$. Also shown in figure 6.2(a) are the new coordinates y_1 and y_2 obtained by the translation, $\mathbf{y} = \mathbf{x} - \mathbf{x}_0$. These coordinates have their origin at the bottom of the quadratic "bowl." In figure 6.2(b) we use the new coordinates z_1 and z_2, which point in the eigenvector directions of A. The level curves in the z coordinates correspond to the equation

$$z_1^2 \lambda_1 + z_2^2 \lambda_2 = C$$

or

$$9z_1^2 + z_2^2 = C$$

or

$$\frac{z_1^2}{1} + \frac{z_2^2}{9} = C',$$

which is the equation for an ellipse. The steepest direction out of the bowl, along the z_1 axis, corresponds to the largest eigenvalue.

In the general two-dimensional case, the level curves of the quadratic form $Q_A(\mathbf{x}) = C$ are of the form

$$z_1^2 \lambda_1 + z_2^2 \lambda_2 = C. \tag{6.28}$$

If $\lambda_1 \lambda_2 > 0$, this equation describes an ellipse, with major and minor axes in the directions of the eigenvectors of A. For $\lambda_1 \lambda_2 < 0$, (6.28) defines a hyperbola. If the eigenvalues differ greatly in magnitude, such as $\lambda_1 \gg \lambda_2$, then the matrix A is said to have a large *eigenvalue disparity*. (This corresponds to the matrix being poorly conditioned.)

Example 6.5.3 Returning to example 6.5.2, we want to make plots of the contours of constant probability, where

$$\mathbf{y}^T R^{-1} \mathbf{y} = C'.$$

Let us write the covariance matrix R as

$$R = U \Lambda U^H.$$

Then R^{-1} has the decomposition

$$R^{-1} = U \Lambda^{-1} U^H,$$

and (6.25) can be written as

$$\mathbf{z}^T \Lambda^{-1} \mathbf{z} = C',$$

since the eigenvalues of R^{-1} are the reciprocals of the eigenvalues of R (see exercise 6.2-10). In two dimensions this is

$$\frac{z_1^2}{\lambda_1} + \frac{z_2^2}{\lambda_2} = C'.$$

When $C' = 1$ this defines an ellipse with major and minor axes $\sqrt{\lambda_1}$ and $\sqrt{\lambda_2}$. Figure 6.3 illustrates the case for

$$R = \begin{bmatrix} 3.88 & 3.84 \\ 3.84 & 6.12 \end{bmatrix}$$

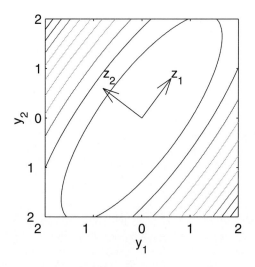

Figure 6.3: Level curves for a Gaussian distribution

6.5 Geometry of Quadratic Forms

(the same as in (6.27)). The level curves are of the form

$$\frac{z_1^2}{9} + \frac{z_2^2}{1} = C'.$$

In this case z_1 points in the direction of *slowest* increase, as it is scaled by the *inverse* of the eigenvalue. Large eigenvalues correspond to large variances, and hence the broad spread in the distribution. □

In higher dimensions the same geometric principle applies.

> Quadratic forms of a matrix A give rise to classical conic sections in two dimensions—ellipses, hyperbolas, and intersecting lines—and multidimensional generalizations of the conic sections for higher dimensions, with orthogonal axis directions determined by the eigenvectors of A.

The quadratic forms of an $m \times m$ matrix with all positive eigenvalues form an ellipsoid in m dimensions. In three dimensions, it helps to envision an American football, an ellipsoid. Figure 6.4(a) shows the locus of points produced by $Q_A(\mathbf{x})$ for $\|\mathbf{x}\| = 1$, the unit ball, where A is a positive definite matrix (all eigenvalues > 0). The l_2 norm of the matrix corresponds to the amount of stretch in the direction that the unit ball is stretched the farthest—the direction of the eigenvector associated with the largest eigenvalue; call it $\mathbf{x}_1$. If we slice the ellipsoid through the largest cross section perpendicular to $\mathbf{x}_1$, as shown in figure 6.4(b), the locus is an ellipse. The largest direction of the ellipse on this plane corresponds to the next largest eigenvalue, and so forth.

The eigenvalues of a self-adjoint matrix can be ordered so that

$$\lambda_1 \geq \lambda_2 \geq \lambda_3 \geq \cdots \geq \lambda_m.$$

With this ordering, let the associated eigenvectors be $\mathbf{x}_1, \mathbf{x}_2, \ldots, \mathbf{x}_m$. It is also convenient to assume that the eigenvectors have been normalized so that $\|\mathbf{x}_i\|_2 = 1, i = 1, 2, \ldots, m$. With this ordering the geometrical reasoning about the ellipsoid can be summarized and generalized to m dimensions by the following theorem.

Theorem 6.5 *(Maximum principle) For a positive-semidefinite self-adjoint matrix A with $Q_A(\mathbf{x}) = \langle A\mathbf{x}, \mathbf{x} \rangle = \mathbf{x}^H A\mathbf{x}$, the maximum*

$$\max_{\|\mathbf{x}\|_2=1} Q_A(\mathbf{x})$$

is λ_1, the largest eigenvalue of A, and the maximizing $\mathbf{x}$ is $\mathbf{x} = \mathbf{x}_1$, the eigenvector corresponding to λ_1.

Furthermore, if we maximize $Q_A(\mathbf{x})$ subject to the constraints that

1. $\langle \mathbf{x}, \mathbf{x}_j \rangle = 0, \ j = 1, 2, \ldots, k-1$, and
2. $\|\mathbf{x}\|_2 = 1$,

then λ_k is the maximized value subject to the constraints and $\mathbf{x}_k$ is the corresponding value of $\mathbf{x}$.

The constraint $\langle \mathbf{x}, \mathbf{x}_j \rangle = 0$ serves to project the search to the space orthogonal to the previously determined eigendirections (e.g., the slice through the ellipsoid).

Proof The proof is carried out by constrained optimization using Lagrange multipliers (see section A.7). We have already seen the first part of the proof in the context of the spectral norm. Form the function

$$J(\mathbf{x}) = \mathbf{x}^H A\mathbf{x} - \lambda \mathbf{x}^H \mathbf{x},$$

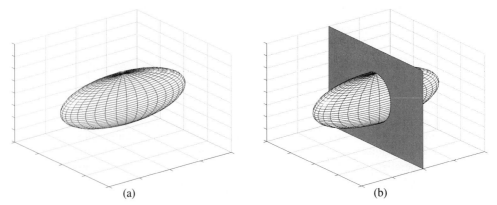

Figure 6.4: The maximum principle. (a) An ellipsoid in three dimensions. (b) The plane orthogonal to the principal eigenvector

where λ is a Lagrange multiplier. Taking the gradient with respect to $\mathbf{x}$ (see section A.6.4) and equating to zero, we obtain

$$\frac{\partial}{\partial \mathbf{x}} J(\mathbf{x}) = A\mathbf{x} - \lambda \mathbf{x} = 0;$$

we see that the maximizing solution[1] $\mathbf{x}^*$ must satisfy

$$A\mathbf{x}^* = \lambda \mathbf{x}^*.$$

Thus $\mathbf{x}^*$ must be an eigenvector of A and λ must be an eigenvalue. For this $\mathbf{x}^*$, we have $Q(\mathbf{x}^*) = (\mathbf{x}^*)^H A \mathbf{x}^* = \lambda (\mathbf{x}^*)^H \mathbf{x}^*$. Maximization of this, subject to the constraint $\|\mathbf{x}^*\| = 1$, requires that we choose λ to be the largest eigenvalue, and $\mathbf{x}^* = \mathbf{x}_1$ to be the eigenvector associated with the largest eigenvalue.

To prove the second part, observe that since the eigenvectors of a self-adjoint matrix are orthogonal, the maximizing solution $\mathbf{x}^*$, subject to the constraints

$$\langle \mathbf{x}^*, \mathbf{x}_j \rangle = 0, \qquad j = 1, 2, \ldots, k-1,$$

must lie in $S_{k,m} = \text{span}\{\mathbf{x}_k, \mathbf{x}_{k+1}, \ldots, \mathbf{x}_m\}$. Let

$$\mathbf{x} = \frac{\mathbf{x}_k + \alpha_{k+1}\mathbf{x}_{k+1} + \alpha_{k+2}\mathbf{x}_{k+2} + \cdots + \alpha_m \mathbf{x}_m}{\|\mathbf{x}_k + \alpha_{k+1}\mathbf{x}_{k+1} + \alpha_{k+2}\mathbf{x}_{k+2} + \cdots + \alpha_m \mathbf{x}_m\|}$$

$$= \frac{\mathbf{x}_k + \alpha_{k+1}\mathbf{x}_{k+1} + \alpha_{k+1}\mathbf{x}_{k+2} + \cdots + \alpha_m \mathbf{x}_m}{\sqrt{1 + |\alpha_{k+1}|^2 + |\alpha_{k+2}|^2 + \cdots + |\alpha_m|^2}}$$

be a normalized candidate solution. Then

$$Q_A(\mathbf{x}) = \lambda_k \frac{1 + |\alpha_{k+1}|^2 \frac{\lambda_{k+1}}{\lambda_k} + \cdots + |\alpha_m|^2 \frac{\lambda_m}{\lambda_k}}{1 + |\alpha_{k+1}|^2 + \cdots + |\alpha_m|^2}. \tag{6.29}$$

Since $\lambda_k \geq \lambda_{k+1} \geq \cdots \geq \lambda_m \geq 0$, $Q_A(\mathbf{x})$ is maximized when $\alpha_{k+1} = \alpha_{k+2} = \cdots = \alpha_m = 0$ (see exercise 6.5-35). Thus, $Q(\mathbf{x})$ has the maximum value λ_k, and $\mathbf{x}_0 = \mathbf{x}_k$. □

The quotient

$$R(\mathbf{x}) = \frac{\mathbf{x}^T A \mathbf{x}}{\mathbf{x}^T \mathbf{x}}$$

---

[1] The symbol * in this case indicates an extremizing value, not an adjoint. There should be little notational ambiguity, since $\mathbf{x}$ is a vector, not an operator.

6.5 Geometry of Quadratic Forms

is known as a *Rayleigh quotient*. From theorem 6.5, we can conclude that

$$\max_{\|\mathbf{x}\|\neq 0} R(\mathbf{x}) = \lambda_1,$$

and that the maximizing value is $\mathbf{x}^* = \mathbf{x}_1$; and that

$$\min_{\|\mathbf{x}\|\neq 0} R(\mathbf{x}) = \lambda_m,$$

where the minimizing value is $\mathbf{x}^* = \mathbf{x}_m$. Some least-squares problems can be couched in terms of Rayleigh quotients, as is shown in section 6.9.

Application of theorem 6.5 requires knowing the first $k-1$ eigenvectors, in order to find the kth eigenvalue and eigenvector. The following theorem provides a means of characterizing the eigenvalues without knowing the first $k-1$ eigenvectors. It is often useful in determining approximate values for the eigenvalues.

Theorem 6.6 (*Courant minimax principle*) *For any self-adjoint $m \times m$ matrix A,*

$$\lambda_k = \min_C \max_{\substack{\|\mathbf{x}\|_2=1 \\ C\mathbf{x}=0}} \langle A\mathbf{x}, \mathbf{x} \rangle,$$

where C is any $(k-1) \times m$ matrix.

Geometrically, the requirement that $C\mathbf{x}=0$ means that $\mathbf{x}$ lies on some $(m-k+1)$-dimensional hyperplane in $\mathbb{R}^m$. We find λ_k by maximizing $Q_A(\mathbf{x})$ for $\mathbf{x}$ lying on the hyperplane, subject to the constraint $\|\mathbf{x}\|_2 = 1$, then move the hyperplane around until the maximum value $Q_A(\mathbf{x})$ is as small as possible. (For example, to find λ_2, think of moving the plane in figure 6.4(b) around.)

Proof For $A = U \Lambda U^H$, we have

$$\langle A\mathbf{x}, \mathbf{x} \rangle = \langle \Lambda \mathbf{y}, \mathbf{y} \rangle = \sum_{i=1}^m \lambda_i |y_i|^2,$$

where $\mathbf{y} = Q^H \mathbf{x}$. Note that $C\mathbf{x} = 0$ if and only if $CQ\mathbf{y} = 0$. Let $B = CQ$. Let

$$\mu = \min_C \max_{\substack{\|\mathbf{x}\|_2=1 \\ C\mathbf{x}=0}} \langle A\mathbf{x}, \mathbf{x} \rangle = \min_B \max_{\substack{\|\mathbf{y}\|_2=1 \\ B\mathbf{y}=0}} \sum_{i=1}^m \lambda_i |y_i|^2.$$

The proof is given by showing that $\mu \leq \lambda_k$ and $\mu \geq \lambda_k$, so that the only possibility is $\mu = \lambda_k$.

It is possible to choose a full-rank B so that $B\mathbf{y} = 0$ implies that $y_1 = y_2 = \cdots = y_{k-1} = 0$. For such a B,

$$\mu \leq \max_{\|\mathbf{y}\|_2=1} \sum_{i=k}^m \lambda_i |y_i|^2 = \lambda_k$$

(where the inequality comes because the minimum over B is not taken).

To get the other inequality, assume that $y_{k+1} = y_{k+2} = \cdots = y_m = 0$. With these $m-k$ constraints, the equation $B\mathbf{y} = 0$ is a system of $k-1$ equations in the k unknowns $y_1, y_2, \ldots, y_k$. This always has nontrivial solutions. Then

$$\mu \geq \min_B \max_{\substack{\|\mathbf{y}\|_2=1 \\ y_{k+1}=\cdots=y_m=0 \\ B\mathbf{y}=0}} \sum_{i=1}^m \lambda_i y_i^2 \geq \min_B \max_{\substack{\|\mathbf{y}\|_2=1 \\ y_{k+1}=\cdots=y_m=0 \\ B\mathbf{y}=0}} \lambda_k \sum_{i=1}^k y_i^2 = \lambda_k,$$

where the first inequality comes by virtue of the extra constraints on the max, and the second inequality follows since λ_k is the smallest of the eigenvalues in the sum. □

6.6 Extremal quadratic forms subject to linear constraints

In the optimization problems of the previous section, we found extrema of quadratic forms, subject to the constraint that the solution be orthogonal to previous solutions. In this section, we modify the constraint somewhat, and consider general linear constraints. Imagine an ellipsoid in three dimensions, as in figure 6.4(a). The axes of the ellipse correspond to the eigenvectors of a matrix, with the length determined by the eigenvalues. Now imagine the ellipsoid sliced by a plane through the origin, as in 6.4(b), but with the plane free to cross at any angle. The intersection of the ellipsoid and the plane forms an ellipse. What are the major and minor axes of this intersecting ellipse? Points on the plane can be described as $\mathbf{x}^T \mathbf{c} = 0$, where $\mathbf{c}$ is the vector orthogonal to the plane. The problem is to determine the stationary points (eigenvectors and eigenvalues) of $\mathbf{x}^H A \mathbf{x}$ (the ellipsoid) subject to the constraints $\mathbf{x}^H \mathbf{x} = 1$ and $\mathbf{x}^H \mathbf{c} = 0$. The problem as stated in three dimensions can obviously be generalized to higher dimensions. Without loss of generality, assume that $\mathbf{c}$ is scaled so that $\|\mathbf{c}\|_2 = 1$. A solution may be found using Lagrange multipliers. Let

$$J(\mathbf{x}) = \mathbf{x}^H A \mathbf{x} - \lambda(\mathbf{x}^H \mathbf{x}) + \mu \mathbf{x}^H \mathbf{c},$$

where λ and μ are Lagrange multipliers. Taking the gradient and equating to zero leads to

$$A\mathbf{x} - \lambda \mathbf{x} + \mu \mathbf{c} = 0. \tag{6.30}$$

Multiplying by $\mathbf{c}^H$ and using $\|\mathbf{c}\|_2 = 1$ leads to $\mu = -\mathbf{c}^H A \mathbf{x}$. Substituting this into (6.30) leads to

$$(I - \mathbf{c}\mathbf{c}^H) A \mathbf{x} = \lambda \mathbf{x}. \tag{6.31}$$

Let $P = I - \mathbf{c}\mathbf{c}^H$. It is apparent that P is a projection matrix, so $P^2 P = P$. Then $PA\mathbf{x} = \lambda \mathbf{x}$ is an eigenvalue problem, but PA may not be Hermitian symmetric (even though both P and A are Hermitian symmetric). Since it is easier to compute eigenvalues for symmetric matrices, it is worthwhile finding a way to symmetricize the problem. Using the fact that the eigenvalues of PA are the same as the eigenvalues of AP we write

$$\lambda(PA) = \lambda(P^2 A) = \lambda(PAP).$$

Let $K = PAP$. Then for an eigenvector $\mathbf{z}$ in

$$K\mathbf{z} = \lambda \mathbf{z},$$

the vector $\mathbf{x} = P\mathbf{z}$ is an eigenvector of PA.

More generally, the eigenproblem may have several constraints,

$$\begin{aligned} C^H \mathbf{x} &= 0, \\ \mathbf{x}^H \mathbf{x} &= 1. \end{aligned} \tag{6.32}$$

If $P = I - C(C^H C)^{-1} C^H$, then the stationary values of $\mathbf{x}^H A \mathbf{x}$ subject to (6.32) are found from the eigenvalues of $K = PAP$ (see exercise 6.6-38).

6.7 The Gershgorin circle theorem

The Gershgorin circle theorem can be used to place bounds on the regions in $\mathbb{C}$ in which the eigenvalues of a matrix A reside. While the regions tend to be large enough that the theorem is generally not useful for computing eigenvalues, it can be used, for example, to determine when a matrix is positive definite.

6.7 The Gershgorin Circle Theorem

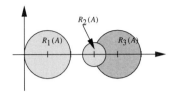

Figure 6.5: Illustration of Gershgorin disks

Let $R_i(A)$ denote the disk in the complex plane defined by

$$R_i(A) = \left\{ x \in \mathbb{C} : |x - a_{ii}| \leq \sum_{\substack{j=1 \\ j \neq i}}^{n} |a_{ij}| \right\}.$$

These disks are referred to as *Gershgorin disks*, and the boundaries are called Gershgorin circles.

Example 6.7.1 Let

$$A = \begin{bmatrix} 1 & 1 & 0 \\ 1/4 & 3 & 1/4 \\ 1/3 & 2/3 & 4 \end{bmatrix}.$$

Then the Gershgorin disks are

$$R_1(A) = \{x \in \mathbb{C} : |x - 1| \leq 1\} \qquad R_2(A) = \{x \in \mathbb{C} : |x - 3| \leq 1/2\}$$

$$R_3(A) = \{x \in \mathbb{C} : |x - 4| \leq 1\}.$$

Figure 6.5 illustrates these regions. The true eigenvalues are

$$\lambda_1 = 0.8914 \qquad \lambda_2 = 2.9328 \qquad \lambda_3 = 4.1758,$$

which fall in the disks. □

Let

$$G(A) = \bigcup_{i=1}^{m} R_i(A).$$

Theorem 6.7 *The eigenvalues of an $m \times m$ matrix A all lie in the union of the Gershgorin disks of A:*

$$\lambda(A) \subset \bigcup_{i=1}^{m} R_i(A) = G(A).$$

Furthermore, if any Gershgorin disk $R_i(A)$ is disjoint from the other Gershgorin disks of A, then it contains exactly one eigenvalue of A. By extension, the union of any k of these disks that do not intersect the remaining $m - k$ circles must contain precisely k of the eigenvalues, counting multiplicities.

Proof The first part of the proof is a straightforward application of the triangle inequality. Let $\mathbf{x} = [x_1, x_2, \ldots, x_m]^T$ be an eigenvector of A with eigenvalue λ, and let $k =$

$\arg\max_i(|x_i|)$. (See box 6.1.) Then the kth component of $A\mathbf{x} = \lambda\mathbf{x}$ is

$$\sum_{j=1}^{m} a_{kj} x_j = \lambda x_k,$$

so that

$$\sum_{\substack{j=1 \\ j \neq k}}^{m} a_{kj} x_j = \lambda x_k - a_{kk} x_k.$$

This leads to

$$|(a_{kk} - \lambda) x_k| = \left| \sum_{\substack{j=1 \\ j \neq k}}^{m} a_{kj} x_j \right| \leq \sum_{\substack{j=1 \\ j \neq k}}^{m} |a_{kj}||x_j|.$$

By the selection of k, $|x_k| \geq |x_j|$ and

$$|(a_{kk} - \lambda) x_k| \leq \sum_{\substack{j=1 \\ j \neq k}}^{m} |a_{kj}| \frac{|x_j|}{|x_k|} \leq \sum_{\substack{j=1 \\ j \neq k}}^{m} |a_{kj}|.$$

Thus $\lambda \in R_k$.

For the second part of the proof, we use a continuity argument. Write

$$A = D + B,$$

where $D = \mathrm{diag}(a_{11}, a_{22}, \ldots, a_{mm})$ and B has zeros on the diagonal. Also let $A_\epsilon = D + \epsilon B$. The eigenvalues of $A_0 = D$ are the diagonal elements of A. As ϵ varies from 0 to 1, the eigenvalues of A_ϵ move from $\lambda(D)$ to $\lambda(A)$. Now we need the following two facts (which we do not prove here—see [142]):

1. The roots of a polynomial depend continuously on the coefficients of the polynomial.
2. As a result of the preceding, the eigenvalues of a matrix are a continuous function of the elements of the matrix.

Thus the locus of eigenvalue locations of A_ϵ traces a continuous curve as ϵ varies. Let $r_i(A_\epsilon)$ be the radius of the ith Gershgorin disk of A_ϵ. Then

$$r_i(A_\epsilon) = r_i(\epsilon B) = \epsilon r_i(A),$$

so $R_i(A_\epsilon) \subset R_i(A)$ for all $\epsilon \in [0, 1]$.

Let $R_1(A)$ be a Gershgorin disk which is disjoint from all other Gershgorin disks of A (in other words, none of the other Gershgorin disks $R_k(A_\epsilon)$ ever intersect $R_1(A)$). $R_1(A)$

Box 6.1: arg max and arg min

The operator $\arg\max(\cdot)$ means "return the argument of the value that maximizes the operand." For example,

$$\arg\max_i (x_i)$$

mean "return the value of i such that x_i is the largest." The function arg min is similarly defined.

has at least one eigenvalue in it; it cannot have more than 1 eigenvalue, however, because the remaining $m - 1$ eigenvalues of A_ϵ start outside $R_1(A)$ and remain outside as ϵ varies over $[0, 1]$. This argument can be extended to regions containing k eigenvalues. □

Definition 6.6 A matrix A is said to be **diagonally dominant** if

$$|a_{ii}| \geq \sum_{\substack{j=1 \\ j \neq i}}^{n} |a_{ij}|$$

for all $i = 1, 2, \ldots, n$. The matrix is **strictly** diagonally dominant if

$$|a_{ii}| > \sum_{\substack{j=1 \\ j \neq i}}^{n} |a_{ij}|$$

for all $i = 1, 2, \ldots, n$. □

It is clear from the geometry of the Gershgorin circles that a *strictly diagonally dominant matrix A cannot have an eigenvalue that is zero, and hence must be invertible.* Note that strict diagonal dominance is required, since otherwise a zero eigenvalue could appear, as in the case of

$$A = \begin{bmatrix} 1 & 1 \\ 1 & 1 \end{bmatrix}.$$

Application of Eigendecomposition methods

6.8 Karhunen–Loève, low-rank approximations, and principal component methods

Let $\mathbf{X}$ be a zero-mean $m \times 1$ random vector, and let $R = E[\mathbf{X}\mathbf{X}^H]$. Let R have the factorization $R = U\Lambda U^H$, where the columns of U are the normalized eigenvectors of R. Let $\mathbf{Y} = U^H \mathbf{X}$. Then $\mathbf{Y}$ is a zero-mean random vector with uncorrelated components:

$$E[\mathbf{Y}\mathbf{Y}^H] = \Lambda.$$

We can thus view the matrix U^H as a "whitening filter." Turning the expression around, we can write

$$\mathbf{X} = U\mathbf{Y} = \sum_{i=1}^{m} \mathbf{u}_i Y_i. \qquad (6.33)$$

This synthesis expression says that we can construct the random variable $\mathbf{x}$ as a linear combination of orthogonal vectors, where the coefficients are uncorrelated random variables. The representation in (6.33) is called the *Karhunen–Loève* expansion of $\mathbf{x}$. In this expansion, the eigenvectors of the correlation matrix R are used as the basis vectors of the expansion.

The Karhunen–Loève expansion could be used to transmit the vector $\mathbf{X}$. If (by some means) the autocorrelation matrix and its eigendecomposition were known at both the transmitter and receiver, then sending the components Y_i would provide, by (6.33), a representation of $\mathbf{X}$. In this representation, m different numbers are needed.

Suppose now that we wanted to provide an approximate representation of $\mathbf{X}$ using fewer components. What is the best representation possible, given the constraint that fewer than m components can be used? Let $\hat{\mathbf{X}} \in \mathbb{C}^m$ be the approximation of $\mathbf{X}$ obtained by

$$\hat{\mathbf{x}} = K\mathbf{x},$$

where K is an $m \times m$ matrix of rank $r < m$. Such a representation is sometimes called a rank-r representation; only r pieces of information are used to approximate $\mathbf{X}$. We desire to determine K so that $\hat{\mathbf{X}}$ is the best approximation of $\mathbf{X}$ in a minimum mean-squared error sense. Such an approximation is sometimes referred to as a *low-rank approximation*. Let $R = E[\mathbf{X}\mathbf{X}^H]$ have eigenvalues $\lambda_1, \lambda_2, \ldots, \lambda_m$ with corresponding eigenvectors $\mathbf{x}_1, \mathbf{x}_2, \ldots, \mathbf{x}_m$. The mean-squared error, as a function of K, is

$$\begin{aligned} e^2(K) &= E[(\mathbf{X} - \hat{\mathbf{X}})^H(\mathbf{X} - \hat{\mathbf{X}})] \\ &= \operatorname{tr} E[(\mathbf{X} - \hat{\mathbf{X}})(\mathbf{X} - \hat{\mathbf{X}})^H] \\ &= \operatorname{tr}(I - K)R(I - K)^H. \end{aligned} \qquad (6.34)$$

Since $e^2(K) = e^2(K^H)$, we may assume that K is Hermitian. We can write K with an orthogonal decomposition,

$$K = \sum_{i=1}^{r} \mu_i \mathbf{u}_i \mathbf{u}_i^H = U M_r U^H, \qquad (6.35)$$

where

$$M_r = \begin{bmatrix} \mu_1 & & & & & & \\ & \mu_2 & & & & & \\ & & \ddots & & & & \\ & & & \mu_r & & & \\ & & & & 0 & & \\ & & & & & \ddots & \\ & & & & & & 0 \end{bmatrix}$$

and U is a unitary matrix. Substituting (6.35) into (6.34) we find (see exercise 6.18-43)

$$e^2(K) = \sum_{i=1}^{r} \mathbf{u}_i^H R \mathbf{u}_i (1 - \mu_i)^2 + \sum_{i=r+1}^{m} \mathbf{u}_i^H R \mathbf{u}_i. \qquad (6.36)$$

To minimize this, clearly we can set $\mu_i = 1, i = 1, 2, \ldots, r$. Then we must minimize

$$\sum_{i=r+1}^{m} \mathbf{u}_i^H R \mathbf{u}_i,$$

subject to the constraints that $\mathbf{u}_i^H \mathbf{u}_j = \delta_{ij}$. But from the discussion of section 6.5, $\mathbf{u}_i, i = r+1, r+2, \ldots, m$ must be the eigenvectors of R corresponding to the $(m - r)$ *smallest* eigenvalues of R. The eigenvectors $\mathbf{u}_i, i = 1, 2, \ldots, r$ which are orthogonal to these form the columns of U, so

$$K = \sum_{i=1}^{r} \mathbf{u}_i \mathbf{u}_i^H = U I_r U^H, \qquad (6.37)$$

where I_r has r ones on the diagonal with the remainder being zeros. The matrix K is a rank-r projection matrix.

The interpretation of this result is as follows: To obtain the best approximation to $\mathbf{X}$ using only r pieces of information, send the values of Y_i corresponding to the r largest eigenvalues of R.

6.8 Low-Rank Approximations

Low-rank approximations and Karhunen–Loève expansions have theoretical application in transform-coding for data compression. A vector **X** is represented by its coefficients in the Karhunen–Loève transform, with the coefficients listed in order of decreasing eigenvalue strength. The first r of these coefficients are quantized and the remaining coefficients are set to zero. The r coefficients provide the representation for the original signal. The corresponding significant eigenvectors of the correlation matrix are assumed somehow to be known. Since the Karhunen–Loève transform provides the optimum low-rank approximation, the reconstructed data should be a good representation of the original data. (However, the Karhunen–Loève transform is rarely used in practice. First, there is the problem of determining R and its eigenvectors for a given signal; second, for each signal set, the eigenvectors selected must somehow be communicated to the decoding side.)

6.8.1 Principal component methods

Related to low-rank approximations are principal component methods. Let **X** be an m-dimensional zero-mean random vector (assumed to be real for convenience), and let $\mathbf{x}_1, \mathbf{x}_2, \ldots, \mathbf{x}_n$ be n observations of **x**. We form the *sample covariance* matrix S by

$$S = \frac{1}{n-1} \sum_{i=1}^{n} \mathbf{x}_i \mathbf{x}_i^T.$$

The principal components of the data are selected so that the ith principal component is the linear combination of the observed data that accounts for the ith largest portion of the variance in the observations [236, page 268]. Let $y_1, y_2, \ldots, y_r$ be the principal components of the data. The first principal component is formed as a linear combination $Y_1 = \mathbf{a}_1^T \mathbf{X}$, where $\mathbf{a}_1$ is chosen so that the sample variance of y_1 is maximized, subject to the constraint that $\|\mathbf{a}_1\| = 1$. The principal component values obtained from the observations are $y_{1,i} = \mathbf{a}_i^T \mathbf{x}_i$, and the sample variance is

$$\sigma_{y_1}^2 = \frac{1}{n-1} \sum_{i=1}^{n} y_{1,i}^2 = \frac{1}{n-1} \sum_{i=1}^{n} \left(\mathbf{a}_1^T \mathbf{x}_i\right)^2 = \mathbf{a}_1^T S \mathbf{a}_1.$$

Maximizing $\mathbf{a}_1^T S \mathbf{a}_1$ subject to $\|\mathbf{a}_1\| = 1$ is a problem we have met before: $\mathbf{a}_1$ is the normalized eigenvector corresponding to the largest eigenvalue of S, λ_1. In this case,

$$\sigma_{y_1}^2 = \mathbf{a}_1^T S \mathbf{a}_1 = \lambda_1 \mathbf{a}_1^T \mathbf{a}_1 = \lambda_1.$$

The second principal component is chosen so that y_2 is uncorrelated with y_1, which leads to the constraint $\mathbf{a}_2^T \mathbf{a}_1 = 0$. Given the discussion in section 6.5, $\mathbf{a}_2$ is the eigenvector corresponding to the second-largest eigenvalue of S, and so forth. The eigenvectors used to compute the principal components are called the principal component directions. If most of the variance of the signal is contained in the principal components, these principal components can be used instead of the data for many statistical purposes.

Example 6.8.1 Figure 6.6 shows 200 sample points from some measured two-dimensional zero-mean data $\mathbf{x}_1, \mathbf{x}_2, \ldots, \mathbf{x}_{200}$. For this data, the covariance matrix is estimated as

$$S = \frac{1}{200 - 1} \sum_{i=1}^{200} \mathbf{x}_i \mathbf{x}_i^T = \begin{bmatrix} 24.1893 & 10.6075 \\ 10.6075 & 6.38059 \end{bmatrix}.$$

Let $\mathbf{s}_1$ and $\mathbf{s}_2$ denote the normalized eigenvalues of S. Then the eigendecomposition of this data is

$$\mathbf{s}_1 = [.9064 \quad .4225]^T \quad \mathbf{s}_2 = [-.4225 \quad .9064]^T,$$
$$\lambda_1 = 29.1343 \quad \lambda_2 = 1.4355.$$

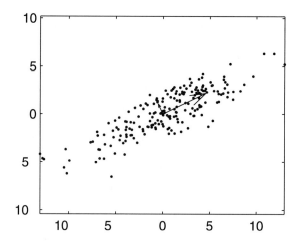

Figure 6.6: Scatter data for principal component analysis

Figure 6.6 also shows a plot of these eigenvectors, the principal component directions of the data, scaled by the square root of the corresponding eigenvalue. The scalar variable

$$y_1 = .9063 X_1 + .4225 X_2$$

accounts for $(100)29.1343/(29.1343 + 1.4355) = 95\%$ of the total variance of the random vector $\mathbf{X} = (X_1, X_2)$, and hence is a good approximation to $\mathbf{X}$ for many statistical purposes. □

6.9 Eigenfilters

Eigenfilters are FIR filters whose coefficients are determined by minimizing or maximizing a quadratic form, subject to some constraint. In this section, two different types of eigenfilter designs are presented. The first is for a random signal in random noise, and the filter is designed in such a way as to maximize the signal-to-noise ratio at the output of the filter. The second is for design of FIR filters with a specified frequency response. As such, they provide an alternative to the standard Parks–McClellan filter design approach.

6.9.1 Eigenfilters for random signals

In the system shown in figure 6.7, let $f[t]$ denote the input signal, which is assumed to be a stationary, zero-mean random process. The input is corrupted by additive white noise $v[t]$ with variance σ^2. The signal then passes through an FIR filter of length m, represented by the vector $\mathbf{h}$, to produce the output $y[t]$. It is desired to design the filter $\mathbf{h}$ to maximize the signal-to-noise ratio. Let

$$\mathbf{f}[t] = \begin{bmatrix} f[t] \\ t[t-1] \\ \vdots \\ f[t-m+1] \end{bmatrix}.$$

6.9 Eigenfilters

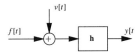

Figure 6.7: Noisy signal to be filtered using an eigenfilter $\mathbf{h}$

Then the filter output can be written as

$$y[t] = \mathbf{h}^H \mathbf{f}[t].$$

The power output due to the input signal is

$$P_0 = E|y[k]|^2 = E\mathbf{h}^H \mathbf{f}[t]\mathbf{f}^H[t]\mathbf{h} = \mathbf{h}^H R \mathbf{h},$$

where $R = E\mathbf{f}[t]\mathbf{f}^H[t]$ is the autocorrelation matrix of f. Let

$$\boldsymbol{\nu}[t] = \begin{bmatrix} \nu[t] \\ \nu[t-1] \\ \vdots \\ \nu[t-m+1] \end{bmatrix}.$$

Then the output of the filter due only to the noise is

$$\mathbf{h}^H \boldsymbol{\nu}[t]$$

and the average noise power output is

$$N_0 = E[\mathbf{h}^H \boldsymbol{\nu}[t] \boldsymbol{\nu}[t]^H \mathbf{h}] = \sigma^2 \mathbf{h}^H \mathbf{h}.$$

The signal-to-noise ratio (SNR) is

$$\text{SNR} = \frac{P_0}{N_0} = \frac{\mathbf{h}^H R \mathbf{h}}{\sigma^2 \mathbf{h}^H \mathbf{h}}.$$

The problem now is to choose the coefficients of the filter $\mathbf{h}$ in such a way as to maximize the SNR. However, this is simply a Rayleigh quotient, which is maximized by taking

$$\mathbf{h} = \mathbf{x}_1,$$

where $\mathbf{x}_1$ is the eigenvector of R corresponding to the largest eigenvalue, λ_1. The maximum SNR is

$$\text{SNR}_{\max} = \frac{\lambda_1}{\sigma^2}.$$

It is interesting to contrast this eigenfilter, which maximizes the SNR for a random input signal, with the matched filter discussed in section 3.19. The operation of the matched filter and the eigenfilter are identical: they both perform an inner-product computation. However, in the case of the matched filter, the filter coefficients are exactly the (conjugate of the) known signal. In the random signal case, the signal can only be known by its statistics. The optimal filter in this case selects that component of the autocorrelation that is most significant.

For this eigenfilter, the important information needed is the eigenvector corresponding to the largest eigenvalue of a Hermitian matrix. Information about the performance of the filter (such as the SNR) may be obtained from the largest eigenvalue. Whereas computing a complete eigendecomposition of a general matrix may be difficult, it is not too difficult to compute the largest eigenvalue and its associated eigenvector. A means of doing this is presented in section 6.15.1.

6.9.2 Eigenfilter for designed spectral response

A variety of filter design techniques exist, the most popular of which is probably the Parks–McClellan algorithm, in which the maximum error between a desired signal spectrum $H_d(e^{j\omega})$ and the filter spectrum $H(e^{j\omega})$ is minimized. In this section we present an alternative filter design technique, which minimizes a quadratic function related to the error $|H_d(e^{j\omega}) - H(e^{j\omega})|^2$. While it does not guarantee to minimize the maximum error, the method does produce good designs, and is of reasonable computational complexity. Furthermore, it is straightforward to impose some constraints on the filter design. The design is exemplified for linear-phase lowpass filters, although it can be extended beyond these restrictions.

It is desired to design a lowpass, linear-phase FIR filter with $m = N + 1$ coefficients that approximates a desired spectral response $H_d(e^{j\omega})$ where N is even. The desired spectral response has a lowpass characteristic,

$$H_d(e^{j\omega}) = \begin{cases} 1 & 0 \leq \omega \leq \omega_p, \\ 0 & \omega_s \leq \omega \leq \pi, \end{cases}$$

with $\omega_p < \omega_s$ (see figure 6.8). The filter is scaled so that the magnitude response at $\omega = 0$ is 1.

The transfer function of the actual (in contrast to the desired) filter is

$$H(z) = \sum_{n=0}^{N} h[n] z^{-n},$$

where the constraint $h[n] = h[N - n]$ is imposed to achieve linear phase. Let $M = N/2$. The frequency response can be written as

$$H(e^{j\omega}) = e^{-jN\omega/2} H_R(\omega), \quad (6.38)$$

where

$$H_R(\omega) = \sum_{n=0}^{M} b_n \cos(\omega n) = \mathbf{b}^T \mathbf{c}(\omega) \quad (6.39)$$

and

$$\mathbf{b} = \begin{bmatrix} b_0 \\ b_1 \\ \vdots \\ b_M \end{bmatrix} = \begin{bmatrix} h(M) \\ 2h(M-1) \\ \vdots \\ 2h(0) \end{bmatrix} \quad \mathbf{c}(\omega) = \begin{bmatrix} 1 \\ \cos \omega \\ \vdots \\ \cos M\omega \end{bmatrix}. \quad (6.40)$$

The squared magnitude response of the filter is

$$|H(e^{j\omega})|^2 = H_R^2(\omega) = \mathbf{b}^T \mathbf{c}(\omega) \mathbf{c}^T(\omega) \mathbf{b}.$$

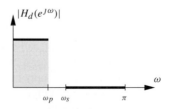

Figure 6.8: Magnitude response specifications for a lowpass filter

6.9 Eigenfilters

Stopband energy

The energy that passes in the stopband, which we want to minimize, is

$$E_s = \frac{1}{\pi} \int_{\omega_s}^{\pi} |(H(e^{j\omega}) - H_d(e^{j\omega})|^2 \, d\omega = \frac{1}{\pi} \mathbf{b}^T \int_{\omega_s}^{\pi} \mathbf{c}(\omega)\mathbf{c}^T(\omega) \, d\omega \mathbf{b}.$$

Let

$$P = \frac{1}{\pi} \int_{\omega_s}^{\pi} \mathbf{c}(\omega)\mathbf{c}^T(\omega) \, d\omega,$$

where the (j, k)th element of P is

$$p_{jk} = \frac{1}{\pi} \int_{\omega_s}^{\pi} \cos(j\omega) \cos(k\omega) \, d\omega.$$

This can be readily computed in closed form.

Passband deviation

The desired DC response $H_d(e^{j0}) = 1$ corresponds to the condition

$$\mathbf{b}^T \mathbf{1} = 1,$$

where **1** is the vector of all 1s. Throughout the passband, we desire the magnitude response to be 1; the deviation from the desired response is

$$1 - \mathbf{b}^T \mathbf{c}(\omega) = \mathbf{b}^T \mathbf{1} - \mathbf{b}^T [\mathbf{1} - \mathbf{c}(\omega)].$$

The square of this deviation can be integrated over the frequencies in the passband as a measure of the quality of the passband error of the filter. Let

$$E_p = \frac{1}{\pi} \int_0^{\omega_p} \mathbf{b}^T [\mathbf{1} - \mathbf{c}(\omega)][\mathbf{1} - \mathbf{c}(\omega)]^T \mathbf{b} \, d\omega = \mathbf{b}^T Q \mathbf{b},$$

with

$$Q = \frac{1}{\pi} \int_0^{\omega_p} [\mathbf{1} - \mathbf{c}(\omega)][\mathbf{1} - \mathbf{c}(\omega)]^T \, d\omega p.$$

This matrix also can be readily computed in closed form.

Overall response functional

Let

$$J(\alpha) = \alpha E_s + (1 - \alpha) E_p$$

be an objective function that trades off the importance of the stopband error with the passband error using the parameter α, where $0 < \alpha < 1$. Combining the errors, we obtain

$$J(\alpha) = \mathbf{b}^T R \mathbf{b},$$

where $R = \alpha P + (1 - \alpha) Q$. Obviously, $J(\alpha)$ can be minimized by setting $\mathbf{b} = \mathbf{0}$. (This corresponds to zero in the passband as well, which matches the deviation requirement, but fails to be physically useful.) To eliminate the trivial filter, we impose the constraint that **b** has unit norm. The final filter coefficients can be scaled from **b** if desired. The design problem thus reduces to:

$$\text{minimize } \mathbf{b}^T R \mathbf{b}$$
$$\text{subject to } \mathbf{b}^T \mathbf{b} = 1.$$

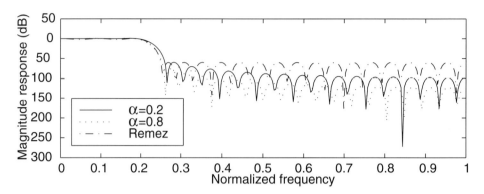

Figure 6.9: Eigenfilter response

This is equivalent to

$$\min_{\|\mathbf{b}\|\neq 0} \frac{\mathbf{b}^T R \mathbf{b}}{\mathbf{b}^T \mathbf{b}},$$

a Rayleigh quotient, which is solved by taking **b** to be the eigenvector corresponding to the *smallest* eigenvalue of the symmetric matrix R.

Figure 6.9 illustrates the magnitude response of a filter design with 45 coefficients, where $\omega_p = 0.2\pi$, $\omega_s = 0.25\pi$. The solid line shows the eigenfilter response when $\alpha = 0.2$, placing more emphasis in the passband. The dotted line shows the eigenfilter response when $\alpha = 0.8$, placing more emphasis in the stopband. For comparative purposes, the response of a 45-coefficient filter designed using the Parks–McClellan algorithm is also shown, with a dash-dot line. The eigenfilter with $\alpha = 0.8$ has better attenuation properties in the stopband, but does not have the equiripple property. MATLAB code that designs the frequency response is shown in algorithm 6.1.

Algorithm 6.1 Eigenfilter design
File: `eigfil.m`
`eigmakePQ.m`

6.9.3 Constrained eigenfilters

One potential advantage of the eigenfilter method over the Parks–McClellan algorithm is that it is fairly straightforward to incorporate a variety of constraints into the design. References on some approaches are given at the end of the chapter. We consider here the problem of adding constraints to fix the response at certain frequencies. Suppose that we desire to specify the magnitude response at r different frequencies, so that

$$H_R(\omega_i) = d_i$$

for $i = 1, 2, \ldots, r$. This can be written as

$$\mathbf{b}^T \mathbf{c}(\omega_i) = d_i$$

for $i = 1, 2, \ldots, r$. Stacking the constraints, we have

$$C^T \mathbf{b} = \mathbf{d}$$

6.9 Eigenfilters

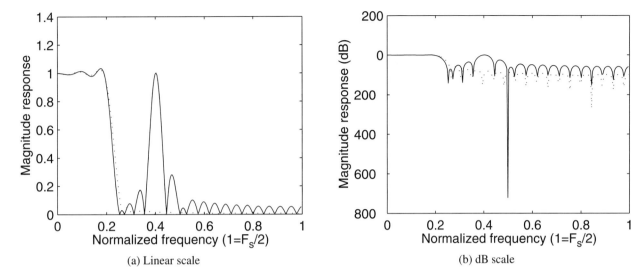

Figure 6.10: Response of a constrained eigenfilter

where

$$C = [\mathbf{c}(\omega_1) \quad \mathbf{c}(\omega_2) \quad \cdots \quad \mathbf{c}(\omega_r)] \qquad \mathbf{d} = \begin{bmatrix} d_1 \\ d_2 \\ \vdots \\ d_r \end{bmatrix}. \tag{6.41}$$

The problem can now be stated as

$$\text{minimize } \mathbf{b}^T R \mathbf{b}$$
$$\text{subject to } C^T \mathbf{b} = \mathbf{d}.$$

A cost functional that includes the r constraints can be written as

$$J(\mathbf{b}) = \mathbf{b}^T R \mathbf{b} + \boldsymbol{\lambda}^T C^T \mathbf{b}, \tag{6.42}$$

where $\boldsymbol{\lambda} = [\lambda_1, \lambda_2, \ldots, \lambda_r]^T$. This leads to the solution (see exercise 6.9-48)

$$\mathbf{b} = R^{-1} C (C^T R^{-1} C)^{-1} \mathbf{d}. \tag{6.43}$$

Algorithm 6.2 shows the code that computes the coefficients. Figure 6.10(a) shows the magnitude response of a 45-coefficient eigenfilter with $\omega_p = 0.2\pi$ and $\omega_S = 0.25\pi$, with constraints so that

$$H_R(e^{j(0)}) = 1 \qquad H_R(e^{j(.4\pi)}) = 1 \qquad H_R(e^{j.5\pi}) = 0 \qquad H_R(e^{j.8\pi}) = 0.$$

Because of the zero outputs, the response is not shown on a dB scale. Figure 6.10(b) shows the dB scale. For comparison, the response of an eigenfilter with the same ω_s and ω_p, but without the extra constraints, is shown with a dotted line.

Algorithm 6.2 Constrained eigenfilter design
File: `eigfilcon.m`

6.10 Signal subspace techniques

In section 1.4.6, we examined methods of determining which sinusoidal signals are present in a signal based upon finding a characteristic equation, then finding its roots. As pointed out in that section, these methods can provide good spectral resolution, but break down quickly in the presence of noise. In this section we continue in that spirit, but account explicitly for the possibility of noise, by breaking the signal out in terms of a *signal subspace* component and a *noise subspace* component.

6.10.1 The signal model

Suppose that a signal $x[t]$ consists of the sum of p complex exponentials in noise,

$$x[t] = \sum_{i=1}^{p} a_i e^{j(2\pi f_i t + \phi_i)},$$

where $f_i \in [0, .5)$ is the frequency (we assume here that all frequencies are distinct), a_i is the amplitude, and ϕ_i is the phase of the ith signal. The phases are assumed to be stationary, statistically independent, and uniformly distributed over $[0, 2\pi)$. The autocorrelation function for $x[t]$ (see exercise 6.10-51) is

$$r_{xx}[k] = E[x[t]\bar{x}[t-k]] = \sum_{i=1}^{p} p_i e^{j2\pi f_i k}, \tag{6.44}$$

where $p_i = a_i^2$. Let

$$\mathbf{x}[t] = \begin{bmatrix} x[t] \\ x[t+1] \\ \vdots \\ x[t+M-1] \end{bmatrix}$$

and let R_{xx} be the $M \times M$ autocorrelation matrix for $x[t]$,

$$R_{xx} = E[\mathbf{x}[t]\mathbf{x}^H[t]] = \begin{bmatrix} r_{xx}[0] & rr_{xx}[-1] & \cdots & r_{xx}[-(M-1)] \\ r_{xx}[1] & r_{xx}[0] & \cdots & r_{xx}[-(M-2)] \\ \vdots & & & \\ r_{xx}[M-1] & r_{xx}[M-2] & \cdots & r_{xx}[0] \end{bmatrix}.$$

The autocorrelation matrix can be written as

$$R_{xx} = \sum_{k=1}^{p} p_k \mathbf{s}_k \mathbf{s}_k^H, \tag{6.45}$$

where

$$\mathbf{s}_i = \begin{bmatrix} 1 \\ e^{j2\pi f_i} \\ e^{j2\pi(2f_i)} \\ \vdots \\ e^{2j\pi(M-1)f_i} \end{bmatrix}.$$

Equation (6.45) can also be written as

$$R_{xx} = SPS^H,$$

6.10 Signal Subspace Techniques

where
$$S = [\mathbf{s}_1 \quad \mathbf{s}_2 \quad \cdots \quad \mathbf{s}_p] \quad \text{and} \quad P = \text{diag}(p_1, p_2, \ldots, p_p). \quad (6.46)$$

The matrix S is a Vandermonde matrix.

The vector space
$$\mathcal{S} = \text{span}\{\mathbf{s}_1, \mathbf{s}_2, \ldots, \mathbf{s}_p\} \quad (6.47)$$

is said to be the *signal subspace* of the signal $x[t]$. This name is appropriate, since every $\mathbf{x}[t]$ can be expressed as a linear combination of the columns of S, hence $\mathbf{x}[t] \in \mathcal{S}$.

It can be shown for $M > p$ that R_{xx} has rank p. Let $\lambda(R_{xx})$ denote the eigenvalues of R_{xx}, ordered so that $\lambda_1 \geq \lambda_2 \geq \cdots \geq \lambda_M$, and let $\mathbf{u}_1, \mathbf{u}_2, \ldots, \mathbf{u}_M$ be the corresponding eigenvectors that are normalized so that $\mathbf{u}_i^T \mathbf{u}_j = \delta_{i-j}$. Then
$$R_{xx}\mathbf{u}_i = \lambda_i \mathbf{u}_i. \quad (6.48)$$

Recall from lemma 6.5 that if R_{xx} has rank p, then $\lambda_{p+1} = \lambda_{p+2} = \cdots = \lambda_M = 0$, so we can write
$$R_{xx} = \sum_{i=1}^{p} \lambda_i \mathbf{u}_i \mathbf{u}_i^H.$$

The eigenvectors $\mathbf{u}_1, \mathbf{u}_2, \ldots, \mathbf{u}_p$ are called the *principal eigenvectors* of R_{xx}.

Lemma 6.6 *The principal eigenvectors of R_{xx} span the signal subspace $\mathcal{S}$,*
$$\text{span}\{\mathbf{u}_1, \mathbf{u}_2, \ldots, \mathbf{u}_p\} = \text{span}\{\mathbf{s}_1, \mathbf{s}_2, \ldots, \mathbf{s}_p\}.$$

Proof Substitute (6.45) into (6.48),
$$\left(\sum_{i=1}^{p} p_i \mathbf{s}_i \mathbf{s}_i^H \right) \mathbf{u}_j = \lambda_j \mathbf{u}_j.$$

Thus,
$$\mathbf{u}_j = \frac{1}{\lambda_j} \sum_{i=1}^{p} p_i \mathbf{s}_i \left(\mathbf{s}_i^H \mathbf{u}_j \right) = \sum_{i=1}^{p} \beta_{i,j} \mathbf{s}_i,$$

where
$$\beta_{i,j} = \frac{1}{\lambda_i} p_k \mathbf{s}_i^H \mathbf{u}_j.$$

Since every $\mathbf{u}_j$ can be expressed as a linear combination of $\{\mathbf{s}_i, i = 1, 2, \ldots, p\}$, and since they are both p-dimensional vector spaces, the spans of both sets are the same. □

Given a sequence of observations of $x[t]$, we can determine (estimate) R_{xx} and find its eigenvectors $\mathbf{u}_i$. Knowing the first p eigenvectors we can determine the space in which the signals reside, even though (at this point) we don't know what the signal frequencies are.

6.10.2 The noise model

Assume that $x[t]$ is observed in noise,
$$y[t] = x[t] + w[t],$$

where $w[t]$ is a stationary, zero-mean, white-noise signal, independent of $x[t]$ with $E[w[t]\overline{w}[t]] = \sigma_w^2$. Then
$$r_{yy}[k] = r_{xx}[k] + \sigma_w^2 \delta[k]$$
$$= \sum_{i=1}^{p} p_i e^{j2\pi f_i k} + \sigma_w^2 \delta[k]. \quad (6.49)$$

Let
$$\mathbf{y}[t] = \begin{bmatrix} y[t] \\ y[t+1] \\ \vdots \\ y[t+M-1] \end{bmatrix} \quad \text{and} \quad \mathbf{w}[t] = \begin{bmatrix} w[t] \\ w[t+1] \\ \vdots \\ w[t+M-1] \end{bmatrix}.$$

Then $R_{yy} = E[\mathbf{y}[t]\mathbf{y}^H[t]]$ can be written as

$$R_{yy} = R_{xx} + \sigma_w^2 I. \tag{6.50}$$

The autocorrelation matrix R_{yy} is full rank because $\sigma_w^2 I$ is full rank. Let $R_{yy}\mathbf{u}_i = \mu_i \mathbf{u}_i$ be the eigenequation for R_{yy} with the eigenvalues sorted as $\mu_1 \geq \mu_2 \geq \cdots \geq \mu_M$. The first p eigenvalues of R_{yy} are related to the first p eigenvalues of R_{xx} by

$$\mu_i = \lambda_i + \sigma_w^2,$$

and the corresponding eigenvectors are equal (see exercise 6.2-8). Furthermore, eigenvalues $\mu_{p+1}, \mu_{p+2}, \ldots, \mu_M$ are all equal to σ_w^2. Thus, we can write

$$R_{yy} = \sum_{i=1}^{p} (\lambda_i + \sigma_w^2)\mathbf{u}_i \mathbf{u}_i^H + \sum_{i=p+1}^{M} \sigma_w^2 \mathbf{u}_i \mathbf{u}_i^H.$$

The space

$$\mathcal{N} = \text{span}(\mathbf{u}_{p+1}, \mathbf{u}_{p+2}, \ldots, \mathbf{u}_M)$$

is called the *noise subspace*. Any vector from the signal subspace is orthogonal to $\mathcal{N}$.

6.10.3 Pisarenko harmonic decomposition

Based on the observation that the signal subspace is orthogonal to the noise subspace, there are various means that can be employed to estimate the signal components in the presence of noise. In the Pisarenko harmonic decomposition (PHD), the orthogonality is exploited directly. Suppose that the number of modes p is known. Then, setting $M = p+1$, the noise subspace is spanned by the single vector $\mathbf{u}_M$, which must be orthogonal to all of the signal subspace vectors:

$$\mathbf{s}_i^H \mathbf{u}_M = 0 \quad i = 1, 2, \ldots, p. \tag{6.51}$$

Letting $\mathbf{u}_M = [u_{M,0}, u_{M,2}, \ldots, u_{M,M-1}]^T$, (6.51) can be written as

$$\sum_{k=0}^{M-1} u_{M,k} e^{-j2\pi f_i k} = 0.$$

This is a polynomial in $e^{-j2\pi f_i}$. The $M - 1 = p$ roots of this polynomial, which lie on the unit circle, correspond to the frequencies of the sinusoidal signal. Once the frequencies are obtained from the roots of the polynomial, the squared amplitudes can be obtained by setting up a system of equations from (6.49) for $m = 1, 2, \ldots, p$:

$$\begin{bmatrix} e^{j2\pi f_1} & e^{j2\pi f_2} & \cdots & e^{j2\pi f_p} \\ e^{j2\pi 2f_1} & e^{j2\pi 2f_2} & \cdots & e^{j2\pi 2f_p} \\ \vdots & \vdots & & \vdots \\ e^{j2\pi pf_1} & e^{j2\pi pf_2} & \cdots & e^{j2\pi pf_p} \end{bmatrix} \begin{bmatrix} p_1 \\ p_2 \\ \vdots \\ p_p \end{bmatrix} = \begin{bmatrix} r_{yy}(1) \\ r_{yy}(2) \\ \vdots \\ r_{yy}(p) \end{bmatrix}. \tag{6.52}$$

The noise strength is obtained from the Mth eigenvalue of R_{yy}. Of course, in practice the correlation matrix R_{yy} must be estimated based on received signals.

6.10 Signal Subspace Techniques

Example 6.10.1 A source $x[t]$ is known to produce $p = 3$ sinusoids. The correlation matrix R_{yy} is estimated to be

$$R_{yy} = \begin{bmatrix} 6.4000 & -2.7361 + 4.6165j & -1.5000 - 3.4410j & 1.7361 + 1.0898j \\ -2.7361 - 4.6165j & 6.4000 & -2.7361 + 4.6165j & -1.5000 - 3.4410j \\ -1.5000 + 3.4410j & -2.7361 - 4.6165j & 6.4000 & -2.7361 + 4.6165j \\ 1.7361 - 1.0898j & -1.5000 + 3.4410j & -2.7361 - 4.6165j & 6.4000 + 0.0000j \end{bmatrix}.$$

Algorithm 6.3 can be used to determine the frequencies of the source. The result of this computation is $\sigma^2 = 0.4$ and $f = [0.2, 0.3, 0.4]^T$. The amplitudes are $\mathbf{p} = [1, 2, 3]^T$. □

Algorithm 6.3 Pisarenko harmonic decomposition
File: `pisarenko.m`

6.10.4 MUSIC

MUSIC stands for MUltiple SIgnal Classification. Like the PHD, it relies on the fact that the signal subspace is orthogonal to the noise subspace. Let

$$\mathbf{s}^T(f) = [1 \quad e^{j2\pi f} \quad e^{j2\pi 2f} \quad \cdots \quad e^{j2\pi(M-1)f}]. \tag{6.53}$$

When $f = f_i$ (one of the input frequencies), then for any vector $\mathbf{x}$ in the noise subspace,

$$\mathbf{s}^H(f)\mathbf{x} = 0,$$

since they are orthogonal. Let

$$M(f) = \sum_{k=p+1}^{M} |\mathbf{s}^H(f)\mathbf{u}_k|^2.$$

Then, theoretically, when $f = f_i$ then $M(f) = 0$, and $1/M(f)$ is infinite. Thus, a plot of $1/M(f)$ should have a tall peak at $f = f_i$ for each of the input frequencies. The function

$$P(f) = \frac{1}{\sum_{k=p+1}^{M} |\mathbf{s}^H(f)\mathbf{u}_k|^2} \tag{6.54}$$

is sometimes referred to as the MUSIC spectrum of f. By locating the peaks, the frequencies can be identified. Knowing the frequencies, the signal strengths can be computed using (6.52), as for the Pisarenko method. The MUSIC spectrum can be computed using Algorithm 6.4.

Example 6.10.2 Using the data from the previous example, compute the spectrum using the MUSIC method.

First, we use the following code to compute the value at a point of the MUSIC spectrum, given the eigenvectors of the autocorrelation matrix.

Algorithm 6.4 Computation of the MUSIC spectrum
File: `musicfun.m`

Then the MUSIC spectrum can be plotted with the following MATLAB code (assuming that R_{yy} is already entered into MATLAB).

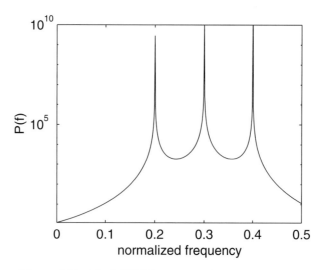

Figure 6.11: The MUSIC spectrum for example 6.10.2

```
[v,u] = eig(Ryy);
f=0:.001:.5;
pf = musicfun(f,3,V);
plot(f,pf);
```

The plot of the MUSIC spectrum is shown in figure 6.11. The peaks are clearly at 0.2, 0.3, and 0.4. Computation of the signal strengths is as in example 6.10.1. □

6.11 Generalized eigenvalues

In addition to the many applications of eigenvalues to signal processing, there has arisen recently an interest in generalized eigenvalue problems of the form

$$A\mathbf{u} = \lambda B\mathbf{u},$$

where A and B are $m \times m$ matrices. The set of matrices $A - \lambda B$ is said to form a *matrix pencil*. The eigenvalues of the matrix pencil, denoted $\lambda(A, B)$, are those values of λ for which

$$\det(A - \lambda B) = 0.$$

For an eigenvalue of the pencil $\lambda \in \lambda(A, B)$, a vector $\mathbf{u} \neq 0$ such that

$$A\mathbf{u} = \lambda B\mathbf{u}$$

is said to be an eigenvector of $A - \lambda B$.

Note that if B is nonsingular then there are n eigenvalues and $\lambda(A, B) = \lambda(B^{-1}A)$. This provides one means of finding the eigenvalues of the matrix pencil. However, it is not particularly well conditioned. An extensive discussion of numerically stable means of computing generalized eigenvalues can be found in [114]. MATLAB can compute the generalized eigenvalues using the `eig` command with two arguments, as `eig(A,B)`.

6.11.1 An application: ESPRIT

One application of generalized eigenvalue decompositions is to sinusoidal estimation using ESPRIT (estimation of signal parameters via rotational invariance techniques). Like MUSIC, the method assumes that there are p sinusoids in white noise, and deals with an eigendecomposition. The same notation as in section 6.10 is employed: $\mathbf{x}[t]$ is a vector of signal samples, $\mathbf{w}[t]$ is a vector of noise samples, and $\mathbf{y}[t] = \mathbf{x}[t] + \mathbf{w}[t]$. We also introduce a new vector of delayed samples, $\mathbf{z}[t] = \mathbf{y}[t+1]$, or

$$\mathbf{z}[t] = \begin{bmatrix} y[t+1] \\ y[t+2] \\ \vdots \\ y[t+M] \end{bmatrix}.$$

As before, we can write

$$R_{yy} = E[\mathbf{y}[t]\mathbf{y}^H[t]] = SPS^H + \sigma_w^2 I.$$

Also (see exercise 6.11-55),

$$R_{yz} = E[\mathbf{y}[t]\mathbf{z}^H[t]] = SP\Phi^H S^H + R_w, \tag{6.55}$$

where

$$R_w = E[\mathbf{w}[t]\mathbf{w}^H[t+1]] = \sigma_w^2 \begin{bmatrix} 0 & 0 & 0 & \cdots & 0 & 0 \\ 1 & 0 & 0 & \cdots & 0 & 0 \\ 0 & 1 & 0 & \cdots & 0 & 0 \\ \vdots & & & & & \\ 0 & 0 & 0 & \cdots & 1 & 0 \end{bmatrix}$$

and Φ represents the phase shift between successive samples,

$$\Phi = \text{diag}(e^{j2\pi f_1}, e^{j2\pi f_2}, \ldots, e^{j2\pi f_p}).$$

If $M > p$, the matrix

$$R_{xx} = R_{yy} - \sigma_w^2 I = SPS^H$$

has rank p. Let

$$C_{yz} = R_{yz} - R_w = SP\Phi^H S^H.$$

Now consider the generalized eigenvalue problem

$$R_{xx}\mathbf{u} = \lambda C_{yz}\mathbf{u}; \tag{6.56}$$

that is,

$$(R_{xx} - \lambda C_{yz})\mathbf{u} = 0.$$

This can be written as

$$SP(I - \lambda \Phi^H)S^H \mathbf{u} = 0.$$

Since Φ is diagonal, it is clear that $\lambda = e^{j2\pi f_i}$ is an eigenvalue of (6.56). From the p generalized eigenvalues that lie on the circle can be obtained the frequencies $f_i, i = 1, 2, \ldots, p$.

Example 6.11.1 For the data-correlation matrix of example 6.10.1, the cross-correlation matrix R_{yz} is determined to be

$$R_{yz} = \begin{bmatrix} -2.7361 - 4.6165i & 6.0000 & -2.7361 + 4.6165i & -1.5000 - 3.4410i \\ -1.1000 + 3.4410i & -2.7361 - 4.6165i & 6.0000 & -2.7361 + 4.6165i \\ 1.7361 - 1.0898i & -1.1000 + 3.4410i & -2.7361 - 4.6165i & 6.0000 \\ -1.5000 + 0.8123i & 1.7361 - 1.0898i & -1.1000 + 3.4410i & -2.7361 - 4.6165i \end{bmatrix}.$$

The following algorithm implements ESPRIT.

Algorithm 6.5 Computation of the frequency spectrum of a signal using ESPRIT
File: `esprit.m`

The results of the computation are $f = [0.2, 0.3, 0.4]$, as before. □

6.12 Characteristic and minimal polynomials

The characteristic polynomial was defined in definition 6.2. In this section, we examine a few of its important properties, as well as the properties of the *minimal* polynomial of a matrix.

6.12.1 Matrix polynomials

We begin with an elementary observation about scalar polynomials.

Lemma 6.7 *When the polynomial $f(x)$ is divided by $x - a$ to form the quotient and remainder,*

$$f(x) = (x - a)q(x) + r(x),$$

where $\deg(r(x)) < \deg(x - a) = 1$, then the remainder term is $f(a)$.

The proof follows simply by evaluating $f(x)$ at $x = 0$ and observing that $r(x)$ must be a constant.

We define a matrix polynomial of degree m by

$$F(x) = F_m x^m + F_{m-1} x^{m-1} + \cdots + F_1 x + F_0,$$

where each F_i is an $m \times m$ matrix, and $F_m \neq \mathbf{0}$. If $\det(F_m) \neq 0$, the matrix polynomial is said to be *regular*. The operations of addition, multiplication, and division of matrix polynomials are as for scalar polynomials, keeping in mind the noncommutativity of matrix multiplication operations. If we divide $F(x)$ by some matrix polynomial $A(x)$ such that

$$F(x) = Q(x)A(x) + R(x),$$

then $Q(x)$ and $R(x)$ are said to be the right quotient and right remainder, respectively. If we divide $F(x)$ such that

$$F(x) = A(x)Q(x) + R(x),$$

then $Q(x)$ and $R(x)$ are the left quotient and left remainder, respectively.

Example 6.12.1 Let $F(x) = x^2 F_2 + x F_1 + F_0$, where

$$F_2 = \begin{bmatrix} 3 & 7 \\ 1 & 6 \end{bmatrix} \quad F_1 = \begin{bmatrix} 23 & 38 \\ 9 & 6 \end{bmatrix} \quad F_0 = \begin{bmatrix} 11 & 22 \\ -5 & -4 \end{bmatrix},$$

and let $A(x) = A_1 x + A_0$, where

$$A_1 = \begin{bmatrix} 1 & 2 \\ 1 & 3 \end{bmatrix} \quad A_0 = \begin{bmatrix} 4 & 5 \\ 6 & 7 \end{bmatrix}.$$

6.12 Characteristic and Minimal Polynomials

Then the right division and left division yield

$$F(x) = \left(\begin{bmatrix} 2 & 1 \\ -3 & 4 \end{bmatrix} x + \begin{bmatrix} 6 & 3 \\ -2 & -1 \end{bmatrix}\right) A(x) + \begin{bmatrix} -31 & -29 \\ 9 & 13 \end{bmatrix}$$

and

$$F(x) = A(x) \left(\begin{bmatrix} 7 & 9 \\ -2 & -1 \end{bmatrix} x + \begin{bmatrix} 53 & 103 \\ -24 & -48 \end{bmatrix}\right) + \begin{bmatrix} -81 & -150 \\ -155 & -286 \end{bmatrix},$$

respectively. □

When $F(x)$ is evaluated at a square matrix value $x = A$, two possible values, generally unequal, result:

$$F_r(A) = F_m A^m + F_{m-1} A^{m-1} + \cdots + F_0$$

and

$$F_l(A) = A^m F_m + A^{m-1} F_{m-1} + \cdots + F_0.$$

The polynomial $F_r(A)$ is the *right* value of F, and $F_l(A)$ is the *left* value of F. Based on this notation, we extend lemma 6.7 as follows.

Lemma 6.8 *(Generalized Bèzout theorem) When $F(x)$ is divided by $xI - A$ on the right, the remainder is $F_r(A)$. When $F(x)$ is divided by $xI - A$ on the left, the remainder is $F_l(A)$.*

Let $\chi_A(\lambda) = \det(\lambda I - A)$ denote the characteristic polynomial of A. Also, let $B_A(\lambda)$ denote the adjugate of $\lambda I - A$. Recall that the adjugate of a matrix X satisfies the property

$$X^{-1} = \text{adj}(X)/\det(X).$$

Thus,

$$X \text{adj}(X) = \det(X) = \text{adj}(X) X. \tag{6.57}$$

Example 6.12.2 For the matrix $\lambda I - A = \begin{bmatrix} \lambda - a_{11} & -a_{12} \\ -a_{21} & \lambda - a_{22} \end{bmatrix}$, the adjugate is

$$\text{adj}(\lambda I - A) = \begin{bmatrix} \lambda - a_{22} & a_{21} \\ a_{12} & \lambda - a_{11} \end{bmatrix}$$

and

$$(\lambda I - A) \text{adj}(\lambda I - A) = \begin{bmatrix} \det(\lambda I - A) & \\ & \det(\lambda I - A) \end{bmatrix} = \text{adj}(\lambda I - A)(\lambda I - A). \quad \square$$

By (6.57),

$$(\lambda I - A) B_A(\lambda) = \chi_A(\lambda) I,$$
$$B_A(\lambda)(\lambda I - A) = \chi_A(\Lambda) I.$$

The matrix polynomial $\chi_A(\lambda) I$ is divisible on both sides without remainder by $A - \lambda I$. Hence we conclude, by lemma 6.8, that $\chi_A(A) = 0$. We have thus proven the following.

Theorem 6.8 *(Cayley–Hamilton theorem)*

> Let $\chi_A(\lambda)$ be the characteristic polynomial of the square matrix A. Then
> $$\chi_A(A) = 0.$$
> In words, a matrix satisfies its own characteristic equation.

Writing
$$\chi_A(\lambda) = \lambda^m + a_{m-1}\lambda^{m-1} + \cdots + a_1\lambda + a_0,$$
we have, by the Cayley–Hamilton theorem,
$$A^m = -a_{m-1}A^{m-1} - a_{m-2}A^{m-2} - \cdots - a_1 A - a_0 I.$$

Example 6.12.3 Let $A = \begin{bmatrix} 2 & 4 \\ 1 & 5 \end{bmatrix}$. Then $\chi_A(\lambda) = \lambda^2 - 7\lambda + 6$, and
$$A^2 - 7A + 6I = \begin{bmatrix} 8 & 28 \\ 7 & 29 \end{bmatrix} - 7\begin{bmatrix} 2 & 4 \\ 1 & 5 \end{bmatrix} + 6\begin{bmatrix} 1 & 0 \\ 0 & 1 \end{bmatrix} = \mathbf{0}.$$

6.12.2 Minimal polynomials

Definition 6.7 A polynomial $f(x)$ is said to be an **annihilating polynomial** of a square matrix A if $f(A) = 0$. □

By the Cayley–Hamilton theorem, the characteristic polynomial $\chi_A(\lambda)$ is an annihilating polynomial. However, for some matrices, there is an annihilating polynomial of lesser degree than the degree of $\chi_A(\lambda)$.

Definition 6.8 The unique monic annihilating polynomial of A of minimal positive degree is called the **minimal polynomial** of A. □

The concept of minimal polynomials is exhaustively studied elsewhere (see, e.g., [98, chapter 6]). We will introduce the concept principally by example.

Example 6.12.4 Let
$$A_1 = \begin{bmatrix} 4 & & \\ & 4 & \\ & & 4 \end{bmatrix} \quad A_2 = \begin{bmatrix} 5 & 1 & 0 \\ 0 & 5 & 0 \\ 0 & 0 & 5 \end{bmatrix} \quad A_3 = \begin{bmatrix} 6 & 1 & 0 \\ 0 & 6 & 1 \\ 0 & 0 & 6 \end{bmatrix}.$$

Matrix A_1 consists of three first-order Jordan blocks. Matrix A_2 consists of a second-order Jordan block and a first-order Jordan block. Matrix A_3 consists of a single Jordan block.

The polynomial $f_1(x) = x - 4$ satisfies $f_1(A_1) = 0$. So the minimal polynomial of A_1 is a degree-one polynomial, even though the characteristic polynomial has degree three.

The polynomial $f_2(x) = (x - 5)^2$ satisfies $f_2(A_2) = 0$. Similarly, the polynomial $f_3(x) = (x - 6)^3$ satisfies $f_3(A_3) = 0$. □

Without presenting all the details, the minimal polynomial of a matrix is obtained as follows: A Jordan block of order l has a minimal polynomial of degree l. The minimal polynomial of a matrix A is the least common multiple of the minimal polynomials of its Jordan blocks. Thus a matrix having m distinct eigenvalues has a minimal polynomial of degree m.

6.13 Moving the eigenvalues around: introduction to linear control

We have seen that the solution of the time-invariant homogeneous differential equation $\dot{\mathbf{x}}(t) = A\mathbf{x}(t)$ is $\mathbf{x}(t) = e^{At}\mathbf{x}_0$, where $\mathbf{x}_0$ is the initial condition. Based on the discussion in section 6.3.1, the solution can be written as
$$\mathbf{x}(t) = Se^{Jt}S^{-1},$$
where J is the Jordan form of A. If the eigenvalues of A are in the left-hand plane, $\text{Re}(\lambda_i) < 0$, then $\mathbf{x}(t) \to 0$ as $t \to \infty$. Similarly, the solution to the time-invariant homogeneous

6.13 Moving the Eigenvalues Around

difference equation $\mathbf{x}[t+1] = A\mathbf{x}[t]$ is $\mathbf{x}[t] = A^t \mathbf{x}_0$, which can be written as

$$\mathbf{x}[t] = SJ^t S^{-1}\mathbf{x}_0.$$

If the eigenvalues of A are inside the unit circle, $|\lambda_i| < 1$, then $\mathbf{x}[t] \to 0$ as $t \to \infty$.

It may happen that for some systems the eigenvalues are not in the left-half plane or inside the unit circle, as necessary for asymptotically stable performance of continuous- or discrete-time systems. Even for stable systems, the eigenvalues may not be suitable locations for some purposes, so that the response is too slow or the overshoot is too high. A principle of control of linear time-invariant systems is that, given the value of the state of the system $\mathbf{x}(t)$ (or $\mathbf{x}[t]$), the eigenvalues of a closed-loop system can be moved to *any* desired location, provided it has the property that it is *controllable*, which will be defined below. We will introduce the notion that eigenvalues may be moved, deferring thorough explanations and applications to the controls literature (see, e.g., [93, 164, 287]). We will develop the concepts in continuous time, with the translation to discrete-time systems being straightforward.

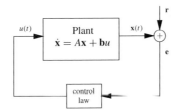

Figure 6.12: Plant with reference input and feedback control

We will consider a system having a state reference input as shown in figure 6.12. A reference state $\mathbf{r}(t)$ is input to the system, and it is desired to have the state of the plant converge to the reference signal. That is, we desire $\mathbf{e}(t) = \mathbf{x}(t) - \mathbf{r}(t)$ to decay such that $\mathbf{e}(t) \to 0$ as $t \to \infty$. Assume that the plant is governed by the equation

$$\dot{\mathbf{x}}(t) = A\mathbf{x}(t) + \mathbf{b}u(t),$$

with a scalar input $u(t)$, where the state variable $\mathbf{x}(t) \in \mathbb{R}^m$. We make two key assumptions:

1. $\dot{\mathbf{r}}(t) = 0$. That is, the system is designed to move to a fixed reference position, and not track a moving reference.
2. $\mathbf{r}$ is in the nullspace of A, so $A\mathbf{r} = 0$. This is less obvious, but is satisfied for many control problems.

(Both of these assumptions are loosened in more sophisticated approaches to design. We present here merely the preliminary concepts.) Based on these assumptions,

$$\dot{\mathbf{e}}(t) = A\mathbf{e}(t) + \mathbf{b}u(t). \tag{6.58}$$

We desire to formulate a plant input $u(t)$ so that $\mathbf{e}(t) \to 0$.

In a linear controller, the control input $u(t)$ is formed as a linear function of the state. We thus assume

$$u(t) = -\mathbf{g}^T \mathbf{e}(t), \tag{6.59}$$

for some gain vector $\mathbf{g}$. By this choice, (6.58) becomes the homogeneous differential equation

$$\dot{\mathbf{e}}(t) = (A - \mathbf{b}\mathbf{g}^T)\mathbf{e}.$$

The control problem is now to choose the gain $\mathbf{g}$ such that $A_c = A - \mathbf{b}\mathbf{g}^T$ has eigenvalues in the desired locations.

Let the characteristic polynomial of A_c be $\chi_{A_c}(s) = s^m + a_{m-1}s^{m-1} + a_1 s + a_0$. Let the desired characteristic polynomial be denoted by

$$d(s) = s^m + d_{m-1}s^{m-1} + \cdots + d_1 s + d_0. \tag{6.60}$$

We desire to choose $\mathbf{g}$ so that $\chi_{A_c}(s) = d(s)$.

Control when A is in first companion form

We begin by assuming that A is in first companion form (see (1.20) and (1.21)), so

$$A = \begin{bmatrix} 0 & 1 & 0 & \cdots & 0 \\ 0 & 0 & 1 & \cdots & 0 \\ \vdots & & & \ddots & \\ 0 & 0 & 0 & \cdots & 1 \\ -a_0 & -a_1 & -a_2 & \cdots & -a_{m-1} \end{bmatrix} \quad \text{and} \quad \mathbf{b} = \begin{bmatrix} 0 \\ 0 \\ \vdots \\ 1 \end{bmatrix}. \tag{6.61}$$

The characteristic polynomial of a matrix in first companion form can be determined simply by looking at the row of coefficients,

$$\chi_A(s) = s^m + a_{m-1}s^{m-1} + \cdots + a_1 s + a_0$$

(see exercise 6.13-60). The closed-loop matrix A_c is

$$A_c = A - \mathbf{b}\mathbf{g}^T = \begin{bmatrix} 0 & 1 & 0 & \cdots & 0 \\ 0 & 0 & 1 & \cdots & 0 \\ \vdots & & & \ddots & \\ 0 & 0 & 0 & \cdots & 1 \\ -a_0 - g_0 & -a_1 - g_1 & -a_2 - g_2 & \cdots & -a_{m-1} - g_{m-1} \end{bmatrix},$$

where

$$\mathbf{g} = [g_0 \quad g_1 \quad \cdots \quad g_{m-1}]^T.$$

Then A_c has the characteristic polynomial

$$\det(sI - A_c) = s^m + (a_{m-1} + g_{m-1})s^{m-1} + \cdots + (a_1 + g_1)s + (a_0 + g_0). \tag{6.62}$$

Equating the desired characteristic polynomial (6.60) to that obtained by the linear feedback (6.62), we see that we must have

$$g_i = d_i - a_i, \quad i = 0, 1, \ldots, m-1.$$

The closed-loop matrix A_c will then have as its eigenvalues the roots of $d(s)$. Writing

$$\mathbf{d} = \begin{bmatrix} d_0 \\ d_1 \\ \vdots \\ d_{m-1} \end{bmatrix} \quad \mathbf{a} = \begin{bmatrix} a_0 \\ a_1 \\ \vdots \\ a_{m-1} \end{bmatrix},$$

we have $\mathbf{g} = \mathbf{d} - \mathbf{a}$.

Control for general *A*

In the more general case, the matrix A is not in first companion form. In this case, the system is transformed by a similarity transformation to a first companion form, the gains are found, and then transformed back. In (6.58), let $\mathbf{z} = T\mathbf{e}$ for a matrix T to be found. Then

$$\dot{\mathbf{z}} = \tilde{A}\mathbf{z} + \tilde{\mathbf{b}}u(t),$$

where $\tilde{A} = TAT^{-1}$ and $\tilde{\mathbf{b}} = T\mathbf{b}$. If T can be found such that $\tilde{A}$ is in first companion form, then a gain vector $\tilde{\mathbf{g}}$ can be found to place the eigenvalues of $\tilde{A}$, and hence of A, in the desired location using the previous technique, since A and $\tilde{A}$ have the same characteristic polynomial. The input control law is then

$$u(t) = -\tilde{\mathbf{g}}^T \mathbf{z} = -\mathbf{g}^T \mathbf{e},$$

from which we find $\mathbf{g} = T^T \tilde{\mathbf{g}}$.

The problem now is to determine the matrix T that transforms a general matrix to a first companion form. This is explored in exercise 6.13-61. It is found that the transformation may be done in two steps:

$$T = VU$$

where U transforms A to a second companion form, and V transforms this to a first companion form. The matrix U is defined by

$$U^{-1} = [\mathbf{b} \quad A\mathbf{b} \quad \cdots \quad A^{m-1}\mathbf{b}].$$

The matrix

$$Q = [\mathbf{b} \quad A\mathbf{b} \quad \cdots \quad A^{m-1}\mathbf{b}] \tag{6.63}$$

is the *controllability test matrix*. If it is not invertible, then there may be sets of eigenvalues that cannot be obtained by the control law (6.59); such a system is said to be not controllable. In the event the system is controllable, the gains are found by

$$\mathbf{g} = [(VU)^T](\mathbf{d} - \mathbf{a}), \tag{6.64}$$

where $\mathbf{d}$ is the vector of desired characteristic equation coefficients, and $\mathbf{a}$ is the vector of open-loop characteristic equation coefficients. Equation (6.64) is known as the *Bass–Gura* formula for pole placement. Pole placement can be accomplished in MATLAB using the `place` command.

6.14 Noiseless constrained channel capacity

In some communications settings there are constraints on the number of runs of symbols. For example, a constraint might be imposed that there can be no runs of more than two repetitions of the same symbol. Thus the sequence 111 would not be allowed, but the sequence 01010 is allowed. A channel in which constraints of this sort are imposed is called a *discrete noiseless channel*. Constraints of this sort arise, for example, in a magnetic recording channel such as a floppy diskette. Data are recorded on a disk by writing no pulse to represent a zero, and writing a pulse to represent a one. The bit string 010001 would correspond to no pulse (blank), followed by a single pulse, followed by three blanks, followed by another pulse. In such a recording channel, there are two constraints that are commonly enforced. First, there is a constraint on the length of a run of zeros. The synchronization of the read/write system requires that pulses must appear sufficiently often to keep the pulse-detection circuitry from drifting too far from the spinning magnetic medium. The second constraint is that ones should be separated by some small number of zeros. If two ones appear next to each other, the magnetic pulses representing them could overlap and lead to degraded performance. To enforce the constraints, a raw (unconstrained) bit stream is passed through a coder which produces a coded sequences that does satisfy the constraints. As the data is read from the disk, the bits are converted back into a raw bit stream.

Example 6.14.1 Consider a channel in which the following constraints must be satisfied:

1. Not more than two zeros in succession.
2. Not more than two ones in succession.

Then the following sequences of length 4 are allowed:

$$\begin{array}{ccccc} 0010 & 0011 & 0100 & 0110 & 1001 \\ 1010 & 0101 & 1100 & 1101 & 1011 \end{array}$$

Since there are 16 possible unconstrained sequences of length 4, and only 10 possible constrained sequences of length 4, there is necessarily a limit on the amount of information that can be put into the coder and have the coder keep up. □

Let $M(k)$ be the number of sequences of length k that satisfy the constraints. In this example, $M(4) = 10$. The average number of bits that the k output symbols can represent is $\frac{1}{k} \log_2 M(k)$. In the constrained code of the example,

$$\frac{1}{4} \log_2 M(4) = 0.8305 \text{ bits/symbol.}$$

As this example demonstrates, a fundamental issue is whether the encoded sequence can convey sufficient information that the coder can keep up with the incoming data stream.

We define the *capacity* of a discrete noiseless channel using the asymptotic value

$$C = \lim_{k \to \infty} \frac{\log_2 M(k)}{k}$$

in units of bits per symbol. The capacity C represents the average number of bits it would take to represent a (long) sequence. For example, if there are no constraints at all, then there are 2^k sequences of length k, and the capacity is one bit per symbol. An analysis problem associated with the discrete noiseless channel is: given the constraints on a channel, determine the capacity of the channel. We shall address this question here, showing that the answer depends upon the largest eigenvalue of a particular matrix. (The more interesting question of how to design coders that approach the capacity is not covered here.)

Our approach will be to find a bound on the number of sequences of length k. The first step is to represent the constraints of the channel using a state diagram, such as that shown in figure 6.13. For example, arriving at state S_0 after a run of ones, a zero must be transmitted. The state diagram can be represented using a *state-transition matrix*. For the

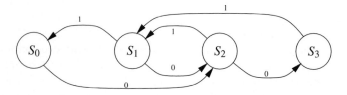

Figure 6.13: State diagram for a constrained channel

6.14 Noiseless Constrained Channel Capacity

state diagram of figure 6.13, the state-transition matrix is

$$T = \begin{bmatrix} 0 & 0 & 1 & 0 \\ 1 & 0 & 1 & 0 \\ 0 & 1 & 0 & 1 \\ 0 & 1 & 0 & 0 \end{bmatrix}. \quad (6.65)$$

A one in position T_{ij} (indexing starting at 0) corresponds to a single path of length 1 from state S_i to state S_j. We can count the number of path segments in this state-transition diagram by summing over the elements in $\sum_{i,j} T_{ij} = 6$. This is an *upper bound* on the number of paths of length 1, since paths emerging from different states may represent the same sequence.

Paths of length 2 in the state-transition diagram can be obtained by computing T^2, and paths of length m can be found from T^m. For example,

$$T^3 = \begin{bmatrix} 1 & 1 & 1 & 0 \\ 1 & 2 & 1 & 1 \\ 1 & 1 & 2 & 1 \\ 0 & 1 & 1 & 1 \end{bmatrix},$$

meaning, for example, that there are two paths of length 3 from state S_2 to state S_2. The total number of paths of length 3 is the sum of the elements in T^3.

If all the sequences have length m, then the number of sequences of length m can be given a upper bound by

$$M(m) \leq \sum_i \sum_j (T^m)_{ij},$$

where there is an upper bound, because not all states are necessarily allowable as starting states for symbol sequences and some sequences may be repeated. The upper bound is sufficient for our purposes.

To find the capacity discrete noiseless channel, we will use the upper bound

$$\lim_{k \to \infty} \frac{1}{k} \log_2 \sum_{ij} (T^m)_{ij}.$$

Let T have the ordered eigenvalues $\lambda_1 > \lambda_2 \geq \cdots \geq \lambda_m$. Factoring $T = S\Sigma S^{-1}$, we note that

$$\sum_{ij}(T^k)_{ij} = \sum_{ij}(S\Lambda^k S^{-1})_{ij})$$

$$= \sum_{ij} \lambda_1^k \alpha_{ij1} + \lambda_2^k \alpha_{ij2} + \cdots + \lambda_m^k \alpha_{ijm}$$

$$= \lambda_1^k \sum_{ij} \alpha_{ij1} + \left(\frac{\lambda_2}{\lambda_1}\right)^k \alpha_{ij2} + \cdots + \left(\frac{\lambda_m}{\lambda_1}\right)^k \alpha_{ijm},$$

where α_{ijl} does not depend upon k. Then

$$C \leq \lim_{k \to \infty} \frac{1}{k} \log_2 \sum_{ij}(T^k)_{ij} = \log_2 \lambda_1.$$

Taking the upper bound as the desired value, we find that the capacity is $\log_2 \lambda_1$ bits/symbol.

Example 6.14.2 For the matrix T of (6.65), the capacity is $\log_2(1.618) = 0.6942$ bits/symbol. An interpretation of this is that 100 bits of this constrained output represent only 69 bits of unconstrained data. □

6.15 Computation of eigenvalues and eigenvectors

The area of numerical analysis dealing with the computation of eigenvalues and eigenvectors is both broad and deep, and we can provide only an introduction here. Because of its importance in eigenfilter and principal component analysis, we discuss means of computing the largest and smallest eigenvalues using the power method. Attention is then directed to the case of symmetric matrices because of their importance in signal processing.

6.15.1 Computing the largest and smallest eigenvalues

A simple iterative algorithm known as the power method can be used to find the largest eigenvalue and its associated eigenvector. Algorithms requiring only the largest eigenvalue can benefit by avoiding the overhead of computing eigenvalues that are not needed. The power method works for both self-adjoint and non–self-adjoint matrices.

Let A be an $m \times m$ diagonalizable matrix with possibly complex eigenvalues ordered as $|\lambda_1| \geq |\lambda_2| \geq \cdots \geq |\lambda_m|$, with corresponding eigenvectors $\mathbf{x}_1, \mathbf{x}_2, \ldots, \mathbf{x}_m$. Let $\mathbf{x}^{[0]}$ be a normalized vector that is assumed to be not orthogonal to $\mathbf{x}_1$. The vector $\mathbf{x}^{[0]}$ can be written in terms of the eigenvectors as

$$\mathbf{x}^{[0]} = a_1 \mathbf{x}_1 + a_2 \mathbf{x}_2 + \cdots + a_m \mathbf{x}_m$$

for some set of coefficients $\{a_i\}$, where $a_1 \neq 0$. We define the power method recursion by

$$\mathbf{x}^{[k+1]} = A\mathbf{x}^{[k]}. \tag{6.66}$$

Then

$$\mathbf{x}^{[1]} = A\mathbf{x}^{[0]} = a_1 \lambda_1 \left(\mathbf{x}_1 + \frac{a_2}{a_1} \frac{\lambda_2}{\lambda_1} + \cdots + \frac{a_m}{a_1} \frac{\lambda_m}{\lambda_1} \right),$$

$$\mathbf{x}^{[2]} = A\mathbf{x}^{[1]} = a_1 \lambda_1^2 \left(\mathbf{x}_1 + \frac{a_2}{a_1} \left(\frac{\lambda_2}{\lambda_1}\right)^2 + \cdots + \frac{a_m}{a_1} \left(\frac{\lambda_m}{\lambda_1}\right)^2 \right),$$

$$\vdots$$

Because of the ordering of the eigenvalues, as $k \to \infty$,

$$\mathbf{x}^{[k]} \to a_1 \mathbf{x}_1,$$

which is the eigenvector of A corresponding to the largest eigenvalue. The eigenvalue itself is found by a Rayleigh quotient,

$$(\mathbf{x}^{[k]})^H A \mathbf{x}^{[k]} / \|\mathbf{x}^{[k]}\| \to \lambda_1.$$

The method is illustrated in algorithm 6.6.

Algorithm 6.6 Computation of the largest eigenvalue using the power method
File: `maxeig.m`

An approach suggested in [236] for finding the second largest eigenvalue and its eigenvector, after knowing the largest eigenvalue, is to form the matrix

$$A_1 = A - \lambda_1 \mathbf{x}_1 \mathbf{x}_1^H,$$

where $\mathbf{x}_1$ is the normalized eigenvector. Algebraically, the largest root of

$$\det(A_1 - \lambda I) = 0$$

6.15 Computation of Eigenvalues and Eigenvectors

is the second largest eigenvalue of A. Computation of the eigenvalue can be obtained by the power method applied to A_1. The result depends on correct computation of λ_1, so there is potential for numerical difficulty. Extending this technique, the ith principal component can be found after the first $i - 1$ are determined by forming

$$A_{i-1} = A - \sum_{j=1}^{i-1} \lambda_i \mathbf{x}_i \mathbf{x}_i^H$$

and using the power method on A_{i-1}. If many eigenvalues and eigenvectors are needed, it may be more efficient computationally and numerically to compute a complete eigendecomposition.

Finding the smallest eigenvalue can be accomplished in at least two ways. If λ_1 is an eigenvalue of A, then $1/\lambda_1$ is an eigenvalue of A^{-1}, and the eigenvectors in each case are the same. The largest eigenvalue λ of A^{-1} is thus the reciprocal of the smallest eigenvalue of A. This method would require inverting A.

Alternatively, we can form $B = \lambda_1 I - A$, which has largest eigenvalue $\lambda_1 - \lambda_m$. The power method can be applied to B to find $\lambda_1 - \lambda_m$, from which λ_m can be obtained. See algorithm 6.7.

Algorithm 6.7 Computation of the smallest eigenvalue using the power method
File: `mineig.m`

6.15.2 Computing the eigenvalues of a symmetric matrix

Finding the full set of eigenvalues and eigenvectors of a matrix has been a matter of considerable study. Thorough discussions are provided in [370] and [114], while some numerical implementations are discussed in [260].

A *real* symmetric matrix A is orthogonally similar to a diagonal matrix Λ,

$$A = Q \Lambda Q^T.$$

The eigenvalues of A are then found explicitly on the diagonal of Λ, and the eigenvectors are found in Q. One strategy for finding the eigenvalues and eigenvectors is to move A toward being a diagonal matrix by a series of orthogonal transformations such as Householder transformations or Givens rotations, which were discussed in conjunction with the QR factorization in section 5.3. One approach to this strategy is to first reduce A to a *tridiagonal* matrix by a series of Householder transformations, then apply a series of Givens rotations that efficiently diagonalize the tridiagonal matrix. This technique has been shown to provide a good mix of computational speed (by means of the tridiagonalization) with numerical accuracy (using the rotations). Throughout the following discussion, MATLAB code is provided to make the presentation concrete. (MATLAB, of course, provides eigenvalues and eigenvectors via the function `eig`.)

Tridiagonalization of A

Let A be an $m \times m$ symmetric matrix. Let

$$Q_1 = \begin{bmatrix} 1 & \\ & H_1 \end{bmatrix}$$

be an orthogonal matrix, where H_1 is a Householder transformation. The transformation is chosen so that $Q_1 A$ has zeros down the first column in positions $3{:}m$. Since A is

symmetric,

$$Q_1 A Q_1^T = \begin{bmatrix} a_{11} & \times & 0 & \cdots & 0 \\ \times & & & & \\ 0 & & & B_1 & \\ \vdots & & & & \\ 0 & & & & \end{bmatrix}, \qquad (6.67)$$

where $\times$ indicates an element which is not zero, and B_1 is an $(m-1) \times (m-1)$ matrix. We continue iteratively, applying Householder transformations to set the subdiagonals and superdiagonals to zero. Then

$$T = Q_{m-2} \cdots Q_2 Q_1 A Q_1^T Q_2^T \cdots Q_{m-2}^T = Q^T A Q$$

is tridiagonal. Algorithm 6.8 illustrated tridiagonalization using Householder transformations. The computation cost can be reduced, by exploiting the symmetry of A (see exercise 6.15-68), to $4m^3/3$ floating operations if the matrix Q is not returned; keeping track of Q requires another $4m^3/3$ floating operations.

Algorithm 6.8 Tridiagonalization of a real symmetric matrix
File: `tridiag.m`

Example 6.15.1 For the matrix

$$A = \begin{bmatrix} 3.15447 & -0.516339 & -0.348363 & -0.0191762 \\ -0.516339 & 2.37035 & 0.318412 & -0.794899 \\ -0.348363 & 0.318412 & 2.29206 & -1.04116 \\ -0.0191762 & -0.794899 & -1.04116 & 2.18312 \end{bmatrix},$$

the tridiagonal forms, T and Q are

$$T = \begin{bmatrix} 3.15447 & 0.623162 & 0 & 0 \\ 0.623162 & 2.56432 & 1.25495 & 0 \\ 0 & 1.25495 & 2.30604 & -0.404962 \\ 0 & 0 & -0.404962 & 1.97516 \end{bmatrix},$$

$$Q = \begin{bmatrix} 1 & 0 & 0 & 0 \\ 0 & -0.828579 & 0.00572494 & -0.559842 \\ 0 & -0.559025 & -0.0634203 & 0.826722 \\ 0 & -0.0307724 & 0.99797 & 0.0557491 \end{bmatrix}.$$

$\square$

6.15.3 The QR iteration

Having found the tridiagonal form, we reduce the matrix further toward a diagonal form using QR iteration. We form the QR factorization of T as

$$T = Q_0 R_0.$$

Then we observe that

$$Q_0^T T Q_0 = Q_0^T Q_0 R_0 Q_0 = R_0 Q_0.$$

6.15 Computation of Eigenvalues and Eigenvectors

Let $T^{[1]} = R_0 Q_0$. We then proceed iteratively, alternating a QR factorization step with a reversal of the product:

$$T^{[0]} = T = Q_0 R_0,$$
$$T^{[1]} = R_0 Q_0$$
$$= Q_1 R_1,$$
$$\vdots$$
$$T^{[k]} = Q_k R_k,$$
$$T^{[k+1]} = R_k Q_k.$$

The key result is provided by the following theorem.

Theorem 6.9 *If the eigenvalues of A (and hence of T) are of different absolute value $|\lambda_i|$, then $T^{[k]}$ approaches a diagonal matrix as $k \to \infty$. In this matrix, the eigenvalues are ordered down the diagonal so that*

$$\left|T_{11}^{[k]}\right| > \left|T_{22}^{[k]}\right| > \cdots > \left|T_{mm}^{[k]}\right|.$$

The proof of the theorem is too length to fit within the scope of this book (see, e.g. [326, chapter 6; 114, chapter 7]). Since $T^{[k]}$ approaches a diagonal matrix, we can read the eigenvalues of A directly off the diagonal of $T^{[k]}$, for k sufficiently large. Since $T^{[0]} = T$ is tridiagonal, $T^{[0]}$ can be converted to upper-triangular form using only $m - 1$ Givens rotations. This is an important reason for tridiagonalizing first, since given proper attention the number of computations can be greatly reduced.

In the proof of theorem 6.9, it is shown that a superdiagonal element of $T^{[k]}$ converges to zero as

$$T_{ij}^{[k]} \approx \left(\frac{\lambda_i}{\lambda_j}\right)^k.$$

Since $\lambda_i < \lambda_j$, this does converge. However, if $|\lambda_i|$ is near to $|\lambda_j|$, convergence is slow. The convergence can be accelerated by means of shifting, which relies on the observation that if λ is an eigenvalue of T, then $\lambda - \tau$ is an eigenvalue of $T - \tau I$. Based on this, we factor

$$T^{[k]} - \tau_k I = Q_k R_k$$

then write

$$T^{[k+1]} = R_k Q_k + \tau_k I.$$

This is known as an *explicit shift* QR iteration. With the shift, the convergence can be shown to be determined by the ratio

$$\frac{\lambda_i - \tau_k}{\lambda_j - \tau_k}.$$

Then the shift τ_k is selected at each k to maximize the rate of convergence. A good choice could be to select τ_k close to the smallest eigenvalue, λ_m; however, this is not generally known in advance. An effective alternative strategy is to compute the eigenvalues of the 2×2 submatrix in the lower right of T, and use that eigenvalue which is closest to $T_{mm}^{[k]}$. This is known as the Wilkinson shift.

While the explicit shift usually works well, subtracting a large τ_k from the diagonal elements can lead to a loss of accuracy for the small eigenvalues. What is preferred is the *implicit QR shift* algorithm. Briefly, how this works is that a Givens rotation matrix is found

so that

$$\begin{bmatrix} c & -s \\ s & c \end{bmatrix} \begin{bmatrix} T_{11}^{[k]} - \tau_k \\ b \end{bmatrix} = \begin{bmatrix} \times \\ 0 \end{bmatrix}.$$

(That is, the rotation zeros out an element below the diagonal of the *shifted* matrix.) However, the shift is never explicitly computed, only the appropriate Givens matrix. Application of the rotation for the shift introduces new elements in the off diagonals. For example, the 5×5 matrix

$$T = \begin{bmatrix} \times & \times & 0 & 0 & 0 \\ \times & \times & \times & 0 & 0 \\ 0 & \times & \times & \times & 0 \\ 0 & 0 & \times & \times & \times \\ 0 & 0 & 0 & \times & \times \end{bmatrix},$$

where $\times$ indicates nonzero elements, when operated on by the Givens rotation G_1 designed to zero out the $(1, 2)$ element of the *shifted* matrix, becomes

$$G_1 T G_1^T = \begin{bmatrix} \times & \times & + & 0 & 0 \\ \times & \times & \times & 0 & 0 \\ + & \times & \times & \times & 0 \\ 0 & 0 & \times & \times & \times \\ 0 & 0 & 0 & \times & \times \end{bmatrix},$$

where $+$ indicates nonzero elements that are introduced. A series of Givens rotations that do not operate on the shifted matrix is then applied to chase these nonzero elements down the diagonal:

$$\begin{bmatrix} \times & \times & + & 0 & 0 \\ \times & \times & \times & 0 & 0 \\ + & \times & \times & \times & 0 \\ 0 & 0 & \times & \times & \times \\ 0 & 0 & 0 & \times & \times \end{bmatrix} \xrightarrow{G_2} \begin{bmatrix} \times & \times & 0 & 0 & 0 \\ \times & \times & \times & + & 0 \\ 0 & \times & \times & \times & 0 \\ 0 & + & \times & \times & \times \\ 0 & 0 & 0 & \times & \times \end{bmatrix} \xrightarrow{G_3} \begin{bmatrix} \times & \times & 0 & 0 & 0 \\ \times & \times & \times & 0 & 0 \\ 0 & \times & \times & \times & + \\ 0 & 0 & \times & \times & \times \\ 0 & 0 & + & \times & \times \end{bmatrix} \xrightarrow{G_4} \begin{bmatrix} \times & \times & 0 & 0 & 0 \\ \times & \times & \times & 0 & 0 \\ 0 & \times & \times & \times & 0 \\ 0 & 0 & \times & \times & \times \\ 0 & 0 & 0 & \times & \times \end{bmatrix}.$$

The steps of introducing the shifted Givens rotation, followed by the Givens rotations which restore the tridiagonal form, are collectively called an *implicit QR shift*. Code which implements this implicit QR shift is shown in algorithm 6.9.

Algorithm 6.9 Implicit QR shift
File: `eigqrshiftstep.m`

Combining the tridiagonalization and the implicit QR shift is shown in algorithm 6.10. Following the initial tridiagonalization, the matrix T is driven toward a diagonal form, with the lower right corner (probably) converging first. The matrix T is split into three pieces,

$$T = \begin{bmatrix} T_1 & & \\ & T_2 & \\ & & T_3 \end{bmatrix},$$

where T_3 is diagonal (as determined by a comparison with a threshold ϵ), and T_1 is also. The implicit QR shift is applied only to T_2. The algorithm iterates until T is fully diagonalized.

6.16 Exercises

Algorithm 6.10 Complete eigenvalue/eigenvector function
File: `neweig.m`

Example 6.15.2 For the matrix

$$A = \begin{bmatrix} 3.15447 & -0.516339 & -0.348363 & -0.0191762 \\ -0.516339 & 2.37035 & 0.318412 & -0.794899 \\ -0.348363 & 0.318412 & 2.29206 & -1.04116 \\ -0.0191762 & -0.794899 & -1.04116 & 2.18312 \end{bmatrix},$$

the statement `[T,X] = neweig(A)` returns the eigenvalues in T and the eigenvectors in X as

$$T = \begin{bmatrix} 4 & 0 & 0 & 0 \\ 0 & 3 & 0 & 0 \\ 0 & 0 & 1 & 0 \\ 0 & 0 & 0 & 2 \end{bmatrix} \quad X = \begin{bmatrix} 0.5 & 0.829341 & -0.182574 & -0.169882 \\ -0.5 & 0.0606835 & -0.365148 & -0.782933 \\ -0.5 & 0.303418 & -0.547723 & 0.598279 \\ 0.5 & -0.46524 & -0.730297 & -0.0147723 \end{bmatrix}.$$ □

6.16 Exercises

6.2-1 (Eigenfunctions and eigenvectors.)

(a) Let $\mathcal{L}$ be the operator that computes the negative of the second derivative, $\mathcal{L}u = -\frac{d^2}{dt^2}u$, defined for functions on $(0, 1)$. Show that

$$u_n(t) = \sin(n\pi t)$$

is an eigenfunction of $\mathcal{L}$, with eigenvalue $\lambda_n = (n\pi)^2$.

(b) In many numerical problems, a differentiation operator is discretized. Show that we can approximate the second derivative operator by

$$\frac{d^2}{dt^2} \approx \frac{u(t+h) + 2u(t) + u(t-h)}{h^2},$$

where h is some small number.

(c) Discretize the interval $[0, 1]$ into $0, t_1, t_2, \ldots, t_N$, where $t_i = i/N$. Let $\mathbf{u} = [u(t_1), \ldots, u(t_{N-1})]^T$, and show that the operator $\mathcal{L}u$ can be approximated by the operator $\frac{1}{N^2}L\mathbf{u}$, where

$$L = \begin{bmatrix} 2 & -1 & & & & \\ -1 & 2 & -1 & & & \\ 0 & -1 & 2 & -1 & & \\ & \ddots & \ddots & \ddots & & \\ & & & -1 & 2 & -1 \\ & & & & 2 & -1 \end{bmatrix}.$$

(d) Show that the eigenvectors of L are

$$\mathbf{x}_n = [\sin(n\pi/N) \quad \sin(2n\pi/N) \quad \cdots \quad \sin((N-1)n\pi/N)]^T \quad n = 1, 2, \ldots, N,$$

where $\lambda_n = 4\sin^2(n\pi/(2N))$. Note that $\mathbf{x}_n$ is simply a sampled version of $x_n(t)$.

6.2-2 Find the eigenvalues of the following matrices:

(a) A diagonal matrix

$$A = \begin{bmatrix} a_{11} & & & \\ & a_{22} & & \\ & & \ddots & \\ & & & a_{nn} \end{bmatrix}.$$

(b) A triangular matrix (either upper or lower; upper in this exercise)

$$A = \begin{bmatrix} a_{11} & a_{12} & a_{13} & \cdots & a_{1n} \\ 0 & a_{22} & a_{23} & \cdots & a_{2n} \\ \vdots & & & \ddots & \\ 0 & 0 & 0 & \cdots & a_{nn} \end{bmatrix}.$$

(c) From these exercises conclude the following.

> The diagonal elements form the eigenvalues of A if A is triangular.

6.2-3 For matrix T in block-triangular form

$$T = \begin{bmatrix} T_{11} & T_{12} \\ 0 & T_{22} \end{bmatrix},$$

show that $\lambda(T) = \lambda(T_{11}) \cup \lambda(T_{22})$.

6.2-4 Show that the determinant of an $n \times n$ matrix is the product of the eigenvalues; that is,

$$\boxed{\det(A) = \prod_{i=1}^{n} \lambda_i}$$

6.2-5 Show that the trace of a matrix is the sum of the eigenvalues,

$$\boxed{\operatorname{tr}(A) = -\sum_{i=1}^{n} \lambda_i}$$

6.2-6 We will use the previous two results to prove a useful inequality (**Hadamard's inequality**). Let A be a symmetric positive-definite $m \times m$ matrix. Then

$$\boxed{\det(A) \leq \prod_{i=1}^{m} a_{ii}}$$

with equality if and only if A is diagonal.

(a) Show that we can write $A = DBD$, where D is diagonal and B has only ones on the diagonal. (Determine D.)

(b) Explain the following equalities and inequalities. (Hint: use the arithmetic–geometric inequality. See equation (A.36). What is λ_i here?)

$$\det(A) = \det(DBD) = \left(\prod_{i=1}^{m} a_{ii}\right) \det(B)$$

$$= \left(\prod_{i=1}^{m} a_{ii}\right) \prod_{i=1}^{m} \lambda_i \leq \left(\prod_{i=1}^{m} a_{ii}\right) \left(\frac{1}{m}\sum_{i=1}^{m} \lambda_i\right)^n$$

$$= \left(\prod_{i=1}^{m} a_{ii}\right) \left(\frac{1}{n}\operatorname{tr}(B)\right)^m = \left(\prod_{i=1}^{m} a_{ii}\right).$$

6.2-7 Suppose A is a rank-1 matrix formed by $A = \mathbf{ab}^T$. Find the eigenvalues and eigenvectors of A. Also show that if A is rank 1, then

$$\det(I + A) = 1 + \operatorname{tr}(A).$$

6.2-8 Show that if λ^* is an eigenvalue of A, then $\lambda^* + r$ is an eigenvalue of $A + rI$, and that A and $A + rI$ have the same eigenvectors.

6.2-9 Show that

> If λ is an eigenvalue of A then λ^n is an eigenvalue of A^n, and A^n has the same eigenvectors as A.

6.2-10 Show that

> If λ is a nonzero eigenvalue of A then $1/\lambda$ is an eigenvalue of A^{-1}.

> The eigenvectors of A corresponding to nonzero eigenvalues are eigenvectors of A^{-1}.

6.2-11 Generalizing the previous two problems, show that if $\lambda_1, \lambda_2, \ldots, \lambda_m$ are the eigenvalues of A, and if $g(x)$ is a scalar polynomial, then the eigenvalues of $g(A)$ are $g(\lambda_1), g(\lambda_2), \ldots, g(\lambda_m)$.

6.2-12 Show that the eigenvalues of a projection matrix P are either 1 or 0.

6.2-13 In this problem you will establish some results on eigenvalues of products of matrices.
  (a) If A and B are both square, show that the eigenvalues of AB are the same as the eigenvalues of BA.
  (b) Show that if the $n \times n$ matrices A and B have a common set of n linearly independent eigenvectors, then $AB = BA$.

  A thorough study of when $AB = BA$, as introduced in this problem, is treated in [245].

6.2-14 Show that a stochastic matrix has $\lambda = 1$ as an eigenvalue, and that $\mathbf{x} = [1, 1, \ldots, 1]^T$ is the corresponding eigenvector. (It can be shown that $\lambda = 1$ is the largest eigenvalue [245].)

6.2-15 (Linear fixed-point problems). Some problems are of the form

$$A\mathbf{x} = \mathbf{x}.$$

If A has an eigenvalue equal to 1, then this problem has a solution. (Conditions guaranteeing that A has an eigenvalue of 1 are described in [227].) Example problems of this sort are the steady-state probabilities for a Markov chain, and the determination of values for a compactly supported wavelet at integer values of the argument.

(a) Let
$$A = \begin{bmatrix} .5 & .3 & .2 \\ .2 & 0 & .7 \\ .3 & .7 & .1 \end{bmatrix}$$
be the state-transition probability matrix for a Markov model. Determine the steady-state probability $\mathbf{p}$ such that $A\mathbf{p} = \mathbf{p}$.

(b) The two-scale equation for a scaling function (3.110) is $\phi(t) = \sum_{k=0}^{N-1} c_k \phi(2t - k)$. Given that we know that $\phi(t)$ is zero for $t \leq 0$ and for $t \geq N - 1$, write an equation of the form
$$\begin{bmatrix} \phi(1) \\ \phi(2) \\ \vdots \\ \phi(N-2) \end{bmatrix} = A \begin{bmatrix} \phi(1) \\ \phi(2) \\ \vdots \\ \phi(N-2) \end{bmatrix},$$
where A is a matrix of wavelet coefficients c_k. Given the set of coefficients $\{c_k\}$, specify A and describe how to solve this equation. Describe how to find $\phi(t)$ at all dyadic rational numbers (numbers of the form $k/2^i$ for integers k and i).

6.3-16 Show that the inertia of a Hermitian matrix A is uniquely determined if the signature and rank of A are known.

6.3-17 (Sylvester's law of inertia) Show that if A and B same inertia, then there is a matrix S such that $A = SBS^H$. Hint: Diagonalize $A = U_A \Lambda_A U_A^H = U_A D_A \Sigma_A D_A U_A^H$, where Σ_A is diagonal with $\{\pm 1, 0\}$ elements. Similarly for B.

6.3-18 Show that if A and B are similar, so that $B = T^{-1}AT$,

(a) A and B have the have the same eigenvalues and the same characteristic equation.

(b) If $\mathbf{x}$ is an eigenvector of A then $\mathbf{z} = T^{-1}\mathbf{x}$ is an eigenvector of B.

(c) If, in addition, C and D are similar with $D = T^{-1}CT$, then $A + C$ is similar to $B + D$.

6.3-19 Determine the Jordan forms of
$$A_1 = \begin{bmatrix} 2 & 1 & 2 \\ 0 & 2 & 3 \\ 0 & 0 & 2 \end{bmatrix}$$
and
$$A_2 = \begin{bmatrix} 2 & 0 & 2 \\ 0 & 2 & 3 \\ 0 & 0 & 2 \end{bmatrix}.$$

6.3-20 Show that (6.15) is true for the 3×3 matrix shown. Then generalize (by induction) to an $m \times m$ Jordan block.

6.3-21 Show that

> A self-adjoint matrix is positive semidefinite if and only if all of its eigenvalues are ≥ 0.

Also show that if all the eigenvalues are positive, then the matrix is positive definite.

6.3-22 Show that if a Hermitian matrix A is positive definite, then so is A^k for $k \in \mathbb{Z}$ (positive as well as negative powers).

6.3-23 Show that if A is nonsingular, then AA^H is positive definite.

6.16 Exercises

6.3-24 Prove theorem 6.2 by establishing the following two steps.
   (a) Show that if A is self-adjoint and U is unitary, then $T = U^H A U$ is also self-adjoint.
   (b) Show that if a self-adjoint matrix is triangular, then it must be diagonal.

6.3-25 Prove lemma 6.5.

6.3-26 A matrix N is **normal** if it commutes with N^H: $N^H N = N N^H$.
   (a) Show that unitary, symmetric, Hermitian, and skew-symmetric and skew-Hermitian matrices are normal. (A matrix A is skew symmetric if $A^T = -A$. It is skew Hermitian if $A^H = -A$.)
   (b) Show that for a normal matrix, the triangular matrix determined by the Schur lemma is diagonal.

6.3-27 Show that for a Hermitian matrix A, if $A^2 = A$ then $\text{rank}(A) = \text{tr}(A)$.

6.3-28 Let

$$F = \begin{bmatrix} 1 & 1 & \cdots & 1 \\ 1 & e^{-j2\pi/N} & \cdots & e^{-j2\pi(N-1)/N} \\ 1 & e^{-j4\pi/N} & \cdots & e^{-j4\pi(N-1)/N} \\ \vdots & & & \vdots \\ 1 & e^{-j\pi(N-1)/N} & \cdots & e^{-j\pi(N-1)^2/N} \end{bmatrix}.$$

The (i, j)th element of this is $f_{ij} = e^{-j2\pi ij/N}$. For a vector $\mathbf{x} = [x_0, \ldots, x_{n-1}]^T$, the product $\mathbf{X} = F\mathbf{x}$ is the DFT of $\mathbf{x}$.
   (a) Prove that the matrix $F/\sqrt{N}$ is unitary. Hint: show the following.

$$\sum_{n=0}^{N-1} e^{j2\pi nk/N} = \begin{cases} N & k \equiv 0 \bmod N \\ 0 & k \not\equiv 0 \bmod N \end{cases}$$

   (b) A matrix

$$C = \begin{bmatrix} c_0 & c_1 & c_2 & \cdots & c_{N-1} \\ c_{N-1} & c_0 & c_1 & \cdots & c_{N-2} \\ \vdots & & & & \\ c_1 & c_2 & c_3 & \cdots & c_0 \end{bmatrix}$$

   is said to be a *circulant matrix*. Show that C is diagonalized by F, $CF = F\Lambda$, where Λ is diagonal. Comment on the eigenvalues and eigenvectors of a circulant matrix.
   (The FFT-based approach to cyclic convolution works essentially by transforming a circulant matrix to a diagonal matrix, where multiplication point-by-point can occur, followed by transformation back to the original space.)

6.4-29 Prove theorem 6.4. Hint: Start with $A = U \Lambda U^H$.

6.4-30 Construct 3×3 matrices according to the following sets of specifications.
   (a) $\lambda_1 = \lambda_2 = 1$, $\lambda_3 = 2$, with invariant subspaces

$$R_1 = \text{span}([1, 2, 1]^T, [-2, 1, 0]^T) \qquad R_2 = \text{span}([-1, -2, 5]^T).$$

   In this case, determine the eigenvalues and eigenvectors of the matrix you construct, and comment on the results.

(b) $\lambda_1 = 1$, $\lambda_2 = 4$, $\lambda_3 = 9$, with corresponding eigenvectors

$$\mathbf{x}_1 = \frac{1}{\sqrt{14}} \begin{bmatrix} 1 \\ 2 \\ 3 \end{bmatrix} \quad \mathbf{x}_2 = \frac{1}{\sqrt{5}} \begin{bmatrix} -2 \\ 1 \\ 0 \end{bmatrix} \quad \mathbf{x}_3 = \frac{1}{\sqrt{70}} \begin{bmatrix} -3 \\ -6 \\ 5 \end{bmatrix}.$$

6.4-31 Let A be $m \times m$ Hermitian with k distinct eigenvalues and spectral representation $A = \sum_{i=1}^{k} \lambda_i P_i$, where P_i is the projector onto the ith invariant subspace. Show that

$$(sI - A)^{-1} = \sum_{i=1}^{k} \frac{P_i}{s - \lambda_i}.$$

6.4-32 [164, page 663] The diagonalization of self-adjoint matrices can be extended to more general matrices. Let A be an $m \times m$ matrix with m linearly independent eigenvectors $\mathbf{x}_1, \mathbf{x}_2, \ldots, \mathbf{x}_m$, and let $S = [\mathbf{x}_1, \mathbf{x}_2, \ldots, \mathbf{x}_m]$. Let $T = S^{-1}$. Then we have $A = S\Lambda T$, where Λ is the diagonal matrix of eigenvalues.

(a) Let $\mathbf{t}_i^T$ be a row of T. Show that $\mathbf{t}_i^T \mathbf{x}_j = \delta_{ij}$.
(b) Show that $A = \sum_{i=1}^{m} \lambda_i \mathbf{x}_i \mathbf{t}_i^T$.
(c) Let $P_i = \mathbf{x}_i \mathbf{t}_i^T$. Show that $P_i P_j = P_i \delta_{ij}$.
(d) Show that $I = \sum_{i=1}^{m} P_i$ (resolution of identity).
(e) Show that

$$(sI - A)^{-1} = \sum_{i=1}^{m} \frac{P_i}{s - \lambda_i}.$$

6.5-33 Let

$$R = \begin{bmatrix} 5.15385 & -3.23077 \\ -3.23077 & 7.84615 \end{bmatrix}.$$

(a) Determine the eigenvalues and eigenvectors of R.
(b) Draw level curves of the quadratic form $Q_R(\mathbf{x})$. Identify the eigenvector directions on the plot, and associate these with the eigenvalues.
(c) Draw the level curves of the quadratic form $Q_{R^{-1}}(\mathbf{x})$, identifying eigenvector directions and the eigenvalues.

6.5-34 Show that if A is a Hermitian $m \times m$ matrix, and if $\mathbf{x}^H A \mathbf{x} \geq 0$ for all vectors $\mathbf{x}$ in a k-dimensional space (with $k \leq m$), then A has at least k nonnegative eigenvalues. Also show that if $\mathbf{x}^H A \mathbf{x} > 0$ for all nonzero $\mathbf{x}$ in a k-dimensional space, then A has at least k positive eigenvalues. Hint: Let S_k be the k-dimensional space, and let $\mathbf{u}_1, \ldots, \mathbf{u}_{n-k}$ span $S_k^\perp$. Let $C = [\mathbf{u}_1, \ldots, \mathbf{u}_{n-k}]$, and use the Courant minimax principle by considering

$$\min_{\substack{\mathbf{x} \neq 0 \\ C\mathbf{x}=0}} \frac{\mathbf{x}^H A \mathbf{x}}{\mathbf{x}^H \mathbf{x}}.$$

6.5-35 In the proof of theorem 6.5,
(a) Show that (6.29) is true.
(b) Show that $Q(\mathbf{x})$ of (6.29) is maximized when $\alpha_{k+1} = \alpha_{k+2} = \cdots = \alpha_m = 0$.

6.5-36 Write and test a MATLAB function plotellipse(A,x0,c) that computes points on the ellipse described by $(\mathbf{x} - \mathbf{x}_0)^T A (\mathbf{x} - \mathbf{x}_0) = c$ suitable for plotting.

6.16 Exercises

6.6-37 Determine stationary values (eigenvalues and eigenvectors) of $\mathbf{x}^T R \mathbf{x}$, subject to $\mathbf{x}^T \mathbf{c} = 0$, where

$$R = \begin{bmatrix} 5.15385 & -3.23077 \\ -3.23077 & 7.84615 \end{bmatrix} \qquad \mathbf{c} = [1, 2]^T.$$

6.6-38 Show that the stationary values of $\mathbf{x}^H R \mathbf{x}$ subject to (6.32) are found from the eigenvalues of PRP, where $P = I - C(C^H C)^{-1} C^H$.

6.7-39 Determine regions in the complex plane where the eigenvalues of A are for

$$A = \begin{bmatrix} 3 & 1 & 1 \\ 1 & 4 & 2 \\ 1 & 2 & 7 \end{bmatrix}.$$

6.7-40 Show that all the eigenvalues of A lie in $G(A) \cap G(A^T)$.

6.7-41 For a real $m \times m$ matrix A with disjoint Gershgorin circles, show that all the eigenvalues of A are real.

6.7-42 [142] In this exercise you will prove a simple version of the *Hoffman–Wielandt* theorem, a theorem frequently used for perturbation studies. Let A and E be $m \times m$ Hermitian matrices, and let A and $A + E$ have eigenvalues λ_i and $\hat{\lambda}_i$, $i = 1, 2, \ldots, m$, respectively, arranged in increasing order,

$$\lambda_1 \leq \lambda_2 \leq \cdots \leq \lambda_m, \qquad \hat{\lambda}_1 \leq \hat{\lambda}_2 \leq \cdots \leq \hat{\lambda}_m.$$

Let $A = Q \Lambda Q^H$, and let $A + E = W \hat{\Lambda} W^H$, where Q and W are unitary matrices.

(a) Starting from $\|E\|_F^2 = \|(A + E) - A\|_F^2 = \|W \hat{\Lambda} W^H - Q \Lambda Q^H\|_F^2$, show that

$$\|E\|_F^2 = \sum_{i=1}^m \left(|\lambda_i|^2 + |\mu_i|^2 \right) - 2 \operatorname{Re} \operatorname{tr}(Z \Lambda Z^H \hat{\Lambda}),$$

where $Z = Q^H W$.

(b) Thus, show that

$$\|E\|_F^2 \geq \sum_{i=1}^m \left(|\lambda_i|^2 + |\mu_i|^2 \right) - 2 \max_{U \text{ unitary}} \operatorname{Re} \operatorname{tr}(U \Lambda U^H \hat{\Lambda}).$$

(It can be shown that the maximum of $\max_{U \text{ unitary}} \operatorname{Re} \operatorname{tr}(U \Lambda U^H \hat{\Lambda})$ occurs when U is a permutation matrix.)

6.8-43 Show using (6.35) that $e^2(K)$ can be written as in (6.36). Hint: Recall how to commute inside a trace.

6.8-44 For a data compression application, it is desired to rotate a set of n-dimensional zero-mean data $Y = \{\mathbf{y}_1, \mathbf{y}_2, \ldots, \mathbf{y}_N\}$ so that it "matches" another set of n-dimensional data $Z = \{\mathbf{z}_1, \mathbf{z}_2, \ldots, \mathbf{z}_M\}$. Describe how to perform the rotation if the match is desired in the dominant q components of the data.

6.8-45 (Computer exercise) This exercise will demonstrate some concepts of principal components.

(a) Construct a symmetric matrix $R \in M_2$ that has unnormalized eigenvectors

$$\mathbf{x}_1 = \begin{bmatrix} 1 \\ 5 \end{bmatrix} \qquad \mathbf{x}_2 = \begin{bmatrix} -5 \\ 1 \end{bmatrix}$$

with corresponding eigenvalues $\lambda_1 = 10$, $\lambda_2 = 2$.

(b) Generate and plot 200 points of zero-mean Gaussian data that have the covariance R.

(c) Form an estimate of the covariance of the generated data, and compute the principal components of the data.

(d) Plot the principal component information over the data, and verify that the principal component vectors lie as anticipated.

6.8-46 ([290]) Low-rank approximation can sometimes be used to obtain a better representation of a noisy signal. Suppose that an m-dimensional zero-mean signal $\mathbf{X}$ with $R_x = E[\mathbf{XX}^H]$ is transmitted through a noisy channel, so that the received signal is

$$\mathbf{R} = \mathbf{X} + \boldsymbol{\nu},$$

as shown in figure 6.14(a). Let $E[\boldsymbol{\nu\nu}^H] = R_\nu = \sigma_\nu^2 I$. The mean-square (MS) error in this signal is

$$e_{\text{direct}}^2 = E[(\mathbf{R} - \mathbf{X})^H(\mathbf{R} - \mathbf{X})] = m\sigma_\nu^2.$$

Alternatively, we can send the signal $\mathbf{X}_1 = U^H\mathbf{X}$, where U is the matrix of eigenvectors of R_x, as in (6.37). The received signal in this case is

$$\mathbf{R}_r = \mathbf{X}_1 + \boldsymbol{\nu},$$

from which an approximation to $\mathbf{X}_r$ is obtained by

$$\hat{\mathbf{X}}_r = UI_r\mathbf{R}_r,$$

where $I_r = \text{diag}(1, 1, \ldots, 1, 0, 0, \ldots, 0)$ (with r ones). Show that

$$e_{\text{indirect}}^2 = E[(\mathbf{X} - \hat{\mathbf{X}}_r)^H(\mathbf{X} - \hat{\mathbf{X}}_r)]$$
$$= \sum_{i=r+1}^{m} \lambda_i + r\sigma_\nu^2.$$

Hence, conclude that for some values of r, the reduced rank method may have lower MS error than the direct error.

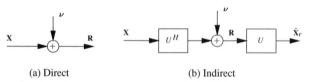

(a) Direct (b) Indirect

Figure 6.14: Direct and indirect transmission through a noisy channel

6.9-47 Consider an input signal with correlation matrix

$$R = \begin{bmatrix} 2 & .3 & .2 \\ .3 & 4 & .1 \\ .2 & .1 & 6 \end{bmatrix}.$$

(a) Design an eigenfilter with three taps that maximizes the SNR at the output of the filter.

(b) Plot the frequency response of this filter.

6.9-48 Show that minimizing (6.42) subject to $C^T\mathbf{b} = \mathbf{d}$ leads to (6.43).

6.9-49 Devise a means of matching a desired response by minimizing $\mathbf{b}^T R\mathbf{b}$, subject to the following constraints:

$$\mathbf{b}^T\mathbf{b} = 1,$$
$$C^T\mathbf{b} = \mathbf{0},$$

6.16 Exercises

where C is as in (6.41). That is, the filter coefficients are constrained in energy, but there are frequencies at which the response should be exactly 0. (Hint: See section 6.6.)

6.9-50 Consider the interpolation scheme shown in figure 6.15. The output can be written as $Y(z) = X(z^L)H(z)$.

(a) Show that if

$$h[Lt] = \begin{cases} c & t = 0, \\ 0 & \text{otherwise,} \end{cases}$$

then $y[lt] = cx[t]$. This means that the input samples are conveyed *without distortion* (but possibly with a scale factor) to the output. Such filters are called *Nyquist* or *Lth band* filters [342].

(b) Explain how to use the eigenfilter design technique to design an optimal mean-square Lth band filter.

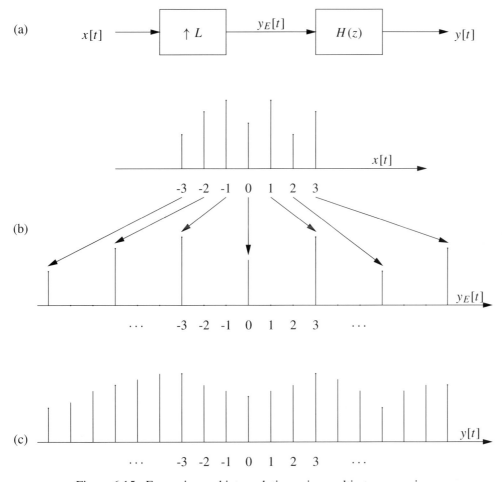

Figure 6.15: Expansion and interpolation using multirate processing

6.10-51 Show that (6.44) is true. Show that (6.45) is true.

6.10-52 Show for $M > p$ that R_{xx} defined in (6.45) has rank p. Hint: See properties of rank on page 266.

6.10-53 Show that every $\mathbf{x}[t] \in \mathcal{S}$, where $\mathcal{S}$ is defined in (6.47).

6.11-54 Numerically compute (using, e.g., MATLAB) the generalized eigenvalues and eigenvectors for
$$A\mathbf{u} = \lambda B\mathbf{u},$$
where
$$A = \begin{bmatrix} 3 & 2 & 4 \\ 1 & 7 & 9 \\ 6 & -2 & 1 \end{bmatrix} \quad B = \begin{bmatrix} 5 & 2 & 1 \\ 5 & 3 & 0 \\ 2 & 1 & 7 \end{bmatrix}.$$

6.11-55 Show that (6.55) is true.

6.12-56 Show that the minimal polynomial is unique. Hint: subtract $f(x) - g(x)$.

6.12-57 Show that the minimal polynomial divides every annihilating polynomial without remainder.

6.12-58 [98] Determine the minimal polynomial of
$$A = \begin{bmatrix} 3 & -3 & 2 \\ -1 & 5 & -2 \\ -1 & 3 & 0 \end{bmatrix}.$$

6.12-59 [164, page 657] (Resolvent identities) Let A be an $m \times m$ matrix with characteristic polynomial
$$\chi_A(s) = \det(sI - A) = s^m + a_{m-1}s^{m-1} + \cdots + a_1 s + a_0.$$
The matrix $(sI - A)^{-1}$ is known as the resolvent of A.

(a) Show that
$$\mathrm{adj}(sI - A) = [I s^{m-1} + (A + a_{m-1}I)s^{m-2} + \cdots +$$
$$(A^{m-1} + a_{m-1}A^{m-2} + \cdots + a_1 I)]. \tag{6.68}$$

Hint: Multiply both sides by $sI - A$, and use the Cayley–Hamilton theorem.

(b) Show that
$$\mathrm{adj}(sI - A) = [A^{m-1} + (s + a_{m-1})A^{m-2} + \cdots +$$
$$(s^{m-1} + a_{m-1}s^{m-2} + \cdots + a_1 I)]. \tag{6.69}$$

(c) Let S_i be the coefficient of s^{m-i} in (6.68). Show that the S_i can be recursively computed as
$$S_1 = I, \quad S_2 = S_1 A + a_{m-1}I, \quad S_3 = S_2 A + a_{m-2}I, \quad \ldots,$$
$$S_m = S_{m-1}A + a_1 I, \quad 0 = S_m + a_0 I.$$

(d) Show that both the coefficients of the characteristic polynomial a_i and the coefficients S_i of the adjoint of the resolvent can be recursively computed as follows:

$$S_1 = I \qquad A_1 = S_1 A \qquad a_{m-1} = -\mathrm{tr}(A_1)$$
$$S_2 = A_1 + a_{m-1}I \qquad A_2 = S_2 A \qquad a_{m-2} = -\frac{1}{2}\mathrm{tr}(A_2)$$
$$S_3 = A_2 + a_{m-2}I \qquad A_3 = S_3 A \qquad a_{m-3} = -\frac{1}{3}\mathrm{tr}(A_3)$$
$$\vdots \qquad \qquad \vdots \qquad \qquad \vdots$$
$$S_{m-1} = A_{m-1} + a_2 I \qquad A_{m-1} = S_{m-1}A \qquad a_1 = -\frac{1}{m-1}\mathrm{tr}(A_{m-1})$$
$$S_m = S_{m-1}A + a_1 I \qquad A_m = S_m A \qquad a_0 = -\frac{1}{m}\mathrm{tr}(A_m).$$

These recursive formulas are known as the Leverrier–Souriau–Faddeeva–Frame formulas [98, p. 88].

Hint: Use the Newton identities [G. Chrystal, *Algebra: An Elementary Textbook* (London: A. C. Black, 1926)]: For the polynomials $p(x) = x^m + a_{m-1}x^{m-1} + \cdots + a_0$, let s_i denote the sum of the ith power of the roots of $p(x)$. Thus s_1 is the sum of the roots, s_2 is the sum of the squares of the roots, and so on. Then

$$s_1 + a_{m-1} = 0,$$
$$s_2 + a_{m-1}s_1 + 2a_{m-2} = 0,$$
$$\vdots$$
$$s_{m-1} + a_{m-1}s_{m-2} + \cdots + (m-1)a_1 = 0.$$

Also use the fact that powers $\lambda(A^k) = \lambda^k(A)$, so that

$$\mathrm{tr}(A^k) = \sum_{i=1}^{m} \lambda_i^k = -s_k.$$

Show that

$$A_k = A^k + a_{m-1}A^{k-1} + \cdots + a_{m-k}A,$$

then take the trace of each side.

6.13-60 Expanding by cofactors, show that the characteristic equation of the matrix in first companion form (6.61) is $s^m + a_{m-1}s^{m-1} + a_1 s + a_0$.

6.13-61 Let A be an $m \times m$ matrix. Refer to figure 6.16.

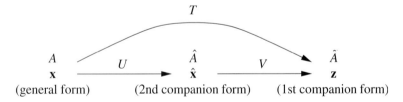

Figure 6.16: Transformation from a general matrix to first companion form

(a) If $\hat{A} = UAU^{-1}$, where

$$U^{-1} = Q = [\mathbf{b} \quad A\mathbf{b} \quad \cdots \quad A^{m-1}\mathbf{b}],$$

then show that $\hat{A}$ has the second companion form

$$\hat{A} = \begin{bmatrix} 0 & 0 & 0 & \cdots & -a_0 \\ 1 & 0 & 0 & \cdots & -a_1 \\ 0 & 1 & 0 & \cdots & -a_2 \\ \vdots & & \ddots & & \\ 0 & 0 & 0 & \cdots & -a_{m-1} \end{bmatrix}. \tag{6.70}$$

Hint: Show that $U^{-1}\hat{A} = AU^{-1}$. Use the Cayley–Hamilton theorem.

(b) Show that $\hat{\mathbf{b}} = U\mathbf{b} = [1, 0, \ldots, 0]^T$.

(c) Show that if $\tilde{A} = V\hat{A}V^{-1}$, where $\hat{A}$ is in second companion form (6.70) and V^{-1} is the Hankel matrix

$$V^{-1} = \begin{bmatrix} a_1 & \cdots & a_{m-3} & a_{m-2} & a_{m-1} & 1 \\ a_2 & \cdots & a_{m-2} & a_{m-1} & 1 & 0 \\ a_3 & \cdots & a_{m-1} & 1 & 0 & 0 \\ \vdots & & & & & \\ 1 & & & & & \end{bmatrix} = W,$$

then $\tilde{A}$ has the first companion form shown in (6.61).

6.13-62 For the system with

$$A = \begin{bmatrix} 0 & 1 & -2 \\ 3 & -2 & 1 \\ 2 & -4 & -5 \end{bmatrix} \qquad b = \begin{bmatrix} 2 \\ 0 \\ 1 \end{bmatrix},$$

determine the gain matrix $\mathbf{g}$ so that the eigenvalues of $A_c = A - \mathbf{b}\mathbf{g}^T$ are at $-3, -2 \pm j3\sqrt{\frac{3}{2}}$.

6.13-63 (Linear observers) In order to do pole placement as described in this section, the state $\mathbf{x}$ must be known. More commonly, only an output y is available, where $y = \mathbf{c}^T\mathbf{x}$. In this case, an *observer* must be constructed to estimate the state.

Assume that the system satisfies $\dot{\mathbf{x}} = A\mathbf{x} + \mathbf{b}u$. Then the observer is of the form (in continuous time)

$$\dot{\hat{\mathbf{x}}} = \hat{A}\hat{\mathbf{x}} + \hat{\mathbf{b}}u + \mathbf{k}y.$$

Let $\mathbf{e} = \mathbf{x} - \hat{\mathbf{x}}$ denote the difference between the true state $\mathbf{x}$ and the estimated state $\hat{\mathbf{x}}$.

(a) Write the differential equation for $\dot{\mathbf{e}}$, and show that in order for the error $\dot{\mathbf{e}}$ to be independent of the state $\mathbf{x}$ and the input u, the following must be the case:

$$\hat{A} = A - \mathbf{k}\mathbf{c}^T \qquad \text{and} \qquad \hat{B} = B.$$

(b) Based on this result, determine a means to place the eigenvalues of the observer matrix $\hat{A}$ at any desired location. Hint: Consider the duality between $A_c = A - \mathbf{b}\mathbf{g}^T$ and $\hat{A} = A - \mathbf{k}\mathbf{c}^T$. You should find that the solution involves a matrix of the form

$$N = \begin{bmatrix} \mathbf{c}^T \\ \mathbf{c}^T A \\ \vdots \\ \mathbf{c}^T A^{m-1} \end{bmatrix},$$

called the *observability test matrix*.

6.13-64 Let

$$Q = [\mathbf{b} \quad A\mathbf{b} \quad A^2\mathbf{b} \quad \cdots \quad A^{m-1}\mathbf{b}] \qquad \text{and} \qquad N = \begin{bmatrix} \mathbf{c}^T \\ \mathbf{c}^T A \\ \vdots \\ \mathbf{c}^T A^{m-1} \end{bmatrix}$$

be the controllability and observability test matrices, respectively, of a system $(A, \mathbf{b}, \mathbf{c})$. Determine the product

$$H = NQ.$$

Note that the elements of H are the Markov parameters introduced in section 1.4. If $\text{rank}(H) = m$, what can you conclude about $\text{rank}(N)$ and $\text{rank}(Q)$?

6.16 Exercises

6.13-65 [164, page 660] (Some properties of companion matrices) Let

$$A = \begin{bmatrix} -a_{m-1} & -a_{m-2} & \cdots & -a_1 & -a_0 \\ 1 & 0 & \cdots & 0 & 0 \\ 0 & 1 & \cdots & 0 & 0 \\ \vdots & & & & \\ 0 & 0 & \cdots & 1 & 0 \end{bmatrix}$$

be an $m \times m$ companion matrix, sometimes called a top companion matrix.

(a) Show that rank$(\lambda_i I - A) \leq m - 1$, where λ_i is an eigenvalue of A.

(b) (Shifting) Show that $\mathbf{e}_i^T A = \mathbf{e}_{i-1}^T$ for $2 \leq i \leq m$, where $\mathbf{e}_i$ is the unit vector with 1 in the ith position. Also $\mathbf{e}_1^T A = [-a_{m-1} \ -a_{m-2} \ \cdots \ -a_0]$.

(c) Show that if A is nonsingular, then A^{-1} is a bottom companion matrix with last row $-[1/a_0, -a_{m-1}/a_0, \ldots, a_1/a_0]$.

(d) (Inverse is companion) Show that A is nonsingular if and only if $a_0 \neq 0$.

6.14-66 Consider a channel constrained to have at least one zero between every one, and runs of zero no longer than 3.

(a) Draw the state-transition diagram.

(b) Determine the state-transition matrix.

(c) Determine the capacity.

6.15-67 The computational routines described in this section apply to *real* matrices. In this problem we examine how to extend real computational routines to Hermitian (complex) matrices. Let A be a Hermitian matrix, and let $A = A_r + jA_i$, where A_r is the real part and A_i is the imaginary part. Let $\mathbf{x} = \mathbf{x}_r + j\mathbf{x}_i$ be an eigenvector of A, with its corresponding real and imaginary parts.

(a) Show that the condition $A\mathbf{x} = \lambda \mathbf{x}$ can be rewritten as

$$\begin{bmatrix} A_r & -A_i \\ A_i & A_r \end{bmatrix} \begin{bmatrix} \mathbf{x}_r \\ \mathbf{x}_i \end{bmatrix} = \lambda \begin{bmatrix} \mathbf{x}_r \\ \mathbf{x}_i \end{bmatrix}.$$

(b) Let $\mathbb{A} = \begin{bmatrix} A_r & -A_i \\ A_i & A_r \end{bmatrix}$. Show that $\mathbb{A}$ is symmetric.

(c) Show that if $[\mathbf{x}_r^T, \mathbf{x}_i^T]^T$ is an eigenvector of $\mathbb{A}$ corresponding to λ, then so is $[\mathbf{x}_i^T, -\mathbf{x}_r^T]^T$.

(d) Conclude that each eigenvalue of $\mathbb{A}$ has multiplicity 2, and that the eigenvalues of A can be obtained by selecting one eigenvector corresponding to each pair of repeated eigenvalues.

6.15-68 ([114]) In the Householder tridiagonalization illustrated in (6.67), the matrix B_1 is operated on by the Householder matrix H_v to produce $H_v B_1 H_v$, for a Householder vector $\mathbf{v}$. Show that $H_v B_1 H_v$ can be computed by

$$H_v B_1 H_v = B - \mathbf{v}\mathbf{w}^T - \mathbf{w}\mathbf{v}^T,$$

where

$$\mathbf{p} = \frac{2}{\mathbf{v}^T \mathbf{v}} B_1 \mathbf{v} \qquad \mathbf{w} = \mathbf{p} - \frac{\mathbf{p}^T \mathbf{v}}{\mathbf{v}^T \mathbf{v}} \mathbf{v}.$$

6.15-69 (Wilkinson shift) If $T = \begin{bmatrix} a_{n-1} & b_{n-1} \\ b_{n-1} & a_n \end{bmatrix}$, show that the eigenvalues are obtained by

$$\mu = a_n + d \pm \sqrt{d^2 + b_{n-1}^2}$$

where $d = (a_{n-1} - a_n)/2$.

6.17 References

The definitive historical work on eigenvalue computations is [370]. A more recent, excellent source on the theory of eigenvalues and eigenvectors is [142]. The Courant minimax theorem is discussed in [55] and [370]. The proof of theorem 6.6 is drawn from [177]. The discussion of the constrained eigenvalue problem of section 6.6 is drawn from [113], where efficient numerical implementation issues are also discussed. Further related discussions are in [317, 97, 91]. Our presentation of the Gershgorin circle theorem is drawn from [142], in which extensive discussion of perturbation of eigenvalue problems is also presented.

The discussion of low-rank approximations is drawn from [132] and [291]. An excellent coverage of principal component methods is found in [236], which also includes a discussion of the asymptotic statistical distribution of the eigenvalues and eigenvectors of correlation matrices; and in [156].

The eigenfilter method for random signals is presented in [132]. The eigenfilter method for the design of FIR filters with spectral requirements is presented in [343]. Additional work on eigenfilters is described in [252, 336, 240, 241]. It also possible to include other constraints, such as minimizing the effect of a known interfering signal, making the response maximally flat, or making the response almost equiripple.

The MUSIC method is due to Schmidt [293] (see [179]). The Pisarenko harmonic decomposition appears in [255]. Considerable work has been done on MUSIC methods since its inception. We cite [331, 328, 329] as representatives; see also [330, 174, 220]. ESPRIT appears in [280, 282]. MUSIC, ESPRIT, and other spectral estimation methods appear in [263].

The noiseless channel coding theorem is discussed in [34], and was originally proposed by Shannon [304]. [202] provides a thorough study of the design of codes for constrained channels, including an explanation of the magnetic recording channel problem. The works of Immink [148, 149] provide an engineering treatment of runlength-limited codes.

Our discussion of characteristic polynomials follows [98]. This source provides an exhaustive treatment of Jordan forms and minimal polynomials. Another excellent source of information about Jordan forms and minimal polynomials is [245]; see also the appendix of [164].

Eigenvalue placement for controls is by now classical; see, for example, [164]. Our discussion of linear controllers, as well as the exercise on linear observers, is drawn from [93].

Our discussion of the computation of eigenvalues and eigenvectors was drawn closely from [114, chapter 8]. The eigenvalue problem is also discussed in [260, 326]. Computation of eigenvectors is reviewed in [150].

Chapter 7

The Singular Value Decomposition

> My First is singular at best:
>   More Plural is my Second:
> My Third is far the pluralest –
> So plural-plural, I protest
>   It scarcely can be reckoned!
>
> — *Lewis Carroll*

The singular value decomposition (SVD) is one of the most important tools of numerical signal processing. It provides robust solution of both overdetermined and underdetermined least-squares problems, matrix approximation, and conditioning of ill-conditioned systems. It is employed in a variety of signal processing applications, such as spectrum analysis, filter design, system identification, model order reduction, and estimation.

7.1 Theory of the SVD

As we have seen in section 6.3.2, a Hermitian matrix A can be factored as $A = U \Lambda U^H$, where U is unitary and Λ is diagonal. The SVD provides a similar factorization for *all* matrices, even matrices that are not square or have repeated eigenvalues. The matrix A can be factored into a product of unitary matrices and a diagonal matrix, as the following theorem explains.

Theorem 7.1 *Every matrix $A \in \mathbb{C}^{m \times n}$ can be factored as*

$$\boxed{A = U \Sigma V^H} \tag{7.1}$$

where $U \in \mathbb{C}^{m \times m}$ is unitary, $V \in \mathbb{C}^{n \times n}$ is unitary, and $\Sigma \in \mathbb{R}^{m \times n}$ has the form

$$\Sigma = \mathrm{diag}(\sigma_1, \sigma_2, \ldots, \sigma_p),$$

where $p = \min(m, n)$.[1]

The diagonal elements of Σ are called the *singular values* of A and are usually ordered so that

$$\sigma_1 \geq \sigma_2 \geq \cdots \geq \sigma_p \geq 0.$$

---

[1] The Greek letter Σ (sigma) employed here is used as a matrix variable, and should not be confused with a summation sign.

If $A \in \mathbb{R}^{m \times n}$ then all matrices in the SVD are real and U and V are orthogonal matrices. Notationally the matrix Σ is written as a diagonal matrix, even though it is not square if $m \neq n$. The elements $\sigma_1, \sigma_2, \ldots, \sigma_p$ are written along the main diagonal, and rows or columns of zeros are appended as necessary to obtain the proper dimension for Σ if it is not square. For example, if $m = 2$ and $n = 3$,

$$\Sigma = \begin{bmatrix} \sigma_1 & 0 & 0 \\ 0 & \sigma_2 & 0 \end{bmatrix}.$$

Also, if $m = 3$ and $n = 2$, then

$$\Sigma = \begin{bmatrix} \sigma_1 & 0 \\ 0 & \sigma_2 \\ 0 & 0 \end{bmatrix}.$$

Before proceeding with a proof of theorem 7.1, it is interesting to observe some properties of the SVD. We note that

$$(A^H A) = V \Sigma^T U^H U \Sigma V^H = V \Sigma^T \Sigma V^H,$$

since U is unitary. Let $\Lambda \in \mathbb{R}^{n \times n}$ be defined by

$$\Lambda = \Sigma^T \Sigma = \operatorname{diag}(\sigma_1^2, \sigma_2^2, \ldots, \sigma_n^2).$$

Then

$$(A^H A) V = V \Lambda.$$

The diagonal elements of Λ are the eigenvalues of $A^H A$. That is, σ_i^2, $i = 1, 2, \ldots, n$ are the eigenvalues of $A^H A$. Since $A^H A$ is symmetric, the eigenvalues must be real and V is formed from the eigenvectors of $A^H A$ (see section 6.3.2). Similarly, computing for AA^H, it follows that

$$AA^H U = U \Lambda$$

and σ_i^2, $i = 1, 2, \ldots, m$ are the eigenvalues of AA^H. The intersecting set σ_i, $i = 1, 2, \ldots, \min(m, n)$ forms the set of singular values of A, and the other singular values must be zero. The SVD simultaneously diagonalizes the inner-product matrix $A^H A$ and the outer-product matrix AA^H.

Proof ([177]) Let

$$A^H A V = V \operatorname{diag}(\lambda_1, \lambda_2, \ldots, \lambda_n) \tag{7.2}$$

be the spectral decomposition of $A^H A$, where the columns of V are eigenvectors,

$$V = [\mathbf{v}_1, \mathbf{v}_2, \ldots, \mathbf{v}_n],$$

and $\lambda_1, \lambda_2, \ldots, \lambda_r > 0$ and $\lambda_{r+1} = \lambda_{r+2} = \lambda_n = 0$, where $r \leq p$. (The existence of the factorization (7.2) is established in theorem 6.2.) For $1 \leq i \leq r$, let

$$\mathbf{u}_i = \frac{A \mathbf{v}_i}{\sqrt{\lambda_i}},$$

and observe that

$$\langle \mathbf{u}_i, \mathbf{u}_j \rangle = \delta_{i,j}. \tag{7.3}$$

The set $\{\mathbf{u}_i, i = 1, 2, \ldots, r\}$ can be extended using the Gram–Schmidt procedure to form an orthonormal basis for $\mathbb{C}^m$. Let

$$U = [\mathbf{u}_1, \mathbf{u}_2, \ldots, \mathbf{u}_m].$$

7.1 Theory of the SVD

Then the $\mathbf{u}_i$ are eigenvectors for AA^H. This is clear for the nonzero eigenvalues of AA^H. For the zero eigenvalues, the eigenvectors must come from the nullspace of AA^H. Since the eigenvectors with zero eigenvalues are, by construction, orthogonal to the eigenvectors with nonzero eigenvectors that are in the range of AA^H, these vectors must, by theorem 4.5, be in the nullspace of AA^H.

Now examine the elements of $U^H AV$. When $i \leq r$, the (i, j)th element of $U^H AV$ is

$$\mathbf{u}_i^H A \mathbf{v}_j = \frac{1}{\sqrt{\lambda_i}} \mathbf{v}_i^H A^H A \mathbf{v}_j = \frac{\lambda_j}{\sqrt{\lambda_i}} \mathbf{v}_i^H \mathbf{v}_j = \sqrt{\lambda_j} \delta_{i,j}.$$

For $i > r$, $AA^H \mathbf{u}_i = 0$. Thus $A^H \mathbf{u}_i$ is in the nullspace of A, and is also, obviously, in the range of A^H. By theorem 4.5, $A^H \mathbf{u}_i = 0$. So $\mathbf{u}_i^H A \mathbf{v}_j = \mathbf{v}_j^H A^H \mathbf{u}_i = 0$. Thus $U^H AV = \Lambda$ is diagonal, where the nonzero elements are $\sqrt{\lambda_j}$, $j = 1, 2, \ldots, r$. $\square$

It is often convenient to break the matrices in the SVD into two parts, corresponding to the nonzero singular values and the zero singular values. Let

$$\Sigma = \begin{bmatrix} \Sigma_1 & \\ & \Sigma_2 \end{bmatrix},$$

where

$$\Sigma_1 = \text{diag}(\sigma_1, \sigma_2, \ldots, \sigma_r) \in \mathbb{R}^{r \times r}$$

and

$$\Sigma_2 = \text{diag}(\sigma_{r+1}, \sigma_{r+2}, \ldots, \sigma_p) = \text{diag}(0, 0, \ldots, 0) \in \mathbb{R}^{(m-r) \times (n-r)}.$$

Then the SVD can be written as

$$A = [U_1 \ U_2] \begin{bmatrix} \Sigma_1 & \\ & \Sigma_2 \end{bmatrix} \begin{bmatrix} V_1^H \\ V_2^H \end{bmatrix} = U_1 \Sigma_1 V_1^H,$$

where U_1 is $m \times r$, U_2 is $m \times (m-r)$, V_1 is $n \times r$, and V_2 is $n \times (n-r)$. The SVD can also be written as

$$A = \sum_{i=1}^{r} \sigma_i \mathbf{u}_i \mathbf{v}_i^H \qquad (7.4)$$

The SVD can also be used to compute two matrix norms:

$$\|A\|_F^2 = \sum_{i=1}^{p} \sigma_i^2 \qquad \text{(Frobenius norm)}, \qquad (7.5)$$

$$\|A\|_2 = \sigma_1 \qquad (l_2 \text{ norm}). \qquad (7.6)$$

Example 7.1.1 In this example, we find the SVD (numerically) of several matrices using the MATLAB command [U,S,V] = svd(A).

A is square symmetric. Let

$$A = \begin{bmatrix} 5 & 6 & 2 \\ 6 & 1 & 4 \\ 2 & 4 & 7 \end{bmatrix}.$$

Then

$$A = U \Sigma U^T,$$

where

$$U = \begin{bmatrix} 0.5923 & -0.6168 & -0.5184 \\ 0.5262 & -0.1911 & 0.8286 \\ 0.6102 & 0.7636 & -0.2114 \end{bmatrix}$$

and
$$\Sigma = \begin{bmatrix} 12.3912 & & \\ & 4.3832 & \\ & & 3.7744 \end{bmatrix}.$$

In this case the SVD is the same as the regular eigendecomposition in (6.18).

A is diagonal

$$A = \begin{bmatrix} 2 & 0 \\ 0 & -4 \end{bmatrix} = \begin{bmatrix} 0 & 1 & 0 \\ -1 & 0 & 0 \\ 0 & 0 & 1 \end{bmatrix} \begin{bmatrix} 4 & 0 \\ 0 & 2 \\ 0 & 0 \end{bmatrix} \begin{bmatrix} 0 & 1 \\ 1 & 0 \end{bmatrix}.$$

In this case, the U and V matrices just shuffle the columns around and change the signs to make the singular values positive.

A is a column vector

$$A = \begin{bmatrix} 2 \\ 3 \\ 4 \end{bmatrix} = \begin{bmatrix} 0.3714 & -0.5571 & -0.7428 \\ 0.5571 & 0.7737 & -0.3017 \\ 0.7428 & -0.3017 & 0.5977 \end{bmatrix} \begin{bmatrix} 5.3852 \\ 0 \\ 0 \end{bmatrix} [1].$$

A is a row vector

$$A = \begin{bmatrix} 2 & 3 & 4 \end{bmatrix} = [1][5.3852 \quad 0 \quad 0] \begin{bmatrix} 0.371391 & -0.928477 & 0 \\ 0.557086 & 0.222834 & -0.8 \\ 0.742781 & 0.297113 & 0.6 \end{bmatrix}. \qquad \square$$

7.2 Matrix structure from the SVD

The rank of a matrix is the number of nonzero singular values. From the notation above,

$$\text{rank}(A) = r.$$

The SVD turns out to be a numerically stable way of computing the rank of a matrix. As pointed out subsequently, it also provides a useful way of determining what the rank of a matrix "almost is," that is, the rank of a matrix of lower rank that is close to A.

The range (column space) of a matrix is

$$\mathcal{R}(A) = \{\mathbf{b} \in \mathbb{C}^m : \mathbf{b} = A\mathbf{x}\}.$$

Substituting in the SVD,

$$\mathcal{R}(A) = \{\mathbf{b}: \mathbf{b} = U\Sigma V^H \mathbf{x}\}$$
$$= \{\mathbf{b}: \mathbf{b} = U\Sigma \mathbf{y}\}$$
$$= \{\mathbf{b}: \mathbf{b} = U_1 \mathbf{y}\} = \text{span}(U_1).$$

The range of a matrix is spanned by the orthogonal set of vectors in U_1, the first r columns of U. The other fundamental spaces of the matrix A can also be determined from the SVD (see exercise 7.2-3):

$$\boxed{\begin{aligned} \mathcal{R}(A) &= \text{span}(U_1) \\ \mathcal{N}(A) &= \text{span}(V_2) \\ \mathcal{R}(A^H) &= \text{span}(V_1) \\ \mathcal{N}(A^H) &= \text{span}(U_2) \end{aligned}} \qquad (7.7)$$

The SVD thus provides an explicit orthogonal basis and a computable dimensionality for each of the fundamental spaces of a matrix.

Stated another way, in the SVD of the $m \times n$ matrix A, the orthogonal matrix U provides a decomposition of $\mathbb{C}^m$ into $\mathbb{C}^m = \mathcal{R}(A) \oplus \mathcal{N}(A^H) = \text{span}(U_1) \oplus \text{span}(U_2)$, and the orthogonal matrix V provides a decomposition of $\mathbb{C}^n$ into $\mathbb{C}^n = \mathcal{R}(A^H) + \mathcal{N}(A) = \text{span}(V_1) \oplus \text{span}(V_2)$. The SVD also reveals explicitly the geometry of the transformation $A\mathbf{x} = U\Sigma V^H \mathbf{x}$. The first transformation $V^H \mathbf{x}$ projects $\mathbf{x}$ onto $\mathcal{R}(A^H)$ and $\mathcal{N}(A)$. The next transformation $\Sigma(V^H \mathbf{x})$ then scales the projection of $\mathbf{x}$ in $\mathcal{R}(A^H)$ and zeros out the projection in $\mathcal{N}(A)$. The final step in the transformation $U(\Sigma V^H \mathbf{x})$ rotates this result into the column space of A.

7.3 Pseudoinverses and the SVD

Consider the solution of the equation $A\mathbf{x} = \mathbf{b}$. If $\mathbf{b}$ does not lie in the column space of A then there is no exact solution. As we have seen, the solution which introduces the minimum amount of error (as measured by the l_2 norm) projects $\mathbf{b}$ onto the range of A by the normal equations $A^H A\hat{\mathbf{x}} = A^H \mathbf{b}$. If $A^H A$ is invertible, this produces the least-squares solution

$$\hat{\mathbf{x}} = (A^H A)^{-1} A^H \mathbf{b}. \tag{7.8}$$

As discussed in section 3.1.1, the matrix $A^H A$ is invertible only if the columns of A are linearly independent. It may occur that the columns of A are not linearly independent, in which case the solution (7.8) cannot be computed, even if $\mathbf{b}$ lies in the column space.

Let $\mathbf{p}$ denote the projection of $\mathbf{b}$ onto the range of A. Then the least-squares equation we are solving is

$$A\hat{\mathbf{x}} = \mathbf{p}. \tag{7.9}$$

If A has dependent columns, then the nullspace of A is not trivial and there is no unique solution. The problem then becomes one of selecting one solution out of the infinite number of possible solutions. As presented in section 3.14, a commonly accepted approach is to select the solution with the smallest length: Solve $A\hat{\mathbf{x}} = \mathbf{p}$ so that $\|\hat{\mathbf{x}}\|$ is minimum. This problem can be solved using the SVD.

To illustrate the concept, consider the solution of the following diagonal problem:

$$\begin{bmatrix} \sigma_1 & 0 & 0 & 0 \\ 0 & \sigma_2 & 0 & 0 \\ 0 & 0 & 0 & 0 \end{bmatrix} \begin{bmatrix} x_1 \\ x_2 \\ x_3 \\ x_4 \end{bmatrix} = \begin{bmatrix} b_1 \\ b_2 \\ b_3 \end{bmatrix}.$$

Since the last component of each column vector is zero, it is clear that there is no solution unless $b_3 = 0$. The vector $\mathbf{p} = [b_1, b_2, 0]^T$ is a projection of $\mathbf{b}$ onto the column space of A; the error $[0, 0, b_3]^T$ is orthogonal to each column of A. Our projected equation is now

$$\begin{bmatrix} \sigma_1 & 0 & 0 & 0 \\ 0 & \sigma_2 & 0 & 0 \\ 0 & 0 & 0 & 0 \end{bmatrix} \begin{bmatrix} \hat{x}_1 \\ \hat{x}_2 \\ \hat{x}_3 \\ \hat{x}_4 \end{bmatrix} = \begin{bmatrix} b_1 \\ b_2 \\ 0 \end{bmatrix}. \tag{7.10}$$

The nullspace of A is spanned by the vectors $[0, 0, 1, 0]^T$ and $[0, 0, 0, 1]^T$. The general solution of (7.10) can be written as

$$\hat{\mathbf{x}} = \begin{bmatrix} b_1/\sigma_1 \\ b_2\sigma_2 \\ 0 \\ 0 \end{bmatrix} + \left(s \begin{bmatrix} 0 \\ 0 \\ 1 \\ 0 \end{bmatrix} + t \begin{bmatrix} 0 \\ 0 \\ 0 \\ 1 \end{bmatrix} \right) = \hat{\mathbf{x}}_r + \hat{\mathbf{x}}_n, \qquad s, t \in \mathbb{R}.$$

Since $\mathbf{x}_r \perp \mathbf{x}_n$, the norm of this solution is (by the Pythagorean theorem)

$$\|\hat{\mathbf{x}}\|^2 = \|\hat{\mathbf{x}}_r\|^2 + \|\hat{\mathbf{x}}_n\|^2.$$

The minimum norm solution is obtained by taking $\mathbf{x}_n = 0$. This can be written as

$$\hat{\mathbf{x}} = \begin{bmatrix} \frac{1}{\sigma_1} & 0 & 0 \\ 0 & \frac{1}{\sigma_2} & 0 \\ 0 & 0 & 0 \end{bmatrix} \begin{bmatrix} b_1 \\ b_2 \\ b_3 \end{bmatrix}.$$

The matrix in this equation is the pseudoinverse of A, which is denoted $A^\dagger$.

More generally, if A is any $m \times n$ diagonal matrix with r nonzero entries

$$A = \begin{bmatrix} \sigma_1 & & & & \\ & \ddots & & & \\ & & \sigma_r & & \\ & & & 0 & \\ & & & & \ddots \end{bmatrix},$$

then the pseudoinverse that provides the minimum length solution is the $n \times m$ diagonal matrix

$$A^\dagger = \begin{bmatrix} 1/\sigma_1 & & & & \\ & \ddots & & & \\ & & 1/\sigma_r & & \\ & & & 0 & \\ & & & & \ddots \end{bmatrix}.$$

Now consider the least-squares solution of a general equation $A\mathbf{x} = \mathbf{b}$. We want to minimize the norm of the error, $\|A\mathbf{x} - \mathbf{b}\|$. Using the SVD we can write

$$\min \|A\mathbf{x} - \mathbf{b}\| = \min \|U \Sigma V^H \mathbf{x} - \mathbf{b}\| = \min \|\Sigma V^H \mathbf{x} - U^H \mathbf{b}\|,$$

where the latter equality follows from the fact that U is unitary, and multiplication by a unitary matrix does not change the length of a vector. Let $\mathbf{v} = V^H \mathbf{x}$ and $\hat{\mathbf{b}} = U^H \mathbf{b}$. Then the least-squares problem can be written as $\min \|\Sigma \mathbf{v} - \hat{\mathbf{b}}\|$. This has reduced the problem to the diagonal form above: the minimum length solution to this problem is

$$\mathbf{v} = \Sigma^\dagger \hat{\mathbf{b}}.$$

The solution for $\hat{\mathbf{x}}$ comes by working backwards:

$$V^H \hat{\mathbf{x}} = \Sigma^\dagger U^H \mathbf{b},$$
$$\hat{\mathbf{x}} = V \Sigma^\dagger U^H \mathbf{b}.$$

The matrix $V \Sigma^\dagger U^H$ is the pseudoinverse of A.

$$\boxed{A^\dagger = V \Sigma^\dagger U^H} \tag{7.11}$$

The pseudoinverse is illustrated in the following commutative diagram (see box 7.1):

$$\begin{array}{ccc} \min \|A\mathbf{x} - \mathbf{b}\| & \xrightarrow{A^\dagger} & \mathbf{x} = A^\dagger \mathbf{b} \\ \downarrow {\scriptsize \mathbf{v}=V^H\mathbf{x},\ \hat{\mathbf{b}}=U^H\mathbf{b}} & & \uparrow {\scriptsize \mathbf{x}=V\mathbf{v},\ \mathbf{b}=U\hat{\mathbf{b}}} \\ \Sigma \mathbf{v} = \hat{\mathbf{b}} & \xrightarrow{\Sigma^\dagger} & \mathbf{v} = \Sigma^\dagger \hat{\mathbf{b}} \end{array}$$

7.4 Numerically Sensitive Problems

> **Box 7.1: Commutative diagrams**
>
> A **commutative diagram** is used to demonstrate equality in different operations. Two (or sometimes more) paths from a starting point to an ending point indicate various ways that an operation can be accomplished. The implication is that whichever path is taken following the arrows, the result is the same. Usually it is used in constructing algebraic identities, but it can also be used to demonstrate the steps to follow to perform some task.
>
> An example of a commutative diagram illustrates their operation. From linear systems theory, it is well known that the "transform of convolution is multiplication." That is, the convolution $f_1(t) * f_2(t)$ can be accomplished by finding the Fourier transform of each function, multiplying the transforms, then computing the inverse transform. This can be represented on a commutative diagram as
>
>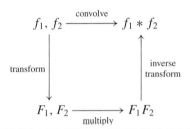

It is enlightening to think about this pseudoinverse in terms of the subspace geometry of the matrix A. In light of the discussion in section 7.2, the multiplication $U^H \mathbf{b}$ decomposes $\mathbf{b}$ into projections into $\mathcal{R}(A)$ and $\mathcal{N}(A^H)$:

$$\begin{bmatrix} U_1^H \\ U_2^H \end{bmatrix} \mathbf{b} \Rightarrow \begin{matrix} U_1^H \mathbf{b} \in \mathcal{R}(A) \\ U_2^H \mathbf{b} \in \mathcal{N}(A^H) \end{matrix}.$$

The multiplication by $\Sigma^\dagger$ scales the projection on $\mathcal{R}(A)$, and zeros out the portion in $\mathcal{N}(A^H)$:

$$\Sigma_1^{-1} U_1^H \mathbf{b} \in \mathcal{R}(A),$$
$$\Sigma_2 U_2^H \mathbf{b} = 0.$$

The multiplication by V transforms the result into an element of $\mathcal{R}(A^H)$:

$$V_1 \Sigma_1^{-1} U_1^H \mathbf{b} \in \mathcal{R}(A^H).$$

Returning to the previous decomposition, if $\hat{\mathbf{x}} = \hat{\mathbf{x}}_r + \hat{\mathbf{x}}_n$ is a least-squares solution where $\hat{\mathbf{x}}_n$ is in $\mathcal{N}(A)$, then the minimum-length solution is $\hat{\mathbf{x}} = \hat{\mathbf{x}}_r$, which is in the **row space of** A. If the nullspace of A is trivial (the columns of A are linearly independent), then there is a unique pseudoinverse.

7.4 Numerically sensitive problems

As discussed in section 4.10, systems of equations that are poorly conditioned are sensitive to small changes in the values. Since, practically speaking, there are always inaccuracies in measured data, the solution to these equations may be almost meaningless. The SVD can help with the solution of ill-conditioned equations by identifying the direction of sensitivity and discarding that portion of the problem.

Example 7.4.1 Recall the example from section 4.10, in which the equation

$$\begin{bmatrix} 1+3\epsilon & 1-3\epsilon \\ 3-\epsilon & 3+\epsilon \end{bmatrix} \begin{bmatrix} x_1 \\ x_2 \end{bmatrix} = \begin{bmatrix} b_1 \\ b_2 \end{bmatrix}$$

or

$$A\mathbf{x} = \mathbf{b}$$

was to be solved. It was shown that when ϵ is small, minor changes in $\mathbf{b}$ lead to drastic changes in the solution. We will reexamine this problem now in light of the SVD.

The SVD of A is

$$A = \frac{1}{\sqrt{10}} \begin{bmatrix} 1 & 3 \\ 3 & -1 \end{bmatrix} \begin{bmatrix} 2\sqrt{5} & \\ & 2\epsilon\sqrt{5} \end{bmatrix} \begin{bmatrix} 1 & 1 \\ 1 & -1 \end{bmatrix} \frac{1}{\sqrt{2}}, \tag{7.12}$$

from which the exact inverse of A is

$$A^{-1} = \sqrt{20} \begin{bmatrix} 1 & 1 \\ 1 & -1 \end{bmatrix} \begin{bmatrix} \frac{1}{2\sqrt{5}} & \\ & \frac{1}{2\epsilon\sqrt{5}} \end{bmatrix} \begin{bmatrix} 1 & 3 \\ 3 & -1 \end{bmatrix} = \frac{1}{20} \begin{bmatrix} 1+3/\epsilon & 3-1/\epsilon \\ 1-3/\epsilon & 3+1/\epsilon \end{bmatrix}.$$

When ϵ is small then A^{-1} has large entries because of the $1/(2\epsilon\sqrt{5})$ that appears in the diagonal matrix, which makes $\mathbf{x} = A^{-1}\mathbf{b}$ unstable: small changes in $\mathbf{b}$ result in large changes in $\mathbf{x}$. Observe from (7.12) that the entry $1/(2\epsilon\sqrt{5})$ multiplies the column $[1, -1]^T$. This is the sensitive direction. The idea of the sensitive direction is shown in figure 7.1. As $\mathbf{b}$ changes slightly, the solution changes in a direction mostly along the sensitive direction.

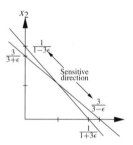

Figure 7.1: Illustration of the sensitive direction

If ϵ is small, then in (7.12) $\sigma_2 = 2\epsilon\sqrt{5}$ may be set to zero to approximate A,

$$A \approx \tilde{A} = \frac{1}{\sqrt{10}} \begin{bmatrix} 1 & 3 \\ 3 & -1 \end{bmatrix} \begin{bmatrix} 2\sqrt{5} & \\ & 0 \end{bmatrix} \begin{bmatrix} 1 & 1 \\ 1 & -1 \end{bmatrix} \frac{1}{\sqrt{2}}.$$

The pseudoinverse is

$$A^\dagger = \frac{1}{\sqrt{20}} \begin{bmatrix} 1 & 1 \\ 1 & -1 \end{bmatrix} \begin{bmatrix} \frac{1}{2\sqrt{5}} & \\ & 0 \end{bmatrix} \begin{bmatrix} 1 & 3 \\ 3 & -1 \end{bmatrix} = \frac{1}{20} \begin{bmatrix} 1 & 3 \\ 1 & 3 \end{bmatrix}.$$

In this case, the multiplier of the sensitive direction vector $[1, -1]$ is zero—no motion in the sensitive direction occurs. Any least-squares solution to the equation $A\mathbf{x} = \mathbf{b}$ is of the form

$$\hat{\mathbf{x}} = A^\dagger \mathbf{b},$$

so that $\hat{\mathbf{x}} = c[1, 1]^T$ for some constant c. As a vector, $\hat{\mathbf{x}}$ points in the direction nearly perpendicular to the sensitive direction of the problem. □

As this example illustrates, the SVD identifies the stable and unstable directions of the problem and, by zeroing the small singular values, eliminates the unstable directions.

7.5 Rank-Reducing Approximations: Effective Rank

The SVD can be used to both illustrate poor conditioning and provide a cure for the ailment. For the equation $A\mathbf{x} = \mathbf{b}$ with solution $\mathbf{x} = A^{-1}\mathbf{b}$, write the solution using the SVD:

$$\mathbf{x} = A^{-1}\mathbf{b} = (U\Sigma V^H)^{-1} = \sum_{i=1}^{n} \mathbf{v}_i \frac{\mathbf{u}_i^H \mathbf{b}}{\sigma_i}.$$

If a singular value σ_i is small, then a small change in $\mathbf{b}$ or a small change in either U or V (resulting in a small change in A) may be amplified into a large change in the solution $\mathbf{x}$. (This amplification does not happen, for example, if $\mathbf{b} \perp \mathbf{u}_i$. However, it is generally desirable to provide a means of solving the problem that will be robust with all right-hand sides.) A small singular value corresponds to a matrix which is nearly singular, and thus more difficult to invert accurately.

The solution that the SVD proposes to the problem of poorly conditioned equations is to zero out the small singular values, then compute a pseudoinverse of the resulting matrix:

1. Compute the SVD of A: $A = U\Sigma V^H$.
2. Examine the singular values of A and zero out any that are "small" to obtain a new (approximate) Σ matrix.
3. Compute the solution by $\hat{\mathbf{x}} = V\Sigma^\dagger U^H \mathbf{b}$.

Determining which singular values are "small" is problem-dependent and requires some judgment. It often happens that there is a clear break in the singular values, with some of them being notably smaller than the largest few singular values.

Computing solutions using the SVD is obviously much more complicated than simply solving a set of equations: it is not the method to be used for every linear problem. There would need to be strong motivation to attempt it for high-speed real-time applications. But for ill-conditioned matrices or noisy data (such as for curve-fitting problems), the SVD is highly recommended.

7.5 Rank-reducing approximations: effective rank

The SVD of a matrix can be used to determine how near (in the l_2-norm) the matrix is to a matrix of lower rank. It can also be used to find the nearest matrix of a given lower rank.

Theorem 7.2 *Let A be an $m \times n$ matrix with $\text{rank}(A) = r$, and let $A = U\Sigma V^H$. Let $k < r$ and let*

$$A_k = \sum_{i=1}^{k} \sigma_i \mathbf{u}_i \mathbf{v}_i^H = U\Sigma_k V^H,$$

where

$$\Sigma_k = \text{diag}(\sigma_1, \sigma_2, \ldots, \sigma_k).$$

Then $\|A - A_k\|_2 = \sigma_{k+1}$, and A_k is the nearest matrix of rank k to A (in the l_2-norm):

$$\min_{\text{rank}(B)=k} \|A - B\| = \|A - A_k\|_2.$$

Proof Since $A - A_k = U \text{diag}(0, 0, \ldots, 0, \sigma_{k+1}, \ldots, \sigma_r, 0, \ldots, 0) V^H$, it follows that $\|A - A_k\| = \sigma_{k+1}$.

The second part of the proof is a "proof by inequality." By the definition of the matrix norm, for any unit vector $\mathbf{z}$,

$$\|A - B\|_2^2 \geq \|(A - B)\mathbf{z}\|_2^2.$$

Let B be a rank-k matrix of size $m \times n$. Then there exist vectors $\{\mathbf{x}_1, \mathbf{x}_2, \ldots, \mathbf{x}_{n-k}\}$ that span $\mathcal{N}(B)$, where $\mathbf{x}_i \in \mathbb{R}^n$. Now consider the vectors from V of the SVD, $\{\mathbf{v}_1, \mathbf{v}_2, \ldots, \mathbf{v}_{k+1}\}$, where $\mathbf{v}_i \in \mathbb{R}^n$. The intersection

$$\text{span}(\mathbf{x}_1, \ldots, \mathbf{x}_{n-k}) \cap \text{span}(\mathbf{v}_1, \ldots, \mathbf{v}_{k+1}) \subset \mathbb{R}^n$$

cannot be only the zero vector, since there are a total of $n+1$ vectors involved. Let $\mathbf{z}$ be a vector from this intersection, normalized so that $\|\mathbf{z}\|_2 = 1$. Then

$$\|A - B\|_2^2 \geq \|(A - B)\mathbf{z}\|_2^2 = \|A\mathbf{z}\|_2^2.$$

Since $\mathbf{z} \in \text{span}(\mathbf{v}_1, \mathbf{v}_2, \ldots, \mathbf{v}_{k+1})$,

$$A\mathbf{z} = \sum_{i=1}^{k+1} \sigma_i \left(\mathbf{v}_i^H \mathbf{z}\right) \mathbf{u}_i.$$

Now

$$\|A - B\|_2^2 \geq \|A\mathbf{z}\|_2^2 = \sum_{i=1}^{k+1} \sigma_i^2 \left(\mathbf{v}_i^H \mathbf{z}\right)^2 \geq \sigma_{k+1}^2, \tag{7.13}$$

where the final inequality is examined in the exercises. The lower bound can be achieved by letting $B = \sum_{i=1}^{k} \sigma_i \mathbf{u}_i \mathbf{v}_i^H$, with $\mathbf{z} = \mathbf{v}_{k+1}$, establishing the theorem. □

A matrix having singular values $\sigma_{k+1}, \ldots, \sigma_p$ that are much smaller than the singular values $\sigma_1, \sigma_2, \ldots, \sigma_k$ is, by this theorem, close to a matrix having only k nonzero singular values. For some numeric problems, even though the rank of the matrix might be greater than k, since the matrix is close to a matrix of rank k, its *effective rank* is only k.

Applications of the SVD

7.6 System identification using the SVD

As discussed in section 1.4, a multi-input, multioutput, LTI discrete-time system in state-space form can be written as

$$\mathbf{x}[t+1] = A\mathbf{x}[t] + B\mathbf{u}[t], \tag{7.14}$$
$$\mathbf{y}[t] = C\mathbf{x}[t]. \tag{7.15}$$

With zero initial conditions, $\mathbf{x}[t] = 0$ for $t < 0$, the response to a unit impulse $\mathbf{u}[t] = \delta[t]I$ is

$$\mathbf{h}[t] = CA^t B.$$

The output sequence $M_t = CA^t B$ is known as the *Markov parameters* of the system. The impulse response of the analogous continuous-time system is $h(t) = Ce^{At}B$, and the Markov parameters of the analogous continuous-time system are computed from

$$M_n = \frac{d^n}{dt^n} h(t)\bigg|_{t=0} \qquad n = 0, 1, \ldots,$$

which yields $M_t = CA^t B, t = 0, 1, \ldots$.

The *order* of the system is the number of elements in the state vector $\mathbf{x}$. In many system identification problems, the order is not known. In this section we consider the

7.6 System Identification Using the SVD

realization problem: given a set of Markov parameters, determine (realize) a system model (A, B, C), including the system order. Because of the importance of this problem, a variety of algorithms have been developed, and we consider only one such approach.

We consider here what is known as the *partial realization* problem: given k observations of the impulse response of the system, determine a system model that matches the observed impulse response. The first step of this method is to form the partial *Hankel* matrix, which consists of the sequence of $k - 1$ impulse responses arranged in shifted rows:

$$\mathcal{H} = \begin{bmatrix} \mathbf{h}[0] & \mathbf{h}[1] & \mathbf{h}[2] & \cdots \\ \mathbf{h}[1] & \mathbf{h}[2] & \mathbf{h}[3] & \cdots \\ \mathbf{h}[2] & \mathbf{h}[3] & \mathbf{h}[4] & \cdots \\ \vdots & & & \\ & \cdots & & \mathbf{h}[k-2] \end{bmatrix}.$$

Lemma 7.1 *Let $\mathcal{H}_{m,n}$ be the Hankel matrix with m block rows and n block columns of Markov parameters. Then, provided that m and n are sufficiently large that $\mathrm{rank}(\mathcal{H}_{m,n}) = \mathrm{rank}(\mathcal{H}_{m+1,n+1})$, the order of the system (A, B, C) is $p = \mathrm{rank}(\mathcal{H}_{m,n})$.*

Proof The proof provides a constructive way of finding the system model (A, B, C). Let $\mathcal{H} = \mathcal{H}_{m,n}$. Using the SVD, write

$$\mathcal{H} = U \Sigma V^H = U \begin{bmatrix} \Sigma_1 & \\ & \Sigma_2 \end{bmatrix} V^H,$$

where Σ_1 is $p \times p$. Then

$$U^H \mathcal{H} V \Sigma^\dagger = \begin{bmatrix} I_p \\ 0 \end{bmatrix},$$

where I_p is the $p \times p$ identity matrix. Let A_{est}, B_{est}, and C_{est} be the indicated submatrices of the products:

$$\begin{aligned} B_{\mathrm{est}} &= U^H \mathcal{H} & (p \times l), \\ C_{\mathrm{est}} &= \mathcal{H} V \Sigma^\dagger & (m \times p), \\ A_{\mathrm{est}} &= U^H \mathrm{shift}(\mathcal{H}, 1) V \Sigma^\dagger & (p \times p), \end{aligned} \quad (7.16)$$

where $\mathrm{shift}(\mathcal{H}, n)$ represents shifting off the n left columns of $\mathcal{H}$,

$$\mathrm{shift}(\mathcal{H}, 1) = \begin{bmatrix} \mathbf{h}[1] & \mathbf{h}[2] & \mathbf{h}[3] & \cdots \\ \mathbf{h}[2] & \mathbf{h}[3] & \mathbf{h}[4] & \cdots \\ \mathbf{h}[3] & \mathbf{h}[4] & \mathbf{h}[5] & \cdots \\ \vdots & & & \\ & \cdots & & \mathbf{h}[k-1] \end{bmatrix}.$$

Then it is straightforward to verify that the Markov parameters

$$C_{\mathrm{est}} A_{\mathrm{est}}^k B_{\mathrm{est}}$$

are equivalent to the Markov parameters $C A^k B$. □

Algorithm 7.1 provides a partial system identification based upon this method.

Algorithm 7.1 System identification using SVD
File: `sysidsvd.m`
`makehankel.m`

In any problem of system identification, it is possible to be misled by the data. For example, suppose that the system has the impulse response

$$h(t) = e^{-t} + \frac{t^{10}}{10!}.$$

Examination of the first few Markov parameters $M_0 = 1, M_1 = -1, M_2 = 1, M_3 = -1, \ldots$, would lead to the wrong conclusion regarding the order of the system. The full order of the system is not revealed until the 10th sample. Such is the risk of the partial realization problem.

Example 7.6.1 A single-input, two-output system has the following impulse response sequence:

$$\left\{ \mathbf{h}_0 = \begin{bmatrix} 11 \\ 9 \end{bmatrix}, \begin{bmatrix} 5 \\ 6 \end{bmatrix} \begin{bmatrix} 2.75 \\ 2.25 \end{bmatrix} \begin{bmatrix} 1.25 \\ 1.5 \end{bmatrix} \begin{bmatrix} 0.6875 \\ 0.5625 \end{bmatrix} \begin{bmatrix} 0.3125 \\ 0.3750 \end{bmatrix} \begin{bmatrix} 0.1719 \\ 0.1406 \end{bmatrix} \begin{bmatrix} 0.0781 \\ 0.0938 \end{bmatrix} = \mathbf{h}_7 \right\}.$$

A Hankel matrix is

$$\mathcal{H} = \begin{bmatrix} 11.0000 & 5.0000 & 2.7500 \\ 9.0000 & 6.0000 & 2.2500 \\ 5.0000 & 2.7500 & 1.2500 \\ 6.0000 & 2.2500 & 1.5000 \end{bmatrix},$$

which has rank 2. Using the method outlined above, a realization of this system, using the following section of code:

```
% the input data:
h0 = [11 9]'; h1 = [5 6]'; h2 = [2.75 2.25]'; h3 = [1.25 1.5]';
h4 = [.6875 .5625]'; h5 = [.3125 .375]'; h6 = [.1719 .1406]';
h7 = [.0781 .0938]';
y = {h0,h1,h2,h3,h4,h5,h6,h7};
[Aest,Best,Cest] = sysidsvd(y);
```

is given as

$$A_{\text{est}} = \begin{bmatrix} 0.513234 & -0.216652 \\ 0.0618908 & -0.513234 \end{bmatrix} \quad \mathbf{b}_{\text{est}} = \begin{bmatrix} 16.7001 \\ -0.73854 \end{bmatrix} \quad C_{\text{est}} = \begin{bmatrix} 0.640217 & -0.417493 \\ 0.570751 & 0.719762 \end{bmatrix}.$$

By comparison, the original data was produced by the system

$$A = \begin{bmatrix} 0 & 0.25 \\ 1 & 0 \end{bmatrix} \quad \mathbf{b} = \begin{bmatrix} 1 \\ 4 \end{bmatrix} \quad C = \begin{bmatrix} 3 & 2 \\ 5 & 1 \end{bmatrix}.$$

Even though the form of the estimated system is not identical to the original form, the modes of the system are correctly identified: the eigenvalues of A_{est} are ± 0.5, the same as the eigenvalues of the original system, so the identified system is similar to the original system. □

Data measured from systems is usually observed in the presence of noise. If a Hankel matrix is formed from a sequence of outputs measured in noise, the Hankel matrix is unlikely to be rank deficient, so the true system order cannot be determined as the rank of the Hankel matrix. However, if the noise power is small enough, the singular values of the Hankel matrix can be examined to determine an effective rank of the system. By setting the smallest singular values to zero, a matrix that is near to the original matrix can be constructed. Unfortunately, the new matrix will not necessarily have the Hankel structure, so the factorization just proposed for system realization does not apply. What is desired is a reduced-rank Hankel matrix that is near to the original matrix. Some approaches to this problem are referenced in section 7.12; another approach is described in section 15.5.3.

Example 7.6.2 Suppose the impulse response data from the previous example are observed in additive white Gaussian noise with $\sigma^2 = 0.01$, and the following data are observed:

$$\left\{ \mathbf{h}_0 = \begin{bmatrix} 11.175 \\ 8.968 \end{bmatrix}, \begin{bmatrix} 4.986 \\ 6.062 \end{bmatrix}, \begin{bmatrix} 2.848 \\ 2.139 \end{bmatrix}, \begin{bmatrix} 1.195 \\ 1.504 \end{bmatrix}, \begin{bmatrix} 0.439 \\ 0.678 \end{bmatrix}, \begin{bmatrix} 0.210 \\ 0.490 \end{bmatrix}, \begin{bmatrix} 0.093 \\ 0.204 \end{bmatrix}, \begin{bmatrix} 0.160 \\ 0.076 \end{bmatrix} = \mathbf{h}_7 \right\}.$$

The 4×3 Hankel matrix constructed from this has singular values 18.83, 1.73, and 0.065; it is near to a rank-2 matrix and so the system can be assumed to be a second-order system. □

7.7 Total least-squares problems

In the least-squares problems considered up to this point, the solution minimizing $\|A\mathbf{x} - \mathbf{b}\|$ has been sought, with the tacit assumption that the matrix A is correct, and any errors in the problem are in $\mathbf{b}$. The vector $\mathbf{b}$ is projected onto the range of A to find the solution. By the assumption, any changes needed to find a solution must come by modifying only $\mathbf{b}$. However, in many problems, the matrix A is determined from measured data, and hence may have errors also. Thus, it may be of use to find a solution to the problem $A\mathbf{x} = \mathbf{b}$ which allows for the fact that both A and $\mathbf{b}$ may be in error. Problems of this sort are known as *total least-squares problems* (TLS).

The LS problem finds a vector $\hat{\mathbf{x}}$ such that

$$\|A\hat{\mathbf{x}} - \mathbf{b}\|_2 = \min,$$

which is accomplished by finding some perturbation $\mathbf{r}$ of the right-hand side of minimum norm, accomplished and finding the solution of the perturbed equation

$$A\mathbf{x} = \mathbf{b} + \mathbf{r}.$$

In the TLS problem, both the right-hand side $\mathbf{b}$ and the data matrix A are assumed to be in error, and solution of the perturbed equation

$$(A + E)\mathbf{x} = (\mathbf{b} + \mathbf{r}) \tag{7.17}$$

is sought, so that $(\mathbf{b} + \mathbf{r}) \in \mathcal{R}(A + E)$ and the norm of the perturbations is minimized. The right-hand side is "bent" toward the left-hand side while the left-hand side is "bent" toward the right-hand side. In usual applications of TLS, the number of equations exceeds the number of unknowns. However, it may also be applied when the number of unknowns exceeds the number of equations. In this case, an infinite number of solutions may exist, and the TLS solution method selects the one with minimum norm.

Some motivation for this problem comes from considering least-squares problems with a single variable. Data are acquired fitting a model

$$y_i = ax_i \qquad i = 1, 2, \ldots, N.$$

For a least-squares fit, a line is found that minimizes the *total vertical distance* from the measured points (x_i, y_i) to the line, as shown in figure 7.2(a). The data x_i are not modified in fitting to the line. For total least squares, a line is found that minimizes the *total distance*

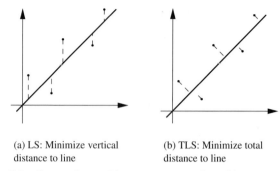

(a) LS: Minimize vertical distance to line

(b) TLS: Minimize total distance to line

Figure 7.2: Comparison of least-squares and total least-squares fit

from the points to the line, as shown in figure 7.2(b). In this case, modifications are made to both the x_i and the y_i data.

Example 7.7.1 An observed signal consists of M complex sinusoids in AWGN,

$$x[t] = \sum_{k=1}^{M} a_k e^{j2\pi f_k t} + w[t], \quad n = 0, 1, \ldots, N-1. \tag{7.18}$$

As discussed in sections 1.4.6 and section 8.1, if there were no noise the signal could be modeled as a zero-input autoregressive signal

$$\sum_{i=1}^{M} c_i x_{M+n-i} = x_{M+n}, \quad n = 0, 1, \ldots, N-M-1, \tag{7.19}$$

for some set of coefficients $\{c_i\}$ to be found. If a solution for the c_i can be obtained, then the frequencies can be obtained by finding the zeros of the characteristic polynomial

$$z^M - c_1 z^{M-1} - \cdots - c_M = 0.$$

Equation (7.19) can be written as

$$\begin{bmatrix} x_0 & x_1 & \cdots & c_{M-1} \\ x_1 & x_2 & \cdots & x_M \\ \vdots & \vdots & & \vdots \\ x_{N-M-1} & x_{N-M} & \cdots & x_{N-1} \end{bmatrix} \begin{bmatrix} c_M \\ c_{M-1} \\ \vdots \\ c_1 \end{bmatrix} = \begin{bmatrix} x_M \\ x_{M+1} \\ \vdots \\ x_{N-1} \end{bmatrix}. \tag{7.20}$$

If there is noise in the observation model (7.19), then (7.20) should be regarded as an approximate equation. Since the observations x_i are on both the left-hand side and right-hand side of the equation, both sides can be considered to be approximate. This equation is thus an ideal candidate for TLS solution. □

Example 7.7.2 Suppose a system can be modeled by the equation

$$y[t] + a_1 y[t-1] + a_2 y[t-2] = b_1 f[t-1] + b_2 f[t-2] + b_3 f[t-3].$$

Let $\boldsymbol{\theta} = [a_1, a_2, b_1, b_2, b_3]^T$ denote the vector of parameters in this system and let $\mathbf{h}[t] = [-y[t-1], -y[t-2], f[t-1], f[t-2], f[t-3]]^T$ denote previous outputs and inputs. Then the system equation can be written as

$$y[k] = \mathbf{h}[t]^T \boldsymbol{\theta}.$$

Let $\mathbf{y} = [y[0], y[1], \ldots, y[N]]^T$, and let

$$H = \begin{bmatrix} \mathbf{h}[0]^T \\ \mathbf{h}[1]^T \\ \vdots \\ \mathbf{h}[N]^T \end{bmatrix}.$$

7.7 Total Least-Squares Problems

Then an equation for the unknown parameters is

$$H\theta = \mathbf{y}.$$

It is desired to estimate θ from noisy observations of the input and output. A least-squares solution to the problem could be found, but that would ignore the fact that there are errors also in the matrix H. The TLS finds a solution which takes into account errors on both sides of the equation. □

Let A be an $m \times n$ matrix. To find a solution to the TLS problem, write (7.17) in the homogeneous form

$$[A + E | \mathbf{b} + \mathbf{r}] \begin{bmatrix} \mathbf{x} \\ -1 \end{bmatrix} = 0$$

or

$$([A|\mathbf{b}] + [E|\mathbf{r}]) \begin{bmatrix} \mathbf{x} \\ -1 \end{bmatrix} = 0, \quad (7.21)$$

where $x|y$ indicates that x and y are stacked side-by-side. Let $C = [A|\mathbf{b}] \in \mathbb{R}^{m \times (n+1)}$ and let $\Delta = [E|\mathbf{r}]$ be the perturbation of the data. In order for (7.21) to have a solution, the augmented vector $[\mathbf{x}^T, -1]^T$ must lie in the nullspace of $C + \Delta$, and in order for the solution not to be trivial, the perturbation Δ must be such that $C + \Delta$ is rank deficient. The TLS solution finds the Δ with smallest norm that makes $C + \Delta$ rank deficient.

The SVD can be used to find the TLS solution. Write the SVD of C as

$$C = U \mathrm{diag}(\sigma_1, \sigma_2, \ldots, \sigma_{n+1}) V^H = \sum_{k=1}^{n+1} \sigma_k \mathbf{u}_k \mathbf{v}_k^H$$

with

$$U = [\mathbf{u}_1, \mathbf{u}_2, \ldots, \mathbf{u}_m] \quad \text{and} \quad V = [\mathbf{v}_1, \mathbf{v}_2, \ldots, \mathbf{v}_{n+1}].$$

Initially we will assume that the singular values are ordered with a *unique* smallest singular value:

$$\sigma_1 \geq \sigma_2 \geq \cdots \geq \sigma_n > \sigma_{n+1} > 0.$$

From theorem 7.2, the reduced-rank matrix closest to C is the matrix

$$C + \Delta = \tilde{C} = \sum_{k=1}^{n} \sigma_i \mathbf{u}_i \mathbf{v}_i^H.$$

The perturbation is therefore of the form

$$\Delta = -\sigma_i \mathbf{u}_{n+1} \mathbf{v}_{n+1}^H.$$

Since $\mathrm{span}(C + \Delta)$ does not contain the vector $\mathbf{v}_{n+1}$, the solution (7.21) must be a multiple of $\mathbf{v}_{n+1}$:

$$\begin{bmatrix} \mathbf{x} \\ -1 \end{bmatrix} = \alpha \mathbf{v}_{n+1}. \quad (7.22)$$

If the last component of $\mathbf{v}_{n+1}$ is not equal to zero, then the desired solution can be found as

$$\mathbf{x} = -\frac{\mathbf{v}_{n+1}(1:n)}{\mathbf{v}_{n+1}(n+1)}.$$

If the last component of $\mathbf{v}_{n+1}$ is zero, there is no solution.

The next level of generality of the TLS solution comes by assuming that the smallest singular value is repeated,

$$\sigma_1 \geq \sigma_2 \geq \cdots \geq \sigma_k > \sigma_{k+1} = \sigma_{k+2} = \cdots = \sigma_{n+1}.$$

Let $\tilde{V} = [\mathbf{v}_{k+1}, \mathbf{v}_{k+2}, \ldots, \mathbf{v}_{n+1}]$, and let
$$S_c = \mathcal{R}(\tilde{V}) = \text{span}(\mathbf{v}_{k+1}, \mathbf{v}_{k+2}, \ldots, \mathbf{v}_{n+1}). \tag{7.23}$$

Then the solution $\begin{bmatrix} \mathbf{x} \\ -1 \end{bmatrix} \in S_c$. We desire to find the solution such that $\|\mathbf{x}\|_2$ is minimized. One approach is explored in exercise 7.7-13. Another approach (based upon exercise 5.3.4-19) is to finde a Householder matrix $\tilde{Q}$ such that

$$\tilde{V}\tilde{Q} = \begin{bmatrix} W & \mathbf{y} \\ \mathbf{0} & \alpha \end{bmatrix}.$$

By the properties of the Householder matrix, the vector $\mathbf{z} = \begin{bmatrix} \mathbf{y} \\ \alpha \end{bmatrix}$ is the vector in $\mathcal{R}(\tilde{V})$ such that the last component of $\tilde{V}\mathbf{z}$ is maximized, subject to the constraint $\|\mathbf{z}\|_2 = 1$. Thus $\mathbf{x} = -\mathbf{y}/\alpha$ is the minimum-norm TLS solution. Algorithm 7.2 computes a TLS solution based upon this concept.

Algorithm 7.2 Total least squares
File: `tls.m`

Example 7.7.3 In this example, the TLS method is applied to the identification of a system from measurements of its inputs and state variables. Suppose we know that a system is a second order system,

$$\mathbf{x}[t+1] = \begin{bmatrix} x_1[t+1] \\ x_2[t+1] \end{bmatrix} = \begin{bmatrix} a_{11} & a_{12} \\ a_{21} & a_{22} \end{bmatrix} \mathbf{x}[t] + \begin{bmatrix} b_1 \\ b_2 \end{bmatrix} u[t] + \mathbf{w}[t].$$

If we are able to measure the input and the state variables, a system of equations can be set up as

$$x_1[1] = a_{11}x_1[0] + a_{12}x_2[0] + b_1 u[0] + w_1[0],$$
$$x_2[1] = a_{21}x_1[0] + a_{22}x_2[0] + b_2 u[0] + w_2[0],$$
$$x_1[2] = a_{11}x_1[1] + a_{12}x_2[1] + b_1 u[1] + w_1[1],$$
$$x_2[2] = a_{21}x_1[1] + a_{22}x_2[1] + b_2 u[1] + w_2[1],$$
$$\vdots$$

These equations can be written in matrix form as

$$\begin{bmatrix} x_1[0] & x_2[0] & & & u[0] & \\ & & x_1[0] & x_2[0] & & u[0] \\ x_1[1] & x_2[1] & & & u[1] & \\ & & x_1[1] & x_2[1] & & u[1] \\ \vdots & & & & & \end{bmatrix} \begin{bmatrix} a_{11} \\ a_{12} \\ a_{21} \\ a_{22} \\ b_1 \\ b_2 \end{bmatrix} \approx \begin{bmatrix} x_1[1] \\ x_2[1] \\ x_1[2] \\ x_2[2] \\ \vdots \end{bmatrix}.$$

It is expected that noise would exist in both the state measurements and the input measurements. Since components on both sides of the equation may be corrupted by noise, a TLS approach is appropriate for the solution.

Suppose we measure the following sequences of state variables

$$\left\{ \mathbf{x}_0 = \begin{bmatrix} 0.0689 \\ 0.8926 \end{bmatrix}, \begin{bmatrix} 0.2926 \\ 0.3016 \end{bmatrix}, \begin{bmatrix} 0.5401 \\ 1.9890 \end{bmatrix}, \begin{bmatrix} 0.5360 \\ 0.6189 \end{bmatrix}, \begin{bmatrix} 0.8275 \\ 3.3191 \end{bmatrix}, \begin{bmatrix} 1.7336 \\ 4.5681 \end{bmatrix}, \begin{bmatrix} 1.5060 \\ 2.8487 \end{bmatrix} = \mathbf{x}_6 \right\}$$

and inputs

$$\{u_0 = 0.0351, 0.5010, 0.1561, 0.7533, 0.8993, 0.2287\}.$$

7.7 Total Least-Squares Problems

The corresponding set of equations is

$$\begin{bmatrix} 0.0689 & 0.8926 & 0 & 0 & 0.0351 & 0 \\ 0 & 0 & 0.0689 & 0.8926 & 0 & 0.0351 \\ 0.2926 & 0.3016 & 0 & 0 & 0.501 & 0 \\ 0 & 0 & 0.2926 & 0.3016 & 0 & 0.501 \\ 0.5401 & 1.989 & 0 & 0 & 0.1561 & 0 \\ 0 & 0 & 0.5401 & 1.989 & 0 & 0.1561 \\ 0.536 & 0.6189 & 0 & 0 & 0.7533 & 0 \\ 0 & 0 & 0.536 & 0.6189 & 0 & 0.7533 \\ 0.8275 & 3.3191 & 0 & 0 & 0.8993 & 0 \\ 0 & 0 & 0.8275 & 3.3191 & 0 & 0.8993 \\ 1.7336 & 4.5681 & 0 & 0 & 0.2287 & 0 \\ 0 & 0 & 1.7336 & 4.5681 & 0 & 0.2287 \end{bmatrix} \begin{bmatrix} a_{11} \\ a_{12} \\ a_{21} \\ a_{22} \\ b_1 b_2 \end{bmatrix} = \begin{bmatrix} 0.2926 \\ 0.3016 \\ 0.5401 \\ 1.989 \\ 0.536 \\ 0.6189 \\ 0.8275 \\ 3.3191 \\ 1.7336 \\ 4.5681 \\ 2.8487 \\ 2.8487 \end{bmatrix}.$$

Then the best estimate of the solution is

$$[a_{11} \quad a_{12} \quad a_{21} \quad a_{22} \quad b_1 \quad b_2]^T =$$
$$[1.38308 \quad 0.0747001 \quad 0.87168 \quad 0.0594366 \quad 0.219796 \quad 3.9033]. \quad \square$$

7.7.1 Geometric interpretation of the TLS solution

It is interesting to examine what the TLS represents geometrically. From the definition of the 2-norm of a matrix,

$$\frac{\|C\mathbf{v}\|}{\|\mathbf{v}\|} \geq \sigma_{n+1}.$$

Equality holds here if and only if $\mathbf{v} \in S_c$, where S_c is defined by (7.23). The TLS problem amounts to finding a vector $\mathbf{x}$ such that

$$\frac{\left\| [A|\mathbf{b}] \begin{bmatrix} \mathbf{x} \\ -1 \end{bmatrix} \right\|}{\left\| \begin{bmatrix} \mathbf{x} \\ -1 \end{bmatrix} \right\|} = \sigma_{n+1},$$

or, squaring everywhere,

$$\min_{\mathbf{x}} \frac{\left\| [A|\mathbf{b}] \begin{bmatrix} \mathbf{x} \\ -1 \end{bmatrix} \right\|^2}{\left\| \begin{bmatrix} \mathbf{x} \\ -1 \end{bmatrix} \right\|^2} = \min_{\mathbf{x}} \sum_{i=1}^{m} \frac{|\mathbf{a}_i^H \mathbf{x} - b_i|^2}{\mathbf{x}^H \mathbf{x} + 1}, \quad (7.24)$$

where $\mathbf{a}_i^T = (a_{i1}, a_{i2}, \ldots, a_{in})$ is the ith row of A. The quantity

$$\frac{|\mathbf{a}_i^H \mathbf{x} - b_i|^2}{\mathbf{x}^H \mathbf{x} + 1}$$

is the square of the distance from the point $\begin{bmatrix} \mathbf{a}_i \\ b_i \end{bmatrix} \in \mathbb{C}^{m+1}$ to the nearest point on the hyperplane P_i defined by

$$P_i = \left\{ \begin{bmatrix} \mathbf{a} \\ b \end{bmatrix} \mid \mathbf{a} \in \mathbb{C}^n, b \in \mathbb{C}, b = \mathbf{x}^H \mathbf{a} \right\}.$$

So the TLS problem amounts to finding the closest hyperplane to the set of points

$$\begin{bmatrix} \mathbf{a}_1 \\ b_1 \end{bmatrix}, \begin{bmatrix} \mathbf{a}_2 \\ b_2 \end{bmatrix}, \ldots, \begin{bmatrix} \mathbf{a}_m \\ b_m \end{bmatrix}.$$

This is the geometry that was suggested by figure 7.2(b).

7.8 Partial total least squares

In the total least-squares method, all of the elements on both sides of the equation $A\mathbf{x} = \mathbf{b}$ are assumed to be noisy. In the oxymoronically named partial total least-squares (PTLS) method, those columns of A or $\mathbf{b}$ that are known to be accurate are not modified. This generalization is useful, for example, in system identification studies in which the input variables may be known precisely, while the output variables are available only through noisy measurements. It is also useful in regression problems, as the following example illustrates.

Example 7.8.1 Consider again the problem of fitting a line to a set of measured points (x_i, y_i), $i = 1, 2, \ldots, N$, according to the model $y_i = ax_i + b$, where a and b are to be determined. The equations can be set up according to

$$\begin{bmatrix} 1 & x_1 \\ 1 & x_2 \\ \vdots & \vdots \\ 1 & x_N \end{bmatrix} \begin{bmatrix} b \\ a \end{bmatrix} = \begin{bmatrix} y_1 \\ y_2 \\ \vdots \\ y_N \end{bmatrix}.$$

The least-squares method assumes that only the right-hand side is noisy. The total least-squares method assumes that both the right-hand and the left-hand sides are noisy. However, the constant column of 1 on the left-hand is *not* in error, so the total least-squares solution is not appropriate here.

As an extension of the total least-squares concept, suppose that somehow some of the data are measured precisely. Suppose, for example, that one measurement (x_1, y_1) is made precisely, while the next two are corrupted by noise,

$$\begin{bmatrix} 1 & x_1 \\ \cdots\cdots \\ 1 & x_2 \\ 1 & x_3 \end{bmatrix} \begin{bmatrix} b \\ a \end{bmatrix} = \begin{bmatrix} y_1 \\ \cdots \\ y_2 \\ y_3 \end{bmatrix}$$

The first row should be solved exactly, while the other rows should be solved in a least-squares sense. □

We will begin with the case in which all the rows are treated equally, but some columns are exact. As with the total least-squares problem, we form an $m \times (n+1)$ matrix $C = [A \quad \mathbf{b}]$ consisting of both sides of the equation, then determine a way to reduce the rank of C so that we have the homogeneous equation

$$[A \quad \mathbf{b}] \begin{bmatrix} \mathbf{x} \\ -1 \end{bmatrix} = 0.$$

We reduce the rank only by modifying certain columns of C. Let C be partitioned as $C = [C_1 \quad C_2]$. (It is assumed that the columns are permuted so that only C_2 is modified.) We wish to find a matrix $\hat{C}_2$ such that $\hat{C} = [C_1 \quad \hat{C}_2]$ has rank n and is closest to C.

Theorem 7.3 *[110] Let C be partitioned as $C = [C_1 \quad C_2]$, where C_1 has l columns and $k = \mathrm{rank}(C_1)$. If $l \leq r$, then $\hat{C}_2$ can be determined so that $[C_1 \quad \hat{C}_2]$ has rank r by forming*

$$\hat{C}_2 = Q_1 R_{12} + Q_2 \hat{R}_{22,r-l},$$

where Q_1, R_{12}, and R_{22} come from the QR factorization of C,

$$[C_1 \quad C_2] = [Q_1 \quad Q_2 \quad Q_3] \begin{bmatrix} R_{11} & R_{12} \\ 0 & R_{22} \\ 0 & 0 \end{bmatrix} = QR, \qquad (7.25)$$

7.8 Partial Total Least Squares

and $R_{22,r-l}$ is the nearest rank $(r - l)$ approximation to R_{22}. In (7.25), R_{11} and R_{22} are upper triangular, and R_{11} has k columns.

Proof Assume that $k = l$; if this is not the case, then C_1 can be replaced with a matrix having l independent columns, which can be restored after the appropriate approximation to C_2 is found.

We need to determine a modification to the second half of the R matrix,

$$\begin{bmatrix} Q_1^H \\ Q_2^H \\ Q_3^H \end{bmatrix} [C_1 \quad \hat{C}_2] = \hat{R} = \begin{bmatrix} R_{11} & \hat{R}_{12} \\ 0 & \hat{R}_{22} \\ 0 & \hat{R}_{32} \end{bmatrix}, \quad (7.26)$$

so that $\hat{R}$ has rank r in such a way that $\hat{R}$ is close to R. We can take $\hat{R}_{12} = R_{12}$, then find $(\hat{R}_{22}, \hat{R}_{23})$ in such a way that they have rank $r - l$. The nearest matrix to $(\hat{R}_{22}, \hat{R}_{23})$ of rank $r - l$ can be determined using the SVD. Denote the solution as $\hat{R}_{22,r-l}$. Then from (7.26), we have

$$\hat{C}_2 = Q_1 R_{12} + Q_2 \hat{R}_{22,r-l}.$$

□

To apply this theorem to the PTLS problem, we want C to be rank deficient, so we let $r = n$ (since C has $n + 1$ columns). Code implementing the PTLS solution to $A\mathbf{x} = \mathbf{b}$ is shown in algorithm 7.3.

Algorithm 7.3 Partial total least squares, part 1
File: `ptls1.m`

The next level of sophistication in the PTLS is achieved by partitioning the data so that the first k_1 rows and the first k_2 columns are assumed to be accurate. The equation $A\mathbf{x} = \mathbf{b}$ can be written in homogeneous form as

$$[A \quad -\mathbf{b}] \begin{bmatrix} \mathbf{x} \\ 1 \end{bmatrix} = \begin{bmatrix} W & X \\ Y & Z \end{bmatrix} \begin{bmatrix} \mathbf{x} \\ 1 \end{bmatrix} = 0,$$

where W is a $k_1 \times k_2$ matrix, corresponding to those rows and columns in the original problem that are to left unmodified. Let

$$T = \begin{bmatrix} W & X \\ Y & Z \end{bmatrix}.$$

Then (as explored in the exercises), T can be written as

$$\tilde{T} = UTV^H = \begin{bmatrix} W_{11} & 0 & 0 & X_{11} & X_{12} \\ 0 & 0 & 0 & X_{21} & 0 \\ 0 & 0 & 0 & 0 & 0 \\ Y_{11} & Y_{12} & 0 & Z_{11} & Z_{12} \\ Y_{21} & 0 & 0 & Z_{21} & Z_{22} \end{bmatrix},$$

where U and V are unitary, and W_{11}, X_{21}, and Y_{12} are either square and nonsingular, or null. By row and column operations, $\tilde{T}$ can be converted into

$$\tilde{\tilde{T}} = \begin{bmatrix} W_{11} & 0 & 0 & 0 & 0 \\ 0 & 0 & 0 & X_{21} & 0 \\ 0 & 0 & 0 & 0 & 0 \\ 0 & Y_{12} & 0 & 0 & 0 \\ 0 & 0 & 0 & 0 & Z_{22} - Y_{21} W_{11}^{-1} X_{12} \end{bmatrix}$$

if W_{11} is not null. If W_{11} is null, then Z_{22} appears in the lower right corner of $\tilde{\tilde{T}}$. The rank of $\tilde{\tilde{T}}$ can be observed to be

$$\text{rank}(\tilde{\tilde{T}}) = \text{rank}(W_{11}) + \text{rank}(X_{21}) + \text{rank}(Y_{12}) + \text{rank}(Z_{22} - Y_{21} W_{11}^{-1} X_{12}).$$

Using the SVD, a matrix ΔZ_{22} can be found so that

$$Z_{22} - Y_{21} W_{11}^{-1} X_{12} + \Delta Z_{22}$$

is rank deficient. Let $\hat{Z}_{22} = Z_{22} + \Delta Z_{22}$, and let

$$\hat{T} = \begin{bmatrix} W_{11} & 0 & 0 & X_{11} & X_{12} \\ 0 & 0 & 0 & X_{21} & 0 \\ 0 & 0 & 0 & 0 & 0 \\ Y_{11} & Y_{12} & 0 & Z_{11} & Z_{12} \\ Y_{21} & 0 & 0 & Z_{21} & \hat{Z}_{22} \end{bmatrix}.$$

Transforming back by

$$\begin{bmatrix} \hat{W} & \hat{X} \\ \hat{Y} & \hat{Z} \end{bmatrix} = U^H \hat{T} V,$$

we obtain the equations

$$\begin{bmatrix} \hat{W} & \hat{X} \\ \hat{Y} & \hat{Z} \end{bmatrix} \begin{bmatrix} \mathbf{x}_1 \\ \mathbf{x}_2 \end{bmatrix} = \mathbf{0},$$

where $\mathbf{x}_1 \in \mathbb{R}_1^k$. From the first row we find

$$\mathbf{x}_1 = -\hat{W}^\dagger \hat{X} \mathbf{x}_2,$$

which, when substituted in the second row, yields the equation

$$(-\hat{Y}\hat{W}^\dagger X + \hat{Z})\mathbf{x}_2 = 0.$$

Since the matrix is constrained to be rank deficient, there is a nontrivial nullspace. From this nullspace, an element $\mathbf{x}_2$ of minimum norm is selected having the last component equal to 1. Based on this value for $\mathbf{x}_2$, the solution $\mathbf{x}_1$ can be found.

The procedure as described here appears in algorithm 7.4.

Algorithm 7.4 Partial total least squares, part 2
File: `ptls2.m`

Example 7.8.2 For the problem of identifying the parameters in the linear model $y_i = ax_i + b$, with unknowns a and b, assume that the first data point is accurately measured. Figure 7.3 illustrates ten data points and lines fit to the data using the PTLS method and the LS method. Observe that the line passes exactly through one point, because of the constraint. □

7.9 Rotation of Subspaces

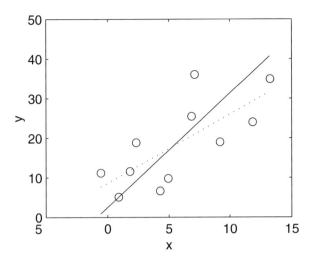

Figure 7.3: PTLS linear parameter identification (solid line is PTLS solution)

7.9 Rotation of subspaces

Example 7.9.1 Figure 7.4 illustrates point data from two data sets. (For example, the data might be salient feature locations from an image). The points in the first data set are indicated with ×; in the second, with ∘. Each point in the second data set is rotated relative to each point in the first data set by some angle $\theta = \theta_0 + \theta_n$, where θ_0 is the average angle, and θ_n is some small random number. The plot illustrates rays to the first data point of each data set. The problem is to estimate the angle of rotation from the first data set to the second data set. □

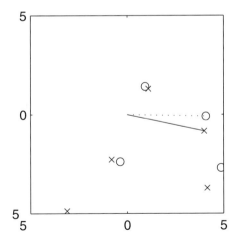

Figure 7.4: A data set rotated relative to another data set

This example motivates the problem known as the *orthogonal Procrustes problem*. Given $A \in \mathbb{C}^{m \times n}$ and $B \in \mathbb{C}^{m \times n}$, find the minimizing rotation of B into A:

$$\text{minimize } \|A - BQ\|_F \quad \text{subject to } Q^H Q = I,$$

where the constraint ensures that the transformation Q is unitary. If Q is unitary, this can be written as

$$\text{minimize } \|A - BQ\|_F = \text{minimize } (\text{tr}(A^H A) + \text{tr}(B^H B) - 2\,\text{tr}(Q^H B^H A)).$$

(See (4.9).) Since the first two terms do not depend upon Q, this is equivalent to the problem

$$\text{maximize tr}(Q^H B^H A), \quad Q \text{ unitary} \quad \text{subject to } Q^H Q = I. \quad (7.27)$$

The maximizing Q can be found by means of the SVD of $B^H A$.

Theorem 7.4 *If*

$$B^H A = U \Sigma V,$$

then the maximizing matrix Q for the orthogonal Procrustes problem (7.27) is

$$Q = U V^T.$$

Proof Let Z be the unitary matrix $Z = V^H Q^H U$, for Q to be determined. Then

$$\begin{aligned}
\text{tr}(Q^H B^H A) &= \text{tr}(Q^H U \Sigma V^H) \\
&= \text{tr}(Z \Sigma) \quad &\text{(operations commute in the trace)} \\
&= \sum_{i=1}^{p} z_{ii} \sigma_i \quad &\text{(definition of the trace)} \\
&\leq \sum_{i=1}^{p} \sigma_i \quad &\text{(see exercise 7.9-16).}
\end{aligned}$$

The upper bound is obtained in the case when the $z_{ii} = 1$, which occurs when $Q = U V^H$: in this case $Z = I$. This is another example of an optimization problem where the conditions satisfying an inequality give rise to the desired optimizer. □

7.10 Computation of the SVD

While the proof of theorem 7.1 is a constructive proof, providing an explicit recipe for finding the SVD using the eigendecomposition of $A^H A$ and AA^H, this method of computing is not recommended. Since these computations involve squaring components of the matrix, small elements of the matrix get smaller, and large elements become even larger. If the condition number of the matrix A is $\kappa(A)$, then the condition of $A^H A$ is $\kappa(A)^2$: the resulting matrix will have worse condition.

In preference to computing eigendecompositions of $A^H A$ and AA^H to find the SVD, we proceed in a manner similar to the computation of the eigenvalues of a symmetric matrix, as discussed in section 6.15. Since A is not, in general, a symmetric matrix, some modifications are necessary. In what follows we assume that A is real; for the SVD of complex matrices see exercise 7.10-17. The outline of the computational approach is as follows:

1. By a series of Householder transformations, determine a matrix B that is orthogonally equivalent to A and that is upper bidiagonal. That is, we find orthogonal matrices U_B and V_B such that

$$U_B^T A V_B = \begin{bmatrix} B \\ 0 \end{bmatrix},$$

7.10 Computation of the SVD

where B has the upper-bidiagonal form

$$B = \begin{bmatrix} d_1 & f_1 & 0 & 0 & \cdots & 0 \\ 0 & d_2 & f_2 & 0 & \cdots & 0 \\ & \ddots & \ddots & \ddots & & \\ \vdots & & & & & f_{n-1} \\ 0 & 0 & 0 & \cdots & 0 & d_n \end{bmatrix}.$$

2. B is now diagonalized using what is known as the Golub–Kahan step. The Golub–Kahan step implicitly creates the tridiagonal matrix $T = B^T B$. Since T is symmetric and tridiagonal, its eigenvalues can be stably computed using the same implicit QR shift as was used in the eigenvalue computation of algorithm 6.10. Specifically, we form the Wilkinson shift by finding the eigenvalue λ of the lower-right 2×2 submatrix of $T = B^T B$ that is closer to $T(n, n)$. Then a Givens rotation G_1 is found that would place a zero in the second position of the shifted $T - I\lambda$,

$$G_1 \begin{bmatrix} d_1^2 - \lambda \\ d_1 f_1 \end{bmatrix} = \begin{bmatrix} \times \\ 0 \end{bmatrix}.$$

However, this Givens rotation is applied to B (on the right), and not to the shifted T. This places a nonzero entry in B. Illustrating for $n = 4$, we have

$$BG_1 = \begin{bmatrix} \times & \times & 0 & 0 \\ + & \times & \times & 0 \\ 0 & \times & \times & 0 \\ 0 & 0 & 0 & \times \end{bmatrix},$$

where $+$ indicates a newly nonzero entry. The new nonzero entry is chased down the diagonal using a series of Givens rotations, U_i applied on the left, and V_i applied on the right:

$$U_1^T(BG_1) = \begin{bmatrix} \times & \times & + & 0 \\ 0 & \times & \times & 0 \\ 0 & 0 & \times & \times \\ 0 & 0 & 0 & \times \end{bmatrix},$$

$$U_1^T(BG_1)V_2 = \begin{bmatrix} \times & \times & 0 & 0 \\ 0 & \times & \times & 0 \\ 0 & + & \times & \times \\ 0 & 0 & 0 & \times \end{bmatrix},$$

$$U_2^T U_1^T(BG_1)V_2 = \begin{bmatrix} \times & \times & 0 & 0 \\ 0 & \times & \times & + \\ 0 & 0 & \times & \times \\ 0 & 0 & 0 & \times \end{bmatrix},$$

and so forth. Then an updated B is obtained by

$$B \leftarrow \left(U_{n-1}^T \cdots U_1^T\right)(BG_1)(V_2 V_3 \cdots V_{n-1}).$$

The process is then repeated until the off-diagonal elements converge essentially to zero.

The Golub–Kahan step operates on a matrix which is nonzero on both its diagonal and its superdiagonal.

The other consideration for the computation of the SVD is that, as the algorithm progresses, the matrix B can in general be partitioned into three pieces,

$$B = \begin{bmatrix} B_{11} & & \\ & B_{22} & \\ & & B_{33} \end{bmatrix},$$

where B_{33} is diagonal and B_{22} is nonzero on its above-diagonal. An additional partitioning can be obtained if B_{22} has a zero (or zeros) on its diagonal. If $(B_{22})_{k,k} = 0$, then the kth row of B_{22} is zeroed out by a succession of Givens rotations, and the Golub–Kahan step is not performed. Combining these ideas, we have algorithm 7.5. This algorithm can compute the singular values alone (saving computations), or the entire SVD, depending on the return arguments.

Algorithm 7.5 Computing the SVD
File: `newsvd.m`
`bidiag.m`
`golubkahanstep.m`
`zerorow.m`

The number of computations to compute the SVD of an $m \times n$ matrix is approximately $4mn^2 - 4n^3/3$ if only the singular values are needed, and $4mn^2 + 8mn^2 + 9n^3$ if it is also necessary to accumulate U and V. A variation on the SVD algorithm (known as the R-bidiagonalization) first triangularizes $A = QR$, then bidiagonalizes the upper-triangular matrix R, before proceeding to the Golub–Kahan step. It has been suggested that the R-bidiagonalization is more efficient if $m \geq 5n/3$. In particular, when there are many more rows than columns, the R-bidiagonalization requires fewer computations.

7.11 Exercises

7.1-1 From the proof of theorem 7.1, show that (7.3) is true.

7.1-2 Using the definition of the matrix 2-norm and the Frobenius norm, verify that (7.5) and (7.6) are true.

7.2-3 Show that the components of the SVD can be used to determine the fundamental subspaces of a matrix as shown in (7.7).

7.3-4 Let
$$\begin{bmatrix} 1 & 4 & 5 & 6 \\ 6 & 7 & 2 & 1 \end{bmatrix} \quad \mathbf{b} = \begin{bmatrix} 48 \\ 30 \end{bmatrix}.$$
One solution to $A\mathbf{x} = \mathbf{b}$ is $\mathbf{x} = [1, 2, 3, 4]^T$. Compute the least-squares solution using the SVD, and compare. Why was the solution chosen?

7.3-5 Show that the minimum squared error when computing the LS $\|A\mathbf{x} - \mathbf{b}\|_2^2$ solution is
$$E_{\min} = \|U_2^H \mathbf{b}\|_2^2.$$
Interpret this result in light of the four fundamental subspaces.

7.11 Exercises

7.4-6 Let
$$A = \begin{bmatrix} 1.04 & .96 \\ 3.99 & 4.01 \end{bmatrix}.$$

(a) If $\mathbf{b} = [1, 1]^T$, determine the solution to $A\mathbf{x} = \mathbf{b}$.

(b) Now let $\mathbf{b} = [1.1, 1]^T$ and solve $A\mathbf{x} = \mathbf{b}$. Comment on how the solution changed.

(c) Determine the SVD of A and the pseudoinverse of A obtained by setting the smallest singular value to zero. Comment on the form of $A^\dagger$. Find the solution to $A\mathbf{x} = \mathbf{b}$ for the two values of $\mathbf{b}$ from the previous part.

7.5-7 Show that the inequalities in (7.13) are correct. Show that the stated conditions for achieving the lower bound are correct.

7.5-8 ([114, page 74]) Prove that the largest singular value of a real $m \times n$ matrix A satisfies
$$\sigma_{\max}(A) = \max_{\mathbf{y} \in \mathbb{R}^m, \mathbf{x} \in \mathbb{R}^n} \frac{\mathbf{y}^T A \mathbf{x}}{\|\mathbf{x}\|_2 \|\mathbf{y}\|_2}.$$

7.5-9 For the 2×2 matrix
$$A = \begin{bmatrix} w & x \\ y & z \end{bmatrix},$$
derive expressions for $\sigma_{\max}(A)$ and $\sigma_{\min}(A)$ in terms of w, x, y, and z.

7.5-10 Using theorem 7.2, prove that the set of full-rank matrices is open.

7.6-11 Show that the matrices from (7.16) produce the Markov parameters in $C_{\text{est}} A_{\text{est}}^k B_{\text{est}}$.

7.7-12 Let P be the plane orthogonal to the normal vector $\mathbf{n} \in \mathbb{R}^{n+1}$,
$$P = \{\mathbf{r} \in \mathbb{R}^{n+1} : \mathbf{r}^T \mathbf{n} = 0\},$$
and let $\mathbf{n}$ have the particular form
$$\mathbf{n} = \begin{bmatrix} \mathbf{x} \\ -1 \end{bmatrix}.$$
Let $\mathbf{p} = \begin{bmatrix} \mathbf{a} \\ b \end{bmatrix}$ be a point in $\mathbb{R}^{n+1}$. Show that the shortest distance from $\mathbf{p}$ to the plane P is
$$\frac{(\mathbf{x}^T \mathbf{a} - b)^2}{\mathbf{x}^T \mathbf{x} + 1}.$$

7.7-13 Let $\tilde{V} = [\mathbf{v}_k, \mathbf{v}_{k+1}, \ldots, \mathbf{v}_{n+1}] \in \mathbb{R}^{(m+1) \times (n-k+2)}$. In the TLS problem, we sought an element $\mathbf{y} \in \mathcal{R}(\tilde{V})$ such that $\mathbf{x} = \tilde{I}\mathbf{y}$ is of minimum norm, where
$$\tilde{I} = \begin{bmatrix} I_m & 0 \end{bmatrix}$$
picks out the first m elements of $\mathbf{y}$, and also such that $\mathbf{y}_{m+1} = -1$. Formulate the problem of finding the minimum-norm $\mathbf{x}$ as a constrained optimization problem, and determine the solution.

7.7-14 (Partial total least squares) Let $T = \begin{bmatrix} W & X \\ Y & Z \end{bmatrix}$.

(a) [71] Show that T can be written as

$$\tilde{T} = UTV^H = \begin{bmatrix} W_{11} & 0 & 0 & X_{11} & X_{12} \\ 0 & 0 & 0 & X_{21} & 0 \\ 0 & 0 & 0 & 0 & 0 \\ Y_{11} & Y_{12} & 0 & Z_{11} & Z_{12} \\ Y_{21} & 0 & 0 & Z_{21} & Z_{22} \end{bmatrix},$$

where U and V are unitary, and W_{11}, X_{21}, and Y_{12} are either square and nonsingular, or null. Hint:

$$\begin{bmatrix} U_3 & \\ & I \end{bmatrix} \begin{bmatrix} W & X \\ Y & Z \end{bmatrix} \begin{bmatrix} V_3^T & \\ & I \end{bmatrix} = \begin{bmatrix} W_{11} & 0 & X_1 \\ 0 & 0 & X_2 \\ Y_1 & Y_2 & Z \end{bmatrix}$$

$$U_4 X_2 V_4^T = \begin{bmatrix} X_{21} & 0 \\ 0 & 0 \end{bmatrix} \qquad U_5 Y_2 V_5^T = \begin{bmatrix} Y_{12} & 0 \\ 0 & 0 \end{bmatrix}.$$

Then stack.

(b) Show that $\tilde{T}$ can be converted into

$$\tilde{\tilde{T}} = \begin{bmatrix} W_{11} & 0 & 0 & 0 & 0 \\ 0 & 0 & 0 & X_{21} & 0 \\ 0 & 0 & 0 & 0 & 0 \\ 0 & Y_{12} & 0 & 0 & 0 \\ 0 & 0 & 0 & 0 & Z_{22} - Y_{21} W_{11}^{-1} X_{12} \end{bmatrix}$$

by row and column operations.

7.9-15 The data

$$A = \begin{bmatrix} 4.84726 & -2.46786 & 0.549036 & -4.31815 \\ 0.813505 & -3.46221 & -0.251525 & 2.97353 \end{bmatrix}$$

are believed to be a rotation of the data

$$B = \begin{bmatrix} 4.51076 & -2.18889 & 0.58951 & -4.715 \\ 1.95213 & -3.64499 & -0.131083 & 2.29282 \end{bmatrix}.$$

Determine the amount of rotation between the data sets.

7.9-16 Let Z be a unitary matrix and let $\sigma_i \geq 0, i = 1, 2, \ldots, p$. Show that

$$\sum_{i=1}^{p} z_{ii} \sigma_i \leq \sum_{i=1}^{p} \sigma_i,$$

where the maximum is achieved when $Z = I$. Hint: Consider the norm of $\mathbf{s} = [\sigma_1^{1/2}, \ldots, \sigma_p^{1/2}]$.

7.10-17 Let $A = B + jC$, where B and C are real. Determine a means of finding the SVD of A in terms of the SVD of

$$\begin{bmatrix} B & -C \\ C & B \end{bmatrix}.$$

See also exercise 6.15-67.

7.10-18 Show that the SVD decomposition of an $n \times m$ matrix A, with $n > m$, can be found by first computing the QR factorization

$$A = Q \begin{bmatrix} R \\ 0 \end{bmatrix},$$

then computing the SVD of the $m \times m$ upper triangular matrix R.

7.12 References

Excellent discussion of the SVD for solution of linear equations is provided in [333]. Our proof of theorem 7.1 comes from [177].

Linear system theory, including Markov parameters and Hankel matrices, is discussed in [164]. A more recent treatment with an excellent discussion of realization theory is [284]. The realization technique described in section 7.6 is based on Ho's algorithm [139]; application of the SVD to Ho's algorithm is suggested in [385]. An interesting recursive approach to the partial realization problem is provided in [274, 277]. Finding a lower-rank approximation with the appropriate Hankel matrix structure is examined in [191, 194, 193]. Additional approaches to system identification using the SVD are presented in [233].

Several other applications of the SVD have been found in numerical analysis that have potential for the solution of problems arising in signal processing. (Among these problems are: the intersection of nullspaces of two matrices, computation of the angle between subspaces, and the intersection of subspaces.) An excellent treatment is found in [114]. The SVD is also useful in quadratic optimization problems with quadratic constraints, as discussed in [91, 317, 97].

The TLS technique is discussed in [112] and references therein. A derivation of the TLS with more geometric insight is presented in [390], along with some applications to array processing. In [344], the TLS technique is applied to system identification problems; and in [269], TLS is used for frequency estimation. Efficient algorithms for computing only that portion of the SVD necessary for the TLS solution are described in [347]. Further discussion of the TLS problems is given in [346]. Our discussion of the PTLS technique is drawn from [110] and [71] (see also [348, 345]). A summary of TLS, directed toward signal-processing applications, appears in [291]. An examination of TLS methods applied to ESPRIT modal analysis appears in [280].

Our discussion of the computation of the SVD follows that of [114]. The Golub–Kahan step was originally proposed in [111]. Increasing interest in the SVD as a computational tool has led to additional approaches to the algorithm. The R-bidiagonalization is presented in [51]. More recently, variations on the Golub–Kahan step have been developed that are more accurate and often faster when computing smaller singular values; see [70, 223, 69]. Recent work has produced efficient SVD computation routines using the Cholesky factorization [86], while other work has led to fast computation of the singular vectors [123].

A brief sampling of some of the many other applications of the SVD to signal processing would include:

1. Spectral analysis [187] and modal analysis [188, 189, 340].
2. System identification [234, 183].
3. Image reconstruction [310].
4. Noise reduction [290].
5. Signal estimation [75].

Chapter 8

Some Special Matrices and Their Applications

Some particular matrix forms arise fairly often in signal processing, in the description and analysis of algorithms (such as in linear prediction, filtering, etc.). This chapter provides an overview of some of the more common special matrix types, along with some applications to signal processing.

8.1 Modal matrices and parameter estimation

The exponential signal model, in which a signal is modeled as a sum of (possibly) damped exponentials, or modes, is frequently encountered. The model arises in a variety of settings, such as frequency estimation, spectrum estimation, system identification, or direction finding. If the modes are real and simple (that is, no repeated roots in the characteristic equation), we can model the signal in discrete (sampled) time as

$$x[t] = \sum_{i=1}^{q} \tilde{a}_i \rho_i^t \cos(2\pi f_i[t] + \phi_i), \qquad (8.1)$$

where $\tilde{a}_i$ is the amplitude, ρ_i is the damping factor, f_i is (discrete) frequency, and ϕ_i is the phase. In the event that $f_i = 0$, it is customary to let the phase be zero. The real signal (8.1) can be written in terms of complex exponentials as

$$x[t] = \sum_{i=1}^{p} \alpha_i z_i^t. \qquad (8.2)$$

In going from (8.1) to (8.2), we have $z_i = \rho_i e^{j2\pi f_i}$ and $\alpha_i = \frac{\tilde{a}_i e^{j\phi_i}}{2}$, with the pair (α_i, z_i) matched by its conjugate pair (α_i^*, z_i^*), unless $f_i = 0$, in which case

$$z_i = \rho_i \qquad \alpha_i = \tilde{a}_i.$$

If we let q_0 be the number of frequencies for which $f_i = 0$, then the number of modes in the complex model is $p = 2(q - q_0) + q_0$. The signal (8.2) is the solution of a homogeneous pth-order difference equation with characteristic polynomial

$$A(z) = (1 - z_1 z^{-1})(1 - z_2 z^{-1}) \cdots (1 - z_p z^{-1}) = \sum_{i=0}^{p} a_i z^{-i}, \qquad (8.3)$$

where $a_0 = 1$, subject to initial conditions that determine the amplitude and phase of the modes. The polynomial $A(z)$ has as its roots the modes z_i, so that $A(z_i) = 0, i = 1, 2, \ldots, p$.

8.1 Modal Matrices and Parameter Estimation

Furthermore,
$$z_i^l A(z_i) = 0 \tag{8.4}$$
for any l.

More generally, the difference equation with characteristic polynomial having repeated roots
$$A(z) = (1 - z_1 z^{-1})^{m_1}(1 - z_2 z^{-1})^{m_2} \cdots (1 - z_r z^{-1})^{m_r},$$
where $m_1 + m_2 + \cdots + m_r = p$, is a pth-order system having a solution that is an exponential signal,
$$x[t] = \sum_{k=1}^{r} \sum_{m=1}^{m_k} \alpha_{k,m} t^{m-1} z_k^t. \tag{8.5}$$

Equation (8.5) is a general form of a pth-order exponential signal. Most often we can neglect the problem of repeated roots and we can use (8.2). In many applications, the problem is to estimate the parameters of the exponential model $\{\alpha_i, z_i, m_i, r\}$ given a set of samples $\{x_0, x_1, x_2, \ldots, x_{N-1}\}$. We will not explore all of the aspects of this problem, but we introduce a fruitful matrix formulation and demonstrate how several important matrix forms arise from this.

Let
$$\mathbf{x} = [x_0 \quad x_1 \quad \cdots \quad x_{N-1}]^T$$
be a vector of data. For the simple exponential signal (8.2)—no repeated modes—we can write
$$\mathbf{x} = V\boldsymbol{\alpha}, \tag{8.6}$$
where
$$V = \begin{bmatrix} 1 & 1 & \cdots & 1 \\ z_1 & z_2 & \cdots & z_p \\ \vdots & & & \\ z_1^{N-1} & z_2^{N-1} & \cdots & z_p^{N-1} \end{bmatrix} \qquad \boldsymbol{\alpha} = \begin{bmatrix} \alpha_1 \\ \alpha_2 \\ \vdots \\ \alpha_p \end{bmatrix}.$$

The matrix V is said to be a *Vandermonde* matrix; several important aspects about Vandermonde matrices are presented in section 8.4. From the coefficients of $A(z)$ in (8.3), we may form the $(N - p) \times N$ matrix A^T by
$$A^T = \begin{bmatrix} a_p & a_{p-1} & \cdots & a_1 & 1 & 0 & 0 & \cdots & 0 \\ 0 & a_p & a_{p-1} & \cdots & a_1 & 1 & 0 & \cdots & 0 \\ \vdots & & & & & & & & \\ 0 & 0 & \cdots & 0 & a_p & a_{p-1} & \cdots & a_1 & 1 \end{bmatrix},$$
which is a Toeplitz matrix. By (8.4), it is straightforward to show that
$$A^T V = 0.$$

The $N - p$ rows of A are linearly independent (why?), and the p columns of V are linearly independent, since $z_i \neq z_j$ for $i \neq j$. (If there were some $z_i = z_j$ for $i \neq j$, this would be a repeated mode, and the signal from (8.5) would have to be used.) So V and A span orthogonal subspaces; the space $\mathcal{R}(V)$ is said to be the signal subspace and the space $\mathcal{R}(A)$ is said to be the orthogonal subspace. The orthogonal projectors onto these subspaces are $P_V = V(V^H V)^{-1} V^H$ and $P_A = A(A^T A)^{-1} A$, and $P_V + P_A = I$.

Since $A^T V = 0$ and $\mathbf{x} = V\alpha$, we have $A^T \mathbf{x} = 0$, or

$$\begin{bmatrix} a_p & a_{p-1} & \cdots & a_1 & 1 & 0 & 0 & \cdots & 0 \\ 0 & a_p & a_{p-1} & \cdots & a_1 & 1 & 0 & \cdots & 0 \\ \vdots & & & & & & & & \\ 0 & 0 & \cdots & 0 & a_p & a_{p-1} & \cdots & a_1 & 1 \end{bmatrix} \begin{bmatrix} x_0 \\ x_1 \\ \vdots \\ x_{N-1} \end{bmatrix} = \mathbf{0},$$

which also can be written as

$$\begin{bmatrix} x_0 & x_1 & \cdots & x_p \\ x_1 & x_2 & \cdots & x_{p+1} \\ \vdots & & & \\ x_{N-1-p} & x_{N-2-p} & \cdots & x_{N-1} \end{bmatrix} \begin{bmatrix} a_p \\ a_{p-1} \\ \vdots \\ a_1 \\ 1 \end{bmatrix} = 0 \quad (8.7)$$

or

$$X\mathbf{a} = 0.$$

This corresponds to the covariance data matrix of section 3.9 for a linear predictor (AR model) of order p (where now we write X instead of A). A matrix X constant on the reverse diagonals is said to be a *Hankel* matrix.

By (8.7), rank$(X) < p + 1$. The following lemma shows that if the modes are distinct and the matrix X is sufficiently large, then rank$(X) = p$.

Lemma 8.1 *Let N samples of x_n be generated according to a pth-order model as in (8.5), and form the $(N-m) \times (m+1)$ Hankel matrix X by*

$$X = \begin{bmatrix} x_0 & x_1 & \cdots & x_m \\ x_1 & x_2 & \cdots & c_{m+1} \\ \vdots & & & \\ x_{N-m-1} & x_{N-m-2} & \cdots & x_{N-1} \end{bmatrix}.$$

Then rank$(X) = \min(p, m, N - m)$.

Proof Clearly, if $m < p$ or $N - m < p$ then the rank is determined by $\min(m, N - m)$. We will assume that $N - m \geq p$ and $m \geq p$ and show that the rank is p, which establishes the result. If there were a set of $q < p$ coefficients such that a linear combination of q columns of X were equal to zero, then there would be a polynomial $B(z)$ of degree q that would be a characteristic polynomial for the difference equation for $x[t]$. This would have to have roots z_i, $i = 1 \ldots, p$. However, a polynomial of degree $q < p$ cannot have p roots. $\square$

The coefficient vector $\mathbf{a}$ in (8.7) can be found by moving the last column to the right-hand side:

$$\begin{bmatrix} x_0 & x_1 & \cdots & x_{p-1} \\ x_1 & x_2 & \cdots & x_{p-2} \\ \vdots & & & \\ x_{N-1-p} & x_{N-2-p} & \cdots & x_{N-2} \end{bmatrix} \begin{bmatrix} a_p \\ a_{p-1} \\ \vdots \\ a_1 \end{bmatrix} = \begin{bmatrix} -x_p \\ -x_{p-1} \\ \vdots \\ -x_{N-1} \end{bmatrix}. \quad (8.8)$$

If $N \geq 2p + 1$ then (8.8) can be solved—directly, or in a least-squares sense, or in a total least-squares sense. The direct solution ($N = 2p + 1$) is known as Prony's method. While very straightforward to compute, Prony's method is sensitive to noise, and superior modal methods are available. However, the effect of noise can be mitigated somewhat using methods described in section 15.3. Also, least-squares or total-least squares solutions can be employed.

Once the coefficient vector **a** is found, the roots of $A(z)$ can be found to obtain $z_1, z_2, \ldots, z_p$. The amplitudes α can then be found from (8.6).

Somewhat more can be said if the signal is known to consist of real undamped sinusoids. Then $z_i = e^{j2\pi f_i}$, and every z_i is accompanied by its conjugate $\bar{z}_i$. If $A(z)$ is the characteristic polynomial, then $A(z_i) = 0$ and $A(\bar{z}_i) = 0$. Furthermore, let $\tilde{A}(z) = z^p A(z^{-1})$. $\tilde{A}(z)$ is known as the *reciprocal polynomial*. If z_i is a root of $A(z)$, then $1/z_i$ is a root of $\tilde{A}(z)$ (see exercise 8.1-3). Furthermore, if the roots lie on the unit circle, as they do for undamped sinusoids, and they occur in complex conjugate pairs, then the roots of $\tilde{A}(z)$ are exactly the same as the roots of $A(z)$. Since the coefficients of $\tilde{A}(z)$ are simply those of $A(z)$ written in reverse order, (8.7) is also true with the coefficients in reverse order,

$$\begin{bmatrix} x_0 & x_1 & \cdots & x_p \\ x_1 & x_2 & \cdots & x_{p+1} \\ \vdots & & & \\ x_{N-1-p} & x_{N-2-p} & \cdots & x_{N-1} \end{bmatrix} \begin{bmatrix} 1 \\ a_1 \\ a_2 \\ \vdots \\ a_p \end{bmatrix} = 0,$$

which in turn can be written as

$$\begin{bmatrix} x_p & x_{p-1} & \cdots & x_0 \\ x_{p+1} & x_p & \cdots & x_1 \\ \vdots & & & \\ x_{N-1} & x_{N-2} & \cdots & x_{N-1-p} \end{bmatrix} \begin{bmatrix} a_p \\ a_{p-1} \\ \vdots \\ a_1 \\ 1 \end{bmatrix} = 0. \tag{8.9}$$

The matrix in (8.9) is now a Toeplitz matrix. Stacking (8.9) and (8.7) provides twice as many equations for N measurements, which when solved using least-squares methods will have lower variance.

8.2 Permutation matrices

Permutation matrices are simple matrices that are used to interchange rows and columns of a matrix.

Definition 8.1 A permutation matrix P is an $m \times m$ matrix with all elements either 0 or 1, with exactly one 1 in each row and column. □

The matrix

$$P = \begin{bmatrix} 0 & 0 & 0 & 1 \\ 0 & 1 & 0 & 0 \\ 1 & 0 & 0 & 0 \\ 0 & 0 & 1 & 0 \end{bmatrix}$$

is a permutation matrix. Let A be a matrix. Then PA is a **row-permuted** version of A, and AP is a **column-permuted** version of A. Permutation matrices are orthogonal: if P is a permutation, then $P^{-1} = P^T$. The product of permutation matrices is another permutation matrix. The determinant of a permutation is ± 1.

Example 8.2.1 Let

$$A = \begin{bmatrix} 1 & 2 & 3 \\ 4 & 5 & 6 \\ 7 & 8 & 9 \end{bmatrix} \quad \text{and} \quad P = \begin{bmatrix} 0 & 1 & 0 \\ 0 & 0 & 1 \\ 1 & 0 & 0 \end{bmatrix}.$$

Then
$$PA = \begin{bmatrix} 4 & 5 & 6 \\ 7 & 8 & 9 \\ 1 & 2 & 3 \end{bmatrix} \qquad AP = \begin{bmatrix} 3 & 1 & 2 \\ 6 & 4 & 5 \\ 9 & 7 & 8 \end{bmatrix}.$$
□

Permutation operations are best implemented, avoiding an expensive multiplication, using an index vector. For example the permutation P of the previous example could be represented in column ordering as the index $i_c = [2, 3, 1]$. Then $PA = A(i_c, :)$. In row ordering, P can be represented as $i_r = [3, 1, 2]$, so $AP = A(:, i_r)$.

It can be shown ([142], Birkhoff's theorem) that every doubly stochastic matrix can be expressed as a convex sum of permutation matrices.

8.3 Toeplitz matrices and some applications

Example 8.3.1 Consider filtering a causal signal $x[t]$ with a filter $h = \{1, 2, 3, -2, -1\}$, using linear (as opposed to circular) convolution. The filtering relationship can be expressed as

$$\mathbf{y} = \begin{bmatrix} y[0] \\ y[1] \\ y[2] \\ \vdots \end{bmatrix} = \begin{bmatrix} 1 & & & & \\ 2 & 1 & & & \\ 3 & 2 & 1 & & \\ -2 & 3 & 2 & 1 & \\ -1 & -2 & 3 & 2 & 1 \\ & -1 & -2 & 3 & 2 & 1 \end{bmatrix} \begin{bmatrix} x[0] \\ x[1] \\ x[2] \\ \vdots \end{bmatrix}.$$

Observe that the elements on the diagonals of the matrix are all the same, the elements of h shifted down and across. Finding $x[t]$ given $y[t]$ would require solving a set of linear equations involving this matrix. □

Definition 8.2 An $m \times m$ matrix R is said to be a **Toeplitz** matrix if the entries are constant along each diagonal. That is, R is Toeplitz if there are scalars $s_{-m+1}, \ldots, s_0, \ldots, s_{m-1}$ such that $t_{ij} = s_{j-i}$ for all i and j. The matrix

$$R = \begin{bmatrix} s_0 & s_1 & s_2 & s_3 \\ s_{-1} & s_0 & s_1 & s_2 \\ s_{-2} & s_{-1} & s_0 & s_1 \\ s_{-3} & s_{-2} & s_{-1} & s_0 \end{bmatrix} \qquad (8.10)$$

is Toeplitz. □

Toeplitz matrices arise in both minimum mean-squared error estimation and least-squares estimation, as the Grammian matrix. For example, in the linear prediction problem, a signal $x[t]$ is predicted based upon its m prior values. Letting $\hat{x}[t]$ denote the predicted value, we have

$$\hat{x}[t] = -\sum_{i=1}^{m} a_{f,i} x[t-i],$$

where $\{-a_{f,i}\}$ are the forward prediction coefficients (see exercise 3.11-17). The forward prediction error is $f_m[t] = x[t] - \hat{x}[t]$. The equations for the prediction coefficients are

$$\begin{bmatrix} r[0] & \bar{r}[1] & \bar{r}[2] & \cdots & \bar{r}[m-1] \\ r[1] & r[0] & \bar{r}[1] & \cdots & \bar{r}[m-2] \\ r[2] & r[1] & r[0] & \cdots & \bar{r}[m-3] \\ \vdots & & & & \vdots \\ r[m-1] & r[m-2] & r[m-3] & \cdots & r[0] \end{bmatrix} \begin{bmatrix} w_{f,1} \\ w_{f,2} \\ w_{f,3} \\ \vdots \\ w_{f,m} \end{bmatrix} = \begin{bmatrix} r[1] \\ r[2] \\ r[3] \\ \vdots \\ r[m] \end{bmatrix}, \qquad (8.11)$$

8.3 Toeplitz Matrices and Some Applications

where $w_{f,i} = -a_{f,i}$. Equation (8.11) can be written as

$$R\mathbf{w} = \mathbf{r}, \tag{8.12}$$

where $R = E[\mathbf{x}[t-1]\mathbf{x}^H[t-1]]$ and $\mathbf{r} = E[x[t]\mathbf{x}[t-1]]$, with

$$\mathbf{x}[t] = \begin{bmatrix} \overline{x}[t] \\ \overline{x}[t-1] \\ \vdots \\ \overline{x}[t-m+1] \end{bmatrix}$$

(note the conjugates) and $r[j] = E[x[t]\overline{x}[t-j]]$. Equation (8.11) is known as the *Yule–Walker* equation.

Before proceeding with the study of Toeplitz matrices, it is useful to introduce a related class of matrices

Definition 8.3 An $m \times m$ matrix B is said to be **persymmetric** if it is symmetric about its northeast–southwest diagonal. That is, $b_{ij} = b_{m-j+1,m-i+1}$. This is equivalent to $B = JB^T J$, where J is the permutation matrix

$$J = \begin{bmatrix} 0 & 0 & \cdots & 0 & 1 \\ 0 & 0 & \cdots & 1 & 0 \\ \vdots & & & & \\ 0 & 1 & \cdots & 0 & 0 \\ 1 & 0 & \cdots & 0 & 0 \end{bmatrix}.$$

□

The matrix J (also denoted J_m, if the dimension is important) is sometimes referred to as the *counteridentity*.

Example 8.3.2 The matrix

$$B = \begin{bmatrix} 1 & 2 & 3 & 4 \\ 6 & 7 & 8 & 3 \\ 9 & 10 & 7 & 2 \\ 11 & 9 & 6 & 1 \end{bmatrix}$$

is persymmetric. □

Persymmetric matrices have the property that the inverse of a persymmetric matrix is persymmetric:

$$B^{-1} = J(B^{-1})^T J. \tag{8.13}$$

Toeplitz matrices are persymmetric.

We first approach the study of the solution of Toeplitz systems of equations in the context of the linear prediction problem (8.12), which will lead us to the solution of Hermitian Toeplitz equations using an algorithm known as Durbin's algorithm. Following the formulation of Durbin's algorithm, we will examine some of the implications of this solution with respect to the linear prediction problem. We will detour slightly to introduce the notation of lattice forms of filters, followed by connections between lattice filters and the solution of the optimal linear predictor equation. After this detour, we will return to the study of Toeplitz equations, this time with a general right-hand side.

To abbreviate the notation somewhat, let R_m denote the $m \times m$ matrix

$$R_m = \begin{bmatrix} r_0 & \bar{r}_1 & \bar{r}_2 & \cdots & \bar{r}_{m-1} \\ r_1 & r_0 & \bar{r}_1 & \cdots & \bar{r}_{m-2} \\ r_2 & r_1 & r_0 & \cdots & \bar{r}_{m-3} \\ \vdots & & & & \\ r_{m-1} & r_{m-2} & r_{m-3} & \cdots & r_0 \end{bmatrix}$$

and let

$$\mathbf{r}_m = \begin{bmatrix} r_1 \\ r_2 \\ \vdots \\ r_m \end{bmatrix}.$$

Observe that

$$R_{m+1} = \begin{bmatrix} R_m & J_m \bar{\mathbf{r}}_m \\ \mathbf{r}_m^T J_m & r_0 \end{bmatrix}, \tag{8.14}$$

where J_m is the $m \times m$ counteridentity.

8.3.1 Durbin's algorithm

We are solving the equation $R_m \mathbf{w}_m = \mathbf{r}_m$, where R_m is the Toeplitz matrix formed by elements of $\mathbf{r}$ as in (8.14), and $\mathbf{w}_m$ is now the vector of unknowns. We proceed inductively. Assume we have a solution for $R_k \mathbf{w}_k = \mathbf{r}_k$, $1 \leq k \leq m-1$. We want to use this solution to find R_{k+1}. Given that we have solved the kth-order Yule–Walker system $R_k \mathbf{w}_k = \mathbf{r}_k$, where $\mathbf{r}_k = [r_1, r_2, \ldots, r_k]^T$, we write the $(k+1)$st Yule–Walker equation as

$$\begin{bmatrix} R_k & J_k \bar{\mathbf{r}}_k \\ \mathbf{r}_k^T J_k & r_0 \end{bmatrix} \begin{bmatrix} \mathbf{z}_k \\ \alpha_k \end{bmatrix} = \begin{bmatrix} \mathbf{r}_k \\ r_{k+1} \end{bmatrix}, \tag{8.15}$$

where J_k is the $k \times k$ counteridentity. The desired solution is

$$\mathbf{w}_{k+1} = \begin{bmatrix} \mathbf{z}_k \\ \alpha_k \end{bmatrix}.$$

Multiplying out the first set of equations in (8.15), we see that

$$\mathbf{z}_k = R_k^{-1}(\mathbf{r}_k - \alpha_k J_k \bar{\mathbf{r}}_k) = \mathbf{w}_k - \alpha_k R_k^{-1} J_k \bar{\mathbf{r}}_k,$$

by the inductive hypothesis. Since R_k^{-1} is persymmetric,

$$R_k^{-1} J_k = J_k R_k^{-1},$$

hence

$$\mathbf{z}_k = \mathbf{w}_k - \alpha_k J_k \bar{\mathbf{w}}_k. \tag{8.16}$$

We observe that the first k elements of $\mathbf{w}_{k+1}$ are obtained as a correction by $\alpha_k J_k \bar{\mathbf{w}}_k$ of the original elements $\mathbf{w}_k$. From the second set of equations in (8.15), we have

$$\alpha_k = \frac{1}{r_0}\left(r_{k+1} - \mathbf{r}_k^T J_k \mathbf{z}_k\right), \tag{8.17}$$

which, by substituting for $\mathbf{z}_k$ from (8.16), gives

$$\alpha_k = \frac{\left(r_{k+1} - \mathbf{r}_k^T J_k \mathbf{w}_k\right)}{r_0 - \mathbf{r}_k^T \bar{\mathbf{w}}_k} = \frac{\left(r_{k+1} - \mathbf{r}_k^T J_k \mathbf{w}_k\right)}{\beta_k} \tag{8.18}$$

8.3 Toeplitz Matrices and Some Applications

where
$$\beta_k = r_0 - \mathbf{r}_k^T \overline{\mathbf{w}}_k. \tag{8.19}$$

For future reference, observe that
$$\alpha_k \beta_k = r_{k+1} - \mathbf{r}_k^T J_k \mathbf{w}_k. \tag{8.20}$$

The parameter α_k is known as the kth *reflection coefficient*.

At this point, sufficient information is available to write an algorithm to recursively solve (8.15). However, some simplifications can be made in the computation of β_k:

$$\begin{aligned}
\beta_k &= r_0 - \mathbf{r}_k^T \overline{\mathbf{w}}_k = r_0 - \begin{bmatrix} \mathbf{r}_{k-1}^T r_k \end{bmatrix} \begin{bmatrix} \overline{\mathbf{z}}_{k-1} \\ \overline{\alpha}_{k-1} \end{bmatrix} \\
&= r_0 - \begin{bmatrix} \mathbf{r}_{k-1}^T r_k \end{bmatrix} \begin{bmatrix} \overline{\mathbf{w}}_{k-1} - \overline{\alpha}_{k-1} J_{k-1} \mathbf{w}_{k-1} \\ \overline{\alpha}_{k-1} \end{bmatrix} \\
&= r_0 - \mathbf{r}_{k-1}^T \overline{\mathbf{w}}_{k-1} + \overline{\alpha}_{k-1} \left(\mathbf{r}_{k-1}^T J_{k-1} \mathbf{w}_{k-1} - r_k \right) \\
&= \beta_{k-1} + \overline{\alpha}_{k-1}(-\alpha_{k-1} \beta_{k-1}) \\
&= \beta_{k-1}(1 - |\alpha_{k-1}|^2), \tag{8.21}
\end{aligned}$$

where the penultimate equality follows from (8.20). This procedure [77] can be summarized as shown in algorithm 8.1.

Algorithm 8.1 Durbin's algorithm
File: `durbin.m`

The complexity of the algorithm is $O(2n^2)$.

8.3.2 Predictors and lattice filters

In this section we examine some signal-processing oriented results of Durbin's algorithm. The reflection coefficients α_k that arose in the derivation of Durbin's algorithm have a useful interpretation in the context of linear prediction. Let $x[t]$ be a stationary stochastic process, and let $\mathbf{x}_m[t-1] = [x[t-1], x[t-2], \ldots, x[t-m]]^T$. (For the present discussion, the vectors are taken as real for convenience; it is straightforward to generalize to complex vectors.) The optimum (MSE) mth-order forward linear predictor is of the form
$$\hat{x}[t] = \mathbf{w}_{f,m}^T \mathbf{x}_m[t-1],$$
where
$$R_m \mathbf{w}_{f,m} = \mathbf{r}_m \tag{8.22}$$
and
$$R_m = E[\mathbf{x}_m[t-1] \mathbf{x}_m^T[t-1]] \quad \text{and} \quad \mathbf{r}_m = E[x[t] \mathbf{x}_m[t-1]].$$

Equation (8.22) is, of course, the Yule–Walker equation solved by the Durbin algorithm. We explore the minimum mean-squared error in light of the Durbin algorithm parameters in the following theorem.

Theorem 8.1 *The minimum mean-squared error for the mth-order forward predictor is*
$$\sigma_{f,m}^2 = \sigma_{f,m-1}^2 \left(1 - \alpha_{m-1}^2\right) = \beta_m,$$
where α_{m-1} is the reflection coefficient from the Durbin algorithm.

Proof (By induction) For the zeroth-order predictor, the error is
$$\sigma_{f,0}^2 = E[x[t]x[t]] = r_0 = \beta_0.$$
For the first-order predictor,
$$\sigma_{f,1}^2 = E[(\hat{x}[t] - x[t])^2] = r_0(1 - \alpha_0^2) = \beta_1 \tag{8.23}$$
(see exercise 8.3-15). Assuming the theorem to be true for the $(k-1)$st order predictor, we write
$$\sigma_{f,k-1}^2 = E[(\hat{x}[t] - x[t])^2] = E\left[\left(\mathbf{w}_{f,k-1}^T \mathbf{x}[t-1] - \mathbf{x}[t]\right)^2\right]$$
$$= r_0 - \mathbf{r}_{k-1}^T \mathbf{w}_{f,k-1}.$$

Now writing $\mathbf{w}_{f,k}$ in terms of its solution in the Durbin algorithm, we obtain
$$\sigma_{f,k}^2 = r_0 - \begin{bmatrix}\mathbf{r}_{k-1}^T & r_k\end{bmatrix} \begin{bmatrix}\mathbf{z}_{k-1} \\ \alpha_{k-1}\end{bmatrix}$$
$$= r_0 - \begin{bmatrix}\mathbf{r}_{k-1}^T & r_k\end{bmatrix} \begin{bmatrix}\mathbf{w}_{f,k-1} - \alpha_{k-1} J_{k-1} \mathbf{w}_{f,k-1} \\ \alpha_{k-1}\end{bmatrix}$$
$$= r_0 - \mathbf{r}_{k-1}^T \mathbf{w}_{f,k-1} + \alpha_{k-1}(\mathbf{r}_{k-1} J_{k-1} \mathbf{w}_{f,k-1} + r_k)$$
$$= \left(r_0 - \mathbf{r}_{k-1}^T \mathbf{w}_{f,k-1}\right) \left(1 + \alpha_{k-1} \frac{\mathbf{r}_{k-1}^T J_{k-1} \mathbf{w}_{f,k-1} - r_k}{r_0 - \mathbf{r}_{k-1}^T \mathbf{w}_{f,k-1}}\right)$$
$$= \left(r_0 - \mathbf{r}_{k-1}^T \mathbf{w}_{f,k-1}\right)\left(1 - \alpha_{k-1}^2\right) \tag{8.24}$$
$$= \beta_k, \tag{8.25}$$
where (8.24) follows from (8.18), and (8.25) follows from (8.19) and (8.21). □

Since, as shown in the following $|\alpha_k| \leq 1$, as the order m grows, there will be less error in the predictor as the number of stages increases, until the predictor is able to predict everything about the signal that is predictable. The prediction error at that point will be white noise.

To motivate the concept of the lattice filters, consider now the problem of "growing" a predictor from kth order to $(k+1)$st order, up to a final predictor of order m, starting from a first-order predictor. The first-order predictor is
$$\hat{x}[t] = -a_{1,1} x[t-1].$$
The second-order predictor is
$$\hat{x}[t] = -a_{2,1} x[t-1] - a_{2,2} x[t-2].$$

According to the recursion (8.16), all of the coefficients in the second-order predictor $\{a_{2,1}, a_{2,2}\}$ are in general different from the coefficients in the first-order predictor $\{a_{1,1}\}$. In the general case, if we desire to extend a kth-order filter to an $(k+1)$st-order filter, all of the coefficients will have to change. We will develop a filter structure, known as a *lattice filter*, to which new filter stages may be added without having to recompute the coefficients for the old filter.

We begin by reviewing some basic notation for predictors. Let
$$f_k[t] = x[t] - \hat{x}_k[t] = \sum_{i=0}^{m} a_{k,i} x[t-i], \qquad a_{k,0} = 1 \tag{8.26}$$
denote the *forward* prediction error of a kth-order predictor. The optimal (MMSE) forward

8.3 Toeplitz Matrices and Some Applications

predictor coefficients satisfy $-R\mathbf{a}_k = \mathbf{r}$, where

$$\mathbf{a}_k = [a_{k,1}, a_{k,2}, \ldots, a_{k,k}]^T.$$

In exercise 3.11-17, the concept of a *backward* predictor, in which $x[t-k]$ is "predicted" using $x[t], x[t-1], \ldots, x[t-k+1]$, was presented. The backward predictor is

$$\hat{x}_b[t-k] = -\sum_{i=0}^{k-1} b_{k,i} x[t-i].$$

Let

$$g_k[t] = x[t-k] - \hat{x}_b[t-k] = \sum_{i=0}^{k} b_{k,i} x[t-i], \qquad b_{k,k} = 1 \qquad (8.27)$$

denote the backward prediction error. As shown in exercise 3.11-17, the optimal (MMSE) backward prediction coefficients satisfy

$$-R\mathbf{b}_k = J_k \bar{\mathbf{r}},$$

where J_k is the $k \times k$ counteridentity, hence the optimal forward predictor coefficients are related to the optimal backward predictor coefficients by

$$\mathbf{a}_k = J\bar{\mathbf{b}}_k. \qquad (8.28)$$

That is, the backward prediction coefficients are the forward prediction coefficients conjugated and in reverse order.

We will now develop the lattice filter by building up a sequence of steps. The first-order forward and backward prediction errors are

$$\begin{aligned} f_1[t] &= x[t] + a_{1,1} x[t-1], \\ g_1[t] &= x[t-1] + b_{1,0} x[t]. \end{aligned} \qquad (8.29)$$

In light of (8.28), the second equation can be written as

$$g_1[t] = x[t-1] + \bar{a}_{1,1} x[t].$$

Now consider the filter structure shown in figure 8.1(a). This structure is known as a *lattice filter*. The outputs of that filter structure can be written as

$$\begin{aligned} f_1[t] &= f_0[t] + \kappa_1 g_0[t-1] = x[t] + \kappa_1 x[t-1], \\ g_1[t] &= g_0[t-1] + \bar{\kappa}_1 f_0[t] = x[t-1] + \bar{\kappa}_1 x[t], \end{aligned} \qquad (8.30)$$

hence, by equating $\kappa_1 = a_{1,1}$, we find that this first-order lattice filter computes both the forward and the backward prediction error for first-order predictors.

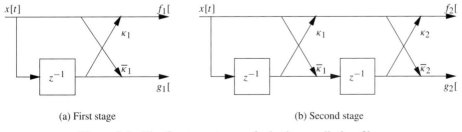

(a) First stage (b) Second stage

Figure 8.1: The first two stages of a lattice prediction filter

Second-order forward and backward predictors satisfy

$$f_2[t] = x[t] + a_{2,1}x[t-1] + a_{2,2}x[t-2],$$
$$g_2[t] = x[t-2] + b_{2,0}x[t] + b_{2,1}x[t-1]$$
$$= x[t-2] + \overline{a}_{2,2}x[t] + \overline{a}_{2,1}x[t-1]. \quad (8.31)$$

For the lattice structure in figure 8.1(b), the output can be written as

$$f_2[t] = x[t] + \kappa_1 x[t-1] + \kappa_2 g_1[t-1]$$
$$= x[t] + (\kappa_1 + \overline{\kappa}_1 \kappa_2)x[t-1] + \kappa_2 x[t-2] \quad (8.32)$$

and, similarly,

$$g_2[t] = x[t-2] + (k_1\overline{k}_2 + \overline{k}_1)x[t-1] + \overline{k}_2 x[t]. \quad (8.33)$$

By equating (8.31) and (8.32), we obtain $a_{2,1} = \kappa_1 + \overline{\kappa}_1 k_2$ and $a_{2,2} = \kappa_2$; again, we have the lattice filter computing both the forward and backward prediction error.

We now generalize to predictors of order k. The forward and backward predictors of (8.26) and (8.27) can be written using the Z-transform as

$$F_k(z) = A_k(z)X(z),$$
$$G_k(z) = B_k(z)X(z), \quad (8.34)$$

where $A_k(z) = \sum_{i=0}^{k} a_{k,i} z^{-i}$. Because of the relationship (8.28), we can write

$$B_k(z) = z^{-k}\overline{A}_k(z^{-1});$$

that is, the polynomial with the coefficients conjugated and in reverse order. The kth-order lattice filter stage shown in figure 8.2 satisfies the equations

$$F_k(z) = F_{k-1}(z) + \kappa_k z^{-1} G_{k-1}(z) \quad k = 1, 2, \ldots, m,$$
$$G_k(z) = \overline{\kappa}_k Fk - 1(z) + z^{-1} G_{k-1}(z) \quad k = 1, 2, \ldots, m. \quad (8.35)$$

Dividing both sides of (8.35) by $X(z)$, we obtain

$$A_k(z) = A_{k-1}(z) + \kappa_k z^{-1} B_{k-1}(z), \quad (8.36)$$
$$B_k(z) = \overline{k}_k A_{k-1}(z) + z^{-1} B_{m-1}(z). \quad (8.37)$$

Equation (8.36) can be written in terms of its coefficients as

$$a_{k,0} = 1,$$
$$a_{k,i} = a_{k-1,i} + \kappa_k \overline{a}_{k-1,k-i}, \quad i = 1, 2, \ldots, k-1, \quad (8.38)$$
$$a_{k,k} = \kappa_k.$$

When iterated from $k = 0, 1, \ldots, m$, (8.38) converts from lattice filter coefficients $\kappa_1, \kappa_2, \ldots, \kappa_m$ to the direct-form filter predictor coefficients $a_{m,1}, a_{m,2}, \ldots, a_{m,m}$. MATLAB code implementing this conversion is shown in algorithm 8.2.

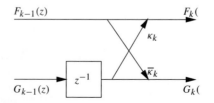

Figure 8.2: The kth stage of a lattice filter

8.3 Toeplitz Matrices and Some Applications

Algorithm 8.2 Conversion of lattice FIR to direct-form
File: `refltodir.m`

To convert from a direct-form implementation to the lattice implementation, we write (8.36) and (8.37) as

$$A_{k-1}(z) = \frac{A_k(z) - \kappa_k B_k(z)}{1 - |\kappa_k|^2}. \tag{8.39}$$

Recalling that $\kappa_k = a_{k,k}$ and writing (8.39) in terms of the coefficients, we obtain the following down-stepping recursion for finding the reflection coefficients from the direct-form filter coefficients. For $k = m, m-1, \ldots, 1$,

$$\kappa_k = a_{k,k},$$
$$a_{k-1,i} = \frac{a_{k,i} - \kappa_k \overline{a}_{k,m-i}}{1 - |\kappa_k|^2}.$$

This recursion works, provided that $|\kappa_k| \neq 1$.

Algorithm 8.3 Conversion of direct-form FIR to lattice
File: `dirtorefl.m`

Example 8.3.3 Suppose we know that $\kappa_1 = 2/3, \kappa_2 = 4/5$, and $\kappa_3 = 1/5$. Then invoking `refltodir` with the argument $\mathbf{k} = [2/3, 4/5, 1/5]$, we obtain $\mathbf{a} = [1 \quad 1.36 \quad 1.04 \quad 0.2]^T$, corresponding to the filter

$$A_3(z) = 1 + 1.36z^{-1} + 1.04z^{-2} + .2z^{-3}.$$

Supplying $\mathbf{a}$ as an argument to `dirtorefl`, we obtain

$$\mathbf{k} = [0.666667 \quad 0.8 \quad 0.2],$$

as expected. □

8.3.3 Optimal predictors and Toeplitz inverses

The lattice representation of an FIR filter applies to *any* FIR filter that is normalized so that the leading filter coefficient is 1. However, for the case of optimal linear predictors, there is a useful relationship between the conversions from the direct-form realization to the lattice realization, and vice versa. Recall that for the solution of the Yule–Walker equation, the update step to go from the kth-order predictor to the $(k+1)$st-order predictor (see (8.16)) is

$$\mathbf{z}_k = \mathbf{w}_k - \alpha_k J_k \overline{\mathbf{w}}_k, \tag{8.40}$$

where $\mathbf{w}_k$ is the solution to the $(k-1)$st Yule–Walker equation. Contrast this with the update equation for converting from lattice to direct form from (8.36) and (8.38),

$$a_{k,i} = a_{k-1,i} + \kappa_k \overline{a}_{k-1,k-i}, \quad i = 1, 2, \ldots, k-1. \tag{8.41}$$

The comparison between (8.40) and (8.41) may be made more direct by writing (8.40) in terms of its components, recalling that $\mathbf{w}_k(1{:}k-1) = \mathbf{z}_{k-1}$. Then (8.40) becomes

$$z_{k,i} = z_{k-1,i} - \alpha_k \bar{z}_{k-1,k-i}, \qquad i = 1, 2, \ldots, k-1. \tag{8.42}$$

Comparison of (8.41) and (8.42) reveals that

$$\alpha_k = -\bar{\kappa}_k. \tag{8.43}$$

Thus, the MMSE predictor error $x[t] - \hat{x}_m[t]$ is precisely computed by the lattice filter with coefficients $-\bar{\alpha}_k$, $k = 1, 2, \ldots, m$. Furthermore, at each stage the forward prediction error $f_k[t] = x[t] - \hat{x}_k[t]$ and the backward prediction error $b_k[t] = x[t-m] - \hat{x}_{m,k}[t]$ are produced by the lattice filter.

Consider now the problem of choosing the lattice coefficient α_k to minimize the MSE at the output of the kth stage of the lattice filter (instead of in the direct-form filter), as in

$$f_k[t] = f_{k-1}[t] + \alpha_k g_{k-1}[t].$$

Minimizing $\langle f_k[t], f_k[t] \rangle = E[|f_k|^2]$ with respect to α_k yields

$$\alpha_k = -\frac{E[f_{k-1}[t]\bar{g}_{k-1}[t-1]]}{E[|g_{k-1}[t-1]|^2]}.$$

By the Cauchy–Schwartz inequality, $-1 \le \alpha_k \le 1$. In light of theorem 8.1, increasing the order of the predictor cannot increase the prediction error power.

By the properties of optimal linear predictors—"the error is orthogonal to the data," where the data is $x[t-i]$, $i = 1, 2, \ldots, m$ for the forward predictor and $x[t-i]$, $i = 0, 1, \ldots, m-1$ for the backward predictor—we can obtain immediately the following orthogonality relationships, where $\langle x, y \rangle = E[x\bar{y}]$:

$$\langle f_m[t], x[t-i] \rangle = 0, \qquad i = 1, 2, \ldots, m,$$
$$\langle g_m[t], x[t-i] \rangle = 0, \qquad i = 0, 1, \ldots, m-1,$$
$$\langle f_m[t], x[t] \rangle = \sigma_{f,m}^2,$$
$$\langle f_i[t], f_j[t] \rangle = \sigma_{f,\max(i,j)}^2,$$
$$\langle g_m[t], g_j[t] \rangle = \begin{cases} 0 & 0 \le j \le m-1, \\ \sigma_{b,m}^2 & j = m. \end{cases}$$

Thus, the backward prediction error is a white sequence having the same span as the input data.

8.3.4 Toeplitz equations with a general right-hand side

We now generalize the solution of Toeplitz systems of equations to equations having a right-hand side that is not formed from components of the left-hand side matrix. This gives the *Levinson algorithm*. In the equation

$$R_m \mathbf{y} = \mathbf{b}, \tag{8.44}$$

R_m is a Toeplitz matrix and $\mathbf{b}$ is an arbitrary vector. As before, the solution is found inductively, but in this case, the update step requires keeping track of both the solution to the Yule–Walker equation

$$R_k \mathbf{w}_k = \mathbf{r}_k$$

(using the same approach as for the Durbin algorithm), and also the equation

$$R_k \mathbf{y}_k = \mathbf{b}_k,$$

8.4 Vandermonde Matrices

which is the one we really want to solve. Assuming that the $\mathbf{w}_k$ and $\mathbf{y}_k$ are known for step k, the solution to the $(k+1)$st step requires solving

$$\begin{bmatrix} R_k & J_k \bar{\mathbf{r}}_k \\ \mathbf{r}_k^T J_k & r_0 \end{bmatrix} \begin{bmatrix} \mathbf{v}_k \\ \mu_k \end{bmatrix} = \begin{bmatrix} \mathbf{b}_k \\ b_{k+1} \end{bmatrix},$$

where

$$\mathbf{y}_{k+1} = \begin{bmatrix} \mathbf{v}_k \\ \mu_k \end{bmatrix}.$$

Using the solutions from time k,

$$\mathbf{v}_k = R_k^{-1}(\mathbf{b}_k - \mu_k J_k \bar{\mathbf{r}}_k) = \mathbf{y}_k - \mu J_k \bar{\mathbf{x}}_k.$$

Then, proceeding as before,

$$\mu_k = \frac{b_{k+1} - \mathbf{r}_k^T J_k \mathbf{y}_k}{r_0 - \mathbf{r}_k^T \bar{\mathbf{w}}_k}.$$

The algorithm that solves for the general right-hand side is attributed to Levinson [199].

Algorithm 8.4 Levinson's algorithm
File: `levinson.m`

8.4 Vandermonde matrices

Definition 8.4 An $m \times m$ **Vandermonde matrix** V has the form

$$V = \begin{bmatrix} 1 & 1 & \cdots & 1 \\ z_0 & z_1 & \cdots & z_{m-1} \\ z_0^2 & z_1^2 & \cdots & z_{m-1}^2 \\ \vdots & & & \\ z_0^{m-1} & z_1^{m-1} & \cdots & z_{m-1}^{m-1} \end{bmatrix}. \quad (8.45)$$

This may be written as

$$V = V(z_0, z_1, \ldots, z_{m-1}).$$

$\square$

Example 8.4.1 Vandermonde matrices arise, for example, in polynomial interpolation. Suppose that the m points $(x_1, y_1), (x_2, y_2), \ldots, (x_m, y_m)$ are to be fitted exactly to a polynomial of degree $m-1$, so that

$$p(x_i) = \sum_{k=0}^{m-1} a_k x_i^k = y_i, \quad i = 1, 2, \ldots, n.$$

This provides the system of equations

$$\begin{bmatrix} 1 & x_1 & x_1^2 & \cdots & x_1^{m-1} \\ 1 & x_2 & x_2^2 & \cdots & x_2^{m-1} \\ \vdots & & & & \\ 1 & x_m & x_m^2 & \cdots & x_m^{m-1} \end{bmatrix} \begin{bmatrix} a_0 \\ a_1 \\ \vdots \\ a_{m-1} \end{bmatrix} = \begin{bmatrix} y_1 \\ y_2 \\ \vdots \\ y_m \end{bmatrix}$$

or

$$V^T \mathbf{a} = \mathbf{y},$$

where V is of the form (8.45).

$\square$

The determinant of a Vandermonde matrix (8.45) is

$$\det(V) = \prod_{\substack{i,j=1 \\ i>j}}^{n} (z_i - z_j) \qquad (8.46)$$

From this it is clear that if $z_i \neq z_j$ for $i \neq j$ then the determinant is nonzero and the matrix is invertible.

Efficient algorithms for solution of Vandermonde systems of equations

$$V\mathbf{x} = \mathbf{b}$$

and

$$V^T\mathbf{x} = \mathbf{b}$$

have been developed.

8.5 Circulant matrices

Definition 8.5 A **circulant matrix** C is of the form

$$C = \begin{bmatrix} c_1 & c_2 & \cdots & c_m \\ c_m & c_1 & \cdots & c_{m-1} \\ c_{m-1} & c_m & \cdots & c_{m-2} \\ \vdots & & & \\ c_2 & c_3 & \cdots & c_1 \end{bmatrix},$$

where each row is obtained by cyclically shifting to the right the previous row. This is also denoted as

$$C = \text{circulant}(c_1, c_2, \ldots, c_m). \qquad \square$$

Example 8.5.1 Let $h = \{1, 2, 3, 4\} = \{h_0, h_1, h_2, h_3\}$ denote the impulse response that is to be cyclically convolved with a sequence $x = \{x_0, x_1, x_2, x_3\}$. The output sequence

$$y = h \circledast x$$

may be computed in matrix form as

$$\mathbf{y} = \begin{bmatrix} 1 & 4 & 3 & 2 \\ 2 & 1 & 4 & 3 \\ 3 & 2 & 1 & 4 \\ 4 & 3 & 2 & 1 \end{bmatrix} \begin{bmatrix} x_0 \\ x_1 \\ x_2 \\ x_3 \end{bmatrix}.$$

Every cyclic convolution corresponds to multiplication by a circulant matrix. $\square$

It can be shown that a matrix A is circulant if and only if $A\Pi = \Pi A$, where $\Pi = \text{circulant}(0, 1, 0, \ldots, 0)$ is a permutation matrix. It is also the case that if C is a circulant matrix, then C^H is a circulant matrix. A matrix $\text{circulant}(c_1, c_2, \ldots, c_m) = \text{circulant}(\mathbf{c})$ can be represented as

$$\text{circulant}(c_1, c_2, \ldots, c_m) = c_1 I + c_2 \Pi + \cdots + c_m \Pi^{m-1}. \qquad (8.47)$$

Let $p_\mathbf{c}(z) = c_1 + c_2 z + \cdots + c_m z^{m-1}$. The power series $p_\mathbf{c}(z^{-1})$ is the z-transform of the sequence of circulant elements. From (8.47), the matrix can be written as

$$C = \text{circulant}(\mathbf{c}) = p_\mathbf{c}(\Pi).$$

8.5 Circulant Matrices

Lemma 8.2

1. If C_1 and C_2 are circulant matrices of the same size, then $C_1 C_2 = C_2 C_1$ (circulant matrices commute).
2. Circulants are normal matrices. (A normal matrix is a matrix C such that $CC^H = C^H C$.)

Proof Write $C_1 = p_{\mathbf{c}_1}(\Pi)$ and $C_2 = p_{\mathbf{c}_2}(\Pi)$. Then
$$C_1 C_2 = p_{\mathbf{c}_1}(\Pi) p_{\mathbf{c}_2}(\Pi),$$
which is just a polynomial in the matrix Π. But polynomials in the same matrix commute,
$$p_{\mathbf{c}_1}(\Pi) p_{\mathbf{c}_2}(\Pi) = p_{\mathbf{c}_2}(\Pi) p_{\mathbf{c}_1}(\Pi),$$
so
$$C_1 C_2 = C_2 C_1.$$
Since C and C^H are both circulants, it follows from part 1 that $CC^H = C^H C$, or C is normal. □

Diagonalization of circulant matrices is straightforward using the Fourier transform matrix. Let
$$F = \begin{bmatrix} 1 & 1 & 1 & \cdots & 1 \\ 1 & \omega & \omega^2 & \cdots & \omega^{m-1} \\ 1 & \omega^2 & \omega^4 & \cdots & \omega^{2(m-1)} \\ \vdots & & & & \\ 1 & \omega^{m-1} & \omega^{2(m-1)} & \cdots & \omega^{(m-1)(m-1)} \end{bmatrix}, \quad (8.48)$$
where $\omega = e^{-j2\pi/m}$. Note that F is a Vandermonde matrix and that
$$FF^H = mI.$$

Theorem 8.2 *If C is an $n \times n$ circulant, with $C = p_\mathbf{c}(\Pi)$, then it is diagonalized by F:*
$$C = \frac{1}{n} F \Lambda F^H$$
where
$$\Lambda = \mathrm{diag}(p_\mathbf{c}(1), p_\mathbf{c}(\omega), \ldots, p_\mathbf{c}(\omega^{n-1})).$$
Conversely, if $\Lambda = \mathrm{diag}(\lambda_1, \lambda_2, \ldots, \lambda_n)$ then
$$C = F \Lambda F^H$$
is circulant.

Proof See exercise 8.5-20. □

Based upon this theorem, we make the following observations:

1. The eigenvalues of circulant$(c_0, c_1, \ldots, c_{m-1})$ are
$$\lambda_i = \sum_{k=0}^{m-1} c_k \exp(-j 2\pi i j / m).$$
That is, the eigenvalues are obtained from the DFT of the sequence $\{c_0, c_1, \ldots, c_{m-1}\}$.

2. The normalized eigenvectors $\mathbf{x}_i$ are

$$\mathbf{x}_i = \frac{1}{\sqrt{m}} \begin{bmatrix} 1 \\ e^{-j2\pi i/m} \\ e^{-j2\pi(2i)/m} \\ \vdots \\ e^{-j2\pi(m-1)i/m} \end{bmatrix}.$$

The eigenvectors of every $m \times m$ circulant matrix are the same. This fact makes circulant matrices particularly easy to deal with: inverses, products, sums, and factors of circulant matrices are also circulant.

The diagonalization of C has a natural interpretation in terms of fast convolution. Write the cyclical convolution

$$\mathbf{y} = C\mathbf{x}$$

as

$$\mathbf{y} = \frac{1}{n}(F^H \Lambda F)\mathbf{x} = \frac{1}{n} F^H \Lambda (F\mathbf{x}). \tag{8.49}$$

Then $F\mathbf{x}$ is the discrete Fourier transform of F. The filtering operation is accomplished by multiplication of the diagonal matrix (element-by-element scaling), then $\frac{1}{m}F^H$ computes the inverse Fourier transform. If the DFT is computed using a fast algorithm (an FFT), then (8.49) represents the familiar fast convolution algorithm.

The diagonalization of C has implications in the solution of equations with circulant matrices. To solve $C\mathbf{x} = \mathbf{b}$, we can write

$$\frac{1}{m} F \Lambda F^H \mathbf{x} = \mathbf{b},$$

which also can be written as

$$\Lambda \mathbf{y} = m F^H \mathbf{b} = \mathbf{d},$$

where $\mathbf{y} = F^H \mathbf{x}$ is the DFT of $\mathbf{x}$, and $\mathbf{d}$ is the scaled DFT of $\mathbf{b}$. Then the solution is

$$y_i = \frac{1}{\lambda_i} d_i.$$

If there are frequency bins at which λ_i becomes small, then there may be amplification of any noise present in the signal.

8.5.1 Relations among Vandermonde, circulant, and companion matrices

Companion matrices were introduced in section 6.13. The following theorem relates Vandermonde, circulant, and companion matrices.

Theorem 8.3 *Let*

$$C = \begin{bmatrix} 0 & 1 & 0 & \cdots & 0 \\ 0 & 0 & 1 & \cdots & 0 \\ \vdots & & & & \\ 0 & 0 & 0 & \cdots & 1 \\ c_0 & c_1 & c_2 & \cdots & c_{m-1} \end{bmatrix}$$

be the companion matrix to the polynomial

$$p(x) = x^m - c_{m-1}x^{m-1} - c_{m-2}x^{m-2} - \cdots - c_1 x - c_0,$$

and let $x_1, x_2, \ldots, x_m$ be the roots of $p(x)$. Let $V = V(x_1, x_2, \ldots, x_m) = V(\mathbf{x})$ be a Vandermonde matrix, and let $D = \text{diag}(\mathbf{x})$ be a diagonal matrix. Then

$$VD = CV.$$

Proof The first $m - 1$ rows can be verified by direct computation. The (m, j)th element of VD is x_j^m. The (m, j)th element of CV is

$$c_0 + c_1 x_j + c_2 x_j^2 + \cdots + c_{m-1} x_j^{m-1} = x_j^m - p(x_j) = x_j^m. \qquad \square$$

8.5.2 Asymptotic equivalence of the eigenvalues of Toeplitz and circulant matrices

There is interest in examining the eigenvalue structure of Toeplitz matrices formed from autocorrelation values, because this provides information about the power spectrum of a stochastic process. Obtaining exact analytical expressions for eigenvalues of a general Toeplitz matrix is difficult. However, because of the similarity between circulant and Toeplitz matrices, and the simple eigenstructure of circulant matrices, there is some hope of obtaining approximate or asymptotic eigenvalue information about a Toeplitz matrix from a circulant matrix that is close to the Toeplitz matrix.

Consider the autocorrelation sequence $\mathbf{r} = \{r_{-m}, r_{-m+1}, \ldots, r_{-1}, r_0, r_1, \ldots, r_m\}$, where $r_k = 0$ for $k < -m$ or $k > m$. The spectrum of the sequence $\mathbf{r}$,

$$S(\omega) = \sum_{k=-m}^{m} r_k e^{-jk\omega},$$

is the power spectrum of some random process. The autocorrelation values can be recovered by the inverse Fourier transform,

$$r_k = \frac{1}{2\pi} \int_0^{2\pi} S(\omega) e^{jk\omega} \, d\omega.$$

Let R_n be the banded $n \times n$ Toeplitz matrix of autocorrelation values

$$R_n = \begin{bmatrix} r_0 & r_{-1} & r_{-2} & \cdots & r_{-m} & & & & & \\ r_1 & r_0 & r_{-1} & r_{-2} & \cdots & r_{-m} & & & & \\ \vdots & & & \ddots & & & & & & \\ r_m & r_{m-1} & r_{m-2} & \cdots & r_0 & r_{-1} & \cdots & r_{-m} & & \\ & r_m & & \cdots & r_1 & r_0 & r_{-1} & \cdots & r_{-m} & \\ & & \ddots & & & \ddots & & & \ddots & \\ & & & r_m & \cdots & r_1 & r_0 & r_{-1} & \cdots & r_{-m} \\ \vdots & & & & \ddots & & & & & \\ & & & & & r_m & \cdots & r_1 & r_0 & r_{-1} \\ & & & & & & r_m & \cdots & r_1 & r_0 \end{bmatrix}. \qquad (8.50)$$

We say that R_n is an mth-order Toeplitz matrix. Except for the upper-right and lower-left corners, R_n has the structure of a circulant matrix. The key to our result is that, as n gets large, the contributions of the elements in the corners become relatively negligible, and the eigenvalues can be approximated using the eigenvalues of the related circulant matrix, which can be found from the DFT of the autocorrelation sequence.

We define an $n \times n$ circulant matrix C_n with the same elements, but with the proper circulant structure,

$$C_n = \text{circulant}(r_0, r_{-1}, \ldots, r_{-m}, 0, \ldots, 0, r_m, r_{m-1}, \ldots, r_1). \tag{8.51}$$

We now want to determine the relationship between the eigenvalues of R_n and the eigenvalues of C_n as $n \to \infty$. To do this, we need to introduce the concept of asymptotic equivalence, and show the relationship between the eigenvalues of asymptotically equivalent matrices.

Definition 8.6 Two sequence of matrices A_n and B_n are said to by **asymptotically equivalent** if the following properties hold.

1. The matrices in each sequence are bounded:

$$\|A_n\|_2 \leq M < \infty, \qquad \|B_n\|_2 \leq M < \infty,$$

   for some finite bound M.

2. $\frac{1}{\sqrt{n}}\|A_n - B_n\|_F \to 0$ as $n \to \infty$. □

Note that the boundedness is stated using the spectral norm, while the convergence is stated in the Frobenius norm. We shall employ the different properties of these two norms below.

Theorem 8.4 ([118]) Let A_n and B_n be asymptotically equivalent matrices with eigenvalues $\lambda_{n,k}$ and $\mu_{n,k}$, respectively. If, for every positive integer l,

$$\lim_{n\to\infty} \frac{1}{n} \sum_{k=0}^{n-1} (\lambda_{n,k})^l < \infty \quad \text{and} \quad \lim_{n\to\infty} \frac{1}{n} \sum_{k=0}^{n-1} (\mu_{n,k})^l < \infty$$

(that is, if the so-called eigenvalue moments exist), then

$$\lim_{n\to\infty} \frac{1}{n} \sum_{k=0}^{n-1} (\lambda_{n,k})^l = \lim_{n\to\infty} \frac{1}{n} \sum_{k=0}^{n-1} (\mu_{n,k})^l. \tag{8.52}$$

That is, the eigenvalue moments of A_n and B_n are asymptotically equal.

Proof Let $A_n = B_n + D_n$. Since the eigenvalues of A_n^l are $(\lambda_{n,k})^l$, we can write

$$\lim_{n\to\infty} \frac{1}{n} \sum_{k=0}^{n-1} (\lambda_{n,k})^l$$

as

$$\lim_{n\to\infty} \frac{1}{n} \operatorname{tr} A_n^l.$$

Let $\Delta_n = A_n^l - B_n^l$. Then (8.52) can be written as

$$\lim_{n\to\infty} \frac{1}{n} \operatorname{tr} \Delta_n = 0.$$

The matrix Δ_n can be written as a finite number of terms, each of which is a product of D_n and B_n, each term containing at least one D_n. For a term such as $D_n^\alpha B_n^\beta$, $\alpha > 0, \beta \geq 0$, using the inequality (4.57), we obtain

$$\left\|D_n^\alpha B_n^\beta\right\|_F \leq M \|D_n\|_F,$$

8.5 Circulant Matrices

where $\|B_n\|_2 \leq M$. But since A_n and B_n are asymptotically equivalent, $\|D_n\|_F \to 0$, establishing the result. □

With this definition and theorem, we are now ready to state the main result of this section.

Theorem 8.5 *[118] The Toeplitz matrix R_n of (8.50) and the circulant matrix C_n of (8.51) are asymptotically equivalent.*

Proof We first establish the boundedness of R_n and C_n. By the definition of the 2-norm,

$$\|R_n\|_2^2 = \max_{\mathbf{x} \neq 0} \frac{\mathbf{x}^H R_n^H R_n \mathbf{x}}{\mathbf{x}^H \mathbf{x}}$$

$$= \frac{\sum_{i=0}^{n-1} \sum_{k=0}^{n-1} r_{i-k} x_i \overline{x}_k}{\sum_{k=0}^{n-1} |x_k|^2}$$

$$= \left(\frac{1}{2\pi} \int_0^{2\pi} \left| \sum_{k=0}^{n-1} x_k e^{-jk\omega} \right|^2 S(\omega) \, d\omega \right) \left(\frac{1}{2\pi} \int_0^{2\pi} \left| \sum_{k=0}^{n-1} x_k e^{-jk\omega} \right|^2 d\omega \right)^{-1} \quad (8.53)$$

$$\leq \max_\omega |S(\omega)| \leq \sum_{k=-m}^{m} |r_k| < \infty. \quad (8.54)$$

The norm $\|C_n\|$ depends upon the largest eigenvalue of C_n; it is straightforward to show that C_n is bounded.

Now, to compute $\|R_n - C_n\|_F$, simply count how many times elements appear in C_n that do not appear in R_n. Then

$$\frac{1}{n} \|R_n - C_n\|_F^2 = \sum_{k=0}^{m} k(|r_k|^2 + |r_{-k}|^2)$$

$$\leq \frac{1}{n} m \sum_{k=0}^{m} (|r_k|^2 + r_{-k})^2.$$

As $n \to \infty$ (with m bounded), $\frac{1}{\sqrt{n}} \|R_n - C_n\|_F \to 0$. □

By theorems 8.5 and 8.4, the eigenvalues $\lambda_{n,k}$ of R_n and the eigenvalues $\mu_{n,k}$ of C_n have the same asymptotic moments:

$$\lim_{n \to \infty} \frac{1}{n} \sum_{k=0}^{n-1} (\lambda_{n,k})^l = \lim_{n \to \infty} \frac{1}{n} \sum_{k=0}^{n-1} (\mu_{n,k})^l,$$

for integer $l \geq 0$. This leads us to the following asymptotic relationship between the eigenvalues of R_n and the power spectrum $S(\omega)$.

Theorem 8.6 *Let R_n be an mth-order Toeplitz matrix, and let $S(\omega)$ denote the Fourier transform of the coefficients of R_n. Let C_n and R_n be asymptotically equivalent. If the eigenvalues of R_n are $\lambda_{n,k}$, and the eigenvalues of C_n are $\mu_{n,k}$, then*

$$\lim_{n \to \infty} \frac{1}{n} \sum_{k=0}^{n-1} (\lambda_{n,k})^l = \frac{1}{2\pi} \int_0^{2\pi} S(\omega)^l \, d\omega$$

for every positive integer l. Furthermore, if R_n is Hermitian, then for any function g continuous on the appropriate interval,

$$\lim_{n\to\infty} \frac{1}{n} \sum_{k=0}^{n-1} g(\lambda_{n,k}) = \frac{1}{2\pi} \int_0^{2\pi} g(S(\omega))\, d\omega.$$

Proof By the preceding discussion, the eigenvalues of C_n are $\mu_{n,i} = \sum_{k=-m}^{m} r_k e^{-j2\pi ik/n} = S(2\pi i/n)$. By the asymptotic equivalence of the moments of the eigenvalues,

$$\lim_{n\to\infty} \frac{1}{n} \sum_{k=0}^{n-1} (\lambda_{n,k})^l = \lim_{n\to\infty} \frac{1}{n} \sum_{k=0}^{n-1} (\mu_{n,k})^l = \lim_{n\to\infty} \frac{1}{n} \sum_{k=0}^{n-1} S(2\pi/n)^l.$$

Now let $\Delta\omega = 2\pi/n$ and $\omega_k = 2\pi k/n$. Then

$$\lim_{n\to\infty} \frac{1}{n} \sum_{k=0}^{n-1} (\lambda_{n,k})^l = \lim_{n\to\infty} \sum_{k=0}^{n-1} S(\omega_k)^l \Delta\omega/2\pi = \frac{1}{2\pi} \int_0^{2\pi} S(\omega)^l\, d\omega. \qquad (8.55)$$

For any polynomial p, by (8.55) we also have

$$\lim_{n\to\infty} \frac{1}{n} \sum_{k=0}^{n-1} p(\lambda_{n,k}) = \frac{1}{2\pi} \int_0^{2\pi} p(S(\omega))\, d\omega.$$

If R_n is Hermitian, its eigenvalues are real. By the Weierstrass theorem, any continuous function g operating on the real interval can be uniformly approximated by a polynomial p. Since the eigenvalues of R_n are real, we can apply the Stone–Weierstrass theorem and conclude that

$$\lim_{n\to\infty} \frac{1}{n} \sum_{k=0}^{n-1} g(\lambda_{n,k}) = \frac{1}{2\pi} \int_0^{2\pi} g(S(\omega))\, d\omega. \qquad \square$$

This theorem is sometimes referred to as *Szegö's theorem*.

What the theorem says, roughly, is that the eigenvalues of R_n have the same distribution as does the spectrum $S(\omega)$. The theorem is somewhat difficult to interpret: whereas there is a definite order to the spectrum $S(\omega)$, the eigenvalues of R_n have no intrinsic order; they are often computed to appear in sorted order. Nevertheless, it can be observed that if R_n has a large eigenvalue disparity (high condition number) then the spectrum $S(\omega)$ will have a large spectral spread: some frequencies will have large $S(\omega)$, while other frequencies have small spectral spread.

Example 8.5.2 Let $r_i = e^{-2|i|}$, $i = -8, -7, \ldots, 8$ be the autocorrelation function for some sequence. Figure 8.3 shows the spectrum $S(\omega)$, and the eigenvalues of R_n for $n = 30$ and $n = 100$. To make the plot, the eigenvalues were computed (in sorted order), then the best match of the eigenvalues to $S(\omega)$ was determined. The improvement in match is apparent as n increases, although even for $n = 30$ there is close agreement between the spectrum and the eigenvalues of R_n. $\qquad \square$

8.6 Triangular matrices

An *upper-triangular matrix* is a matrix of the form

$$T = \begin{bmatrix} t_{11} & t_{12} & t_{13} & \cdots & t_{1n} \\ 0 & t_{22} & t_{23} & \cdots & t_{2n} \\ 0 & \ddots & t_{33} & \cdots & t_{3n} \\ & & & & \\ 0 & 0 & 0 & \cdots & t_{nn} \end{bmatrix}.$$

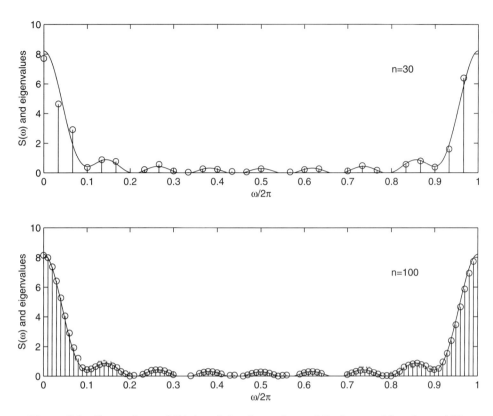

Figure 8.3: Comparison of $S(\omega)$ and the eigenvalues of R_n for $n = 30$ and $n = 100$

A lower-triangular matrix is a matrix such that its transpose is triangular. Triangular matrices arise in conjunction with the LU factorization (see section 5.1) and the QR factorization (see section 5.3). Triangular matrices have the following properties:

1. The product of two upper-triangular matrices is upper triangular; the product of two lower-triangular matrices is lower triangular.
2. The inverse of an upper-triangular matrix is upper triangular; the inverse of a lower-triangular matrix is lower triangular.

Triangular matrices are frequently seen in solving systems of equations using the LU factorization. There are also system realizations that are built on triangular matrices.

8.7 Properties preserved in matrix products

With regard to the varieties of matrices with special structures that we have encountered throughout this book, it is valuable to know when their properties are preserved under matrix multiplication. That is, if A and B are matrices possessing some special structure, when does $C = AB$ also posses this structure? What follows is a simple list.

Matrix properties preserved under matrix multiplication

1. Unitary
2. Circulant

3. Nonsingular
4. Lower (or upper) triangular

Matrix properties not preserved under matrix multiplication

1. Hermitian
2. Positive definite
3. Toeplitz
4. Vandermonde
5. Normal
6. Stable (e.g., eigenvalues inside unit circle)

8.8 Exercises

8.1-1 Consider a third-order exponential signal with repeated modes, $r = 2, m_1 = 2, m_2 = 1$.
   (a) Write an explicit expression for $x[t]$ using (8.5).
   (b) Determine the form of V in equation (8.6) for this signal with repeated modes. Is it still a Vandermonde matrix?

8.1-2 Let $A(z) = 1 + 2z + 3z^2$. Find the reciprocal polynomial $\tilde{A}(z)$. What is the relation of the coefficients of $A(z)$ to those of $\tilde{A}(z)$?

8.1-3 Show that if $\gamma \neq 0$ is a root of a polynomial $A(z)$, then $1/\gamma$ is a root of the reciprocal polynomial $\tilde{A}(z)$.

8.1-4 Show, by finding a counterexample, that the symmetry of coefficients is necessary but not sufficient for the roots of a polynomial to lie on the unit circle.

8.1-5 (**Computer experiment**) Using MATLAB, generate a signal with two real modes, having roots of the characteristic equation at $0.95e^{j\pm\pi/5}$ and $0.92e^{j\pm\pi/3}$, and explore Prony's method. Let the signal amplitudes be $\tilde{a}_1 = 1, \tilde{a}_2 = 0.5$.
   (a) Generate sufficient data to use Prony's method, solve for the coefficients, and plot the pole locations in the Z-plane.
   (b) Now add noise to the signal and determine how the Prony's method deteriorates as a function of SNR. Try SNR = 10 dB, 5 dB, 0 dB, -3 dB. Measure the SNR relative to the stronger signal.
   (c) Repeat the previous two steps using least-squares and total least-squares Prony's methods, varying the number of equations employed.

8.1-6 A useful way of interpreting the exponential model is as the impulse response of an ARMA$(p, p-1)$ model with transfer function

$$H(z) = \frac{B(z)}{A(z)} = \frac{\sum_{i=0}^{p-1} b_i z^{-1}}{1 + \sum_{i=1}^{p} a_i z^{-1}}. \tag{8.56}$$

In the case of simple modes, this can be written using partial fraction expansion as

$$H(z) = \sum_{i=1}^{p} \frac{\alpha_i}{1 - z_i z^{-1}},$$

from which the relationship

$$x[t] = \mathcal{Z}^{-1}[H(z)] = \sum_{i=1}^{p} \alpha_i z_i^{pt}$$

8.8 Exercises

is obvious. By writing out the difference equation implied by (8.56), develop a set of equations

$$\tilde{A}\tilde{\mathbf{x}} = \tilde{\mathbf{b}},$$

where $\tilde{A}$ is a Toeplitz matrix with coefficients from $A(z)$, $\tilde{\mathbf{x}}$ has time samples, and $\tilde{\mathbf{b}}$ has coefficients from $B(z)$. From this equation, the coefficients of $B(z)$ can be found without finding the roots of $A(z)$.

8.2-7 Show that the determinant of a permutation matrix is ± 1.

8.2-8 The "bit-reverse shuffle" of the FFT algorithm is a permutation. Table 8.1 illustrates a bit-reverse shuffle for an 8-point DFT. Determine a permutation matrix which permutes an incoming column vector according to the bit-reverse shuffle.

Table 8.1: Bit-reverse shuffle

n	(binary)	(bit reverse)	bit-reversed n
0	000	000	0
1	001	100	4
2	010	010	2
3	011	110	6
4	100	001	1
5	101	101	5
6	110	011	3
7	111	111	7

8.2-9 The $m \times m$ permutation matrices form a group. Determine the number of members in the group of $m \times m$ permutation matrices. Determine a power k such that all 3×3 permutation matrices P satisfy $P^k = I$.

8.3-10 Show that if R is persymmetric, then $B^{-1}J = J(B^{-1})^T$, where J is the counteridentity.

8.3-11 The $m \times m$ matrices

$$B = \begin{bmatrix} 0 & 1 & 0 & \cdots & 0 \\ 0 & 0 & 1 & \cdots & 0 \\ 0 & 0 & 0 & \ddots & 0 \\ \vdots & & & & \\ 0 & 0 & 0 & \cdots & 1 \\ 0 & 0 & 0 & \cdots & 0 \end{bmatrix} \quad \text{and} \quad F = \begin{bmatrix} 0 & 0 & 0 & \cdots & 0 & 0 \\ 1 & 0 & 0 & \cdots & 0 & 0 \\ 0 & 1 & 0 & \cdots & 0 & 0 \\ 0 & 0 & 0 & \ddots & 0 & 0 \\ 0 & 0 & 0 & \cdots & 1 & 0 \end{bmatrix}$$

are called *backward-shift* and *forward-shift* matrices, respectively.

(a) Let $\mathbf{a} = [1, 2, 3, 4]^T$. Compute $B\mathbf{a}$ and $F\mathbf{a}$ for 4×4 backward- and forward-shift matrices. Compute $B^2\mathbf{a}$ and $F^2\mathbf{a}$. Comment on the name of the matrices.

(b) Show that an $m \times m$ matrix of the form in (8.10) can be written as

$$R = \sum_{k=1}^{m} s_{-k} F^k + \sum_{k=0}^{m} s_k B^k.$$

8.3-12 Let $r_0, r_{\pm 1}, r_{\pm 2}, \ldots, r_{\pm m}$ denote the autocorrelation sequence of a stationary stochastic process, and let

$$S(\omega) = \sum_{n=-m}^{m} e^{-j\omega n} r_n$$

be its power spectral density. Show that if $S(\omega) \geq 0$ then the Toeplitz matrix R with elements $R_{ij} = r_{i-j}$ is positive semidefinite.

8.3-13 The algorithms durbin and levinson are designed for a symmetric Toeplitz matrix. Develop similar algorithms suitable for nonsymmetric matrices. Hint: Propagate two solutions.

8.3-14 Show that for a Hermitian Toeplitz matrix R_{k+1},

$$\begin{bmatrix} I & J_k \mathbf{w}_k \\ \mathbf{0} & -1 \end{bmatrix}^H R_{k+1} \begin{bmatrix} I & J_k \mathbf{w}_k \\ \mathbf{0} & -1 \end{bmatrix} = \begin{bmatrix} R_k & 0 \\ 0 & r_0 - \mathbf{w}_k^H \mathbf{r}_k \end{bmatrix}.$$

Hence conclude that if R_{k+1} is positive definite, then $r_0 - \mathbf{r}_k^T \overline{\mathbf{w}}_k$ in (8.18) is not zero.

8.3-15 Show that the variance of the first-order forward prediction error filter (8.23) is correct. Hint: Show that $\alpha_0 = r_1/r_0$.

8.4-16 Show that the formula (8.46) for the determinant of the Vandermonde matrix is correct. Hint: Use row operations, induction, and the cofactor expansion.

8.4-17 Determine a polynomial interpolating the points

$$\{(1, 2), (1.2, 4), (4.1, -3), (2.2, 7)\}.$$

8.5-18 Show that circulant$(1, 1, 1, -1)$ is a Hadamard matrix; that is, that if $H = $ circulant$(1, 1, 1, -1)$ then $HH^T = 4I$ (see section 9.2). It is believed that this is the only circulant Hadamard matrix [64].

8.5-19 Show that the matrix F defined in (8.48) satisfies $FF^H = mI$.

8.5-20 Prove theorem 8.2.

8.5-21 (Some properties of circulant matrices.)

(a) Show that if A and B are circulant matrices of the same size, then AB is circulant.

(b) Show that if A is a circulant matrix, then for any fixed $r > 0$,

$$\sum_{k=0}^{r} a_k A^k$$

is circulant, where the a_k are scalars.

(c) Show that the inverse of an $m \times m$ circulant matrix A is

$$A^{-1} = mF^H \Lambda^{-1} F.$$

(d) Show that the determinant of an $m \times m$ circulant matrix is $A = $ circulant$(\mathbf{c})$ is

$$\det(\text{circulant}(\mathbf{c})) = \prod_{j=1}^{m} p_\mathbf{c}(\omega^{j-1}),$$

where $\omega = e^{-j2\pi/m}$.

(e) Show that the pseudoinverse of a circulant matrix C is
$$C^\dagger = m F^H \Lambda^\dagger F.$$

8.5-22 Justify (8.53) and (8.54).

8.5-23 Using theorem 8.6, show that
$$\lim_{n \to \infty} [\det(R_n)]^{1/n} = \exp\left(\frac{1}{2\pi} \int_0^{2\pi} \ln S(\omega)\, d\omega\right).$$

Hint: $g(\cdot) = \ln$.

8.5-24 Show that a circulant matrix is Toeplitz, but a Toeplitz matrix is not necessarily circulant.

8.7-25 For each of the properties listed in section 8.7 that *fail* to be preserved under matrix multiplication, find an example to demonstrate this failure.

8.9 References

Exponential signal models are discussed in, for example, [291, 48, 174, 327]. Prony's method has a considerable history, dating to 1795 [65]. The least-squares Prony method appears in [137]. The observation about a real undamped signal having a symmetric characteristic polynomial appears in [189].

The basic algorithms for solution of Toeplitz equations come from [114]. Signal processing interpretations of Toeplitz matrices are found, for example, in [132] and [263]. Inversion for block Toeplitz matrices is discussed in [3]. An interesting survey article is [167]. Related articles discussing solutions of equations with Hankel and Toeplitz matrices appear in [276, 275].

Diagonalization of circulant matrices is also discussed in [147], where it is shown that a block circulant matrix (a matrix that circulates blocks of matrices) can be diagonalized by a two-dimensional DFT. Application of this to image processing is discussed in the survey [9].

Our discussion of the asymptotic equivalence of Toeplitz and circulant matrices is drawn closely from [118], which in turn draws from [119] and [363]. Asymptotic equivalence of the power spectrum and the eigenvalues of autocorrelation function (by means of the Karhunen–Loève representation) are discussed in [96, chapter 8]. Eigenvalues of Toeplitz matrices are also discussed in [16], while (John Makhoul, "On the Eigenvectors of Symmetric Toeplitz Matrices," *IEEE Trans. Acoust., Speech, Signal Proc.*, 29(4): 868–871, 1981) and [272] treat the eigenvectors of symmetric Toeplitz matrices.

Information on a variety of special matrices is in [142]. Though brief, [64] has a great deal of material on circulant matrices, Toeplitz matrices, block matrices, permutations, and pseudoinverses, among other things. The summary of matrix properties preserved under multiplication comes from [342].

Chapter 9

Kronecker Products and the vec Operator

The Kronecker product has recently been used in the expression and development of some fast signal-processing algorithms. In this chapter, we present first the basic theory of Kronecker products, then demonstrate some application areas. In addition, we present the vec operator, which is useful in restructuring matrix equations by turning matrices into vectors. In many problems, this restructuring provides notational leverage by allowing the use, for example, of the formulas for derivatives with respect to vectors.

9.1 The Kronecker product and Kronecker sum

Definition 9.1 Let A be an $n \times p$ matrix and B be an $m \times q$ matrix. The $mn \times pq$ matrix

$$A \otimes B = \begin{bmatrix} a_{11}B & a_{12}B & \cdots a_{1p}B \\ a_{21}B & a_{22}B & \cdots & a_{2p}B \\ \vdots & & & \\ a_{n1}B & a_{n2}B & \cdots & a_{np}B \end{bmatrix}$$

is called the **Kronecker product** of A and B. It is also called the direct product or the tensor product. □

Example 9.1.1 Let

$$A = \begin{bmatrix} 1 & 2 \\ 3 & 4 \end{bmatrix} \qquad B = \begin{bmatrix} 7 & 7 & 7 \\ 9 & 9 & 9 \\ 11 & 11 & 11 \end{bmatrix};$$

then

$$A \otimes B = \begin{bmatrix} 7 & 7 & 7 & 14 & 14 & 14 \\ 9 & 9 & 9 & 18 & 18 & 18 \\ 11 & 11 & 11 & 22 & 22 & 22 \\ 21 & 21 & 21 & 28 & 28 & 28 \\ 27 & 27 & 27 & 36 & 36 & 36 \\ 33 & 33 & 33 & 44 & 44 & 44 \end{bmatrix}.$$

□

9.1 The Kronecker Product and Kronecker Sum

The Kronecker product has the following basic properties (the indicated operations are assumed to be defined).

1. $A \otimes B \neq B \otimes A$, in general.
2. For a scalar x,
$$(xA) \otimes B = A \otimes (xB) = x(A \otimes B). \tag{9.1}$$
3. Distributive properties:
$$(A + B) \otimes C = (A \otimes C) + (B \otimes C), \tag{9.2}$$
$$A \otimes (B + C) = (A \otimes B) + (A \otimes C). \tag{9.3}$$
4. Associative property:
$$(A \otimes B) \otimes C = A \otimes (B \otimes C). \tag{9.4}$$
5. Transposes:
$$(A \otimes B)^T = A^T \otimes B^T, \tag{9.5}$$
$$(A \otimes B)^H = A^H \otimes B^H. \tag{9.6}$$
6. Trace (for square A and B):
$$\operatorname{tr}(A \otimes B) = \operatorname{tr}(A)\operatorname{tr}(B). \tag{9.7}$$
7. If A is diagonal and B is diagonal, then $A \otimes B$ is diagonal.
8. Determinant, where A is $m \times m$ and B is $n \times n$:
$$\det(A \otimes B) = \det(A)^n \det(B)^m. \tag{9.8}$$
9. The Kronecker product theorem:
$$(A \otimes B)(C \otimes D) = (AC) \otimes (BD), \tag{9.9}$$
provided that the matrices are shaped such that the indicated products are allowed.
10. Inverses: If A and B are nonsingular, then $A \otimes B$ is nonsingular, and
$$(A \otimes B)^{-1} = A^{-1} \otimes B^{-1}. \tag{9.10}$$
11. There is a permutation matrix P such that
$$B \otimes A = P^T (A \otimes B) P. \tag{9.11}$$

In proving properties involving Kronecker products, it is often helpful to consider the (i, j)th block of the result. For example, to prove (9.2), we note that the (i, j)th block of $(A + B) \otimes C$ is $(a_{ij} + b_{ij})C$, and that the (i, j)th block of $A \otimes C + B \otimes C$ is $a_{ij}C + b_{ij}C = (a_{ij} + b_{ij})C$.

Theorem 9.1 *Let A be an $m \times m$ matrix with eigenvalues $\lambda_1, \lambda_2, \ldots, \lambda_m$ and corresponding eigenvectors $\mathbf{x}_1, \mathbf{x}_2, \ldots, \mathbf{x}_m$, and let B be an $n \times n$ matrix with eigenvalues $\mu_1, \mu_2, \ldots, \mu_n$ and corresponding eigenvectors $\mathbf{y}_1, \mathbf{y}_2, \ldots, \mathbf{y}_n$. Then the mn eigenvalues of $A \otimes B$ are $\lambda_i \mu_j$ for $i = 1, 2, \ldots, m$, $j = 1, 2, \ldots, n$, and the corresponding eigenvectors are $\mathbf{x}_i \otimes \mathbf{y}_j$.*

Proof [117, page 27] Let $A\mathbf{x}_i = \lambda_i \mathbf{x}_i$ and $B\mathbf{y}_j = \mu_j \mathbf{y}_j$. By (9.9),

$$(A \otimes B)(\mathbf{x}_i \otimes \mathbf{y}_j) = (A\mathbf{x}_i) \otimes (B\mathbf{y}_j),$$
$$= \lambda_i \mu_j (\mathbf{x}_i \otimes \mathbf{y}_j).$$
□

Example 9.1.2 For the matrices

$$A = \begin{bmatrix} 4 & 5 \\ 2 & 1 \end{bmatrix} \quad B = \begin{bmatrix} 7 & -3 \\ -3 & 7 \end{bmatrix},$$

$$A \otimes B = \begin{bmatrix} 28 & -12 & 35 & -15 \\ -12 & 28 & -15 & 35 \\ 14 & -6 & 7 & -3 -6 & 14 & -3 & 7 \end{bmatrix}.$$

The eigenvalues of A are $6, -1$ and the eigenvalues of B are $10, 4$. The eigenvalues of $A \otimes B$ are $60, 24, 4, -10$. □

Definition 9.2 The **Kronecker sum** of an $m \times m$ matrix A and an $n \times n$ matrix B is

$$A \oplus B = (I_n \otimes A) + (I_m \otimes B), \tag{9.12}$$

where I_m denotes the $m \times m$ identity matrix. □

Example 9.1.3 For A and B of example 9.1.1,

$$A \oplus B = I_3 \otimes A + I_2 \otimes B$$

$$= \begin{bmatrix} 1 & 2 & 0 & 0 & 0 & 0 \\ 3 & 4 & 0 & 0 & 0 & 0 \\ 0 & 0 & 1 & 2 & 0 & 0 \\ 0 & 0 & 3 & 4 & 0 & 0 \\ 0 & 0 & 0 & 0 & 1 & 2 \\ 0 & 0 & 0 & 0 & 3 & 4 \end{bmatrix} + \begin{bmatrix} 7 & 7 & 7 & 0 & 0 & 0 \\ 9 & 9 & 9 & 0 & 0 & 0 \\ 11 & 11 & 11 & 0 & 0 & 0 \\ 0 & 0 & 0 & 7 & 7 & 7 \\ 0 & 0 & 0 & 9 & 9 & 9 \\ 0 & 0 & 0 & 11 & 11 & 11 \end{bmatrix}$$

$$= \begin{bmatrix} 8 & 9 & 7 & 0 & 0 & 0 \\ 12 & 13 & 9 & 0 & 0 & 0 \\ 11 & 11 & 12 & 2 & 0 & 0 \\ 0 & 0 & 3 & 11 & 7 & 7 \\ 0 & 0 & 0 & 9 & 10 & 11 \\ 0 & 0 & 0 & 11 & 14 & 15 \end{bmatrix}.$$
□

As for the Kronecker product, the eigenvalues of the Kronecker sum are determined by the eigenvalues of the constituent parts, as stated in the following theorem.

Theorem 9.2 Let A be $m \times m$ with eigenvalues $\lambda_1, \lambda_2, \ldots, \lambda_m$ and corresponding eigenvectors $\mathbf{x}_1, \mathbf{x}_2, \ldots, \mathbf{x}_m$, and let B be $n \times n$ with eigenvalues $\mu_1, \mu_2, \ldots, \mu_n$ and corresponding eigenvectors $\mathbf{y}_1, \mathbf{y}_2, \ldots, \mathbf{y}_n$. Then the mn eigenvalues of the Kronecker sum (9.12) are $\lambda_i + \mu_j$, and the eigenvectors are $\mathbf{y}_j \otimes \mathbf{x}_i$.

Proof [117, page 30] Let $A\mathbf{x}_i = \lambda_i \mathbf{x}_i$ and $B\mathbf{y}_j = \mu_j \mathbf{y}_j$. Then

$$(A \oplus B)(\mathbf{x}_i \otimes \mathbf{y}_j) = (A \otimes I)(\mathbf{x}_i \otimes \mathbf{y}_j) + (I \otimes B)(\mathbf{x}_i \otimes \mathbf{y}_j)$$
$$= (A\mathbf{x}_i \otimes \mathbf{y}_j) + (\mathbf{x}_i \otimes B\mathbf{y}_j)$$
$$= \lambda_i (\mathbf{x}_i \otimes \mathbf{y}_j) + \mu_j (\mathbf{x}_i \otimes \mathbf{y}_j)$$
$$= (\lambda_i + \mu_j)(\mathbf{x}_i \otimes \mathbf{y}_j).$$
□

9.2 Some applications of Kronecker products

9.2.1 Fast Hadamard transforms

Definition 9.3 An $n \times n$ **Hadamard matrix** is a matrix H_n whose elements are ± 1 such that

$$H_n H_n^T = nI. \qquad \square$$

The 2×2 Hadamard matrix is

$$H_2 = \begin{bmatrix} 1 & 1 \\ 1 & -1 \end{bmatrix}.$$

One way of constructing Hadamard matrices, the so-called Sylvester construction, is to form

$$H_{2n} = H_2 \otimes H_n. \qquad (9.13)$$

Based upon this construction,

$$H_4 = \begin{bmatrix} 1 & 1 & 1 & 1 \\ 1 & -1 & 1 & -1 \\ 1 & 1 & -1 & -1 \\ 1 & -1 & -1 & 1 \end{bmatrix}.$$

The Hadamard transform matrix (after normalization) is a unitary matrix, and hence is of interest in some transform coding problems. The Hadamard transform is also used in conjunction with error-correction algorithms for Reed–Muller codes. Because of the relationship (9.13), fast algorithms can be developed for computing the Hadamard transform.

Example 9.2.1 Let $\mathbf{x} = [x_1, x_2, x_3, x_4]^T$, and let $\mathbf{z} = H_4 \mathbf{x}$ be the 4-point Hadamard transform of $\mathbf{x}$. Then

$$\mathbf{z} = \begin{bmatrix} x_1 + x_2 + x_3 + x_4 \\ x_1 - x_2 + x_3 - x_4 \\ x_1 + x_2 - x_3 - x_4 \\ x_1 - x_2 - x_3 + x_4 \end{bmatrix}.$$

In the general case of an n-point transform, $n(n-1)$ additions/subtractions are required. However, the computations can be ordered differently. If we compute

$$\mathbf{y}_1 = \begin{bmatrix} y_{11} \\ y_{12} \end{bmatrix} = H_2 \begin{bmatrix} x_1 \\ x_2 \end{bmatrix} = \begin{bmatrix} x_1 + x_2 \\ x_1 - x_2 \end{bmatrix} \qquad \mathbf{y}_2 = \begin{bmatrix} y_{21} \\ y_{22} \end{bmatrix} = H_2 \begin{bmatrix} x_3 \\ x_4 \end{bmatrix} = \begin{bmatrix} x_3 + x_4 \\ x_3 - x_4 \end{bmatrix},$$

then

$$\mathbf{z} = \begin{bmatrix} y_{11} + y_{21} \\ y_{12} + y_{22} \\ y_{11} - y_{21} \\ y_{12} - y_{22} \end{bmatrix},$$

which can be computed using another pair of 2-point Hadamard transforms. The processing is outlined in figure 9.1. The number of additions/subtractions in this case is 8. In general, an n-point fast Hadamard transform can be computed in $n \log n$ additions/subtractions. $\square$

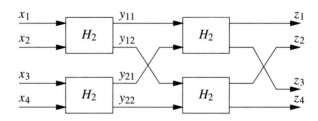

Figure 9.1: 4-point fast Hadamard transform

An $n = 2^m$-point Hadamard transform can be computed using the following decomposition.

Lemma 9.1 *[214] If H_n is a Sylvester-type Hadamard matrix, where $n = 2^m$, then*

$$H_n = M_n^{(1)} M_n^{(2)} \cdots M_n^{(m)}$$

where $M_n^{(i)} = I_{2^{i-1}} \otimes H_2 \otimes I_{2^{m-i}}$.

Proof (By induction) When $m = 1$, the result is obvious. Assume that the result is true for m. Then for $1 \leq i \leq m$,

$$M_{2^{m+1}}^{(i)} = I_{2^{i-1}} \otimes H_2 \otimes I_{2^{m+1-i}} = I_{2^{i-1}} \otimes H_2 \otimes I_{2^{m-i}} \otimes I_2$$
$$= M_{2^m}^{(i)} \otimes I_2,$$

and $M_{2^{m+1}}^{(m+1)} = I_{2^m} \otimes H_2$. Then

$$M_{2^{m+1}}^{(1)} M_{2^{m+1}}^{(2)} \cdots M_{2^{m+1}}^{(m+1)} = \left(M_{2^m}^{(1)} \otimes I_2\right)\left(M_{2^m}^{(2)} \otimes I_2\right) \cdots \left(M_{2^m}^{(m)} \otimes I_2\right)\left(I_{2^m} \otimes H_2\right)$$
$$= \left(M_{2^m}^{(1)} M_{2^m}^{(2)} \cdots M_{2^m}^{(m)} I_{2^m}\right) \otimes H_2$$
$$= H_{2^{m+1}} \otimes H_2 = H_{2^{m+1}}. \qquad \square$$

For example,

$$M_4^{(1)} = \begin{bmatrix} 1 & 0 & 1 & 0 \\ 0 & 1 & 0 & 1 \\ 1 & 0 & -1 & 0 \\ 0 & 1 & 0 & -1 \end{bmatrix} \qquad M_4^{(2)} = \begin{bmatrix} 1 & 1 & 0 & 0 \\ 1 & -1 & 0 & 0 \\ 0 & 0 & 1 & 1 \\ 0 & 0 & 1 & -1 \end{bmatrix}.$$

It is straightforward to verify that $M_4^{(1)} M_4^{(2)} = H_4$. In use, the fast Hadamard transform first computes in succession

$$\mathbf{y}_1 = M_n^{(m)} \mathbf{x},$$
$$\mathbf{y}_2 = M_n^{(m-1)} \mathbf{y}_1,$$
$$\vdots$$
$$\mathbf{z} = M_n^{(1)} \mathbf{y}_{m-1}.$$

Since each row of $M_n^{(i)}$ has only two nonzero elements, each stage can be computed in n additions/subtractions.

9.2.2 DFT computation using Kronecker products

The Kronecker product can be used in the definition of fast algorithms (e.g., reduced numbers of multiplications) for FFTs and convolution, by building large DFT algorithms from small ones. The computational advantage of these algorithms stems from the fact that as

9.2 Some Applications of Kronecker Products

the matrices in the decomposition become smaller, the number of computations decreases. Fast large-n DFTs can be built up by carefully implementing a set of small-n DFTs, then putting the pieces together using the Kronecker product.

Suppose an n-point DFT is to be computed. If n is a power of 2, the familiar Cooley–Tukey algorithm can be used. However, in this section we assume that n is not a power of 2, but can be factored as $n = n_L n_{l-1} \cdots n_1$, where the factors are relatively prime, $(n_i, n_j) = 1$ for $i \neq j$. We will demonstrate the principal of stitching together small DFTs to make larger ones, using a 6-point DFT.

Let

$$\mathbf{y} = \begin{bmatrix} y_0 \\ y_1 \\ y_2 \\ y_3 \\ y_4 \\ y_5 \end{bmatrix} = \begin{bmatrix} 1 & 1 & 1 & 1 & 1 & 1 \\ 1 & w_6^1 & w_6^2 & w_6^3 & w_6^4 & w_6^5 \\ 1 & w_6^2 & w_6^4 & w_6^6 & w_6^8 & w_6^{10} \\ 1 & w_6^3 & w_6^6 & w_6^9 & w_6^{12} & w_6^{15} \\ 1 & w_6^4 & w_6^8 & w_6^{12} & w_6^{16} & w_6^{20} \\ 1 & w_6^5 & w_6^{10} & w_6^{15} & w_6^{20} & w_6^{25} \end{bmatrix} \begin{bmatrix} x_0 \\ x_1 \\ x_2 \\ x_3 \\ x_4 \\ x_5 \end{bmatrix} = F_6 \mathbf{x} \quad (9.14)$$

be the 6-point DFT of the vector $\mathbf{x}$, where $w_N = e^{-j2\pi/N}$. We wish to formulate an $N = 6$-point DFT in terms of the 2-point and 3-point DFTs represented by

$$F_2 = \begin{bmatrix} 1 & 1 \\ 1 & w_6^3 \end{bmatrix} \qquad F_3 = \begin{bmatrix} 1 & 1 & 1 \\ 1 & w_6 & w_6^2 \\ 1 & w_6^2 & w_6^4 \end{bmatrix}.$$

We find that

$$F = F_2 \otimes F_3 = \begin{bmatrix} 1 & 1 & 1 & 1 & 1 & 1 \\ 1 & w_6^2 & w_6^4 & 1 & w_6^2 & w_6^4 \\ 1 & w_6^4 & w_6^8 & 1 & w_6^4 & w_6^8 \\ 1 & 1 & 1 & w_6^3 & w_6^3 & w_6^3 \\ 1 & w_6^2 & w_6^4 & w_6^3 & w_6^5 & w_6^7 \\ 1 & w_6^4 & w_6^8 & w_6^3 & w_6^7 & w_6^{11} \end{bmatrix}. \quad (9.15)$$

Unfortunately, F is not the same as F_6 defined in (9.14). However, careful comparison of F and F_6 reveals that the DFT can be computed by reordering the input and output sequence. The DFT can be computed as

$$\begin{bmatrix} y_0 \\ y_4 \\ y_2 \\ y_3 \\ y_1 \\ y_5 \end{bmatrix} = \begin{bmatrix} 1 & 1 & 1 & 1 & 1 & 1 \\ 1 & w_6^2 & w_6^4 & 1 & w_6^2 & w_6^4 \\ 1 & w_6^4 & w_6^8 & 1 & w_6^4 & w_6^8 \\ 1 & 1 & 1 & w_6^3 & w_6^3 & w_6^3 \\ 1 & w_6^2 & w_6^4 & w_6^3 & w_6^5 & w_6^7 \\ 1 & w_6^4 & w_6^8 & w_6^3 & w_6^7 & w_6^{11} \end{bmatrix} \begin{bmatrix} x_0 \\ x_2 \\ x_4 \\ x_3 \\ x_5 \\ x_1 \end{bmatrix}. \quad (9.16)$$

The index ordering of the input is [0, 2, 4, 3, 5, 1] and the index ordering of the output is [0, 4, 2, 3, 1, 5]. The details behind this particular index scheme will not be treated here, as they require some background in number theory; the sources cited in the references section provide details.

Figure 9.2 illustrates the butterfly signal flow diagram used to compute this 6-point DFT. Interestingly, the "twiddle factors" familiar from traditional Cooley–Tukey FFT algorithms

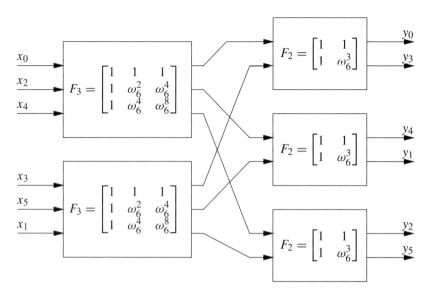

Figure 9.2: 6-point DFT using Kronecker decomposition

are absent between the blocks of the diagram. More generally, we can compute the n-point DFT, where $n = n_1 n_2 \cdots n_L$, using the Kronecker product of

$$F = F_L \otimes \cdots \otimes F_2 \otimes F_1 \tag{9.17}$$

where each F_i is an $n_i \times n_i$ matrix. By correct reordering of the input and output, a DFT equivalent to that obtained by F_n may be obtained.

9.3 The vec operator

Definition 9.4 For an $m \times n$ matrix $A = [\mathbf{a}_1, \mathbf{a}_2, \ldots, \mathbf{a}_n]$, the vec operator converts the matrix to a column vector by stacking the columns of A,

$$\mathrm{vec}(A) = \begin{bmatrix} \mathbf{a}_1 \\ \mathbf{a}_2 \\ \vdots \\ \mathbf{a}_n \end{bmatrix},$$

to obtain a vector of mn elements. □

The vec operation can be computed in MATLAB by indexing with a single colon; for vec(A) type A(:). A vector can be reshaped using the reshape function.

Example 9.3.1 Let

$$A = \begin{bmatrix} 1 & 2 & 3 \\ 4 & 5 & 6 \end{bmatrix}.$$

Then

$$\mathrm{vec}(A) = \begin{bmatrix} 1 \\ 4 \\ 2 \\ 5 \\ 3 \\ 6 \end{bmatrix}.$$

□

9.3 The vec Operator

The vec representation can be used to rewrite a variety of operations. For example,
$$\text{tr}(AB) = (\text{vec}(A^T))^T \text{vec}(B). \tag{9.18}$$
(see exercise 9.3-12).

Theorem 9.3
$$\text{vec}(AYB) = (B^T \otimes A) \text{vec } Y. \tag{9.19}$$

Proof Let B be $m \times n$. Observe that the kth column of (AYB) can be written (see section C.1 for notation) as

$$(AYB)_{:k} = \sum_{j=1}^{m} b_{jk} AY_{:j} = [b_{1k}A \; b_{2k}A \; \cdots \; b_{nk}A] \begin{bmatrix} Y_{:1} \\ Y_{:2} \\ \vdots \\ Y_{:n} \end{bmatrix} \quad k = 1, 2, \ldots, n.$$

This in turn can be written as
$$(AYB)_{:k} = (B'_{:k} \otimes A) \text{vec}(Y).$$

Stacking the columns together, we obtain the desired result:

$$\text{vec}(AYB) = \begin{bmatrix} (AYB)_{:1} \\ (AYB)_{:2} \\ \vdots \\ (AYB)_{:n} \end{bmatrix} = \begin{bmatrix} (B'_{:1} \otimes A) \text{vec}(Y) \\ (B'_{:2} \otimes A) \text{vec}(Y) \\ \vdots \\ (B'_{:n} \otimes A) \text{vec}(Y) \end{bmatrix} = \begin{bmatrix} B'_{:1} \otimes A \\ B'_{:2} \otimes A \\ \vdots \\ B'_{:n} \otimes A \end{bmatrix} \text{vec}(Y)$$
$$= (B^T \otimes A) \text{vec}(Y). \qquad \square$$

Example 9.3.2 The vec operator can be used to convert matrix equations to vector equations. The equation
$$\begin{bmatrix} a_{11} & a_{12} \\ a_{21} & a_{22} \end{bmatrix} \begin{bmatrix} x_{11} & x_{12} \\ x_{21} & x_{22} \end{bmatrix} = \begin{bmatrix} c_{11} & c_{12} \\ c_{21} & c_{22} \end{bmatrix}$$

can be vectorized by writing
$$AXI = C$$
so that
$$\text{vec}(AXI) = \text{vec}(C),$$
or, by (9.19),
$$(I \otimes A) \text{vec}(X) = \text{vec}(C).$$

This is equivalent to
$$\begin{bmatrix} a_{11} & a_{12} & 0 & 0 \\ a_{21} & a_{22} & 0 & 0 \\ 0 & 0 & a_{11} & a_{12} \\ 0 & 0 & a_{21} & a_{22} \end{bmatrix} \begin{bmatrix} x_1 \\ x_2 \\ x_3 \\ x_4 \end{bmatrix} = \begin{bmatrix} c_{11} \\ c_{21} \\ c_{12} \\ c_{22} \end{bmatrix}. \qquad \square$$

Example 9.3.3 Suppose we desire to solve the equation
$$AXB = C \tag{9.20}$$
for the matrix X. If A and B are invertible, one method of solution is simply
$$X = A^{-1}CB^{-1}. \tag{9.21}$$

Another approach is to rewrite (9.20) as

$$Y\mathbf{x} = \mathbf{c},$$

where $Y = B^T \otimes A$, $\mathbf{x} = \text{vec}(X)$, and $\mathbf{c} = \text{vec}(C)$.

Generalizing the problem, suppose we desire to solve

$$A_1 X B_1 + A_2 X B_2 + \cdots + A_s X B_s = C$$

for X. In this case, simple matrix inversion as in (9.21) will not suffice. However, it can be vectorized as we have seen, where

$$Y = B_1^T \otimes A_1 + B_2^T \otimes A_2 + \cdots + B_s^T \otimes A_s.$$ □

Definition 9.5 A linear operator A is said to be *separable* if $A = A_1 \otimes A_2$ for some A_1 and A_2. □

Operations involving separable linear operators can be reduced in complexity by the use of (9.19). For example, suppose that A is $m^2 \times m^2$. Computation of the product

$$\mathbf{b} = A\mathbf{x} \qquad (9.22)$$

will require $O(N^4)$ operations. If $A = A_1 \otimes A_2$, where each A_i is $m \times m$, then (9.22) can be written as

$$B = A_2 X A_1^T, \qquad (9.23)$$

where B and X are $m \times m$. The two matrix multiplications in (9.23) require a total of $2O(N^3)$ operations.

Example 9.3.4 The matrix

$$A = \begin{bmatrix} 2 & 3 & 4 & 6 \\ -5 & 6 & -10 & 12 \\ 10 & 15 & 14 & 21 \\ -25 & 30 & -35 & 42 \end{bmatrix}$$

is separable,

$$A = \begin{bmatrix} 1 & 2 \\ 5 & 7 \end{bmatrix} \otimes \begin{bmatrix} 2 & 3 \\ -5 & 6 \end{bmatrix}.$$ □

Another vectorizing problem that occurs in some minimization problems is as follows: given an $m \times n$ matrix X, determine $\text{vec}(X^T)$ in terms of $\text{vec}(X)$. The transpose shuffles the columns around, so it may be anticipated that

$$\text{vec}(X^T) = P \text{vec}(X),$$

where P is a permutation matrix.

Example 9.3.5 Let

$$X = \begin{bmatrix} x_{11} & x_{12} \\ x_{21} & x_{22} \end{bmatrix}.$$

Then

$$\text{vec}(X) = \begin{bmatrix} x_{11} \\ x_{21} \\ x_{12} \\ x_{22} \end{bmatrix} \quad \text{and} \quad \text{vec}(X^T) = \begin{bmatrix} x_{11} \\ x_{12} \\ x_{12} \\ x_{22} \end{bmatrix} = \begin{bmatrix} 1 & 0 & 0 & 0 \\ 0 & 0 & 1 & 0 \\ 0 & 1 & 0 & 0 \\ 0 & 0 & 0 & 1 \end{bmatrix} \text{vec}(X).$$ □

The permutation matrix can be determined using element matrices. Observe that X can be written in terms of unit element matrices (see section C.1) as

$$X = \sum_{r=1}^{m}\sum_{s=1}^{n} x_{rs} E_{rs},$$

where the unit element matrix E_{rs} is $m \times n$. Then X^T can be written as

$$X^T = \sum_{r=1}^{m}\sum_{s=1}^{n} x_{rs} E_{sr},$$

where in this case E_{sr} is $n \times m$. It is straightforward to show (see exercise 9.3-15) that the right-hand side can be written as

$$X^T = \sum_{r=1}^{m}\sum_{s=1}^{n} E_{sr} X E_{sr},$$

with E_{sr} of size $n \times m$. Then, using (9.19),

$$\text{vec}(X^T) = \text{vec}\sum_{r=1}^{m}\sum_{s=1}^{n} E_{sr} X E_{sr}$$

$$= \sum_{r=1}^{m}\sum_{s=1}^{n} (E_{rs} \otimes E_{sr}) \text{vec}(X),$$

so that

$$P = \sum_{r=1}^{m}\sum_{s=1}^{n} E_{rs} \otimes E_{sr}, \qquad (9.24)$$

with the unit element matrices suitably sized.

9.4 Exercises

9.1-1 Prove each of the eleven listed properties of the Kronecker product, with the exception of the associative property. Hints: To prove the determinant property, use theorem 9.1. To prove 9.11, use theorem 9.3 with $X = AYB^T$ and $X^T = BYA^T$.

9.1-2 If $A = U_A S_A V_A^H$ and $B = U_B S_B V_B^H$ are the SVDs of A and B, show that $A \otimes B = (U_A \otimes U_B)(S_A \otimes S_B)(V_A \otimes V_B)^H$.

9.1-3 Let

$$B = \begin{bmatrix} A+2I & -I & & & & \\ -I & A+2I & & & & \\ & -I & A+2I & & & \\ & & & \ddots & & \\ & & & & & -I \\ & & & & -I & A+2I \end{bmatrix},$$

where

$$A = \begin{bmatrix} 2 & -1 & & & \\ -1 & 2 & -1 & & \\ & -1 & 2 & -1 & \\ & & \ddots & \ddots & \\ & & & & -1 \\ & & & -1 & 2 \end{bmatrix}.$$

(a) Show that $B = (A \otimes I) + (I \otimes A)$.

(b) Find the eigenvalues and eigenvectors of B.

Hints: See exercise **6.2-1** and use Kronecker addition.

9.1-4 The equation

$$AX - XB = C,$$

where A, B, and C are $n \times n$ and known and X is $n \times n$ and is to be determined, is sometimes called *Sylvester's equation*.

(a) Show that $AX - XB = C$ can be written as a set of n^2 equations, using the Kronecker sum, as

$$[(I \otimes A) + (-B^T \otimes I)]\mathbf{x} = \mathbf{c}$$

where

$$\mathbf{x} = \begin{bmatrix} \mathbf{x}_1 \\ \mathbf{x}_2 \\ \vdots \\ \mathbf{x}_n \end{bmatrix} \quad \text{and} \quad \mathbf{c} = \begin{bmatrix} \mathbf{c}_1 \\ \mathbf{c}_2 \\ \vdots \\ \mathbf{c}_n \end{bmatrix},$$

and where $\mathbf{x}_i$ is the ith column of X and $\mathbf{c}_i$ is the ith column of C.

(b) Show that Sylvester's equation has a unique solution if and only if A and B have no common eigenvalues.

9.1-5 Show that if A and B are both

(a) Normal (i.e., $A^H A = AA^H$),

(b) Hermitian,

(c) Positive definite,

(d) Positive semidefinite,

(e) Unitary,

then $A \otimes B$ has the corresponding property.

9.1-6 Show that:

(a) $(I \otimes A)^k = (I \otimes A^k)$.

(b) If A is an $m \times m$ matrix,

$$e^{I \otimes A} = I \otimes e^A$$

and

$$e^{A \otimes I} = e^A \otimes I.$$

9.2-7 Show that if a Hadamard matrix of order n exists, then $n = 1$, or $n = 2$, or $n \equiv 0 \mod 4$ (i.e., n is a multiple of 4).

9.2-8 Show that if A and B are Hadamard matrices of order m and n, respectively, then $A \otimes B$ is a Hadamard matrix.

9.2-9 Show that

$$H_{2^{n+1}} = (H_{2^n} \otimes I_2)(I_{2^n} \otimes H_2).$$

9.2-10 Show that the DFT scheme shown in figure 9.2 computes the 6-point DFT.

9.5 References

9.2-11 For the 10-point DFT:

(a) Write down F_{10}.

(b) Write down F_2, F_5, and $F = F_2 \otimes F_5$, in terms of w_{10}.

(c) Determine the shuffling of the input and output so that F computes a 10-point DFT.

(d) Draw a block diagram indicating how to compute the 10-point DFT using F.

9.3-12 Show that
$$\text{tr}(AB) = (\text{vec}(A^T))^T \text{vec}(B).$$

9.3-13 Show that for $n \times n$ matrices A and B
$$\text{vec}(AB) = (I \otimes A) \text{vec } B,$$
$$\text{vec}(AB) = (B^T \otimes A) \text{vec } I.$$

9.3-14 Find the solution X to the equation
$$A_1 X B_1 + A_2 X B_2 = C,$$
where
$$A_1 = \begin{bmatrix} 4 & 2 \\ 1 & 2 \end{bmatrix} \quad B_1 = \begin{bmatrix} 0 & 1 \\ 1 & 1 \end{bmatrix},$$
$$A_2 = \begin{bmatrix} 5 & 2 \\ 1 & 0 \end{bmatrix} \quad B_2 = \begin{bmatrix} 2 & 0 \\ 0 & 2 \end{bmatrix},$$
$$C = \begin{bmatrix} 1 & 2 \\ 3 & 4 \end{bmatrix}.$$

9.3-15 Show that for an $m \times n$ matrix X,
$$\sum_{r=1}^{m} \sum_{s=1}^{n} x_{rs} E_{sr} = \sum_{r=1}^{m} \sum_{s=1}^{n} E_{sr} X E_{sr}.$$
Hint: Show that $E_{ij} X E_{kl} = x_{jk} E_{il}$.

9.3-16 Let A be a separable matrix:
$$A = \begin{bmatrix} 24 & 8 & -12 & -4 \\ 4 & 12 & -2 & -6 \\ 12 & 4 & -6 & -2 \\ 2 & 6 & -1 & -3 \end{bmatrix}.$$

(a) Determine A_1 and A_2 so that $A = A_1 \otimes A_2$.

(b) Let $\mathbf{x} = [1, 2, 3, 4]^T$. Compute the product $A\mathbf{x}$ both directly and using (9.19).

9.5 References

For a wealth of information about Kronecker products, the vec operator, and gradients, see [117]. The Kronecker product is also discussed in [245].

Hadamard transforms are discussed in [128] and [152]. Applications to error-correction coding appear in [361] and [214].

There are many FFT algorithms based upon Kronecker products; our presentation barely scratches the surface. A classic reference on the topic is [242]. Excellent coverage is also provided in [33] and [78]. Other sources include [296, 133]. The use of the Kronecker product in image processing is presented in [152].

Sylvester's equation as explored in exercise 9.1-4(b) arises in some controls problems; see, for example, [93]. The problem is examined from a numerical standpoint in [114]. Additional treatments are in [14, 196, 245 (chapter 6), 114 (chapter 7)].

The description of the vec operator is taken from [117], and the exercises are drawn from examples and exercises found there.

Part III

Detection, Estimation, and Optimal Filtering

In this part, we undertake a study of the problem of signal processing in the presence of random noise. The first focus of study is the making of decisions in noise, known as *detection theory*. This forms a foundation discipline for a variety of application areas, including pattern recognition, radar processing, and digital communications.

Next, we consider the discipline that studies the estimation of parameters in the presence of noise, *estimation theory*. We have already seen some aspects of this theory in the context of vector spaces. Estimation theory encompasses a variety of applications, including tracking, spectral estimation, and synchronization.

Optimal filtering addesses the design and implementation of linear filters for estimating and predicting stochastic processes in the presence of noise. We examine in particular the Kalman and its application filter.

Chapter 10

Introduction to Detection and Estimation, and Mathematical Notation

> The mind's deepest desire, even in its most elaborate operations, parallels man's unconscious feeling in the face of his universe: it is an insistence upon familiarity, an appetite for clarity. Understanding the world for a man is reducing it to the human, stamping it with his seal.
>
> *Albert Camus*
> The Myth of Sisyphus

In this chapter, we introduce statistical decision making as an instance of a game (in the mathematical sense), then formalize the elements of the problem. We then present basic notation and concepts related to random samples, followed by some basic theory that will be of use in our study:

1. Conditional expectations,
2. Transformations of random variables,
3. Sufficient statistics,
4. Exponential families.

10.1 Detection and estimation theory

Observations of signals in physical systems are frequently made in the presence of noise, and effective processing of these signals often relies upon techniques drawn from the statistical literature. These statistical techniques are generally applied to two different kinds of problems, illustrated in the following examples.

Example 10.1.1 *Detection.* Let
$$x(t) = A \cos(2\pi f_c t), \qquad t \in [0, T],$$
where A takes on one of two values, $A \in \{1, -1\}$. The signal $x(t)$ is observed in noise,
$$y(t) = x(t) + n(t),$$
where $n(t)$ is a random process. An example of a *detection problem* is the choice between the two

values of A (the signal amplitude), given the observation $y(t), t \in [0, T)$. This problem arises in the transmission of binary data over a noisy channel.

Estimation. The signal $y(t) = x(t)\cos(2\pi f_c t + \theta) + n(t)$ is measured at a receiver, where θ is an unknown phase. An example of an *estimation problem* is the determination of the phase, based upon observation of the signal over some interval of time. □

Detection theory involves making a choice over some countable (usually finite) set of options, while estimation involves making a choice over a continuum of options.

10.1.1 Game theory and decision theory

Taking a broader perspective, the component of statistical theory that we are concerned with fits in an even larger mathematical construct, that of game theory. Therefore, to establish these connections, to introduce some notation, and to provide a useful context for future development, we will begin our discussion of this topic with a brief detour into the general area of mathematical games.

In a two-person game, each "person" (either of whom may be Nature) has options open to them, and each attempts to make a choice that appears to help them achieve their goal (e.g., of "winning"). In a *zero-sum* game, one person's loss is another person's gain. More formally, we have the following.

Definition 10.1 A two-person, zero-sum mathematical game, which we will refer to from now on simply as a **game**, consists of three basic components:

1. A nonempty set, Θ_1, of possible actions available to Player 1.
2. A nonempty set, Θ_2, of possible actions available to Player 2.
3. A loss function, $L: \Theta_1 \times \Theta_2 \mapsto \mathbb{R}$, representing the loss incurred by Player 1 (which, under the zero-sum condition, corresponds to the gain obtained by Player 2).

Any such triple (Θ_1, Θ_2, L) defines a game.

The losses are expressed with respect to player 1; a negative loss is interpreted as a gain for player 2. □

Here is a simple example [85, page 2].

Example 10.1.2 (Odd or even) Two contestants simultaneously put up either one or two fingers. Player 1 wins if the sum of the digits showing is odd, and Player 2 wins if the sum of the digits showing is even. The winner in all cases receives in dollars the sum of the digits showing, this being paid to him by the loser.

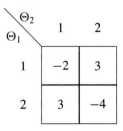

Figure 10.1: Loss function (or matrix) for "odd or even" game

10.1 Detection and Estimation Theory

To create a triple (Θ_1, Θ_2, L) for this game, we define $\Theta_1 = \Theta_2 = \{1, 2\}$ and define a loss function by

$$L(1, 1) = -2,$$
$$L(1, 2) = 3,$$
$$L(2, 1) = 3,$$
$$L(2, 2) = -4.$$

It is customary to arrange the loss function into a *loss matrix* as depicted in figure 10.1. □

An important class of games are those in which one player is able to obtain information relating to the choice made by their opponent, before committing to their own choice. To illustrate: suppose, with the "odd or even" game, that Player 2 is able to observe data regarding the action to be taken by Player 1, but that these data are subject to error. This modification is a significant complication of the original game, which must now be expanded to account for this additional structure. One way to incorporate this additional information is for Player 2 to model the observation in terms of probability theory.

The characterization of uncertain information in terms of probability theory provides a powerful addition to the basic game-theoretic structure provided by definition 10.1. This addition is of great value in the context in which we concentrate our attention—that of decision and estimation theory. We view decision and estimation theory as a two-person game between Nature, in the role of Player 1, and a decision-making or computational agent, in the role of Player 2. The "choices" available to nature are represented as elements of a set Θ. The decisions that the agent makes are represented as a element of set Δ. In addition, the agent has at its disposal samples of a random variable, or vector, X. As with the original two-person game, there is a loss function L.

A **statistical game** is a game represented by the triple (Θ, Δ, L), coupled with a random observable, X, defined over a **sample space**, or **observation space**, $\mathcal{X}$, whose distribution depends on the state $\theta \in \Theta$ chosen by nature. Assosiated with this random variable is a decision function, ϕ, that maps the observed value of X into the decision space.

1. $\Theta \subset \mathbb{R}^k$ is a nonempty set of possible states of nature, or parameter. Θ is sometimes referred to as the **parameter space**. An element of Θ is denoted θ (for a scalar parameter) or $\boldsymbol{\theta}$ (for a vector parameter).

2. Δ is a nonempty set of possible decisions available to the agent, sometimes called the **decision space**. An element of Δ is represented as δ.

3. $L: \Theta \times \Delta \mapsto \mathbb{R}$ is a **loss function** or **cost function**.

4. $X: \mathcal{X} \mapsto \mathbb{R}^n, n \geq 1$, is a random variable or vector whose cumulative distribution function is $F_X: \mathcal{X} \times \Theta \mapsto [0, 1]$. We represent this cumulative distribution function as $F_X(x \mid \theta)$. That is, the distribution of X is governed by the parameters $\theta \in \Theta$.

5. $\phi: \mathcal{X} \mapsto \Delta$ is a **decision rule**, alternatively termed a **strategy**, **decision function**, or **test**, that provides the coupling between the observations (and therefore the state of nature through $F_X(\cdot \mid \theta)$), and the decisions.

In the detection or estimation statistical game, nature chooses a point $\theta \in \Theta$, and an observation $X = x \in \mathcal{X}$ is generated at random according to the distribution $F_X(x \mid \theta)$. The agent, using the observation x but without other explicit knowledge of nature's choice, chooses an action $\phi(x) = \delta \in \Delta$. As a consequence of these choices, the agent experiences a loss $L(\theta, \delta)$. The elements of this structure are represented in figure 10.2.

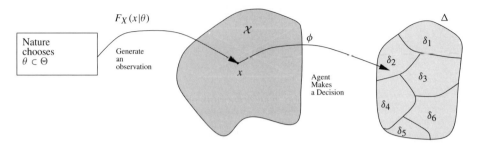

Figure 10.2: Elements of the statistical decision game

Θ \ D	ϕ_1	ϕ_2	ϕ_3	ϕ_4
0	0	λ_0	$1-\lambda_0$	1
1	1	λ_1	$1-\lambda_1$	0

Figure 10.3: A simple binary communications channel

Example 10.1.3 In this example we modify the concepts from the game in example 10.1.2 to apply to a communication channel. Consider the binary channel shown in figure 10.3. The bits zero or one can be chosen (where the transmitter takes the role of Player 1, or Nature). The parameter space is thus $\Theta = \{0, 1\}$. As the transmitted bits pass through the channel, they are corrupted. The receiver is to decide whether a 0 or a 1 was sent; thus the decision space is $\Delta = \{0, 1\}$. In a communication problem, a common cost structure is to impose a cost of 1 on incorrect decisions, and a cost of 0 on correct decisions. Thus

$$L(\theta, \delta) = \begin{cases} 0 & \text{if } \theta = \delta, \\ 1 & \text{if } \theta \neq \delta. \end{cases}$$
□

Some of the issues that arise in our exploration are described in the following list.

1. Determination of the decision rule, ϕ, by which the agent makes a decision: a common approach is to choose ϕ such that the average loss is as small as possible. This decision rule is fundamental to the detection or estimation problem, as it indicates, given an observation, which action (estimate, decision) should be made. We shall denote the space of all possible decision rules as D. Thus, the design problem is to select some $\phi \in D$ such that the goals of the agent are met.

2. Evaluation of the quality of the decision rule: for a detection problem, the quality might be measured, for example, in terms of probability of error, probability of conditional error, or cost of false alarm. For an estimation problem, the quality of the decision rule, and its resulting estimate, is examined in terms of the bias and variance of the estimate.

3. In some problems, the question of invariance: how may the detector or estimator be developed in such a way that it is insensitive (invariant) to transformations on the data.

10.1.2 Randomization

We introduced the decision rule, or strategy, ϕ, as a single function mapping observations into the decision space. Such a function is termed a **pure strategy**, or **nonrandomized decision rule**. We may generalize the notion of a decision rule, however, by specifying a *probability distribution*, φ, over the space of all possible nonrandomized rules. Such a decision rule is called a **mixed strategy**, or **randomized decision rule**. Let D denote the space of all nonrandomized decision rules, and let D^* denote the space of all randomized decision rules. Then $\varphi: D \to [0, 1]$ is a probability distribution that specifies the probability of selecting the elements of D.

If D contains countably many elements, $\{\phi_1, \phi_2, \ldots\}$, let $\varphi = (\pi_1, \pi_2 \ldots)$, $\pi_i \geq 0$, $i = 1, 2, \ldots, \sum_i \pi_i = 1$, with the understanding that we invoke decision rule ϕ_i with probability π_i. For example, suppose there are two distinct pure strategies, so $D = \{\phi_1, \phi_2\}$. Define

$$\varphi_\pi = (\pi, 1 - \pi), 0 \leq \pi \leq 1,$$

with $\pi = P(\phi_1) = 1 - P(\phi_2)$, where $P(\phi_i)$ is the probability that rule ϕ_i, $i = 1, 2$, is invoked. Clearly $D^* = \{\varphi_\pi, \pi \in [0, 1]\}$. Applying the randomized rule φ_π means that the nonrandomized rule ϕ_1 will be selected with probability π, and ϕ_2 will be selected with probability $1 - \pi$. A nonrandomized rule may be viewed as a degenerate randomized rule, such that all of the probability mass of the randomized rule is applied to a single pure strategy. For example, letting $\pi = 1$ means that ϕ_1 will be selected with probability one.

Randomized decision rules constitute an important mathematical concept, necessary for certain fundamental results, such as the Neyman–Pearson lemma (see section 11.2) and the minimax theorem (see section 11.10.5). Although the mathematical treatment of randomized rules is above reproach, the actual application of randomized rules is a topic worthy of considerable debate. For an interesting discussion of this concept see, for example, [208].

10.1.3 Special cases

The preceding framework provides a formalism for much of the statistical analysis we do in this text. However, only a part of statistics is represented by this formalism. We do not discuss such topics as the choice of experiments, the design of experiments, or sequential analysis. In each case, however, additional structure could be added to the basic framework to include these topics, and the problem could be reduced again to a simple game. Most of the body of statistical decision making involves three special cases of the general game formulation presented above.

1. Δ *consists of two points,* $\Delta = \{\delta_0, \delta_1\}$. Corresponding to each decision is a hypothesis. By choosing decision δ_0 the agent accepts hypothesis H_0 (thereby rejecting hypothesis H_1), and by choosing decision δ_1 the agents accepts hypothesis H_1 (thereby rejecting hypothesis H_0). In this case with only two decisions, the problem is called a *binary hypothesis testing problem*.

   As a specific example of a hypothesis testing problem, suppose that a radar signal is examined at a receiver to determine whether a target is present. Further suppose that the observed return is of the form

   $$X = \theta + N,$$

   where θ represents the reflected energy of a radar signal, and N is receiver noise. If θ is sufficiently small, then we conclude that there is no reflected signal (and hence no target). Although Θ may take on a continuum of values, these represent only two states of nature, which are summarized in the following two hypotheses:

   H_0: no target present, $\theta \leq \theta_0$,

   H_1: target present, $\theta > \theta_0$.

   In statistical parlance, H_0 is termed the **null** hypothesis, and H_1 the **alternative** hypothesis. The agent makes, on the basis of observing X, a decision (and its associated action) about what the state of nature is. With this simple problem, four outcomes are possible.

H_0 true, choose δ_0: Target not present, decide target not present: correct decision.

H_1 true, choose δ_1: Target present, decide target present: correct decision.

H_1 true, choose δ_0: Target present, decide target not present: **missed detection**. This type of error is also known as a **Type I error**.

H_0 true, choose δ_1: Target not present, decide target present: **false alarm**. This type of error is also known as a **Type II error**.

With this structure in place, the problem is to determine the decision function $\phi(x)$, which makes a selection out of Δ based on the observed value of X. In chapter 11, we present two ways of developing the decision function:

- The Neyman–Pearson test, in which the test is designed for maximum probability of detection for a fixed probability of false alarm.
- The Bayes test, in which an average cost is minimized. By appropriate selection of costs, this is equivalent to minimizing the probability of error, but other more general costs and decision structures can be developed.

In each case, the test can be expressed in terms of a *likelihood ratio* function.

2. Δ *consists of M points,* $\Delta = \{\delta_1, \delta_2, \ldots, \delta_M\}$, $M \geq 3$. These problems are called *multiple decision problems*, or *multiple hypothesis testing problems*.

    Multiple hypothesis testing problems arise in digital communications in which the signal constellations have more than 2 points; and in pattern recognition problems, in which one of M classes of data is to be distinguished.

3. Δ *consists of the real line,* $\Delta = \mathbb{R}$. Such decision problems are referred to as *point estimation of a real parameter*. Point estimation problems appear in a variety of contexts: target bearing estimation, frequency estimation, model parameter estimation, state estimation, phase estimation, and symbol timing, to name but a few.

    Consider the case where $\Theta = \mathbb{R}$ and the loss function is given by

$$L(\theta, \delta) = c(\theta - \delta)^2,$$

where c is some positive constant. A decision function, d, is in this case a real-valued function defined on the sample space, and is often called an *estimate* of the true unknown state of nature, θ. It is the agent's desire to choose the function d to minimize average loss.

10.2 Some notational conventions

Having briefly introduced the focus of part III, we must now pause in our development to ensure that we have the necessary tools to understand and apply these concepts.

We use notation of the form $F_X(x \mid \theta)$ to indicate a probability distribution for the random variable X. We also commonly refer to the probability density function (pdf) $f_X(x \mid \theta)$ or the probability mass function (pmf) $f_X(x \mid \theta)$. (The same notation is employed for both pmf and pdf, with the context of the problem determining what is intended.)

The symbol θ may be regarded as a parameter, or it may be regarded as a random variable. In the former case, the notation $f_X(x \mid \theta)$ simply represents θ as a parameter, even though it appears as if it were a conditioning variable. In the case where θ is regarded as a random variable—for Bayes detection or estimation—a perhaps more appropriate notation is $f_{X \mid \theta}(x \mid \vartheta)$, where the conditioning is demonstrated explicitly by including θ

10.2 Some Notational Conventions

(here denoting a random variable) in the subscript, with ϑ denoting the value assumed by θ. While we have attempted to be consistent in our usage, the notational awkwardness of $f_{X|\theta}(x \mid \vartheta)$ discourages its use in many sources.

We now introduce some other notation that is commonly employed in the literature. When referring to a distribution, pdf, or pmf, it is common to suppress the subscript indicating the random variable. For example, $F_X(x \mid \theta)$ may be represented simply using $F(x \mid \theta)$. The argument x may even be replaced by another variable, so that only the problem context provides an indication of the random variable intended.

Another commonly used notation is to denote the distribution function $F_X(x \mid \theta)$ by the abbreviated form $F_\theta(x)$. (This can be downright confusing because, by common convention, the subscript is used to indicate the random variable.) For discrete random variables, the pmf denoted by $f_X(x \mid \theta)$ or $f_\theta(x)$ and, similarly, for continuous random variables, the pdf is also denoted by $f_X(x \mid \theta)$ or $f_\theta(x)$.

We are also required to take the mathematical expectation of various random variables. As usual, we let $E(\cdot)$ denote the expectation operator (with or without parentheses, depending upon the chances of confusion). When we write EX it is understood that this expectation is performed using the distribution function of X, but when the distribution function for X is parameterized by θ, we must augment this notation by writing $E_\theta X$:

$$E_\theta[X] = \int x f_{X|\theta}(x \mid \theta)\,dx = \int x f_\theta(x)\,dx.$$

We also employ the notation $P_\theta[s]$ to denote the probability of the event s under the condition that θ is the true parameter.

We (mostly) use a bold capital font to represent random vectors. For example,

$$\mathbf{X} = \begin{bmatrix} X_1 \\ X_2 \\ \vdots \\ X_n \end{bmatrix}.$$

The pmf or pdf $f_{X_1, X_2, \ldots, X_n}(x_1, x_2, \ldots, x_n)$ is represented interchangeably with $f_\mathbf{X}(\mathbf{x})$. The same holds true for conditional pmfs or pdfs.

10.2.1 Populations and statistics

As we have described, the problem of estimation is, essentially, to obtain a set of data, or observations, and use this information to fashion a guess for the value of an unknown parameter (the parameter may be a vector). One of the ways to achieve this goal is through the method of random sampling.

Let X be a random variable known as the **population random variable**. The **distribution of the population** is the distribution of X. The population is discrete or continuous accordingly as X is discrete or continuous.

By "sampling" we mean that we repeat a given experiment a number of times; the ith repetition involves the creation, mathematically, of a replica, or copy, of the population on which a random variable X_i is defined. The distribution of the random variable X_i is the same as the distribution of X, the population random variable. (It is possible that sampling will change the distribution of the population.) The random variables $X_1, X_2, \ldots$, are called **sample random variables** or, sometimes, the **sample values** of X.

The act of sampling can take many forms. Perhaps the simplest sampling procedure is that of *sampling with replacement*, where the distribution of the population is unchanged by the sampling.

In decision making, we frequently take a collection of observations, and compute some function of it.

Definition 10.2 A function of the sample values of a random variable X is called a **statistic** of X. □

Example 10.2.1 Let X be a random variable with unknown mean value. Suppose we have a collection of *independent* samples of X, which we will denote by $X_1, \ldots, X_n$. The sample mean, written as the random variable $\overline{X}$, is given by

$$\overline{X} = \frac{1}{n} \sum_{i=1}^{n} X_i.$$

This is a function of the random variables, and is thus an example of a statistic. □

Before continuing with this discussion, it is important to make a distinction between random variables and the values they may take. Once the observations have been taken, the sample values become evaluated at the points $X_i = x_i$, and the array $(x_1, \ldots, x_n)$ is a collection of *real numbers*, not random variables. After the observations, therefore, the sample mean may be evaluated as

$$\overline{x} = \frac{1}{n} \sum_{i=1}^{n} x_i.$$

The real number $\overline{x}$ is *not* a random variable, nor are the quantities $x_1, \ldots x_n$. When we talk about quantities such as the mean or variance, they are associated with random variables, and not the values these assume. We can certainly talk about the average of the numbers $x_1, \ldots, x_n$, but this average is not the mathematical expectation of the random variable X. The only way we can think of $\overline{x}$ as a random variable is in a degenerate sense, where all of the mass is located at the number $\overline{x}$. Outside this context, it is meaningless to speak of the mean or variance of $\overline{x}$, but it *is* highly relevant to speak of the mean and variance of the random variable $\overline{X}$.

10.3 Conditional expectation

As we shall see, conditional expectation forms one of the key mathematical concepts in estimation theory, allowing us to form estimates based upon some given, *conditioning* information. It is therefore important to have firm understanding of conditional expectation and to provide a notation for it. In this section we point out the main properties of conditional expectation. (Note: a thorough explanation of conditional expectation requires measure theory, which is beyond the scope of this text. Interested readers are referred to [30].)

For our purposes, we will use the following definition of conditional expectation, which assumes the existence of $f_{X|Y}(x \mid y)$.

Definition 10.3 (Continuous distributions) Let X and Y be random variables with conditional pdf $f_{X|Y}(x \mid y)$. Then the **conditional expectation** of X given Y is

$$E(X \mid Y) = \int x f_{X|Y}(x \mid y)\, dx. \tag{10.1}$$

(Discrete distributions) Let X and Y be random variables with conditional pmf $f_{X|Y}(x \mid Y)$. Then the conditional expectation of X given Y is

$$E(X \mid Y) = \sum x f_{X|Y}(x \mid y).$$

□

The properties of conditional expectation that will be of use to us are given in the following theorem.

Theorem 10.1 *(Properties of conditional expectations)*

1. $E(X \mid Y) = EX$ if X and Y are independent.
2. $E(X \mid Y)$ is a function of Y: $E(X \mid Y) = h(Y)$.
3. $EX = E[E(X \mid Y)]$.
4. $E[g(Y)X \mid Y] = g(Y)E(X \mid Y)$, where $g(\cdot)$ is a function.
5. $E(c \mid Y) = c$ for any constant c.
6. $E[g(Y) \mid Y] = g(Y)$.
7. $E[(cX + dZ) \mid Y] = cE(X \mid Y) + dE(Z \mid Y)$ for any constants c and d.

Proof The proof is given for continuous distributions; similar methods apply for discrete distributions.

1. If X and Y are independent, then $f_{X \mid Y}(x \mid y) = f_X(x)$, so
$$E[X \mid Y] = \int x f_X(x) \, dx = EX.$$

2. By the definition,
$$E[X \mid Y = y] = \int f_{X \mid Y}(x \mid Y = y) \, dx,$$
which is a function of y.

3. $E(X \mid Y)$ is a function of Y, so in $E[E(X \mid Y)]$, the outer expectation is with respect to Y:
$$EE(X \mid Y) = \int \left(\int x f_{X \mid Y}(x \mid Y = y) \, dx \right) f_Y(y) \, dy$$
$$= \int x \left(\int f_{X \mid Y}(x \mid y) f_Y(y) \, dy \right) dx = \int x f_X(x) \, dx$$
$$= E(X).$$

4. $E[g(y)X \mid Y = y] = \int g(y) f_{X \mid Y}(x \mid Y = y) \, dx = g(y)E[X \mid Y = y]$, since $g(y)$ is a constant in the integral.

The last three properties follow similarly. □

10.4 Transformations of random variables

We review here transformations of random variables, which subsequently will be useful to us.

Theorem 10.2 *Let X and Y be continuous random variables with $Y = g(X)$. Suppose g is one-to-one, and both g and its inverse function, g^{-1}, are continuously differentiable. Then*

$$f_Y(y) = f_X[g^{-1}(y)] \left| \frac{dg^{-1}(y)}{dy} \right|. \tag{10.2}$$

Proof Since g is one-to-one, it is either increasing or decreasing; suppose it is increasing. Let a and b be real numbers such that $a < b$; we have

$$P[Y \in (a, b)] = P[g(X) \in (a, b)] = P[X \in (g^{-1}(a), g^{-1}(b))].$$

But
$$P[Y \in (a, b)] = \int_a^b f_Y(y)\,dy$$
and
$$P[X \in (g^{-1}(a), g^{-1}(b))] = \int_{g^{-1}(a)}^{g^{-1}(b)} f_X(x)\,dx$$
$$= \int_a^b f_X[g^{-1}(y)] \left|\frac{dg^{-1}(y)}{dy}\right| dy.$$

Thus, for all intervals (a, b), we have

$$\int_a^b \left[f_Y(y)\,dy - f_X[g^{-1}(y)] \left|\frac{dg^{-1}(y)}{dy}\right| \right] dy = 0. \tag{10.3}$$

Suppose that (10.2) is not true, so that there exists some y^* such that equality does not hold; but then, by the continuity of the density functions f_X and f_Y, (10.3) must be nonzero for some open interval containing y^*. This yields a contradiction, so (10.2) is true if g is increasing. To show that it holds for decreasing g, we simply note that the change of variable will also reverse the limits as well as the sign of the slope. Thus, the absolute value will be required. □

Example 10.4.1 Suppose that a random variable X has the density

$$f_X(x) = \frac{1}{2\sigma^2} e^{-x/2\sigma^2}$$

(X is a χ_2^2 random variable). Let
$$R = \sqrt{X};$$
that is, $R = g(X)$, where $g(X) = \sqrt{X}$, so $g^{-1}(R) = X^2$. Then

$$f_R(r) = f_X(g^{-1}(r)) \frac{\partial g^{-1}(r)}{\partial r}$$
$$= \left(\frac{1}{2\sigma^2} e^{-(r^2)/2\sigma^2} \right) (2r)$$
$$= \frac{r}{\sigma} e^{-r^2/2\sigma^2}.$$

The random variable R is said to be a *Rayleigh* random variable. □

Theorem 10.3 *Let* $\mathbf{X}$ *and* $\mathbf{Y}$ *be continuous n-dimensional random vectors with* $\mathbf{Y} = \mathbf{g}(\mathbf{X})$. *Suppose* $\mathbf{g} \colon \mathbb{R}^n \to \mathbb{R}^n$ *is one-to-one, and both* $\mathbf{g}$ *and its inverse function,* $\mathbf{g}^{-1}$, *are continuously differentiable. Then*

$$f_Y(\mathbf{y}) = f_X(\mathbf{g}^{-1}(\mathbf{y})) \left| \frac{\partial \mathbf{g}^{-1}(\mathbf{y})}{\partial \mathbf{y}} \right|, \tag{10.4}$$

where $\left|\frac{\partial \mathbf{g}^{-1}(\mathbf{y})}{\partial \mathbf{y}}\right|$ is the absolute value of the Jacobian determinant.

The proof of this theorem is similar to the proof for the univariate case, and we do not repeat it here.

10.5 Sufficient statistics

The question of sufficiency addresses an important issue in decision theory: much information must be retained from sample data in order to make valid decisions (e.g., in detection or estimation problems). The notion of sufficiency arises frequently in the work that follows.

10.5 Sufficient Statistics

Suppose that we have a collection of samples $\{X_1, X_2, \ldots, X_n\}$, to be used for parameter estimation or other decision-making purposes. (For example, the statistic

$$\overline{X} = \frac{1}{n} \sum_{i=1}^{n} X_i$$

is one of many possible statistics (or functions) to be obtained from the samples $X_1, \ldots, X_n$.) Suppose our objective in collecting the observations is to estimate the mean value of the random variable X. Let us ask ourselves, "What is the best estimate of the mean value of X that we can make on the basis of the sample values alone?" This question is not yet mathematically meaningful, since the notion of "best" has not been defined. Yet, with the preceding example, there is a strong compulsion to suppose that the random variable, $\overline{X}$, captures everything that there is to learn about the expectation of X from the random variables $X_1, \ldots, X_n$. As we show, the random variable $\overline{X}$ contains some special properties that qualify it as a *sufficient statistic* for the mean of the random variable X.

Definition 10.4 Let X be a random variable whose distribution depends on a parameter θ. A real-valued function T of X is said to be **sufficient** for θ if the conditional distribution of X, given $T = t$, is independent of θ. That is, T is sufficient for θ if

$$F_{X|T}(x \mid t, \theta) = F_{X|T}(x \mid t). \qquad \square$$

This definition remains unchanged if X, θ, and T are vector-valued, rather than scalar-valued.

Example 10.5.1 A coin with unknown probability of heads p, $0 \leq p \leq 1$, is tossed independently n times. Let X_i be zero if the outcome of the ith toss is tails and one if the outcome is heads. The random variables $X_1, \ldots, X_n$ are independent and identically distributed (i.i.d.), with common probability mass function

$$f_X(x_i \mid p) = P(X_i = x_i \mid p) = p^{x_i}(1-p)^{1-x_i} = \begin{cases} p & x = 1 \\ 1-p & x = 0 \end{cases} \quad \text{for } x_i = 0, 1. \qquad (10.5)$$

The random variable X known as a *Bernoulli* random variable. We will indicate this as $X \sim \mathcal{B}(p)$.

If we are looking at the outcome of this sequence of tosses in order to make a guess of the value of p, it is clear that the important thing to consider is the total number of heads and tails. It is hard to see how the information concerning the order of heads and tails can help us once we know the total number of heads. In fact, if we let T denote the total number of heads, $T = \sum_{i=1}^{n} X_i$, then intuitively the conditional distribution of $X_1, \ldots, X_n$, given $T = j$, is uniform over the $\binom{n}{j}$ n-tuples that have j ones and $n - j$ zeros; that is, given that $T = j$, the distribution of $X_1, \ldots, X_n$ may be obtained by choosing completely at random the j places in which ones go and putting zeros in the other locations. *This may be done without knowing p.* Thus, once we know the total number of heads, being given the rest of the information about $X_1, \ldots, X_n$ is like being told the value of a random variable whose distribution does not depend on p at all. In other words, the total number of heads carries all the information the sample has to give about the unknown parameter p. We claim that the total number of heads is a sufficient statistic for p.

To prove this fact, we need to show that the conditional distribution of $\{X_1, \ldots, X_n\}$, given $T = t$, is independent of p. This conditional distribution is

$$f_{X_1,\ldots X_n|T}(x_1, \ldots, x_n \mid t, p) = \frac{P(X_1 = x_1, \ldots, X_n = x_n, T = t \mid p)}{P(T = t \mid p)}. \qquad (10.6)$$

The denominator of this expression is the binomial probability

$$P(T = t \mid p) = \binom{n}{t} p^t (1-p)^{n-t}. \qquad (10.7)$$

We now examine the numerator. Since t represents the sum of the values X_i takes, we must set the probability that $X_1 + \cdots + X_n \neq t$ to zero, otherwise we will have an inconsistent probability. Thus, the numerator is zero except when $x_1 + \cdots + x_n = t$, and each $x_i = 0$ or 1, and then

$$\begin{aligned} P(X_1 = x_1, \ldots, X_n = x_n, T = t \mid p) &= P(X_1 = x_1, \ldots, X_n = x_n \mid p) \\ &= p^{x_1}(1-p)^{1-x_1} \cdots p^{x_n}(1-p)^{1-x_n} \\ &= p^{\sum x_i}(1-p)^{n - \sum x_i}. \end{aligned} \quad (10.8)$$

But $t = \sum x_i$; thus, substituting (10.7) and (10.8) into (10.6), we obtain

$$f_{X_1,\ldots,X_n \mid T}(x_1, \ldots, x_n \mid t, p) = \binom{n}{t}^{-1},$$

where $t = \sum x_i$ and each $x_i = 0$ or 1. This distribution is independent of p for all $t = 0, 1, \ldots, n$, which proves the sufficiency of T. □

The results of this example are likely not surprising; it makes intuitive sense without requiring a rigorous mathematical proof. We do learn from this example, however, that the notion of sufficiency is central to the study of statistics. But it would be tedious to establish sufficiency by essentially proving a new theorem for every application. Fortunately, we won't have to do so. The factorization theorem gives us a convenient mechanism for testing the sufficiency of a statistic. We state and prove this theorem for the discrete variables, and sketch a proof for continuous random variables as well.

Theorem 10.4 (Factorization theorem) *Let $\mathbf{X} = [X_1, X_2, \ldots, X_n]^T$ be a discrete random vector whose probability mass function $f_\mathbf{X}(\mathbf{x} \mid \theta)$ depends on a parameter $\theta \in \Theta$. The statistic $\mathbf{T} = \mathbf{t}(\mathbf{x})$ is sufficient for θ if, and only if, the probability mass function factors into a product of a function of $\mathbf{t}(\mathbf{x})$ and θ and a function of $\mathbf{x}$ alone; that is,*

$$f_\mathbf{X}(\mathbf{x} \mid \theta) = b(\mathbf{t}(\mathbf{x}), \theta) a(\mathbf{x}). \quad (10.9)$$

Proof (Discrete variables) Suppose $\mathbf{T} = \mathbf{t}(\mathbf{X})$, and note that by consistency we must have

$$f_{\mathbf{X},\mathbf{T}}(\mathbf{x}, \mathbf{t}(\mathbf{x}) \mid \theta) = \begin{cases} f_\mathbf{X}(\mathbf{x} \mid \theta) & T = \mathbf{t}(\mathbf{x}), \\ 0 & \text{otherwise.} \end{cases}$$

Assume that $\mathbf{T}$ is sufficient for θ. Then the conditional distribution of $\mathbf{X}$ given $\mathbf{T}$ is independent of θ, and we may write

$$\begin{aligned} f_\mathbf{X}(\mathbf{x} \mid \theta) &= f_{\mathbf{X},\mathbf{T}}(\mathbf{x}, \mathbf{t}(\mathbf{x}) \mid \theta) \\ &= f_{\mathbf{X} \mid \mathbf{T}}(\mathbf{x} \mid \mathbf{t}(\mathbf{x}), \theta) f_\mathbf{T}(\mathbf{t}(\mathbf{x}) \mid \theta) \\ &= f_{\mathbf{X} \mid \mathbf{T}}(\mathbf{x} \mid \mathbf{t}(\mathbf{x})) f_\mathbf{T}(\mathbf{t}(\mathbf{x}) \mid \theta), \end{aligned}$$

provided the conditional probability is well defined. Hence, we define $a(\mathbf{x})$ and $b(\mathbf{t}(\mathbf{x}), \theta)$ by

$$a(\mathbf{x}) = f_{\mathbf{X} \mid \mathbf{T}}(\mathbf{x} \mid \mathbf{t}(\mathbf{x})),$$
$$b(\mathbf{t}(\mathbf{x}), \theta) = f_\mathbf{T}(\mathbf{t}(\mathbf{x}) \mid \theta),$$

(there are other possible assignments) and the factorization is established.

Conversely, suppose $f_\mathbf{X}(\mathbf{x} \mid \theta) = b(\mathbf{t}(\mathbf{x}), \theta) a(\mathbf{x})$. Let $\mathbf{t}_0$ be chosen such that $f_\mathbf{T}(\mathbf{t}_0 \mid \theta) > 0$ for some $\theta \in \Theta$. Then

$$f_{\mathbf{X} \mid \mathbf{T}}(\mathbf{x} \mid \mathbf{t}_0, \theta) = \frac{f_{\mathbf{X},\mathbf{T}}(\mathbf{x}, \mathbf{t}_0 \mid \theta)}{f_\mathbf{T}(\mathbf{t}_0 \mid \theta)}. \quad (10.10)$$

The numerator is zero for all θ whenever $\mathbf{t}(\mathbf{x}) \neq \mathbf{t}_0$, and when $\mathbf{t}(\mathbf{x}) = \mathbf{t}_0$, the numerator is simply $f_\mathbf{X}(\mathbf{x} \mid \theta)$, by our previous argument. The denominator may be written using the

10.5 Sufficient Statistics

factorization as

$$f_T(t_0 \mid \theta) = \sum_{x: t(x)=t_0} f_X(x \mid \theta)$$
$$= b(t_0, \theta) \sum_{x: t(x)=t_0} a(x). \quad (10.11)$$

Substituting (10.9) and (10.11) into (10.10), we obtain

$$f_{X \mid T}(x \mid t_0, \theta) = \begin{cases} 0 & \text{if } t(x) \neq t_0, \\ \dfrac{a(x)}{\sum_{x': t(x')=t_0} a(x')} & \text{if } t(x) = t_0. \end{cases}$$

Thus, $f_{X \mid T}(x \mid t_0)$ is independent of θ, for all t_0 and θ for which it is defined. $\square$

The proof of the factorization with continuous random variables relies upon transformations of random variables.

Proof (Continuous random variables) Let X be an n-dimensional random vector. Denote the dimensionality of $T = t(x)$ by r. In general, $r < n$, so there is no invertible mapping from T to X. We therefore adjoin to $t(X)$ the auxiliary statistic $u(X)$, so that $w(X) = [t(X), u(X)]$ is continuous, one-to-one, of dimension n, and the inverse mapping w^{-1} is continuous. (The existence of this mapping depends upon the inverse function theorem.*) Let $Y = w(X) = [T, U]$. Then by theorem 10.3,

$$f_Y(y \mid \theta) = f_X(w^{-1}(y) \mid \theta) \left| \frac{\partial w^{-1}(y)}{\partial y} \right|.$$

If we can write $f_X(x \mid \theta) = b(t(x), \theta) a(x)$, then

$$f_Y(y \mid \theta) = b(t(w^{-1}(y)), \theta) a(w^{-1}(y)) \left| \frac{\partial w^{-1}(y)}{\partial y} \right|,$$

so that f_Y also factors. The density for T is obtained by integrating the density for $Y = [T, U]$ over U:

$$f_T(t \mid \theta) = \int f_Y(t, u, \mid \theta) \, du = b(t, \theta) \left(\int a(w^{-1}(t, u)) \left| \frac{\partial w^{-1}(y)}{\partial y} \right| \right) du. \quad (10.12)$$

The conditional density $f_{X \mid T}(x \mid t, \theta)$ is therefore obtained by

$$f_{X \mid T}(x \mid t, \theta) = \frac{f_{X,T}(x, t \mid \theta)}{f_T(t \mid \theta)} = \frac{f_X(x \mid \theta)}{f_T(t \mid \theta)}$$
$$= \frac{a(x)}{\int a(w^{-1}(t, u)) \left| \frac{\partial w^{-1}(y)}{\partial y} \right| du},$$

which is independent of θ.

Conversely, we observe that

$$f_X(x \mid \theta) = f_{X \mid T}(x \mid t, \theta) f_T(t \mid \theta).$$
$\square$

---

*The inverse function theorem says that if $G(X): \mathbb{R}^n \to \mathbb{R}^n$ is continuous at x_0 and the Jacobian of G is invertible at x_0, then there is a neighborhood U and x_0 in which an inverse G^{-1} exists.

From (10.11) and (10.12), we observe that the distribution of $f_\mathbf{T}(\mathbf{t}\,|\,\theta)$ is proportional to $b(\mathbf{t}(\mathbf{x}), \theta)$, where the constant of proportionality depends upon $a(\mathbf{x})$—that is, upon $\mathbf{x}$ and $\mathbf{t}$—but not on θ.

10.5.1 Examples of sufficient statistics

Example 10.5.2 (Bernoulli random variables) Let $\mathbf{X} = [X_1, X_2, \ldots, X_n]^T$ be a random vector, where the X_i are independent Bernoulli random variables. Then from (10.5), with probability parameter $\theta = p$,

$$f_\mathbf{X}(\mathbf{x}\,|\,\theta) = \prod_{i=1}^{n} p^{x_i}(1-p)^{1-x_i} = p^t(1-p)^{n-t},$$

where $t = \sum_{i=1}^{n} x_i$. Identifying $a(\mathbf{x}) = 1$ and

$$b(t, \theta) = p^t(1-p)^{n-t},$$

we note that t is sufficient for p.

The random variable $T = \sum_{i=1}^{n} X_i$ has pmf

$$f_T(t) = c p^t (1-p)^{n-t},$$

where c is chosen to make $f_T(t)$ sum to 1.

The distribution of the sufficient statistic is found using (10.11), as

$$f_T(t\,|\,\theta) = b(t, \theta) \sum_{\mathbf{x}:\sum x_i = t} 1 = p^t(1-p)^{n-t} \binom{n}{t}.$$

$\square$

Example 10.5.3 Let $\mathbf{X} = [X_1, \ldots, X_n]^T$, where each X_i is from $\mathcal{N}(\mu, \sigma^2)$. The joint density of $\mathbf{X}$ is

$$f_\mathbf{X}(\mathbf{x}\,|\,\mu, \sigma) = (2\pi\sigma^2)^{-\frac{n}{2}} \exp\left[-(2\sigma^2)^{-1} \sum_{i=1}^{n}(x_i - \mu)^2\right]. \tag{10.13}$$

There are three different subsets of parameters than can be taken as unknown.

1. If σ^2 is known and μ is unknown, then $\theta = \mu$. From the factorization theorem, it is straightforward to show that

$$\overline{X} = t(\mathbf{X}) = \frac{1}{n} \sum_{i=1}^{n} X_i$$

is sufficient for μ and is Gaussian distributed:

$$\overline{X} \sim \mathcal{N}(\mu, \sigma^2/n).$$

2. If μ is known but σ^2 is unknown, then $\theta = \sigma^2$. Then it can be shown that

$$t(\mathbf{X}) = \sum_{i=1}^{n} (X_i - \mu)^2$$

is sufficient for σ^2.

3. If both μ and σ^2 are unknown, then $\boldsymbol{\theta} = (\mu, \sigma^2)$.
Let $\overline{x} = \frac{1}{n}\sum_{i=1}^{n} x_i$ and $s^2 = \frac{1}{n}\sum_{i=1}^{n}(x_i - \overline{x})^2$ (the normalization by $1/n$ is only for convenience). The density (10.13) may be written

$$f_\mathbf{X}(\mathbf{x}\,|\,\mu, \sigma) = (2\pi\sigma^2)^{-\frac{n}{2}} \exp[-ns^2/2\sigma^2] \cdot \exp[-n(\overline{x} - \mu)^2/2\sigma^2].$$

The pair $(\overline{X}, S^2)$ is a sufficient statistic for (μ, σ^2). (We adopt the notation that $\overline{X}$ and S^2 are the random variables corresponding to the realizations $\overline{x}$ and s^2.) $\square$

10.5 Sufficient Statistics

Example 10.5.4 Consider a sample $X_1, \ldots, X_n$ from the uniform distribution over the interval $[\alpha, \beta]$. The joint density of $X_1, X_2, \ldots, X_n$ is

$$f_\mathbf{X}(\mathbf{x} \mid \alpha, \beta) = (\beta - \alpha)^{-n} \prod_{i=1}^{n} I_{[\alpha,\beta]}(x_i),$$

where I_A is the indicator function:

$$I_A(x) = \begin{cases} 1 & \text{if } x \in A, \\ 0 & \text{if } x \notin A. \end{cases}$$

This joint density may be rewritten as

$$f_\mathbf{X}(\mathbf{x} \mid \alpha, \beta) = (\beta - \alpha)^{-n} I_{[\alpha,\infty)}(\min x_i) I_{(-\infty,\beta]}(\max x_i),$$

since we must have all of the $\{x_i\} \geq \alpha$. (Hence, the smallest of them must be $\geq \alpha$, and we must also have all of the $\{x_i\} \leq \beta$, hence the largest of them must be $\leq \beta$.)

We examine three cases. First, if α is known, then $\max X_i$ is a sufficient statistic for β; second, if β is known, then $\min X_i$ is a sufficient statistic for α; and if both α and β are unknown, then $(\min X_i, \max X_i)$ is a sufficient statistic for (α, β). □

10.5.2 Complete sufficient statistics

As we have seen, the concept of a sufficient statistic leads to economy in the design of algorithms to compute estimates, and may simplify the requirements for data acquisition and storage, since only the sufficient statistic needs to be retained for purposes of estimation. Clearly, not all sufficient statistics are created equal. As an extreme case, the mapping $\mathbf{T}_1 = \mathbf{t}_1(X_1, \ldots, X_n) = (X_1, \ldots, X_n)$, which retains all of the data, is always a sufficient statistic for the mean, but no reduction in information is obtained. At the other extreme, if the random variables X_i are i.i.d., then, as we have seen, a sufficient statistic for the mean is $T_2 = t_2(X_1, \ldots, X_n) = \overline{X}$, and it is hard to see how the data could be reduced further. What about the vector-valued statistic $\mathbf{T}_3 = \mathbf{t}_3(X_1, \ldots, X_n) = (\sum_{i=1}^{n-1} X_i, X_n)$? This statistic is also sufficient for the mean. Obviously, T_2 would require the least bandwidth to transmit, the least memory to store, and would be simplest to use, but all three are sufficient for the mean. In fact, it easy to see that $\mathbf{T}_3$ can be expressed as a function of $\mathbf{T}_1$ but not vice versa, and that T_2 can be expressed as a function of $\mathbf{T}_3$ (and, consequently, of $\mathbf{T}_1$). This leads to a useful definition.

Definition 10.5 A sufficient statistic for a parameter $\theta \in \Theta$ that is a function of all other sufficient statistics for θ is said to be a **minimal sufficient statistic**, or *necessary and sufficient statistic*, for θ. Such a sufficient statistic represents the smallest amount of information that is still sufficient for the parameter. □

There are a number of questions one might ask about minimal sufficient statistics: (a) Does one always exist: (b) If so, is it unique? (c) If it exists, how do I find it? Rather than try to answer these questions directly, we defer instead to a related concept, that of *completeness*, and use this to approach the question of minimality.

Definition 10.6 Let $\mathbf{T}$ be a sufficient statistic for a parameter $\theta \in \Theta$, and let $w(\mathbf{T})$ be any real-valued function of $\mathbf{T}$. $\mathbf{T}$ is said to be **complete** if

$$E_\theta(w(\mathbf{T})) = 0 \tag{10.14}$$

for all $\theta \in \Theta$ implies that

$$P_\theta[w(T) = 0] = 1 \quad \forall \theta \in \Theta.$$

That is, $w(T) = 0$ with probability 1 (i.e., except on a set of measure zero) for every possible value of θ. □

Example 10.5.5 Let $X_1, \ldots, X_n$ be a sample from the uniform distribution over the interval $[0, \theta]$, $\theta > 0$. Then $T = \max_j X_j$ is sufficient for θ. We may compute the density of T as follows. For any real number t, the event $[\max_i X_i \leq t]$ occurs if and only if $[X_i \leq t], \forall i = 1, \ldots, n$. Thus, using the independence of the X_i, we have

$$P_\theta[T \leq t] = P_\theta(x_1 \leq t, x_2 \leq t, \ldots, x_n \leq t] = \prod_{i=1}^{n} P_\theta[X_i \leq t] = \begin{cases} 0 & \text{if } t \leq 0, \\ \frac{t^n}{\theta^n} & \text{if } 0 \leq t \leq \theta, \\ 1 & \text{if } \theta < t. \end{cases}$$

Taking derivatives, we find that the density is

$$f_T(t \mid \theta) = n \frac{t^{n-1}}{\theta^n} I_{[0,\theta]}(t).$$

Now let w be a function of T. Then

$$E_\theta w(T) = n\theta^{-n} \int_0^\theta w(t) t^{n-1} dt.$$

If this is identically zero for all $\theta > 0$, we must have that $\int_0^\theta w(t)t^{n-1} dt = 0$ for all $\theta > 0$. This implies that $w(t) = 0$ for all $t > 0$, except for a set of measure zero. At all points of continuity, the fundamental theorem of calculus shows that $w(t)$ is zero. Hence, $P_\theta[w(T) = 0] = 1$ for all $\theta > 0$, so that T is a complete sufficient statistic. □

We present two of the most important properties of complete sufficient statistics. We precede these properties by an important definition.

Definition 10.7 Let **X** be a random variable whose sample values are used to estimate a parameter θ of the distribution of **X**. We will use the notation $\hat{\theta}(\mathbf{X})$ to indicate that $\hat{\theta}$ is a function of **X** which returns an estimate of θ. An estimate $\hat{\theta}(\mathbf{X})$ of a θ is said to be **unbiased** if, when θ is the true value of the parameter, the mean of the distribution of $\hat{\theta}(\mathbf{X})$ is θ; that is,

$$E_\theta \hat{\theta}(\mathbf{X}) = \theta \quad \forall \theta.$$ □

Theorem 10.5 (Lehmann–Scheffé). *Let T be a complete sufficient statistic for a parameter $\theta \in \Theta$, and let w be a function of T that produces an unbiased estimate of θ; then w is unique with probability 1.*

Proof Let w_1 and w_2 be two functions of T that produce unbiased estimates of θ. Thus,

$$E_\theta w_1(T) = E_\theta w_2(T) = \theta \quad \forall \theta \in \Theta.$$

But then

$$E_\theta[w_1(T) - w_2(T)] = 0 \quad \forall \theta \in \Theta.$$

We note, however, that $w_1(T) - w_2(T)$ is a function of T, so by the completeness of T, we must have $w_1(T) - w_2(T) = 0$ with probability 1 for all $\theta \in \Theta$. □

Now the notion of completeness will allow us to make a determination about minimality:

Theorem 10.6 *A complete sufficient statistic for a parameter $\theta \in \Theta$ is minimal.*

Proof The proof relies on the properties of conditional expectations from section 10.3. Let T be a complete sufficient statistic and let S be another sufficient statistic, and suppose that S is minimal. By property 3 of theorem 10.1, we know that $ET = E[E(T|S)]$. By property 2, we know that the conditional expectation $E(T|S)$ is a function of S. But, because S is minimal, we also know that S is a function of T. Thus, the random variable $T - E(T|S)$ is a function of T, and this function has zero expectation for all $\theta \in \Theta$. Therefore, since T is complete, it follows that $T = E(T|S)$ with probability 1. This makes T a function of S, and since S is minimal, T is therefore a function of all other sufficient statistics, and T is itself minimal. □

10.6 Exponential families

Complete sufficient statistics, with their desirable qualities, do not always exist. We have seen that for the family of normal distributions, the two-dimensional statistic $(\sum X_i, \sum X_i^2)$ (or, equivalently, the sample mean and the sample variance) is sufficient for (μ, σ^2), and it is at least intuitively obvious that this statistic is also minimal. This motivates us to look for properties of the distribution that would be conducive to completeness and, hence, to minimality.

The exponential family is a family of distributions with surprisingly broad coverage, for which it is straightforward to determine complete sufficient statistics. This family covers many of the familiar distributions, including Gaussian, Poisson, and binomial. In addition, it is straightforward to determine the distribution of the sufficient statistics for distributions in the exponential family.

Definition 10.8 A family of distributions with probability mass function or density $f_X(x \mid \theta)$ is said to be a k-parameter **exponential family** if $f_X(x \mid \theta)$ has the form

$$f_X(x \mid \theta) = c(\theta) a(x) \exp\left[\sum_{i=1}^{k} \pi_i(\theta) t_i(x)\right]. \qquad (10.15)$$

In this definition, θ may be either a scalar or vector of parameters. □

Because $f(x \mid \theta)$ is a probability mass function or density function of a distribution, the function $c(\theta)$ is determined by the functions $a(x)$, $\pi_i(\theta)$, and $t_i(x)$, by means of the formulas

$$c(\theta) = \frac{1}{\sum_x a(x) \exp\left[\sum_{i=1}^{k} \pi_i(\theta) t_i(x)\right]}$$

in the discrete case and

$$c(\theta) = \frac{1}{\int_x a(x) \exp\left[\sum_{i=1}^{k} \pi_i(\theta) t_i(x)\right] dx}$$

in the continuous case.

If $f(x \mid \theta)$ is in the exponential family, and $X_1, X_2, \ldots, X_n$ are independent samples of X, then the joint distribution $\mathbf{X} = (X_1, X_2, \ldots, X_n)$ is also in the exponential family:

$$f_\mathbf{X}(\mathbf{x} \mid \theta) = f_{X_1,\ldots,X_n}(x_1, \ldots, x_n)$$

$$= c^n(\theta) \prod_{j=1}^{n} a(x_j) \exp\left[\sum_{i=1}^{k} \pi_i(\theta) \sum_{j=1}^{n} t_i(x_j)\right]$$

$$= \tilde{c}(\theta) a(\mathbf{x}) \exp\left[\sum_{i=1}^{k} \pi_i(\theta) t_i(\mathbf{x})\right],$$

where

$$a(\mathbf{x}) \triangleq \prod_{j=1}^{n} a(x_j) \qquad t_i(\mathbf{x}) = \sum_{j=1}^{n} t_i(x_j).$$

It is straightforward to identity sufficient statistics for the exponential family. Now let $X_1, \ldots, X_n$ be a sample of size n from an exponential family of distributions with either mass or density function given by (10.15). Then the joint probability mass or density is

$$f_{\mathbf{X}}(\mathbf{x} \mid \theta) = f_{X_1, \ldots, X_n}(x_1, \ldots, x_n \mid \theta) \qquad (10.16)$$

$$= c^n(\theta) \left(\prod_{j=1}^{n} a(x_j) \right) \exp\left[\sum_{i=1}^{k} \pi_i(\theta) \sum_{j=1}^{n} t_i(x_j) \right]. \qquad (10.17)$$

Clearly, this can be factored as

$$f_{\mathbf{X}}(\mathbf{x} \mid \theta) = \left(\prod_{j=1}^{n} a(x_j) \right) \left(c^n(\theta) \exp\left[\sum_{i=1}^{k} \pi_i(\theta) \sum_{j=1}^{n} t_i(x_j) \right] \right),$$

$$= a(\mathbf{x}) b(\mathbf{t}(\mathbf{x}), \theta)$$

so that by the factorization theorem it is clear that

$$\mathbf{T} = [T_1, \ldots, T_k]^T = \left[\sum_{j=1}^{n} t_1(X_j), \ldots, \sum_{j=1}^{n} t_k(X_j) \right]^T$$

is a sufficient statistic. We will denote

$$t_i(\mathbf{X}) = \sum_{j=1}^{n} t_i(X_j).$$

The distribution function for the sufficient statistic is determined by the following theorem.

Theorem 10.7 *Let $X_1, \ldots, X_n$ be a sample from the exponential family (10.15), either continuous or discrete. (We assume, in the continuous case, that a density exists.) Then the distribution of the sufficient statistic $\mathbf{T} = [T_1, \ldots, T_k]^T$ has the form*

$$f_{\mathbf{T}}(\mathbf{t} \mid \theta) = c(\theta) a_0(\mathbf{t}) \exp\left[\sum_{i=1}^{k} \pi_i(\theta) t_i \right], \qquad (10.18)$$

where $\mathbf{t} = [t_1, \ldots, t_k]^T$.

Proof (Continuous case) From the proof of the factorization theorem (see (10.12)), we may write the marginal distribution of $\mathbf{T}$ as

$$f_{\mathbf{T}}(\mathbf{t} \mid \theta) = \int f_{\mathbf{Y}}(\mathbf{t}, \mathbf{u}, \mid \theta) \, d\mathbf{u} = b(\theta, \mathbf{t}) \left(\int a(\mathbf{w}^{-1}(\mathbf{t}, \mathbf{u})) \left| \frac{\partial \mathbf{w}^{-1}(\mathbf{t}, \mathbf{u})}{\partial (\mathbf{t}, \mathbf{u})} \right| \right) d\mathbf{u}.$$

Also, by the factorization theorem, we know that

$$b(\mathbf{t}(\mathbf{x}), \theta) = \frac{f_{\mathbf{X}}(\mathbf{x} \mid \theta)}{a(\mathbf{x})}$$

and, when f_X is exponential, we may write

$$b(\mathbf{t}(\mathbf{x}), \theta) = \frac{c(\theta) a(\mathbf{x}) \exp\left[\sum_{i=1}^{k} \pi_i(\theta) t_i(\mathbf{x}) \right]}{a(\mathbf{x})},$$

10.6 Exponential Families

so, substituting this into the marginal for **T**, we obtain

$$f_{\mathbf{T}}(\mathbf{t}\mid\theta) = c(\theta)\left[\int a[\mathbf{w}^{-1}(\mathbf{t},\mathbf{u})]\left|\frac{\partial\mathbf{w}^{-1}(\mathbf{t},\mathbf{u})}{\partial(\mathbf{t},\mathbf{u})}\right|d\mathbf{u}\right]\exp\left[\sum_{i=1}^{k}\pi_i(\theta)t_i\right],$$

which is of the desired form if we set

$$a_0(\mathbf{t}) = \int a[\mathbf{x}(\mathbf{t},\mathbf{u})]\left|\frac{\partial\mathbf{w}^{-1}(\mathbf{t},\mathbf{u})}{\partial(\mathbf{t},\mathbf{u})}\right|d\mathbf{u}. \qquad \square$$

We are now in a position to state a key result, which in large measure justifies our attention to exponential families of distributions.

Theorem 10.8 *For a k-parameter exponential family, the sufficient statistic*

$$\mathbf{T} = \left[\sum_{j=1}^{n}t_1(X_j),\ldots,\sum_{j=1}^{n}t_k(X_j)\right]^T$$

is complete, and therefore a minimal sufficient statistic.

Proof To establish completeness, we need to show that, for any function w of **T**, the condition $E_\theta w(\mathbf{T}) = 0$, $\forall \theta \in \Theta$ implies $P_\theta[w(\mathbf{T}) = 0] = 1$. But the expectation is

$$E_\theta w(\mathbf{T}) = \int w(\mathbf{t})c(\theta)a(\mathbf{t})\exp\left[\sum_{i=1}^{k}\pi_i(\theta)t_i\right]d\mathbf{t},$$

and we observe that this is the k-dimensional Laplace transform of $w(\mathbf{t})c(\theta)a(\mathbf{t})$. By the uniqueness of the Laplace transform, we must have $w(\mathbf{t}) = \mathbf{0}$ for almost all $\mathbf{t}$ (that is, all $\mathbf{t}$ except possibly on a set of measure zero). $\qquad \square$

Example 10.6.1 The pmf for the binomial distribution of the number of successes in m independent trials, when θ is the probability of success at each trial, is

$$f_X(x\mid\theta) = \binom{m}{x}\theta^x(1-\theta)^{m-x} = (1-\theta)^m\binom{m}{x}\exp\{x[\log\theta - \log(1-\theta)]\},$$

for $x = 0, 1, \ldots, m$—so this family of distributions is a one-parameter exponential family with

$$c(\theta) = (1-\theta)^m,$$
$$a(x) = \binom{m}{x}\qquad \pi_1(\theta) = \log\theta - \log(1-\theta),$$
$$t_1(x) = x.$$

Hence, for sample size n, $\sum_{j=1}^{n}X_j$ is sufficient for θ. $\qquad \square$

Example 10.6.2 The pmf for the Poisson distribution of the number of events that occur in a unit-time interval, when the events are occurring in a Poisson process at rate $\theta > 0$ per unit time, is

$$f_X(x) = \frac{\theta^x}{x!}e^{-\theta} = e^{-\theta}\frac{1}{x!}e^{(\log\theta)x},$$

for $x = 0, 1, \ldots$. This is a one-parameter exponential family with

$$c(\theta) = e^{-\theta},$$
$$a(x) = \frac{1}{x!},$$
$$\pi_1(\theta) = \log\theta,$$
$$t_1(x) = x.$$

Hence, the number of events that occur during the specified time interval is a sufficient statistic for θ. □

Example 10.6.3 The normal pdf is

$$f_X(x) = \frac{1}{\sqrt{2\pi}\sigma} \exp\left[\frac{-(x-\mu)^2}{2\sigma^2}\right] = \frac{1}{\sqrt{2\pi}\sigma} \exp\left[-\frac{\mu^2}{2\sigma^2}\right] \exp\left[\frac{-1}{2\sigma^2}x^2 + \frac{\mu}{\sigma^2}x\right].$$

This is a two-parameter exponential family with

$$c(\theta) = \frac{1}{\sqrt{2\pi}\sigma} \exp\left[-\frac{\mu^2}{2\sigma^2}\right],$$
$$a(x) = 1,$$
$$\pi_1(\mu, \sigma^2) = -\frac{1}{2\sigma^2},$$
$$\pi_2(\mu, \sigma^2) = \frac{\mu}{\sigma^2},$$
$$t_1(x) = x^2,$$
$$t_2(x) = x.$$

Hence, for sample size n, $(\sum_{i=1}^{n} X_i, \sum_{i=1}^{n} X_i^2)$ are sufficient for (μ, σ^2). □

Example 10.6.4 An important family of distributions that is *not* exponential is the family of uniform distributions. (We already have identified a complete sufficient statistic for that distribution.) □

10.7 Exercises

10.1.1-1 Consider the well-known game of Prisoner's Dilemma. Two agents, denoted X_1 and X_2, are accused of a crime. They are interrogated separately, but the sentences that are passed are based upon the joint outcome. If they both confess, they are both sentenced to a jail term of three years. If neither confesses, they are both sentenced to a jail term of one year. If one confesses and the other refuses to confess, then the one who confesses is set free and the one who refuses to confess is sentenced to a jail term of five years. This payoff matrix is illustrated in figure 10.4. The first entry in each quadrant of the payoff matrix corresponds to X_1's payoff, and the second entry corresponds to X_2's payoff. This particular game represents slight extension to our original definition, since it is not a zero-sum game.

When playing such a game, a reasonable strategy is for each agent to make a choice such that, once chosen, neither player would have an incentive to depart unilaterally from the outcome. Such a decision pair is called a *Nash equilibrium* point. In other words, at the Nash equilibrium point, both players can only hurt themselves by departing from their decision. What is the Nash equilibrium point for the Prisoner's Dilemma game? Explain why this problem is considered a "dilemma."

	X_2	
X_1	silent	confesses
silent	1,1	5,0
confesses	0,5	3,3

Figure 10.4: A typical payoff matrix for the Prisoner's Dilemma game

10.7 Exercises

10.5-2 Let $\mathbf{X} = (\mathbf{X}_1, \mathbf{X}_2, \ldots, \mathbf{X}_n)$ denote a random sample of an m-dimensional Gaussian random vector X_i, where $\mathbf{X}_i \sim \mathcal{N}(\boldsymbol{\mu}, R)$. Show that the statistics

$$\mathbf{m} = \frac{1}{n} \sum_{i=1}^{n} \mathbf{x}_i$$

and

$$S^2 = \sum_{i=1}^{n} (\mathbf{x}_i - \mathbf{m})(\mathbf{x}_i - \mathbf{m})^T$$

are sufficient for $(\boldsymbol{\mu}, R)$. The matrix S^2 is called the *scatter matrix* of the data. Hint:

$$\sum_i (\mathbf{x}_i - \boldsymbol{\mu})^T R^{-1} (\mathbf{x}_i - \boldsymbol{\mu}) = \text{tr}\left[R^{-1} \sum_i (\mathbf{x}_i - \boldsymbol{\mu})(\mathbf{x}_i - \boldsymbol{\mu})^T \right].$$

10.5-3 A **Poisson random variable** has pmf

$$f_X(x \mid \theta) = e^{-\theta} \frac{1}{x!} \theta^x,$$

where $\theta > 0$ is the parameter of the distribution. We write

$$X \sim \mathcal{P}(\theta).$$

The Poisson distribution models the distribution of the number of events that occur in the unit interval $(0, 1)$ when the events are occurring at an average rate of θ events per unit time.

Let $\mathbf{X} = [X_1, X_2, \ldots, X_n]^T$ be a random sample, where each X_i is Poisson distributed.

(a) Show that if $X \sim \mathcal{P}(\theta)$ then $EX = \theta$ and $\text{var}(X) = \theta$.

(b) Show that

$$K = \sum_{i=1}^{n} X_i$$

is sufficient for θ.

(c) Determine the distribution of K.

10.5-4 Let $\mathbf{X}$ be a random vector with density $f_\mathbf{X}(\mathbf{x} \mid \theta)$, and let

$$\mathbf{Y} = \mathbf{w}(\mathbf{X})$$

be an invertible transformation. Suppose that $\mathbf{s}(\mathbf{Y})$ is a sufficient statistic for θ in $f_\mathbf{Y}(\mathbf{y} \mid \theta)$. Show that

$$T(\mathbf{x}) = \mathbf{s}(\mathbf{w}(\mathbf{x}))$$

is a sufficient statistic for θ in $f_\mathbf{X}(\mathbf{x} \mid \theta)$.

10.5-5 The **binomial distribution** has pmf

$$f(x \mid p) = \binom{n}{x} p^x (1-p)^{n-x},$$

where n is a positive integer. The notation $X \sim \mathcal{B}(n, p)$ means that X has a binomial distribution with parameters n and p. The binomial $\mathcal{B}(n, p)$ represents the distribution of the total number of successes in n independent Bernoulli $\mathcal{B}(p)$ trials, where the probability of success in each trial is p.

(a) Show that the mean of $\mathcal{B}(n, p)$ is np and the variance is $np(1-p)$.

(b) Let $X_1, X_2, \ldots, X_n$ be n independent Bernoulli random variables with $P(X_i = 1) = p$.

Show that
$$X = X_1 + X_2 + \cdots + X_n$$
is $\mathcal{B}(n, p)$.

(c) If $X_1, X_2, \ldots, X_n$ are $\mathcal{B}(n_i, \theta)$, $i = 1, 2, \ldots, n$, show that:

i. $\sum_{i=1}^{n} X_i$ is sufficient for θ.

ii. $(\sum_{i=1}^{n} X_i) \sim \mathcal{B}(\sum_{i=1}^{n} n_i, \theta)$; that is, the distribution of the sufficient statistic is itself binomially distributed.

10.5-6 It is interesting to contemplate the use of sufficient statistics for data compression. Let $X_i, i = 1, 2, \ldots, n$ be Bernoulli random variables. Compare the number of bits required to represent the sufficient statistic
$$t = \sum_{i=1}^{n} x_i$$
with the number of bits required to code the sequence $(x_1, x_2, \ldots, x_n)$.

10.5-7 Let $\mathbf{X} \sim \mathcal{N}(H\theta, R)$, where H is $m \times p$ and θ is $p \times 1$. Show that if H and R are known, then $H^T R^{-1} \mathbf{x}$ is sufficient for θ. Determine the distribution of the random variable $H^T R^{-1} \mathbf{X}$.

10.5-8 [29] Let $X_1, X_2, \ldots, X_n$ be a sample from a population with density
$$f_X(x \mid \theta) = \begin{cases} \dfrac{1}{\sigma} \exp[-(x - \mu)/\sigma] & x \geq \mu, \\ 0 & \text{otherwise.} \end{cases}$$
The parameters are $\theta = (\mu, \sigma)$, where $\mu \in \mathbb{R}$ and $\sigma > 0$.

(a) Show that $\min(X_1, X_2, \ldots, X_n)$ is sufficient for μ when σ is known.

(b) Find a one-dimensional sufficient statistic for σ when μ is known.

(c) Find a two-dimensional sufficient statistic for θ.

10.6-9 Let T be a sufficient statistic that is distributed as $T \sim \mathcal{B}(2, \theta)$. Show that T is a complete sufficient statistic.

10.6-10 Show that each of the following statistics is not complete, by finding a nonzero function w such that $E[w(T)] = 0$.

(a) $T \sim \mathcal{U}(-\theta, \theta)$ (T is uniformly distributed from $-\theta$ to θ).

(b) $T \sim \mathcal{N}(0, \theta)$.

10.6-11 Let X_i have pmf $f_X(x \mid \theta) = \theta^x(1 - \theta)^{1-x}$, $x = 0, 1$, for $i = 1, 2, \ldots, n$. Show that $T = \sum_{i=1}^{n} X_i$ is a complete sufficient statistic for θ. Also, find a function of T that is an unbiased estimator of θ.

10.6-12 Express the following pdfs or pmfs as members of the exponential family and determine the sufficient statistics.

(a) Exponential (pdf): $f_X(x \mid \theta) = \theta e^{-\theta x}$, $x \geq 0$.

(b) Rayleigh (pdf): $f_X(x \mid \theta) = 2\theta e^{-\theta x^2}$, $x \geq 0$.

(c) Gamma (pdf): $f_X(x \mid \theta_1, \theta_2) = \dfrac{\theta_2^{\theta_1+1}}{\Gamma(\theta_1+1)} x^{\theta_1} e^{-\theta_2 x}$, $x \geq 0$.

(d) Poisson (pmf): $f_X(x \mid \theta) = (\theta^x/x!)e^{-\theta}$, $x = 0, 1, 2, \ldots$.

(e) Multinomial (pmf): $f_X(\mathbf{x} \mid \theta_1, \theta_2, \ldots, \theta_d) = (\prod_{i=1}^{d} \theta_i^{x_i}) m! / \prod_{i=1}^{d} x_i!$, $x_i = 0, 1, 2, \ldots$, and $\sum_{i=1}^{d} x_i = m$ and $\sum_{i=1}^{d} \theta_i = 1$ with $\theta_i > 0$.

(f) Geometric (pmf): $f_X(x \mid \theta) = (1 - \theta)^x \theta$.

10.8 References

10.6-13 Let $X_i \sim \mathcal{B}(p), i = 1, 2, \ldots, n$, and let $T = \sum_{i=1}^{n} X_i$. Show that T is a complete minimal statistic.

10.6-14 Let $X_1, \ldots, X_n$ be a sample from the exponential family (10.15), either continuous or discrete. Show that the distribution of the sufficient statistic $\mathbf{T} = [T_1, \ldots, T_k]^T$ has the form

$$f_\mathbf{T}(\mathbf{t} \mid \theta) = c(\theta) a_0(\mathbf{t}) \exp\left[\sum_{i=1}^{k} \pi_i(\theta) t_i\right], \qquad (10.19)$$

where $\mathbf{t} = [t_1, \ldots, t_k]^T$.

10.6-15 ([291]) Let $(\mathbf{X}_0, \mathbf{X}_1, \ldots, \mathbf{X}_{M-1})$ denote a random sample of N-dimensional random vectors $\mathbf{X}_n$, each of which has mean value $\mathbf{m}$ and covariance matrix R. Show that the sample mean

$$\hat{\mathbf{m}}_t = \frac{1}{t+1} \sum_{n=0}^{t} \mathbf{x}_n$$

and the sample covariance

$$S_t(\hat{\mathbf{m}}_t) = \frac{1}{t+1} \sum_{n=0}^{t} (\mathbf{x}_n - \hat{\mathbf{m}}_t)(\mathbf{x}_n - \hat{\mathbf{m}}_t)^T$$

may be written recursively as

$$\hat{\mathbf{m}}_t = \frac{t}{t+1} \hat{\mathbf{m}}_{t-1} + \frac{1}{t+1} \mathbf{x}_t, \qquad \hat{\mathbf{m}}_0 = \mathbf{x}_0,$$

and

$$S_t(\hat{\mathbf{m}}_t) = Q_t - \hat{\mathbf{m}}_t \hat{\mathbf{m}}_t^T,$$

where

$$Q_t = \frac{t}{t+1} Q_{t-1} + \frac{1}{t+1} \mathbf{x}_t \mathbf{x}_t^T.$$

10.8 References

The consideration of decision making in terms of games, and the special cases presented here, are promoted in [85]. A solid analytical coverage of measure theory and conditional expectation. For those interested in general game theory, [208] is a reasonable introduction. Another work on games, with connections to linear programming, is [171].

Both [29] and [141] provide a good background to the introductory material on transformations of variables, conditional expectations, exponential families, and sufficient statistics; material and insight has also been drawn from [291].

Chapter 11

Detection Theory

> I often say that when you can measure what you are speaking about and express it in numbers you know something about it; but when you cannot measure it, when you cannot express it in numbers, your knowledge of it is of a meagre and unsatisfactory kind.
>
> — *William Thompson, Lord Kelvin*

11.1 Introduction to hypothesis testing

In the detection problem, an observation of a random variable (or signal) x is used to make decisions about a finite number of outcomes. More specifically, in an M-ary hypothesis testing problem, it is assumed that the parameter space $\Theta = \Theta_0 \cup \Theta_1 \cup \cdots \cup \Theta_{M-1}$, where the Θ_i are mutually disjoint. Corresponding to each of these sets are choices—or hypotheses—denoted as

$$H_0: \theta \in \Theta_0,$$
$$H_1: \theta \in \Theta_1,$$
$$\vdots$$
$$H_{M-1}: \theta \in \Theta_{M-1}.$$

The parameter θ determines the distribution of a random variable x that takes values in a space $\mathcal{X}$ according to the distribution function $F_X(x \mid \theta)$. Based on the observation $x = x$, a decision is made by a decision-making agent. In the simplest case, the decision space is $\Delta = \{\delta_0, \delta_2, \ldots, \delta_{M-1}\}$, with one choice corresponding to each hypothesis, such that δ_i represents the decision to "accept hypothesis H_i^n (thereby rejecting the others). There are two major approaches to detection.

Bayesian approach. In the Bayesian approach, the emphasis is on *minimizing loss*. With the Bayesian approach, we assume that the parameters are actually random variables, governed by a prior probability. A loss function is established for each possible outcome and each possible decision, and decisions are made to minimize the average loss. The Bayesian approach can be applied well to the M-ary detection problem, for $M \geq 2$.

Neyman–Pearson approach. The Neyman–Pearson approach is used primarily for the binary detection problem. In this approach, the probability of false alarm is fixed at some value, and the decision function is found which maximizes the probability of detection. In each case, the tests are reduced to comparisons of ratios of probability density or probability mass, forming what is called a *likelihood ratio test*.

11.1 Introduction to Hypothesis Testing

The theory presented here has had great utility for detection using radar signals, and some of the terminology used in that context has permeated the general field. (Notions such as false alarm, missed detection, receiver operating characteristic, and so forth, owe their origins to radar.) Statistics has coined its own vocabulary for these concepts, however, and we will find it desirable to become familiar with both the engineering and statistics terminology. The fact that more than one discipline has embraced these concepts is a testimony to their great utility.

We will begin our investigation of hypothesis testing with the *binary* detection problem. Despite its apparent simplicity, there is a considerable body of theory associated with the problem. In the classical binary decision problem, as we have seen, H_0 is called the *null hypothesis* and H_1 is the *alternative hypothesis*. Only one of these disjoint hypotheses is true, and the job of the decision maker is to detect (guess) which hypothesis is true. The decision space is $\Delta = \{\delta_0 \text{ (accept } H_0), \delta_1 \text{ (reject } H_0)\}$.

Example 11.1.1 (Digital communications) One of two possible signals is sent. We can take the parameter space as $\Theta = \{1, -1\}$. The receiver decides between H_0: $\theta = 1$ and H_1: $\theta = -1$, based upon the observation of a random variable.

In a common signal model (additive Gaussian noise channel), the received signal is modeled as

$$R = S + N,$$

where S is the transmitted signal and N is a random variable. If $N \sim \mathcal{N}(0, \sigma^2)$ and $S = \theta a$, for some amplitude a, then the distribution for R, conditioned upon knowing the transmitted signal θ, is

$$f_R(r \mid \theta) = \frac{1}{\sqrt{2\pi}\sigma} e^{(r-a\theta)^2/2\sigma^2}$$

$\square$

Example 11.1.2 (Optical communications) Another signal model more appropriate for an optical channel is to assume that

$$R = X,$$

where X is a Poisson random variable whose rate depends upon θ:

$$f_X(x \mid \theta) = \begin{cases} \exp[-\lambda_1]\frac{\lambda_1^x}{x!} & \theta = 0, \\ \exp[-\lambda_2]\frac{\lambda_2^x}{x!} & \theta = 1. \end{cases}$$

This models, for example, the rate of received photons, where different photon intensities are used to represent the two possible values of θ. $\square$

Example 11.1.3 (Radar detection) Assume that a received signal is represented as

$$R = \theta + N,$$

where N is a random variable representing the noise, and θ is a random variable indicating the presence or absence of some target. The two hypotheses can now be described as

H_0: target is absent: $\theta \leq \theta_0$,

H_1: target is present: $\theta > \theta_0$.

$\square$

Based upon the observation x, we must make a decision regarding which hypothesis to accept. We divide the space $\mathcal{X}$ into two disjoint regions, $\mathcal{R}$ and $\mathcal{A}$, with $\mathcal{X} = \mathcal{R} \cup \mathcal{A}$. We formulate our decision function $\phi(x)$ as

$$\phi(x) = \begin{cases} 1 & \text{if } x \in \mathcal{R}, \\ 0 & \text{if } x \in \mathcal{A}. \end{cases} \tag{11.1}$$

We interpret this decision rule as follows: If $x \in \mathcal{R}$ (reject) we take action δ_1 (accepting H_1, rejecting H_0); and if $x \in \mathcal{A}$ (accept) we take action δ_0 (accepting H_0, rejecting H_1). The decision regions that are chosen depend upon the particular structure present in the problem.

11.2 Neyman–Pearson theory

In the Neyman–Pearson approach to detection, the focus is on the conditional probabilities. In particular, it is desired to maximize the probability of choosing H_1 when, in fact, H_1 is true, while not exceeding a fixed probability of choosing H_1 when it is not true. That is, we want to maximize the probability of detection, while not exceeding a standard for the probability of false alarm.

11.2.1 Simple binary hypothesis testing

We first look at the case where H_0 and H_1 are *simple*.

Definition 11.1 A test for $\theta \in \Theta_i$, $i = 0, 1, \ldots, k-1$ is said to be **simple** if each Θ_i consists of exactly one element. If any Θ_i has more than one point, a test is said to be **composite**. □

Example 11.2.1 Let $\Theta = \{0, 1\}$. The test

$$H_0: \theta = 0$$
$$H_1: \theta = 1$$

is a simple test.

Now let $\Theta = \mathbb{R}^+$. The test

$$H_0: \theta = 0$$
$$H_1: \theta > 0$$

is a composite test. □

For a binary hypothesis test, the decision space consists of two points, $\Delta = \{\delta_0, \delta_1\}$, corresponding to accepting H_0 and H_1. Then, if $\theta = \theta_0$ is the true value of the parameter, we prefer to take action δ_0, whereas if θ_1 is the true value we prefer δ_1.

Definition 11.2 The probability of rejecting the null hypothesis H_0 when it is true is called the **size** of the rule ϕ, and is denoted α. This is called a *type I error*, or *false alarm*. □

For the simple binary hypothesis test,

$$\alpha = P[\text{decide } H_1 \mid H_0 \text{ is true}] = P[\phi(X) = 1 \mid \theta_0]$$
$$= E_{\theta_0} \phi(X)$$
$$= P_{FA}.$$

The notation P_{FA} is standard for the probability of a false alarm. This latter terminology stems from radar applications, where a pulsed electromagnetic signal is transmitted. If a return signal is reflected from the target, we say a target is detected. But due to receiver noise, atmospheric disturbances, spurious reflections from the ground and other objects, and other signal distortions, it is not possible to determine with absolute certainty whether a target is present.

Definition 11.3 The **power**, or **detection probability**, of a decision rule ϕ is the probability of correctly accepting the alternative hypothesis, H_1, when it is true, and is denoted by β.

11.2 Neyman–Pearson Theory

The probability of accepting H_0 when H_1 is true is $1 - \beta$, resulting in a **type II error**, or **missed detection**. □

We thus have

$$\beta = P[\text{decide } H_1 \mid H_1 \text{ is true}] = P[\phi(X) = 1 \mid \theta_1]$$
$$= E_{\theta_1}\phi(X)$$
$$= P_D.$$

The notation P_D is standard for the probability of a detection, and

$$P_{MD} = 1 - P_D$$

is the probability of a missed detection.

Definition 11.4 A test ϕ is said to be **best of size** α for testing H_0 against H_1 if $E_{\theta_0}\phi(X) = \alpha$ and if, for every test ϕ' for which $E_{\theta_0}\phi'(X) \leq \alpha$, we have

$$\beta = E_{\theta_1}\phi(X) \geq E_{\theta_1}\phi'(X) = \beta';$$

that is, a test ϕ is best of size α if, out of all tests with P_{FA} not greater than α, ϕ has the largest probability of detection. □

11.2.2 The Neyman–Pearson lemma

We now give a general method for finding the best tests of a simple hypothesis against a simple alternative. The test will take the following form:

$$\phi(x) = \begin{cases} 0 & \text{condition 1} \\ \gamma & \text{condition 2} \\ 1 & \text{condition 3}. \end{cases}$$

where the three conditions are mutually exclusive. If condition 1 is satisfied, then the test chooses decision 0 (selects H_0). If condition 3 is satisfied, then the test chooses decision 1 (selects H_1). However, if condition 2 is satisfied and γ is chosen, what this means is that a random selection takes place. Decision 1 is chosen with probability γ, and decision 0 is chosen with probability $1 - \gamma$. The instantiation of condition 3 is an example of a randomized decision rule.

The best test of size α is provided by the following important lemma.

Lemma 11.1 *(Neyman–Pearson lemma) Suppose that $\Theta = \{\theta_0, \theta_1\}$ and that the distributions of X have densities (or mass functions) $f_X(x \mid \theta)$. Let $\nu > 0$ be a threshold.*

1. *Any test $\phi(X)$ of the form*

$$\phi(x) = \begin{cases} 1 & \text{if } f_X(x \mid \theta_1) > \nu f_X(x \mid \theta_0) \\ \gamma & \text{if } f_X(x \mid \theta_1) = \nu f_X(x \mid \theta_0) \\ 0 & \text{if } f_X(x \mid \theta_1) < \nu f_X(x \mid \theta_0) \end{cases} \quad (11.2)$$

for some $0 \leq \gamma \leq 1$, is best of its size for testing H_0: $\theta = \theta_0$ against H_1: $\theta = \theta_1$. Corresponding to $\nu = \infty$, the test

$$\phi(x) = \begin{cases} 1 & \text{if } f_X(x \mid \theta_0) = 0 \\ 0 & \text{if } f_X(x \mid \theta_0) > 0 \end{cases} \quad (11.3)$$

is best of size 0 for testing H_0 against H_1.

2. (Existence) *For every α, $0 \leq \alpha \leq 1$, there exists a test of the form above with γ a constant, for which $E_{\theta_0}\phi(X) = \alpha$.*

3. (Uniqueness) *If ϕ' is a best test of size α for testing H_0 against H_1, then it has the form given by (11.2), except perhaps for a set of x with probability 0 under H_0 and H_1.*

Proof The proof that follows is for the continuous case; the discrete case is left to the reader, and may be proven by replacing integrals everywhere with summations.

1. Choose any $\phi(x)$ of the form (11.2) and let $\phi'(x)$, $0 \leq \phi'(x) \leq 1$, be any test whose size is not greater than the size of $\phi(x)$, that is, for which

$$E_{\theta_0}\phi'(x) \leq E_{\theta_0}\phi(x).$$

We must show that $E_{\theta_1}\phi'(x) \leq E_{\theta_1}\phi(x)$; that is, that the power of $\phi'(x)$ is not greater than the power of $\phi(x)$. Note that

$$\int [\phi(x) - \phi'(x)][f_X(x \mid \theta_1) - \nu f_X(x \mid \theta_0)]\,dx$$

$$= \int_{A_+} [1 - \phi'(x)][f_X(x \mid \theta_1) - \nu f_X(x \mid \theta_0)]\,dx$$

$$+ \int_{A_-} [0 - \phi'(x)][f_X(x \mid \theta_1) - \nu f_X(x \mid \theta_0)]\,dx$$

$$+ \int_{A_0} [\gamma - \phi'(x)][f_X(x \mid \theta_1) - \nu f_X(x \mid \theta_0)]\,dx,$$

where

$$A_+ = \{x: f_X(x \mid \theta_1) - \nu f_X(x \mid \theta_0) > 0\},$$
$$A_- = \{x: f_X(x \mid \theta_1) - \nu f_X(x \mid \theta_0) < 0\},$$
$$A_0 = \{x: f_X(x \mid \theta_1) - \nu f_X(x \mid \theta_0) = 0\}.$$

Since $\phi'(x) \leq 1$, the first integral is nonnegative. Also, the second integral is nonnegative by inspection, and the third integral is identically zero. Thus,

$$\int [\phi(x) - \phi'(x)][f_X(x \mid \theta_1) - \nu f_X(x \mid \theta_0)]\,dx \geq 0. \tag{11.4}$$

This implies that

$$E_{\theta_1}\phi(X) - E_{\theta_1}\phi'(X) \geq \nu E_{\theta_0}\phi(X) - \nu E_{\theta_0}\phi'(X) \geq 0,$$

where the last inequality is a consequence of the hypothesis that $E_{\theta_0}\phi'(X) \leq E_{\theta_0}\phi(X)$. This proves that $\phi(X)$ is more powerful than $\phi'(X)$, that is,

$$\beta - \beta' \geq \nu(\alpha - \alpha').$$

For the case $\nu = \infty$, any test ϕ' of size $\alpha = 0$ must satisfy

$$\alpha = \int \phi'(x)f_X(x \mid \theta_0)\,dx = 0, \tag{11.5}$$

hence $\phi'(x)$ must be zero almost everywhere on the set $\{x: f_X(x, \mid \theta_0) > 0\}$. Thus,

11.2 Neyman–Pearson Theory

using this result and (11.3),

$$E_{\theta_1}[\phi(X) - \phi'(X)] = \underbrace{\int_{\{x: f_X(x|\theta_0)>0\}} (\phi(x) - \phi'(x)) f_X(x|\theta_1)\, dx}_{=0}$$

$$+ \int_{\{x: f_X(x|\theta_0)=0\}} (\phi(x) - \phi'(x)) f_X(x|\theta_1)\, dx$$

$$= \int_{\{x: f_X(x|\theta_0)=0\}} (1 - \phi'(x)) f_X(x|\theta_1)\, dx \geq 0,$$

since $\phi(x) = 1$ whenever the density $f_X(x|\theta_0) = 0$ by (11.3), and $\phi'(x) \leq 1$. This completes the proof of the first part.

2. Since a best test of size $\alpha = 0$ is given by (11.3), we may restrict attention to $0 < \alpha \leq 1$. The size of the test (11.2), is

$$E_{\theta_0}\phi(X) = P_{\theta_0}[f_X(X|\theta_1) > \nu f_X(X|\theta_0)] + \gamma P_{\theta_0}[f_X(X|\theta_1) = \nu f_X(X|\theta_0)]$$

$$= 1 - P_{\theta_0}[f_X(X|\theta_1) \leq \nu f_X(X|\theta_0)] + \gamma P_{\theta_0}[f_X(X|\theta_1) = \nu f_X(X|\theta_0)].$$
(11.6)

For fixed α, $0 < \alpha \leq 1$, we are to find ν and γ so that $E_{\theta_0}\phi(X) = \alpha$; or, equivalently, using the representation (11.6),

$$1 - P_{\theta_0}[f_X(X|\theta_1) \leq \nu f_X(X|\theta_0)] + \gamma P_{\theta_0}[f_X(X|\theta_1) = \nu f_X(X|\theta_0)] = \alpha$$

or

$$P_{\theta_0}[f_X(X|\theta_1) \leq \nu f_X(X|\theta_0)] - \gamma P_{\theta_0}[f_X(X|\theta_1) = \nu f_X(X|\theta_0)] = 1 - \alpha. \quad (11.7)$$

If there exists a ν_0 for which $P_{\theta_0}[f_X(X|\theta_1) \leq \nu f_X(X|\theta_0)] = 1 - \alpha$, we take $\gamma = 0$ and $\nu = \nu_0$. If not, then there is a discontinuity in $P_{\theta_0}[f_X(X|\theta_1) \leq \nu f_X(X|\theta_0)]$, when viewed as a function of ν, that brackets the particular value $1 - \alpha$; that is, there exists a ν_0 such that

$$P_{\theta_0}[f_X(X|\theta_1) < \nu_0 f_X(X|\theta_0)] < 1 - \alpha \leq P_{\theta_0}[f_X(X|\theta_1) \leq \nu_0 f_X(X|\theta_0)]. \quad (11.8)$$

Figure 11.1 illustrates this situation. Using (11.7) for $1 - \alpha$ in (11.8), and solving the

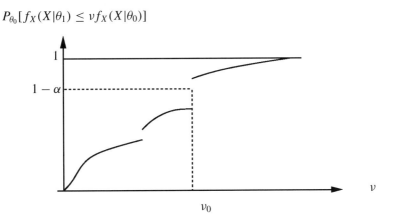

Figure 11.1: Illustration of threshold for Neyman–Pearson test

equation
$$1 - \alpha \leq P_{\theta_0}[f_X(X \mid \theta_1) \leq \nu_0 f_X(X \mid \theta_0)]$$
for γ, yields
$$\gamma = \frac{P_{\theta_0}[f_X(X \mid \theta_1) \leq \nu_0 f_X(X \mid \theta_0)] - (1 - \alpha)}{P_{\theta_0}[f_X(X \mid \theta_1) = \nu_0 f_X(X \mid \theta_0)]}. \tag{11.9}$$

Since this satisfies (11.7) and $0 \leq \gamma \leq 1$, letting $\nu = \nu_0$, the second part is proved.

3. If $\alpha = 0$, the argument in part 1 shows that $\phi(x) = 0$ almost everywhere on the set $\{x: f_{\theta_0}(x) > 0\}$. If ϕ' has a minimum probability of the second kind of error, then $1 - \phi'(x) = 0$ almost everywhere on the set $\{x: f_{\theta_1}(x) > 0\} \sim \{x: f_{\theta_0}(x) > 0\}$. Thus, ϕ' differs from the ϕ of (11.3) by a set of probability 0 under either hypothesis.

If $\alpha > 0$, let ϕ be the best test of size α of the form (11.2). Then, because $E_{\theta_i}\phi(X) = E_{\theta_i}\phi'(X), i = 0, 1$, the integral (11.4) must be equal to zero. But because this integral is nonnegative it must be zero almost everywhere; that is to say, on the set for which $f_X(x \mid \theta_1) \neq f_X(x \mid \theta_0)$, we have $\phi(x) = \phi'(x)$ almost everywhere. Thus, except for a set of probability 0, $\phi'(x)$ has the same form as (11.2) with the same value for ν as $\phi(x)$, thus the function $\phi(x)$ satisfies the uniqueness requirement. □

11.2.3 Application of the Neyman–Pearson lemma

The Neyman–Pearson lemma provides a general decision rule for a simple hypothesis versus a simple alternative. We would apply it as follows:

1. For a given binary decision problem, determine which is to be the null hypothesis and which is to be the alternative. This choice is at the discretion of the analyst. As a practical issue, it would be wise to choose as the null hypothesis the one that has the most serious consequences if rejected, because the analyst is able to choose the size of the test, which enables control of the probability of rejecting the null hypothesis when it is true.

2. Select the size of the test. It seems to be the tradition for many applications to set $\alpha = 0.05$ or $\alpha = 0.01$, which correspond to common "significance levels" used in statistics. The main issue, however, is to choose the size relevant to the problem at hand. For example, in a radar target detection problem, if the null hypothesis is "no target present," setting $\alpha = 0.05$ means that we are willing to accept a 5% chance that a target will not be there when our test tells us that a target is present. The smaller the size, in general, the smaller also is the power, as will be made more evident in the discussion of the receiver operating characteristic.

3. Calculate the threshold, ν. The way to do this is not obvious from the theorem. Clearly, ν must be a function of the size, α, but until specific distributions are used, there is no obvious formula for determining ν. That is one of the tasks examined in the examples to follow.

The structure of the test when $\gamma \neq 0$ deserves some discussion. If this equality condition obtains, then there is a nonzero probability that $f_X(x \mid \theta_1) = \nu f_X(x \mid \theta_0)$. The parameter, γ, defined in the proof of the Neyman–Pearson lemma has a natural interpretation as the probability of setting $\phi(x) = 1$ when the equality condition obtains. Accordingly, we may define the randomized decision rule, $\varphi_\gamma = (\gamma, 1 - \gamma)$, when $\gamma = P(\phi_1) = 1 - P(\phi_2)$ (the probability of choosing rule ϕ_1), and

$$\phi_1(x) = \begin{cases} 1 & \text{if } f_X(x \mid \theta_1) \geq \nu f_X(x \mid \theta_0), \\ 0 & \text{if } f_X(x \mid \theta_1) < \nu f_X(x \mid \theta_0), \end{cases} \tag{11.10}$$

11.2 Neyman–Pearson Theory

and

$$\phi_2(x) = \begin{cases} 1 & \text{if } f_X(x \mid \theta_1) > \nu f_X(x \mid \theta_0), \\ 0 & \text{if } f_X(x \mid \theta_1) \leq \nu f_X(x \mid \theta_0). \end{cases} \quad (11.11)$$

11.2.4 The likelihood ratio and the receiver operating characteristic (ROC)

The key quantities in the Neyman–Pearson theory are the density functions $f_X(x \mid \theta_1)$ and $f_X(x \mid \theta_0)$. These quantities are sometimes viewed as the conditional pdfs (or pmfs) of X given θ. The concept of conditioning, however, requires that the quantity θ be a random variable. But nothing in the Neyman–Pearson theory requires θ to be so viewed; in fact, the Neyman–Pearson approach is often considered to be an alternative to the Bayesian approach, in which θ *is* viewed as a random variable. Since the purists insist that the Neyman–Pearson not be confused with the Bayesian approach, they have coined the term *likelihood function* for $f_X(x \mid \theta_1)$ and $f_X(x \mid \theta_0)$. To maintain tradition, we respect this convention and call these likelihood functions, or likelihoods, when required.

The inequality

$$f_X(x \mid \theta_1) \lessgtr \nu f_X(x \mid \theta_0)$$

has emerged as a natural expression in the statement and proof of the Neyman–Pearson lemma. Using the ratio

$$\ell(x) = \frac{f_X(x \mid \theta_1)}{f_X(x \mid \theta_0)}, \quad (11.12)$$

known as the **likelihood ratio**, the Neyman–Pearson test can be expressed as one of the three comparisons in

$$\ell(x) \lessgtr \nu.$$

The test (11.2) may be rewritten as a **likelihood ratio test** (LRT):

$$\phi(x) = \begin{cases} 1 & \text{if } \ell(x) > \nu, \\ \gamma & \text{if } \ell(x) = \nu, \\ 0 & \text{if } \ell(x) < \nu. \end{cases} \quad (11.13)$$

For many distributions, it is convenient to use the logarithm of the likelihood function. Accordingly, we define (where appropriate)

$$\Lambda(x) = \log \ell(x) = \log \frac{f_X(x \mid \theta_1)}{f_X(x \mid \theta_0)}. \quad (11.14)$$

The function $\Lambda(x)$ (or some multiple of it, as convenient) is known as the *log-likelihood ratio*. Since the log function is monotonically increasing, we can rewrite the test (11.2) as

$$\phi(x) = \begin{cases} 1 & \text{if } \Lambda(x) > \log \nu, \\ \gamma & \text{if } \Lambda(x) = \log \nu, \\ 0 & \text{if } \Lambda(x) < \log \nu. \end{cases} \quad (11.15)$$

Since log-likelihood functions are common, we will find it convenient to introduce a new threshold variable for our test,

$$\eta = \log \nu.$$

You may have noticed in the proof of the lemma expressions such as $f_X(X \mid \theta_1)$, where we have used the random variable X as an argument of the density function. When we do this, the function $f_X(X \mid \theta_1)$ is, of course, a random variable since it becomes a function of a random variable. The likelihood ratio $\ell(X)$ is also a random variable, as is the log-likelihood ratio $\Lambda(X)$.

A false alarm (accepting H_1 when H_0 is true) occurs if $\ell(x) > \nu$ when $\theta = \theta_0$ and $X = x$. Let $f_\ell(l \mid \theta_0)$ denote the density of ℓ given $\theta = \theta_0$; then

$$\alpha = P_{FA} = P_{\theta_0}[\ell(X) > \nu] = \int_\nu^\infty f_\ell(l, \mid \theta_0)\, dl.$$

Thus, if we could compute the density of ℓ given $\theta = \theta_0$, we would have a method of computing the value of the threshold, ν. Or, in terms of the log-likelihood, we can write

$$\alpha = P_{FA} = P_{\theta_0}[\Lambda(X) > \eta] = \int_\eta^\infty f_\Lambda(l \mid \theta_0)\, dl,$$

where $f_\Lambda(l, \mid \theta_0)$ is the density of the random variable $\Lambda(X)$.

The probability of detection can similarly be found:

$$\beta = P_D = P_{\theta_1}[\ell(X) > \nu] = \int_\nu^\infty f_\ell(l, \mid \theta_1)\, dl$$
$$= P_{\theta_1}[\Lambda(X) > \eta] = \int_\eta^\infty f_\Lambda(l \mid \theta_1)\, dl.$$

In practice, we are often interested in comparing how P_{FA} varies with P_D. For a Neyman–Pearson test, the size and power, as specified by P_{FA} and P_D, completely specify the test performance. We can gain some valuable insight by crossplotting these parameters for a given test; the resulting plot is called the *Receiver operating characteristic*, or ROC curve, borrowing from radar terminology. ROC curves are perhaps the most useful single method for evaluating the performance of a binary detection system. We present some examples of ROCs in what follows.

11.2.5 A Poisson example

We wish to design a Neyman–Pearson detector for the Poisson random variable introduced in Example 11.1.2. The two hypotheses are

$$H_0: X \sim \exp[-\lambda_0]\frac{\lambda_0^x}{x!},$$
$$H_1: X \sim \exp[-\lambda_1]\frac{\lambda_1^x}{x!}.$$

The likelihood ratio for the problem is

$$\ell(x) = \exp[\lambda_0 - \lambda_1]\left(\frac{\lambda_1}{\lambda_0}\right)^x,$$

and (11.2) becomes, after simplification,

$$\phi(x) = \begin{cases} 1 & \text{if } x > \frac{\log \nu + \lambda_1 - \lambda_0}{\log \lambda_1 - \log \lambda_0}, \\ \gamma & \text{if } x = \frac{\log \nu + \lambda_1 - \lambda_0}{\log \lambda_1 - \log \lambda_0}, \\ 0 & \text{if } x < \frac{\log \nu + \lambda_1 - \lambda_0}{\log \lambda_1 - \log \lambda_0}. \end{cases} \qquad (11.16)$$

For a fixed size, α, we must compute the threshold, ν. The probability of a false alarm (deciding the $\theta = \lambda_1$ when $\theta = \lambda_0$ is true) is equal to the probability, under the null

11.2 Neyman–Pearson Theory

hypothesis, that $\ell(X) > \nu$; that is,

$$P_{FA} = P_{\theta_0}(\ell(X) > \nu) + \gamma P_{\theta_0}(\ell(X) = \nu).$$

Let $q(\nu) = \frac{\log \nu + \lambda_1 - \lambda_0}{\log \lambda_1 - \log \lambda_0}$ be such that it always takes integer values (by appropriate selection of ν). Then

$$P_{FA} = P_{\theta_0}(X > q(\nu)) + \gamma P_{\theta_0}(X = q(\nu))$$

$$= \sum_{k=q(\nu)+1}^{\infty} \frac{\exp[-\lambda_0]\lambda_0^k}{k!} + \gamma P(X = q(\nu))$$

$$= 1 - \sum_{k=0}^{q(\nu)} \frac{\exp[-\lambda_0]\lambda_0^k}{k!} + \gamma \exp[\lambda_0] \frac{\lambda_0^{q(\nu)}}{q(\nu)!}.$$

If α is such that there exists an integer q' that satisfies

$$1 - \alpha = \sum_{k=0}^{q'} \frac{\exp[-\lambda_0]\lambda_0^k}{k!}, \tag{11.17}$$

then we may take $\gamma = 0$ and

$$\nu = \exp\left[q' \log \frac{\lambda_1}{\lambda_0} - \lambda_1 + \lambda_0\right]. \tag{11.18}$$

In general, however, there will not be an q' that solved (11.17) with $\gamma = 0$, and we must set q' in (11.18) equal to

$$q' = \arg \min_{n \in \mathbb{Z}} \left\{ \sum_{k=0}^{n} \frac{\exp[-\lambda_0]\lambda_0^k}{k!} > 1 - \alpha \right\},$$

and apply (11.9) to yield

$$\gamma = \frac{\sum_{k=0}^{q'} \frac{\exp[-\lambda_0]\lambda_0^k}{k!} - (1 - \alpha)}{\frac{\exp[-\lambda_0]\lambda_0^{q'}}{q'!}}.$$

As a simple numerical example, let $\lambda_0 = 1$, $\lambda_1 = e$, and $\alpha = 0.1$. Straightforward calculation yields $q' = 2$, $\nu = 1.325$, and $\gamma = 0.1071$. Thus, if the observed value, $X = x$ is greater than 2, decide that $\theta = e$; if the value is less than 2, decide that $\theta = 1$; and if the observed value equals 2, make a random selection with the probability of choosing $\theta = 1$ being equal to 0.1071. This decision rule assures that the probability of detection will be maximized while holding the probability of a false alarm to exactly 0.1.

11.2.6 Some Gaussian examples

In this section we present several examples and implications of Neyman–Pearson detection where the observations are governed by random variables. Not only do these examples illustrate several important aspects of the theory, but they arise frequently in practice. We present a sequence of problems, more-or-less in order of increasing difficulty.

1. Scalar Gaussian detection, with different means and common variances:

$$H_0: X \sim \mathcal{N}(\mu_0, \sigma^2),$$
$$H_1: X \sim \mathcal{N}(\mu_1, \sigma_2).$$

We compute P_{FA} and P_D by introducing the Q function.

2. Vector Gaussian detection, with different means and common covariances:

$$H_0: X \sim \mathcal{N}(\mathbf{m}_0, R),$$
$$H_1: X \sim \mathcal{N}(\mathbf{m}_1, R).$$

We demonstrate detector architectures and performance.

3. Vector Gaussian detection, with common means and different covariances. Without loss of generality, we assume the means to be zero:

$$H_0: X \sim \mathcal{N}(\mathbf{0}, \sigma_0^2 I),$$
$$H_1: X \sim \mathcal{N}(\mathbf{0}, \sigma_1^2 I).$$

Analysis of performance in this case will require introduction of the χ^2 distribution.

Scalar Gaussian detection; different means, common variance

As a physical motivation for this problem, let us assume that, under hypothesis H_1, a source output is a constant voltage μ_1, and that under H_0 the source output is a constant voltage μ_0. Before observation, the voltage is corrupted by an additive noise; the sample random variables are

$$X = \theta + Z, \tag{11.19}$$

where $\theta \in \{\theta_0, \theta_1\}$ with $\theta_0 = \mu_0$ and $\theta_1 = \mu_1$. The random variables Z are zero-mean Gaussian random variables with known variance σ^2, and are also independent of the source output, θ. We desire to formulate a test to discriminate between the two hypotheses. We have

$$H_0: X = Z + \mu_0 \qquad H_1: X = Z + \mu_1,$$

with

$$f_Z(z) = \frac{1}{\sqrt{2\pi}\sigma} \exp\left[-\frac{z^2}{2\sigma^2}\right].$$

The probability densities of X under each hypothesis are

$$f_X(x \mid \theta_0) = \frac{1}{\sqrt{2\pi}\sigma} \exp\left[-\frac{(x-\mu_0)^2}{2\sigma^2}\right],$$

$$f_X(x \mid \theta_1) = \frac{1}{\sqrt{2\pi}\sigma} \exp\left[-\frac{(x-\mu_1)^2}{2\sigma^2}\right].$$

The problem can also be stated as

$$H_0: X \sim \mathcal{N}(\mu_0, \sigma^2), \qquad H_1: X \sim \mathcal{N}(\mu_1, \sigma^2).$$

The likelihood ratio is

$$\ell(x) = \frac{f_X(x \mid \theta_1)}{f_X(x \mid \theta_0)} = \frac{\frac{1}{\sqrt{2\pi}\sigma} \exp\left[-\frac{(x-\mu_1)^2}{2\sigma^2}\right]}{\frac{1}{\sqrt{2\pi}\sigma} \exp\left[\frac{-(x-\mu_0)^2}{2\sigma^2}\right]}.$$

After canceling common terms and taking the logarithm, we have

$$\Lambda(x) = \log \ell(x) = \frac{1}{\sigma^2} x(\mu_1 - \mu_0) + \frac{1}{2\sigma^2}(\mu_0^2 - \mu_1^2). \tag{11.20}$$

11.2 Neyman–Pearson Theory

The log-likelihood ratio test then becomes

$$\phi(x) = \begin{cases} 1 & \text{if } \Lambda(x) > \eta, \\ \gamma & \text{if } \Lambda(x) = \eta, \\ 0 & \text{if } \Lambda(x) < \eta, \end{cases} \quad (11.21)$$

where $\eta = \log \nu$. Since $\Lambda(x) = \eta$ with probability 0 (because the pdf is continuous), the middle choice in the test can be removed with no effect on the probability of error. Also, letting

$$\tau = \frac{\mu_1 + \mu_0}{2} + \sigma^2 \frac{\eta}{\mu_1 - \mu_0},$$

we see that the test can be written as

$$\phi(x) = \begin{cases} 1 & \text{if } x \geq \tau, \\ 0 & \text{if } x < \tau. \end{cases} \quad (11.22)$$

Figure 11.2 illustrates a block diagram of this test. The test simply becomes a matter of testing against a threshold.

In order to quantify the performance of this test, we need to determine the distribution of the log-likelihood function. We observe that $\Lambda(X)$ is a linear function of the random variable X, so that $\Lambda(X)$ is itself a Gaussian random variable, with mean and covariance, under hypotheses H_0 and H_1, of

$$\mu_{\theta_0} = E_{\theta_0}\Lambda(X) = -\frac{1}{2\sigma^2}(\mu_0 - \mu_1)^2 \quad \sigma^2_{\theta_0} = \text{var}_{\theta_0}\Lambda(X) = \frac{1}{\sigma^2}(\mu_1 - \mu_0)^2, \quad (11.23)$$

$$\mu_{\theta_1} = E_{\theta_1}\Lambda(X) = \frac{1}{2\sigma^2}(\mu_0 - \mu_1)^2 \quad \sigma^2_{\theta_1} = \text{var}_{\theta_1}\Lambda(X) = \frac{1}{\sigma^2}(\mu_1 - \mu_0)^2. \quad (11.24)$$

Thus, the log-likelihood function has the distributions

$$f_L(l \mid \theta_0) \sim \mathcal{N}(\mu_{\theta_0}, \sigma^2_{\theta_0}) \qquad f_L(l \mid \theta_1) \sim \mathcal{N}(\mu_{\theta_1}, \sigma^2_{\theta_1}).$$

Then

$$\alpha = P_{FA} = P_{\theta_0}(\Lambda(X) > \eta) = P\big(\mathcal{N}(\mu_{\theta_0}, \sigma^2_{\theta_0}) > \eta\big) = \int_\eta^\infty \frac{1}{\sqrt{2\pi}\sigma_{\theta_0}} e^{-(y-\mu_{\theta_0})^2/2\sigma^2_{\theta_0}}\, dy. \quad (11.25)$$

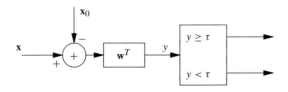

Figure 11.2: Scalar Gaussian detection of the mean

Figure 11.3 illustrates the normal curves for the two hypotheses under consideration, showing P_{FA} as the area under the curve $f_X(x \mid \theta_0)$ to the right of the threshold τ, and P_D as the area under $f_X(x \mid \theta_1)$ to the right of the threshold.

Based on the definition of the Q function (see box 11.1), (11.25) can be written

$$\alpha = Q\left(\frac{\eta + \frac{1}{2\sigma^2}(\mu_0 - \mu_1)^2}{\frac{1}{\sigma}(\mu_1 - \mu_0)}\right) = Q(\sigma\eta/(\mu_1 - \mu_0) + |\mu_0 - \mu_1|/(2\sigma)).$$

If we let $d = |\mu_0 - \mu_1|$ be the distance between the means, we have

$$\alpha = Q(\sigma\eta/d + d/(2\sigma)). \qquad (11.26)$$

Box 11.1: The Q function

The Q function is frequently used to determine the probability of error analysis in communications problems. If $Z \sim \mathcal{N}(0, 1)$ (that is, Z is a unit Gaussian random variable), then

$$Q(x) = P(Z > x) = \int_x^\infty \frac{1}{\sqrt{2\pi}} e^{-y^2/2}\, dy$$

If $W \sim \mathcal{N}(\mu, \sigma^2)$, it is straightforward to show by a change of variables that

$$P(W > x) = Q\left(\frac{x - \mu}{\sigma}\right).$$

It is also straightforward to show that $Q(x) = 1 - Q(-x)$. The plot below illustrates the Q-function for $x \geq 0$.

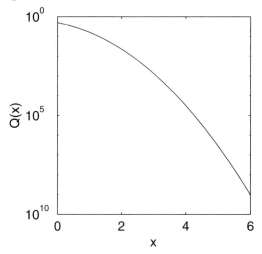

See also the bounds in exercise 11.2-18.

The Q function is related to the complementary error function common in statistics. It may be computed in MATLAB using the following code.

```
function p = qf(x)
% function p = qf(x)
% compute the Q function:
% p = 1/sqrt(2pi)int_x^infty exp(-t^2/2)dt
p = 0.5*erfc(x/sqrt(2));
```

11.2 Neyman–Pearson Theory

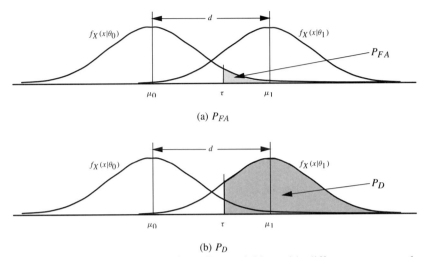

(a) P_{FA}

(b) P_D

Figure 11.3: Error probabilities for Gaussian variables with different means and equal variances

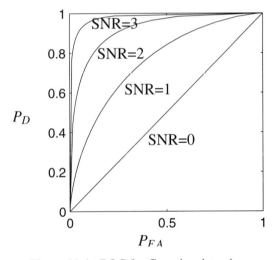

Figure 11.4: ROC for Gaussian detection

We can also write this as
$$\alpha = Q(z), \qquad (11.27)$$
where
$$z = \sigma\eta/d + d/(2\sigma). \qquad (11.28)$$
Similarly, the probability of detection is obtained from
$$\beta = P_D = Q\left(\frac{\eta - \frac{1}{2\sigma^2}(\mu_0 - \mu_1)^2}{\frac{1}{\sigma}(\mu_1 - \mu_0)}\right) = Q(\sigma\eta/d - d/(2\sigma)) = Q(z - d/\sigma). \qquad (11.29)$$

A plot of the ROC is shown in figure 11.4. The plot shows the performance for various values of the SNR, which is defined here as
$$SNR = \frac{d}{\sigma} = \frac{|\mu_0 - \mu_1|}{\sigma}. \qquad (11.30)$$
As the SNR increases, it is possible to obtain greater power for a given size.

Vector Gaussian detection; different means, common variance

Let $m_i(\theta)$, $i = 1, 2, \ldots, n$ be samples of a signal, parameterized by some parameter θ. Suppose that the signal is observed in noise, producing a measurement

$$X_i = m_i(\theta) + Z_i, \qquad i = 1, 2, \ldots, n,$$

where the Z_i are $\mathcal{N}(0, \sigma^2)$ and are independent. Then, because the Z_i are independent, the joint pdf of $\mathbf{X} = (X_1, X_2, \ldots, X_n)$ is simply the product of the individual pdfs:

$$f_{X_1, X_2, \ldots, X_n}(x_1, x_2, \ldots, x_n \mid \theta) = f_{\mathbf{X}}(\mathbf{x} \mid \theta)$$

$$= \prod_{i=1}^{n} \frac{1}{\sqrt{2\pi}\sigma} \exp\left[-\frac{(x_i - m_i(\theta))^2}{2\sigma^2}\right]$$

$$= \frac{1}{(2\pi)^{n/2}\sigma^n} \exp\left[-\frac{1}{2\sigma^2}(\mathbf{x} - \mathbf{m}(\theta))^T (\mathbf{x} - \mathbf{m})\right],$$

where $\mathbf{m}(\theta) = [m_1(\theta), m_2(\theta), \ldots, m_n(\theta)]^T$. Using this model, we can consider detection problems such as the determination of which signal was sent.

Example 11.2.2 (on-off signaling) Suppose that there are two possible signals, $\Theta = \{0, 1\}$, corresponding to the hypotheses

$$H_0: \mathbf{m} = \mathbf{0} \qquad (\theta = 0),$$
$$H_1: \mathbf{m} = \mathbf{m}_1 \qquad (\theta = 1).$$

That is, the signal is either absent, or it is present and the observed vector $\mathbf{X}$ has mean $\mathbf{m}_1$. □

We can generalize the detection problem to samples that are not independent. Consider the simple binary Gaussian detection problem

$$H_0: \mathbf{X} \sim \mathcal{N}(\mathbf{m}_0, R),$$
$$H_1: \mathbf{X} \sim \mathcal{N}(\mathbf{m}_1, R).$$

Then

$$f_{\mathbf{X}}(\mathbf{x} \mid \theta_0) = \frac{1}{(2\pi)^{n/2} |R|^{1/2}} \exp\left[-\frac{1}{2}(\mathbf{x} - \mathbf{m}_0)^T R^{-1} (\mathbf{x} - \mathbf{m}_0)\right],$$

$$f_{\mathbf{X}}(\mathbf{x} \mid \theta_1) = \frac{1}{(2\pi)^{n/2} |R|^{1/2}} \exp\left[-\frac{1}{2}(\mathbf{x} - \mathbf{m}_1)^T R^{-1} (\mathbf{x} - \mathbf{m}_1)\right].$$

As we did for the scalar Gaussian detection case, we determine the likelihood ratio

$$\ell(\mathbf{X}) = \frac{f_X(\mathbf{x} \mid \theta_1)}{f_X(\mathbf{x} \mid \theta_0)}$$

$$= \exp\left[(\mathbf{m}_1 - \mathbf{m}_0)^T R^{-1} \mathbf{x} + \frac{1}{2}(\mathbf{m}_0 + \mathbf{m}_1)^T R^{-1}(\mathbf{m}_0 - \mathbf{m}_1)\right]$$

and log-likelihood ratio,

$$\Lambda(\mathbf{x}) = (\mathbf{m}_1 - \mathbf{m}_0)^T R^{-1}(\mathbf{x} - \mathbf{x}_0),$$

where

$$\mathbf{x}_0 = \frac{1}{2}(\mathbf{m}_1 + \mathbf{m}_0).$$

Letting $\mathbf{w} = R^{-1}(\mathbf{m}_1 - \mathbf{m}_0)$, we can write

$$\Lambda(\mathbf{x}) = \mathbf{w}^T (\mathbf{x} - \mathbf{x}_0).$$

The set of points where $\Lambda(\mathbf{x}) = 0$ forms a plane orthogonal to $\mathbf{w}$, passing through $\mathbf{x}_0$.

11.2 Neyman–Pearson Theory

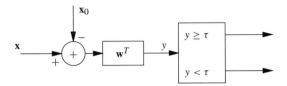

Figure 11.5: Test for vector Gaussian random variables with different means

The decision based upon the log-likelihood ratio is

$$\Lambda(\mathbf{x}) = \begin{cases} 1 & \mathbf{w}^T(\mathbf{x} - \mathbf{x}_0) >= \eta, \\ 0 & \mathbf{w}^T(\mathbf{x} - \mathbf{x}_0) > \eta. \end{cases}$$

Figure 11.5 illustrates the block diagram for this test.

The performance for this vector Gaussian case is straightforward to determine. We observe that $\Lambda(\mathbf{X})$ is a scalar Gaussian random variable, with

$$\mu_{\theta_0} = E_{\theta_0}\Lambda(\mathbf{X}) = -\frac{1}{2}(\mathbf{m}_1 - \mathbf{m}_0)^T R^{-1}(\mathbf{m}_1 - \mathbf{m}_0) = -\frac{1}{2}\mathbf{w}^T R \mathbf{w}, \quad (11.31)$$

$$\operatorname{var}_{\theta_0}\Lambda(\mathbf{X}) = (\mathbf{m}_1 - \mathbf{m}_0)^T R^{-1}(\mathbf{m}_1 - \mathbf{m}_0) = \mathbf{w}^T R \mathbf{w}, \quad (11.32)$$

and similarly

$$\mu_{\theta_1} = \frac{1}{2}\mathbf{w}^T R \mathbf{w}, \qquad \operatorname{var}_{\theta_1} = \mathbf{w}^T R \mathbf{w}.$$

Let

$$s^2 = \mathbf{w}^T R \mathbf{w}. \quad (11.33)$$

Then, under H_0,

$$\Lambda(X) \sim \mathcal{N}(-s^2/2, s^2)$$

and under H_1,

$$\Lambda(X) \sim \mathcal{N}(s^2/2, s^2).$$

The performance of the detector is

$$P_{FA} = P_{\theta_0}(\Lambda(X) \geq \eta) = Q\left(\frac{\eta + s^2/2}{s}\right) = Q(z)$$

and

$$P_D = P_{\theta_1}(\Lambda(X) \geq \eta) = Q\left(\frac{\eta - s^2/2}{s}\right) = Q(z - s),$$

where $z = \eta/s - s/2$. By comparison with (11.26), the quantity s is directly analogous to d/σ, which we defined as the signal-to-noise ratio. Thus, the ROC for the vector Gaussian case is identical to that of the scalar Gaussian case, when plotted as a function of SNR $= s$.

Simplifications when $R = I$

It is interesting to examine certain detector structures under the frequently encountered circumstance that $R = I$, that is, that the samples of the signal are independent. Then

$$\Lambda(X) = (\mathbf{m}_1 - \mathbf{m}_0)^T(\mathbf{x} - \mathbf{x}_0).$$

The quantity s^2 defined in (11.33) is simply

$$s^2 = \frac{1}{\sigma^2}\|\mathbf{m}_1 - \mathbf{m}_0\|^2 = \frac{d^2}{\sigma^2},$$

where $d = \|\mathbf{m}_1 - \mathbf{m}_0\|$. (Note: the l_2 (Euclidean) norm is used here, and throughout the discussion of Gaussian detection. It is a natural norm to use for problems associated with Gaussian problems.)

An additional simplification occurs when $\|\mathbf{m}_1\| = \|\mathbf{m}_0\|$. Then the log-likelihood ratio is

$$\Lambda(\mathbf{x}) = \frac{1}{\sigma^2} (\mathbf{m}_1^T - \mathbf{m}_0^T)\mathbf{x} + c,$$

where c is a constant that does not depend upon $\mathbf{x}$. Absorbing the constant and the factor σ^2 into the threshold, the detector computes $(\mathbf{m}_1 - \mathbf{m}_0)^T \mathbf{x}$ and compares this inner product to the (modified) threshold). In this case, the detector determines, on the basis of the angle between the signals, which signal the received vector is most similar.

Example 11.2.3 The detection problem applies directly to digital communications, where signals $\mathbf{m}_0$ and $\mathbf{m}_1$ are sent, and we desire to distinguish between them at the receiver. Suppose that $\mathbf{m}_0$ or $\mathbf{m}_1$ are sent with equal probability. Most commonly we choose the threshold so that there is the same probability of error given that a zero is sent as there is given that a one is sent. That is, we set

$$\alpha = 1 - \beta,$$

which corresponds to the case that $\eta = 0$. Let $P(\mathcal{E} \mid \mathbf{m}_i)$ be the probability of error given that $\mathbf{m}_i$ was sent. Then the probability of error, denoted $P(\mathcal{E})$, is

$$P(\mathcal{E}) = P(\mathbf{m}_0)P(\mathcal{E} \mid \mathbf{m}_0) + P(\mathbf{m}_1)P(\mathcal{E} \mid \mathbf{m}_1) = \frac{1}{2}\alpha + \frac{1}{2}(1-\beta) = Q(f/2).$$

This can be written as

$$\boxed{P(\mathcal{E}) = Q\left(\frac{d}{2\sigma}\right)}$$

where $d = \|\mathbf{m}_1 - \mathbf{m}_0\|$. In a digital communications setting, the probability of error for binary communications ultimately *depends upon the distance between signals d relative to the noise energy*. This is why the SNR is such an important measure in communications. Figure 11.6 illustrates the probability of error for BPSK signaling as a function of SNR, in dB. The SNR is given (for reasons which become clear subsequently) as $SNR = E_b/N_0$.

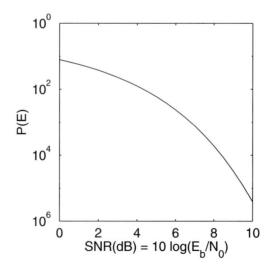

Figure 11.6: Probability of error for BPSK signaling

11.2 Neyman–Pearson Theory

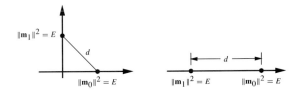

(a) Orthogonal constellation (b) Antipodal constellation

Figure 11.7: An orthogonal and antipodal binary signal constellation

Consider now the two binary signal constellations shown in figure 11.7. In each constellation, the signals have equal energy,

$$\|\mathbf{m}_0\| = \|\mathbf{m}_1\| = 1.$$

In the *orthogonal signal constellation*, in which $\mathbf{m}_0^T \mathbf{m}_1 = 0$, the distance between the signals is

$$d = \sqrt{2E}.$$

In the *antipodal signal constellation*, the distance between the signals is

$$d = 2E.$$

In comparing the two distances, the antipodal signaling has a 3 dB advantage in SNR over orthogonal signaling. □

Vector Gaussian; same means, different covariance

Let us now consider a different kind of detection problem, in which the means are the same but the covariances are different. We assume for convenience that the means are equal to zero. We wish to examine the detection problem

$$H_0: \mathbf{X} \sim \mathcal{N}(\mathbf{0}, \sigma_0^2 I),$$
$$H_1: \mathbf{X} \sim \mathcal{N}(\mathbf{0}, \sigma_1^2 I),$$

in which $\mathbf{X}$ is an n-dimensional random vector. The log-likelihood is

$$\Lambda(\mathbf{X}) = \log \frac{\sigma_0}{\sigma_1} + \mathbf{x}^T \mathbf{x} \left(\frac{1}{2\sigma_0^2} - \frac{1}{2\sigma_1^2} \right).$$

Since the first term does not depend on the data, we will discard it and write

$$\Lambda(\mathbf{x}) = \mathbf{x}^T \mathbf{x} \left(\frac{1}{2\sigma_0^2} - \frac{1}{2\sigma_1^2} \right). \tag{11.34}$$

Let us denote $\gamma^2 = (\frac{1}{2\sigma_0^2} - \frac{1}{2\sigma_1^2})$, so that $\Lambda(\mathbf{x}) = \gamma^2 \mathbf{x}^T \mathbf{x}$. The Neyman–Pearson test becomes

$$\phi(\mathbf{x}) = \begin{cases} 1 & \gamma^2 \mathbf{x}^T \mathbf{x} \geq \eta, \\ 0 & \gamma^2 \mathbf{x}^T \mathbf{x} < \eta, \end{cases} \tag{11.35}$$

for some threshold η.

In the evaluation of the performance of this test, it must be recognized that $\Lambda(\mathbf{X})$, being a *quadratic* function of a Gaussian vector, is no longer Gaussian distributed. We must examine a new distribution to determine the power and size of this test.

> **Box 11.2: The Γ function**
>
> The Γ function is defined by the integral
> $$\Gamma(x) = \int_0^\infty t^{x-1} e^{-t}\, dt.$$
> Using integration by parts, it is straightforward to show that for $x > 0$,
> $$\Gamma(x+1) = x\Gamma(x),$$
> so that for an integer k, $\Gamma(k) = (k-1)!$.
> Two useful special values of the Γ function are
> $$\Gamma(1/2) = \sqrt{\pi},$$
> $$\Gamma(m+1/2) = \frac{1 \cdot 3 \cdot 5 \cdots (2m-1)}{2^m}\sqrt{\pi} \qquad m = 1, 2, 3, \ldots.$$

χ^2 random variables

To analyze the performance of the detector in (11.34), we need to introduce a new distribution. Suppose that
$$Z = \sum_{i=1}^n Y_i^2,$$
where the random variables Y_i, $i = 1, \ldots, n$, are independent and $\mathcal{N}(0, 1)$. The random variable Z is said to be (central) chi-squared with n degrees of freedom, denoted as $Z \sim \chi_n^2$.

Theorem 11.1 *If $Z \sim \chi_n^2$, then*
$$f_Z(z) = \frac{1}{\Gamma(n/2) 2^{n/2}} z^{n/2-1} e^{-z/2}. \tag{11.36}$$

The gamma function $\Gamma(\cdot)$ is described in box 11.2.

Proof Let $Y_1 \sim \mathcal{N}(0, 1)$, and let $Z_1 = Y_1^2$. Then
$$P(Z_1 \leq z) = P(-\sqrt{z} \leq Y_1 \leq \sqrt{z}) = 2\int_0^{\sqrt{z}} \frac{1}{\sqrt{2\pi}} e^{-x^2/2}\, dx$$
$$= \int_0^z \frac{1}{\sqrt{2\pi}} e^{-x/2} x^{-1/2}\, dx.$$

By taking the derivative with respect to z, we obtain
$$f_{Z_1}(z) = \frac{1}{\sqrt{2\pi}} e^{-z/2} z^{-1/2}, \qquad z \geq 0.$$

The characteristic function of Z_1 is
$$\phi_{Z_1}(\omega) = E[e^{j\omega Z_1}] = \frac{1}{(1 - 2j\omega)^{1/2}}. \tag{11.37}$$

Now let
$$Z = \sum_{i=1}^n Y_i^2,$$
where each $Y_i \sim \mathcal{N}(0, 1)$ independently. Then $\phi_Z(\omega)$ is the n-fold product of $\phi_{Z_1}(\omega)$,
$$\phi_Z(\omega) = \frac{1}{(1 - 2j\omega)^{n/2}}. \tag{11.38}$$

The inverse Fourier transform of this function is

$$f_Z(z) = \frac{1}{\Gamma(n/2)2^{n/2}} z^{n/2-1} e^{-z/2}. \tag{11.39}$$

$\square$

A result that we will need shortly relates to quadratic forms of Gaussian random variables, a generalization of χ_n^2 random variables.

Theorem 11.2 *[291] Let* $\mathbf{X} \sim \mathcal{N}(0, R)$ *be n-dimensional, and let*

$$Q = \mathbf{X}^T P \mathbf{X},$$

where P is symmetric. If $PR = RP$, *then the characteristic function of Q is*

$$\phi_Q(\omega) = \frac{1}{|I - 2j\omega RP|^{1/2}}.$$

Hence if RP is a projection matrix with r nonzero eigenvalues, then Q is a χ_r^2 *random variable.*

Proof

$$\phi_Q(\omega) = E e^{j\omega Q} = \frac{1}{(2\pi)^{n/2}|R|^{1/2}} \int \exp\left[-\frac{1}{2}\mathbf{x}^T R^{-1} \mathbf{x}\right] \exp[j\omega \mathbf{x}^T P \mathbf{x}] \, d\mathbf{x}$$

$$= \frac{1}{(2\pi)^{n/2}|R|^{1/2}} \int \frac{|I - 2j\omega RP|^{1/2}}{|I - 2j\omega RP|^{1/2}} \exp\left[-\frac{1}{2}\mathbf{x}^T R^{-1}(I - 2j\omega RP)\mathbf{x}\right] d\mathbf{x}$$

$$= \frac{1}{|I - 2j\omega RP|^{1/2}} \frac{1}{(2\pi)^{n/2}|R(I - 2j\omega RP)^{-1}|^{1/2}|}$$

$$\times \int \exp\left[-\frac{1}{2}\mathbf{x}^T R^{-1}(I - 2j\omega RP)\mathbf{x}\right] d\mathbf{x}$$

$$= \frac{1}{|I - 2j\omega RP|^{1/2}}.$$

Now, suppose RP is a rank-r projection matrix. Since the eigenvalues of RP are either 0 or 1, the diagonalization of RP using the orthogonal eigenvector matrix U is

$$U^T R P U = \text{diag}(1, 1, \ldots, 1, 0, 0, \ldots, 0),$$

where there are r ones on the diagonal. In this case,

$$\phi_Q(\omega) = \frac{1}{(1 - 2j\omega)^r},$$

which is the characteristic function for a χ_r^2 random variable. $\square$

Performance of detectors when covariances differ

We return now to analyzing the performance of the detector (11.35). Under H_0,

$$\frac{\Lambda(\mathbf{X})}{\gamma^2 \sigma_0^2} \sim \chi_n^2,$$

and under H_1,

$$\frac{\Lambda(\mathbf{X})}{\gamma^2 \sigma_1^2} \sim \chi_n^2.$$

Then

$$P_{FA} = P_{\theta_0}(\Lambda(\mathbf{X}) > \eta)$$
$$= P\big(\Lambda(\mathbf{X})/(\gamma^2\sigma_0^2) > \eta/(\gamma^2\sigma_0^2)\big)$$
$$= P\big(\chi_n^2 > \tau/(\gamma^2\sigma_0^2)\big)$$
$$= \int_{\eta/(\gamma^2\sigma_0^2)}^{\infty} \frac{1}{\Gamma(n/2)2^{n/2}} z^{n/2-1} e^{-z/2}\, dz. \tag{11.40}$$

Similarly,

$$P_D = \int_{\eta/(\gamma^2\sigma_1^2)}^{\infty} \frac{1}{\Gamma(n/2)2^{n/2}} z^{n/2-1} e^{-z/2}\, dz. \tag{11.41}$$

In the case of general n, (11.40) and (11.41) must be computed numerically. However, as the next example illustrates, the ROC is readily obtained when $n = 2$.

Example 11.2.4 The computations in (11.40) and (11.41) are readily accomplished when $n = 2$, since the density of a χ_2^2 random variable Y is

$$f_Y(z) = \frac{1}{2} e^{-y/2}.$$

Letting $\epsilon = \eta/\gamma^2$, we have

$$P_{FA} = e^{-\epsilon/\sigma_0^2} \quad \text{and} \quad P_D = e^{-\epsilon/\sigma_1^2}.$$

Given a size α, the threshold for the test that uses

$$\mathbf{X}^T \mathbf{X}$$

as the statistic can be determined from

$$\epsilon = -\sigma_0^2 \log P_{FA}.$$

Furthermore, the ROC can be obtained readily since

$$P_D = P_{FA}^{\sigma_0^2/\sigma_1^2}.$$

Figure 11.8 illustrates this ROC for

$$\rho = \frac{\sigma_1^2}{\sigma_0^2} = 1, 2, 3, 4, 5.$$

As expected, there is improved performance as the ratio between the variances increases. □

We consider briefly the problem

$$H_0\colon \mathbf{X} \sim \mathcal{N}(\mathbf{0}, R_0),$$
$$H_1\colon \mathbf{X} \sim \mathcal{N}(\mathbf{0}, R_1).$$

Developing the likelihood ratio test is straightforward (see exercise 11.2-17). However, quantifying the performance is more difficult, because the pdf of $\Lambda(\mathbf{X})$ can only be obtained by numerical integration.

11.2.7 Properties of the ROC

Property 1. *All likelihood ratio tests have ROC curves that are concave.*

Proof Suppose the ROC has a segment that is convex. To be specific, suppose (P_{FA}^a, P_D^a) and (P_{FA}^b, P_d^b) are points on the ROC curve, but the curve is convex between these two

11.2 Neyman–Pearson Theory

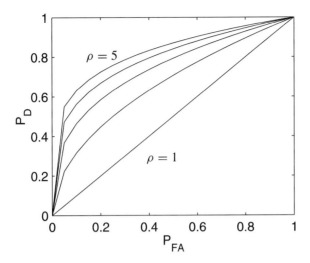

Figure 11.8: ROC: normal variables with equal means and unequal variances

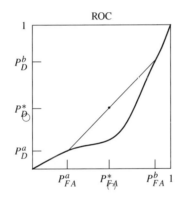

Figure 11.9: Demonstration of the concave property of the ROC

points, as illustrated in figure 11.9. Let $\phi_a(x)$ and $\phi_b(x)$ be the decision rules obtained for the corresponding sizes and powers, as given by the Neyman–Pearson lemma. Now form a new rule by choosing ϕ_a with probability q and ϕ_b with probability $1-q$, for any $0 < q < 1$; that is,

$$\phi^*(x) = \begin{cases} \phi_a(x) & \text{with probability } q, \\ \phi_b(x) & \text{with probability } 1-q. \end{cases}$$

This is a randomized rule, under which the decision maker would take action corresponding to ϕ_a with probability q, and otherwise would take action corresponding to rule ϕ_b. The probability of detection, P_D^*, for this randomized rule is

$$P_D^* = q P_D^a + (1-q) P_D^b,$$

a *convex combination* of P_D^a and P_D^b. The set of all such convex combinations must lie on the line connecting P_D^a and P_D^b, hence the rule $\phi^*(x)$ of size P_{FA}^*, has greater power than the rule provided by the Neyman–Pearson test, thus contradicting the optimality of the Neyman–Pearson test. Thus, the ROC curve cannot be concave. □

Property 2. *All continuous likelihood ratio tests have ROC curves that are above the $P_D = P_{FA}$ line.*

This property is just a special case of property 1, because the points $(0, 0)$ and $(1, 1)$ are contained on all ROC curves.

Property 3. *The slope of the ROC curve at any differentiable point is equal to the value of the threshold ν required to achieve the P_D and P_{FA} of that point, using the original likelihood ratio (not the log-likelihood ratio).*

Proof Let ℓ be the likelihood ratio, and suppose that ν is a given threshold. Then

$$P_D = \int_\nu^\infty f_\ell(l \mid \theta_1) \, dl,$$

$$P_{FA} = \int_\nu^\infty f_\ell(l \mid \theta_0) \, dl.$$

Let δ be a small perturbation in the threshold; then

$$\Delta P_D = \int_\nu^{\nu+\delta} f_\ell(l \mid \theta_1) \, dl,$$

$$\Delta P_{FA} = \int_\nu^{\nu+\delta} f_\ell(l \mid \theta_0) \, dl,$$

represent the changes in P_D and P_{FA}, respectively, as a result of the change in threshold. Then the slope of the ROC curve is given by

$$\lim_{\delta \to 0} \frac{\Delta P_D}{\Delta P_{FA}} = \lim_{\delta \to 0} \frac{\delta f_\ell(\nu \mid \theta_1)}{\delta f_\ell(\nu \mid \theta_0)} = \frac{f_\ell(\nu \mid \theta_1)}{f_\ell(\nu \mid \theta_0)}. \tag{11.42}$$

To establish that this ratio equals ν, we observe that, in general,

$$E_{\theta_1} \ell^n(X) = \int \ell^n(x) f_X(x \mid \theta_1) \, dx$$

$$= \int \frac{f_X^n(x \mid \theta_1)}{f_X^n(x \mid \theta_0)} f_X(x \mid \theta_1) \, dx$$

$$= \int \frac{f_X^{n+1}(x \mid \theta_1)}{f_X^{n+1}(x \mid \theta_0)} f_X(x \mid \theta_0) \, dx$$

$$= \int \ell^{n+1}(X) f_X(x \mid \theta_0) \, dx$$

$$= E_{\theta_0} \ell^{n+1}(X).$$

But the condition $E_{\theta_1} \ell^n = E_{\theta_0} \ell^{n+1}$ requires that

$$\int l^n f_\ell(l \mid \theta_1) \, dl = \int l^{n+1} f_\ell(l \mid \theta_0) \, dl$$

must hold for all n, which implies that

$$f_\ell(l \mid \theta_1) = l f_\ell(l \mid \theta_0) \tag{11.43}$$

must hold for all values of l. Thus, applying (11.43) to (11.42), we obtain the desired result:

$$\frac{dP_D}{dP_{FA}} = \frac{f_\ell(v \mid \theta_1)}{f_\ell(v \mid \theta_0)} = v.$$

□

11.3 Neyman–Pearson testing with composite binary hypotheses

Thus far, we have dealt with the simplest form of binary hypothesis testing: a simple hypothesis versus a simple alternative. We now generalize our thinking to composite hypotheses. As mentioned in definition 11.1, a hypothesis $H_0: \theta \in \Theta_0$ is said to be *composite* if Θ_0 consists of at least two elements. We are interested in testing a composite hypothesis $H_0: \theta \in \Theta_0$ against a composite alternative $H_1: \theta \in \Theta_1$. Before pursuing the development of a theory for composite hypotheses, we need to generalize the notions of size and power for this situation.

Definition 11.5 A test ϕ of $H_0: \theta \in \Theta_0$ against $H_1: \theta \in \Theta_1$ is said to have **size** α if

$$\sup_{\theta \in \Theta_0} E_\theta \phi(X) = \alpha.$$

□

Definition 11.6 A test ϕ_0 is said to be **uniformly most powerful (UMP) of size** α for testing $H_0: \theta \in \Theta_0$ against $H_1: \theta \in \Theta_1$ if ϕ_0 is of size α and if, for any other test ϕ of size at most α,

$$E_\theta \phi_0(X) \geq E_\theta \phi(X)$$

for each $\theta \in \Theta_1$.

□

For a test to be UMP, it must maximize the power $E_\theta \phi(X)$ for each $\theta \in \Theta_1$. This is a very stringent condition, and the existence of a uniformly most powerful test is not guaranteed in all cases. For example, although the Neyman–Pearson lemma tells us that there exists a most powerful test of size α for fixed $\theta_1 \in \Theta_1$, there is no reason why this same test should also be most powerful of size α for $\theta_2 \neq \theta_1$, with $\theta_2 \in \Theta_1$. Our goal in this section is to arrive at conditions for which the existence of a UMP test can indeed be guaranteed. That is, we want to establish conditions under which there exists a test such that the probability of false alarm is less than a given α for all $\theta \in \Theta_0$, but at the same time has maximum probability of detection for all $\theta \in \Theta_1$.

We approach this development through an example; this result motivates the characterization of the conditions for the existence of a UMP test.

Example 11.3.1 Let $X \sim \mathcal{N}(\theta, 1)$. Let $\Theta_0 = (-\infty, \theta_0]$, and let $\Theta_1 = (\theta_0, \infty)$. We wish to test $H_0: \theta \in \Theta_0$ against $H_1: \theta \in \Theta_1$. We desire the test to be uniformly most powerful out of the class of all tests ϕ for which

$$E_\theta \phi(X) \leq \alpha \quad \forall \theta \leq \theta_0. \tag{11.44}$$

To solve this problem we first solve a related problem, and seek the best test ϕ_0 of size α for testing the simple hypothesis $H_0': \theta = \theta_0$ against the simple alternative $H_1': \theta = \theta_1$, where $\theta_1 > \theta_0$. By the Neyman–Pearson lemma, this test is of the form

$$\phi_0(x) = \begin{cases} 1 & \text{if } \frac{1}{\sqrt{2\pi}} \exp[-(x-\theta_1)^2/2] > \frac{v}{\sqrt{2\pi}} \exp[-(x-\theta_0)^2/2], \\ \gamma & \text{if } \frac{1}{\sqrt{2\pi}} \exp[-(x-\theta_1)^2/2] = \frac{v}{\sqrt{2\pi}} \exp[-(x-\theta_0)^2/2], \\ 0 & \text{if } \frac{1}{\sqrt{2\pi}} \exp[-(x-\theta_1)^2/2] < \frac{v}{\sqrt{2\pi}} \exp[-(x-\theta_0)^2/2]. \end{cases}$$

After taking logarithms and rearranging, this test assumes an equivalent form

$$\phi_0(x) = \begin{cases} 1 & \text{if } x > \nu', \\ 0 & \text{otherwise}, \end{cases} \qquad (11.45)$$

where

$$\nu' = \frac{(\theta_1^2/2 - \theta_0^2/2) + \eta}{(\theta_1 - \theta_0)}.$$

(We may set $\gamma = 0$ since the probability that $X = \nu'$ is zero.) With this test, we see that

$$P_{\theta_0}[X > \nu'] = \int_{\nu'}^{\infty} \frac{1}{\sqrt{2\pi}} \exp[-(x - \theta_0)^2/2] \, dx$$

$$= \int_{\nu' - \theta_0}^{\infty} \frac{1}{\sqrt{2\pi}} \exp[-x^2/2] \, dx$$

$$= Q(\nu' - \theta_0) = \alpha$$

implies that

$$\nu' = \theta_0 + Q^{-1}(\alpha). \qquad (11.46)$$

It is important to note that ν' depends only on θ_0 and α, but *not otherwise on* θ_1. In fact, exactly the same test as given by (11.45), with ν' determined by (11.46), is best, according to the Neyman–Pearson lemma, for *all* $\theta_1 \in (\theta_0, \infty)$. Thus, ϕ_0 given by (11.45) is UMP out of the class of all tests for which

$$E_{\theta_0}\phi(X) \leq \alpha.$$

We have thus established that ϕ_0 is UMP for H_0: $\theta = \theta_0$ (simple) and H_1: $\theta > \theta_0$ (composite). To complete the development, we need to extend the discussion to permit H_0: $\theta \leq \theta_0$ (composite). We may do this by establishing that ϕ_0 satisfies the condition given by (11.44). Fix ν' by (11.46) for the given α. Now examine

$$E_{\theta}\phi_0(X) = P_{\theta}[X > \nu']$$

$$= \int_{\nu'}^{\infty} \frac{1}{\sqrt{2\pi}} \exp[-(x - \theta)^2/2] \, dx = Q(\nu' - \theta_0),$$

and note that this quantity is an increasing function of θ (ν' being fixed). Hence,

$$E_{\theta}\phi_0(X) < E_{\theta_0}\phi_0(X) \leq \alpha \qquad \forall \theta \leq \theta_0$$

and, consequently,

$$\sup_{\theta \in (-\infty, \theta_0]} E_{\theta}\phi_0(X) \leq \alpha.$$

Hence, ϕ_0 is uniformly best out of all tests satisfying (11.44); in other words, it is UMP. □

Summarizing, we have established that there does indeed exist a uniformly most powerful test of the hypotheses H_0: $\theta \leq \theta_0$ against the alternatives H_1: $\theta > \theta_0$, for any θ_0 where θ_0 is the mean of a normal random variable X with known variance. Such a test is said to be *one-sided*, and has a very simple form: reject H_0 if $X > \nu'$ and accept H_0 if $X \leq \nu'$, where ν' is chosen to make the size of the test equal to α.

We now turn attention to the issue of determining what conditions on the distribution are sufficient to guarantee the existence of a UMP.

Definition 11.7 A real parameter family of distributions is said to have **monotone likelihood ratio** if densities (or probability mass functions) $f(x \mid \theta)$ exist such that, whenever

11.4 Bayes Decision Theory

$\theta_1 < \theta_2$, the likelihood ratio

$$\ell(x) = \frac{f(x \mid \theta_2)}{f(x \mid \theta_1)}$$

is a nondecreasing function of x in the set of its existence (that is, for x in the set of points for which at least one of $f(x \mid \theta_1)$ and $f(x \mid \theta_2)$ is positive). If $f(x \mid \theta_1) = 0$ and $f(x \mid \theta_2) > 0$, the likelihood ratio is defined as $+\infty$. □

Thus, if the distribution has monotone likelihood ratio, the larger that x is, the more likely that the alternative, H_1, is true.

Theorem 11.3 *(Karlin and Rubin) If the distribution of X has monotone likelihood ratio, then any test of the form*

$$\phi(x) = \begin{cases} 1 & \text{if } x > x_0, \\ \gamma & \text{if } x = x_0, \\ 0 & \text{if } x < x_0, \end{cases} \quad (11.47)$$

has nondecreasing power. Any test of the form (11.47) is UMP of its size for testing H_0: $\theta \leq \theta_0$ against H_1: $\theta > \theta_0$ for any $\theta_0 \in \Theta$, provided its size is not zero. For every $0 < \alpha \leq 1$ and every $\theta_0 \in \Theta$, there exist numbers $-\infty < x_0 < \infty$ and $0 \leq \gamma \leq 1$ such that the test (11.47) is UMP of size α for testing H_0: $\theta \leq \theta_0$ against H_1: $\theta > \theta_0$.

Proof Let θ_1 and θ_2 be any points of Θ with $\theta_1 < \theta_2$. By the Neyman–Pearson lemma, any test of the form

$$\phi(x) = \begin{cases} 1 & \text{if } f_X(x \mid \theta_2) > \nu f_X(x \mid \theta_1), \\ \gamma & \text{if } f_X(x \mid \theta_2) = \nu f_X(x \mid \theta_1), \\ 0 & \text{if } f_X(x \mid \theta_2) < \nu f_X(x \mid \theta_1), \end{cases} \quad (11.48)$$

for $0 \leq \nu < \infty$, is best of its size for testing $\theta = \theta_1$ against $\theta = \theta_2$. Because the distribution has monotone likelihood ratio, any test of the form (11.47) is also of the form (11.48). To see this, note that if $x' < x_0$, then $\ell(x') \leq \ell(x_0)$. For any ν in the range of ℓ there exists an x_0 such that if $\ell(x) = \nu$, then $x = x_0$. Thus, (11.47) is best of size $\alpha > 0$ for testing $\theta = \theta_1$ against $\theta = \theta_2$. The remainder of the proof is essentially the same as the proof for the normal distribution, and will be omitted. □

Example 11.3.2 The one-parameter exponential family of distributions with density (or probability mass function)

$$f(\mathbf{x} \mid \theta) = c(\theta) a(\mathbf{x}) \exp[\pi(\theta) t(\mathbf{x})]$$

has a monotone likelihood ratio provided that both π and t are nondecreasing. To see this, simply write, with $\theta_1 < \theta_2$,

$$\frac{f(\mathbf{x} \mid \theta_2)}{f(\mathbf{x} \mid \theta_1)} = \frac{c(\theta_2)}{c(\theta_1)} \exp\{[\pi(\theta_2) - \pi(\theta_1)] t(\mathbf{x})\},$$

which is nondecreasing in x. □

11.4 Bayes decision theory

Thus far, in our treatment of decision theory, we have considered the parameter as an unknown quantity, but not a random variable, and formulated a decision rule on the basis of maximizing the probability of correct detection (the power) while attempting to keep the probability of false alarm (the size) to an acceptably low level. The result was the likelihood ratio test and ROC.

Decision theory is nothing more than the art of guessing and, as with any art, there is no absolute, or objective, measure of quality. In fact, we are free to invent any principle we like with which to govern our choice of decision rule. In our study of Neyman–Pearson theory, we have seen one attempt at the invention of a principle by which to order decision rules, namely, the notions of power and size. The Bayesian approach constitutes another approach.

11.4.1 The Bayes principle

The Bayes theory requires that the parameter θ be viewed as a random variable (or random vector), rather than just an unknown quantity. This assumption is a major leap, and should not be glossed over lightly. It requires us to accept the premise that nature has specified a particular probability distribution, called the *prior*, or *a priori*, distribution of θ. Furthermore, strictly speaking, Bayesianism requires that we know what this distribution is. These are large pills for some people to swallow, particularly for those of the so-called "objectivist" school that includes those of the Neyman–Pearson persuasion. Bayesianism has been subjected to much criticism from this quarter over the years. But the more modern school of subjective probability has gone a long way towards the development of a rationale for Bayesianism.

Briefly, subjectivists argue that it is not necessary to believe that nature actually chooses a state according to a prior distribution; rather, the prior distribution is viewed merely as a reflection of the belief of the decision-making agent about where the true state of nature lies, and the acquisition of new information, usually in the form of observations, acts to modify the agent's belief about the state of nature. In fact, it can be shown that, in general, every really good decision rule is essentially a Bayes rule with respect to some prior distribution.

In the interest of distinguishing the random variable from the values it assumes, we adopt the notational convention that θ denotes the state of nature viewed as a random variable, and ϑ denotes the values assumed by θ (that is, $\vartheta \in \Theta$, where Θ is the parameter space). Thus we write $[\theta = \vartheta]$ to mean the event that the random variable θ takes on the parameter value ϑ, similar to the way we write $[X = x]$ to mean the event that the random variable X takes on the value x.

To characterize θ as a random variable, we must be able to define the joint distribution of X and θ. Let this distribution be represented by

$$F_{X,\theta}(x, \vartheta).$$

We assume, for our treatment, that such a joint distribution exists, and recall that

$$F_{X,\theta}(x, \vartheta) = F_{X|\theta}(x \mid \vartheta) F_\theta(\vartheta) = F_{\theta|X}(\vartheta \mid x) F_X(x).$$

Note a slight notational change here. With the Neyman–Pearson approach, we did not explicitly include the θ in the subscript of the distribution function, we merely carried it along as a parameter in the argument list of the function. While that notation was suggestive of conditioning, it was not required that we interpret it in that light. Within the Bayesian context, however, we wish to emphasize that the parameter is viewed as a random variable and $F_{X|\theta}$ is a conditional distribution, so we are careful to carry it in subscript of the distribution function as well as in its argument list.

Definition 11.8 The distribution of the random variable θ is called the **prior**, or **a priori** distribution. The set of all possible prior distributions is denoted by the set Θ^*. We make two assumptions about this set of prior distributions: (a) it contains all finite distributions,

11.4 Bayes Decision Theory

that is, all distributions that give all their mass to a finite number of points of Θ; and (b) it is convex, that is, if $\tau_1 \in \Theta^*$ and $\tau_2 \in \Theta^*$, then $a\tau_1 + (1-a)\tau_2 \in \Theta^*$, for all $0 \le a \le 1$ (this is the set of so-called convex combinations). $\square$

11.4.2 The risk function

As we have seen, a strategy (nonrandomized), or decision rule, or decision function, $\phi: \mathcal{X} \to \Delta$ is a rule for deciding $\delta = \phi(x)$ after having observed $X = x$. In the Neyman–Pearson approach, the decision function was chosen in light of the conditional probabilities α and β. In the Bayes approach, a cost is associated with each decision for each state of nature, and an attempt is made to choose a decision function that minimizes the cost.

Recall that in section 10.1.1 we introduced the concept of statistical games. As part of the game, we introduced the cost function $L: \Theta \times \Delta \to \mathbb{R}$, so that $L(\vartheta, \delta)$ is the cost of making decision δ when θ is the true state of nature. If the agent uses decision function $\delta = \phi(X)$, then the loss becomes $L(\vartheta, \phi(X))$ which, for fixed $\vartheta \in \Theta$, is a random variable (i.e., it is a function of the random variable X).

Definition 11.9 The expectation of the loss $L(\vartheta, \phi(X))$, where the expectation is with respect to X, is called the **risk function** $R: \Theta \times D \to \mathbb{R}$, denoted $R(\vartheta, \phi)$:

$$R(\vartheta, \phi) = EL(\vartheta, \phi(X)).$$
$\square$

To ensure that risk is well defined, we must restrict the set of nonrandomized decision rules, D, to only those funtions $\phi: \mathcal{X} \to \mathbb{R}$ for which $R(\vartheta, \phi)$ exists and is finite for all $\vartheta \in \Theta$.

If a pdf $f_{X|\theta}(x \mid \vartheta)$ exists, then the risk function may be written as

$$R(\vartheta, \phi) = \int_{-\infty}^{\infty} L(\vartheta, \phi(x)) f_{X|\theta}(x \mid \vartheta) \, dx.$$

If the probability is purely discrete with pmf $f_{X|\theta}(x_k \mid \vartheta)$, then the risk function may be expressed as

$$R(\vartheta, \phi) = \sum_{k=1}^{N} L(\vartheta, \phi(x_k)) f_{X|\theta}(x_k \mid \vartheta).$$

The risk represents the average loss to the agent when the true state of nature is ϑ and the agent uses the decision rule ϕ. Application of the Bayes principle, however, permits us to view $R(\theta, \phi)$ as a random variable, since it is a function of the random variable θ.

Example 11.4.1 (Odd or even) The game of "odd or even," mentioned in example 10.1.2, may be extended to a statistical decision problem. Suppose that before the game is played, the agent is allowed to ask nature how many fingers it intends to put up and that nature must answer truthfully with probability 3/4 (hence untruthfully with probability 1/4). (This models, for example, a noisy observation.) The agent observes a random variable X (the answer nature gives), taking the values of 1 or 2. If $\theta = 1$ is the true state of nature, the probability that $X = 1$ is 3/4; that is, $P(X = 1 \mid \theta = 1) = 3/4$. Similarly, $P(X = 1 \mid \theta = 2) = 1/4$. The observation space in this case is $\mathcal{X} = \{1, 2\}$. The choice of nature, and the observation produced, can be represented as shown in figure 11.10. The decision space is $\Delta = \{1, 2\}$. Recall that the loss function is

$$L(1, 1) = -2,$$
$$L(1, 2) = 3,$$
$$L(2, 1) = 3,$$
$$L(2, 2) = -4.$$

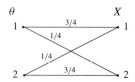

Figure 11.10: Illustration of even–odd observations

We first examine the possible decision functions. For small decision problems such as this, it is possible to exhaustively enumerate all decision functions. There are exactly four possible functions from $\mathcal{X}$ into Δ, so $D = \{\phi_1, \phi_2, \phi_3, \phi_4\}$, where

$$
\begin{aligned}
\phi_1(1) &= 1 & \phi_1(2) &= 1, \\
\phi_2(1) &= 1 & \phi_2(2) &= 2, \\
\phi_3(1) &= 2 & \phi_3(2) &= 1, \\
\phi_4(1) &= 2 & \phi_4(2) &= 2.
\end{aligned}
\tag{11.49}
$$

Rules ϕ_1 and ϕ_4 ignore the value of X. Rule ϕ_2 reflects the agent's belief that nature is telling the truth, and rule ϕ_3, that nature is not telling the truth.

Let us now examine the risk function $R(\theta, \phi)$ for this game. For example,

$$R(\theta, \phi_1) = \sum_{x=1}^{2} L(\theta, \phi_1(x)) f_X(x \mid \theta).$$

When $\theta = 1$ we have

$$R(1, \phi_1) = \sum_{x=1}^{2} L(1, \phi_1(1)) f_X(1 \mid 1) + L(1, \phi_1(2)) f_X(2 \mid 1) = (-2)(3/4) + (-2)(1/4) = -2.$$

The risk matrix, given in figure 11.11, characterizes this statistical game. □

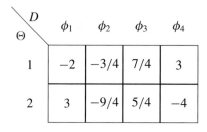

Figure 11.11: Risk function for statistical odd or even game

Example 11.4.2 (Binary channel) Consider now the problem of transmission in a binary channel with crossover probabilities λ_0 and λ_1, as shown in figure 11.12. As for the odd or even game, four possible decision functions exist:

$$
\begin{aligned}
\phi_1(0) &= 0 & \phi_1(1) &= 0, \\
\phi_2(0) &= 0 & \phi_2(1) &= 1, \\
\phi_3(0) &= 1 & \phi_3(1) &= 0, \\
\phi_4(0) &= 1 & \phi_4(1) &= 1.
\end{aligned}
\tag{11.50}
$$

where the first and last decision functions ignore the measured value, and the third decision function

11.4 Bayes Decision Theory

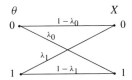

Figure 11.12: A binary channel

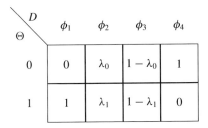

Figure 11.13: Risk function for binary channel

reflects a belief that the observed value is incorrect. If we assume the cost structure

$$L(x, y) = \delta_{x-y}$$

(that is, the cost of making bit errors), then the risk function for this "game" is shown in figure 11.13. □

With the introduction of the risk function, R, and the class of decision functions, D, we may replace the original game (Θ, Δ, L) by a new game which we will denote by the triple (Θ, D, R), in which the space D and the function R have an underlying structure, depending on Δ and L and the distribution of X, whose exploitation is the main objective of decision theory. Sometimes the triple (Θ, D, R) is called a statistical game.

11.4.3 Bayes risk

We might suppose that a reasonable decision criterion would be to choose the decision rule ϕ such that the risk is minimized, but generally this is not possible, since the value θ assumes is *unknown*—we cannot unilaterally minimize the risk as long as the loss function depends on θ (and that takes in just about all interesting cases). A natural way to deal with this situation in the Bayesian context is to compute the average risk and then find a decision rule that minimizes this average risk. Under the assumption that θ is a random variable, we can now introduce the concept of Bayes risk.

Definition 11.10 The **Bayes risk function** with respect to a prior distribution, F_θ, denoted $r(F_\theta, \phi)$, is given by $r(F_\theta, \phi) = ER(\theta, \phi)$, where the expectation is taken over the space Θ of values that θ may assume:

$$r(F_\theta, \phi) = \int_\Theta R(\vartheta, \phi) f_\theta(\vartheta) \, d\vartheta$$

when F_θ has a pdf $f_\theta(\vartheta)$, and

$$r(F_\theta, \phi) = \sum_{\vartheta \in \Theta} R(\vartheta_i, \phi) f_\theta(\vartheta)$$

when F_θ has a pmf $f_\theta(\vartheta)$. □

We note that the risk R is defined as the average of the loss function, obtained by averaging over all values $X = x$ for a fixed θ; the Bayes risk r, however, is the average value of the loss function obtained by averaging over all values $X = x$ and $\theta = \vartheta$. For example, when both X and θ are continuous,

$$r(F_\theta, \phi) = EL[\theta, \phi(X)]$$
$$= \int_\Theta R(\vartheta, \phi) f_\theta(\vartheta) \, d\vartheta$$
$$= \int_\Theta \int_\mathcal{X} L[\vartheta, \phi(x)] f_{X|\theta}(x \mid \vartheta) f_\theta(\vartheta) \, dx \, d\vartheta. \quad (11.51)$$

If X is continuous and θ is discrete, then

$$r(F_\theta, \phi) = EL[\theta, \phi(X)]$$
$$= \sum_{\vartheta \in \Theta} R(\vartheta, \phi) f_\theta(\vartheta)$$
$$= \sum_{\vartheta \in \Theta} \int_\mathcal{X} L[\vartheta, \phi(x)] f_{X|\theta}(x \mid \vartheta) f_\theta(\vartheta) \, dx. \quad (11.52)$$

The remaining constructions when X is discrete are also easily obtained.

We may extend the definition of Bayes risk to randomized decision rules by taking the expectation of Bayes risk with respect to the rule. Let $\varphi \in D^*$ be a randomized rule over the set D of nonrandomized rules. Then the Bayes risk with respect to the prior θ and the randomized decision rule φ is

$$r(F_\theta, \varphi) = E_\varphi r(F_\theta, \phi). \quad (11.53)$$

For example, if $D = \{\phi_1, \ldots, \phi_k\}$, and $\varphi = (\pi_1, \ldots, \pi_k)$, then

$$r(F_\theta, \varphi) = \sum_{i=1}^k r(F_\theta, \phi_i) \pi_i. \quad (11.54)$$

11.4.4 Bayes tests of simple binary hypotheses

Let $\Theta = \{\vartheta_0, \vartheta_1\}$, corresponding to the hypotheses H_0 and H_1, respectively, and let $\Delta = (\delta_0, \delta_1)$, correspond respectively. We desire to fashion a decision rule, or $\phi \colon \mathcal{X} \to \mathbb{R}$ such that, when $X = x$ is observed,

$$\phi(x) = \begin{cases} 1 & \text{if } x \in \mathcal{R} \text{ (reject } H_0), \\ 0 & \text{if } x \in \mathcal{A} \text{ (accept } H_0), \end{cases} \quad (11.55)$$

where $\mathcal{R}$ and $\mathcal{A}$ are disjoint subsets of $\mathcal{X}$, and $\mathcal{X} = \mathcal{R} \cup \mathcal{A}$. We interpret this decision rule as follows: If $x \in \mathcal{R}$ we take action δ_1 (that is, choose H_1), and if $x \in \mathcal{A}$ we take action δ_0 (choose H_0). In order to establish ϕ, we must determine the sets $\mathcal{R}$ and $\mathcal{A}$. The risk function for rule (11.55) is

$$R(\theta, \phi) = \int_\mathcal{A} L(\theta, \phi(x)) f_X(x \mid \theta) \, dx + \int_\mathcal{R} L(\theta, \phi(x)) f_X(x \mid \theta) \, dx$$
$$= L(\theta, \delta_0) P(\mathcal{A} \mid \theta) + L(\theta, \delta_1) P(\mathcal{R} \mid \theta)$$
$$= [1 - P(\mathcal{R} \mid \theta)] L(\theta, \delta_0) + P(\mathcal{R} \mid \theta) L(\theta, \delta_1)$$
$$= L(\theta, \delta_0) + P(\mathcal{R} \mid \theta)[L(\theta, \delta_1) - L(\theta, \delta_0)],$$

11.4 Bayes Decision Theory

where by $P(\mathcal{R}\,|\,\theta)$ we mean the conditional probability that X will take values in $\mathcal{R}$, given θ. For our particular choice of decision rule, we observe that the conditional expectation of $\phi(X)$ given θ is

$$E[\phi(X)\,|\,\theta] = 1 \cdot P(\mathcal{R}\,|\,\theta) + 0 \cdot [1 - P(\mathcal{R}\,|\,\theta)]$$
$$= P(\mathcal{R}\,|\,\theta),$$

so we may write

$$R(\theta, \phi) = L(\theta, \delta_0) + E[\phi(X)\,|\,\theta][L(\theta, \delta_1) - L(\theta, \delta_0)].$$

For the case of binary alternatives and simple binary hypotheses, there are four types of cost that we might incur:

1. The cost of deciding H_0, given that H_0 is correct, denoted L_{00}.
2. The cost of deciding H_1, given that H_1 is correct, denoted L_{11}.
3. The cost of deciding H_0, given that H_1 is correct, denoted L_{10}.
4. The cost of deciding H_1, given that H_0 is correct, denoted L_{01}.

More generally, L_{ij} indicates the cost of choosing H_j, given that H_i is correct.

The risk function becomes

$$R(\theta, \phi) = \begin{cases} L_{00}P(\mathcal{A}\,|\,\theta_0) + L_{01}P(\mathcal{R}\,|\,\theta_1) & \theta = \vartheta_0, \\ L_{10}P(\mathcal{A}\,|\,\theta_0) + L_{11}P(\mathcal{R}\,|\,\theta_1) & \theta = \vartheta_1. \end{cases} \quad (11.56)$$

We also introduce the probability notation

$$P(\mathcal{A}\,|\,\theta_0) = \text{probability of correct acceptance} = 1 - \alpha = P_{00},$$
$$P(\mathcal{R}\,|\,\theta_0) = \text{probability of false alarm} = \alpha = P_{01},$$
$$P(\mathcal{A}\,|\,\theta_1) = \text{probability of missed detection} = 1 - \beta = P_{10},$$
$$P(\mathcal{R}\,|\,\theta_1) = \text{probability of detection} = \beta = P_{11}.$$

On this basis, we can write

$$R(\theta, \phi) = \begin{cases} L_{00}P_{00} + L_{01}P_{01} & \theta = \vartheta_0, \\ L_{10}P_{10} + L_{11}P_{11} & \theta = \vartheta_1, \end{cases}$$

$$= \begin{cases} (L_{01} - L_{00})P_{01} + L_{00} & \theta = \vartheta_0, \\ (L_{10} - L_{11})P_{10} + L_{11} & \theta = \vartheta_1. \end{cases} \quad (11.57)$$

From (11.57), we observe that, no matter what decision we make, there is a constant cost L_{00} associated with the case $\theta = \vartheta_0$, and similarly a constant cost L_{11} associated with $\theta = \vartheta_1$. It is customary to assume that $L_{00} = L_{11} = 0$, making adjustments to L_{01} and L_{10} as necessary. We then have

$$R(\theta, \phi) = \begin{cases} L_{01}P_{01} = L_{01}\alpha & \theta = \vartheta_0, \\ L_{10}P_{10} = L_{10}(1 - \beta) & \theta = \vartheta_1. \end{cases} \quad (11.58)$$

We now introduce the number p as the prior probability,

$$p = f_\theta(\vartheta_1) = P[\theta = \vartheta_1],$$
$$1 - p = f_\theta(\vartheta_0) = P[\theta = \vartheta_0]. \quad (11.59)$$

Although the preceding development involves only nonrandomized rules, we may easily extend to randomized rules by replacing ϕ with φ in all cases (recall that nonrandomized rules may be viewed as degenerate randomized rules, where all of the probability mass is placed on one nonrandomized rule). As p represents the distribution F_θ, we will write the Bayes risk $r(F_\theta, \varphi)$ as $r(p, \varphi)$ (see (11.53)). The Bayes risk is

$$r(p, \varphi) = (1 - p)R(\vartheta_0, \varphi) + pR(\vartheta_1, \varphi). \tag{11.60}$$

Any (randomized) decision function that, for fixed p, minimizes the Bayes risk is said to be **Bayes with respect to p**, and will be denoted φ_p, which satisfies

$$\varphi_p = \arg \min_{\varphi \in D^*} r(p, \phi). \tag{11.61}$$

The usual intuitive meaning associated with (11.60) is the following. Suppose that you know (or believe) that the unknown parameter θ is in fact a random variable with specified prior probabilities of p and $1 - p$ of taking values ϑ_1 and ϑ_0, respectively. Then for any decision function φ, the "global" expected loss will be given by (11.60), and hence it will be reasonable to use the decision function φ_p that minimizes $r(p, \varphi)$.

We now proceed to find the decision function φ_p which minimizes the Bayes risk. We assume that the two conditional distributions of X, for $\theta = \vartheta_0$ and $\theta = \vartheta_1$, are given in terms of density functions $f_{X|\theta}(x \mid \vartheta_0)$ and $f_{X|\theta}(x \mid \vartheta_1)$. Then, from (11.58) and (11.60), we have

$$r(p, \phi) = pL_{10}(1 - E[\phi(X) \mid \theta = \vartheta_1]) + (1 - p)L_{01}E[\phi(X) \mid \theta = \vartheta_0]$$

$$= pL_{10}\left(1 - \int_{\mathcal{X}} f_{X|\theta}(x \mid \vartheta_1)\phi(x)\,dx\right) + (1 - p)L_{01}\int_{\mathcal{X}} f_{X|\theta}(x \mid \vartheta_0)\phi(x)\,dx$$

$$= pL_{10} + \int_{\mathcal{X}} (-pL_{10}f_{X|\theta}(x \mid \vartheta_1) + (1 - p)L_{01}f_{X|\theta}(x \mid \vartheta_0))\phi(x)\,dx. \tag{11.62}$$

This last expression is minimized by minimizing the integrand for each x; that is, by defining $\phi(x)$ to be

$$\phi(x) = \begin{cases} 1 & \text{if } (1 - p)L_{01}f_{X|\theta}(x \mid \vartheta_0) < pL_{10}f_{X|\theta}(x \mid \vartheta_1), \\ 0 & \text{if } (1 - p)L_{01}f_{X|\theta}(x \mid \vartheta_0) > pL_{10}f_{X|\theta}(x \mid \vartheta_1), \\ \text{arbitrary} & \text{if } (1 - p)L_{01}f_{X|\theta}(x \mid \vartheta_0) = pL_{10}f_{X|\theta}(x \mid \vartheta_1). \end{cases}$$

For this binary problem, the Bayes risk is unaffected by the equality condition $(1 - p)L_{01}f_{X|\theta}(x \mid \vartheta_0) = pL_{10}f_{X|\theta}(x \mid \vartheta_1)$ and, therefore, without loss of generality, we may place all of the probability mass of the randomized decision rule on the nonrandomized rule

$$\phi_p(x) = \begin{cases} 1 & \text{if } (1 - p)L_{01}f_{X|\theta}(x \mid \vartheta_0) < pL_{10}f_{X|\theta}(x \mid \vartheta_1), \\ 0 & \text{otherwise.} \end{cases} \tag{11.63}$$

We may define the sets $\mathcal{R}$ and $\mathcal{A}$ as

$$\mathcal{R} = \{x \colon (1 - p)L_{01}f_{X|\theta}(x \mid \vartheta_0) < pL_{10}f_{X|\theta}(x \mid \vartheta_1)\},$$

$$\mathcal{A} = \{x \colon (1 - p)L_{01}f_{X|\theta}(x \mid \vartheta_0) \geq pL_{10}f_{X|\theta}(x \mid \vartheta_1)\};$$

then (11.62) becomes

$$r(p, \phi_p) = pL_{10}\left(1 - \int_{\mathcal{X}} f_{X|\theta}(x \mid \vartheta_1)I_{\mathcal{R}}(x)\,dx\right) + (1 - p)L_{01}\int_{\mathcal{X}} f_{X|\theta}(x \mid \vartheta_0)I_{\mathcal{R}}(x)\,dx$$

$$= pL_{10}\int_{\mathcal{X}} f_{X|\theta}(x \mid \vartheta_1)I_{\mathcal{A}}(x)\,dx + (1 - p)L_{01}\int_{\mathcal{X}} f_{X|\theta}(x \mid \vartheta_0)I_{\mathcal{R}}(x)\,dx.$$

$$\tag{11.64}$$

11.4 Bayes Decision Theory

Since we decide $\theta = \vartheta_1$ if $x \in \mathcal{R}$ and $\theta = \vartheta_0$ if $x \in \mathcal{A}$, we observe that by setting $L_{01} = L_{10} = 1$, the Bayes risk (11.64) becomes the total probability of error:

$$r(p, \phi_p) = \underbrace{P[\mathcal{R} \mid \theta = \vartheta_0]}_{P_{FA}} P[\theta = \theta_0] + \underbrace{P[\mathcal{A} \mid \theta = \vartheta_1]}_{P_{MD}} P[\theta = \theta_1]. \qquad (11.65)$$

Observe that $\phi_p(x)$ of (11.63) may be written as a likelihood ratio test:

$$\phi_p(x) = \begin{cases} 1 & \text{if } \dfrac{f_{X\mid\theta}(x \mid \vartheta_1)}{f_{X\mid\theta}(x \mid \vartheta_0)} > \dfrac{L_{01} P(H_0)}{L_{10} P(H_1)}, \\ 0 & \text{otherwise.} \end{cases} \qquad (11.66)$$

It is important to note that for binary decision problems under simple hypotheses, this test is identical in form to the solution to the Neyman–Pearson test; only the threshold is changed. Whereas the threshold for the Neyman–Pearson test was determined by the size of the test, the Bayesian formulation provides the threshold as a function of the prior distribution on θ, and the costs associated with the decisions.

Example 11.4.3 (Binary channel) Consider again the binary channel of example 11.4.2. We want to devise a Bayes test for this channel.

$$f_X(1 \mid 1) = P(X = 1 \mid \theta = 1) = 1 - \lambda_1 \qquad f_X(0 \mid 1) = P(X = 0 \mid \theta = 1) = \lambda_1,$$
$$f_X(1 \mid 0) = P(X = 1 \mid \theta = 0) = \lambda_0 \qquad f_X(0 \mid 0) = P(X = 0 \mid \theta = 0) = 1 - \lambda_0.$$

We can write a likelihood ratio

$$\ell(x) = \frac{f_X(x \mid 1)}{f_X(x \mid 0)} = \begin{cases} \dfrac{1 - \lambda_1}{\lambda_0} & x = 1, \\ \dfrac{\lambda_1}{1 - \lambda_0} & x = 0. \end{cases}$$

If costs are appropriate for communications, $L_{01} = L_{10} = 1$, then the decision rule is

$$\phi_p(y) = \begin{cases} 1 & \ell(y) > \dfrac{1-p}{p}, \\ 0 & \text{otherwise.} \end{cases} \qquad (11.67)$$

For example, when $p = 1/2$, the decision rule is

$$\phi(1) = \begin{cases} 1 & 1 - \lambda_1 \geq \lambda_0, \\ 0 & \text{otherwise,} \end{cases}$$

$$\phi(0) = \begin{cases} 1 & \lambda_1 \geq 1 - \lambda_0, \\ 0 & \text{otherwise.} \end{cases} \qquad \square$$

Example 11.4.4 Let $\Theta = \{\theta_0, \theta_1\} = \{\mathbf{m}_0, \mathbf{m}_1\}$. Let us assume that, under hypothesis H_1, $\mathbf{X} \sim \mathcal{N}(\mathbf{m}_1, R)$, and under hypothesis H_0 that $\mathbf{X} \sim \mathcal{N}(\mathbf{m}_0, R)$, where $\mathbf{X}$ is an n-dimensional random vector. Denoting the mean of the distribution by $\mathbf{m}$, we assume that we have the following prior information:

$$P[\mathbf{m} = \mathbf{m}_1] = p,$$
$$P[\mathbf{m} = \mathbf{m}_0] = 1 - p.$$

The likelihood ratio is

$$\ell(\mathbf{x}) = \frac{f_{\mathbf{X}}(\mathbf{x} \mid \theta_1)}{f_{\mathbf{X}}(\mathbf{x} \mid \theta_0)}$$

$$= \frac{\exp\left[-\tfrac{1}{2}(\mathbf{x} - \mathbf{m}_1)^T R^{-1} (\mathbf{x} - \mathbf{m}_1)\right]}{\exp\left[-\tfrac{1}{2}(\mathbf{x} - \mathbf{m}_0)^T R^{-1} (\mathbf{x} - \mathbf{m}_0)\right]}.$$

After canceling common terms and taking the logarithm, we have the log-likelihood ratio

$$\Lambda(\mathbf{x}) = \log \ell(\mathbf{x}) = (\mathbf{m}_1 - \mathbf{m}_0)^T R^{-1}(\mathbf{x} - \mathbf{x}_0), \tag{11.68}$$

where $\mathbf{x}_0 = \frac{1}{2}(\mathbf{m}_1 + \mathbf{m}_0)$.

Suppose the problem now is to detect the mean of $\mathbf{X}$, and the only criterion is correctness of the decision. Then, for a cost function we can take $L_{01} = L_{10} = 1$. Based on the decision rule from (11.66), we have

$$\phi_p(\mathbf{x}) = \begin{cases} 1 & \text{if } \Lambda(\mathbf{x}) > \log \frac{1-p}{p} = \eta, \\ 0 & \text{otherwise.} \end{cases}$$

As before, we could compute the probabilities of error $P_{FA} = P(\phi_p = 1 \mid H_0)$, and the probability of missed detection $P_{MD} = P(\phi_t = 0 \mid H_1)$. In this case,

$$P_{FA} = P_{\theta_0}(\Lambda(\mathbf{X}) > \eta)$$
$$= Q((\eta + s^2/2)/s) = Q(z),$$

where

$$s^2 = \mathbf{w}^T R \mathbf{w} \quad \text{and} \quad \mathbf{w} = R^{-1}(\mathbf{m}_1 - \mathbf{m}_0)$$

and

$$z = \eta/s + s/2.$$

Also,

$$P_D = P_{\theta_1}(\Lambda(\mathbf{X}) > \eta)$$
$$= Q((\eta - s^2/2)/s) = Q(z - s).$$

Given our model of the prior distribution, we can also compute the total probability of error:

$$P(\mathcal{E}) = P(\phi_t = 1 \mid H_0)P(H_0) + P(\phi_t = 0 \mid H_1)P(H_1)$$
$$= pQ((s^2 - \log((1-p)/p))/s) + (1-p)Q((\log((1-p)/p) + s^2)/s)$$
$$= p(1 - Q(z-s)) + (1-p)Q(z). \tag{11.69}$$

□

11.4.5 Posterior distributions

If the distribution of the parameter θ before observations are made is called the prior distribution, then it is natural to consider defining a posterior distribution as the distribution of the parameter after observations are taken and processed.

We first consider the case for X and θ both continuous. Assuming we can reverse the order of integration in (11.51), we obtain

$$r(F_\theta, \phi) = \int_\Theta \int_\mathcal{X} L[\vartheta, \phi(x)] f_{X|\theta}(x \mid \vartheta) f_\theta(\vartheta) \, dx \, d\vartheta$$
$$= \int_\mathcal{X} \int_\Theta L[\vartheta, \phi(x)] \underbrace{f_{X|\theta}(x \mid \vartheta) f_\theta(\vartheta)}_{f_{X\theta}(x, \vartheta)} \, d\vartheta \, dx$$
$$= \int_\mathcal{X} \left\{ \int_\Theta L[\vartheta, \phi(x)] f_{\theta|X}(\vartheta \mid x) \, d\vartheta \right\} f_X(x) \, dx, \tag{11.70}$$

where we have used the fact that

$$f_{X|\theta}(x \mid \vartheta) f_\theta(\vartheta) = f_{X\theta}(x, \vartheta) = f_{\theta|X}(\vartheta \mid x) f_X(x).$$

11.4 Bayes Decision Theory

In other words, a choice of θ by the marginal distribution $f_\theta(\vartheta)$, followed by a choice of X from the conditional distribution $f_{X|\theta}(x \mid \vartheta)$, determines a joint distribution of θ and X, which in turn can be determined by first choosing X according to its marginal distribution $f_X(x)$ and then choosing θ according to the conditional distribution of θ, $f_{\theta|X}(\vartheta \mid x)$, given $X = x$.

With this change in the order of integration, some very useful insight may be obtained. We see that we may minimize the Bayes risk given in (11.70) by finding a decision function $\phi(x)$ that *minimizes the inside integral separately for each x*; that is, we may find for each x a rule that minimizes

$$\int_\Theta L[\vartheta, \phi(x)] f_{\theta|X}(\vartheta \mid x) \, d\vartheta. \tag{11.71}$$

Definition 11.11 The conditional distribution of θ, given X, denoted $f_{\theta|X}(\vartheta \mid x)$, is called the **posterior**, or **a posteriori**, distribution of θ. It is frequently written using Bayes theory as

$$f_{\theta|X}(\vartheta_1 \mid x) = \frac{f_{X|\theta}(x \mid \vartheta_1) f_\theta(\vartheta_1)}{\int f_{X|\theta}(x \mid \vartheta) f_\theta(\vartheta) \, d\vartheta}.$$

□

The expression (11.71) is the expected loss given that $X = x$, and we may, therefore, interpret a Bayes decision rule as one that *minimizes the posterior conditional expected loss, given the observation*.

The above results need be modified only in notation for the case where X and θ are discrete. For example, if θ is discrete, say $\Theta = \{\vartheta_1, \ldots, \vartheta_k\}$, we reverse the order of summation and integration in (11.52) to obtain

$$r(p, \phi) = \sum_{i=1}^k \int_\mathcal{X} L[\vartheta_i, \phi(x)] f_{X|\theta}(x \mid \vartheta_i) f_\theta(\vartheta_i) \, dx$$

$$= \int_\mathcal{X} \sum_{i=1}^k L[\vartheta_i, \phi(x)] f_{X|\theta}(x \mid \vartheta_i) f_\theta(\vartheta_i) \, dx$$

$$= \int_\mathcal{X} \left\{ \sum_{i=1}^k L[\vartheta_i, \phi(x)] f_{\theta|X}(\vartheta_i \mid x) \right\} f_X(x) \, dx. \tag{11.72}$$

Suppose that there are only two hypotheses, $\Theta = \{\vartheta_0, \vartheta_1\}$, and decisions corresponding to each of these. Then

$$\sum_{i=1}^2 L[\vartheta_i, \phi(x)] f_{\theta|X}(\vartheta_i \mid x) = \begin{cases} L_{00} f_{\theta|X}(\vartheta_0 \mid x) + L_{10} f_{\theta|X}(\vartheta_1 \mid x) & \phi(x) = 0, \\ L_{01} f_{\theta|X}(\vartheta_0 \mid x) + L_{11} f_{\theta|X}(\vartheta_1 \mid x) & \phi(x) = 1. \end{cases}$$

Determination of $\phi(x)$ on the basis of minimum risk can be stated as follows: set $\phi(x) = 1$ if

$$L_{01} f_{\theta|X}(\vartheta_0 \mid x) + L_{11} f_{\theta|X}(\vartheta_1 \mid x) < L_{00} f_{\theta|X}(\vartheta_0 \mid x) + L_{10} f_{\theta|X}(\vartheta_1 \mid x),$$

which leads to the likelihood ratio test

$$\phi(x) = \begin{cases} 1 & \frac{f_{\theta|X}(\vartheta_1 \mid x)}{f_{\theta|X}(\vartheta_0 \mid x)} > \frac{L_{01} - L_{00}}{L_{10} - L_{11}}, \\ 0 & \text{otherwise.} \end{cases} \tag{11.73}$$

It is interesting to contrast this rule with that derived in (11.66), which is reproduced here:

$$\phi_p(x) = \begin{cases} 1 & \text{if } \dfrac{f_{X|\theta}(x\,|\,\vartheta_1)}{f_{X|\theta}(x\,|\,\vartheta_0)} > \dfrac{L_{01}(1-p)}{L_{10}p}\,\eta, \\ 0 & \text{otherwise.} \end{cases}$$

In (11.73), the threshold is determined only by the Bayes costs, and the ratio is a ratio of *posterior* densities (strictly speaking, not a likelihood ratio).

Example 11.4.5 Let us consider the simple hypothesis versus simple alternative problem formulation, and let $\Theta = \{\vartheta_0, \vartheta_1\}$ and $\Delta = \{0, 1\}$. Assume that we observe a random variable X taking values in $\{x_0, x_1\}$, with the following conditional distributions:

$$f_{X|\theta}(x_1\,|\,\vartheta_0) = P[X = x_1\,|\,\theta = \vartheta_0] = \frac{3}{4} \qquad f_{X|\theta}(x_0\,|\,\vartheta_0) = P[X = x_0\,|\,\theta = \vartheta_0] = \frac{1}{4},$$

$$f_{X|\theta}(x_1\,|\,\vartheta_1) = P[X = x_1\,|\,\theta = \vartheta_1] = \frac{1}{3} \qquad f_{X|\theta}(x_0\,|\,\vartheta_1) = P[X = x_0\,|\,\theta = \vartheta_1] = \frac{2}{3}.$$

We will take the loss function for this problem as given by the matrix in figure 11.14. (This example could be thought of as a generalization of the binary channel, with crossover probabilities 3/4 and 2/3, and with different costs associated with the different kinds of error.)

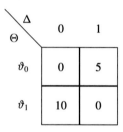

Figure 11.14: Loss function

Let $P[\theta = \vartheta_1] = p$ and $P[\theta = \vartheta_0] = 1 - p$ be the prior distribution for θ, for $0 \le p \le 1$. We will address this problem by solving for the *a posteriori* pmf. The posterior pmf is given, via Bayes theorem, as

$$f_{\theta|X}(\vartheta_1\,|\,x) = \frac{f_{X|\theta}(x\,|\,\vartheta_1)f_\theta(\vartheta_1)}{f_{X|\theta}(x\,|\,\vartheta_0)f_\theta(\vartheta_0) + f_{X|\theta}(x\,|\,\vartheta_1)f_\theta(\vartheta_1)} \tag{11.74}$$

$$= \begin{cases} \dfrac{\frac{1}{3}p}{\frac{3}{4}(1-p) + \frac{1}{3}p} & \text{if } x = x_1, \\[2mm] \dfrac{\frac{2}{3}p}{\frac{1}{4}(1-p) + \frac{2}{3}p} & \text{if } x = x_0. \end{cases} \tag{11.75}$$

Note that

$$f_{\theta|X}(\vartheta_0\,|\,x) = 1 - f_{\theta|X}(\vartheta_1\,|\,x).$$

After the value $X = x$ has been observed, a choice must be made between the two actions $\delta = 0$ and $\delta = 1$. The Bayes decision rule is

$$\phi_p(x) = \arg\min_{\phi}\{L(\vartheta_1, \phi)f_{\theta|X}(\vartheta_1\,|\,x) + L(\vartheta_0, \phi)f_{\theta|X}(\vartheta_0\,|\,x)\}$$

$$= \begin{cases} \arg\min_\phi \left\{ L(\vartheta_1,\phi)\dfrac{\frac{1}{3}p}{\frac{3}{4}(1-p)+\frac{1}{3}p} + L(\vartheta_0,\phi)\dfrac{\frac{3}{4}(1-p)}{\frac{3}{4}(1-p)+\frac{1}{3}p} \right\} & \text{if } x = x_1, \\[3mm] \arg\min_\phi \left\{ L(\vartheta_1,\phi)\dfrac{\frac{2}{3}p}{\frac{1}{4}(1-p)+\frac{2}{3}p} + L(\vartheta_0,\phi)\dfrac{\frac{1}{4}(1-p)}{\frac{1}{4}(1-p)+\frac{2}{3}p} \right\} & \text{if } x = x_0, \end{cases} \tag{11.76}$$

11.4 Bayes Decision Theory

for $\phi \in \{0, 1\}$. Consider the case when $x = x_1$: the risk function is either

$$10\frac{\frac{1}{3}p}{\frac{3}{4}(1-p) + \frac{1}{3}p} \quad \text{or} \quad 5\frac{\frac{3}{4}(1-p)}{\frac{3}{4}(1-p) + \frac{1}{3}p}, \tag{11.77}$$

depending upon whether $\phi = 0$ or $\phi = 1$. A plot of these two risk functions is shown on the left hand side of figure 11.15. Equating the two risk functions in (11.77) to find the point of intersection, we find that

$$\phi_p(x_1) = \begin{cases} 0 & \text{if } p \leq \frac{9}{17}, \\ 1 & \text{if } p > \frac{9}{17}. \end{cases}$$

The right-hand side of figure 11.15 similarly shows the risk function when $x = x_0$. Again, the threshold can be found, and the decision rule in this case is

$$\phi_p(x_0) = \begin{cases} 0 & \text{if } p \leq \frac{3}{19}, \\ 1 & \text{if } p > \frac{3}{19}. \end{cases}$$

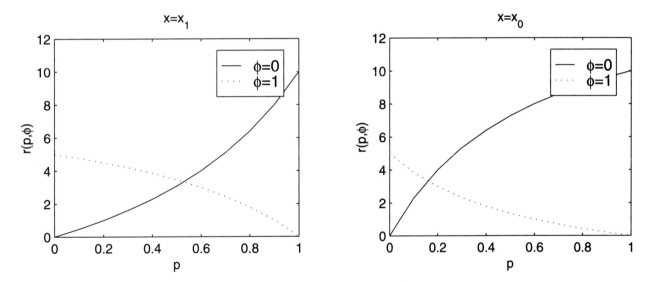

Figure 11.15: Bayes risk for a decision

We may compute the Bayes risk function as follows. If $0 \leq p < \frac{3}{19}$, then it follows that $\phi_p(x) \equiv 0$ will be the Bayes rule, whatever the value of x. The corresponding Bayes risk is $0 \cdot (1-p) + 10p = 10p$. If $\frac{3}{19} \leq p \leq \frac{9}{17}$, then $\phi_p(x_0) = 1$ and $\phi_p(x_1) = 0$ is the Bayes decision function, and the corresponding risk is

$$r(p, \phi_p) = pR(\vartheta_1, \phi_p) + (1-p)R(\vartheta_0, \phi_p)$$

$$= p\left[10 \cdot \frac{1}{3} + 0 \cdot \frac{2}{3}\right] + (1-p)\left[0 \cdot \frac{3}{4} + 5 \cdot \frac{1}{4}\right]$$

$$= \frac{10}{3}p + \frac{5}{4}(1-p) = \frac{25}{12}p + \frac{5}{4}.$$

If $\frac{9}{17} < p \leq 1$, then $\phi_p(x) \equiv 1$ is the Bayes rule, and the Bayes risk is $5(1-p)$. □

11.4.6 Detection and sufficiency

We have seen that for binary tests in both the Neyman–Pearson and Bayes detectors, the decision function can be expressed in terms of the likelihood ratio,

$$\ell(\mathbf{x}) = \frac{f_\mathbf{X}(\mathbf{x} \mid \theta_1)}{f_\mathbf{X}(\mathbf{x} \mid \theta_0)}.$$

If $\mathbf{t}(\mathbf{x})$ is sufficient for θ, so that $f_\mathbf{X}(\mathbf{x} \mid \theta_i) = b(\mathbf{t}(\mathbf{x}), \theta_i)a(\mathbf{x})$, the likelihood ratio becomes

$$\ell(\mathbf{x}) = \frac{b(\mathbf{t}(\mathbf{x}), \theta_1)}{b(\mathbf{t}(\mathbf{x}), \theta_0)}. \tag{11.78}$$

The ratio in (11.78) is equivalent to the ratio of density functions,

$$\ell(\mathbf{x}) = \frac{f_\mathbf{T}(\mathbf{t} \mid \theta_1)}{f_\mathbf{T}(\mathbf{t} \mid \theta_0)},$$

which is naturally denoted as $\ell(\mathbf{t})$. On this basis, the decision function for a binary test becomes a function only of the sufficient statistic, not of the entire set of observed data:

$$\phi(\mathbf{t}) = \begin{cases} 1 & \ell(\mathbf{t}) > \nu, \\ \gamma & \ell(\mathbf{t}) = \nu, \\ 0 & \ell(\mathbf{t}) < \nu, \end{cases}$$

for some suitably chosen threshold ν. In the Neyman–Pearson test, the threshold is selected to produce the desired size α for the test. In the Bayes test, it is selected for minimum risk.

Example 11.4.6 Suppose $X_i \sim \mathcal{P}(\lambda)$ (that is, X_i is Poisson distributed), for $i = 1, 2, \ldots, n$. We desire to test H_0: $\lambda = \lambda_0$ versus H_1: $\lambda = \lambda_1$, where $\lambda_1 > \lambda_0$. The random variable

$$T = \sum_{i=1}^{n} X_i$$

is sufficient for λ. T is Poisson distributed,

$$T \sim \mathcal{P}(n\lambda).$$

For a given threshold ν, the probability of false alarm is

$$\alpha = \sum_{k > \nu}^{\infty} \frac{e^{-n\lambda_0}(n\lambda_0)^k}{k!} + \gamma \frac{e^{-n\lambda_0}(n\lambda_0)^\nu}{\nu!}. \qquad \square$$

11.4.7 Summary of binary decision problems

The following observations summarize the results we have obtained for the binary decision problem.

1. Using either a Neyman–Pearson or a Bayes criterion, we see that the optimum test is a likelihood ratio test. (If the distribution is not continuous, a randomized test may be necessary for the Neyman-Pearson decision.) Thus, regardless of the dimensionality of the observation space, the test consists of comparing a scalar variable $\ell(\mathbf{x})$ with a threshold.

2. In many cases, construction of the likelihood ratio test can be simplified by using a sufficient statistic.

3. A complete description of the likelihood ratio test performance can be obtained by plotting the conditional probabilities P_D versus P_{FA} as the threshold is varied. The

11.5 Some *M*-ary Problems

resulting ROC curve can be used to calculate either the power for a given size (and vice versa) or the Bayes risk (the probability of error).

4. A Bayes rule minimizes the expected loss under the posterior distribution.

11.5 Some *M*-ary problems

Up to this point in the chapter, all of the tests have been binary. We now generalize to M-ary tests. Suppose there are $M \geq 2$ possible outcomes, each of which corresponds to one of the M hypotheses. We observe the output and are required to decide which source was used to generate it. In terms of the radar detection problem previously discussed, suppose there are M different target possibilities, and we not only have to detect the presence of a target, but to classify it as well. For example, we may be required to choose between three alternatives: H_0: no target present, H_1: target is present and hostile, H_2: target is present and friendly. Another common example is digital communication, in which there are more than two points in the signal constellation.

Formally, the parameter space Θ is of the form $\Theta = \{\vartheta_0, \vartheta_1, \ldots, \vartheta_{M-1}\}$. Let $H_0: \theta = \vartheta_0, H_1: \theta = \vartheta_1, \ldots, H_{M-1}: \theta = \vartheta_{M-1}$ denote the M hypotheses to test. We will employ the Bayes criterion to address this problem, and assume that $\mathbf{p} = [p_0, \ldots, p_{M-1}]^T$ is the corresponding *a priori* probability vector, where

$$p_j = f_\theta(\vartheta_j).$$

(In other words, $\mathbf{p}$ represents F_θ.) We will denote the cost of each course of action as L_{ji}, where the subscript i signifies that the ith hypothesis is chosen, and the subscript j signifies that the jth hypothesis is true. L_{ji} is the cost of choosing H_i when H_j is true.

We observe a random variable $\mathbf{X}$ taking values in $\mathcal{X} \subset \mathbb{R}^k$. We wish to generalize the notion of a threshold test, which was so useful for the binary case. Our approach will be to compute the posterior conditional expected loss for $\mathbf{X} = \mathbf{x}$.

The natural generalization of the binary case is to partition the observation space into M disjoint regions $S_0, \ldots, S_{M-1}$, that is, $\mathcal{X} = S_0 \cup \cdots \cup S_{M-1}$, and to invoke a decision rule of the form

$$\phi(\mathbf{x}) = n \quad \text{if } \mathbf{x} \in S_n, \quad n = 0, \ldots, M-1. \tag{11.79}$$

The loss function then assumes the form

$$L[\vartheta_j, \phi(\mathbf{x})] = \sum_{i=0}^{M-1} L_{ji} I_{S_i}(\mathbf{x}),$$

where $I_{S_i}(\mathbf{x})$ is the indicator function, equal to 1 if $\mathbf{x} \in S_i$. From (11.72), the Bayes risk is

$$r(F_\theta, \phi) = r(\mathbf{p}, \phi) = \int_\mathcal{X} \left\{ \sum_{j=0}^{M-1} L[\vartheta_j, \phi(\mathbf{x})] f_{\theta|\mathbf{X}}(\vartheta_j \mid \mathbf{x}) \right\} f_\mathbf{X}(\mathbf{x}) \, d\mathbf{x}$$

$$= \int_\mathcal{X} \left\{ \sum_{j=0}^{M-1} \sum_{i=0}^{M-1} L_{ji} I_{S_i}(\mathbf{x}) f_{\theta|\mathbf{X}}(\vartheta_i \mid \mathbf{x}) \right\} f_\mathbf{X}(\mathbf{x}) \, d\mathbf{x},$$

and we may minimize this quantity by minimizing the quantity in braces for each $\mathbf{x}$. It suffices to minimize the posterior conditional expected loss,

$$r'(\mathbf{p}, \phi) = \sum_{j=0}^{M-1} \sum_{i=0}^{M-1} L_{ji} I_{S_i}(\mathbf{x}) f_{\theta|\mathbf{X}}(\vartheta_i \mid \mathbf{x}). \tag{11.80}$$

The problem reduces to determining the sets S_i, $i = 0, \ldots, M - 1$, that result in the minimization of r'.

From Bayes rule, we have

$$f_{\theta|\mathbf{X}}(\vartheta_j \mid \mathbf{x}) = \frac{f_{\mathbf{X}|\theta}(\mathbf{x} \mid \vartheta_j) f_\theta(\vartheta_j)}{f_{\mathbf{X}}(\mathbf{x})}. \tag{11.81}$$

We will denote the prior probabilities $f_\theta(\vartheta_j)$ as

$$p_j = f_\theta(\vartheta_j).$$

Then (11.81) becomes

$$f_{\theta|\mathbf{X}}(\vartheta_j \mid \mathbf{x}) = \frac{f_{\mathbf{X}|\theta}(\mathbf{x} \mid \vartheta_j) p_j}{f_{\mathbf{X}}(\mathbf{x})},$$

which, when substituted into (11.80), yields

$$r'(\mathbf{p}, \phi) = \sum_{j=0}^{M-1} \sum_{i=0}^{M-1} L_{ji} I_{S_i}(\mathbf{x}) \frac{f_{\mathbf{X}|\theta}(\mathbf{x} \mid \vartheta_j) p_j}{f_{\mathbf{X}}(\mathbf{x})}.$$

We now make an important observation: Given $\mathbf{X} = \mathbf{x}$, we can minimize the posterior conditional expected loss by minimizing

$$\sum_{j=0}^{M-1} \sum_{i=0}^{M-1} L_{ji} I_{S_i}(x) f_{X|\theta}(x \mid \vartheta_j) p_j;$$

that is, $f_{\mathbf{X}}(\mathbf{x})$ is simply a scale factor for this minimization problem, since $\mathbf{x}$ is assumed to be fixed. Since

$$\sum_{j=0}^{M-1} \sum_{i=0}^{M-1} L_{ji} I_{S_i}(\mathbf{x}) f_{\mathbf{X}|\theta}(\mathbf{x} \mid \vartheta_j) p_j = \sum_{i=0}^{M-1} I_{S_i}(\mathbf{x}) \sum_{i=0}^{M-1} L_{ji} f_{\mathbf{X}|\theta}(\mathbf{x} \mid \vartheta_j) p_j,$$

we may now ascertain the structure of the sets S_i that result in the Bayes decision rule $\phi(\mathbf{x})$ given by (11.79).

$$S_k = \left\{ \mathbf{x} \in \mathcal{X} : \sum_{j=0}^{M-1} L_{jk} f_{\mathbf{X}|\theta}(\mathbf{x} \mid \vartheta_j) p_j \leq \sum_{j=0}^{M-1} L_{ji} f_{\mathbf{X}|\theta}(x \mid \vartheta_j) p_j \quad \forall i \neq k \right\}. \tag{11.82}$$

The decision determined by the sets in (11.82) can be written another way. We set our estimate $\hat{\vartheta}$ equal to the value ϑ_k that minimizes

$$\sum_{j=0}^{M-1} L_{jk} f_{\mathbf{X}|\theta}(\mathbf{x} \mid \vartheta_j) p_j.$$

That is, $\hat{\vartheta} = \vartheta_k$ if

$$k = \arg\min_k \sum_{j=0}^{M-1} L_{jk} f_{\mathbf{X}|\theta}(\mathbf{x} \mid \vartheta_j) p_j. \tag{11.83}$$

The general structure of these decision regions is rather messy to visualize and lengthy to compute, but we can learn almost all there is to know about this problem by simplifying it a bit. In the important case of digital communication, it is appropriate to consider a decision cost that depends upon incorrect decisions only. Thus we set

$$L_{ii} = 0,$$
$$L_{ji} = 1, \quad i \neq j.$$

11.5 Some *M*-ary Problems

Then

$$S_k = \left\{ \mathbf{x} \in \mathcal{X} : \sum_{j, j \neq k} f_{\mathbf{X}|\theta}(\mathbf{x} \mid \vartheta_j) p_j \leq \sum_{j, j \neq i} f_{\mathbf{X}|\theta}(\mathbf{x} \mid \vartheta_j) p_j, \quad \forall i \neq k \right\}.$$

This is equivalent to

$$S_k = \left\{ \mathbf{x} \in \mathcal{X} : f_{\mathbf{X}|\theta}(\mathbf{x} \mid \vartheta_k) p_k \geq \max_{i \neq k} f_{\mathbf{X}|\theta}(\mathbf{x} \mid \vartheta_i) p_i \right\}.$$

Stated in terms of decisions, the best decision is $\hat{\vartheta} = \vartheta_k$, where

$$k = \arg\max_k f_{\mathbf{X}|\theta}(\mathbf{x} \mid \vartheta_k) p_k. \tag{11.84}$$

Stated in words, the best (Bayes) decision is that which *maximizes* the posterior probability $f_{\mathbf{X}|\theta}(\mathbf{x} \mid \vartheta_k) p_k$. Such a test is sometimes called the *maximum* a posteriori *test*, or the MAP test.

Example 11.5.1 (Detection in Gaussian noise) Given

$$H_0 : \mathbf{X} \sim \mathcal{N}(\mathbf{m}_0, \sigma^2 I),$$
$$H_1 : \mathbf{X} \sim \mathcal{N}(\mathbf{m}_1, \sigma^2 I),$$
$$H_2 : \mathbf{X} \sim \mathcal{N}(\mathbf{m}_2, \sigma^2 I),$$

with prior probabilities p_0, p_1, p_2, we consider the boundaries between decision regions. Let us consider the boundary between the decision region for H_0 and H_1. Forming the likelihood ratio, we have

$$\frac{f_{X|\theta}(\mathbf{x} \mid \vartheta_1)}{f_{X|\theta}(\mathbf{x} \mid \vartheta_0)} \underset{H_0}{\overset{H_1}{\gtrless}} \frac{p_0}{p_1}$$

After some simplification, we find the test

$$(\mathbf{m}_1 - \mathbf{m}_0)^T (\mathbf{x} - \mathbf{x}_0) \underset{H_0}{\overset{H_1}{\gtrless}} \log \frac{p_0}{p_1},$$

where $\mathbf{x}_0 = \frac{1}{2}(\mathbf{m}_1 + \mathbf{m}_0)$. The boundary between the decision regions occurs where

$$(\mathbf{m}_1 - \mathbf{m}_0)^T (\mathbf{x} - \mathbf{x}_0) = \log \frac{p_0}{p_1}. \tag{11.85}$$

Equation (11.85) is the equation of a plane orthogonal to $\mathbf{m}_1 - \mathbf{m}_0$. In the comparison between H_0 and H_1, if $\mathbf{x}$ falls on the side of the plane nearest $\mathbf{m}_0$, then H_0 is selected, and if $\mathbf{x}$ falls on the side of the plane nearest $\mathbf{m}_1$, then H_1 is selected. We can get a better understanding of the separating plane by letting

$$\mathbf{d} = \left(\log \frac{p_0}{p_1} \right) \frac{(\mathbf{m}_1 - \mathbf{m}_0)}{\|\mathbf{m}_1 - \mathbf{m}_0\|^2},$$

so that $(\mathbf{m}_1 - \mathbf{m}_0)^T \mathbf{d} = \log p_0/p_1$. Then the equation for the separating plane of (11.85) can be written

$$(\mathbf{m}_1 - \mathbf{m}_0)^T (\mathbf{x} - \tilde{\mathbf{x}}) = 0, \tag{11.86}$$

where $\tilde{\mathbf{x}} = \mathbf{x}_0 + \mathbf{d}$. Equation (11.86) represents a plane orthogonal to the vector $\mathbf{m}_1 - \mathbf{m}_0$ between the means, that passes through the point $\tilde{\mathbf{x}}$ (see figure 11.16). The situation is even more clear when $p_0 = p_1$. Then $\tilde{\mathbf{x}} = \mathbf{x}_0$, the point midway between the means, so the separating plane lies midway between $\mathbf{m}_1$ and $\mathbf{m}_0$.

Similar separating planes can be found between each pair of means; they divide space up into decision regions. □

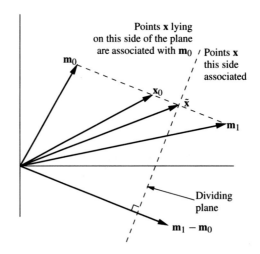

Figure 11.16: Geometry of the decision space for multivariate Gaussian detection

Example 11.5.2 Now consider the quaternary detection problem with the means

$$\mathbf{m}_0 = \begin{bmatrix} 4 \\ 4 \end{bmatrix} \qquad \mathbf{m}_1 = \begin{bmatrix} -2 \\ -2 \end{bmatrix} \qquad \mathbf{m}_2 = \begin{bmatrix} 2 \\ 5 \end{bmatrix} \qquad \mathbf{m}_3 = \begin{bmatrix} 1 \\ 1 \end{bmatrix}.$$

Figure 11.17 illustrates the decision regions for this problem. The dashed lines are the lines between the means, the heavy solid lines indicate the boundaries of the decision regions, and the light solid lines are portions of the decision lines that do not contribute to the decision boundaries. In figure 11.17(a), each selection is equally probable, and in figure 11.17(b), $\mathbf{m}_0$ occurs with probability 0.99, the remaining probability split equally between the others. The effect of this change in probability is to make the decision region for H_0 larger. □

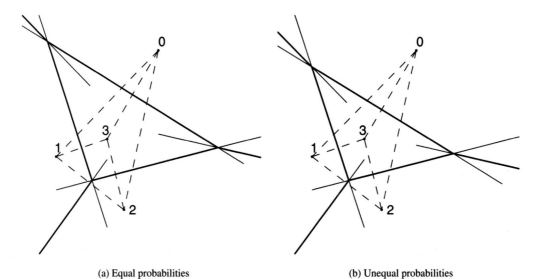

(a) Equal probabilities (b) Unequal probabilities

Figure 11.17: Decision boundaries for a quaternary decision problem

The Bayes risk, or probability of error, for the M-ary classifier can be stated in terms of the probability of making a *correct* decision,

$$P(\mathcal{E}) = 1 - P(\mathcal{C}),$$

where

$$P(\mathcal{C}) = \sum_{j=1}^{m} P(\mathcal{C} \mid \vartheta_j) p_j$$

and

$$P(\mathcal{C} \mid \vartheta_j) = \int_{S_j} f_{\mathbf{x}|\theta}(\mathbf{x} \mid \vartheta_j) \, d\mathbf{x}.$$

In the general case it can be difficult to compute these probabilities exactly. In some special cases—such as decision regions having rectangular boundaries with Gaussian observations—the computation is straightforward. A recent result [313] provides extension of probability computations to more complicated polygonal regions.

11.6 Maximum-likelihood detection

The decision criterion specified in (11.84) requires knowledge of the functions $f_{\mathbf{x}|\theta}(\mathbf{x} \mid \vartheta_k)$ and the prior probabilities. In many circumstances, the prior probabilities are all equal,

$$p_0 = p_1 = \cdots = p_{M-1} = \frac{1}{M}$$

(or, lacking information to the contrary, they are assumed to be equal). A decision made on this basis can be stated as: Set $\hat{\vartheta} = \vartheta_k$ if

$$k = \arg\max_k f_{\mathbf{x}|\theta}(\mathbf{x} \mid \vartheta_k).$$

A decision made on the basis of this criterion is said to be a *maximum-likelihood* estimate, and the conditional probability $f_{\mathbf{x}|\theta}(\mathbf{x} \mid \vartheta)$ is said to be the *likelihood function*, being viewed (usually) as a function of ϑ.

11.7 Approximations to detection performance: the union bound

As has been observed, obtaining exact expressions for the probability of error for M-ary detection in Gaussian noise can be difficult. However, it is straightforward to obtain an *upper bound* on the probability of error, using what is known as the union bound.

Consider the problem of computing the probability of error for the union of two events A and B. This probability can be expressed as

$$P(A \cup B) = P(A) + P(B) - P(A \cap B),$$

where the subtracted term prevents the event in the intersection from being counted twice (see figure 11.18). Since every probability ≥ 0, we must have

$$P(A \cup B) \leq P(A) + P(B).$$

By this bound, the probability of a complicated event $(A \cup B)$ is bounded by the probabilities of more elementary events (A and B).

Now consider the problem of finding the probability of error for a PSK signal constellation, such as that shown in figure 11.19. (Assume that all signals are sent with equal

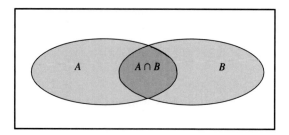

Figure 11.18: Venn diagram for the union of two sets

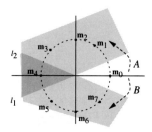

Figure 11.19: Bound on the probability of error for PSK signaling

probability.) Suppose that $\mathbf{m}_0$ is sent; the received signal will be correctly detected only if it falls in the white wedge. Looked at from another point of view, the signal will be detected if either event A occurs, which is the event that the received signal lies above the line l_1, or event B occurs, which is the event that the received signal lies below the line l_2. It is also possible for both events to occur (the darkly shaded wedge). Using the union bound, we have

$$P(\mathcal{E} \mid \mathbf{m}_0) = P(A \cup B) \leq P(A) + P(B).$$

But $P(A)$ is the *binary* probability of error between the signals $\mathbf{m}_0$ and $\mathbf{m}_1$, and $P(B)$ is the binary probability of error between the signals $\mathbf{m}_0$ and $\mathbf{m}_7$, so that

$$P(A) = P(B) = Q(d/2\sigma),$$

where d is the distance between adjacent signals in the PSK constellation. Thus

$$P(\mathcal{E}) \leq 2Q(d/2\sigma). \tag{11.87}$$

As the SNR increases, the probability of falling in the darkly shaded wedge becomes smaller, and the bound (11.87) becomes increasingly tight.

11.8 Invariant tests

The goal of an invariant test is to provide a function that eliminates unavoidable and unknown transformations on data, leaving the data invariant with respect to such a transformation. Rather than formally develop the theory of invariant tests, we present a few examples of invariant transformations that illustrate the theory.

Example 11.8.1 Consider the detection problem

$$H_0: X_i \sim \mathcal{N}(\mu_0 + c, \sigma^2),$$
$$H_1: X_i \sim \mathcal{N}(\mu_1 + c, \sigma^2),$$

11.8 Invariant Tests

where $i = 1, 2, \ldots, n$, and c is an unknown constant. We can formulate a new variable **Y** that is *invariant* with respect to any value of c,

$$Y_1 = X_1 - X_n \qquad Y_2 = X_2 - X_n \qquad \cdots \qquad Y_{n-1} = X_1 - X_n,$$

and develop a test based on **Y**. □

Example 11.8.2 More generally, we desire to distinguish

$$H_0: \mathbf{X} \sim \mathcal{N}(\mathbf{m}_0 + \gamma \mathbf{c}, \sigma^2 I),$$
$$H_1: \mathbf{x} \sim \mathcal{N}(\mathbf{m}_1 + \gamma \mathbf{c}, \sigma^2 I),$$

where **c** is a fixed vector and $\gamma \in \mathbb{R}$. Figure 11.20 illustrates the concept. We desire to make a test that is invariant with respect to any value of γ. A little thought will reveal that if we project **X** in the direction *orthogonal* to **c** we will obtain a variable that is invariant to $\gamma \mathbf{c}$. Accordingly, we define

$$\mathbf{Y} = P_{c^\perp} \mathbf{X},$$

where $P_{c^\perp}$ is the projection matrix

$$P_{c^\perp} = I - \mathbf{c}\mathbf{c}^T / (\mathbf{c}^T \mathbf{c}).$$

Then we have the detection problem

$$H_0: \mathbf{Y} \sim \mathcal{N}\left(P_{c^\perp} \mathbf{m}_0, \sigma^2 P_{c^\perp}\right),$$
$$H_1: \mathbf{Y} \sim \mathcal{N}\left(P_{c^\perp} \mathbf{m}_1, \sigma^2 P_{c^\perp}\right).$$

Clearly, if $\mathbf{m}_0$ and $\mathbf{m}_1$ lie in span(**c**), then there are no useful tests for this new detection problem. □

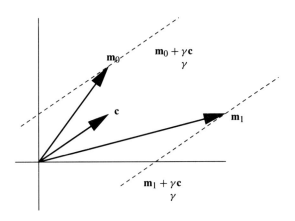

Figure 11.20: A test biased by $\gamma \mathbf{c}$

Example 11.8.3 [291] Suppose that $\mathbf{X} \sim \mathcal{N}(\mu \mathbf{m}, \sigma^2 I)$ is an n-dimensional random vector and we desire to test

$$H_0: \mu \leq 0,$$
$$H_1: \mu > 0.$$

However, rather than observing **X**, suppose that we are only able to observe a corrupted version, **Y**

$$\mathbf{Y} = \gamma Q_{m^\perp} \mathbf{X},$$

where γ is an unknown positive gain, and $Q_{m\perp}$ is a rotation by an unknown amount in the space orthogonal to $\mathbf{m}$, as shown in figure 11.21. Our first challenge is to express such a rotation algebraically.

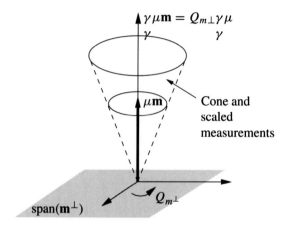

Figure 11.21: Channel gain and rotation

Let $S = \text{span}(\mathbf{m})$ and let $T = S^\perp$, the orthogonal complement of S in $\mathbb{R}^n$. The rotation $Q_{m\perp}$ combines a rotation in T with a projection onto S. Let P_T be a projection matrix onto T. Also let P_S be a projection onto S. Then

$$Q_{m\perp} = P_T Q P_T + P_S,$$

where Q is a rotation matrix: $QQ^T = I$. Writing a vector $\mathbf{x}$ in terms of its components in the two spaces,

$$\mathbf{x} = P_S \mathbf{x} + P_T \mathbf{x},$$

it is straightforward to show that $Q_{m\perp}$ leaves the component of $\mathbf{x}$ in S unchanged, while rotating the component in T:

$$(P_T Q P_T + P_S)(P_S \mathbf{x}) = P_S \mathbf{x},$$
$$(P_T Q P_T + P_S)(P_T \mathbf{x}) = P_T Q P_T \mathbf{x}.$$

With $\mathbf{Y} = \gamma Q_{m\perp} \mathbf{X}$, our problem becomes

$$H_0: Y \sim \mathcal{N}\left(\mu \gamma Q_{m\perp} \mathbf{m}, \gamma^2 I\right), \quad \mu \leq 0,$$
$$H_1: Y \sim \mathcal{N}\left(\mu \gamma Q_{m\perp} \mathbf{m}, \gamma^2 I\right), \quad \mu > 0,$$

since $Q_{m\perp} Q_{m\perp}^T = I$.

A test invariant with respect to scaling by γ and rotation by $Q_{m\perp}$ must combat both of these transformations. The following statistic,

$$t = \frac{\mathbf{m}^T \mathbf{y}}{\|\mathbf{m}\| \|\mathbf{y}\|}, \tag{11.88}$$

works—as may be verified by substituting $\mathbf{y} = \gamma Q_{m\perp} \mathbf{x}$. □

Example 11.8.4 (t statistic) Let us examine another invariant test, this one involving an unknown variance. Suppose that $X \sim \mathcal{N}(\mu \mathbf{m}, \sigma^2 I)$, where $\mu = 0$ under H_0 and $\mu > 0$ under H_1, and σ^2 is not known. We can write $\mathbf{X} = \sigma \mathbf{Y}$, where $\mathbf{Y} \sim \mathcal{N}(\mu/\sigma, I)$. We desire a test that is invariant to changes in σ. A useful test statistic is

$$T = \frac{(\mathbf{m}^T P_{\mathbf{m}} \mathbf{X})/(\sigma \|\mathbf{m}\|)}{[\mathbf{X}^T (I - P_{\mathbf{m}}) \mathbf{X}/(\sigma(n-1))]^{1/2}}, \tag{11.89}$$

where the matrix $P_\mathbf{m}$ projects onto span($\mathbf{m}$). Note that σ cancels out of both numerator and denominator, so the statistic can be formed without explicit knowledge of σ.

The numerator of (11.89) is an $\mathcal{N}(\mu/\sigma, 1)$ random variable. The random variable

$$\frac{1}{\sigma^2}(I - P_\mathbf{m})\mathbf{X}$$

in the denominator is $\mathcal{N}(\mathbf{0}, (I - P_\mathbf{m}))$, which is independent of the variable $P_\mathbf{m}\mathbf{x}$ in the numerator. Based on theorem 11.2, the denominator

$$\frac{1}{\sigma^2}\mathbf{X}^T(I - P_\mathbf{m})\mathbf{x}$$

is a χ^2_{n-1} random variable, which is independent of the Gaussian in the numerator.

Thus (see box 11.3), under H_0, the statistic T in (11.89) is a t distributed random variable. The random variable T is invariant to σ^2. On this basis, we can formulate a decision as

$$\phi(t) = \begin{cases} 1 & t > t_0, \\ 0 & t \leq t_0, \end{cases}$$

for some threshold t_0 which may be chosen (for example) to set the probability of false alarm:

$$P_{FA} = P(T > t_0 \mid H_0) = \int_{t_0}^{\infty} f_T(t)\, dt.$$

□

11.8.1 Detection with random (nuisance) parameters

Example 11.8.5 Let $\mathbf{X} = (X_1, X_2) \in \mathbb{R}^2$, and consider the testing problem

$$\begin{aligned} H_0 &: X_1 = N_1 & X_2 &= N_2, \\ H_1 &: X_1 = A\cos\theta + N_1 & X_2 &= A\sin\theta + N_2, \end{aligned} \quad (11.92)$$

where $N_1 \sim \mathcal{N}(0, \sigma^2)$ and $N_2 \sim \mathcal{N}(0, \sigma^2)$ are independent and θ is the (unknown) uniformly distributed random "phase"

$$\theta \sim \mathcal{U}[0, 2\pi),$$

independent of N_1 and N_2. We desire to determine when the component $(A\cos\theta, A\sin\theta)$ is present. In this problem, there is a nuisance parameter θ which, although we do not need to know it explicitly, stands in the way of performing the desired detection. □

Box 11.3: The t distribution

The ratio $T = Z/\sqrt{Y/r}$, where $Z \sim \mathcal{N}(0, 1)$ and $Y \sim \chi^2_r$, is said to have a (central) t distribution. This is also known as the Student's t distribution. The pdf for the t distributed random variable (see exercise 11.8-24) is

$$f_T(t) = \frac{\Gamma((r+1)/2)}{\Gamma(r/2)\sqrt{\pi r}} \frac{1}{(1 + t^2/r)^{(r+1)/2}}, \quad t \in \mathbb{R}. \quad (11.90)$$

If, in $T = Z/\sqrt{Y/r}$, the numerator is distributed as $Z \sim \mathcal{N}(\mu, 1)$, then T is said to be a noncentral t distribution. The pdf in this case is

$$f_T(t) = \frac{r^{r/2}}{\sqrt{\pi}\,\Gamma(r/2)2^{(r-1)/2}(t^2 + r)^{(r+1)/2}} e^{-r\mu^2/(2t^2 + r)}$$

$$\times \int_0^\infty \exp\left[-\frac{1}{2}\left(x - \frac{\mu t}{\sqrt{t^2 + r}}\right)^2\right] x^r\, dx. \quad (11.91)$$

The situation illustrated in the preceding example can described in more general terms, as follows. First, we suppose that the set of parameters is split into two subsets,

$$\Theta = \{\Theta_1, \Theta_2\}.$$

Θ_1 is the set of parameters that we desire to detect. For example, for a binary detection problem,

$$\Theta_1 = \{\vartheta_0, \vartheta_1\}.$$

The set Θ_2 is a set of parameters that enter into the problem, but do not distinguish between the cases to be detected. Then the hypothesis testing problem (for the binary case) can be expressed as

$$H_0: (\theta_1, \theta_2) = \{\vartheta_0, \theta\} \text{ for some } \theta \in \Theta_2,$$
$$H_1: (\theta_1, \theta_2) = \{\vartheta_1, \theta\} \text{ for some } \theta \in \Theta_2.$$

We assume that θ is a random variable governed by some distribution function $f_\theta(\theta)$. That is, we wish to distinguish between the two cases $\theta_1 = \vartheta_0$ and $\theta_1 = \vartheta_1$. The parameter $\theta_2 = \theta$ is common to both problems, and thus does not assist in the detection. Indeed, the presence of θ_2 may prevent the formulation of a useful test. Because of this, θ_2 is sometimes regarded as a *nuisance parameter*. We wish somehow to make a test that is invariant with respect to θ_2.

Example 11.8.6 The detection problem of the previous example can be expressed in this notation as

$$\Theta = \{\Theta_1, \Theta_2\}$$

where

$$\Theta_1 = \{0, A\} \quad \text{and} \quad \Theta_2 = [0, 2\pi).$$

We wish to determine if the signal is present—that is, to determine which $\vartheta_1 \in \Theta_1$ has occurred—but we do not know θ. We need to find a test that is invariant with respect to θ. (Our intuition suggests that we take the magnitude of the signal; this is precisely what we do.) □

When there is an unknown parameter, a commonly used approach can be stated as follows.

> To make the distribution invariant with respect to the unknown random parameter, we average over the unknown parameter.

Then

$$f_{x|\theta}(x|\theta_1) = E f_{x|\theta}(x|\theta_1, \theta_2)$$
$$= \int_{\theta_2 \in \Theta_2} f_{x|\theta}(x|\theta_1, \theta_2) f_\theta(\theta_2) \, d\theta_2. \tag{11.93}$$

This distribution can be used in a likelihood ratio test.

Example 11.8.7 Returning to example 11.8.5, under H_0 we have no need to worry about the nuisance parameter θ_1,

$$f_{\mathbf{x}|\theta}(\mathbf{x}|\theta_1 = \vartheta_0) = f_{\mathbf{x}|\theta}(x_1, x_2 | \theta_1 = \vartheta_0) = \frac{1}{2\pi\sigma^2} \exp\left[-\frac{1}{2\sigma^2}\left(x_1^2 + x_2^2\right)\right].$$

Under H_1,

$$f_{\mathbf{x}|\theta}(\mathbf{x}|\theta_1 = \vartheta_1, \theta_2) = \frac{1}{2\pi\sigma^2} \exp\left[-\frac{1}{2\sigma^2}[(x_1 - A\cos\theta_2)^2 + (x_2 - A\sin\theta_2)^2]\right].$$

11.8 Invariant Tests

Averaging out the nuisance parameter θ_2 according to (11.93), we obtain

$$f_{\mathbf{X}|\theta}(\mathbf{x}|\theta_1 = \vartheta_1) = \frac{1}{2\pi}\frac{1}{2\pi\sigma^2}\int_0^{2\pi}\exp\left[-\frac{1}{2\sigma^2}[(x_1 - A\cos\theta_2)^2 + (x_2 - A\sin\theta_2)^2]\right]d\theta_2$$

$$= \frac{1}{4\pi^2\sigma^2}\exp\left[-\frac{1}{2\sigma^2}(x_1^2 + x_2^2 + A^2)\right]\int_0^{2\pi}\exp\left[\frac{A}{\sigma^2}(x_1\cos\theta_2 + x_2\sin\theta_2)\right]d\theta_2.$$

Now we introduce the change of variables

$$x_1 = r\cos\phi \qquad x_2 = r\sin\phi,$$
$$r = \sqrt{x_1^2 + x_2^2} \qquad \phi = \tan^{-1}\frac{x_2}{x_1}.$$

Then in the exponent in the integral, we have

$$x_1\cos\theta_2 + x_2\sin\theta_2 = r(\cos\phi\cos\theta_2 + \sin\phi\sin\theta_2) = r\cos(\theta_2 - \phi).$$

By the definition of the zeroth-order modified Bessel function (see box 11.4),

$$f_{\mathbf{X}|\theta}(\mathbf{x}|\theta_1 = \vartheta_1) = \frac{1}{4\pi^2\sigma^2}\exp\left[-\frac{1}{2\sigma^2}(x_1^2 + x_2^2 + A^2)\right]\int_0^{2\pi}\exp\left[\frac{Ar}{\sigma^2}\cos(\theta_2 - \phi)\right]d\theta_2,$$

$$= \frac{1}{4\pi^2\sigma^2}\exp\left[-\frac{1}{2\sigma^2}(x_1^2 + x_2^2 + A^2)\right]I_0\left(\frac{A}{\sigma^2}\sqrt{x_1^2 + x_2^2}\right).$$

The likelihood ratio can now be expressed as

$$\ell(x_1, x_2) = \frac{f_{\mathbf{X}|\theta}(\mathbf{x}|\theta_1 = \vartheta_1)}{f_{\mathbf{X}|\theta}(\mathbf{x}|\theta_1 = \vartheta_0)} = \exp[-A^2/2\sigma^2]I_0\left(A/\sigma^2\left(\sqrt{x_1^2 + x_2^2}\right)\right),$$

leading to the decision rule

$$\phi(x_1, x_2) = \begin{cases} 1 & \exp[-A^2/2\sigma^2]I_0\left(A/\sigma^2\left(\sqrt{x_1^2 + x_2^2}\right)\right) > \nu, \\ 0 & \text{otherwise} \end{cases}$$

for some threshold ν. In light of the monotonic nature of I_0, this can be rewritten in such a way that it not necessary to compute either the square root or the Bessel function:

$$\phi(x_1, x_2) = \begin{cases} 1 & x_1^2 + x_2^2 > \eta, \\ 0 & \text{otherwise,} \end{cases}$$

where

$$\eta = \left[\frac{\sigma^2}{A}I_0^{-1}\left(\nu e^{A^2/2\sigma^2}\right)\right]^2.$$

Thus the optimum decision depends upon the quantity $r^2 = x_1^2 + x_2^2$—the squared magnitude of the received signal components. The detector variable r^2 is proportional to a χ_2^2 random variable under H_0. □

Example 11.8.8 In this example, we extend the results of example 11.8.7. Two signals, $\mathbf{s}_0(\theta)$ and $\mathbf{s}_1(\theta)$, are transmitted in the presence of noise,

$$H_0: \mathbf{X} = \mathbf{s}_0(\theta) + \mathbf{N},$$
$$H_1: \mathbf{X} = \mathbf{s}_1(\theta) + \mathbf{N},$$

where $\mathbf{N} \sim \mathcal{N}(0, \sigma^2 I)$, and where θ is a nuisance parameter, uniformly distributed on $[0, 2\pi)$. The

> **Box 11.4: The function $I_0(x)$**
>
> The modified Bessel function of zeroth-order $I_0(x)$ is defined as an integral,
>
> $$I_0(x) = \frac{1}{2\pi} \int_0^{2\pi} \exp[x \cos(\theta)] \, d\theta, \qquad x \in \mathbb{R}, \quad x \geq 0.$$
>
> The plot below illustrates $I_0(x)$.
>
>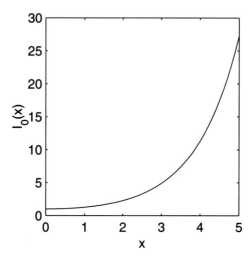
>
> Some useful properties of $I_0(x)$:
>
> 1. $I_0(0) = 1$.
> 2. $I_0(x)$ is monotonically increasing.
> 3. A series for $I_0(x)$ is
>
> $$I_0(x) = \sum_{k=0}^{\infty} \frac{(x^2/4)^k}{(k!)^2}.$$
>
> 4. Asymptotically, as $x \to \infty$,
>
> $$I_0(x) \sim \frac{e^x}{\sqrt{2\pi x}}.$$

signals are

$$\mathbf{s}_0(\theta) = [s_{00}(\theta), s_{01}(\theta), \ldots, s_{0,n-1}(\theta)]^T \qquad \mathbf{s}_1(\theta) = [s_{10}(\theta), s_{11}(\theta), \ldots, s_{1,n-1}(\theta)]^T,$$

where

$$s_{ki}(\theta) = a_i \sin(i\omega_k T_s + \theta)$$

with $a_0, a_1, \ldots, a_{n-1}$ a known amplitude sequence. In other words, we desire to distinguish between two sinusoidally modulated signals, with each signal having its own frequency, where the phase of the sinusoid is unknown. Such a detection problem, with unknown phase, is referred to as *incoherent* detection. We can write the parameter set as

$$\Theta = \{\Theta_1, \Theta_2\},$$

where

$$\Theta_1 = \{\vartheta_0 = 0, \vartheta_1 = 1\} \qquad \text{and} \qquad \Theta_2 = [0, 2\pi).$$

11.8 Invariant Tests

The conditional density of $\mathbf{X}$ given $\theta_1 = \vartheta_0$ is

$$f_{\mathbf{X}|\theta}(\mathbf{x}|\theta_1 = \vartheta_0, \theta_2 = \theta) = \frac{1}{(2\pi)^{n/2}\sigma^n} \exp\left[-\frac{1}{2\sigma^2}[(\mathbf{x}-\mathbf{s}_0(\theta))^T(\mathbf{x}-\mathbf{s}_0(\theta))]\right]$$

$$= \frac{1}{(2\pi)^{n/2}\sigma^n} \exp\left[-\frac{1}{2\sigma^2}(\mathbf{x}^t\mathbf{x} - 2\mathbf{x}^T\mathbf{s}_0(\theta) + \mathbf{s}_0^T(\theta)\mathbf{s}_0(\theta))\right]. \quad (11.94)$$

We can write the cross term in the exponent of (11.94) as

$$\mathbf{x}^T \mathbf{s}_0(\theta) = \sum_{i=0}^{n-1} x_i a_i \sin(i\omega_0 T_c + \theta)$$

$$= \sum_{i=0}^{n-1} x_i a_i \cos(i\omega_0 T_s) \sin(\theta) + \sum_{i=0}^{n-1} x_i a_i \sin(i\omega_0 T_s) \cos(\theta)$$

$$= x_{c0} \sin(\theta) + x_{s0} \cos(\theta),$$

where the identity $\sin(a+b) = \sin(a)\cos(b) + \sin(b)\cos(a)$ has been used, and where

$$x_{c0} = \sum_{i=0}^{n-1} x_i \cos(i\omega_0 T_s),$$

$$x_{s0} = \sum_{i=0}^{n-1} x_i \sin(i\omega_0 T_s).$$

We can write the term $\mathbf{s}_0^T(\theta)\mathbf{s}_0(\theta)$ from the exponent of (11.94),

$$\sum_{i=0}^{n-1} s_i^2(\theta) = \sum_{i=0}^{n-1} a_i^2 \sin^2(i\omega_0 T_s + \theta)$$

$$= \frac{1}{2}\left(\sum_{i=0}^{n-1} a_i^2 - \sum_{i=0}^{n-1} a_i^2 \cos(2i\omega_0 T_s + 2\theta)\right). \quad (11.95)$$

If the signal amplitude $\{a_0, a_1, \ldots, a_{n-1}\}$ is slowly varying (or better, constant) in comparison with $\omega_0 T_s$, then the double-frequency term on the right-hand side of (11.95) is approximately zero. We assume that this is, in fact, the case and write

$$\sum_{i=0}^{n-1} s_i^2(\theta) = \frac{1}{2} \sum_{i=0}^{n-1} a_i^2 = E_a.$$

Based on these notations, we can write

$$f_{\mathbf{X}|\theta}(\mathbf{x}|\theta_1 = \vartheta_0, \theta_2 = \theta) = \frac{1}{(2\pi)^{n/2}\sigma^n} \exp\left[-\frac{1}{2\sigma^2}(\mathbf{x}^T\mathbf{x} + E_a)\right]$$

$$\times \exp\left[-\frac{1}{2\sigma^2}(x_{c0}\sin\theta + x_{s0}\cos\theta)\right], \quad (11.96)$$

and by integrating,

$$f_{\mathbf{X}|\theta}(\mathbf{x}|\theta_1 = \vartheta_0) = \frac{1}{2\pi}\int_0^{2\pi} f_{\mathbf{X}|\theta}(\mathbf{x}|\theta_1 = \vartheta_0, \theta_2 = \theta)\,d\theta_2 \quad (11.97)$$

$$= \frac{1}{(2\pi)^{n/2}\sigma^n} \exp\left[-\frac{1}{2\sigma^2}(\mathbf{x}^T\mathbf{x} + E_a)\right] I_0(r_0/2\sigma^2), \quad (11.98)$$

where $r_0 = \sqrt{x_{c0}^2 + x_{s0}^2}$. Proceeding similarly for the case $\theta_1 = \vartheta_1$, we have

$$f_{X_1, X_2|\theta}(x_1, x_2|\theta_1 = \vartheta_1) = \frac{1}{(2\pi)^{n/2}\sigma^n} \exp\left[-\frac{1}{2\sigma^2}(\mathbf{x}^T\mathbf{x} + E_a)\right] I_0(r_1/2\sigma^2)$$

where $r_1 = \sqrt{x_{c1}^2 + x_{s1}^2}$, and

$$x_{c1} = \sum_{i=0}^{n-1} x_i \cos(i\omega_1 T_s),$$

$$x_{s1} = \sum_{i=0}^{n-1} x_i \sin(i\omega_1 T_s).$$

The decision rule is based on the likelihood ratio test,

$$\phi(\mathbf{x}) = \begin{cases} 1 & \frac{I_0(r_1 2\sigma^2)}{I_0(r_0/2\sigma^2)} > \nu, \\ 0 & \text{otherwise,} \end{cases}$$

for some threshold ν. If $\nu\tau = 1$, which occurs when Bayes costs are based on probability of error and the hypotheses are equally likely, then based on the monotonic nature of I_0, the test can be written

$$\phi(\mathbf{x}) = \begin{cases} 1 & r_1^2/r_0^2 > \eta, \\ 0 & \text{otherwise,} \end{cases}$$

for a threshold η.

A block diagram of the test (assuming these latter simplifications) is shown in figure 11.22. Characterization of the performance P_{FA} and P_D for incoherent detectors is discussed in [262, section 5-4-4] and [23, appendix A]. □

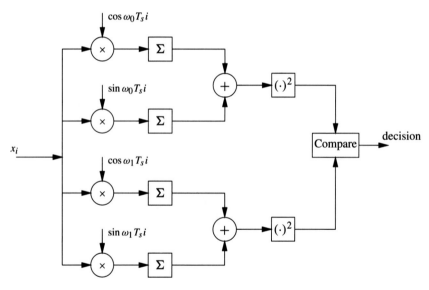

Figure 11.22: Incoherent binary detector

11.9 Detection in continuous time

Our presentation of detection theory up to this point has assumed that decisions are made based upon observations of a random variable X or a random vector $\mathbf{X}$. In many practical applications, decisions are based upon the observation of a continuous-time random process $X(t), t \in [0, T)$ for some T. This is the continuous-time detection problem.

Given our experience with vector spaces, in which functions are represented as points therein, it should come as no surprise that in many instances, we can represent $X(t)$ as a

11.9 Detection in Continuous Time

generalized Fourier series

$$X(t) = \sum_{i=1}^{\infty} X_i \psi_i(t), \qquad (11.99)$$

for some appropriate set of basis functions $\Psi = \{\psi_i(t), i = 1, 2, \ldots, \}$. If an adequate representation (11.99) exists, then we can use the discrete random variables $\{X_i, i = 1, 2, \ldots\}$ in place of the function $X(t)$ as the basis for decisions, for either Neyman–Pearson or Bayes detectors. Given that $X(t)$ is a random process, the existence of the set Ψ is a proposition that should be examined carefully. Our approach, however, is more cavalier, so that we arrive more quickly at some applications. We present some special cases that work without difficulty, and serve in a variety of practical applications, then point the direction to some of the technical difficulties without actually discussing the solution. (A more technical treatment of these ideas appears in [257].)

We treat the detection problem in which

$$X(t) = s_\theta(t) + N(t), \qquad (11.100)$$

where $s_\theta(t)$ is a nonstochastic signal, selected from a finite set of signals

$$\{s_0(t), s_1(t), \ldots, s_{M-1}(t)\},$$

corresponding to a set $\Theta = \{\vartheta_0 = 1, \vartheta_1 = 2, \ldots, \vartheta_{M-1} = M\}$. The signal $N(t)$ is a stationary *Gaussian white-noise random process*.

Definition 11.12 A random process $N(t)$ is said to be a stationary **Gaussian white-noise** process if:

1. $N(t)$ is wide-sense stationary.
2. All random vectors formed from samples of $N(t)$ are jointly Gaussian distributed; that is, the vector $\mathbf{N} = [N(t_1), N(t_2), \ldots, N(t_n)]^T$ is Gaussian distributed for all sample times $t_1, t_2, \ldots, t_n$.
3. The autocorrelation function is

$$R_N(t-s) = E[N(t)N(s)] = \frac{N_0}{2}\delta(t-s), \qquad (11.101)$$

where $\delta(t)$ is the Dirac δ function, and $N_0 < \infty$. □

The PSD of a random process N with autocorrelation given by (11.101) is

$$S_N(\omega) = \frac{N_0}{2},$$

constant for all frequencies. (This implies that the power in the process is infinite, and hence is nonphysical. However, it is an extremely valuable model. A discussion of this is provided in [257].) We assume, for convenience, that the $N(t)$ is zero mean:

$$EN(t) = 0 \quad \forall t.$$

Let us *assume* that an orthonormal sequence of functions $\{\psi_i(t)\}$ exists such that

$$X(t) = \sum_{i=1}^{\infty} X_i \psi_i(t), t \in [0, T).$$

By the orthonormality of the basis functions,

$$X_i = \langle X(t), \psi_i(t) \rangle = \int_0^T X(t) \psi_i(t) \, dt.$$

(We are assuming for convenience that all signals are real.) Using (11.100), we have

$$X_i = \int_0^T s_\theta(t)\psi_i(t)\,dt + \int_0^T N(t)\psi_i(t)\,dt$$
$$= s_{\theta,i} + N_i.$$

Let us characterize the random variables N_i. Because they were obtained by a linear operation on a Gaussian random process, the N_i are Gaussian. The mean and covariance are

$$EN_i = E\int_0^T N(t)\psi_i(t)\,dt = \int_0^T EN(t)\psi_i(t)\,dt = 0,$$

$$\text{cov}(N_i, N_j) = E\int_0^T N(t)\psi_i(t)\,dt \int_0^T N(s)\psi_j(s)\,ds$$
$$= \int_0^T\int_0^T E[N(t)N(s)]\psi_i(t)\psi_j(s)\,dt\,ds = \frac{N_0}{2}\int_0^T \psi_i(t)\psi_j(t)\,dt$$
$$= \begin{cases} N_0/2 & i = j, \\ 0 & i \neq j. \end{cases}$$

Let us assume that we have a finite number N of samples. Because the variables X_i are independent, the density of $\mathbf{X} = (X_1, X_2, \ldots, X_N)$, conditioned upon knowing which signal is sent, is

$$f_{\mathbf{X}|\theta}(\mathbf{x}|\theta) = \prod_{i=1}^N \frac{1}{\sqrt{\pi N_0}} \exp\left[-\frac{1}{N_0}(X_i - s_{\theta,i})^2\right].$$

The likelihood ratio for comparing $f_{\mathbf{X}|\theta}(x_1, \ldots, x_N | \vartheta_j)$ to $f_{\mathbf{X}|\theta}(x_1, \ldots, x_N | \vartheta_k)$ is

$$\ell(x_1, \ldots, x_N) = \frac{\prod_{i=1}^N \exp\left[-\frac{1}{N_0}(X_i - s_{j,i})^2\right]}{\prod_{i=1}^N \exp\left[-\frac{1}{N_0}(X_i - s_{k,i})^2\right]}$$

$$= \frac{\exp\left[-\frac{1}{N_0}\sum_{i=1}^N (X_i - s_{j,i})^2\right]}{\exp\left[-\frac{1}{N_0}\sum_{i=1}^N (X_i - s_{k,i})^2\right]}$$

In the limit as $N \to \infty$, the exponent of the numerator (see exercise 11.9-27) approaches

$$-\frac{1}{N_0}\int_0^T (X(t) - s_j(t))^2\,dt$$

and the exponent of the denominator approaches

$$-\frac{1}{N_0}\int_0^T (X(t) - s_k(t))^2\,dt,$$

so that the likelihood ratio for $\mathbf{x} = x_1, x_2, \ldots,$ is

$$\ell(x_1, \ldots) = \frac{\exp\left[-\frac{1}{N_0}\int_0^T (X(t) - s_j(t))^2\,dt\right]}{\exp\left[-\frac{1}{N_0}\int_0^T (X(t) - s_k(t))^2\,dt\right]} \qquad (11.102)$$

11.9 Detection in Continuous Time

We call the function

$$g_{X|\theta}(\mathbf{x}, \theta) = \exp\left[-\frac{1}{N_0}\int_0^T (X(t) - s_\theta(t))^2\, dt\right]$$

the *likelihood function* for X given θ, and view it most commonly as a function of θ.

Example 11.9.1 A signal $X(t) = s_i(t) + N(t)$ is observed over $[0, T)$, where $N(t)$ is white Gaussian noise, and we desire to detect which signal is sent. The decision function, based on the likelihood ratio in (11.102), is

$$\phi(X) = \begin{cases} 1 & \dfrac{\exp\left[-\dfrac{1}{N_0}\int_0^T (X(t) - s_1(t))^2\, dt\right]}{\exp\left[-\dfrac{1}{N_0}\int_0^T (X(t) - s_0(t))^2\, dt\right]} > \nu, \\ 0 & \text{otherwise}, \end{cases}$$

for some threshold ν. Suppose that $s_0(t)$ and $s_1(t)$ have energies

$$E_0 = \int_0^T s_0^2(t)\, dt \quad \text{and} \quad E_1 = \int_0^T s_1^2(t)\, dt.$$

Then the decision function simplifies to

$$\phi(X) = \begin{cases} 1 & \int_0^T x(t) s_1(t)\, dt - E_1/2 > \eta\left(\int_0^T x(t) s_0(t)\, dt - E_0/2\right) \\ 0 & \text{otherwise}, \end{cases}$$

where $\eta = N_0 \log \nu$. Note that this detector is simply a matched filter, as discussed in section 3.19.

Let $r_1 = \int_0^T x(t) s_1(t)\, dt - E_1/2$ and $r_0 = \int_0^T x(t) s_0(t)\, dt - E_0/2$. We will quantify the behavior of the detector under the simplifying assumption that $\eta = 1$. The probability of false alarm is

$$P_{FA} = P(r_1 > r_0 \mid \theta = 0)$$

or

$$P_{FA} = P(r_1 - r_0 > 0 \mid \theta = 0).$$

Also, in this case,

$$P_D = 1 - P_{MD} = 1 - P_{FA}.$$

Since $r_1 - r_0$ is Gaussian, it is straightforward to show (see exercise 11.9-26) that

$$P_{FA} = Q(d/2\sigma),$$

where $d^2 = \int_0^T (s_0(t) - s_1(t))^2\, dt$ is the squared distance between the signals, and

$$\sigma^2 = \frac{N_0}{2}.$$

When $s_0(t) = -s_1(t)$, the signaling scheme is said to be *binary phase-shift keying* (BPSK). In this case, $d^2 = 4E_b$, and the probability of false alarm becomes

$$P_{FA} = Q(\sqrt{2E_b/N_0}).$$

The quantity E_b is sometimes referred to as the bit energy—the energy required to send one signal conveying one bit of information—and the ratio E_b/N_0 is referred to as the signal-to-noise ratio. A plot of the probability of error $Q(\sqrt{2E_b/N_0})$ is shown in figure 11.23, as a function of

$$(\text{SNR})_{dB} = 10 \log(E_b/N_0).$$

□

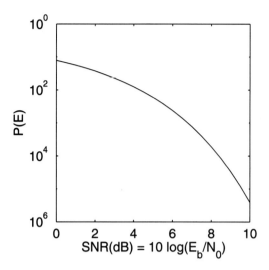

Figure 11.23: Probability of error for BPSK

11.9.1 Some extensions and precautions

The preceding treatment of the detection of a deterministic signal in noise covers many cases of practical interest. It can be readily extended, for example, to cover signals with random phase, analogous to those of example 11.8.8, and all the familiar signal constellations that we have seen, such as PSK, QAM, orthogonal, and so on. However, there are several technical issues to address that did not arise in the examples of the previous section. We go just far enough to show some of the problems that might arise, without actually presenting the solutions.

The first issue is that of the existence of the set of orthonormal basis functions $\Psi = \{\psi_i(t), i = 1, 2, \ldots\}$, which can be used to characterize the random process $N(t)$. Actually, before becoming too engrossed in the solution of this problem, it is worth pointing out that for detection of *deterministic* functions in white Gaussian noise, it is not necessary to fully characterize the noise.

Suppose that the deterministic signal $s_\theta(t)$ can be represented in terms of the m basis functions $\phi_i(t)$, $i = 1, 2, \ldots, m$, as

$$s_\theta(t) = \sum_{i=1}^{m} a_{\theta,i} \phi_i(t),$$

where (for convenience) we assume that the $\{\phi_i\}$ are orthonormal, so that

$$s_{\theta,i} = \langle s_\theta(t), \phi_i(t) \rangle.$$

Suppose also that there is a set of basis functions $\{\psi_i(t)\}$, $i = 1, 2, \ldots$, which can be used to represent the noise. In general, $\{\phi_i(t), i = 1, 2, \ldots, m, \psi_i(t), i = 1, 2, \ldots\}$ will not form an orthogonal set. Using the Gram–Schmidt process, we form an orthogonal set that we denote as $\{\phi_i(t), i = 1, 2, \ldots, m, \hat{\psi}_i(t), i = m+1, \ldots\}$. We let

$$s_{\theta,i} = \langle s_\theta(t), \hat{\psi}_i(t) \rangle \qquad i = m+1, m+2, \ldots.$$

Then the likelihood function for samples $x_1, x_2, \ldots, x_N$ of $X(t)$ (assuming $N > m$), given

11.9 Detection in Continuous Time

$s_\theta(t)$, is

$$\ell(x_1, x_2, \ldots, x_N) = \prod_{i=1}^{N} \exp\left[-\frac{1}{N_0}(X_i - s_{\theta,i})^2\right]$$

$$= \prod_{i=1}^{m} \exp\left[-\frac{1}{N_0}(X_i - s_{\theta,i})^2\right] \prod_{i=m+1}^{N} \exp\left[-\frac{1}{N_0}(X_i - s_{\theta,i})^2\right]$$

$$= \prod_{i=1}^{m} \exp\left[-\frac{1}{N_0}(X_i - s_{\theta,i})^2\right] \prod_{i=m+1}^{N} \exp\left[-\frac{1}{N_0}X_i^2\right],$$

so that

$$\prod_{i=1}^{m} \exp\left[-\frac{1}{N_0}(X_i - s_{\theta,i})^2\right]$$

is *sufficient* for detecting $s_{\theta,i}$. We can write

$$\prod_{i=1}^{m} \exp\left[-\frac{1}{N_0}(X_i - s_{\theta,i})^2\right] = \exp\left[-\frac{1}{N_0}\int_0^T (\tilde{X}(t) - s_\theta(t))^2\, dt\right],$$

where $\tilde{X}(t)$ is the projection of $X(t)$ onto the space spanned by $\phi_i(t)$, $i = 1, 2, \ldots, m$:

$$\tilde{X}(t) = \sum_{i=1}^{m} X_i \phi_i(t)$$

with $X_i = \langle X(t), \phi_i(t)\rangle$.

Thus, for all practical purposes, when detecting deterministic signals in noise, the noise that is orthogonal to the signal space is *irrelevant*. (We have just proven what is sometimes called the theorem of irrelevance.) We can call the function

$$\ell(X(t)) = \exp\left[-\frac{1}{N_0}\int_0^T (\tilde{X}(t) - s_\theta(t))^2\, dt\right]$$

the projected likelihood function. We will denote $\Lambda(X(t)) = \log \ell(X(t))$. Expanding $\Lambda(X(t))$, we have

$$\Lambda(X(t)) = -\frac{1}{N_0}\int_0^T \tilde{X}^2(t)\, dt - 2\int_0^T \tilde{X}(t)s_\theta(t) + \int_0^T s_\theta^2(t)\, dt$$

$$= -\frac{1}{N_0}\int_0^T \tilde{X}^2(t)\, dt + \frac{2}{N_0}\int_0^T \tilde{X}(t)s_\theta(t)\, dt - \frac{1}{N_0}\int_0^T s_\theta^2(t)\, dt$$

The first term does not depend upon θ, and the third term can be written in terms of the energy $E_\theta = \int_0^T s_\theta^2(t)$. The decision can thus be based upon

$$\Lambda'(X(t)) = \int_0^T \tilde{X}(t)s_\theta(t)\, dt - \frac{E_\theta}{2}. \qquad (11.103)$$

The first term of (11.103) can be written as

$$\int_0^T \tilde{X}(t)s_\theta(t)\, dt = \sum_{i=1}^{m} X_i s_{\theta,i},$$

where $X_i = \langle X(t), \phi_i(t)\rangle$. For a known signal $s_\theta(t)$, the coefficients $s_{\theta,i}$ can be precomputed. The received signal is correlated with the basis functions $\phi_i(t)$, and these statistics are used to make decisions, as shown in figure 11.24. Geometrically, this amounts to projecting $X(t)$ onto span$\{\phi_i(t)\}$, then finding the signal point closest to the projection.

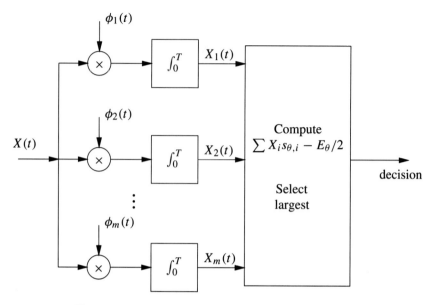

Figure 11.24: A projection approach to signal detection

We now return to the question of finding the functions $\{\psi_i(t)\}$ that span the space of noise signals. We do not assume (now) that the signal is necessarily white. One approach to obtaining these functions is by means of the Karhunen–Loève representation, first presented in section 6.7 for discrete-time random variables. Let $Z(t), t \in [0, T)$ be a zero-mean random process with $EZ^2(t) < \infty$ for all $t \in [0, T)$, and let

$$C_Z(t, u) = \text{cov}(Z(t), Z(u)) \tag{11.104}$$

be the autocovariance function for Z. In the equation

$$\lambda \psi(t) = \int_0^T C_Z(t, u) \psi(u) \, du, \tag{11.105}$$

the quantity λ is an *eigenvalue* and $\psi(t)$ is an *eigenfunction*, analogous to eigenvalues and eigenvectors of a symmetric matrix.

Theorem 11.4 *Let $Z(t)$ be a zero-mean random process, with $EZ^2(t) < \infty$ for all t and covariance $C_Z(t, u)$ as in (11.104), having eigenvalues $\{\lambda_1, \lambda_2, \ldots,\}$ and eigenfunctions $\{\psi_1(t), \psi_2(t), \ldots,\}$. Then:*

1. *The eigenvalues are real.*
2. *The eigenfunctions are orthogonal.*
3. *(Mercer's theorem) The covariance function can be written as*

$$C_Z(t, u) = \sum_{k=1}^{\infty} \lambda_k \psi_k(t) \psi_k(u), \tag{11.106}$$

where the convergence of the sum is uniform.

4. *(Karhunen–Loève representation) The random process $Z(t)$ can be represented as*

$$Z(t) = \sum_{k=1}^{\infty} Z_k \psi_k(t)$$

11.9 Detection in Continuous Time

where

$$Z_k = \int_0^T \psi_k(t) Z(t)\, dt.$$

Proof (1) and (2) are just as for eigenvalues and eigenvectors, as discussed in chapter 6, since $C_Z(t, s) = C_Z(s, t)$.

Property (11.106) is directly analogous to (6.33), but more difficult to prove because there may be an infinite number of dimensions. A thorough proof of Mercer's theorem is presented in [207]. We simply present a substantiating observation. Let

$$R_n(t, u) = C_Z(t, u) - \sum_{i=1}^n \lambda_i \psi_i(t) \psi_i(u).$$

Then, by the definition of the eigenfunctions, for any eigenfunction $\psi_j(u)$,

$$\langle R_n(t, u), \psi_j(u) \rangle = 0. \tag{11.107}$$

Thus the eigenfunctions of $C_Z(t, u)$ are all orthogonal to the difference. What remains to be shown is that all functions (in some appropriate space) are orthogonal to this difference, so the difference must be zero.

Proof of (4) follows from Mercer's theorem. Let

$$Z_n(t) = \sum_{i=1}^n Z_n \psi_i(t),$$

and let

$$e_n(t) = E[(Z(t) - Z_n(t))^2]$$

denote the mean-squared error. We will show that $e_n(t) \to 0$ as $n \to \infty$.

$$e_n(t) = EZ^2(t) - 2EZ(t) \sum_{i=1}^n Z_i \psi_i(t) + E \sum_{i=1}^n \sum_{j=1}^n Z_i Z_j \psi_i(t) \psi_j(t)$$

$$= C_Z(t, t) - 2EZ(t) \int_0^T Z(u) \psi_i(u)\, du\, \psi_i(t) + \sum_{i=1}^n \lambda_i \psi_i^2(t)$$

$$= C_Z(t, t) - 2 \sum_{i=1}^n \int_0^T C_Z(t, u) \psi_i(u)\, du\, \psi_i(t) + \sum_{i=1}^n \lambda_i \psi_i^2(t)$$

$$= C_Z(t, t) - \sum_{i=1}^n \lambda_i \psi_i^2(t),$$

where the second inequality is established in exercise 11.9-27. Applying Mercer's theorem to the last cquality, $e_n(t) \to 0$ as $n \to \infty$. □

Theorem 11.4 addresses the existence of the desired series representation. Application of the theorem also allows the detection of known signals in noise that is not white. However, there are still other potential technical problems. One of these is *singular detection*. A singular detection problem is one that, theoretically, has a detector that could operate with no error. Consider, for example, the problem of detecting a step function $u(t)$ in bandlimited noise $N(t)$. Given $X(t) = u(t) + N(t)$, there should be no possibility of detection error, since any discontinuity in $X(t)$ must arise due to the discontinuity in $u(t)$, and not due to a discontinuity in $N(t)$ (since $N(t)$ is bandlimited, it can have no discontinuities). A set of basis functions $\{\psi_i(t)\}$, derived via the Karhunen–Loève representation for $N(t)$, will be inadequate for representation of $X(t)$, since it will not represent the discontinuities. Functions represented by directions that are *not* represented by the set $\{\psi_i(t)\}$ are necessary to accomplish the desired detection. (Continuity considerations do not arise in discrete-time systems, so singular detection problems have not been previously apparent.)

Singular detection can also be a problem when we desire to detect *random* signals in noise. Again, there is a potential representation problem: that the random signal to be detected, $S(t)$, may not lie in the space spanned by $\{\psi_i(t)\}$, and that those directions that are not present are precisely those in which the distinguishing information about $S(t)$ must lie.

A thorough treatment of detection in continuous time requires some background in measure theory. The discussion, while important, would take us too far afield. The interested student is referred to [257], which provides an engineering-oriented discussion of these issues. Another excellent applied discussion of detection in nonwhite noise appears in [313].

11.10 Minimax Bayes decisions

Thus far, we have taken two approaches to prior probabilities: assuming nothing (focusing on conditional probabilities of error, as in the Neyman–Pearson test), or making certain assumptions (focusing on minimal risk, as in Bayes theory). In this section, we return to Bayes decision theory, but address the problem of finding decision functions when the prior probabilities are unknown. This will lead us to the minimax Bayes decision procedure. We gain some understanding of the minimax problem by means of the Bayes envelope function; we then introduce the minimax principle in the context of multiple hypothesis testing.

11.10.1 Bays envelope function

For binary hypothesis testing, we can introduce the Bayes envelope function. Suppose that, rather than invoking a rule to assign a specific action δ for a given observation x, we instead invoke a randomized rule. Let $\phi \in D$ and $\phi' \in D$ be two nonrandomized rules, and let $\varphi_\pi \in D^*$ be the randomized decision rule corresponding to using rule ϕ with probability π, where $\pi \in [0, 1]$, and using rule ϕ' with probabilty $1 - \pi$. To compute the risk function corresponding to this randomized rule we must take the expectation with respect to the rule itself, in addition to taking the expectation with respect to X. This yields (see (11.54))

$$r(\vartheta, \varphi_\pi) = \pi r(\vartheta, \phi) + (1 - \pi) r(\vartheta, \phi').$$

Definition 11.13 The function $\rho(\cdot)$ defined by

$$\rho(p) = r(p, \varphi_p) = \min_{\varphi \in D^*} r(p, \varphi) \qquad (11.108)$$

is called the **Bayes envelope function**. It represents the minimal global expected loss attainable by any decision function, when θ is a random variable with *a priori* distribution $P[\theta = \vartheta_1] = p$ and $P[\theta = \vartheta_0] = 1 - p$. □

We observe that, for $p = 0$, $\rho(p) = 0$, and also for $p = 1$, $\rho(p) = 0$. Furthermore, it easy to see that $\rho(p)$ must be concave downward; if it were not, we could construct a randomized rule that would improve performance, in a manner analogous to the way that we analyzed the construction of a randomized rule in the ROC curve context. Figure 11.25 is an example of a Bayes envelope function (which is the parabolically shaped curve in the figure).

11.10 Minimax Bayes Decisions

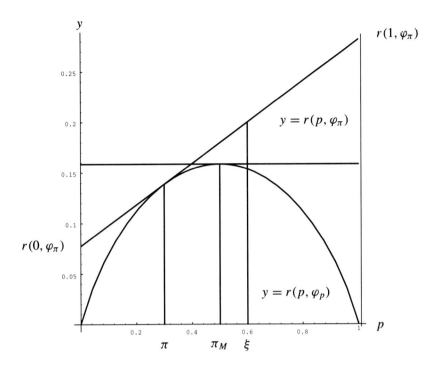

Figure 11.25: Bayes envelope function

Theorem 11.5 (Concavity of Bayes envelope function) *For any distributions of θ p_1 and p_2, and for any number q such that $0 \leq q \leq 1$,*

$$\rho(qp_1 + (1-q)p_2) \geq q\rho(p_1) + (1-q)\rho(p_2).$$

Proof Since the Bayes risk defined in (11.60) is linear in p, it follows that for any decision rule ϕ,

$$r(qp_1 + (1-q)p_2, \phi) = qr(p_1, \phi) + (1-q)r(p_2, \phi).$$

To obtain the Bayes envelope, we must minimize this expression over all decision rules ϕ. But the minimum of the sum of two quantities can never be smaller than the sum of their individual minima, hence

$$\min_{\phi} r(qp_1 + (1-q)p_2, \phi) = \min_{\phi}[qr(p_1, \phi) + (1-q)r(p_2, \phi)]$$
$$\geq \min_{\phi} qr(p_1, \phi) + \min_{\phi}(1-q)r(p_2, \phi). \qquad \square$$

Now consider the function defined by

$$y_\pi(p) = pR(\vartheta_1, \varphi_\pi) + (1-p)R(\vartheta_0, \varphi_\pi)$$
$$= r(p, \varphi_\pi).$$

As a function of p, $y_\pi(p)$ is a straight line from $y(0) = R(\vartheta_0, \varphi_\pi)$ to $y(1) = R(\vartheta_1, \varphi_\pi)$. We see that, for each fixed π, the curve $\rho(p)$ lies entirely below the straight line $y_\pi(p) = r(p, \varphi_\pi)$. The quantity $y_\pi(p)$ may be regarded as the expected loss incurred by assuming that $P[\theta = \vartheta_1] = \pi$ and hence using the decision rule φ_π, when in fact $P[\theta = \vartheta_1] = p$;

the excess of $y_\pi(p)$ over $\rho(p)$ is the cost of the error in incorrectly estimating the true value of the *a priori* probability $p = P[\theta = \vartheta_1]$ (see figure 11.25).

The minimax estimator addresses the question, What if the prior probability p is unknown? What is best detector rule Φ_π that we can use to minimize the maximum cost of the decision?

Example 11.10.1 Consider again the problem of detecting

$$H_0: \mathbf{X} \sim \mathcal{N}(\mathbf{m}_0, \sigma^2 I), \qquad H_1: \mathbf{X} \sim \mathcal{N}(\mathbf{m}_1, \sigma^2 I).$$

As we have seen, the probability of error is

$$P(\mathcal{E}) = (1-p)Q(\sigma/d \log((1-p)/p) + d/(2\sigma)) + pQ(d/(2\sigma) - \sigma/d \log((1-p)/p)).$$

Setting $L_{01} = L_{10} = 1$, the Bayes risk is the total probability of error. Figure 11.26 illustrates the corresponding Bayes envelope functions for various values of $d = \|\mathbf{m}_0 - \mathbf{m}_1\|$. □

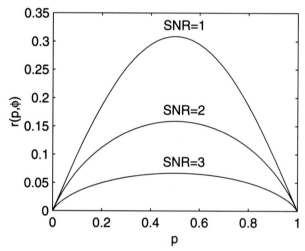

Figure 11.26: Bayes envelope function: normal variables with unequal means and equal variances

Example 11.10.2 Consider again the detection problem of example 11.4.5, where the risk function corresponding to the optimal decision rule was found to be

$$r(p, \phi_p) = \begin{cases} 10p & 0 \le p < \frac{3}{19}, \\ \frac{25}{12}p + \frac{5}{4} & \frac{3}{19} \le p < \frac{9}{17}, \\ 5(1-p) & \frac{9}{17} \le p \le 1. \end{cases}$$

A plot of the Bayes envelope function is provided in figure 11.27. □

Example 11.10.3 Consider the binary channel of example 11.10.3, and assume that $\lambda_0 = 1/4$ and $\lambda_1 = 1/3$. The Bayes risk functions for each decision function are

$$r(p, \phi_1) = p,$$
$$r(p, \phi_2) = (1-p)\lambda_0 + p\lambda_1,$$
$$r(p, \phi_3) = (1-p)(1-\lambda_0) + p(1-\lambda_1),$$
$$r(p, \phi_4) = 1 - p.$$

Figure 11.28 shows the Bayes risk function $r(p, \phi_i)$ for each of the possible decision functions. Also shown (in the darker line) is the minimum Bayes risk function—the Bayes envelope. □

11.10 Minimax Bayes Decisions

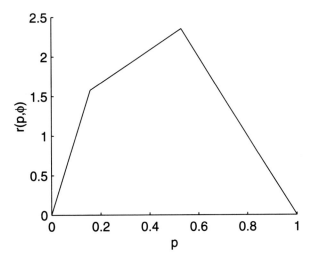

Figure 11.27: Bayes envelope function for example 11.4.5

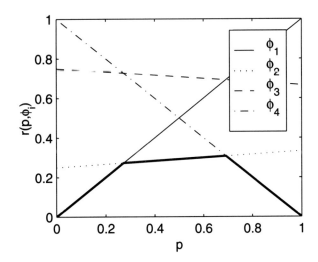

Figure 11.28: Bayes envelope for binary channel

11.10.2 Minimax rules

An interesting approach to decision making is to consider ordering decision rules according to the worst that could happen. Consider the value $p = \pi_M$ on the Bayes envelope plot given in figure 11.25. At this value, we have that

$$r(0, \varphi_{\pi_M}) = r(1, \varphi_{\pi_M}) = \max_p \rho(p).$$

Thus, for $p = \pi_M$, the maximum possible expected loss due to ignorance of the true state of nature is minimized by using φ_{π_M}. This observation motivates the introduction of the so-called minimax decision rules.

Definition 11.14 We say that a decision rule φ_1 is **preferred** to rule φ_2 if

$$\max_{\vartheta \in \Theta} R(\vartheta, \varphi_1) < \max_{\vartheta \in \Theta} R(\vartheta, \varphi_2).$$

□

This notion of preference leads to a linear ordering of the rules in D^*. A rule that is most preferred in this ordering is called a *minimax* decision rule. That is, a rule φ_0 is said to be *minimax* if

$$\max_{\vartheta \in \Theta} R(\vartheta, \varphi_0) = \min_{\varphi \in D^*} \max_{\vartheta \in \Theta} R(\vartheta, \varphi). \qquad (11.109)$$

The value on the right-hand side of (11.109) is called the *minimax value*, or *upper value* of the game.

In words, (11.109) means, essentially, that if we first find, for each rule $\varphi \in D^*$, the value of ϑ that maximizes the risk, and then find the rule $\varphi_0 \in D^*$ that minimizes the resulting set of risks, we have the minimax decision rule. This rule corresponds to an attitude of "cutting our losses." We first determine what state nature would take if we were to use rule φ and it were perverse, then we take the action that minimizes the amount of damage that nature can do to us.

A paranoid agent would be inclined toward a minimax rule. But, as they say, "Just because I'm paranoid doesn't mean they're *not* out to get me," and indeed nature may have it in for a decision-making agent. In such a situation, nature would search through the family of possible prior distributions, and would choose one that does the agent the most damage, even if the agent adopts a minimax stance.

Definition 11.15 A distribution $p_0 \in \Theta^*$ is said to be a **least favorable prior** (lfp) if

$$\min_{\varphi \in D^*} r(p_0, \varphi) = \max_{p \in \Theta^*} \min_{\varphi \in D^*} r(p, \varphi). \qquad (11.110)$$

The value on the right-hand side of (11.110) is called the *maximin value*, or *lower value* of the Bayes risk. □

The terminology, "least favorable," derives from the fact that, if I were told which prior nature was using, I would like least for it to be a distribution p_0 satisfying (11.110), because that would mean that nature had taken a stance that would allow me to cut my losses by the least amount.

11.10.3 Minimax Bayes in multiple-decision problems

In developing the solution to the minimax decision problem, we will generalize beyond the binary hypothesis test to the M-ary decision problem. Suppose that Θ consists of $M \geq 2$ points, $\Theta = \{\vartheta_1, \ldots, \vartheta_M\}$. The general decision problem is to determine a test to select among these M options.

Suppose that the prior distribution on Θ is

$$P(\theta = \vartheta_1) = p_1, P(\theta = \vartheta_2) = p_2, \ldots, P(\theta = \vartheta_M) = p_M.$$

We can represent the vector of priors as

$$\mathbf{p} = [p_1, p_2, \ldots, p_M]^T.$$

As in the binary case, we can talk about the risk and the Bayes risk, where risk is denoted as $R(\vartheta_i, \varphi)$, and Bayes risk as

$$r(\mathbf{p}, \varphi) = \sum_{i=1}^{M} p_i R(\vartheta_i, \varphi).$$

Using the notation $\mathbf{y}(\varphi) = [R(\vartheta_1, \varphi), \ldots, R(\vartheta_M, \varphi)]^T$, we have

$$r(\mathbf{p}, \varphi) = \mathbf{p}^T \mathbf{y}.$$

11.10 Minimax Bayes Decisions

Definition 11.16 The **risk set** $S \subset \mathbb{R}^M$ is the set of the form

$$S = \{R(\vartheta_1, \varphi), \ldots, R(\vartheta_M, \varphi)\},$$

where φ ranges through D^*, the set of all randomized decision rules. In other words, S is the set of all M-tuples $(y_1, \ldots, y_M)$, such that $y_i = R(\vartheta_i, \varphi)$, $i = 1, \ldots, M$, for some $\varphi \in D^*$. □

The risk set will be fundamental to our understanding of minimax tests.

Theorem 11.6 *The risk set S is a convex subset of $\mathbb{R}^M$.*

Proof Let $\mathbf{y} = [y_1, \ldots, y_M]^T$ and $\mathbf{y}' = [y'_1, \ldots, y'_M]^T$ be arbitrary points in S. According to the definition of S, there exist decision rules φ and φ' in D^* for which $y_i = R(\vartheta_i, \varphi)$ and $y'_i = R(\vartheta_i, \varphi')$, for $i = 1, \ldots, M$. Let π be arbitrary such that $0 \leq \pi \leq 1$, and consider the decision rule φ_π that chooses rule φ with probability π and rule φ' with probability $1 - \pi$. Clearly, $\varphi_\pi \in D^*$, and

$$R(\vartheta_i, \varphi_\pi) = \pi R(\vartheta_i, \varphi) + (1 - \pi) R(\vartheta_i, \varphi')$$

for $i = 1, \ldots, M$. If $\mathbf{z}$ denotes the point whose ith coordinate is $R(\vartheta_i, \varphi_\pi)$, then $\mathbf{z} = \pi \mathbf{y} + (1 - \pi) \mathbf{y}'$, thus $\mathbf{z} \in S$. □

A prior distribution for nature is an M-tuple of nonnegative numbers $(p_1, \ldots, p_M)$ such that $\sum_{i=1}^M p_i = 1$, with the understanding that p_i represents the probability that nature chooses ϑ_i. Let $\mathbf{p} = [p_1, \ldots, p_M]^T$. For any point $\mathbf{y} \in S$ determined by some rule φ, the Bayes risk is then the inner product

$$r(\mathbf{p}, \varphi) = \mathbf{p}^T \mathbf{y} = \sum_{i=1}^M p_i y_i = \sum_{i=1}^M p_i R(\vartheta_i, \varphi).$$

We make the following observations:

1. There may be multiple points with the same Bayes risk (for example, suppose one or more entries in $\mathbf{p}$ is zero). Consider the set of all vectors $\mathbf{y}$ that satisfy, for a given $\mathbf{p}$, the relationship

$$\mathbf{p}^T \mathbf{y} = b \tag{11.111}$$

   for any real number b. All of these points (and the corresponding decision rules) are equivalent.

2. The set of points $\mathbf{y}$ that satisfy (11.111) lie in a hyperplane; this plane is perpendicular to the vector from the origin to the point $(p_1, \ldots, p_M)$. To see this, consider figure 11.29, where, for $M = 2$, the risk set and sets of equivalent points are displayed (the concepts carry over to the general case for $M > 2$). If $\mathbf{y}$ is such that $\mathbf{p}^T \mathbf{y} = b$, then for a vector $\mathbf{x} \perp \mathbf{p}$,

$$\mathbf{p}^T (\mathbf{y} + \mathbf{x}) = b.$$

3. The quantity b can be visualized by noting that the point of intersection of the diagonal line $y_1 = \cdots = y_M$ with the plane $\mathbf{p}^T \mathbf{y} = \sum_i p_i y_i = b$ must occur at $[b, \ldots, b]^T$.

4. To find the Bayes rules, we find the minimum of those values of b, call it b_0, for which the plane $\mathbf{p}^T \mathbf{y} = b_0$ intersects the set S. Decision rules corresponding to points in this intersection are Bayes with respect to the prior $\mathbf{p}$.

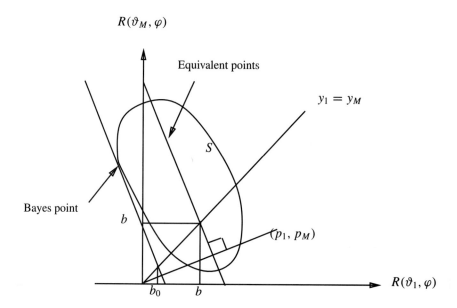

Figure 11.29: Geometrical interpretation of the risk set

The minimax problem can thus be visualized using the risk set: as **p** varies, how does the point b_0 of minimum risk vary? The point of minimum risk is the minimax risk, as we now explore.

The maximum risk for a fixed rule φ is given by

$$\max_i R(\vartheta_i, \varphi).$$

All points $\mathbf{y} \in S$ that yield this same value of $\max_i y_i$ are equivalent with respect to the minimax principle. Thus, all points on the boundary of the set

$$Q_c = \{(y_1, \ldots, y_M): y_i \leq c \quad \text{for } i = 1, \ldots, M\}$$

for any real number c are equivalent. To find the minimax rules, we find the minimum of those values of c, call it c_0, such that the set Q_{c_0} intersects S. Then we observe the following:

Any decision rule φ whose associated risk point $[R(\vartheta_1, \varphi) \cdots R(\vartheta_M, \varphi)]^T$ is an element of $Q_{c_0} \cap S$ is a minimax decision rule.

Figure 11.30 depicts a minimax rule for $M = 2$.

Thus, for a minimax rule, we must have risk equalization. For the minimax rule φ,

$$R(\vartheta_1, \varphi) = R(\vartheta_2, \varphi) = \cdots = R(\vartheta_M, \varphi). \tag{11.112}$$

Due to the equal risk, at the point of minimax risk,

$$r(\mathbf{p}, \varphi) = R(\vartheta_1, \varphi)(p_0 + p_1 + \cdots + p_M) = R(\vartheta_1, \varphi),$$

so that the *Bayes risk is independent of the prior*: any attempts by nature to find a less favorable prior are neutralized.

11.10 Minimax Bayes Decisions

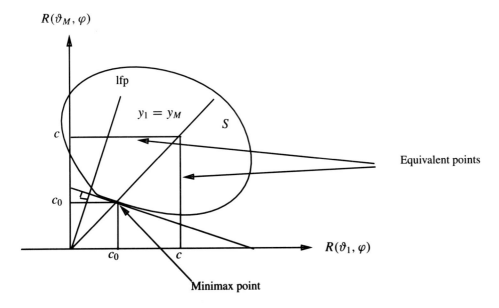

Figure 11.30: Geometrical interpretation of the minimax rule

Figure 11.30 also depicts the least favorable prior, which is visualized as follows. As we have seen, a strategy for nature is a prior distribution $\mathbf{p} = [p_1, \ldots, p_M]^T$ that represents the family of planes perpendicular to $\mathbf{p}$. In using a Bayes rule to minimize the risk, we must find the plane from this family that is tangent to and below S. Because the minimum Bayes risk is b_0, where $[b_0, \ldots, b_0]^T$ is the intersection of the line $y_1 = \cdots = y_M$ and the plane, tangent to and below S and perpendicular to $\mathbf{p}$, a least favorable prior distribution is the choice of $\mathbf{p}$ that makes the intersection as far up the line as possible. Thus the least favorable prior (lfp) is a Bayes rule whose risk is $b_0 = c_0$.

Example 11.10.4 We can be more explicit about the risk set S in $M = 2$ dimensions. From (11.64),

$$r(p, \phi_p) = (1-p)R(\vartheta_0, \phi) + pR(\vartheta_1, \phi), \quad (11.113)$$

$$= (1-p)L_{01}\alpha + pL_{10}(1-\beta). \quad (11.114)$$

Let ϕ be a Neyman–Pearson test associated with the binary hypothesis problem. The ROC, associated with the Neyman–Pearson test, is a plot of β versus α for the test ϕ. Let $\hat{\phi} = 1 - \phi$ denote the test that is *conjugate* to ϕ. Let P_{FA} and $\hat{P}_{FA}$ denote the probability of choosing decision 1, given that $\theta = \theta_0$, for ϕ and $\hat{\phi}$, respectively, and let P_D and $\hat{P}_D$ be defined similarly. Then, from table 11.1, we note that $\hat{\phi}$ has $\hat{P}_{FA} = 1 - \alpha$ and $\hat{P}_D = 1 - \beta$. A plot of the ROC for ϕ and $\hat{\phi}$ is shown in figure 11.31(a). There are no tests outside the shape shown, since such points would violate the Neyman–Pearson lemma. Figure 11.31(b) shows the boundaries of the risk set, found by plotting $L_{01}P_{FA}$ and $L_{10}(1 - P_D)$ for each of the two sets ($L_{01} = 0.4$ and $L_{10} = 1.5$ in this figure). □

Table 11.1: Probabilities for Neyman–Pearson and conjugate Neyman–Pearson tests

		$\theta = \theta_0$	$\theta = \theta_1$
$\phi = 0$	$\hat{\phi} = 1$	$1 - \alpha = \hat{P}_{FA}$	$1 - \beta = \hat{P}_D$
$\phi = 1$	$\hat{\phi} = 0$	$\alpha = P_{FA}$	$\beta = P_D$

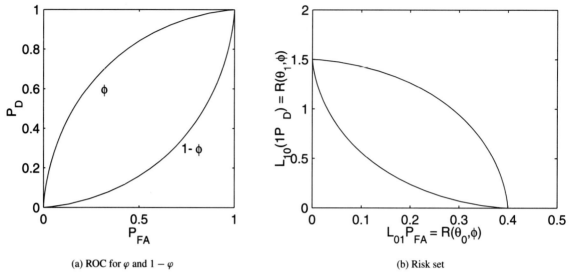

(a) ROC for φ and $1-\varphi$ (b) Risk set

Figure 11.31: The risk set and its relation to the Neyman–Pearson test

11.10.4 Determining the least favorable prior

Given that a minimax solution is found, it may be of interest to determine the least-favorable prior. As discussed, the probability vector $\mathbf{p} = [p_0, p_1, \ldots, p_M]^T$ is orthogonal to the boundary of the risk set at the point where $R(\vartheta_0, \phi) = R(\vartheta_1, \phi) = \cdots = R(\vartheta_M, \phi)$. Determining the least favorable prior requires finding a vector tangent to the boundary of the risk set, then finding a vector normal to that surface, normalized to be a probability vector.

Let the boundary B of S that intersects Q_c be a surface parameterized by some parameter $\mathbf{q} \in \mathbb{R}^{M-1}$, and assume that B is a differentiable function of the components of $\mathbf{q}$ for some $\mathbf{q}$ in an open neighborhood of $\mathbf{q}_0$. That is, the point $[R_{\mathbf{q}}(\vartheta_0, \phi), R(\vartheta_1, \phi), \ldots, R_{\mathbf{q}}(\vartheta_{M-1}, \phi)]$ is a point on B, and we take the point $\mathbf{q} = \mathbf{q}_0$ as that value of parameter which is the minimax risk. Then the vectors

$$\begin{bmatrix} \dfrac{\partial R_{\mathbf{q}}(\vartheta_0, \phi)}{\partial q_1} \\ \dfrac{\partial R_{\mathbf{q}}(\vartheta_1, \phi)}{\partial q_1} \\ \vdots \\ \dfrac{\partial R_{\mathbf{q}}(\vartheta_{M-1}, \phi)}{\partial q_1} \end{bmatrix} \begin{bmatrix} \dfrac{\partial R_{\mathbf{q}}(\vartheta_0, \phi)}{\partial q_2} \\ \dfrac{\partial R_{\mathbf{q}}(\vartheta_1, \phi)}{\partial q_2} \\ \vdots \\ \dfrac{\partial R_{\mathbf{q}}(\vartheta_{M-1}, \phi)}{\partial q_2} \end{bmatrix} \cdots \begin{bmatrix} \dfrac{\partial R_{\mathbf{q}}(\vartheta_0, \phi)}{\partial q_{M-1}} \\ \dfrac{\partial R_{\mathbf{q}}(\vartheta_1, \phi)}{\partial q_{M-1}} \\ \vdots \\ \dfrac{\partial R_{\mathbf{q}}(\vartheta_{M-1}, \phi)}{\partial q_{M-1}} \end{bmatrix},$$

evaluated at $\mathbf{q} = \mathbf{q}_0$, are tangent to B at the minimax risk point. A vector which is orthogonal to all of these vectors, normalized to be a probability vector, is thus a least favorable prior probability.

In two dimensions, the least favorable prior (p_0, p_1) can be determined with less sophistication. At the point of equal risk, the minimax test is a Bayes test with likelihood

11.10 Minimax Bayes Decisions

ratio test threshold

$$v = \frac{p_0 L_{01}}{p_1 L_{10}}. \tag{11.115}$$

If the threshold v can be determined, then (11.115) can be solved for p_0.

11.10.5 A minimax example and the minimax theorem

Example 11.10.5 We now can develop solutions to the "odd or even" game introduced earlier in the chapter. As you recall, nature and yourself simultaneously put up either one or two fingers. Nature wins if the sum of the digits showing is odd, and you win if the sum of the digits showing is even. The winner in all cases receives in dollars the sum of the digits showing, this being paid by the loser. Before the game is played, you are allowed to ask nature how many fingers it intends to put up and nature must answer truthfully with probability 3/4 (hence untruthfully with probability 1/4). You therefore observe a random variable X (the answer nature gives) taking the values of 1 or 2. If $\theta = 1$ is the true state of nature, the probability that $X = 1$ is 3/4; that is, $P_{\theta=1}(1) = 3/4$. Similarly, $P_{\theta=1}(2) = 1/4$. The risk matrix, given in figure 11.32, characterizes this statistical game.

$\Theta \backslash D$	d_1	d_2	d_3	d_4
1	-2	$-3/4$	$7/4$	3
2	3	$-9/4$	$5/4$	-4

Figure 11.32: Risk function for statistical odd or even game

The risk set for this example is given in figure 11.33, which must contain all of the lines between any two of the points $(-2, 3)$, $(-3/4, -9/4)$, $(7/4, 5/4)$, $(3, -4)$. According to our earlier analysis, the minimax point corresponds to the point indicated in the figure, which is on the line L connecting $(R(1, \phi_1), R(2, \phi_1))$ with $(R(1, \phi_2), R(2, \phi_2))$. The parametric equation for this line is

$$(y_1, y_2) = q(-2, 3) + (1-q)(-3/4, -9/4),$$

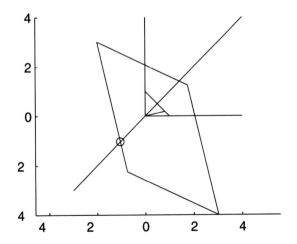

Figure 11.33: Risk set for odd or even game

as q ranges over the interval $[0, 1]$, which can be written as

$$y_1 = -\frac{5}{4}q - \frac{3}{4},$$

$$y_2 = \frac{21}{4}q - \frac{9}{4}.$$

This line intersects the line $y_1 = y_2$ at $\frac{5}{4}q - 2 = -\frac{21}{4}q + 3$, that is, when $q = \frac{3}{13}$. The minimax risk is

$$-\frac{5}{4}\frac{3}{13} - \frac{3}{4} = -\frac{27}{26}.$$

The randomized decision rule is: Use rule d_1 with probability $q = \frac{3}{13}$, and use d_2 with probability $1 - q = \frac{10}{13}$.

We may compute the least favorable prior as follows. Let nature take action $\vartheta = 1$ with probability p and $\vartheta = 2$ with probability $1 - p$. The vector $\mathbf{p} = [p, 1 - p]^T$ is perpendicular to the surface of S, which in this case is the line L previously parameterized. The tangent to this line has slope

$$\frac{\frac{dy_2}{dq}}{\frac{dy_1}{dq}} = -21/5.$$

By the orthogonality of the least favorable prior vector, we require

$$\frac{1-p}{p} = \frac{5}{21}$$

or $p = \frac{21}{26}$.

Thus, if nature chooses to hold up one finger with probability $21/26$, it will maintain your expected loss to at least $-\frac{27}{26}$, and if you select decision rule d_1 with probability $\frac{3}{13}$, you will restrict your average loss to no more than $-\frac{27}{26}$. It seems reasonable to call $-\frac{27}{26}$ the *value* of the game. If a referee were to arbitrate this game, it would seem fair to require nature to pay you $\frac{27}{26}$ dollars in lieu of playing the game.

We should point out what is achieved in the least favorable prior. The selection is only the probability p of choosing some particular outcome. What is *not* changed is the conditional probability upon which measurements are made, $f_{X|\theta}(x \mid \vartheta)$. □

The preceding example demonstrates a situation in which the best you can do in response to the worst nature can do yields the same expected loss as if nature did its worst in response to the best you can do. This result is summarized in the following theorem (which we do not prove).

Theorem 11.7 (Minimax theorem). *If, for a given decision problem (Θ, D, R) with finite $\Theta = \{\vartheta_1, \ldots, \vartheta_k\}$, the risk set S is bounded below, then*

$$\min_{\varphi \in D^*} \max_{p \in \Theta^*} r(p, \varphi) = \max_{p \in \Theta^*} \min_{\varphi \in D^*} r(p, \varphi),$$

and there exists a least favorable distribution p_0.

This condition is called the *saddlepoint condition*. More on saddlepoint optimality is presented in section 18.8.

Example 11.10.6 Consider again the binary channel of example 11.10.3, and take $\lambda_0 = 1/4$ and $\lambda_1 = 1/3$. Figure 11.34 illustrates the risk set for this case. The line of the minimax solution lies on the risk for ϕ_2 and ϕ_4; it is parameterized by

$$(y_1, y_2) = q(1/4, 1/3) + (1-q)(1, 0),$$

so that the minimax solution, when $y_1 = y_2$, occurs when $q = 12/13$. That is, ϕ_2 should be

11.10 Minimax Bayes Decisions

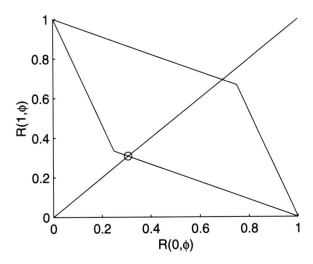

Figure 11.34: Risk set for the binary channel

employed with probability 12/13. The corresponding minimum Bayes risk is 4/13—this is the minimax probability of error. The least favorable prior is found by finding the slope of the risk function

$$\frac{\frac{dy_2}{dq}}{\frac{dy_1}{dq}} = -\frac{4}{9}.$$

The LFP has perpendicular slope:

$$\frac{9}{4} = \frac{p}{1-p},$$

so that $p = 9/13$ is least favorable.

It is interesting to compare these results with the Bayes envelope of figure 11.28. The LFP and minimax Bayes risk are both apparent in this figure. □

Example 11.10.7 Given $\Theta = \{\vartheta_0, \vartheta_1\}$ corresponding, respectively, to the hypotheses H_0 and H_1 defined by

$$H_0: \mathbf{x} \sim \mathcal{N}(\mathbf{m}_0, \sigma^2 I),$$

$$H_1: \mathbf{x} \sim \mathcal{N}(\mathbf{m}_1, \sigma^2 I),$$

we form the log-likelihood

$$\Lambda(\mathbf{x}) = \frac{1}{\sigma^2}(\mathbf{m}_1 - \mathbf{m}_0)^T \left(\mathbf{x} - \frac{1}{2}(\mathbf{m}_1 + \mathbf{m}_0)\right).$$

The log-likelihood function is Gaussian distributed: under H_0, $\Lambda(\mathbf{X}) \sim \mathcal{N}(-f^2/2, f^2)$, and under H_1, $\Lambda(\mathbf{X}) \sim \mathcal{N}(f^2/2, f^2)$, where

$$f^2 = \frac{1}{\sigma^2}\|\mathbf{m}_1 - \mathbf{m}_0\|.$$

The decision is

$$\phi(\mathbf{x}) = \begin{cases} 1 & \Lambda(\mathbf{x}) > \log \nu = \eta, \\ 0 & \Lambda(\mathbf{x}) < \eta. \end{cases}$$

The size and power are

$$\alpha = Q(\mu),$$
$$\beta = Q(\mu - f), \qquad (11.116)$$

where

$$\mu = \eta/f + f/2. \qquad (11.117)$$

Now suppose that we impose the costs $L_{00} = L_{11} = 0$ and $L_{01} = cL_{10}$. That is, the cost of a false alarm is k times more than the cost of a missed detection. We desire to determine the threshold η which minimizes the risk against all possible priors. The risks are

$$R(\vartheta_0, \phi) = L_{01}\alpha,$$

$$R(\vartheta_1, \phi) = L_{10}(1 - \beta).$$

By (11.112), the minimax rule must satisfy

$$L_{01}\alpha = L_{10}(1 - \beta)$$

or

$$cL_{10}\alpha = L_{10}(1 - \beta),$$

so $\beta = 1 - c\alpha$. From (11.116), we must have

$$Q(\mu - f) = 1 - cQ(\mu). \tag{11.118}$$

Determination of μ and (from (11.117)) the log-likelihood threshold η, can be accomplished by numerical solution of (11.118), which can be done by iterating

$$\mu^{[k+1]} = Q^{-1}((1 - Q(\mu^{[k]} - f))/c),$$

starting from some initial $\mu^{[0]}$.

Once μ is found, the least favorable prior is found. We can describe the boundary of S using the ROC curve as a function of the threshold,

$$R_\mu(\vartheta_0, \phi) = L_{01}\alpha = L_{01}Q(\mu),$$

$$R_\mu(\vartheta_1, \phi) = L_{10}(1 - \beta) = L_{10}(1 - Q(\mu - f)).$$

Then the tangent vector has slope

$$dR_\mu(\vartheta_1, \phi)/dR_\mu(\vartheta_0, \phi) = -\frac{L_{10}g(\mu - f)}{L_{01}g(\mu)} = \frac{g(\mu - f)}{cg(\mu)},$$

where $g(x) = \frac{1}{\sqrt{2\pi}} e^{-x^2/2}$, and the orthogonal vector $[p, 1-p]$ must satisfy

$$\frac{1 - p}{p} = c\frac{g(\mu)}{g(\mu - f)}.$$

MATLAB code that computes μ, ϵ, the minimax value, and the least favorable prior is shown in algorithm 11.1. ☐

Algorithm 11.1 Example Bayes minimax calculations
File: `bayes3.m`

11.11 Exercises

11.2-1 Consider the test
$$H_0: Y \sim \mathcal{U}(0, 1), \qquad H_1: Y \sim \mathcal{U}(0, 2).$$

(a) Set up the likelihood ratio test and determine the decision regions as a function of the threshold.

(b) Find P_F and P_D.

11.11 Exercises

11.2-2 [11] Consider the test

$$H_0: Y = N,$$
$$H_1: Y = S + N.$$

where $S \sim \mathcal{U}(-1, 1)$ and $N \sim \mathcal{U}(-2, 2)$, and S and N are statistically independent.

(a) Set up the likelihood ratio test and determine the decision regions when (i) $\nu = 1/4$; (ii) $\nu = 2$; (iii) $\nu = 1$.

(b) Find P_{FA} and P_D for each of these values of ν.

(c) Sketch the ROC.

11.2-3 Show that the means and variances in (11.23) and (11.24) are correct.

11.2-4 Show that the mean and variance in (11.31) and (11.32) are correct. Hint: use the fact that $\int_0^\infty x^{\nu-1} e^{-\mu x} dx = \Gamma(\nu)/\mu^\nu$.

11.2-5 Show that the inverse Fourier transform of the characteristic function in (11.38) is (11.39).

11.2-6 [11] Consider two hypotheses:

$$H_0: f_R(r \mid H_0) = \frac{1}{\sqrt{2\pi}} \exp\left(-\frac{1}{2}r^2\right),$$

$$H_1: f_R(r \mid H_1) = \frac{1}{2} \exp(-|r|).$$

(a) Plot the density functions.

(b) Find the likelihood ratio and the log-likelihood ratio. Plot the log-likelihood ratio as a function of r for $\nu = \sqrt{\pi/2}$.

(c) Compute the decision regions as a function of ν.

(d) Determine expressions for α and β.

11.2-7 The random variable X is normal, zero-mean, and has unit variance. It is passed through one of two nonlinear transformations:

$$H_0: Y = X^2,$$
$$H_1: Y = X^3.$$

Find the LRT.

11.2-8 **Poisson characteristic function**. Let $X \sim P(\lambda)$, that is, X is Poisson-distributed with parameter λ. Then

$$P(X = k) = \frac{\lambda^k}{k!} e^{-\lambda}, \quad k \geq 0.$$

Show that the characteristic function of X is

$$\Phi_X(\omega) = \sum_{k=0}^{\infty} P(X = k) e^{j\omega k} = e^{\lambda(e^{j\omega} - 1)}.$$

11.2-9 **Detection of Poisson random variables** Let X_i, $i = 1, 2, \ldots, n$, be independent Poisson random variables with rate λ. We will say that $X_i \sim P(\lambda)$.

(a) Show that $X = \sum_{i=1}^n X_i$ is $P(n\lambda)$.

(b) Find the likelihood ratio of a test for $\lambda_1 > \lambda_0$.

(c) Determine how to find the threshold for a test of size $\alpha = 0.01$ in a Neyman–Pearson test of $H_0: \lambda = 2$ versus $H_1: \lambda = 4$. (The intermediate test γ will be required.)

11.2-10 [291] In the optical communication channel using on/off signalling, suppose that a leaky detector is used. When a pulse is sent, photons arrive at the detector at a rate λ_1, and when no pulse is sent, only background photons arrive, at a rate $\lambda_0 < \lambda_1$. In the leaky detector, photons arrive at the receiver according the probability law $P(X(t) = k) = e^{-\lambda t}(\lambda t)^k/k! \sim P(\lambda t)$, but each photon is detected with probability p. Let the output of the detector be $Y(t)$.

(a) Show that $P(Y(t) = k \mid X(t) = n)$ is $\mathcal{B}(n, p)$. (See exercise 10.5-5.)

(b) Show that $P(Y(t) = k)$ is $P(p\lambda t)$.

(c) Find the Neyman–Pearson version of this detector.

(d) Compute and plot the ROC when $\lambda_1 = 2$, $\lambda_0 = 1$, and $p = 0.99$.

11.2-11 (Coherent FSK) A signal vector

$$\mathbf{s}_i = [s_{i0}, s_{i1}, \ldots, s_{i,n-1}]^T$$

is obtained by $s_{ij} = \cos 2\pi f_i j/n$, $j = 0, 1, \ldots, n-1$ for an integer frequency f_i, $i = 0, 1$. The received signal is $\mathbf{Y} = \mathbf{s} + \mathbf{N}$, where $\mathbf{N} \sim \mathcal{N}(0, \sigma^2 I)$.

(a) Determine an optimal Neyman–Pearson detector.

(b) Draw a block diagram of the detector structure.

11.2-12 By integration by parts, show that the Γ function, introduced in box 11.2 as

$$\Gamma(x) = \int_0^\infty t^{x-1} e^{-t} \, dt$$

satisfies $\Gamma(x+1) = x\Gamma(x)$ for $x > 0$.

11.2-13 For the detection problem

$$H_0: \mathbf{X} \sim \mathcal{N}(\mathbf{0}, R_0),$$
$$H_1: \mathbf{X} \sim \mathcal{N}(\mathbf{0}, R_1),$$

develop a likelihood ratio test. Express the test in terms of the "signal-to-noise ratio" $S = R_0^{-1/2} R_1 R_0^{-1/2}$. Simplify as much as possible.

11.2-14 **Bounds and approximations to the Q function**

(a) Show that

$$\sqrt{2\pi} Q(x) = \frac{1}{x} e^{-x^2/2} - \int_x^\infty \frac{1}{y^2} e^{-y^2/2} \, dy \qquad x > 0.$$

Hint: integrate by parts.

(b) Show that

$$0 < \int_x^\infty \frac{1}{y^2} e^{-y^2/2} \, dy < \frac{1}{x^3} e^{-x^2/2}.$$

(c) Conclude from (b) that

$$\frac{1}{\sqrt{2\pi} x} e^{-x^2/2} (1 - 1/x^2) < Q(x) < \frac{1}{\sqrt{2\pi} x} e^{-x^2/2} \qquad x > 0.$$

(d) Plot these lower and upper bounds on a plot with $Q(x)$ (use a log scale).

(e) Another useful bound is $Q(x) \leq \frac{1}{2} e^{-x^2/2}$. Derive this bound. Hint: Identify $[Q(\alpha)]^2$ as the probability that the zero-mean unit-Gaussian random variables lie in the shaded region shown on the left in figure 11.35, (the region $[\alpha, \infty) \times [\alpha, \infty)$). This probability is exceeded by the probability that (x, y) lies in the shaded region shown on the right (extended to ∞). Evaluate this probability.

11.11 Exercises

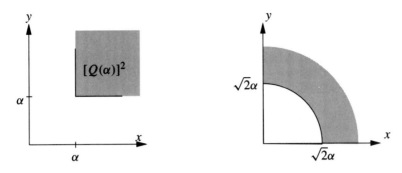

Figure 11.35: Regions for bounding the Q function

11.4-15 [11] For the hypothesis testing problem

$$H_0: f_Y(y \mid H_1) = e^{-y}, \quad y > 0 \qquad H_1: Y \sim \mathcal{U}(0, 2)$$

(a) Set up the likelihood ratio test and determine the decision regions as a function of the threshold.

(b) Find the minimum probability of error when (i) $p_0 = 1/2$; (ii) $p_0 = 2/3$; (iii) $p_0 = 1/3$.

11.4-16 [373] One of two signals $s_0 = -1$ or $s_1 = 1$ is transmitted over the channel shown in figure 11.36(a), where the noises N_1 and N_2 are independent Laplacian noise with pdf

$$f_N(\alpha) = \frac{1}{2} e^{-|\alpha|}.$$

(a) Show that the optimum decision regions for equally likely messages are as shown in figure 11.36(b).

(b) Determine the probability of error for this detector.

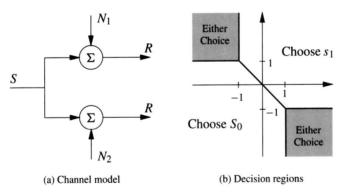

(a) Channel model (b) Decision regions

Figure 11.36: Channel with Laplacian noise and decision region

11.4-17 (**Computer exercise:** signal space simulation.) In this exercise you will simulate several different digital communications signal constellations and their detection. Suppose that an M-ary transmission scheme is to be simulated, where $M = 2^k$. The following is the general algorithm to estimate the probability of error:

Generate k random bits
Map the bits into the M-ary constellation to produce the signal **S**. (This is one symbol.)
Generate a Gaussian random number (noise) with variance $\sigma^2 = N_0/2$
    in each signal component direction.
Add the noise to the signal constellation point: $\mathbf{R} = \mathbf{S} + \mathbf{N}$.

Perform a detection on the received signal **R**.
Map the detected point $\hat{\mathbf{x}}$ back to bits.
Compare the detected bits with the transmitted bits, and count bits in error.

Repeat this until many (preferably at least 100) bits in error have been counted. The estimated *bit error* probability is

$$P_b \approx \frac{\text{number of bits in error}}{\text{total number of bits generated}}.$$

The estimated *symbol error* probability is

$$P_E \approx \frac{\text{number of symbols in error}}{\text{total number of symbols generated}}.$$

In general, $P_b \neq P_E$, since a symbol in error may actually have several bits in error.

The process above should be repeated for values of SNR (E_b/N_0) in the range from 0–10 dB.

(a) Plot the theoretical probability of error for BPSK detection with equal probabilities as a function of SNR (in dB) versus P_b (on a log scale). Your plot should look like figure 11.23.

(b) By simulation, estimate the probability of error for BPSK transmission, using the method just outlined. Plot the results on the same axes as the theoretical plot. (They should be very similar.)

(c) Plot the theoretical probability of *symbol* error for QPSK. Simulate using QPSK, and plot the estimated symbol error probability.

(d) Plot the upper bound for the probability of 8-PSK. Simulate using 8-PSK, and plot the estimated error probability.

(e) Repeat parts (a) and (b) using unequal prior probabilities,

$$P(\mathbf{m}_0) = 0.8 \qquad P(\mathbf{m}_1) = 0.2.$$

(f) Compare the theoretical and experimental plots, and comment.

11.5-18 For some distributions of means, the probability of classification error is straightforward to compute. For the set of points representing means shown in figure 11.37, compute the probability of error, assuming that each hypothesis occurs with equal probability, and that the noise is $\mathcal{N}(0, \sigma^2 I)$. (These sets of means could represent signal constellations in a digital communications setting.) In each constellation, the distance between nearest signal points is d.

Also, compute average energy E_s of the signal constellation as a function of d. If the means are at $\mathbf{m}_i$, then the average energy is

$$E = \frac{1}{M} \sum_{i=1}^{M} \|\mathbf{m}_i\|^2.$$

For example, for the 4-PSK constellation,

$$E = \frac{1}{4}(4)[(d/2)^2 + (d/2)^2] = d^2/2.$$

For each constellation, express the probability of error as a function of E_s.

11.11 Exercises

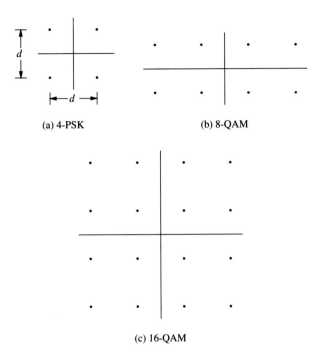

Figure 11.37: Some signal constellations

11.5-19 Let $M = 2^k$, where k is an even number. Determine the probability of error for a signal constellation with M points arranged in a square centered at the origin, with minimum distance between points equal to d and noise variance σ^2. Assume the noise is Gaussian. Express this as a function of E_s, the average signal energy for the constellation.

11.7-20 In an M-dimensional orthogonal detection problem, there are M hypotheses, $H_i \colon \mathbf{X} \sim \mathcal{N}(\mathbf{m}_i, \sigma^2 I)$, where
$$\mathbf{m}_i \perp \mathbf{m}_j \qquad i \ne j.$$
Assume that $E_s = \|\mathbf{m}_i\|^2$ for $i = 1, 2, \ldots, M$. Let $M = 2^k$, and assume that these M orthogonal signals are used to send k bits of information.

(a) Show that the minimum distance between signals is $d = \sqrt{2E_s}$. Also show that E_b, the energy per bit, is $E_b = E_s/k$.

(b) By the union bound, show that the probability of symbol error is bounded by $P(\mathcal{E}) \le (M-1)Q(d/2\sigma)$.

(c) Using the upper bound on the Q function $Q(x) \le \tfrac{1}{2} e^{-x^2/2}$, show that the error approaches zero as $k \to \infty$, provided that $E_b/\sigma^2 > 4\ln 2$.

11.7-21 For points in the signal constellation shown in figure 11.38 (where the basis functions are orthonormal), determine an upper bound on the probability of error using the union bound. Assume that the noise is AWGN with variance $\sigma^2 = 0.1$. Express your answer in terms of the Q function.

11.8-22 [291] Suppose that a signal is of the form
$$\mathbf{s} = H\boldsymbol{\theta},$$
where H is a known $m \times p$ matrix, but $\boldsymbol{\theta}$ is not known. That is, the signal is known to

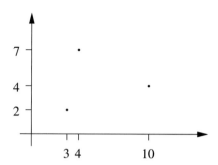

Figure 11.38: Signal constellation with three points

lie in $\mathcal{R}(H)$, but the particular point in that space is not known. Let $\mathbf{X} = \mu H\boldsymbol{\theta} + \mathbf{N}$, where $\mathbf{N} \sim \mathcal{N}(\mathbf{0}, \sigma^2 I)$. It is desired to distinguish between H_0: $\mu = 0$ (signal absent) and H_1: $\mu > 0$ (signal present). However, it is not possible to observe $\mathbf{X}$ directly. Instead, we observe the output $\mathbf{Y}$ of a channel that introduces some bias $\mathbf{v} \perp \mathcal{R}(H)$, and also rotates $\mathbf{X}$ in $\mathcal{R}(H)$. Let Q indicate the rotation in $\mathcal{R}(H)$.

(a) Show that $Q = U_H \tilde{Q} U_H + P_{H^\perp}$, where $P_{H^\perp}$ is a projection onto $[\mathcal{R}(H)]^\perp$, and $P_H = U_H U_H^T$ is a projection onto $\mathcal{R}(H)$, and Q is an orthogonal matrix.

(b) Show that the rotation of $\mu H \boldsymbol{\theta}$ is $\mu H \boldsymbol{\theta}'$ for some $\boldsymbol{\theta}'$.

(c) Show that the statistic
$$Z = \mathbf{Y}^T P_H \mathbf{Y}$$
is invariant with respect to the offset $\mathbf{v}$ and any rotation Q.

(d) Show that under H_0, Z/σ^2 is distributed as χ_m^2.

11.8-23 [291] Let $\mathbf{X} \sim \mathcal{N}(\mu H\boldsymbol{\theta}, \sigma^2 I)$, where H is a known $m \times p$ matrix, but σ^2 is unknown. Assume that the signal is biased by a vector $\mathbf{v} \perp \mathcal{R}(H)$ and rotated in $\mathcal{R}(H)$ to produce the measurement $\mathbf{Y}$.

(a) Show that the statistic
$$F = \frac{\mathbf{y}^T P_H \mathbf{y}/(\sigma^2 p)}{\mathbf{y}^T(I - P_H)\mathbf{y}/(\sigma^2(m-p))}$$
is invariant with respect to $\mathbf{v}$ and Q and independent of σ^2.

(b) Explain why F is the ratio of independent χ^2 random variables.

(c) The distribution of F is called the F-distribution. It is known to have a monotone likelihood ratio. Based on this fact, write down a uniformly most powerful test.

11.8-24 (t **distribution**) Let $T = Z/\sqrt{Y/r}$, where $Z \sim \mathcal{N}(0, 1)$ and $Y \sim \chi_r^2$. Let $(t, u) = W(z, y)$ be an invertible transformation, where
$$t = z/\sqrt{y/r} \qquad u = y.$$

(a) Show that the Jacobian of the transformation is
$$J = \det \begin{bmatrix} \dfrac{\partial z}{\partial t} & \dfrac{\partial z}{\partial u} \\ \dfrac{\partial y}{\partial t} & \dfrac{\partial y}{\partial u} \end{bmatrix} = \sqrt{u/r}.$$

(b) Hence, show that the joint density $f_{TU}(t, u)$ is
$$f_{TU}(t, u) = \sqrt{u/r} \, \frac{1}{\sqrt{2\pi}\,\Gamma(r/2) 2^{r/2}} u^{r/2-1} e^{-u/2}.$$

(c) Finally, integrate out u to derive the density (11.90). Use $\int_0^\infty x^{\nu-1} e^{-\mu x} dx = \Gamma(\nu)/\mu^\nu$.

11.11 Exercises

11.9-25 Show that
$$\lim_{N \to \infty} \sum_{i=1}^{N} (X_i - s_{j,i})^2 = \int_0^T (X(t) - s_j(t))^2 \, dt.$$

11.9-26 For the binary detector in Gaussian noise of example 11.9.1, verify that
$$P_{FA} = P_{MD} = Q(d/2\sigma).$$

11.9-27 In the proof of theorem 11.4, we used the fact that
$$E \sum_{i=1}^{n} \sum_{j=1}^{n} Z_i Z_j \psi_i(t) \psi_j(t) = \sum_{i=1}^{n} \lambda_i \psi_i^2(t).$$
Show that this is true.

11.9-28 Show that (11.107) is true.

11.9-29 Draw the block diagram for an incoherent detector for the problem
$$H_0: X(t) = s_0(t) + N(t),$$
$$H_1: X(t) = s_1(t) + N(t),$$
where
$$s_i(t) = a \sin(\omega_i t + \theta), \qquad 0 \le t \le T,$$
and where θ is uniformly distributed as $\mathcal{U}(0, 2\pi)$ and $N(t)$ is Gaussian white noise.

11.10-30 Consider the binary channel represented by the accompanying diagram.

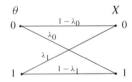

(a) Determine the likelihood ratio test.

(b) Determine the threshold ν required to obtain a test of size α when $\lambda_0 = \lambda_1 = \lambda$ as a function of λ.

(c) If $\lambda_0 = \lambda_1 = \lambda$, determine and plot the ROC for a Neyman–Pearson test on the channel for $\lambda = 1/8$, $\lambda = 1/4$, $\lambda = 3/8$, and $\lambda = 1/2$.

(d) Determine the Bayes decision rule when the prior probabilities $p_0 = P(\theta = 0)$ and $p_1 = P(\theta = 1)$ are equal and the costs are uniform.

(e) Plot the Bayes envelope function when $\lambda_0 = 0.1$ and $\lambda_1 = 0.2$.

24 Consider two boxes A and B, each of which contains both red and green balls. It is known that, in one of the boxes, $\frac{1}{2}$ of the balls are red and $\frac{1}{2}$ are green; and that, in the other box, $\frac{1}{4}$ of the balls are red and $\frac{3}{4}$ are green. Let the box in which $\frac{1}{2}$ of the balls are red be denoted box W, and suppose $P(W = A) = \xi$ and $P(W = B) = 1 - \xi$. Suppose you may select one ball at random from either box and that, after observing its color, you must decide whether $W = A$ or $W = B$. Prove that, if $\frac{1}{2} < \xi < \frac{2}{3}$, to maximize the probability of making a correct decision, you should select the ball from box B. Prove also that if $\frac{2}{3} \le \xi \le 1$, then it does not matter from which box the ball is selected.

24 A wildcat oilman must decide how to finance the drilling of a well. It costs \$100,000 to drill the well. The oilman has three options available:

H_0: finance the drilling himself and retain all the profits,

H_1: accept \$70,000 from investors in return for paying them 50% of the profits,

H_2: accept \$120,000 from investors in return for paying them 90% of the profits.

The profits will be 3θ, where θ is the number of barrels of oil in the well.
From past data, it is believed that $\theta = 0$ with probability 0.9, and the density for $\theta > 0$ is

$$g(\vartheta) = \frac{0.1}{300{,}000} e^{-\vartheta/300{,}000} I_{(0,\infty)}(\vartheta).$$

A seismic test is performed to determine the likelihood of oil in the given area. The test tells which type of geological structure, x_1, x_2, or x_3, is present. It is known that the probabilities of the x_i given θ are

$$f_{X|\theta}(x_1 \mid \vartheta) = 0.8 e^{-\vartheta/100{,}000},$$
$$f_{X|\theta}(x_2 \mid \vartheta) = 0.2,$$
$$f_{X|\theta}(x_3 \mid \vartheta) = 0.8(1 - e^{-\vartheta/100{,}000}).$$

(a) For monetary loss, what is the Bayes decision rule if $X = x_1$ is observed?

(b) For monetary loss, what is the Bayes decision rule if $X = x_2$ is observed?

(c) For monetary loss, what is the Bayes decision rule if $X = x_3$ is observed?

11.10-31 [?] Suppose that a device has been created that can classify blood as type A, B, AB, or O. The device measures a quantity X, which has density

$$f_{X|\theta}(x \mid \vartheta) = e^{-(x-\vartheta)} I_{(\vartheta,\infty)}(x).$$

If $0 < \theta < 1$, the blood is of type AB; if $1 < \theta < 2$, the blood is of type A; if $2 < \theta < 3$, the blood is of type B; and if $\theta > 3$ the blood is of type O. In the population as a whole, θ is distributed according to the density

$$f_\theta(\vartheta) = e^{-\vartheta} I_{(0,\infty)}(\vartheta).$$

The loss in misclassifying the blood is given by the following table.

		Classification			
		AB	A	B	O
	AB	0	1	1	2
True	A	1	0	2	2
type	B	1	2	0	2
	O	3	3	3	0

If $X = 4$ is observed, what is the Bayes decision rule?

11.10-32 For a binary channel, take $\lambda_0 = 1/3$ and $\lambda_1 = 1/4$. Determine:

(a) The risk set.

(b) The minimax Bayes risk.

(c) The optimum decision rule.

(d) The least favorable prior.

11.10-33 In these last two exercises, we introduce briefly some other topics in detection theory. This problem deals with **detection of change**. Suppose that a signal changes its mean at some unknown time n_0, and the problem is to detect the change. We set up the following hypothesis test:

$$H_0: X_i \sim \mathcal{N}(m_0, \sigma^2) \quad i = 1, 2, \ldots, n,$$
$$H_1: X_i \sim \mathcal{N}(m_0, \sigma^2) \quad i = 1, 2, \ldots, n_0 - 1,$$
$$X_i \sim \mathcal{N}(m_1, \sigma^2) \quad i = n_0, n_0 + 1, \ldots, n,$$

where means $m_1 > m_0$ are assumed to be known, as is σ^2. Assume that n_0 is known.

(a) Based upon a likelihood ratio test, show that a test for the change is: Decide H_1 if

$$T(\mathbf{x}) = \frac{1}{n - n_0 + 1} \sum_{i=n_0}^{n} (x_i - m_0) > \eta$$

for some threshold η.

(b) Determine the distribution of $T(\mathbf{x})$ under the two hypotheses, and determine an expression for P_{FA} as a function of the threshold η.

11.10-34 For the detection of change problem of the previous exercise, assume now that we don't know n_0. Forming the likelihood ratio

$$\ell(n_0, \mathbf{x}) = \frac{f_\mathbf{x}(\mathbf{x} \mid H_1)}{f_\mathbf{x}(\mathbf{x} \mid H_0)},$$

we choose the maximum likelihood estimate of n_0 to be that value which maximizes $\ell(n_0, \mathbf{x})$. Show that this reduces to

$$\max_{n_0} \sum_{i=n_0}^{n-1} \left(x_i - m_0 - \frac{m_1 - m_0}{2} \right).$$

11.12 References

For the results on Neyman–Pearson detection, we have drawn heavily on [291 (chapter 4), 85 (chapter 5), 349]. Also, [257] is useful reading. Discussion of the philosophy of Bayesian decision making is found in [145]. For this development we rely on [291, 85, 67].

The application of decision theory to the detection of signals is a mainstay of digital communications, in which several different signal sets are characterized by their detector structures and their probability of error performance. Many excellent books on communications exist, of which we cite [261, 23, 198].

The game theory touched on in the examples is but the tip of a very large body of research, first formalized in [357]. The connections between games and linear programming are explored in [171]. An interesting discussion of the Prisoner's dilemma game appears in [140] and [7]. The concept of a minimax point—minimizing the maximum loss—has seen application in a variety of areas besides game theory, among them the minimax filter approximation approach [249, 250].

A discussion of detection in continuous time is provided in [349] and [257]. An excellent discussion of detection in nonwhite Gaussian noise also appears in [313]. See also the survey article [166].

The detection of change problems introduced in exercises 11.10-35 and 11.10-36 are thoroughly discussed in [19, 17]. See also [175 (chapter 12), 18, 138, 10, 371].

In section 11.3, we introduced the notion of a uniformly most powerful test for composite hypotheses. A significantly more thorough coverage of tests for composite hypotheses appears in [175 (chapter 6)].

Chapter 12

Estimation Theory

> HYLAS.   You still take things in a strict literal sense: that is not fair, Philonous.
> PHILONOUS.   I am not for imposing any sense on your words: you are at liberty to explain them as you please. Only, I beseech you, make me understand something by them.
> — *George Berkeley*
> The First Dialogue Between Hylas and Philonous

Estimation is the process of making decisions over a continuum of parameters. We have seen that there are two major philosophies to detection: the Neyman–Pearson approach, in which no prior probabilities are assumed on the parameters; and the Bayes approach, in which a prior probability is assumed. The same dichotomy exists with estimation, since we may view the unknown parameter as either an unknown (but deterministic) quantity, or as a random variable. Consequently, there are multiple schools of thought regarding estimation. On the one hand, when no prior distribution is assumed, the estimation is commonly based upon the principle of *maximum likelihood*. When a prior distribution for the parameter is assumed, a *Bayes* estimate is formed.

12.1   The maximum-likelihood principle

The essential feature of the principle of maximum likelihood (ML), as it applies to estimation theory, is that is requires one to choose, as an estimate of a parameter, that value for which the probability of obtaining an actually observed sample is as large as possible. That is, having obtained observations, one "looks back" and computes the probability, from the point of view of one about to perform the experiment, that the given sample values will be observed. This probability will in general depend on the parameter, which is then given that value for which this probability is maximized. This is reminiscent of the story of the crafty politician who, once he observes which way the crowd is going, hurries to the front of the group as if to lead the parade.

Suppose that the random variable X has a probability distribution that depends on a parameter θ. The parameter θ must lie in a space of possible parameters Θ. Let $f_X(x \mid \theta)$ denote either a pmf or pdf of X. We suppose that the form of f_X is known, but not the value of the parameter θ. The joint pmf of m sample random variables evaluated at the sample points $x_1, \ldots, x_m$ is

$$\ell(\theta, x_1, \ldots, x_m) = \ell(\theta, \mathbf{x}) = f_{\mathbf{X}}(\mathbf{x} \mid \theta) = \prod_{i=1}^{m} f_X(x_i \mid \theta). \qquad (12.1)$$

12.1 The Maximum-Likelihood Principle

This function is also known as the *likelihood function* of the sample; we are particularly interested in it as a function of θ when the sample values $x_1, \ldots, x_m$ are fixed. The principle of maximum likelihood requires us to choose as an estimate of the unknown parameter that value of θ for which the likelihood function assumes its largest value.

If the parameter θ is a vector, say $\boldsymbol{\theta} = [\theta_1, \ldots, \theta_k]^T$, then the likelihood function will be a function of all of the components of $\boldsymbol{\theta}$. Thus, we are free to regard $\boldsymbol{\theta}$ as a vector in (12.1), and the maximum-likelihood estimate of θ is then the vector of numbers that render the likelihood function a maximum.

Example 12.1.1 *(A maximum-likelihood detector)* Suppose you are given a coin and told that it is biased, with one side four times as likely to turn up as the other; you are allowed three tosses and must then guess whether it is biased in favor of head or in favor of tails.

Let θ be the probability of heads (denoted H, with T corresponding to tails) on a single toss. Define the random variable, $X: \{H, T\} \to \{0, 1\}$; $X(H) = 1$ and $X(T) = 0$. The pmf for X is given by

$$f_X(0 \mid 4/5) = 1/5 \qquad f_X(1 \mid 4/5) = 4/5,$$
$$f_X(0 \mid 1/5) = 4/5 \qquad f_X(1 \mid 1/5) = 1/5.$$

Suppose you throw the coin three times, resulting in the samples HTH. The sample values are $x_1 = 1$, $x_2 = 0$, $x_3 = 1$. The likelihood function is

$$\begin{aligned} \ell(\theta, x_1, x_2, x_3) &= f_{X_1 X_2 X_3}(x_1, x_2, x_3 \mid \theta) \\ &= f_{X_1 X_2 X_3}(1, 0, 1 \mid \theta) \\ &= f_{X_1}(1 \mid \theta) f_{X_2}(0 \mid \theta) f_{X_3}(1 \mid \theta), \end{aligned}$$

or

$$\ell(4/5, 1, 0, 1) = (4/5)(1/5)(4/5) = 16/125,$$
$$\ell(1/5, 1, 0, 1) = (1/5)(4/5)(1/5) = 4/125.$$

Clearly, $\theta = 4/5$ yields the larger value of the likelihood function, so by the likelihood principle we are compelled to decide that the coin is biased in favor of heads. □

Although, as this example demonstrates, the principle of maximum likelihood may be applied to discrete decision problems, it has found greater utility for problems where the distribution is continuous and differentiable in θ. The reason for this is that we will usually be taking derivatives in order to find maxima. But it is important to remember that general decision problems can, in principle, be addressed via the principle of maximum likelihood. Notice, for this example, that neither cost functions nor *a priori* knowledge of the distribution of the parameters is needed to fashion a maximum-likelihood estimate.

Example 12.1.2 *(Empiric distributions)* Let X be a random variable of unknown distribution, and $X_1, \ldots, X_m$ be sample random variables from the population of X. Suppose we are required to estimate the distribution function of X. There are many ways to approach this problem. One would be to assume some general structure, such as an exponential family, and try to estimate the parameters of this family. But then one has the simultaneous problems of (a) estimating the parameters and (b) justifying the structure. Although there are many ways of solving both of these problems, this approach is not easy. The maximum-likelihood method gives us a fairly simple approach that, if for no other reason, would be valuable as a baseline for evaluating other, more sophisticated approaches.

To apply the principle of maximum likelihood to this problem, we must first define the parameters. We do this by setting

$$\theta_i = P[X_i = x_i], \quad i = 1, \ldots, m.$$

The event

$$[X_1 = x_1, \ldots, X_m = x_m]$$

is observed and, according to the maximum-likelihood principle, we wish to choose the values of θ_i that maximize the probability that this event will occur. Since the events $[X_i = x_i]$, $i = 1, \ldots, m$, are independent, we have

$$P[X_1 = x_1, \ldots, X_m = x_m] = \prod_{i=1}^{m} P[X_i = x_i] = \prod_{i=1}^{m} \theta_i.$$

We wish to maximize subject to the constraint $\sum_{i=1}^{m} \theta_i = 1$, which we shall do via Lagrange multipliers. Let

$$J = \prod_{i=1}^{m} \theta_i + \lambda \left(\sum_{i=1}^{m} \theta_i - 1 \right),$$

and set the gradient of J with respect to θ_i, $i = 1, \ldots, m$, and with respect to λ, to zero:

$$\frac{\partial J}{\partial \theta_j} = \prod_{i \neq j} \theta_i + \lambda = 0, \quad j = 1, \ldots, m,$$

$$\frac{\partial J}{\partial \lambda} = \sum_{i=1}^{m} \theta_i - 1 = 0.$$

But the only way that all of the products $\prod_{i \neq j} \theta_i$ can be equal is if $\theta_1 = \cdots = \theta_m$, and the constraint therefore requires that $\theta_i = 1/m$, $i = 1, \ldots, m$.

We define the maximum-likelihood estimate for the distribution as follows. Let $\tilde{X}$ be a random variable, called the *empiric random variable*, whose distribution function is

$$F_{\tilde{X}}(x) = P[\tilde{X} \leq x] = \frac{1}{m} \sum_{i=1}^{m} I_{[x_i, \infty)}(x).$$

Figure 12.1 illustrates the structure of the empiric distribution function. For large samples, it is convenient to quantize the observations and construct the empiric density function by building a histogram.

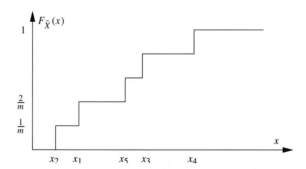

Figure 12.1: Empiric distribution function

Thus, the empiric distribution is precisely that distribution for which the influence of the sample values actually observed is maximized at the expense of other possible values of X. Of course, the actual utility of this distribution is limited, since the number of parameters may be very large. But it is a maximum-likelihood estimate of the distribution function. □

It must be stressed that the likelihood function $\ell(\theta, x)$ is to be viewed as a function of θ, with x being a fixed quantity, rather than a variable. This is in contradistinction to the way we view the density function $f_X(x \mid \theta)$, where θ is a fixed quantity and x is viewed as

12.1 The Maximum-Likelihood Principle

a variable. So remember, even though we may write $\ell(\theta, x) = f_X(x \mid \theta)$, we view the roles of x and θ in the two expressions entirely differently.

It is actually more convenient, for many applications, to consider the logarithm of the likelihood function, which we denote

$$\Lambda(\theta, \mathbf{x}) = \log f_{\mathbf{X}}(\mathbf{x} \mid \theta),$$

and call the *log-likelihood function*. Since the logarithm is a monotonic function, the maximization of the likelihood and log-likelihood functions is equivalent, that is, θ_{ML} maximizes the likelihood function if and only if it also maximizes the log-likelihood function. Thus, in this development we will deal mainly with the log-likelihood function.

The major issue before us is to find a way to maximize the likelihood function. If the maximum is interior to the range of θ, and $\Lambda(\theta, \mathbf{x})$ has a continuous first derivative, then a necessary condition for $\hat{\theta}_{ML}$ to be the maximum-likelihood estimate for θ is that

$$\left. \frac{\partial \Lambda(\theta, \mathbf{x})}{\partial \theta} \right|_{\theta = \hat{\theta}_{ML}} = 0. \tag{12.2}$$

In the case of vector parameters $\boldsymbol{\theta}$, we write this as

$$\left. \frac{\partial \Lambda(\boldsymbol{\theta}, \mathbf{x})}{\partial \boldsymbol{\theta}} \right|_{\boldsymbol{\theta} = \hat{\boldsymbol{\theta}}_{ML}} = 0. \tag{12.3}$$

Equation (12.2) or (12.3) is called the *likelihood equation*. We now give some examples to illustrate the maximization process.

Example 12.1.3 This example shows that, while the likelihood equation is frequently useful, more general principles of optimization can be used to obtain maximum-likelihood estimates even when the maximum may not occur in the interior of the set of possible values.

Let $X_1, \ldots, X_m$ denote a random sample of size m from a uniform distribution over $[0, \theta]$. We wish to find the maximum-likelihood estimate of θ. Let $I_A(x) = \begin{cases} 1 & x \in A \\ 0 & x \notin A \end{cases}$ be the indicator function for the set A. The likelihood function is

$$\ell(\theta, x_1, \ldots, x_m) = \theta^{-m} \prod_{i=1}^{m} I_{(0,\theta)}(x_i)$$

$$= \theta^{-m} \prod_{i=1}^{m} I_{(0,\max_i\{x_i\})}(\min_i x_i) I_{(\min_i\{x_i\},\theta)}(\max_i\{x_i\})$$

$$= \theta^{-m} \prod_{i=1}^{m} I_{(\min_i\{x_i\},\theta)}(\max_i\{x_i\})$$

$$= \theta^{-m} \prod_{i=1}^{m} I_{(\max_i\{x_i\},\infty)}(\theta).$$

Since the maximum of this quantity does not occur on the interior of the range of θ, we can't take derivatives and set to zero. But we don't need to do that for this example, since θ^{-m} is monotonically decreasing in θ. Consequently, the likelihood function is maximized at

$$\hat{\theta}_{ML} = \max_i \{x_i\}.$$

Intuitively, we should expect the range of a uniformly distributed sample to be determined by the largest value that is observed. □

Example 12.1.4 Let $X_1, \ldots, X_m$ denote a random sample of size m from the normal distribution $\mathcal{N}(\mu, \sigma^2)$. We wish to find the maximum-likelihood estimates for μ and σ^2. The density function is

$$f_\mathbf{X}(\mathbf{x} \mid \mu, \sigma) = \prod_{i=1}^m \frac{1}{\sqrt{2\pi}\sigma} \exp\left[-\frac{(x_i - \mu)^2}{2\sigma^2}\right],$$

and the log-likelihood function is then

$$\Lambda(\mu, \sigma, \mathbf{x}) = -m \log \sqrt{2\pi} - m \log \sigma - \frac{1}{2\sigma^2} \sum_{i=1}^m (x_i - \mu)^2.$$

Taking the gradient and equating to zero yields

$$\frac{\partial \Lambda}{\partial \mu} = \frac{1}{\sigma^2} \sum_{i=1}^m (x_i - \mu) = 0$$

$$\Rightarrow \hat{\mu}_{ML} = \frac{1}{m} \sum_{i=1}^m x_i,$$

and

$$\frac{\partial \Lambda}{\partial \sigma} = -\frac{m}{\sigma} + \sigma^{-3} \sum_{i=1}^m (x_i - \mu)^2 = 0$$

$$\Rightarrow \hat{\sigma}^2_{ML} = \frac{1}{m} \sum_{i=1}^m (x_i - \mu)^2.$$

When the mean is not known, we can write

$$\hat{\sigma}_{ML} = \frac{1}{m} \sum_{i=1}^m (x_i - \hat{\mu}_{ML})^2.$$

It is satisfying that these estimates coincide with what our intuition would suggest.

If we view the estimators as random variables,

$$\hat{\mu}_{ML} = \frac{1}{m} \sum_{i=1}^m X_i \qquad \hat{\sigma}^2_{ML} = \frac{1}{m} \sum_{i=1}^m (X_i - \hat{\mu}_{ML})^2,$$

we can examine their means:

$$E\hat{\mu}_{ML} = \mu, \tag{12.4}$$

$$E\hat{\sigma}^2_{ML} = \sigma^2 \frac{m-1}{m}. \tag{12.5}$$

We note that $\hat{\mu}_{ML}$ is an unbiased estimator, and $\hat{\sigma}^2_{ML}$ is a biased estimator. So a maximum-likelihood estimate is not necessarily an unbiased estimate. (However, as $m \to \infty$ the estimate becomes unbiased.)

We can also examine the variance of the estimators. For example, it can be shown that

$$\operatorname{var} \hat{\mu}_{ML} = \frac{\sigma^2}{m}, \tag{12.6}$$

so that the variance decreases the larger the number of samples used to determine the estimate. □

Before we get too euphoric over the simplicity and seemingly magical powers of the maximum-likelihood approach, consider the following example.

12.2 ML Estimates and Sufficiency

Example 12.1.5 Let $X_1 \sim \mathcal{N}(\theta, 1)$ and $X_2 \sim \mathcal{N}(-\theta, 1)$, and define

$$Y = \begin{cases} X_1 & \text{with probability } 1/2, \\ X_2 & \text{with probability } 1/2; \end{cases}$$

then

$$f_Y(y \mid \theta) = \frac{1}{2} \frac{1}{\sqrt{2\pi}} e^{-\frac{1}{2}(y-\theta)^2} + \frac{1}{2} \frac{1}{\sqrt{2\pi}} e^{-\frac{1}{2}(y+\theta)^2}.$$

Now let $Y = y'$ be a given sample value. According to our procedure, we would evaluate the likelihood function at y', yielding

$$l(\theta, y') = \frac{1}{2} \frac{1}{\sqrt{2\pi}} e^{-\frac{1}{2}(y'-\theta)^2} + \frac{1}{2} \frac{1}{\sqrt{2\pi}} e^{-\frac{1}{2}(y'+\theta)^2},$$

and choose, as the maximum-likelihood estimate of θ, that value that maximizes $l(\theta, y')$. But this function does not have a unique maximum, so there is no unique estimate. Both $\hat{\theta}_{ML} = y'$ and $\hat{\theta}_{ML} = -y'$ qualify as maximum-likelihood estimates for θ. □

12.2 ML estimates and sufficiency

If $t(\mathbf{x})$ is a sufficient statistic for θ in $f_\mathbf{X}(\mathbf{x}|\theta)$, then

$$f_\mathbf{X}(\mathbf{x} \mid \theta) = b(\mathbf{t}(\mathbf{x}), \theta) a(\mathbf{x})$$

and

$$f_\mathbf{T}(\mathbf{t} \mid \theta) = b(\mathbf{t}, \theta) \left(\int a(\mathbf{w}^{-1}(\mathbf{t}, \mathbf{u})) \left| \frac{\partial \mathbf{w}^{-1}(\mathbf{y})}{\partial \mathbf{y}} \right| \right) d\mathbf{u},$$

so that $f_\mathbf{X}(\mathbf{x} \mid \theta)$ is proportional to $f_\mathbf{T}(\mathbf{t} \mid \theta)$ and the constant of proportionality depends upon $\mathbf{x}$ but not on θ. Thus the log-likelihood can be written as

$$\Lambda(\theta, \mathbf{x}) = C(\mathbf{x}) + \log f_T(\mathbf{t} \mid \theta),$$

where $C(\mathbf{x})$ does not depend upon θ. For purposes of maximizing the log likelihood function, we can ignore the constant $C(\mathbf{x})$ and write

$$\Lambda(\theta, \mathbf{t}) = \log f_T(\mathbf{t} \mid \theta).$$

Example 12.2.1 Let $X_1, X_2, \ldots, X_n$ be independent Bernoulli random variables $\mathcal{B}(p)$, where p is the unknown parameter. Then

$$f_\mathbf{X}(\mathbf{x} \mid p) = \prod_{i=1}^n p^{x_i} (1-p)^{1-x_i}.$$

A sufficient statistic for this distribution is

$$k = \sum_{i=1}^n x_i,$$

which is binomial $\mathcal{B}(n, p)$ distributed:

$$f_K(k \mid p) = \binom{n}{k} p^k (1-p)^{n-k}. \tag{12.7}$$

In finding a maximum-likelihood estimate for p, we may maximize the distribution of the sufficient statistic in (12.7). The log likelihood based on this distribution is

$$\Lambda(p, k) = \log f_K(k \mid p) = \log \binom{n}{k} + k \log p + (n-k) \log(1-p). \tag{12.8}$$

The constant $\log \binom{n}{k}$ does not depend upon p and can be ignored. Taking the derivative of (12.8) with respect to p and equating to zero, we obtain

$$\hat{p} = \frac{k}{n}.$$

□

12.3 Estimation quality

Estimation theorists are sometimes consumed, not only with devising and understanding various algorithms for estimation, but with evaluations of how reliable they are. We usually ask the question in the superlative: "What is the best estimate?"

We might be tempted to answer that the best estimate is the one closest to the true value of the parameter to be estimated. But every estimate is a function of the sample values, and thus is the observed value of some random variable. There is no means of predicting just what the individual values are to be for any given experiment, so the accuracy of an estimate cannot be judged reliably from individual values. As we repeatedly sample the population, however, we may form statistics, such as the sample mean and variance, whose distributions we may calculate. If we are able to form estimators from these statistics, then the best we can hope for is that the bulk of the mass in the distribution is concentrated in some small neighborhood of the true value. In such circumstances, there is a high probability that the estimate will only differ from the true value by a small amount. From this point of view, we may order the quality of estimators as a function of how the sample distribution is concentrated about the true value.

This intuitive notion is embodied by the Chebyshev inequality, which states that for a random variable Y with mean μ_y and variance σ_y^2, the probability that an observation $Y = y$ differs from the mean by ϵ is

$$P[|y - \mu_y| > \epsilon] \leq \frac{\sigma^2}{\epsilon^2}.$$

The smaller σ^2 is, the less the probability that an observation is far from the mean. In the context of estimation, we want the estimate of a parameter to be close to the mean—the true value—so we want the variance of the estimate to be as small as possible.

Extending this simple concept to vector parameters, for vector parameters $\boldsymbol{\theta}$ with a vector estimate $\hat{\boldsymbol{\theta}}$, a good estimator is one for which the covariance

$$C = E(\hat{\boldsymbol{\theta}} - E\boldsymbol{\theta})(\hat{\boldsymbol{\theta}} - E\boldsymbol{\theta})^T$$

is as small as possible. "Small" here means the following: Let A and B be Hermitian matrices. Then we say that $A < B$ if $B - A$ is positive definite: $\mathbf{x}^T(B - A)\mathbf{x} > 0$ for all non-zero vectors $\mathbf{x}$.

While there are other measures of the quality of an estimate, most estimation techniques exclusively use the variance (or covariance for multidimensional parameters) as a means of evaluating the quality of the estimate. This choice is motivated strongly by the important case when the sampling distributions of the estimates are at least approximately normal, since the second-order moment is then the unique measure of dispersion.

12.3.1 The score function

The maximum-likelihood method of estimation does not provide, as a byproduct of calculating the estimate, any measure of the concentration (that is, the variance) of the estimation error. Although the variance can be calculated for many important examples, it is difficult

12.3 Estimation Quality

for others. Rather than approach the problem of calculating the variance for an estimate directly, therefore, we will first calculate a lower bound for the variance of the estimation error for *any unbiased estimator*, then we will see how the variance of the maximum-likelihood estimation error compares with this lower bound. Before stating the main result of this section, we need to establish some new notation and terminology, and prove some preliminary results.

Definition 12.1 Let $\mathbf{X} = [X_1, \ldots, X_n]^T$ denote an n-dimensional random vector, and $\boldsymbol{\theta} = [\theta_1, \ldots, \theta_p]^T$ denote a p-dimensional parameter vector. The **score function** $\mathbf{s}(\boldsymbol{\theta}, \mathbf{X})$ is the gradient of the log-likelihood function:

$$\mathbf{s}(\boldsymbol{\theta}, \mathbf{X}) = \frac{\partial \Lambda(\boldsymbol{\theta}, \mathbf{X})}{\partial \boldsymbol{\theta}} = \frac{1}{\ell(\boldsymbol{\theta}, \mathbf{X})} \frac{\partial \ell(\boldsymbol{\theta}, \mathbf{X})}{\partial \boldsymbol{\theta}}. \tag{12.9}$$

□

We see that at an ML estimate $\boldsymbol{\theta}_{ML}$ (on the interior of the range of $\boldsymbol{\theta}$),

$$s(\boldsymbol{\theta}_{ML}, \mathbf{X}) = 0;$$

the ML estimate is a zero of the score function. We also say that good scores are those with values near zero. It is important to notice that since the score is related to the gradient, the results of this method apply to estimates $\dot{\boldsymbol{\theta}}$ lying on the interior of Θ

Before continuing, we prove some useful facts about the score function. We begin with the following theorem.

Theorem 12.1 *If* $\mathbf{s}(\boldsymbol{\theta}, \mathbf{X})$ *is the score of a likelihood function* $\ell(\boldsymbol{\theta}, \mathbf{X})$ *and if* $\mathbf{t}$ *is any vector-valued function of* $\mathbf{X}$ *and* $\boldsymbol{\theta}$, *then*

$$E\mathbf{s}(\boldsymbol{\theta}, \mathbf{X})\mathbf{t}^T(\boldsymbol{\theta}, \mathbf{X}) = \frac{\partial}{\partial \boldsymbol{\theta}} E\mathbf{t}^T(\boldsymbol{\theta}, \mathbf{X}) - E\frac{\partial}{\partial \boldsymbol{\theta}} \mathbf{t}^T(\boldsymbol{\theta}, \mathbf{X}). \tag{12.10}$$

Before embarking on the proof, we note that if $\mathbf{t}(\boldsymbol{\theta}) = [t_1(\boldsymbol{\theta}), t_2(\boldsymbol{\theta}), \ldots, t_k(\boldsymbol{\theta})]^T$, then

$$\frac{\partial}{\partial \boldsymbol{\theta}} \mathbf{t}^T = \begin{bmatrix} \frac{\partial t_1}{\partial \theta_1} & \frac{\partial t_2}{\partial \theta_1} & \cdots & \frac{\partial t_k}{\partial \theta_1} \\ & & & \\ \vdots & & & \\ \frac{\partial t_1}{\partial \theta_p} & \frac{\partial t_2}{\partial \theta_p} & \cdots & \frac{\partial t_k}{\partial \theta_p} \end{bmatrix}.$$

Proof We have

$$E\mathbf{t}^T(\boldsymbol{\theta}, \mathbf{X}) = \int \mathbf{t}^T(\boldsymbol{\theta}, \mathbf{x}) f_{\mathbf{X}}(\mathbf{x} \mid \boldsymbol{\theta}) \, d\mathbf{x}$$

$$= \int \mathbf{t}^T(\boldsymbol{\theta}, \mathbf{x}) \ell(\boldsymbol{\theta}, \mathbf{x}) \, d\mathbf{x}.$$

Upon differentiating both sides with respect to $\boldsymbol{\theta}$, and taking the differentiation under the integral sign in the right-hand side (assuming differentiability conditions as appropriate), we obtain

$$\frac{\partial}{\partial \boldsymbol{\theta}} E \mathbf{t}^T(\boldsymbol{\theta}, \mathbf{X}) = \int \ell(\boldsymbol{\theta}, \mathbf{x}) \frac{\partial}{\partial \boldsymbol{\theta}} \log \ell(\boldsymbol{\theta}, \mathbf{x}) \mathbf{t}^T(\boldsymbol{\theta}, \mathbf{x}) \, d\mathbf{x} + \int \ell(\boldsymbol{\theta}, \mathbf{x}) \frac{\partial}{\partial \boldsymbol{\theta}} \mathbf{t}(\boldsymbol{\theta}, \mathbf{x}) \, d\mathbf{x}$$
$$= E \mathbf{s}(\boldsymbol{\theta}, \mathbf{x}) \mathbf{t}^T(\boldsymbol{\theta}, \mathbf{x}) + E \frac{\partial}{\partial \boldsymbol{\theta}} \mathbf{t}^T(\boldsymbol{\theta}, \mathbf{x}). \tag{12.11}$$

The result follows from simplifying and rearranging this expression. □

We may quickly obtain two useful corollaries of this theorem.

Corollary 12.1 *If* $\mathbf{s}(\boldsymbol{\theta}, \mathbf{X})$ *is the score corresponding to a (differentiable) likelihood function* $\ell(\boldsymbol{\theta}, \mathbf{X})$, *then*

$$E \mathbf{s}(\boldsymbol{\theta}, \mathbf{X}) = \mathbf{0}. \tag{12.12}$$

Proof Choose $\mathbf{t}$ as any constant vector. Then, since $\mathbf{t}$ is not a function of $\boldsymbol{\theta}$, its derivative vanishes; so by (12.10),

$$E \mathbf{s}(\boldsymbol{\theta}, \mathbf{X}) \mathbf{t}^T = E[\mathbf{s}(\boldsymbol{\theta}, \mathbf{X})] \mathbf{t}^T = \mathbf{0},$$

which can happen for arbitrary $\mathbf{t}$ only if $E[\mathbf{s}(\boldsymbol{\theta}, \mathbf{X})] = \mathbf{0}$. □

We note that (12.12) can be written as

$$E \mathbf{s}(\boldsymbol{\theta}, \mathbf{X}) = E \frac{\partial}{\partial \boldsymbol{\theta}} \log \ell(\boldsymbol{\theta}, \mathbf{X})$$
$$= \frac{\partial}{\partial \boldsymbol{\theta}} \int \ell(\boldsymbol{\theta}, \mathbf{x}) \, dx = 0. \tag{12.13}$$

(This is also a manifestation of the trivial observation that $\frac{\partial}{\partial \boldsymbol{\theta}} \int f_\mathbf{X}(\mathbf{x} \mid \boldsymbol{\theta}) d\mathbf{x} = \frac{\partial}{\partial \boldsymbol{\theta}} 1 = 0$, since $f_\mathbf{X}(\cdot \mid \boldsymbol{\theta})$ is a density function.)

Corollary 12.2 *If* $\mathbf{s}(\boldsymbol{\theta}, \mathbf{X})$ *is the score corresponding to a differentiable likelihood function* $\ell(\boldsymbol{\theta}, \mathbf{X})$ *and* $\mathbf{t}(\mathbf{X})$ *is any unbiased estimator of* $\boldsymbol{\theta}$, *then*

$$E[\mathbf{s}(\boldsymbol{\theta}, \mathbf{X}) \mathbf{t}^T(\mathbf{X})] = \mathbf{I}. \tag{12.14}$$

Proof Since the estimate is unbiased, we have $E \mathbf{t}(\mathbf{X}) = \boldsymbol{\theta}$, and since $\mathbf{t}$ is not a function of $\boldsymbol{\theta}$, we have $\frac{\partial}{\partial \boldsymbol{\theta}} \mathbf{t}^T = 0$, thus by (12.10),

$$E[\mathbf{s}(\boldsymbol{\theta}, \mathbf{X}) \mathbf{t}^T(\mathbf{X})] = \frac{\partial}{\partial \boldsymbol{\theta}} \boldsymbol{\theta}^T = \mathbf{I}.$$
□

12.3.2 The Cramér–Rao lower bound

Definition 12.2 The covariance matrix of the score function is the **Fisher information matrix**, denoted $J(\boldsymbol{\theta})$. Since by (12.12) the score function is zero-mean, we have

$$J(\boldsymbol{\theta}) = E \mathbf{s}(\boldsymbol{\theta}, \mathbf{X}) \mathbf{s}^T(\boldsymbol{\theta}, \mathbf{X}) = E \left(\frac{\partial}{\partial \boldsymbol{\theta}} \log \ell(\boldsymbol{\theta}, \mathbf{x}) \right) \left(\frac{\partial}{\partial \boldsymbol{\theta}} \log \ell(\boldsymbol{\theta}, \mathbf{x}) \right)^T. \tag{12.15}$$

□

Lemma 12.1 *The Fisher information* $J(\boldsymbol{\theta})$ *of (12.15) can be written as*

$$J(\boldsymbol{\theta}) = -E \frac{\partial}{\partial \boldsymbol{\theta}} \left(\frac{\partial}{\partial \boldsymbol{\theta}} \log \ell(\boldsymbol{\theta}, \mathbf{x}) \right)^T. \tag{12.16}$$

12.3 Estimation Quality

Proof Letting $\mathbf{t} = \mathbf{s}$ in theorem 12.1, we obtain

$$E\mathbf{s}(\theta, \mathbf{X})\mathbf{s}^T(\theta, \mathbf{X}) = \frac{\partial}{\partial \theta} E\mathbf{s}^T(\theta, \mathbf{X}) - E\frac{\partial}{\partial \theta}\mathbf{s}^T(\theta, \mathbf{X}) \quad (12.17)$$

$$= -E\frac{\partial}{\partial \theta}\mathbf{s}^T(\theta, \mathbf{X}), \quad (12.18)$$

since $E\mathbf{s}^T(\theta, \mathbf{X}) = 0$, by corollary 12.12. □

Theorem 12.2 (Cramér–Rao). *If $\mathbf{t}(\mathbf{X})$ is any unbiased estimator of θ based on a differentiable likelihood function, then*

$$E[\mathbf{t}(\mathbf{X}) - \theta][\mathbf{t}(\mathbf{X}) - \theta]^T \geq J^{-1}(\theta), \quad (12.19)$$

where $J(\theta)$ is the Fisher information matrix.

That is, $J(\theta)$ provides a lower bound on the covariance of *any* unbiased estimator of θ.

The proof of this theorem is yet another application of the Cauchy–Schwartz inequality. Before proving the Cramér–Rao lower bound, we introduce an auxiliary result that is needed in the proof.

Lemma 12.2 *Let J be a positive-definite matrix, and let $\mathbf{a}$ be a fixed vector. The maximum of $\mathbf{a}^T \mathbf{c}$, subject to the constraint*

$$\mathbf{c}^T J \mathbf{c} = 1, \quad (12.20)$$

is attained at

$$\mathbf{c} = \frac{J^{-1}\mathbf{a}}{(\mathbf{a}^T J^{-1} \mathbf{a})^{\frac{1}{2}}}.$$

The proof of this lemma is explored in the exercises.

Proof (Cramér–Rao theorem) Let $\mathbf{a}$ and $\mathbf{c}$ be two p-dimensional vectors and let $\mathbf{s}(\theta, \mathbf{X})$ be the score function. Form the two random variables $\alpha = \mathbf{a}^T \mathbf{t}(\mathbf{X})$ and $\beta = \mathbf{c}^T \mathbf{s}(\theta, \mathbf{X})$. Since the correlation coefficient,

$$\rho_{\alpha\beta} = \frac{E\alpha\beta}{\sqrt{\text{var}(\alpha)\,\text{var}(\beta)}},$$

is bounded in magnitude by 1, we have

$$\frac{E^2(\alpha\beta)}{\text{var}(\alpha)\,\text{var}(\beta)} \leq 1. \quad (12.21)$$

But since the score function is zero mean, it is immediate that

$$\text{var}(\beta) = E\mathbf{c}^T \mathbf{s}(\theta, \mathbf{X})\mathbf{s}^T(\theta, \mathbf{X})\mathbf{c}$$
$$= \mathbf{c}^T \text{var}(\mathbf{s}(\theta, \mathbf{X}))\mathbf{c}$$
$$= \mathbf{c}^T J(\theta)\mathbf{c}.$$

Also,

$$\text{var}(\alpha) = \mathbf{a}^T \text{cov}(\mathbf{t})\mathbf{a}.$$

Furthermore, by (12.14), we have

$$E\alpha\beta = \mathbf{a}^T E[\mathbf{t}(\mathbf{X})\mathbf{s}^T(\theta, \mathbf{X})]\mathbf{c}$$
$$= \mathbf{a}^T I \mathbf{c}$$
$$= \mathbf{a}^T \mathbf{c}.$$

Substituting these expressions into (12.21) and squaring,

$$\frac{[E(\alpha\beta)]^2}{\text{var}(\alpha)\,\text{var}(\beta)} = \frac{(\mathbf{a}^T\mathbf{c})^2}{\mathbf{a}^T\text{cov}(\mathbf{t})\mathbf{a}\mathbf{c}^T J(\theta)\mathbf{c}} \leq 1. \qquad (12.22)$$

In the interest of finding the largest value of this expression (and with the assistance of considerable hindsight), let us substitute the result of lemma 12.2 into (12.22):

$$\frac{(\mathbf{a}^T\mathbf{c})^2}{\mathbf{a}^T\text{cov}(\mathbf{t})\mathbf{a}\mathbf{c}^T J(\theta)\mathbf{c}} \leq \frac{\left(\frac{\mathbf{a}^T J^{-1}(\theta)\mathbf{a}}{\sqrt{\mathbf{a}^T J^{-1}(\theta)\mathbf{a}}}\right)^2}{\mathbf{a}^T \text{var}(\mathbf{t})\mathbf{a}}$$

$$= \frac{\mathbf{a}^T J^{-1}(\theta)\mathbf{a}}{\mathbf{a}^T\text{cov}(\mathbf{t})\mathbf{a}} \leq 1. \qquad (12.23)$$

We observe that this inequality must hold for all $\mathbf{a}$, so

$$\mathbf{a}^T J^{-1}(\theta)\mathbf{a} \leq \mathbf{a}^T \text{cov}(\mathbf{t})\mathbf{a}$$

or

$$\mathbf{a}^T(\text{cov}(\mathbf{t}) - J^{-1}(\theta))\mathbf{a} \geq 0$$

for all $\mathbf{a}$, which is equivalent to (12.19). □

The inverse of the Fisher information matrix is a lower bound on the variance that may be attained by any unbiased estimator of the parameter θ, given the observations $\mathbf{X}$. It is interesting to determine conditions under which, the Cramér–Rao lower bound may be achieved. From (12.21), we see that equality is possible if

$$E^2(\alpha\beta) = \text{var}(\alpha)\text{var}(\beta),$$

or

$$E(\alpha\beta) = \sqrt{\text{var}(\alpha)}\sqrt{\text{var}(\beta)}.$$

But from the Cauchy–Schwartz inequality, equality is possible if and only if α and β are linearly related, that is, if

$$\mathbf{t}(\mathbf{X}) = k(\theta)\mathbf{s}(\theta, \mathbf{X})$$

for some function $k(\theta)$.

12.3.3 Efficiency

Definition 12.3 An estimator is said to be **efficient** if it is unbiased and the covariance of the estimation error achieves the Cramér–Rao lower bound. That is, if $\hat{\theta} = \mathbf{t}(\mathbf{X})$ is an estimator for θ, then $\hat{\theta}$ is efficient if

$$E\hat{\theta} = \theta$$
$$E[\hat{\theta} - \theta][\hat{\theta} - \theta]^T = J^{-1}(\theta).$$
□

Theorem 12.3 (Efficiency) *An unbiased estimator $\hat{\theta}$ is efficient if and only if*

$$J(\theta)(\hat{\theta} - \theta) = \mathbf{s}(\theta, \mathbf{X}). \qquad (12.24)$$

Furthermore, any unbiased efficient estimator is a maximum-likelihood estimator.

Proof Suppose $J(\theta)(\hat{\theta} - \theta) = \mathbf{s}(\theta, \mathbf{X})$. Then, from the definition,

$$J(\theta) = E\mathbf{s}(\theta, \mathbf{X})\mathbf{s}^T(\theta, \mathbf{X})$$
$$= J(\theta) E[\hat{\theta} - \theta][\hat{\theta} - \theta]^T J(\theta).$$

But this result implies $E[\hat{\theta} - \theta][\hat{\theta} - \theta]^T J(\theta) = \mathbf{I}$, which yields efficiency.

12.3 Estimation Quality

Conversely, suppose $\hat{\theta}$ is efficient. From (12.12) and (12.14), it follows that

$$E\mathbf{s}(\theta, \mathbf{X})(\hat{\theta} - \theta)^T = \mathbf{I},$$

so by the Cauchy–Schwartz inequality,

$$\mathbf{I} = \{E\mathbf{s}(\theta, \mathbf{X})(\hat{\theta} - \theta)^T\}^2 \leq E[\mathbf{s}(\theta, \mathbf{X})\mathbf{s}^T(\theta, \mathbf{X})]E[(\hat{\theta} - \theta)(\hat{\theta} - \theta)^T]$$
$$= J(\theta)E(\hat{\theta} - \theta)(\hat{\theta} - \theta)^T = \mathbf{I} \quad \text{(by efficiency assumption)}$$

Equality can hold with the Cauchy–Schwartz inequality if and only if

$$\mathbf{s}(\theta, \mathbf{X}) = K(\theta)(\hat{\theta} - \theta)$$

for some constant $K(\theta)$. Multiplying both sides of this expression by $(\hat{\theta} - \theta)^T$ and taking expectations yields $K(\theta) = J(\theta)$.

To show that any unbiased efficient estimator is a maximum-likelihood estimator, let $\hat{\theta}$ be efficient and unbiased, and let $\tilde{\theta}$ be a maximum-likelihood estimate of θ. Evaluating (12.24) at $\theta = \tilde{\theta}$ yields

$$J(\tilde{\theta})(\hat{\theta} - \tilde{\theta}) = \mathbf{s}(\tilde{\theta}, \mathbf{X});$$

but the score function is zero when evaluated at the maximum-likelihood estimate, consequently,

$$\hat{\theta} = \tilde{\theta}.$$

□

12.3.4 Asymptotic properties of maximum likelihood estimators

Unfortunately, it is the exception rather than the rule that an unbiased efficient estimator can be found for problems of practical importance. This fact motivates us to analyze just how close we can get to the ideal of an efficient estimate. Our approach will be to examine the large-sample properties of maximum-likelihood estimates.

In our preceding development we have considered the size of the sample as a fixed integer $m \geq 1$. Let us now suppose that an unbiased estimate can be defined for all m, and consider the asymptotic behavior of $\hat{\theta}_{ML}$ as m tends to infinity. In this section we present (without proof) three key results (which are subject to sufficient regularity of the distributions):

1. Maximum-likelihood estimates are consistent.
2. Maximum-likelihood estimates are asymptotically normally distributed.
3. Maximum-likelihood estimates are asymptotically efficient.

In the interest of clarity, we treat only the case for scalar θ. We assume, in the statement of the following three theorems, that all of the appropriate regularity conditions are satisfied. The outlines of proofs presented here are quite technical, and students may wish to skip them on a first reading. More thorough proofs of these results can be found in [115].

Definition 12.4 Let $\hat{\theta}_m$ be an estimator based on m samples of a random variable. The sequence $\{\hat{\theta}_m, m = 1, \ldots, \infty\}$ is said to be a **consistent** (also known as **strongly consistent**) sequence of estimators of θ if $\lim_{m \to \infty} \hat{\theta}_m = \theta$ almost surely (that is, with probability 1), written

$$\hat{\theta}_m \xrightarrow{a.s.} \theta.$$

□

Theorem 12.4 (Consistency) *Let $\hat{\theta}_m$ designate the maximum-likelihood estimate of θ based on m i.i.d. random variables $X_1, \ldots, X_m$. Then, if θ_0 is the true value of the parameter, $\hat{\theta}_m$ converges almost surely to θ_0.*

Proof Although this theorem is true in a very general setting, its rigorous proof is beyond the scope of our preparation. Consequently, we content ourselves with a heuristic demonstration. We proceed through all of the major steps of the proof, but assume sufficient regularity and other nice properties, when necessary, to make life bearable. We also assume that θ is a scalar parameter.

To simplify things, let $\mathbf{x} = (x_1, \ldots, x_m)$, and introduce the following notation:

$$f_m(\mathbf{x} \mid \theta) \stackrel{\text{def}}{=} f_{X_1, \ldots, X_m}(x_1, \ldots, x_m \mid \theta).$$

We can get away with this since the quantities $x_1, \ldots, x_m$ do not change throughout the proof (rather, the parameter that changes is θ).

From theorem 12.1 and its corollaries,

$$E\left[\frac{\partial \log f_m(\mathbf{X} \mid \theta)}{\partial \theta}\right] = 0, \tag{12.25}$$

where $\mathbf{X} = \{X_1, \ldots, X_m\}$.

Suppose the true value of the parameter θ is θ_0. Now let us expand $\frac{\partial \log f_m(\mathbf{x} \mid \theta)}{\partial \theta}$ in a Taylor series about θ_0, to obtain

$$\left.\frac{\partial \log f_m(\mathbf{x} \mid \theta)}{\partial \theta}\right|_{\theta=\theta'} = \left.\frac{\partial \log f_m(\mathbf{x} \mid \theta)}{\partial \theta}\right|_{\theta=\theta_0} + \left.\frac{\partial^2 \log f_m(\mathbf{x} \mid \theta)}{\partial \theta^2}\right|_{\theta=\theta^*} (\theta' - \theta_0), \tag{12.26}$$

where θ^* is chosen to force equality. Let $\hat{\theta}_m$ be the maximum-likelihood estimate based on $X_1, \ldots, X_m$, which consequently satisfies

$$\left.\frac{\partial \log f_m(\mathbf{x} \mid \theta)}{\partial \theta}\right|_{\theta=\hat{\theta}_m} = 0.$$

Hence, evaluating (12.26) at $\theta' = \hat{\theta}_m$, we obtain

$$\left.\frac{\partial^2 \log f_m(\mathbf{x} \mid \theta)}{\partial \theta^2}\right|_{\theta=\theta^*} (\hat{\theta}_m - \theta_0) = -\left.\frac{\partial \log f_m(\mathbf{x} \mid \theta)}{\partial \theta}\right|_{\theta=\theta_0}. \tag{12.27}$$

Since $X_1, \ldots, X_m$ are i.i.d., we have, with $f_X(x \mid \theta)$ the common density function,

$$\frac{\partial \log f_m(\mathbf{x} \mid \theta)}{\partial \theta} = \frac{\partial}{\partial \theta} \log \prod_{i=1}^{m} f_X(x_i \mid \theta)$$

$$= \frac{\partial}{\partial \theta} \sum_{i=1}^{m} \log f_X(x_i \mid \theta)$$

$$= \sum_{i=1}^{m} \frac{\partial \log f_X(x_i \mid \theta)}{\partial \theta}.$$

By a similar argument,

$$\frac{\partial^2 \log f_m(\mathbf{x} \mid \theta)}{\partial \theta^2} = \sum_{i=1}^{m} \frac{\partial^2 \log f_X(x_i \mid \theta)}{\partial \theta^2}.$$

12.3 Estimation Quality

From the strong law of large numbers[1] it follows that

$$\frac{1}{m}\sum_{i=1}^{m}\frac{\partial \log f_X(x_i\mid\theta)}{\partial \theta} \xrightarrow{a.s} E\left[\frac{\partial \log f_X(X\mid\theta)}{\partial \theta}\right] = 0, \quad (12.28)$$

where the equality holds from (12.25). Similarly,

$$\frac{1}{m}\sum_{i=1}^{m}\frac{\partial^2 \log f_X(x_i\mid\theta)}{\partial \theta^2}\bigg|_{\theta=\theta^*} \xrightarrow{a.s} E\left[\frac{\partial^2 \log f_X(X\mid\theta)}{\partial \theta^2}\right]_{\theta=\theta^*}. \quad (12.29)$$

We now make the assumption that

$$E\left[\frac{\partial^2 \log f_X(X\mid\theta)}{\partial \theta^2}\right]_{\theta=\theta^*} \neq 0.$$

This assumption is essentially equivalent to the condition that the likelihood function is a concave function for all values of θ. We might suspect that most of the common distributions we use would satisfy this condition—but we will not expend the effort to prove it. Given the above assumption and substituting (12.28) and (12.29) into (12.27), we obtain that

$$(\hat{\theta}_m - \theta_0) \xrightarrow{a.s.} \frac{E\left[\frac{\partial \log f_X(X\mid\theta)}{\partial \theta}\right]_{\theta=\theta_0}}{E\left[\frac{\partial^2 \log f_X(X\mid\theta)}{\partial \theta^2}\right]_{\theta=\theta^*}} = 0. \quad (12.30)$$

□

The preceding theorem shows that, as $m \to \infty$, the maximum-likelihood estimate $\hat{\theta}_m$ tends to θ_0 with probability 1, the true value of the parameter. The next theorem shows us that, for large m, the values of $\hat{\theta}_m$ from different trials are clustered around θ_0 with a normal distribution.

Theorem 12.5 (Asymptotic normality) *Let $\hat{\theta}_m$ designate the maximum-likelihood estimate of θ based on m i.i.d. random variables $X_1, \ldots, X_m$. Let $\hat{\theta}_m$ be in the interior of Θ, and assume that $f_m(\mathbf{x}\mid\theta)$ is atleast twice differentiable and $J(\theta)$ is nonsingular. Then if θ_0 is the true value of the parameter, $\hat{\theta}_m$ converges in law (also called convergence in distribution) to a normal random variable; that is,*

$$\sqrt{m}(\hat{\theta}_m - \theta_0) \xrightarrow{law} Y,$$

where

$$Y \sim \mathcal{N}(0, J^{-1}(\theta_0))$$

and $J(\theta)$ is the Fisher information.

Proof Due to the complexity of the proof of this result, we again content ourselves with a heuristic demonstration.

First, we form a Taylor expansion about the true parameter value, θ_0:

$$\frac{\partial \log f_m(\mathbf{x}\mid\theta)}{\partial \theta}\bigg|_{\theta=\hat{\theta}_m} = \frac{\partial \log f_m(\mathbf{x}\mid\theta)}{\partial \theta}\bigg|_{\theta=\theta_0} + \frac{\partial^2 \log f_m(\mathbf{x}\mid\theta)}{\partial \theta^2}\bigg|_{\theta=\theta_0}(\hat{\theta}_m - \theta_0) \quad (12.31)$$
$$+ O\left((\hat{\theta}_m - \theta_0)^2\right).$$

---

[1] The strong law of large numbers says that for $\{X_i\}$, a sequence of i.i.d. random variables with common expectation μ, $\frac{1}{n}\sum_{i=1}^{n} x_i \xrightarrow{a.s} \mu$.

Since $\hat{\theta}_m \xrightarrow{a.s.} \theta_0$, we assume sufficient regularity to neglect the higher order terms. Also, since $\hat{\theta}_m$ is the maximum-likelihood estimate, the left-hand side of (12.31) is zero, and therefore

$$\frac{1}{\sqrt{m}} \left.\frac{\partial \log f_m(\mathbf{x}\mid\theta)}{\partial \theta}\right|_{\theta=\theta_0} = -\frac{\sqrt{m}}{m} \left.\frac{\partial^2 \log f_m(\mathbf{x}\mid\theta)}{\partial \theta^2}\right|_{\theta=\theta_0} (\hat{\theta}_m - \theta_0). \qquad (12.32)$$

But from the strong law of large numbers,

$$\frac{1}{m} \frac{\partial^2 \log f_m(\mathbf{x}\mid\theta)}{\partial \theta^2} \xrightarrow{a.s.} E\left[\frac{\partial^2 \log f_X(X\mid\theta)}{\partial \theta^2}\right], \qquad (12.33)$$

and from Lemma 12.1, we obtain

$$E\left[\frac{\partial \log f_X(X\mid\theta)}{\partial \theta}\right]^2 = J(\theta). \qquad (12.34)$$

We have thus established that the random variable

$$\left.\frac{\partial \log f_X(X_i\mid\theta)}{\partial \theta}\right|_{\theta=\theta_0}$$

is a zero-mean random variable with variance $J(\theta_0)$. Thus, by the central limit theorem[2], the left-hand side of (12.32) converges to a normal random variable; that is,

$$\frac{1}{\sqrt{m}} \sum_{i=1}^{m} \left.\frac{\partial \log f_X(X_i\mid\theta)}{\partial \theta}\right|_{\theta=\theta_0} \xrightarrow{law} W,$$

where $W \sim \mathcal{N}[0, J(\theta_0)]$. Consequently, the right-hand side of (12.32) also converges to W; that is,

$$\sqrt{m} J(\theta_0)(\hat{\theta}_m - \theta_0) \xrightarrow{law} W.$$

Finally, it is evident, therefore, that

$$\sqrt{m}(\hat{\theta}_m - \theta_0) \xrightarrow{law} \frac{1}{J(\theta_0)} W \sim \mathcal{N}[0, J^{-1}(\theta_0)]. \qquad (12.35)$$

$\square$

Theorem 12.6 (Asymptotic efficiency) *Within the class of consistent uniformly asymptotically normal estimators, $\hat{\theta}_m$ is asymptotically efficient in the sense that asymptotically it attains the Cramér–Rao lower bound as $m \to \infty$.*

Proof This result is an immediate consequence of the previous theorem and the Cramér–Rao lower bound. $\square$

This theorem is of great practical significance, since it shows that the maximum-likelihood estimator makes efficient use of all the available data for large samples.

12.3.5 The multivariate normal case

Because of its general importance in engineering, we develop the maximum-likelihood estimate for the mean and covariance of the multivariate normal distribution.

---

[2]The version of the central limit theorem we need is as follows: Let $\{X_n\}$ be a sequence of i.i.d. random variables with common expectation μ and common variance σ^2. Let $Z_n = \frac{X_1+\cdots+X_n-n\mu}{\sqrt{n}\sigma}$. Then $Z_n \to Z$ where Z is distributed $\mathcal{N}(0, 1)$. Stated another way, let $W_n = \frac{X_1+\cdots+X_n}{\sqrt{n}}$. Then $W_n \to W$, where W is $\mathcal{N}(\mu, \sigma^2)$.

12.3 Estimation Quality

Suppose $\mathbf{X}_1, \ldots, \mathbf{X}_m$ is a random n-dimensional sample from $\mathcal{N}(\mathbf{m}, R)$, where $\mathbf{m}$ is an n-vector and R is an $n \times n$ covariance matrix. The likelihood function for this sample is

$$\ell(\mathbf{m}, R, \mathbf{X}_1, \ldots, \mathbf{X}_m) = (2\pi)^{-\frac{mn}{2}} |R|^{-\frac{m}{2}} \exp\left\{-\frac{1}{2} \sum_{i=1}^{m} (\mathbf{x}_i - \mathbf{m})^T R^{-1} (\mathbf{x}_i - \mathbf{m})\right\}, \tag{12.36}$$

and, taking logarithms,

$$\Lambda(\mathbf{m}, R, \mathbf{X}_1, \ldots, \mathbf{X}_m) = -\frac{mn}{2} \log(2\pi) - \frac{m}{2} \log |R| - \frac{1}{2} \sum_{i=1}^{m} (\mathbf{x}_i - \mathbf{m})^T R^{-1} (\mathbf{x}_i - \mathbf{m}). \tag{12.37}$$

Equation (12.37) can be simplified as follows. First, let

$$\bar{\mathbf{x}} = \frac{1}{m} \sum_{i=1}^{m} \mathbf{x}_i.$$

We then write

$$\begin{aligned}(\mathbf{x}_i - \mathbf{m})^T R^{-1} (\mathbf{x}_i - \mathbf{m}) &= (\mathbf{x}_i - \bar{\mathbf{x}} + \bar{\mathbf{x}} - \mathbf{m})^T R^{-1} (\mathbf{x}_i - \bar{\mathbf{x}} + \bar{\mathbf{x}} - \mathbf{m}) \\ &= (\mathbf{x}_i - \bar{\mathbf{x}})^T R^{-1} (\mathbf{x}_i - \bar{\mathbf{x}}) + (\bar{\mathbf{x}} - \mathbf{m})^T R^{-1} (\bar{\mathbf{x}} - \mathbf{m}) \\ &\quad + 2 (\bar{\mathbf{x}} - \mathbf{m})^T R^{-1} (\mathbf{x}_i - \bar{\mathbf{x}}).\end{aligned}$$

Summing over the index $i = 1, \ldots, m$, the final term on the right-hand side vanishes, and we are left with

$$\sum_{i=1}^{m} (\mathbf{x}_i - \mathbf{m})^T R^{-1} (\mathbf{x}_i - \mathbf{m}) = \sum_{i=1}^{m} (\mathbf{x}_i - \bar{\mathbf{x}})^T R^{-1} (\mathbf{x}_i - \bar{\mathbf{x}}) + m (\bar{\mathbf{x}} - \mathbf{m})^T R^{-1} (\bar{\mathbf{x}} - \mathbf{m}). \tag{12.38}$$

Since each term of $(\mathbf{x}_i - \bar{\mathbf{x}})^T R^{-1} (\mathbf{x}_i - \bar{\mathbf{x}})$ is a scalar, it equals the trace of itself. Hence, since the trace of the product of matrices is invariant under any cyclic permutation of the matrices,

$$(\mathbf{x}_i - \bar{\mathbf{x}})^T R^{-1} (\mathbf{x}_i - \bar{\mathbf{x}}) = \operatorname{tr} R^{-1} (\mathbf{x}_i - \bar{\mathbf{x}}) (\mathbf{x}_i - \bar{\mathbf{x}})^T. \tag{12.39}$$

Summing (12.39) over the index i and substituting into (12.38) yields

$$\sum_{i=1}^{m} (\mathbf{x}_i - \mathbf{m})^T R^{-1} (\mathbf{x}_i - \mathbf{m}) = \operatorname{tr} R^{-1} \left\{\sum_{i=1}^{m} (\mathbf{x}_i - \bar{\mathbf{x}})(\mathbf{x}_i - \bar{\mathbf{x}})^T\right\} + m (\bar{\mathbf{x}} - \mathbf{m})^T R^{-1} (\bar{\mathbf{x}} - \mathbf{m}). \tag{12.40}$$

Now define

$$S = \frac{1}{m} \sum_{i=1}^{m} (\mathbf{x}_i - \bar{\mathbf{x}})(\mathbf{x}_i - \bar{\mathbf{x}})^T;$$

using (12.40) in (12.37) gives

$$\begin{aligned}\Lambda(\mathbf{m}, R, \mathbf{X}_1, \ldots, \mathbf{X}_m) = &-\frac{mn}{2} \log(2\pi) - \frac{m}{2} \log |R| - \frac{m}{2} \operatorname{tr} R^{-1} S \\ &+ \frac{m}{2} (\bar{\mathbf{x}} - \mathbf{m})^T R^{-1} (\bar{\mathbf{x}} - \mathbf{m}).\end{aligned} \tag{12.41}$$

Calculation of the score function

To facilitate the calculation of the score function, it is convenient to parameterize the log-likelihood equation in terms of $V = R^{-1}$, yielding

$$\Lambda(\mathbf{m}, V, \mathbf{X}_1, \ldots, \mathbf{X}_m) = -\frac{mn}{2}\log(2\pi) + \frac{m}{2}\log|V| - \frac{m}{2}\operatorname{tr} VS$$
$$- \frac{m}{2}\operatorname{tr} V(\bar{\mathbf{x}} - \mathbf{m})(\bar{\mathbf{x}} - \mathbf{m})^T. \quad (12.42)$$

To calculate the score function, we must evaluate $\frac{\partial}{\partial \mathbf{m}} \Lambda$ and $\frac{\partial}{\partial V} \Lambda$. Referring to the gradient formulas of appendix E, we have

$$\frac{\partial}{\partial \mathbf{m}} \Lambda = \frac{m}{2} \frac{\partial}{\partial \mathbf{m}} (\bar{\mathbf{x}} - \mathbf{m})^T V (\bar{\mathbf{x}} - \mathbf{m})$$
$$= mV(\bar{\mathbf{x}} - \mathbf{m}). \quad (12.43)$$

To calculate $\frac{\partial}{\partial V}\Lambda$, we need to compute gradients of the form $\frac{\partial}{\partial V}\log|V|$ and $\frac{\partial}{\partial V}\operatorname{tr}(VS)$. Using the results from appendix E, we have

$$\frac{\partial}{\partial V}\log|V| = 2R - \operatorname{diag} R, \quad (12.44)$$

$$\frac{\partial}{\partial V}\operatorname{tr} VS = 2S - \operatorname{diag}(S), \quad (12.45)$$

and

$$\frac{\partial}{\partial V}(\bar{\mathbf{x}} - \mathbf{m})^T V(\bar{\mathbf{x}} - \mathbf{m}) = \frac{\partial}{\partial V}\operatorname{tr} V(\bar{\mathbf{x}} - \mathbf{m})(\bar{\mathbf{x}} - \mathbf{m})^T$$
$$= 2(\bar{\mathbf{x}} - \mathbf{m})(\bar{\mathbf{x}} - \mathbf{m})^T - \operatorname{diag}(\bar{\mathbf{x}} - \mathbf{m})(\bar{\mathbf{x}} - \mathbf{m})^T. \quad (12.46)$$

Combining (12.44), (12.45), and (12.46), we obtain

$$\frac{\partial \Lambda}{\partial V} = \frac{m}{2}(2M - \operatorname{diag} M), \quad (12.47)$$

where

$$M = R - S - (\bar{\mathbf{x}} - \mathbf{m})(\bar{\mathbf{x}} - \mathbf{m})^T. \quad (12.48)$$

To find the maximum-likelihood estimate of $\mathbf{m}$ and R, we must solve

$$\frac{\partial \Lambda}{\partial \mathbf{m}} = \mathbf{0} \qquad \frac{\partial \Lambda}{\partial V} = \mathbf{0}. \quad (12.49)$$

From (12.43) we see that the maximum-likelihood estimate of $\mathbf{m}$ is

$$\hat{\mathbf{m}}_{ML} = \bar{\mathbf{x}}. \quad (12.50)$$

To obtain the maximum-likelihood estimate of R we require, from (12.47), that $M = 0$, which yields

$$\hat{R}_{ML} = S + (\bar{\mathbf{x}} - \mathbf{m})(\bar{\mathbf{x}} - \mathbf{m})^T;$$

but since the solutions for $\mathbf{m}$ and S must satisfy (12.49), we must have $\mathbf{m} = \hat{\mathbf{m}}_{ML} = \bar{\mathbf{x}}$, hence we obtain

$$\hat{R}_{ML} = S. \quad (12.51)$$

12.3 Estimation Quality

Computation of the Cramér–Rao bound

In the case when R is known, the Cramér–Rao bound is easily computed. From (12.43), the score function is $s(\mathbf{m}, \mathbf{X}) = mV(\bar{\mathbf{x}} - \mathbf{m})$. Using the expression from lemma 12.1,

$$J(\mathbf{m}) = -E\frac{\partial}{\partial \mathbf{m}}s(\mathbf{m}, \mathbf{x}) = EmV = mV = mR^{-1}.$$

It is straightforward to show that

$$\text{cov}(\hat{\mathbf{m}}_{ML}) = R/m = J^{-1}(\mathbf{m}).$$

Thus the estimate $\hat{\mathbf{m}}_{ML}$ is an unbiased estimate of the mean that achieves the Cramér–Rao lower bound—it is efficient.

12.3.6 Minimum-variance unbiased estimators

We have identified several desirable properties of estimators. In chapter 10, we introduced the concept of sufficiency to encapsulate the notion that it may be possible to reduce the complexity of an estimate by combining the observations in various ways; we introduced the ideas of completeness and minimality in recognition of the fact that there are ways to formulate sufficient statistics that reduce the complexity of the statistics to a minimum; and we introduced unbiasedness to express the concept of the average value of the estimator being equal to the parameter being sought. To these concepts we have now added efficiency, a notion involving the covariance of the estimate. Intuitively, the smaller the covariance, the higher the probability that an (unbiased) estimate will lie near its mean value. Thus, the covariance of the estimate provides a convenient means of evaluating the quality of the estimator. It is therefore desirable to choose an estimator with a covariance that it as small as possible.

Definition 12.5 An estimator $\hat{\theta}$ is said to be a **minimum-variance unbiased** estimate of θ if

(a) $E_\theta \hat{\theta}(\mathbf{X}) = \theta$,

(b) $\sigma_{\hat{\theta}}^2 = \min_{\tilde{\theta} \in \hat{\Theta}}\{E_\theta(\hat{\theta}(\mathbf{X}) - \theta)^2\}$, where $\hat{\Theta}$ is the set of all possible unbiased estimates, given X, of θ. □

The notion of minimum variance is a conceptually powerful one. From our familiarity with Hilbert space, we know that variance has a valid interpretation as squared distance; a minimum variance estimate possesses the property, therefore, that this measure of distance between the estimate and the true parameter is minimized. This appears to be desirable. To explore this in more detail, we begin by eatablishing the Rao–Blackwell theorem.

Theorem 12.7 (Rao–Blackwell) Let Y be a random variable such that $E_\theta Y = \theta \; \forall \theta \in \Theta$ and $\sigma_Y^2 = E_\theta(Y - \theta)^2$. Let Z be a random variable that is sufficient for θ, and let $g(Z)$ be the conditional expectation of Y given Z,

$$g(Z) = E(Y \mid Z).$$

Then

(a) $Eg(Z) = \theta$, and

(b) $E(g(Z) - \theta)^2 \leq \sigma_Y^2$.

Proof The proof of (a) is immediate from property 2 of conditional expectation:
$$Eg(Z) = E[E(Y \mid Z)] = EY = \theta.$$
To established (b), we write
$$\sigma_Y^2 = E_\theta(Y-\theta)^2 = E_\theta[Y - g(Z) + g(Z) - \theta]^2$$
$$= \underbrace{E_\theta[Y-g(Z)]^2}_{\gamma^2 \geq 0} + \underbrace{E[g(Z)-\theta]^2}_{\sigma^2_{g(Z)}} + 2E_\theta[Y-g(Z)][g(Z)-\theta].$$

We next examine the term $E_\theta[Y - g(Z)][g(Z) - \theta]$, and note that, by properties 2 and 4 of conditional expectation,
$$E_\theta[Y - g(Z)][g(Z) - \theta] = E_\theta(E\{[Y-g(Z)]\underbrace{[g(Z)-\theta]}_{\text{fcn of } Z} \mid Z\})$$
$$= E_\theta[g(Z) - \theta]E[Y - g(Z) \mid Z])$$
$$= 0$$
since $E[Y - g(Z) \mid Z] = g(Z) = 0$. Thus,
$$\sigma_Y^2 = \gamma^2 + \sigma^2_{g(Z)},$$
which establishes (b). □

The relevance of this theorem is as follows: Let $\mathbf{X} = \{X_1, \ldots, X_n\}$ be sample values of a random variable X whose distribution is parameterized by $\theta \in \Theta$, and let $Z = t(\mathbf{X})$ be a sufficient statistic for θ. Let $Y = \tilde{\theta}$ be any unbiased estimator of $\tilde{\theta}$. The Rao–Blackwell theorem states that the estimate $E[\tilde{\theta} \mid t(\mathbf{X})]$ is unbiased and has variance at least as small as that of the estimate, $\tilde{\theta}$.

Let's review what we have accomplished with all of our analysis. We started with the assumption of minimum-variance unbiasedness as our criterion for optamility. The Rao–Blackwell theorem showed us that the minimum-variance estimate was based upon a sufficient statistic. We recognized, completely justifiably, that if we are going base our estimate on a sufficient statistic, then we should use a complete sufficient statistic. But the Lehmann–Scheffé theorem tells us that there is at most one unbiased estimate based on a completely sufficient statistic. So we have established that the set of optimal estimates, according to our criterion, contains *at most one* member. Thus, if we find an unbiased estimate based on a complete suffcient statistic, not only is it the best one, in terms of being of minimum variance, it is the only one.

Example 12.3.1 ([85]) This example illustrates the dubious optimality of minimum-variance unbiasedness. Suppose a telephone operator who, after working 10 minutes, wonders if he would be missed if he took 20-minute break. He assumes that calls are coming in to his switchboard as a Poisson process at the unknown rate of λ calls per 10 minutes. Let X denote the number of calls received within the first 10 minutes. As we have seen, X is a sufficient statistic for λ. On the basis of observing X, the operator wishes to estimate the probability that no calls will be received within the next 20 minutes. Since the probability of no calls in any 10-minute interval is $f_X(0) = \frac{\lambda^0}{0!}e^{-\lambda}$, the probability of no calls in a 20-minute interval is $\theta = e^{-2\lambda}$. If operator is enamored with unbiased estimates, he will look for an estimate $\hat{\theta}(X)$ for which
$$E_\lambda \hat{\theta}(X) = \sum_{x=0}^{\infty} \hat{\theta}(x) \frac{e^{-\lambda}\lambda^x}{x!} \equiv e^{-2\lambda}.$$

After multiplying both sides by e^λ and expanding $e^{-\lambda}$ in a power series, he would obtain
$$\sum_{x=0}^{\infty} \hat{\theta}(x) \frac{\lambda^x}{x!} \equiv \sum_{x=0}^{\infty} (-1)^x \frac{\lambda^x}{x!}.$$

Two convergent power series can be equal only if corresponding coefficient are equal. The only unbiased estimate of $\theta = e^{-2\lambda}$ is $\hat{\theta}(x) = (-1)^x$. Thus he would estimate the probability of receiving no calls in the next 20 minutes as $+1$ if he received an even number of calls in the last 10 minutes, and as -1 if he received an odd number of calls in the last 10 minutes. This ridiculous estimate nonetheless is a minimum-variance unbiased estimate. □

12.3.7 The linear statistical model

Suppose now that we have observations $\mathbf{X}_i \sim \mathcal{N}(H\boldsymbol{\theta}, R)$, where $\mathbf{X}_i$ is an n-dimensional vector, H is a known vector, and $\boldsymbol{\theta}$ is an unknown parameter vector. Then the likelihood function based on m independent observations is

$$f_\mathbf{X}(\mathbf{x} \mid \boldsymbol{\theta}) = (2\pi)^{-mn/2} |R|^{-m/2} \exp\left[-\frac{1}{2} \sum_{i=1}^{m} (\mathbf{x}_i - H\boldsymbol{\theta})^T R^{-1} (\mathbf{x}_i - H\boldsymbol{\theta})\right],$$

from which the score function for $\boldsymbol{\theta}$ (see exercise 12.3-4) can be obtained as

$$s(\boldsymbol{\theta}, \mathbf{X}) = H^T R^{-1} \sum_{i=1}^{m} (\mathbf{x}_i - H\boldsymbol{\theta}). \tag{12.52}$$

Based on this, the maximum-likelihood estimate of $\boldsymbol{\theta}$ is

$$\hat{\boldsymbol{\theta}} = (H^T R^{-1} H)^{-1} H^T R^{-1} \hat{\mathbf{m}}, \tag{12.53}$$

where

$$\hat{\mathbf{m}} = \frac{1}{m} \sum_{i=1}^{m} \mathbf{x}_i.$$

The form of the solution in (12.53) is revealing: it is precisely the form of solution obtained for the weighted least-squares solution. Thus, least-squares solutions (and weighted least-squares solutions) correspond to maximum-likelihood estimates of the mean, when the observations are made in Gaussian noise. The score function is

$$J = -E \frac{\partial}{\partial \boldsymbol{\theta}} s(\boldsymbol{\theta}, \mathbf{x}) = m H^T R^{-1} H.$$

We observe that we can write the score function as

$$s(\boldsymbol{\theta}, \mathbf{X}) = J(\hat{\boldsymbol{\theta}} - \boldsymbol{\theta}),$$

so that, by theorem 12.3, the estimate $\hat{\boldsymbol{\theta}}$ must be efficient. Furthermore,

$$E\hat{\boldsymbol{\theta}} = \boldsymbol{\theta},$$

so that the estimate is unbiased.

12.4 Applications of ML estimation

Maximum-likelihood estimation is employed in a variety of signal processing applications, for which we can only provide a taste here. Selection of these few applications was on the basis of relevance, and on the way that they demonstrate some of the concepts developed for vector-space problems.

12.4.1 ARMA parameter estimation

We have encountered several times the problem of estimating the parameters of an AR process; this was a major theme of chapter 3. In this section we resume this examination, obtaining the same results as before, and also add some additional details to address the case of ARMA parameter estimation.

Let

$$H(z) = \frac{\sum_{i=0}^{p-1} b_i z^{-i}}{1 + \sum_{i=1}^{p} a_i z^{-i}},$$

and let the sequence $\{x_0, x_1, \ldots, x_{n-1}\}$ be n samples of the impulse response of the system. Then we have

$$x_t = \begin{cases} 0 & t < 0, \\ -\sum_{i=1}^{p} a_i x_{t-i} + b_t & 0 \le t < p, \\ -\sum_{i=1}^{p} a_i x_{t-i} & t \ge p, \end{cases}$$

which may be written in matrix form as

$$\begin{bmatrix} 1 & & & & \\ a_1 & 1 & & & \\ \vdots & & & & \\ a_{p-1} & \cdots & a_1 & 1 & \\ a_p & & \cdots & a_1 & 1 \\ & & & & \\ & a_p & & \cdots & a_1 & 1 \end{bmatrix} \begin{bmatrix} x_0 \\ x_1 \\ \vdots \\ x_{n-1} \end{bmatrix} = \begin{bmatrix} b_0 \\ b_1 \\ \vdots \\ b_{p-1} \\ 0 \\ \vdots \\ 0 \end{bmatrix}.$$

Let this be written in terms of vectors and matrices as

$$K^{-1}\mathbf{x} = \begin{bmatrix} \mathbf{b} \\ \mathbf{0} \end{bmatrix},$$

where we decompose K^{-1} as

$$K^{-1} = \begin{bmatrix} 1 & & & \\ a_1 & 1 & & \\ & & \ddots & \\ a_{p-1} & \cdots & a_1 & 1 \\ & & A^T & \end{bmatrix}$$

with A^T the $(n-p) \times n$ banded matrix

$$A^T = \begin{bmatrix} a_p & \cdots & a_1 & 1 & & \\ & a_p & & \cdots & a_1 & 1 \\ \vdots & & & & & \\ & & & a_p & \cdots & a_1 & 1 \end{bmatrix}.$$

Solving for $\mathbf{y}$, we obtain

$$\mathbf{x} = K \begin{bmatrix} \mathbf{b} \\ \mathbf{0} \end{bmatrix}.$$

Now let $H = K(:, 1:p)$ (the first p columns of K), and explicitly indicate the dependence of H upon the AR coefficients as $H = H(\mathbf{a})$. Then we can write

$$\mathbf{x} = H(\mathbf{a})\mathbf{b}. \tag{12.54}$$

If the impulse response observations are made in the presence of noise, then, we have

$$\mathbf{y} = \mathbf{x} + \mathbf{n},$$

12.4 Applications of ML Estimation

where **n** is the noise vector. (For convenience, and following a notation that is fairly standard, we indicate the random vectors using lower-case letters.) Substituting from (12.54), we have

$$\mathbf{y} = H(\mathbf{a})\mathbf{b} + \mathbf{n}. \tag{12.55}$$

If we assume an appropriate noise model for **n**, then we can attempt maximum-likelihood estimation of the model parameters.

If we assume that $\mathbf{n} \sim \mathcal{N}(\mathbf{0}, \sigma^2 I)$, then the log-likelihood function for the parameters **a** and **b** (after eliminating constants that do not depend upon the parameters) is

$$L(\mathbf{a}, \mathbf{b}, \mathbf{y}) = -\frac{1}{2\sigma^2}(\mathbf{y} - H(\mathbf{a})\mathbf{b})^T (\mathbf{y} - H(\mathbf{a})\mathbf{b}).$$

The maximum-likelihood estimate of the parameters satisfies

$$(\hat{\mathbf{a}}, \hat{\mathbf{b}}) = \arg \min_{(\mathbf{a},\mathbf{b})} (\mathbf{y} - H(\mathbf{a})\mathbf{b})^T (\mathbf{y} - H(\mathbf{a})\mathbf{b}).$$

This is a nonlinear optimization problem, due to the product of $H(\mathbf{a})\mathbf{b}$. (It is for this reason that we have not treated the ARMA case previously in this text, restricting ourselves to the easier AR problem for which linear optimization problems can be formulated.)

If we know the value of $H(\mathbf{a})$, then the ML estimate of **b** is

$$\hat{\mathbf{b}} = (H(\mathbf{a})H(\mathbf{a}))^T H(\mathbf{a})^T \mathbf{y}, \tag{12.56}$$

the least-squares solution. Assuming that this value is available, we substitute it into the log-likelihood function to obtain

$$L(\mathbf{a}, \hat{\mathbf{b}}, \mathbf{y}) = -\frac{1}{2\sigma^2} \mathbf{y}^T (I - P(\mathbf{a})) \mathbf{y}, \tag{12.57}$$

where $P(\mathbf{a}) = H(\mathbf{a})(H(\mathbf{a})^T H(\mathbf{a}))^{-1} H^T(\mathbf{a})$ is a projection matrix that projects onto the column space of H, and thus

$$I - P(\mathbf{a})$$

is a projector on the space orthogonal to the column space of $H(\mathbf{a})$. The likelihood function of (12.57), obtained by substituting an estimate of one parameter back into the likelihood function, is known as a *compressed* likelihood function. (A similar technique can also be used in combining maximum-likelihood detection and estimation, in which an ML estimate is substituted back into a likelihood function, which then is used for detection. This is called a *generalized likelihood ratio test*.)

From (12.57), we see that the ML estimate of **a** is obtained from

$$\hat{\mathbf{a}} = \arg \min_{\mathbf{a}} \mathbf{y}^T (I - P(\mathbf{a})) \mathbf{y}. \tag{12.58}$$

We now address the problem of finding values for $\hat{\mathbf{a}}$ from (12.58). Interestingly enough, an estimate for **a** can be found that does not depend upon knowledge the value of **b**. Writing the equation $K^{-1} K \begin{bmatrix} \mathbf{b} \\ \mathbf{0} \end{bmatrix}$ as

$$\begin{bmatrix} 1 & & & \\ a_1 & 1 & & \\ & & \ddots & \\ a_{p-1} & \cdots & a_1 & 1 \\ & & A^T & \end{bmatrix} [H(\mathbf{a}) \ \times] \begin{bmatrix} \mathbf{b} \\ \mathbf{0} \end{bmatrix} = \begin{bmatrix} 1 & & & \\ a_1 & 1 & & \\ & & \ddots & \\ a_{p-1} & \cdots & a_1 & 1 \\ & & A^T & \end{bmatrix} H(\mathbf{a})\mathbf{b} = \begin{bmatrix} \mathbf{b} \\ \mathbf{0} \end{bmatrix},$$

(where × indicates columns of no significance), we observe that

$$\begin{bmatrix} 1 & & & \\ a_1 & 1 & & \\ \vdots & & \ddots & \\ a_{p-1} & \cdots & a_1 & 1 \end{bmatrix} H(\mathbf{a}) = I$$

and

$$A^T H(\mathbf{a}) = 0. \tag{12.59}$$

From (12.59), we note that the subspace span(A) is orthogonal to the subspace span($H(\mathbf{a})$). The projector onto the orthogonal space $I - P(\mathbf{a})$ is therefore the projector onto the column space of A:

$$P_A = A(A^T A)^{-1} A^T.$$

Thus (12.58) can be written as

$$\hat{\mathbf{a}} = \arg\min_{\mathbf{a}} \mathbf{y}^T P_A \mathbf{y} = \arg\min_{\mathbf{a}} \mathbf{y}^T A(A^T A)^{-1} A^T \mathbf{y}. \tag{12.60}$$

The factor $A^T \mathbf{y}$ that arises from (12.60) can be written as

$$A^T \mathbf{y} = \begin{bmatrix} a_p & \cdots & a_1 & 1 & & \\ & \ddots & & \ddots & \ddots & \\ & & a_p & \ddots & a_1 & a_1 \end{bmatrix} \begin{bmatrix} y_0 \\ y_1 \\ \vdots \\ y_{n-1} \end{bmatrix}.$$

This can be rewritten as

$$A^T \mathbf{y} = \begin{bmatrix} y_0 & y_1 & \cdots & y_p \\ y_1 & y_2 & \cdots & y_{p+1} \\ \vdots & & & \\ y_{n-1-p} & y_{n-2-p} & \cdots & y_{n-1} \end{bmatrix} \begin{bmatrix} a_p \\ a_{p-1} \\ \vdots \\ a_1 \\ 1 \end{bmatrix} = Y^T \mathbf{a},$$

where Y is a Hankel matrix of the impulse response. Using this representation, (12.60) can be written as

$$\hat{\mathbf{a}} = \arg\min_{\mathbf{a}} \mathbf{a}^T Y (A^T A)^{-1} Y^T \mathbf{a}. \tag{12.61}$$

The advantage of the representation (12.61) is that it is directly in terms of the measured parameters Y and the desired parameters A, and does not require forming $H(\mathbf{a})$, which would involve a matrix inverse.

An iterative solution to (12.61) can be obtained as follows. Start from an initial solution $\hat{\mathbf{a}}^{[0]}$. Let $A(\hat{\mathbf{a}}^{[k]})$ denote the matrix A formed from the coefficients in $\hat{\mathbf{a}}^{[k]}$. Then iterate the following: Minimize $\hat{\mathbf{a}}^{[k+1]} Y(A^T(\hat{\mathbf{a}}^{[k]})A(\hat{\mathbf{a}}^{[k]}))^{-1} Y^T \hat{\mathbf{a}}^{[k+1]}$ with respect to $\hat{\mathbf{a}}^{[k+1]}$, subject to the constraint that

$$(\hat{\mathbf{a}}^{[k+1]})^T \mathbf{e}_{p+1} = 1,$$

where $\mathbf{e}_p$ is the vector that is in the $(p+1)$st element and zero elsewhere (see exercise 12.3-16). Repeat until convergence.

Once $\hat{\mathbf{a}}$ is obtained, then $\hat{\mathbf{b}}$ may be obtained from (12.56). Code that implements this solution is shown in algorithm 12.1.

12.4 Applications of ML Estimation

Algorithm 12.1 Maximum-likelihood ARMA estimation
File: `kissarma.m`

12.4.2 Signal subspace identification

Suppose that we have the linear signal model $\mathbf{x}_t = H\boldsymbol{\theta}$, where H is an $n \times p$ matrix that is *unknown*. All of the signals lie in span(H), which is a p-dimensional subspace of $\mathbb{R}^n$. Assuming $p < n$, we develop an ML method for identifying span(H)—the subspace in which the signals lie—based upon a sequence of measurements $\{\mathbf{y}_1, \mathbf{y}_2, \ldots, \mathbf{y}_m\}$ that are measurements of the linear model made in the presence of noise,

$$\mathbf{y}_t = \mathbf{x}_t + \mathbf{n}_t$$
$$= H\boldsymbol{\theta} + \mathbf{n}_t.$$

Let $A = [\mathbf{a}_1 \quad \mathbf{a}_2 \quad \ldots \quad \mathbf{a}_{n-p}]$ be a matrix orthogonal to H: $A^T H = 0$, so that

$$A^T \mathbf{x}_t = 0, \qquad t = 1, 2, \ldots, n. \tag{12.62}$$

Assuming that the noise vectors are distributed as $\mathbf{n}_t \sim \mathcal{N}(0, \sigma^2 I)$, the log-likelihood function (neglecting inessential constants) is

$$L(H, \boldsymbol{\theta}, \mathbf{y}_1, \ldots, \mathbf{y}_n) = -\frac{1}{2\sigma^2} \sum_{i=1}^{} (\mathbf{y}_i - H\boldsymbol{\theta})^T (\mathbf{y}_i - H\boldsymbol{\theta}).$$

We want to maximize the log-likelihood function subject to the constraint (12.62). A cost functional incorporating the constraints by means of Lagrange multipliers is

$$J(\mathbf{x}) = \sum_{i=1}^{m} (\mathbf{y}_i - \mathbf{x}_i)^T (\mathbf{y}_i - \mathbf{x}_i) + \sum_{i=1}^{m} \boldsymbol{\lambda}_i^T A^T \mathbf{x}_i,$$

where $\boldsymbol{\lambda}_i \in \mathbb{R}^{n-p}$. Taking the gradient with respect to $\mathbf{x}_t$ and equating to zero, we obtain

$$-2\mathbf{y}_t + 2\mathbf{x}_t + A\boldsymbol{\lambda}_t = 0,$$

from which

$$\mathbf{x}_t = \mathbf{y}_t - \frac{1}{2} A\boldsymbol{\lambda}_t. \tag{12.63}$$

To find $\boldsymbol{\lambda}_t$ and enforce the constraint (12.62), we multiply by A^T:

$$A^T \mathbf{x}_t = A\left(\mathbf{y}_t - \frac{1}{2} A\boldsymbol{\lambda}_t\right) = 0,$$

from which $\boldsymbol{\lambda}_t = 2(A^T A)^{-1} A^T \mathbf{y}_t$. Substituting this into (12.63), we obtain

$$\mathbf{x}_t = (I - A(A^T A)^{-1} A^T)\mathbf{y}_t. \tag{12.64}$$

We recognize that $P_A = A(A^T A)^{-1} A^T$ is the orthogonal projector onto the column space of A, and $I - P_A$ is the projector onto the orthogonal space. This provides a reasonable solution.

However, A is unknown, so the solution is not complete. We substitute the solution (12.64) into the likelihood function to obtain the compressed likelihood

$$L = -\frac{1}{2\sigma^2} \sum_{t=1}^{m} \mathbf{y}_t^T P_A \mathbf{y}_t.$$

It can be shown (see exercise 12.3-13) that L can be written as

$$L = -\frac{m}{2\sigma^2} \operatorname{tr}(P_A S), \quad (12.65)$$

where

$$S = \frac{1}{m} \sum_{t=1}^{m} \mathbf{y}_t \mathbf{y}_t^T$$

is the sample correlation matrix.

Let us factor S and P_A as

$$S = U \Lambda U^T \qquad P_A = V I_{n-p} V^T,$$

where Λ is the diagonal matrix of eigenvalues of S $\lambda_1, \lambda_2, \ldots, \lambda_n$ and I_{n-p} is the portion of an identity matrix with $n - p$ ones along the diagonal. Then

$$\operatorname{tr}(P_A S) = \operatorname{tr}(V I_{n-p} V^T U \Lambda U^T) = \operatorname{tr}(I_{n-p} \Lambda)$$

$$= \sum_{n-p} \lambda_i, \quad (12.66)$$

where the sum over $n - p$ denotes summing over some set of $n - p$ values. To maximize L in (12.65), we must minimize the sum in (12.66), which can be accomplished by summing over the $n - p$ *smallest* eigenvalues of S (since all the eigenvalues of S are nonnegative). If we order the eigenvalues λ_i and their corresponding eigenvectors $\mathbf{u}_i$ as

$$\lambda_1 \geq \lambda_2 \geq \cdots \geq \lambda_n,$$

then the minimum of $\operatorname{tr}(P_A S)$ is obtained when

$$P_A = \sum_{i=p+1}^{n} \mathbf{u}_i \mathbf{u}_i^T.$$

Thus the subspace $\operatorname{span}(H)$ is the space orthogonal to P_A:

$$\operatorname{span}(H) = \operatorname{span}(\mathbf{u}_1, \mathbf{u}_2, \ldots, \mathbf{u}_p).$$

12.4.3 Phase estimation

In this section we introduce the important problem of estimation of phase. We also introduce the ML problem for a continuous-time random process. Suppose that a signal is $s(t; \theta) = A \cos(\omega_c t + \theta)$, where ω_c is known and θ is an unknown phase to be determined. The signal $s(t; \theta)$ is observed over an interval $t \in [0, T)$ in stationary white Gaussian noise to produce the signal

$$y(t) = s(t; \theta) + n(t),$$

where $n(t)$ has the autocorrelation function

$$r_n(\tau) = \frac{N_0}{2} \delta(\tau).$$

As discussed in section 11.9, a log-likelihood function can be written in this case (neglecting inessential constants) as

$$\Lambda(\theta, y(t)) = -\frac{1}{N_0} \int_0^T (y(t) - s(t; \theta))^2 \, dt. \quad (12.67)$$

Expanding this, we obtain

$$\Lambda(\theta, y(t)) = -\frac{1}{N_0} \int_0^T (y^2(t) - 2y(t)s(t; \theta) + s^2(t; \theta)) \, dt.$$

12.4 Applications of ML Estimation

The first term $y^2(t)$ is independent of θ. The third term $s^2(t;\theta)$ integrates to the signal energy over T seconds, which is independent of θ for the given signal model. We can therefore simplify the log-likelihood function to that term which depends upon θ, as

$$L'(\theta, y(t)) = 2\frac{1}{N_0} \int_0^T y(t) s(t;\theta)\, dt. \qquad (12.68)$$

We desire to maximize (12.68) with respect to $\theta \in [0, 2\pi)$. Taking the derivative with respect to θ and equating to zero, we obtain the equation

$$\int_0^T y(t) \sin(\omega_c t + \theta)\, dt \Big|_{\theta=\hat{\theta}_{ML}} = 0. \qquad (12.69)$$

We examine two different approaches to the solution of the problem. The first is an explicit solution, and the second is a device for tracking the phase—the phase-locked loop (PLL).

The explicit solution is obtained by expanding (12.69) using the trigonometric identity $\sin(a+b) = \sin(a)\cos(b) + \sin(b)\cos(a)$, and solving for $\hat{\theta}_{ML}$, which gives

$$\tan \hat{\theta}_{ML} = -\frac{\int_0^T y(t) \sin(\omega_c t)\, dt}{\int_0^T y(t) \cos(\omega_c t)\, dt}.$$

A block diagram illustrating the solution is shown in figure 12.2.

The phase-locked loop is obtained by building a feedback device that enforces the maximum-likelihood equation (12.69). Rather than integrating simply over $[0, T]$, we assume that we are "integrating" over a period of time using a lowpass filter. The block diagram of figure 12.3 illustrates the basic concept. A voltage-controlled oscillator (VCO) is used to generate a reference signal $\sin(\omega_c t + \hat{\theta})$, which is correlated with the incoming signal $y(t)$. When $\hat{\theta} = \theta$, by (12.67) the correlation value is zero and the maximum-likelihood equation (12.69) is satisfied. Otherwise, some output voltage is produced that drives the VCO to either advance or delay the phase.

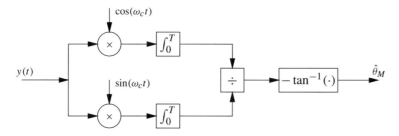

Figure 12.2: Explicitly computing the estimate of the phase

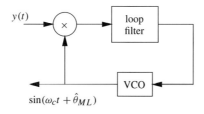

Figure 12.3: A phase-locked loop

12.5 Bayes estimation theory

Suppose you observe a random variable X, whose distribution depends on a parameter θ. The maximum-likelihood approach to estimation says that you should take as your estimate of an unknown parameter that value that is the most likely, out of all possible values of the parameter, to have given rise to the observed data. Before observations are taken, therefore, the maximum-likelihood method is silent—it makes no predictions about either the value of the parameter or the values future observations will take. Instead, the attitude of a rabid "max-like" enthusiast is: "Wait until all of the data are collected, give them to me, be patient, and soon I will give you an estimate of what the values of the parameters were that generated the data." If you were to ask him for his best guess, before you collected the data, as to what values would be assumed by either the data or the parameters, his response would simply be: "Don't be ridiculous."

On the other hand, a Bayesian would be all too happy to give you estimates, both before and after the data have been obtained. Before the observation, she would give you, perhaps, the mean value of the *a priori* distribution of the parameter, and after the data were collected she would give you the mean value of the *a posteriori* distribution of the parameter. She would offer, as predicted values of the observations, the mean value of the conditional distribution of X given the expected value of θ (based on the *a priori* distribution).

Some insight into how the prior distribution affects the problem of estimation may be gained through the following example.

Example 12.5.1 Let $X_1, \ldots, X_m$ denote a random sample of size m from the normal distribution $\mathcal{N}(\theta, \sigma^2)$. Suppose σ is known, and we wish to estimate θ. We are given the prior density $\theta \sim \mathcal{N}(\vartheta_0, \sigma_\theta^2)$, that is,

$$f_\theta(\vartheta) = \frac{1}{\sqrt{2\pi}\sigma_\theta} \exp\left[-\frac{(\vartheta - \vartheta_0)^2}{2\sigma_\theta^2}\right].$$

Before getting involved in deep Bayesian principles, let's just think about ways we could use this prior information.

1. We could consider computing the maximum-likelihood estimate of θ (which we saw earlier is just the sample average) and then simply averaging this result with the mean value of the prior distribution, yielding

$$\hat{\theta}_a = \frac{\vartheta_0 + \hat{\theta}_{ML}}{2}.$$

   This naive approach, while it factors in the prior information, gives equal weight to the prior information as compared to *all* of the direct observations. Such a result might be hard to justify, especially if the data quality is high.

2. We could treat ϑ_0 as one extra "data" point and average it in with all of the other x_i's, yielding

$$\hat{\theta}_b = \frac{\vartheta_0 + \sum_{i=1}^m x_i}{m+1}.$$

   This approach has a very nice intuitive appeal; we simply treat the *a priori* information in exactly the same way as we do the real data. $\hat{\theta}_b$ is therefore perhaps more reasonable than $\hat{\theta}_a$, but it still suffers a drawback: it is treated as being exactly equal in informational content to each of the x_i's, whether or not σ_θ^2 equals σ^2.

3. We could take a weighted average of the *a priori* mean and the maximum-likelihood estimate, each weighted inversely proportional to the variance, yielding

$$\hat{\theta}_c = \frac{\frac{\vartheta_0}{\sigma_\theta^2} + \frac{\hat{\theta}_{ML}}{\sigma_{ML}^2}}{\frac{1}{\sigma_\theta^2} + \frac{1}{\sigma_{ML}^2}},$$

where σ_{ML}^2 is the variance of $\hat{\theta}_{ML}$, and is given by

$$\sigma_{ML}^2 = E\left[\frac{1}{m}\sum_{i=1}^{m} X_i - \theta\right]^2.$$

To calculate the above expectation, we temporarily take off our Bayesian hat and put on our ML hat, view θ as simply an unknown parameter, and take the expectation with respect to the random variables X_i only. In so doing, it follows after some manipulations that $\sigma_{ML}^2 = \sigma^2/m$. Consequently,

$$\hat{\theta}_c = \frac{\sigma^2/m}{\sigma_\theta^2 + \sigma^2/m}\vartheta_0 + \frac{\sigma_\theta^2}{\sigma_\theta^2 + \sigma^2/m}\hat{\theta}_{ML}. \qquad (12.70)$$

The estimate $\hat{\theta}_c$ seems to incorporate all of the information, both *a priori* and *a posteriori*, that we have about θ. We see that, as m becomes large, the *a priori* information is forgotten, and the maximum-likelihood portion of the estimator dominates. We also see that if $\sigma_\theta^2 \ll \sigma^2$, then the *a priori* information tends to dominate.

The estimate provided by $\hat{\theta}_c$ appears to be, of the three we have presented, the one most worthy of our attention. We shall eventually see that it is indeed a Bayesian estimate. $\square$

12.6 Bayes risk

The starting point for Bayesian estimation, as for Bayesian detection, is the specification of a loss function and the calculation of the Bayes risk. Recall that the cost function is a function of the state of nature and the decision function; that is, it is of the general form $L[\theta, \phi(X)]$. For our development in Bayes estimation theory, we restrict the structure of the loss function to be the function of the *difference*, that is, to be of the form $L[\theta - \phi(X)]$. Although this restricts us to only a small subset of all possible loss functions, we see that it still leads us to some very interesting and useful results. We examine three different cost functionals:

1. Squared error,
2. Absolute error, and
3. Uniform cost.

Of these, the squared-error criterion will emerge as being the most important and deserving of study.

Recall from section 11.4 that the risk function $R: \Theta \times \Delta \to \Delta$ is the average loss, where the average is with respect to X,

$$R(\theta, \phi) = E[L(\theta)\phi(X)] = \int_{-\infty}^{\infty} L(\theta, \phi(x)) f_{X|\theta}(x \mid \theta)\, dx.$$

The Bayes risk is the expectation of the risk with respect to an assumed prior distribution on θ,

$$r(F_\theta, \phi) = E[R(\theta, \phi)] = \int_\Theta R(\vartheta, \phi) f_\theta(\vartheta)\, d\vartheta.$$

We saw earlier that, under appropriate regularity conditions, we may reverse the order of integration in the calculation of the Bayes risk function to obtain

$$r(\tau, \phi) = \int_\mathcal{X} \left\{\int_\Theta L[\vartheta, \phi(x)] f_{\theta|X}(\vartheta \mid x)\, d\vartheta\right\} f_X(x)\, dx,$$

and noted that we could minimize the Bayes risk by minimizing the inner integral *for each x separately*; that is, we may find, for each x, the action, call it $\phi(x)$, that minimizes

$$\int L[\vartheta, \phi(x)] f_{\theta|X}(\vartheta \mid x) \, d\vartheta.$$

In other words, **the Bayes decision rule minimizes the posterior conditional expected loss, given the observations**.

Let us now examine the structure of the Bayes rule under the three cost functionals we have defined.

Squared-error loss

Let us first consider squared-error loss, and introduce the concept via the following example.

Example 12.6.1 Consider the estimation problem in which $\Theta = \Delta = (0, \infty)$ and

$$L(\theta, \delta) = (\theta - \delta)^2.$$

Our problem is to estimate the value of θ. That is, the "decision" $\delta \in \Delta$ is our estimate of θ, so we can write $\hat{\theta} = \delta$.

Suppose we observe the value of a random variable X having a uniform distribution on the interval $(0, \theta)$ with density

$$f_{X|\theta}(x \mid \vartheta) = \begin{cases} 1/\vartheta & \text{if } 0 < x < \vartheta, \\ 0 & \text{otherwise.} \end{cases}$$

Note that we may write

$$f_{X|\theta}(x \mid \vartheta) = \frac{1}{\vartheta} I_{(0,\vartheta)}(x) = \frac{1}{\vartheta} I_{(x,\infty)}(\vartheta).$$

We are to find a Bayes rule with respect to the prior distribution. Assume, for some reason, that we know (or suspect) that the parameter θ is distributed according to an exponential density

$$f_\theta(\vartheta) = \begin{cases} \vartheta e^{-\vartheta} & \text{if } \vartheta > 0, \\ 0 & \text{otherwise.} \end{cases}$$

(This is a significant point of departure from maximum-likelihood estimation; at this point we have no physical or mathematical justification for this assumption. For now, this density simply appears in the development.) The joint density of X and θ is, therefore,

$$f_{X\theta}(x, \vartheta) = f_{X|\theta}(x \mid \vartheta) f_\theta(\vartheta),$$

and the marginal distribution of X has the density

$$f_X(x) = \int_{-\infty}^{\infty} f_{X\theta}(x, \vartheta) \, d\vartheta = \begin{cases} e^{-x} & \text{if } x > 0, \\ 0 & \text{otherwise.} \end{cases}$$

Hence, the posterior distribution of θ, given $X = x$, has the density

$$f_{\theta|X}(\vartheta \mid x) = \frac{f_{X\theta}(x, \vartheta)}{f_X(x)} = \begin{cases} e^{x-\vartheta} & \text{if } \vartheta > x, \\ 0 & \text{otherwise,} \end{cases}$$

where $x > 0$. (Again, we see a significant difference between Bayesian estimation and ML estimation; in ML estimation there was no concept of a posterior, because there was no concept of a prior.) The posterior expected loss, given $X = x$, is

$$E[L(\theta, \delta) \mid X = x] = e^x \int_x^\infty (\vartheta - \delta)^2 e^{-\vartheta} \, d\vartheta.$$

12.6 Bayes Risk

To find the δ that minimizes this expected loss, we may set the derivative with respect to δ to zero:

$$\frac{d}{d\delta} E[L(\theta, \delta) \mid X = x] = -2e^x \int_x^\infty (\vartheta - \delta) e^{-\vartheta} d\vartheta = 0.$$

This implies

$$\phi(x) = \delta = \frac{\int_x^\infty \vartheta e^{-\vartheta} d\vartheta}{\int_x^\infty e^{-\vartheta} d\vartheta} = \frac{(x+1)e^{-x}}{e^{-x}} = x + 1.$$

This, therefore, is a Bayes decision rule with respect to F_θ: if $X = x$ is observed, then the estimate of θ is $x + 1$. □

The problem of point estimation of a real parameter, using quadratic loss, occurs so frequently in engineering applications that it is worthwhile to make the following observation. The posterior expected loss, given $X = x$, for a quadratic loss function at δ is the second moment about δ of the posterior distribution of θ given x. That is,

$$EL(\theta, \delta \mid X = x) = \int_{-\infty}^\infty (\vartheta - \delta)^2 f_{\theta \mid X}(\vartheta \mid x) d\vartheta.$$

Theorem 12.8 *The posterior expected loss given $X = x$,*

$$E[L(\theta, \delta) \mid X = x] = \int_{-\infty}^\infty (\vartheta - \delta)^2 f_{\theta \mid X}(\vartheta \mid x) d\vartheta, \qquad (12.71)$$

is minimized by taking δ as the mean of the distribution, that is,

$$\phi(x) = \delta = E(\theta \mid X = x).$$

Proof Taking the derivative of (12.71) with respect to δ and simplifying, we obtain

$$\int \vartheta f_{\theta \mid X}(\vartheta \mid x) d\vartheta = \delta \int f_{\theta \mid X}(\vartheta \mid x) d\vartheta.$$

On the left-hand side, we recognize $E(\theta \mid X = x)$, and on the right-hand side we have simply δ. □

(Note: strictly speaking, δ is a *function*, and so a first variation, not a derivative, should be employed here. However, the derivation works because for every $X = x$, δ is a constant independent of the variable of integration.) The estimate of θ given by this theorem is the minimum *mean-square* estimate of θ, and is denoted $\hat{\theta}_{MS}$.

Absolute-error loss

Another important loss function is absolute value of the difference,

$$L(\theta, \delta) = |\theta - \delta|.$$

The Bayes risk is minimized by minimizing

$$EL(\theta, \delta \mid X = x) = \int_{-\infty}^\infty |\vartheta - \delta| f_{\theta \mid X}(\vartheta \mid x) d\vartheta. \qquad (12.72)$$

The minimization here is more awkward than for the squared-error loss, since the absolute value function is not differentiable everywhere. Our approach is to consider two cases, and take derivatives of each piece.

1. When $\vartheta > \delta$, then

$$\frac{\partial}{\partial \delta} \int |\theta - \vartheta| f_{\theta \mid X}(\vartheta \mid x) d\vartheta = \int (-1) f_{\theta \mid X}(\vartheta \mid x) d\vartheta.$$

2. When $\vartheta < \delta$, then

$$\frac{\partial}{\partial \delta} \int |\theta - \vartheta| f_{\theta|X}(\vartheta \mid x) \, d\vartheta = \int (1) f_{\theta|X}(\vartheta \mid x) \, d\vartheta.$$

Combining these two by means of the limits of integration, and setting the derivative with respect to δ equal to zero, we obtain

$$\int_{\infty}^{\delta} f_{\theta|X}(\vartheta \mid x) \, d\vartheta - \int_{\delta}^{\infty} f_{\theta|X}(\vartheta \mid x) \, d\vartheta = 0$$

or

$$\int_{\infty}^{\delta} f_{\theta|X}(\vartheta \mid x) \, d\vartheta = \int_{\delta}^{\infty} f_{\theta|X}(\vartheta \mid x) \, d\vartheta.$$

That is, the integral under the density to the left of δ is the same as that to the right of δ. We have thus proven the following theorem.

Theorem 12.9 $EL(\theta, \delta \mid X = x) = \int_{-\infty}^{\infty} |\vartheta - \delta| f_{\theta|X}(\vartheta \mid x) \, d\vartheta$ *is minimized by taking*

$$\phi(x) = \delta = \text{median } f_{\theta|X}(\vartheta \mid x).$$

That is, a Bayes rule corresponding to the absolute error criterion is to take δ as the median of the posterior distribution of θ, given $X = x$.

Uniform cost

The loss function associated with uniform cost is defined as

$$L(\vartheta, \delta) = \begin{cases} 0 & \text{if } |\vartheta - \delta| \leq \epsilon/2, \\ 1 & \text{if } |\vartheta - \delta| > \epsilon/2. \end{cases}$$

In other words, an error less than $\epsilon/2$ is as good as no error, and if the error is greater than $\epsilon/2$, we assign a uniform cost. The Bayes risk is minimized by minimizing

$$\int_{-\infty}^{\infty} L(\vartheta, \delta) f_{\theta|X}(\vartheta \mid x) \, d\vartheta = \int_{-\infty}^{\delta-\epsilon/2} f_{\theta|X}(\vartheta \mid x) \, d\vartheta + \int_{\delta+\epsilon/2}^{\infty} f_{\theta|X}(\vartheta \mid x) \, d\vartheta$$

$$= 1 - \int_{\delta-\epsilon/2}^{\delta+\epsilon/2} f_{\theta|X}(\vartheta \mid x) \, d\vartheta.$$

Consequently, the Bayes risk is minimized when the integral

$$\int_{\delta-\epsilon/2}^{\delta+\epsilon/2} f_{\theta|X}(\vartheta \mid x) \, d\vartheta$$

is maximized. When ϵ is sufficiently small and $f_{\theta|X}(\vartheta|x)$ is continuous in ϑ,

$$\int_{\delta-\epsilon/2}^{\delta+\epsilon/2} f_{\theta|X}(\vartheta \mid x) \, d\vartheta \approx 2\epsilon f_{\theta|X}(\delta \mid x).$$

In this case, it is evident that the integral is maximized when ϑ assumes the value at which the posterior density $f_{\theta|X}(\vartheta \mid x)$ is maximized.

Definition 12.6 The **mode** of a distribution is that value that maximizes the probability density function. □

We have proven the following.

12.6 Bayes Risk

Theorem 12.10 *The Bayes risk with uniform cost is minimized when the integral*

$$\int_{\delta-\epsilon/2}^{\delta+\epsilon/2} f_{\theta|X}(\vartheta \mid x)\, d\vartheta$$

is minimized. As $\epsilon \to 0$, *the minimum value is obtained by choosing δ to be the mode—the maximizing value—of $f_{\theta|X}(\vartheta \mid x)$.*

12.6.1 MAP estimates

Definition 12.7 The value of ϑ that maximizes the *a posteriori* density (that is, the mode of the posterior density) is called the **maximum a posteriori probability** estimate of θ. □

If the posterior density of θ given X is unimodal and symmetric, then it is easy to see that the MAP estimate and the mean squared estimate coincide, for then the posterior density attains its maximum value at its expectation. Furthermore, under these circumstances, the median also coincides with the mode and the expectation. Thus, if we are lucky enough to be dealing with such distributions, the various estimates all tend toward the same value.

Although in the development of maximum-likelihood estimation theory we eschewed the characterization of θ as random, we may gain some valuable understanding of the maximum-likelihood estimate by considering θ to be a random variable whose prior distribution is so dispersed (that is, has such a large variance) that the information provided by the prior is vanishingly small. If the theory is consistent, we would have a right to expect that the maximum-likelihood estimate would be the limiting case of such a Bayesian estimate.

Let θ be considered as a random variable distributed according to the *a priori* density $f_\theta(\vartheta)$. The *a posteriori* distribution for θ, then, is given by

$$f_{\theta|X}(\vartheta \mid x) = \frac{f_{X|\theta}(x \mid \vartheta) f_\theta(\vartheta)}{f_X(x)}. \tag{12.73}$$

If the logarithm of the *a posteriori* density is differentiable with respect to θ, then the MAP estimate is given by the solution to

$$\left. \frac{\partial \log f_{\theta|X}(\vartheta \mid x)}{\partial \vartheta} \right|_{\vartheta = \hat{\theta}_{MAP}} = 0. \tag{12.74}$$

This equation is called the *MAP equation*.

Taking the logarithm of (12.73) yields

$$\log f_{\theta|X}(\vartheta \mid x) = \log f_{X|\theta}(x \mid \vartheta) + \log f_\theta(\vartheta) - \log f_X(x),$$

and since $f_X(x)$ is not a function of θ, the MAP equation becomes

$$\frac{\partial \log f_{\theta|X}(\vartheta \mid x)}{\partial \vartheta} = \frac{\partial \log f_{X|\theta}(x \mid \vartheta)}{\partial \vartheta} + \frac{\partial \log f_\theta(\vartheta)}{\partial \vartheta}. \tag{12.75}$$

Comparing (12.75) to the standard maximum-likelihood equation

$$\left. \frac{\partial L(\theta, x)}{\partial \theta} \right|_{\theta = \hat{\theta}_{ML}} = 0,$$

we see that the two expressions differ by $\frac{\partial \log f_\theta(\vartheta)}{\partial \vartheta}$. If $f_\theta(\vartheta)$ is sufficiently "flat," (that is, if the variance is very large) its logarithm will also be flat, so the gradient of the logarithm will be nearly zero, and the *a posteriori* density will be maximized, in the limiting case, at the maximum-likelihood estimate.

Example 12.6.2 Let $X_1, \ldots, X_m$ denote a random sample of size m from the normal distribution $\mathcal{N}(\theta, \sigma^2)$. Suppose σ is known, and we wish to find the MAP estimate for the mean, θ. The joint density function for $X_1, \ldots, X_m$ is

$$f_{X_1,\ldots,X_m}(x_1, \ldots, x_m \mid \theta) = \prod_{i=1}^{m} \frac{1}{\sqrt{2\pi}\sigma} \exp\left[-\frac{(x_i - \vartheta)^2}{2\sigma^2}\right],$$

Suppose θ is distributed $\mathcal{N}(0, \sigma_\theta^2)$, that is,

$$f_\theta(\vartheta) = \frac{1}{\sqrt{2\pi}\sigma_\theta} \exp\left[-\frac{\vartheta^2}{2\sigma_\theta^2}\right].$$

Straightforward manipulation yields

$$\frac{\partial \log f_{\theta \mid X}(\vartheta \mid x)}{\partial \vartheta} = \frac{1}{\sigma^2} \sum_{i=1}^{m} (x_i - \vartheta) - \frac{\vartheta}{\sigma_\theta^2}.$$

Equating this expression to zero and solving for ϑ yields

$$\hat{\theta}_{MAP} = \frac{\sigma_\theta^2}{\sigma_\theta^2 + \frac{\sigma^2}{m}} \frac{1}{m} \sum_{i=1}^{m} x_i.$$

Now, it is clear that as $\sigma_\theta^2 \to \infty$, the limiting expression is the maximum-likelihood estimate $\hat{\theta}_{ML}$. It is also true that, as $m \to \infty$, the MAP estimate asymptotically approaches the ML estimate. Thus, as the knowledge about θ from the prior distribution tends to zero, or as the amount of data becomes overwhelming, the MAP estimate converges to the maximum-likelihood estimate. □

12.6.2 Summary

From the preceding results, we have seen that the Bayes estimate of ϑ based upon the measurement of a random variable X depends upon the posterior density $f_{\theta \mid X}(\vartheta \mid x)$. The conversion of the prior information about θ represented by $f_\theta(\vartheta)$ to the posterior density is accomplished via the expression

$$f_{\theta \mid X}(\vartheta \mid x) = \frac{f_{X \mid \theta}(x \mid \vartheta)}{f_X(x)} f_\theta(\vartheta). \tag{12.76}$$

The posterior density $f_{\theta \mid X}(\vartheta \mid x)$ represents our state of knowledge after the measurement of X. It is on the posterior density that we base our estimate and, for Bayesian purposes, it contains all the information necessary for estimation. On the basis of the posterior, estimates can be obtained in several ways:

1. For a minimum variance (quadratic loss function),

$$\hat{\vartheta} = E[\theta \mid X].$$

2. To minimize $|\vartheta - \hat{\vartheta}|$, set $\hat{\vartheta}$ to the median of $f_{\theta \mid X}(\vartheta \mid x)$.

3. To maximize the probability that $\hat{\vartheta} = \vartheta$, set $\hat{\vartheta}$ to the mode (maximum value) of $f_{\theta \mid X}(\vartheta \mid x)$.

12.6.3 Conjugate prior distributions

In general, the marginal density $f_X(x)$ and the posterior density $f_{\theta \mid X}(\vartheta \mid x)$ are not easily calculated. We are interested in establishing conditions on the structure of the distributions involved that ensure tractability in the calculation of the posterior distribution. We introduce in the following the idea of sequential estimation, in which a Bayesian estimate is updated after each observation in a sequence. In order to have tractable sequential observations, we must be able to propagate one posterior density to the next by means of an update step. This is most tractable if the distributions involved belong to a conjugate family.

12.6 Bayes Risk

Definition 12.8 Let $\mathcal{F}$ denote a class of conditional density functions $f_{X|\theta}$, indexed by ϑ as ϑ ranges over all the values in Θ. A class $\mathcal{P}$ of distributions is said to be a **conjugate family** for $\mathcal{F}$ if the posterior $f_{\theta|X} \in \mathcal{P}$ for all $f_{X|\theta} \in \mathcal{F}$, and all priors $f_\theta \in \mathcal{P}$. In other words, a family of distributions is a conjugate family if it contains both the prior f_θ and the posterior $f_{\theta|X}$ for all possible conditional densities. A conjugate family is said to be *closed under sampling*. □

We give some examples of conjugate families. For more examples and analysis, the interested reader is referred to [67].

Example 12.6.3 Suppose that $X_1, \ldots, X_m$ is a random sample from a Bernoulli distribution with parameter $0 \leq \theta \leq 1$ that has density

$$f_{X|\theta}(x \mid \vartheta) = \begin{cases} \vartheta^x (1-\vartheta)^{1-x} & x \in \{0, 1\}, \\ 0 & \text{otherwise.} \end{cases}$$

Suppose also that the prior distribution of θ is a β distribution with parameters $\alpha > 0$ and $\beta > 0$, with density (see box 12.1)

$$f_\theta(\vartheta) = \begin{cases} \dfrac{\Gamma(\alpha+\beta)}{\Gamma(\alpha)\Gamma(\beta)} \vartheta^{\alpha-1}(1-\vartheta)^{\beta-1} & 0 < \vartheta < 1, \\ 0 & \text{otherwise.} \end{cases}$$

Then the joint distribution of θ and $\mathbf{X} = [X_1, X_2, \ldots, X_m]^T$ is

$$\begin{aligned} f_{\mathbf{X},\theta}(\mathbf{x}, \vartheta) &= f_{\mathbf{X}|\theta}(\mathbf{x} \mid \theta) f_\theta(\vartheta) \\ &= \frac{\Gamma(\alpha+\beta)}{\Gamma(\alpha)\Gamma(\beta)} \theta^{\alpha+y-1}(1-\theta)^{\beta+m-y-1}, \end{aligned}$$

where $y = \sum_{i=1}^m x_i$. The posterior distribution of θ given $\mathbf{X}$ is

$$f_{\theta|\mathbf{X}}(\vartheta \mid \mathbf{x}) = \frac{f_{\mathbf{X},\theta}(\mathbf{x}, \vartheta)}{f_{\mathbf{X}}(\mathbf{x})} = \frac{f_{\mathbf{X},\theta}(\mathbf{x}, \vartheta)}{\int f_{\mathbf{X},\theta}(\mathbf{x}, \vartheta)\, d\vartheta}.$$

It can be shown (see exercise 12.6-22) that

$$f_{\theta|\mathbf{X}}(\vartheta \mid \mathbf{x}) = \frac{\Gamma(\overline{\alpha}+\overline{\beta})}{\Gamma(\overline{\alpha})\Gamma(\overline{\beta})} \vartheta^{\overline{\alpha}-1}(1-\vartheta)^{\overline{\beta}-1} \sim \beta(\overline{\alpha}, \overline{\beta}). \tag{12.77}$$

where $\overline{\alpha} = \alpha + y$ and $\overline{\beta} = \beta + m - y$. Thus, both f_θ and $f_{\theta|X}$ have a β distribution. □

Example 12.6.4 Suppose that $X_1, \ldots, X_m$ is a random sample from a Poisson distribution with parameter $\theta > 0$ that has pmf

$$f_{X|\theta}(x \mid \vartheta) = \begin{cases} \dfrac{e^{-\vartheta} \vartheta^x}{x!} & x = 0, 1, 2, \ldots, \\ 0 & \text{otherwise.} \end{cases}$$

Suppose also that the prior distribution of θ is a Γ distribution with parameters $\alpha > 0$ and $\beta > 0$ (see

---

Box 12.1: The β distribution

The β pdf is given by

$$f_X(x) = \frac{\Gamma(\alpha+\beta)}{\Gamma(\alpha)\Gamma(\beta)} x^{\alpha-1}(1-x)^{\beta-1},$$

for $0 \leq x \leq 1$, where α and β are parameters. This is denoted by saying $X \sim \beta(\alpha, \beta)$. The mean and variance are $\mu = \frac{\alpha}{\alpha+\beta}$ and $\sigma^2 = \frac{\alpha\beta}{(\alpha+\beta)^2(\alpha+\beta+1)}$.

Box 12.2: The Γ distribution

The Γ pdf is parameterized by two parameters α and β, having pdf
$$f_X(x) = \frac{1}{\beta^\alpha \Gamma(\alpha)} x^{\alpha-1} e^{-x/\beta}$$
for $x > 0$. This is denoted by $X \sim \Gamma(\alpha, \beta)$. The mean and variance are $\mu = \alpha\beta$ and $\sigma^2 = \alpha\beta^2$.

box 12.2), with density
$$f_\theta(\vartheta) = \begin{cases} \dfrac{\beta^\alpha}{\Gamma(\alpha)} \vartheta^{\alpha-1} e^{\beta\vartheta} & \vartheta > 0, \\ 0 & \text{otherwise.} \end{cases}$$

Then (see exercise 12.6-23) the posterior distribution of θ when $X_i = x_i, i = 1, \ldots, m$, is a $\Gamma(\alpha + y, 1/(1/\beta + m))$ distribution, where $y = \sum_{i=1}^m x_i$. □

Example 12.6.5 Suppose that $X_1, \ldots, X_m$ is a random sample from an exponential distribution with parameter $\theta > 0$ that has density
$$f_{X|\theta}(x \mid \vartheta) = \begin{cases} \theta e^{-\theta x} & x > 0, \\ 0 & \text{otherwise.} \end{cases}$$

Suppose also that the prior distribution of θ is a Γ distribution with parameters $\alpha > 0$ and $\beta > 0$, that have density
$$f_\theta(\vartheta) = \begin{cases} \dfrac{1}{\beta^\alpha \Gamma(\alpha)} \vartheta^{\alpha-1} e^{-\beta/\vartheta} & \vartheta > 0, \\ 0 & \text{otherwise.} \end{cases}$$

Then (see exercise 12.6-24) the posterior distribution of θ when $X_i = x_i, i = 1, \ldots, m$, is a $\Gamma(\alpha+m, 1/(1/\beta + y))$ distribution, where $y = \sum_{i=1}^m x_i$. □

Example 12.6.6 Suppose that $X_1, \ldots, X_m$ is a random sample from a Gaussian distribution with unknown mean θ and known variance σ^2. Suppose also that the prior distribution of θ is a Gaussian distribution with mean ϑ_0 and variance σ_θ^2. Then the posterior distribution of θ when $X_i = x_i$, $i = 1, \ldots, m$, is a Gaussian distribution with mean

$$\hat{\theta}_c = \frac{\dfrac{\vartheta_0}{\sigma_\theta^2} + \dfrac{\bar{x}}{\sigma_m^2}}{\dfrac{1}{\sigma_\theta^2} + \dfrac{1}{\sigma_m^2}} \tag{12.78}$$

and variance

$$\sigma_{\hat{\theta}}^2 = \frac{\sigma_m^2 \sigma_\theta^2}{\sigma_m^2 + \sigma_\theta^2}, \tag{12.79}$$

where
$$\bar{x} = \frac{1}{m} \sum_{i=1}^m x_i \quad \text{and} \quad \sigma_m^2 = \sigma^2/m.$$

Due to its importance, we provide a demonstration of the above claim. For $-\infty < \vartheta < \infty$, the conditional density of $X_1, \ldots, X_m$ satisfies

$$f_{\mathbf{X}|\theta}(\mathbf{x} \mid \vartheta) = \prod_{i=1}^m \frac{1}{\sqrt{2\pi}\sigma} \exp\left[-\frac{(x_i - \vartheta)^2}{2\sigma^2}\right]$$

$$= (2\pi)^{-\frac{m}{2}} \sigma^{-m} \exp\left[-\frac{1}{2\sigma^2} \sum_{i=1}^m (x_i - \bar{x})^2\right] \exp\left[\frac{-m}{2\sigma^2}(\vartheta - \bar{x})^2\right]. \tag{12.80}$$

12.6 Bayes Risk

The prior density of θ satisfies

$$f_\theta(\vartheta) = \frac{1}{\sqrt{2\pi}\sigma_\theta} \exp\left[-\frac{(\vartheta - \vartheta_0)^2}{2\sigma_\theta^2}\right], \tag{12.81}$$

and the posterior density function of θ will be proportional to the product of (12.80) and (12.81). Letting the symbol $\propto$ denote proportionality, we have

$$f_{\theta|\mathbf{X}}(\vartheta \mid \mathbf{x}) \propto \exp\left[-\frac{m}{2\sigma^2}(\vartheta - \overline{x})^2\right] \exp\left[-\frac{(\vartheta - \vartheta_0)^2}{2\sigma_\theta^2}\right]$$

$$= \exp\left[-\frac{(\vartheta - \overline{x})^2}{2\sigma_m^2} - \frac{(\vartheta - \vartheta_0)^2}{2\sigma_\theta^2}\right].$$

Simplifying the exponent, we obtain

$$\frac{(\vartheta - \overline{x})^2}{\sigma_m^2} + \frac{(\vartheta - \vartheta_0)^2}{\sigma_\theta^2} = \frac{\sigma_m^2 + \sigma_\theta^2}{\sigma_m^2 \sigma_\theta^2}(\vartheta - \hat{\theta}_c)^2 + \frac{1}{\sigma_m^2 + \sigma_\theta^2}(\overline{x} - \vartheta_0)^2,$$

where $\hat{\theta}_c$ is given by (12.78). Thus,

$$f_{\theta|\mathbf{X}}(\vartheta \mid \mathbf{x}) \propto \exp\left[-1/2 \frac{\sigma_m^2 + \sigma_\theta^2}{\sigma_m^2 \sigma_\theta^2}(\vartheta - \hat{\theta}_c)^2\right].$$

Consequently, we see that the posterior density of θ given $X_1, \ldots, X_m$, suitably normalized, is normal with mean given by (12.78) and variance given by (12.79).

Upon rearranging (12.78), we see that

$$\hat{\theta}_c = \frac{\sigma_m^2}{\sigma_\theta^2 + \sigma_m^2}\vartheta_0 + \frac{\sigma_\theta^2}{\sigma_\theta^2 + \sigma_m^2}\overline{x},$$

which is exactly the same as the estimate given by (12.70). Thus, the weighted average, proposed as a reasonable way to incorporate prior information into the estimate, turns out to be exactly a Bayes estimate for the parameter given that the prior is a member of the normal conjugate family. □

As this example shows, the conjugate prior for a Gaussian distribution is a Gaussian distribution—yet another reason for engineering interest in these distributions.

We will see subsequently that conjugate classes of distributions are useful in sequential estimation, in which a posterior density at one stage of computation is used as a prior for the next stage.

12.6.4 Connections with minimum mean-squared estimation

In chapter 3, considerable effort was devoted to explaining and exploring minimum mean-squared estimation. In that context, an estimate $\hat{x}$ of a signal x, where $\hat{x}$ is a linear combination of some set of data

$$\hat{x} = c_1 p_1 + c_2 p_2 + \cdots c_m p_m,$$

was determined so that the average of the squared error, where

$$e = x - \hat{x},$$

is minimized. That is, $E[e^2] = E[(x - \hat{x})^2]$ is minimized.

Now, recall from theorem 12.8 that for a Bayes estimator using a quadratic loss function, the best estimate of a random parameter $\hat{\theta}$ given a measurement X is the conditional expectation

$$\hat{\theta} = E(\theta \mid X),$$

and that the Bayes cost $E[L(\theta, \delta) \mid X]$ was termed the mean-squared error of $\hat{\theta}$. Thus the conditional mean is the estimator that minimizes the mean-squared error.

Obviously there must be some connection between the two techniques, since both of them rely on a minimum mean-squared error criterion. We make some observations in this regard. Our comparison is aided by using a notation that is similar for each case. For the first case, we will write our estimate as

$$\hat{\theta} = c_1 X_1 + c_2 X_2 + \cdots + c_m X_m; \qquad (12.82)$$

that is, we are estimating the parameter θ as a linear combination of the m random variables $X_1, X_2, \ldots, X_m$. We will refer to this as a linear estimator. In the second case, we might actually have several observations, so our estimator will be of the form

$$\hat{\theta} = E(\theta \mid X_1, X_2, \ldots, X_m). \qquad (12.83)$$

We will refer to this as a conditional mean estimator.

By the formulation of the linear estimator (12.82), we have restricted our attention to only that class of estimators that are *linear functions* of the observations. The conditional mean estimator (12.83) has no such restrictions: the conditional mean may not be a linear function of the observations. The conditional mean estimator may, in fact, be a nonlinear function of the observations. The conditional mean estimate thus guarantees minimum mean-squared error across all possible estimates. However, for some distributions the resulting nonlinearity may make the computation intractable.

Fortunately, in the case of estimating the mean of a Gaussian distribution, the conditional mean estimate *is linear*, as we see in the next section, so that the linear estimator and the conditional mean estimator coincide. This is yet another reason why the Gaussian distribution is of practical interest.

12.6.5 Bayes estimation with the Gaussian distribution

We have encountered throughout this text the Gaussian distribution in a variety of settings. We consider again the problem of jointly distributed Gaussian random variables, such as (X, Y), or random vectors, such as $(\mathbf{X}, \mathbf{Y})$. Since the distribution of Gaussian random variables is unimodal and symmetric, and since the conditional distribution $f_{X|Y}$ is also Gaussian, this conditional distribution provides what is needed for estimating the random variable X for a variety of cost functions:

1. For a squared-error loss function, the best estimate is the conditional mean.
2. For an absolute-error loss function, the best estimate is the median, which for a Gaussian is the same as the mean.
3. For a uniform cost function, the best estimate is the mode, which for a Gaussian is the same as the mean.

Thus, determining the conditional distribution and identifying the mean provides the necessary estimates for the most common Bayes loss functions. (It should be noted that in this section we denote the object of our interest in estimation as the random variable $\mathbf{X}$, rather than the random variable $\boldsymbol{\theta}$. This provides a notational transition toward considering $\mathbf{X}$ as a state variable to be estimated, as is done in following sections.)

Recall that in section 4.12 we computed the distribution of the conditional random variable $X \mid Y$, using the formulas for the inverse of a partitioned matrix. These results will now be put to work.

12.6 Bayes Risk

In example 4.12.1, the distribution of the random variable $\mathbf{Z} = (\mathbf{X}, \mathbf{Y})$, where $\mathbf{X} \in \mathbb{R}^m$ and $\mathbf{Y} \in \mathbb{R}^n$, $\mathbf{X} \sim \mathcal{N}(\mu_x, R_{xx})$ and $\mathbf{Y} \sim \mathcal{N}(\mu_y, R_{yy})$, is found to be

$$f_{\mathbf{Z}}(\mathbf{z}) = f_{\mathbf{Z}}(\mathbf{x}, \mathbf{y}) = \frac{1}{(2\pi)^{p/2} \det(R_{zz})} \exp\left[-\frac{1}{2}(\mathbf{z} - \mu_z)^T R_{zz}^{-1}(\mathbf{z} - \mu_z)\right]$$

where $p = m + n$, and

$$R_{zz} = \text{cov}(\mathbf{Z}) = \begin{bmatrix} R_{xx} & R_{xy} \\ R_{yx} & R_{yy} \end{bmatrix}.$$

We now consider the estimation problem. Given a measurement of $\mathbf{Y}$, we want to estimate $\mathbf{X}$. This requires finding $f_{\mathbf{X}|\mathbf{Y}}(\mathbf{x}|\mathbf{y})$. However, we have already dealt with this problem in example 4.12.1: $f_{\mathbf{X}|\mathbf{Y}}(\mathbf{x}|\mathbf{y})$ was shown to be Gaussian with mean

$$\mu_{x|y} = E[\mathbf{X} | \mathbf{Y} = \mathbf{y}] = \mu_x + R_{xy} R_y^{-1}(\mathbf{y} - \mu_y) \quad (12.84)$$

and covariance

$$\text{cov}(\mathbf{X} | \mathbf{Y} = \mathbf{y}) = R_{xx} - R_{xy} R_{yy}^{-1} R_{yx} = P. \quad (12.85)$$

The quantity $\hat{\mathbf{x}} = \mu_{x|y}$ is the Bayes estimate of $\mathbf{x}$, given the measurement $\mathbf{Y}$, in the sense of being the mean, mode, and median of the distribution. It can be interpreted as follows: Prior to any measurements, the best estimate of $\mathbf{X}$ is obtained via the prior density $f_\mathbf{X}(\mathbf{x})$ to be μ_x, the mean of $\mathbf{X}$. By means of the measurement, the prior distribution $f_\mathbf{X}(\mathbf{x})$ "evolves" into the posterior distribution by

$$f_{\mathbf{X}|\mathbf{Y}}(\mathbf{x}|\mathbf{y}) = \frac{f_{\mathbf{Y}|\mathbf{X}}(\mathbf{y}|\mathbf{x})}{f_\mathbf{Y}(\mathbf{y})} f_\mathbf{X}(\mathbf{x}). \quad (12.86)$$

On the basis of the posterior density, the prior estimate is modified by an amount proportional to how far the measurement $\mathbf{y}$ is from its expected value. The proportionality depends upon the how strongly X and Y are correlated (by means of R_{xy}) and inversely on the variance of R_y^{-1}: measurements with high variance are not accorded as much weight as measurements with low variance.

Let us examine the estimator $\hat{\mathbf{x}} = \mu_{x|y}$ further.

1. The estimator is unbiased:

$$E\hat{\mathbf{x}} = E\mu_x + R_{xy} R_y^{-1} E(\mathbf{Y} - \mu_y) = \mu_x.$$

2. The estimator error $\mathbf{e} = \mathbf{x} - \hat{\mathbf{x}}$ is uncorrelated with $\hat{\mathbf{x}} - \mu_x$:

$$E\mathbf{e}(\hat{\mathbf{x}} - \mu_x)^T = 0. \quad (12.87)$$

3. The error is uncorrelated with the $\mathbf{y} - \mu_y$:

$$E\mathbf{e}(\mathbf{y} - \mu_y)^T = 0. \quad (12.88)$$

("The error is orthogonal to the data.")

4. The covariance of $\hat{\mathbf{x}}$ is

$$\text{cov}(\hat{\mathbf{x}}) = E[\mathbf{e}\mathbf{e}^T] = R_{xx} - R_{xy} R_{yy}^{-1} R_{yx}.$$

Thus this has "smaller" covariance than the *a priori* covariance R_{xx}.

In the case of the linear model

$$\mathbf{Y} = H\mathbf{X} + \nu, \quad (12.89)$$

where ν is a zero-mean random variable with $\text{cov}(\nu) = R$, then

$$R_{xy} = R_{xx} H^T \quad \text{and} \quad R_{yy} = R \quad \text{and} \quad \mu_y = H\mu_x.$$

Then (12.84) can be written as

$$\mu_{x|y} = \mu_x + R_{xx} H^T R^{-1}(\mathbf{y} - \mu_y). \tag{12.90}$$

It will be convenient to write $K = R_{xx} H^T R^{-1}$, where K is called the *Kalman gain*. Then

$$\mu_{x|y} = \mu_x + K(\mathbf{y} - H\mu_x).$$

12.7 Recursive estimation

We now examine the problem of estimating the state of a system using observations of the system, where the state evolves in the presence of noise and the observations are made sequentially in the presence of noise. We will use the notation $\mathbf{x}_t$ to indicate the parameter to be estimated (instead of θ), and use $\mathbf{y}_t$ to indicate the observation data. Our problem is to estimate the state $\mathbf{x}_t$, $t = 0, 1, \ldots$, based on a sequence of observations $\mathbf{y}_t$, $t = 0, 1, \ldots$. In this development, we assume that the state sequence $\mathbf{x}_t$ is a Markov random process, that is, for any random variable z that is a function of $\mathbf{x}_s$, $s \geq t$,

$$f(z \mid \mathbf{x}_t, \mathbf{x}_{t-1}, \ldots, \mathbf{x}_0) = f(z \mid \mathbf{x}_t). \tag{12.91}$$

In particular, we have

$$f(\mathbf{x}_{t+1} \mid \mathbf{x}_t, \ldots, \mathbf{x}_0) = f(\mathbf{x}_{t+1} \mid \mathbf{x}_t). \tag{12.92}$$

Also, we will assume that the observation $\mathbf{y}_{t+1}$ depends upon $\mathbf{x}_{t+1}$ and possibly on random noise that is independent from sample to sample, but is conditionally independent of prior observations, given $\mathbf{x}_{t+1}$; that is,

$$f(\mathbf{y}_{t+1} \mid \mathbf{x}_{t+1}, \mathbf{y}_t, \ldots, \mathbf{y}_0) = f(\mathbf{y}_{t+1} \mid \mathbf{x}_{t+1}). \tag{12.93}$$

Notation: The vector $\mathbf{x}_t$ is a *random vector*, as is $\mathbf{y}_t$. In making the change to lower case (rather than upper case, as previously in this part), we are following a notational convention now decades old. In statistics, the standard notation for a random variable is to use a capital symbol, and we have retained that usage up to this point, mainly to reinforce the concept that we are dealing with random variables and not their actual values. But we will now depart from the traditional notation of statistics.

We are headed in this development in the direction of the Kalman filter, an important recursive estimator. This is presented in detail in chapter 13, building upon the concepts presented here.

We employ the following notation. The set of measurements $\{y_0, y_1, \ldots, y_t\}$ is denoted as $\mathcal{Y}_t$. The notation $\hat{\mathbf{x}}_{t|\tau}$ is used to denote the Bayes estimate of $\mathbf{x}_t$, given the data $\mathbf{y}_0, \mathbf{y}_1, \ldots, \mathbf{y}_\tau = \mathcal{Y}_\tau$. For example, the estimate $\hat{\mathbf{x}}_{t|t-1}$ indicates the estimate of $\mathbf{x}_t$, using the data $\mathcal{Y}_{t-1}$. We denote the covariance of the estimate of $\hat{\mathbf{x}}_{t|\tau}$ as $P_{t|\tau}$,

$$P_{t|\tau} = E(\hat{\mathbf{x}}_{t|\tau} - E\hat{\mathbf{x}}_{t|\tau})(\hat{\mathbf{x}}_{t|\tau} - E\hat{\mathbf{x}}_{t|\tau})^T. \tag{12.94}$$

For notational convenience we also eliminate the subscript notation on the density functions for now, using the arguments to indicate the random variables, as

$$f(\mathbf{x}_{t+1} \mid \mathcal{Y}_{t+1}) = f_{\mathbf{x}_{t+1} \mid \mathcal{Y}_{t+1}}(\mathbf{x}_{t+1} \mid \mathcal{Y}_{t+1}).$$

Starting from a prior density $f(\mathbf{x}_0)$, the first observation $\mathbf{y}_0$ is used to compute a posterior distribution using Bayes theorem (12.76), as

$$f(\mathbf{x}_0 \mid \mathbf{y}_0) = \frac{f(\mathbf{y}_0 \mid \mathbf{x}_0)}{f(\mathbf{y}_0)} f(\mathbf{x}_0).$$

12.7 Recursive Estimation

Based on $f(\mathbf{x}_0 \mid \mathbf{y}_0)$, an estimate $\hat{\mathbf{x}}_{0|0}$ is obtained. This is the "update" step. This density is now "propagated" ahead in time (by some means using the state update equation for $\mathbf{x}_t$) to obtain $f(\mathbf{x}_1 \mid \mathbf{y}_0)$, from which the estimate $\hat{\mathbf{x}}_{1|0}$ is obtained.

We now want to generalize this first step, to the updating of an estimate conditioned on $\mathcal{Y}_t$ to one conditioned on $\mathcal{Y}_{t+1}$. From the point of view of Bayesian estimation, the problem now is to determine the posterior density $f(\mathbf{x}_{t+1} \mid \mathcal{Y}_{t+1})$ recursively from the posterior density $f(\mathbf{x}_t \mid \mathcal{Y}_t)$. That is, we wish to find a function $\mathcal{F}$ such that

$$f(\mathbf{x}_{t+1} \mid \mathcal{Y}_{t+1}) = \mathcal{F}[f(\mathbf{x}_t \mid \mathcal{Y}_t), \mathbf{y}_{t+1}].$$

Identification of the function $\mathcal{F}$ will provide the desired posterior density, from which the estimate may be obtained. Let us begin by writing the desired result using Bayes theory:

$$f(\mathbf{x}_{t+1} \mid \mathcal{Y}_{t+1}) = f(\mathbf{x}_{t+1} \mid \mathcal{Y}_t, \mathbf{y}_{t+1})$$
$$= \frac{f(\mathbf{y}_{t+1} \mid \mathbf{x}_{t+1}, \mathcal{Y}_t)}{f(\mathbf{y}_{t+1} \mid \mathcal{Y}_t)} f(\mathbf{x}_{t+1} \mid \mathcal{Y}_t). \qquad (12.95)$$

We now observe that

$$f(\mathbf{y}_{t+1} \mid \mathbf{x}_{t+1}, \mathcal{Y}_t) = f(\mathbf{y}_{t+1} \mid \mathbf{x}_{t+1}). \qquad (12.96)$$

To be explicit about why this is true, note that we can write

$$f(\mathbf{y}_{t+1} \mid \mathbf{x}_{t+1}, \mathcal{Y}_t) = f(\mathbf{y}_{t+1} \mid \mathbf{x}_{t+1}, \ldots, \mathbf{x}_0, \mathbf{y}_t, \ldots, \mathbf{y}_0)$$
$$= f(\mathbf{y}_{t+1} \mid \mathbf{x}_{t+1}, \ldots, \mathbf{x}_0) \quad \text{(by (12.91))}$$
$$= f(\mathbf{y}_{t+1} \mid \mathbf{x}_{t+1}) \quad \text{(by (12.93))}.$$

Substituting (12.96) into (12.95) yields

$$\underbrace{f(\mathbf{x}_{t+1} \mid \mathcal{Y}_{t+1})}_{\text{posterior}} = \frac{f(\mathbf{y}_{t+1} \mid \mathbf{x}_{t+1})}{f(\mathbf{y}_{t+1} \mid \mathcal{Y}_t)} \underbrace{f(\mathbf{x}_{t+1} \mid \mathcal{Y}_t)}_{\text{"prior"}}. \qquad (12.97)$$

Equation (12.97) is directly analogous to (12.76), with the following identification. The prior probability $f_\theta(\vartheta)$ is identified as $f(\mathbf{x}_{t+1} \mid \mathcal{Y}_t)$, and $f(\vartheta \mid x)$ is identified as the posterior $f(\mathbf{x}_{t+1} \mid \mathcal{Y}_{t+1})$. We may call (12.97) the "update" step.

Computation of the update step requires finding

$$f(\mathbf{x}_{t+1} \mid \mathcal{Y}_t).$$

The density $f(\mathbf{x}_{t+1} \mid \mathcal{Y}_t)$ is the "propagation" step. This step can be written as

$$f(\mathbf{x}_{t+1} \mid \mathcal{Y}_t) = \int f(\mathbf{x}_{t+1} \mid \mathbf{x}_t, \mathcal{Y}_t) f(\mathbf{x}_t \mid \mathcal{Y}_t) \, d\mathbf{x}_t$$
$$= \int f(\mathbf{x}_{t+1} \mid \mathbf{x}_t) f(\mathbf{x}_t \mid \mathcal{Y}_t) \, d\mathbf{x}_t, \qquad (12.98)$$

where the equality in (12.98) follows by the Markov property of $\mathbf{x}_t$ and the fact that $\mathbf{y}_t$ depends upon $\mathbf{x}_t$.

The two steps represented by (12.97) (update) and (12.98) (propagate) are illustrated in figure 12.4. The prior distribution $f(\mathbf{x}_0)$ is updated by means of (12.97) to produce the posterior $f(\mathbf{x}_0 \mid \mathbf{y}_0)$, from which the estimate $\hat{\mathbf{x}}_{0|0}$ is obtained. The density is propagated by (12.98) to $f(\mathbf{x}_1 \mid \mathbf{y}_0)$, which is then used as the prior for the next stage and from which $\hat{\mathbf{x}}_{1|0}$ is obtained. Iterating these two equations provides for an update of the Bayes estimate as new data arrive. In going from one stage to the next, the conditional $f(\mathbf{x}_{t+1} \mid \mathcal{Y}_t)$ becomes the prior for the next stage. In order to preserve the computational structure from one stage to the next, it is expedient to have the conditional density be of the same type as the prior density; this means that they should be members of a conjugate class. (In practice, however, it is only Gaussian random variables that admit finite-dimensional filtering implementations.)

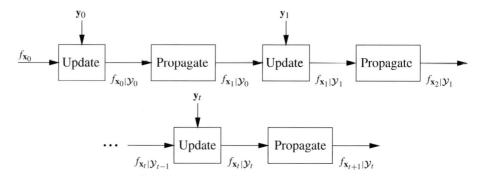

Figure 12.4: Illustration of the update and propagate steps in sequential estimation

12.7.1 An example of non-Gaussian sequential Bayes

We will demonstrate the concept of sequential estimation with a simple probability structure. The key equations are the Bayes update equation (12.97) and the propagate equation (12.98), which are used to provide the evolution of the distributions as new information is obtained. For convenience, an example with a discrete distribution has been selected, so that all integrals are replaced by summations.

In this example [41, page 385], the state of the scalar signal $X_t \in \{0, 1\}$ is governed by a Bernoulli distribution with pmf

$$P(X_t = x_t) = (1 - q_0)\delta(x_t) + q_0\delta(1 - x_t), \tag{12.99}$$

where

$$\delta(x) = \begin{cases} 1 & x = 0, \\ 0 & x \neq 0. \end{cases}$$

This distribution holds for all t. Let N_t be a scalar Markov Bernoulli sequence with

$$P(N_0 = n_0) = (1 - a_0)\delta(n_0) + a_0\delta(1 - n_0), \quad a_0 < 1/2. \tag{12.100}$$

Suppose also that $P(N_t)$ evolves according to

$$P(N_{t+1} = n_{t+1} \mid N_t = n_t) = \left(1 - a_t - \frac{n_t}{2}\right)\delta(n_{t+1}) + \left(a_t + \frac{n_t}{2}\right)\delta(1 - n_{t+1}).$$

This conditional update tends to favor the reoccurrence of a 1: if $N_t = 1$, then N_{t+1} is more likely to be so.

The measurement equation is

$$Y_t = X_t \vee N_t,$$

where $\vee$ indicates the maximum value of its arguments. Based on a sequence of observations $y_0, y_1, \ldots$, we desire to estimate $x_0, x_1, \ldots$, and $n_0, n_1, \ldots$. These equations represent a simple (but imperfect) model of a detection system, in which the state x_t indicates the presence of a target—occurring in isolated samples—and the noise n_t represents blocking of the signal by some large body that gives a false indication of the target (if the blocking was present at the last measurement, it will be more likely to appear in the next measurement). For example, the system might apply to an infrared detection system in which clouds might block the view and give a false signal.

12.7 Recursive Estimation

From the prior probabilities in (12.99) and (12.100), updated pmfs based upon the observation y_0 can be computed from

$$P(N_0 = n_0 \mid Y_0 = y_0) = \frac{P(Y_0 = y_0 \mid N_0 = n_0)}{P(Y_0 = y_0)} P(N_0 = n_0), \qquad (12.101)$$

$$P(X_0 = x_0 \mid Y_0 = y_0) = \frac{P(Y_0 = y_0 \mid X_0 = x_0)}{P(Y_0 = y_0)} P(X_0 = x_0), \qquad (12.102)$$

where $P(Y_0 = y_0)$ is obtained from explicit enumeration:

$$P(Y_0 = 0) = P(X_0 = 0, N_0 = 0) = (1 - a_0)(1 - q_0),$$
$$P(Y_0 = 1) = P(X_0 = 0, N_0 = 1) + P(X_0 = 1, N_0 = 0) + P(X_0 = 1, N_0 = 1)$$
$$= a_0(1 - q_0) + q_0(1 - a_0) + a_0 q_0.$$

Then from (12.101) we have

$$P(N_0 = 1 \mid Y_0 = y_0) = a_{0\mid 0} = \begin{cases} \frac{P(Y_0=1\mid N_0=1)}{P(Y_0=1)} P(N_0 = 1) & \text{if } Y_0 = 1 \\ \frac{P(Y_0=0\mid Y_0=1)}{P(Y_0=0)} P(N_0 = 1) & \text{if } Y_0 = 0 \end{cases} = \frac{a_0 \delta(y_0 - 1)}{a_0 + q_0 - a_0 q_0} \qquad (12.103)$$

and, similarly,

$$P(X_0 = 1 \mid Y_0 = y_0) = q_{0\mid 0} = \frac{q_0 \delta(y_0 - 1)}{a) + q_0 - a_0 q_0}.$$

The updated densities then can be written as

$$P(N_0 = n_0 \mid Y_0 = y_0) = (1 - a_{0\mid 0})\delta(n_0) + a_{0\mid 0}\delta(1 - n_0),$$
$$P(X_0 = x_0 \mid Y_0 = y_0) = (1 - q_{0\mid 0})\delta(x_0) + q_{0\mid 0}\delta(1 - x_0),$$

which are of the same form as the original pmfs in (12.100) and (12.99), except that the probabilities have changed.

The update step is straightforward, using (12.98):

$$P(N_1 = n_1 \mid Y_0 = y_0) = \sum_{n_0} P(N_1 = n_1 \mid N_0 = n_0) P(N_0 = n_0 \mid Y_0 = y_0)$$
$$= (1 - a_{1\mid 0})\delta(n_1) + a_{1\mid 0}\delta(1 - n_1),$$

where

$$a_{1\mid 0} = a + \frac{a_{0\mid 0}}{2}, \qquad (12.104)$$

Also, $P(X_1 = x_1 \mid Y_0 = y_0) = P(X_1 = x_1)$.

Letting $\mathcal{Y}_t = \{y_0, \ldots, y_t\}$, we have

$$P(N_t = n_t \mid \mathcal{Y}_t) = (1 - a_{t\mid t})\delta(n_t) + a_{t\mid t}\delta(1 - n_t) \qquad (12.105)$$

where

$$a_{t\mid t} = \frac{a_{t\mid t-1}\delta(1 - y_t)}{(1 - a_{t\mid t-1})(1 - q)\delta(y_t) + (a_{t\mid t-1} + q - a_{t\mid t-1}q)\delta(1 - y_t)},$$

and

$$P(N_{t+1} = n_{t+1} \mid \mathcal{Y}_t) = (1 - a_{t+1\mid t})\delta(n_{t+1}) + a_{t+1\mid t}\delta(1 - n_{t+1})$$

where

$$a_{t+1\mid t} = a + \frac{a_{t\mid t}}{2}.$$

Similarly, we have
$$P(X_t = x_t \mid \mathcal{Y}_t) = (1 - q_{t|t})\delta(x_t) + q_{t|t}\delta(1 - x_t) \qquad (12.106)$$
where
$$x_{t|t} = \frac{q\delta(1 - y_t)}{(1 - a_{t|t-1})(1 - q)\delta(y_t) + (a_{t|t-1} + q - a_{t|t-1}q)\delta(1 - y_t)}$$
and
$$P(X_{t+1} = x_{t+1} \mid \mathcal{Y}_t) = P(X_{t+1} = x_{t+1})$$

Suppose $a = 1/4$ and $q = 1/6$. If the sequence $\mathcal{Y}_3 = \{0, 0, 0, 1\}$ is observed, then
$$P(X_3 = 1 \mid \mathcal{Y}_3) = .4444 \qquad P(N_3 = 1 \mid \mathcal{Y}_3) = .6667.$$
If the sequence $\mathcal{Y}_3 = \{0, 1, 1, 1\}$ is observed, then
$$P(X_3 = 1 \mid \mathcal{Y}_3) = .2230 \qquad P(N_3 = 1 \mid \mathcal{Y}_3) = .9324.$$

As an interpretation, consider $P(x_3 = 1 \mid \mathcal{Y}_3)$ as the probability that a target is present (as opposed to the blocking). Comparing the first case ($P = .4444$) with the second case ($P = .2230$), there is more probability that the target is present in the first case—the sequence of ones is suggestive of the blocking.

Because the distributions in this example were chosen to be discrete, this estimation problem can be interpreted as a detection problem.

12.8 Exercises

12.1-1 Show that the means in (12.4) and (12.5) are correct.

12.1-2 Show that (12.6) is correct.

12.1-3 [291] Let $X_1, X_2, \ldots, X_m$ be a random sample, where $X_i \sim \mathcal{U}(0, \theta)$ (uniform).
  (a) Show that $\hat{\theta}_{ML} = \max X_i$.
  (b) Show that the density of $\hat{\theta}_{ML}$ is $f_\theta(x) = \frac{m}{\theta^m} x^{m-1}$.
  (c) Find the expected value of $\hat{\theta}_{ML}$.
  (d) Find the variance of $\hat{\theta}_{ML}$.

12.3-4 Justify (12.11) and show how it leads to (12.10).

12.3-5 Show that (12.52) is correct.

12.3-6 (Linear statistical model) Consider an m-dimensional Gaussian random vector $\mathbf{Y}$ with mean value $\mathbf{c}\theta$ (where $\mathbf{c}$ is a constant m-dimensional vector) and covariance matrix R (an $m \times m$ known matrix).
  (a) Show that the maximum likelihood estimate of θ is
  $$\hat{\theta} = (\mathbf{c}^T R^{-1} \mathbf{c})^{-1} \mathbf{c}^T R^{-1} \mathbf{Y}.$$
  (b) Find the mean and variance of $\hat{\theta}$.
  (c) Find the Fisher information matrix, and show that $\hat{\theta}$ is efficient.

12.3-7 Consider the system presented in exercise 12.3-6, but with R having the special form $R = \sigma^2 I$, where σ^2 is to be estimated. Show that the maximum likelihood estimators for θ and σ^2 are
$$\hat{\theta} = (\mathbf{c}^T \mathbf{c})^{-1} \mathbf{c}^T \mathbf{Y},$$
$$\hat{\sigma}^2 = (1/m)(\mathbf{Y} - \mathbf{c}\hat{\theta})^T (\mathbf{Y} - \mathbf{c}\hat{\theta}).$$

12.8 Exercises

12.3-8 (Linear statistical model) Consider N independent observations of an m-dimensional random vector $\{\mathbf{Y}_k, k \in (1, 2, \ldots, N)\}$, such that each $\mathbf{Y}_k$ has a Gaussian distribution with mean $\mathbf{c}_k \theta$ and common covariance R.

(a) Show that a necessary condition for $\hat{\theta}$ and $\hat{R}$ to be maximum-likelihood estimators of θ and R, respectively, is that they simultaneously satisfy

$$\hat{\theta} = \left[\sum_{k=1}^{N} \mathbf{c}_k^T \hat{R}^{-1} \mathbf{c}_k\right]^{-1} \sum_{k=1}^{N} \mathbf{c}_k^T \hat{R}^{-1} \mathbf{Y}_k, \tag{12.107}$$

$$\hat{R} = \frac{1}{N} \sum_{k=1}^{N} (\mathbf{Y}_k - \mathbf{c}_k \hat{\theta})(\mathbf{Y}_k - \mathbf{c}_k \hat{\theta})^T. \tag{12.108}$$

(To establish this result, you may need some of the matrix differentiation identities from appendix E.)

(b) Equations (12.107) and (12.108) do not have simple closed-form solutions. However, they can be solved by a *relaxation* algorithm as follows:

i. Pick any value of $\hat{R}$ (say $\mathbf{I}$).
ii. Solve (12.107) for $\hat{\theta}$ using $\hat{R}$.
iii. Solve (12.108) for $\hat{R}$ using $\hat{\theta}$.
iv. Stop if converged, otherwise go to (b).

Unfortunately, no proof of global convergence of the above relaxation algorithm is known. Computational studies, however, indicate that it works well in practice. What *can* be shown is that regardless of the value of $\hat{R}$, the estimate $\hat{\theta}$ given by (12.107) is an unbiased estimate of θ. Prove this fact.

12.3-9 (Another proof of the Cramér–Rao lower bound) let $\mathbf{t}(\mathbf{X})$ be an unbiased estimator of θ. From the $2p \times 1$ vector

$$\mathbf{v} = \begin{bmatrix} \mathbf{t}(\mathbf{X}) - \theta \\ \mathbf{s}(\theta, \mathbf{X}) \end{bmatrix}$$

which, since $\mathbf{t}(\mathbf{X})$ is unbiased and by (12.12), is zero-mean.

(a) Show that

$$\text{cov}(\mathbf{v}) = E \begin{bmatrix} \mathbf{t}(\mathbf{X}) - \theta \\ \mathbf{s}(\theta, \mathbf{X}) \end{bmatrix} [(\mathbf{t}(\mathbf{X}) - \theta)^T \quad \mathbf{s}^T(\theta, \mathbf{X})] = \begin{bmatrix} \text{cov}(\mathbf{t}(\mathbf{X})) & I \\ I & J(\theta) \end{bmatrix}.$$

(b) Argue that $\text{cov}(\mathbf{v})$ is positive definite.

(c) Show that there is a matrix B such that

$$B^T \text{cov}(\mathbf{v}) B = \begin{bmatrix} \text{cov}(\mathbf{t}(\mathbf{X})) - J(\theta)^{-1} & 0 \\ 0 & J(\theta) \end{bmatrix}.$$

Hence we must have that $\text{cov}(\mathbf{t}(\mathbf{X})) - J^{-1}(\theta)$ is positive semidefinite:

$$\text{cov}(\mathbf{t}(\mathbf{X})) \geq J(\theta)^{-1}.$$

12.3-10 Prove lemma 12.2.

12.3-11 If $\hat{\mathbf{m}}_{ML} = \bar{\mathbf{x}}$ in the Gaussian case, show that

$$\text{cov}(\hat{\mathbf{m}}_{ML}) = R/m.$$

12.3-12 For the Gaussian random sample $\mathbf{X}$, let $\mathbf{s}(\mathbf{m}, \mathbf{X}) = m R^{-1}(\bar{\mathbf{x}} - \mathbf{m})$, where $\bar{\mathbf{x}} = \frac{1}{m} \sum_{i=1}^{m} \mathbf{x}_i$. Compute the Fisher information matrix $J(\mathbf{m})$, both by $E \mathbf{s}(\mathbf{m}, \mathbf{X}) \mathbf{s}^T(\mathbf{m}, \mathbf{X})$ and by $-E \frac{\partial}{\partial \mathbf{m}} \mathbf{s}(\mathbf{m}, \mathbf{X})$, and show that they are the same.

12.3-13 Let $X_1, X_2, \ldots, X_m$ be a random sample, where each X_i has an exponential density with parameter θ,

$$f_X(x) = \frac{1}{\theta} e^{-x/\theta}, \qquad x \geq 0.$$

(a) Determine the ML estimate of θ.

(b) Determine $E\hat{\theta}_{ML}$. Is the estimate unbiased?

(c) Find the Fisher information matrix. Is the estimate minimum variance? efficient?

12.4-14 Show that (12.65) is correct.

12.4-15 Show that for $\mathbf{a} \in \mathbb{R}^{p+1}$, the minimum of $\mathbf{a}^T Q \mathbf{a}$, subject to the constraint that $a_{p+1} = 1$, is obtained when

$$\mathbf{a} = \frac{1}{\mathbf{e}_{p+1}^T Q^{-1} \mathbf{e}_{p+1}} Q^{-1} \mathbf{e}_{p+1},$$

where $\mathbf{e}_{p+1}$ is the unit vector with 1 in element $p+1$.

12.4-16 [291] Let R be a real circulant matrix (see section 8.5)

$$R = \begin{bmatrix} r_0 & r_1 & \cdots & r_{n-1} \\ r_{n-1} & r_0 & \cdots & r_{n-2} \\ \vdots & & & \\ r_1 & r_2 & \cdots & r_0 \end{bmatrix}.$$

Recall that DFT vectors diagonalize R:

$$R\mathbf{u}_k = \lambda_k \mathbf{u}_k$$

where

$$\sqrt{n}\mathbf{u}_k = \begin{bmatrix} 1 & e^{-j2\pi k/n} & e^{-j2\pi 2k/n} & \cdots & e^{-j2\pi(n-1)k/n} \end{bmatrix}^T$$

and

$$\lambda_k = \sum_{i=0}^{n-1} r_n e^{-j2\pi ik/n}.$$

Also, let $\mathbf{X} = (\mathbf{X}_1, \mathbf{X}_2, \ldots, \mathbf{X}_m)$ be a random sample of $\mathcal{N}(0, R)$ random variables, where R is a real, symmetric, nonnegative definite, $n \times n$ circulant matrix. Note that since R is symmetric, $\lambda_k = \lambda_{n-k}$.

(a) Find ML estimates of λ_k.

(b) Are the ML estimates unbiased?

(c) Find CR bounds on unbiased estimates of λ_k.

(d) Are the ML estimates of λ_k minimum variance? efficient?

(e) Find ML estimates of $\{r_n\}$.

(f) Are the ML estimates of r_n unbiased? minimum variance? efficient?

12.4-17 For the discrete-time signal

$$s_t = \cos(\omega t + \theta), \qquad t = 1, 2, \ldots, m$$

with known frequency and unknown phase θ, determine a maximum-likelihood estimate of θ, assuming that s_t is observed in Gaussian noise,

$$y_t = s_t + n_t,$$

where $n_t \sim \mathcal{N}(0, \sigma^2)$.

12.4-18 (Acoustic level) A vertical post has two speakers separated by a known distance d, with speaker 1 emitting $s_1(t) = \cos(\omega_c t)$ and speaker 2 emitting $s_2(t) = \cos(2\omega_c t)$. These signals travel through the air until they are picked up by a microphone (see figure 12.5). Develop

12.8 Exercises

a maximum-likelihood framework for detecting the difference in phase between the two received signals, and from that describe how to use this as a level. (Assume that the background noise is white and Gaussian.)

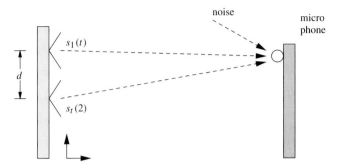

Figure 12.5: Acoustic level framework

12.6-19 Suppose that $X \sim P(\lambda_1)$ (Poisson) and $N \sim P(\lambda_2)$, independently. Let $Y = X + N$ (signal plus noise). (This might model an optical communications problem, where the received photon counts Y are modeled as the signal photon counts X plus some background photon counts N.)

(a) Find the distribution of Y.

(b) Find the conditional pmf for X given Y.

(c) Find the minimum mean-squared error estimator of X.

(d) Compute the mean and mean-squared error for your MMSE estimator. Is the estimate unbiased?

12.6-20 [291] (Imperfect Geiger counter) A radioactive source emits n radioactive particles. We assume that the particle generation is governed by a Poisson distribution with parameter λ:

$$f_N(n) = P(N = n) = \frac{\lambda^n}{n!} e^{-\lambda} \qquad n \geq 0.$$

The n particles emitted are detected by an imperfect Geiger counter, which detects with probability p. Of the n particles emitted, the imperfect Geiger counter detects $k \leq n$ of them. The problem we examine is estimating n from the measurement k, using Bayesian methods.

(a) Show that k (the number of detected particles) is conditionally distributed as

$$P[k|n] = \binom{n}{k} p^k (1-p)^{n-k}$$

(binomial distribution).

(b) Show that the joint distribution is

$$P[k, n] = \binom{n}{k} p^k (1-p)^{n-k} e^{-\lambda} \frac{\lambda^n}{n!}.$$

(c) Show that k is distributed as

$$P[k] = \frac{(\lambda p)^n}{k!} e^{-\lambda p},$$

(Poisson with parameter λp).

(d) Compute the posterior distribution $P[n \mid k]$.

(e) Show that the conditional mean (the minimum mean-square estimate) is

$$E[n|k] = k + \lambda(1 - p).$$

Also show that the conditional variance (the variance of the estimate) is

$$E[(n - E[n|k])^2|k] = \lambda(1 - p).$$

12.6-21 [291] Let $X_1, X_2, \ldots, X_n$ each be i.i.d. $P(\lambda)$ (Poisson distributed) with parameter $\theta = \lambda$,

$$f_{X|\lambda}(x \mid \lambda) = \frac{\lambda^x}{x!} e^{-\lambda}, \qquad x \in \mathbb{Z}^+, \qquad \lambda \geq 0.$$

Suppose that we have a known prior on λ that is exponential,

$$f_\lambda(\lambda) = ae^{-a\lambda}, \qquad \lambda \geq 0; \qquad a > 0.$$

(a) Show that $t = \sum_{i=1}^{n} x_i$ is sufficient for λ.
(b) Show that the marginal density for X is

$$f_X(x) = \frac{a}{\prod_{i=1}^{n} x_i!} \frac{\Gamma(t+1)}{(n+a)^{t-1}}.$$

(c) Show that the conditional (posterior) density for λ, given x, is

$$f_{\lambda|x}(\lambda \mid x) = e^{-(n+a)\lambda} \lambda^t \frac{(n+a)^{t-1}}{\Gamma(t+1)}.$$

This is a Γ density with parameters $t + 1$ and $n + a$.

(d) Show that the conditional mean (Bayes estimate) of λ is

$$\hat{\lambda} = \frac{t+1}{N+a}.$$

(e) Show that the conditional variance of $\hat{\lambda}$ is $(t+2)/(n+a)^2$.

12.6-22 Show that (12.77) is true.

12.6-23 Show that the posterior density $f_{\theta|X}(\vartheta \mid x)$ of example 12.6.4 is a $\Gamma(\alpha + y, 1/(1/\beta + m))$ density.

12.6-24 Show that the posterior density $f_{\theta|X}(\vartheta \mid x)$ of example 12.6.5 is a $\Gamma(\alpha + m, 1/(1/\beta + y))$ density.

12.6-25 Show that if $X_1 \sim \Gamma(p, \lambda)$ and $X_2 \sim \Gamma(q, \lambda)$ independently, then
(a) $Y = X_1 + X_2$ is distributed as $\Gamma(p + q, \lambda)$ (sums of gammas are gammas).
(b) $Z = X_1/(X_1 + X_2)$ is distributed as $\beta(p, q)$.

12.6-26 [291] There are other ways to consider the joint distribution model that are useful in developing intuition about the problem. In this exercise we explore some of these. In each case, $\mathbf{X}$ and $\mathbf{Y}$ are jointly distributed Gaussian random variables with mean and covariance $\boldsymbol{\mu}_x$, R_x and $\boldsymbol{\mu}_y$, R_y, respectively. They can be regarded as being generated by the diagram shown in figure 12.6(a).

(a) Show that conditioned upon measuring $\mathbf{X}$, for the random variable $\mathbf{Y}$,

$$\mathbf{Y} \mid \mathbf{X} \sim \mathcal{N}(\boldsymbol{\mu}_y + R_{yx} R_x^{-1} (\mathbf{x} - \boldsymbol{\mu}_x), Q)$$

where

$$Q = R_y - R_{yx} R_x^{-1} R_{xy}.$$

This interpretation is as shown in figure 12.6(b) for zero mean variables.

12.8 Exercises

(a) Jointly distributed **X** and **Y**

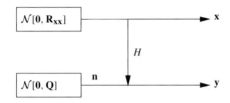

(b) Marginally distributed **X** and conditionally distributed **Y**

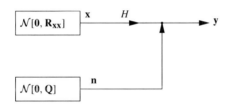

(c) Channel model: linearly transformed **X** plus noise **n**

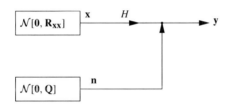

(d) Signal-plus-noise model

Figure 12.6: Equivalent representations for the Gaussian estimation problem

(b) Show that an equivalent way of generating **X** and **Y** with equivalent joint distribution is to model this as a signal-plus-noise model,

$$\mathbf{Y} = H\mathbf{X} + \mathbf{N},$$

where $H = R_{yx} R_x^{-1}$, $\mathbf{X} \sim (\boldsymbol{\mu}_x, R_x)$, and $\mathbf{N} \sim \mathcal{N}(0, Q)$. This model is illustrated in figure 12.6(c).

(c) Conditioning now on a measurement of **Y**, show that an equivalent representation for the joint distribution is as shown in figure 12.6(c), where

$$G = R_{xy} R_y^{-1}.$$

That is,

$$\tilde{X} = \boldsymbol{\mu}_x + G(\mathbf{Y} - \boldsymbol{\mu}_y) + \mathbf{N}$$

has the same distribution as **X**.

12.6-27 Show that (12.87) and (12.88) are correct.

12.7-28 Show that (12.103) is correct.

12.7-29 Write MATLAB code to compute the probabilities $P(X_t = x_t \mid \mathcal{Y}_t)$ and $P(N_t = n_t \mid \mathcal{Y}_t)$ of (12.105) and (12.106), given a sequence of observations and the initial probabilities a and q.

12.7-30 In the example of section 12.7.1, suppose that the observation equation is
$$y_t = x_t \oplus n_t,$$
where $\oplus$ represents addition modulo 2, with everything else as before. Derive the update and propagation equations in this case.

12.9 References

Maximum-likelihood estimation is discussed in [291, 349, 58, 40]. The discussion of the Cramér–Rao bound is drawn from [219]. You may contrast this development with the more conventional proofs given in [291, 349]. The proof of consistency is drawn from [115].

Our discussion of the ARMA parameter identification closely follows [291, 186], in which the observation is made that this ARMA identification is equivalent to the Steiglitz–McBride algorithm [319]. A derivation of the Fisher information matrix for this estimation problem is also provided in [291].

The problem of phase estimation is treated in a variety of sources, for example, [261]. This source also provides an excellent discussion of joint estimation and detection, such as estimating phase and discrete amplitude for digital communication. It also provides a good introduction to phase-locked loops, including a linear model of the response of the PLL.

Bayes estimation is also discussed in [291, 349, 58, 40].

Chapter 13

The Kalman Filter

> One would always like to settle oneself, get braced, say, "Now I am going to begin"—and then begin. But as the necessary quiet seems about to descend, a hand is felt at one's back, shoving. And that is the way with the river when a current is running: once the connection with the shore is broken, the journey has begun.
>
> — *Wendell Berry*
> Recollected Essays 1965–1980

The Kalman filter is a recursive estimator used to estimate the state of a linear time-varying state equation, in which the states are driven by noise and observations are made in the presence of noise. Like most significant ideas in mathematics, there is more than one way to derive the Kalman filter, each with its own criterion of optimality. In this chapter we detail two derivations. The first is based upon a recursive Bayes approach—an application of the recursive Bayes estimation of the previous chapter—and the second upon the innovations approach—building upon the principle of orthogonality explored in chapter 3. These two derivations will highlight the connection pointed out in section 12.6.4: the linear estimator we derive is identical to the Bayesian estimator derived for the Gaussian noise case. However, it is commonly used in a variety of applications in which the noise is *not* necessarily Gaussian, as being simply the optimum linear minimum mean-squared error estimator. As a homework problem, the Kalman filter is also derived using calculus-based techniques. These derivations illustrate the interplay between orthogonality, minimality, and optimality.

Following the basic development using both techniques, other practical aspects of Kalman filtering are considered, such as filtering continuous-time signals and linearizing non-linear systems. We end the chapter with an introduction to smoothing—making estimates of the state using both future and past data.

13.1 The state-space signal model

The Kalman filter is an application of the general results of sequential estimation from the previous chapter to a *discrete-time* state-variable system[1] driven by noise, with noisy observations

$$\begin{aligned} \mathbf{x}_{t+1} &= A_t \mathbf{x}_t + \mathbf{w}_t, \\ \mathbf{y}_t &= C_t \mathbf{x}_t + \boldsymbol{\nu}_t. \end{aligned} \quad (13.1)$$

---

[1] A continuous-time Kalman filter also exists, but we do not examine it here.

As indicated, the state-variable system may be time-varying.

The following assumptions are made about this system:

1. For convenience, all processes are assume to be *real*.
2. The state noise process $\mathbf{w}_t$ is zero-mean, with covariance
$$E\mathbf{w}_t\mathbf{w}_\tau^T = Q_t \delta_{t,\tau}.$$
The noise is uncorrelated among samples. For the Bayesian approach to the Kalman filter, we assume that $\mathbf{w}_t$ is Gaussian.
3. The observation noise $\boldsymbol{\nu}_t$ is zero-mean, with covariance
$$E\boldsymbol{\nu}_t\boldsymbol{\nu}_\tau^T = R_t \delta_{t,\tau}. \tag{13.2}$$
The noise is uncorrelated among samples. For the Bayesian approach, we assume that $\boldsymbol{\nu}_t$ is Gaussian.
4. Initially, we assume that the state noise and the observation noise have correlation
$$E\mathbf{w}_t\boldsymbol{\nu}_\tau^T = M_t \delta_{t,\tau} \qquad \forall t, \tau.$$
For the first derivation we assume that the state and observation noise are uncorrelated: $M_t \equiv 0$ for all t. We subsequently lift this restriction.
5. There is some *initial condition*, or *a priori* density, on the random variable $\mathbf{x}_0$ with mean $\boldsymbol{\mu}_x[0]$ and covariance Π_0. Again, for the Bayesian case we assume $\mathbf{x}_0$ is Gaussian. The derivations are simplified notationally if it is assumed that $\boldsymbol{\mu}_x[0] = \mathbf{0}$.

The Kalman filtering problem can be stated as follows: given a sequence of measurements $\mathbf{y}_0, \mathbf{y}_1, \ldots$, determine a sequence of estimates of the state of the system $\mathbf{x}_t$ in a computationally feasible, recursive manner.

13.2 Kalman filter I: The Bayes approach

We first derive the Kalman filter from the Bayesian point of view. This is a natural outgrowth of the results in section 12.7. For the Bayesian approach, we assume that the noise processes are Gaussian distributed. Then the Bayes estimate of $\mathbf{x}_t$ amounts to finding the conditional mean of $\mathbf{x}_t$, given the observations.

The key equations in the Bayes derivation are the propagate step (see (12.98)),
$$f(\mathbf{x}_{t+1} \mid \mathcal{Y}_t) = \int f(\mathbf{x}_{t+1} \mid \mathbf{x}_t) f(\mathbf{x}_t \mid \mathcal{Y}_t) \, d\mathbf{x}_t, \tag{13.3}$$
from which the estimate is propagated using the state update equation into the future; and
$$f(\mathbf{x}_{t+1} \mid \mathcal{Y}_{t+1}) = \frac{f(\mathbf{y}_{t+1} \mid \mathbf{x}_{t+1})}{f(\mathbf{y}_{t+1} \mid \mathcal{Y}_t)} f(\mathbf{x}_{t+1} \mid \mathcal{Y}_t), \tag{13.4}$$
which is the measurement update step from (12.97).

We will begin by finding explicit formulas for the propagate step in (13.3).

1. The density $f(\mathbf{x}_t \mid \mathcal{Y}_t)$ corresponds to the estimate of $\mathbf{x}_t$, given measurements up to time t. Under the assumption of Gaussian noise and using the notation just introduced, the random variable $\mathbf{x}_t$ conditioned upon $\mathcal{Y}_t$ is Gaussian,
$$\mathbf{x}_t \mid \mathcal{Y}_t \sim \mathcal{N}(\hat{\mathbf{x}}_{t|t}, P_{t|t}). \tag{13.5}$$

13.2 Kalman Filter I: The Bayes Approach

2. The density $f(\mathbf{x}_{t+1} \mid \mathbf{x}_t)$ is obtained by noting from (13.1) that, conditioned upon $\mathbf{x}_t$, $\mathbf{x}_{t+1}$ is distributed as

$$\mathbf{x}_{t+1} \mid \mathbf{x}_t \sim \mathcal{N}(A_t \mathbf{x}_t, Q_t). \tag{13.6}$$

Inserting (13.5) and (13.6) into (13.3) and performing the integration (which involves expanding and completing the square), we find that $\mathbf{x}_{t+1} \mid \mathcal{Y}_t$ is Gaussian, with mean

$$\hat{\mathbf{x}}_{t+1 \mid t} = A_t \hat{\mathbf{x}}_{t \mid t} \tag{13.7}$$

and covariance

$$P_{t+1 \mid t} = A_t P_{t \mid t} A_t^T + Q_t. \tag{13.8}$$

Equation (13.7) provides a means to propagate the estimate ahead in time, in the absence of measurements, and (13.8) shows that, without measurements, the estimate covariance grows in time.

Let us now examine the update step in (13.4). This is a Bayes update of a Gaussian random variable. The mean of $\mathbf{x}_{t+1} \mid \mathcal{Y}_{t+1}$ is obtained analogous to (12.84), in which the mean of the prior is updated:

$$\hat{\mathbf{x}}_{t+1 \mid t+1} = E[\mathbf{x}_{t+1} \mid \mathcal{Y}_t] = \hat{\mathbf{x}}_{t+1 \mid t} + R_{xy, \mathcal{Y}_t} R_{yy, \mathcal{Y}_t}^{-1} (\mathbf{y}_{t+1} - E[\mathbf{y}_{t+1} \mid \mathcal{Y}_t]). \tag{13.9}$$

We now examine the pertinent components of this mean value.

1. The notation $R_{xy, \mathcal{Y}_t}$ is used to denote the correlation, conditioned upon $\mathcal{Y}_t$:

$$R_{xy, \mathcal{Y}_t} = E[(\mathbf{x}_{t+1} - E[\mathbf{x}_{t+1}])(\mathbf{y}_{t+1} - E[\mathbf{y}_{t+1}])^T \mid \mathcal{Y}_t].$$

Then we have

$$R_{xy, \mathcal{Y}_t} = E[(\mathbf{x}_{t+1} - \hat{\mathbf{x}}_{t+1 \mid t})(C_{t+1}(\mathbf{x}_{t+1} - \hat{\mathbf{x}}_{t+1 \mid t}) + \boldsymbol{\nu}_{t+1})^T \mid \mathcal{Y}_t]$$
$$= P_{t+1 \mid t} C^T \tag{13.10}$$

by the definition of $P_{t+1 \mid t+1}$ in (12.94).

2. The notation $R_{yy, \mathcal{Y}_t}$ denotes the covariance of $\mathbf{y}_t$, conditioned upon $\mathcal{Y}_t$:

$$R_{yy, \mathcal{Y}_t} = E[(\mathbf{y}_{t+1} - E[\mathbf{y}_{t+1}])(\mathbf{y}_{t+1} - E[\mathbf{y}_{t+1}])^T \mid \mathcal{Y}_t] \tag{13.11}$$
$$= E[(C_{t+1}(\mathbf{x}_{t+1} - \hat{\mathbf{x}}_{t+1 \mid t}) + \boldsymbol{\nu}_{t+1})(C_{t+1}(\mathbf{x}_{t+1} - \hat{\mathbf{x}}_{t+1 \mid t}) + \boldsymbol{\nu}_{t+1})^T \mid \mathcal{Y}_t]$$
$$= C_{t+1} P_{t+1 \mid t} C_{t+1}^T + R_{t+1} \tag{13.12}$$

3. The mean $E[\mathbf{y}_{t+1} \mid \mathcal{Y}_t]$ is equal to $C_{t+1} \hat{\mathbf{x}}_{t+1 \mid t}$.

Putting these pieces together, we have the following update step:

$$\hat{\mathbf{x}}_{t+1 \mid t+1} = \hat{\mathbf{x}}_{t+1 \mid t} + P_{t+1 \mid t} C_{t+1}^T \left(C_{t+1} P_{t+1 \mid t} C_{t+1}^T + R_{t+1} \right)^{-1} (\mathbf{y}_{t+1} - C_{t+1} \hat{\mathbf{x}}_{t+1 \mid t}). \tag{13.13}$$

It will be convenient to let

$$K_{t+1} = P_{t+1 \mid t} C_{t+1}^T \left(C_{t+1} P_{t+1 \mid t} C_{t+1}^T + R_{t+1} \right)^{-1} \tag{13.14}$$

so that the mean update can be written as

$$\hat{\mathbf{x}}_{t+1 \mid t+1} = \hat{\mathbf{x}}_{t+1 \mid t} + K_{t+1}(\mathbf{y}_{t+1} - C_{t+1} \hat{\mathbf{x}}_{t+1 \mid t}). \tag{13.15}$$

The quantity K_t is called the *Kalman gain*.

Let us now consider the covariance of $\mathbf{x}_{t+1} \mid \mathcal{Y}_{t+1}$, which is the variance of the estimator error $\tilde{\mathbf{x}}_{t+1 \mid t+1} = \hat{\mathbf{x}}_{t+1} - \hat{\mathbf{x}}_{t+1 \mid t+1}$. This covariance can be found by identification from the conditional Gaussian model examined in example 4.12.1. In that example, we found that the conditional density $\mathbf{X} \mid \mathbf{Y}$ had covariance

$$\operatorname{cov}(\mathbf{X} \mid \mathbf{Y}) = R_{xx} - R_{xy} R_{yy}^{-1} R_{yx}. \tag{13.16}$$

To apply this result, we identity the random variable $\mathbf{X}$ in (13.16) with the estimate $\hat{\mathbf{x}}_{t+1|\mathcal{Y}_t}$; the observation $\mathbf{Y}$ with the observation $\mathbf{y}_{t+1}$. The covariance R_{xx} of (13.16) is thus analogous to $P_{t+1|t}$. The matrix R_{xy} is analogous to $R_{xy,\mathcal{Y}}$, and R_{yy} is analogous to $R_{yy,\mathcal{Y}_t}$. Substituting from (13.10) and (13.12), we have

$$\begin{aligned} P_{t+1|t+1} &= P_{t+1|t} - P_{t+1|t} C_{t+1}^T \left(C_{t+1} P_{t+1|t} C_{t+1}^T + R_{t+1} \right)^{-1} C_{t+1} P_{t+1|t} \\ &= \left(I - K_{t+1} C_{t+1}^T \right) P_{t+1|t}. \end{aligned} \tag{13.17}$$

This completes the derivation of the Kalman filter. In summary we have the following: Starting from an initial estimate $\hat{\mathbf{x}}_{0|-1}$, with an initial covariance of $P_{-1|-1}$, for each observation $\mathbf{y}_t$, $t = 0, 1, \ldots$, the estimate of the state is updated using the following steps:

1. **State estimate extrapolation:** $\hat{\mathbf{x}}_{t+1|t} = A_t \hat{\mathbf{x}}_{t|t}$.
2. **Error covariance extrapolation:** $P_{t+1|t} = A_t P_{t|t} A_t^T + Q_t$.
3. **Kalman gain:** $K_{t+1} = P_{t+1|t} C_{t+1}^T \left(C_{t+1} P_{t+1|t} C_{t+1}^T + R_{t+1} \right)^{-1}$.
4. **State estimate update:** $\hat{\mathbf{x}}_{t+1|t+1} = \hat{\mathbf{x}}_{t+1|t} + K_{t+1}(\mathbf{y}_{t+1} - C_{t+1} \hat{\mathbf{x}}_{t+1|t})$.
5. **Error covariance update:** $P_{t+1|t+1} = (I - K_{t+1} C_{t+1}) P_{t+1|t}$.

In the interest of reducing computations, the Kalman gain and error covariance update are sometimes written (see exercise 13.2-4) as

$$K_{t+1} = P_{t+1|t+1} C_{t+1}^T R_{t+1}^{-1}, \tag{13.18}$$
$$P_{t+1|t+1}^{-1} = P_{t+1|t}^{-1} + C_{t+1}^T R_{t+1}^{-1} C_{t+1}. \tag{13.19}$$

Algorithm 13.1 illustrates an implementation of the steps we have listed.

Algorithm 13.1 Kalman filter I
File: `kalman1.m`

Example 13.2.1 In our first example of Kalman filtering, we will take a simple setup, explaining subsequently (in section 13.5) our rationale for the structure of the example. We consider simple kinematic motion in two coordinates, $(x_{1,t}, x_{2,t})$, with a motion update according to

$$x_{i,t+1} = x_{i,t} + \Delta \dot{x}_{i,t} + w_{i,t},$$

where $\dot{x}_{i,t}$ is the velocity in the ith direction and Δ represents some sampling time. We also assume that the velocity is subject to random fluctuations,

$$\dot{x}_{i,t+1} = \dot{x}_{i,t} + \dot{w}_t.$$

Stacking up the state vector as $\mathbf{x}_t = [x_{1,t}, \dot{x}_{1,t}, x_{2,t}, \dot{x}_{2,t}]^T$, we have the state equation

$$\mathbf{x}_{t+1} = \begin{bmatrix} 1 & \Delta & 0 & 0 \\ 0 & 0 & 1 & \Delta \\ 0 & 1 & 0 & 0 \\ 0 & 0 & 0 & 1 \end{bmatrix} \mathbf{x}_t + \mathbf{w}_t,$$

13.3 Kalman Filter II: The Innovations Approach

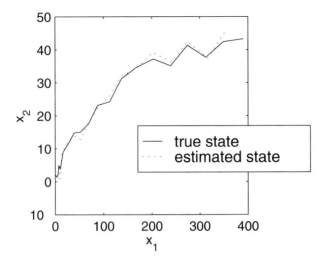

Figure 13.1: Illustration of Kalman filter

where $\mathbf{w}_t$ is zero-mean and Gaussian. For this example, the covariance is chosen to be

$$\text{cov}(\mathbf{w}_t) = Q = 2I.$$

We assume, simply, that we observe the position variables in noise,

$$\mathbf{y}_t = \begin{bmatrix} 1 & 0 & 0 & 0 \\ 0 & 1 & 0 & 0 \end{bmatrix} \mathbf{x}_t + \boldsymbol{\nu}_t,$$

where $\boldsymbol{\nu}_t$ is zero-mean, Gaussian, and with covariance (for this example)

$$\text{cov}(\boldsymbol{\nu}_t) = R = 3I.$$

The code in algorithm 13.2 illustrates the simulation of this dynamical system and the estimate of its state using a Kalman filter based upon the observation $\mathbf{y}_t$. Figure 13.1 illustrates the tracking ability of the filter, where the position variables of the true state and the estimated state are shown. □

Algorithm 13.2 Kalman filter example
File: `kalex1.m`

13.3 Kalman filter II: The innovations approach

In this section we obtained the Kalman filter based upon the minimum mean-squared error principle obtained in chapter 3, as embodied by the orthogonality principle. As we have observed, the best estimator of the state $\mathbf{x}_t$ would be the conditional expectation $E[\mathbf{x}_t \mid \mathcal{Y}_t]$. However, unless the noise is Gaussian, this conditional expectation may not be a linear function of the observations, and hence may be intractable. We therefore restrict our attention to *linear* minimum mean-squared error filters, imposing the structure that the estimate must be a linear function of the observations. Our problem then becomes one of determining the coefficients.

As before, our intent is to develop a *recursive* formulation of the estimate. To do this, we will introduce the notion of innovations, which are a sequence of orthogonal data derived from the observations, which span the same space as the observations. The orthogonality of the innovations makes the recursive update straightforward.

13.3.1 Innovations for processes with linear observation models

We introduce the innovations concept by assuming that there is a random variable $\mathbf{x}$ that we wish to estimate using a linear combination of the vectors in the sequence of vectors $\mathcal{Y}_t = \{\mathbf{y}_0, \mathbf{y}_1, \ldots, \mathbf{y}_t\}$. For now, $\mathbf{x}$ is a fixed vector—we will include the state update concept further on. We assume that $\mathbf{y}_t$ is an observation of the form in (13.1), that is,

$$\mathbf{y}_t = C_t \mathbf{x} + \boldsymbol{\nu}_t, \tag{13.20}$$

where $\boldsymbol{\nu}_t$ is an uncorrelated noise sequence.

We denote the estimate of $\mathbf{x}$ using $\mathcal{Y}_t$ as $\mathbf{x}_{|t}$, and restrict our attention to *linear* estimators, so that

$$\hat{\mathbf{x}}_{|t} = \sum_{i=0}^{t} a_i \mathbf{y}_i$$

for some set of coefficients $\{a_i\}$. Clearly, the estimate $\hat{\mathbf{x}}_{|t}$ must lie in span($\mathcal{Y}_t$). If we select the coefficients on the basis of minimizing the average squared length of the error vector—the minimum mean-squared error criterion—then the error $\tilde{\mathbf{x}} = \mathbf{x} - \hat{\mathbf{x}}_{t|t}$ must be orthogonal to the data, where the orthogonality in this case is naturally defined using the expectation:

$$E(\mathbf{x} - \hat{\mathbf{x}}_{t|})\mathbf{y}_i^T = 0, \qquad i = 0, 1, \ldots, t.$$

On the basis of these equations, we could set up a system of equations to find the coefficients. However, as we observed in chapter 3, determining coefficients in linear representations is easier if the vectors comprising the space are orthogonal. Thus we seek a sequence of *orthogonal* vectors $\{\epsilon_0, \epsilon_1, \ldots, \epsilon_t\}$ that collectively span the same space as $\mathcal{Y}_t$:

$$\text{span}(\epsilon_0, \epsilon_1, \ldots, \epsilon_t) = \text{span}(\mathbf{y}_0, \mathbf{y}_1, \ldots, \mathbf{y}_t).$$

The vectors ϵ_i which have this property are said to be **innovations** of $\mathbf{y}_i$, and the random process $\{\epsilon_i\}$ is said to be an innovations process (or innovations sequence).

The innovations sequence can be determined by means of the Gram–Schmidt process (section 2.15). In the current notation we can write this as follows:

1. Set $\epsilon_0 = \mathbf{y}_0$.
2. Subtract from $\mathbf{y}_1$ its projection onto the space spanned by ϵ_0:

$$\epsilon_1 = \mathbf{y}_1 - \left\langle \mathbf{y}_1, \frac{\epsilon_0}{\|\epsilon_0\|} \right\rangle \frac{\epsilon_0}{\|\epsilon_0\|}.$$

3. Iterate, subtracting from $\mathbf{y}_t$ its projection onto the space spanned by $\mathcal{Y}_{t-1}$:

$$\epsilon_t = \mathbf{y}_t - \sum_{j=0}^{t-1} \langle \mathbf{y}_t, \epsilon_j \rangle \frac{\epsilon_j}{\|\epsilon_j\|^2}.$$

Let us denote the projection of $\mathbf{y}_t$ onto the space spanned by $\mathcal{Y}_{t-1}$ as $\hat{\mathbf{y}}_{|t-1}$. Then we have

$$\epsilon_t = \mathbf{y}_t - \hat{\mathbf{y}}_{|t-1}.$$

We have similarly that $\hat{\mathbf{x}}_{|t-1}$ and $\hat{\boldsymbol{\nu}}_{|t-1}$ are the projection of $\mathbf{x}_t$ and $\boldsymbol{\nu}_t$, respectively, onto $\mathcal{Y}_{t-1}$.

13.3 Kalman Filter II: The Innovations Approach

If we assume an observation model of the form (13.20), specifically

$$\mathbf{y} = C\mathbf{x} + \nu,$$

then, by projecting onto $\mathcal{Y}_{t-1}$, we have

$$\hat{\mathbf{y}}_{|t-1} = C_t \hat{\mathbf{x}}_{|t-1} + \hat{\nu}_{|t-1}.$$

But since $\hat{\nu}_{|t-1}$ is a linear combination of the elements in $\mathcal{Y}_{t-1}$, and the noise ν_t is uncorrelated with (orthogonal to) previous samples, $\hat{\nu}_{|t-1} = 0$. Thus we have

$$\hat{\mathbf{y}}_{|t-1} = C_i \hat{\mathbf{x}}_{|t-1}.$$

Let

$$\mathcal{E}_t = \{\epsilon_0, \epsilon_1, \ldots, \epsilon_t\}.$$

Then by construction, $\text{span}(\mathcal{E}_t) = \text{span}(\mathcal{Y}_t)$. Also,

$$\text{cov}(\epsilon_i, \epsilon_j) = \begin{cases} R_{\epsilon_i \epsilon_i} & i = j, \\ 0 & i \neq j, \end{cases}$$

since the vectors are orthogonal.

For the purposes of linear estimation, the innovations process contains *exactly* the same information as the initial process $\{\mathbf{y}_t\}$. It may seem counterintuitive that a white-noise process could contain information, but because of our construction—relying upon the Gram–Schmidt orthogonalization procedure—the computation of the innovations process from the original process is both causal and causally invertible: it is an invertible mapping, with no extra ("future") data required, and must convey the same information when inverted.

13.3.2 Estimation using the innovations process

The minimum linear mean-squared error estimate of $\mathbf{x}$ using $\mathcal{Y}_t = \mathcal{E}_t$, which we denote as $\mathbf{x}_{|t}$, is

$$\hat{\mathbf{x}}_{|t} = \sum_{i=0}^{t} a_i \mathbf{y}_i$$

for some set of coefficients $\{a_i\}$. Since $\mathcal{Y}_t$ and $\mathcal{E}_t$ span the same space, there is a set of coefficients $\{b_i\}$ such that

$$\hat{\mathbf{x}}_{|t} = \sum_{i=0}^{t} b_i \epsilon_i,$$

where, by the orthogonality of the basis functions,

$$b_i = R_{x, \epsilon_i} R_{\epsilon_i, \epsilon_i}^{-1} \epsilon_i,$$

where $R_{x, \epsilon_i} = E\mathbf{x}\epsilon_i^T$.

By means of the innovations structure, recursive updates to the estimate $\mathbf{x}_{|t}$ are straightforward. If an additional data sample $\mathbf{y}_{t+1}$ becomes available, then

$$\hat{\mathbf{x}}_{|t+1} = \sum_{i=0}^{t+1} b_i \epsilon_i$$

$$= \hat{\mathbf{x}}_{|t} + \text{estimate of } \mathbf{x} \text{ given } \epsilon_{t+1}$$

$$= \hat{\mathbf{x}}_{|t} + R_{x, \epsilon_{t+1}} R_{\epsilon_{t+1} \epsilon_{t+1}}^{-1} \epsilon_{t+1}, \qquad (13.21)$$

where

$$\epsilon_{t+1} = \mathbf{y}_{t+1} - C_t \hat{\mathbf{x}}_{|t}.$$

13.3.3 Innovations for processes with state-space models

Now let us return to the problem of estimating $\mathbf{x}_t$, where $\mathbf{x}_t$ has the state-update equation (13.1). Our first step is to project $\mathbf{x}_{t+1}$ onto span($\mathcal{E}_t$). We thus obtain

$$\hat{\mathbf{x}}_{t+1|t} = A_t \hat{\mathbf{x}}_{t|t} + \hat{\mathbf{w}}_{t|t}.$$

The projection $\hat{\mathbf{w}}_{t|t}$ is obtained from

$$\hat{\mathbf{w}}_{t|t} = \sum_{j=0}^{t} R_{w_t, \epsilon_j} R_{\epsilon_j \epsilon_j}^{-1} \epsilon_j$$

$$= \sum_{j=0}^{t} E \mathbf{w}_t \epsilon_j^T R_{\epsilon_j \epsilon_j}^{-1} \epsilon_j.$$

But

$$E \mathbf{w}_t \epsilon_j^T = E \mathbf{w}_t (C_j \mathbf{x}_j + \boldsymbol{\nu}_j - C_j \hat{\mathbf{x}}_{j|j-1})^T = M_t \delta_{jt}$$

by the definition of the innovations, and because

$$E \mathbf{w}_t \mathbf{x}_j = 0 \quad \text{for } j \leq t,$$
$$E \mathbf{w}_t \boldsymbol{\nu}_j^T = M_t \delta_{tj};$$

hence,

$$\hat{\mathbf{w}}_{t|t} = M_t R_{\epsilon_t \epsilon_t}^{-1} \epsilon_t,$$

so that

$$\hat{\mathbf{x}}_{t+1|t} = A_t \hat{\mathbf{x}}_{t|t} + M_t R_{\epsilon_t \epsilon_t}^{-1} \epsilon_t. \tag{13.22}$$

This is the "propagate step." (Equation 13.22 should be compared with (13.7); the difference is due to the fact that here we have assumed that there is some correlation between the state and observation noise, whereas previously the correlation was assumed to be zero.)

Let us now incorporate a new measurement $\mathbf{y}_{t+1}$ into the estimate using the innovations update formula (13.21), where the random variable $\mathbf{x}$ is interpreted as $\mathbf{x}_{t+1}$, giving

$$\hat{\mathbf{x}}_{t+1|t+1} = \hat{\mathbf{x}}_{t+1|t} + E\left[\mathbf{x}_{t+1} \epsilon_{t+1}^T\right] R_{\epsilon_{t+1} \epsilon_{t+1}}^{-1} \epsilon_{t+1}. \tag{13.23}$$

Equations (13.22) and (13.23) constitute the heart of the Kalman filter. It is interesting to note the (relative) ease with which we reached this point. For the Bayesian approach, the optimal estimator with the squared-error cost function led to conditional estimation, which can be very tricky. In the present development, the need for conditional expectation has been eliminated. In essence, by projecting all of the computations onto the space of $\mathcal{Y}_t$ (or, equivalently, $\mathcal{E}_t$), all of the "conditioning" takes place automatically. This is a potent idea: When estimating with Gaussian random variables, *conditional expectation is projection onto the space of the variables that do the conditioning*.

What remains in our development of the Kalman filter is to determine explicit, recursive representations for the expectations appearing in these formulas,

$$R_{\epsilon_{t+1} \epsilon_{t+1}} \quad \text{and} \quad E\left[\mathbf{x}_{t+1} \epsilon_{t+1}^T\right].$$

To find these expectations, it will be useful to introduce some new notation. Let

$$\tilde{\mathbf{x}}_{t|t-1} = \mathbf{x}_t - \hat{\mathbf{x}}_{t|t-1}$$

be the *predicted state-estimation error*. The covariance of this random sequence, denoted

$$P_{t|t-1} = E \tilde{\mathbf{x}}_{t|t-1} \tilde{\mathbf{x}}_{t|t-1}^T,$$

13.3 Kalman Filter II: The Innovations Approach

is called the predicted state-estimation error covariance, or more briefly the *estimation error covariance matrix*.

The innovation ϵ_t can be written as

$$\epsilon_t = \mathbf{y}_t - C_t \hat{\mathbf{x}}_{t|t-1} = C_t \mathbf{x}_t + \boldsymbol{\nu}_t - C_t \hat{\mathbf{x}}_{t|t-1}$$
$$= C_t \tilde{\mathbf{x}}_{t|t-1} + \boldsymbol{\nu}_t. \tag{13.24}$$

Then we can express

$$R_{\epsilon_t \epsilon_t} = E\epsilon_t \epsilon_t^T = C_t \underbrace{E\tilde{\mathbf{x}}_{t|t-1}\tilde{\mathbf{x}}_{t|t-1}^T}_{P_{t|t-1}} C_t^T + C_t E\tilde{\mathbf{x}}_{t|t-1}\boldsymbol{\nu}_t^T + E\boldsymbol{\nu}_t \tilde{\mathbf{x}}_{t|t-1}^T C_t^T + \underbrace{E\boldsymbol{\nu}_t \boldsymbol{\nu}_t^T}_{R_t}.$$

But

$$E\tilde{\mathbf{x}}_{t|t-1}\boldsymbol{\nu}_t^T = \mathbf{0},$$

since $\boldsymbol{\nu}_t$ is orthogonal to both $\mathbf{x}_t$ and $\hat{\mathbf{x}}_{t|t-1}$. Consequently,

$$R_{\epsilon_t,\epsilon_t} = C_t P_{t|t-1} C_t^T + R_t. \tag{13.25}$$

Also,

$$E\mathbf{x}_t \epsilon_t^T = E\mathbf{x}_t \left(\tilde{\mathbf{x}}_{t|t-1}^T C_t^T + \boldsymbol{\nu}_t^T \right)$$
$$= E\mathbf{x}_t \tilde{\mathbf{x}}_{t|t-1}^T C_t^T + \underbrace{E\mathbf{x}_t \boldsymbol{\nu}_t^T}_{0}$$
$$= E[\hat{\mathbf{x}}_{t|t-1} + \tilde{\mathbf{x}}_{t|t-1}]\tilde{\mathbf{x}}_{t|t-1}^T C_t^T$$
$$= \underbrace{E\hat{\mathbf{x}}_{t|t-1}\tilde{\mathbf{x}}_{t|t-1}^T}_{0} C_t^T + \underbrace{E\tilde{\mathbf{x}}_{t|t-1}\tilde{\mathbf{x}}_{t|t-1}^T}_{P_{t|t-1}} C_t^T.$$

Thus,

$$E\mathbf{x}_t \epsilon_t^T = P_{t|t-1} C_t^T. \tag{13.26}$$

The remaining step in the derivation is to find a recursive expression for $P_{t|t-1}$, on which both (13.25) (13.26) depend.

13.3.4 A recursion for $P_{t|t-1}$

To find $P_{t|t-1}$, the estimation error covariance matrix, it will be convenient to introduce two more covariance matrices. Let Π_t denote the covariance of $\mathbf{x}_t$:

$$\Pi_t = E\mathbf{x}_t \mathbf{x}_t^T.$$

Let $\Sigma_{t|t-1}$ denote the covariance of $\hat{\mathbf{x}}_{t|t-1}$:

$$\Sigma_{t|t-1} = E\hat{\mathbf{x}}_{t|t-1}\hat{\mathbf{x}}_{t|t-1}^T.$$

Now observe that, since $\hat{\mathbf{x}}_{t|t-1}$ and $\tilde{\mathbf{x}}_{t|t-1}$ are orthogonal, we have an *orthogonal decomposition* of $\mathbf{x}_t$:

$$\mathbf{x}_t = \hat{\mathbf{x}}_{t|t-1} + \tilde{\mathbf{x}}_{t|t-1}.$$

Consequently, taking the variance of both sides of this expression (assuming all random variables are zero-mean), we obtain

$$\Pi_t = E\mathbf{x}_t \mathbf{x}_t^T = E[\hat{\mathbf{x}}_{t|t-1} + \tilde{\mathbf{x}}_{t|t-1}][\hat{\mathbf{x}}_{t|t-1} + \tilde{\mathbf{x}}_{t|t-1}]^T$$
$$= \underbrace{E\hat{\mathbf{x}}_{t|t-1}\hat{\mathbf{x}}_{t|t-1}^T}_{\Sigma_{t|t-1}} + \underbrace{E\hat{\mathbf{x}}_{t|t-1}\tilde{\mathbf{x}}_{t|t-1}^T}_{0} + \underbrace{E\tilde{\mathbf{x}}_{t|t-1}\hat{\mathbf{x}}_{t|t-1}^T}_{0} + \underbrace{E\tilde{\mathbf{x}}_{t|t-1}\tilde{\mathbf{x}}_{t|t-1}^T}_{P_{t|t-1}},$$

or,

$$\Pi_t = \Sigma_{t|t-1} + P_{t|t-1}. \tag{13.27}$$

Taking the initial condition $\Sigma_{0|-1} = \mathbf{0}$, we have $P_{0|-1} = \Pi_0$. Equation (13.27) can be rearranged as

$$P_{t|t-1} = \Pi_t - \Sigma_{t|t-1},$$

or, replacing $i \to i + 1$,

$$P_{t+1|t} = \Pi_{t+1} - \Sigma_{t+1|t}.$$

We now find update formulas for Π_{t+1} and $\Sigma_{t+1|t}$.

- Π_{t+1}:

$$\Pi_{t+1} = E\mathbf{x}_{t+1}\mathbf{x}_{t+1}^T = E\left[A_t\mathbf{x}_t + \mathbf{w}_t\right]\left[A_t\mathbf{x}_t + \mathbf{w}_t\right]^T$$
$$= A_t \underbrace{E\mathbf{x}_t\mathbf{x}_t^T}_{\Pi_t} A_t^T + A_t E\mathbf{x}_t\mathbf{w}_t^T + E\mathbf{w}_t\mathbf{x}_t^T A_t^T + \underbrace{E\mathbf{w}_t\mathbf{w}_t^T}_{Q_t}.$$

Because, by the modeling assumptions, we have $E\mathbf{x}_t\mathbf{w}_t^T = \mathbf{0}$,

$$\Pi_{t+1} = A_t \Pi_t A_t^T + Q_t, \quad \Pi_0 \text{ given.} \tag{13.28}$$

- $\Sigma_{t+1|t}$ (using (13.22)):

$$\Sigma_{t+1|t} = E\hat{\mathbf{x}}_{t+1|t}\hat{\mathbf{x}}_{t+1|t}^T = E(A_t\hat{\mathbf{x}}_{t|t} + M_t R_{\epsilon_t\epsilon_t}^{-1}\epsilon_t)(A_t\hat{\mathbf{x}}_{t|t} + M_t R_{\epsilon_t\epsilon_t}^{-1}\epsilon_t)^T$$
$$= A_t E\hat{\mathbf{x}}_{t|t}\hat{\mathbf{x}}_{t|t}^T A_t^T + A_t E\hat{\mathbf{x}}_{t|t}\epsilon_t^T R_{\epsilon_t\epsilon_t}^{-1} M_t + M_t R_{\epsilon_t\epsilon_t}^{-1} E\epsilon_t\hat{\mathbf{x}}_{t|t}^T + M_t R_{\epsilon_t\epsilon_t}^{-1} M_t.$$

In this expression, $E[\hat{\mathbf{x}}_{t|t}\hat{\mathbf{x}}_{t|t}^T]$ must be found. Using (13.23), we find

$$E\hat{\mathbf{x}}_{t|t}\hat{\mathbf{x}}_{t|t}^T = \Sigma_{t|t-1} + E\left[\mathbf{x}_t\epsilon_t^T\right] R_{\epsilon_t\epsilon_t}^{-1} E\left[\epsilon_t\mathbf{x}_t^T\right]. \tag{13.29}$$

Substitution of (13.29) into (13.3.4) and rearrangement gives

$$\Sigma_{t+1|t} = A_t \Sigma_{t|t-1} A_t^T + \left(A_t E\left[\mathbf{x}_t\epsilon_t^T\right] + M_t\right) R_{\epsilon_t\epsilon_t}^{-1} \left(A_t E\left[\mathbf{x}_t\epsilon_t^T\right] + M_t\right)^T. \tag{13.30}$$

From (13.3.4), and applying (13.28) and (13.30), we have

$$P_{t+1|t} = \Pi_{t+1} - \Sigma_{t+1|t}$$
$$= A_t \Pi_t A_t^T + Q_t - A_t \Sigma_{t|t-1} A_t^T - W_t R_{\epsilon_t\epsilon_t}^{-1} W_t^T, \tag{13.31}$$

where we define

$$W_t = A_t P_{t|t-1} C_t^T + M_t. \tag{13.32}$$

Substituting (13.32), (13.25), and (13.26) into (13.31), we obtain

$$P_{t+1|t} = A_t P_{t|t-1} A_t^T + Q_t$$
$$- \left[A_t P_{t|t-1} C_t^T + M_t\right]\left[C_t P_{t|t-1} C_t^T + R_t\right]^{-1}\left[A_t P_{t|t-1} C_t^T + M_t\right]^T \tag{13.33}$$

with

$$P_{0|-1} = \Pi_0. \tag{13.34}$$

Equation (13.33) is known as a matrix Riccati difference equation, after the mathematician who first analyzed nonlinear differential equations of similar form. This difference equation is nonlinear, but can be solved easily by recursive means.

13.3.5 The discrete-time Kalman filter

With the development of the matrix Riccati equation (13.33), we have completed the steps needed for the celebrated Kalman filter. We present one method—the time-update measurement form—leaving the presentation of another form for the exercises.

Since both the state estimate and the associated error covariance need to be updated, we derive time- and measurement-update equations for both of these quantities. First, we consider the time-update equation for the state. Substitution of (13.25) into (13.22) yields the time-update equation:

$$\hat{\mathbf{x}}_{t+1|t} = A_t \hat{\mathbf{x}}_{t|t} + M_t \left[C_t P_{t|t-1} C_t^T + R_t \right]^{-1} \underbrace{[\mathbf{y}_t - C_t \hat{\mathbf{x}}_{t|t-1}]}_{\epsilon_t}. \quad (13.35)$$

Also, substitution of (13.25) and (13.26) into (13.23) yields the measurement-update equation:

$$\hat{\mathbf{x}}_{t+1|t+1} = A_t \hat{\mathbf{x}}_{t+1|t} + P_{t+1|t} C_{t+1}^T \left[C_{t+1} P_{t+1|t} C_{t+1}^T + R_{t+1} \right]^{-1} \underbrace{[\mathbf{y}_{t+1} - C_{t+1} \hat{\mathbf{x}}_{t+1|t}]}_{\epsilon_{t+1}}. \quad (13.36)$$

The covariance matrix, $P_{t|t-1}$, is obtained via (13.33) and (13.34).

These equations for the Kalman filter should be compared with those of (13.7) and (13.13). There are two principle differences: the time update (13.35) depends upon the innovation ϵ_i, and the measurement update (13.36) must have the covariance matrix $P_{i+1|i}$ computed as the solution to the matrix Ricatti equation (13.33). Both of these differences are due to an assumption that was made in the innovations approach that was not made in the previous Bayes approach. It was a assumed that the measurement noise and the state-update noise are correlated. Removing this assumption by setting $M_t \equiv \mathbf{0}$ will simplify our expressions. Let us see the result in this case.

We already have introduced the estimation error covariance matrix, $P_{t|t-1} = E\tilde{\mathbf{x}}_{t|t-1}\tilde{\mathbf{x}}_{t|t-1}^T$. What we also need to develop is an expression for the filtered state-estimation error covariance, $P_{t|t} = E\tilde{\mathbf{x}}_{t|t}\tilde{\mathbf{x}}_{t|t}^T$, where $\tilde{\mathbf{x}}_{t|t} = \mathbf{x}_t - \hat{\mathbf{x}}_{t|t}$.

Define the *Kalman gain* matrix

$$\begin{aligned} K_t &= P_{t|t-1} C_t^T R_{\epsilon_t \epsilon_t}^{-1} \\ &= P_{t|t-1} C_t^T \left[C_t P_{t|t-1} C_t^T + R_t \right]^{-1}. \end{aligned} \quad (13.37)$$

From the measurement-update equation (13.23), we have

$$\begin{aligned} \hat{\mathbf{x}}_{t+1|t+1} &= \hat{\mathbf{x}}_{t+1|t} + P_{t+1|t} C_{t+1}^T R_{\epsilon_t \epsilon_t}^{-1} \epsilon_{t+1} \\ &= \hat{\mathbf{x}}_{t+1|t} + K_{t+1} \epsilon_{t+1} \\ &= \hat{\mathbf{x}}_{t+1|t} + K_{t+1} (\mathbf{y}_{t+1} - C_{t+1} \hat{\mathbf{x}}_{t+1|t}). \end{aligned} \quad (13.38)$$

Now let us formulate the *filtered state-estimation error covariance* matrix

$$P_{t|t} = E[\mathbf{x}_t - \hat{\mathbf{x}}_{t|t}][\mathbf{x}_t - \hat{\mathbf{x}}_{t|t}]^T = E\left[\tilde{\mathbf{x}}_{t|t}\tilde{\mathbf{x}}_{t|t}^T\right]. \quad (13.39)$$

Substituting (13.38) into (13.39), we obtain

$$\begin{aligned} P_{t|t} &= E[\tilde{\mathbf{x}}_{t|t-1} - K_t \epsilon_t][\tilde{\mathbf{x}}_{t|t-1} - K_t \epsilon_t]^T \\ &= \underbrace{E\tilde{\mathbf{x}}_{t|t-1}\tilde{\mathbf{x}}_{t|t-1}^T}_{P_{t|t-1}} - E\tilde{\mathbf{x}}_{t|t-1}\epsilon_t^T K_t^T - K_t E\epsilon_t \tilde{\mathbf{x}}_{t|t-1}^T + K_t E\epsilon_t \epsilon_t^T K_t^T. \end{aligned} \quad (13.40)$$

But using the fact that $\epsilon_t = \mathbf{y}_t - C_t \hat{\mathbf{x}}_{t|t-1} = C_t(\mathbf{x}_t - \hat{\mathbf{x}}_{t|t-1}) + \boldsymbol{\nu}_t$, we have

$$E\tilde{\mathbf{x}}_{t|t-1}\epsilon_t^T = \underbrace{E\tilde{\mathbf{x}}_{t|t-1}\tilde{\mathbf{x}}_{t|t-1}^T}_{P_{t|t-1}} C_t^T + \underbrace{E\tilde{\mathbf{x}}_{t|t-1}\boldsymbol{\nu}_t^T}_{0}. \quad (13.41)$$

Inserting (13.41) and (13.25) into (13.40), expanding, and simplifying, we obtain

$$P_{t|t} = P_{t|t-1} - P_{t|t-1}C_t^T K_t^T - K_t C_t P_{t|t-1} + K_t \left[C_t P_{t|t-1} C_t^T + R_t \right] K_t^T$$

$$= P_{t|t-1} - P_{t|t-1}C_t^T \left[C_t P_{t|t-1} C_t^T + R_t \right]^{-1} C_t P_{t|t-1} \qquad (13.42)$$

$$= [I - K_t C_t] P_{t|t-1}. \qquad (13.43)$$

We will complete the time-update/measurement-update structure for the Riccati equation by obtaining an expression for $P_{t+1|t}$ in terms of $P_{t|t}$. Rearranging (13.33) with $M_t \equiv 0$,

$$P_{t+1|t} = A_t \underbrace{\left[P_{t|t-1} - P_{t|t-1}C_t^T \left[C_t P_{t|t-1} C_t^T + R_t \right]^{-1} C_t P_{t|t-1} \right]}_{P_{t|t}} A_t^T + Q_t. \qquad (13.44)$$

Summarizing from (13.35), (13.44), (13.37), (13.38), and (13.43), we have the following formulation of the Kalman filter:

Time update

$$\hat{x}_{t+1|t} = A_t \hat{x}_{t|t},$$
$$P_{t+1|t} = A_t P_{t|t} A_t^T + Q_t. \qquad (13.45)$$

Kalman gain

$$K_{t+1} = P_{t+1|t} C_{t+1}^T \left[C_{t+1} P_{t+1|t} C_{t+1}^T + R_{t+1} \right]^{-1}. \qquad (13.46)$$

Measurement update

$$\hat{x}_{t+1|t+1} = \hat{x}_{t+1|t} + K_{t+1}[y_{t+1} - C_{t+1}\hat{x}_{t+1|t}]$$
$$P_{t+1|t+1} = (I - K_{t+1}C_{t+1})P_{t+1|t}. \qquad (13.47)$$

These are identical to the Kalman filter equations derived using Bayesian methods in section 13.2.

13.3.6 Perspective

We see that the Kalman filter is a solution to the general problem of estimating the state of a linear system. Such restrictions as stationarity or time-invariance are not important to the derivation. What is important, however, is the assumption that the noise processes are uncorrelated. Also, we do not need to know the complete distribution of the noise—only its first and second moments. This is a major simplification to the problem, and one of the nice things about *linear* estimation theory, and is not true of general nonlinear systems.

There are many ways to derive the Kalman filter, of which two have been presented: the Bayesian approach and the innovations approach. Each of them has conceptual value. The innovations approach is closest in spirit to the important concept introduced in chapter 2— namely, orthogonal projections. This is essentially the way that Kalman first derived the filter. By way of comparison, we note the following:

- The orthogonal projections approach does not rely on anything more than knowledge of the first and second moments of the distributions of all the random processes involved. If we do in fact have complete knowledge of all distributions involved, we should perhaps wonder if there might be a way to do better than we can partial knowledge. This is a realistic question to address, and the answer, for linear *Gaussian* systems, is that we do not gain anything! The reason succinctly, is that the first and second moments completely specify the Gaussian distribution. The Bayesian approach relies upon transformations of density functions. Since a Gaussian density is entirely characterized by its mean and covariance, the Kalman filter is optimal.

- A minimum-variance approach might also be considered (see, in fact, exercise 13.3-15). It might be hoped that such an approach might yield a different—perhaps even better—estimator. However, the Kalman filter admits interpretations as both a minimum-variance and maximum-likelihood estimator.

The fact is that, under fairly wide and applicable conditions, the least-squares, conditional expectations, maximum-likelihood, minimum-variance, and other interpretations of the Kalman filter are all equivalent. This is quite remarkable—even mystifying—with the reason for equivalence, perhaps, lying in the basic structure of linear systems combined with the basic mathematical concept of orthogonality.

13.3.7 Comparison with the RLS adaptive filter algorithm

Let us return to the least-squares filtering problem introduced in section 3.9 and the RLS filter introduced in section 4.11.1. We desired to find a set of FIR filter coefficients **h** so that when the sequence $\{f[t]\}$ is passed through the filter, the output matches as closely as possible some desired sequence $d[t]$.

The output of the filter is

$$y[t] = \sum_{i=0}^{m-1} h[i] f[t-i].$$

Assuming real inputs and filter coefficients for simplicity, we can write

$$\mathbf{q}[t] = \begin{bmatrix} x[t] \\ x[t-1] \\ \vdots \\ x[t-m+1] \end{bmatrix} \quad C = \begin{bmatrix} h[0] \\ h[1] \\ \vdots \\ h[m-1] \end{bmatrix};$$

thus, we have

$$y[t] = \mathbf{q}^T[t] C$$

In the adaptive filter application, the filter coefficients are chosen adaptively to minimize the sum of the squares of $e[t] = d[t] - y[t]$.

In our Kalman filter interpretation, we regard the filter coefficient vector **h** as the "state" of some system, and assume that the state is fixed. The adaptive filter adapts by trying to find the fixed state, so that estimating the adaptive filter coefficients becomes the problem of estimating the state of the system. We can regard

$$d[t] = \mathbf{q}^T[t] \mathbf{h} + e[t] \tag{13.48}$$

as the state-space observation equation. Since the state is fixed, the state-update matrix A_t is the identity matrix, and there is no state noise. For the observation equation (13.48), we will denote the estimate of the state as $\mathbf{h}(t)$—the adaptive filter coefficients at time t. Thus, from the point of view of a Kalman filter, the problem of finding the best (least-squares error) filter coefficients has become the problem of estimating the state of the system.

While the original least-squares problem does not have a probability model for the observation error $e[t]$, it is still useful to apply a Kalman filter to this problem. We note that since the state update has $A_t = I$, we must have $P_{t+1|t} = P_{t|t}$. We make the following identifications:

	state space	RLS
State-update matrix	A_t	I
State-noise variance	Q_t	0
Observation matrix	C_t	$\mathbf{q}^T[t]$
Observation-noise variance	R_t	σ_e^2
Observation	$\mathbf{y}[t]$	$d[t]$
Estimation-error covariance	$P_{t+1|t}$	$P(t)$
State estimate	$\hat{\mathbf{x}}_{t|t-1}$	$\mathbf{h}(t)$

Denoting the filter weight estimate at time t as $\mathbf{h}(t)$, based upon (13.46) and (13.47) we find that an updated filter weight is obtained by

$$\mathbf{k}_{t+1} = P_{t+1|t}\mathbf{q}_{t+1}\left(\mathbf{q}^T[t]P_{t+1|t}\mathbf{q}[t] + \sigma_e^2\right)^{-1} = \frac{P_{t+1|t}\mathbf{q}_{t+1}}{\sigma_e^2 + \mathbf{q}^T[t]P_{t+1|t}\mathbf{q}[t]},$$

$$\mathbf{h}(t+1) = \mathbf{h}(t) + \mathbf{k}_{t+1}(d[t+1] - \mathbf{q}^T[t+1]\mathbf{h}(t)),$$

$$P_{t+1|t+1} = (I - \mathbf{k}_{t+1})P_{t|t}.$$

Comparison with the RLS algorithm of section 4.11.1 reveals that the algorithms are essentially identical. The only difference is that in section 4.11.1, the Kalman gain was given (equivalently) by

$$\mathbf{k}_{t+1} = \frac{P_{t+1|t}\mathbf{q}[t]}{1 + \mathbf{q}_{t+1}^T P_{t+1|t}\mathbf{q}_{t+1}},$$

so that the term σ_e^2 does not appear in the denominator of the Kalman gain. However, this can be canceled out if it is assumed that the initial error covariance is $P_{0|-1} = \sigma_e^2 \delta^{-1} I$, for some small positive δ. Since δ was assumed to be arbitrary, this does not represent a significant modification.

We conclude that for the estimation of a constant state, the RLS filter (involving a least-squares criterion) is equivalent to a Kalman filter (in which the estimation error is modeled as observation noise).

13.4 Numerical considerations: Square-root filters

In the early days of its deployment, the Kalman filter was often implemented using fixed-point arithmetic. It was found that the Kalman filter equations require a wide dynamic range of numerical coefficients, and that the equations tend to be somewhat poorly conditioned numerically, especially for fixed-point implementations. To address these problems, algorithms were developed that have better numerical conditioning. Even today, with floating-point numeric processing readily available, the conditioning of the problem may suggest that numerically stable algorithms be employed for the Kalman filter. Using these algorithms is analogous to using the QR algorithm for solution of the least-squares problem, in which the Grammian is never explicitly computed and hence is never explicitly inverted. We present one of these numerically stable algorithms for the Kalman filter. The references at the end of the chapter provide suggestions for others.

The algorithm we develop is known as the *square-root* Kalman filter. It relies upon computing and propagating the Cholesky factorization of $P_{t+1|t}$, as

$$P_{t+1|t} = S_{t+1}S_{t+1}^T,$$

where S_{t+1} is the "square root" of $P_{t+1|t}$. The Cholesky factorization is done in such a way

13.4 Numerical Considerations: Square-Root Filters

that S_{t+1} is a lower-triangular matrix. In formulating the algorithm, it is helpful to place the Kalman filter in its one-step predictor form. We can write

$$W_t = R_t + C_t P_{t|t-1} C_t^T, \qquad (13.49)$$

$$\hat{x}_{t+1|t} = A_t \hat{x}_{t|t-1} + A_t P_{t|t-1} C_t^T W_t^{-1} (y_t - C_t \hat{x}_{t|t-1}), \qquad (13.50)$$

$$P_{t+1|t} = A_t \left(I - P_{t|t-1} C_t^T W_t^{-1} C_t \right) P_{t|t-1} A_t^T + Q_t. \qquad (13.51)$$

The computation involved in the square-root Kalman filter is summarized as follows.

1. Form the matrix

$$X_t = \begin{bmatrix} R_t^{1/2} & C_t S_t & 0 \\ 0 & A_t S_t & Q_t^{1/2} \end{bmatrix}$$

where, for example, $Q_t^{1/2}$ is a matrix such that $Q_t^{1/2} (Q_t^{1/2})^T = Q_t$ (a Cholesky factor).

2. Triangularize X_t using an orthogonal matrix U_t. That is, find an orthogonal matrix U_t such that $Y_t = X_t U_t$ is lower triangular, and identify the following components:

$$X_t U_t = \begin{bmatrix} R_t^{1/2} & C_t S_t & 0 \\ 0 & A_t S_t & Q_t^{1/2} \end{bmatrix} U_t = \begin{bmatrix} W_k^{1/2} & 0 & 0 \\ G_t & S_{t+1} & 0 \end{bmatrix} = Y_t. \qquad (13.52)$$

The orthogonal matrix U_t can be found using Householder transformations (see section 5.3.4).

3. Update the estimate as

$$\hat{x}_{t+1|t} = A_t \hat{x}_{t|t-1} + G_t W_t^{-1/2} (y_t - C_t \hat{x}_{t|t-1}). \qquad (13.53)$$

The operation of this algorithm is revealed by computing

$$X_t X_t^T = (X_t U_t)(X_t U_t)^T = \begin{bmatrix} R_t + C_t S_t S_t^T C_t & C_t S_t S_t^T A_t^T \\ A_t S_t S_t^T C_t^T & A_t S_t S_t^T A_t^T + Q_t \end{bmatrix}$$

$$= \begin{bmatrix} R_t + C_t P_{t|t-1} C_t^T & C_t P_{t|t-1} A_t^T \\ A_t P_{t|t-1} C_t^T & A_t P_{t|t-1} A_t^T + Q_t \end{bmatrix}, \qquad (13.54)$$

and by computing, from the right-hand side of (13.52),

$$Y_t Y_t^T = \begin{bmatrix} W_t & W_t^{1/2} G_t^T \\ G_t W_t^{1/2} & G_t G_t^T + S_{t+1} S_{t+1}^T \end{bmatrix}. \qquad (13.55)$$

Equating (13.54) and (13.55), we must have

$$G_t W_t^{1/2} = A_t P_{t|t-1} C_t^T \qquad (13.56)$$

so that $G_t W_t^{-1/2} = A_t P_{t|t-1} C_t^T W_t^{-1}$, giving the desired update gain matrix as required from (13.50). Also, from equating (13.54) and (13.55), we have

$$G_t G_t^T + S_{t+1} S_{t+1}^T = A_t P_{t|t-1} A_t^T + Q_t, \qquad (13.57)$$

from which

$$S_{t+1} S_{t+1}^T = A_t P_{t|t-1} A_t^T + Q_t - G_t G_t^T.$$

In light of (13.56), this can be written as

$$S_{t+1} S_{t+1}^T = A_t P_{t|t-1} A_t^T + Q_t - A_t P_{t|t-1} C_t^T W_t^{-1} C_t P_{t|t-1} A_t^T,$$

so that S_{t+1} is, in fact, a factor of $P_{t+1|t}$ as shown in (13.51).

13.5 Application in continuous-time systems

In this section, we describe how a continuous-time kinematic model in state space is converted to a discrete-time state-space form, suitable for state estimation using a Kalman filter. We will demonstrate the discretization process for a four-state linear system with a two-dimensional observations vector, corresponding to two-dimensional equations of the motion of a vehicle. The observations consist of noisy samples of the vehicle position.

13.5.1 Conversion from continuous time to discrete time

Suppose that we are given the following time-invariant, continuous-time system,

$$\dot{\mathbf{x}}(t) = \mathcal{F}\mathbf{x}(t) + \mathcal{G}\mathbf{u}(t),$$

where $\mathbf{u}(t)$ is a zero-mean, continuous-time, white-noise random process with

$$E\mathbf{u}(t)\mathbf{u}^T(s) = \mathcal{Q}\delta(t-s).$$

From this continuous-time system, samples are taken every Δ seconds. We wish to write the dynamics of the corresponding discrete-time system,

$$\mathbf{x}_{k+1} = A\mathbf{x}_k + \mathbf{w}_k,$$

where $\mathbf{w}_k$ is a random process with covariance Q. (We use k as the time index here, to avoid confusion between the continuous-time process and the discrete-time process). Conversion from continuous time to discrete time requires finding A and Q. The key to the conversion is the solution of the continuous-time differential equation, presented in section 1.4.3. Applying (1.35), we have

$$\mathbf{x}(t+\Delta) = e^{\mathcal{F}\Delta}\mathbf{x}(t) + \int_t^{t+\Delta} e^{\mathcal{F}(t+\Delta-\lambda)}\mathcal{G}\mathbf{u}(\lambda)\,d\lambda.$$

By identifying the discrete-time process with the continuous time process as $\mathbf{x}_k \equiv \mathbf{x}(t_k)$, where $t_k = k\Delta t$, we see that

$$A = e^{\mathcal{F}\Delta} \qquad (13.58)$$

and

$$\mathbf{w}_k = \int_{t_k}^{t_{k+1}} e^{\mathcal{F}(t_{k+1}-\lambda)}\mathcal{G}\mathbf{u}(\lambda)\,d\lambda.$$

The covariance of $\mathbf{w}_k$ can be found from

$$Q = \text{cov}(\mathbf{w}_k) = \int_{t_k}^{t_{k+1}}\int_{t_k}^{t_{k+1}} e^{\mathcal{F}(t_{k+1}-\lambda)}\mathcal{G}E[\mathbf{u}(\lambda)\mathbf{u}^T(\sigma)]\mathcal{G}^T e^{\mathcal{F}^T(t_{k+1}-\sigma)}\,d\lambda d\sigma$$

$$= \int_{t_k}^{t_{k+1}} e^{\mathcal{F}(t_{k+1}-\lambda)}\mathcal{G}\mathcal{Q}\mathcal{G}^T e^{\mathcal{F}^T(t_{k+1}-\lambda)}\,d\lambda. \qquad (13.59)$$

13.5.2 A simple kinematic example

Assume the following continuous-time system, in which the acceleration is applied via the random noise

$$\underbrace{\begin{bmatrix}\dot{x}(t)\\\dot{y}(t)\\\ddot{x}(t)\\\ddot{y}(t)\end{bmatrix}}_{\dot{\mathbf{x}}(t)} = \underbrace{\begin{bmatrix}0 & 0 & 1 & 0\\0 & 0 & 0 & 1\\0 & 0 & 0 & 0\\0 & 0 & 0 & 0\end{bmatrix}}_{\mathcal{F}} \underbrace{\begin{bmatrix}x(t)\\y(t)\\z(t)\end{bmatrix}}_{\mathbf{x}(t)} + \underbrace{\begin{bmatrix}0 & 0\\0 & 0\\1 & 0\\0 & 1\end{bmatrix}}_{\mathcal{G}} \underbrace{\begin{bmatrix}u_{xt}\\u_{yt}\\u_{zt}\end{bmatrix}}_{\mathbf{u}(t)}, \qquad (13.60)$$

13.6 Extensions of Kalman Filtering to Nonlinear Systems

where the system matrices $\mathcal{F}$ and $\mathcal{G}$ are defined in (13.60), and $\mathbf{u}_t$ is a continuous-time white noise with covariance

$$E\mathbf{u}(t)\mathbf{u}^T(s)) = \mathcal{Q}\delta(t-s) = \begin{bmatrix} q_x^2 & \\ & q_y^2 \end{bmatrix}\delta(t-s).$$

The matrix exponential is

$$\Phi(t) = \exp\{\mathcal{F}t\} = \exp\left\{\begin{bmatrix} 0 & 0 & 1 & 0 \\ 0 & 0 & 0 & 1 \\ 0 & 0 & 0 & 0 \\ 0 & 0 & 0 & 0 \end{bmatrix} t\right\} = \begin{bmatrix} 1 & 0 & t & 0 \\ 0 & 1 & 0 & t \\ 0 & 0 & 1 & 0 \\ 0 & 0 & 0 & 1 \end{bmatrix}, \quad (13.61)$$

so that from (13.58) we have

$$A = \exp(\mathcal{F}\Delta) = \begin{bmatrix} 1 & 0 & \Delta & 0 \\ 0 & 1 & 0 & \Delta \\ 0 & 0 & 1 & 0 \\ 0 & 0 & 0 & 1 \end{bmatrix}.$$

To compute Q, we substitute the values for $\Phi(t)$ and $\mathcal{G}$ from (13.61) and (13.60) into (13.59), to obtain

$$Q = E\mathbf{w}_i\mathbf{w}_i^T$$

$$= \int_{t_i}^{t_{i+1}} \begin{bmatrix} t_{i+1}-t & 0 \\ 0 & t_{i+1}-t \\ 1 & 0 \\ 0 & 1 \end{bmatrix} \begin{bmatrix} q_x^2 & 0 \\ 0 & q_y^2 \end{bmatrix} \begin{bmatrix} t_{i+1}-t & 0 & 1 & 0 \\ 0 & t_{i+1}-t & 0 & 1 \end{bmatrix} dt$$

$$= \int_{t_i}^{t_{i+1}} \begin{bmatrix} q_x^2(t_{i+1}-t)^2 & 0 & q_x^2(t_{i+1}-t) & 0 \\ 0 & q_y^2(t_{i+1}-t)^2 & 0 & q_y^2(t_{i+1}-t) \\ q_x^2(t_{i+1}-t) & 0 & q_x^2 & 0 \\ 0 & q_y^2(t_{i+1}-t) & 0 & q_y^2 \end{bmatrix} dt$$

$$= \begin{bmatrix} q_x^2 \frac{\Delta^3}{3} & 0 & q_x^2 \frac{\Delta^2}{2} & 0 \\ 0 & q_y^2 \frac{\Delta^3}{3} & 0 & q_y^2 \frac{\Delta^2}{2} \\ q_x^2 \frac{\Delta^2}{2} & 0 & q_x^2 \Delta & 0 \\ 0 & q_y^2 \frac{\Delta^2}{2} & 0 & q_y^2 \Delta \end{bmatrix}.$$

In computing the discrete-time Q matrix, the process noise induced on the position components is due to the acceleration error accumulation over the integration interval. Even though the continuous-time system had no noise components on the position components, there are noise components on the position components in the discrete-time model.

13.6 Extensions of Kalman filtering to nonlinear systems

Consider a general nonlinear discrete-time system of the form

$$\mathbf{x}_{t+1} = A(\mathbf{x}_t, t) + \mathbf{w}_t, \quad (13.62)$$
$$\mathbf{y}_t = C(\mathbf{x}_t, t) + \mathbf{\nu}_t, \quad (13.63)$$

for $t = 0, 1, \ldots$, with $\{\mathbf{w}_t, t = 0, 1, \ldots\}$ and $\{\mathbf{\nu}_t, t = 0, 1, \ldots\}$ representing uncorrelated, zero-mean state-noise and observation-noise sequences, respectively. The general nonlinear estimation problem is extremely difficult, and no general solution to the general nonlinear filtering problem is available. One reason the linear problem is easy to solve is that when the process noise, observation noise, and initial conditions, $\mathbf{x}_0$, are Gaussian distributed, the state $\mathbf{x}_t$ is Gaussian, and so is the conditional expectation $\hat{\mathbf{x}}_{t|\tau}$. But if A is nonlinear, then the state is no longer guaranteed to be Gaussian distributed, and if either A or C is nonlinear, then the conditional expectation $\hat{\mathbf{x}}_{t|\tau}$ is not guaranteed to be Gaussian distributed. Thus, we cannot, in general, obtain the estimate as a function of only the first two moments of the conditional distribution. The general solution would require the propagation of the entire conditional distribution. Thus, we cannot easily get an exact solution, and we resort to the time-honored method of obtaining an approximate solution by means of linearization.

In what follows, we present the general method for linearizing dynamical systems, which requires the use of a nominal trajectory. Subsequently, we use the Kalman estimate of the state as the nominal trajectory, producing what is known as the extended Kalman filter.

Linearization of dynamical systems

Nonlinear dynamical systems can be linearized using a Taylor series about some operating point. We introduce the linearization for discrete-time systems; the method is similar for continuous-time systems. To linearize the nonlinear dynamics and observation equation of the form

$$\mathbf{x}[t+1] = \mathbf{f}(\mathbf{x}[t], \mathbf{u}[t], t),$$
$$\mathbf{y}[t] = \mathbf{h}(\mathbf{x}[t], \mathbf{u}[t], t),$$

where $\mathbf{f}: \mathbb{R}^p \times \mathbb{R}^l \times \mathbb{Z} \to \mathbb{R}^p$ and $\mathbf{h}: \mathbb{R}^p \times \mathbb{R}^l \times \mathbb{Z} \to \mathbb{R}^m$, we assume the existence of a *nominal* input $\mathbf{u}_0[t]$, a nominal state trajectory $\mathbf{x}_0[t]$, and a nominal output trajectory $\mathbf{y}_0[t]$. We assume that the input $\mathbf{u}[t]$ and state $\mathbf{x}[t]$ are close to the nominal input and state, and create the linearization by truncating the Taylor series expansion for $\mathbf{f}$. Let

$$\mathbf{u}[t] = \mathbf{u}_0[t] + \mathbf{u}_\delta[t] \qquad \mathbf{x}[t] = \mathbf{x}_0[t] + \mathbf{x}_\delta[t] \qquad \mathbf{y}[t] = \mathbf{y}_0[t] + \mathbf{y}_\delta[t],$$

where $\mathbf{u}_\delta[t]$ is a vector with components $u_{\delta,1}[t], u_{\delta,2}[t]$, and so forth. Then the nonlinear dynamics equation can be written as

$$\mathbf{x}_0[t+1] + \mathbf{x}_\delta[t+1] = f(\mathbf{x}_0[t] + \mathbf{x}_\delta[t], \mathbf{u}_0[t] + \mathbf{u}_\delta[t], t).$$

Now, assuming that $f \in C^1$, we expand f in a Taylor series about $\mathbf{x}_0[t]$ and $\mathbf{u}_0[t]$. Since $\mathbf{x}_\delta[t]$ and $\mathbf{u}_\delta[t]$ are assumed small, we retain only the first-order term of the series. The ith component of $\mathbf{f}$ can be expanded as

$$f_i(\mathbf{x}_0 + \mathbf{x}_\delta, \mathbf{u}_0 + \mathbf{u}_\delta, t) \approx f_i(\mathbf{x}_0, \mathbf{u}_0, t) + \frac{\partial f_i}{\partial x_1}(\mathbf{x}_0, \mathbf{u}_0, t) x_{\delta,1} + \cdots + \frac{\partial f_i}{\partial x_p}(\mathbf{x}_0, \mathbf{u}_0, t) x_{\delta,p}$$
$$+ \frac{\partial f_i}{\partial u_1}(\mathbf{x}_0, \mathbf{u}_0, t) u_{\delta,1} + \cdots + \frac{\partial f_i}{\partial u_l}(\mathbf{x}_0, \mathbf{u}_0, k) u_{\delta,l}.$$

Let

$$\frac{\partial \mathbf{f}}{\partial \mathbf{x}} = \begin{bmatrix} \frac{\partial f_1}{\partial x_1} & \frac{\partial f_1}{\partial x_2} & \cdots & \frac{\partial f_2}{\partial x_p} \\ \frac{\partial f_2}{\partial x_1} & \frac{\partial f_2}{\partial x_1} & \cdots & \frac{\partial f_2}{\partial x_p} \\ \vdots & & & \\ \frac{\partial f_p}{\partial x_1} & \frac{\partial f_p}{\partial x_1} & \cdots & \frac{\partial f_p}{\partial x_p} \end{bmatrix}.$$

13.6 Extensions of Kalman Filtering to Nonlinear Systems

The $p \times p$ matrix $\frac{\partial \mathbf{f}}{\partial \mathbf{x}}$ is called the *Jacobian* of $\mathbf{f}$ with respect to $\mathbf{x}$. Similarly, let $\frac{\partial \mathbf{f}}{\partial \mathbf{u}}$ denote the $p \times l$ Jacobian of $\mathbf{f}$ with respect to $\mathbf{u}$. Then the Taylor series for the linearization can be written as

$$\mathbf{x}_0[t+1] + \mathbf{x}_\delta[t+1] = \mathbf{f}(\mathbf{x}_0[t], \mathbf{u}_0[t], t) + \frac{\partial \mathbf{f}}{\partial \mathbf{x}}(\mathbf{x}_0[t], \mathbf{u}_0[t], t)\mathbf{x}_\delta[t] \\ + \frac{\partial \mathbf{f}}{\partial \mathbf{u}}(\mathbf{x}_0[t], \mathbf{u}_0[t], t)\mathbf{u}_\delta[t]. \tag{13.64}$$

The nominal solution satisfies the original dynamics equation

$$\mathbf{x}_0[t+1] = \mathbf{f}(\mathbf{x}_0[t], \mathbf{u}_0[t], t).$$

Subtracting this equation from (13.64), we obtain the time-varying linear dynamics equation

$$\mathbf{x}_\delta[t+1] = \frac{\partial \mathbf{f}(\mathbf{x}_0[k], \mathbf{u}_0[k], k)}{\partial \mathbf{u}} \mathbf{x}_\delta[t] + \frac{\partial \mathbf{f}(\mathbf{x}_0[k], \mathbf{u}_0[k], k)}{\partial \mathbf{u}} \mathbf{u}_\delta[t]. \tag{13.65}$$

It is convenient to let

$$A(t) = \frac{\partial \mathbf{f}}{\partial \mathbf{x}}(\mathbf{x}_0[t], \mathbf{u}_0[t], t) \quad \text{and} \quad B(t) = \frac{\partial \mathbf{f}}{\partial \mathbf{u}}(\mathbf{x}_0[t], \mathbf{u}_0[t], t).$$

Then we obtain

$$\mathbf{x}_\delta[t+1] = A(t)\mathbf{x}_\delta[t] + B(t)\mathbf{u}_\delta[t].$$

The observation equation can be similarly linearized by expanding $\mathbf{h}[t]$ in a Taylor series. We obtain thereby

$$\mathbf{y}_\delta[t] = C(t)\mathbf{x}_\delta[t] + D(t)\mathbf{u}_\delta[t],$$

where $C(t)$ and $D(t)$ are respectively the $m \times p$ and $m \times l$ matrices defined by

$$C(t) = \frac{\partial \mathbf{h}}{\partial \mathbf{x}}(\mathbf{x}_0[t], \mathbf{u}_0[t], t) \qquad D(k) = \frac{\partial \mathbf{h}}{\partial \mathbf{x}}(\mathbf{x}_0[t], \mathbf{u}_0[t], t).$$

Example 13.6.1 Determine a linearization of the system

$$x_1[t+1] = 2x_1[t](1 - x_2[t]) + u[t],$$
$$x_2[t+1] = x_2[t] - 3x_1^2[t] + u[t]x_1[t],$$

about the nominal trajectory

$$x_{01}[t+1] = 0,$$
$$x_{02}[t+1] = 0,$$

where $u_0[t] = 0$. The Jacobian of $\mathbf{f}$ with respect to $\mathbf{x}$ is

$$A(t) = \frac{\partial \mathbf{f}}{\partial \mathbf{x}}(\mathbf{x}_0[t], \mathbf{u}_0[t], t) = \begin{bmatrix} 2(1 - x_{02}[t]) & -2x_{01}[t] \\ 6x_{01}[t] & 1 \end{bmatrix}.$$

Evaluated along the nominal trajectory we have

$$A(k) = \begin{bmatrix} 2 & 0 \\ 0 & 1 \end{bmatrix},$$

so the linearized system is

$$\mathbf{x}_\delta[k+1] = \begin{bmatrix} 2 & 0 \\ 0 & 1 \end{bmatrix} \mathbf{x}_\delta[k]. \qquad \square$$

One of the more difficult aspects of determining a linearization is finding an appropriate nominal trajectory. Very commonly, the nominal trajectory is taken about a point of *equilibrium*, which is a point such that

$$\mathbf{f}(\mathbf{x}, \mathbf{u}, k) = \mathbf{0}.$$

At such points,

$$\mathbf{x}[k+1] = \mathbf{x}[k].$$

The equilibrium points are *fixed points* of the dynamical system. The linearization of the system about those points thus provides an indication of how the system behaves in a neighborhood of equilibrium. For continuous-time systems, the equilibrium points are those at which

$$\dot{\mathbf{x}}(t) = \mathbf{0}.$$

Linearization about a nominal trajectory

In the preceding pages, to linearize a nonlinear state-space system, we postulated a nominal, or reference, trajectory about which a Taylor series was expanded. For our purposes here, we denote the nominal trajectory at time t by $\bar{\mathbf{x}}_t$, and the deviation $\delta\mathbf{x}_t$ as

$$\delta\mathbf{x}_t = \mathbf{x}_t - \bar{\mathbf{x}}_t.$$

Then, based upon the techniques of section 13.6, the nonlinear dynamics equation (13.62) can be approximated by

$$\delta\mathbf{x}_t \approx A_t \delta\mathbf{x}_t + \mathbf{w}_t,$$

where A is the matrix of partials

$$A_t = \left. \frac{\partial A(\mathbf{x}, t)}{\partial \mathbf{x}} \right|_{\mathbf{x}=\bar{\mathbf{x}}_t}. \tag{13.66}$$

We assume that the approximation is close enough that we can write

$$\delta\mathbf{x}_{t+1} = A\delta\mathbf{x}_t + \mathbf{w}_t. \tag{13.67}$$

Similarly, the observation equation (13.63) can be written as the approximation equation (replacing approximation, $\approx$, with equality, $=$, from here on)

$$\delta\mathbf{y}_t = C_t \delta\mathbf{x}_t + \nu_t, \tag{13.68}$$

where

$$\delta\mathbf{y}_t = \mathbf{y}_t - \bar{\mathbf{y}}_t$$

and

$$C_t = \left. \frac{\partial C(\mathbf{x}, t)}{\partial \mathbf{x}} \right|_{\mathbf{x}=\bar{\mathbf{x}}_t}. \tag{13.69}$$

Once the linearized dynamics and observations equations given by (13.67) and (13.68) are obtained, we may apply the Kalman filter to this system in $\delta\mathbf{x}_t$ in the standard way. The algorithm consists of the following steps:

1. Obtain a reference trajectory $\{\bar{\mathbf{x}}_t, t = 0, 1, \ldots, T\}$.
2. Evaluate the partials of A and C at $\bar{\mathbf{x}}_t$; identify these quantities as A_t and C_t, respectively.
3. Compute the reference observations, $\bar{\mathbf{y}}_t$, and calculate $\delta\mathbf{y}_t$.

4. Apply the Kalman filter to the linearized model

$$\delta \mathbf{x}_{t+1} = A_t \delta \mathbf{x}_t + \mathbf{w}_t,$$
$$\delta \mathbf{y}_t = C_t \delta \mathbf{x}_t + \nu_t,$$

to obtain the deviation estimates $\delta \hat{\mathbf{x}}_{t|t}$.

5. Add the deviation estimates to the nominal trajectory, to obtain the trajectory estimates:

$$\hat{\mathbf{x}}_{t|t} = \overline{\mathbf{x}}_t + \delta \hat{\mathbf{x}}_{t|t}.$$

The approach thus outlined is called *global linearization*, and it presents several potential problems. First and foremost, it assumes that a reliable nominal trajectory is available, so that the A_t and C_t matrices are valid. Furthermore, as the system is time-varying, these A_t and C_t matrices must be stored for every time step. But many important estimation problems do not enjoy the luxury of having foreknowledge sufficient to generate a reference trajectory. Also, even if the A_t and C_t matrices are not grossly in error, the approach is predicated on the assumption that higher-order terms in the Taylor expansion may be safely ignored. It would be highly fortuitous if the nominal trajectory were of such high quality that neither of these concerns were manifest.

In the general case, the development of a nominal trajectory is problematic. In some special cases it may be possible to generate such a trajectory via computer simulations; in other cases, experience and intuition may guide in development. Often, however, one may simply have to rely on guesses and hope for the best. The estimates may diverge, but even if they do not, the results may be suspect because of the sensitivity of the results to the operating point. Of course, one could perturb the operating point and evaluate the sensitivity of the estimates to this perturbation, but that would be a tedious procedure, certainly not feasible in real-time applications.

The extended kalman filter

Global linearization about a predetermined reference trajectory is not the only way to approach the linearization problem. Another approach is to calculate a local nominal trajectory "on the fly," and update it as information becomes available.

We wish to construct a recursive estimator; and regardless of its linearity properties, we are under obligation to provide the estimator with initial conditions in the form of $\hat{\mathbf{x}}_{0|-1}$ and $P_{0|-1}$, the *a priori* state estimate and covariance. The state $\mathbf{x}_{0|-1}$ represents the best information we have concerning the value $\mathbf{x}_0$, so it makes sense to use this value as the first point in the nominal trajectory; that is, to define

$$\overline{\mathbf{x}}_0 = \hat{\mathbf{x}}_{0|-1},$$

and use this value to compute the C_0 matrix as

$$C_0 = \left. \frac{\partial C(\mathbf{x}, 0)}{\partial \mathbf{x}} \right|_{\mathbf{x} = \hat{\mathbf{x}}_{0|-1}}$$

and the deviation observation equation as

$$\delta \mathbf{y}_0 = \mathbf{y}_0 - C(\overline{\mathbf{x}}_0, 0) = \mathbf{y}_0 - C(\hat{\mathbf{x}}_{0|-1}, 0).$$

Using these values, we may process $\delta \mathbf{y}_0$ using a standard Kalman filter applied to (13.67) and (13.68). The resulting measurement update is

$$\delta \hat{\mathbf{x}}_{0|0} = \delta \hat{\mathbf{x}}_{0|-1} + K_0 [\delta \mathbf{y}_0 - C_0 \delta \hat{\mathbf{x}}_{0|-1}], \qquad (13.70)$$
$$P_{0|0} = [\mathbf{I} - K_0 C_0] P_{0|-1}, \qquad (13.71)$$

where $K_0 = P_{0|-1}C_0^T[C_0 P_{0|-1}C_0^T + R_0]^{-1}$. But note that $\hat{\mathbf{x}}_{0|-1}$ fulfills two roles: (a) it is the initial value of the state estimate, and (b) it is the nominal trajectory about which we linearize, namely $\bar{\mathbf{x}}_0$. Consequently,

$$\delta\hat{\mathbf{x}}_{0|-1} = \hat{\mathbf{x}}_{0|-1} - \bar{\mathbf{x}}_0 = \mathbf{0}.$$

Furthermore,

$$\delta\hat{\mathbf{x}}_{0|0} = \hat{\mathbf{x}}_{0|0} - \bar{\mathbf{x}}_0 = \hat{\mathbf{x}}_{0|0} - \hat{\mathbf{x}}_{0|-1},$$

so (13.70) becomes

$$\hat{\mathbf{x}}_{0|0} = \hat{\mathbf{x}}_{0|-1} + K_0[\mathbf{y}_0 - C(\hat{\mathbf{x}}_{0|-1}, 0)]. \tag{13.72}$$

Consequently, (13.72) and (13.71) constitute the measurement-update equations at time $t = 0$.

The next order of business is to predict to the time of the next observation, and then update. We need to compute the predicted state, $\hat{\mathbf{x}}_{1|0}$, and the predicted covariance, $P_{1|0}$. To predict the state, we simply apply the nonlinear dynamics equation:

$$\hat{\mathbf{x}}_{1|0} = A(\hat{\mathbf{x}}_{0|0}, 0). \tag{13.73}$$

To predict the covariance, we need to obtain a linear model, which will enable us to predict the covariance as

$$P_{1|0} = A_0 P_{0|0} A_0^T + Q_0. \tag{13.74}$$

The question is, what should we use as a nominal trajectory at which to evaluate (13.66)? According to our philosophy, we should use the best information we currently have about $\mathbf{x}_0$, and this is our filtered estimate. Thus, we take, for the calculation of A_0, the value $\bar{\mathbf{x}}_0 = \hat{\mathbf{x}}_{0|0}$. Using this value, the prediction step at time $t = 0$ is given by (13.73) and (13.74).

The next step, of course, is to perform the time update at time $t = 1$, yielding

$$\delta\hat{\mathbf{x}}_{1|1} = \delta\hat{\mathbf{x}}_{1|0} + K_1[\delta\mathbf{y}_1 - C_1 \delta\hat{\mathbf{x}}_{1|0}],$$
$$P_{1|1} = [\mathbf{I} - K_1 C_1]P_{1|0},$$

which requires us to employ a reference trajectory $\bar{\mathbf{x}}_1$. Following our philosophy, we simply use the best information we have at time $t = 1$, namely, the predicted estimate, so we set $\bar{\mathbf{x}}_1 = \hat{\mathbf{x}}_{1|0}$. Consequently, $\delta\hat{\mathbf{x}}_{1|0} = \hat{\mathbf{x}}_{1|0} - \bar{\mathbf{x}}_1 = \mathbf{0}$, and $\delta\hat{\mathbf{x}}_{1|1} = \hat{\mathbf{x}}_{1|1} - \hat{\mathbf{x}}_{1|0}$, which yields

$$\hat{\mathbf{x}}_{1|1} = \hat{\mathbf{x}}_{1|0} + K_1[\mathbf{y}_1 - C(\hat{\mathbf{x}}_{1|0}, 1)],$$

where

$$K_1 = P_{1|0}C_1^T \left[C_1 P_{1|0} C_1^T + R_1\right]^{-1},$$

with

$$C_1 = \left.\frac{\partial C(\mathbf{x}, 1)}{\partial \mathbf{x}}\right|_{\mathbf{x}=\hat{\mathbf{x}}_{1|0}}.$$

The pattern should now be quite clear. The resulting algorithm is called the *extended Kalman filter*, summarized as follows.

Measurement update

$$\hat{\mathbf{x}}_{t+1|t+1} = \hat{\mathbf{x}}_{t+1|t} + K_{t+1}[\mathbf{y}_{t+1} - C(\hat{\mathbf{x}}_{t+1|t}, t)], \tag{13.75}$$

$$P_{t+1|t+1} = [\mathbf{I} - K_{t+1} C_{t+1}]P_{t+1|t}, \tag{13.76}$$

where

$$K_t = P_{t+1|t}C_{t+1}^T \left[C_{t+1} P_{t+1|t} C_{t+1}^T + R_{t+1}\right]^{-1}, \tag{13.77}$$

with

$$C_t = \left.\frac{\partial C(\mathbf{x}, t)}{\partial \mathbf{x}}\right|_{\mathbf{x}=\hat{\mathbf{x}}_{t+1|t}}. \tag{13.78}$$

Time update

$$\hat{\mathbf{x}}_{t+1|t} = A(\hat{\mathbf{x}}_{t|t}, t), \tag{13.79}$$

$$P_{t+1|t} = A_t P_{t|t} A_t^T + Q_t, \tag{13.80}$$

where

$$A_t = \left.\frac{\partial A(\mathbf{x}, t)}{\partial \mathbf{x}}\right|_{\mathbf{x}=\hat{\mathbf{x}}_{t|t}}. \tag{13.81}$$

The extended Kalman filter is initialized in exactly the same way as the standard Kalman filter; namely, by supplying the *a priori* estimate and covariance, $\mathbf{x}_{0|-1}$ and $P_{0|-1}$, respectively.

13.7 Smoothing

The Kalman filter provides a state estimate conditioned on the past and present observations, and so is a causal estimator. Such an estimator is appropriate for real-time operation, but in applications, it often is possible to delay the calculation of the estimate until future data are obtained. In such a postprocessing environment, we ought to consider constructing a *smoothed*, or noncausal, estimator that uses the future, as well as the past, data.

In our discussions of filtering, we have employed a double-subscript notation of the form $\hat{\mathbf{x}}_{j|k}$ to denote the estimate of the state $\mathbf{x}_j$ given data up to time k, where we have assumed that the data set is of the form $\{\mathbf{y}_0, \mathbf{y}_1, \ldots, \mathbf{y}_k\}$. For the ensuing discussions, however, it is convenient, though a bit cumbersome, to modify this notation as follows: Let the estimate $\hat{\mathbf{x}}_{j|i:k}$, $i \leq k$, denote the estimate of $\mathbf{x}_j$ given data $\{\mathbf{y}_i, \mathbf{y}_{i+1}, \ldots, \mathbf{y}_k\}$. In this notation, the filtered estimate $\hat{\mathbf{x}}_{j|k}$ becomes $\hat{\mathbf{x}}_{j|0:k}$. The estimation error covariance for these estimates will be denoted by $P_{j|i:k} = E[\mathbf{x}_j - \hat{\mathbf{x}}_{j|i:k}][\mathbf{x}_j - \hat{\mathbf{x}}_{j|i:k}]^T$.

We assume that, in general, the entire set of data available are the samples $\mathcal{Y}_T = \{\mathbf{y}_0, \mathbf{y}_1, \ldots, \mathbf{y}_T\}$. There are three general smoothing situations.

Fixed-lag smoothing. In fixed-lag smoothing, an estimate of $\mathbf{x}_i$ is obtained using N points of future observation, producing the estimate denoted $\hat{\mathbf{x}}_{i|0:i+N}$.

Fixed-point smoothing. In fixed-point smoothing, the state is estimated at one fixed time only, using all of the data available. For fixed t_0, the fixed-point smoother is denoted $\hat{\mathbf{x}}_{t_0|0:T}$, where $0 \leq t_0 \leq T$.

Fixed-interval smoothing. Given the set of data $\mathcal{Y}_T$, the fixed-interval smoother provides estimates $\hat{\mathbf{x}}_{k|0:T}$ for all k in the range $0 \leq k \leq T$.

Fixed-lag and fixed-point smoothing are specialized applications that are found in various texts and are not developed here. Fixed-point smoothing may actually be viewed as a special case of fixed-interval smoothing.

13.7.1 The Rauch–Tung–Striebel fixed-interval smoother

There are at least three approaches to the development of the fixed-interval smoother: (a) the forward–backward smoother, (b) the two-point boundary-value approach, and (c) the Rauch–Tung–Striebel smoother. We present only the Rauch–Tung–Striebel approach.

Assume that for each time t the filtered estimate and covariance, $\hat{\mathbf{x}}_{t|0:t}$ and $P_{t|0:t}$, and predicted estimate and covariance, $\hat{\mathbf{x}}_{t+1|0:t}$ and $P_{t+1|0:t}$, have been computed, using a Kalman filter. We want to use these quantities to obtain a recursion for the fixed-interval smoothed estimate and covariance, $\hat{\mathbf{x}}_{t|0:T}$ and $P_{t|0:T}$.

We assume that $\mathbf{x}_t$ and $\mathbf{x}_{t+1}$ are jointly normal, given $\{\mathbf{y}_0, \ldots, \mathbf{y}_T\}$. We consider the conditional joint density function

$$f_{\mathbf{x}_t, \mathbf{x}_{t+1}|\mathbf{y}_0,\ldots,\mathbf{y}_T}(\mathbf{x}_t, \mathbf{x}_{t+1}|\mathbf{y}_0, \ldots, \mathbf{y}_T),$$

and seek the values of $\mathbf{x}_t$ and $\mathbf{x}_{t+1}$ that maximize this joint conditional density, resulting in the maximum-likelihood estimates for $\mathbf{x}_t$ and $\mathbf{x}_{t+1}$, given all of the data available over the full extent of the problem. (We eventually show that the maximum-likelihood estimate is indeed the orthogonal projection of the state onto the space spanned by all of the data, although we do not attack the derivation initially from that point of view.) For the remainder of this derivation, we suspend the subscripts, and let the reader infer the structure of the densities involved from the argument list.

We write

$$\begin{aligned} f(\mathbf{x}_t, \mathbf{x}_{t+1}|\mathbf{y}_0, \ldots \mathbf{y}_T) &= \frac{f(\mathbf{x}_t, \mathbf{x}_{t+1}, \mathbf{y}_0, \ldots \mathbf{y}_T)}{f(\mathbf{y}_0, \ldots, \mathbf{y}_T)} \\ &= \frac{f(\mathbf{x}_t, \mathbf{x}_{t+1}, \mathbf{y}_0, \ldots, \mathbf{y}_t, \mathbf{y}_{t+1}, \ldots, \mathbf{y}_T)}{f(\mathbf{y}_0, \ldots, \mathbf{y}_T)} \\ &= \frac{f(\mathbf{x}_t, \mathbf{x}_{t+1}, \mathbf{y}_{t+1}, \ldots, \mathbf{y}_T|\mathbf{y}_0, \ldots, \mathbf{y}_t)f(\mathbf{y}_0, \ldots, \mathbf{y}_t)}{f(\mathbf{y}_0, \ldots, \mathbf{y}_T)} \\ &= f(\mathbf{y}_{t+1}, \ldots, \mathbf{y}_T|\mathbf{x}_t, \mathbf{x}_{t+1}, \mathbf{y}_0, \ldots, \mathbf{y}_t) \\ &\quad \times f(\mathbf{x}_t, \mathbf{x}_{t+1}|\mathbf{y}_0, \ldots, \mathbf{y}_t)\frac{f(\mathbf{y}_0, \ldots, \mathbf{y}_t)}{f(\mathbf{y}_0, \ldots, \mathbf{y}_T)}. \end{aligned} \quad (13.82)$$

But, conditioned on $\mathbf{x}_{t+1}$, the distribution of $\{\mathbf{y}_0, \ldots, \mathbf{y}_T\}$ is independent of all previous values of the state and the observations, so

$$f(\mathbf{y}_{t+1}, \ldots, \mathbf{y}_T|\mathbf{x}_t, \mathbf{x}_{t+1}, \mathbf{y}_0, \ldots, \mathbf{y}_t) = f(\mathbf{y}_t, \ldots, \mathbf{y}_T|\mathbf{x}_{t+1}). \quad (13.83)$$

Furthermore,

$$\begin{aligned} f(\mathbf{x}_t, \mathbf{x}_{t+1}|\mathbf{y}_0, \ldots, \mathbf{y}_t) &= f(\mathbf{x}_t|\mathbf{x}_{t+1}, \mathbf{y}_0, \ldots, \mathbf{y}_t)f(\mathbf{x}_t|\mathbf{y}_0, \ldots, \mathbf{y}_t) \\ &= f(\mathbf{x}_t|\mathbf{x}_{t+1})f(\mathbf{x}_t|\mathbf{y}_0, \ldots, \mathbf{y}_t), \end{aligned} \quad (13.84)$$

where the last equality obtains since $\mathbf{x}_{t+1}$ conditioned on $\mathbf{x}_t$ is independent of all previous observations. Substituting (13.83) and (13.84) into (13.82) yields

$$f(\mathbf{x}_t, \mathbf{x}_{t+1}|\mathbf{y}_0, \ldots, \mathbf{y}_T) = f(\mathbf{x}_{t+1}|\mathbf{x}_t)f(\mathbf{x}_t|\mathbf{y}_0, \ldots, \mathbf{y}_t) \\ \times \underbrace{\frac{f(\mathbf{y}_{t+1}, \ldots, \mathbf{y}_T|\mathbf{x}_{t+1})f(\mathbf{y}_0, \ldots, \mathbf{y}_t)}{f(\mathbf{y}_0, \ldots, \mathbf{y}_T)}}_{\text{independent of } \mathbf{x}_t}. \quad (13.85)$$

Our approach to the smoothing problem will be to assume that $\hat{\mathbf{x}}_{t+1|0:T}$ is available, and maximize (13.85) given this assumption. (This assumption provides a boundary condition, as we shall see.) Assuming normal distributions, the densities $f(\mathbf{x}_{t+1} | \mathbf{x}_t)$ and $f(\mathbf{x}_t | \mathbf{y}_0, \ldots, \mathbf{y}_t)$ are

$$f(\mathbf{x}_{t+1} | \mathbf{x}_t) = \mathcal{N}(A_t\mathbf{x}_t, Q_t),$$
$$f(\mathbf{x}_t | \mathbf{y}_0, \ldots, \mathbf{y}_t) = \mathcal{N}(\hat{\mathbf{x}}_{t|0:t}, P_{t|0:t}).$$

The problem of maximizing the conditional probability density function,

$$f(\mathbf{x}_t, \mathbf{x}_{t+1} | \mathbf{y}_0, \ldots, \mathbf{y}_T),$$

13.7 Smoothing

with respect to $\mathbf{x}_t$ assuming $\mathbf{x}_{t+1}$ is given as the smoothed estimate at time $t+1$ is equivalent to the problem of minimizing

$$J(\mathbf{x}_t) = \frac{1}{2}[\mathbf{x}_{t+1} - A_t\mathbf{x}_t]^T Q_t^{-1}[\mathbf{x}_{t+1} - A_t\mathbf{x}_t] + \frac{1}{2}[\mathbf{x}_t - \hat{\mathbf{x}}_{t|0:t}]^T P_{t|0:t}^{-1}[\mathbf{x}_{t+1} - \hat{\mathbf{x}}_{t|0:t}]$$

evaluated at $\mathbf{x}_{t+1} = \hat{\mathbf{x}}_{t+1|0:T}$. By taking the gradient of $J(\mathbf{x}_t)$ with respect to $\mathbf{x}_t$, and equating it to zero, the solution can be shown to be

$$\hat{\mathbf{x}}_{t|0:T} = \left[P_{t|0:t}^{-1} + A_t Q_t^{-1} A_t\right]^{-1} \left[P_{t|0:t}^{-1}\hat{\mathbf{x}}_{t|0:t} + A_t Q_t^{-1} A_t \hat{\mathbf{x}}_{t+1|0:T}\right]. \quad (13.86)$$

Using the matrix-inversion identity (4.33) and its corollary (4.34), it can be shown that

$$\hat{\mathbf{x}}_{t|0:T} = \hat{\mathbf{x}}_{t|0:t} + S_t(\hat{\mathbf{x}}_{t+1|0:T} - A_t\hat{\mathbf{x}}_{t|0:t}) \quad (13.87)$$

where

$$\begin{aligned} S_t &= P_{t|0:t} A_t \left[A_t P_{t|0:t} A_t^T + Q_t\right]^{-1} \\ &= P_{t|0:t} A_t^T P_{t+1|0:t}^{-1}. \end{aligned} \quad (13.88)$$

Equation (13.87) is the Rauch–Tung–Streibel smoother. The smoother operates in backward time with $\mathbf{x}_{T|0:T}$, the final filtered estimate, as the initial condition for the smoother. For example, the order of computations is as follows.

1. Initialize: Given the data $\mathcal{Y}_T$, compute $\hat{\mathbf{x}}_{T|T}$ using the Kalman filter. Set $t = T - 1$.
2. Compute $\hat{\mathbf{x}}_{t|0:t}$ using the Kalman filter, then $\hat{\mathbf{x}}_{t|0:T}$ using (13.87).
3. Let $t = t - 1$ and repeat from step 2 as necessary.

We next seek an expression for the covariance of the smoothing error, $\tilde{\mathbf{x}}_{t|0:T} = \mathbf{x}_t - \hat{\mathbf{x}}_{t|0:t}$:

$$P_{t|0:T} = E\tilde{\mathbf{x}}_{t|0:T}\tilde{\mathbf{x}}_{t|0:T}^T.$$

From (13.87),

$$\tilde{\mathbf{x}}_{t|0:T} = \mathbf{x}_t - \hat{\mathbf{x}}_{t|0:T} = \mathbf{x}_t - \hat{\mathbf{x}}_{t|0:t} - S_t\left(\hat{\mathbf{x}}_{t+1|0:T} - A_t\hat{\mathbf{x}}_{t|0:t}\right),$$

so that

$$\tilde{\mathbf{x}}_{t|0:T} + S_t\hat{\mathbf{x}}_{t+1|0:T} = \tilde{\mathbf{x}}_{t|0:t} + S_t A_t\hat{\mathbf{x}}_{t+1|0:t}.$$

Multiplying both sides by the transpose and taking expectations yields

$$\begin{aligned} E\tilde{\mathbf{x}}_{t|0:T}\tilde{\mathbf{x}}_{t|0:T}^T + E\tilde{\mathbf{x}}_{t|0:T}\hat{\mathbf{x}}_{t+1|0:T}^T S_t^T + \\ S_t E\hat{\mathbf{x}}_{t+1|0:T}\tilde{\mathbf{x}}_{t|0:T}^T + S_t E\hat{\mathbf{x}}_{t+1|0:T}\hat{\mathbf{x}}_{t+1|0:T}^T S_t^T = \\ E\tilde{\mathbf{x}}_{t|0:t}\tilde{\mathbf{x}}_{t|0:t}^T + E\tilde{\mathbf{x}}_{t|0:t}\hat{\mathbf{x}}_{t|0:t}^T A_t^T S_t^T + \\ S_t A_t E\hat{\mathbf{x}}_{t|0:t}\tilde{\mathbf{x}}_{t|0:t}^T + S_t A_t E\hat{\mathbf{x}}_{t|0:t}\hat{\mathbf{x}}_{t|0:t}^T A_t^T S_t^T. \end{aligned} \quad (13.89)$$

Examining the cross terms of these expressions yields, for example,

$$\begin{aligned} E\tilde{\mathbf{x}}_{t|0:T}\hat{\mathbf{x}}_{t+1|0:T}^T &= E\tilde{\mathbf{x}}_{t|0:T}[A_t\hat{\mathbf{x}}_{t|0:T} + \hat{\mathbf{w}}_{t|0:T}]^T = E\tilde{\mathbf{x}}_{t|0:T}\hat{\mathbf{x}}_{t|0:T}^T A_t^T \\ &= E\left\{E\left[\tilde{\mathbf{x}}_{t|0:T}\hat{\mathbf{x}}_{t|0:T}^T | \mathbf{y}_0, \ldots, \mathbf{y}_T\right]\right\}A_t^T \\ &= E\left\{E[\tilde{\mathbf{x}}_{t|0:T}|\mathbf{y}_0, \ldots, \mathbf{y}_T]\hat{\mathbf{x}}_{t|0:T}^T\right\}A_t^T \\ &= E\left\{\left(\underbrace{E[\mathbf{x}_t|\mathbf{y}_0, \ldots, \mathbf{y}_T]}_{\hat{\mathbf{x}}_{t|0:T}} - \hat{\mathbf{x}}_{t|0:T}\right)\hat{\mathbf{x}}_{t|0:T}^T\right\}A_t^T \\ &= \mathbf{0}. \end{aligned}$$

By a similar argument (or from previous orthogonality results),
$$E\tilde{\mathbf{x}}_{t|0:t}\hat{\mathbf{x}}_{t|0:t}^T = \mathbf{0},$$
and so all cross terms in (13.89) vanish, leaving the expression
$$P_{t|0:T} + S_t E\hat{\mathbf{x}}_{t+1|0:T}\hat{\mathbf{x}}_{t+1|0:T}^T S_t^T = P_{t|0:t} + S_t A_t E\hat{\mathbf{x}}_{t|0:t}\hat{\mathbf{x}}_{t|0:t}^T A_t^T S_t^T. \quad (13.90)$$
An important byproduct of the above derivations is the result
$$E\tilde{\mathbf{x}}_{t|0:T}\hat{\mathbf{x}}_{t|0:T} = \mathbf{0}. \quad (13.91)$$
This result establishes the fact that the smoothed estimation error is orthogonal to the smoothed estimate, which is equivalent to the claim that the smoothed estimate is the projection of the state onto the space spanned by the entire set of observations. Thus smoothing preserves orthogonality.

Next, we compute the term from (13.90)
$$E\hat{\mathbf{x}}_{t+1|0:T}\hat{\mathbf{x}}_{t+1|0:T}^T.$$
To solve for this term, we use the just-established fact that
$$\mathbf{x}_{t+1} = \hat{\mathbf{x}}_{t+1|0:T} + \tilde{\mathbf{x}}_{t+1|0:T}$$
is an orthogonal decomposition, so
$$E\mathbf{x}_{t+1}\mathbf{x}_{t+1}^T = E\hat{\mathbf{x}}_{t+1|0:T} E\hat{\mathbf{x}}_{t+1|0:T}^T + E\tilde{\mathbf{x}}_{t+1|0:T} E\tilde{\mathbf{x}}_{t+1|0:T}^T$$
$$= E\hat{\mathbf{x}}_{t+1|0:T} E\hat{\mathbf{x}}_{t+1|0:T}^T + P_{t+1|0:T}.$$
Similarly,
$$E\mathbf{x}_t\mathbf{x}_t^T = E\hat{\mathbf{x}}_{t|0:t} E\hat{\mathbf{x}}_{t|0:t}^T + P_{t|0:t}.$$
We have previously found that
$$E\mathbf{x}_{t+1}\mathbf{x}_{t+1}^T = A_t E\mathbf{x}_t\mathbf{x}_t^T A_t^T + Q_t.$$
Substituting these results into (13.90) yields
$$P_{t|0:T} + S_t[E\mathbf{x}_{t+1}\mathbf{x}_{t+1} - P_{t+1|0:T}]^T S_t^T$$
$$= P_{t|0:t} + S_t A_t [E\mathbf{x}_t\mathbf{x}_t^T - P_{t|0:t}] A_t^T S_t^T, \quad (13.92)$$
or
$$P_{t|0:T} + S_t[E\mathbf{x}_t\mathbf{x}_t + Q_t - P_{t+1|0:T}]^T S_t^T$$
$$= P_{t|0:t} + S_t A_t [E\mathbf{x}_t\mathbf{x}_t^T - P_{t|0:t}] A_t^T S_t^T, \quad (13.93)$$
which simplifies to
$$P_{t|0:T} = P_{t|0:t} + S_t [P_{t+1|0:T} - Q_t - A_t P_{t|0:t} A_t^T]^T S_t^T$$
$$= P_{t|0:t} + S_t [P_{t+1|0:T} - P_{t+1|0:t}]^T S_t^T. \quad (13.94)$$

13.8 Another approach: H_∞ smoothing

Recently, other approaches to Kalman filtering have been developed based upon other criteria. The H_∞ approach can be modeled as a two-person game, in which the filter (the first player) attempts to prepare the best performance (minimum estimation error) against the worst strategy that the other play (nature) can produce. That is, the filter is designed for uniformly small error for any meaurement noise that nature produces, whether Gaussian or not. Rather than discuss this method in detail, we refer the interested reader to the references at the end of the chapter.

13.9 Exercises

13.1-1 Given the assumptions on our model, verify each of the following relationships:

$$E\mathbf{x}_j \boldsymbol{\nu}_k^T = \mathbf{0} \quad k \geq j,$$
$$E\mathbf{x}_j \mathbf{w}_k^T = \mathbf{0} \quad k \geq j,$$
$$E\mathbf{y}_j \boldsymbol{\nu}_k^T = \mathbf{0} \quad k > j,$$
$$E\mathbf{y}_j \mathbf{w}_k^T = \mathbf{0} \quad k > j,$$
$$E\mathbf{y}_k \boldsymbol{\nu}_k^T = C_k^T,$$
$$E\mathbf{y}_k \mathbf{w}_k^T = R_k.$$

13.2-2 Show by using the definition

$$P_{t+1|t+1} = E\left[\tilde{\mathbf{x}}_{t+1|t+1} \tilde{\mathbf{x}}_{t+1|t+1}^T\right],$$

where $\tilde{\mathbf{x}}_{t+1|t+1} = \tilde{\mathbf{x}}_{t+1} - \hat{\mathbf{x}}_{t+1|t+1}$ that (13.17) is correct.

13.2-3 Show that (13.18) and (13.19) are correct. (Use the matrix inversion lemma.)

13.2-4 Show that the optimal estimate and its error are orthogonal:

$$E\left[\hat{\mathbf{x}}_{t+1|t+1} (\mathbf{x}_t - \hat{\mathbf{x}}_{t+1|t+1})^T\right] = 0.$$

13.3-5 Suppose that y_t is a scalar zero-mean discrete-time stationary random process modeled by

$$y_t - \rho y_{t-1} = n_t$$

(an $AR(1)$ process), where n_t is a zero-mean, uncorrelated random process with $E[n_t^2] = 1 - \rho^2$. Suppose also that y_t is normalized so that $Ey_0 = 0$ and $Ey_0^2 = 1$. Show that this model has a correlation function of the form $Ey_t y_\tau = \rho^{|t-\tau|}$.

13.3-6 Let y_t be a scalar zero-mean discrete-time stationary random process with

$$Ey_t y_\tau = \rho^{|t-\tau|}.$$

Show that $\epsilon_t = y_t - \rho y_{t-1}$ is an innovations process, and that $E\epsilon_t^2 = 1 - \rho^2$. (This is the whitening filter for the process in the previous exercise.)

13.3-7 Let y_t be a scalar random process, and let ϵ_t be its innovation process. Then there is a linear relationship

$$\boldsymbol{\epsilon} = W\mathbf{y}$$

for an invertible matrix W, where $\boldsymbol{\epsilon} = [\epsilon_0, \epsilon_1, \ldots, \epsilon_n]^T$ and $\mathbf{y} = [y_0, y_1, \ldots, y_n]^T$. Let $\hat{\mathbf{x}}$ be the linear MMSE estimate based upon $\mathbf{y}$:

$$\hat{\mathbf{x}} = R_{xy} R_{yy}^{-1} \mathbf{y}.$$

In addition, let L be a Cholesky factor of R_{yy}, as

$$R_{yy} = LL^T.$$

(a) Show that $R_{yy} = W^{-1} R_{\epsilon\epsilon} W^{-T}$.
(b) Let $\overline{W} = L^{-1}$. Show that

$$\overline{W} = R_{\epsilon\epsilon}^{-1/2} W,$$

where $R_{\epsilon\epsilon}^{-1/2}$ is a matrix such that $(R_{\epsilon\epsilon}^{-1/2})^T R_{\epsilon\epsilon}^{-1/2} = R_{\epsilon\epsilon}^{-1}$. Also show that $R_{yy}^{-1} = \overline{W}^T \overline{W}$.

(c) Using these results, show that

$$\hat{\mathbf{x}} = R_{x\epsilon} R_{\epsilon\epsilon}^{-1} \epsilon,$$

so that the estimate $\hat{\mathbf{x}}$ is a linear function of the innovations.

13.3-8 The Kalman filter as we have derived it has been for state equations of the following form (for convenience we suppress the time dependence):

$$\begin{aligned} \mathbf{x}_{t+1} &= A\mathbf{x}_t + \mathbf{w}_t, \\ \mathbf{y}_t &= C\mathbf{x}_t + \nu_t, \end{aligned} \qquad (13.95)$$

where

$$\operatorname{cov}(\mathbf{w}_t) = Q \qquad \operatorname{cov}(\nu_t) = R.$$

Frequently, the model is specified as

$$\begin{aligned} \mathbf{x}_{t+1} &= A\mathbf{x}_t + B\mathbf{w}_t, \\ \mathbf{y}_t &= C\mathbf{x}_t + D\nu. \end{aligned} \qquad (13.96)$$

Explain how the Kalman filter derived for (13.95) can be employed to provide an estimate for the state in (13.96).

13.3-9 Show that the Kalman filter can be put into the *one-step predictor* form

$$\hat{\mathbf{x}}_{t+1|t} = A_t \hat{\mathbf{x}}_{t|t-1} + A_t P_{t|t-1} C_t^T \left[C_t P_{t|t-1} C_t^T + R_t \right]^{-1} [\mathbf{y}_t - C_t \hat{\mathbf{x}}_{t|t-1}]$$

with $P_{t|t-1}$ given by (13.33) and (13.34).

13.3-10 A random process $\{y_0, y_1, \ldots\}$ is defined by the following recursive procedure: Let y_0 be a random variable uniformly distributed over $(0, 1)$, and define y_t as the fractional part of $2y_{t-1}, t = 1, 2, \ldots$.

(a) Show that $Ey_t = 0.5$, $\operatorname{cov}(y_t, y_\tau) = \frac{2^{-|t-\tau|}}{12}$.

(b) Show that $\hat{y}_{t|t-1} = \frac{1}{4} + \frac{1}{2} y_{t-1}$, is the *linear* least-squares predictor of y_t given $\{y_0, \ldots, y_{t-1}\}$. Demonstrate that $E(y_t - \hat{y}_{t|t-1})^2 = \frac{1}{16}$.

Note: If $y_0 = 0.a_1 a_2 a_3 \cdots$, observe that the $\{a_k\}$ are independent random variables taking values $\{0, 1\}$, each with probability $\frac{1}{2}$, and that we have

$$y_k = 0.a_k a_{k+1} \cdots = \sum_{i=1}^{\infty} \frac{a_{k+i}}{2^i}.$$

13.3-11 Consider a process $\{\mathbf{y}_t\}$ with a state-space model

$$\begin{aligned} \mathbf{x}_{t+1} &= A\mathbf{x}_t + B\mathbf{w}_t \quad t \geq 0, \\ \mathbf{y}_t &= C\mathbf{x}_t + \nu_t, \end{aligned}$$

with

$$E \begin{bmatrix} \mathbf{w}_i \\ \nu_i \\ \mathbf{x}_0 \end{bmatrix} \begin{bmatrix} \mathbf{w}_j^T & \nu_j^T & \mathbf{x}_0^T \end{bmatrix} = \begin{bmatrix} Q & H & 0 \\ H^T & R & 0 \\ 0 & 0 & \Pi_0 \end{bmatrix} \delta_{ij},$$

where δ_{ij} is the Kronecker delta function. Define $\Pi_k = E\mathbf{x}_k \mathbf{x}_k^T$. Show that we can write

$$E\mathbf{y}_i \mathbf{y}_j^T = \begin{cases} CA^{i-j} N_j + R\delta_{ij} & i \geq j, \\ N_i^T A^{j-i} C^T & i < j, \end{cases}$$

13.9 Exercises

where
$$\mathbf{N}_j = \Pi_j C^T + BH.$$

13.3-12 A process $\{y_k\}$ is called wide-sense Markov if the linear least-squares estimate of y_{k+j}, $j > 0$, given $\{y_i, i \leq k\}$, depends only upon the value of y_k. Show that a process is wide-sense Markov if and only if
$$f(i,k) = f(i,j)f(j,k), \quad i \leq j \leq k,$$
where
$$f(i,j) \stackrel{def}{=} \frac{r(i,j)}{r(j,j)},$$
$$r(i,j) \stackrel{def}{=} Ey_i y_j.$$

13.3-13 In this problem, yet another derivation of the Kalman filter is presented, which relies upon a linear minimum-variance criterion. Let us impose the linear measurement-update model
$$\hat{\mathbf{x}}_{t+1|t+1} = D_{t+1}\hat{\mathbf{x}}_{t+1|t} + K_{t+1}\mathbf{y}_{t+1},$$
where D_{t+1} and K_{t+1} are to be found. Let us also use the following notation for the estimation error:
$$\tilde{\mathbf{x}}_{t+1|t+1} = \mathbf{x}_{t+1} - \hat{\mathbf{x}}_{t+1|t+1}, \quad \tilde{\mathbf{x}}_{t+1|t} = \mathbf{x}_{t+1} - \hat{\mathbf{x}}_{t+1|t}.$$

(a) Show that if $\hat{\mathbf{x}}_{t+1|t+1}$ is to be unbiased, then
$$D_{t+1} = I - K_{t+1}C_{t+1}. \tag{13.97}$$

(You may assume, as a recursive step, that $E[\tilde{\mathbf{x}}_{t+1|t}] = 0$.)

(b) On the basis of (13.97), show that
$$\hat{\mathbf{x}}_{t+1|t+1} = \hat{\mathbf{x}}_{t+1|t} + K_{t+1}(\mathbf{y}_{t+1} - C_{t+1}\hat{\mathbf{x}}_{t+1|t}).$$

(c) Show that
$$P_{t+1|t+1} = E\left[\tilde{\mathbf{x}}_{t+1|t+1}\tilde{\mathbf{x}}^T_{t+1|t+1}\right]$$
may be written as
$$P_{t+1|t+1} = (I - K_{t+1}C_{t+1})P_{t+1|t}(I - K_{t+1}C_{t+1})^T + K_{t+1}R_{t+1}K^T_{t+1}.$$

(d) A reasonable performance criterion is to minimize
$$E\left[\hat{\mathbf{x}}^T_{t+1|t+1}\hat{\mathbf{x}}_{t+1|t+1}\right] = \text{tr}(P_{t+1|t+1}).$$

Using the gradient formulas in appendix E, show that the minimum can be achieved by taking the gradient with respect to K_{t+1}, to obtain
$$K_{t+1} = P_{t+1|t}C^T_{t+1}\left(C_{t+1}P_{t+1|t}C^T_{t+1} + R_{t+1}\right)^{-1},$$
and that the corresponding error covariance is
$$P_{t+1|t+1} = (I - K_{t+1})C_{t+1}P_{t+1|t}.$$

(e) Compare these formulas with those for the Kalman filter derived via Bayesian and innovations methods.

13.6-14 For the following homogeneous (zero-input) systems, find the equilibrium points and a linearized system about the equilibrium points.

(a) $\dot{x}_1 = x_2$,
$\dot{x}_2 = -x_1 + x_1^3/8 - x_2$.

(b) $\dot{x}_1 = -x_1^3 + 2x_2$,
$\dot{x}_2 = x_1 - x_2^3$.

(c) $\dot{x}_1 = 2x_2 \cos x_1$,
$\dot{x}_2 = 2 \sin x_1$.

13.7-15 Show that (13.87) and (13.88) are correct.

13.7-16 Write a MATLAB function `xsmooth = smooth(X,t,A,C,Q,R,P0)` that provides an estimate of $\mathbf{x}_t$ using the data in X (stored in columns).

13.10 References

The Kalman filter was introduced in [169] and [170], initially for continuous time. Since the time of its introduction, a variety of presentations have been made, including [41, 154, 4, 102, 291]. The innovations approach has been thoroughly examined in a series of papers starting in [163] and extending through [104] (and references therein). The minimum-variance approach is promoted in [102].

The comparison of the Kalman filter and the RLS filter is made in [289]. The smoothing algorithm of section 13.7 is described in [271]. Square-root algorithms are summarized in [356], which also provides a comparison of their numeric conditioning and computational complexity.

The problem of initializing the Kalman filter with uncertain initial values has been addressed using set-valued Kalman filters, described in [235], with an application to multitarget tracking in [232].

H_∞ filter design is discussed in [8, 122, 302, 237, 125, 129, 306, 377, 15, 307]. A thorough discussion of the H_∞ viewpoint is provided in [386].

Part IV

Iterative and Recursive Methods in Signal Processing

Iterative algorithms are those for which a new solution is obtained as a refinement of a previous solution. Recursive systems provide an update of a solution as new data become available. We begin this part with an introduction to the kinds of behaviors that can occur in iterative systems, then present some basic results relating to convergence. This is followed by an introduction to several iterative algorithms: Newton's method, steepest descent, composite mapping, clustering, iterative solution of linear systems, and iterative maximum likelihood (the EM algorithm). By this array of choices, we demonstrate a broad (but not exhaustive) cross-section of iterative techniques. Applications of iterative algorithms are legion, of which we illustrate a few: LMS adaptive filtering, multilayer perceptrons (neural networks), bandlimited signal reconstruction, vector quantization, training hidden Markov models, and others.

Chapter 14

Basic Concepts and Methods of Iterative Algorithms

"...et procédant ainsi infiniment, l'on approche infiniment plus près au requis".

[...and proceeding in this way unendingly, one approaches infinitely closer to the required value.]

— S. Stevin
La pratique d'arithmetique

In digital computation in general, there are very few algorithms that are not at least partly iterative in nature. LU decomposition and Cholesky factorization of matrices are examples of algorithms that iterate through several steps, row by row, until completion. However, iteration of this sort is not what is intended as the current focus of study. Iterative methods, as described in this part, may be loosely categorized as those for which at each stage a solution exists that is approximately correct, and for which successive iterations of the algorithm may improve the quality of the solution, either by incorporating more data into the solution or by simply improving the solution already found. By this loose definition, we are thus encompassing algorithms that are truly *iterative*, and also algorithms that are *recursive* in nature. A recursive algorithm is one for which a new solution is computed as new data become available, where the new value is specifically obtained by updating the old value, as opposed to computing the new result from scratch (starting at the beginning of the data set). An example of an interative algorithm is minimization by steepest descent: at each pass through the algorithm, the solution approaches closer to a final minimizing solution. Another example that we have already encountered—which is, strictly speaking, data-recursive—is the RLS algorithm, in which a least-squares filtering problem is iteratively solved, with more data used at each step of the algorithm.

Iterative algorithms are employed in signal processing for a variety of reasons. One of the most important is that closed-form solutions of some equations may not be obtainable, but it may be possible to approach an acceptable solution by degrees. This is often the case for nonlinear equations. It is often possible to introduce into iterative algorithms constraints that would be difficult to incorporate into a closed-form solution.

Another reason for iterative methods is adaptivity: instead of fixing a permanent solution, the system evolves as new data comes in. (Again, more precisely, such algorithms should be termed recursive.) This is particularly important for real-time systems. Adaptive filters are examples of algorithms of this sort. Another advantage is computational expediency: it may require substantially fewer computations to get a close solution than one that is "exact." In many applications the need for speed exceeds the need for precision.

When a solution is approached by stages, one of the issues that must be addressed is convergence: if the algorithm were to run sufficiently long, would it converge to a final solution? The speed of convergence is also an issue. It is desirable to approach the final answer as quickly as possible, where the speed may be measured either in the amount of data it takes to obtain the solution, or the number of computational cycles. An important part of the study of iterative algorithms is the determination of convergence properties. In the methods presented in this section, issues of convergence are discussed at a variety of levels. For some algorithms, convergence is easy to establish. For other algorithms, convergence can only be determined empirically, on the basis of computerized tests. While the latter may not be as rigorously satisfying as a theoretical proof, it is by no means uncommon or unacceptable. In this case, enough tests must be run for some degree of confidence in the convergence to be established, for a variety of conditions.

There are a variety of iterative algorithms that have been developed in response to diverse needs. The methods presented in this part may be regarded as important representatives of iterative methods, but by no means an exhaustive survey. The field is so broad that for each type of algorithm, entire books, or even careers, have been dedicated to their understanding.

In this chapter, we examine the basic definitions and theorems associated with iterative methods. We examine the qualitative behavior that may be seen at fixed points of an iterated system. The important contraction mapping theorem is introduced. Following this, some simple but important iterative methods are introduced: Newton's method and steepest descent, together with variations and applications.

14.1 Definitions and qualitative properties of iterated functions

In this section we introduce some important definitions for iterated functions. While not all iterative methods rely on iterated functions, the concepts presented here are important to keep in mind when dealing with iterative methods and their convergence.

In iterative methods, we often evaluate a function on the result of a previous computation. Suppose we have a function f that maps a space S into itself, $f: S \to S$. Starting from a point $\mathbf{x} \in S$, we can compute $f(\mathbf{x})$. Since this also lies in S, we can apply $f(\mathbf{x})$ again to obtain $f(f(\mathbf{x}))$, and so on. We can designate the result of the nth operation as $\mathbf{x}^{[n]}$, with $\mathbf{x}^{[0]} = \mathbf{x}$, the starting point. Thus

$$\mathbf{x}^{[0]} = \mathbf{x},$$
$$\mathbf{x}^{[1]} = f(\mathbf{x}),$$
$$\mathbf{x}^{[2]} = f(f(\mathbf{x})),$$

and so forth. The set of points $\mathbf{x}^{[n]}$, $n = 0, 1, 2, \ldots$, is called the **orbit** of the point $\mathbf{x}$ under the transformation f. (Strictly speaking, it is the forward orbit.) To denote the repeated application of the function f, we use the notation $f^{\circ n}(\mathbf{x})$, which mean f is applied successively n times, so that

$$f^{\circ n}(\mathbf{x}) = f(f^{\circ n-1}(\mathbf{x})).$$

A point $\mathbf{x}$ such that $\mathbf{x} = f(\mathbf{x})$, that is, one in which application of f does not change the value, is called a **fixed point** of the mapping. If there is a point $\mathbf{x}$ such that $\mathbf{x} = f^{\circ nk}(\mathbf{x})$ for some $n \in \mathbb{Z}^+$, then $\mathbf{x}$ is said to be a **periodic point** with period k. A fixed point is periodic

14.1 Definitions and Qualitative Properties of Iterated Functions

with period 1. The result of finding $\mathbf{x}^{[n+1]}$ by application of a transformation $f(\mathbf{x}^{[n]})$ is sometimes referred to as **successive approximation**.

The result of repeated applications of functions over the real numbers may be visualized as shown in figure 14.1. Starting from a point $x^{[0]}$, the function is evaluated to produce $f(x^{[0]})$. This y value must become the x value for the next iteration. Moving from $f(x)$ horizontally to the line $x = y$, we find graphically the location of the next x value, $x^{[1]} = f(x^{[0]})$. Repeating this, we obtain a succession of values. In figure 14.1, the result of repeated application is that a fixed point is eventually reached. This point is called an **attractor**, or *attractive fixed point*, because x values that are not initially on the fixed point are eventually drawn to it by repeated application of f. The sequence $\{x^{[0]}, x^{[1]}, x^{[2]}, \ldots\}$ (the orbit of $x^{[0]}$) converges to the fixed point. Note that the fixed point is the point where the line $y = x$ crosses $f(x)$. A little thought reveals that this must be the case. Figure 14.2 illustrates another possibility. The iteration begins at a value near the fixed point, which is near $x = 3.8$. However, rather than moving toward the fixed point, successive iterations

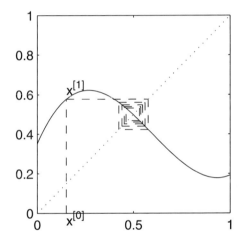

Figure 14.1: Illustration of an orbit of a function with an attractive fixed point

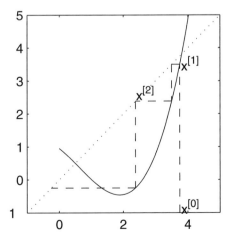

Figure 14.2: Illustration of an orbit of a function with a repelling fixed point

move *away* from the fixed point. If the iteration had started exactly at the fixed point, then successive iterations would have stayed there. However, for points not at that point of intersection, the orbit diverges. Such a fixed point is a *nonattractive*, or *repelling*, or *unstable* fixed point.

14.1.1 Basic theorems of iterated functions

The following theorems provide results concerning the existence and nature of fixed points in an iterated function system for scalar variables. Generalizations to vector variables exist in most cases.

Theorem 14.1 *([74, page 13]) Let $I = [a, b] \subset \mathbb{R}$ be an interval and let $f: I \to I$ be continuous. Then f has at least one fixed point in I.*

Proof Let $h(x) = f(x) - x$. Suppose that $f(a) > a$ and $f(b) < b$. (Otherwise, either a or b is fixed and the theorem is established.) Then $h(a) > 0$ and $h(b) < 0$. By the intermediate value theorem (see section A.5), there is some value c, $a < c < b$, such that $h(c) = 0$. This means that $f(c) = c$. □

Definition 14.1 An **attractive fixed point** x^* of a mapping f is a point such that $f(x^*) = x^*$ with an open neighborhood U surrounding it, such that points in U tend toward x^* under iteration of f. □

Theorem 14.2 *Let $I = [a, b]$ be an interval and let $f: I \to I$ be continuous. If $|f'(x)| < 1$ for all $x \in I$, then there exists a unique fixed point for f in I. Also, $|f(x) - f(y)| < |x - y|$.*

The condition that $|f(x) - f(y)| < |x - y|$ is defined in the proof as a "contraction mapping." Contraction mappings are discussed in more detail in section 14.2.

Proof For $x, y \in I$, with $x \neq y$, there is, by the mean value theorem (see section A.6.1), a $c \in [x, y]$ so that

$$|f(x) - f(y)| = |f'(c)||x - y| < |x - y|.$$

Thus, f is established as a contraction mapping.

By theorem 14.1, f has at least one fixed point. It remains to establish uniqueness. Assume that there are two fixed points x and y. By the mean value theorem, there is a $c \in [x, y]$ so that

$$f'(c) = \frac{f(y) - f(x)}{y - x}.$$

But if x and y are fixed points, we have

$$\frac{f(y) - f(x)}{y - x} = \frac{y - x}{y - x} = 1,$$

which is a violation of $|f'(c)| < 1$. □

Theorem 14.3 *If $f(x)$ is a C^1 function with fixed point x^*, such that $|f'(x^*)| < 1$, then x^* is an attractive fixed point.*

Proof Since f is C^1, there is an $\epsilon > 0$ such that for $x \in [x^* - \epsilon, x^* + \epsilon]$, $|f'(x)| < 1$. By application of the mean value theorem, we obtain

$$\frac{|f(x) - x^*|}{x - x^*} = \frac{|f(x) - x^*|}{x - x^*} = f'(c)$$

14.1 Definitions and Qualitative Properties of Iterated Functions

for $c \in [x, x^*]$. But because, therefore, $c \in [x^* - \epsilon, x^* + \epsilon]$, it follows that $|f'(c)| < 1$ and
$$|f(x) - x^*| < |x - x^*|.$$
By the same argument, $|f^{\circ n}(x) - x^*| \leq |f'(c)|^n |x - x^*|$, so $f^{\circ n}(x) \to x^*$ as $n \to \infty$. □

14.1.2 Illustration of the basic theorems

In the interest of understanding the types of behavior that may result from iterated function systems, we explore some of the behavior of iterations of a simple quadratic function known as the *logistic function*. This simple function demonstrates a variety of behaviors, including attractive fixed points, repelling fixed points, periodic cycles and, perhaps most interestingly, chaos. These differing behaviors are obtained by the variation of a single parameter of the system. The logistic map is defined by the function
$$f(x) = \lambda x (1 - x),$$
where x is in the range $0 \leq x \leq 1$ and λ is a parameter of the mapping. Figure 14.3(a)

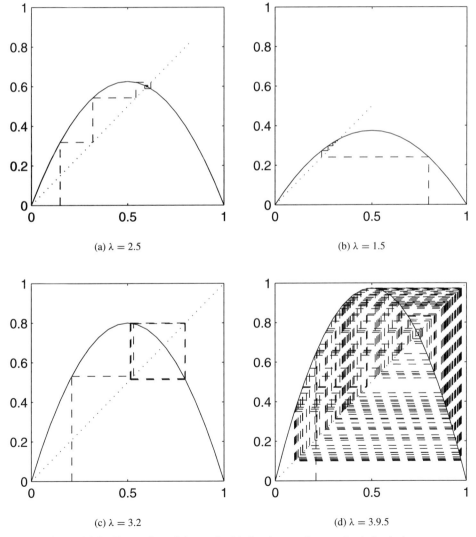

(a) $\lambda = 2.5$ (b) $\lambda = 1.5$

(c) $\lambda = 3.2$ (d) $\lambda = 3.9.5$

Figure 14.3: Examples of dynamical behavior on the quadratic logistic map

shows the trajectory orbit with $x^0 = 0.15$ and $\lambda = 2.5$, and was produced with the code in algorithm 14.1.

Algorithm 14.1 Logistic function orbit
File: `ifs3a.m`
             `logistic.m`

Observe that for this value of λ there is a fixed point at $x^* = 0.6$. The location of the fixed point $f(x) = x$ is

$$x^* = \frac{\lambda - 1}{\lambda}.$$

The derivative at the fixed point is $f'(0.6) = -0.5$. In the plot, the fixed point is located where the graph of $y = f(x)$ intersects the graph of $y = x$. The fixed point x^* is an attractive fixed point because values of x nearby are drawn toward x^* through successive iterations, as predicted by theorem 14.2. Observe that the derivative condition $|f'(x)| < 1$ is a sufficient but not necessary, condition to ensure convergence, since there are values of x such that this condition is not satisfied but that still approach the fixed point.

The logistic function maps $[0, 1] \to [0, 1]$ as long as $1 < \lambda < 4$.

Figure 14.3(b) shows the trajectory with $\lambda = 1.5$ and $x^0 = 0.8$. Again there is a single fixed point, $x = .3333$, where the derivative satisfies $f'(x^*) = 0.5$. In both of these figures there is also a fixed point at $x = 0$, but it is a repelling fixed point since $f'(0) = \lambda > 1$.

Figure 14.3(c) shows a different behavior when $\lambda = 3.2$. (The starting value for the figure is at $x^0 = 0.21$.) The trajectory is converging to oscillate between two values, one at $x = .5130$ and the other at $x = 0.7994$. These points are periodic with period 2. The reason for the presence of two fixed points may be understood from figure 14.4, which shows

$$g(x) = f(f(x)) = \lambda^2 x(1-x)(1-\lambda(1-x)x),$$

along with the line $y = x$. There are four intersections of the graph of $y = g(x)$ with $y = x$.

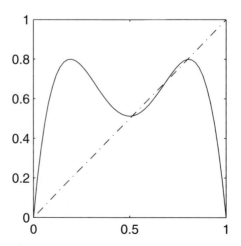

Figure 14.4: Illustration of $g(x) = f(f(x))$ when $\lambda = 3.2$

At the periodic points,
$$g'(0.513) = 0.159 \qquad g'(0.799) = 0.170,$$
so these points are attractive in $g(x)$. Since $g(x)$ is $f(f(x))$, the points are periodic with period 2. There is actually another fixed point of $g(x)$ at $x = 0.6875$, but $g'(.6875) = 1.44$, so this is a repelling fixed point. While only period-two attractive points are shown, by changing the value of λ, periodic points with different periods can be found. (Can you find, for example, a value of λ leading to period-4 behavior?)

Finally, figure 14.3(d) shows iterations of the logistic map when $\lambda = 3.9$, in which something fundamentaly appears. To accentuate the phenomenon, 100 iterations of the function are shown. In this case, there is no fixed or periodic point. The trajectory bounces all over the place. What is evidenced by this is **chaos**: there is no simple periodic attractor. Chaotic behavior such as this is actually fairly common in nonlinear dynamical systems. Chaotic dynamical systems have been used (with varying success) as noise generators.

In summary, several types of behavior are possible with iterated functions:

1. There may be a simple attractive fixed point. From the point of view of the convergence of an iterated algorithm, this is what is usually desired.

2. There may be a fixed point that is not an attractor. From the point of view of most iterative algorithms, this is not useful, because the only way to converge to that fixed point is to *start* on it: any other point in a neighborhood of the fixed point will iterate away from it. This leads (usually) to divergence.

3. There may be periodic attractors. From the point of view of the convergence of an iterative algorithm, this type of behavior is probably not desired, as it leads to what is sometimes called "limit cycle" behavior.

4. There may be no attractors whatsoever. As iterations are computed, the result may not diverge (in the sense of answers that become numerically larger and larger), but neither does it converge, in the sense of successive iterations that become closer to a final answer.

14.2 Contraction mappings

The contraction mapping theorem is a powerful tool that is often used to prove convergence of iterative algorithms. We begin with a definition and demonstration of what a contraction mapping is.

Definition 14.2 Let S be a subspace of a normed space X, and let $T: S \to S$ be a transformation from S into S. Then T is a **contraction mapping** if there is an α with $0 \le \alpha < 1$ such that $\|T(\mathbf{x}) - T(\mathbf{x})\| \le \alpha \|\mathbf{x} - \mathbf{y}\|$ for all $\mathbf{x}, \mathbf{y} \in S$. $\qquad \square$

The effect of a contraction mapping is to bring points in S closer together. What happens when the function is iterated several times?

Example 14.2.1 Let $S = \mathbb{R}^2$, and let
$$T(\mathbf{x}) = \begin{bmatrix} 0.8800 & -0.0800 \\ 0.1800 & 0.8800 \end{bmatrix} \mathbf{x} + \begin{bmatrix} .1 \\ .02 \end{bmatrix}.$$
A mapping of the form $T(\mathbf{x}) = A\mathbf{x} + \mathbf{b}$ is called an **affine transformation**. To see the effect of this transformation, let $\mathcal{X} = [0, 1] \times [0, 1] \subset \mathbb{R}$ be the unit square shown in figure 14.5, considered as a set of points. Let $T(\mathcal{X})$ denote the application of $T(\mathbf{x})$ to every point in $\mathcal{X}$. Then the orbit of the points in $\mathcal{X}$ is a set of increasingly smaller square regions. From the figure, it is clear that the squares

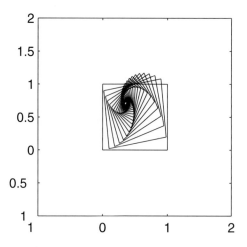

Figure 14.5: Iterations of an affine transformation, acting on a square

are converging to a single point. In fact, for this mapping, any point in $\mathbb{R}^2$ will converge to the same limit point. □

This example demonstrates the following important theorem.

Theorem 14.4 *(Contraction mapping theorem) If T is a contraction mapping on a convex set S of a Banach space, there is a* unique *vector* $\mathbf{x}^* \in S$ *such that* $\mathbf{x}^* = T(\mathbf{x}^*)$.

The contraction mapping theorem states that every contraction mapping on a closed space has a unique fixed point. In addition, that it is possible to reach the fixed point by repeated iteration of the transformation T.

Proof Let $\mathbf{x}_0 \in S$, and let $\mathbf{x}_n = T^{\circ n}(\mathbf{x}_0)$. Then

$$\|\mathbf{x}_{n+1} - \mathbf{x}_n\| = \|T(\mathbf{x}_n) - T(\mathbf{x}_{n-1})\| \leq \alpha \|\mathbf{x}_n - \mathbf{x}_{n-1}\|.$$

Inductively, we see that

$$\|\mathbf{x}_{n+1} - \mathbf{x}_n\| \leq \alpha^n \|\mathbf{x}_1 - \mathbf{x}_0\|.$$

We will show that the sequence $\{\mathbf{x}_n\}$ converges, by showing that it is a Cauchy sequence (see section 2.1.2); then, since S is in a Banach space, the limit point must exist. To see that $\{\mathbf{x}_n\}$ forms a Cauchy sequence, note that for any $p > 0$,

$$\|\mathbf{x}_{n+p} - \mathbf{x}_n\| = \|\mathbf{x}_{n+p} - \mathbf{x}_{n+p-1} + \mathbf{x}_{n+p-1}\| - \mathbf{x}_n\|$$

$$\leq \|\mathbf{x}_{n+p} - \mathbf{x}_{n+p-1}\| + \|\mathbf{x}_{n+p-1} - \mathbf{x}_n\|.$$

Repeating this process, we find that

$$\|\mathbf{x}_{n+p} - \mathbf{x}_n\| \leq \|\mathbf{x}_{n+p} - \mathbf{x}_{n+p-1}\| + \|\mathbf{x}_{n+p-1} - \mathbf{x}_{n+p-2}\| + \cdots + \|\mathbf{x}_{n+1} - \mathbf{x}_n\|$$

$$\leq (\alpha^{n+p-1} + \alpha^{n+p-2} + \cdots + \alpha^n) \|\mathbf{x}_1 - \mathbf{x}_0\|$$

$$\leq \alpha^n \sum_{k=0}^{\infty} \alpha^k \|\mathbf{x}_1 - \mathbf{x}_0\| = \frac{\alpha^{n-1}}{1-\alpha} \|\mathbf{x}_1 - \mathbf{x}_0\|.$$

Thus, for n sufficiently large, $\|\mathbf{x}_{n+p} - \mathbf{x}_n\|$ is arbitrarily small for any p and, since S is closed, there must be a limit element $\mathbf{x}^*$ such that $\mathbf{x}_n \to \mathbf{x}^*$.

To show that the limit satisfies $\mathbf{x}^* = T(\mathbf{x}^*)$, observe that

$$\|\mathbf{x}^* - T(\mathbf{x}^*)\| \le \|\mathbf{x}^* - \mathbf{x}_n\| + \|\mathbf{x}_n - T(\mathbf{x}_n)\|$$
$$\le \|\mathbf{x}^* - \mathbf{x}_n\| + \alpha^n \|\mathbf{x}_1 - \mathbf{x}_0\|.$$

Since $\mathbf{x}_n \to \mathbf{x}^*$, both terms $\to 0$ as $n \to \infty$. Thus $\mathbf{x}^* = T(\mathbf{x}^*)$.

Uniqueness of the limit may be shown by assuming that there are two distinct limits, $\mathbf{x}^*$ and $\mathbf{y}^*$. Then

$$\|\mathbf{x}^* - \mathbf{y}^*\| = \|T(\mathbf{x}^*) - T(\mathbf{y}^*)\| \le \alpha \|\mathbf{x}^* - \mathbf{y}^*\|.$$

But since $\alpha < 1$, this means that $\mathbf{x}^* = \mathbf{y}^*$. □

By theorem 14.2, if $|T'(\mathbf{x})| < 1$ for all $\mathbf{x} \in S$, then T is a contraction mapping.

Example 14.2.2 *Jacobi iteration* provides an iterative means of solving the linear equation $A\mathbf{x} = \mathbf{b}$, without explicitly inverting A. The solution to $A\mathbf{x} = \mathbf{b}$ is the solution to

$$\mathbf{x} = (I - A)\mathbf{x} + \mathbf{b}.$$

Let $T(\mathbf{x}) = (I - A)\mathbf{x} + \mathbf{b}$. This is seen to be an affine transformation. The fixed point (if there is one) is the solution $\mathbf{x} = A^{-1}\mathbf{b}$. Convergence of the method may be established by showing that $T(\mathbf{x})$ is a contraction mapping. Since

$$\|T(\mathbf{x}) - T(\mathbf{y})\| \le \|A - I\| \|\mathbf{x} - \mathbf{y}\|,$$

$T(\mathbf{x})$ is a contraction mapping provided that

$$\|A - I\| < 1$$

for some matrix norm $\|\cdot\|$. Further details of Jacobi iterations are found in section 16.2. □

Example 14.2.3 ([209, page 275]) The integral equation

$$x(t) = f(t) + \lambda \int_a^b K(t,s) x(s) \, ds,$$

with bounded kernel

$$\int_a^b \int_a^b K^2(s,t) \, dt \, ds = \beta^2 < \infty,$$

defines a bounded linear operator on $L_2[a,b]$ with norm $\le \beta$. Define the mapping

$$T(x) = f(x) + \lambda \int_a^b K(t,s) x(s) \, ds.$$

Then $T(x)$ is a contraction mapping, provided that $|\lambda| < 1/\beta$. For λ in this range, a unique solution $x(t)$ can be determined by iteration of $T(x)$. □

14.3 Rates of convergence for iterative algorithms

For iterative algorithms approaching a final fixed point $\mathbf{x}^{[n]} \to \mathbf{x}^*$, it is of interest to qualitatively determine how fast the limit is approached.

Definition 14.3 Let the sequence $\{\mathbf{x}^{[n]}\}$ converge to $\mathbf{x}^*$. The **order** of convergence of $\{\mathbf{x}^{[n]}\}$ is defined as the supremum of the nonnegative integers p that satisfy

$$0 \le \limsup_{n \to \infty} \frac{\|\mathbf{x}^{[n+1]} - \mathbf{x}^*\|}{\|\mathbf{x}^{[n]} - \mathbf{x}^*\|^p} < \infty.$$

□

This definition applies as $n \to \infty$: in the rate of convergence, how the sequence starts out may have little impact on the final rate of convergence. The order of convergence, then, is a measure of how fast the *tail* of a sequence decreases—that part of the sequence that is arbitrarily far out.

If p is the order of convergence of a sequence, then the higher p is, the faster the convergence, since the distance from $\mathbf{x}^{[n]}$ to the limit is reduced by a factor of p in a single step. In other words, if there is a finite, nonzero β such that

$$\beta = \lim_{n \to \infty} \frac{\|\mathbf{x}^{[n+1]} - \mathbf{x}^*\|}{\|\mathbf{x}^{[n]} - \mathbf{x}^*\|^p},$$

then, asymptotically,

$$\|\mathbf{x}^{[n+1]} - \mathbf{x}^*\| = \beta \|\mathbf{x}^{[n]} - \mathbf{x}^*\|^p.$$

Example 14.3.1 We consider two simple examples.

1. Suppose $x^{[n]} = a^n$, for $0 < a < 1$. Then $x^{[n]} \to 0$. The ratio of successive terms is

$$\frac{x^{[n+1]}}{x^{[n]}} = a,$$

so that the order of convergence is 1.

2. Suppose $x^{[n]} = a^{2^n}$ for $0 < a < 1$. Then the ratio of successive terms is

$$\frac{x^{[n+1]}}{x^{[n]}} = a^{2^n},$$

so that $x^{[n+1]}/(x^{[n]})^2 = 1$. Hence, the sequence converges to zero with order 2. □

Definition 14.4 If the sequence $\{\mathbf{x}^{[n]}\}$ converges to $\mathbf{x}^*$ in such a way that

$$\lim_{n \to \infty} \frac{\|\mathbf{x}^{[n+1]} - \mathbf{x}^*\|}{\|\mathbf{x}^{[n]} - \mathbf{x}^*\|} = \beta, \tag{14.1}$$

for some $\beta < 1$, then the sequence is said to converge **linearly**, with ratio β. If $\beta = 0$ in (14.1), then the convergence is said to be *superlinear*. □

The tail of a linearly convergent sequence decays at a rate $c\beta^n$ for some constant c; linear convergence thus decays (oddly enough) at a geometric rate.

Example 14.3.2 Let $x^{[n]} = 1/n$. Then the convergence is of order 1, but it is *not* linear, since

$$\lim_{n \to \infty} \frac{x^{[n+1]}}{x^{[n]}} = 1,$$

so that β is not less than 1. □

14.4 Newton's method

Newton's method is an iterative method for finding a solution of an equation of the form

$$f(x) = 0.$$

We introduce this first for scalar variables, then generalize to vector variables.*

---

*For a brief biography of this phenomenal innovator, please see box 14.1.

14.4 Newton's Method

> **Box 14.1: Isaac Newton (1642–1727)**
>
> The "method" that bears his name is only the smallest of the many contributions made by Newton during his lifetime. He is still regarded by many as the greatest mathematician of all time.
>
> Newton was born the son of a farmer, but his mechanical cleverness led to continuation of schooling, and he entered Trinity College at Cambridge University at age eighteen. It was at this time that he directed his attention to mathematics. When he was twentythree, the schools were closed for two years due to the bubonic plague. These years were spent in productive effort; during that time Newton developed the binomial theorem, differential calculus, and his theory of gravitation; and explored the nature of color. It is retrospectively remarkable that he was as successful as he was with calculus without the benefit of a consistent concept of limits. He also made (at first, by trial and error) discoveries about infinite series.
>
> Major works published during his life include *Principia* (1687), *Opticks* (1704), *Arithmetica universalis* (1707), and *Analysis per Series, Fluxiones, etc.* (1711). In 1669, he was offered the Lucasian professorship at Cambridge (the seat most recently held by Steven Hawking), where he remained for eighteen years. In 1696 he was made Warden of the Mint (which appears to be have been a sinecure), and in 1703 he was elected President of the Royal Society, a position he retained until his death.

To derive Newton's method, we write a Taylor series for $f(x)$ about a starting point $x^{[n]}$,

$$f(x) = f(x^{[n]}) + (x - x^{[n]})f'(x^{[n]}) + \text{h.o.t.}$$

(where h.o.t. denotes "higher-order terms"). The linear approximation is obtained by throwing away the higher-order terms. We obtain an approximate solution to $f(x) = 0$ by setting this linear approximation to zero and solving,

$$f(x) \approx f(x^{[n]}) + (x - x^{[n]})f'(x^{[n]}) = 0,$$

which leads to

$$x = x^{[n]} - \frac{f(x^{[n]})}{f'(x^{[n]})}.$$

We assume, of course, that $f'(x^{[n]}) \neq 0$. The solution point x can now be used as the starting point for the next iteration. The algorithm can be stated as

$$\boxed{x^{[n+1]} = x^{[n]} - \frac{f(x^{[n]})}{f'(x^{[n]})}} \qquad (14.2)$$

Example 14.4.1 Newton's method can be used to find roots of numbers. To illustrate, we compute the cube root of a number. Solution of the equation $f(x) = x^3 - a = 0$ will compute the cube root of a. By Newton's method,

$$x^{[n]} = x^{[n]} - \frac{(x^{[n]})^3 - a}{3x^{[n]}} = \frac{1}{3}\left(2x^{[n]} + \frac{2}{(x^{[n]})^2}\right).$$

As a specific example, let $a = 8$. Starting from $x^{[0]} = 1.5$, the following sequence of solutions is obtained:

n	$x^{[n]}$	$(x^{[n]})^3$
0	1.5000	3.3750
1	1.2963	2.1783
2	1.2609	2.0048
3	1.2599	2.0000

□

The initial point $x^{[0]}$ can determine which root of $f(x)$ is found, or whether the method converges at all.

Example 14.4.2 Let $f(x) = x\sin(x) - \cos(x)$, as shown in figure 14.6. Running Newton's method with $x^{[0]} = 1$, the final value is $x^* = 0.8603$, so a root near to the initial value is located. Running Newton's method with $x^{[0]} = 2$, the final value is $x^* = -.8603$, which is not the nearest root to $x^{[0]}$. Because the derivative at $x = 2$ is small enough, $x^{[1]} = -.2658$, which causes the next root below zero to be found. When $x^{[0]} = 2.3$, which is on a falling edge of the plot, the solution found is at $x^* = 59.707$, far away from the initial value. □

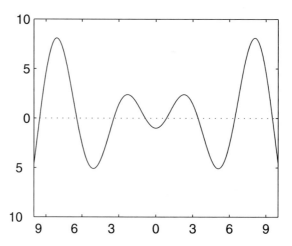

Figure 14.6: Illustration of Newton's method

The point of this example is that a good estimate of the root location can be a crucial condition for ending up where you expect (or desire). Newton's method in general is *not guaranteed* to converge. However, if the initial point is sufficiently close to a root, convergence often occurs. In order to get sufficiently close to a root, other methods are often used, with Newton's method used to refine the solution.

One result on the convergence of Newton's method can be obtained using the contraction mapping principle. Newton's method can be expressed as a transformation

$$T(x) = x - \frac{f(x)}{f'(x)}.$$

Then, iterating Newton's method can be viewed as successive approximation, and the solution as a fixed point of $T(x)$. The derivative of the transformation is

$$T'(x) = \frac{f(x)f'(x)}{f''(x)}.$$

14.4 Newton's Method

If the functions are bounded as

$$\left\|\frac{1}{f'(x)}\right\| < \beta \qquad \|f''(x)\| < \kappa \qquad \left\|\frac{f(x)}{f'(x)}\right\| < \eta$$

and have $h = \beta\kappa\beta$, we have $\|T'(x)\| < h$. If $h < 1$ for every point in the region of interest, we would expect convergence. In fact, it can be shown [209, page 279] that if $h < \frac{1}{2}$ at $x = x^{[0]}$, then $h < \frac{1}{2}$ for all points in the iteration and, thus, by the contraction mapping theorem, converges.

When Newton's method converges, the rate of convergence can be determined as follows. Write the Taylor series expansion about $x^{[n]}$ as

$$f(x) = f(x^{[n]}) = f'(x^{[n]})(x - x^{[n]}) + \frac{1}{2}f''(\eta)(x - x^{[n]})^2$$

where $\eta \in [x, x^{[n]}]$. At the solution x^*, $f(x^*) = 0$ and

$$0 = f(x^*) = f(x^{[n]}) + f'(x^{[n]})(x^* - x^{[n]}) + \frac{1}{2}f''(\eta)(x^* - x^{[n]})^2.$$

When this is divided by $f'(x^{[n]})$, it can be rearranged as

$$x^* - \left(x^{[n]} - \frac{f(x^{[n]})}{f'(x^{[n]})}\right) = \frac{f''(\eta)}{2f'(x^{[n]})}(x^* - x^{[n]})^2.$$

The left-hand side is simply $x^* - x^{[n+1]}$, so we obtain

$$x^* - x^{[n+1]} = C_n(x^* - x^{[n]})^2,$$

where $C_n = \frac{f''(\eta)}{2f'(x^{[n]})}$. From this we observe that the error $x^* - x^{[n+1]}$ decays *quadratically* with n, provided that $f'(x^{[n]}) \neq 0$. This rapid convergence makes Newton's method attractive computationally. However, the quadratic convergence is typically observed only when the solution is quite close.

Newton's method is often used to minimize functions. Let $f(x)$ be a C^2 (twice differentiable) function, and let

$$F(x) = f'(x).$$

Then there is an extremum of $f(x)$ at a point x^* where $F(x^*) = 0$. If $F'(x^*) = f''(x^*) > 0$ at the point of extremum, then x^* is a minimizer of $f(x)$. The update for Newton's method is

$$x^{[n+1]} = x^{[n]} - \frac{f'(x^{[n]})}{f''(x^{[n]})}.$$

As before, Newton's method for minimization exhibits quadratic convergence.

Newton's method can be used to find zeros of transformations in higher dimensions, or to minimize scalar functions of multiple variables. Let $\mathbf{f}(\mathbf{x}): \mathbb{R}^n \to \mathbb{R}^n$, and let

$$\mathbf{f}'(\mathbf{x}) = \begin{bmatrix} \frac{\partial f_1}{\partial x_1} & \frac{\partial f_1}{\partial x_2} & \cdots & \frac{\partial f_1}{\partial x_n} \\ \frac{\partial f_2}{\partial x_1} & \frac{\partial f_2}{\partial x_2} & \cdots & \frac{\partial f_2}{\partial x_n} \\ \vdots & & & \\ \frac{\partial f_n}{\partial x_1} & \frac{\partial f_n}{\partial x_2} & \cdots & \frac{\partial f_n}{\partial x_n} \end{bmatrix}.$$

A step in Newton's method in the multidimensional case is

$$\mathbf{x}^{[n+1]} = \mathbf{x}^{[n]} - [\mathbf{f}'(\mathbf{x}^{[n]})]^{-1}\mathbf{f}(\mathbf{x}^{[n]}).$$

Newton's method can also be employed to minimize functions of multiple variables. Let $g(\mathbf{x}): \mathbb{R}^n \to \mathbb{R}$. Then an extremum of $g(\mathbf{x})$ exists where $\frac{\partial g}{\partial \mathbf{x}} = 0$. That is, we let $\mathbf{f}(\mathbf{x}) = \frac{\partial g}{\partial \mathbf{x}}$, and find the zero of $\mathbf{f}$. To derive Newton's method for this minimization problem, write a Taylor expansion for $g(\mathbf{x})$ about a point $g(\mathbf{x}^{[n]})$:

$$g(\mathbf{x}) = g(\mathbf{x}^{[n]}) + (\mathbf{x} - \mathbf{x}^{[n]})\nabla g + \frac{1}{2}(\mathbf{x} - \mathbf{x}^{[n]})^T \nabla^2 g (\mathbf{x} - \mathbf{x}^{[n]}) + \text{h.o.t.},$$

where

$$\nabla g = \frac{\partial g}{\partial \mathbf{x}} = \begin{bmatrix} \frac{\partial g}{\partial x_1} \\ \frac{\partial g}{\partial x_2} \\ \vdots \\ \frac{\partial g}{\partial x_n} \end{bmatrix} \quad (14.3)$$

and

$$\nabla^2 g = \begin{bmatrix} \frac{\partial^2 g}{\partial x_1 \partial x_1} & \frac{\partial^2 g}{\partial x_1 \partial x_2} & \cdots & \frac{\partial^2 g}{\partial x_1 \partial x_n} \\ \frac{\partial^2 g}{\partial x_2 \partial x_1} & \frac{\partial^2 g}{\partial x_2 \partial x_2} & \cdots & \frac{\partial^2 g}{\partial x_2 \partial x_n} \\ \vdots & & & \\ \frac{\partial^2 g}{\partial x_n \partial x_1} & \frac{\partial^2 g}{\partial x_n \partial x_2} & \cdots & \frac{\partial^2 g}{\partial x_n \partial x_n} \end{bmatrix}.$$

The matrix $\nabla^2 g$ is the **Hessian** of g. If the Hessian is positive definite, the extremum is a (local) minimum. We obtain a quadratic approximation to $g(\mathbf{x})$ by throwing out the higher-order terms in (14.3), then obtain the Newton step by minimizing the resulting quadratic. Taking the gradient of the quadratic

$$Q = g(\mathbf{x}^{[n]}) + (\mathbf{x} - \mathbf{x}^{[n]})\nabla g(\mathbf{x}^{[n]}) + \frac{1}{2}(\mathbf{x} - \mathbf{x}^{[n]})^T \nabla^2 g(\mathbf{x}^{[n]})(\mathbf{x} - \mathbf{x}^{[n]})$$

with respect to $(\mathbf{x} - \mathbf{x}^{[n]})$, and equating to zero, leads to the Newton equation

$$\nabla^2 g(\mathbf{x}^{[n]})(\mathbf{x} - \mathbf{x}^{[n]}) + \nabla g(\mathbf{x}^{[n]}) = 0.$$

Solving $\mathbf{x}$ as a minimizing value, we obtain

$$\mathbf{x}^{[n+1]} = \mathbf{x}^{[n]} - (\nabla^2 g(\mathbf{x}^{[n]}))^{-1} \nabla g(\mathbf{x}^{[n]}).$$

Example 14.4.3 The function $g(\mathbf{x}) = 100(x_2 - x_1^2)^2 + (1 - x_1)^2$ is known as Rosenbrock's function, and is a favorite among numerical analysts for testing minimization algorithms. It has a minimum at $(1, 1)$, approached by a fairly flat "valley." The gradient and Hessian are

$$\nabla g(\mathbf{x}) = \begin{bmatrix} -400x_1(x_2 - x_1^2) + 2(x_1 - 1) \\ 200(x_2 - x_1^2) \end{bmatrix} \quad \nabla^2 g(\mathbf{x}) = \begin{bmatrix} 1200x_1^2 - 400x_2 + 2 & -400x_1 \\ -400x_1 & 200 \end{bmatrix}.$$

Figure 14.7 shows the results of Newton's algorithm, starting with $\mathbf{x}^{[0]} = [-2, 5]^T$. Observe from this example that it is possible for points on the iteration to be *higher* than the starting point. Newton's method does not necessarily always take a "downhill" direction. □

14.5 Steepest Descent

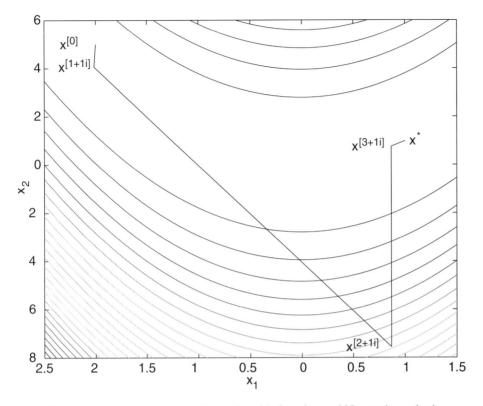

Figure 14.7: Contour plots of Rosenbrock's function and Newton's method

The results on convergence via contraction mapping, and quadratic convergence in the univariable Newton's method, extend to the multivariable Newton's method as well.

The quadratic convergence of Newton's method, when it converges, makes it a valuable tool. However, because there is rarely any guarantee of convergence, it must be used with care.

14.5 Steepest descent

> I wish I had a nickel for every time I heard about gradient descent.
> — J. Clarke Stevens

One method for optimizing a function—in this case, to minimize it—is to iterate in such a way that $f(\mathbf{x}^{[n+1]}) < f(\mathbf{x}^{[n]})$, unless $f(\mathbf{x}^{[n+1]}) = f(\mathbf{x}^{[n]})$, in which case a minimum point (or other extremum) is reached. One general framework for accomplishing this is to update the point $\mathbf{x}^{[n]}$ by

$$\mathbf{x}^{[n+1]} = \mathbf{x} + \alpha_n \mathbf{p}_n, \tag{14.4}$$

where α_n is a scalar, which denotes a step size, and $\mathbf{p}_n$ is a direction of motion, selected so that the successive steps decrease f. Depending on the vector $\mathbf{p}_n$ selected, we can get steepest descent, conjugate-gradient descent, or other successive-approximation algorithms.

It is worth pointing out that iterating (14.4) will *not* necessarily reach the global minimum of a function. For example, if the function is as shown in figure 14.8, starting from

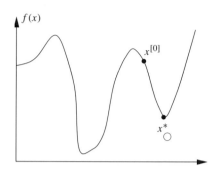

Figure 14.8: A function with local and global minima

$x^{[0]}$, then the best that can be hoped for is to reach the local minimum at x^*. From the given starting point, the global minimum will never be reached by a strict descent method.

A widely used method for minimizing a functional f is *steepest descent*, which is applicable to differentiable functionals defined on a Hilbert space. It is based on the following fact:

Theorem 14.5 *Let $f: \mathbb{R}^m \to \mathbb{R}$ be a differentiable function in some open set D. The gradient $\frac{\partial f}{\partial \mathbf{x}}$ points in the direction of the maximum increase of f at the point $\mathbf{x}$.*

Proof We can write a linear approximation to f (see theorem A.3) where $\Delta \mathbf{x} = \lambda \beta$, with β a unit-length direction vector, as

$$f(\mathbf{x} + \lambda \beta) = f(\mathbf{x}) + \lambda \left(\frac{\partial f}{\partial \mathbf{x}}\right)^T \beta + R, \tag{14.5}$$

where $R = O(\lambda^2)$. Then

$$\frac{f(\mathbf{x} + \lambda \beta) - f(\mathbf{x})}{\lambda} = (\nabla f(\mathbf{x}))^T \beta + \frac{R}{\lambda}.$$

The term R/λ is negligible as $\lambda \to 0$. The maximum change $f(\mathbf{x} + \lambda \beta) - f(\mathbf{x})$ therefore occurs when $(\nabla f(\mathbf{x}))^T \beta$ is maximized. Using the Cauchy–Schwarz inequality, it is clear that this occurs when β is proportional to $\nabla f(\mathbf{x})$. So the maximum-change direction is the direction in which the gradient vector points. □

Since the gradient points in the direction of maximum increase, the negative of the gradient points in the direction of maximum decrease. We thus obtain the method of steepest descent:

$$\mathbf{x}^{[n+1]} = \mathbf{x}^{[n]} - \alpha_n \nabla f(\mathbf{x}^{[n]}). \tag{14.6}$$

The parameter α_n determines how far we move at step n. Frequently, steepest descent algorithms use $\alpha_n = \alpha$, for some constant α.

For functions defined on two variables, such as $f(x_1, x_2)$, the idea of steepest descent is straightforward. The contour lines of the plot indicate the locus of constant function value. The gradient of the function at a point is *orthogonal* to the contour line, pointing in the direction of steepest increase, so the negative of the gradient points in the direction of steepest decrease. (Think of walking across a steep slope so that your altitude remains constant. This is the direction of the contour line. To one side, the slope increases steeply; that is the direction in which the gradient points. To the other side the slope deceases steeply; that is the direction of the negative of the gradient.)

14.5 Steepest Descent

Because the update direction is determined by the gradient, steepest-descent algorithms are also known as *gradient-descent* algorithms.

Example 14.5.1 Figure 14.9 shows contours of the function

$$f(\mathbf{x}) = \mathbf{x}^T R \mathbf{x} - 2\mathbf{b}\mathbf{x},$$

where

$$R = \begin{bmatrix} 105 & 95 \\ 95 & 105 \end{bmatrix}, \qquad \mathbf{b} = \begin{bmatrix} 200 \\ 200 \end{bmatrix},$$

and $\lambda(R) = \{10, 200\}$. The eigenvectors of R point in the directions of

$\begin{bmatrix} 1 \\ 1 \end{bmatrix}$ (with eigenvalue $\lambda = 200$) and $\begin{bmatrix} 1 \\ -1 \end{bmatrix}$ (with eigenvalue $\lambda = 10$).

The minimum of $f(\mathbf{x})$ is at $\mathbf{x}^* = R^{-1}\mathbf{b} = [1, 1]^T$. In figure 14.9(a), 50 iterations of steepest descent are shown for $\alpha = 0.004$. The steepest-descent direction does *not* point toward the minimum of the function, so there is some oscillation as the algorithm converges. Figure 14.9(b) demonstrates that convergence is not guaranteed. In this case, with $\alpha = 0.0051$, the algorithm rocks higher and higher, diverging away from the minimum value but passing over it (in the limit) with each iteration. □

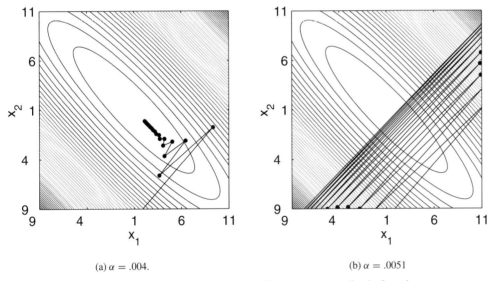

(a) $\alpha = .004$. (b) $\alpha = .0051$

Figure 14.9: Convergence of steepest descent on a quadratic function

Some valuable information about the convergence of steepest descent can be obtained by studying its application to the problem of the last example,

$$f(\mathbf{x}) = \mathbf{x}^T R \mathbf{x} - 2\mathbf{b}^T \mathbf{x}, \qquad (14.7)$$

where R is symmetric positive definite and $\mathbf{x} \in \mathbb{R}^m$. Even though an analytic solution to this minimization problem is available, insight that will benefit more complicated problems may be gained by studying this simple problem, since near a minimizing value a function can be well approximated by truncating its Taylor series at the quadratic term. Equation (14.7) can thus be viewed as an approximation to a general function expanded about its optimizing value. Recall from section 6.4 that in two dimensions the contour lines of $f(\mathbf{x})$ form ellipses in the plane, and in higher dimensions the level surfaces are hyperplanes, with the axes of the ellipse determined by the eigenvectors of R. Let

$$\mathbf{r} = \mathbf{b} - R\mathbf{x}$$

denote the residual. Note that $\nabla f(\mathbf{x}) = 2R\mathbf{x} - 2\mathbf{b} = -2\mathbf{r}$. Steepest descent applied to f with a fixed step size α yields

$$\mathbf{x}^{[n+1]} = \mathbf{x}^{[n]} + 2\alpha \mathbf{r}_n,$$

with initial value $\mathbf{x}^{[0]}$. Let $\mathbf{x}^*$ denote the solution to $R\mathbf{x} = \mathbf{b}$. Shifting coordinates centered around $\mathbf{x}^*$, and letting $\mu = 2\alpha$ for notational convenience, we obtain

$$\mathbf{x}^{[n+1]} - \mathbf{x}^* = \mathbf{x}^{[n]} - \mathbf{x}^* + \mu(\mathbf{b} - R\mathbf{x}^{[n]}).$$

Substituting $\mathbf{y}^{[n]} = \mathbf{x}^{[n]} - \mathbf{x}^*$, we obtain

$$\mathbf{y}^{[n+1]} = (I - \mu R)\mathbf{y}^{[n]}, \tag{14.8}$$

from which it follows inductively that

$$\mathbf{y}^{[n]} = (I - \mu R)^n \mathbf{y}^{[0]}. \tag{14.9}$$

Convergence of this equation from any initial point $\mathbf{y}^{[0]} = \mathbf{x}^{[0]} + \mathbf{x}^*$ requires that

$$\|I - \mu R\| < 1$$

in some norm. Geometrically the L_2 norm is convenient. Let

$$\Lambda = QRQ^T,$$

where Q is the orthogonal matrix composed of eigenvectors of R, and Λ is the diagonal matrix of eigenvalues. Let $\mathbf{z} = Q\mathbf{y}$. This change of variables has the effect of rotating the coordinate system so that the elements of $\mathbf{z}$ are aligned with the axes of the ellipsoid. Under this change of variables, (14.8) becomes

$$\mathbf{z}^{[n+1]} = (QQ^T - \mu QRQ^T)\mathbf{z}^{[n]},$$

which leads to the solution

$$\mathbf{z}^{[n]} = (I - \mu \Lambda)^n \mathbf{z}^{[0]}. \tag{14.10}$$

Since the matrix $I - \mu\Lambda$ is diagonal, (14.10) can be expressed as the set of decoupled equations

$$z_1^{[n]} = (1 - \mu\lambda_1)^n z_1^{[0]},$$

$$z_2^{[n]} = (1 - \mu\lambda_2)^n z_2^{[0]},$$

$$\vdots$$

$$z_m^{[n]} = (1 - \mu\lambda_m)^n z_m^{[0]}.$$

It is clear from these that convergence can occur from any starting point $\mathbf{z}^{[0]}$ only if

$$|1 - \mu\lambda_i| < 1 \qquad i = 1, 2, \ldots, m,$$

that is, if

$$0 < \mu < \frac{2}{\lambda_i} \qquad i = 1, 2, \ldots, m.$$

14.5 Steepest Descent

Since a separate μ is not provided for each direction, we must take the μ satisfying all of the constraints. We must therefore take

$$\boxed{0 < \mu < \frac{2}{\lambda_{\max}}} \qquad (14.11)$$

Example 14.5.2 Returning to the steepest-descent results of example 14.5.1, convergence was observed when $\mu = 0.004 < \frac{1}{200}$, where 200 is the largest eigenvalue of R. With $\mu = 0.0051 > \frac{1}{200}$, the steepest descent diverged. □

There are rates of convergence in each z-coordinate direction that depend on the eigenvalue associated with that coordinate. The z-coordinate directions in which $|1 - \mu\lambda_i|$ is smallest converge the fastest. In figure 14.10(a), the direction along the eigenvector $\begin{bmatrix} 1 \\ 1 \end{bmatrix}$ converges fastest, while convergence along the eigenvector $\begin{bmatrix} 1 \\ -1 \end{bmatrix}$ direction is much slower. The error in each z-coordinate direction is shown in figure 14.10, where part (a) shows the error when $\alpha = 0.004$, and part (b) shows the error when $\alpha = 0.0051$. The solid line is the error component z_1 and the dashed line is the error component z_2. Since the value of μ must be chosen according to (14.11), there is no way to significantly speed convergence in each eigenvector direction if one or more eigenvalues are large in comparison to other eigenvalues. The difference in convergence rates due to the difference in magnitude of the eigenvalues is referred to as being due to *eigenvalue disparity*, or the eigenvalue spread. The eigenvalue spread is sometimes expressed as a ratio,

$$\frac{\lambda_{\max}}{\lambda_{\min}}.$$

In section 4.10, we learned that a matrix is poorly conditioned if the eigenvalue discrepancy is large. Here we see that a poorly conditioned matrix also has slow convergence properties when it is used in a steepest-descent algorithm.

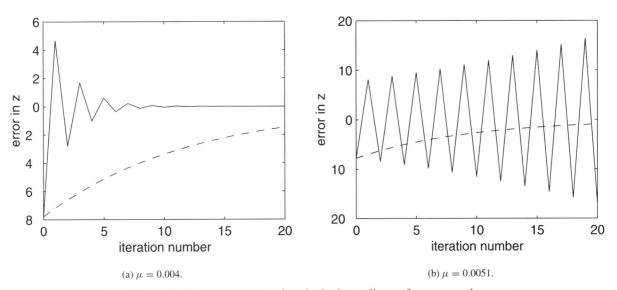

(a) $\mu = 0.004$.   (b) $\mu = 0.0051$.

Figure 14.10: Error components in principal coordinates for steepest descent

14.5.1 Comparison and discussion; other techniques

It is interesting to compare Newton's method to gradient descent. Newton's method is used to find roots of an equation $\mathbf{f}(\mathbf{x}) = 0$, while gradient descent is used to minimize a scalar function $f(\mathbf{x})$. But Newton's method can be applied to the minimization problem by finding a zero of $\nabla f(\mathbf{x})$. Also, gradient descent can be used to find a zero of a function by finding the minimum value of $g(\mathbf{x}) = (f(\mathbf{x}))^2$. Thus either method can be applied to a variety of problems.

One of the characteristics to compare is convergence rate. When Newton's method converges, it converges quadratically, which means that the number of places of precision doubles at every step (at least, near the convergent point). When gradient descent converges, the convergence is geometric, but the overall convergence rate is limited by the eigenvalue disparity of the Hessian matrix of the function.

Convergence of both methods requires that the initial point be somewhat near the final solution. Usually, other methods are used to arrive at a good initial guess. Among these methods, the **bisection method** and the **secant method** are commonly employed, particularly for one-dimensional problems. The bisection method is used to find roots of functions. It begins with two points x_1 and x_2 such that $f(x_1)f(x_2) < 0$, that is, the two points bracket a zero. Then a midpoint between x_1 and x_2 is selected, and a new pair of points is selected so that one of the initial points and the new midpoint again bracket the solution. Convergence of the bisection method is linear, with the relative error decreasing by a factor of 0.5 every iteration.

In the secant method, two points are again used which bracket the root. These two points are used to construct a linear approximation of the function $f(x)$, and the point of intersection of this line is the new estimate of the solution. Convergence is superlinear with a rate $r = \frac{1}{2}(1 + \sqrt{2})$ (the Golden ratio). Proof of the convergence rate is found in [246].

Both of these methods can be generalized to multiple dimensions. For the bisection method, we start with two points $\mathbf{x}_1$ and $\mathbf{x}_2$ such that $f(\mathbf{x}_1)f(\mathbf{x}_2) < 0$. Then a midpoint on the line

$$\alpha \mathbf{x}_1 + (1 - \alpha)\mathbf{x}_2 \qquad 0 \leq \alpha \leq 1$$

is selected by taking $\alpha = 0.5$ (that is, the new point is $(\mathbf{x}_1 + \mathbf{x}_2)/2$), and the algorithm proceeds as before. The secant method is generalized similarly. In the multidimensional setting, finding a zero by searching along a single dimension is sometimes referred to as a *line search*.

Both Newton's method and gradient descent require the use of derivative information. For minimization, Newton's method requires a second derivative (the Hessian for multivariate problems). For some problems, obtaining the Hessian can be difficult or impossible, so a variety of methods have been developed for estimating the Hessian. Another problem is that the Hessian may not be positive definite; some of the eigenvalues may be zero. The iterates cannot converge along the eigendirections with zero eigenvalues. For these circumstances, it has been proposed to replace the Hessian with a nearby matrix that is positive definite. For a discussion of such algorithms, see [260].

Since these methods both require the use of derivatives, they are not applicable for functions that are not differentiable. A commonly employed approach to minimization that does not require derivatives is the Nelder–Mead simplex algorithm. A **simplex** is a convex shape with $m + 1$ vertices in m-dimensional space. In two-space, the simplex is triangular; in three-space it is the tetrahedron. In the Nelder–Mead algorithm, a simplex is formed in m dimensions whose vertices lies on the surface of the function to be minimized, and the vertex $\mathbf{x}_k$ for which $f(\mathbf{x}_k)$ is minimum is taken as an estimate of the direction of descent. The points of the simplex are moved in the direction of the minimum by the algorithm, and the simplex more-or-less rolls or crawls down the surface to be minimized, then collapses to the convergent point at the bottom. Descriptions of the Nelder–Mead method are provided in [260, 239].

Gradient descent can descend only to the bottom of the bowl that the initial point lies within and, hence, may fail to find a global minimum. Newton's method also fails to find a global minimum, and if it is close enough to a solution it can find only that solution. The problem of locating the global minimum for any function is still unsolved. Some progress has been made with stochastic optimization methods; at least two of these methods have become widely used. The first is *simulated annealing*. Simulated annealing may be viewed as randomly "shaking" the error surface while a descent algorithm slides toward a minimum. Because of the random shaking, from time to time the solution may be dislodged from its current basin of attraction to another one. By starting the algorithm with vigorous shaking (i.e., with a lot of "thermal motion" and a "hot" system) a lot of basins are explored. As the algorithm progresses and the candidate solution is deeper in its basin, it becomes more and more difficult to dislodge the solution from the basin. The "temperature" is lowered so that there is less randomness, and the solution eventually converges to a minimum. On the computer, of course, the shaking is produced by random number generators, and the decrease of randomness simply corresponds to lowering the variance as the algorithm progresses. Even for simulated annealing, however, there is no guarantee of convergence to a global minimum. Simulated annealing is described, for example, in [260, 1].

Another stochastic optimization method of recent interest is *genetic algorithms*. These algorithms attempt to exploit the principles of evolutionary adaptation observed in nature. Multiple solutions are coded as bit strings that are regarded as a "gene pool." Each solution is evaluated as to its fitness according to some function to be optimized, then a succeeding generation is selected randomly according to the fitness of the solutions. Bit elements from selected strings are crossed over in an operation that simulates meiosis, and occasional transcription errors are also introduced, simulating mutation. A readable introduction to genetic algorithms is [108]. Many applications have been developed, including adaptive filters (see, e.g., [338]).

Some applications of basic iterative methods

14.6 An application of steepest descent: LMS adaptive filtering

Let us consider again the problem of minimum mean-squared error filtering, introduced in chapter 3. (For convenience, we consider real signals). The signal $y[t]$ is the output of a filter with finite impulse response $h[t]$ and input $f[l]$, so that

$$y[t] = \sum_{l=0}^{m-1} h[l] f[t-l] = \mathbf{f}[t]^T \mathbf{h}$$

where

$$\mathbf{f}[t] = [f[t] \quad f[t-1] \quad \ldots \quad f[t-m+1]]^T \qquad \mathbf{h} = [h[0] \quad h[1] \quad \ldots \quad h[m-1]]^T.$$

Let $d[t]$ be a desired signal. Then we can formulate a cost function as

$$J(\mathbf{h}) = E[(d[t]-y[t])^2] = E[(d[t]-\mathbf{f}[t]^T\mathbf{h})^2]$$
$$= \sigma_d^2 - \mathbf{p}^T\mathbf{h} - \mathbf{h}^T\mathbf{p} + \mathbf{h}^T R \mathbf{h}, \qquad (14.12)$$

where

$$R = E[\mathbf{f}[t]\mathbf{f}^T[t]] \qquad \mathbf{p} = E[\mathbf{f}[t]\mathbf{d}[t]].$$

To minimize this with respect to the filter coefficients $\mathbf{h}$, using gradient descent, we propose to update it by

$$\mathbf{h}[t+1] = \mathbf{h}[t] - \frac{1}{2}\mu\frac{\partial J(\mathbf{h}[t])}{\partial \mathbf{h}[t]}.$$

Using the partial derivative formulas in appendix E, we find

$$\frac{\partial J(\mathbf{h})}{\partial \mathbf{h}} = -2\mathbf{p} + 2R\mathbf{h}.$$

Thus, a steepest-descent rule is

$$\mathbf{h}[t+1] = \mathbf{h}[t] + \mu(\mathbf{p} - R\mathbf{h}[t]). \tag{14.13}$$

This provides an iterative update to the weights that satisfies the minimum mean-squared error criterion. Since the original problem (14.12) is quadratic, iteration of (14.13) will converge to the optimum solution in m steps.

While (14.13) provides an iterative update, it is not yet in an *adaptive*—or practical—form. In fact, if we know R and $\mathbf{p}$, as we must to compute the gradient, then we know enough to find the solution for the optimal filter, since $\mathbf{h}_{\text{opt}} = R^{-1}\mathbf{p}$ minimizes $J(\mathbf{h})$ in (14.12). To make the LMS adaptive filter more practical, our next step is to eliminate—by an approximation—the need to know R and $\mathbf{p}$.

We can form a (very) rough approximation of $R = E\mathbf{f}[t]\mathbf{f}^T[t]$ and $\mathbf{p} = E\mathbf{f}[t]d[t]$ simply by *throwing away* the expectation. Thus we use one of the sample values as an estimate of the expected value. That is, we take

$$\hat{R}[t] = \mathbf{f}[t]\mathbf{f}^T[t] \tag{14.14}$$

and

$$\hat{\mathbf{p}}[t] = \mathbf{f}[t]d[t]. \tag{14.15}$$

While this appears to rest on shaky grounds theoretically, we observe that after making several steps, the algorithm will employ several samples in the computation of $\mathbf{h}[t]$, whose effect will tend to average together. Furthermore, since the algorithm is always updating the filter weights, any imprecision introduced by the approximation hopefully will be (on average) corrected sooner or later by some other data.

Substituting (14.14) and (14.15) into (14.13), we obtain the LMS adaptive filter weight update:

$$\mathbf{h}[t+1] = \mathbf{h}[t] + \mu\mathbf{f}[t](d[t] - \mathbf{x}^T[t]\mathbf{h}[t]). \tag{14.16}$$

The quantity $e[t] = d[t] - \mathbf{x}^T[t]\mathbf{h}[t]$ is the error between the filter output and the desired output. The LMS update is often written as

$$\mathbf{h}[t+1] = \mathbf{h}[t] + \mu\mathbf{f}[t]e[t].$$

If the desired output matches the filter output exactly at some step, so that $e[t] = 0$, then no update is made to the filter weights at that step. The update step (14.16) is sometimes referred to as *stochastic* gradient descent, since the gradient is only approximately known. The LMS adaptive filter is of very low computational complexity. Once the filter output is obtained (in m multiply/adds), then the filter weights are updated in another m multiply/adds.

The update step size μ must be chosen carefully. If it is too small, the convergence is too slow. On the other hand, if it is too large, then the LMS algorithm does not converge.

14.6 LMS Aadaptive Filtering

As we show, the LMS algorithm converges in mean if

$$0 < \mu < \frac{2}{\lambda_{\max}},$$

where $\lambda_{\max}$ is the largest eigenvalue of R. Since this is often unknown (as there is now no need, with the adaptation algorithm, to explicitly compute R), the trace of R is sometimes taken as a conservative estimate, so that μ is bounded by

$$0 < \mu < \frac{2}{\text{tr}(R)}.$$

But $\text{tr}(R) = mr(0)$, since R is Toeplitz, and $r(0)$ is the power of the input signal $f[t]$. Thus, μ bounded by

$$0 < \mu < \frac{2}{mr(0)}$$

leads to an LMS algorithm that converges in mean.

An implementation of the LMS adaptive filter is shown in algorithm 14.2.

Algorithm 14.2 LMS adaptive filter
File: lms.m
         lmsinit.m

14.6.1 An example LMS application

In this section, we again work through the application presented for the RLS filter, that of channel equalization. A set of binary ± 1 data are passed through an unknown channel, as shown in figure 4.9. The problem is to find an adaptive filter to equalize the channel response. The desired signal is the channel input (somehow provided at the adaptive filter for training purposes). The LMS adaptive filter algorithm was employed at the equalizer. Shown in figure 14.11 is the squared filter error, averaged over 200 iterations, for $\mu = 0.075$ and

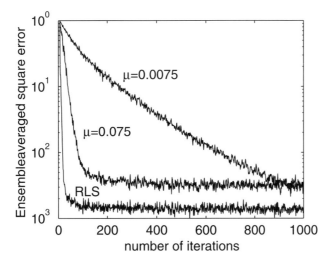

Figure 14.11: Error in the LMS algorithm for $\mu = 0.075$ and $\mu = 0.0075$, compared with the RLS algorithm, for an adaptive equalizer problem

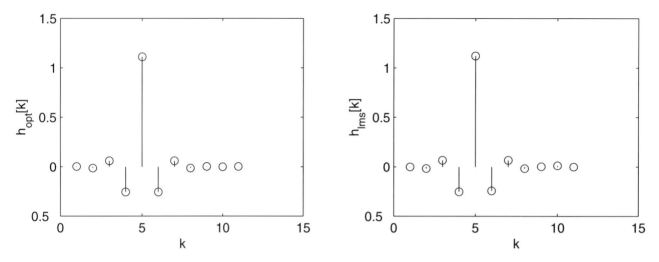

Figure 14.12: Optimal equalizer coefficients and adaptive equalizer coefficients

$\mu = 0.0075$. Clearly, the algorithm converges faster for the larger value of μ. Also shown, for comparison, is the error for the RLS algorithm. The RLS algorithm converges much faster, but requires more computations. Figure 14.12 shows the optimal Wiener filter equalizer coefficients, compared with the filter coefficients obtained using the LMS algorithm with $\mu = 0.075$.

14.6.2 Convergence of the LMS algorithm

Let $\epsilon[t] = \mathbf{h}[t] - \mathbf{h}_0$, where $\mathbf{h}_0$ is the optimal-weight vector for the minimum mean-squared error filter, $\mathbf{h}_0 = R^{-1}\mathbf{p}$. That is, $\epsilon[t]$ is the weight-error vector. Also let $e_0[t] = d[t] - \mathbf{h}_0\mathbf{f}[t]$ denote the estimation error produced in the optimum minimum mean-squared solution. We will examine convergence of the LMS algorithm in the *mean-square* sense. That is, we will examine

$$E[\|\epsilon[t]\|^2].$$

Despite the simplicity of the LMS algorithm, **proof of its convergence is**, in all generality, quite difficult. In order to make the **proof tractable, several approximations** and assumptions are made. While these weaken the proof, **the results that are obtained may be validated**, and lead to useful insight in practice.

Independence assumptions

In order to make the analysis more tractable, we invoke several *independence assumptions*. While strictly speaking these assumptions are not true, they are commonly used in adaptive filter analysis for at least two reasons. First, as mentioned, they do simplify the analysis. Second, simulations of the convergence rates of adaptive filter algorithms show that the theoretical results obtained on the basis of these assumptions track fairly well (but not perfectly) with experimental results. The assumptions are as follows.

1. The inputs $\mathbf{f}[0], \mathbf{f}[1], \ldots$, are statistically independent. (This is obviously not true, since the components of $\mathbf{f}[1]$ are mostly composed of components of $\mathbf{f}[0]$.)
2. The input $\mathbf{f}[t]$ is statistically independent of $d[t-1], d[t-2], \ldots, d[0]$.

14.6 LMS Adaptive Filtering

3. The desired response $d[t]$ is dependent upon $\mathbf{f}[t]$ (otherwise there is no sense in trying to find a filter relationship between them), but $d[t]$ is statistically independent of $\mathbf{f}[t-1], \mathbf{f}[t-2], \ldots, \mathbf{f}[0]$.

On the basis of these assumptions, we conclude that $\mathbf{h}[t+1]$ is independent of $\mathbf{f}[t+1]$ and $d[t+1]$.

Error weight update

It is straightforward to show that

$$\epsilon[t+1] = (I - \mu \mathbf{f}[t]\mathbf{f}^T[t])\epsilon[t] + \mu \mathbf{f}[t]e_0[t]. \tag{14.17}$$

In the interest of simplifying this equation with respect to variations in the input data, we invoke the *direct-averaging method* assumption: for small values of μ,

$$\mu \mathbf{f}[t]\mathbf{f}^T[t] \approx \mu R.$$

On the basis of this assumption,

$$\epsilon[t+1] \approx (I - \mu R)\epsilon[t] + \mu \mathbf{f}[t]e_0[t], \tag{14.18}$$

and we subsequently regard the approximate equality as an equality. We define the correlation matrix

$$P[t] = E\epsilon[t]\epsilon^T[t].$$

Then, freely invoking the independence assumptions, we find (using (14.18)) that

$$P[t+1] = (I - \mu R)P[t](I - \mu R) + K, \tag{14.19}$$

where $K = \mu^2 E e_0[t]\mathbf{f}[t]\mathbf{f}^T[t]e_0[t]$.

Let us now consider the estimation error $e[t] = d[t] - y[t]$. This can be written as

$$e[t] = d[t] - \mathbf{h}[t]\mathbf{f}[t] = e_0[t] - \epsilon^t[t]\mathbf{f}[t].$$

Then, with $J(t)$ denoting the mean-squared error at time t, we have

$$J(t) = E[|e[t]|^2] = E(e_0[t] - \epsilon^T[t]\mathbf{f}[t])(e_0[t] - \epsilon^T[t]\mathbf{f}[t])^T$$
$$= J_{\min} + E\epsilon^T[t]\mathbf{f}[t]\mathbf{f}^T[t]\epsilon[t], \tag{14.20}$$

where $J_{\min}$ is the minimum mean-squared error. Evaluating the expectation in (14.20), we employ the observation that the quantity is a scalar, and that the trace of a scalar is equal to that scalar, then use the fact that quantities commute within a trace:

$$E\epsilon^T[t]\mathbf{f}[t]\mathbf{f}^T[t]\epsilon[t] = \text{tr}(E\epsilon^T[t]\mathbf{f}[t]\mathbf{f}^T[t]\epsilon[t])$$
$$= \text{tr}(E[\mathbf{f}[t]\mathbf{f}^T[t]\epsilon[t]\epsilon^T[t]]).$$

Again employing the independence assumption, we have

$$E[\mathbf{f}[t]\mathbf{f}^T[t]\epsilon[t]\epsilon^T[t]] = E[\mathbf{f}[t]\mathbf{f}^T[t]]E[\epsilon[t]\epsilon^T[t]] = RP[t].$$

Thus

$$J(t) = J_{\min} + \text{tr}(RP[t]).$$

The quantity $\text{tr}(RP[t])$ is known as the *excess mean-squared error*; it is the amount by which the mean-squared error in the LMS algorithm exceeds the minimum mean-squared error of a Wiener filter. We will denote this excess error as $J_{\text{ex}}(t)$:

$$J_{\text{ex}}(t) = J(t) - J_{\min} = \text{tr}(RP[t]).$$

More can be said about the convergence by diagonalizing the R matrix in $J_{\text{ex}}(t)$ using the by-now familiar transformation

$$Q^T R Q = \Lambda,$$

where Q is an orthogonal matrix and Λ is the diagonal matrix of the eigenvalues of R. Also, we define the new quantity $X[t]$ by

$$Q^T P[t] Q = X[t].$$

Employing this factorization with (14.19), we obtain the recursion

$$X[t+1] = (I - \mu\Lambda) X[t] (I - \mu\Lambda) + Q^T K Q. \tag{14.21}$$

The excess mean-squared error can be expressed as

$$J_{\text{ex}} = \text{tr}(RP[t]) = \text{tr}(Q\Lambda Q^T Q X[t] Q^T) = \text{tr}(\Lambda X[t]) = \sum_{i=1}^{m} \lambda_i x_{ii}[t].$$

The excess mean-squared error thus depends upon the diagonal elements of $X[t]$ only, which we denote as $x_{ii}[t]$. We can rewrite (14.21) in terms of these diagonal elements as

$$x_{ii}[t+1] = (1 - \mu\lambda_i)^2 x_{ii}[t] + (Q^T K Q)_{ii}. \tag{14.22}$$

The solution of the difference equation (14.22) grows exponentially unless $|1 - \mu\lambda_i| < 1$. If this inequality is satisfied, then $x_{ii}[t+1]$ approaches a constant value. Stability is obtained if

$$0 < \mu < \frac{2}{\lambda_i}$$

for $i = 1, 2, \ldots, m$. Thus the same stability condition required for the steepest-descent algorithm holds for the LMS algorithm.

14.7 An application of steepest descent: Neural networks

A neural network is (loosely defined) a collection of interconnected simple processing elements that are "trained" to perform some particular computational task, such as pattern recognition, signal discrimination, nonlinear filtering, and so on. There are a variety of neural network designs described in the current literature; depending on the particular network architecture, the training may be done in a variety of ways. One of the most popular and powerful neural network designs is known as a multilayer perceptron. A multilayer perceptron consists of a set of simple computational devices—neurons—interconnected by weighted connections. Figure 14.13 illustrates this concept. There is an *output layer* at which

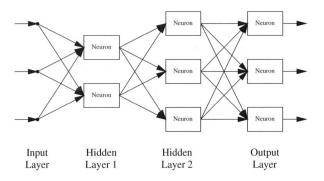

Figure 14.13: Representation of the layers of an artificial neural network

14.7 Neural Networks

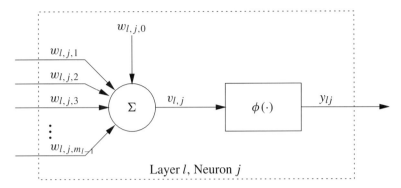

Figure 14.14: An artificial neuron

the outputs are obtained, one or more *hidden layers* that perform the computations, and an *input layer*, which has no neurons and serves simply to distribute the input patterns to the first hidden layer. An example neuron is shown in figure 14.14. It has weighted coefficients that are adjusted to train the algorithm, which are linearly combined, then passed through a nonlinearity ϕ. The nonlinearity is important. Without it, the neural network could only compute linear functions of its inputs. By means of the nonlinearity, a variety of functions can be at least approximately represented. While a multilayer perceptron is an engineering artifice, there are some similarities between artificial neural networks and the way that natural neurons are believed to work. A neuron by itself is a simple cell unit, consisting of a cell body with several long appendages known as axons. An axon divides into several dendrites. (Think of your arms as axons and your fingers as dendrites.) Connection between one neuron and the next is via a *synapse*—a gap between dendrites of the two neurons. Because a neuron has a large number of dendrites that can interconnect with other neurons, a natural neural network consists of an enormous number of interconnected neurons. The strength of the connection between neurons across the synapse is believed to be modified by a learning process. In a similar manner, the operation of the artificial neural network is modified by means of the weighting coefficients in the artificial neurons.

In training the multilayer perceptron, a *supervised training* algorithm is used, in which a set of *known* input/output data combinations are presented to the network. Using a *back-propagation algorithm*, which is simply steepest descent, the network is trained so that the network output matches as closely as possible the desired output, for each input data point.

There are N pieces of training data, consisting of a set of inputs and the corresponding desired outputs. Let the input data be denoted as $\mathbf{x}(n), n = 1, 2, \ldots, N$, and let the corresponding desired output data be $\mathbf{d}(n), n = 1, 2, \ldots, N$. Let the output layer of the neural network corresponding to the input $\mathbf{x}(n)$ be denoted by $\mathbf{y}(n)$. Then the squared error at the output, based upon the data $(\mathbf{x}(n), \mathbf{d}(n))$, is

$$\mathcal{E}(n) = \frac{1}{2} \|\mathbf{d}(n) - \mathbf{y}(n)\|^2. \tag{14.23}$$

The error averaged across all the training data is

$$\mathcal{E}_{\text{av}} = \frac{1}{N} \sum_{n=1}^{N} \mathcal{E}(n). \tag{14.24}$$

The neural network is trained by adjusting the interconnecting weights to minimize $\mathcal{E}_{\text{av}}$.

After the training process, the neural network can be used for its designed purpose. For example, if the training data are samples of pattern-recognition data, then the neural network can be used to recognize patterns by providing a feature vector at the input and using the outputs to classify it.

14.7.1 The backpropagation training algorithm

The backpropagation training algorithm is simply steepest descent applied to the neuron. To present the algorithm we make use of the following notation. (Reference to figure 14.15 will clarify the use of this notation.)

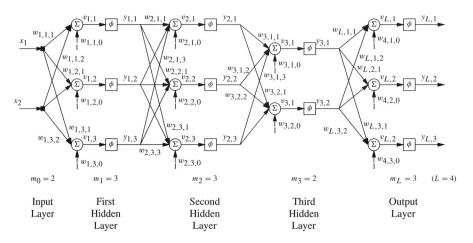

Figure 14.15: Notation for a multilayer neural network

- The index n refers to the input/output training data point number. The training input and desired output data associated with index n are $\mathbf{x}(n)$ and $\mathbf{d}(n)$, respectively.
- The index k (as a superscript) indicates the iteration of training.
- The index l denotes a layer number. The input layer is layer $l = 0$, the output layer is layer $l = L$.
- i will be used to index inputs, j will be used to index neuron number.
- m_l is the number of neurons on layer l. m_L is the number of neurons in the output layer, and m_0 is the number of inputs (the number of nodes in the input layer).
- $w_{l,j,i}$ is the weight on the jth neuron of the lth layer, coming from the ith neuron on the $(l-1)$st (previous) layer. During training, the weights are updated in response to the nth input, and we denote the change in the weight as $\Delta w_{l,j,i}(n)$. The weight at iteration k is

$$w_{l,j,i}^{[k]}.$$

The weights associated with layer l can be viewed as a matrix of size $m_l \times (1 + m_{l-1})$. The reason that the number of columns is $1 + m_{l-1}$ is that a default constant input "1" to every neuron is assumed.

- $y_{l,j}(n)$ is the output of the jth neuron on the lth layer in response to input $\mathbf{x}(n)$. If l is not specified, then the output layer is assumed. We use $y_{l,0} = 1$ to indicate a constant input available at every neuron in the hidden layers. This can be used to provide a constant offset at the next layer (if that is found to be necessary by the training algorithm). We

14.7 Neural Networks

also stack outputs as vectors. For example,

$$\mathbf{y}(n) = \begin{bmatrix} y_1(n) \\ y_2(n) \\ \vdots \\ y_{m_L} \end{bmatrix}$$

denotes the set of neuron outputs on the output layer, and

$$\mathbf{y}_l(n) = \begin{bmatrix} 1 \\ y_{l,1}(n) \\ y_{l,2}(n) \\ \vdots \\ y_{l,m_l}(n) \end{bmatrix}$$

denotes the set of outputs of the neurons on layer l.

- The *activity* of the neuron is the linear combination of the inputs and the connection weights. The activity of neuron j on layer l due to input number n is

$$v_{l,j}(n) = \sum_{i=0}^{m_{l-1}} w_{l,j,i} y_{l-1,i}(n). \tag{14.25}$$

- The output of the neural network is a nonlinear function of the activity, so that the neuron output is

$$y_{l,j} = \phi_{l,j}(v_{l,j}(n)) \tag{14.26}$$

for some function $\phi_{l,j}(\cdot)$. In all generality, different functions could be used in each neuron, and we retain this notation throughout the derivation. In practice, however, it is most common to employ the same function at each neuron, so $\phi_{l,j}(\cdot) = \phi(\cdot)$.

- For the output layer, the desired response can be obtained directly from the training data, and $d_{L,j}(n)$ is denoted as $d_j(n)$ or, in vector form, $\mathbf{d}(n)$. For hidden layers, the desired data must be inferred by backpropagation from the output layer, as discussed in the following.

- The error between the neuron output and the desired output for the neuron at layer l is

$$e_{l,j}(n) = d_{l,j}(n) - y_{l,j}(n). \tag{14.27}$$

For the output layer, this is abbreviated $\mathbf{e}_j(n)$. The error and average error of (14.23) and (14.24) may be written as

$$\mathcal{E}(n) = \mathcal{E}_L(n) = \frac{1}{2} \sum_{j=1}^{m_L} e_{L,j}^2(n) = \frac{1}{2} \sum_{j=1}^{m_L} (d_{L,j}(n) - y_{L,j}(n))^2 \tag{14.28}$$

and

$$\mathcal{E}_{\text{av}} = \frac{1}{N} \sum_{n=1}^{N} \mathcal{E}_L(n).$$

The goal of the training algorithm is to minimize the average squared error (14.24). The error $\mathcal{E}_{\text{av}}$ is a function of all of the weights in the neural network, and we will employ a steepest-descent algorithm for their minimization. Based upon input n, an update for a weight in the neural network, due to the nth input, using the steepest descent, is

$$w_{l,j,i}^{[k+1]} = w_{l,j,i}^{[k]} - \mu \frac{\partial \mathcal{E}(n)}{\partial w_{l,j,i}^{[k]}}, \tag{14.29}$$

where μ is a learning-rate parameter. This update step is computed for each input set $n = 1, 2, \ldots, N$, possibly multiple times, until the weights have converged.

Determining the value of μ to use is somewhat problematic. From our studies on steepest descent, we know that its maximum value is determined by the largest eigenvalue of the Hessian matrix. However, the Hessian is unknown, and would be very difficult to compute. What is frequently done is simply to set μ to a small value, and determine (based on the total squared error) whether it is small enough for the neural network to converge.

The key to the training algorithm, based upon the steepest-descent approach to weight adjustment, is finding the partial derivatives

$$\frac{\partial \mathcal{E}(n)}{\partial w_{l,j,i}^{[k]}}.$$

There is a difference in how the derivatives are computed at the output layer and the hidden layers, since the desired output $d_j(n)$ is explicitly available at the output layer, but not for the hidden layers. Accordingly, we treat each of those cases separately, starting with the output layer. (It may be helpful for the reader to review the generalization of the chain rule to composite functions, as found in section 18.2.)

Derivative at the output layer

Using the chain rule when $l = L$, we have

$$\frac{\partial \mathcal{E}(n)}{\partial w_{L,j,i}^{[k]}} = \frac{\partial \mathcal{E}(n)}{\partial e_{L,j}(n)} \frac{\partial e_{L,j}(n)}{\partial y_{L,j}(n)} \frac{\partial y_{L,j}(n)}{\partial v_{L,j}(n)} \frac{\partial v_{L,j}(n)}{\partial w_{L,j,i}^{[k]}}. \tag{14.30}$$

We now examine each partial derivative in turn:

- From (14.23) and the definition of (14.28),

$$\frac{\partial \mathcal{E}(n)}{\partial e_{L,j}(n)} = e_{L,j}(n).$$

- From (14.27), we find

$$\frac{\partial e_{L,j}(n)}{\partial y_{L,j}(n)} = -1.$$

- From (14.26), we have

$$\frac{\partial y_{L,j}(n)}{\partial v_{L,j}(n)} = \phi'_{L,j}(v_{L,j}(n)),$$

where the "prime" symbol, $'$, indicates differentiation with respect to the argument. This is left in this form until a particular function $\phi_{l,j}(\cdot)$ is chosen.

- From (14.25), we have

$$\frac{\partial v_{L,j}(n)}{\partial w_{L,j,i}^{[k]}} = y_{L-1,i}(n),$$

since the activity $v_{L,j}(n)$ at this instant depends on the present value of $w_{L,j,i}$, which is $w_{L,j,i}^{[k]}$.

Combining these partial derivatives, we have

$$\frac{\partial \mathcal{E}(n)}{\partial w_{L,j,i}^{[k]}} = -e_{L,j}(n)\phi'_{L,j}(v_{L,j}(n))y_{L-1,i}(n). \tag{14.31}$$

14.7 Neural Networks

In what follows, it will be helpful to use the following definition. Let us write the chain rule of (14.30) as

$$\frac{\partial \mathcal{E}(n)}{\partial w_{l,j,i}^{[k]}} = \left[\frac{\partial \mathcal{E}(n)}{\partial e_{l,j}(n)} \frac{\partial e_{l,j}(n)}{\partial y_{l,j}(n)} \frac{\partial y_{l,j}(n)}{\partial v_{l,j}(n)}\right] \frac{\partial v_{l,j}(n)}{\partial w_{l,j,i}^{[k]}}$$

$$= \delta_{l,j}(n) \frac{\partial v_{l,j}(n)}{\partial w_{l,j,i}^{[k]}}, \qquad (14.32)$$

where $\delta_{l,j}$ is called the *local gradient*, and is defined by

$$\delta_{l,j}(n) = \frac{\partial \mathcal{E}(n)}{\partial v_{l,j}(n)}. \qquad (14.33)$$

Then, in terms of the local gradient, the derivative (14.31) can be written as

$$\frac{\partial \mathcal{E}(n)}{\partial w_{L,j,i}(n)} = \delta_{L,j}(n) y_{L-1,i}(n).$$

The local gradient at the output layer is

$$\delta_{L,j}(n) = -e_{L,j}(n) \phi'_{L,j}(v_{L,j}(n)). \qquad (14.34)$$

Derivative at a hidden layer

We will write the derivative, using the local gradient, as

$$\frac{\partial \mathcal{E}(n)}{\partial w_{l,j,i}^{[k]}} - \delta_{l,j}(n) \frac{\partial v_{l,j}(n)}{\partial w_{l,j,i}^{[k]}} = \delta_{l,j}(n) y_{l-1,l}(n). \qquad (14.35)$$

The local gradient in (14.35) can be computed as

$$\delta_{l,j}(n) = \frac{\partial \mathcal{E}(n)}{\partial v_{l,j}(n)} = \sum_{p=1}^{m_{l+1}} \frac{\partial \mathcal{E}(n)}{\partial v_{l+1,p}(n)} \frac{\partial v_{l+1,p}(n)}{\partial v_{l,j}(n)} = \sum_{p=1}^{m_{l+1}} \delta_{l+1,p}(n) \frac{\partial v_{l+1,p}(n)}{\partial v_{l,j}(n)}. \qquad (14.36)$$

Since

$$v_{l+1,p}(n) = \sum_{i=0}^{m_l} w_{l+1,p,i} \phi_{l,i}(v_{l,j}(n)),$$

the partial derivative in (14.36) can be written as

$$\frac{\partial v_{l+1,p}(n)}{\partial v_{l,j}(n)} = w_{l+1,p,j} \phi'_{l,j}(v_{l,j}(n)).$$

Substitution of this result into (14.36) gives

$$\delta_{l,j}(n) = \phi'_{l,j}(v_{l,j}(n)) \sum_{p=1}^{m_{l+1}} \delta_{l+1,p}(n) w_{l+1,p,j}(n). \qquad (14.37)$$

14.7.2 The nonlinearity function

The function $\phi_{l,j}(\cdot)$ provides the nonlinearity necessary for the neural network. In general, any monotonically increasing function can be used. One is commonly used function is the *sigmoidal* nonlinearity, defined by

$$\phi(v) = \frac{1}{1 + e^{-v}}$$

and plotted in figure 14.16. This function has the property that the derivative is easy to compute. If $y = \phi(v)$, then

$$\phi'(v) = y(1 - y).$$

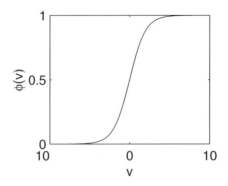

Figure 14.16: The sigmoidal nonlinearity

The sigmoid function has a range from 0 to 1. The desired outputs of the neural network must be scaled so that they are in the same range. It has been found that scaling the outputs so that they are in the range $0 + \epsilon$ to $1 - \epsilon$ for some small ϵ speeds the convergence of the training algorithm, since the weights are moved to drive the outputs to their limiting values.

14.7.3 The forward–backward training algorithm

In the training of the neural network, an input $\mathbf{x}(n)$ is presented, and the outputs $y_{l,j}(n)$ are computed for each neuron in the network. This is the forward step. Then, starting at the output layer, the weights are updated and the local gradients are computed. Since the local gradients at layer l depend upon those at layer $l + 1$ (see (14.37)), and since the local gradients at the output layer can be computed explicitly (see (14.34)), the update rule can be propagated backward through the network.

In implementation, it is common to select the training data $(\mathbf{x}(n), \mathbf{d}(n))$ at random from the pool of training data. This tends to make the search more stochastic, avoiding limit cycles in the learning process.

Learning continues until some stopping criterion is reached. Ideally, the learning would stop when the gradient of $\mathcal{E}_{\text{av}}$ with respect to each of the weights was zero, but this would require extra computation. This criterion is approximated by stopping when the change in the total squared error $\mathcal{E}_{\text{av}}$ is sufficiently small.

14.7.4 Adding a momentum term

The steepest-descent weight update equation

$$w_{l,j,i}^{[k+1]} = w_{l,j,i}^{[k]} + \Delta w_{l,j,i}(n)$$

with

$$\Delta w_{l,j,i}(n) = -\mu \frac{\partial \mathcal{E}(n)}{\partial w_{l,j,i}^{[k]}}$$

may be modified to produce an update of the form

$$\Delta w_{l,j,i}(n) = \alpha \Delta w_{l,j,i}(n-1) - \mu \frac{\partial \mathcal{E}(n)}{\partial w_{l,j,i}^{[k]}}. \tag{14.38}$$

The modification $\alpha \Delta w_{l,j,i}(n-1)$ is called a *momentum* term, and α is called the *momentum constant*. When $\alpha \neq 0$, the update (14.38) is known as the *generalized delta rule*. When $\alpha = 0$, the steepest-descent algorithm is obtained. When $\alpha \neq 0$, the momentum term has

14.7 Neural Networks

been shown to increase the learning rate of the neural network in some cases. When the derivative term $\frac{\partial \mathcal{E}(n)}{\partial w_{l,j,i}^{[k]}}$ has the same sign as $\frac{\partial \mathcal{E}(n)}{\partial w_{l,j,i}^{[k-1]}}$, then the presence of the momentum term accelerates descent in these directions. On the other hand, when the consecutive derivatives have opposite signs, the momentum term provides a drag and tends to minimize oscillations, thereby providing a stabilizing influence.

14.7.5 Neural network code

Algorithm 14.3 implements the forward computations of a neural network. It requires as its arguments the input to the network and the weights (as MATLAB cells). The code automatically determines from the weights the number of layers and the number of neurons on each layer.

Example 14.7.1 We demonstrate the weights for a neural network with one input layer, two hidden layers, and one output layer, using the notation $w\{l\}$ to indicate the weights for layer l. The input layer has two nodes. The first hidden layer has 3 neurons, and the second hidden layer has 2 neurons. The output layer has 1 neuron. The weights chosen are arbitrary, but the shapes of the matrices are not.

$$w\{1\} = \begin{bmatrix} 1 & 2 & 3 \\ 4 & 5 & 6 \\ 7 & 8 & 9 \end{bmatrix} \qquad \text{first hidden layer,}$$

$$w\{2\} = \begin{bmatrix} 10 & 11 & 12 & 13 \\ 14 & 15 & 16 & 17 \end{bmatrix} \qquad \text{second hidden layer,}$$

$$w\{3\} = [18 \quad 19 \quad 20] \qquad \text{output layer.}$$

The first set of weights $w\{1\}$ is 3×3, which is equivalent to (number of weights in layer 1) × (number of inputs + 1). The second set of weights is 2×4, which is equivalent to (number of weights in layer 2) × (number of weights in layer 1 + 1), and so forth. □

Algorithm 14.3 Neural network forward-propagation algorithm
File: `nn1.m`

The training process is shown in algorithm 14.4. The input is provided in the matrices x and d, where the input/output data vectors are stored in columns.

Algorithm 14.4 Neural network backpropagation training algorithm
File: `nntrain1.m`
`nnrandw.m`

Now consider training the neural network for an example pattern-recognition problem. A neural network is to be trained to recognize the difference between the × and ○ data shown in figure 14.17(a). The dashed line indicates the boundary between the data classes—essentially the line that the neural network needs to learn. We employ a neural network with two outputs (one output would suffice, but two are used for demonstration purposes), with

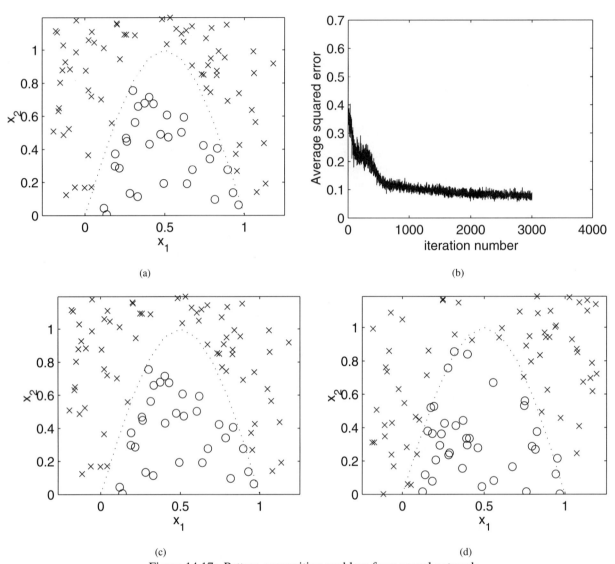

Figure 14.17: Pattern-recognition problem for a neural network

one hidden layer and one output layer. There are two inputs and five neurons on the hidden layer. Code illustrating how the data are obtained is shown in algorithm 14.5

Algorithm 14.5 Neural network test example
File: `testnn10.m`

In the training process, 100 points of × data are generated at random over the region of interest, and the corresponding desired outputs are chosen. These are chosen so that $d_1(n) = .8$ for those data in class ×, and $d_1(n) = 0.2$ for those data in class ○, with $d_2(n) = 1 - d_1(n)$.

Figure 14.17(b) illustrates the learning curve for an experiment with $\mu = .1$, showing $\mathcal{E}_{av}$ as a function of the iteration number, where each iteration consists of presenting

14.7 Neural Networks

all $N = 100$ data points to the training algorithm in random order. Even though steepest descent is employed as the training algorithm, the decrease in $\mathcal{E}_{av}$ is not monotonic. This is because the weight update (14.29) is based upon the error $\mathcal{E}(n)$, not the average error $\mathcal{E}_{av}$. A change in weights that decreases $\mathcal{E}(n)$ may not, in general, be a change that decreases $\mathcal{E}_{av}$ (see exercise 14.7-16). Nevertheless, on the ensemble the error does decrease. The implementation of the algorithm retains the weights corresponding to the minimum attained value of $\mathcal{E}_{av}$ and returns that value. (A variation on the training is to reject those weight updates that increase $\mathcal{E}_{av}$. This is done simply by removing the comment on the line `else w = wmin; end;` in algorithm 14.4. However, this leaves the network overly trained for some particular order of presentation, and generally leads to inferior pattern recognition ability.)

Figure 14.17(c) shows the results of classifying the training data using the neural network. The results show (almost) no misclassifications. Figure 14.17(d) shows the results of classifying some new random data using the trained neural network. Again, the neural network performs very well, classifying almost all points correctly. Figure 14.18 illustrates the desired output $d_1(n)$ (either 0.8 or 0.2) as a function of the data set number n. Superimposed on this is the corresponding output for the neural network, which we call $\hat{d}_1(n)$ (dashed line). While the neural network output does not correspond exactly to the desired output, because some kind of thresholding takes place in the pattern recognition problem, the error has little impact on the recognition capability of this neural network.

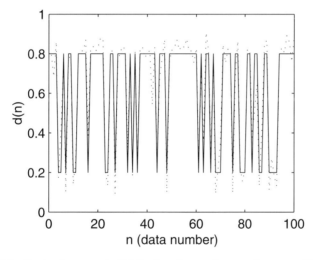

Figure 14.18: Desired output (solid line) and neural network output (dashed line)

Figure 14.19 illustrates the effect of varying the parameters. Part (a) shows the effect of varying the momentum constant α when $\mu = 0.1$: larger α tends to lead to quicker convergence. Part (b) shows the same comparison with $\mu = 0.5$. The improvement in convergence due to a larger α is slight, because the convergence is already faster due to the larger μ. Parts (c) and (d) present the same data in a different format, comparing the rates of convergence for fixed α and varying μ. As μ increases, the rate of convergence improves (for this problem).

It cannot be concluded from figure 14.19 that larger values of μ or α are necessarily always desirable. Obviously, if μ is made too large, then the steepest-descent algorithm will not converge at all. And if the momentum term is too large, the training might overshoot the minimum. Also, larger μ might lead to a larger misadjustment after training. In practice, it

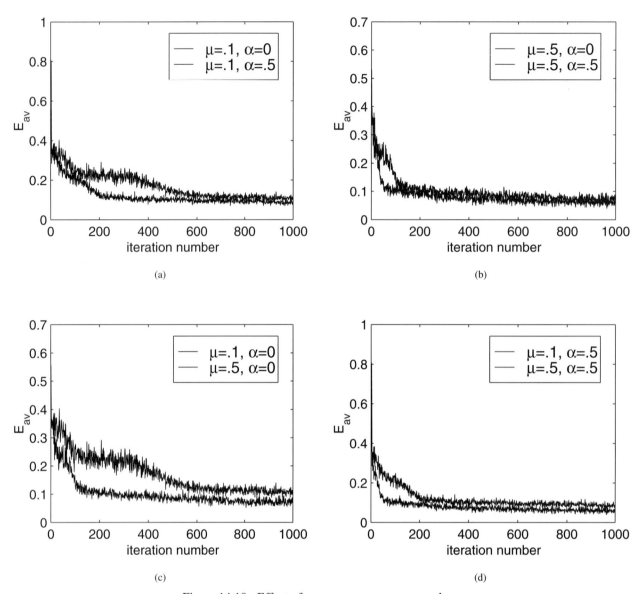

Figure 14.19: Effect of convergence rate on μ and α

is common to experiment with several values of μ and α to explore convergence rates for a given problem.

14.7.6 How many neurons?

One of the open-ended practical questions for this technique concerns the number of neurons that need to be employed in each layer, and the number of layers that are necessary. There are no hard and fast answers, but some guidelines are available from the literature.

The neural network can be viewed as a function approximation device: a mapping from the inputs to the outputs. Thus, the question of how many layers and neurons a neural network needs can be posed in terms of how many layers and neurons are needed to approximate (with arbitrary reliability) an arbitrary function. A theoretical result in this regard [60] is that a *single* hidden layer is sufficient to uniformly approximate *any* continuous function with support over [0, 1] on each input, and similar range on each output. However, a single layer may not be optimum, and a function may require an infinite number of neurons for representation.

From a more practical point of view, it has been found that having two hidden layers is often very effective. The first hidden layer acts as a feature extraction layer, and the second examines the global interactions of these features.

Another issue in the selection of the number of neurons concerns what is known as *overtraining*. Think of the training process as curve-fitting: the process of training provides the curve fit, and when the neural network is used, function values not specifically trained in are obtained by interpolation. If the network is overtrained, input values very similar to those used for training provide desired outputs, while input values different from those used for training may give answers completely wrong: a poor interpolation between training data is obtained. The network is said to have poor generalization capability. Overtraining can be a result of many factors: too many neurons, too much training time, or training with a sophisticated algorithm on an insufficiently rich training set.

It is difficult to be precise about how many neurons are required for a general application. It is common practice to train a network, then evaluate its performance. If the performance is inadequate, then the number of neurons is increased.

14.7.7 Pattern recognition: ML or NN?

In chapter 11, we demonstrated pattern recognition using a maximum-likelihood or Bayes criterion. When the data are Gaussian distributed, the surfaces dividing classes are always planar. Frequently (whether fully justified or not), the distributions are *assumed* to be Gaussian, and the parameters (mean and covariance) are estimated from the data, often using a clustering algorithm (see section 16.1). To the extent that the data are not truly Gaussian distributed, or the parameters are imprecisely identified, there will be some degradation of the performance of the pattern-recognition algorithm.

In contrast, the neural network offers several advantages. First, the separating surfaces need not be planar. As the example in the previous section shows, the dividing surfaces can, in principle, take any shape, provided that there are sufficient data to effectively train the neural network. Second, there is no explicit assumption on the type of density (at least, as we have presented neural networks). This avoids the problem of having to estimate the parameters of a distribution whose form was (usually) assumed in the first place.

Offsetting these advantages of the neural network is the requirement to train, which requires a significant amount of data and processing power. Also, while the example shows that good recognition may be obtained, there is no claim to optimality—the training algorithm does not indicate whether there might be a better classifier. Also, computationally, neural networks tend to be more involved than Gaussian classifiers.

In summary, there are no hard and fast answers about when a formal detection approach should be employed in comparison to a neural network approach for any given problem. If, as happens in many communications problems, there are known distribution functions on the noise, employing detection theory is encouraged. For many other pattern-recognition problems, neural networks are frequently used. Both techniques are useful and should be available in the signal processor's toolbox.

14.8 Blind source separation

Consider the scenario pictured in figure 14.20, in which the vector signal $\mathbf{s}[t] = [s_1[t], s_2[t], \ldots, s_n[t]]^T$, is passed through a system represented by the matrix A, and the vector output $\mathbf{x}[t] = [x_1[t], x_2[t], \ldots, x_n[t]]$, is observed, where

$$\mathbf{x}[t] = A\mathbf{s}[t].$$

The source-separation problem is to identify the input sources $s_1[t], s_2[t], \ldots, s_n[t]$ from the outputs. This is also called the "cocktail party problem"—the problem of selecting and understanding individual speakers in a room full of people speaking at the same time. The *blind* source-separation problem is the problem of determining $s_i[t]$ when neither A nor the input signals are known. This is an example of a general class of problems in signal processing, in which unknown information must be extracted from the signal before the signal of interest can be determined. Other examples include blind equalization (training an adaptive equalizer without a training sequence) and blind image enhancement (processing an image to remove artifacts such as blurring without an explicit model for the artifacts).

In this section, we consider the blind source-separation problem. Ideally, we would determine a matrix W such that

$$\mathbf{y}[t] = W\mathbf{x}[t] = \Lambda P \mathbf{s}[t],$$

where Λ is a diagonal matrix of amplitudes, and P is a permutation matrix. That is, W unmixes the sources, returning the original signal components, up to a permutation and scaling of amplitude. Rather than find such a matrix W directly, we formulate the problem as a neural network problem, and train the neural network to separate the signals. Clearly, in order to accomplish this, additional data or assumptions of some sort are necessary. One assumption that has led to some success is that the sources are stochastic processes that are statistically *independent*. Then the neural network is determined so that the components of $\mathbf{y}[t]$ are also statistically independent. Because of this criterion, blind source separation is frequently associated with independent component analysis [49, 54]. Another approach [21] makes use of the information-theoretic concept of *mutual information* [56]. The neural network is selected to maximize the mutual information between $\mathbf{y}[t]$ and $\mathbf{x}[t]$. As we show in the following, these two criteria are strongly related.

14.8.1 A bit of information theory

In order to pursue separation according to a criterion of maximum mutual information, we need to introduce a little information theory. Let $\mathbf{X}$ be a discrete random vector (or variable), that has outcomes $\mathbf{x}$ in a set $\mathcal{X}$ that occur with probability $f_X(\mathbf{x})$. The **entropy** of $\mathbf{X}$ depends

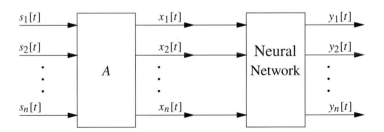

Figure 14.20: The blind source-separation problem

14.8 Blind Source Separation

on the probabilities (not the outcomes); it is written as

$$H(\mathbf{X}) = -\sum_{\mathbf{x} \in \mathcal{X}} f_X(\mathbf{x}) \log f_X(\mathbf{x}) = -E[\log f_X(\mathbf{x})]. \quad (14.39)$$

(Even though it is written as if it were a function of $\mathbf{X}$, the entropy is not a random variable—it is a function of the pmf of $\mathbf{X}$.) The entropy of $\mathbf{X}$ represents the amount of uncertainty about $\mathbf{X}$ that is resolved when $\mathbf{X}$ is observed. If the logarithm in (14.39) is base 2, then the units of entropy are bits; if the logarithm is base e, then the units are *nats* (natural units).

Example 14.8.1 Let X be a Bernoulli random variable with probability p. Then

$$f_X(1) = p \qquad f_X(0) = 1 - p$$

and

$$H(X) = -(p \log p + (1-p) \log(1-p)).$$

The entropy in this case is often denoted (more accurately) as $H(p)$, and is called the binary entropy function. A plot of $H(p)$ is shown in figure 14.21. When $p = 0$ or $p = 1$, there is no uncertainty about X to be resolved (the outcome is sure one way or the other), and the entropy is 0. When $p = 0.5$, a full bit of uncertainty is resolved when X is observed. □

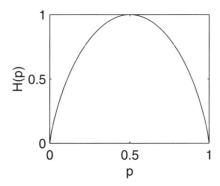

Figure 14.21: The binary entropy function $H(p)$

Let $\mathbf{X}$ and $\mathbf{Y}$ be jointly distributed random discrete variables taking values in the sets $\mathcal{X}$ and $\mathcal{Y}$, respectively, with joint pmf $f_{\mathbf{XY}}(\mathbf{x},\mathbf{y})$. The **joint entropy** is defined by

$$H(\mathbf{X}, \mathbf{Y}) = -\sum_{\mathbf{x}\in\mathcal{X},\mathbf{y}\in\mathcal{Y}} f_{XY}(\mathbf{x},\mathbf{y}) \log f_{XY}(\mathbf{x},\mathbf{y}).$$

This is simply a modification of the definition in (14.39).

The next information-theoretic quantity we introduce is the **conditional entropy**,

$$H(\mathbf{Y}\mid\mathbf{X}) = \sum_{\mathbf{x}\in\mathcal{X}} f_X(\mathbf{x}) H(\mathbf{Y}\mid\mathbf{X}=\mathbf{x}) = -\sum_{\mathbf{x}\in\mathcal{X},\mathbf{y}\in\mathcal{Y}} f_{XY}(\mathbf{x},\mathbf{y}) \log f_{Y\mid X}(\mathbf{y}\mid\mathbf{x}). \quad (14.40)$$

This represents how much uncertainty remains in $\mathbf{Y}$ when $\mathbf{X}$ is known.

Example 14.8.2 Let X be a random variable that passes through a deterministic system to produce an output Y. Then $H(Y\mid X) = 0$, since when X is known, there is no uncertainty about Y. □

It is straightforward to show that

$$H(\mathbf{X}, \mathbf{Y}) = H(\mathbf{X}) + H(\mathbf{Y}\mid\mathbf{X}). \quad (14.41)$$

This has the intuitively appealing interpretation that the combined uncertainty in **X** and **Y** is the sum of the uncertainty about **X** alone and the uncertainty remaining about **Y** when **X** is known.

The last information-theoretic quantity we present is the **mutual information** between **X** and **Y**. This is defined by

$$I(\mathbf{X}, \mathbf{Y}) = H(\mathbf{Y}) - H(\mathbf{Y} \mid \mathbf{X}) = H(\mathbf{X}) - H(\mathbf{X} \mid \mathbf{Y}). \tag{14.42}$$

Intuitively, $I(X, Y)$ indicates the reduction in uncertainty of **Y** due to knowledge of **X** (or vice versa). It can be shown that

$$H(\mathbf{X}, \mathbf{Y}) = H(\mathbf{X}) + H(\mathbf{Y}) - I(\mathbf{X}, \mathbf{Y}). \tag{14.43}$$

Also, mutual information has the property that

$$I(\mathbf{X}, \mathbf{Y}) \geq 0, \tag{14.44}$$

with equality if and only if **X** and **Y** are statistically independent random vectors (or variables).

14.8.2 Applications to source separation

Returning now to the source-separation problem, for the moment let us model the input signal as a random vector **S**, and the measured signal as $\mathbf{X} = A\mathbf{S}$. Consider a single-stage, multiple-input, multiple-output neural network in which the processing that takes place is

$$\mathbf{Y} = \mathbf{g}(\mathbf{U}) \quad \text{where} \quad \mathbf{U} = W\mathbf{X} + \mathbf{w}_0,$$

where $\mathbf{g}(\mathbf{U})$ indicates that a function g is applied separately to each component of **U**,

$$\mathbf{g}(\mathbf{U}) = \begin{bmatrix} g(U_1) \\ g(U_2) \\ \vdots \\ g(U_n) \end{bmatrix}.$$

We take g as the logistic function commonly used in neural network processing,

$$g(u) = \frac{1}{1 + e^{-u}}.$$

Our optimization criterion now can be stated as: determine the neural network parameters $(W, \mathbf{w}_0)$ so that the mutual information between **Y** and **X** is maximized. Since

$$I(\mathbf{Y}, \mathbf{X}) = H(\mathbf{Y}) - H(\mathbf{Y} \mid \mathbf{X}),$$

and since $H(\mathbf{Y} \mid \mathbf{X}) = 0$ (knowing **X** reduces all uncertainty about **Y** because there is a deterministic system from **X** to **Y**), our criterion becomes equivalently: determine the neural network parameters $(W, \mathbf{w}_0)$ to maximize $H(\mathbf{Y})$.

This maximum-entropy criterion has another interpretation. If **Y** consists, for example, of two components Y_1 and Y_2, then (see (14.43))

$$H(Y_1, Y_2) = H(Y_1) + H(Y_2) - I(Y_1, Y_2).$$

Maximizing the joint entropy $H(Y_1, Y_2)$ is accomplished by maximizing the individual entropies while minimizing the mutual information $I(Y_1, Y_2)$. When $I(Y_1, Y_2)$ is zero then, as observed previously, Y_1 and Y_2 are statistically independent. Thus, the maximum mutual information optimization criterion is often effectively the same as a minimum statistical dependence (independent component) criterion (see [21, page 1141]).

14.8 Blind Source Separation

To maximize

$$H(\mathbf{Y}) = -\sum_{\mathbf{y}} f_Y(\mathbf{y}) \log f_Y(\mathbf{y}) = -E[\log f_Y(\mathbf{y})] \quad (14.45)$$

we use an iterative approach, updating the neural network weights at time k according to

$$W^{[k+1]} = W^{[k]} + \mu \Delta W^{[k]},$$
$$\mathbf{w}_0^{[k+1]} = \mathbf{w}_0^{[k]} + \mu \Delta \mathbf{w}^{[k]},$$

where μ is a learning rate parameter. We employ a steepest ascent approach, in which the change in weights is the gradient of the criterion function,

$$\Delta W^{[k]} = \frac{\partial}{\partial W} H(\mathbf{Y}) \bigg|_{\mathbf{Y}=\mathbf{g}(W^{[k]}\mathbf{X}+\mathbf{w}_0^{[k]})}, \quad (14.46)$$

$$\Delta \mathbf{w}_0^{[k]} = \frac{\partial}{\partial \mathbf{w}_0} H(\mathbf{Y}) \bigg|_{\mathbf{Y}=\mathbf{g}(W^{[k]}\mathbf{X}+\mathbf{w}_0^{[k]})}. \quad (14.47)$$

We turn our attention now to evaluating the gradients. Assuming that the neural nonlinearity g is monotonically increasing, the density of $\mathbf{Y}$ can be written as

$$f_Y(\mathbf{y}) = f_X(\mathbf{x}) \frac{1}{|J|} \quad \text{when} \quad \mathbf{y} = \mathbf{g}(W\mathbf{x} + \mathbf{w}_0), \quad (14.48)$$

where $|J|$ is the absolute value of the Jacobian,

$$J = \det \begin{bmatrix} \frac{\partial y_1}{\partial x_1} & \cdots & \frac{\partial y_1}{\partial x_n} \\ \vdots & & \vdots \\ \frac{\partial y_n}{\partial x_1} & \cdots & \frac{\partial y_n}{\partial x_1} \end{bmatrix}.$$

Substituting (14.48) into (14.45), we have

$$H(\mathbf{Y}) = -E[\log f_X(\mathbf{x})/|J|] = E[\log |J|] - E[\log f_X(\mathbf{x})]$$
$$= E[\log |J|] + H(\mathbf{X}).$$

Since $H(\mathbf{X})$ is unaffected by the parameters $(W, \mathbf{w}_0)$, we need only examine

$$\frac{\partial}{\partial W} E[\log |J|]$$

and

$$\frac{\partial}{\partial \mathbf{w}_0} E[\log |J|],$$

where $\mathbf{y} = \mathbf{g}(W^{[k]}\mathbf{x} + \mathbf{w}_0^{[k]})$.

We now make a stochastic gradient approximation, as we did for the LMS filter. That is, we approximate the expected value by a single point in the ensemble. We thus approximate the expectation by

$$E[\log |J|] \approx \log |J|,$$

on the basis that across an ensemble of training data the set of $\mathbf{x}$ data approximates the density $f_X(\mathbf{x})$. Assuming this approximation as accurate, we set

$$\Delta W^{[k]} = \frac{\partial}{\partial W} \log |J| \quad \text{where} \quad \mathbf{y} = \mathbf{g}\big(W^{[k]}\mathbf{x} + \mathbf{w}_0^{[k]}\big),$$

$$\Delta \mathbf{w}_0^{[k]} = \frac{\partial}{\partial \mathbf{w}_0} \log |J| \quad \text{where} \quad \mathbf{y} = \mathbf{g}\big(W^{[k]}\mathbf{x} + \mathbf{w}_0^{[k]}\big).$$

These derivatives are given by the following lemma.

Lemma 14.1 *The change in weights satisfies*

$$\frac{\partial}{\partial W} \log |J| = W^{-T} + (\mathbf{1} - 2\mathbf{y})\mathbf{x}^T,$$

$$\frac{\partial}{\partial \mathbf{w}_0} \log |J| = \mathbf{1} - 2\mathbf{y},$$

where $\mathbf{1}$ is a vector of all ones, and the log is assumed to be natural. The vector $\mathbf{y} = \mathbf{g}(W\mathbf{x} + \mathbf{w}_0)$.

Proof Let $\mathbf{u} = W\mathbf{x} + \mathbf{w}_0$. Then

$$J = \det \begin{bmatrix} \frac{\partial y_1}{\partial x_1} & \cdots & \frac{\partial y_1}{\partial x_n} \\ \vdots & & \vdots \\ \frac{\partial y_n}{\partial x_1} & \cdots & \frac{\partial y_n}{\partial x_n} \end{bmatrix} = \det \begin{bmatrix} \frac{\partial y_1}{\partial u_1}\frac{\partial u_1}{\partial x_1} & \cdots & \frac{\partial y_1}{\partial u_n}\frac{\partial u_n}{\partial x_n} \\ \vdots & & \vdots \\ \frac{\partial y_n}{\partial u_n}\frac{\partial u_n}{\partial x_1} & \cdots & \frac{\partial y_n}{\partial u_n}\frac{\partial u_n}{\partial x_n} \end{bmatrix} = \det(W) \prod_{l=1}^{n} y_l'.$$

It follows that

$$\frac{\partial}{\partial W} \log |J| = \frac{\partial}{\partial W} \log |\det(W)| + \frac{\partial}{\partial W} \log \prod_{i=1}^{n} |y_i'|.$$

Then, from example E.2.2, we have

$$\frac{\partial}{\partial W} \log |\det(W)| = W^{-T}.$$

Also, since g is increasing, we can remove the absolute value condition from $|y_i'|$. We also use the fact that $y_l' = y_l(1 - y_l)$. Then, the components of the partial derivative are found from

$$\frac{\partial}{\partial w_{ij}} \log \prod_{l=1}^{n} y_l' = \sum_{l=1}^{n} \frac{\partial}{\partial w_{ij}} \log y_l(1 - y_l)$$

$$= \sum_{l=1}^{n} \frac{1}{y_l(1 - y_l)} \left[(1 - y_l)\frac{\partial y_l}{\partial w_{ij}} + y_l \frac{\partial(1 - y_l)}{\partial w_{ij}} \right] = x_j(1 - 2y_i).$$

Putting the components together, we find that

$$\Delta W = W^{-T} + (\mathbf{1} - 2\mathbf{y})\mathbf{x}^T.$$

The approach is similar (but substantially easier) to find $\Delta \mathbf{w}_0$. □

14.8.3 Implementation aspects

In the case of blind source separation, the observed vectors $\mathbf{x}[t]$ are modeled as a sequence of instances of random vectors $\mathbf{X}$. In the training algorithm that we now present, the data $\mathbf{x}[t]$ are randomly permuted, with a different permutation used for each pass through the data. The matrix update $\Delta W^{[k]}$ is computationally intensive, since it involves the inverse of a matrix. To reduce the computation, b updates are batched together to produce a single update. Code to test the blind source separation is shown in algorithm 14.6.

Algorithm 14.6 Blind source separation test
File: `bss1.m`
`permutedata.m`

Example 14.8.3 The code in algorithm 14.6 is tested by recording approximately 10 seconds of three sources, using a microphone and a sound card in an office environment and 8-bit quantization at 8000 samples/second. The three sources are a female speaker `ag`, a male speaker `tkm`, and vocal ensemble music `vm`. These sources are mixed using the *A* matrix shown in algorithm 14.6, and the mixed signals are processed using the algorithm with $\mu = 0.01$ by iterating 15 times through the data, where the data vector for processing is chosen at random (without replacement) to get a statistical representation of the distribution. The initial conditions are $W^{[0]} = I$ and $\mathbf{w}_0^{[0]} = \mathbf{0}$. The results show good separation, with the most dramatic improvement evident for the speakers. The matrix *WA* is

$$WA = \begin{bmatrix} 0.202 & -0.0343 & \underline{2.2048} \\ -0.1126 & \underline{-2.4561} & -0.0722 \\ \underline{1.6283} & 0.0849 & -0.0164 \end{bmatrix},$$

with the underlined components being quite strong, compared to the other elements. (The data files can be found on the included media in `.wav` format, with `bssini.wav` being the input files, and `bssouti.wav` being the output (separated) files.) □

14.9 Exercises

14.1-1 What is the orbit of the logistic map if $x_0 < 0$? If $x_0 > 1$?

14.1-2 Explore the behavior of the logistic map for various values of λ. Determine the fixed points (if any), and whether they are attractive or repelling. Try the following values of λ: 0.5, 3.44, 3.5, 3.631, 3.7, 3.831, 3.99.

14.2-3 (Designing affine transformations) In example 14.2.1, an affine transformation $A\mathbf{x} + \mathbf{b}$ was given that mapped a square to a square. In this exercise, you will devise a method for finding such a transformation.

Suppose that a polygonal set in $\mathbb{R}^2$ is defined by its vertices, $\{\mathbf{x}_0^{[0]}, \mathbf{x}_1^{[0]}, \ldots, \mathbf{x}_k^{[0]}\}, \mathbf{x}_i^{[0]} \in \mathbb{R}^2$. We desire to find an affine transformation that maps this region to the polygonal set with vertices $\{\mathbf{x}_0^{[1]}, \mathbf{x}_1^{[1]}, \ldots, \mathbf{x}_k^{[1]}\}$ by the transformation $x_i^{[1]} = Ax_i^{[0]} + b$.

(a) Determine the number of vertices k necessary to uniquely define the affine transformation.

(b) What happens if fewer than k vertices are available? What happens if more than k vertices are available?

(c) Determine a means of finding A and $\mathbf{b}$ for the case of a unique transformation.

(d) Find an affine transformation that will map the vertices

$$\left\{ \begin{bmatrix} 1 \\ 1 \end{bmatrix}, \begin{bmatrix} 2 \\ 1 \end{bmatrix}, \begin{bmatrix} 2 \\ 2 \end{bmatrix} \right\}$$

to the vertices

$$\left\{ \begin{bmatrix} 1.2 \\ 1 \end{bmatrix}, \begin{bmatrix} 1.7 \\ 1.2 \end{bmatrix}, \begin{bmatrix} 1.5 \\ 1.7 \end{bmatrix} \right\},$$

respectively. Draw the polygonal region before and after.

(e) Write MATLAB code that iterates your affine transformation to produce a diagram such as figure 14.5.

14.2-4 For the Jacobi iteration, show that for a matrix A with $a_{ii} = 1$, then $\|A - I\| < 1$ if A is diagonally dominant, where the norm is the max norm.

14.2-5 The affine mapping $\mathbf{x}^{[k+1]} = A\mathbf{x}^{[k]} + \mathbf{b}$ with

$$A = \begin{bmatrix} \cos\theta & -\sin\theta \\ \sin\theta & \cos\theta \end{bmatrix} \begin{bmatrix} e & \lambda \\ 0 & 1 \end{bmatrix}$$

has been proposed as a means of generating points for a specialized codebook for a data compression application.

(a) Show that the kth point of the orbit is

$$\mathbf{x}^{[k]} = \sum_{j=0}^{k-1} A^j \mathbf{b} + A^k \mathbf{x}^{[0]},$$

which can be expressed as

$$\mathbf{x}^{[k]} = (A - I)^{-1}(A^k - I) + A^k \mathbf{x}^{[0]}.$$

(b) If $\|A\|_2 = 1$, the orbit lies on an ellipse. Determine the center point of the ellipse, $\mathbf{x}_0$.

(c) For $\|A\|_2 = 1$, the ellipse can be expressed as

$$(\mathbf{x} - \mathbf{x}_0)^T U (\mathbf{x} - \mathbf{x}_0) = c$$

for any $\mathbf{x}$ in the orbit, where the matrix U is related to A and $\mathbf{b}$. If A and $\mathbf{b}$ are known, specify a means of finding the matrix U.

(d) Using MATLAB, examine the orbit of the transformation for various values of e, λ, and θ.

14.3-6 Show that $x^{[n]} = (1/n)^n$ is of order 1 and has superlinear convergence.

14.4-7 Determine a Newton's method for computing the square root of a number. Code it in a high-level language, and determine how many iterations are required for convergence to six decimal places for a variety of starting points.

14.4-8 Show that if Newton's method is used to minimize $f(x)$ when $f(x)$ is a quadratic function, then it requires only one step.

14.4-9 Sketch an example of a function with an initial condition for which Newton's method will not converge.

14.5-10 A steepest-descent problem on $f(\mathbf{x}) = \mathbf{x}^T R \mathbf{x} - \mathbf{b}^T \mathbf{x}$ with variable step-size α_n can be written as

$$\mathbf{x}^{[n+1]} = \mathbf{x}^{[n]} + \alpha_n \mathbf{r}_n,$$

where $\mathbf{r}_n = 2R\mathbf{x}^{[n]} - \mathbf{b}$. Show that

$$\alpha_n = \frac{\mathbf{r}_n^T \mathbf{r}_n}{\mathbf{r}_n^T R \mathbf{r}_n}$$

minimizes $f(\mathbf{x}^{[n+1]})$ at each step.

14.5-11 Write MATLAB algorithms that find the root of a function $f\colon \mathbb{R}^n \to \mathbb{R}$ using (a) the bisection method and (b) the secant method. As parameters to the function, pass in f, x1, and x2, the function name, and the points x_1 and x_2 that bracket the solution.

14.9 Exercises

14.6-12 Show that (14.19) is true.

14.6-13 The constant K appearing in (14.19) is equal to $K = \mu^2 E e_0[t] \mathbf{f}[t] \mathbf{f}^T[t] e_0[t]$. Using the independence assumption, and also assuming that the input vector $\mathbf{f}[t]$ and the desired response $d[t]$ are mutually Gaussian distributed, show that

$$K = \mu^2 J_{\min} R,$$

where $J_{\min}$ is the minimum mean-squared error,

$$J_{\min} = E(e_0[t])^2,$$

and $R = E\mathbf{f}[t]\mathbf{f}^T[t]$. Note: Use the *Gaussian moment factoring theorem*: if $x_1, x_2, x_3,$ and x_4 are complex Gaussian, then

$$E[\bar{x}_1 \bar{x}_2 x_3 x_4] = E[\bar{x}_1 x_3] E[\bar{x}_2 x_4] + E[\bar{x}_2 x_3] E[\bar{x}_1 x_4].$$

14.6-14 An adaptive filter related to the LMS algorithm is the *normalized LMS* algorithm. This is obtained as the solution to a constrained optimization problem, which can be expressed as follows: Given an input vector $\mathbf{f}[t]$ and a desired response $d[t]$, determine the tap weight vector $\mathbf{h}[t+1]$ to minimize $\|\Delta \mathbf{h}[t+1]\|$, where

$$\Delta \mathbf{h}[t+1] = \mathbf{h}[t+1] - \mathbf{h}[t],$$

subject to the constraint $\mathbf{f}^T[t]\mathbf{h}[t+1] = d[t]$. Show that the adaptive filter that satisfies these constraints has the update equation

$$\mathbf{h}[t+1] = \mathbf{h}[t] + \frac{1}{\|\mathbf{f}[t]\|^2} \mathbf{f}[t] e[t],$$

where $e[t] = d[t] - \mathbf{f}T[t]\mathbf{h}[t]$.

In order to provide more flexibility in the normalized LMS algorithm, the update is often written as

$$\mathbf{h}[t+1] = \mathbf{h}[t] + \frac{\mu}{\|\mathbf{f}[t]\|^2 + a} \mathbf{f}[t] \mathbf{e}[t],$$

where μ is a constant in the range $0 < \mu < 2$, and a is some small positive constant meant to ensure numerical stability when $\mathbf{f}[t]$ becomes small.

14.6-15 (Computer experiment) A white-noise signal $f[t]$ is passed through a system with impulse response $\{.5, -1, -2, 1, .5\}$. Program a MATLAB simulation to identify the system using an LMS algorithm. Try your experiment where the variance of $f[t]$ is $\sigma_f^2 = 0.1$, and the following variations:

(a) The LMS filter has five coefficients.

(b) The LMS filter has ten coefficients.

(c) The LMS filter has three coefficients.

Determine experimentally the range of μ for which the adaptive filter converges.

14.7-16 The steepest-descent neural network training algorithm has the following modification to *batch mode* processing.

(a) Show that a steepest-descent weight update based upon $\mathcal{E}_{av}$ can be written as

$$\mu \Delta w_{l,j,i} = -\mu \frac{\partial \mathcal{E}_{av}}{\partial w_{l,j,i}}$$

$$= -\frac{\mu}{N} \sum_{n=1}^{N} e_j(n) \frac{\partial e_j(n)}{\partial w_{l,j,i}}.$$

(b) Modify the code in algorithm 14.4 to implement this training rule, and test your results.

14.7-17 Show that the generalized delta rule in (14.38) can be written as

$$\Delta w_{l,j,i}(n) = -\mu \sum_{t=0}^{n} \alpha^{n-t} \frac{\partial \mathcal{E}(t)}{\partial w_{l,j,i}(t)}.$$

Argue that for stability, α must be within the range $0 \leq |\alpha| < 1$.

14.7-18 Another nonlinearity employed in the multilayer perceptron is

$$\phi(v) = a \tanh(bv)$$

for some constant a and b, where in applications $a \approx 1.7$ and $b = 2/3$ are used. Describe the modifications to the backpropagation algorithm when this nonlinear function is employed.

14.8-19 Show that (14.41) is true.

14.8-20 Show that (14.43) is true.

14.8-21 Show that (14.44) is true. (Hint: Use the "information inequality" $\ln x \leq x - 1$.)

14.8-22 Show that $I(X, Y) \leq \min(H(X), H(Y))$.

14.8-23 Consider a scalar system with time delay, where the output is processed from the measured signal $x(t)$ by

$$y(t) = g(wx(t - d)).$$

Determine a rule (for steepest ascent) for learning the delay d. Interpret the result.

14.8-24 Other functions $g(s)$ are possible. For the following functions, show that the slopes $y'_i = \partial y_i / \partial u_i$ and the gradient term $\frac{\partial}{\partial w_{ij}} \ln |y'_i|$ are corrected as indicated:

(a) $y_i = \tanh(u_i)$ $y'_i = 1 - y_i^2$ $\frac{\partial}{\partial w_{ij}} \ln |y'_i| = 2 x_j y_i$.

(b) $y_i = \arctan(u_i)$ $y'_i = 1/(1 + u_i^2)$ $\frac{\partial}{\partial w_{ij}} = -(2 x_j u_j)/(1 + u_i^2)$.

14.10 References

The discussion of the properties of iterated functions in section 14.1 is really an introduction to nonlinear dynamical systems. An excellent introduction to nonlinear dynamical systems, leading to a good discussion of chaos, is found in [74]. Iterated function systems are described in relationship to fractals in [12]. Another starting point in this area is [332], in which the orbits of the logistic function are presented.

The contraction mapping theorem is presented well in [209], and is common in books on analysis as well, such as [283]. A description (with insightful figures) of contraction mappings is in [12].

Our discussion in section 14.3 of the rate of convergence closely follows that of [210], which in turn draws from [246].

Newton's method is described in all books on numerical analysis; see, for example, [181]. Some of the results on the convergence of Newton's method were drawn from [209].

The LMS adaptive filter introduced in section 14.6 has been the object of an enormous amount of research, both in applications and in the theory of convergence. An early but still insightful examination of the LMS algorithm is [364] (see also [365, 367, 99]). An excellent summary in both areas is provided in [132, chapter 9]. Among the applications covered in that text are: adaptive equalization (also introduced here), system identification, adaptive line enhancement, beam nulling, and noise cancellation. Convergence of the LMS algorithm

14.10 References

continues to be studied, as does, adaptive filtering based on *blocks* of data—providing that improved estimates of R and **p** are presented, as in algorithms that adapt in the frequency domain. [212] and [213] provide an excellent starting point for analyses that weaken the independence assumption. An early and important work on the convergence of stochastic gradient algorithms is [278]. In some LMS algorithms, a variable step size is employed to speed convergence and reduce excess error. The normalized LMS algorithm presented in exercise 14.6-14 is one such algorithm. Other examples of variable step-size algorithms are in [222].

The literature on neural networks is also very broad, and the introduction presented in this chapter only scratches the surface. A historically important work, previous and preparatory to the explosion of research, is [228], which examines single-layer perceptrons with step functions as their nonlinearities. A groundbreaking text, covering the multilayer perceptron and a myriad of other topics, is [285]. A recent text, which provides an excellent coverage and includes fairly recent research, is [131]; another text, which takes a rather more generalized view of what constitutes a neural network, is [192]. [184] presents neural networks in the context of dynamical systems, and provides an interesting history. A tutorial introduction to several neural network varieties is [204].

A very useful discussion of iterative optimization methods appears in [260], where it is made clear that steepest descent, though conceptually easy, is often very bad in practice. Several powerful alternatives are presented there. See also the discussion on conjugate-gradient methods in section 16.5.

An introduction to the general problem of blind signal-processing problems appears in [47, 190, 130, 132]. Some blind equalization problems are also introduced in [9]. The principles of blind source separation are summarized in [49] and treated in more depth in [50]. The material in this chapter was largely drawn from [21]; see also [54] for more information on independent component analysis.

Chapter 15

Iteration by Composition of Mappings

> Tell me, what will it be, and O, where will it end?
> Say, if you have permission to tell;
> Is there any fixed point into which prospects tend?
> Does a focus belong to pell-mell?
>
> — *Eliza R. Snow,*
> 20 August 1842, *The Wasp*, Nauvoo, Ill.

15.1 Introduction

Many signals or the data structures associated with them have known properties. These properties might include, for example, signal attributes such as being real, even, positive, bandlimited, or possessing some kind of symmetry. Data matrices formed from signals might have properties such as being symmetric; having Toeplitz, Vandermonde or Hankel structure; having a known rank; and so on. However, in acquiring a signal for processing, desired theoretical properties may not be evident if the data are corrupted by the measurement process. The measured signal might be modified, for example, by to the need to obtain a finite-length sample of the data, or by measurement noise. As a result, the measured data may not satisfy the theoretical properties that the original data is known (or believed) to.

The iterative methods described in this chapter provide a means of enhancing measured signal data by finding the nearest signal that satisfies desired properties. The general technique is known as iteration by composition of mappings.

The topic may be introduced by means of an example. Suppose a function $f(t)$ with Fourier transform $F(\omega)$ is known to be bandlimited, that is,

$$F(\omega) = 0 \quad \text{for } |w| > b.$$

Let $g(t)$ be the measurement of $f(t)$ obtained over a time interval $-T < t < T$,

$$g(t) = \begin{cases} f(t) & |t| < T, \\ 0 & \text{otherwise.} \end{cases}$$

Since the length in time of $g(t)$ is finite, its Fourier transform cannot be bandlimited. Given $g(t)$, we desire to reconstruct an approximation $\hat{f}(t)$ to $f(t)$ that satisfies two

15.2 Alternating Projections

constraints:

1. For $|t| < T$, the reconstructed signal $\hat{f}$ should be consistent with the measured signal, $\hat{f}(t) = g(t)$ for $|t| < T$.
2. $\hat{f}(t)$ should be bandlimited: $\hat{F}(\omega) = 0$ for $|\omega| > b$.

This problem is sometimes called the bandlimited reconstruction problem. As we show in this chapter, an approach to the bandlimited reconstruction problem is to successively produce a function satisfying the first, then the second property, then the first again, and so forth. Each of these operations is a mapping, and the reconstructed signal is obtained by repeated composition of these mappings.

More generally, a signal may possess several properties, each of which has its own associated mapping. Satisfaction of these separate properties may be accomplished by iterating the associated mappings. In this chapter, we explore this basic technique, as well as some observations about convergence of the iterations.

Historically, the bandlimited reconstruction problem was addressed without the broader context, and the first proof of convergence did not rely on the theorems presented here [247]. A technique based on satisfying two constraints by composition of mappings was later examined, and the bandlimited reconstruction method was shown to be an application [380]. This was subsequently generalized to methods involving multiple constraints, in which the constraint sets are convex [379]. This gives rise to the method of *projection on convex sets* (POCS). A later generalization eliminated the need for the constraint sets to be convex. We refer to this version as the *composite mapping* (CM) method.

15.2 Alternating projections

We begin the formal tour of composite mapping methods by finding a means of satisfying two properties, as in the bandlimited reconstruction problem in the previous section. Let $\mathcal{H}$ be a Hilbert space with inner product $\langle x, y \rangle$. All of the signals are members of $\mathcal{H}$. In the context of these iterative methods, signals with particular properties are identified as members of subsets of $\mathcal{H}$. Requiring membership in a set is equivalent to requiring satisfaction of a certain constraint. For example, the set of bandlimited signals form a set, and requiring a signal to lie in this set enforces the constraint that a signal is to be bandlimited. In the alternating projection algorithm, we require that the constraint sets be closed linear manifolds (CLMs) in the Hilbert space $\mathcal{H}$.

Definition 15.1 A set of points $\mathcal{M}$ in $\mathbb{R}^n$ is a **manifold** if each point $f \in \mathcal{M}$ has an open neighborhood that has a one-to-one map onto an open set of $\mathbb{R}^n$.

A **linear manifold** $\mathcal{M}$ in a Hilbert (or Banach) space $\mathcal{H}$ is a subspace of $\mathcal{H}$. That is, the linear combination of elements in $\mathcal{M}$ is again in $\mathcal{M}$. If $\mathcal{M}$ is closed (that is, it is itself a Hilbert or Banach space), then $\mathcal{M}$ is said to be a closed linear manifold. □

The key idea of a manifold is that locally it "looks" like $\mathbb{R}^n$. An example manifold is the surface of a sphere: each sufficiently small neighborhood looks like a piece of $\mathbb{R}^2$, even though globally a sphere cannot be mapped exactly to $\mathbb{R}^2$.

Let $\mathcal{P}$ be a closed linear manifold of the Hilbert space, and let $\mathcal{P}^\perp$ be its orthogonal complement. We will say that a set such as $\mathcal{P}$ is synonymous with the property that defines the set: every signal $f \in \mathcal{P}$ has that particular property that determines inclusion in the set, and every signal that possesses the property is in $\mathcal{P}$.

Let P be the orthogonal projection operator onto $\mathcal{P}$. We can define P operationally as follows: Let $f \in \mathcal{H}$. Then the projection of f onto $\mathcal{P}$ is the unique point $Pf = g \in \mathcal{P}$ that is nearest to f in some appropriate norm (not necessarily an induced norm). Thus

$$\inf_{g \in \mathcal{P}} \|f - g\| = \|f - Pf\|. \tag{15.1}$$

(Recall that for finite-dimensional subspaces with known spanning vectors, the projection operator can be explicitly written. The definition here extends the concept to more general manifolds.) The projection of a signal f onto $\mathcal{P}$ by Pf means that we find the signal in $\mathcal{P}$ that is nearest to f; that is, the nearest signal that satisfies the constraints represented by the set $\mathcal{P}$. The uniqueness of the projection Pf is guaranteed since a CLM is convex (see exercise 15.2-1). The idea of a projection onto a convex set is shown in figure 15.1(a). Note for future reference that when a set is not convex, as in 15.1(b), there may be more than one point in the set nearest to f. We restrict our attention for the moment to sets that are CLMs and appreciate their convexity, then remove this restriction in section 15.3.

Let Q be the orthogonal projection operator onto $\mathcal{P}^\perp$, $Q = I - P$ where I is the identity operator. By the projection theorem (see theorem 2.8), every element $f \in \mathcal{H}$ has the unique decomposition

$$f = x + y \tag{15.2}$$

where $x \in \mathcal{P}$ and $y \in \mathcal{P}^\perp$. The orthogonal decomposition (15.2) gives rise to the decomposition

$$\mathcal{H} = \mathcal{P} \oplus \mathcal{P}^\perp.$$

The components of f in (15.2) can be obtained as $x = Pf$ and $y = Qf = (I - P)f$.

Now suppose that there are two CLMs, $\mathcal{P}_a \subset \mathcal{H}$ and $\mathcal{P}_b \subset \mathcal{H}$, representing different properties. The problem addressed by the method of alternating projections is this: An element $f \in \mathcal{H}$ is known to belong to the CLM $\mathcal{P}_b$ (representing known properties) but, due to difficulties in measurement, the available data g are in $\mathcal{P}_a$, which is the projection of f onto $\mathcal{P}_a$. We desire to find a way to compute f (the signal with desired properties) from g (the available signal). Let P_a, Q_a, P_b, and Q_b denote the projection operators onto $\mathcal{P}_a$, $\mathcal{P}_a^\perp$, $\mathcal{P}_b$, and $\mathcal{P}_b^\perp$, respectively. Then the measured data (assumed to be available free of noise) is

$$g = P_a f.$$

Since $f = P_b f$ (because P_b is a projection and $f \in \mathcal{P}_b$), we obtain

$$g = P_a P_b f = (I - Q_a) P_b f = f - Q_a P_b f = Af, \tag{15.3}$$

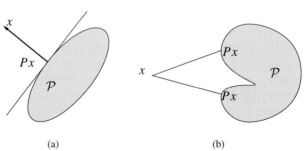

Figure 15.1: Illustration of a projection on a set. (a) When the set is convex, the projection is unique. (b) When the set is not convex, there may be more than one nearest point

15.2 Alternating Projections

where

$$A = I - Q_a P_b$$

and I is the identity operator. The operator A has the nullspace

$$\mathcal{N}(A) = \mathcal{P}_b \cap \mathcal{P}_a^\perp,$$

since every $x \in \mathcal{P}_b \cap \mathcal{P}_a^\perp$ satisfies $Ax = 0$, and conversely (by the orthogonal decomposition), every element in the nullspace of A must be in $\mathcal{P}_b \cap \mathcal{P}_a^\perp$.

If there is an inverse operator $T = A^{-1}$, then we can find

$$f = Tg.$$

Not only is the existence of the inverse important, but also its "condition"—that is, whether it is stable—because usually we don't know g exactly, but rather a noise-perturbed version $g + n$. When the available measurement is $g = P_a f + n$, the reconstructed signal is

$$\hat{f} = Tg = T(P_a f + n),$$

and the difference between the true f and $\hat{f}$ is therefore

$$\Delta f = Tn.$$

We hope that $\|\Delta f\|$ is small when $\|n\|$ is small. More precisely, we desire to impose the condition that

$$\sup_{n \in \mathcal{H}} \frac{\|Tn\|}{\|n\|} < \infty,$$

which is to say $\|T\| < \infty$.

Rather than attempt to find the inverse operator T explicitly, we formulate an operator whose iterates converge to T. From the equation $f = g + Q_a P_b f$ obtained from (15.3), we propose the recursive algorithm

$$f^{[k+1]} = g + Q_a P_b f^{[k]} \qquad (15.4)$$

with initial condition $f^{[0]} = g$. Equation (15.4) is *the method of alternating projections*.

Figure 15.2 provides some geometric insight into the algorithm in a two-dimensional space. The vector $f \in \mathcal{P}_b$ is desired. We know $g = \mathcal{P}_a f$. In figure 15.2(a), one iteration of the algorithm is illustrated. We first project onto $\mathcal{P}_b$. Since the projection of this vector

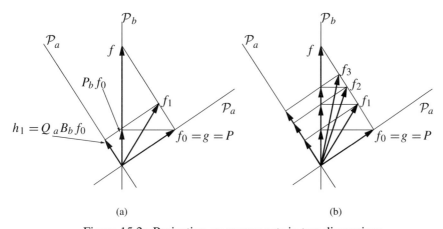

Figure 15.2: Projection on convex sets in two dimensions

back onto $\mathcal{P}_a$ does not match the starting vector g, we therefore compute a correction by projecting $P_b\mathbf{g}$ onto $\mathcal{P}_a^\perp$ to form

$$h^{[1]} = Q_a P_b f^{[0]}.$$

Finally, we obtain an update to our estimate as

$$f^{[1]} = g + h^{[1]} = g + Q_a P_b f^{[0]}.$$

Figure 15.2(b) shows additional iterates $f^{[2]}$ and $f^{[3]}$. These are clearly approaching the desired final value f.

The condition that establishes the convergence of (15.4) is based on the angle between the subspaces $\mathcal{P}_a$ and $\mathcal{P}_b$. For points $f \in \mathcal{P}_a$ and $g \in \mathcal{P}_b$, we define

$$\cos\psi(f, g) = \frac{|\langle f, g\rangle|}{\|f\|\|g\|}.$$

We define the angle between subspaces $\mathcal{P}_a$ and $\mathcal{P}_b$ as

$$\psi(\mathcal{P}_a, \mathcal{P}_b) = \inf_{\substack{f \in \mathcal{P}_a \\ g \in \mathcal{P}_b}} \psi(f, g),$$

or, equivalently,

$$\cos\psi(\mathcal{P}_a, \mathcal{P}_b) = \sup_{\substack{f \in \mathcal{P}_a \\ g \in \mathcal{P}_b}} \frac{|\langle f, g\rangle|}{\|f\|\|g\|}.$$

Geometrically, if $\mathcal{P}_a$ and $\mathcal{P}_b$ are orthogonal, then from the projection $g = P_b f$, the information necessary to recover f is lost. This is formalized in the following theorem.

Theorem 15.1 *[380] Let $\mathcal{P}_a$ and $\mathcal{P}_b$ be any two closed linear manifolds in a Hilbert space $\mathcal{H}$, with P_a, P_b, Q_a, and Q_b the projectors onto these respective spaces and their complements. Let $f \in \mathcal{P}_b$.*

1. *f is uniquely determined by its projection $P_a f$ onto $\mathcal{P}_a$ if and only if*

$$\mathcal{P}_b \cap \mathcal{P}_a^\perp = 0.$$

2. *The inverse operator $T = (I - Q_a P_b)^{-1}$ has a bounded inverse if and only if*

$$\psi\left(\mathcal{P}_b, \mathcal{P}_a^\perp\right) > 0.$$

This angle constraint is satisfied if and only if

$$\|Q_a P_b\| < 1.$$

Due to its length, the proof of this theorem is not presented here (but may be found in [379]). However, after the uniqueness and existence of the inverse are established, the key point of the proof is that the iterates in (15.4) can be written as

$$f^{[k+1]} = \sum_{i=0}^{k}(Q_a P_b)^i g = \sum_{i=0}^{k}(Q_a P_b)^i(f - Q_a P_b f) = f - (Q_a P_b)^{k+1} f.$$

It can be shown that $\|Q_a P_b\| = \cos\psi(\mathcal{P}_b, \mathcal{P}_a^\perp)$ (this is the hard part of the proof), so that under the hypotheses of the theorem

$$\|Q_a P_b\| < 1.$$

15.2.1 An application: bandlimited reconstruction

Hence, in the limit,
$$\lim_{k \to \infty} (Q_a P_b)^k = h_0,$$
where $h_0 \in \mathcal{P}_b \cap \mathcal{P}_a^\perp$. But by the conditions of the theorem, $h_0 = 0$, so that $f^{[k+1]} \to f$.

15.2.1 An application: bandlimited reconstruction

Let $\mathcal{H}$ be the Hilbert space L_2, and let $\mathcal{P}_T \subset \mathcal{H}$ denote the set of time-limited functions that vanish for $|t| > T$. It is clear that $\mathcal{P}_T$ is a CLM, since the linear combination of any two functions time limited to T is clearly time limited, and any sequence of time-limited functions is time limited. Also, let $\mathcal{P}_b \subset \mathcal{H}$ denote the set of functions bandlimited to b; that is, for $f(t) \in \mathcal{P}_b$ with $f(t) \leftrightarrow F(\omega)$, $F(\omega) = 0$ for $|\omega| > b$. This space is also a CLM. The projection operator $P_a: \mathcal{H} \to \mathcal{P}_T$ is defined by

$$P_T f = \begin{cases} f(t) & |t| < 0, \\ 0 & \text{otherwise.} \end{cases}$$

For convenience, we define the function $p_T(t)$ by

$$p_T(t) = \begin{cases} 1 & |t| < T, \\ 0 & \text{otherwise.} \end{cases}$$

Then $P_T f = p_T f$. The projection operator $P_b: \mathcal{H} \to \mathcal{P}_b$ is defined by

$$P_b f \leftrightarrow p_b(\omega) F(\omega),$$

that is, by truncating the Fourier transform. We can write explicitly

$$P_b f = \int_{-b}^{b} F(\omega) e^{j\omega t}.$$

We can write the iteration (15.4) as

$$f^{[k+1]} = g + (1 - p_T(t)) \mathcal{F}^{-1}[f^{[k]}(\omega) p_b(\omega)]. \tag{15.5}$$

A section of code to implement bandlimited reconstruction, using discrete Fourier transforms for projection onto $\mathcal{P}_b$, is shown in algorithm 15.1

Algorithm 15.1 Bandlimited reconstruction using alternating projections
File: `bliter1.m`

As a specific example, let $f(t) = \sin(\pi t)/(\pi t)$. This is a signal bandlimited to π radians/sec. Let $g(t)$ be the signal obtained over $-0.5 < t < 0.5$. The results of the reconstruction when the data are obtained by sampling every 0.02 seconds are shown in figure 15.3, where (a) shows the true signal $f(t)$ and its reconstruction, and (b) shows the spectrum of the reconstructed signal. In this example, $N = 1024$ points were used in the FFT.

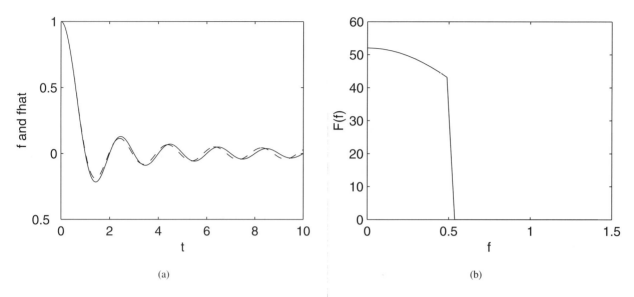

Figure 15.3: Results of the bandlimited reconstruction algorithm. (a) solid line $= f(t)$; dashed line $= \hat{f}(t)$. (b) Spectrum of $\hat{f}(t)$

15.3 Composite mappings

We now generalize the idea of alternating projections to more than two projections. We also allow for the constraint sets to be nonconvex. Also, we remove the need for a Hilbert space; closure of the space is retained, but for the convergence results stated here no inner product is necessary. Instead, we consider simply a metric space (X, d).

Suppose that there are m properties that we desire to enforce. Corresponding to these properties are sets $\mathcal{P}_1, \mathcal{P}_2, \ldots, \mathcal{P}_m$, each $\mathcal{P}_k \subset X$,

$$\mathcal{P}_k = \{x \in X : x \text{ possesses property } k\}.$$

These sets are not necessarily convex, although in many applications they turn out to be convex. The set of elements in X that possess all of the properties is the intersection of the property sets

$$\mathcal{P} = \mathcal{P}_1 \cap \mathcal{P}_2 \cap \cdots \cap \mathcal{P}_m.$$

The set $\mathcal{P}$ is assumed to be nonempty, or else there is no solution to the problem. The concept is illustrated in figure 15.4. Let the original measured signal be g. The problem of finding a signal f nearest to g that satisfies all of the constraints may be expressed as

$$\hat{f} = \arg \inf_{f \in \mathcal{P}} d(g, f). \tag{15.6}$$

As illustrated in figure 15.1, there may, in general, be multiple solutions to (15.6) if $\mathcal{P}$ is not convex, so we refer to a solution *set*, rather than a unique solution.

Direct minimization in (15.6) is often difficult, due to the need to satisfy multiple constraints and minimization over possibly nonconvex functions. Instead of solving the optimization problem subject to all the constraints simultaneously, the approach taken by the composite mapping algorithm is look for a solution to the optimization problems taking each constraint separately, by solving

$$\inf_{f \in \mathcal{P}_k} d(g, f). \tag{15.7}$$

15.4 Closed Mappings and The Global Convergence Theorem

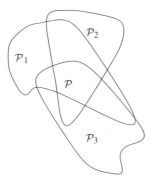

Figure 15.4: Property sets in X and their intersection $\mathcal{P}$

We denote by $P_k g$ the solution set to (15.7). It will be useful to think of $P_k g$ as a mapping that acts on g. For each property there is a mapping, and we obtain the set of mappings $P_1, P_2, \ldots, P_m$.

We now form a composite mapping by sequential application of the individual property mappings specified by

$$P = P_m \cdots P_2 P_1$$

or

$$Pg = P_m(\cdots(P_2(P_1(g)))\cdots).$$

The composite mapping P will in general be a point-to-set mapping. The ordering of mappings does make a difference, since the property mappings do not in general commute.

In order to say something about the convergence of the method, we take a slight but important detour to present a general convergence theorem for closed point-to-set mappings.

15.4 Closed mappings and the global convergence theorem

The utility of point-to-set mappings, as opposed to simple point-to-point mappings, is that they allow for some ambiguity in representation. For example, provided that convergence can be proven for a point-to-set mapping, numeric details from one computer to another may have little bearing on the convergence of any particular implementation, as long as the results obtained on any computer can all be said to fall within some set in which convergence is observed.

Example 15.4.1 For $x \in \mathbb{R}$, define the point-to-set mapping

$$P(x) = [-|x|/2, |x|/2].$$

If we start from $x_0 = 10$, any of the following sequences could be in sets that are produced by the mapping P:

$$\{10, -4.5, 2, -.75, .25, -.1, .05, -.01, .005, \ldots\},$$
$$\{10, 5, 2.5, 1.25, 0.625, .313, .156, \ldots\},$$
$$\{10, -1, 0.5, .25, .125, .0625, .0313, \ldots\}.$$

In each sequence the numbers are different, but each sequence is converging to 0. □

In this example, the numbers diminish with every iteration. It will be useful to introduce a function to measure how "small" the elements in the metric space are, and to introduce the idea of a solution set, from which no additional decrease in size is possible.

Definition 15.2 Let X be a metric space, let $Z: X \to \mathbb{R}$ be a function, and let $\Gamma \subset X$ be a set associated with Z and a mapping P, which may be set valued (map to sets). The function Z is said to be a **descent function** for Γ and P if it satisfies the following properties.

1. If $x \notin \Gamma$ and $y \in Px$ then $Z(y) < Z(x)$. That is, an iteration of x under P leads to a set whose elements y decrease Z, as long as x is not already in the solution set.
2. If $x \in \Gamma$ and $y \in P(x)$ then $Z(y) \leq Z(x)$. That is, if $x \in \Gamma$, then an iteration of x under P may not lead to a further decrease in Z. □

Let $x_k \to x$ indicate that the sequence of elements x_k is converging to x in (X, d); that is,

$$d(x_k, x) \to 0 \qquad \text{as } k \to \infty.$$

Definition 15.3 [210, 384] Let P be a point-to-set mapping of the set X into itself. P is said to be **closed** at $x \in X$ if

$$x_k \to x \quad \text{with } x_k \in X,$$
$$y_k \to y \quad \text{with } y_k \in P(x_k),$$

imply $y \in P(x)$. The point-to-set mapping P is closed on a set $A \subset X$ if it is closed on each $x \in A$. □

Example 15.4.2 Consider again the mapping in example 15.4.1. Let

$$x_k = 2 - \frac{1}{k},$$

and note that $x_k \to 2 = x$. Let

$$y_k = x_k/2 = (2 - 1/k)/2$$

and note that $y_k \in P(x_k)$ and $y_k \to 1 \in P(x)$. Thus $P(x)$ is a closed mapping.

To contrast, define the mapping

$$G(x) = [-|x|/2, |x|/2).$$

Using the same sequences x_k and y_k, note that $y_k \to 1$, but $1 \notin G(x)$. Thus $G(x)$ is *not* a closed mapping. □

A mapping $P(x)$ can be seen to be closed from the graph of the map, which is the set $\{(x, y): x \in X, y \in P(x)\}$. If X is closed then P is closed throughout X if and only if the graph is a closed set.

Compositions of closed mappings are defined as follows.

Definition 15.4 Let $A: X \to Y$ and $B: Y \to Z$ be point-to-set mappings. The composite mapping $C = BA$ is defined as the point-to-set mapping $C: X \to Z$ with

$$C(x) = \bigcap_{y \in A(x)} B(y).$$

□

The mappings in this definition are illustrated in figure 15.5. The issue of closure of the composite mapping C is determined by the following theorem.

Theorem 15.2 *[210, page 186] Let $A: X \to Y$ and $B: Y \to Z$ be point-to-set mappings. If the following properties are true:*

1. *A is closed at x and B is closed on $A(x)$,*

15.4 Closed Mappings and The Global Convergence Theorem

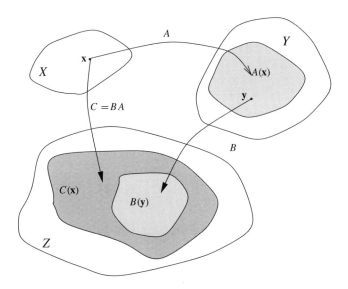

Figure 15.5: Illustration of the composition of point-to-set mappings

2. $x_k \to x$ and $y_k \in A(x_k)$, and
3. there is a y such that for a subsequence $\{y_{k_i}\}$ $y_{k_i} \to y$,

then the composite mapping $C = BA$ is closed at x.

Proof Let $x_k \to x$ and $z_k \to z$ with $z_k \in C(x_k)$. Closure is established if it can be shown that $z \in C(x)$. Let $y_k \in A(x_k)$ be such that $z_k \in B(y_k)$ and, according to the hypothesis of the theorem, let y and $\{y_{k_i}\}$ be such that $y_{k_i} \to y$. Since A is closed at x, it follows that $y \in A(x)$.

Likewise, let $z_{k_i} \in B(y_{k_i})$. Since $y_{k_i} \to y$ and B is closed at y, it follows that $z \in B(y) \subset BA(x) = C(x)$. □

Corollary 15.1 *If $A: X \to Y$ and $B: Y \to Z$ are point-to-set mappings with A closed at x, B closed on $A(x)$, and Y compact, then the composite mapping $C = BA$ is closed at x.*

This follows from the theorem since for compact Y, every convergent subsequence y_{k_i} converges in Y. As a second corollary, we have the following.

Corollary 15.2 *Let $A: X \to Y$ be a point-to-point mapping and $B: Y \to Z$ be a point-to-set mapping. If A is continuous at x and B is closed at $A(x)$, then the composite mapping $C = BA$ is closed at x.*

Since A is point-to-point, the continuity of A at x guarantees the closure of the mapping A.

With this nomenclature, we can now state and prove the global convergence theorem, which is a powerful result used to establish convergence for iterations of mappings. It can be used to prove convergence of the composite mapping algorithm.

Theorem 15.3 *(Global convergence theorem) [210, page 187] Let P be a mapping onto X, and suppose that for an initial point $f^{[0]}$ the sequence $\{f^{[k]}\}$ is generated according to*

$$f^{[k+1]} = P(f^{[k]}).$$

Assume that there is some continuous objective function $Z: X \to \mathbb{R}$ that is a descent function with respect to F and a solution set Γ. Also assume that the mapping F is closed at points outside Γ. Then the limit of any convergent subsequence of $\{x_k\}$ is a solution.

Although the proof is somewhat technical, it is presented to reinforce some of the analytical concepts first presented in chapter 2.

Proof Let the convergent subsequence $\{f_k\}$, $k \in \mathcal{K}$ converge to the limit f, where $\mathcal{K}$ is some index set (such as $\mathbb{Z}$). Since Z is continuous, it follows that for $k \in \mathcal{K}$, $Z(f_k) \to Z(f)$, so that Z is convergent with respect to the subsequence $\mathcal{K}$. We show that it is convergent with respect to the entire sequence. By the monotonicity of Z on the sequence $\{\mathbf{x}_k\}$, we have $Z(f_k) - Z(f_K) \leq 0$ for $k > K$. By the convergence of Z on the subsequence, for every $\epsilon > 0$ there is a $K \in \mathcal{K}$ such that $Z(f_k) - Z(f) < \epsilon$ for $k > K$ with $k \in \mathcal{K}$. For any $k > K$, consider the following:

$$Z(f_k) - Z(f) = (Z(f_k) - Z(f_K)) + (Z(f_k) - Z(f)).$$

By monotonicity, $(Z(f_k) - Z(f_K)) < 0$, and by convergence of the subsequence, $(Z(f_k) - Z(f)) < \epsilon$. Thus,

$$Z(f_k) - Z(f) < \epsilon$$

and we conclude that $Z(f_k) \to Z(f)$.

It remains to show that the convergent value f is in the solution set Γ. Suppose, to the contrary, that $f \notin \Gamma$. Consider the subsequence $\{f_{k+1}\}$ for $k \in \mathcal{K}$. Since all members of the sequence $\{f_k\}$ are in a compact set, there is a subsequence $\overline{\mathcal{K}} \subset \mathcal{K}$ such that $\{f_{k+1}\}$ for $k \in \overline{\mathcal{K}}$ converges to some limit $\bar{f}$. We thus have

$$f_k \to f \qquad \text{for } k \in \overline{\mathcal{K}},$$
$$f_{k+1} \to \bar{f} \qquad \text{for } k \in \overline{\mathcal{K}}.$$

Since F is closed at f, it is also true that $\bar{f} \in A(f)$. But by the results above, $Z(\bar{f}) = Z(f)$, regardless of the subsequence chosen, which contradicts the fact that Z is a descent function outside of the solution set Γ. □

The requirement that the mapping F be closed is perhaps the most important condition, and failure to satisfy that condition may prevent an algorithm from converging.

15.5 The composite mapping algorithm

Let (X, d) be a metric space. To apply the global convergence theorem to the composite mapping algorithm, we let $f_r \in X$ be a reference signal satisfying all of the properties, $f_r \in \mathcal{P}$, and let the objective function Z be the metric distance between the signal $f \in X$ and the reference signal,

$$Z(f) = d(f, f_r).$$

Convergence of the composite mapping

$$f^{[k+1]} = P(f^{[k]})$$

follows from the global convergence theorem if the mapping P is closed and distance-reducing relative to Z. Because P may be a point-to-set mapping, the sequence gen-

15.5 The Composite Mapping Algorithm

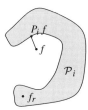

Figure 15.6: Projection onto a non-convex set may actually increase the distance to the desired point

erated by it may not be unique, but we are assured by the theorem that the sequence generated converges to a signal in $\mathcal{P}$.

Since nonconvex sets may arise in some applications, the restriction that the mapping P be distance-reducing relative to Z may not be trivial. While any projection P_i mapping to $\mathcal{P}_i$ will reduce (more precisely, not increase) the distance to $\mathcal{P}_i$, this mapping could, if the sets are not convex, actually increase the distance to the intersection set, $d(P_i f, f_r)$, so that the composition P could be non–distance-reducing. An illustration of this increase in distance in shown in figure 15.6. However, for all the properties described in the remainder of this section, convergence is obtained. It is further shown (in section 15.6) that when the property sets *are* convex, the composition of operators is distance-reducing relative to Z. In addition, it is shown in section 15.6 that over-relaxed operators can also be used, which can potentially speed the rate of convergence. We now examine several different properties and examine mappings onto their constraint sets. In these examples, we point out constraint sets that are convex.

15.5.1 Bandlimited reconstruction, revisited

The bandlimited reconstruction problem examined in section 15.2 can be expressed in terms of composite mappings. Let $\mathcal{P}_b$ be the set of functions bandlimited to b. Let $\mathcal{T}_T$ denote the set of functions that are time limited to T, that is, that vanish for $|t| > T$. Let $\mathcal{T}_T^\perp$ denote its orthogonal complement. Also, let $\mathcal{P}_g$ denote the set of functions whose projections onto $\mathcal{T}_T$ is equal to $g(t)$. Then

$$\mathcal{P}_g = g + \mathcal{T}_T^\perp,$$

which is a linear variety. Combining projections onto $\mathcal{P}_b$ and $\mathcal{P}_g$, we obtain the reconstruction algorithm (15.5).

15.5.2 An example: Positive sequence determination

Given a sequence of data, we desire to find a sequence near to it with the property that it has a real, nonnegative Fourier transform. This problem has application in filter synthesis, data windowing, and spectral analysis. More precisely stated, the problem is this: Given a sequence of data $\{x_k\}$, determine the nearest sequence that has a positive (and hence real) Fourier transform. In order for the transform to be real, it is necessary that $x_k = \overline{x}_{-k}$, that is, that x has conjugate even symmetry. A sequence having the property that its Fourier transform is real is said (in this context) to be a positive sequence.

We employ the composite mapping property to find a sequence of length $2q + 1$ with positive Fourier transform. For computational purposes, we use the DFT. To enforce the

conjugate even symmetry, let

$$\mathbf{x} = [x_0 \; x_1 \; \cdots \; x_q \; 0 \; 0 \; \cdots \; 0 \; \overline{x}_q \; \overline{x}_{q-1} \; \cdots \; \overline{x}_1]'$$

be a conjugate symmetric vector of length $N > 2q + 1$. There are $N - 2q - 1$ zeros in the middle. Let

$$\mathbf{X} = F_N \mathbf{x}$$

be the N-point DFT of $\mathbf{x}$. There are two properties to enforce:

1. π_1: the sequence $\mathbf{x} \in \mathbb{R}^N$ is of length $2q + 1$. That is, there must be $N - 2q - 1$ zeros in the middle of the final sequence. The set $\mathcal{P}_1$ of vectors in $\mathbb{R}^N$ with $N - 2q - 1$ zeros in the middle is convex.
2. π_2: the Fourier transform of the sequence is positive. The set $\mathcal{P}_2$ of vectors with positive Fourier transform is convex.

The mapping P_1 to the set of vectors satisfying property π_1 is found as follows. The conjugate symmetric vector $\mathbf{y}$ with $2q + 1$ nonzero elements that is closest to an arbitrary conjugate symmetric vector $\mathbf{x}$ of length N is found by the projection mapping

$$\mathbf{y} = P_1 \mathbf{x} = [x_0 \; x_1 \; \cdots \; x_q \; 0 \; 0 \; \cdots \; 0 \; \overline{x}_q \; \overline{x}_{q-1} \; \cdots \; \overline{x}_1]'.$$

In other words, the $N - 2q - 1$ elements in the middle are simply set to zero. The mapping P_2 to the set of vectors satisfying property π_2 is found as follows. Let $[x]^+$ denote the operator

$$[x]^+ = \begin{cases} x & x \geq 0, \\ 0 & x < 0. \end{cases} \tag{15.8}$$

Then

$$P_2 \mathbf{x} = F_N^{-1} [F_N \mathbf{x}]^+.$$

That is, we compute the Fourier transform, clip any negative values at zero, then compute the inverse Fourier transform. Then the composite mapping iteration step is defined by

$$\mathbf{x}^{[k+1]} = F \mathbf{x}^{[k]} = P_1 P_2 \mathbf{x}^{[k]}.$$

Example 15.5.1 The Hamming window

$$x[k] = 0.54 + 0.46 \cos(\pi k / M) \qquad -M \leq k \leq M$$

is not a positive sequence. The code in algorithm 15.2 executes the composite mapping algorithm on the data to produce a nearby positive sequence $\tilde{x}[k]$.

Algorithm 15.2 Mapping to a positive sequence
File: `compmap2.m`

Figure 15.7(a) shows the Fourier transform of $x[k]$ and of $\tilde{x}[k]$. Figure 15.7(b) shows both $x[k]$ and $\tilde{x}[k]$. □

15.5 The Composite Mapping Algorithm

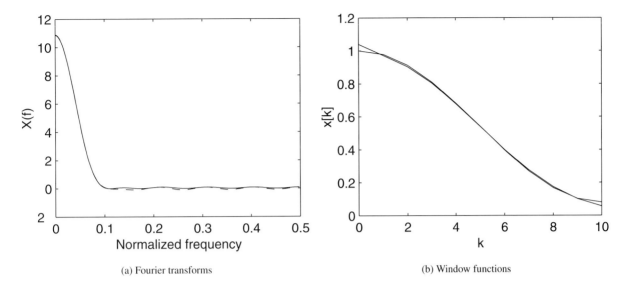

(a) Fourier transforms (b) Window functions

Figure 15.7: Producing a positive sequence from the Hamming window. Solid line: Hamming window; dashed line: positive sequence

15.5.3 Matrix property mappings

A variety of problems in signal processing give rise to matrices that have a particular theoretical structure—such as Toeplitz, Hankel, stochastic, and so forth,—or a specification on the rank. While the theoretical structure of a matrix might be known, a matrix obtained from noisy data might fail to have the anticipated properties. Using composite mappings, it may be possible to find a nearest matrix that approximates the noisy matrix and has the desired property. We discuss a variety of properties and mappings to obtain them. Other properties and mappings can often be similarly defined.

Nearest matrix with nonnegative elements. Let A be a matrix. A mapping that determines the nearest matrix to A that has all nonnegative elements is to be determined. The Frobenius is an appropriate norm, because it works on an element-by-element basis. The appropriate mapping is

$$P_1(A) = [A]^+;$$

that is, the operator defined in (15.8), applied element by element to A. The operator is closed. Note, however, that an operator that mapped to strictly positive elements would not be closed.

The set of matrices with nonnegative elements is convex.

Nearest matrix of given rank. As discussed in section 7.5, the matrix nearest to A (using the Frobenius norm) that has rank $r \leq \text{rank}(A)$ is

$$\sum_{k=1}^{r} \sigma_k \mathbf{u}_k \mathbf{v}_k^H,$$

where σ_k, $\mathbf{u}_k$ and $\mathbf{v}_k$ come from the SVD of A. An operator which maps A to the nearest matrix of rank r is therefore

$$P_2(A) = \sum_{k=1}^{r} \sigma_k \mathbf{u}_k \mathbf{v}_k.$$

The set of matrices with a given rank is *not* convex, as may be seen by counterexample. Let

$$A = \begin{bmatrix} 1 & 2 & 1 \\ 2 & 3 & 2 \\ 4 & 8 & 4 \end{bmatrix} \quad \text{and} \quad B = \begin{bmatrix} 12 & 14 & 9 \\ 7 & 8 & 5 \\ 5 & 6 & 4 \end{bmatrix}.$$

Then

$$\text{rank}(A) = 2 \quad \text{rank}(B) = 2 \quad \text{but} \quad \text{rank}(A+B) = 3.$$

Nearest symmetric matrix. The mapping to the symmetric matrix nearest to A is given by

$$P_3(A) = \frac{A + A^T}{2}.$$

The set of symmetric matrices is convex.

Nearest positive-semidefinite matrix. Let X be a Hermitian $n \times n$ matrix, represented as

$$X = \sum_{k=1}^{n} \lambda_k \mathbf{u}_k \mathbf{u}_k^H. \tag{15.9}$$

Arrange the eigenvalues as $\lambda_1 \geq \lambda_2 \geq \cdots \geq \lambda_n$, with the first p being positive. A positive-semidefinite matrix has eigenvalues ≥ 0. The mapping

$$P_4(X) = \sum_{k=1}^{p} \lambda_k \mathbf{u}_k \mathbf{u}_k^H \tag{15.10}$$

produces a matrix that is positive semidefinite and that is closest in Frobenius norm to X. Verification of this fact comes by recognition that the representation (15.9) is equivalent to the SVD $X = U \Sigma U^H$ for a symmetric matrix X. The distance from a positive-semidefinite matrix X^+ to X is $\|X - X^+\|_F = \|\Sigma - \tilde{X}^+\|_F$, where $\tilde{X} = U^H X^+ U$. The positive-semidefinite matrix $\tilde{X}$ closest to Σ is that matrix that is diagonal, with elements matching the positive values of Σ and zeros elsewhere. This corresponds to the mapping (15.10). The set of positive-semidefinite matrices is convex.

Matrix with rows summing to 1. (**stochastic matrix**) The matrix B, nearest in Frobenius norm to an $m \times n$ matrix A, satisfying

$$B\mathbf{1} = \mathbf{1},$$

that is, whose rows sum to 1, can be written as

$$B = P_5(A) = \begin{bmatrix} \mathbf{a}_1^T - \frac{\mathbf{a}_1^T \mathbf{1} - 1}{n} \mathbf{1}^T \\ \mathbf{a}_2^T + \frac{\mathbf{a}_2^T \mathbf{1} - 1}{n} \mathbf{1}^T \\ \vdots \\ \mathbf{a}_m^T + \frac{\mathbf{a}_m^T \mathbf{1} - 1}{n} \mathbf{1}^T \end{bmatrix}. \tag{15.11}$$

In (15.11), A is represented in terms of its rows,

$$A = \begin{bmatrix} \mathbf{a}_1^T \\ \mathbf{a}_2^T \\ \vdots \\ \mathbf{a}_m^T \end{bmatrix},$$

and $\mathbf{1}$ is an $n \times 1$ vector of all ones (see exercise 15.5-4). The set of matrices with rows summing to 1 is convex.

15.5 The Composite Mapping Algorithm

Nearest linear-structured matrix. To introduce linear-structured matrices, consider the Toeplitz matrix

$$A = \begin{bmatrix} a_2 & a_1 \\ a_3 & a_2 \\ a_4 & a_3 \end{bmatrix}.$$

This matrix has only four independent elements. The vectorized version of A (see section 9.3) is

$$\mathbf{a} = \text{vec}(A) = \begin{bmatrix} a_2 & a_3 & a_4 & a_1 & a_2 & a_3 \end{bmatrix}^T.$$

This can be written in terms of a minimal vector of parameters

$$\mathbf{v} = \begin{bmatrix} a_1 \\ a_2 \\ a_3 \\ a_4 \end{bmatrix},$$

by means of the *linear-structured matrix* S:

$$\mathbf{a} = S\mathbf{v} = \begin{bmatrix} 0 & 1 & 0 & 0 \\ 0 & 0 & 1 & 0 \\ 0 & 0 & 0 & 1 \\ 1 & 0 & 0 & 0 \\ 0 & 1 & 0 & 0 \\ 0 & 0 & 1 & 0 \end{bmatrix} \begin{bmatrix} a_1 \\ a_2 \\ a_3 \\ a_4 \end{bmatrix}. \quad (15.12)$$

More generally, a linear-structured $m \times n$ matrix A is one whose vectorization $\mathbf{a}$ can be written as

$$\text{vec}(A) = \mathbf{a} = S\mathbf{v},$$

where the length of $\mathbf{v}$ is less than the length of $\mathbf{a}$. The original matrix A thus has a linear dependence between its elements that may be exploited in the vectorized notation. Toeplitz, Hankel, and symmetric matrices are examples of linear-structured matrices.

Given a matrix X, we can find the matrix A that is closest to it and that has a desired linear structure, in three steps.

1. Vectorize X to form $\mathbf{x}$.
2. Find the vector $\mathbf{a} = S\mathbf{v}$ that is closest to $\mathbf{x}$. Using the Euclidean norm, the minimization is

$$\min_{\mathbf{v}} \|\mathbf{x} - S\mathbf{v}\|_2.$$

   The least-squares solution is

$$\mathbf{v} = S(S^T S)^{-1} S^T \mathbf{x}.$$

3. Unvectorize $\mathbf{a}$ to form A with the desired linear structure property.

The three steps can be combined into a single operator as

$$P_6(X) = \text{vec}^{-1}(S(S^T S)^{-1} S^T \text{vec}(X)).$$

Computation of the operator $S(S^T S)^{-1} S^T = SS^\dagger$ should rarely be done directly. The linear-structured matrix S may be very large, and directly computing its pseudoinverse would be slow. Instead, it is worthwhile to examine what the operator is actually doing. Let X be a 3×2 matrix

$$X = \begin{bmatrix} x_{11} & x_{12} \\ x_{21} & x_{22} \\ x_{31} & x_{32} \end{bmatrix},$$

and consider finding the nearest Toeplitz matrix

$$A = \begin{bmatrix} a_2 & a_1 \\ a_3 & a_2 \\ a_4 & a_3 \end{bmatrix}.$$

Using S from (15.12), we find that the vectorized solution is

$$\mathbf{v} = S(S^T S)^{-1} S \mathbf{x} = \begin{bmatrix} .5 & 0 & 0 & 0 & .5 & 0 \\ 0 & .5 & 0 & 0 & 0 & .5 \\ 0 & 0 & 1 & 0 & 0 & 0 \\ 0 & 0 & 0 & 1 & 0 & 0 \\ .5 & 0 & 0 & 0 & 0.5 & 0 \\ 0 & .5 & 0 & 0 & 0 & .5 \end{bmatrix} \mathbf{x}.$$

The component equations can be written as

$$v_2 = \frac{1}{2}(x_{11} + x_{22}),$$

$$v_3 = \frac{1}{2}(x_{21} + x_{32}),$$

$$v_1 = x_{12},$$

$$v_4 = x_{32}.$$

The elements of A are obtained by simple averages of those elements of X that correspond. The set of matrices with a given linear structure is convex.

A few examples of the use of combinations of these properties should demonstrate their utility.

Example 15.5.2 We desire to devise an algorithm to find the nearest stochastic matrix to a matrix A. A stochastic matrix has all of its elements nonnegative and all of its rows sum to 1. To satisfy this, two mapping are necessary, P_1—which finds the nearest positive matrix—and P_5—which finds the nearest matrix whose rows sum to 1. The composite map is $P_1 P_5$. A MATLAB routine to perform this function is shown in algorithm 15.3.

Algorithm 15.3 Mapping to the nearest stochastic matrix
File: `tostoch.m`

When

$$A = \begin{bmatrix} 0.2190 & 0.6793 & 0.5194 \\ 0.0470 & 0.9347 & 0.8310 \\ 0.6789 & 0.3835 & 0.0346 \end{bmatrix}$$

is input, the resulting output is

$$A = \begin{bmatrix} 0.0797 & 0.5401 & 0.3802 \\ 0 & 0.5519 & 0.4481 \\ 0.6466 & 0.3512 & 0.0023 \end{bmatrix},$$

which can be seen to satisfy the properties. □

15.5 The Composite Mapping Algorithm

Example 15.5.3 Suppose X is an $m \times n$ matrix, and we desire to find the nearest $m \times n$ Hankel matrix A of rank $p < \min(m, n)$. In this case, the mappings are P_2 with a linear structure map of the form P_6. An $m \times n$ Hankel matrix has $m + n - 1$ parameters. An example 3×2 Hankel matrix A,

$$A = \begin{bmatrix} a_1 & a_2 \\ a_2 & a_3 \\ a_3 & a_4 \end{bmatrix},$$

can be expressed in vectorized form using a linear-structured matrix S as

$$\mathbf{a} = S\mathbf{v} = \begin{bmatrix} 1 & 0 & 0 & 0 \\ 0 & 1 & 0 & 0 \\ 0 & 0 & 1 & 0 \\ 0 & 1 & 0 & 0 \\ 0 & 0 & 1 & 0 \\ 0 & 0 & 0 & 1 \end{bmatrix} \begin{bmatrix} a_1 \\ a_2 \\ a_3 \\ a_4 \end{bmatrix}.$$

The MATLAB code that enforces these two properties is shown in algorithm 15.4

Algorithm 15.4 Mapping to a Hankel matrix of given rank
File: `tohankel.m`

When

$$X = \begin{bmatrix} 0.2190 & 0.6793 \\ 0.0470 & 0.9347 \\ 0.6789 & 0.3835 \end{bmatrix}$$

is used as an argument to `A = tohankel(X,1)`, the answer is

$$A = \begin{bmatrix} 0.3463 & 0.4311 \\ 0.4311 & 0.5367 \\ 0.5367 & 0.6680 \end{bmatrix},$$

which is Hankel and has rank 1. □

Example 15.5.4 Let $x[t]$ be modeled as the sum of $p/2$ real sinusoids in noise,

$$x[t] = \sum_{i=1}^{p/2} a_i \cos(\omega_i t + \theta_n) + e[t].$$

Let N samples of data be taken. If there were no noise on the data, then, as discussed in section 8.1, for $m > p$ the $(N - m) \times (m + 1)$ Toeplitz data matrix

$$X_f = \begin{bmatrix} x[m+1] & x[m] & \cdots & x[1] \\ x[m+2] & x[m+1] & \cdots & x[2] \\ \vdots & & & \\ x[N] & x[N-1] & \cdots & x[N-m] \end{bmatrix}$$

would satisfy

$$X_f \mathbf{a} = 0 \tag{15.13}$$

for some vector $\mathbf{a}$. Furthermore, the $(N-m) \times (m+1)$ Hankel matrix

$$X_b = \begin{bmatrix} x[1] & x[2] & \cdots & x[m+1] \\ x[2] & x[3] & \cdots & x[m+2] \\ \vdots & & & \\ x[N-m] & x[N-m+1] & \cdots & x[N] \end{bmatrix}$$

would, in the absence of noise on the x data, also satisfy

$$X_b \mathbf{a} = 0, \tag{15.14}$$

where $\mathbf{a}$ in (15.14) is the same as in (15.13). The stack of X_f and X_b thus satisfies

$$X\mathbf{a} = \begin{bmatrix} X_f \\ X_b \end{bmatrix} \mathbf{a} = 0.$$

In the absence of noise, the rank of X is p. The matrix X formed from the data should therefore have both the Toeplitz/Hankel block property and the rank-p property. Since X is formed from real, noisy data, it will not necessarily have these properties. Using composite mappings, we form a matrix $\hat{X}$ that is nearest to X and that has both properties. The method is similar to that of the previous example, with two linear-structured matrices employed to accommodate both structured matrix types. The linear-structured matrices S_T and S_H are large, each being $((N-m)(m+1)) \times N$, so that

$$S = \begin{bmatrix} S_T \\ S_H \end{bmatrix}$$

is $(2(N-m)(m+1)) \times N$. Algorithm 15.5 demonstrates this mapping.

Algorithm 15.5 Mapping to a Toeplitz/Hankel matrix stack of given rank
File: `tohanktoep.m`

A numerical experiment is performed to enhance a signal with two sinusoids in noise. The signal generated is

$$x[t] = 15\cos(0.15\pi t) + 10\cos(0.68\pi t) + e[t],$$

where $e[t]$ is a Gaussian white-noise sequence with variance $\sigma^2 = 112.5$, so that the SNR relative to the largest signal is 0 dB. A total of $N = 128$ points were taken. Figure 15.8(a) shows the spectrum of the clean data (no noise). The two spectral lines are clearly observed. Figure 15.8(b) shows the spectrum of the signal with the noise included. The noisy data is, formed into a Toeplitz/Hankel block matrix X. Then the nearest matrix $\hat{X}$ satisfying the two properties (linear structure and rank 4) is found, from which a resulting time series $\hat{\mathbf{x}}$ is extracted. Figure 15.8(c) shows $\|X^{[k+1]} - X^{[k]}\|$ as an indication of the convergence rate of the algorithm. After approximately five iterations the algorithm has mostly converged. Figure 15.8(d) shows the cleaned data, which shows a close approximation to the original clean data. □

15.6 Projection on Convex Sets

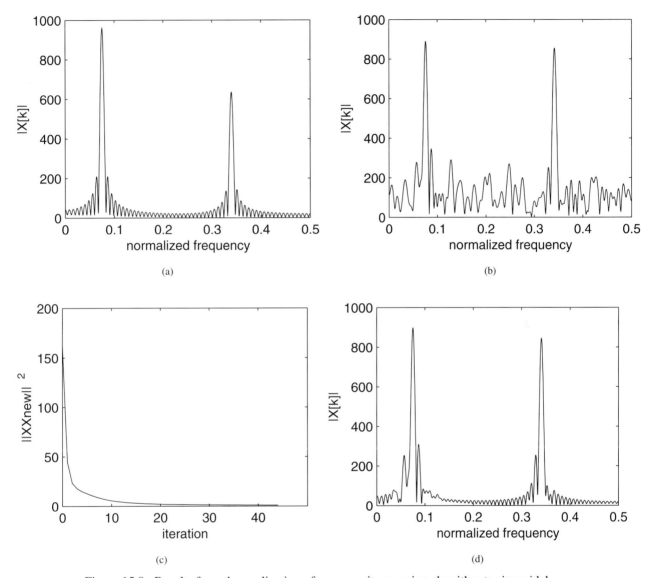

Figure 15.8: Results from the application of a composite mapping algorithm to sinusoidal data. (a) Spectrum of original data, (b) spectrum of noisy data, (c) rate of convergence, (d) spectrum of cleaned data

15.6 Projection on convex sets

Projection on convex sets (POCS) is a special case of the composite mapping algorithm that has been widely used in a variety of settings such as tomography and image restoration. Because the constraint sets are convex, it is easier to guarantee convergence. The basic method can be extended to the use of an over-relaxed operator that might speed convergence; discussion of this extension justifies our presentation.

As with the composite mapping algorithm, we assume that there are several sets $\mathcal{P}_1, \mathcal{P}_2, \ldots, \mathcal{P}_m$, determined by specific properties. We assume that these sets are *convex*. This is the major restriction that POCS has in comparison to composite mapping. The algorithm is examined in the Hilbert space $\mathcal{H}$. We assume that the desired signal f lies in the intersection of the convex sets (similar to our assumption for composite mapping),

$$f \in \bigcap_{i=1}^{m} \mathcal{P}_i.$$

We let P_i denote the projection operator onto the set $\mathcal{P}_i$.

The following two lemmas lay the groundwork for the convergence of this iterative method, by leading to the concept of a nonexpansive operator.

Lemma 15.1 *Let $\mathcal{P}$ be a closed convex subset of a Hilbert space $\mathcal{H}$. A necessary and sufficient condition that a point $p_0 \in \mathcal{P}$ be the projection of x onto $\mathcal{P}$ is that*

$$\operatorname{Re} \langle x - p_0, p - p_0 \rangle \leq 0, \tag{15.15}$$

for every $p \in \mathcal{P}$.

Proof Let $p_0 = Px$, the projection of x onto $\mathcal{P}$. Suppose that there is a $p_1 \in \mathcal{P}$ such that

$$\operatorname{Re} \langle x - p_0, p_1 - p_0 \rangle = \epsilon > 0.$$

Since $\mathcal{P}$ is convex, every point of the form $p_\alpha = (1 - \alpha) p_0 + \alpha p_1$ for $0 \leq \alpha \leq 1$ is in $\mathcal{P}$. Then

$$\|x - p_\alpha\|^2 = \|(1 - \alpha)(x - p_0) + \alpha (x - p_1)\|^2,$$
$$b = (1 - \alpha)^2 \|x - p_0\|^2 + 2\alpha(1 - \alpha) \operatorname{Re} \langle x - p_0, x - p_1 \rangle + \alpha^2 \|x - p_1\|^2.$$

Taking the derivative of $\|x - p_\alpha\|$ with respect to α and evaluating at $\alpha = 0$, we obtain

$$\left. \frac{d}{d\alpha} \|x - p_\alpha\|^2 \right|_{\alpha=0} = -2\|x - p_0\|^2 + 2\operatorname{Re} \langle x - p_0, x - p_1 \rangle$$
$$= -2 \operatorname{Re} \langle x - p_0, p_1 - p_0 \rangle = -2\epsilon < 0.$$

Since the derivative is negative at $\alpha = 0$ and the function is continuous in α, it must follow for some small positive α that $\|x - p_\alpha\| < \|x - p_0\|$, which contradicts the fact that p_0 is the projection of x onto $\mathcal{P}$, the nearest point.

Conversely, suppose the $p_0 \in \mathcal{P}$ is such that $\operatorname{Re} \langle x - p_0, p - p_0 \rangle \leq 0$ for all $p \in \mathcal{P}$. Then for $p \neq p_0$,

$$\|x - p\|^2 = \|x - p_0 + p_0 - p\|^2$$
$$= \|x - p_0\|^2 + 2\operatorname{Re} \langle x - p_0, p_0 - k \rangle + \|x - p\|^2 > \|x - p_0\|.$$

Thus p_0 is the minimizing vector. □

In a real Hilbert space, the inequality (15.15) can be written as

$$\langle x - Px, y - Px \rangle \leq 0,$$

so that the angle between the two vectors $x - Px$ and $y - Px$ is greater than 90°. A diagram illustrating this lemma is shown in figure 15.9.

Lemma 15.2 *Let $\mathcal{P}$ be a closed convex set in a Hilbert space $\mathcal{H}$ and let P be a projection operator onto $\mathcal{P}$. Then for $x, y \in \mathcal{H}$,*

$$\|Px - Py\| \leq \operatorname{Re} \langle x - y, Px - Py \rangle. \tag{15.16}$$

Proof Application of (15.15) to both Px and Py yields

$$\operatorname{Re} \langle x - Px, Py - Px \rangle \leq 0$$

15.6 Projection on Convex Sets

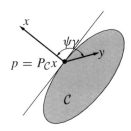

Figure 15.9: Geometric properties of convex sets

and
$$\text{Re}\,\langle y - Py, Px - Py\rangle \leq 0.$$

Adding these and separating the terms yields
$$\text{Re}\,[\langle Px, Px\rangle + \langle Py, Py\rangle - \langle Px, Py\rangle - \langle Py, Px\rangle]$$
$$\leq \text{Re}\,[\langle x, Px\rangle + \langle y, Py\rangle - \langle x, Py\rangle - \langle y, Px\rangle],$$

which is equivalent to (15.16). □

The projection operator P and the projection operator onto the complement $Q = I - P$ can be used to decompose a signal x as

$$x = Px + Qx. \tag{15.17}$$

Definition 15.5 An operator P such that $\|Px - Py\| \leq \|x - y\|$ is said to be **nonexpansive.** □

A nonexpansive operator is similar to a contraction operator, as defined in definition 14.2, except that the distance between two points can be preserved (left equal), rather than being strictly decreased as in a contraction map.

Lemma 15.3
$$\|Px - Py\| \leq \|x - y\|, \tag{15.18}$$

with equality if and only if $x \in \mathcal{P}$ and $y \in \mathcal{P}$.

Proof Using the decomposition (15.17), we can write
$$\|Px - Py\| = \|x - Qx - y + Qy\|$$
$$\leq \|x - y\| + \|Qx - Qy\|.$$

If $x \in \mathcal{P}$ and $y \in \mathcal{P}$ then $Qx = 0$ and $Qy = 0$, in which case
$$\|Px - Py\| = \|x - y\|.$$

Otherwise, we have
$$\|Px - Py\| < \|x - y\|.$$
□

By this lemma, for points $x, y \notin \mathcal{P}$, P is a contraction operator, and the results of the contraction mapping apply. Looked at from another point of view, let $y \in \mathcal{C}$; then $d(x, y) = \|x - y\|$ is a descent function for $\mathcal{P}$ and P, and the theory of global convergence can be applied.

As we did for the composite mapping algorithm, we now assume that there are m convex sets $\mathcal{P}_1, \mathcal{P}_2, \ldots, \mathcal{P}_m$, representing certain signal properties (e.g., constraints on positivity, matrix structure, symmetry, etc.). We desire to take a point $x \in \mathcal{H}$ representing measured data and find a point f that is near x and that is in the intersection of the convex sets: it satisfies all the properties. Let

$$\mathcal{P} = \bigcap_{i=1}^{m} \mathcal{P}_i.$$

Then the desired point $f \in \mathcal{H}$ satisfies

$$f \in \mathcal{P}.$$

Associated with each convex set $\mathcal{P}_i$ we determine a projection operator P_i such that

$$\inf_{f \in \mathcal{P}_i} \|x - f\| = \|x - P_i x\|,$$

and we introduce the composition operator

$$P = P_m P_{m-1} \cdots P_1.$$

While P may not be the projection operator onto $\mathcal{P}$ for every $x \in \mathcal{H}$, every point in $\mathcal{P}$ is a fixed point of every P_i and hence of P. Also, every fixed point of P is an element of $\mathcal{P}$. We introduce the recursive update on x as

$$x^{[k+1]} = P x^{[k]}.$$

Figure 15.10 illustrates the concept geometrically. We first project onto the set $\mathcal{P}_1$, then from there onto $\mathcal{P}_2$, then back to $\mathcal{P}_1$, and so forth. Since the set $\mathcal{P} = \mathcal{P}_1 \cap \mathcal{P}_2$ is not an empty set, repeated application of the algorithm leads to the point at the intersection.

In figure 15.10, the first few iterations have the most dramatic effect on the nearness to the final solution. As the algorithm proceeds the rate of convergence slows due to the near-tangency of the boundaries of the convex sets. One procedure that may improve the convergence rate is to define an "over-relaxed" operator that extends the projection beyond the boundaries of the sets by

$$T_i = 1 + \lambda_i (P_i - 1) \qquad i = 1, 2, \ldots, m, \tag{15.19}$$

and then define the composition operator

$$T = T_m T_{m-1} \cdots T_1.$$

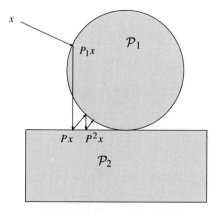

Figure 15.10: Projection onto two convex sets

Under fairly general conditions, successive approximation using T converges. If it can be shown that T is a closed mapping and is distance-reducing relative to $\mathcal{P}$, then the global convergence theorem can be invoked to prove convergence. The key issue is to establish that the over-relaxed operator and its compositions are nonexpansive; then, for points x not in $\mathcal{P}$, the operator will be distance-reducing.

Lemma 15.4 *The operator $T = T_m T_{m-1} \cdots T_1$ is nonexpansive.*

Proof To show that T_i as defined in (15.19) is nonexpansive, we consider first the case $0 < \lambda_i < 1$. In this case, nonexpansiveness is straightforward. When $\lambda_i > 1$ then $1 - \lambda_i < 0$, and we have for any $x, y \in \mathcal{H}$,

$$\begin{aligned}
\|T_i x - T_i y\|^2 &= \|(1-\lambda_i)(x-y) + \lambda_i(T_i x - T_i y)\|^2 \\
&= (1-\lambda_i)^2 \|x-y\|^2 + 2\lambda_i(1-\lambda_i)\operatorname{Re}(x-y, T_i x - T_i y) \\
&\quad + \lambda_i^2 \|T_i x - T_i y\|^2 \\
&\leq (1-\lambda_i)^2 \|x-y\|^2 + \left(\lambda_i^2 + 2\lambda_i(1-\lambda_i)\right)\|T_i x - T_i y\| \\
&\leq (\lambda_i(2-\lambda_i) + (1-\lambda_i)^2)\|x-y\|^2 = \|x-y\|^2.
\end{aligned}$$

Composition of nonexpansive operators is nonexpansive. We show this for $m = 2$; extension to larger m is similar. Let T_1 and T_2 be nonexpansive and let $T = T_2 T_1$. Then

$$\|Tx - Ty\| = \|T_2 T_1 x - T_2 T_1 y\| = \|T_2 x_1 - T_2 y_1\| \leq \|x_1 - y_1\|,$$

where $x_1 = T_1 x$ and $y_1 = T_1 y$. But since T_1 is nonexpansive, $\|x_1 - y_1\| \leq \|x - y\|$, and thus $\|Tx - Ty\| \leq \|x - y\|$. □

15.7 Exercises

15.2-1 Show that a closed linear manifold is convex.

15.2-2 Show for two CLMs $\mathcal{P}_a$ and $\mathcal{P}_b$ that if $\psi(\mathcal{P}_a, \mathcal{P}_b) > 0$ then the intersection $\mathcal{P}_a \cap \mathcal{P}_b$ contains only the zero vector.

15.2-3 Let P_a and P_b be the projection operators onto $\mathcal{P}_a$ and $\mathcal{P}_b$, respectively. Show that

$$\|P_a P_b\| = \cos \psi(\mathcal{P}_a, \mathcal{P}_b) = \|P_b P_a\|.$$

15.2-4 Prove part (1) of theorem 15.1.

15.5-5 Let A be a matrix, and let B be a matrix whose rows sum to 1. Show that the B nearest to A in the Frobenius norm is given in (15.11).

15.5-6 Show the linear-structured matrix for a 3×3 symmetric matrix.

15.5-7 Show the linear-structured matrix for a 3×3 Hermitian matrix.

15.5-8 Let A be a linear-structured matrix, and $\mathbf{a} = \operatorname{vec}(A)$. Suppose a real linear-structured matrix for $\mathbf{a}$ has the form

$$\mathbf{a} = \begin{bmatrix} S_1 \\ S_2 \\ \vdots \\ S_K \end{bmatrix} \mathbf{v},$$

where $\mathbf{v}$ is the set of parameters in A. For a vector $\mathbf{x}$, show that the nearest vector $\mathbf{y}$ to $\mathbf{x}$ with the linear structure of **a** is

$$\mathbf{y} = S \left[\sum_{k=1}^{K} S_k^T S_k \right]^{-1} \begin{bmatrix} S_1^T & S_2^T & \cdots & S_K^T \end{bmatrix} \mathbf{x}.$$

15.5-9 The $M \times M$ matrix R_{yy} 6.50 used in the MUSIC algorithm is a theoretical Hermitian–Toeplitz matrix, and the smallest $M - p$ eigenvalues correspond to noise power, where p is the number of complex sinusoids in the received signal. It may be possible to enhance the MUSIC algorithm by finding a nearby matrix with the proper structure, but which has rank p. Write and test a MATLAB function that enforces the linear structure and rank-p properties of a matrix R.

15.5-10 A Vandermonde matrix does not have a linear structure. Nevertheless, it is still possible to define a mapping from a general $n \times n$ matrix X to a nearest Vandermonde matrix.

    (a) Given a sequence $\{x_1, x_2, \ldots, x_n\}$, determine a mapping to the nearest sequence $\{ca, ca^2, \ldots, ca^n\}$ under some appropriate norm.

    (b) Find a mapping from a matrix X to the nearest Vandermonde matrix.

    (c) Code and test your algorithm.

15.5-11 Given an $m \times 1$ vector **a** and a vector **b** assumed to be a permutation of **a**,

$$\mathbf{b} = P\mathbf{a},$$

determine an algorithm to find the best (in the minimum least-squares sense) permutation matrix P. Code your algorithm in MATLAB. (At first blush, this seems like a difficult problem, possibly involving a search over all $m!$ permutations. This example demonstrates the power of composite mapping methods.

15.6-12 Show that the projection onto a convex set as defined by (15.1) is unique.

15.6-13 Show that for $0 < \lambda < 1$, the operator

$$T = 1 + \lambda(P - 1)$$

is nonexpansive if P is nonexpansive.

15.8 References

The method of alternating on convex sets is described in [380], and our description in section 15.6 follows closely the development there. Generalizations and regularization are also discussed in this excellent paper. The method of projection on convex sets can be found in [379], with an example given in [301]. Excellent material on projections is also found in [209].

    The composite mapping algorithm is discussed in [48], where an application to bearing estimation using MUSIC is also presented. The concepts relating to closed mappings come from [210] and [384]. The problem of finding a near Hankel matrix of specified rank was explored by alternative methods in [194], for the purpose of system identification. The reconstruction of bandlimited functions was proposed by Papoulis [247], where convergence is proved by expansion using prolate spheroidal functions. The proof is specialized but nevertheless interesting and valuable.

    Projection on convex sets is introduced with examples for image restoration in [379]. A valuable discussion of convergence is also provided there. A more recent example of applications of this nature is in [389].

    The local and global properties of manifolds, as referred to in definition 15.1, are discussed at length in [297].

Chapter 16

Other Iterative Algorithms

In this chapter, we present two classes of iterative algorithms that have been of particular interest to the signal processing community. The first class of iterative algorithms concerns clustering: finding structure in data. This technique is commonly used in pattern-recognition problems, as well as data compression. Algorithms in the other class are iterative methods for solving systems of linear equations. These algorithms move toward a solution by improving on a previous solution, and have application for both sparse systems and adaptive processing.

16.1 Clustering

Clustering algorithms provide a means of representing a set of N data points by a set of M data points, where $M < N$, by finding a set of M points that is somehow representative of the entire set. Most clustering algorithms are iterative, in which an initial set of cluster representatives is refined by processes such as splitting, agglomeration, or averaging, to produce a set of cluster representatives that is better in some measure than before.

In this section we present two examples of the use of clustering: vector quantization for data compression, and pattern recognition.

16.1.1 An example application: Vector quantization

In this section we present briefly the concept of vector quantization (VQ) as an application of clustering. Vector quantization is often used in lossy data compression. Because the reconstructed signal is not an exact representation of the original signal, significantly higher compression is obtainable than for lossless data compression. The design issue is to obtain the maximum amount of compression with the minimum amount of distortion between the input and output. In practice, VQ provides only part of the solution of an effective data compression technique.

The process of (scalar) quantization takes a scalar variable x (which may, for example, be a continuous voltage) and assigns it a quantization index $i = Q(x)$ (which may, for example, be an eight-bit number). Let R denote the number of bits required to uniquely identify the quantization index. This is the *rate*. The number of quantization values is thus $M = 2^R$. As a general rule, quantization results in a loss of information. Usually it is not possible to reconstruct the variable x given $Q(x)$, but only to reconstruct an approximation $\hat{x} = P(Q(x))$.

In the design of quantizers, the goal is to design Q and P in such a way that the difference between x and $\hat{x}$ is small according to some norm. In the analysis of quantization systems it is customary to regard x as a random variable with some distribution, and to measure the quality of the quantization system by an average $D = E\|x - \hat{x}\|$, where E is the expectation operator. The quantity D is called the *distortion*. Clearly, the more values that $Q(x)$ takes on, the better the quantizer can perform and the lower the distortion can be for a well-designed quantizer.

Suppose that we have a sequence of variables $x_0, x_1, x_2, \ldots$ to be quantized. In such a case, Shannon's rate-distortion theory applies, (see, for example [56]), which states that there is a theoretical lower bound on the distortion D as a function of R. Rate-distortion theory indicates that in order to achieve the lowest possible distortion, variables should be quantized as *vectors*. That is, we form n-dimensional vectors by stacking elements, such as

$$\mathbf{x}_0 = \begin{bmatrix} x_0 \\ x_1 \\ \vdots \\ x_{n-1} \end{bmatrix} \quad \mathbf{x}_1 = \begin{bmatrix} x_n \\ x_{n+1} \\ \vdots \\ x_{2n-2} \end{bmatrix} \quad \ldots$$

We then pass the vector $\mathbf{x}_i$ through a quantizer function that returns an index $i = Q(\mathbf{x})$. The quantizer has $M = 2^R$ different representable values. An approximation of the original variable is reconstructed using a function P,

$$\hat{\mathbf{x}} = P(i) = P(Q(\mathbf{x})).$$

The most common coding technique is to have a set of representative vectors $\mathbf{y}_0, \mathbf{y}_1, \ldots, \mathbf{y}_{M-1}$. These vectors constitute the *codebook*. The quantization function Q encodes $\mathbf{x}$ by selecting the codebook vector $\mathbf{y}_i$ that is nearest to $\mathbf{x}$ in some distortion measure $d(\mathbf{x}, \mathbf{y}_i) = \|\mathbf{x} - \mathbf{y}_i\|$. The index of the codebook vector is the quantization index,

$$i = Q(\mathbf{x}) = \arg \min_{i=0,1,\ldots,M-1} d(\mathbf{x}, \mathbf{y}_i).$$

The value of i is used to represent the data. It is assumed that both the source of the data and the sink (the place where the data gets used) are stored in a fixed location in the codebook, so that the codebook does not need to be sent. The reconstructed vector $\hat{\mathbf{x}}$ is simply

$$\hat{\mathbf{x}} = P(i) = \mathbf{y}_i.$$

Example 16.1.1 Suppose that the scalar data x is represented with 8-bit numbers, and that $n = 16$ dimensional vectors are used. If the codebook has 1024 elements, then $R = 10 = \log_2(1024)$ bits are used to represent each vector. Thus for each $(8)(16) = 128$ bits input into the quantizer, 10 bits are output, and the compression ratio is

$$\frac{128}{10}.$$

Higher compression is achievable by larger dimensionality or smaller codebooks. □

A design issue in VQ is the selection of the codebook vectors $\mathbf{y}_0, \ldots, \mathbf{y}_{M-1}$ so that these vectors provide a good representation of the data $\mathbf{x}_i$. The method commonly used is a set of training data that is similar to the types of data that are expected to be used in practice. Based on the training data, a set of M vectors are chosen that represent the training data by having a low overall distortion. The vectors are chosen according to a clustering procedure.

Example 16.1.2 Figure 16.1(a) shows 500 data points that have been selected according to the distribution $\mathcal{N}(0, R)$, where

$$R = \begin{bmatrix} 110 & 90 \\ 90 & 110 \end{bmatrix}.$$

16.1 Clustering

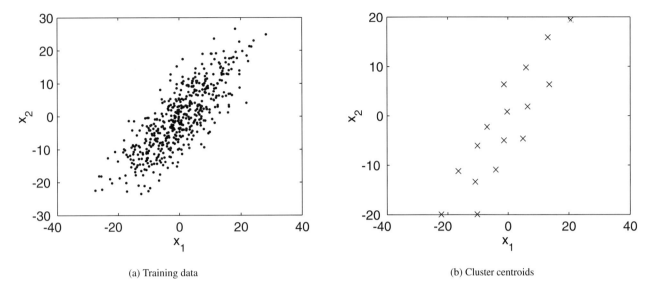

(a) Training data (b) Cluster centroids

Figure 16.1: Demonstration of clustering.

This data is passed into the clustering algorithm lgb, described in section 16.1.3, and 16 representative vectors are found, which are shown in figure 16.1(b). Observe that the clusters do give a fair representation of the original data. □

16.1.2 An example application: Pattern recognition

In chapter 11, the problem of pattern recognition was addressed from the point of view of decision theory. Suppose that there are C classes to distinguish among, each characterized by a likelihood function $f(x|i)$ and prior probability $p_i, i = 1, 2, \ldots, C$. Given an observation vector x, the optimum (minimum probability of error) decision is to choose that class with the highest posterior probability,

$$\hat{c} = \arg\max_{i \in \{1,2,\ldots,C\}} f(x|i) p_i.$$

In the case that the prior probabilities are not known, then a maximum-likelihood decision can be made under the assumption that each class is equally probable:

$$\hat{c} = \arg\max_{i \in \{1,2,\ldots,C\}} f(x|i).$$

For practical application of the theory, there must be some means of determining what the probability densities are. Exploring this question has generated a tremendous amount of research (a good starting source is [76]). In this section, we present one method based on clustering.

Often, it can be assumed that the data in each class are Gaussian distributed. Under this assumption, it is only necessary to determine the mean and the variance of the data in class i to obtain the density $f(x|i)$. One way of doing this is by clustering.

Example 16.1.3 In a particular speech-pattern recognition problem, two-dimensional feature vectors representing (say) the ratio of high-frequency energy to low-frequency energy, and the scaled formant frequency, are to be used to distinguish among five different classes. Figure 16.2(a) shows 300 data vectors. The problem is to identify the densities represented by this data. Figure 16.2(b) shows

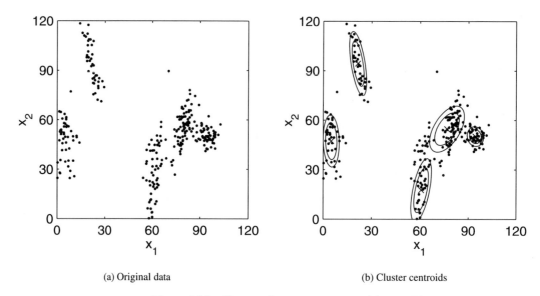

(a) Original data (b) Cluster centroids

Figure 16.2: Clusters for a pattern recognition problem.

the results of running a clustering algorithm on the data. The cluster centroids are shown as ×, and the covariance directions from each cluster are indicated using ellipses of constant probability contours. The five centroids were reasonably placed, and the centroid directions are more-or-less as expected. □

16.1.3 *k*-means Clustering

The *k*-means algorithm for clustering in n dimensions produces k mean vectors that represent k classes of data. It has a long history; a relevant reference within the context of signal processing is [203]. The algorithm presented there (and here) is often referred to as the LGB algorithm, using the initials of the authors of that paper.

The algorithm relies on a distortion measure $d(\mathbf{x}, \mathbf{y})$ between points in $\mathbb{R}^n$. A variety of norms can be used, such as L_1, L_2, L_∞, or others specific to the problem.

Let the set of training data be $X = \{\mathbf{x}_1, \mathbf{x}_2, \ldots, \mathbf{x}_N\}$. Given a cluster point $\mathbf{y}_i$ (a **centroid**), the set of points in X that is closer under the distortion measure d to $\mathbf{y}_i$ than to any other centroid is called the **Voronoi region** of $\mathbf{y}_i$. We will denote the Voronoi region for $\mathbf{y}_i$ as $\mathcal{V}_i$:

$$\mathcal{V}_i = \{\mathbf{x} \in X : d(\mathbf{x}, \mathbf{y}_i) < d(\mathbf{x}, \mathbf{y}_j), i \neq j\}.$$

The number of vectors in a Voronoi region is denoted as $|\mathcal{V}_i|$. The centroid of vectors in a Voronoi region is given by

$$\frac{1}{|\mathcal{V}_i|} \sum_{\mathbf{x} \in \mathcal{V}_i} \mathbf{x}.$$

The centroid of the Voronoi region is used as the representative of all the data in the region. Let $\mathbf{y}(\mathbf{x})$ denote the centroid $\mathbf{y}_i$ that is closest to $\mathbf{x}$. The average distortion across the entire set of data X is

$$d(X) = \frac{1}{|X|} \sum_{\mathbf{x} \in X} d(\mathbf{x}, \mathbf{y}(\mathbf{x})).$$

16.1 Clustering

With this notation, the k-means algorithm can be stated as follows:

1. Choose an initial set of centroids $\mathbf{y}_1, \mathbf{y}_2, \ldots, \mathbf{y}_k$.
2. Determine the Voronoi region for each $\mathbf{y}_i$.
3. Compute the centroid of each Voronoi region.
4. If the algorithm has not converged, go to step 2. Otherwise stop.

There are some issues left to be explained in this description. First is the stopping criterion. One stopping criterion is to compute the average distortion, and to stop if the distortion is small enough. Another criterion is to compute the change in distortion from one iteration to the next, and to stop if the change in distortion is small enough.

Another issue is choosing the initial set of centroids. There are a variety of approaches to this choice. The initial selection is important, because even though the algorithm will converge, there is *no guarantee* that it will converge to a global minimum. The final configuration of centroids is highly dependent on the initial choice. One way to get the initial set of centroids is to choose points at random in the region of $\mathbb{R}^n$ where the data points are. Another is to choose M points out of the training data (without replacement) as the initial data vectors. A more sophisticated technique is to use splitting. In splitting, the number of clusters is increased from 1 to the desired number k. When there is one cluster, the centroid of the data is found. This centroid is then split into two by choosing two centroids as random perturbations of this initial centroid. Clustering is again performed, and each region is split again. This proceeds until the desired number of centroids is obtained. A variation on this technique is to split, not randomly, but in the direction of maximum variation, as determined by the principle components of the data in each Voronoi region.

MATLAB code that computes the k-means algorithm is shown in algorithm 16.1. It uses the basic l_2 norm between vectors, random selection to get the initial set of centroids, and change in total distortion as a stopping criterion.

Algorithm 16.1 k-means clustering (LGB algorithm)
    File: `lgb.m`
              `initcluster.m`

For the purpose of making the transition to the clustering described in the next section, it is useful to introduce some new notation. Let $u_{ij} \in \{0, 1\}$ be an indicator of the membership of a data point $\mathbf{x}_j$ to a cluster i, where

$$u_{ij} = \begin{cases} 1 & \text{if } \mathbf{x}_j \text{ is in cluster } i, \\ 0 & \text{otherwise.} \end{cases}$$

The fact that u_{ij} can take on only the values 0 or 1 stems from the fact that for the algorithm just described, a data point $\mathbf{x}_j$ is either in cluster i or it is not; there is no middle ground. Since every element must belong to a cluster, the following constraints hold for u_{ij}.

$$\sum_{i=1}^{k} u_{ij} = 1 \qquad j = 1, 2, \ldots, N,$$

$$\sum_{i=1}^{k} \sum_{j=1}^{N} u_{ij} = N.$$

The number of elements in the ith cell (its Voronoi region) is

$$|\mathcal{V}_i| = \sum_{j=1}^{N} u_{ij}.$$

16.1.4 Clustering using fuzzy k-means

The basic k-means algorithm has been extended in a variety of ways over the years. In this section we generalize it by expanding the concept of cluster membership. In the k-means algorithm just described, a vector $\mathbf{x}$ either belongs to the cluster described by the centroid $\mathbf{y}_i$, or it does not. There is no provision for data to belong partly to one cluster and partly to another. However, in many pattern-classification problems, this "crispness" is unwarranted by the data. We can generalize this concept by utilizing a membership function that indicates the *degree* to which a vector belongs to a cluster. It is the fact that the degree of membership to a set is explicitly represented that leads to the word *fuzzy* in describing this kind of clustering.

We generalize the concept of the indicator u_{ij} (as introduced in the last section) to allow for

$$u_{ij} \in [0, 1].$$

Under this specification, u_{ij} indicates the *degree* to which $\mathbf{x}_j$ is in cluster i and is said to be a *set membership* function. To determine a clustering algorithm based on this generalization, we introduce the weighted-criterion function

$$J(U, Y) = \sum_{j=1}^{N} \sum_{i=1}^{k} (u_{ij})^m d^2(\mathbf{x}_j, \mathbf{y}_i), \qquad (16.1)$$

where $U = [u_{ij}]$ is the matrix of membership functions, $Y = [\mathbf{y}_1, \mathbf{y}_2, \ldots, \mathbf{y}_k]$ is the set of centroids, and m is a weighting exponent, $1 \leq m < \infty$. The distance measure is taken, for analytical purposes, to be

$$d^2(\mathbf{x}_j, \mathbf{y}_i) = (\mathbf{x}_j - \mathbf{y}_i)^T (\mathbf{x}_j - \mathbf{y}_i). \qquad (16.2)$$

The functional $J(U, V)$ measures the penalty for representing k fuzzy clusters indicated by U with centroids represented by Y. The goal is to minimize (16.1), subject to the constraints

$$\sum_{i=1}^{k} u_{ij} = 1, \qquad j = 1, 2, \ldots, N \qquad (16.3)$$

$$u_{ij} \geq 0 \qquad i = 1, 2, \ldots, k, \qquad j = 1, 2, \ldots, N. \qquad (16.4)$$

When $m = 1$ in (16.1), the algorithm turns out to be simply the k-means algorithm. When $m > 1$, it is straightforward to set up a constrained optimization problem (see exercise 16.1-1) to determine that

$$u_{ij} = \frac{1}{(d(\mathbf{x}_j, \mathbf{y}_i))^{2/(m-1)} \sum_{l=1}^{k} \frac{1}{(d(\mathbf{x}_j, \mathbf{y}_l))^{2/(m-1)}}} \qquad i = 1, 2, \ldots, k \quad j = 1, 2, \ldots, N, \qquad (16.5)$$

$$\mathbf{y}_i = \frac{\sum_{j=1}^{N} (u_{ij})^m \mathbf{x}_j}{\sum_{j=1}^{N} (u_{ij})^m}. \qquad (16.6)$$

Observe that (16.6) is a centroid computation; and that, in the limit, as $m \downarrow 1$, u_{ij} takes on only the values in $\{0, 1\}$. Clustering by fuzzy k-means is accomplished by choosing an initial set of means, then iterating (16.5) and (16.6) until convergence.

16.2 Iterative methods for computing inverses of matrices

The most costly part of many signal-processing problems is the solution of a set of linear equations of the form $A\mathbf{x} = \mathbf{b}$. Where the structure of the matrix A can be exploited to reduce the number of computations required, as compared with a general problem solver, it is often expedient to do so. Also, if the system is changing over time, so that we have $A\mathbf{x}(n) = \mathbf{b}(n)$ at time n, we want to use as much information about the solution at time n as possible to reduce the computations at time $n + 1$. One approach to solving this problem is by means of iterative methods for computing inverses of matrices. These methods are commonly employed for solving sparse systems of matrices, for which a solution method such as the LU factorization would be inefficient, since it would turn a single sparse matrix into two dense matrices. In the context of signal processing, iterative methods can also be used to develop adaptive algorithms that provide updated solutions as the data changes.

In iterative methods for the solution of $A\mathbf{x} = \mathbf{b}$, we form an approximate solution $\mathbf{x}^{[k]}$, then update the solution to refine it. The update for many iterative methods can be expressed as

$$\mathbf{x}^{[k+1]} = B\mathbf{x}^{[k]} + \mathbf{c}, \tag{16.7}$$

for a matrix B and vector $\mathbf{c}$ that depend upon the type of iterative method employed. The Jacobi and Gauss–Seidel can be expressed in this way, while a third method—successive over-relaxation—can be expressed as a generalization of this form. Conditions for convergence of these methods are given by the following theorem.

Theorem 16.1 *[181, page 229] For the iteration*

$$\mathbf{x}^{[k+1]} = B\mathbf{x}^{[k]} + \mathbf{c}$$

to produce a sequence converging to $(I - B)^{-1}\mathbf{c}$ for any starting vector $\mathbf{x}^{[0]}$, it is necessary and sufficient that the spectral radius of B satisfy $\rho(B) < 1$.

Proof Take $\rho(B) < 1$. Then, by theorem 4.4, there is a matrix norm such that $\|B\| < 1$. The kth term of the iteration can be written as

$$\mathbf{x}^{[k]} = B^k \mathbf{x}^{[0]} + \sum_{j=0}^{k-1} B^j \mathbf{c}. \tag{16.8}$$

The norm of the first term goes to zero as $k \to \infty$:

$$\|B^k \mathbf{x}^{[0]}\| \leq \|B^k\| \|\mathbf{x}^{[0]}\| \leq \|B\|^k \|\mathbf{x}^{[0]}\| \to 0$$

as $k \to \infty$. For the summation in (16.8), we use the Neumann identity (see section 4.2.2)

$$\sum_{j=0}^{\infty} B^j \mathbf{c} = (I - B)^{-1} \mathbf{c}.$$

Thus, as $k \to \infty$ in (16.8), we obtain

$$\lim_{k \to \infty} \mathbf{x}^{[k]} = (I - B)^{-1} \mathbf{c}.$$

To prove the converse, assume that $\rho(B) \geq 1$, and let $B\mathbf{u} = \lambda \mathbf{u}$, where λ is an eigenvalue, with $|\lambda| \geq 1$. Let $\mathbf{c} = \mathbf{u}$ and $\mathbf{x}^{[0]}$. Then, by (16.8),

$$\mathbf{x}^{[k]} = \sum_{j=0}^{k-1} \lambda^j \mathbf{u},$$

which, for $\lambda > 1$, diverges as $k \to \infty$. $\square$

It is straightforward to show that if A is row diagonally dominant or column diagonally dominant then the conditions of the theorem are satisfied and an iterative method of the form (16.7) will converge.

16.2.1 The Jacobi method

The Jacobi method can be introduced by an example. Consider the 2×2 system of equations

$$\begin{bmatrix} a_{11} & a_{12} \\ a_{21} & a_{22} \end{bmatrix} \begin{bmatrix} x_1 \\ x_2 \end{bmatrix} = \begin{bmatrix} b_1 \\ b_2 \end{bmatrix},$$

where $a_{ii} \neq 0$. The two equations can be solved for x_1 and x_2, respectively, as

$$x_1 = \frac{1}{a_{11}}(b_1 - a_{12}x_2),$$

$$x_2 = \frac{1}{a_{22}}(b_2 - a_{21}x_1).$$

We form an iterative solution by using elements of $\mathbf{x}^{[k]}$ on the right-hand side to obtain updates on the left-hand side:

$$x_1^{[k+1]} = \frac{1}{a_{11}}\left(b_1 - a_{12}x_2^{[k]}\right),$$

$$x_2^{[k+1]} = \frac{1}{a_{22}}\left(b_2 - a_{21}x_1^{[k]}\right).$$

More generally, for an $n \times n$ system, we obtain updates as

$$x_i^{[k+1]} = \frac{1}{a_{ii}}\left(b_i - \sum_{j \neq i} a_{ij}x_j^{[k]}\right), \quad i = 1, 2, \ldots, n. \tag{16.9}$$

Equation (16.9) is the Jacobi method. This element-by-element update can be expressed in matrix form as follows. Decompose the matrix A as

$$A = L + D + U,$$

where L is lower triangular, D is diagonal, and U is upper triangular. Then the Jacobi method can be expressed as

$$\mathbf{x}^{[k+1]} = -D^{-1}(L + U)\mathbf{x}^{[k]} + D^{-1}\mathbf{b}.$$

For the 2×2 example, the decomposition is

$$L = \begin{bmatrix} 0 & 0 \\ a_{21} & 0 \end{bmatrix} \quad D = \begin{bmatrix} a_{11} & 0 \\ 0 & a_{22} \end{bmatrix} \quad U = \begin{bmatrix} 0 & a_{12} \\ 0 & 0 \end{bmatrix}.$$

A MATLAB code for a Jacobi iteration is shown in algorithm 16.2.

16.2 Iterative Methods for Computing Inverses of Matrices

Algorithm 16.2 Jacobi iteration
File: `jacobi.m`

16.2.2 Gauss–Seidel iteration

In the Jacobi method, all of the values of $\mathbf{x}^{[k]}$ are held fixed while $\mathbf{x}^{[k+1]}$ is computed. Gauss–Seidel iteration is similar, except that as soon as a new update to a component in $\mathbf{x}^{[k+1]}$ is computed, it is used for later components of $\mathbf{x}^{[k+1]}$. To illustrate for the 2×2 case, the first update equation is

$$x_1^{[k+1]} = \frac{1}{a_{11}}\left(b_1 - a_{12}x_2^{[k]}\right).$$

Now we use the updated $x_1^{[k+1]}$ to obtain the updated $x_2^{[k+1]}$:

$$x_2^{[k+1]} = \frac{1}{a_{22}}\left(b_2 - a_{21}x_1^{[k+1]}\right).$$

For the general $n \times n$ matrix, we can write

$$x_i^{[k+1]} = \frac{1}{a_{ii}}\left(b_i - \sum_{j<i} a_{ij}x_j^{[k+1]} - \sum_{j>i} a_{ij}x_j^{[k]}\right). \tag{16.10}$$

The computation proceeds sequentially through the elements: it is not possible to compute $x_2^{[k+1]}$ until $x_1^{[k+1]}$ has been computed. This is in contrast to the Jacobi method, for which there is no mixture between updated prior components. This means that updates on components for the Jacobi method can be assigned to separate processors in a parallel computing environment, but not for the Gauss–Seidel method. On the other hand, since the information is used as soon as it becomes available, the Gauss–Seidel method tends to converge more quickly. The ordering of elements within the vector can affect the rate of convergence; a good ordering improves the rate of convergence.

In matrix form, the Gauss–Seidel update can be written as

$$\mathbf{x}^{[k+1]} = -(D+L)^{-1}(U\mathbf{x}^{[k]} - \mathbf{b}),$$

where $A = L + D + U$, as before. Observe that the matrix $(D+L)^{-1}$ is the inverse of a lower-triangular system, which can be solved by backsubstitution. A routine for Gauss–Seidel iteration is shown in algorithm 16.3.

Algorithm 16.3 Gauss–Seidel iteration
File: `gaussseid.m`

For nonsparse $n \times n$ matrices, both the Jacobi and Gauss–Seidel methods require $O(n^2)$ flops per iteration. Provided that an LU factorization of a matrix A is obtained, an iteration is thus comparable to a backsubstitution step. Of course, there would usually be an extra overhead for the LU factorizations. However, for sparse matrices (say, with roughly $r \ll n$ nonzero elements per row of A), $O(rn)$ flops are required. This could be a significant savings, particularly if n is large. Also, the Jacobi iteration may be numerically unstable unless a relaxation parameter is used.

Another application in which these iterative methods might be useful is tracking, in which A and possibly also $\mathbf{b}$ change with each iteration, and an approximate solution that attempts to track the true solution is acceptable.

Example 16.2.1 For the system of time-varying equations

$$A[t]\mathbf{x}[t] = \mathbf{b}[t],$$

let

$$A[t] = \begin{bmatrix} 4 & 2 & -1 \\ 1 & 6 & -2 \\ 4 & -3 & 9 \end{bmatrix} + \begin{bmatrix} \cos \omega_1 t & \cos \omega_2 t & \cos \omega_3 t \\ \cos \omega_1 t & \cos \omega_2 t & \cos \omega_3 t \\ \cos \omega_1 t & \cos \omega_2 t & \cos \omega_3 t \end{bmatrix} \qquad \mathbf{b}[t] = \begin{bmatrix} 2 \\ 4 \\ 7 \end{bmatrix} + \begin{bmatrix} \cos \omega_1 t \\ \cos \omega_2 t \\ \cos \omega_3 t \end{bmatrix}.$$

Figure 16.3(a) shows the true solution $\mathbf{x}[t] = A[t]^{-1}\mathbf{b}[t]$. Figure 16.3(b) shows the approximate solution $\hat{\mathbf{x}}_J[t]$ obtained by using one iterate of the Jacobi update for each time step. The approximate

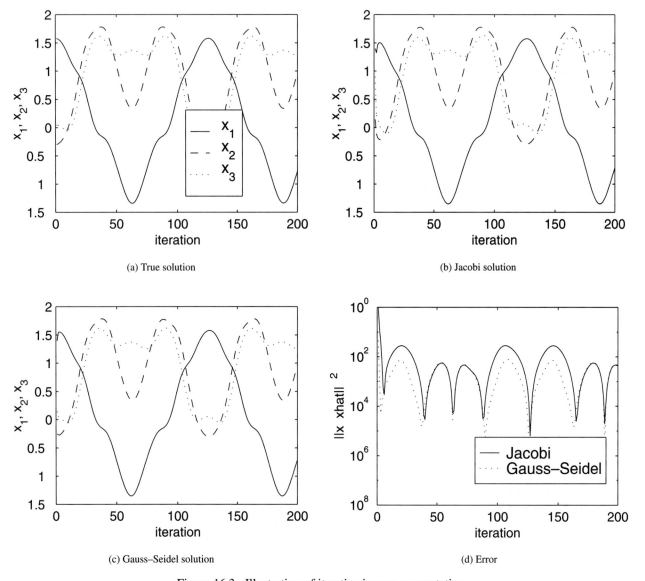

Figure 16.3: Illustration of iterative inverse computation.

16.2 Iterative Methods for Computing Inverses of Matrices

solution tracks the true solution closely. Figure 16.3(c) shows the approximation solution $\hat{\mathbf{x}}_{GS}[t]$ obtained by one iterate of the Gauss–Seidel solution per time step, again with close tracking. Figure 16.3(d) shows the error $\|\mathbf{x}-\hat{\mathbf{x}}_J\|^2$ and $\|\mathbf{x}-\hat{\mathbf{x}}_{GS}\|^2$. The Gauss–Seidel approximation generally has lower error, due to the fact that it uses more information for each update step. □

16.2.3 Successive over-relaxation (SOR)

Consider now the more generalized update

$$Q\mathbf{x}^{[k+1]} = (Q - A)\mathbf{x}^{[k]} + \mathbf{b},$$

where Q is an invertible matrix known as a **splitting matrix**. The iterative methods previously introduced can be shown to be of this form (see exercise 16.2-6). The iterative method converges, provided that

$$\|I - Q^{-1}A\| < 1.$$

The successive over-relaxation method can be used to solve the equation $A\mathbf{x} = \mathbf{b}$, where A is a Hermitian, positive definite matrix. In the SOR method, the splitting matrix is chosen to be of the form

$$Q = \alpha D - C, \tag{16.11}$$

where $\alpha > \frac{1}{2}$, D is a positive definite Hermitian matrix, and C is any matrix that satisfies

$$C + C^H = D - A. \tag{16.12}$$

Then, for a positive-definite Hermitian matrix A, the SOR iteration converges for any starting vector $\mathbf{x}^{[0]}$. To see this, we will show that

$$B = I - Q^{-1}A$$

has $\rho(B) < 1$. To see this, let λ be an eigenvalue of B, with corresponding eigenvector $\mathbf{x}$. Then let $\mathbf{y}$ be defined by

$$\mathbf{y} - \mathbf{x} = B\mathbf{x} = \mathbf{x} - \lambda\mathbf{x} = Q^{-1}A\mathbf{x} \tag{16.13}$$

and

$$Q - A = (\alpha D - C) - (D - C - \overline{C}) = (\alpha - 1)D + \overline{C}. \tag{16.14}$$

From (16.13), we have

$$(\alpha D - C)\mathbf{y} = Q\mathbf{y} = A\mathbf{x}. \tag{16.15}$$

Combining (16.13) and (16.14), we obtain

$$((\alpha - 1)D + \overline{C})\mathbf{y} = A(\mathbf{x} - \mathbf{y}) = AB\mathbf{x}. \tag{16.16}$$

Let us now take inner products with $\mathbf{y}$, using (16.15) and (16.16) to obtain

$$\alpha\langle D\mathbf{y}, \mathbf{y}\rangle - \langle C\mathbf{y}, \mathbf{y}\rangle = \langle A\mathbf{x}, \mathbf{y}\rangle, \tag{16.17}$$
$$(\alpha - 1)\langle \mathbf{y}, D\mathbf{y}\rangle + \langle \mathbf{y}, \overline{C}\mathbf{y}\rangle = \langle \mathbf{y}, AG\mathbf{x}\rangle. \tag{16.18}$$

Adding (16.17) and (16.18), and using the fact that $\langle \mathbf{y}, D\mathbf{y}\rangle = \langle D\mathbf{y}, \mathbf{y}\rangle$ (since D is Hermitian) we obtain

$$(2\alpha - 1)\langle D\mathbf{y}, \mathbf{y}\rangle = \langle A\mathbf{x}, \mathbf{y}\rangle + \langle \mathbf{y}, AB\mathbf{x}\rangle. \tag{16.19}$$

Now we use the fact that $\mathbf{y} = (1 - \lambda)\mathbf{x}$ in (16.19) to write

$$(2\alpha - 1)|1 - \lambda|^2\langle D\mathbf{x}, \mathbf{x}\rangle = (1 - |\lambda|^2)\langle A\mathbf{x}, \mathbf{x}\rangle. \tag{16.20}$$

If $|\lambda| \neq 1$, then the left-hand side of (16.20) is positive since D is positive definite; hence, the right-hand side must be positive, so that $|\lambda| < 1$. If, on the other hand, $\lambda = 1$, then $\mathbf{y} = 0$; so that, from (16.15), $A\mathbf{x} = 0$. But this contradicts the fact that A is positive definite.

The choice of α determines the convergence rate. Frequently α appears in the literature as $\alpha = 1/\omega$, where $0 < \omega < 2$. Selection of a value for ω is difficult, in general, with some discussion appearing in [381, 350].

There is great flexibility in the SOR method in the choice of D and C, the only constraints being (16.11) and (16.12). Frequently, D is chosen as the diagonal of A, and C is chosen as the negative lower-triangular part of A.

Having shown that the SOR method converges, we summarize it as follows, using the aforementioned convention. Let A be divided as

$$A = D + C_L + C_U,$$

where D is diagonal and $C_L = C_U^H$ is strictly lower triangular. Let $C = -C_L$. For some ω in the range $0 < \omega < 2$:

$$Q = (D + \omega C_L)/\omega, \tag{16.21}$$

$$(D + \omega C_L)\mathbf{x}^{[k+1]} = [(1 - \omega)D - \omega C_U]\mathbf{x}^{[k]} + \omega\mathbf{b}. \tag{16.22}$$

Solving (16.22) for $\mathbf{x}^{[k+1]}$ is easily accomplished, since the matrix on the left-hand side is lower triangular and backsubstitution can be employed. For a given matrix A, the matrix factors should be precomputed. Algorithm 16.4 illustrates the SOR update technique using a backsubstitution step.

Algorithm 16.4 Successive over-relaxation
File: `sor.m`

16.3 Algebraic reconstruction techniques (ART)

Some linear problems involve very large matrices that are sparse. For example, in $A\mathbf{x} = \mathbf{b}$, the $m \times n$ matrix A might have only approximately r nonzero elements per row, where $r \ll n$. Such large sparse problems arise, for example, in projective tomography reconstruction problems, where the elements of $\mathbf{x}$ correspond to tissue density, the elements of $\mathbf{b}$ correspond to detector measurements, and the elements of A model the projection of tissue density onto the detectors in the tomographic process.

Iterative techniques have been developed for such large linear problems, which require only a small number of computations per inverse. In the tomographic literature they are known as *algebraic reconstruction techniques*. It has been shown that these methods converge when constraints are added to the solution (such as enforcing the requirement that each element of $\mathbf{x}$ be positive, or be in the range $0 < x_i < 1$). Such constraints are helpful in the tomographic regime, where tissue density must be positive, and where imposition of the constraints helps the problem to be less poorly conditioned.

We present an iterative solution to the (possibly overdetermined) linear equation

$$A\mathbf{x} = \mathbf{b}.$$

16.3 Algebraic Reconstruction Techniques (ART)

Let $\mathbf{x}^{[0]}$ be an initial solution. Also, let A be written in terms of its rows as

$$A = \begin{bmatrix} \mathbf{a}_0^T \\ \mathbf{a}_1^T \\ \vdots \\ \mathbf{a}_{m-1}^T \end{bmatrix}.$$

We interpret $\mathbf{a}_i^T$ as the transpose of the ith row. Also, let $\mathbf{b} = [b_0, b_1, \ldots, b_{m-1}]^T$, $m \geq n$. Then a partial update to the solution is obtained by

$$\tilde{\mathbf{x}}^{[k+1]} = \mathbf{x}^{[k]} + \frac{b_{(k+1) \bmod m} - \mathbf{a}_{(k+1) \bmod m}^T \mathbf{x}^{[k]}}{\|\mathbf{a}_{(k+1) \bmod m}\|^2} \mathbf{a}_{(k+1) \bmod m} \qquad m = 1, 2, \ldots.$$

We note that $\tilde{\mathbf{x}}^{[k+1]}$ satisfies the $(k+1)$st row (mod m) of $A\mathbf{x} = \mathbf{b}$:

$$\mathbf{a}_{(k+1) \bmod m}^T \tilde{\mathbf{x}}^{[k+1]} = \mathbf{a}_{(k+1) \bmod m}^T \mathbf{x}^{[k]}$$
$$+ \frac{b_{(k+1) \bmod m} - \mathbf{a}_{(k+1) \bmod m}^T \mathbf{x}^{[k]}}{\|\mathbf{a}_{(k+1) \bmod m}\|^2} \mathbf{a}_{(k+1) \bmod m}^T \mathbf{a}_{(k+1) \bmod m}$$
$$= b_{(k+1) \bmod m}.$$

Then, $\tilde{\mathbf{x}}^{[k+1]}$ is mapped into $\mathbf{x}^{[k+1]}$ using various conditions of constraint:

Unconstrained. Set $\mathbf{x}^{[k+1]} = \tilde{\mathbf{x}}^{[k+1]}$. It has been shown that if the set $\mathcal{X}_1 = \{\mathbf{x} | A\mathbf{x} = \mathbf{b}\}$ is nonempty, then the iterative algorithm converges to the element of $\mathcal{X}_1$ nearest to $\mathbf{x}^{[0]}$. (Hence, a good initial solution is very desirable.)

Partially constrained. Set

$$\mathbf{x}_i^{[k+1]} = \begin{cases} 0 & \tilde{\mathbf{x}}_i^{[k+1]} < 0, \\ \tilde{\mathbf{x}}_i^{[k+1]} & \text{otherwise.} \end{cases}$$

Fully constrained. Set

$$\mathbf{x}_i^{[k+1]} = \begin{cases} 0 & \tilde{\mathbf{x}}_i^{[k+1]} < 0, \\ \tilde{\mathbf{x}}_i^{[k+1]} & 0 < \tilde{\mathbf{x}}_i^{[k+1]} < 1, \\ 1 & \tilde{\mathbf{x}}_i^{[k+1]} > 1. \end{cases}$$

Algorithm 16.5 illustrates an implementation of this technique (without constraints) that processes each row of $\mathbf{x}$ once.

Algorithm 16.5 Algebraic reconstruction technique
File: `art1.m`

The ART algorithm tends to converge somewhat slowly, as the following example illustrates.

Example 16.3.1 Let

$$A = \begin{bmatrix} 1 & 2 & 3 \\ -4 & -2 & -5 \\ 2 & 7 & 10 \\ 4 & 3 & -2 \end{bmatrix} \qquad \mathbf{b} = \begin{bmatrix} 14 \\ -23 \\ 46 \\ 4 \end{bmatrix}.$$

The true solution is $\mathbf{x} = [1, 2, 3]^T$. Iteratively calling `art1` (starting from the initial value of $\mathbf{x}^{[0]} = [1, 1, 1]^T$) yields, after 10 iterations, the approximate solution

$$\mathbf{x}^{[200]} = \begin{bmatrix} 1.01463 \\ 1.98554 \\ 3.00758 \end{bmatrix}, \quad \text{with} \quad A\mathbf{x}^{[10]} = \begin{bmatrix} 14.0085 \\ -23.0675 \\ 46.0038 \\ 4 \end{bmatrix}.$$

(As previously observed, the last row updated has an exact solution.) The solution is close to $\mathbf{b}$, but there is still a residual. Figure 16.4 illustrates the error $\|\mathbf{b} - A\mathbf{x}^{[k]}\|$ as a function of iteration. Note the slow convergence after the initial improvement. □

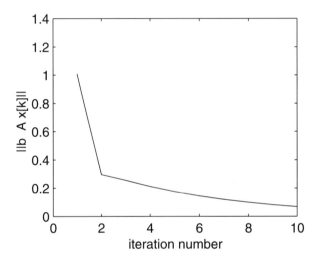

Figure 16.4: Residual error in the ART algorithm as a function of iteration.

16.4 Conjugate-direction methods

Conjugate-direction methods can be used to solve the system of equations

$$Q\mathbf{x} = \mathbf{b}$$

when Q is a symmetric positive definite $n \times n$ matrix. Conjugate-direction methods can also be used as an iterative minimization technique that is generally faster than Newton's method, but without the need to compute a Hessian matrix. Unlike in the steepest-descent method, the minimum of a quadratic form in n dimensions can be obtained in n steps using exact arithmetic with conjugate-direction methods.

Let Q be a positive-definite matrix, and consider minimizing the quadratic problem

$$f(\mathbf{x}) = \frac{1}{2}\mathbf{x}^T Q \mathbf{x} - \mathbf{b}^T \mathbf{x}.$$

By taking the gradient with respect to $\mathbf{x}$, it is clear that this is equivalent to solving the equation

$$Q\mathbf{x} = \mathbf{b}.$$

Conversely, suppose the equation $Q\mathbf{x} = \mathbf{b}$ is to be solved. Clearly, a solution exists at the point where $f(\mathbf{x})$ has a minimum. There is thus a duality between solving the symmetric linear equation and solving the minimization problem.

16.4 Conjugate-Direction Methods

In section 2.8, the inner product with respect to a matrix Q was defined as

$$\langle \mathbf{x}, \mathbf{y} \rangle_Q = \mathbf{x}^T Q \mathbf{y}. \tag{16.23}$$

(We assume for convenience that all matrices and vectors are real.) Vectors $\mathbf{x}$ and $\mathbf{y}$ are orthogonal with respect to this inner product if $\mathbf{x}^T Q \mathbf{y} = 0$. In the context of the conjugate-direction literature, this orthogonality is given another name.

Definition 16.1 For a symmetric matrix Q, vectors $\mathbf{x}$ and $\mathbf{y}$ are said to be ***Q*-orthogonal** or **conjugate with respect to** Q or ***Q*-conjugate** if $\mathbf{x}^T Q \mathbf{y} = 0$.

A set of vectors $\mathbf{d}_0, \mathbf{d}_1, \ldots, \mathbf{d}_k$ is said to be a ***Q*-orthogonal** set if $\mathbf{d}_i^T Q \mathbf{d}_j = 0$ for $i \neq j$. □

An important fact about Q-orthogonal sets is given in the following lemma, whose proof is given as an exercise.

Lemma 16.1 *If the vectors $\mathbf{d}_0, \mathbf{d}_1, \ldots, \mathbf{d}_k$ are all nonzero and form a Q-orthogonal set for a positive definite Q, then the vectors are linearly independent.*

Let $\mathbf{d}_0, \mathbf{d}_1, \ldots, \mathbf{d}_{n-1}$ be a given Q-orthogonal set. Then, since these n vectors must span n-space, the solution to the equation $Q\mathbf{x} = \mathbf{b}$ can be written as a linear combination of the vectors in the Q-orthogonal set:

$$\mathbf{x} = \alpha_0 \mathbf{d}_0 + \alpha_1 \mathbf{d}_1 + \cdots + \alpha_{n-1} \mathbf{d}_{n-1}.$$

The coefficients α_i can be found by premultiplying both sides by $\mathbf{d}_i^T Q$. Using the Q-orthogonality property and solving for α_i, we obtain

$$\alpha_i = \frac{\mathbf{d}_i^T Q \mathbf{x}}{\mathbf{d}_i^T Q \mathbf{d}_i} = \frac{\mathbf{d}_i^T \mathbf{b}}{\mathbf{d}_i^T Q \mathbf{d}_i}.$$

By using the Q-orthogonality, there is no need to compute an expression for $Q\mathbf{x}$, which is fortunate because we don't yet know $\mathbf{x}$. Instead, we can express the coefficients in terms of $\mathbf{b}$. We can write the expansion for $\mathbf{x}$ as

$$\mathbf{x} = \sum_{i=1}^{n} \frac{\mathbf{d}_i^T \mathbf{b}}{\mathbf{d}_i^T Q \mathbf{d}_i} \mathbf{d}_i.$$

This construction of the solution for $\mathbf{x}$ can be viewed as an iterative algorithm in which the solution can be built up from any starting point. The following theorem clarifies this point.

Theorem 16.2 *(Conjugate-direction theorem) Let $\{\mathbf{d}_0, \mathbf{d}_1, \ldots, \mathbf{d}_{n-1}\}$ be a set of nonzero Q-orthogonal vectors. For any $\mathbf{x}^{[0]} \in \mathbb{R}^n$, the sequence $\{\mathbf{x}^{[k]}\}$ generated by*

$$\mathbf{x}^{[k+1]} = \mathbf{x}^{[k]} + \alpha_k \mathbf{d}_k \qquad k \geq 0 \tag{16.24}$$

with

$$\alpha_k = -\frac{\mathbf{g}_k^T \mathbf{d}_k}{\mathbf{d}_k^T Q \mathbf{d}_k}$$

and

$$\mathbf{g}_k = Q\mathbf{x}^{[k]} - \mathbf{b}$$

converges to the unique solution $\mathbf{x}^*$ of $Q\mathbf{x} = \mathbf{b}$ after n steps.

It is interesting and pertinent to note that

$$\mathbf{g}_k = Q\mathbf{x}^{[k]} - \mathbf{b}$$

is the gradient of $f(\mathbf{x})$ evaluated at $\mathbf{x} = \mathbf{x}^{[k]}$.

Proof By applying the iterative process (16.24) from $\mathbf{x}^{[0]}$ to $\mathbf{x}^{[k]}$,

$$\mathbf{x}^{[k]} - \mathbf{x}^{[0]} = \alpha_0 \mathbf{d}_0 + \alpha_1 \mathbf{d}_1 + \cdots + \alpha_{k-1}\mathbf{d}_{k-1}.$$

By the Q-orthogonality of the $\mathbf{d}_k$, it follows that

$$\mathbf{d}_k^T Q(\mathbf{x}^{[k]} - \mathbf{x}^{[0]}) = 0. \tag{16.25}$$

Since the $\mathbf{d}_i$ are linearly independent, we can write

$$\mathbf{x}^* - \mathbf{x}^{[0]} = \alpha_0 \mathbf{d}_0 + \alpha_1 \mathbf{d}_1 + \cdots + \alpha_{n-1}\mathbf{d}_{n-1}$$

for some α_i. As before, we can find the coefficient α_i by premultiplying by $\mathbf{d}_i^T Q$, to obtain

$$\alpha_i = \frac{\mathbf{d}_i^T Q(\mathbf{x}^* - \mathbf{x}^{[0]})}{\mathbf{d}_i^T Q \mathbf{d}_i} = \frac{\mathbf{d}_i^T Q(\mathbf{x}^* - \mathbf{x}^{[k]} + \mathbf{x}^{[k]} - \mathbf{x}^{[0]})}{\mathbf{d}_i^T Q \mathbf{d}_i}$$

$$= \frac{\mathbf{d}_i^T Q(\mathbf{x}^{[k]} - \mathbf{x}^{[0]})}{\mathbf{d}_i^T Q \mathbf{d}_i} + \frac{\mathbf{d}_i^T Q(\mathbf{x}^* - \mathbf{x}^{[k]})}{\mathbf{d}_i^T Q \mathbf{d}_i}. \tag{16.26}$$

Now we substitute (16.25) into the first term of (16.26) to obtain

$$\alpha_i = \frac{\mathbf{d}_i^T Q(\mathbf{x}^* - \mathbf{x}_k)}{\mathbf{d}_i^T Q \mathbf{d}_i} = -\frac{\mathbf{g}_i^T \mathbf{d}_i}{\mathbf{d}_i^T Q \mathbf{d}_i}.$$

We thus obtain a series in $\mathbf{d}_i$ for $\mathbf{x}^*$, and the theorem is proved. $\square$

Of course, to make the conjugate-direction method useful for solving $Q\mathbf{x} = \mathbf{b}$, it is necessary to have a Q-orthogonal set $\{\mathbf{d}_i\}$. This could be found (for example) by using the Gram–Schmidt process on a sequence with respect to the inner product (16.23). However, this is numerically and computationally suspicious. In addition, minimizing a quadratic function f is too easy to justify all this effort. However, we now extend the method so that it can apply to functions that are not exactly quadratic, and so that the Q-orthogonal set is computed as the algorithm proceeds.

16.5 Conjugate-gradient method

In the conjugate gradient method, the conjugate direction vectors $\{\mathbf{d}_i\}$ are not computed beforehand, but are computed as the method progresses. The algorithm proceeds as follows: Let $\mathbf{x}^{[0]} \in \mathbb{R}^n$ be an arbitrary starting vector, and let $\mathbf{d}_0 = -\mathbf{g}_0 = \mathbf{b} - Q\mathbf{x}_0$. The conjugate gradient algorithm proceeds by iterating

$$\mathbf{x}^{[k+1]} = \mathbf{x}^{[k]} + \alpha_k \mathbf{d}_k, \tag{16.27}$$

$$\alpha_k = -\frac{\mathbf{g}_k^T \mathbf{d}_k}{\mathbf{d}_k^T Q \mathbf{d}_k}, \tag{16.28}$$

$$\mathbf{g}_{k+1} = Q\mathbf{x}^{[k+1]} - \mathbf{b}, \tag{16.29}$$

$$\mathbf{d}_{k+1} = -\mathbf{g}_{k+1} + \beta_k \mathbf{d}_k, \tag{16.30}$$

$$\beta_k = \frac{\mathbf{g}_{k+1}^T Q \mathbf{d}_k}{\mathbf{d}_k^T Q \mathbf{d}_k}. \tag{16.31}$$

Observe that the first step starts moving in the direction of the negative gradient, hence the method starts like steepest descent. However, unlike the traditional steepest descent, the step-size parameter (here denoted as α_k) changes at each step. Each successive move is a weighted combination of the current gradient $\mathbf{g}_k$ and the previous direction of motion.

16.5 Conjugate-Gradient Method

Algorithm 16.6 illustrates the conjugate gradient solution of $Q\mathbf{x} = \mathbf{b}$ for a symmetric matrix Q.

Algorithm 16.6 Conjugate-gradient solution of a symmetric linear equation
File: `conjgrad1.m`

Example 16.5.1 Figure 16.5 shows the contours of $\|R\mathbf{x} - \mathbf{b}\|$ and the results of the conjugate-gradient algorithm for

$$R = \begin{bmatrix} 105 & 95 \\ 95 & 105 \end{bmatrix} \quad \text{and} \quad \mathbf{b} = \begin{bmatrix} 200 \\ 200 \end{bmatrix},$$

starting from $\mathbf{x}^{[0]} = [1 \ -9]^T$. Observe that, unlike the steepest-descent example for these same parameters (see example 14.5.1), the conjugate-gradient method converges in exactly two steps for a purely quadratic problem. The values of α are

$$\alpha_1 = .005 \qquad \alpha_2 = .0998.$$

Note that α_1 is exactly the largest value that a steepest descent-step can be and still satisfy (14.11): it steps just far enough in the steepest direction to reach the end of the valley. The next step heads straight into the solution. □

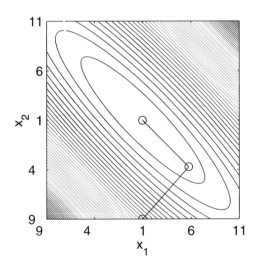

Figure 16.5: Convergence of conjugate gradient on a quadratic function.

Example 16.5.2 Using the conjugate-gradient algorithm, solve

$$\begin{bmatrix} 1 & 3 & 7 \\ 3 & 2 & 4 \\ 7 & 4 & 6 \end{bmatrix} \mathbf{x} = \begin{bmatrix} 33 \\ 30 \\ 60 \end{bmatrix}.$$

Proceeding as in the steps outlined above yields the solution

$$\mathbf{x} = \begin{bmatrix} 4 \\ 5 \\ 2 \end{bmatrix}.$$

Starting from the initial point $\mathbf{x}^{[0]} = [1\ 0\ 0]^T$, the conjugate-direction vectors are

$$\mathbf{d}_0 = \begin{bmatrix} 32 \\ 27 \\ 53 \end{bmatrix} \quad \mathbf{d}_1 = \begin{bmatrix} -4.7226 \\ -0.4416 \\ 3.7624 \end{bmatrix} \quad \mathbf{d}_2 = \begin{bmatrix} 0.1539 \\ -0.6634 \\ 0.2549 \end{bmatrix}.$$

It is straightforward to verify that these vectors are Q-orthogonal. □

In order to validate that the algorithm is, in fact, a conjugate-direction algorithm, it must be determined that the vectors $\{\mathbf{d}_k\}$ are Q-orthogonal. This condition, and more, is established by the following theorem.

Theorem 16.3 *If the conjugate gradient algorithm (16.27)–(16.31) does not terminate at $\mathbf{x}_k$, then:*

(a) $\operatorname{span}\{\mathbf{g}_0, \mathbf{g}_1, \ldots, \mathbf{g}_k\} = \operatorname{span}\{\mathbf{g}_0, Q\mathbf{g}_0, \ldots, Q^k\mathbf{g}_0\}$.
(b) $\operatorname{span}\{\mathbf{d}_0, \mathbf{d}_1, \ldots, \mathbf{d}_k\} = \operatorname{span}\{\mathbf{g}_0, Q\mathbf{g}_0, \ldots, Q^k\mathbf{g}_0\}$.
(c) $\mathbf{d}_k^T Q \mathbf{d}_i = 0$ *for* $i < k$.

Proof The proof is by induction simultaneously on parts (a), (b), and (c). For $k = 0$ the result is clear. We assume that (a), (b), and (c) are true for k and show that they are true for $k + 1$. The gradient update can be written as

$$\mathbf{g}_{k+1} = \mathbf{g}_k + \alpha_k Q \mathbf{d}_k.$$

By the induction hypothesis and part (a),

$$\mathbf{g}_k \in \operatorname{span}\{\mathbf{g}_0, Q\mathbf{g}_0, \ldots, Q^k\mathbf{g}_0\} \subset \operatorname{span}\{\mathbf{g}_0, Q\mathbf{g}_0, \ldots, Q^k\mathbf{g}_0, Q^{k+1}\mathbf{g}_0\},$$

and by part (b),

$$\mathbf{d}_k \in \operatorname{span}\{\mathbf{g}_0, Q\mathbf{g}_0, \ldots, Q^k\mathbf{g}_0\} \subset \operatorname{span}\{\mathbf{g}_0, Q\mathbf{g}_0, \ldots, Q^k\mathbf{g}_0, Q^{k+1}\mathbf{g}_0\}.$$

Hence, $\mathbf{g}_{k+1} \in \operatorname{span}\{\mathbf{g}_0, Q\mathbf{g}_0, \ldots, Q^k\mathbf{g}_0, Q^{k+1}\mathbf{g}_0\}$. But $\mathbf{g}_{k+1}$ cannot be in $\operatorname{span}\{\mathbf{g}_0, Q\mathbf{g}_0, \ldots, Q^k\mathbf{g}_0\} = \operatorname{span}\{\mathbf{d}_0, \mathbf{d}_1, \ldots, \mathbf{d}_k\}$ because, for a conjugate-direction method, $\mathbf{g}_{k+1} \perp \mathbf{d}_i$, $i = 0, 1, \ldots, k$ (see exercise 16.5-11), and this would make $\mathbf{g}_{k+1} = 0$. We conclude that

$$\operatorname{span}\{\mathbf{g}_0, \mathbf{g}_1, \ldots, \mathbf{g}_{k+1}\} = \operatorname{span}\{\mathbf{g}_0, Q\mathbf{g}_0, \ldots, Q^{k+1}\mathbf{g}_0\},$$

establishing part (a). To prove part (b), we note that

$$\mathbf{d}_{k+1} = -\mathbf{g}_{k+1} + \beta_k \mathbf{d}_k$$

from which (b) follows from (a).

To prove (c), we write

$$\mathbf{d}_{k+1}^T Q \mathbf{d}_i = -\mathbf{g}_{k+1}^T Q \mathbf{d}_i + \beta_k \mathbf{d}_k^T Q \mathbf{d}_i.$$

When $i = k$, the right-hand side is zero, by the definition of β_k. When $i < k$, the second term vanishes, by the induction hypothesis (c). The first term vanishes since the expanding subspace theorem (see exercise 16.5-10) guarantees that $\mathbf{g}_{k+1} \perp \operatorname{span}\{\mathbf{d}_0, \mathbf{d}_1, \ldots, \mathbf{d}_{i+1}\}$, and since

$$Q\mathbf{d}_i \in \operatorname{span}\{\mathbf{d}_0, \mathbf{d}_1, \ldots, \mathbf{d}_{i+1}\}.$$

By (c), the $\mathbf{d}_i$ are Q-orthogonal, which establishes that this is a conjugate-direction method. □

It should be noted that even though the algorithm should converge in n steps for the quadratic functional $f(\mathbf{x})$, because of numerical roundoff, there may not be exact convergence. This failure to stop in n steps led to disinterest in the conjugate-gradient algorithm. More recently, however, interest in the algorithm has been rejuvenated as an *iterative* algorithm suited for some sparse matrices, in which the number of iterates to convergence is hoped to be $< n$.

16.6 Nonquadratic problems

An unconstrained problem

$$\min_{\mathbf{x} \in \mathbb{R}^n} f(\mathbf{x})$$

can be approached using conjugate-gradient methods. In the conjugate-gradient algorithm, we identify $\mathbf{g}_k$ as $\nabla f(\mathbf{x}_k)$, and Q as the Hessian

$$\frac{\partial^2 f(\mathbf{x})}{\partial x_i \partial x_j}.$$

However, since the problem is not exactly a quadratic problem, convergence in n steps is not expected. What is commonly done is to proceed through n steps of the conjugate-gradient algorithm, then restart with a pure gradient step. MATLAB code for computing this is shown in algorithm 16.7.

Algorithm 16.7 conjugate-gradient solution for unconstrained minimization
File: `conjgrad2.m`

One problem with this conjugate-gradient solution is the need to compute the Hessian, which is often a very expensive calculation. This can be alleviated by taking two extra steps. First, it is possible (see exercise 16.5-11) to write

$$\beta_{k+1} = \frac{\mathbf{g}_{k+1}^T \mathbf{g}_{k+1}}{\mathbf{g}_k^T \mathbf{g}_k},$$

which eliminates the Hessian from β_k. For the step (16.27), instead of computing an α_k, the line $\mathbf{x}^{[k]} + \mu \mathbf{d}_k$ is searched to find the minimizer of $f(\mathbf{x}^{[k]} + \mu \mathbf{d}_k)$. This line search is often less expensive than computation of the Hessian.

16.7 Exercises

16.1-1 For the fuzzy k-means algorithm, we need to minimize (16.1) subject to (16.3). Set up a cost functional incorporating Lagrange multipliers and derive (16.5). Using the distance measure (16.2), derive (16.6).

16.1-2 Write a MATLAB function that does fuzzy clustering on a set of data.

16.2-3 Show that the update in the Jacobi iteration can be obtained by

$$x_i^{[k+1]} = \tilde{b}_i - \sum_{j \neq i} \tilde{a}_{ij} x_j^{[k]}$$

(no divisions), by suitable normalization of A and $\mathbf{b}$ prior to beginning the Jacobi iteration.

16.2-4 Show that if A is diagonally dominant, so

$$|a_{ii}| > \max_i \sum_{j \neq i} |a_{ij}|,$$

then the Jacobi method converges.

16.2-5 Show that:

(a) If an iterative update is of the form

$$Q\mathbf{x}^{[k+1]} = (Q - A)\mathbf{x}^{[k]} + \mathbf{b}, \qquad (16.32)$$

where Q is an invertible matrix, then the iterative method converges if $\|I - Q^{-1}A\| < 1$, for some matrix norm.

(b) Show that the Jacobi method can be written in the form of (16.32).

(c) Show that the Gauss–Seidel method can be written in the form of (16.32).

(d) (Richardson's method) Show that if $Q = I$, then the iteration (16.32) has a fixed point equal to $A^{-1}\mathbf{b}$ if $\|I - A\| < 1$.

16.2-6 Show that if A is diagonally dominant, then

$$|a_{ii}| > \max_i \sum_{j \neq i} |a_{ij}/a_{ii}|,$$

and the Gauss–Seidel method converges.

16.2-7 (Acceleration) Consider an iterative update of the form

$$\mathbf{x}^{[k+1]} = B\mathbf{x}^{[k]} + \mathbf{c}.$$

We define a new update formula by

$$\mathbf{x}^{[k+1]} = \gamma(B\mathbf{x}^{[k+1]} + \mathbf{c}) + (1 - \gamma)\mathbf{x}^{[k]} = B_\gamma \mathbf{x}^{[k]} + \mathbf{c},$$

where $B_\gamma = \gamma B + (1 - \gamma)I$. In this problem, you will examine how to find an optimal value of γ to speed convergence in some cases.

(a) Show that if the eigenvalues of G are known to lie in the interval $[a, b]$, then every eigenvalue λ of G_γ must be in the range

$$\gamma a + 1 - \gamma \leq \lambda \leq \gamma b + 1 - \gamma. \qquad (16.33)$$

Hint: See exercise 6.2-11.

(b) Now, assuming that $b < 1$, let $d = 1 - b$. Show when $\gamma = 2/(2 - a - b)$ that

$$-1 + \gamma d \leq \lambda \leq 1 - \gamma d.$$

(c) Conclude that, in this case, the $\rho(G_\gamma) \leq 1 - \gamma d$.

(d) Argue, that $\gamma = 2/(2 - a - b)$ is the optimum value of γ.

(e) Repeat steps (b)–(d), assuming that $a > 1$, using $d = a - 1$.

16.2-8 Show that the Gauss–Seidel iteration is a special case of SOR.

16.5-9 Prove lemma 16.1.

16.5-10 (The expanding subspace theorem) Let $\mathcal{B}_k$ be the subspace spanned by $\{\mathbf{d}_0, \mathbf{d}_1, \ldots, \mathbf{d}_{k-1}\}$, and let $f(\mathbf{x}) = \frac{1}{2}\mathbf{x}^T Q\mathbf{x} - \mathbf{b}^T\mathbf{x}$.

(a) Show that $\mathbf{x}^{[k]} = \mathbf{x}^{[k-1]} + \alpha_{k-1}\mathbf{d}_{k-1}$ lies in the linear variety $\mathbf{x}_0 + \mathcal{B}_k$.

(b) Show that since f is a convex function, the point $\mathbf{x} \in \mathbf{x}_0 + \mathcal{B}_k$ that minimizes f is the point at which

$$\mathbf{g}(\mathbf{x}) = Q\mathbf{x} - \mathbf{b}$$

is orthogonal to $\mathcal{B}_k$. Note that $\mathbf{g}(\mathbf{x})$ is the gradient of f.

(c) Show for the recursion given in theorem 16.2 that $\mathbf{g}_k \perp \mathcal{B}_k$, and hence $f(\mathbf{x})$ is minimized in $\mathbf{x}_0 + \mathcal{B}_k$. (Hint: Use induction and the Q-orthogonality of the $\mathbf{d}_i$.)

(d) Hence argue that $\mathbf{x}_n$ must be the overall minimum of f.

16.5-11 Show that for the conjugate-gradient algorithm,

$$\beta_k = \frac{\mathbf{g}_{k+1}^T \mathbf{g}_{k+1}}{\mathbf{g}_k^T \mathbf{g}_k}.$$

16.5-12 Given a sequence $\mathbf{p}_0, \mathbf{p}_1, \ldots, \mathbf{p}_{n-1}$, a Q-orthogonal set $\mathbf{d}_0, \mathbf{d}_1, \ldots, \mathbf{d}_{n-1}$ can be produced using the Gram-Schmidt process using the Q-weighted inner product. In the particular case when the $\mathbf{p}_k$ are generated by $\mathbf{p}_k = Q^k \mathbf{p}_0$, show that $\mathbf{d}_{k+1}$ can be generated by a three-term recursion involving $Q\mathbf{d}_k$, $\mathbf{d}_k$, and $\mathbf{d}_{k-1}$.

16.5-13 Let $f(\mathbf{x}) = \frac{1}{2}\mathbf{x}^T Q\mathbf{x} - \mathbf{b}^T\mathbf{x}$ for $\mathbf{x} \in \mathbb{R}^n$ and positive-definite Q. Let $\mathbf{x}_1$ be a minimizer of f over a subspace of $\mathbb{R}^n$ containing the vector $\mathbf{d}$, and let $\mathbf{x}_2$ be a minimizer over another subspace containing $\mathbf{d}$. Suppose that $f(\mathbf{x}_1) < f(\mathbf{x}_2)$. Show that $\mathbf{x}_1 - \mathbf{x}_2$ is Q-conjugate to $\mathbf{d}$.

16.5-14 Let Q be a symmetric matrix. Show that any two vectors of Q corresponding to distinct eigenvalues are Q-conjugate.

16.5-15 Let $\mathbf{d}_0, \mathbf{d}_1, \ldots, \mathbf{d}_{n-1}$ be Q-conjugate for symmetric Q. Describe how to find a matrix E such that $E^T Q E$ is diagonal.

16.5-16 Modify algorithm 14.4 to use conjugate-gradient methods to train a neural network, implementing the line search instead of (16.27).

16.8 References

There is a very large literature on clustering in association with multidimensional statistics, pattern recognition, and data classification. An excellent survey of some classical techniques is provided in [76, Chapter 6].

The fuzzy k-means was introduced in [26]. It was later generalized to fuzzy k-varieties, which generalized the shape of clusters that can be found. To a large extent, the clusters produced by the generic k-means algorithm are essentially circular in the geometry determined by the distance function. Data with linear or planar clusters might therefore be misclassified, as the ends of a set of data lying on a line might fall into different clusters. This problem is addressed by clustering into linear varieties; clustering into fuzzy k-varieties are discussed in [27, 28], and generalized further to a piecewise regression problem in [124]. More recent related work is in [185, 101].

The question of cluster validity is still largely open. A statistically based discussion appears in [153]. Research addressing the question for fuzzy clustering is presented in [376].

The methods discussed for iterative solution of linear systems are widely reported (see, for example, [114, 181]). Sparse systems of equations are described in [381, 350]. The method of successive over-relaxation, which provides a weighted average of prior values with a Gauss–Seidel update, is also discussed in [13], along with a variety of nonstationary methods including conjugate-gradient, generalized minimal residual, biconjugate-gradient, conjugate-gradient squared, and Chebyshev methods. Our discussion of SOR comes from [181, chapter 4].

The algebraic reconstruction technique discussed here was proposed in [159], with additional discussions being found in [116, 105]. See also the discussion in [152].

The conjugate-gradient method was introduced in [136], and is discussed in a variety of sources including [209, 181]. The presentation here, and some of the exercises, loosely follow [210]. This source also provides an interesting discussion of the treatment of functions with penalty terms, in which the eigenvalues may be divided into two groups. By a partial conjugate-gradient method, the eigenvalue disparity can be neatly avoided. Iterative approaches that are similar in spirit to conjugate gradient but applicable to general (nonsymmetric) matrices are reviewed in [13]. As an example of the way that conjugate-gradient methods are applied in signal processing, [157] discusses image restoration. Conjugate-gradient techniques are also considered for training neural networks; see, for example, [192]. Preconditioning of the conjugate gradient to improve the numerical stability is covered well in [181].

Chapter 17

The EM Algorithm in Signal Processing

> It is no paradox to say that in our most theoretical moods we may be nearest to our most practical applications.
>
> — *A.N. Whitehead*

In this chapter, we introduce a means of maximum-likelihood estimation of parameters that is applicable in many cases when direct access to the data necessary to make the estimates is impossible, or when some of the data are missing. Such inaccessible data are present, for example, when an outcome is a result of an accumulation of simpler outcomes, or when outcomes are clumped together (e.g., in a binning or histogram operation). There may also be data dropouts or clustering such that the number of underlying data points is unknown (censoring and/or truncation). The EM (expectation–maximization) algorithm is ideally suited to problems of this sort, in that it produces maximum-likelihood (ML) estimates of parameters when there is a many-to-one mapping from an underlying distribution to the distribution governing the observation. The EM algorithm consists of two primary steps: an expectation step, followed by a maximization step. The expectation is obtained with respect to the unknown underlying variables, using the current estimate of the parameters and conditioned upon the observations. The maximization step then provides a new estimate of the parameters. These two steps are iterated until convergence. The concept is illustrated in figure 17.1.

The EM algorithm was discovered and employed independently by several different researchers; Dempster [72] brought their ideas together, proved convergence, and coined the term "EM algorithm." Since this seminal work, hundreds of papers employing the EM algorithm in many areas have been published. A typical application area of the EM algorithm is genetics, where the observed data (the phenotype) is a function of the underlying, unobserved gene pattern (the genotype); see, for example, [155]. Another area is estimating parameters of mixture distributions, as in [273]. The EM algorithm has also been widely used in econometric, clinical, and sociological studies that have unknown factors affecting the outcomes ([292]). Some applications to the theory of statistical methods are found in [205].

In the area of signal processing applications, the largest area of interest in the EM algorithm is maximum-likelihood tomographic image reconstruction (see, for example, [309, 315]). Another commonly cited application is the training of hidden Markov models (see section 1.7), especially for speech recognition, as in [266].

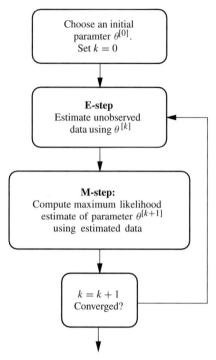

Figure 17.1: An overview of the EM algorithm. After initialization, the E-step and the M-step are alternated until the parameter estimate has converged (no more change in the estimate).

Other signal processing and engineering applications began appearing in the mid-1980s. These include: parameter estimation ([299, 382]), ARMA modeling ([151, 388]), image modeling, reconstruction, and processing ([195, 173]), simultaneous detection and estimation ([5, 83, 168]), pattern recognition and neural network training ([46, 158, 335]), direction finding ([226]), noise suppression ([351]), signal enhancement ([360]), spectroscopy, signal and sequence detection ([103]), time-delay estimation ([6]), and specialized developments of the EM algorithm itself ([300]). The EM algorithm is also related to algorithms used in information theory to compute channel capacity and rate-distortion functions ([31, 59]), since the expectation step in the algorithm produces a result similar to entropy. The EM algorithm is philosophically similar to ML detection in the presence of unknown phase (incoherent detection) or other unknown parameters: the likelihood function is averaged with respect to the unknown quantity (i.e., the expected value of the likelihood function is computed) before detection, which is a maximization step (see, for example, [261, chapter 5]).

The algorithm is presented first by means of an extended example, then formally in section 17.2. The convergence properties of the algorithm are discussed in section 17.3. Several signal processing algorithms are then discussed. Some concluding remarks appear in section 17.9.

17.1 An introductory example

The following problem, although somewhat contrived, illustrates most of the principles of the EM algorithm. In many aspects it is similar to a problem that is of practical interest—the emission tomography problem discussed in section 17.5.

17.1 An Introductory Example

Suppose that in an image-pattern recognition problem, there are two general classes to be distinguished: a class of dark objects and a class of light objects. The class of dark objects may be further subdivided into two shapes: round and square. We desire to determine the probability of a dark object. For the sake of the example, assume that we know the objects to be trinomially distributed. Let the random variable X_1 represent the number of round dark objects, X_2 represent the number of square dark objects, and X_3 represent the number of light objects, and let $[x_1, x_2, x_3]^T = \mathbf{x}$ be the vector of values the random variables take for some image. The general trinomial distribution is

$$P(X_1 = x_1, X_2 = x_2, X_3 = x_3) = \left(\frac{n!}{x_1! x_2! x_3!}\right) p_1^{x_1} p_2^{x_2} p_3^{x_3},$$

where $n = x_1 + x_2 + x_3$ and $p_1 + p_2 + p_3 = 1$. The parameters of this distribution are p_1, p_2, and p_3. However, in this problem we assume that enough is known about the probabilities of the different classes so that the probability may be written as

$$P(X_1 = x_1, X_2 = x_2, X_3 = x_3 \mid p) = \left(\frac{n!}{x_1! x_2! x_3!}\right) \left(\frac{1}{4}\right)^{x_1} \left(\frac{1}{4} + \frac{p}{4}\right)^{x_2} \left(\frac{1}{2} - \frac{p}{4}\right)^{x_3}$$

$$= f_\mathbf{X}(x_1, x_2, x_3 \mid p), \tag{17.1}$$

where p is now the single unknown parameter of the distribution. Recall that (in general) we use the symbol f to indicate either a pdf or a pmf.

Suppose that for some reason it is not possible to distinguish each of the classes of objects separately. For the sake of the example, we assume that a feature extractor is employed that can distinguish which objects are light and which are dark, but cannot distinguish shape. Let $[y_1, y_2]^T = \mathbf{y}$ be the number of dark objects and number of light objects detected, respectively, so that $y_1 = x_1 + x_2$ and $y_2 = x_3$, and let the corresponding random variables be Y_1 and Y_2. There is a many-to-one mapping between $\{x_1, x_2\}$ and y_1. For example, if $y_1 = 3$, there is no way to tell from the measurements whether $x_1 = 1$ and $x_2 = 2$, or $x_1 = 2$ and $x_2 = 1$. The EM algorithm is specifically designed for problems with such many-to-one mappings. Based on observations y_1 and y_2, we desire to determine a maximum likelihood estimate of p, the parameter of the distribution.

The random variable $Y_1 = X_1 + X_2$ is binomially distributed (see appendix F),

$$P(Y_1 = y_1 \mid p) = \binom{n}{y_1} \left(\frac{1}{2} + \frac{p}{4}\right)^{y_1} \left(\frac{1}{2} - \frac{p}{4}\right)^{n-y_1}.$$

$$= g(y_1 \mid p).$$

(The symbol g is used to indicate the probability function for the observed data.)

In this case, it would be possible to compute an ML estimate of p by solving

$$p_{\text{ML}} = \arg\max_p g(Y_1 = y_1 \mid p).$$

In more interesting problems, however, such straightforward estimation is not always possible. In the interest of introducing the EM algorithm, we do not take the direct approach to the ML estimate. The key idea behind the EM algorithm is that, even when we do not know x_1 and x_2, knowledge of the form of the underlying distribution $f_\mathbf{X}(x_1, x_2, x_3 \mid p)$ can be used to determine an estimate for p. This is done by first estimating the underlying data—in this case, x_1 and x_2—then using these data to update our estimate of the parameter. This is repeated until convergence. Let $p^{[k]}$ indicate the estimate of p after the kth iteration,

$k = 1, 2, \ldots$. An initial parameter value $p^{[0]}$ is assumed. The algorithm consists of two primary steps:

Expectation step (E-step). Compute the expected value of the x data using the current estimate of the parameter and the observed data.

The expected value of x_1, given the measurement y_1 and based upon the current estimate of the parameter, may be computed as

$$x_1^{[k+1]} = E[x_1 \mid y_1, p^{[k]}].$$

Using the results of appendix F,

$$x_1^{[k+1]} = y_1 \frac{\frac{1}{4}}{\frac{1}{2} + \frac{p^{[k]}}{4}}. \tag{17.2}$$

Similarly,

$$x_2^{[k+1]} = E[x_2 \mid y_1, p^{[k]}] = y_1 \frac{\frac{1}{4} + \frac{p^{[k]}}{4}}{\frac{1}{2} + \frac{p^{[k]}}{4}}. \tag{17.3}$$

In the current example, x_3 is known explicitly and does not need to be estimated.

Maximization step (M-step). Use the data from the expectation step as if it were actually measured data, to determine an ML estimate of the parameter. This estimated data is sometimes called "imputed" data.

In this example, with $x_1^{[k+1]}$ and $x_2^{[k+1]}$ imputed and x_3 available, the ML estimate of the parameter is obtained by taking the derivative of $\log f_\mathbf{X}(x_1^{[k+1]}, x_2^{[k+1]}, x_3 \mid p)$ with respect to p, equating it to zero, and solving for p,

$$0 = \frac{d}{dp} \log f_\mathbf{X}\left(x_1^{[k+1]}, x_2^{[k+1]}, x_3 \mid p\right)$$

$$\Rightarrow p^{[k+1]} = \frac{2x_2^{[k+1]} - x_3}{x_2^{[k+1]} + x_3}. \tag{17.4}$$

The estimate $x_1^{[k+1]}$ is not used in (17.4) and so, for this example, need not be computed. The EM algorithm consists of iterating (17.3) and (17.4) until convergence. Intermediate computation and storage may be eliminated by substituting (17.3) into (17.4) to obtain a one-step update:

$$p^{[k+1]} = \frac{p^{[k]}(4y_1 - 2x_3) + 2y_1 - 2x_3}{p^{[k]}(2y_1 + 2x_3) + y_1 + 2x_3}. \tag{17.5}$$

As a numerical example, suppose that the true parameter is $p = 0.5$, and $n = 100$ samples are drawn, with $y_1 = 63$. (The true values of x_1 and x_2 are 25 and 38, respectively, but the algorithm does not know this.) The code segment in algorithm 17.1 illustrates the computations.

Algorithm 17.1 EM algorithm example computations
File: `em1.m`

The results of these computations are shown in table 17.1, starting from $p^{[0]} = 0$. The final estimate $p^* = 0.52$ is in fact the ML estimate of p that would have been obtained by maximizing (17.1) with respect to p, had the x data been available.

Table 17.1: Results of the EM algorithm for an example using trinomial data

k	$x_1^{[k]}$	$x_2^{[k]}$	$p^{[k]}$
1	31.500000	31.500000	0.379562
2	26.475460	36.524540	0.490300
3	25.298157	37.701843	0.514093
4	25.058740	37.941260	0.518840
5	25.011514	37.988486	0.519773
6	25.002255	37.997745	0.519956
7	25.000441	37.999559	0.519991
8	25.000086	37.999914	0.519998
9	25.000017	37.999983	0.520000
10	25.000003	37.999997	0.520000

17.2 General statement of the EM algorithm

Let $\mathcal{Y}$ denote the sample space of the observations, and let $\mathbf{y} \in \mathbb{R}^m$ denote an observation from $\mathcal{Y}$. Let $\mathcal{X}$ denote the underlying space and let $\mathbf{x} \in \mathbb{R}^n$ be an outcome from $\mathcal{X}$, with $m < n$. The data $\mathbf{x}$ is referred to as the **complete data**. The complete data $\mathbf{x}$ are not observed directly, but only by means of $\mathbf{y}$, where $\mathbf{y} = \mathbf{y}(\mathbf{x})$, and $\mathbf{y}(\mathbf{x})$ is a many-to-one mapping. An observation $\mathbf{y}$ determines a subset of $\mathcal{X}$, which is denoted as $\mathcal{X}(\mathbf{y})$. Figure 17.2 illustrates the mapping.

The pdf of the complete data is $f_X(\mathbf{x} \mid \boldsymbol{\theta}) = f(\mathbf{x} \mid \boldsymbol{\theta})$, where $\boldsymbol{\theta} \in \Theta \subset \mathbb{R}^r$ is the set of parameters of the density. The pdf (or pmf) f is assumed to be a continuous function of $\boldsymbol{\theta}$ and appropriately differentiable. The ML estimate of $\boldsymbol{\theta}$ is assumed to lie within the region Θ. The pdf of the incomplete data is

$$g(\mathbf{y} \mid \boldsymbol{\theta}) = \int_{\mathcal{X}(\mathbf{y})} f(\mathbf{x} \mid \boldsymbol{\theta}) \, d\mathbf{x}.$$

Let

$$l_y(\boldsymbol{\theta}) = g(\mathbf{y} \mid \boldsymbol{\theta})$$

denote the likelihood function, and let

$$L_y(\boldsymbol{\theta}) = \log g(\mathbf{y} \mid \boldsymbol{\theta})$$

denote the log-likelihood function.

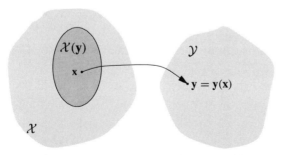

Figure 17.2: Illustration of a many-to-one mapping from $\mathcal{X}$ to $\mathcal{Y}$. The point $\mathbf{y}$ is the image of $\mathbf{x}$, and the set $\mathcal{X}(\mathbf{y})$ is the inverse map of $\mathbf{y}$.

The basic idea behind the EM algorithm is that we would like to find $\boldsymbol{\theta}$ to maximize $\log f(\mathbf{x} \mid \boldsymbol{\theta})$, but we do not have the data $\mathbf{x}$ to compute the log-likelihood. So, instead, we maximize the expectation of $\log f(\mathbf{x} \mid \boldsymbol{\theta})$ given the data $\mathbf{y}$ and our current estimate of $\boldsymbol{\theta}$. This can be accomplished in two steps. Let $\boldsymbol{\theta}^{[k]}$ be our estimate of the parameters at the kth iteration.

E-step. Compute

$$Q(\boldsymbol{\theta} \mid \boldsymbol{\theta}^{[k]}) = E[\log f(\mathbf{x} \mid \boldsymbol{\theta}) \mid \mathbf{y}, \boldsymbol{\theta}^{[k]}]. \tag{17.6}$$

It is important to distinguish between the first and second arguments of the Q functions. The second argument is a conditioning argument to the expectation and is regarded as fixed and known at every E-step. The first argument conditions the likelihood of the complete data.

M-step. Let $\boldsymbol{\theta}^{[k+1]}$ be that value of $\boldsymbol{\theta}$ that maximizes $Q(\boldsymbol{\theta} \mid \boldsymbol{\theta}^{[k]})$:

$$\boldsymbol{\theta}^{[k+1]} = \arg\max_{\boldsymbol{\theta}} Q(\boldsymbol{\theta} \mid \boldsymbol{\theta}^{[k]}). \tag{17.7}$$

It is important to note that the maximization is with respect to the first argument of the Q function, the conditioner of the complete-data likelihood.

The EM algorithm consists of choosing an initial $\boldsymbol{\theta}^{[k]}$, then performing the E-step and the M-step successively until convergence. Convergence may be determined by observing when the parameters stop changing; for example, when $\|\boldsymbol{\theta}^{[k]} - \boldsymbol{\theta}^{[k-1]}\| < \epsilon$ for some ϵ and some appropriate distance measure $\|\cdot\|$.

The general form of the EM algorithm as stated in (17.6) and (17.7) may be specialized and simplified somewhat by restriction to distributions in the *exponential family* (see section 10.6). These are pdfs (or pmfs) of the form

$$f(\mathbf{x} \mid \boldsymbol{\theta}) = a(\mathbf{x})c(\boldsymbol{\theta}) \exp[\boldsymbol{\pi}(\boldsymbol{\theta})^T \mathbf{t}(\mathbf{x})], \tag{17.8}$$

where $\boldsymbol{\theta}$ is a vector of parameters for the family, and where

$$\mathbf{t}(\mathbf{x}) = [t_1(\mathbf{x}), \ldots, t_q(\mathbf{x})]^T$$

is the vector of sufficient statistics for $\boldsymbol{\theta}$. For exponential families, the E-step can be written as

$$Q(\boldsymbol{\theta} \mid \boldsymbol{\theta}^{[k]}) = E[\log a(\mathbf{x}) \mid \mathbf{y}, \boldsymbol{\theta}^{[k]}] + \boldsymbol{\pi}(\boldsymbol{\theta})^T E[\mathbf{t}(\mathbf{x}) \mid \mathbf{y}, \boldsymbol{\theta}^{[k]}] + \log c(\boldsymbol{\theta}).$$

Let $\mathbf{t}^{[k+1]} = E[\mathbf{t}(\mathbf{x}) \mid \mathbf{y}, \boldsymbol{\theta}^{[k]}]$. Because a conditional expectation is an estimator, $\mathbf{t}^{[k+1]}$ is an estimate of the sufficient statistic.[1] In light of the fact that the M-step will be maximizing

$$E[\log a(\mathbf{x}) \mid \mathbf{y}, \boldsymbol{\theta}^{[k]}] + \boldsymbol{\pi}(\boldsymbol{\theta})^T \mathbf{t}^{[k+1]} + \log c(\boldsymbol{\theta})$$

with respect to $\boldsymbol{\theta}$ and that $E[\log a(\mathbf{x}) \mid \mathbf{y}, \boldsymbol{\theta}^{[k]}]$ does not depend upon $\boldsymbol{\theta}$, it is sufficient to write the following.

E-step. Compute

$$\mathbf{t}^{[k+1]} = E[\mathbf{t}(\mathbf{x}) \mid \mathbf{y}, \boldsymbol{\theta}^{[k]}]. \tag{17.9}$$

---

[1] The EM algorithm is sometimes called the estimation–maximization algorithm because, for exponential families, the first step is an estimator. It has also been called the expectation–modification algorithm [266].

M-step. Compute
$$\theta^{[k+1]} = \arg\max_{\theta} \pi(\theta)^T \mathbf{t}^{[k+1]} + \log c(\theta). \quad (17.10)$$

The EM algorithm may be diagrammed starting from an initial guess of the parameter $\theta^{[0]}$ as follows:

$$\theta^{[0]} \xrightarrow{\text{E-step}} \mathbf{t}^{[1]} \xrightarrow{\text{M-step}} \theta^{[1]} \xrightarrow{\text{E-step}} \mathbf{t}^{[2]} \xrightarrow{\text{M-step}} \cdots.$$

The EM algorithm has the advantage of being simple, at least in principle; actually computing the expectations and performing the maximizations may be computationally (or intellectually!) taxing. Unlike other optimization techniques, it does not require the computation of gradients or Hessians, nor is it necessary to worry about setting step-size parameters, as algorithms such as gradient descent require.

17.3 Convergence of the EM algorithm

For every iterative algorithm, the question of convergence must be addressed: does the algorithm come finally to a solution, or does it iterate *ad nauseam*, ever learning but never coming to a knowledge of the truth? For the EM algorithm, convergence may be stated simply: at every iteration of the algorithm, a value of the parameter is computed so that the likelihood function of $\mathbf{y}$ does not decrease. That is, at every iteration, the estimated parameter provides an increase in the likelihood function until a local maximum is achieved, at which point the likelihood function cannot increase (but will not decrease).

We present a proof of this general concept as follows. Let

$$k(\mathbf{x} \mid \mathbf{y}, \theta) = \frac{f(\mathbf{x} \mid \theta)}{g(\mathbf{y} \mid \theta)}, \quad (17.11)$$

and note that $k(\mathbf{x} \mid \mathbf{y}, \theta)$ may be interpreted as a conditional density. Then the log-likelihood function $L_y(\theta) = \log g(\mathbf{y} \mid \theta)$ may be written

$$L_y(\theta) = \log f(\mathbf{x} \mid \theta) - \log k(\mathbf{x} \mid \mathbf{y}, \theta).$$

Define

$$H(\theta' \mid \theta) = E[\log k(\mathbf{x} \mid \mathbf{y}, \theta') \mid \mathbf{y}, \theta].$$

Let $M: \theta^{[k]} \to \theta^{[k+1]}$ represent the mapping defined by the EM algorithm in (17.6) and (17.7), so that $\theta^{[k+1]} = M(\theta^{[k]})$.

Theorem 17.1 $L_y(M(\theta^{[k+1]})) \geq L_y(\theta)$, with equality if and only if

$$Q(M(\theta) \mid \theta) = Q(\theta \mid \theta)$$

and

$$k(\mathbf{x} \mid \mathbf{y}, M(\theta)) = k(\mathbf{x} \mid \mathbf{y}, \theta).$$

That is, the likelihood function increases at each iteration of the EM algorithm, until the conditions for equality are satisfied and a fixed point of the iteration is reached. If θ^* is an ML parameter estimate, so that $L_y(\theta^*) \geq L_y(\theta)$ for all $\theta \in \Theta$, then $L_y(M(\theta^*)) = L_y(\theta^*)$. In other words, ML estimates are fixed points of the EM algorithm. Since the likelihood function is bounded (for distributions of practical interest), the sequence of parameter estimates $\theta^{[0]}, \theta^{[1]}, \ldots, \theta^{[k]}$ yields a bounded nondecreasing sequence $L_y(\theta^{[0]}) \leq L_y(\theta^{[1]}) \leq \cdots \leq L_y(\theta^{[k]})$, which must converge as $k \to \infty$.

Proof

$$L_y(M(\theta)) - L_y(\theta) = Q(M(\theta) \mid \theta) - Q(\theta \mid \theta) + H(\theta \mid \theta) - H(M(\theta) \mid \theta). \quad (17.12)$$

By the definition of the M-step, it must be the case that

$$Q(M(\theta) \mid \theta) \geq Q(\theta \mid \theta)$$

for every $\theta \in \Theta$. For any pair $(\theta', \theta) \in \Theta \times \Theta$, it is the case that

$$H(\theta' \mid \theta) \leq H(\theta \mid \theta).$$

This can be proven with Jensen's inequality (see section A.3), which states: If $f(x)$ is a concave function, then $E[f(x)] \leq f(E[x])$, with equality if and only if x is constant (nonrandom). This inequality may be employed as follows.

$$H(\theta' \mid \theta) - H(\theta \mid \theta) = E\left[\log \frac{k(\mathbf{x} \mid \mathbf{y}, \theta')}{k(\mathbf{x} \mid \mathbf{y}, \theta)} \,\bigg|\, \mathbf{y}, \theta\right]$$

$$\leq \log E\left[\log \frac{k(\mathbf{x} \mid \mathbf{y}, \theta')}{k(\mathbf{x} \mid \mathbf{y}, \theta)} \,\bigg|\, \mathbf{y}, \theta\right] \quad (17.13)$$

$$= \log \int_{\mathcal{X}} \frac{k(\mathbf{x} \mid \mathbf{y}, \theta')}{k(\mathbf{x} \mid \mathbf{y}, \theta)} k(\mathbf{x} \mid \mathbf{y}, \theta) \, d\mathbf{x} \quad (17.14)$$

$$= \log \int_{\mathcal{X}} k(\mathbf{x} \mid \mathbf{y}, \theta') \, d\mathbf{x}$$

$$= 0.$$

Equation (17.13) follows from Jensen's inequality, with $f(x) = \log(x)$, which is concave; and (17.14) is true since $k(\mathbf{x} \mid \mathbf{y}, \theta)$ is a conditional density.

Examination of (17.12) in light of the M-step and the conditions for equality in Jensen's inequality reveals that equality in the theorem can only hold for the stated conditions. □

The theorem falls short of proving that the fixed points of the EM algorithm are in fact ML estimates. The latter is true, under rather general conditions, but the proof is somewhat involved and is not presented here (see [375]).

Despite the convergence attested in theorem 17.1, there is no guarantee that the convergence will be to a *global* maximum. For likelihood functions with multiple maxima, convergence will be to a local maximum that depends on the initial starting point $\theta^{[0]}$.

17.3.1 Convergence rate: Some generalizations

The convergence rate of the EM algorithm is also of interest. Based on mathematical and empirical examinations, it has been determined that the convergence rate is usually slower than the quadratic convergence typically available with a Newton-type method [273]. However, as observed by Dempster [72], the convergence near the maximum (at least for exponential families) depends upon the eigenvalues of the Hessian of the update function M, so that rapid convergence may be possible. More precisely [72], let

$$F_y = -\nabla^2 L_y(\hat{\theta})$$

and

$$F_x = E[-\nabla^2 L_x(\hat{\theta}) \mid \mathbf{y}, \hat{\theta}].$$

Then the asymptotic rate of convergence is proportional to the largest eigenvalue of $(F_x - F_y)F_y^{-1}$. It has also been shown that the monotic rate of convergence is equal to $\|F_x^{-1/2}(F_x - F_y)F_x^{-1/2}\|$, where the norm is the matrix l_2 norm [135]. On this basis, we observe that as F_x approaches F_y (i.e., as the complete data X becomes less informative), the speed of convergence of the EM algorithm increases. However, as a general rule, the M-step becomes more difficult as X becomes less informative. This tradeoff can be eased by using "hidden data" sets [87] that are less informative than complete data sets, and that can vary at each iteration. This algorithm carries the acronym SAGE—"space alternating expectation maximization"—and is characterized by updating only small groups of the parameters at each iteration. This algorithm is similar to the "expectation conditional maximization either" (ECME) algorithm [206]. A further generalization appears in the "alternating expectation conditional maximization" (AECM) algorithm [225].

Regardless of the specific form of the algorithm employed, there are general advantages to EM algorithms over Newton-type algorithms. In the first place, no Hessian needs to be computed. Also, there is no chance of "overshooting" the target or diverging from the maximum. The EM algorithm is guaranteed to be stable and to converge to an ML estimate. Further discussion of convergence appears in [375, 39].

Example applications of the EM algorithm

We now present several applications of the EM algorithm to problems of signal-processing interest, to illustrate the computations required in the steps of the algorithm. The diversity of the applications illustrates the breadth of the EM algorithm's utility. The example of section 17.5 and the introductory example of section 17.1 illustrate the case in which the densities are members of the exponential family. The examples of sections 17.6, 17.7, and 17.8 treat densities that are not in the exponential family, so the more general statement of the EM algorithm must be applied. The focus of the examples is on the algorithm; assumptions and details of the systems involved are therefore not presented. The interested reader is encouraged to examine the references for details.

17.4 Introductory example, revisited

The multinomial distribution of the introductory example is a member of the exponential family with $\mathbf{t}(\mathbf{x}) = \mathbf{x}$. The E-step consists simply of estimating the underlying data, given the current estimate and the accessible data. This is followed by a straightforward maximization.

17.5 Emission computed tomography (ECT) image reconstruction

In ECT [309], tissues within the body are stimulated to emit photons by means of a radioactive tracer that is administered to the body. These emitted photons are measured by detectors surrounding the tissue. For purposes of computation, the body is divided into B boxes. We model the photons generated in each box by a random variable $N(b)$, with a particular realization at box b having $n(b)$ photons, where $b = 1, 2, \ldots, B$. There are D detectors around the body, and we model the measurement of the detector as a random variable $Y(d)$, with a particular realization denoted by $y(d)$, where $d = 1, 2, \ldots, D$. The measurement configuration is diagrammed in figure 17.3. Let $\mathbf{y} = [y(1), y(2), \ldots, y(D)]^T$ denote a vector of observations.

In a commonly assumed model, the generation of the photons from box b can be described as a Poisson process $N(b)$ with mean $\lambda(b)$. We use the notation $f(n(b) \mid \lambda(b))$

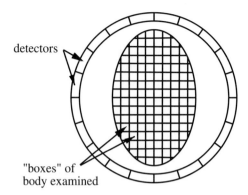

Figure 17.3: Representation of emission tomography. There are B boxes in the body and D detectors surrounding the body.

to indicate the pmf, that is,

$$f(n(b) \mid \lambda(b)) = P(N(b) = n \mid \lambda(b)) = e^{-\lambda(b)} \frac{\lambda(b)^n}{n!}.$$

The parameter $\lambda(b)$ is a function of the uptake of radioactive tracer in the tissue, so that by estimating the parameters $\lambda(b)$ in each box it is possible to construct an image of the body on the basis of the tracer density in the tissue. The boxes are assumed (in the simplest development) to be independent. Let the set of unknown parameters be denoted by

$$\boldsymbol{\lambda} = \{\lambda(1), \lambda(2), \ldots, \lambda(B)\}.$$

A photon emission from box b is detected in detector d with a probability that we denote as $p(d \mid b)$. We assume in this development that all emitted photons are detected by some detector, so that

$$\sum_{d=1}^{D} p(d \mid b) = 1 \quad \forall\, b. \tag{17.15}$$

The transition probability $p(d \mid b)$ depends upon the geometry of the detectors relative to the body under examination, and the physics of the excitation process. A description of a fairly realistic model of $p(d \mid b)$ for emission tomography is given in [309]. The simulations presented in the following present a simplified model.

The detector variables $y(d)$ are Poisson distributed,

$$f(y(d) \mid \lambda(d)) = P(Y(d) = y(d)) = e^{-\lambda(d)} \frac{\lambda(d)^{y(d)}}{y(d)!},$$

where

$$\lambda(d) = E[y(d)] = \sum_{b=1}^{B} \lambda(b) p(d \mid b).$$

Let $x(b, d)$ be the sample of the number of emissions from box b detected in detector d, an observation of the random variable $X(b, d)$, and let $\mathbf{x} = \{x(b, d), b = 1, \ldots, B, d = 1, \ldots, D\}$. For any given set of detector data $\{y(d)\}$, there are many different ways that the photons could have been generated. There is thus a many-to-one mapping from $x(b, d)$ to $y(d)$, and $\mathbf{x}$ constitutes the complete data set. Each variable of the complete data $x(b, d)$ is Poisson with mean

$$\lambda(b, d) = \lambda(b) p(d \mid b). \tag{17.16}$$

17.5 Emission Computed Tomography (ECT) Image Reconstruction

Assuming that each box generates photons independently of every other box and that the detectors operate independently, the likelihood function of the complete data is

$$l_x(\lambda) = f(\mathbf{x} \mid \lambda) = \prod_{b=1}^{B} \prod_{d=1}^{D} e^{-\lambda(b,d)} \frac{\lambda(b,d)^{x(b,d)}}{x(b,d)!} \qquad (17.17)$$

and, using (17.16), the log-likelihood function is

$$L_x(\lambda) = \log l_x(\lambda)$$
$$= \sum_{b=1}^{B} \sum_{d=1}^{D} -\lambda(b) p(d \mid b) + x(b,d) \log \lambda(b) + x(b,d) \log p(d \mid b) - \log x(b,d)!. \qquad (17.18)$$

Application of the EM algorithm is now straightforward. What is assumed to be known is the set of detector measurements $\mathbf{y}$, which are Poisson distributed, and the transition probabilities $p(d \mid b)$. Poisson distributions are in the exponential family. The sufficient statistics for the distribution are the data, $\mathbf{t}(\mathbf{x}) = \mathbf{x}$. Let $\lambda^{[k]}$ be the estimate of the parameters at the kth iteration and let $x^{[k]}(b,d)$ be the estimate of the complete data. For the E-step, compute

$$x^{[k+1]}(b,d) = E[x(b,d) \mid \mathbf{y}, \lambda^{[k]}] = E[x(b,d) \mid y(d), \lambda^{[k]}],$$

where the latter equality follows since each box is independent. Since $x(b,d)$ is Poisson with mean $\lambda^{[k]}(b,d)$ and $y(d) = \sum_{b=1}^{B} x(b,d)$ is Poisson with mean $\lambda^{[k]}(d) = \sum_{b=1}^{B} \lambda^{[k]}(b,d)$, the conditional expectation may be computed (using techniques similar to those in appendix F)

$$x^{[k+1]}(b,d) = \frac{y(d) \lambda^{[k]}(b,d)}{\sum_{b'=1}^{B} \lambda^{[k]}(b',d)}, \quad b = 1, 2, \ldots, B, \quad d = 1, 2, \ldots, D. \qquad (17.19)$$

For the M-step, $x^{[k+1]}(b,d)$ is used in the likelihood function (17.18), which is maximized with respect to $\lambda(b)$. We will compute (to avoid confusing indices) the derivative with respect to $\lambda(\beta)$, and equate the result to zero:

$$0 = \frac{\partial}{\partial \lambda(\beta)} \sum_{b=1}^{B} \sum_{d=1}^{D} -\lambda(b) p(d \mid b) + x^{[k+1]}(b,d) \log \lambda(b)$$
$$+ x^{[k+1]}(b,d) \log p(d \mid b) - \log x^{[k+1]}(b,d)!$$
$$\rightarrow \lambda^{[k+1]}(b) = \sum_{d=1}^{D} x^{[k+1]}(b,d), \qquad (17.20)$$

where (17.15) has been used.

Equations (17.19) and (17.20) may be iterated until convergence. The overhead of storing $x^{[k+1]}(b,d)$ at each iteration may be eliminated by substituting (17.19) into (17.20) using (17.16), much as was done in the introductory example. This gives

$$\lambda^{[k+1]}(b) = \lambda^{[k]}(b) \sum_{d=1}^{D} \frac{y(d) p(d \mid b)}{\sum_{b'=1}^{B} \lambda^{[k]}(b') p(d \mid b')}. \qquad (17.21)$$

Example 17.5.1 In this example, we demonstrate the principles of tomographic reconstruction with an artificial data set. Both the detector geometry and the conditional probability model are

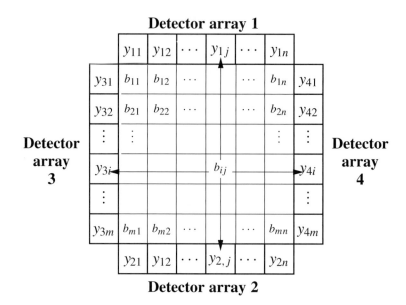

Figure 17.4: Detector arrangement for tomographic reconstruction example.

significantly simplified from physical measurement systems for this example. Consider the grid shown in figure 17.4. Each box corresponds to a pixel. Around the periphery of the box are detectors, with a detector in each row and column. For purposes of simulation, it is convenient to represent four arrays of detectors, where array 1 represents the data across the top of the image, array 2 represents the data along the bottom of the image, and array 3 and array 4 represent the left-hand and right-hand side of the image, respectively. The detectors are numbered by banks, so that y_{ij}, $i = 1, 2, 3, 4$, $j = 1, 2, \ldots, n_i$, represents the jth detector in detector bank i, and $n_1 + n_2 + n_3 + n_4 = D$, the total number of detectors. The detectors are constrained so that they measure a projection of the data *only* in their own row or column. Thus, each pixel projects onto exactly four detectors, as suggested by the arrows in the diagram. As a simple measurement model, we assume that the probability $p(d \mid b)$ is linearly proportional to the distance between the detector d and the pixel b. By this scheme, an $m \times n$ pixel array projects onto $2(m + n)$ detectors. The problem is highly underconstrained, so that perfect image reconstruction should not be expected. For the 60×60 image below, only 240 measurements are available, from which the original image is to be reconstructed as closely as possible.

Figure 17.5(a) shows an artificially generated image. Figure 17.5(c) shows the array of detector outputs for the bank of detectors along the top of the image, and figure 17.5(d) shows the array of detector outputs along the bottom of the output. (The noise-like variation is due to the fact that the detectors measure samples of Poisson-distributed random variables, not the actual mean values of the random variables.)

Figure 17.5(b) shows the result of applying the EM algorithm embodied in (17.21) on the observed data for two iterations. The reconstruction conveys the general characteristics of the data, but due to the horizontal/vertical measurement symmetry imposed by the structure of the detectors, there are significant striping artifacts. (Other measurement geometries with more radial symmetry would reduce these artifacts somewhat.)

The code in algorithm 17.2 demonstrates the processing that takes place. The algorithm expects an image in the variable `im`. Based upon the geometry just mentioned, the transfer probabilities $p(d \mid b)$ are computed, then the detector outputs. This is followed by the reconstruction algorithm.

Also shown in algorithm 17.2 is the code for the Poisson random number generator used in the simulation. □

17.6 Active Noise Cancellation (ANC)

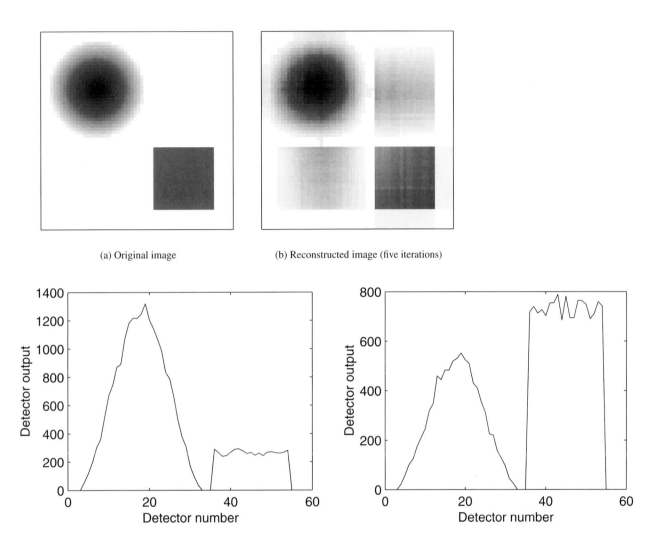

(a) Original image

(b) Reconstructed image (five iterations)

(c) Detector array 1 outputs

(d) Detector array 2 outputs

Figure 17.5: Example emission tomography reconstruction.

Algorithm 17.2 Simulation and reconstruction of emission tomography
File: `testet.m`
          `et1.m`
          `poisson.m`
          `initpoisson.m`

17.6 Active noise cancellation (ANC)

Active noise cancellation is accomplished by measuring a noise signal and then using a speaker driven out of phase with the noise to cancel it. In many traditional ANC techniques, two microphones are used in conjunction with an adaptive filter to provide cancellation

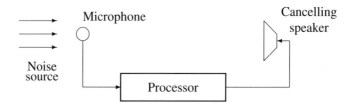

Figure 17.6: Single-microphone ANC system.

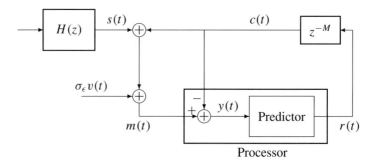

Figure 17.7: Processor block diagram of the ANC system.

(see, for example, [368, 324]). Using the EM algorithm, ANC may be achieved with only one microphone [84]. The physical system is depicted in figure 17.6, with a block diagram for the ANC in figure 17.7.

The signal to be canceled is modeled as the output of an all-pole filter,

$$s[t] = -\sum_{k=1}^{p} a_k s[t-k] + \sigma_s u[t]$$
$$= -\mathbf{s}_{p-1}^T[t-1]\mathbf{a} + \sigma_s u[t],$$

where

$$\mathbf{s}_p[t] = [s[t-p], s[t-p+1], \ldots, s[t]]^T,$$
$$\mathbf{a} = [a_{p-1}, a_{p-2}, \ldots, a_1]^T,$$

and $u[t]$ is a discrete-time, white, unit-variance, zero-mean Gaussian process. The signal $r[t]$ is generated by the processor and corresponds to the input of the speaker; the delay z^{-M} is the delay from the speaker to the microphone. The signal $\sigma_\epsilon v[t]$ models the measurement error at the microphone. According to figure 17.7, the input to the processor can be written

$$y[t] = s[t] + \sigma_\epsilon v[t];$$

we assume that $v[t]$ is a unit-variance, white Gaussian process. The set of unknown parameters is

$$\boldsymbol{\theta} = \left[\mathbf{a}^T, \sigma_s^2, \sigma_\epsilon^2\right]^T.$$

A block of N measurements is used for processing. The observed data vector is

$$\mathbf{y} = [y[1], y[2], \ldots, y[N]]^T;$$

these observations span a set of autoregressive samples given by

$$\mathbf{s} = [s[1-p], s[2-p], \ldots, s[N]]^T.$$

17.6 Active Noise Cancellation (ANC)

The complete data set is $\mathbf{x} = [\mathbf{y}^T, \mathbf{s}^T]^T$. If we knew $\mathbf{s}$, estimation of the AR parameters would be straightforward using familiar spectrum estimation techniques.

The likelihood function for the complete data is

$$f(\mathbf{x} \mid \boldsymbol{\theta}) = f(\mathbf{y}, \mathbf{s} \mid \boldsymbol{\theta}) = f(\mathbf{y} \mid \mathbf{s}, \boldsymbol{\theta}) f(\mathbf{s} \mid \boldsymbol{\theta}). \qquad (17.22)$$

The conditioning step provides important leverage because it is straightforward to determine $f(\mathbf{y} \mid \mathbf{s}, \boldsymbol{\theta})$. The conditioning can be further broken down as

$$f(\mathbf{x} \mid \boldsymbol{\theta}) = f(\mathbf{y} \mid \mathbf{s}, \boldsymbol{\theta}) f(s[1], s[2], \ldots, s[N] \mid \mathbf{s}_{p-1}[0], \boldsymbol{\theta}) f(\mathbf{s}_{p-1}[0] \mid \boldsymbol{\theta}).$$

Then

$$f(\mathbf{y} \mid \mathbf{s}, \boldsymbol{\theta}) = \frac{1}{(2\pi \sigma_\epsilon^2)^{N/2}} \exp\left[-\frac{1}{2\sigma_\epsilon^2} \sum_{t=1}^{N} (y[t] - s[t])^2\right]$$

and

$$f(s[1], s[2], \ldots, s[N] \mid \mathbf{s}_{p-1}[0], \boldsymbol{\theta}) = \frac{1}{(2\pi \sigma_s^2)^{(N-p)/2}} \exp\left[-\frac{1}{2\sigma_s^2} \sum_{t=1}^{N} (s[t] + \mathbf{a}^T \mathbf{s}_{p-1}[t])^2\right]$$

(see [174, page 187]). The E-step may be computed as

$$E[\log f(\mathbf{x} \mid \boldsymbol{\theta}) \mid \mathbf{y}, \boldsymbol{\theta}^{[k]}] = \log f(\mathbf{s}_{p-1}[0] \mid \boldsymbol{\theta}) - N \log \sigma_s - N \log \sigma_\epsilon$$

$$- \frac{1}{2\sigma_s^2} \sum_{t=1}^{N} [E[s^2[t] \mid \mathbf{y}, \boldsymbol{\theta}^{[k]}] + 2\mathbf{a}^T E[\mathbf{s}_{p-1}[t-1]s[t] \mid \mathbf{y}, \boldsymbol{\theta}]$$

$$+ \mathbf{a}^T E[\mathbf{s}_{p-1}[t-1]\mathbf{s}_{p-1}[t-1] \mid \mathbf{y}, \boldsymbol{\theta}^{[k]}]\mathbf{a}]$$

$$- \frac{1}{2\sigma_\epsilon^2} \sum_{t=1}^{N} [y^2[t] - 2y[t]E[s[t] \mid \mathbf{y}, \boldsymbol{\theta}^{[k]}] + E[s^2[t] \mid \mathbf{y}, \boldsymbol{\theta}^{[k]}]].$$

Taking the gradient with respect to $\mathbf{a}$, and derivatives with respect to σ_s and σ_ϵ to maximize, yields

$$\mathbf{a}^{[k+1]} = -\left[\sum_{t=1}^{N} E\left[\mathbf{s}_{p-1}[t-1]\mathbf{s}_{p-1}^T[t-1] \mid \mathbf{y}, \boldsymbol{\theta}^{[k]}\right]\right]^{-1} \sum_{t=1}^{N} E[\mathbf{s}_{p-1}[t-1]s[t] \mid \mathbf{y}, \boldsymbol{\theta}^{[k]}], \qquad (17.23)$$

$$(\sigma_s^2)^{[k+1]} = \frac{1}{N} \sum_{t=1}^{N} E[s^2[t] \mid \mathbf{y}, \boldsymbol{\theta}^{[k]}] + (\mathbf{a}^{[k]})^T \sum_{t=1}^{N} E[\mathbf{s}_{p-1}[t-1]s[t] \mid \mathbf{y}, \boldsymbol{\theta}^{[k]}], \qquad (17.24)$$

$$(\sigma_\epsilon^2)^{[k+1]} = \frac{1}{N} \sum_{t=1}^{N} [y^2[t] - 2y[t]E[s[t] \mid \mathbf{y}, \boldsymbol{\theta}^{[k]}] + E[s^2[t] \mid \mathbf{y}, \boldsymbol{\theta}^{[k]}]]. \qquad (17.25)$$

The expectations in (17.23), (17.24), and (17.25) are first and second moments of Gaussians, conditioned upon observation, which may be computed using a Kalman smoother. The variable $\mathbf{s}_p$ may be put into state-space form by

$$\mathbf{s}_p[t] = \Phi \mathbf{s}_p[t-1] + \mathbf{g}u[t],$$
$$y[t] = \mathbf{h}^T \mathbf{s}_p[t] + \sigma_\epsilon v[t],$$

where

$$\Phi = \begin{bmatrix} 0 & I \\ 0 & -\mathbf{a}^T \end{bmatrix},$$

$$\mathbf{g}^T = [0, 0, \ldots, 0, \sigma_s],$$

and

$$\mathbf{h}^T = [0, \ldots, 0, 1].$$

With an estimate of the parameters, the canceling signal $c[t+M]$ is obtained by estimating $\mathbf{s}[t+M]$ using $E[\mathbf{s}_p[t] \mid \mathbf{y}, \boldsymbol{\theta}]$ and $\boldsymbol{\theta}^{[t]}$.

17.7 Hidden Markov models

The hidden Markov model is a stochastic model of a process that exhibits features that change over time. It has been applied in a broad variety of sequential pattern recognition problems such as speech recognition and handwriting recognition [266, 314]. Detailed descriptions of HMMs and their application are given in [265, 68, 266]; see also the introduction in section 1.7. In this section we introduce one method of training the parameters of an HMM, known as the *forward–backward* algorithm or the *Baum–Welch* method. Another method of training the HMM, based upon the Viterbi algorithm, is presented in section 19.6.

A Markov model is a stochastic model of a system that is capable of being in a finite number of states $\{1, 2, \ldots, S\}$. The state of the system is a random variable $S[t]$, governed by an underlying Markov process. The particular value of the state at a time t is denoted by $s[t]$. (We will use s to indicate the state here, unlike in section 1.7, reserving the symbol $\mathbf{x}$ to represent the complete data). The probability of transition from a state at the current (discrete) time t to any other state at time $t+1$ depends only on the current state, and not on any prior states:

$$P(S[t+1] = i \mid S[t] = j, S[t-1] = j_1, \ldots) = P(S[t+1] = i \mid S[t] = j).$$

It is common to express the transition probabilities as a matrix A with elements $P(s[t+1] = i \mid s[t] = j) = a_{i,j}$,

$$A = \begin{bmatrix} P(1 \mid 1) & P(1 \mid 2) & \cdots & P(1 \mid S) \\ P(2 \mid 1) & P(2 \mid 2) & \cdots & P(2 \mid S) \\ \vdots & & & \\ P(S \mid 1) & P(S \mid 2) & \cdots & P(S \mid S) \end{bmatrix},$$

where $P(i \mid j)$ is used as an abbreviation for $P(S[t+1] = i \mid S[t] = j)$. The initial state $S[1]$ is chosen according to the probability

$$\boldsymbol{\pi} = [P(S[1] = 1), \ldots, P(S[1] = S)]^T = [\pi_1, \ldots, \pi_S]^T.$$

In each state at time t, a (possibly vector) random variable $\mathbf{Y}[t] \in \mathbb{R}^m$ is selected according to the density (or pmf) $f_{Y \mid S}(\mathbf{y}[t] \mid S[t] = i)$. The variable $\mathbf{y}[t]$ is observed as an instance of $\mathbf{Y}[t]$, but the underlying state is not, hence the name *hidden* Markov model. We will denote the density by which $\mathbf{Y}[t]$ is selected when in state s by $f(\mathbf{y}[t] \mid s)$ or, more simply, by $f_s(\mathbf{y}[t])$, and denote the set of densities $\{f_1, f_2, \ldots, f_S\}$ by $f_{\{s\}}$. The probability of a sequence, such as $f(\mathbf{y}[1], \mathbf{y}[2] \mid s[1], s[2])$, we will denote as $f(\mathbf{y} \mid \mathbf{s})$.

As introduced in section 1.7, the HMM operates as follows. An initial state $S[1] = s[1]$ is chosen according to the probability law $\boldsymbol{\pi}$, and an output $\mathbf{y}[1]$ is generated from that state. At succeeding times $t = 2, 3 \ldots$, the state $S[t]$ is selected according to the Markov probability transition matrix A. At each time instant $t = 2, 3, \ldots$, an output $\mathbf{Y}[t] = \mathbf{y}[t]$

17.7 Hidden Markov Models

is generated according to $f_{s[t]}$. Then a new state is chosen, and the process continues. It is common to assume that the distributions in each state have the same form, but not necessarily the same parameters—for example discrete or Gaussian—so that the problem of finding parameters for each state is identical.

The triple $\mathcal{M} = (A, \pi, f_{\{s\}})$ defines the HMM. (In section 1.7, the densities were discrete, and the family of densities was represented by the matrix C.) The parameters of the HMM are the initial state probability π, the components of the state transition matrix A, and the parameters of the densities in each state. Let the elements of the HMM be parameterized by θ, that is, there is a mapping $\theta \to (A(\theta), \pi(\theta), f_{\{s\}}(\cdot \mid \theta))$. The mapping is assumed to be appropriately smooth. In training the HMM, the densities are viewed as being parameterized by θ. For example, the output generation probability $f_s(\mathbf{y}) = f(\mathbf{y} \mid S = s)$, viewed as a function of the parameters, is written as $f_s(\mathbf{y} \mid \theta)$, and the state transition probability $a_{i,j}$ is written as $a_{i,j}(\theta)$. The parameter estimation problem for an HMM is this: given a sequence of observations, $\mathbf{y} = \{\mathbf{y}^T[1], \mathbf{y}^T[2], \ldots, \mathbf{y}^T[T]\}$, determine the parameter θ which maximizes the likelihood function

$$l_y(\theta) = f(\mathbf{y}[1], \mathbf{y}[2], \ldots, \mathbf{y}[T] \mid \theta)$$

$$= \sum_{s[1],\ldots,s[T]=1}^{S} \pi_{s[1]}(\theta) f_{s[1]}(\mathbf{y}[1] \mid \theta) a_{s[2]s[1]}(\theta) f_{s[2]}(\mathbf{y}[2] \mid \theta) a_{s[3],s[2]}(\theta)$$

$$\times f_{s[3]}(\mathbf{y}[3] \mid \theta) \cdots a_{s[T],s[T-1]}(\theta) f_{s[T]}(\mathbf{y}[T] \mid \theta). \quad (17.26)$$

From the complicated structure of (17.26), it is clear that this is a difficult maximization problem. The EM algorithm, however, provides the leverage necessary.

We introduce some notation that is useful in what follows. Let $\mathbf{y}_{t_1}^{t_2}$ represent the sequence

$$\mathbf{y}_{t_1}^{t_2} = \{\mathbf{y}[t_1], \ldots, \mathbf{y}[t_2]\}.$$

Then the *partial forward sequence* of observations at time t is

$$\mathbf{y}_1^t = \{\mathbf{y}[1], \mathbf{y}[2], \ldots, \mathbf{y}[t]\}$$

and the *partial backward sequence* of observations at time t is

$$\mathbf{y}_{t+1}^T = \{\mathbf{y}[t+1], \mathbf{y}[t+2], \ldots, \mathbf{y}[T]\}.$$

The notation $\mathbf{y}$ can also be written as

$$\mathbf{y} = \mathbf{y}_1^T.$$

Let $\mathbf{s} = \{s[1], s[2], \ldots, s[T]\}$ be the sequence of the (unobserved) states. The complete data can be expressed as $\mathbf{x} = (\mathbf{y}, \mathbf{s})$. The pdf of the complete data can be written as

$$f(\mathbf{x} \mid \theta) = f(\mathbf{y}, \mathbf{s} \mid \theta) = f(\mathbf{y} \mid \mathbf{s}, \theta) f(\mathbf{s} \mid \theta). \quad (17.27)$$

This factorization, with the pdf of the observation conditioned upon the unknown state sequence and the distribution of the unknown state sequence, is the key in the application of the EM algorithm.

Because of the Markov structure of the state, the state probabilities in (17.27) may be written

$$f(\mathbf{s} \mid \theta) = \pi_{s[1]}(\theta) \prod_{t=2}^{T} a_{s[t],s[t-1]}(\theta), \quad (17.28)$$

where the explicit dependence of the initial probabilities and the state transition probability upon the parameter θ is indicated. The logarithm of (17.28), used in what follows, is

$$\log f(\mathbf{s} \mid \theta) = \log \pi_{s[1]}(\theta) + \sum_{t=2}^{T} \log a_{s[t],s[t-1]}(\theta). \quad (17.29)$$

For the moment, we will leave the output density $f(\mathbf{y}\,|\,\mathbf{s},\boldsymbol{\theta})$ unspecified, and focus on finding the initial probability π and the state transition probabilities A in terms of this density. We subsequently consider the problem of identifying parameters of the density $f(\mathbf{y}\,|\,\mathbf{s},\boldsymbol{\theta})$ for various families of distributions, but for now we simply make the observation that the pdf of the observations, conditioned upon the unobserved states, factors as

$$f(\mathbf{y}\,|\,\mathbf{s},\boldsymbol{\theta}) = \prod_{t=1}^{T} f(\mathbf{y}[t]\,|\,s[t],\boldsymbol{\theta}). \tag{17.30}$$

The logarithm is

$$\log f(\mathbf{y}\,|\,\mathbf{s},\boldsymbol{\theta}) = \sum_{t=1}^{T} \log f(\mathbf{y}[t]\,|\,s[t],\boldsymbol{\theta}). \tag{17.31}$$

17.7.1 The E- and M-steps

The first step of the parameter estimation is the E-step,

$$Q(\boldsymbol{\theta}\,|\,\boldsymbol{\theta}^{[k]}) = E[\log f(\mathbf{y},\mathbf{s}\,|\,\boldsymbol{\theta})\,|\,\mathbf{y},\boldsymbol{\theta}^{[k]}].$$

Since the expectation is conditioned upon the observations, the only random component comes from the state variable. The E-step can thus be written

$$\begin{aligned}
Q(\boldsymbol{\theta}\,|\,\boldsymbol{\theta}^{[k]}) &= \sum_{\mathbf{s}\in\mathcal{S}} f(\mathbf{s}\,|\,\mathbf{y},\boldsymbol{\theta}^{[k]}) \log[f(\mathbf{y},\mathbf{s}\,|\,\boldsymbol{\theta})] \\
&= \sum_{\mathbf{s}\in\mathcal{S}} f(\mathbf{s}\,|\,\mathbf{y},\boldsymbol{\theta}^{[k]}) \log[f(\mathbf{y}\,|\,\mathbf{s},\boldsymbol{\theta})f(\mathbf{s}\,|\,\boldsymbol{\theta})],
\end{aligned} \tag{17.32}$$

where $\mathcal{S} = \{1,2,\ldots,S\}_1^T$ denotes the set of all possible state sequences of length T. The conditional probability in (17.32) can be written as

$$f(\mathbf{s}\,|\,\mathbf{y},\boldsymbol{\theta}^{[k]}) = \frac{f(\mathbf{y}\,|\,\mathbf{s},\boldsymbol{\theta}^{[k]})f(\mathbf{s}\,|\,\boldsymbol{\theta}^{[k]})}{f(\mathbf{y}\,|\,\boldsymbol{\theta}^{[k]})} = \frac{f(\mathbf{y},\mathbf{s}\,|\,\boldsymbol{\theta}^{[k]})}{f(\mathbf{y}\,|\,\boldsymbol{\theta}^{[k]})}. \tag{17.33}$$

Substituting from (17.29), (17.31), and (17.33), we obtain

$$Q(\boldsymbol{\theta}\,|\,\boldsymbol{\theta}^{[k]}) = \frac{1}{f(\mathbf{y}\,|\,\boldsymbol{\theta}^{[k]})} \sum_{\mathbf{s}\in\mathcal{S}} f(\mathbf{y},\mathbf{s}\,|\,\boldsymbol{\theta}^{[k]})$$

$$\times \left[\sum_{t=1}^{T} \log f(\mathbf{y}[t]\,|\,s[t],\boldsymbol{\theta}) + \log \pi_{s[1]}(\boldsymbol{\theta}) + \sum_{t=2}^{T} \log a_{s[t],s[t-1]}(\boldsymbol{\theta})\right]. \tag{17.34}$$

The updated parameters are then obtained by the M-step. We start with the parameters of the underlying Markov chain. The Markov chain parameters π_i and $a_{i,j}$ may also be obtained by maximizing (17.34) with constraints to preserve the probabilistic nature of the parameters. We will demonstrate the details for $a_{i,j}$:

$$a_{i,j}^{[k+1]} = \arg\max_{a_{i,j}(\boldsymbol{\theta})} Q(\boldsymbol{\theta}\,|\,\boldsymbol{\theta}^{[k]}) \qquad \text{subject to} \sum_{i=1}^{S} a_{i,j}^{[k+1]} = 1,\ a_{i,j} \geq 0.$$

The constrained optimization may be accomplished using Lagrange multipliers. We set up a Lagrange multiplier problem, take the derivative, and equate to zero as follows:

$$\begin{aligned}
0 &= \frac{\partial}{\partial a_{i,j}(\boldsymbol{\theta})} Q(\boldsymbol{\theta}\,|\,\boldsymbol{\theta}^{[k]}) - \lambda \sum_{l=1}^{S} a_{l,m}(\boldsymbol{\theta}) \\
&= \frac{1}{f(\mathbf{y}\,|\,\boldsymbol{\theta}^{[k]})} \sum_{\mathbf{s}\in\mathcal{S}} f(\mathbf{y},\mathbf{s}\,|\,\boldsymbol{\theta}^{[k]}) \sum_{t=2}^{T} \frac{1}{a_{i,j}(\boldsymbol{\theta})} \delta_{s[t],i}\delta_{s[t-1],j} - \lambda,
\end{aligned} \tag{17.35}$$

17.7 Hidden Markov Models

where $\delta_{m,n}$ is the Kronecker delta. The updated parameter $a_{i,j}^{[k+1]}$ is that value of $a_{i,j}(\boldsymbol{\theta})$ that solves (17.35),

$$
\begin{aligned}
a_{i,j}^{[k+1]} &= \frac{1}{\lambda} \sum_{\mathbf{s} \in \mathcal{S}} f(\mathbf{y}, \mathbf{s} \mid \boldsymbol{\theta}^{[k]}) \sum_{t=2}^{T} \delta_{s[t],i} \delta_{s[t-1],j} \\
&= \frac{1}{\lambda} \sum_{t=2}^{T} \sum_{\mathbf{s} \in \mathcal{S}} f(\mathbf{y}, \mathbf{s} \mid \boldsymbol{\theta}^{[k]}) \delta_{s[t],i} \delta_{s[t-1],j} \\
&= \frac{1}{\lambda} \sum_{t=2}^{T} \sum_{\mathbf{s}: s[t]=i, s[t-1]=j} f(\mathbf{y}, \mathbf{s} \mid \boldsymbol{\theta}^{[k]}). \quad (17.36)
\end{aligned}
$$

The value of λ will be chosen to normalize the probability. Let us contemplate what the inner sum in (17.36) means: it is the probability of generating the sequence $\mathbf{y}_1^T$ over any state sequence, provided that $s[t]$ and $s[t-1]$ are constrained to be i and j, respectively. Performing the computations in a brute-force manner would require a very large number of computations: there are S states, which are free for each of $T-2$ time steps, resulting in S^{T-2} terms in the sum. However, this can be simplified considerably by introducing some new notation.

17.7.2 The forward and backward probabilities

Let us define $\alpha(\mathbf{y}_1^t, j)$ as the *forward probability* of generating the forward partial sequence $\mathbf{y}_1^t$ and ending up in state j:

$$\alpha(\mathbf{y}_1^t, j) = P(\mathbf{Y}[1] = \mathbf{y}[1], \ldots, \mathbf{Y}[t] = \mathbf{y}[t], S[t] = j) = P\left(\mathbf{Y}_1^t = \mathbf{y}_1^t, S[t] = j\right).$$

Where the dependence upon the parameters is to be explicitly indicated, we write $\alpha(\mathbf{y}_1^t, j \mid \boldsymbol{\theta})$. Note that $\alpha(\mathbf{y}_1^t, j)$ denotes the probability independent of the state sequence, provided that the final state is j:

$$\alpha(\mathbf{y}_1^t, j) = \sum_{s[0], s[1], \ldots, s[t-1]} P\left(\mathbf{Y}_1^t = \mathbf{y}_1^t, S[0] = s[0], \ldots, S[t-1] = s[t-1], S[t] = j\right).$$

This summing over the states is what is needed in (17.36). Fortunately, the sum over all states is not necessary, since $\alpha(\mathbf{y}_1^t, j)$ can be computed using a simple recursion:

$$\alpha\left(\mathbf{y}_1^{t+1}, j\right) = \sum_{k=1}^{S} \alpha\left(\mathbf{y}_1^t, k\right) a_{j,k} f_j(\mathbf{y}[t+1]), \quad (17.37)$$

with initial value

$$\alpha\left(\mathbf{y}_1^1, j\right) = \pi_j f_j(\mathbf{y}[1]).$$

Let us also define the *backward probability* as the probability of generating the backward partial sequence $\mathbf{y}_{t+1}^T$, given that $s[t] = i$:

$$\beta(\mathbf{y}_{t+1}^T \mid i) = P\left(\mathbf{Y}_{t+1}^T = \mathbf{y}_{t+1}^T \mid s[t] = i\right).$$

As before, when dependence on the parameters is to be explicitly shown we write $\beta(\mathbf{y}_{t+1}^T \mid i, \boldsymbol{\theta})$. A recursive formula for computing $\beta(\mathbf{y}_{t+1}^T \mid i)$ can be written as

$$\beta(\mathbf{y}_{t+1}^T \mid i) = \sum_{k=1}^{S} \beta(\mathbf{y}_{t+2}^T \mid k) \, a_{k,i} \, f_k(\mathbf{y}[t+1]), \quad (17.38)$$

where recursion is initialized by

$$\beta(y_{T+1}^T \mid i) = \begin{cases} 1 & \text{if } i \text{ is a valid final state,} \\ 0 & \text{otherwise.} \end{cases}$$

Now let us return to (17.36), and write the probability in terms of these forward and backward probabilities.

$$a_{i,j}^{[k+1]} = \frac{1}{\lambda} \sum_{t=2}^{T} \alpha\left(\mathbf{y}_1^{t-1}, j \mid \boldsymbol{\theta}^{[k]}\right) a_{i,j}(\boldsymbol{\theta}^{[k]}) f_i(\mathbf{y}[t] \mid \boldsymbol{\theta}^{[k]}) \beta\left(\mathbf{y}_{t+1}^T \mid i, \boldsymbol{\theta}^{[k]}\right). \quad (17.39)$$

The normalizing factor is

$$\lambda = \sum_{i=1}^{S} \sum_{t=2}^{T} \alpha\left(\mathbf{y}_1^{t-1}, j \mid \boldsymbol{\theta}^{[k]}\right) a_{i,j}(\boldsymbol{\theta}^{[k]}) f_i(\mathbf{y}[t] \mid \boldsymbol{\theta}^{[k]}) \beta\left(\mathbf{y}_{t+1}^T \mid i, \boldsymbol{\theta}^{[k]}\right)$$

$$= \sum_{t=2}^{T} \alpha\left(\mathbf{y}_1^{t-1}, j \mid \boldsymbol{\theta}^{[k]}\right) \sum_{i=1}^{S} a_{i,j}(\boldsymbol{\theta}^{[k]}) f_i(\mathbf{y}[t] \mid \boldsymbol{\theta}^{[k]}) \beta\left(\mathbf{y}_{t+1}^T \mid i, \boldsymbol{\theta}^{[k]}\right) \quad (17.40)$$

$$= \sum_{t=2}^{T} \alpha\left(\mathbf{y}_1^{t-1}, j \mid \boldsymbol{\theta}^{[k]}\right) \beta\left(\mathbf{y}_t^T \mid j, \boldsymbol{\theta}^{[k]}\right), \quad (17.41)$$

where the last equality follows from (17.38). Combining (17.39) and (17.41), we obtain the update formula

$$a_{i,j}^{[k+1]} = \frac{\sum_{t=2}^{T} \alpha\left(\mathbf{y}_1^{t-1}, j \mid \boldsymbol{\theta}^{[k]}\right) a_{i,j}(\boldsymbol{\theta}^{[k]}) f_i(\mathbf{y}[t] \mid \boldsymbol{\theta}^{[k]}) \beta\left(\mathbf{y}_{t+1}^T \mid i, \boldsymbol{\theta}^{[k]}\right)}{\sum_{t=2}^{T} \alpha\left(\mathbf{y}_1^{t-1}, j \mid \boldsymbol{\theta}^{[k]}\right) \beta\left(\mathbf{y}_t^T \mid j, \boldsymbol{\theta}^{[k]}\right)}. \quad (17.42)$$

The initial state probabilities can be found as the solution to the constrained optimization problem

$$\pi_i^{[k+1]} = \arg\max_{\pi_i(\boldsymbol{\theta})} Q(\boldsymbol{\theta} \mid \boldsymbol{\theta}^{[k]}) \qquad \text{subject to} \sum_{i=1}^{S} \pi_i = 1, \quad \pi_i \geq 0.$$

Proceeding as for the transition probabilities, the initial probabilities can be found from (17.34):

$$\pi_i^{[k+1]} = \frac{\alpha\left(\mathbf{y}_1^1, i \mid \boldsymbol{\theta}^{[k]}\right) \beta\left(\mathbf{y}_2^T \mid i, \boldsymbol{\theta}^{[k]}\right)}{P\left(\mathbf{y}_1^T \mid \boldsymbol{\theta}^{[k]}\right)}. \quad (17.43)$$

17.7.3 Discrete output densities

Let us suppose, that there are M possible discrete outcomes in each state, so that

$$f_s(y[t]) = P(y[t] = i \mid s[t] = s) = c_{i,s}, \qquad i = 1, 2, \ldots, M.$$

Then

$$Q(\boldsymbol{\theta} \mid \boldsymbol{\theta}^{[k]}) = \frac{1}{f(\mathbf{y} \mid \boldsymbol{\theta}^{[k]})} \sum_{\mathbf{s} \in \mathcal{S}} f(\mathbf{y}, \mathbf{s} \mid \mathbf{s}, \boldsymbol{\theta}^{[k]}) \left[\sum_{t=1}^{T} \log c_{y[t],s[t]}(\boldsymbol{\theta}) + \log f\left(\mathbf{s} \mid \boldsymbol{\theta}^{[k]}\right) \right].$$

Now taking the derivative with respect to $c_{i,j}(\boldsymbol{\theta})$, under the constraint that $\sum_{i=1}^{M} c_{i,j}(\boldsymbol{\theta}) = 1$, and equating to zero, we find

$$c_{i,j}^{[k+1]} = \frac{\sum_{t=1, y[t]=i}^{T} \alpha\left(y_1^t, j \mid \boldsymbol{\theta}^{[k]}\right) \beta\left(y_{t+1}^T \mid j, \boldsymbol{\theta}^{[k]}\right)}{\sum_{t=1}^{T} \alpha\left(y_1^t, j \mid \boldsymbol{\theta}^{[k]}\right) \beta\left(y_{t+1}^T \mid j, \boldsymbol{\theta}^{[k]}\right)}. \quad (17.44)$$

17.7.4 Gaussian output densities

Let us now assume that the density in each state is Gaussian,

$$f_s \sim \mathcal{N}(\boldsymbol{\mu}_s, R_s), \qquad s = 1, 2, \ldots, S,$$

17.7 Hidden Markov Models

with covariance R_s and mean μ_s, $s = 1, 2, \ldots, S$. Then, showing the explicit dependence of the parameters, we have

$$f(\mathbf{y}[t] \mid s[t], \boldsymbol{\theta}) = \frac{1}{(2\pi)^{m/2} \mid R_{s[t]}\mid^{1/2}(\boldsymbol{\theta})\mid} \exp\left[-\frac{1}{2}(\mathbf{y}[t] - \boldsymbol{\mu}_{s[t]}(\boldsymbol{\theta}))^T R_{s[t](\boldsymbol{\theta})}^{-1} (\mathbf{y}[t] - \boldsymbol{\mu}_{s[t]}(\boldsymbol{\theta}))\right]. \tag{17.45}$$

The updates to the mean $\boldsymbol{\mu}_s$ and the covariance R_s can be found by taking the derivative of $Q(\boldsymbol{\theta} \mid \boldsymbol{\theta}^{[k]})$ with respect to $\boldsymbol{\mu}_s(\boldsymbol{\theta})$ and $R_s(\boldsymbol{\theta})$, respectively, then equating the result to zero and solving for $\boldsymbol{\mu}_i^{[k+1]}$ and $R_i^{[k+1]}$ (as those values of $\boldsymbol{\mu}_i(\boldsymbol{\theta})$ and $R_i(\boldsymbol{\theta})$, respectively, solve the derivative equations). By this means we find

$$\boldsymbol{\mu}_i^{[k+1]} = \frac{\sum_{t=1}^T \alpha\left(\mathbf{y}_1^t, i \mid \boldsymbol{\theta}^{[k]}\right) \beta\left(\mathbf{y}_{t+1}^T \mid i, \boldsymbol{\theta}^{[k]}\right) \mathbf{y}[t]}{\sum_{t=1}^T \alpha\left(\mathbf{y}_1^t, i \mid \boldsymbol{\theta}^{[k]}\right) \beta\left(\mathbf{y}_{t+1}^T \mid i, \boldsymbol{\theta}^{[k]}\right)}, \tag{17.46}$$

$$R_i^{[k+1]} = \frac{\sum_{t=1}^T \alpha\left(\mathbf{y}_1^t, i \mid \boldsymbol{\theta}^{[k]}\right) \beta\left(\mathbf{y}_{t+1}^T \mid i, \boldsymbol{\theta}^{[k]}\right) (\mathbf{y}[t] - \boldsymbol{\mu}_i^{[k]})(\mathbf{y}[t] - \boldsymbol{\mu}_i^{[k]})^T}{\sum_{t=1}^T \alpha\left(\mathbf{y}_1^t, i \mid \boldsymbol{\theta}^{[k]}\right) \beta\left(\mathbf{y}_{t+1}^T \mid i, \boldsymbol{\theta}^{[k]}\right)}. \tag{17.47}$$

17.7.5 Normalization

Because formulas for computing the updates for the HMM involve products of probabilities, the numbers involved can become very small, often beyond the range of accurate representation. It is important in practical algorithms to normalize the computations as far as possible.

Let $\hat{\alpha}(\mathbf{y}_1^t, j)$ be the normalized forward probability, defined by

$$\hat{\alpha}\left(\mathbf{y}_1^t, j\right) = \alpha\left(\mathbf{y}_1^t, j\right) c(1)$$

where

$$c(1) = \frac{1}{\sum_{s=1}^S \alpha\left(\mathbf{y}_1^t, s\right)}. \tag{17.48}$$

Now, update (17.37) to compute $\alpha(\mathbf{y}_1^2, j)$, but replace $\alpha(\mathbf{y}_1^1, j)$ with $\hat{\alpha}(\mathbf{y}_1^1, i)$ and call the result $\tilde{\alpha}(\mathbf{y}_1^2, j)$:

$$\tilde{\alpha}\left(\mathbf{y}_1^2, j\right) = \sum_{k=1}^S \hat{\alpha}\left(\mathbf{y}_1^1, k\right) a_{j,k} f_j(\mathbf{y}[2]) = c(1)\alpha\left(\mathbf{y}_1^2, j\right).$$

Again, we normalize to obtain

$$\hat{\alpha}\left(\mathbf{y}_1^2, j\right) = \tilde{\alpha}\left(\mathbf{y}_1^2, j\right) c(2) = \alpha\left(\mathbf{y}_1^2, j\right) c(1)c(2)$$

where

$$c(2) = \frac{1}{\sum_{s=1}^S \tilde{\alpha}\left(\mathbf{y}_1^2, s\right)}.$$

Proceeding inductively, we define

$$\tilde{\alpha}\left(\mathbf{y}_1^t, j\right) = \sum_{k=1}^S \hat{\alpha}\left(\mathbf{y}_1^{t-1}, k\right) a_{j,k} f_j(\mathbf{y}[t]) \tag{17.49}$$

and

$$\hat{\alpha}\left(\mathbf{y}_1^t, j\right) = c(t)\tilde{\alpha}\left(\mathbf{y}_1^t, j\right) = \left(\prod_{l=1}^t c(l)\right) \alpha\left(\mathbf{y}_1^t, j\right), \tag{17.50}$$

where

$$c(t) = \frac{1}{\sum_{s=1}^S \tilde{\alpha}\left(\mathbf{y}_1^t, s\right)}. \tag{17.51}$$

Now, substituting (17.49) into (17.51) and placing the result in (17.50) using (17.49), we find the normalized update

$$\hat{\alpha}\left(\mathbf{y}_1^t, j\right) = \frac{\sum_{k=1}^S \hat{\alpha}\left(\mathbf{y}_1^{t-1}, k\right) a_{j,k} f_j(\mathbf{y}[t])}{\sum_{s=1}^S \sum_{k=1}^S \hat{\alpha}\left(\mathbf{y}_1^{t-1}, k\right) a_{s,k} f_s(\mathbf{y}[t])}. \tag{17.52}$$

It is straightforward to show that

$$\hat{\alpha}\left(\mathbf{y}_1^t, j\right) = \frac{\alpha\left(\mathbf{y}_1^t, j\right)}{\sum_{s=1}^S \alpha\left(\mathbf{y}_1^t, s\right)}.$$

Let us also normalize the backward probability by the same normalization factor, since the forward and backward probabilities are roughly of the same magnitude. That is, define

$$\hat{\beta}\left(\mathbf{y}_{t+1}^T \mid j\right) = \left(\prod_{l=t}^T c(l)\right) \beta\left(\mathbf{y}_{t+1}^T \mid j\right).$$

Using the normalized variables, the transition probability update formula (17.42) can be written using (17.40) as

$$a_{i,j}^{[k+1]} = \frac{\sum_{t=2}^T \alpha\left(\mathbf{y}_1^{t-1}, j \mid \boldsymbol{\theta}^{[k]}\right) a_{i,j}(\boldsymbol{\theta}^{[k]}) f_i(\mathbf{y}[t] \mid \boldsymbol{\theta}^{[k]}) \beta\left(\mathbf{y}_{t+1}^T \mid i, \boldsymbol{\theta}^{[k]}\right)}{\sum_{s=1}^S \sum_{t=2}^T \alpha\left(\mathbf{y}_1^{t-1}, j \mid \boldsymbol{\theta}^{[k]}\right) a_{s,j}(\boldsymbol{\theta}^{[k]}) f_s(\mathbf{y}[t] \mid \boldsymbol{\theta}^{[k]}) \beta\left(\mathbf{y}_{t+1}^T \mid s, \boldsymbol{\theta}^{[k]}\right)}$$

$$= \frac{\sum_{t=2}^T \frac{1}{\prod_{l=1}^{t-1} c(l)} \alpha\left(\mathbf{y}_1^{t-1}, j \mid \boldsymbol{\theta}^{[k]}\right) a_{i,j}(\boldsymbol{\theta}^{[k]}) f_i(\mathbf{y}[t] \mid \boldsymbol{\theta}^{[k]}) \frac{1}{\prod_{l=t}^T c(l)} \beta\left(\mathbf{y}_{t+1}^T \mid i, \boldsymbol{\theta}^{[k]}\right)}{\sum_{s=1}^S \sum_{t=2}^T \frac{1}{\prod_{l=1}^{t-1} c(l)} \alpha\left(\mathbf{y}_1^{t-1}, j \mid \boldsymbol{\theta}^{[k]}\right) a_{s,j}(\boldsymbol{\theta}^{[k]}) f_s(\mathbf{y}[t] \mid \boldsymbol{\theta}^{[k]}) \frac{1}{\prod_{l=t}^T c(l)} \beta\left(\mathbf{y}_{t+1}^T \mid s, \boldsymbol{\theta}^{[k]}\right)}$$

$$= \frac{\sum_{t=2}^T \hat{\alpha}\left(\mathbf{y}_1^{t-1}, j \mid \boldsymbol{\theta}^{[k]}\right) a_{i,j}(\boldsymbol{\theta}^{[k]}) f_i(\mathbf{y}[t] \mid \boldsymbol{\theta}^{[k]}) \hat{\beta}\left(\mathbf{y}_{t+1}^T \mid i, \boldsymbol{\theta}^{[k]}\right)}{\sum_{s=1}^S \sum_{t=2}^T \hat{\alpha}\left(\mathbf{y}_1^{t-1}, j \mid \boldsymbol{\theta}^{[k]}\right) a_{s,j}(\boldsymbol{\theta}^{[k]}) f_s(\mathbf{y}[t] \mid \boldsymbol{\theta}^{[k]}) \hat{\beta}\left(\mathbf{y}_{t+1}^T \mid s, \boldsymbol{\theta}^{[k]}\right)}$$

The initial probability, discrete output probability, and Gaussian parameter probabilities similarly can be written in terms of the scaled parameters, as examined in the exercises.

17.7.6 Algorithms for HMMs

The concepts described above are embodied in the MATLAB code in this section. Several short functions are presented that provide for generation of HMM outputs, computation of likelihoods, and update of the parameters, based upon the scaled forward and backward probabilities. Algorithm 17.3 provides an overview of the code and the data structures employed; for completeness, it also provides a list of appropriate functions for training HMMs using the Viterbi algorithm methods discussed in section 19.6. The functions in algorithm 17.4 provide (scaled) forward and backward probability computations, and likelihood computations (of samples and sequences). Algorithm 17.5 contains the model update functions. The function `hmmupdaten` calls `hmmApiupn` to update the Markov parameters, and `hmmfupdaten` to update the output distribution, which in turn calls functions appropriate for the distributions. Finally, algorithm 17.6 has functions for generating outputs from an HMM, for simulation and testing purposes.

Algorithm 17.3 Overview of HMM data structures and functions
File: `hmmnotes.m`

17.7 Hidden Markov Models

Algorithm 17.4 HMM likelihood computation functions
File: `hmmlpyseqn.m`
`hmmabn.m`
`hmmf.m`
`hmmdiscf.m`
`hmmgausf.m`

Algorithm 17.5 HMM model update functions
File: `hmmupdaten.m`
`hmmApiupn.m`
`hmmfupdaten.m`
`hmmdiscfupn.m`
`hmmgausfupn.m`

Algorithm 17.6 HMM generation functions
File: `hmmgendat.m`
`hmmgendisc.m`
`hmmgengaus.m`

Example 17.7.1 We briefly demonstrate the use of the HMM code. A three-state HMM with Gaussian output densities having means

$$\boldsymbol{\mu}_1 = \begin{bmatrix} 1 \\ 1 \\ 1 \end{bmatrix} \qquad \boldsymbol{\mu}_2 = \begin{bmatrix} -1 \\ -1 \\ -1 \end{bmatrix} \qquad \boldsymbol{\mu}_3 = \begin{bmatrix} 5 \\ 5 \\ 5 \end{bmatrix}$$

and covariance matrices

$$R_1 = .5I \qquad R_2 = .6I \qquad R_3 = .7I,$$

and with initial probability and state-transition probability matrix

$$\pi = \begin{bmatrix} 1 \\ 0 \\ 0 \end{bmatrix} \qquad A = \begin{bmatrix} 0 & .25 & .2 \\ 1 & .25 & .4 \\ 0 & .5 & .4 \end{bmatrix},$$

is represented in MATLAB using the code

```
HMM.A = [0 1/4 .2;   1 1/4 .4;   0 1/2 .4];
HMM.pi = [1;0;0];
HMM.final = [0 0 1];
HMM.f{1} = 2;    % Gaussian distribution
HMM.f{2,1} = [1;1;1];
HMM.f{2,2} = [-1;-1;-1];
```

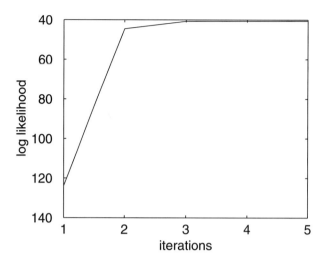

Figure 17.8: $\log P(\mathbf{y}_1^T \mid \theta^{[k]})$ for an HMM.

```
HMM.f{2,3}  =  [5;5;5];
HMM.f{3,1}  =  .5*eye(3);
HMM.f{3,2}  =  .6*eye(3);
HMM.f{3,3}  =  .7*eye(3);
```

A sequence of eight outputs can be generated by

```
[y,ss] = hmmgendat(8,HMM);
```

The code to compute the likelihood, then train an HMM (starting from the initial HMM) is

```
lpy = hmmlpyseqn(y,HMM);            % compute the log-likelihood
lpv = lpy;
hmmnewn = HMM;                       % start with the given HMM
for i=1:4
  hmmnewn = hmmupdaten(y,hmmnewn);   % update the HMM
  lpy = hmmlpyseqn(y,hmmnew);        % compute the likelihood, and
  lpv = [lpv lpy];                   % save it
end
```

Figure 17.8 shows $\log P(\mathbf{y}_1^T \mid \theta^{[k]})$, illustrating how the algorithm converges. Convergence is obtained (in this example) in three iterations. □

17.8 Spread-spectrum, multiuser communication

In direct-sequence spread-spectrum multiple-access (SSMA) communications (also known as code-division multiple access (CDMA)), all users in a channel transmit simultaneously, using quasiorthogonal spreading codes to reduce the interuser interference [354]. The system block diagram is shown in figure 17.9. A signal received in a K-user system through a Gaussian channel may be written as

$$r(t) = S(t, \mathbf{b}) + \sigma N(t),$$

17.8 Spread-Spectrum, Multiuser Communication

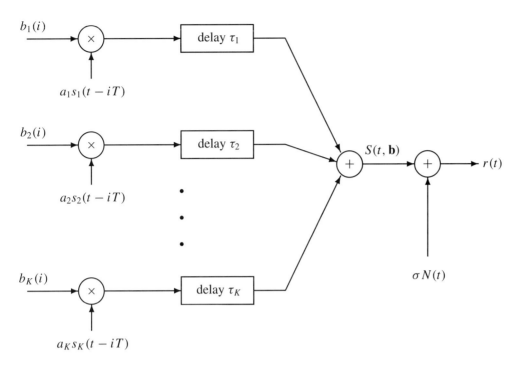

Figure 17.9: Representation of signals in an SSMA system.

where $N(t)$ is unit-variance, zero-mean, white Gaussian noise, and

$$S(t, \mathbf{b}) = \sum_{k=1}^{K} a_k \sum_{i=-M}^{M} b_k(i) s_k(t - iT - \tau_k)$$

is the composite signal from all K transmitters. Here, a_k is the amplitude of the kth transmitted signal (as seen at the receiver), $\mathbf{b}$ represents the symbols of all the users, $b_k(i)$ is the ith bit of the kth user, τ_k is the channel propagation delay for the kth user, and $s_k(t)$ is the signaling waveform of the kth user, including the spreading code. For this example, coherent reception of each user is assumed so that the amplitudes are real.

At the receiver, the signal is passed through a bank of matched filters, with a filter matched to the spreading signal of each of the users, as shown in figure 17.10. (This

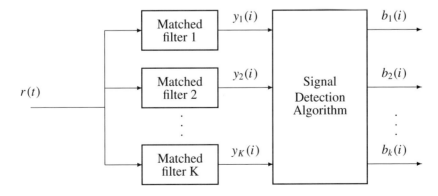

Figure 17.10: Multiple-access receiver matched-filter bank.

assumes that synchronization for each user has been obtained.) The set of matched filter outputs for the ith bit interval is

$$\mathbf{y}(i) = [y_1(i), y_2(i), \ldots, y_K(i)]^T.$$

Because the interference among the users is similar to intersymbol interference, optimal detection requires dealing with the entire sequence of matched filter vectors

$$\mathbf{y} = [\mathbf{y}(-M)^T, \mathbf{y}(-M+1)^T, \ldots, \mathbf{y}(M)^T]^T.$$

For a Gaussian channel, it may be shown that

$$\mathbf{y} = H(\mathbf{b})\mathbf{a} + \mathbf{z}, \qquad (17.53)$$

where $H(\mathbf{b})$ depends upon the correlations between the spreading signals and the bits transmitted, and $\mathbf{z}$ is nonwhite, zero-mean Gaussian noise. The likelihood function for the received sequence may be written (see [258]) as

$$f(\mathbf{y}\,|\,\mathbf{a},\mathbf{b}) = c\exp\left[\frac{1}{2\sigma^2}(2\mathbf{a}^T R(\mathbf{b})\mathbf{y} - \mathbf{a}^T S(\mathbf{b})\mathbf{a})\right], \qquad (17.54)$$

where $R(\mathbf{b})$ and $S(\mathbf{b})$ depend upon the bits and correlations and c is a constant that makes the density integrate to 1. Note that even though the noise is Gaussian, which is in the exponential family, the overall likelihood function is not Gaussian because of the presence of the random bits—it is actually a mixture of Gaussians. For the special case of a single user, the likelihood function becomes

$$f(\mathbf{y}\,|\,\mathbf{a},\mathbf{b}) = c\exp\left[\frac{1}{2\sigma^2}\left(2a_1\sum_{i=-M}^{M} b_1(i)y_1(i) - a_1^2\right)\right].$$

What we ultimately desire from the detector is the set of bits for each user. It has been shown [354] that the interuser interference increases the probability of error very little, provided that sophisticated detection algorithms are employed after the matched filters. However, most of the algorithms that have been developed require knowledge of the amplitudes of each user [211]. Therefore, in order to determine the bits reliably, the amplitude of each user must also be known. Seen from the point of view of amplitude estimation, the bits are unknown nuisance parameters. (Other estimation schemes relying on decision feedback may take a different point of view.)

If the bits were known, an ML estimate of the amplitudes could be easily obtained: $\mathbf{a}_{ml} = S(\mathbf{b})^{-1} R(\mathbf{b})\mathbf{y}$. Lacking the bits, however, more sophisticated tools for obtaining the amplitudes must be applied as a precursor to detecting the bits. One approach to estimating the signal amplitudes is the EM algorithm [258]. For purposes of applying the EM algorithm, the complete data set is $\mathbf{x} = \{\mathbf{y}, \mathbf{b}\}$ and the parameter set is $\boldsymbol{\theta} = \mathbf{a}$. To compute the expectations in the E-step, it is assumed that the bits are independent and equally likely ± 1.

The likelihood function of the complete data is

$$f(\mathbf{x}\,|\,\mathbf{a}) = f(\mathbf{y},\mathbf{b}\,|\,\mathbf{a}) = f(\mathbf{y}\,|\,\mathbf{b},\mathbf{a})f(\mathbf{b}\,|\,\mathbf{a}). \qquad (17.55)$$

This conditioning is similar to that of (17.22) and (17.27): the complete-data likelihood is broken into a likelihood of the observation, conditioned upon the unobserved data, multiplied by a likelihood of the unobserved data. From (17.54), $f(\mathbf{y}\,|\,\mathbf{b},\mathbf{a})$ is Gaussian. To compute the E-step

$$E[\log f(\mathbf{x}\,|\,\mathbf{a})\,|\,\mathbf{y}, \mathbf{a}^{[k]}] = \sum_{\mathbf{b}\in\{\pm 1\}^{(M+1)K}} f(\mathbf{b}\,|\,\mathbf{y}, \mathbf{a}^{[k]}) \log f(\mathbf{x}\,|\,\mathbf{a}),$$

it is necessary to determine the conditional probability $f(\mathbf{b}\,|\,\mathbf{y}, \mathbf{a}^{[k]})$.

It is revealing to consider a single-user system. In this case the log-likelihood function is

$$\log f(\mathbf{x} \mid a_1) = \frac{a_1}{\sigma^2} \sum_{i=-M}^{M} b_1(i) y_1(i) - \frac{a_1^2}{2\sigma^2}(2M+1) + \text{constant},$$

and the E-step becomes

$$E[\log f(\mathbf{x} \mid \mathbf{a}) \mid \mathbf{y}, \mathbf{a}^{[k]}] = \sum_{i=-M}^{M} \sum_{b_1(i) \in \pm 1} f\left(b_1(i) \mid y_1(i), a_1^{[k]}\right) \log f(x_1(i) \mid a_1). \quad (17.56)$$

The conditional probability required for the expectation is

$$f\left(b_1(i) \mid y_1(i), a_1^{[k]}\right) = \frac{f\left(b_1(i), y_1(i) \mid a_1^{[k]}\right)}{f\left(y_1(i) \mid a_1^{[k]}\right)} = \frac{f\left(b_1(i), y_1(i) \mid a_1^{[k]}\right)}{\sum_{b \in \pm 1} f\left(b, y_1(i) \mid a_1^{[k]}\right)} \quad (17.57)$$

$$= \frac{\exp\left[\frac{1}{2\sigma^2}\left(2a_1^{[k]} b_1(i) y_1(i) - a_1^{[k]2}\right)\right]}{\sum_{b \in \pm 1} \exp\left[\frac{1}{2\sigma^2}\left(2a_1^{[k]} b y_1(i) - a_1^{[k]2}\right)\right]}$$

$$= \frac{\exp\left[\frac{1}{\sigma^2}\left(a_1^{[k]} b_1(i) y_1(i)\right)\right]}{\cosh\left(y_1(i) a_1^{[k]}/\sigma^2\right)}. \quad (17.58)$$

Substituting (17.58) into (17.56) yields

$$E\left[\log f(\mathbf{x} \mid a_1) \mid \mathbf{y}, a_1^{[k]}\right] = \frac{a_1}{\sigma^2} \sum_{i=-M}^{M} y_1(i) \tanh\left(a_1^{[k]} y_1(i)/\sigma^2\right) - \frac{a_1^2}{2\sigma^2}(2M+1) + \text{constant}. \quad (17.59)$$

Conveniently, (17.59) is quadratic in a_1, and the M-step is easily computed by differentiating (17.59) with respect to a_1, giving

$$a_1^{[k+1]} = \frac{1}{2M+1} \sum_{i=-M}^{M} y_1(i) \tanh\left(a_1^{[k]} y_1(i)/\sigma^2\right). \quad (17.60)$$

Equation (17.60) gives the update equation for the amplitude estimate, which may be iterated until convergence.

For multiple users, the E-step and M-step are structurally similar, but more involved computationally [258].

17.9 Summary

The EM algorithm may be employed when there is an underlying set with a known distribution function that is observed by means of a many-to-one mapping. If the distribution of the underlying complete data is exponential, the EM algorithm may be specialized as in (17.9) and (17.10). Otherwise, it will be necessary to use the general statement of (17.6) and (17.7). In many cases, the type of conditioning exhibited in (17.22), (17.27), or (17.55) may be used: the observed data is conditioned upon the unobserved data, so that the likelihood function may be computed. In general, if the complete data set is $\mathbf{x} = (\mathbf{y}, \mathbf{z})$ for some unobserved $\mathbf{z}$, then

$$E[\log f(\mathbf{x} \mid \boldsymbol{\theta}) \mid \mathbf{y}, \boldsymbol{\theta}^{[k]}] = \int f(\mathbf{z} \mid \mathbf{y}, \boldsymbol{\theta}^{[k]}) \log f(\mathbf{x} \mid \boldsymbol{\theta}) \, d\mathbf{z},$$

since, conditioned upon $\mathbf{y}$, the only random component of $\mathbf{x}$ is $\mathbf{z}$.

Analytically, the most difficult portion of the EM algorithm is the E-step. This is often also the most difficult computational step: for the general EM algorithm the expectation must be computed over all values of the unobserved variables. There may be, as in the case of the HMM, efficient algorithms to ease the computation, but even these cannot completely eliminate the computational burden.

In most instances where the EM algorithm applies, there are other algorithms that also apply, such as gradient descent (see, for example, [243]). As already observed, however, these may have problems of their own, such as requiring derivatives or setting of convergence-rate parameters. Because of its generality and guaranteed convergence, the EM algorithm is a good choice to consider for many estimation problems.

As noted in section 17.3.1, there have also been more recent developments, such as ECME, SAGE, and AECM, which provide the advantages of the EM algorithm with faster convergence. Work continues in this area.

17.10 Exercises

17.2-1 [273] A mixture density is a density of the form

$$f(y \mid \boldsymbol{\theta}) = \sum_{i=1}^{m} \alpha_i f_i(y \mid \boldsymbol{\theta}_i), \qquad (17.61)$$

which models data from a statistical population that is a mixture of m component densities f_i and mixing proportions α_i. (For example, think of the distribution of weights of a human population, which is a mixture of the weights of males and the weights of females). In (17.61), the mixing proportions satisfy $\sum_{i=1}^{m} \alpha_i = 1$ and $\alpha_i \geq 0$. The total parameter set $\boldsymbol{\theta} = (\boldsymbol{\theta}_1, \ldots, \boldsymbol{\theta}_m, \alpha_1, \ldots, \alpha_m)$. The parameter estimation problem is to determine the mixture parameters α_i, as well as the parameter set $\boldsymbol{\theta}_i$, for the density f_i. Assume that $y_1, y_2, \ldots, y_n$ is an independent sample.

If the data $\mathbf{y} = [y_1, y_2, \ldots, y_n]$ were labeled according to which distribution generated y_i, then the parameters of the distribution f_i could be estimated based upon the data associated with it. Most commonly, however, the data are not labeled.

Let $\mathbf{x} = (\mathbf{y}, \mathbf{c})$ be the complete data, where $\mathbf{c} = (c_1, c_2, \ldots, c_n)$, $c_i \in \{1, \ldots, m\}$, is the set of labels. Then $\alpha_i = P(c_j = i)$, the probability that the ith density is used.

(a) Show that the log-likelihood function is

$$L_y(\boldsymbol{\theta}) = \sum_{j=1}^{n} \log f(y_i \mid \boldsymbol{\theta}).$$

(b) Show that the $Q(\boldsymbol{\theta} \mid \boldsymbol{\theta}^{[k]})$ computed in the E-step is

$$Q(\boldsymbol{\theta} \mid \boldsymbol{\theta}^{[k]}) = \sum_{i_1=1}^{m} \cdots \sum_{i_n=1}^{m} \sum_{j=1}^{n} \log \alpha_{i_j} f_{i_j}(y_j \mid \boldsymbol{\theta}_{i_j}) \prod_{l=1}^{n} \frac{\alpha_{i_l}^{[k]} f_{i_l}\left(y_l \mid \boldsymbol{\theta}_{i_l}^{[k]}\right)}{f(y_l \mid \boldsymbol{\theta}^{[k]})}$$

$$= \sum_{i=1}^{m} \sum_{j=1}^{n} \log \alpha_i f_i(y_j \mid \boldsymbol{\theta}_i) \frac{\alpha_i^{[k]} f_i\left(y_j \mid \boldsymbol{\theta}_i^{[k]}\right)}{f(y_j \mid \boldsymbol{\theta}^{[k]})}$$

$$= \sum_{i=1}^{m} \left[\sum_{j=1}^{n} \frac{\alpha_i^{[k]} f_i\left(y_j \mid \boldsymbol{\theta}_i^{[k]}\right)}{f(y_j \mid \boldsymbol{\theta}^{[k]})} \right] \log \alpha_i$$

$$+ \sum_{i=1}^{m} \sum_{j=1}^{n} \log f_i(y_j \mid \boldsymbol{\theta}_i) \frac{\alpha_i^{[k]} f_i\left(y_j \mid \boldsymbol{\theta}_i^{[k]}\right)}{f(y_j \mid \boldsymbol{\theta}^{[k]})}.$$

(c) Show that the M step provides

$$\alpha_i^{[k+1]} = \frac{1}{n} \sum_{j=1}^{n} \frac{\alpha_i^{[k]} f_i\left(y_j \mid \theta_i^{[k]}\right)}{f(y_j \mid \theta^{[k]})}.$$

(d) Show that

$$\theta_i^{[k+1]} = \arg\max_{\theta_i} \sum_{j=1}^{n} \log f_i(y_j \mid \theta_i) \frac{\alpha_i^{[k]} f_i\left(y_j \mid \theta_i^{[k]}\right)}{f(y_j \mid \theta^{[k]})}.$$

17.5-2 Show, using the update formula (17.21), that for every iteration,

$$\sum_d y(d) = \sum_b \lambda^{[k+1]}(b).$$

17.7-3 Show that for an HMM,

$$P\left(\mathbf{y}_1^T\right) = \sum_{i=1}^{S} \alpha\left(\mathbf{y}_1^t, i\right) \beta\left(\mathbf{y}_{t+1}^T \mid i\right).$$

Hence, show that

$$P\left(\mathbf{y}_1^T\right) = \sum_{\text{all legal final } i} \alpha\left(\mathbf{y}_1^T, i\right).$$

17.7-4 Show that (17.43) is correct.

17.7-5 Show that (17.44) is correct.

17.7-6 Show that (17.46) is correct.

17.7-7 Show that (17.47) is correct.

17.7-8 (Mixture HMM) In some applications of an HMM, a Gaussian mixture model is used.

$$f(\mathbf{y}[t] \mid s) = \sum_{l=1}^{M} c_{sl} \mathcal{N}(\mathbf{y}[t]; \boldsymbol{\mu}_{s,l}, R_{s,l}),$$

where the c_{sl} are the mixture gains satisfying

$$\sum_{l=1}^{m} c_{sl} = 1 \qquad c_{sl} \geq 0,$$

and $\mathcal{N}$ is a Gaussian density function,

$$\mathcal{N}(\mathbf{y}; \boldsymbol{\mu}_{sl}, R_{sl}) = \frac{1}{(2\pi)^{m/2} \mid R_{sl} \mid^{1/2}} \exp\left[-\frac{1}{2}(\mathbf{y} - \boldsymbol{\mu}_{sl})^T R_{sl}^{-1} (\mathbf{y} - \boldsymbol{\mu}_{sl})\right].$$

In the mixture model, the density for each state is parameterized by the means, covariance matrices, and mixture gains.

(a) Show that the update rule for the mean in the mixture model is

$$\boldsymbol{\mu}_{sl}^{[k+1]} = \frac{\sum_{t=1}^{T} \gamma(s, l, t \mid \theta^{[k]}) \mathbf{y}[t]}{\sum_{t=1}^{T} \gamma(s, l, t \mid \theta^{[k]})},$$

where

$$\gamma(s, l, t \mid \boldsymbol{\theta}) = \frac{\alpha\left(\mathbf{y}_1^t, s \mid \boldsymbol{\theta}\right) \beta\left(\mathbf{y}_{t+1}^T \mid s, \boldsymbol{\theta}\right)}{\sum_{t=1}^T \alpha\left(\mathbf{y}_1^t, s \mid \boldsymbol{\theta}\right) \beta\left(\mathbf{y}_{t+1}^T \mid s, \boldsymbol{\theta}\right)} \frac{c_{sl}\mathcal{N}(\mathbf{y}[t], \boldsymbol{\mu}_{sl}(\boldsymbol{\theta}), R_{sl}(\boldsymbol{\theta}))}{\sum_{l=1}^M c_{sl}\mathcal{N}(\mathbf{y}[t], \boldsymbol{\mu}_{sl}(\boldsymbol{\theta}), R_{sl}(\boldsymbol{\theta}))}.$$

(b) Show that

$$R_{sl}^{[k+1]} = \frac{\sum_{t=1}^T \gamma(s, l, t \mid \boldsymbol{\theta}^{[k]})(\mathbf{y}[t] - \boldsymbol{\mu}_{sl}(\boldsymbol{\theta}^{[k]}))(\mathbf{y}[t] - \boldsymbol{\mu}_{sl}(\boldsymbol{\theta}^{[k]}))^T}{\sum_{t=1}^T \gamma(s, l, t \mid \boldsymbol{\theta}^{[k]})}.$$

(c) Show that

$$c_{sl}^{[k+1]} = \frac{\sum_{t=1}^T \gamma(s, l, t \mid \boldsymbol{\theta}^{[k]})}{\sum_{t=1}^T \sum_{l=1}^M \gamma(s, l, t \mid \boldsymbol{\theta}^{[k]})}.$$

(d) Show that the initial probability estimate for an HMM can be written in terms of the scaled forward and backward probabilities as

$$\pi_i^{[k+1]} = \frac{\hat{\alpha}\left(\mathbf{y}_1^1, i \mid \boldsymbol{\theta}^{[k]}\right) \hat{\beta}\left(\mathbf{y}_2^T \mid i, \boldsymbol{\theta}^{[k]}\right)}{\sum_{s=1}^S \hat{\alpha}\left(\mathbf{y}_1^1, s \mid \boldsymbol{\theta}^{[k]}\right) \hat{\beta}\left(\mathbf{y}_2^T \mid s, \boldsymbol{\theta}^{[k]}\right)}.$$

(e) Show that the mean and covariance for the HMM model can be updated using

$$\boldsymbol{\mu}_i^{[k+1]} = \frac{\sum_{t=1}^T \frac{1}{c(t)} \hat{\alpha}\left(\mathbf{y}_1^t, i \mid \boldsymbol{\theta}^{[k]}\right) \hat{\beta}\left(\mathbf{y}_{t+1}^T \mid i, \boldsymbol{\theta}^{[k]}\right) \mathbf{y}[t]}{\sum_{t=1}^T \frac{1}{c(t)} \hat{\alpha}\left(\mathbf{y}_1^t, i \mid \boldsymbol{\theta}^{[k]}\right) \hat{\beta}\left(\mathbf{y}_{t+1}^T \mid i, \boldsymbol{\theta}^{[k]}\right)},$$

$$R_i^{[k+1]} = \frac{\sum_{t=1}^T \frac{1}{c(t)} \hat{\alpha}\left(\mathbf{y}_1^t, i \mid \boldsymbol{\theta}^{[k]}\right) \hat{\beta}\left(\mathbf{y}_{t+1}^T \mid i, \boldsymbol{\theta}^{[k]}\right) (\mathbf{y}[t] - \boldsymbol{\mu}_i^{[k]})(\mathbf{y}[t] - \boldsymbol{\mu}_i^{[k]})^T}{\sum_{t=1}^T \frac{1}{c(t)} \hat{\alpha}\left(\mathbf{y}_1^t, i \mid \boldsymbol{\theta}^{[k]}\right) \hat{\beta}\left(\mathbf{y}_{t+1}^T \mid i, \boldsymbol{\theta}^{[k]}\right)}.$$

17.7-9 Show that for an HMM, the log-likelihood function $\log P(\mathbf{y}_1^T \mid \boldsymbol{\theta})$ can be written as

$$\log P\left(\mathbf{y}_1^T \mid \boldsymbol{\theta}\right) = -\sum_{t=1}^T \log(c(t)) + \log \sum_{\text{valid final } i} \hat{\alpha}\left(\mathbf{y}_1^T, i \mid \boldsymbol{\theta}\right).$$

17.7-10 Write MATLAB code to do the following:

(a) Generate a sequence of eight outputs of a three-state HMM with Gaussian state output distributions and the following parameters:

$$A = [.8 \quad .2 \quad 0.2 \quad .4 \quad .40 \quad .4 \quad .6] \qquad \pi = \begin{bmatrix} .5 \\ .3 \\ .2 \end{bmatrix},$$

$$f_1 \sim \mathcal{N}(3, 1) \qquad f_2 \sim \mathcal{N}(2, 1) \qquad f_3 \sim \mathcal{N}(3, 1).$$

(b) Compute the log-likelihood of this sequence using the current HMM.

(c) Starting from an initial training HMM with uniform prior and state-transition probabilities, and initial-state output probabilities that are $\mathcal{N}(\text{random}, 1)$, train an HMM to match your output sequence until convergence. Determine an appropriate convergence criterion.

(d) Plot the log-likelihood of the sequence, given your training HMM, as a function of iteration.

17.11 References

The following references should be considered in addition to those that have appeared throughout the chapter. This chapter is a modified and extended version of [231]. A seminal reference on the EM algorithm is [72]. A more recent survey article with specific application to mixture-density parameter estimation is [273] (an even more recent survey with more modern techniques is found in [225]). Our presentation was helped by the very short summary appearing in [134]. A recent book dedicated to the EM algorithm is [224].

Our discussion of emission tomography was drawn largely from [309]. The HMM material was drawn largely from Rabiner's excellent tutorial [266], with additional input from [68]; [265] also has valuable information on HMMs.

Part V

Methods of Optimization

Signal-processing design and analysis frequently begin by the statement of a criterion of optimality, followed by an optimization procedure. In this part, we examine constrained optimization, shortest-path optimization on graphs, and linear programming, as well as applications of these concepts.

Chapter 18

Theory of Constrained Optimization

> Life is a constrained optimization problem.
> — Rick Frost

> What e'er thou art, act well thy part.
> — Scottish proverb

The approach to an effective design often begins with a statement of a function to be optimized. We have encountered several design criteria in this book: maximum likelihood, minimum variance, minimum norm, minimum risk, minimum mean-squared error, and so forth. In some cases, a criterion of optimality in a design leads to a problem that is differentiable in the parameters of the problem. We focus on differentiable problems such as these in this chapter. In particular, we wish to address the question of how to incorporate constraints in the optimization problem. Constrained optimization problems have arisen several times throughout this text, and the problem has been treated using Lagrange multipliers, which were introduced in section A.7. This chapter provides geometrical justification for the use of Lagrange multipliers, as well as a treatment of inequality constraints.

The theory of optimization is very broad, a testament to both the importance of the problem, and the theoretical and algorithmic richness necessary to treat it well. This chapter can only serve to introduce a few significant points of the theory.

For the purposes of this chapter, we consider optimization to mean *minimization* of some function f over a domain $\Omega \subset \mathbb{R}^n$. When a function g is to be maximized, we will consider instead the minimization of $f = -g$.

18.1 Basic definitions

Consider the function shown in figure 18.1, where the domain is $\Omega = \{x : x \geq 0\}$. In that function, there are three points at which the function obtains a minimum; the point x_1 is a local minimum, while the point x_2 is a global minimum and x_4 is a boundary point at which a minimum occurs. The concept demonstrated by the figure is formalized in the following definition.

Definition 18.1 A point $\mathbf{x}^* \in \Omega$ is said to be a **local minimum point** (or minimizer, or weak local minimum) of $f : \mathbb{R}^n \to \mathbb{R}$ over the domain Ω if there is an $\epsilon > 0$ such that $f(\mathbf{x}) \geq f(\mathbf{x}^*)$, for all $\mathbf{x} \in \Omega$ where $|\mathbf{x} - \mathbf{x}^*| < \epsilon$. If $f(\mathbf{x}) > f(\mathbf{x}^*)$ for all such $\mathbf{x}$, then the point $\mathbf{x}^*$ is a **strict** (or **strong**) **local minimum**.

A point $\mathbf{x}^* \in \Omega$ is a **global minimum** point of f over Ω if $f(\mathbf{x}) \geq f(\mathbf{x}^*)$ for all $\mathbf{x} \in \Omega$. □

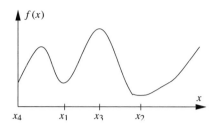

Figure 18.1: Examples of minimizing points.

Consider again the function in figure 18.1. The point x_1 is a strong local minimum. The point x_2 is a weak local minimum. At the points x_1 and x_2, the derivative of the function is zero. At these points, the second derivative is positive. At the point x_3, the derivative is zero, but the second derivative is negative—x_3 is a maximum point. At the point x_4 the derivative is nonzero, even though a minimum occurs there, this is because x_4 is on the boundary of Ω. The zero-derivative condition provides a way to characterize extrema (minimizers, maximizers, or points of inflection) on the interior of a domain, but not on the boundary of the domain. We are looking for a way to generalize the derivative condition to characterize the minimum (or maximum) on the boundary of a domain, and to generalize it in such a way that it can be extended to multiple dimensions.

Let $\mathbf{x} \in \Omega$, and consider points $\mathbf{y} = \mathbf{x} + \alpha \mathbf{d}$ for a scalar α and a displacement vector $\mathbf{d}$. The point $\mathbf{y}$ is said to be a **feasible point** if $\mathbf{y} \in \Omega$, and the displacement vector $\mathbf{d}$ is said to be a **feasible direction** at $\mathbf{x}$ if there is an $\epsilon_0 > 0$ such that $\mathbf{x} + \epsilon \mathbf{d} \in \Omega$ for all ϵ where $0 \leq \epsilon \leq \epsilon_0$.

For the point x_4 in figure 18.1, a feasible direction is $d = 1$, but $d = -1$ is not a feasible direction. For the points x_1 and x_2, both $d = 1$ or $d = -1$ are feasible directions.

We recall the definition of the gradient of a (real) function defined over a subset of $\mathbb{R}^n$, and define the gradient of a C^1 function f by

$$\frac{\partial f(\mathbf{x})}{\partial \mathbf{x}} = \nabla_x f(\mathbf{x}) = \begin{bmatrix} \frac{\partial f(\mathbf{x})}{\partial x_1} \\ \frac{\partial f(\mathbf{x})}{\partial x_2} \\ \vdots \\ \frac{\partial f(\mathbf{x})}{\partial x_n} \end{bmatrix}.$$

In many circumstances, the derivative is taken with respect to variables that are clear from the context. In such cases, we omit the subscript. Thus, we may write $\nabla_x f(x)$ or $\nabla f(x)$.

We define the Hessian (second derivative) of the C^2 function to be the matrix

$$\nabla^2 f(\mathbf{x}) = \begin{bmatrix} \frac{\partial^2 f}{\partial x_1 \partial x_1} & \frac{\partial^2 f}{\partial x_1 \partial x_2} & \cdots & \frac{\partial^2 f}{\partial x_1 \partial x_n} \\ \frac{\partial^2 f}{\partial x_2 \partial x_1} & \frac{\partial^2 f}{\partial x_2 \partial x_2} & \cdots & \frac{\partial^2 f}{\partial x_2 \partial x_n} \\ \vdots & & & \\ \frac{\partial^2 f}{\partial x_n \partial x_1} & \frac{\partial^2 f}{\partial x_n \partial x_2} & \cdots & \frac{\partial^2 f}{\partial x_n \partial x_n} \end{bmatrix}.$$

We are now led to the following theorem.

18.1 Basic Definitions

Theorem 18.1 *(Necessary conditions for minimality [210, pages 169, 174])* Let $\Omega \subset \mathbb{R}^n$, and let $f: \mathbb{R}^n \to \mathbb{R}$ be a C^1 function on $\Omega \subset \mathbb{R}^n$.

1. If $\mathbf{x}^*$ is a local minimum of f over Ω, then for any $\mathbf{d} \in \mathbb{R}^n$ that is a feasible direction at $\mathbf{x}^*$, we have
$$(\nabla f(\mathbf{x}^*))^T \mathbf{d} \geq 0. \tag{18.1}$$

2. If $\mathbf{x}^*$ is an interior point of Ω, then
$$\nabla f(\mathbf{x}^*) = \mathbf{0}.$$

3. If, in addition, $f \in C^2$ and $(\nabla f(\mathbf{x}^*))^T \mathbf{d} = 0$, then
$$\mathbf{d}^T \nabla^2 f(\mathbf{x}^*) \mathbf{d} \geq 0.$$

Proof

1. Let $\mathbf{x}(\epsilon) = \mathbf{x}^* + \epsilon \mathbf{d}$, and define the function $g(\epsilon) = f(\mathbf{x}(\epsilon))$. The function $g(\epsilon)$ has a relative minimum at $\epsilon = 0$ since $\mathbf{x}^*$ is a minimum, and has a Taylor series expansion about $\epsilon = 0$ as
$$g(\epsilon) = g(0) + g'(0)\epsilon + o(\epsilon),$$
so that
$$g(\epsilon) - g(0) = g'(0)\epsilon + o(\epsilon).$$
If $g'(0) < 0$, then for sufficiently small values of ϵ we must have $g(\epsilon) - g(0) < 0$, which contradicts the minimality of $g(0)$. Thus $g'(0) \geq 0$. But we observe (see section 18.2) that
$$g'(0) = \frac{d}{d\epsilon} f(\mathbf{x}(\epsilon))\bigg|_{\epsilon=0} = \sum_{i=1}^n \frac{d}{dx_i} f(\mathbf{x}(\epsilon)) \frac{\partial x_i}{\partial \epsilon}\bigg|_{\epsilon=0}$$
$$= \sum_{i=1}^n \frac{d}{dx_i} f(\mathbf{x}(\epsilon)) d_i \bigg|_{\epsilon=0}$$
$$= (\nabla f(\mathbf{x}^*))^T \mathbf{d}.$$

2. If $\mathbf{x}^*$ is an interior point, then every direction is feasible, so that if $\mathbf{d}$ is feasible then so is $-\mathbf{d}$. Hence, if $(\nabla f(\mathbf{x}))^T \mathbf{d} \geq 0$ and $-(\nabla f(\mathbf{x}))^T \mathbf{d} \geq 0$, we must have $\nabla f(\mathbf{x}) = \mathbf{0}$.

3. In the case that $(\nabla f(\mathbf{x}^*))^T \mathbf{d} = 0$, from the Taylor series for $g(\epsilon)$ we obtain
$$g(\epsilon) - g(0) = \frac{1}{2} g''(0) \epsilon^2 + o(\epsilon^2).$$
If $g''(0) < 0$, then for sufficiently small values of ϵ we have $g(\epsilon) - g(0) < 0$, which contradicts the minimality of $g(0)$. Thus, we must have $g''(0) \geq 0$, and we compute that
$$g''(0) = \mathbf{d}^T \nabla^2 f(\mathbf{x}) \mathbf{d} \geq 0. \qquad \square$$

Observe that if the point $\mathbf{x}^*$ is an interior point of Ω, then every direction is feasible, and so
$$\mathbf{d}^T \nabla^2 f(\mathbf{x}^*) \mathbf{d} \geq 0$$
for any $\mathbf{d}$; that is, the Hessian $\nabla^2 f(\mathbf{x}^*)$ is *positive semidefinite*.

Example 18.1.1 Let

$$f(x_1, x_2, x_3) = 3x_1^2 + 2x_1x_2 + 3x_2^2 - 20x_1 + 4x_2$$

be minimized subject to $x_1 \geq 0$, $x_2 \geq 0$. The derivatives are

$$\frac{\partial f}{\partial x_1} = 6x_1 + 2x_2 - 20,$$

$$\frac{\partial f}{\partial x_2} = 2x_1 + 6x_2 + 4.$$

In the absence of the constraints, the global minimum would be at $(x_1, x_2) = (4, -2)$, where the function takes on the value -44. However, the minimum taking into account the constraints is at $(x_1, x_2) = (10/3, 0)$. At this point, the function takes on the value -33.33, and the derivatives take on the values

$$\frac{\partial f}{\partial x_1} = 6x_1 + 2x_2 - 20 = 0,$$

$$\frac{\partial f}{\partial x_2} = 2x_1 + 6x_2 + 4 = 32/3.$$

Because of the constraint, both derivatives do not vanish at the minimum point, but since any feasible direction has a positive x_2 component, condition (18.1) is satisfied.

Figure 18.2 shows the contours of $f(x_1, x_2)$, where the unconstrained global minimum is indicated with ○ and the constrained minimum is indicated with ×. For future reference, we make the observation that the constraint line $\mathbf{x}_2 = 0$, indicated with the dotted, line is tangent to the contour at the point of minimum. □

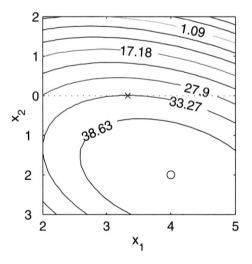

Figure 18.2: Contours of $f(x_1, x_2)$, showing minimum and constrained minimum.

A converse to theorem 18.1 provides sufficient conditions for local minimization when the minimum is in the interior of the domain:

Theorem 18.2 *([210, page 176]) Let $f \in C^2$ be a function defined on a region Ω, and let $\mathbf{x}^*$ be in the interior of Ω. If $\nabla f(\mathbf{x}^*) = 0$ and $\nabla^2 f(\mathbf{x}^*)$ is positive definite, then $\mathbf{x}^*$ is a strict local minimum of f.*

18.2 Generalization of the Chain Rule to Composite Functions

Proof Expanding $f(\mathbf{x}^* + \mathbf{d})$ using Taylor's theorem, we have

$$f(\mathbf{x}^* + \mathbf{d}) = f(\mathbf{x}^*) + (\nabla f(\mathbf{x}^*))^T \mathbf{d} + \frac{1}{2}\mathbf{d}^T \nabla^2 f(\mathbf{x}^*)\mathbf{d} + o(|\mathbf{d}|^2),$$

which, in light of the conditions of the theorem, can be written as

$$f(\mathbf{x}^* + \mathbf{d}) - f(\mathbf{x}^*) = \frac{1}{2}\mathbf{d}^T \nabla^2 f(\mathbf{x}^*)\mathbf{d} + o(|\mathbf{d}|^2). \tag{18.2}$$

Since $\nabla^2 f(\mathbf{x}^*)$ is (by hypothesis) positive definite, then for any $\mathbf{d}$ there is an $\epsilon > 0$ such that

$$\mathbf{d}^T \nabla^2 f(\mathbf{x}^*)\mathbf{d} \geq \epsilon |\mathbf{d}|^2.$$

Employing this in (18.2), we have

$$f(\mathbf{x}^* + \mathbf{d}) - f(\mathbf{x}) \geq \epsilon/2 |\mathbf{d}|^2 + o(|\mathbf{d}|^2).$$

For $|\mathbf{d}|$ sufficiently small, the first term dominates the second, and we have that $f(\mathbf{x}^* + \mathbf{d}) > f(\mathbf{x}^*)$. $\square$

18.2 Generalization of the chain rule to composite functions

In the following development, we need to take derivatives of composite functions, as a generalization of the familiar chain rule

$$\boxed{\frac{d}{dx}(f(g(x)) = f'(g(x))g'(x)}$$

to functions that are composites of multiple functions and multiple variables. We introduce the generalization by an example. Let

$$f(x, y) = x^2 y,$$
$$g(x, y) = 3y^2 x,$$
$$h(x, y) = x - 2y.$$

Also let

$$F(x, y) = f(g(x, y), h(x, y)).$$

We desire to compute $\frac{\partial F}{\partial x}$ and $\frac{\partial F}{\partial y}$. It is helpful to introduce auxiliary variables. Let

$$v = g(x, y) \qquad w = h(x, y),$$

so that

$$u = f(v, w).$$

Figure 18.3 illustrates the relationships among the variables and functions. The first partial derivative may be computed as

$$\frac{\partial F}{\partial x} = \frac{\partial u}{\partial x}$$
$$= \frac{\partial u}{\partial v}\frac{\partial v}{\partial x} + \frac{\partial u}{\partial w}\frac{\partial w}{\partial x}.$$

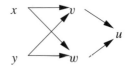

Figure 18.3: Relationships between variables in composite functions.

In our example, we get

$$\frac{\partial u}{\partial v} = 2vw = 6x^2y^2 - 12xy^3,$$

$$\frac{\partial v}{\partial x} = 3y^2,$$

$$\frac{\partial u}{\partial w} = v^2 = 9x^2y^4,$$

$$\frac{\partial w}{\partial x} = 1.$$

Then

$$\frac{\partial F}{\partial x} = 27x^2y^4 - 36xy^5.$$

More generally, suppose that $x_1, x_2, \ldots, x_n$ are the independent variables, and that there are m functions $g_1(x_1, \ldots, x_n), g_2(x_1, \ldots, x_n), \ldots, g_m(x_1, \ldots, x_n)$. If our function of interest is

$$F(x_1, \ldots, x_n) = f(g_1, g_2, \ldots, g_m),$$

then the derivative is

$$\frac{\partial F}{\partial x_j} = \sum_{i=1}^{m} D_i f D_j g_i, \tag{18.3}$$

where the operator D_k is the derivative with respect to the kth argument.

The situation can be more complicated when there are multiple dependencies. For example, let

$$u = f(x, v, w) \qquad v = g(x, w, y) \qquad w = h(x, y).$$

The functional dependencies are shown in figure 18.4. Each path from x to u represents a term that must be included in the derivative. Thus,

$$\frac{\partial u}{\partial x} = \underbrace{\frac{\partial u}{\partial x}}_{(*)} + \underbrace{\frac{\partial u}{\partial v}\frac{\partial v}{\partial x}}_{(**)} + \underbrace{\frac{\partial u}{\partial w}\frac{\partial w}{\partial x}}_{(***)} + \underbrace{\frac{\partial u}{\partial v}\frac{\partial v}{\partial w}\frac{\partial w}{\partial x}}_{(****)}.$$

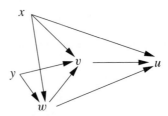

Figure 18.4: Illustration of functional dependencies.

The correspondence between the terms in this partial derivative and the paths in figure 18.4 are

$$(*): x \to u,$$
$$(**): x \to v \to u,$$
$$(***): x \to w \to u,$$
$$(****): x \to w \to v \to u.$$

The notation may be more clear by writing

$$\frac{\partial u}{\partial x} = D_1 f + D_2 f D_1 g + D_3 f D_1 h + D_2 f D_2 g D_1 h.$$

18.3 Definitions for constrained optimization

Constraints may enter into an optimization problem in a variety of ways. For example, there may be a constraint that all the variables in a problem must be integers; this frequently leads to equations known as Diophantine equations, which have their own methods of solution (when they can be solved at all). Or there may be a constraint as to the nature of a signal—for example, the signal may need to be positive—where there is no obvious connection between the constraint and the available parameters. In this chapter, however, we restrict our attention to optimization problems that can be expressed in the following form:

$$\begin{aligned}
\text{minimize} \quad & f(\mathbf{x}) \\
\text{subject to} \quad & h_1(\mathbf{x}) = 0, \\
& h_2(\mathbf{x}) = 0, \\
& \quad \vdots \\
& h_m(\mathbf{x}) = 0, \\
& g_1(\mathbf{x}) \leq 0, \\
& g_2(\mathbf{x}) \leq 0, \\
& \quad \vdots \\
& g_p(\mathbf{x}) \leq 0, \\
& \mathbf{x} \in \Omega,
\end{aligned} \quad (18.4)$$

for $\Omega \subset \mathbb{R}^n$. This can be represented more briefly by letting $\mathbf{h} = (h_1, h_2, \ldots, h_m)$ and $\mathbf{g} = (g_1, g_2, \ldots, g_p)$, so the problem is stated as

$$\begin{aligned}
\text{minimize} \quad & f(\mathbf{x}) \\
\text{subject to} \quad & \mathbf{h}(\mathbf{x}) = \mathbf{0}, \\
& \mathbf{g}(\mathbf{x}) \leq \mathbf{0}, \\
& \mathbf{x} \in \Omega.
\end{aligned} \quad (18.5)$$

The constraints $\mathbf{h}(\mathbf{x}) = \mathbf{0}$ are said to be *equality* constraints, and the constraints $\mathbf{g}(\mathbf{x}) \leq \mathbf{0}$ are *inequality* constraints; combined, they form the functional constraints. A point $\mathbf{x} \in \Omega$ that satisfies all of the functional constraints is said to be *feasible*. The optimization problem can be stated in words as: determine that value of $\mathbf{x}$ that is feasible and that minimizes $f(\mathbf{x})$.

Example 18.3.1 Consider again the optimization problem of example 18.1.1,

$$\begin{aligned}
\text{minimize} \quad & f(x_1, x_2) = 3x_1^2 + 2x_1 x_2 + 3x_2^2 - 20x_1 + 4x_2 \\
\text{subject to} \quad & x_1 \geq 0, \\
& x_2 \geq 0.
\end{aligned}$$

The minimum is at $(x_1, x_2) = (10/3, 0)$. The element of the solution $x_2 = 0$ is on the boundary of the feasible region, making the constraint associated with x_2 *active*. The component of the solution $x_1 = 10/3$ is not on the boundary of the feasible region; the constraint on x_1 is *inactive*. □

In general, a constraint $g_i(\mathbf{x}) \leq 0$ is said to be active at a feasible point $\mathbf{x}$ if $g_i(\mathbf{x}) = 0$, so that $\mathbf{x}$ is on the boundary of the feasible region. Note that a constraint of the form $h_i(\mathbf{x}) = 0$ must always be active in the feasible region. A constraint $g_i(\mathbf{x}) \leq 0$ is inactive if $g_i(\mathbf{x}_i) < 0$. An inactive constraint has no influence on the solution in a sufficiently small neighborhood of $\mathbf{x}$, whereas the active constraints restrict the feasible values of $\mathbf{x}$. All active constraints affect the problem locally as equality constraints. If the set of active constraints could be determined at any point $\mathbf{x}$, then the problem could formulated at that point in terms of equality constraints alone. This observation will guide most of the theoretical development that follows.

18.4 Equality constraints; Lagrange multipliers

We assume for now that the optimization problem of interest has only equality constraints. (Any inequality constraints are either inactive, or are regarded here as belonging to the set of equality constraints.) For $\mathbf{x} \in \mathbb{R}^n$, a set of m constraints

$$h_1(\mathbf{x}) = 0 \quad h_2(\mathbf{x}) = 0 \quad \cdots \quad h_m(\mathbf{x}) = 0,$$

which we also write as

$$\mathbf{h}(\mathbf{x}) = \mathbf{0},$$

determines a hypersurface S of $n - m$ dimensions. In order to apply the calculus-based theory of this chapter, we assume that the surface S is smooth; in practice, we need at least $h_i \in C^1$. (If this smoothness requirement is not met, then more specialized techniques must be applied, which are beyond the scope of this chapter.)

We first argue geometrically that:

> A local extremum $\mathbf{x}^*$ of f, subject to the equality constraints $\mathbf{h}(\mathbf{x}) = \mathbf{0}$, is obtained when *the gradient of f is orthogonal to the tangent plane of* $\mathbf{h}$ at $\mathbf{x}^*$.

We then back up this geometric viewpoint with additional mathematical justification.

Recall that the gradient of a function $f(\mathbf{x})$ points in a direction normal to the contours of the function. At a point $\mathbf{x}$, the gradient of f points "uphill"—toward increasing values of f—so that motion in a direction opposite the gradient is downhill. This fact is the basis of the gradient-descent optimization procedure introduced in section 14.5. Consider the function $f(\mathbf{x})$ as a mountain range. In the absence of constraints, a marble released upon a mountain will roll downhill (in the direction of the negative gradient) until it arrives at the lowest point in the valley in which it initially finds itself.

An equality constraint introduces a surface upon which the final solution must occur. A marble rolling downhill will continue in the negative direction of the gradient until it runs into the constraint surface, and it must remain on this surface (satisfy the equality constraint). At this point, it will slide downhill along the constraint surface until the surface cannot allow any more downhill motion. This occurs when the gradient of f at a point $\mathbf{x}$ is pointing in a direction orthogonal to the constraint $h(\mathbf{x})$. Any other motion along the constraint curve can only increase the function, while motions not along the curve are restricted by the constraint.

As an example, consider the function

$$f(x_1, x_2) = 10\cos(2\pi x_1) + 1)(2x_2 - .7)^2 + 10$$

18.4 Equality Constraints; Lagrange Multipliers

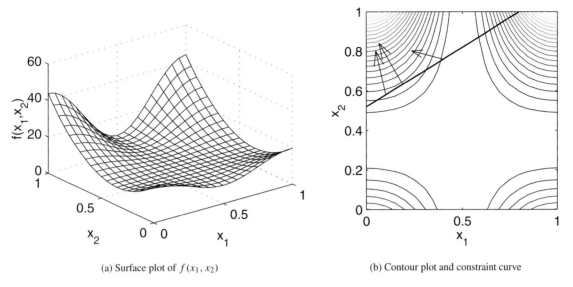

(a) Surface plot of $f(x_1, x_2)$

(b) Contour plot and constraint curve

Figure 18.5: Surface and contour plots of $f(x_1, x_2)$. The arrows in (b) indicate gradient directions.

plotted in figure 18.5(a). The global minimum occurs at $(x_1, x_2) = (0.5, 0.7)$. The contours of this function are shown in figure 18.5(b). Now, consider finding the minimum of this function, subject to the constraint

$$h(x_1, x_2) = 0.6x_1 + 0.52 - x_2 = 0, \tag{18.6}$$

which is the "curve" (actually a line, for plotting convenience) shown on the contours of figure 18.5(b). The gradient vectors of the function are shown at three different points along the constraint line. At the point of constrained minimum, $(x_1 = 0.1888, 0.6333)$, the gradient vector is orthogonal to the constraint line.

In extending this geometric concept to more constraints and higher dimensions, we need to describe the concept of a tangent plane to a surface at a point and to review how planes are described analytically. Figure 18.6 illustrates a curve and a surface, and "planes" tangent to each of them. (The tangent "plane" to a curve is simply a line.) The tangent plane is defined as follows. We define a curve C on a surface S as the set of points $\mathbf{x}(\xi) \in S$ continuously parameterized by $\xi \in \mathbb{R}$ for some $a \leq \xi \leq b$. Figure 18.7 illustrates several curves on a surface S. The derivative of the curve at the point $\mathbf{x}(\xi)$ is

$$\dot{\mathbf{x}}(\xi) = \frac{d}{d\xi}\mathbf{x}(\xi).$$

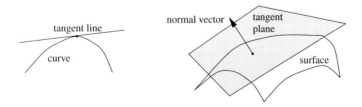

Figure 18.6: Tangent plane to a surface.

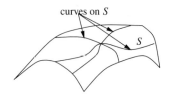

Figure 18.7: Curves on a surface.

(The dot in this case indicates differentiation with respect to the independent parameter ξ.) The derivative is a vector[1] in $\mathbb{R}^n$ that points in the direction of the tangent to the curve C at the point $\mathbf{x}(\xi)$. Different curves defined upon a surface S have different derivatives (different directions) at a point $\mathbf{x}$.

Definition 18.2 The **tangent plane** to a surface S at a point $\mathbf{x}$ on S is the linear span of the derivatives of all of the differentiable curves on S at $\mathbf{x}$. □

Example 18.4.1 The "curve" defined by (18.6) is described by the set

$$S = \{x_1, x_2 \colon h(x_1, x_2) = 0\}.$$

A parameterization of points on the set is

$$x_1(\xi) = \xi, \qquad x_2(\xi) = 0.6\xi + 0.52.$$

The tangent vector at a point (x_1, x_2) points in the direction

$$\frac{\partial}{\partial \xi}[x_1(\xi), x_2(\xi)]^T = [1, .6]^T.$$

(Since the "curve" is actually a straight line, the tangent vector is constant.) □

Example 18.4.2 Consider the constraint curve $h(x_1, x_2) = x_1 \cos(x_2) - 1$. The surface (a curve) described by this constraint is

$$S = \{x_1, x_2 \colon x_1 \cos(x_2) - 1 = 0\}.$$

A parameterization of points on S is

$$x_1 = \frac{1}{\cos \xi} \qquad x_2 = \xi.$$

The tangent to S at a point (x_1, x_2) is

$$\frac{\partial}{\partial \xi}[x_1, x_2]^T = [\sec x_2 \tan x_2, 1]^T.$$

□

At a point $\mathbf{x}^*$, the gradient to a function $h(\mathbf{x})$ is *orthogonal* to the level curves $h(\mathbf{x}) = c$.

Example 18.4.3 The gradient of the constraint function $h(x_1, x_2)$ in (18.6) is

$$\frac{\partial h}{\partial \mathbf{x}} = [.6, -1]^T,$$

---

[1]Strictly speaking, there is a difference between a "vector" that is a gradient and a "vector" that is a tangent vector. In the language of differential geometry, a gradient is a *covariant vector* and a tangent to a curve is a *contravariant vector* (see, for example, [297]). However, we have no need of this distinction here.

18.4 Equality Constraints; Lagrange Multipliers

a vector that is orthogonal to the level curve $h(x_1, x_2) = c$, which has a tangent pointing in the direction $[1, .6]^T$. □

If $\mathbf{x}(\xi)$ is a parameterization of points satisfying $h(x(\xi)) = 0$ (that is, $\mathbf{x}(\xi)$ is a curve on the "surface" $h(\mathbf{x}) = 0$), then, as mentioned, $d/d\xi \mathbf{x}(\xi)$ is *tangent* to the curve. These ideas are explored in exercise 18.4-2.

Now recall that a plane P passing through a point $\mathbf{x}^*$ is defined in terms of the point $\mathbf{x}^*$ and a vector $\mathbf{n}$ normal to the surface at $\mathbf{x}^*$. If $\mathbf{n}$ is a vector normal to the surface at $\mathbf{x}^*$, then the plane can be written as

$$P = \{\mathbf{y} \in \mathbb{R}^n : \mathbf{n}^T(\mathbf{y} - \mathbf{x}^*) = 0\}.$$

We can use this concept to define a tangent plane to a surface at a point $\mathbf{x}^*$: The tangent plane P to a surface S defined by $\mathbf{h}(\mathbf{x}) = 0$ at the point $\mathbf{x}^*$ is given by

$$P = \{\mathbf{y} \in \mathbb{R}^n : \nabla \mathbf{h}(\mathbf{x})^T(\mathbf{y} - \mathbf{x}^*) = 0\},$$

where

$$\nabla \mathbf{h}(\mathbf{x}) = [\nabla h_1(\mathbf{x}) \quad \nabla h_2(\mathbf{x})) \quad \cdots \quad \nabla h_m(\mathbf{x}))] = \begin{bmatrix} \frac{\partial h_1}{\partial x_1} & \frac{\partial h_2}{\partial x_1} & \cdots & \frac{\partial h_m}{\partial x_1} \\ \frac{\partial h_1}{\partial x_2} & \frac{\partial h_2}{\partial x_2} & \cdots & \frac{\partial h_m}{\partial x_2} \\ \vdots & & & \\ \frac{\partial h_1}{\partial x_n} & \frac{\partial h_2}{\partial x_2} & \cdots & \frac{\partial h_m}{\partial x_n} \end{bmatrix}.$$

Example 18.4.4 For $h(x_1, x_2)$ defined in (18.6), we have

$$\nabla h(\mathbf{x}) = [.6 \quad -1]^T,$$

and the tangent plane at a point (x_1, x_2) is described by

$$P = \{(y_1, y_2) : [.6 \quad -1][y_1 - x_1 \quad y_2 - x_2] = 0\}$$

which, given that (x_1, x_2) satisfy (18.6), is equivalent to

$$0.6y_1 - y_2 + 0.52 = 0,$$

as expected. □

For reasons to be clarified, in this specification of the tangent plane, it is important that the gradient vectors

$$\nabla h_1(\mathbf{x}), \nabla h_2(\mathbf{x}), \ldots, \nabla h_m(\mathbf{x})$$

be *linearly independent* at the point $\mathbf{x} = \mathbf{x}^*$.

Definition 18.3 When the gradient vectors $\nabla h_1, \nabla h_2, \ldots, \nabla h_m$ are linearly independent at $\mathbf{x}^*$, the point $\mathbf{x}^*$ is said to be a **regular point**. □

Example 18.4.5 To see the impact of having a point that is irregular, let $h(x_1, x_2) = x_2$. Then the surface $h(\mathbf{x}) = 0$ is the x_1 axis, and

$$\nabla h(\mathbf{x}) = \begin{bmatrix} 0 \\ 1 \end{bmatrix}.$$

The tangent plane is the x_1 axis.

Now consider $\tilde{h}(x_1, x_2) = x_2^2$. The surface $\tilde{h}(\mathbf{x}) = 0$ is the same as before, the x_1 axis, which also has the x_1 as the tangent plane. However, the gradient gives

$$\nabla \tilde{h}(\mathbf{x}) = \begin{bmatrix} 0 \\ 2x_2 \end{bmatrix}.$$

At $(x_1, x_2) = (0, 0)$, the set of points

$$P = \{y \in \mathbb{R}^2 : (\nabla \tilde{h}(\mathbf{x}))^T \mathbf{y} = 0\}$$

is the entire plane $\mathbb{R}^2$, which is not the same as the tangent plane. This is because $\tilde{h}$ is irregular at the origin. □

As this example shows, the attribute of being a regular point depends upon the representation of the function $\mathbf{h}$. We assume that the representation has been chosen so that points have the regularity property as necessary.

Given the notion of a curve on the constraint surface, we are now ready for the following lemma.

Lemma 18.1 *Let $\mathbf{x}(\xi)$ be a parameterization of a curve on the surface $\mathbf{h}(\mathbf{x}) = \mathbf{0}$, where the parameterization is chosen so that $\mathbf{x}^* = \mathbf{x}(0)$ is a constrained local minimum of f. Then*

$$\left. \frac{d}{d\xi} f(\mathbf{x}(\xi)) \right|_{\xi=0} = 0. \tag{18.7}$$

Proof Expand $f(\mathbf{x}(\xi))$ as a Taylor series,

$$f(\mathbf{x}(\xi)) = f(\mathbf{x}(0)) + \xi \frac{d}{d\xi} f(\mathbf{x}(\xi)) + o(\xi),$$

so that

$$f(\mathbf{x}(\xi) - f(\mathbf{x}(0)) = \xi \frac{d}{d\xi} f(\mathbf{x}(\xi)) + o(\xi).$$

If $f(\mathbf{x}(0))$ is a local minimum, then for sufficiently small $|\xi|$ we must have $f(\mathbf{x}(\xi)) - f(\mathbf{x}(0)) \geq 0$. Since ξ can be both positive and negative, we must therefore have $\frac{d}{d\xi} f(\mathbf{x}(\xi)) = 0$ and, furthermore, it follows that $\frac{d^2}{d\xi^2} f(\mathbf{x}(\xi))$ must be positive as well. □

Recall from section 18.2 that we can write

$$\frac{d}{d\xi} f(\mathbf{x}(\xi^*)) = \sum_{i=1}^{n} \frac{\partial f}{\partial x_i^*} \dot{\mathbf{x}}(\xi^*)_i = (\nabla f(\mathbf{x}^*) \dot{\mathbf{x}}(\xi^*)). \tag{18.8}$$

Since $\dot{\mathbf{x}}(\xi^*)$ is a vector tangent to the surface $\mathbf{h}(\mathbf{x}) = \mathbf{0}$, we can state the geometric result of the constrained minimization problem as follows: **at a feasible point $\mathbf{x}^*$ of an extremum of f, the gradient $\nabla f(\mathbf{x}^*)$ is orthogonal to the plane tangent to $\mathbf{h}(\mathbf{x}^*)$**. More formally, we have the following.

Lemma 18.2 *Let $\mathbf{x}^*$ be a regular point of the surface defined by the constraints $\mathbf{h}(\mathbf{x}) = \mathbf{0}$ and a local extremum of f subject to these constraints. For all $\mathbf{y} \in \mathbb{R}^n$ such that*

$$\nabla \mathbf{h}(\mathbf{x}^*)^T \mathbf{y} = 0$$

(that is, for any vector in the tangent plane), then

$$\nabla f(\mathbf{x}^*) \mathbf{y} = 0.$$

Proof Let the coordinate system be translated so that $\mathbf{x}^* = \mathbf{0}$, to simplify our notation. Since $\mathbf{x}^*$ is a regular point, the tangent plane is equivalent to the set of vectors $\mathbf{z}$ satisfying

18.4 Equality Constraints; Lagrange Multipliers

$\nabla \mathbf{h}(\mathbf{x}^*)\mathbf{z} = 0$. Let $\mathbf{y}$ be any particular vector on the tangent plane. Then we have

$$\nabla \mathbf{h}(\mathbf{x}^*)\mathbf{y} = 0.$$

Let $\mathbf{x}(\xi)$ be a smooth curve on the constraint surface such that $\mathbf{x}(0) = \mathbf{x}^*$ and $\dot{\mathbf{x}}(0) = \mathbf{y}$. From (18.7) and (18.8), we have

$$0 = \left.\frac{d}{d\xi} f(\mathbf{x}(\xi))\right|_{\xi=0} = (\nabla f(\mathbf{x}^*))^T \mathbf{y}.$$

$\square$

Another way of looking at the geometry of the problem is apparent from figure 18.5. At a point of constrained extremum $\mathbf{x}^*$, the gradient $\nabla f(\mathbf{x}^*)$ is parallel to each gradient $\nabla h_i(\mathbf{x})$, $i = 1, 2, \ldots, m$, so that there exist constants λ_i such that

$$\nabla f(\mathbf{x}^*) + \lambda_i \nabla h_i(\mathbf{x}^*) = 0, \quad i = 1, 2, \ldots, m.$$

Since these must hold simultaneously, they hold in the sum,

$$\nabla f(\mathbf{x}^*) + \sum_{i=1}^{m} \lambda_i \nabla h_i(\mathbf{x}^*) = 0.$$

Letting

$$\boldsymbol{\lambda} = \begin{bmatrix} \lambda_1 \\ \lambda_2 \\ \vdots \\ \lambda_m \end{bmatrix},$$

we can write this as

$$\nabla f(\mathbf{x}^*) + \nabla \mathbf{h}(\mathbf{x}^*)\boldsymbol{\lambda} = 0.$$

We have thus established the following theorem, which is the "bottom line" for equality-constrained optimization problems.

Theorem 18.3 *(Necessary conditions for equality constraints) Let $\mathbf{x}^*$ be a local extremum point of f subject to the m constraints $\mathbf{h}(\mathbf{x}) = 0$, and let $\mathbf{x}^*$ be a regular point of these constraints. Then there is a $\boldsymbol{\lambda} \in \mathbb{R}^m$ such that*

$$\boxed{\nabla f(\mathbf{x}^*) + \nabla \mathbf{h}(\mathbf{x}^*)\boldsymbol{\lambda} = 0} \qquad (18.9)$$

The quantities λ_i are called **Lagrange multipliers**. In the interest of making the condition clearer it can be rewritten as

$$\boxed{\nabla f(\mathbf{x}^*) + \sum_{i=1}^{m} \lambda_i \nabla h_i(\mathbf{x}^*))^T = 0}$$

In the constrained optimization problem as stated, there are n unknowns in $\mathbf{x}^*$ and m unknowns in $\boldsymbol{\lambda}$. The m equations in $\mathbf{h}(\mathbf{x}^*) = \mathbf{0}$ and the n equations in $\nabla f(\mathbf{x}^*) + \nabla \mathbf{h}(\mathbf{x}^*)\boldsymbol{\lambda} = 0$ give a total of $n + m$ (generally nonlinear) equations, so that the solution is at least locally unique.

In formulating constrained optimization problems, it is common to formulate a *Lagrangian* as

$$L(\mathbf{x}, \boldsymbol{\lambda}) = f(\mathbf{x}) + \mathbf{h}(\mathbf{x})^T \boldsymbol{\lambda}.$$

The necessary conditions for optimization are then expressed as

$$\nabla_x L(\mathbf{x}, \boldsymbol{\lambda}) = 0,$$
$$\nabla_\lambda L(\mathbf{x}, \boldsymbol{\lambda}) = 0.$$

18.4.1 Examples of equality-constrained optimization

Example 18.4.6 We want to maximize

$$f(x_1, x_2) = x_1^2 + x_1 x_2 + x_2 x_3$$

subject to the single constraint

$$x_1 + x_2 = 4.$$

The constraint function is $h(x_1, x_2) = x_1 + x_2 - 4$.

The Lagrangian is

$$L(x_1, x_2, \lambda) = f(x_1, x_2) + \lambda(x_1 + x_2 - 4).$$

The necessary conditions are

$$\nabla_x L(x_1, x_2, \lambda) = \nabla_x f(x_1, x_2) + \lambda \nabla_x (x_1 + x_2 - 4) = \mathbf{0}$$
$$= \begin{bmatrix} 2x_1 + x_2 + \lambda \\ x_1 + x_3 + \lambda \\ x_2 \end{bmatrix}$$

and

$$\nabla_\lambda L(x_1, x_2, \lambda) = \nabla_\lambda (x_1 + x_2 - 4) = \mathbf{0}$$
$$= x_1 + x_2 - 4.$$

The latter condition, as always, is simply a restatement of the original constraints. Solution of these four equations leads to

$$\begin{bmatrix} x_1 \\ x_2 \\ x_3 \\ \lambda \end{bmatrix} = \begin{bmatrix} 4 \\ 0 \\ 4 \\ -8 \end{bmatrix}.$$

We are left with the question of whether this is a minimum, a maximum, or perhaps some kind of saddle point. This question is addressed in the following using second-order conditions. (We see in example 18.5.2 that this is in fact a saddle point.) □

Example 18.4.7 (Maximum entropy I) The entropy of a discrete random variable that takes on the values $x_1, x_2, \ldots, x_n$ with corresponding probabilities $\mathbf{p} = (p_1, p_2, \ldots, p_n)$ is

$$H = -\sum_{i=1}^{n} p_i \log p_i. \tag{18.10}$$

We desire to find the distribution $\mathbf{p}$ that maximizes the entropy. The constraints on the probabilities are

$$\sum_{i=1}^{n} p_i = 1,$$
$$p_i \geq 0.$$

18.4 Equality Constraints; Lagrange Multipliers

We will ignore for the moment the nonnegativity constraints, and return to check that they are satisfied. The Lagrangian is

$$L(\mathbf{p}, \lambda) = -\sum_{i=1}^{n} p_i \log p_i + \lambda \left(\sum_{i=1}^{n} p_i - 1 \right),$$

leading to the necessary conditions

$$-\log p_i - 1 + \lambda = 0 \quad i = 1, 2, \ldots, n.$$

Since this is true for every p_i, we must have $p_i = c = e^{\lambda-1}$, which is constant for all i. This means that $p_i = 1/n$, and the uniform distribution maximizes the entropy. Note that in this case it is not necessary to explicitly determine the Lagrange multiplier; simply observing its effect on the solution suffices. □

Example 18.4.8 (Maximum entropy II) We now want to maximize the entropy (18.10) subject to the constraint that the mean of the random variable is equal to m. The constraints are

$$\sum_{i=1}^{n} p_i = 1,$$

$$\sum_{i=1}^{n} p_i x_i = m.$$

The Lagrangian now incorporates two constraints, each with their own Lagrange multipliers λ_1 and λ_2:

$$L(\mathbf{p}, \lambda_1, \lambda_2) = H(\mathbf{p}) + \lambda_1 \left(\sum_{i=1}^{n} p_i - 1 \right) + \lambda_2 \left(\sum_{i=1}^{n} p_i x_i - m \right).$$

The necessary conditions are

$$-\log p_i - 1 + \lambda_1 + \lambda_2 x_i = 0, \quad i = 1, 2, \ldots, n,$$

so that

$$p_i = \exp[\lambda_1 - 1 + \lambda_2 x_i], \quad i = 1, 2, \ldots, n. \qquad (18.11)$$

These probabilities are all positive, so the nonnegativity constraints are inactive. The probabilities in (18.11) are of the form $p_i = c e^{\lambda_2 x_i}$, which is a discrete exponential density. The parameters c and λ_2 must be chosen (by solving nonlinear equations) to satisfy the constraints. □

Example 18.4.9 (Constrained least squares) Ellipsoidal sets arise in several signal-processing applications, such as set-valued Kalman filtering and set-based estimation. In some problems, it is of interest to determine the point of an ellipsoid nearest to a point outside the set. This is a constrained minimization problem.

Let $\|\cdot\|$ be the L_2 norm, let E be an ellipsoidal set in $\mathbb{R}^n$,

$$E = \{\mathbf{z} \in \mathbb{R}^n : \|L\mathbf{z}\| \leq 1\},$$

and let $\mathbf{y}$ be a point not in the ellipsoid, as shown in figure 18.8. Since it is clear geometrically that the nearest point to $\mathbf{y}$ must be on the boundary of the ellipse, the problem can be stated as

$$\text{minimize } \|\mathbf{y} - \mathbf{z}\|$$

$$\text{subject to } \|L\mathbf{z}\| = 1.$$

Since $\|\mathbf{z}\|^2 = \mathbf{z}^T \mathbf{z}$ is employed, the problem is differentiable, and is in fact a constrained quadratic minimization problem. This problem can be viewed as an example of a class of *constrained*

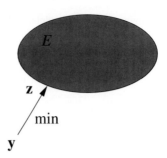

Figure 18.8: Minimizing the distance to an ellipse.

least-squares problems. A Lagrangian can be written as

$$L(\mathbf{z}, \lambda) = (\mathbf{y} - \mathbf{z})^T(\mathbf{y} - \mathbf{z}) + \lambda (L\mathbf{z})^T(L\mathbf{z}),$$

leading to the necessary conditions

$$(I + \lambda L^T L)\mathbf{z} = \mathbf{y}.$$

Since solution of this equation provides a least-squares solution, an equation of this sort for a constrained quadratic problem is sometimes called a *normal equation*. If λ were known, determination of the minimizing $\mathbf{z}$ would be straightforward. Solving for $\mathbf{z}$ in terms of the unknown λ we have

$$\mathbf{z}(\lambda) = (I + \lambda L^T L)^{-1}\mathbf{y}. \tag{18.12}$$

The computations can be simplified significantly using the SVD of L,

$$L = U\Sigma V^T,$$

from which

$$\mathbf{z}(\lambda) = V(I + \lambda \Sigma^2)\mathbf{e}$$

where $\mathbf{e} = V^T\mathbf{y}$. We then have

$$L\mathbf{z}(\lambda) = U\Sigma(I + \lambda \Sigma^2)^{-1}\mathbf{e}, \tag{18.13}$$

from which the constraint $\|L\mathbf{z}\| = 1$ becomes

$$\sum_{j=1}^{n} \frac{e_j^2 \sigma_j^2}{\left(1 + \lambda \sigma_j^2\right)^2} = 1, \tag{18.14}$$

where the σ_i are the singular values of L. An equation of the form (18.14), used to satisfy the constraint in an optimization problem, is sometimes called a *secular equation*. This nonlinear equation in the unknown λ can be readily solved numerically using Newton's method. □

Example 18.4.10 (Least squares with multiple constraints) Consider the following problem:

$$\begin{aligned}
\text{minimize} \quad & \mathbf{b}^T P \mathbf{b} \\
\text{subject to} \quad & \mathbf{b}^T \mathbf{c}_1 = \beta_1, \\
& \mathbf{b}^T \mathbf{c}_2 = \beta_2, \\
& \vdots \\
& \mathbf{b}^T \mathbf{c}_m = \beta_m,
\end{aligned}$$

18.5 Second-Order Conditions

Let

$$C = [\mathbf{c}_1 \; \mathbf{c}_2 \; \cdots \; \mathbf{c}_m] \quad \text{and} \quad \lambda = \begin{bmatrix} \lambda_1 \\ \lambda_2 \\ \vdots \\ \lambda_m \end{bmatrix} \quad \text{and} \quad \beta = \begin{bmatrix} \beta_1 \\ \beta_2 \\ \vdots \\ \beta_m \end{bmatrix}.$$

Then a Lagrangian (incorporating a factor of -2 for later convenience) is

$$\mathbf{b}^T P \mathbf{b} - 2\mathbf{b}^T C \lambda$$

with the constraint

$$\mathbf{b}^T C = \beta^T. \tag{18.15}$$

The necessary conditions are

$$2P\mathbf{b} - 2C\lambda = 0.$$

A solution (assuming that P is invertible) is

$$\mathbf{b} = P^{-1} C \lambda.$$

If P is not invertible, a pseudoinverse can be used. Incorporating the constraints using (18.15), we find (again, assuming invertibility) that

$$\lambda = (C^T P C)^{-1} \beta,$$

from which

$$\mathbf{b} = P^{-1} C (C^T P C)^{-1} \beta. \qquad \square$$

18.5 Second-order conditions

Second-order conditions provide information about whether an extremum is a minimum or maximum (or neither). Second-order conditions for constrained optimizations exist that are similar to those for unconstrained optimization.

We let $F(\mathbf{x})$ be the $n \times n$ Hessian of $f(\mathbf{x})$,

$$F(\mathbf{x}) = \left[\frac{\partial^2 f(\mathbf{x})}{\partial x_i \partial x_j}\right],$$

and $H(\mathbf{x}, \lambda)$ be the $n \times n$ matrix defined by

$$H(\mathbf{x}, \lambda) = \sum_{k=1}^{m} \frac{\partial^2 \mathbf{h}_k(\mathbf{x})}{\partial x_i \partial x_j} \lambda_k.$$

We introduce the matrix

$$\mathbf{L}(\mathbf{x}) = F(\mathbf{x}) + H(\mathbf{x}, \lambda)$$

as the matrix of second partial derivatives of the Lagrangian L.

Theorem 18.4 *(Second-order conditions, [210, page 306])*
Let f and $\mathbf{h}$ be in C^2.

1. *(Necessity)* Suppose there is a point $\mathbf{x}^*$ that is a local minimum of f subject to $\mathbf{h}(\mathbf{x}) = 0$, and that $\mathbf{x}^*$ is a regular point of $\mathbf{h}$. Then there is a vector λ such that

$$\nabla f(\mathbf{x}^*) + \nabla \mathbf{h}(\mathbf{x}^*) \lambda = 0.$$

Let P be the tangent plane of $\mathbf{h}(\mathbf{x})$ at $\mathbf{x}^*$, that is,
$$P = \{\mathbf{y}: \nabla \mathbf{h}(\mathbf{x}^*)\mathbf{y} = 0\};$$
then the matrix
$$\mathbf{L}(\mathbf{x}^*) = F(\mathbf{x}^*) + H(\mathbf{x}^*, \boldsymbol{\lambda}) \qquad (18.16)$$
is positive semidefinite for values of $\mathbf{y} \in P$. (We say that $\mathbf{L}(\mathbf{x}^*)$ is positive semidefinite on P.)

2. (Sufficiency) If $\mathbf{x}^*$ satisfies $\mathbf{h}(\mathbf{x}^*) = \mathbf{0}$, and there is a $\boldsymbol{\lambda}$ such that
$$\nabla f(\mathbf{x}^*) + \nabla \mathbf{h}(\mathbf{x}^*)\boldsymbol{\lambda} = \mathbf{0},$$
and there is a matrix $\mathbf{L}(\mathbf{x}^*)$ defined as in (18.16) that is positive definite on P, then $\mathbf{x}^*$ is a strict local minimum of f subject to $\mathbf{h}(\mathbf{x}^*) = 0$.

Proof

1. Let $\mathbf{x}(\xi)$ be a curve on the constraint surface S through $\mathbf{x}^*$ with $\mathbf{x}(0) = \mathbf{x}^*$. Then (see exercise 18.4-5),
$$\left. \frac{d^2}{d\xi^2} f(\mathbf{x}(\xi)) \right|_{\xi=0} \geq 0. \qquad (18.17)$$

We also have (see exercise 18.4-4)
$$\frac{d^2}{d\xi^2} f(\mathbf{x}(\xi)) = \dot{\mathbf{x}}(\xi)^T F \dot{\mathbf{x}} + \nabla f(\mathbf{x})^T \ddot{\mathbf{x}}(\xi), \qquad (18.18)$$

and (see exercise 18.5-6)
$$\left. \frac{d^2}{d\xi^2} \sum_{k=1}^m \mathbf{h}(\mathbf{x}(\xi))\lambda_k \right|_{\xi=0} = \dot{\mathbf{x}}(0)^T H(\mathbf{x}^*, \boldsymbol{\lambda})\dot{\mathbf{x}}(0) + \ddot{\mathbf{x}}(0)^T \nabla \mathbf{h} \boldsymbol{\lambda} = 0. \qquad (18.19)$$

Adding (18.19) and (18.18) using (18.17), we obtain
$$\dot{\mathbf{x}}(0)\mathbf{L}(\mathbf{x}^*)\dot{\mathbf{x}}(0) \geq 0.$$

Since $\dot{\mathbf{x}}(0)$ is arbitrary in P, $\mathbf{L}(\mathbf{x}^*)$ must be positive semidefinite on P.

2. Let $\mathbf{x}(\xi)$ be a curve on the surface $\mathbf{h}(\mathbf{x}) = 0$, with $\mathbf{x}(0) = \mathbf{x}^*$. We write out the Taylor series for $f(\mathbf{x}(\xi))$ to the quadratic term
$$f(\mathbf{x}(\xi)) = f(\mathbf{x}(0)) + \xi(\nabla f(\mathbf{x}^*))^T \mathbf{x}(0)$$
$$+ \frac{\xi^2}{2}[\dot{\mathbf{x}}(0) F \dot{\mathbf{x}}(0) + (\nabla f(\mathbf{x}(0)))^T \ddot{\mathbf{x}}(0)] + o(\xi^2), \qquad (18.20)$$

and the Taylor series for $h_i(\mathbf{x}(\xi))$, for $i = 1, 2, \ldots, m$, as
$$h_i(\mathbf{x}(\xi)) = h_i(\mathbf{x}(0)) + \xi(\nabla h_i(\mathbf{x}^*))^T \dot{\mathbf{x}}(0)$$
$$+ \frac{\xi^2}{2}[\dot{\mathbf{x}}^T(0) H_i \dot{\mathbf{x}}(0) + (\nabla h_i(\mathbf{x}^*))^T \ddot{\mathbf{x}}(0)] + o(\xi^2). \qquad (18.21)$$

We note that $h_i(\mathbf{x}(\xi))\big|_{\xi=0} = h_i(\mathbf{x}^*) = 0$. Now, for each $i = 1, 2, \ldots, m$, multiply (18.21) by λ_i, then add all these with m times (18.20). Using (18.16), we obtain
$$f(\mathbf{x}(\xi)) - f(\mathbf{x}^*) = \frac{\xi^2}{2}\dot{\mathbf{x}}^T(0)\mathbf{L}(\mathbf{x}^*)\dot{\mathbf{x}}(0) + o(\xi^2).$$

18.5 Second-Order Conditions

Since $\dot{\mathbf{x}}(0) \in P$, and since $\mathbf{L}(\mathbf{x}^*)$ is positive definite on P, we conclude, for sufficiently small ξ, that

$$f(\mathbf{x}(\xi)) - f(\mathbf{x}^*) > 0,$$

so that $f(\mathbf{x}^*)$ is a local minimum. ☐

Obviously, if $\mathbf{x}^*$ is a local maximum, then a negative definite $\mathbf{L}$ matrix on P results.

Example 18.5.1 Consider the problem

$$\text{maximize } x_1 x_2 + x_2 x_3 + x_1 x_3$$
$$\text{subject to } x_1 + x_2 + x_3 = 3.$$

Solution of this leads to

$$\mathbf{L} = \begin{bmatrix} 0 & 1 & 1 \\ 1 & 0 & 1 \\ 1 & 1 & 0 \end{bmatrix}.$$

$\mathbf{L}$ has eigenvalues $-1, -1$, and 2, so it is neither positive nor negative definite. The tangent plane is $P = \{\mathbf{y}: y_1 + y_2 + y_3 = 0\}$. To determine if a maximum or minimum is obtained, we must find out if $\mathbf{x}^T L \mathbf{x} > 0$ for every $\mathbf{x} \in P$. For this particular problem, using $\mathbf{x} = [y_1, y_2, -(y_1 + y_2)] \in P$, it is straightforward to show that

$$\mathbf{x}^T L \mathbf{x} = -\left(y_1^2 + y_2^2 + y_3^2\right),$$

so L is negative definite, and a maximizing point is obtained. ☐

Example 18.5.2 Returning to example 18.4.6, we find that

$$\mathbf{L} = F + \lambda H = \begin{bmatrix} 2 & 1 & 0 \\ 1 & 0 & 1 \\ 0 & 1 & 0 \end{bmatrix} + \mathbf{0} = \begin{bmatrix} 2 & 1 & 0 \\ 1 & 0 & 1 \\ 0 & 1 & 0 \end{bmatrix}.$$

There are two positive eigenvalues and one negative eigenvalue of $\mathbf{L}$, so the problem is at this point indeterminate.

Let us therefore restrict attention to the tangent plane

$$P = \left\{ \mathbf{y}: [1, 1, 0] \begin{bmatrix} y_1 \\ y_2 \\ y_3 \end{bmatrix} = 0 \right\} = \{\mathbf{y}: y_1 + y_2 = 0\}.$$

Let $\mathbf{x} = [y_1, -y_1, y_3]^T$ be a point in P. Then

$$\mathbf{x}^T L \mathbf{x} = -2 y_1 y_3.$$

This is neither positive definite nor negative definite. The solution actually obtained in example 18.4.6 is a saddle point. ☐

In determining the definiteness of $\mathbf{L}$, it is necessary to consider specifically the restriction of points to P. Problems will not, in general, be as straightforward as these examples, since P may be more difficult to describe.

The restriction of $\mathbf{L}$ to P may be understood geometrically as follows. Let $\mathbf{y} \in P$. The vector $\mathbf{Ly}$ might not be in P. We therefore project $\mathbf{Ly}$ back into P. The operator that first computes $L\mathbf{y}$ then projects into P we will call L_P (see figure 18.9). We desire to find an explicit matrix representation for L_P so that its eigenvalues can be found, and hence the nature of the solution. This may be done by finding a basis for P. Let $\mathbf{e}_1, \mathbf{e}_2, \ldots, \mathbf{e}_{n-m}$ be an orthonormal basis for P, and let $E = [\mathbf{e}_1, \mathbf{e}_2, \ldots, \mathbf{e}_{n-m}]$. Any point $\mathbf{y} \in P$ can be written

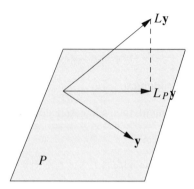

Figure 18.9: The projection of **Ly** into P to form L_P.

as $\mathbf{y} = E\mathbf{z}$ for some $\mathbf{z} \in \mathbb{R}^{n-m}$. Then the projection of $\mathbf{Ly} = \mathbf{L}E\mathbf{z}$ into P is

$$E(E^T E)^{-1} E^T \mathbf{L} E \mathbf{z} = E E^T \mathbf{L} E \mathbf{z}.$$

Thus $E^T \mathbf{L} E$ gives the coordinates of the projection in terms of the basis E. The matrix $E^T \mathbf{L} E$ is the matrix representation of **L** restricted to P. The positive or negative definiteness of **L** can be determined by finding the eigenvalues of $E^T \mathbf{L} E$.

Example 18.5.3 Consider again L and P of example 18.5.1. P is the null space of the matrix [1 1 1]. A basis for the null space may be found using the SVD (see section 7.2); an orthogonal basis is found to be

$$\mathbf{e}_1 = \frac{1}{\sqrt{6}} \begin{bmatrix} -2 \\ 1 \\ 1 \end{bmatrix} \qquad \mathbf{e}_2 = \frac{1}{\sqrt{2}} \begin{bmatrix} 0 \\ -1 \\ 1 \end{bmatrix}.$$

Then

$$E^T \mathbf{L} E = \begin{bmatrix} -1 & 0 \\ 0 & -1 \end{bmatrix},$$

which is negative definite. □

Example 18.5.4 Returning to example 18.5.2, P is the nullspace of [1 1 0] for which a basis is found to be

$$\mathbf{e}_1 = \begin{bmatrix} 1 \\ -1 \\ 0 \end{bmatrix} \qquad \mathbf{e}_2 = \begin{bmatrix} 0 \\ 0 \\ 1 \end{bmatrix}.$$

Then

$$E^T \mathbf{L} E = \begin{bmatrix} 0 & -1 \\ -1 & 0 \end{bmatrix},$$

which has eigenvalues 1 and -1. □

18.6 Interpretation of the Lagrange multipliers

A question that may arise in the context of constrained minimization is: What do the Lagrange multipliers mean in a physical sense? We will show that, in some regard, the Lagrange multipliers provide an indication of how much the constraints "cost," in the

18.6 Interpretation of the Lagrange Multipliers

neighborhood of the optimum solution. Let us allow for the constraint equations to take on different values $\mathbf{c}$,

$$\begin{aligned}\text{minimize } & f(\mathbf{x}) \\ \text{subject to } & \mathbf{h}(\mathbf{x}) = \mathbf{c}.\end{aligned} \qquad (18.22)$$

Let $\mathbf{x}^*$ be the solution to (18.22) when $\mathbf{c} = \mathbf{0}$ (as before), and let $\boldsymbol{\lambda}^*$ be the vector of Lagrange multipliers at this solution. For $\mathbf{c}$ in a sufficiently small neighborhood of $\mathbf{0}$ there is a solution to (18.22) that depends continuously upon $\mathbf{c}$. Let this solution be denoted by $\mathbf{x}(\mathbf{c})$, with $\mathbf{x}(\mathbf{0}) = \mathbf{x}^*$.

Our development is straightforward, but involves considerable manipulation of derivatives via the chain rule.

Computation of $\nabla_c \mathbf{h}(\mathbf{x})$ We first note from (18.22) that

$$\nabla_c \mathbf{h}(\mathbf{x}(\mathbf{c})) = I. \qquad (18.23)$$

We also have

$$\nabla_c \mathbf{h}(\mathbf{x}(\mathbf{c})) = [\nabla_c h_1(\mathbf{x}(\mathbf{c})) \quad \nabla_c h_2(\mathbf{x}(\mathbf{c})) \quad \cdots \quad \nabla_c h_m(\mathbf{x}(\mathbf{c}))]$$

where, by the chain rule, the ith column is of the form

$$\nabla_c h_i(\mathbf{x}(\mathbf{c})) = \begin{bmatrix} \sum_{k=1}^{m} \frac{\partial}{\partial x_k} h_i(\mathbf{x}(\mathbf{c})) \frac{\partial x_k}{\partial c_1} \\ \sum_{k=1}^{m} \frac{\partial}{\partial x_k} h_i(\mathbf{x}(\mathbf{c})) \frac{\partial x_k}{\partial c_2} \\ \vdots \\ \sum_{k=1}^{m} \frac{\partial}{\partial x_k} h_i(\mathbf{x}(\mathbf{c})) \frac{\partial x_k}{\partial c_m} \end{bmatrix} = \begin{bmatrix} \Delta_{c_1}(\mathbf{x}(\mathbf{c}))^T \nabla_x h_i(\mathbf{x}(\mathbf{c})) \\ \Delta_{c_2}(\mathbf{x}(\mathbf{c}))^T \nabla_x h_i(\mathbf{x}(\mathbf{c})) \\ \vdots \\ \Delta_{c_m}(\mathbf{x}(\mathbf{c}))^T \nabla_x h_i(\mathbf{x}(\mathbf{c})) \end{bmatrix},$$

in which we introduce the notation

$$\Delta_{c_i}(\mathbf{x}(\mathbf{c}))^T = \begin{bmatrix} \frac{\partial x_1}{\partial c_i} & \frac{\partial x_2}{\partial c_i} & \cdots & \frac{\partial x_n}{\partial c_i} \end{bmatrix}.$$

Stacking these together, we have

$$\nabla_c \mathbf{h}(\mathbf{x}(\mathbf{c})) = \begin{bmatrix} \Delta_{c_1}\mathbf{x}(\mathbf{c})^T \nabla_x h_1(\mathbf{x}(\mathbf{c})) & \cdots & \Delta_{c_1}\mathbf{x}(\mathbf{c})^T \nabla_x h_m(\mathbf{x}(\mathbf{c})) \\ \Delta_{c_2}\mathbf{x}(\mathbf{c})^T \nabla_x h_1(\mathbf{x}(\mathbf{c})) & \cdots & \Delta_{c_2}\mathbf{x}(\mathbf{c})^T \nabla_x h_m(\mathbf{x}(\mathbf{c})) \\ \vdots & & \\ \Delta_{c_m}\mathbf{x}(\mathbf{c})^T \nabla_x h_1(\mathbf{x}(\mathbf{c})) & \cdots & \Delta_{c_m}\mathbf{x}(\mathbf{c})^T \nabla_x h_m(\mathbf{x}(\mathbf{c})) \end{bmatrix}$$

$$= \Delta_c \mathbf{x}(\mathbf{c})^T \nabla_x \mathbf{h}(\mathbf{x}(\mathbf{c})), \qquad (18.24)$$

where

$$\Delta_c \mathbf{x}(\mathbf{c}) = [\Delta_{c_1}\mathbf{x}(\mathbf{c}) \quad \Delta_{c_2}\mathbf{x}(\mathbf{c}) \quad \cdots \quad \Delta_{c_m}\mathbf{x}(\mathbf{c})].$$

Combining (18.23) and (18.24), we have

$$\Delta_{c_i}\mathbf{x}(\mathbf{c})^T \nabla_x \mathbf{h}_j(\mathbf{x}(\mathbf{c}))|_{\mathbf{c}=\mathbf{0}} = \delta_{i,j}, \qquad (18.25)$$

where $\delta_{i,j}$ is the Kronecker delta function.

Computation of $\nabla_c f(\mathbf{x(c)})$. The ith component of $\nabla_c f(\mathbf{x(c)})$ is

$$[\nabla_c f(\mathbf{x(c)})]_i = \sum_{k=1}^n \frac{\partial}{\partial x_i} f(\mathbf{x(c)}) \frac{\partial x_k}{\partial c_i} = \Delta_{c_i} \mathbf{x(c)}^T \nabla_x f(\mathbf{x(c)}).$$

Using the necessary conditions for the constrained solution, we write

$$[\nabla_c f(\mathbf{x(c)})]_i = -\Delta_{c_i} \mathbf{x(c)}^T \nabla_x \mathbf{h(x)} \boldsymbol{\lambda}.$$

Stacking these and using (18.25), we obtain

$$\nabla_c f(\mathbf{x(c)})|_{\mathbf{c}=0} = - \begin{bmatrix} \Delta_{c_1} \mathbf{x(c)}^T \nabla \mathbf{h(x(c))} \boldsymbol{\lambda} \\ \Delta_{c_2} \mathbf{x(c)}^T \nabla \mathbf{h(x(c))} \boldsymbol{\lambda} \\ \vdots \\ \Delta_{c_m} \mathbf{x(c)}^T \nabla \mathbf{h(x(c))} \boldsymbol{\lambda} \end{bmatrix} = \begin{bmatrix} \mathbf{e}_1^T \\ \mathbf{e}_2^T \\ \vdots \\ \mathbf{e}_m^T \end{bmatrix} \boldsymbol{\lambda}, \tag{18.26}$$

where $\mathbf{e}_i = [0 \;\; \cdots \;\; 1 \;\; \cdots \;\; 0]^T$ is the vector with 1 in the ith place and ϕ elsewhere. Thus, from (18.26), we obtain

$$\nabla_c f(\mathbf{x(c)})|_{\mathbf{c}=0} = -\boldsymbol{\lambda}.$$

In other words, $-\boldsymbol{\lambda}$ indicates how f changes at (or near) the optimum solution as the constraint values are changed.

Example 18.6.1

$$\text{Minimize } f(x_1, x_2) = x_1^2 + x_2^2$$
$$\text{subject to } x_1 + x_2 - 1 = c.$$

To demonstrate the principle, we first find a direct solution without Lagrange multipliers. By symmetry, $x_1 = x_2$. From the constraint, $x_1 = (1+c)/2$. Then

$$f(x_1, x_2) = (1+c)^2/2,$$

and

$$\frac{\partial f}{\partial c} = (1+c).$$

Now we solve the problem using Lagrange multipliers. Using the necessary conditions (18.9), we obtain

$$\begin{bmatrix} 2 & 0 & 1 \\ 0 & 2 & 1 \\ 1 & 1 & 0 \end{bmatrix} \begin{bmatrix} x_1 \\ x_2 \\ \lambda \end{bmatrix} = \begin{bmatrix} 0 \\ 0 \\ 1+c \end{bmatrix},$$

which has solution $\lambda = -(1+c)$. The interpretation of this is that an incremental increase ϵ in c leads to an increase in the value of f by approximately $\epsilon(1+c)$.

If the constraint $x_1 + x_2 - 2 = c$ is used, then we find that $\lambda = -(2+c)$, and $df/dc = (2+c)$. In this case, there is more change in f as c is changed. □

As this example illustrates, the Lagrange multipliers can be interpreted as providing a measure of the *sensitivity* of the minimum of f to changes in the constraints. Large absolute values of the Lagrange multipliers mean that a small change in the constraint leads to a large change in the value of f: the constraint costs more in that case.

18.7 Complex constraints

The theory of optimization over real vectors as just presented can be easily generalized to optimization over complex vectors. For the problem

$$\text{minimize } f(\mathbf{x})$$
$$\text{subject to } \mathbf{h}(\mathbf{x}) = 0,$$

where $\mathbf{x}$ is now *complex* and

$$f: \mathbf{C}^n \to \mathbb{R} \qquad \mathbf{h}: \mathbf{C}^n \to \mathbf{C}^m,$$

we consider the real and imaginary parts of the constraints separately, forming the Lagrangian as

$$L = f(\mathbf{x}) + \text{Re}[\mathbf{h}(\mathbf{x})^T]\lambda_1 + \text{Im}[\mathbf{h}(\mathbf{x})^T]\lambda_2, \qquad (18.27)$$

where $\lambda_1 \in \mathbb{R}^m$ and $\lambda_2 \in \mathbb{R}^m$. Now, letting $\lambda = \lambda_1 + j\lambda_2$, we can rewrite (18.27) as

$$L = f(\mathbf{x}) + \text{Re}[\mathbf{h}(\mathbf{x})^T \overline{\lambda}] = 0.$$

Minimization can be accomplished by taking the derivative with respect to $\overline{\mathbf{x}}$.

Example 18.7.1 Let $\mathbf{x}$ be complex, and minimize

$$f(\mathbf{x}) = \mathbf{x}^H R \mathbf{x},$$

where R is Hermitian symmetric, subject to the linear constraint

$$\mathbf{x}^H \mathbf{s} = g.$$

The Lagrangian is

$$\mathbf{x}^H R \mathbf{x} + \text{Re}(\mathbf{x}^H \mathbf{s} \overline{\lambda}),$$

where $\lambda \in \mathbb{C}$. Taking the gradient, using the methods in section A.6.4, we find

$$\frac{\partial}{\partial \overline{\mathbf{x}}} L = R\mathbf{x} + \frac{1}{2}\mathbf{s}\lambda = 0; \qquad (18.28)$$

or, by taking the conjugate transpose, we have

$$\mathbf{x}^H R + \frac{1}{2}\overline{\lambda}\mathbf{s}^H = 0,$$

Multiplying by R^{-1} and then by $\mathbf{s}$, we obtain

$$\mathbf{x}^H \mathbf{s} + \frac{1}{2}\overline{\lambda}\mathbf{s}^H R^{-1}\mathbf{s} = 0,$$

from which the Lagrange multiplier may be found:

$$\lambda = -2\frac{\overline{g}}{\mathbf{s}^H R^{-1}\mathbf{s}}.$$

Substituting back into (18.28), the solution is found to be

$$\mathbf{x} = \frac{\overline{g} R^{-1}}{\mathbf{s}^H R^{-1}\mathbf{s}}.$$

□

18.8 Duality in optimization

A principle of some importance in many optimization problems is that of duality. Essentially what the duality principle states is that for a minimization problem, there is a corresponding maximization problem, such that the minimizer of the former problem is the maximizer of the latter.

Example 18.8.1 The problem of finding the shortest distance to a line l is a constrained minimization problem: min $\|\mathbf{x}\|$ such that $\mathbf{x}$ lies on l. The dual to this is the problem of finding the maximum distance to planes that pass through the line l. □

Example 18.8.2 Let K be a convex set. A minimization problem is to find the point in K nearest to the origin. The dual problem is to find the largest distance to a plane separating the convex set from the origin (see figure 18.10). □

We introduce the notion of duality by studying a problem with a quadratic cost function and linear constraints. Generalizations are possible and are discussed in the references cited at the end of the chapter. Consider the problem

$$\text{minimize } f(\mathbf{x}) = \frac{1}{2}\mathbf{x}C^{-1}\mathbf{x} - \mathbf{b}^T\mathbf{x} \qquad \text{subject to } A^T\mathbf{x} = \mathbf{c}, \qquad (18.29)$$

where C is a symmetric positive-definite matrix. This problem is simply a generalization of the "nearest distance to the line" problem, where $f(\mathbf{x})$ measures the squared distance (in some norm), and the "line" is actually the plane $A\mathbf{x} = \mathbf{b}$. We determine the dual problem by means of Lagrange multipliers. The Lagrangian is

$$L(\mathbf{x}, \boldsymbol{\lambda}) = f(\mathbf{x}) + (A^T\mathbf{x} - \mathbf{c})^T\boldsymbol{\lambda}. \qquad (18.30)$$

Taking derivatives with respect to $\mathbf{x}$ and $\boldsymbol{\lambda}$ and equating to zero, we find

$$C^{-1}\mathbf{x} + A\boldsymbol{\lambda} = \mathbf{b},$$
$$A^T\mathbf{x} = \mathbf{c}.$$

From the first of these we obtain $\mathbf{x} = C(\mathbf{b} - A\boldsymbol{\lambda})$. Substituting of $\mathbf{x}$ into (18.30) and simplifying, we obtain

$$L(\mathbf{x}, \boldsymbol{\lambda}) = g(\boldsymbol{\lambda}) = -\frac{1}{2}(\mathbf{b} - A\boldsymbol{\lambda})^T C(\mathbf{b} - A\boldsymbol{\lambda}) - \boldsymbol{\lambda}^T\mathbf{c}. \qquad (18.31)$$

Because of the minus sign, this function is concave. The function $g(\boldsymbol{\lambda})$ is the dual function, the one to maximize to obtain the optimum solution. Thus

$$\begin{array}{c}\text{minimize } f(\mathbf{x})\\ \text{subject to } A^T\mathbf{x} = \mathbf{c}\end{array} \qquad \text{is equivalent to} \qquad \text{maximize } g(\boldsymbol{\lambda})$$

in the sense that they both have the same value, and describe, from two different points of view, the same problem. The first optimization problem is called the "primal" problem, in comparison to the dual problem. Note that whereas the primal problem in this case is constrained, the dual problem is unconstrained, the constraint having been absorbed by the

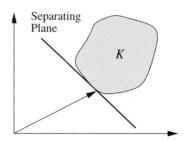

Figure 18.10: Duality: the nearest point to K is the maximum distance to a separating hyperplane.

18.8 Duality in Optimization

process of constructing $g(\lambda)$. Since unconstrained problems are often easier to solve, or may involve fewer variables, it may be of benefit in certain cases to determine the dual problem.

Example 18.8.3 We desire to maximize

$$f(\mathbf{x}) = 2x_1^2 + x_2^2 - 2x_1 - 6x_2$$

subject to $3x_1 - 2x_2 = -1$. We identify

$$C^{-1} = \begin{bmatrix} 4 & 0 \\ 0 & 2 \end{bmatrix} \quad \mathbf{b} = \begin{bmatrix} 1 \\ 6 \end{bmatrix} \quad A = \begin{bmatrix} 3 \\ -2 \end{bmatrix} \quad c = -1$$

and, using (18.31),

$$g(\lambda) = -\frac{(1-3\lambda)^2}{8} - \frac{(6+2\lambda)^2}{4} + \lambda.$$

A plot of this function is shown in figure 18.11. As is apparent, this achieves a maximum value at $\lambda = 1$: an unconstrained maximization problem. Knowing λ, x_1 and x_2 can be obtained from $\mathbf{x} = C(\mathbf{b} - A\lambda)$ as $x_1 = 1$, $x_2 = 2$. □

Another aspect of the constrained optimization problem remains to be examined. The Lagrangian $L(\mathbf{x}, \lambda)$ is *convex* in $\mathbf{x}$, since $f(\mathbf{x})$ is a convex function, and *concave* in λ. As shown in exercise 18.8-11, for every $\mathbf{x}$ satisfying the constraint $A^T\mathbf{x} = \mathbf{c}$,

$$g(\lambda) \leq f(\mathbf{x}).$$

Since $g(\lambda)$ is unconstrained and does not depend upon $\mathbf{x}$, $g(\lambda)$ forms a lower bound on $f(\mathbf{x})$; similarly, $f(\mathbf{x})$ forms an upper bound on $g(\lambda)$. The constrained optimization essentially consists of *minimizing $f(\mathbf{x})$* while *maximizing $g(\lambda)$*. At the point of optimality, a saddle point is achieved, where both functions achieve their extrema:

$$g(\lambda) = f(\mathbf{x}),$$

subject to the constraints.

At the point of optimality,

$$\min_{\mathbf{x}} \max_{\lambda} L(\mathbf{x}, \lambda) = \max_{\lambda} \min_{\mathbf{x}} L(\mathbf{x}, \lambda). \tag{18.32}$$

(To envision what is being described, it is helpful to think of a saddle surface, as shown in figure 18.12. While not an exact portrayal of the current circumstance in which L is linear in λ, it conveys the notion of the geometry.) To show that the minimization and maximization

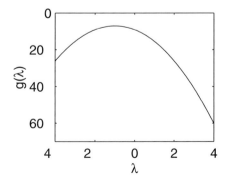

Figure 18.11: The dual function $g(\lambda)$.

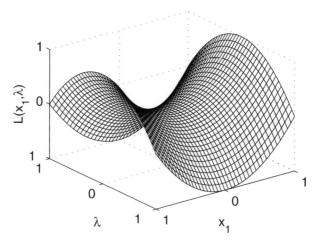

Figure 18.12: Saddle surface for minimax optimization.

can be done in either order for the particular case of quadratic cost with linear constraints, consider first the maximum with respect to λ:

$$\max_{\lambda} L(\mathbf{x}, \boldsymbol{\lambda}) = \max_{\lambda} f(\mathbf{x}) + (A^T \mathbf{x} - \mathbf{c})^T \boldsymbol{\lambda}.$$

If the constraint is satisfied so that $A^T \mathbf{x} = \mathbf{c}$, then the maximum value is $f(\mathbf{x})$. If the constraint is not satisfied, then the maximum value is ∞, since $\mathbf{x}$ can be chosen arbitrarily large with the appropriate sign so that $(A^T \mathbf{x} - \mathbf{c})^T \boldsymbol{\lambda}$ is infinitely large. Thus, maximization over $\boldsymbol{\lambda}$ followed by minimization over $\mathbf{x}$ leads to the condition that $A^T \mathbf{x} = \mathbf{c}$. In this case, we have

$$\min_{\mathbf{x}} \max_{\boldsymbol{\lambda}} L(\mathbf{x}, \boldsymbol{\lambda}) = \min_{A^T \mathbf{x} = \mathbf{c}} f(\mathbf{x}).$$

In the other case, taking the minimum first with respect to $\mathbf{x}$ by computing $\partial L / \partial \mathbf{x} = 0$, we obtain the expression $\mathbf{x} = C(\mathbf{b} - A\boldsymbol{\lambda})$. Substituting this into $L(\mathbf{x}, \boldsymbol{\lambda})$ gives

$$\min_{\mathbf{x}} L(\mathbf{x}, \boldsymbol{\lambda}) = -\frac{1}{2} (\mathbf{b} - A\mathbf{x})^T C (\mathbf{b} - A\mathbf{x}) - \mathbf{x}^T \mathbf{c} = g(\boldsymbol{\lambda}).$$

Now the maximum is over $g(\boldsymbol{\lambda})$, which we observed is upper-bounded by $f(\mathbf{x})$.

The min/max interchange expressed in (18.32), which we proved for quadratic costs with linear constraints, is true in more general cases. In fact, we have the following generalization.

Theorem 18.5 *Let $L(\mathbf{x}, \boldsymbol{\lambda})$ be a real-valued continuous function that is convex in $\mathbf{x}$ for each value of $\boldsymbol{\lambda}$, and concave in $\boldsymbol{\lambda}$ for each value of $\mathbf{x}$. If C and D are closed bounded concave sets, then*

$$\min_{\mathbf{x} \in C} \max_{\boldsymbol{\lambda} \in D} L(\mathbf{x}, \boldsymbol{\lambda}) = \min_{\mathbf{x} \in C} \max_{\boldsymbol{\lambda} \in D} L(\mathbf{x}, \boldsymbol{\lambda}).$$

The proof of this theorem is found in [171, page 28].

More insight into Lagrangian duality can be obtained by studying duality for linear programming problems, exercise 20.5-11 should be valuable in this regard.

18.9 Inequality constraints: Kuhn–Tucker conditions

We return to the problem with inequality constraints,

$$\text{minimize } f(\mathbf{x})$$
$$\text{subject to } \mathbf{h}(\mathbf{x}) = \mathbf{0}, \quad (18.33)$$
$$\mathbf{g}(\mathbf{x}) \leq \mathbf{0},$$

where there are m equality constraints and p inequality constraints. In the context of constrained minimization, when we say that a vector is less than zero, such as

$$\mathbf{g} < \mathbf{0},$$

we mean that each component of $\mathbf{g}$ is less than zero, and similarly for other comparisons. We say that a point $\mathbf{x}^*$ is regular if the gradient vectors $\nabla_x h_i(\mathbf{x}^*)$ and the gradient vectors $\nabla_x g_j(\mathbf{x}^*)$ are linearly independent for $i = 1, 2, \ldots, m$ and for all j such that g_j is an active constraint.

Inequality constraints can make solution of these problems more difficult, since it is not clear in advance which constraints are active, and only active constraints directly affect the solution. The fundamental result for the optimization is provided by the following theorem.

Theorem 18.6 *(Kuhn–Tucker conditions) Let $\mathbf{x}^*$ be a local minimum to (18.33), and suppose that the constraints are regular at $\mathbf{x}^*$ (considering both the equality and inequality constraints). Then there is a Lagrange multiplier vector $\boldsymbol{\lambda} \in \mathbb{R}^m$ and a Lagrange multiplier vector $\boldsymbol{\mu} \in \mathbb{R}^p$ with*

$$\boldsymbol{\mu} \geq \mathbf{0}$$

such that

$$\mathbf{g}(\mathbf{x}^*)^T \boldsymbol{\mu} = 0, \quad (18.34)$$
$$\nabla f(\mathbf{x}^*) + \nabla \mathbf{h}(\mathbf{x}^*)^T \boldsymbol{\lambda} + \nabla \mathbf{g}(\mathbf{x}^*)^T \boldsymbol{\mu} = \mathbf{0}. \quad (18.35)$$

The condition that $(\mathbf{x}^*)\boldsymbol{\mu} = \mathbf{0}$ is sometimes called the complementarity condition: the Lagrange multiplier is zero when the constraint is not (that is, when the constraint is inactive).

Proof For those constraints that are active, the problem becomes (relative to those constraints) a Lagrange multiplier problem, so that (18.35) can be seen to hold using these Lagrange multipliers by setting $\mu_i = 0$ for every $g_i(\mathbf{x})$ that is nonzero (i.e., not an active constraint). On this basis, (18.34) also holds.

Suppose $g_i(\mathbf{x})$ is not an active constraint, that is, $g_i(\mathbf{x}) < 0$. Then, by (18.34) and $\boldsymbol{\mu} \geq \mathbf{0}$, we must have $\mu_i = 0$. Stated another way, a component μ_i may be nonzero only if $g_i(\mathbf{x})$ is active so that $g_i(\mathbf{x}) = 0$.

We now show that $\boldsymbol{\mu} \geq \mathbf{0}$. Our thinking is demonstrated first with a one-dimensional function, following which the concept is extended to higher dimensions. Let $\mathbf{x}^*$ be a local minimum for which $g_k(\mathbf{x}^*)$ is active, and let $S_{\bar{k}}$ denote the surface defined by all active constraints *other* than g_k. Let $\mathbf{x}(\xi)$ be a curve on $S_{\bar{k}}$ passing through $\mathbf{x}^*$ at $\xi = 0$. Figure 18.13(a) illustrates $f(\mathbf{x}(\xi))$ and the point $g_k(\mathbf{x}(\xi)) = 0$. The directions of the gradients $\nabla_x f$ and $\nabla_x g_k$ are also shown as arrows. If, as shown, $g_k(\mathbf{x}(\xi)) > 0$ in the direction of decreasing f, then no further decrease in f is possible, since this would violate the constraint $g_k(\mathbf{x}) \leq 0$. Since the gradient directions (on the constrained surface) point in opposite directions, there is a *positive* μ_k such that

$$\nabla f + \mu_k \nabla g_k = 0.$$

On the other hand, if (as in figure 18.13(b)), $g_k(\mathbf{x}(\xi)) < 0$ in the direction of decreasing f, then the point $\mathbf{x}^*$ cannot be an optimum, because further decrease in the value of f would

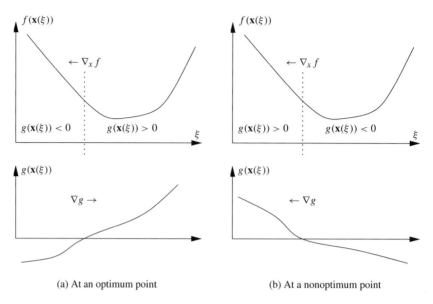

Figure 18.13: Illustration of the Kuhn–Tucker condition in a single dimension.

be possible by decreasing g_k without violating the inequality constraint. If the direction of increase of f is the same as the direction of increase of g_k, then further decrease of f is possible by moving f downhill, rendering g_k inactive. More generally, let $S_{\bar{k}}$ be as before, and let $P_{\bar{k}}$ be the tangent plane of $S_{\bar{k}}$ at $\mathbf{x}^*$. We show by contradiction that $\boldsymbol{\mu} \geq \mathbf{0}$. For any $\mathbf{z} \in P_{\bar{k}}$,

$$(\nabla \mathbf{h}(\mathbf{x}^*))^T \mathbf{z} = \mathbf{0} \tag{18.36}$$

and

$$\nabla g_i(\mathbf{x}^*)^T \mathbf{z} = 0 \tag{18.37}$$

for an active constraint g_i, with $i \neq k$. Let $\mathbf{y} \in P_{\bar{k}}$ be a vector such that $\nabla g_k(\mathbf{x}^*)^T \mathbf{y} > 0$. (By the regularity assumption, there is such a $\mathbf{y}$.) This vector points in the general direction of increase of $g_k(\mathbf{x})$ at the point $\mathbf{x}^*$. Let $\mathbf{x}(\xi)$ be a curve on $S_{\bar{k}}$ passing through $\mathbf{x}^*$ at $\xi = 0$, parameterized such that $\dot{\mathbf{x}}(0) = \mathbf{y}$. Then

$$\left.\frac{df(\mathbf{x}(\xi))}{d\xi}\right|_{\xi=0} = \nabla f(\mathbf{x}^*)^T \dot{\mathbf{x}}(0) = \nabla f(\mathbf{x}^*)^T \mathbf{y}$$

$$= -\boldsymbol{\lambda}^T \nabla \mathbf{h}(\mathbf{x}^*) \mathbf{y} - \boldsymbol{\mu}^T \nabla \mathbf{g}(\mathbf{x}^*) \mathbf{y} \tag{18.38}$$

$$= -\boldsymbol{\mu}^T \nabla \mathbf{g}(\mathbf{x}^*) \mathbf{y} \tag{18.39}$$

$$= -\mu_k (\nabla g_k(\mathbf{x}^*))^T \mathbf{y} < 0, \tag{18.40}$$

where (18.38) comes from (18.35), (18.39) comes from (18.36), and (18.40) follows from (18.37). Since at a point of minimum we should have $df(\mathbf{x}(\xi))/d\xi > 0$, we have a contradiction. □

The Kuhn–Tucker conditions may be applied as follows. The necessary conditions are established via (18.35). Then solutions are found using various combinations of active constraints, and that solution that gives $\boldsymbol{\mu} \geq \mathbf{0}$ is selected.

18.9 Inequality Constraints: Kuhn–Tucker Conditions

Example 18.9.1

$$\text{minimize } f(x_1, x_2) = 3x_1^2 + 4x_2^2 + 6x_1x_2 - 8x_2$$
$$\text{subject to } g_1(x_1, x_2) = x_1^2 + x_2^2 - 9 \leq 0,$$
$$g_2(x_1, x_2) = 2x_1 - x_2 - 4 \leq 0.$$

From (18.35), the necessary first-order conditions are

$$6x_1 + 6x_2 - 6 + 2\mu_1 x_1 + 2\mu_2 = 0,$$
$$6x_1 + 8x_2 - 8 + 2\mu_1 x_2 - \mu_2 = 0,$$
$$\mu_1(x_1^2 + x_2^2 - 9) + \mu_2(2x_1 - x_2 - 4) = 0,$$

$$\mu_1 \geq 0 \qquad \mu_2 \geq 0.$$

We try various combinations of no, one, or two active constraints, and check the signs of the Lagrange multipliers μ_i.

1. (g_1 is active, g_2 is inactive) This has $\mu_2 = 0$ and leads to the equations

$$6x_1 + 6x_2 - 6 + 2\mu_1 x_1 = 0,$$
$$6x_1 + 8x_2 - 8 + 2\mu_1 x_2 = 0,$$

$$x_1^2 + x_2^2 = 9.$$

There are four solutions (obtained, for example, using MATHEMATICA), all of which have $\mu_1 < 0$. For example, one solution is

$$\mu_1 = -8.20525 \qquad x_1 = -1.90977 \qquad x_2 = -2.31361.$$

2. (g_2 is active and g_1 is inactive) This leads to the equations

$$6x_1 + 6x_2 - 6 + 2\mu_2 = 0,$$
$$6x_1 + 8x_2 - 8 - \mu_2 = 0,$$

$$2x_1 - x_2 = 4.$$

The solution is

$$x_1 = 1.77419 \qquad x_2 = -0.451613 \qquad \mu_2 = -0.967742.$$

3. (Both constraints are active) In this case, we find the points of intersection of the two constraints (the intersection of the circle and the line) at

$$x_1 = 0.522967 \qquad x_2 = -2.95407$$

or

$$x_1 = 2.67703 \qquad x_2 = 1.35407.$$

The first solution has $(\mu_1, \mu_2) = (-7.20275, 14.0601)$, while the second solution has $(\mu_1, \mu_2) = (-5.19725, 4.8199)$.

4. (Each of the solutions to this point fails to satisfy the constraint $\mu \geq 0$) We therefore make all inequality constraints inactive, obtaining $\mu = 0$, and find the minimizing point at $(x_1, x_2) = (0, 1)$. □

Example 18.9.2 [56] In this example, we illustrate the application of Kuhn–Tucker in a more complicated problem that demonstrates typical issues that arise when inequality constraints are employed.

The problem is that of a gambler placing repeated bets at horse races. (It also has bearing on certain communication models.) We first present a background to the problem, then proceed with the optimization.

In a racing scenario there are m horses. The odds on the horses are o_i, so that a gambler receives o_i dollars for every bet placed on horse i if that horse wins. In this problem, we assume that the race is "subfair," meaning that track gets a take (a portion) on every bet. This means that

$$\sum_{i=1}^{m} \frac{1}{o_i} > 1.$$

The gambler knows (somehow!) the win probabilities for each horse; we will call these probabilities p_i. If play occurs only once, it is reasonable to place all the money on the best horse. However, when play is repeated and the winnings are "reinvested" in the next match, it is better to distribute the money across all of the options. The gambler places his bets on the horses using the proportions b_i, where b_i is the proportion bet on the ith horse. Since there is a track take, it may make sense for the gambler to hold some back. Let b_0 represent the amount held back.

The money the gambler has after a race is

$$S(i) = b_0 + b_i o_i$$

if horse i wins. Then, the wealth after N matches is

$$S_n = \prod_{k=1}^{n} S(i_k) S_0,$$

where i_k is the winner on the kth race and S_0 is the initial money. The rate of growth of the gamblers stake after many races is

$$\lim_{n \to \infty} \frac{1}{n} \log S_n = E[\log S(x)] = \sum_{k=1}^{m} p_k \log(b_0 + b_k o_k),$$

where E is the expectation (average) operator. The problem can now be stated as follows: How can the gambler distribute his money so as to maximize the rate of growth

$$\sum_{k=1}^{m} p_k \log(b_0 + b_k o_k),$$

subject to the constraints that he cannot use more or less money than he has,

$$\sum_{k=0}^{m} b_k = 1,$$

and that each bet placed cannot be negative, $b_k \geq 0$. We form the Lagrangian

$$L = \sum_{k=1}^{m} p_k \log(b_0 + b_k o_k) + \lambda \sum_{k=0}^{m} b_k + \sum_{k=1}^{m} \mu_k b_k.$$

(Since $b_0 > 0$, there is no need to introduce the Lagrange multiplier μ_0.) We obtain a necessary condition by

$$\frac{\partial L}{\partial b_0} = \sum_{k=1}^{m} \frac{p_k}{b_0 + b_k o_k} + \lambda = 0. \qquad (18.41)$$

To apply the Kuhn–Tucker condition, we let B denote the set of indices of outcomes for which $b_k > 0$

18.9 Inequality Constraints: Kuhn–Tucker Conditions

(the constraints are inactive), and let B' denote the complement of that set. Then we have the following two cases:

1. For $i \in B$, we must have $\mu_i = 0$, and

$$\frac{\partial L}{\partial b_i} = \frac{p_i o_i}{b_0 + b_i o_i} + \lambda = 0. \tag{18.42}$$

2. For $i \in B'$, we must have $\mu_i < 0$ (since the constraint is that $b_i \geq 0$), and we obtain

$$\frac{\partial L}{\partial b_i} = \frac{p_i o_i}{b_0 + b_i o_i} + \lambda + \mu_i = 0.$$

Since, in this case, $b_i = 0$ and $\mu_i \leq 0$, the preceding can be rewritten as

$$\frac{p_i o_i}{b_0} = -\lambda - \mu_i \leq -\lambda. \tag{18.43}$$

We first use the equalities (18.41) and (18.42). From (18.42), we obtain

$$b_i = -\frac{p_i}{\lambda} - \frac{b_0}{o_i}, \quad b_i \in B. \tag{18.44}$$

From the constraint

$$b_0 + \sum_{k \in B} b_k = 1$$

and (18.44), we obtain

$$b_0 = \frac{\lambda + \pi_B}{\lambda - \lambda \sigma_B}, \tag{18.45}$$

where

$$\pi_B = \sum_{k \in B} p_k \qquad \sigma_B = \sum_{k \in B} \frac{1}{o_k}.$$

Substituting (18.45) into (18.41), we obtain

$$\lambda = -1.$$

Using this value of λ, let us denote the proportion of the bet withheld as a function of the set B of bets placed, as

$$b_{0,B} = \frac{1 - \pi_B}{1 - \sigma_B}. \tag{18.46}$$

From (18.44), we find

$$b_i = p_i - \frac{b_{0,B}}{o_i} \qquad i \in B. \tag{18.47}$$

Now making using of the inequality, (18.43) can be written as

$$p_i o_i \leq b_{0,B} \qquad i \in B'. \tag{18.48}$$

We note that since $b_0 \geq 0$, from (18.46) we must have $\sigma_B < 1$. It remains to determine the set of active bets B. This must be done so that the inequality (18.48) and other constraints are satisfied. This is most conveniently done by reordering the indices so that the expected returns $p_i o_i$ are ordered, $p_i o_i \geq p_{i+1} o_{i+1}$, and bets are placed starting (potentially) with the highest probable return $p_1 o_1$. Let t indicate how many bets are placed, and let

$$\pi_t = \sum_{i=1}^{t} p_i \qquad \sigma_t = \sum_{i=1}^{t} \frac{1}{o_i}$$

and

$$b_{0,t} = \frac{1 - \pi_t}{1 - \sigma_t} \quad \text{with} \quad b_{0,0} = 1.$$

We make the observation that (for these sorted returns) if

$$p_{t+1} o_{t+1} < b_{0,t} \quad \text{then} \quad b_{0,t+1} > b_{0,t}, \qquad (18.49)$$

provided that we maintain $o_t < 1$. In particular, if $p_1 o_1 < 1$, then $b_{0,t}$ increases with t and no bets should be placed. We can find the minimum value of $b_{0,t}$ subject to the constraints by increasing t until the minimum positive value of $b_{0,t}$ is obtained. After the minimum value of $b_{0,t}$ is found, then

$$b_i = \max(p_i - b_{0,t}/o_i, 0).$$

As a particular numerical example, suppose that the probabilities and odds are

$$\mathbf{p} = \begin{bmatrix} \tfrac{1}{3} & \tfrac{1}{3} & \tfrac{1}{3} \end{bmatrix} \qquad \mathbf{o} = \begin{bmatrix} 4 & 2 & 1 \end{bmatrix}$$

(already sorted). Then the minimum value of $b_{0,t}$ is found when $t = 1$, and

$$b_0 = \frac{8}{9} \qquad b_1 = \frac{1}{9} \qquad b_2 = 0 \qquad b_3 = 0.$$

This result is computed using the example code shown in algorithm 18.1.

Algorithm 18.1 A constrained optimization of a racing problem
File: `gamble.m`

□

Example 18.9.3 The capacity of a channel is a measure of how much information (in bits per channel use) can be transmitted reliably through the channel. When a message passing through a channel is corrupted with additive white Gaussian noise, the capacity is known to be

$$\frac{1}{2} \log \left(1 + \frac{P}{N} \right),$$

where P is the signal power and N is the variance (average power) of the noise.

Suppose that there are n channels over which we can send information, and that a total amount of power P is available. Each channel has its own variance $N_1, N_2, \ldots, N_n$. We desire to apportion the power into each of the n channels to maximize the total transmission capacity without exceeding the total power availability. We must also enforce the inequality constraint $P_i \geq 0$. The Lagrangian is

$$L = \sum_{k=1}^{n} \frac{1}{2} \log \left(1 + \frac{P_k}{N_k} \right) + \lambda \sum_{k=1}^{n} P_k + \sum_{k=1}^{n} \mu_k P_k.$$

We now have two cases.

1. If $P_i > 0$ (an inactive constraint) then, by the Kuhn–Tucker condition, $\mu_i = 0$ and

$$\frac{\partial L}{\partial P_i} = \frac{1}{2} \frac{1}{N_i + P_i} + \lambda = 0,$$

from which

$$P_i = \kappa - N_i, \qquad (18.50)$$

where $\kappa = -1/2\lambda$.

2. On the other hand, if $P_i = 0$ (an active constraint) then, by the Kuhn–Tucker condition, $\mu_i \leq 0$. We obtain

$$\frac{\partial L}{\partial P_i} = \frac{1}{2} \frac{1}{N_i + P_i} + \lambda + \mu_i = 0.$$

18.9 Inequality Constraints: Kuhn–Tucker Conditions

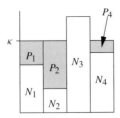

Figure 18.14: Illustration of "waterfilling" solution.

This can be written as

$$P_i \geq \kappa - N_i.$$

In conjunction with this, we also have the constraint $P_i = 0$.

Combining (18.50) with the nonnegativity constraint, we obtain

$$P_i = \max(\kappa - N_i, 0)$$

where κ is chosen so that

$$\sum_{k=1}^{n} P_i = P.$$

A graphical illustration of the solution is shown in figure 18.14. This solution is commonly called a *waterfilling* solution. The different noise levels are shown as vertical levels. As the available power P increases, power is first put into the channels with lowest noise, then the next highest, and so forth. The power fills channels until the total power expended is equal to the power available. If the noise in a channel is too high (as N_3 in the figure), then no power is expended in that channel. □

18.9.1 Second-order conditions for inequality constraints

The second-order conditions for problems with inequality constraints, used to determine whether a maximum or minimum is obtained, are very similar to the results for problems with equality constraints. We present the following results without proof.

Theorem 18.7 *Let $\mathbf{x}^*$ be a regular point of $\mathbf{h}$ and the active constraints in $\mathbf{g}$, and let P be the tangent space of the active constraints at $\mathbf{x}^*$. Then, if $\mathbf{x}^*$ is a local minimum of f,*

$$\mathbf{L}(\mathbf{x}^*) = F(\mathbf{x}^*) + \sum_{k=1}^{m} \frac{\partial^2 \mathbf{h}_k(\mathbf{x}^*)}{\partial x_i \partial x_j} \lambda_k + \sum_{k=1}^{p} \frac{\partial^2 \mathbf{g}_k(\mathbf{x}^*)}{\partial x_i \partial x_j} \mu_k \qquad (18.51)$$

is positive semidefinite on the tangent subspace of the active constraints.

(Almost) conversely, if there exist a $\boldsymbol{\lambda}$ and $\boldsymbol{\mu}$ satisfying (18.34) and (18.35) at a point $\mathbf{x}^*$, and if $\mathbf{L}(\mathbf{x}^*)$ defined in (18.51) is positive definite on the tangent subspace, then $\mathbf{x}^*$ is a strict local minimum of f.

18.9.2 An extension: Fritz John conditions

A minor extension of the Kuhn–Tucker conditions is found the *Fritz John conditions* [218, 94]. Under these conditions, a necessary condition for the inequality-constrained problem (18.33) to have a solution $\mathbf{x}^*$ is that there exists a vector $\boldsymbol{\lambda} \in \mathbb{R}^m$, a vector $\boldsymbol{\mu} \in \mathbb{R}^{p+1}$, and a scalar μ_0, such that

$$\mu_0 \nabla f(\mathbf{x}^*) + \nabla \mathbf{h}(\mathbf{x}^*)\boldsymbol{\lambda} + \nabla \mathbf{g}(\mathbf{x}^*)\boldsymbol{\mu} = \mathbf{0}, \tag{18.52}$$

$$\mathbf{g}(\mathbf{x}^*)^T \boldsymbol{\mu} = 0, \tag{18.53}$$

$$\begin{bmatrix} \mu_0 \\ \boldsymbol{\mu} \end{bmatrix} \geq \mathbf{0}, \tag{18.54}$$

$$\begin{bmatrix} \mu_0 \\ \boldsymbol{\mu} \\ \boldsymbol{\lambda} \end{bmatrix} \neq \mathbf{0}. \tag{18.55}$$

18.10 Exercises

18.2-1 Let
$$f(x, y) = x^2 - y,$$
$$h(x, y) = y + 2xy,$$
$$g(x, y) = x + y,$$

and let
$$F(x, y) = f(g(x, y), h(x, y)).$$

Determine
$$\frac{\partial F}{\partial x} \quad \text{and} \quad \frac{\partial F}{\partial y}.$$

18.4-2 This problem explores the orthogonality of gradients and tangents.
  (a) Let $h(\mathbf{x}) = h(x_1, x_2) = x_1 - x_2 + 1$.
      i. Make a plot of the "surface" $h(\mathbf{x}) = 0$.
      ii. Determine a parameterization $\mathbf{x}(\xi)$ of points on the surface between, say, the points $(x_1, x_2) = (1, 2)$ and $(3, 4)$. Plot the vector $\mathbf{x}(\xi)$ at the point $(2, 3)$.
      iii. Compute $d\mathbf{x}/d\xi$ and $\nabla h(\mathbf{x})$ and show that they are orthogonal.
  (b) Let $h(\mathbf{x}) = x_1^2 - x_2$.
      i. Make a plot of the "surface" $h(\mathbf{x}) = 0$.
      ii. Determine a parameterization $\mathbf{x}(\xi)$ of points on the surface.
      iii. Plot the vector $\mathbf{x}(\xi)/d\xi$ and the vector $\nabla h(\mathbf{x})$ at the point $\mathbf{x} = (2, 4)$. Show that these vectors are orthogonal.
  (c) More generally, let $h(\mathbf{x}) = 0$ be a curve in $\mathbb{R}^n$, and let $\mathbf{x}(\xi)$ be a parameterization of the curve for some $a \leq \xi \leq b$. Show that
  $$\frac{d}{d\xi}\mathbf{x}(\xi) \quad \text{and} \quad \nabla h(\mathbf{x}) = 0$$
  are orthogonal at every point $\mathbf{x}$ of the parameterization.

18.4-3 Work out the details of the development from equation (18.12) through (18.14).

18.4-4 Show that
$$\frac{d^2}{d\xi^2} f(\mathbf{x}(\xi)) = \sum_{i=1}^n \frac{\partial f}{\partial x_i} \frac{d^2 x_i(\xi)}{d\xi^2} + \sum_{i=1}^n \sum_{j=1}^n \frac{\partial^2 f}{\partial x_i \partial x_j} \frac{dx_i}{d\xi} \frac{dx_j}{d\xi}$$
$$= \dot{\mathbf{x}}(\xi)^T F \dot{\mathbf{x}} + (\nabla f(\mathbf{x}))^T \ddot{\mathbf{x}}(\xi),$$

where F is the Hessian matrix of f.

18.10 Exercises

18.4-5 Extend lemma 18.1 by showing that if $\mathbf{x}(\xi)$ is a curve on the surface $\mathbf{h}(\mathbf{x}) = \mathbf{0}$ and $\mathbf{x}(0) = \mathbf{x}^*$ is a constrained local minimum of f, then

$$\left.\frac{d^2}{d\xi^2} f(\mathbf{x}(\xi))\right|_{\xi=0} \geq 0.$$

Use a Taylor series argument.

18.5-6 Show that

$$\left.\frac{d^2}{d\xi^2} \sum_{k=1}^{m} \mathbf{h}(\mathbf{x}(\xi))\lambda_k \right|_{\xi=0} = \dot{\mathbf{x}}(0)^T H(\mathbf{x}^*, \boldsymbol{\lambda})\dot{\mathbf{x}}(0) + \ddot{\mathbf{x}}(0)^T \nabla \mathbf{h} \boldsymbol{\lambda} = 0.$$

18.5-7 Consider the problem

$$\text{maximize } 2x_1 x_2 + x_2 x_3 + x_1 x_3$$
$$\text{subject to } x_1 + x_2 + x_3 = 3.$$

Determine the solution (x_1, x_2, x_3, λ). Compute the matrix $\mathbf{L}$, and determine if your solution is a minimum, maximum, or neither.

18.5-8 In example 18.4.7, let $n = 4$. Examine the application of the second-order conditions to the maximum-entropy solution found there, by finding $E^T L E$, where E is a matrix of basis functions for the nullspace representing the tangent plane.

18.5-9 Verify that $\lambda = -1$ satisfies (18.41) when (18.45) is substituted in.

18.5-10 Show that (18.49) is true.

18.8-11 Let $f(\mathbf{x})$ be the quadratic function in (18.29), and let $g(\boldsymbol{\lambda})$ be as in (18.31), where C is symmetric positive definite. Prove that

$$g(\boldsymbol{\lambda}) \leq f(\mathbf{x})$$

for all $\mathbf{x}$ and $\boldsymbol{\lambda}$ for which the constraint $A^T \mathbf{x} = \mathbf{c}$ is satisfied. This is known as *weak duality*.

18.8-12 Consider the problem

$$\text{minimize } f(\mathbf{x}) = 4x_1^2 + 3x_2^2 + 6x_1 x_2 - x_1 + x2_x$$
$$\text{subject to } 2x_1 - x_2 = 1.$$

(a) Identify C^{-1}, A, $\mathbf{b}$, and c as in (18.29).
(b) Find the solution using Lagrange multipliers.
(c) Determine the dual function $g(\lambda)$, and determine its maximizing λ.

18.8-13 Consider the problem

$$\text{minimize } f(\mathbf{x}) = x_1^2 + 2x_2^2 + 4x_3^2 - x_1 + x2_x + 3x_3$$
$$\text{subject to } 3x_1 - 2x_2 + x_3 = 4.$$

(a) Identify C^{-1}, A, $\mathbf{b}$, and c as in (18.29).
(b) Find the solution using Lagrange multipliers.
(c) Determine the dual function $g(\lambda)$, and determine its maximizing λ.

18.8-14 The first (easy) half of theorem 18.5 can be stated as follows: If $L(\mathbf{x}, \boldsymbol{\lambda})$ is a real-valued function on some domain $C \times D$ (not necessarily closed, bounded, or convex), then

$$\inf_{\boldsymbol{\lambda} \in D} \sup_{\mathbf{x} \in C} L(\mathbf{x}, \mathbf{y}) \geq \sup_{\mathbf{x} \in C} \inf_{\boldsymbol{\lambda} \in D} L(\mathbf{x}, \mathbf{y}).$$

Show that this is true.

18.9-15 (Kuhn–Tucker) Determine conditions for minimizing the function

$$J(\mathbf{d}) = \sum_{i=1}^{m} \frac{1}{2} \ln \frac{\sigma_i^2}{d_i},$$

subject to the constraints that $\sum_{i=1}^{m} d_i = d$ (where d is known and fixed) and also to the constraints that

$$d_i \leq \sigma_i^2.$$

18.9-16 ([210, page 321]) Maximize $14x - x^2 + 6y - y^2 + 7$ subject to $x + y \leq 2$, $2 + 2y \leq 3$.

18.9-17 By expressing the constrained optimization problem (18.5) as the min/max problem

$$\min_{\mathbf{x}} \max_{\boldsymbol{\lambda},\boldsymbol{\mu}} f(\mathbf{x}) + [\boldsymbol{\lambda}^T \boldsymbol{\mu}^T] \begin{bmatrix} \mathbf{h}(\mathbf{x}) \\ \mathbf{g}(\mathbf{x}) \end{bmatrix},$$

argue that the complementarity condition evident in the Kuhn–Tucker conditions must be true; that is,

$$\mu_i g_i(\mathbf{x}) = 0.$$

18.11 References

An excellent source for both the theory and practice of optimization is [210], from which the material of this chapter is largely drawn. Other good sources on optimization include [89], [106], [254], and [260]. The constrained least-squares problem of example 18.4.9 is discussed in greater generality in [97] and [91]. The gambling problem is discussed in [56] and [178]. The multichannel capacity problem is discussed in [56].

The concept of duality introduced in section 18.8 is of considerable importance in game theory; in this regard, see [171]. A general statement related to Lagrangian optimization appears in [217, 218]. Our presentation of duality was drawn from [332].

Chapter 19

Shortest-Path Algorithms and Dynamic Programming

> The Viterbi algorithm is to sequence estimation what the FFT is to convolution.
> — *Darryl Morrell*

A variety of problems can be expressed in terms of finding a shortest path through a graph. Some examples, to name a few: hidden Markov model training, optimum sequence detection, dynamic time warping in speech recognition, convolutional and trellis code decoding, constrained optimal bit allocation, and a variety of problems in discrete-time optimum control. We present in the following section the basic concepts behind forward and backward (Viterbi) dynamic programming algorithms. After the initial discussion, specific applications are discussed.

19.1 Definitions for graphs

Many types of problems, particularly those that involve sequences of steps, may be represented using graphs. We begin with nomenclature relating to graphs to facilitate discussion.

Definition 19.1 A **graph** $G = (V, E)$ is a collection of **vertices** V and **edges** E. A **vertex** is a point in space, and an edge is a connection between two vertices. We will also refer to a vertex as a *node* and, in the context of trellises, as a *state*. We will refer to a **branch** as a single edge. The edge between the vertices a and b is denoted as (a, b).

A **path** from a vertex a to a vertex b is a list in which successive vertices are connected by edges in the graph; it is a list of adjacent branches. A **directed graph** is a graph in which travel along an edge is allowed in only one direction: an edge may go from say, a to b, but not the other way. Often the edges in a directed graph are indicated by arrows.

A **weighted** graph is one in which numerical weights are assigned to each edge. The **cost** of a path is the sum of the weights along the path. □

Example 19.1.1 Figure 19.1(a) shows a graph with four vertices and four edges. The vertex set is $V = \{a, b, c, d\}$, and the edge set is $E = \{(a, d), (a, c), (a, b), (b, c)\}$. Figure 19.1(b) shows a weighted directed graph. The cost of the path passing through vertices a, b, and c is 13. □

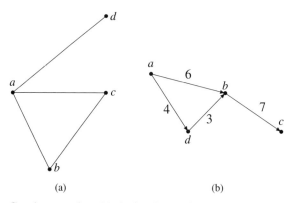

Figure 19.1: Graph examples. (a) A simple graph. (b) A weighted directed graph.

Representations of graphs in software take a variety of forms; we mention just a few. In these representations, it is most convenient to label the nodes numerically, rather than alphabetically as in the previous examples.

Adjacency matrix. If $G = (V, E)$ has m vertices labeled $1, 2, \ldots, m$, then the adjacency matrix A of the graph is the $m \times m$ matrix in which the i, jth element is the number of edges joining vertex i to vertex j. If the graph is undirected, then the adjacency matrix is symmetric. For the graph of figure 19.1(a), the adjacency matrix is

$$A = \begin{bmatrix} 0 & 1 & 1 & 1 \\ 1 & 0 & 1 & 0 \\ 1 & 1 & 0 & 0 \\ 1 & 0 & 0 & 0 \end{bmatrix}.$$

For the directed graph of figure 19.1(b), the adjacency matrix is

$$A = \begin{bmatrix} 0 & 1 & 0 & 1 \\ 0 & 0 & 1 & 0 \\ 0 & 0 & 0 & 0 \\ 0 & 1 & 0 & 0 \end{bmatrix},$$

where a corresponds to $i = 1$, b corresponds to $i = 2$, and so forth. For a weighted graph, a cost adjacency matrix is sometimes used, in which the elements indicate the weights (using a weight of ∞ to indicate the absence of a branch). For the graph of figure 19.1(b), a cost adjacency matrix is

$$A = \begin{bmatrix} \infty & 6 & \infty & 4 \\ \infty & \infty & 7 & \infty \\ \infty & \infty & \infty & \infty \\ \infty & 3 & \infty & \infty \end{bmatrix}.$$

Cost adjacency matrices are also sometimes used in conjunction with adjacency matrices to avoid the presence of ∞.

Incidence matrix. If there are n labeled edges of a graph, then the incidence matrix M is the $m \times n$ matrix in which the i, jth element is 1 if vertex i is incident to edge j, and 0 otherwise. In figure 19.1(a), let the edges be labeled, respectively, as

$$\{(a, d), (a, c), (a, b), (b, c)\} = \{1, 2, 3, 4\}$$

(these are labels, not branch weights). Then an incidence matrix for this graph is

$$M = \begin{bmatrix} 1 & 1 & 1 & 0 \\ 0 & 0 & 1 & 1 \\ 0 & 1 & 0 & 1 \\ 1 & 0 & 0 & 0 \end{bmatrix}.$$

Next-node lists. In some cases, a graph is conveniently represented using next-node lists, with corresponding weight lists. For the graph of figure 19.1(b), the next-node lists could be

next(1) = {2, 4}   next(2) = {3}   next(3) = {}   next(4) = {2},

where next(i) is the list of successor nodes to node i. Corresponding to these are the branch weights,

weight(1) = {6, 4}   weight(2) = {7}   weight(3) = {}   weight(4) = {3}.

19.2 Dynamic programming

The fundamental problem we address in this section can be stated: given a weighted, directed graph G with a vertex a designated as an initial node, find the best path to some terminating vertex. In some cases, the terminating vertex is explicitly specified; in other cases, the terminating vertex must be among a specified set of vertices. Depending on the metric of the problem, the "best" path may be interpreted as either "lowest cost" or "highest return" path. Throughout our discussion we assume that a shortest (lowest cost) path is desired. If the path cost is to be maximized, the problem can be converted to one of minimizing the path cost by replacing each path cost by its negative.

Consider the graph shown in figure 19.2. A directed graph $G = (V, E)$ of this form, in which the vertices can be partitioned into disjoint sets $V_1, V_2, \ldots, V_k$, such that if (u, v) is an edge in E then $u \in V_i$ and $v \in V_{i+1}$, is said to be a **multistage** graph. The sets V_i are **stage sets**. For the discussion of dynamic programming, we restrict our attention to multistage graphs.

One way to find the shortest path through a graph is to enumerate all possible paths. However, for graphs of even modest size, the number of possible paths can be too large for practical use. Consider the graph in figure 19.2. At the first stage, there are four choices. At the second stage most vertices have two choices (we will approximate the complexity by assuming that all vertices have the same number of edges), and each choice may be made independently of the first choice. At the third stage, there are (again assuming equal

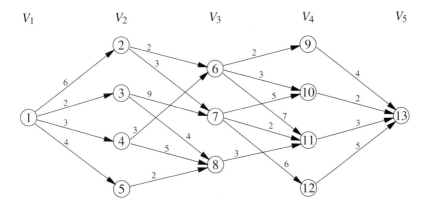

Figure 19.2: A multistage graph.

structure from each node) three choices, followed by a final choice at the fourth stage. The total number of paths is thus upper bounded by

$$N = 4 \cdot 2 \cdot 3 \cdot 1 = 24$$

paths. More generally, for an m-stage multistage graph with n_i possible choices at the ith stage, the number of paths grows as

$$N = \prod_{i=1}^{m} n_i.$$

If each stage has the same number of choices n (as is most frequently the case), then the number of paths grows as $N = n^m$, exponentially.

While the total complexity grows quickly, the number of paths that must actually be examined can often be dramatically reduced by employing the *principle of optimality*, which is introduced now by example.

In the graph of figure 19.2, we will find the shortest path by starting at the terminating node ⑬ and working toward the initial node ①. Let $F(i, j)$ denote the cost to go from vertex number j at stage i to the terminating node. At the penultimate stage, determining $F(4, j)$ requires no computation:

$$F(4, 9) = 4,$$
$$F(4, 10) = 2,$$
$$F(4, 11) = 3,$$
$$F(4, 12) = 5.$$

Observe that from ⑥ the shortest path can be computed as

$$F(3, 6) = \min\{2 + F(4, 9), 3 + F(4, 10), 7 + F(4, 11)\} = 5$$

where the path passing through ⑩ is selected. Now observe that any shortest path to ⑬ that passes through ⑥ must also pass through ⑩. If not, then a shorter path could be obtained by choosing a path that does go through ⑩. Similarly, the other paths in V_3 to ⑬ can be obtained as

$$F(3, 7) = \min\{5 + F(4, 10), 2 + F(4, 11), 6 + F(4, 12)\} = 5 \ (\text{through } ⑪),$$
$$F(3, 8) = \min\{3 + F(4, 11)\} = 6 \ (\text{through } ⑪).$$

This example demonstrates what has come to be known as Bellman's principal of optimality [22]: *An optimal sequence of decisions has the property that whatever the initial state and initial decision are, the remaining decisions must constitute an optimal policy with regard to the state resulting from the first decision.* As applied to this example, for any path through V_4, the decisions from V_4 toward the final node must be optimal.

Continuing to work backward, we can find the shortest path. For vertices in V_2, we have

$$F(2, 2) = \min\{2 + F(3, 6), 3 + F(3, 7)\} = 7 \ (\text{through } ⑥),$$
$$F(2, 3) = \min\{9 + F(3, 7), 4 + F(3, 8)\} = 10 \ (\text{through } ⑧),$$
$$F(2, 4) = \min\{3 + F(3, 6), 5 + F(3, 8)\} = 8 \ (\text{through } ⑥),$$
$$F(2, 5) = \min\{2 + F(3, 8)\} = 8 \ (\text{through } ⑧),$$
$$F(1, 1) = \min\{6 + F(2, 2), 2 + F(2, 3), 3 + F(2, 4), 4 + F(2, 5)\} = 11 \ (\text{through } ④).$$

The optimal path from ① to ⑬ is (1, 4, 6, 10, 13), with a total cost of 11.

19.3 The Viterbi Algorithm

This process of finding the optimal path is known as *forward* dynamic programming: the decision at stage k is made in terms of the optimal decisions for stages $k+1, k+2, \ldots, r$, looking *forward*. Confusingly, in forward dynamic programming, the path is formed by working backward from the terminal vertex.

More generally, the cost at stage i and vertex j can be obtained as follows. Let $\mathcal{S}_j$ denote the successor vertices to node j (which, for multistage graphs, will be in stage $i+1$), and let $c(j, k)$ denote the cost of the path from vertex j to vertex k. Then, for each $j \in V_i$,

$$F(i, j) = \min_{k \in \mathcal{S}_j}\{c(j, k) + F(i+1, k)\}.$$

An implementation of forward dynamic programming on multistage graphs is shown in algorithm 19.1. This function requires that the n vertices are enumerated by stages, as shown in figure 19.2. The parameters G and W are a list of next-state vertices and edge costs. For the graph of figure 19.2, these parameters are

```
G{1} = [2,3,4,5];
G{2} = [6,7];      G{3} = [7,8];      G{4} = [6,8];      G{5} = 8;
G{6} = [9,10,11];  G{7} = [10,11,12]; G{8} = 11;
G{9} = 13;         G{10} = 13;        G{11} = 13;        G{12} = 13;
W{1} = [6,2,3,4];
W{2} = [2,3];      W{3} = [9,4];      W{4} = [3,5];      W{5} = 2;
W{6} = [2,3,7];    W{7} = [5,2,6];    W{8} = 3;
W{9} = 4;          W{10} = 2;         W{11} = 3;         W{12} = 5;
```

For example, this says that the states that are successors to ① are ②, ③, ④, and ⑤, with path costs 6, 2, 3, and 4, respectively.

Algorithm 19.1 Forward dynamic programming
File: `fordyn.m`

19.3 The Viterbi algorithm

It is also possible to find the shortest path through a graph using *reverse* dynamic programming which, perversely, starts at the initial node and works toward a terminating node.

In figure 19.2, consider the shortest path to ⑥, which has two paths to it from ①, one path through ② and the other path through ④. The key observation is: *any optimum path that is an extension of a path from node ① to node ⑥ must be an extension of the optimum path from ① to ⑥*. Thus only the shortest path to ⑥ needs to be retained, which in this case is ①—④—⑥.

Let $B(i, j)$ be the cost from the initial vertex ① to vertex j at stage i. For the graph of figure 19.2 the shortest costs are obtained as

$$B(2, 2) = 6,$$
$$B(2, 3) = 2,$$
$$B(2, 4) = 3,$$
$$B(2, 5) = 4,$$

$$B(3, 6) = \min\{2 + B(2, 2), 3 + B(2, 4)\} = 6 \; (\text{through } ④),$$
$$B(3, 7) = \min\{3 + B(2, 2), 9 + B(2, 3)\} = 9 \; (\text{through } ②),$$
$$B(3, 8) = \min\{4 + B(2, 3), 5 + B(2, 4), 2 + B(2, 5)\} = 6 \; (\text{through } ③ \text{ or } ⑤),$$
$$B(4, 9) = \min\{2 + B(3, 6)\} = 8,$$
$$B(4, 10) = \min\{3 + B(3, 6), 5 + B(3, 7)\} = 9 \; (\text{through } ⑥),$$
$$B(4, 11) = \min\{7 + B(3, 6), 2 + B(3, 7), 3 + B(3, 8)\} = 9 \; (\text{through } ⑧),$$
$$B(4, 12) = \min\{7 + B(3, 7)\} = 15,$$
$$B(5, 13) = \min\{4 + B(4, 9), 2 + B(4, 10), 3 + B(4, 11), 5 + B(4, 12)\}$$
$$= 11 \; (\text{through } ⑩).$$

The best path can be obtained by working backward through the graph to obtain (1, 4, 6, 10, 13), as before.

The Viterbi algorithm (VA) is a reverse dynamic programming algorithm. The graphs that most commonly occur in conjunction with applications of the VA are **trellises**, which are multistage graphs that have the same edge structure between each stage, with the possible exception of some initial and terminal stages. We present examples that illustrate why the trellis structure is of particular interest. A trellis is shown in figure 19.3. The graph is directed with movement along the edges always toward the right, so that arrowheads are not portrayed along the edges. The trellis may be viewed as proceeding from left to right in time, and a time index $k = 0, 1, 2, \ldots$, is indicated.

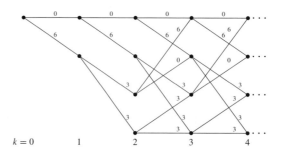

Figure 19.3: A trellis diagram.

The vertices at a given stage in a trellis are frequently referred to as states—this is because a trellis is often used to represent the time behavior of a state machine. For example, the trellis of figure 19.3 represents the set of possible sequences and outputs of the state machine with four states shown in figure 19.4. The branch weights of the trellis correspond to outputs of the state machine. In this context, finding the shortest path through the trellis corresponds to finding the best sequence of states in the state machine. In the VA in general, it is not known in advance which vertices will be in the final optimum path, so the best path to each vertex must be computed at each time. A *survivor path* is maintained to each vertex at each time. The Viterbi algorithm can be summarized as follows.

1. Find the path metric for each path entering each vertex at time k by adding the path metric of the survivor path to each vertex at time $k - 1$ to the branch metric (weight) for the branch from the vertex set at time $k - 1$.

19.3 The Viterbi Algorithm

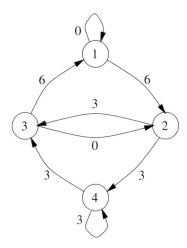

Figure 19.4: State machine corresponding to a trellis.

2. For each path into vertices at time k, determine the path with the best metric. Call this path the survivor path.
3. Store the survivor path and metric for each vertex at time k.
4. Increment k and repeat.

Example 19.3.1 We examine some particular aspects of a VA implementation with an artificial example. Suppose a source produces a sequence of outputs, according to the state machine in figure 19.4, where the numbers along the branches are machine outputs, not branch weights. The outputs of the source are observed after passing through a noisy channel, as shown in figure 19.5. The problem is to determine (in an optimal fashion) the sequence of outputs (and hence the sequence of states) produced by the source. Since the output at time k depends upon the state of the source at that time, the outputs are not independent. Given a sequence of measured values $\mathbf{r} = \{r_0, r_1, \ldots\}$, the branch metric (weight) along a branch at time k labeled with a value a_k is obtained as $w = (r_k - a_k)^2$; the path metric is the sums of the squares of the distances along the path. □

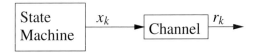

Figure 19.5: State-machine output observed after passing through a noisy channel.

For many shortest-path problems on trellises, including this example, the number of branches may be indefinitely long. In implementing the VA, therefore, some provision must be made to produce outputs before the end of the observed sequence, since the observed sequence may continue indefinitely. A common approach is to preserve the paths in the VA over a finite-length window. When the length of the path has grown to the window length, the state with the lowest cost is selected. The path leading to this best state is searched backward, and the first branch in the window is produced as the output. At the next time step, the window "slides over" by one: the paths at the end of the window are each extended, the best path is selected and traversed backward, and again a single output is produced. If the window is sufficiently wide, this approach almost always produces the optimum path. However, since the optimum path is defined from the beginning to the end of the sequence, it is still possible that windowing the VA will occasionally produce an incorrect output.

Example 19.3.2 The windowed VA for the state-machine example is illustrated in figure 19.6. The window width has been selected to be seven branches. (A width of seven is actually shorter than is typical of most applications, but has been chosen to make the graphical representation easier.) In the sequence of paths shown, the sequence

$$\mathbf{r} = \{6, 3, 6, 6, 3, 3, 0, 5, 1, 3\}$$

is observed. We desire to determine the best estimate of the outputs and the corresponding states.

The numbers shown at the right of the paths in figure 19.6 are the path metrics. The paths grow until they are the width of the window, with no outputs produced, as shown in the first six frames. At the seventh frame, the state with the shortest path to it is chosen, and that path is traversed back (as shown by the thick line) to the first time frame in the window, where the first state in the surviving path is shown with a $\diamond$. The path segment which begins at the state $\diamond$ is on the least-cost path. The

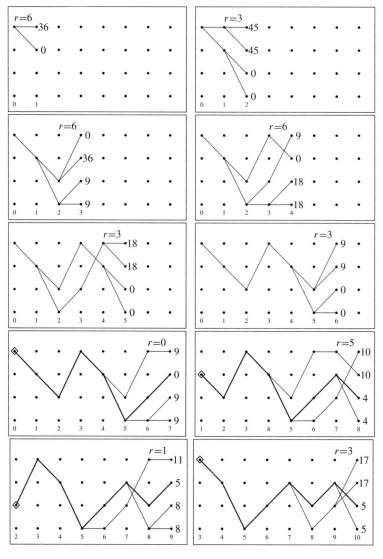

Figure 19.6: Steps in the Viterbi algorithm.

paths are shown after step 3 of the VA outlined previously, in which the survivor paths to each node have already been selected. □

It may be observed that as the algorithm progresses, the survivor paths usually have the first few branches in common. For example, for the paths shown in frame 6 in figure 19.6, the first four branches of all paths are common. This could be used to provide an alternate way of producing output: any branches common to all survivor paths can be output. However, determining this condition is usually more difficult than searching backward from the state with lowest path cost and, provided that the survivor paths do in fact merge within the window width, provides the same output.

It may also be observed that there may be tie conditions, both in selecting the best path to a state and in selecting the state with lowest cost. The tie may be handled by selecting one of the tied values arbitrarily.

If the input sequence is of finite length, at the end of the input sequence the branches on the best path can be "flushed" out of the window to get the rest of the data by simply finding the best path and working backwards, without processing any more inputs.

19.4 Code for the Viterbi algorithm

Code implementing the VA is shown in algorithm 19.2. Before calling this code, the data structures must be set up by calling `initvit1`, which is shown in algorithm 19.3. The description of the trellis is provided by the variables `trellis` and `branchweight`. The `trellis` variable indicates the successor states. For example, writing in MATLAB notation,

```
trellis{1} = [1 3]; trellis{2} = [3 4];
trellis{3} = [1 2]; trellis{4} = [3 4];
```

indicates that the successor states to state 1 are 1 and 3; the successor states to state 2 are 3 and 4; and so forth. This therefore describes the trellis shown in figure 19.3. The `branchweight` variable describes the state-machine outputs on each branch. Provision is made for vector-valued outputs by the use of MATLAB cells. The `branchweight` for the trellis of this example is

```
tbranchweight{1,1} = 0;    % node 1, branch 1
tbranchweight{1,2} = 6;    % node 1, branch 2
tbranchweight{2,1} = 3;    % node 2, branch 1
tbranchweight{2,2} = 3;    % node 2, branch 2
tbranchweight{3,1} = 6;    % node 3, branch 1
tbranchweight{3,2} = 0;    % node 3, branch 2
tbranchweight{4,1} = 3;    % node 4, branch 1
tbranchweight{4,2} = 3;    % node 4, branch 2
```

The third argument to `initvit1` is the length of the window, and the fourth is the norm function used. This can be a simple function, such as:

```
function d = vitsqnorm(branch,input,state,nextstate)
% function d = vitsqnorm(branch,input,state,nextstate)
%
% Compute the square norm of the difference between inputs
% This function may be feval'ed for use with the Viterbi algorithm
% (state and nextstate are not used here)
d = norm(branch-input)^2;
```

A routine to flush the window is `vitflush`, shown in algorithm 19.4, which takes as an argument the desired target state (or states), or a 0 to indicate that the state with lowest path cost should be selected.

In this implementation, the survivor path list is maintained in the array `savepath`, which indicates the previous state. Time indexing into `savepath` is done modulo the path length (a circular buffer), to produce a shifting window with no need to physically shift the data. The initialization function takes the trellis and weight description and a string representing the norm function, and sets up the appropriate data structures. The most important computation is to determine the data in `priorstate`. This is used to set up the `savepath` data by recording the predecessor of every state.

Algorithm 19.2 The Viterbi algorithm
File: `viterbi1.m`

Algorithm 19.3 Initializing the Viterbi algorithm
File: `initvit1.m`

Algorithm 19.4 Flushing the shortest path in the VA
File: `vitflush.m`

Example 19.4.1 The following code produces the paths shown in figure 19.6.

```
% demonstrate the VA

ttrellis{1} = [1 2]; ttrellis{2} = [3 4];
ttrellis{3} = [1 2]; ttrellis{4} = [3 4];
tbranchweight{1,1} = 0; tbranchweight{1,2} = 6;
tbranchweight{2,1} = 3; tbranchweight{2,2} = 3;
tbranchweight{3,1} = 6; tbranchweight{3,2} = 0;
tbranchweight{4,1} = 3; tbranchweight{4,2} = 3;

initvit1(ttrellis,tbranchweight,7,'vitsqnorm');

rlist = [6, 3, 6, 6, 3, 3, 0, 5, 1, 3];
plist = [];
% Go through the inputs one at a time
for r = rlist
  p = viterbi1(r);
  if(p)
plist = [plist p];
  end
end
% Now flush the rest out
plist = [plist vitflush(0)];
```

19.4 Code for the Viterbi Algorithm

The list of states produced by this example code (the shortest path sequence) is

```
plist = [1 2 3 1 2 4 3 2 3 2 3]
```
□

The function `viterbi1` has provisions for more than regular trellises with scalar-valued branches.

Example 19.4.2 The irregular trellis shown in figure 19.7, with different numbers of branches emerging from each state, can be represented by the following.

```
ttrellis{1} = [1 2 3];      ttrellis{2} = [4];
ttrellis{3} = [1 2 3];      ttrellis{4} = [3 4];

tbranchweight{1,1} = 1;     tbranchweight{1,2} = 2;
tbranchweight{1,3} = 3;

tbranchweight{2,1} = 4;

tbranchweight{3,1} = 5;     tbranchweight{3,2} = 6;
tbranchweight{3,3} = 7;

tbranchweight{4,1} = 8;     tbranchweight{4,2} = 9;
```
□

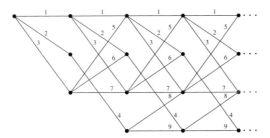

Figure 19.7: A trellis with irregular branches.

Example 19.4.3 A trellis with vector outputs, such as might be used for a convolutional code, can be set using a vector branchweight such as:

```
branchweight{1,1} = [0 0]; branchweight{1,2} = [1 1];
branchweight{2,1} = [0 1]; branchweight{2,2} = [1 0];
branchweight{3,1} = [1 1]; branchweight{3,2} = [0 0];
branchweight{4,1} = [1 0]; branchweight{4,2} = [0 1];
```

The trellis represented by this assignment is shown in figure 19.8. For a multioutput trellis such as this, a norm function such as the following can be used.

```
function d = convnorm(branch,input,state,nextstate)
%
% Compute the Hamming distance between the branchweights and
% the input
% This function may be feval'ed for use with the Viterbi algorithm
% (state and nextstate are not used here)
d = sum(r ~= branch)
```
□

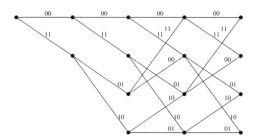

Figure 19.8: A trellis with multiple outputs.

19.4.1 Related algorithms: Dijkstra's and Warshall's

The VA as just described provided the shortest path through a multistage graph. A minor modification of this, known as Dijkstra's algorithm, is a shortest-path algorithm that works for a general weighted directed graph. Given a starting node, the algorithm finds the shortest path to every vertex that can be reached from the starting node. The key idea is as in the VA: if multiple paths from the starting vertex converge at a given node, then only the shortest path to that vertex is retained as the surviving path. The outline of the algorithm is as follows.

1. Initialization: Denote the initial node as a. Let S indicate the set of vertices to which the shortest paths have been found. For a vertex w, let $d(w)$ be the length of the shortest path starting at a and ending at vertex w. The initial value of $d(w)$ is set to $c(a, w)$, where $c(a, w)$ is the branch cost ($c(a, w)$ may be set to ∞ if there is no edge between a and w).

2. New vertex: Choose the vertex u that has minimum distance $d(u)$ among all those vertices not in S; u now becomes a vertex in S.

3. Distance update: Update the shortest path to all vertices w not in S. If the distance does change, it is because there must be a shorter path starting at a passing through u and going to w.

4. Repeat from step 2 until all vertices have been examined.

Algorithm 19.5 illustrates an implementation.

Algorithm 19.5 Dijkstra's shortest-path algorithm
File: `dijkstra.m`

Example 19.4.4 Consider the accompanying graph and its associated cost adjacency matrix (not shown in the graph for drawing convenience is the fact that the weight of the path from any node to itself is 0).

$$A = \begin{bmatrix} 0 & \infty & \infty & \infty & \infty & \infty & \infty & \infty \\ 3 & 0 & \infty & \infty & \infty & \infty & \infty & \infty \\ 10 & 8 & 0 & \infty & \infty & \infty & \infty & \infty \\ \infty & \infty & 12 & 0 & \infty & \infty & \infty & \infty \\ \infty & \infty & \infty & 15 & 0 & 2.5 & 9 & \infty \\ \infty & \infty & \infty & 10 & \infty & 0 & \infty & 14 \\ \infty & \infty & \infty & \infty & \infty & \infty & 0 & 10 \\ 17 & \infty & \infty & \infty & \infty & \infty & \infty & 0 \end{bmatrix}$$

19.4 Code for the Viterbi Algorithm

The results of calling `dijkstra` with this cost adjacency matrix and $a = 5$ are the set of shortest distances

$$\mathbf{d} = [33.5 \quad 32.5 \quad 24.5 \quad 12.5 \quad 0 \quad 2.5 \quad 9 \quad 16.5]$$

and the set of previous nodes

$$\mathbf{p} = [8 \quad 3 \quad 4 \quad 6 \quad 5 \quad 5 \quad 5 \quad 6].$$

As an example, the shortest path to ① is shown with a bold line. □

Definition 19.2 The **transitive closure** of a graph G is the directed graph G_c that contains an edge (a, b) if there is a way to get from a to b in the graph G in one or more branches. □

In other words, the transitive closure indicates which vertices are reachable from which other vertices, perhaps by traversing multiple branches. One way to determine the transitive closure would be to run `dijkstra` starting from each node, and determine those nodes to which there is a finite-cost path. Another way is using Warshall's algorithm, which is based on the observation that if there is a way to get from a to b, and a way to get from b to c, then there must be a way to get from a to c. This concept is embodied in the following algorithm.

Algorithm 19.6 Warshall's transitive closure algorithm
File: `warshall.m`

Example 19.4.5 Calling `warshall` with the adjacency matrix for the graph of example 19.4.4,

$$A = \begin{bmatrix} 0 & 0 & 0 & 0 & 0 & 0 & 0 & 0 \\ 1 & 0 & 0 & 0 & 0 & 0 & 0 & 0 \\ 1 & 1 & 0 & 0 & 0 & 0 & 0 & 0 \\ 0 & 0 & 1 & 0 & 0 & 0 & 0 & 0 \\ 0 & 0 & 0 & 1 & 0 & 1 & 1 & 0 \\ 0 & 0 & 0 & 1 & 0 & 0 & 0 & 1 \\ 0 & 0 & 0 & 0 & 0 & 0 & 0 & 1 \\ 1 & 0 & 0 & 0 & 0 & 0 & 0 & 0 \end{bmatrix} \quad \text{yields} \quad \tilde{A} = \begin{bmatrix} 0 & 0 & 0 & 0 & 0 & 0 & 0 & 0 \\ 1 & 0 & 0 & 0 & 0 & 0 & 0 & 0 \\ 1 & 1 & 0 & 0 & 0 & 0 & 0 & 0 \\ 1 & 1 & 1 & 0 & 0 & 0 & 0 & 0 \\ 1 & 1 & 1 & 1 & 0 & 1 & 1 & 1 \\ 1 & 1 & 1 & 1 & 0 & 0 & 0 & 1 \\ 1 & 0 & 0 & 0 & 0 & 0 & 0 & 1 \\ 1 & 0 & 0 & 0 & 0 & 0 & 0 & 0 \end{bmatrix}$$

as the adjacency matrix for the graph of the transitive closure. □

19.4.2 Complexity comparisons of Viterbi and Dijkstra

Let us consider the complexity of the Dijkstra and Viterbi algorithms using the O notation introduced in section A.4. Starting with the Dijkstra algorithm, we see that the line `for nn=2:n` must run $n - 1$ times, where n is the number of vertices, and that finding the minimum distance to all paths not in S requires $O(n)$ operations. Updating S then requires $O(n)$ operations, so the overall complexity is $O(n^2)$.

If Dijkstra's algorithm were used to find the minimal spanning tree, then the complexity would be multiplied by the number of vertices, resulting in $O(n^3)$ complexity. By contrast, Warshall's algorithm requires $O(n)$ iterations through the outer loop `for b=1:n` and $O(n)$ iterations through the inner loop `for a=1:n`, resulting in an overall complexity of $O(n^2)$.

The Viterbi algorithm cost per iteration is computed as follows. Each state must be examined in the loop `for state=1:numstate`. Then each succeeding state must be examine in each state in the loop `for nextstate=trellis{state}`. If we let n denote the number of states, and m denote the number of connections to next states, then the complexity per trellis stage is $O(mn)$.

Applications of path search algorithms

We now illustrate a variety of applications in which shortest-path problems arise.

19.5 Maximum-likelihood sequence estimation

We have already met the problem of maximum-likelihood sequence estimation in the introduction to the VA associated with figure 19.3. More formally, we have the following: A state machine produces a sequence of outputs $\mathbf{x} = \{\mathbf{x}_1, \mathbf{x}_2, \ldots\}$ as a function of the sequence of states in the state machine $\mathbf{s} = \{s_1, s_2, \ldots\}$. The outputs are observed after passing through a noisy channel, to obtain

$$\mathbf{r}_k = f(\mathbf{x}_k),$$

for some function f. In the case of an additive noise channel, for example,

$$\mathbf{r}_k = \mathbf{x}_k + \mathbf{n}_k, \qquad k = 1, 2, \ldots.$$

The MLSE problem is to determine the most likely sequence of symbols $\mathbf{x}$, given the observations $\mathbf{r}$. The Viterbi algorithm provides a computational approach to solving the sequence-estimation problem. In this case, the branch weights are distances (using an appropriate metric) between the observed values $\mathbf{r}_k$ and the state machine output $\mathbf{x}_k$. The vertices in the graph underlying the problem are associated with the states of the state machine. Application of the VA in practice depends on being able to identify that there is an underlying state "machine" in the process.

If the noise is zero-mean AWGN with noise independent from sample to sample and variance σ^2, then the likelihood function for the sequence (assuming, for convenience, that the samples are scalar valued) can be written as

$$f(r_1, r_2, \ldots, r_K \mid x_1, x_2, \ldots, x_K) = \prod_{k=1}^{K} \frac{1}{\sqrt{2\pi}\sigma} \exp\left[-\frac{(r_k - x_k)^2}{2\sigma^2}\right].$$

The distance function is then the log-likelihood function (throwing away terms that are independent of the r_i),

$$D(\mathbf{r}, \mathbf{x}) = \sum_{k=1}^{k} (r_k - x_k)^2.$$

We now illustrate several circumstances in which MLSE problems arise.

19.5.1 The intersymbol interference (ISI) channel

A linearly-modulated signal may be represented as

$$v(t) = \sum_{n} I_n g(t - nT),$$

where I_n (possibly complex) represents the signal amplitude that depends upon the bits to be transmitted, and T is the symbol time. An example waveform is shown in figure 19.9(a),

19.5 Maximum-Likelihood Sequence Estimation

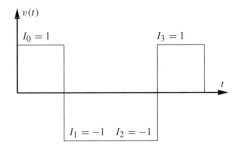

(a) Transmitted waveforms

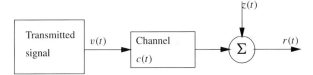

(b) Channel with noise

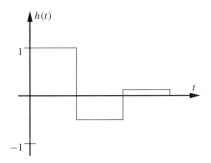

(c) Channel response $h(t)$

Figure 19.9: MLSE detection in ISI.

where $g(t)$ is a simple flat-topped pulse and the amplitudes are binary, $I_n \in \{\pm 1\}$. If this signal is transmitted through a channel with impulse response $c(t)$, then corrupted with additive Gaussian noise, as shown in figure 19.9(b), then the received signal can be expressed as

$$r(t) = \sum_n I_n h(t - nT) + z(t),$$

where

$$h(t) = \int_{-\infty}^{\infty} g(\tau) c(t - \tau)\, d\tau$$

and $z(t)$ is the noise.

Example 19.5.1 As an example, suppose that $c(t)$ has the simple impulse response

$$c(t) = \delta(t) - 0.5\delta(t - T) + 0.1\delta(t - 2T).$$

Then, still using the flat-topped pulse for $g(t)$, the signal $h(t)$ is as shown in figure 19.9(c). □

The filter action of the channel causes the output symbols to overlap in time. Intuitively, if the amplitudes of the signals overlapping the kth symbol could be determined and canceled out, then the kth symbol could be reliably determined. Of course, in order to determine reliably the symbols overlapping the kth symbol, we must first determine the symbols overlapping the $k - 1$st symbol, and so forth. Thus the optimal detector must detect the entire sequence, which requires the use of the Viterbi algorithm.

Using the techniques of section 11.9, the likelihood function for the received signal, conditioned upon the sequence $\mathbf{I} = \{\ldots, I_{-1}, I_0, I_1, \ldots\}$, can be written as

$$\ell(\mathbf{I}, r(t)) = f(r(t) \mid \mathbf{I}) = C \exp\left[-\frac{1}{N_0}\int_{-\infty}^{\infty}\left|r(t) - \sum_n I_n h(t - nT)\right|^2 dt\right].$$

Expanding the logarithm of the likelihood function and eliminating those terms that do not depend upon the I_k, we have

$$\Lambda(\mathbf{I}, r(t)) = 2\operatorname{Re}\sum_n \bar{I}_n \int_{-\infty}^{\infty} r(t)\bar{h}(t-nT)\,dt - \sum_n \sum_m \bar{I}_n I_m \int_{-\infty}^{\infty} \bar{h}(t-nT)h(t-mT)\,dt. \tag{19.1}$$

Let

$$y_k = y(kT) = \int_{-\infty}^{\infty} r(t)\bar{h}(t - kT)\,dt; \tag{19.2}$$

this is just the output of a filter matched to the signal $h(t)$. Also let

$$x_k = x(kT) = \int_{-\infty}^{\infty} \bar{h}(t)h(t + kT)\,dt; \tag{19.3}$$

this is the autocorrelation function of $h(t)$, and provides a measure of the channel dispersion. In a transmission system, the ISI will affect only a finite number of symbols, so that $x_k = 0$ for $|k| > M$ for some M.

Example 19.5.2 For the signal $h(t)$ shown in figure 19.9(c), we have (with $T = 1$)

$$x_k = \begin{cases} 1.26 & k = 0, \\ -.55 & |k| = 1, \\ -.01 & |k| = 2. \end{cases}$$

In this case $M = 2$. □

Using (19.2) and (19.3), we can write the log-likelihood function (19.1) as

$$\Lambda(\mathbf{I}, r(t)) = 2\operatorname{Re}\sum_n \bar{I}_n y_n - \sum_n \sum_{k=-M}^{M} \bar{I}_n I_{n-k} x_k. \tag{19.4}$$

The log-likelihood function (19.4) forms the path metric—the total cost of detecting the sequence $\{y_n\}$. In order to form a branch metric for the kth branch, we determine those terms of the log-likelihood that depend upon $\bar{I}_k$ and the M previous values—this makes computation of the branch metric causal. Letting $b(y_k, I_k \mid \sigma)$ denote the branch metric corresponding to the matched filter output y_k and the symbol I_k at the state σ_k, we see that

$$b(y_n, I_n \mid \sigma_k) = y_k \bar{I}_k + x_0 |I_k|^2 + 2\operatorname{Re}\sum_{i=1}^{M} \bar{I}_k I_{k-i} x_i. \tag{19.5}$$

19.5 Maximum-Likelihood Sequence Estimation

In order to compute this, we must know $\{I_k, I_{k-1}, \ldots, I_{k-M}\}$. The previous inputs form the state:

$$\sigma_k = \{I_{k-1}, I_{k-2}, \ldots, I_{k-M}\}.$$

As various branches are taken depending on the value of I_k used, the state changes accordingly.

Example 19.5.3 A trellis diagram corresponding to the case when $M = 2$ is shown in figure 19.10(a). □

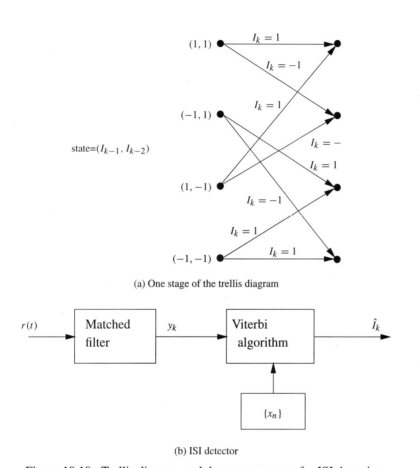

(a) One stage of the trellis diagram

(b) ISI detector

Figure 19.10: Trellis diagram and detector structure for ISI detection.

The detector consists of a matched filter followed by a Viterbi algorithm. The overall MLSE detector is shown in figure 19.10(b).

While the MLSE just described provides an optimal (in an ML sense) detector, it can be very computationally complex. If there are S symbols in the alphabet ($S = 2$ for binary transmission), and the channel impulse response lasts for $M + 1$ symbol times, then there are S^M states necessary in the trellis. In many practical channels, the impulse response can last for more than ten symbol times. Even for simple binary transmission, this corresponds to more than 1000 states. For this many states, real-time detection is difficult to achieve. What is commonly employed in these circumstances is a suboptimal detector of some sort, such as those described briefly in section 19.7.

19.5.2 Code-division multiple access

Closely related to the ISI channel (at least in terms of detection algorithms) is code-division multiple access. In CDMA, K users transmit simultaneously. The signal sequence of the kth user $v_k(t)$ is multiplied by a spreading sequence $c_k(t)$, which is chosen to reduce the correlation between the users. In traveling from the transmitter to the receiver, the kth user's signal is delayed by τ_k. The received signal is

$$r(t) = \sum_{k=1}^{K} v_k(t - \tau_k) c_k(t - \tau_k) + z(t), \qquad (19.6)$$

where $z(t)$ is noise and where

$$v_k(t) = \sum_n A_k I_{k,n} g(t - nT) \qquad (19.7)$$

is the kth users signal. A system model is shown in figure 19.11(a).

Ideally, the correlation between spreading sequences of different users would be zero, so that a receiver can detect the signal from any user without any interference from any of the other users. However, in practice there is correlation between users that depends on the relative delay of the signals. The optimum receiver takes into account the correlation between all of the received signals, just as for the ISI channel. Intuitively, if the sequence of transmitted symbols for all users that interfere with the mth symbol of user k were known, then this interference could be subtracted and the symbol could be reliably detected. Of course, determining the interfering symbols again requires knowing the symbols from all the other users, and so the entire sequence must be detected. Writing down the likelihood function leads, after some straightforward analysis, to a Viterbi-type sequence estimation problem, where the path cost is similar to that for the ISI problem in (19.4), and where the state consists of the symbols of the previous $K - 1$ users.

To be more specific, consider the case with $K = 3$ users. Assume that the users are transmitting nonsynchronously, so that each user has a delay between transmission and reception. Denote the delay for the kth user as τ_k, and assume that the users are ordered so that $\tau_1 \leq \tau_2 \leq \tau_3$. The received signal is the sum of the overlapping signals shown in figure 19.11(b). Let the symbols be indexed sequentially as $\nu = 0, 1, 2, \ldots$, where

$$\nu = Kn + ((k - 1) \bmod K).$$

As an example, consider decoding the symbol indexed by $\nu = 4$, shown shaded in figure 19.11(b). Optimal decoding of this would require decoding the overlapping symbols, those with indices $\nu = 3$ and $\nu = 2$ (the $K - 1$ previous symbols) and with indices $\nu = 5$ and $\nu = 6$ (the $K - 1$ following symbols), each of which in turn is optimally decoded by knowing what symbols overlap with them. This is an ideal case for the Viterbi algorithm. The state of the Viterbi algorithm is given by the $K - 1$ prior symbols, and the branch metric is that portion of the log-likelihood function that can be causally computed given the state, much as was done for the ISI case. The branch metric requires knowledge of the cross-correlations among the users,

$$h_{jk}(m) = \int c_j(t) \bar{c}_k(t - mT) \, dt,$$

19.5 Maximum-Likelihood Sequence Estimation

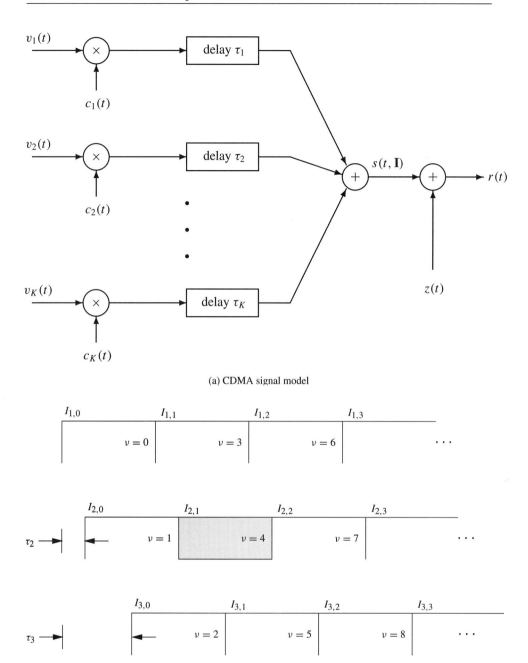

(a) CDMA signal model

(b) Overlapping CDMA signals

Figure 19.11: CDMA signal model.

just as ISI decoding requires knowing the correlation of the channel output signal x_n. A trellis diagram for this three-user system, where each user employs binary signaling, is shown in figure 19.12(a), where I_ν indicates the symbol $I_{k,n}$ for the value of k, n corresponding to ν. The branch metric is a function of the correlation between the overlapping signals, which in turn depends upon the relative delays between signals and the spreading codes selected

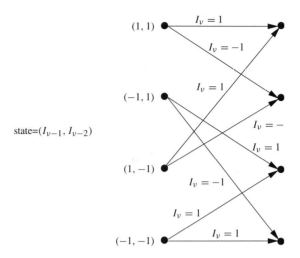

(a) One stage of the trellis for CDMA decoding

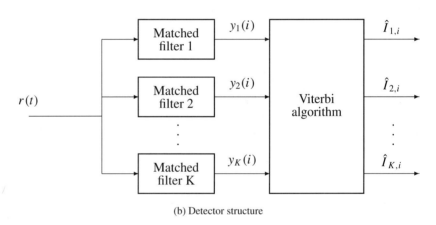

(b) Detector structure

Figure 19.12: CDMA detection.

for each user, and the outputs of filters matched to each of the user's signals,

$$y_k(n) = \int r(t) c_k(t - nT).$$

The general receiver structure is illustrated in figure 19.12(b).

The number of states in the trellis is 2^{K-1} for binary transmission. Since the decoding complexity is proportional to the number of states, computational issues impose limitations on the number of users when optimal decoding is employed. For this reason, a variety of suboptimal decoding strategies that have been developed are discussed in the following sections.

19.5.3 Convolutional decoding

Error-correction coding is commonly employed in the transmission of digital data to decrease the probability of error for a given signal transmission energy. Error-correction codes operate by adding redundancy to a signal in such a way that some errors that occur on the

19.5 Maximum-Likelihood Sequence Estimation

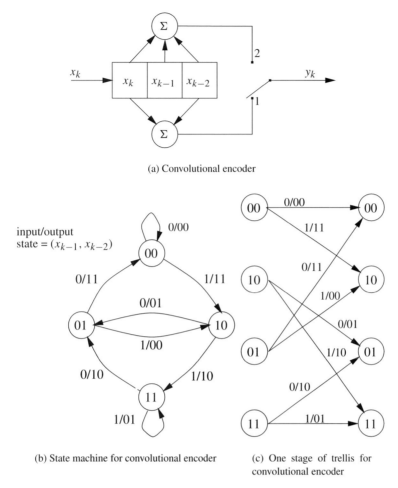

(a) Convolutional encoder

(b) State machine for convolutional encoder

(c) One stage of trellis for convolutional encoder

Figure 19.13: Convolutional coding.

signal can be eliminated at the decoder. One major family of error-correction codes is the *convolutional codes*. A convolutional encoder is (in general) a multi-input, multioutput, linear time-invariant filter. The number of inputs is denoted by k; the number of outputs is denoted by n, and in order to obtain the necessary redundancy, it is always the case that $n > k$.

An example of a convolutional encoder is shown in figure 19.13(a), with $k = 1$ and $n = 2$. A single input bit x_k is presented at the input, and a pair of bits is obtained from each of the outputs in succession, the output y_k. The finite memory in the encoder gives rise to a finite-state machine, as shown in figure 19.13(b), and a trellis that describes the sequence. One stage of the trellis associated with the convolutional encoder of figure 19.13(a) is shown in figure 19.13(c).

What allows the convolutional encoder to correct errors is that only certain paths through the trellis are permitted. If a sufficiently small number of errors occur, then the shortest path through the trellis corresponds to the original path through the trellis. This is illustrated in the following example.

Example 19.5.4 The sequence of bits $\{1, 1, 0, 0, 1, 0, 1, 0\}$ is passed through the convolutional encoder of figure 19.13. The following table shows the results of the coding operation, assuming that the encoder starts in state 00.

time k	input x_k	state x_{k-1}, x_{k-2}	output y_k
0	1	00	11
1	1	10	10
2	0	11	10
3	0	01	11
4	1	00	11
5	0	10	01
6	1	01	00
7	0	10	01

The output sequence is

$$\mathbf{y} = 11\ 10\ 10\ 11\ 11\ 01\ 00\ 01.$$

Now suppose that this sequence passes through a binary symmetric channel that introduces two errors, so that the received sequence is

$$\mathbf{r} = 11\ 10\ \underline{0}0\ 1\underline{0}\ 11\ 01\ 00.$$

The received bits in error are underlined, although the receiver does not know this yet.

The received sequence is passed through a Viterbi algorithm that uses as the branch metric the Hamming distance (number of bits different) between the received bits for that branch and the trellis bits for that branch. The result is that the best path through the trellis corresponds to the bit sequence

$$\hat{\mathbf{r}} = 11\ 10\ 10\ 11\ 11\ 01\ 00\ 01,$$

the original sequence! Two errors have been corrected. From this sequence, knowing the structure of the state machine, the original input sequence x_k can be reconstructed. □

The branch metric employed in decoding the signal depends upon the assumed channel model. For the binary symmetric channel model of the previous example, the appropriate metric is the Hamming distance, which counts the number of differing bits between the received sequence and a path through the trellis. If the channel is assumed to be AWGN, a Euclidean distance metric is employed.

19.6 HMM likelihood analysis and HMM training

The hidden Markov model is described in section 1.7, with additional information and a maximum likelihood training procedure presented in section 17.7. In this section, we present a Viterbi algorithm for determining the most likely sequence of states that an HMM traverses, based on an observed output from the HMM. This sequence of states can be used to compute the likelihood of the sequence, given the HMM model, in what is known as the "best path" approach. Using this technique, the state corresponding to a given output can be estimated—the states are no longer "hidden." On the basis of knowing which outputs correspond to which states, a modified training algorithm is proposed.

Let $\mathcal{M} = (A, \pi, f_{\{S\}})$ be a given HMM, and let $\mathbf{y} = \{\mathbf{y}_1, \mathbf{y}_2, \ldots, \mathbf{y}_T\}$ be a sequence of observations generated by the HMM. Let $S[t]$ denote the state (random variable) at time t, and let $\mathbf{S} = \{s[1], s[2], \ldots, s[T]\}$ denote a sequence of state outcomes. Since the outputs of the HMM are conditionally independent, given the states, the probability of observing

19.6 HMM Likelihood Analysis and HMM Training

the sequence of outputs is

$$P(\mathbf{y}\,|\,\mathbf{s}, \mathcal{M}) = f(\mathbf{y}[1]|s[1])f(\mathbf{y}[2]|s[2])\cdots f(\mathbf{y}[T]|s[T])$$
$$= f_{s[1]}(\mathbf{y}[1])f_{s[2]}(\mathbf{y}[2])\cdots f_{s[T]}(\mathbf{y}[T]). \tag{19.8}$$

Given the Markov structure, the probability of the state sequence $\mathbf{s}$ is

$$P(\mathbf{s}\,|\,\mathcal{M}) = P(S[1] = s[1])P(S[2] = s[2]|S[1] = s[1])$$
$$\cdots P(S[T] = s[T]|S[T-1] = s[T-1])$$
$$= \pi(s[1])a(s[2], s[1])\cdots a(s[T], s[T-1]). \tag{19.9}$$

Multiplying (19.8) and (19.9), we obtain

$$P(\mathbf{y}, \mathbf{s}\,|\,\mathcal{M})$$
$$= f_{s[1]}(\mathbf{y}[1])f_{s[2]}(\mathbf{y}[2])\cdots f_{s[t]}(\mathbf{y}[T])\pi(s[1])a(s[2], s[1])\cdots a(s[T], s[T-1]). \tag{19.10}$$

The maximum-likelihood state sequence is that sequence of states that maximizes (19.10):

$$\mathbf{s}^* = \arg\max_{\mathbf{s}} P(\mathbf{y}, \mathbf{s}\,|\,\mathcal{M}).$$

It will be convenient to deal with the negative log-likelihood function instead of the likelihood function. One reason is that the products in (19.10) tend to become very small; another reason is that it converts the multiplicative cost into an additive cost. We will want to *minimize* the negative log-likelihood function

$$L(\mathbf{s}) = -\log P(\mathbf{y}, \mathbf{s}|\mathcal{M}) = -\log \pi(s[1]) - \sum_{t=1}^{T} \log a(s[t], s[t-1]) - \log f_{s[t]}(\mathbf{y}[t]). \tag{19.11}$$

The log-likelihood function of (19.11) can be used as the path metric in a maximum-likelihood sequence estimation problem, where the ML sequence corresponds to the shortest path through the trellis derived from the state diagram underlying the HMM, where the path weight is

$$b(y[k], s[k]|\sigma_k) = -\log f_{s[k]}(\mathbf{y}[k]) - \log a(s[k], s[k-1]),$$

and where the state in this case is

$$\sigma_k = s[k-1].$$

The initial path cost is set according to $-\log \pi(s[k])$.

Using the VA code of section 19.4 requires computation of an appropriate branch metric, as well as setting up the trellis parameters based on the parameters of the HMM. Code to accomplish these is shown in algorithm 19.7, which has the functions `hmmnorm` to compute the norm, and `hmminitvit` to initialize the trellis information.

Algorithm 19.7 Norm and initialization for Viterbi HMM computations
    File: `hmmnorm.m`
              `hmminitvit.m`

Using the VA, the best-path cost is the maximum negative log-likelihood

$$-\log P(\mathbf{y}, \mathbf{s}\,|\,\mathcal{M}),$$

which can be used to measure how closely the sequence **y** "fits" the HMM. In section 1.7, an algorithm was provided for computing $\log P(\mathbf{y} \mid \mathcal{M})$. This is referred to as the "any path" likelihood, since any (and every possible) path that contributes to the likelihood is considered. Using the VA, only the best (in the ML sense) path is used to compute the likelihood. This is referred to as the "best path" likelihood. Code that computes the best-path likelihood, by finding the shortest path through the trellis associated with the HMM, is given in algorithm 19.8.

Algorithm 19.8 Best-path likelihood for the HMM
  File: `hmmlpyseqv.m`
     `vitbestcost.m`

Once a state sequence $\{s[1], s[2], \ldots, s[T]\}$ is available, an update of the transition probabilities and the output density parameters is possible conditioned upon this sequence. The transition probabilities can be estimated as

$$a(i, j) = \frac{\text{number of times the transition from } i \text{ to } j \text{ occurs}}{\text{number of times transition from } i \text{ occurs}}.$$

Note that if several observation sequences are available, they can be pooled together to estimate these transition numbers.

The output density parameters associated with each state can also be updated if the state sequence is known. Let $\mathbf{y}_s$ be the set of output data associated with state s, $s = 1, 2, \ldots, S$. The density parameters can be updated as follows.

Discrete outputs. The probability $b_{i,s} = P(Y[t] = i \mid S[t] = s)$ can be estimated from

$$b_{i,s} = \frac{\text{number of times output } i \text{ occurs in } \mathbf{y}_s}{\text{number of outputs from state } j}.$$

Gaussian distributions. The mean and covariance of the Gaussian distribution associated with state s is simply the sample mean and sample covariance of the data $\mathbf{y}_s$.
MATLAB code that provides an update of the HMM parameters is shown in algorithm 19.9.

Algorithm 19.9 HMM training using Viterbi methods
  File: `hmmupdatev.m`
     `hmmupfv.m`

With these methods, there is some flexibility in training and evaluation. We now have two training methods—based on the EM algorithm and the Viterbi algorithm—as well as two methods of computing the likelihood—the any-path method and the best-path method. The following example illustrates the use of the Viterbi methods associated with the HMM, as well as providing a comparison among the four possible choices of training crossed with evaluation.

Example 19.6.1 The code shown in algorithm 19.10 performs the following:

1. Data for a test HMM are established, and a sequence of eight data samples **y** is generated.

19.6 HMM Likelihood Analysis and HMM Training

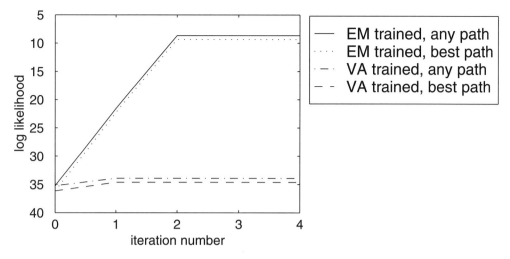

Figure 19.14: Comparing HMM training algorithms.

2. The results of updating the HMM based on this observation sequence are presented for both methods of training, and both methods of computing likelihood. Training continues for four iterations.

Figure 19.14 illustrates the results of the four possible outcomes. The following observations may be drawn for this particular example with a small amount of test data:

1. The Viterbi methods converge after only one iteration.
2. The any-path likelihood is larger than the best-path likelihood, since the any-path likelihood includes the best path, as well as others. However, they tend to follow each other closely. □

Algorithm 19.10 Use of the Viterbi methods with HMMs
File: `hmmtest2vb.m`

19.6.1 Dynamic warping

Let A represent a sequence of feature vectors,

$$A = \{\mathbf{a}_1, \mathbf{a}_2, \ldots, \mathbf{a}_M\}.$$

This sequence of vectors is to be compared with another sequence of vectors

$$B = \{\mathbf{b}_1, \mathbf{b}_2, \ldots, \mathbf{b}_N\}.$$

Ideally, we could make a comparison vector by vector, and add up the distortion, as in

$$d(A, B) = \sum_{i=1}^{M} \|\mathbf{a}_i - \mathbf{b}_i\|.$$

However, if vector $\mathbf{a}_i$ does not "line up" with $\mathbf{b}_i$, then this simple cumulative distance may not work. In particular, the sequence A and the sequence B might not even be of the same length, so that $N \neq M$. In this case, it may be necessary to introduce an association function between the vectors in A and the vectors in B. This is illustrated in figure 19.15, in which the

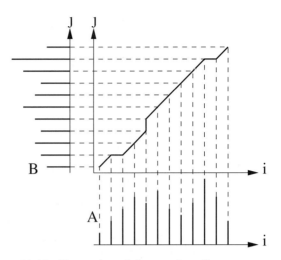

Figure 19.15: Illustration of the warping alignment process.

horizontal axis represents vectors in A indexed by i and the vertical axis represents vectors in B indexed by j. (The vectors are drawn as scalar quantities for the sake of representation.) The correspondence between the data is given by C, where

$$\mathbf{C} = \{c_1, c_2, \ldots, c_K\}$$

with

$$c_k = (i_k, j_k), \qquad k = 1, 2, \ldots, K.$$

With each correspondence there is a discrepancy (distance) between the associated points in A and B, which is

$$d[c_k] = d(a_{i_k}, b_{j_k}) = \|a_{i_k} - b_{j_k}\|.$$

The correspondence C determines an overall distance between A and B as

$$D(C; A, B) = \sum_{k=1}^{K} d[c_k].$$

The goal in dynamic warping is to determine the best correspondence C, so that the distance $D(C; A, B)$ is minimized. This is a "warping" of A onto B so that they are as similar as possible. Some reasonable constraints imposed on C are the following:

1. The function must be monotonic (proceed from right to left and from bottom to top):

$$i_k \geq i_{k-1} \qquad j_k \geq j_{k-1}.$$

2. The function must match the endpoints of A and B:

$$i_1 = j_1 = 1,$$
$$i_K = M,$$
$$j_K = N.$$

3. The function must not skip any points:

$$i_k - i_{k-1} \leq 1 \qquad j_k - j_{k-1} \leq 1.$$

4. A global limit on the maximum amount of warping must also be imposed:

$$|i_k - j_k| \leq Q.$$

The computational task of finding the best corresponding path is reduced by the principle of optimality: the best path from the starting point $(1, 1)$ to any other point (i, j) is independent of what happens beyond that point. Hence, the total cost to the point (i_k, j_k) is determined by the cost of the best path to a predecessor to that point, plus the cost of that point. If $\mathbf{C}_k$ denotes the sequence of correspondences $C_k = \{c_1, c_2, \ldots, c_k\}$, then

$$D(C_k) = d[c(k)] + \min_{\text{legal } c_{k-1}} [D(C_{k-1})],$$

where "legal c_{k-1}" denotes all allowable predecessors of the point c_k, which are $(i, j - 1)$, $(i - 1, j - 1)$, and $(i - 1, j)$.

A common application of this warping algorithm is in speech recognition, where the vectors in A and B are feature vectors, and it is desired to match an observed word to a sequence of stored templates for purposes of recognition. Due to the temporal variation in speech, the alignment of the speech vectors may vary. In such an application, the shortest-path algorithm is referred to as *dynamic time warping*.

Another application is for spectral matching. Given a measured spectrum (computed, for example, using FFT-based techniques), it is desired to determine which other measured spectrum best matches. If there are shifts in the some of the frequency peaks, then simply computing the difference spectrum may be inappropriate. In this case, it may be more appropriate to first align the peaks using this warping technique, then use the distance between the aligned peaks as the degree of spectral match. For this kind of matching, we coin the term *dynamic frequency warping*.

MATLAB code that performs dynamic warping is shown in algorithm 19.11 (following [251]).

Algorithm 19.11 Warping code
File: `warp.m`

19.7 Alternatives to shortest-path algorithms

Despite the fact that optimal solutions to a variety of sequential problems can be theoretically computed, the algorithms are not always employed because of the computational complexity involved in computing the true optimum. We discuss briefly some alternatives to the path-search algorithms thus far described.

Reduced-width search. Instead of maintaining and propagating a path to each state, only the best m paths are propagated at each time. Provided that m is sufficiently large, then with high probability the correct path will be propagated. An examination of this and related algorithms can be found in [229].

State reduction. In some instances the number of states, hence the computational complexity of the VA, can be reduced by dividing the set of states into subsets of states that are used in the VA search. An exploration of this concept is provided in [81, 82].

Linear and adaptive equalization. This alternative is very common for ISI communication and CDMA detection. Instead of an attempt to perform optimal detection, the received signal is passed through a channel that models (in some sense) the inverse of the channel that gives rise to the interference. At the output of this channel, the signal is passed through a detector. In some applications, the filter is determined adaptively. The minimum mean-squared error and least-squares error filter designs discussed in chapter 3 are often employed. A thorough discussion of these techniques is provided in [261, chapters 10 and 11].

Decision feedback equalization (DFE). This method is also appropriate as a suboptimal approach to detection in ISI or CDMA channels. In DFE, decisions made at the output of the detector are fed back and subtracted from the incoming signals. While there is always the probability that an incorrect decision will be subtracted (leading to potentially worse performance), DFE tends to work fairly well in practice, in the presence of modest noise and interference. For ISI channels, this is discussed in [261, chapter 11].

Sequential algorithms. Sequential algorithms are search algorithms related to the VA. However, unlike the VA, they do not rely on a trellis structure—they are applicable for searching trees—and may not find the true optimum solution. Also, because there is some forward and backward searching on the tree, the number of computations made as the computation proceeds through the tree is a random variable. This means that buffers employed in the implementation of the algorithm must be large, so that they do not overfill (or only do so with very low probability).

The basic concept of sequential algorithms is to trace a path through the tree, observing how the path metric grows. If the path metric starts to grow too quickly, the algorithm will back up through the last few frames of data and explore alternate paths. The particular rules for computing the path metric appropriate for the search and for backing up are determined by the specific type of sequential algorithm. A window of some width b is maintained, and as the path exceeds b branches in the window, the oldest branch is output, and the window is shifted, much as for the VA.

An example of a sequential algorithm is the stack algorithm. The stack algorithm extends a number q of paths at each frame. The stack algorithm maintains three pieces of information on the stack for every path it is considering: the path length, the path description, and the path discrepancy (path metric). Starting from an empty path, the following steps are used:

1. Extend the path at the top of the stack (the best path on the stack) to its q successors, computing the metric for each new path.
2. Sort the stack according to the new path metrics.
3. If the top path is at the end of the tree (or window width is achieved) output a branch decision.
4. Return to step 1.

Since the path metrics are for paths of different lengths, in order to make a valid comparison in step 2 (the sorting step), the metric must be normalized in some sense by the path length. A thorough discussion of the stack algorithm, including an appropriate metric for binary sequences, is provided in [201]. An alternative sequential algorithm that is slower but requires less memory is the Fano algorithm, also described in [201].

19.8 Exercises

19.1-1 For the graph of figure 19.2, determine an adjacency matrix, a cost adjacency matrix, and an incidence matrix.

19.2-2 [254] A performance measure P is to be minimized, where

$$P = \sum_{i=1}^{3} f_i(x_i),$$

in which the functions f_i are the piecewise linear functions shown in the following table.

$f_i(x_i)$	$x_i = 0$	$1 \leq x_i \leq 5$	$5 \leq x_i \leq 15$
$f_1(x_1) =$	0	$5 + 10x_1$	$15 + 8x_1$
$f_2(x_2) =$	0	$2 + 8x_2$	$-18 + 12x_2$
$f_3(x_3) =$	0	$1 + 9x_3$	$-4 + 10x_3$

P is to be minimized with integer values of x_i, subject to the constraint

$$\sum_{i=1}^{3} x_i = b$$

for an integer b.

(a) Determine the optimum solution (x_1, x_2, x_3) and optimum cost $P_{\min}$ when $b = 14$.

(b) Repeat when $b = 10$ and $b = 1$.

19.2-3 [143] Consider a warehouse with a storage capacity of B units and an initial stock of v units. Each month, y_i units are sold, where $i = 1, 2, \ldots, n$. The unit selling price is P_i in month i. (The price may vary from month to month.) Let x_i be the quantity purchased in month i, at a buying price of c_i. At the end of each month, the stock on hand must be no more than B, so that

$$v + \sum_{i=1}^{j}(x_i - y_i) \leq B \qquad j = 1, 2, \ldots, n.$$

The amount sold each month cannot be more than the stock at the end of the previous month (new stock arrives at the end of each month), so that $y_i \leq v + \sum_{i=1}^{j-1}(x_j - y_j)$ for $i = 1, 2, \ldots, n$. Also, x_i and y_i must be nonnegative integers. The total profit is

$$P_n = \sum_{j=1}^{n}(p_j y_j - c_j x_j).$$

The problem is to determine x_j and y_j so that P_n is maximized. Let $f_i(v_i)$ represent the maximum profit that can be earned in months $i + 1, i + 2, \ldots, n$, starting with v_i units of stock at the end of month i. Then $f_0(v)$ is the maximum value of P_n.

(a) Obtain the dynamic programming recurrence for $f_i(v_i)$ in terms of $f_{i+1}(v_i)$.

(b) What is $f_n(v_i)$?

(c) Show that $f_i(v_i) = a_i x_i + b_i v_i$ for some constants a_i and b_i.

(d) Show that an optimal P_n is obtained using the following strategy: (i) If $p_i \geq c_i$ and

$b_{i+1} \geq c_i$	then	$y_i = v_i$	and	$x_i = B$,
$b_{i+1} \leq c_i$	then	$y_i = v_i$	and	$x_i = 0$.

(ii) If $p_i \leq c_i$ and

$$b_{i+1} \geq c_i \quad \text{then} \quad y_i = 0 \quad \text{and} \quad x_i = B - v_i,$$
$$b_{i+1} \leq p_i \quad \text{then} \quad y_i = v_i \quad \text{and} \quad x_i = 0,$$
$$p_i \leq b_{i+1} \leq c_i \quad \text{then} \quad y_i = 0 \quad \text{and} \quad x_i = 0.$$

(e) Assume $B = 100$ and $v = 60$. For the following price/cost quantities, determine an optimal decision sequence:

i	1	2	3	4	5	6	7	8,
p_i	8	8	2	3	4	3	2	5,
c_i	3	6	7	1	4	5	1	3.

19.3-4 In a certain network, the probability of failure of the links between the branches is shown in figure 19.16, where each link fails independently of the other links. Determine the most reliable network path between nodes ① and ⑦.

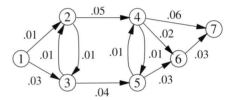

Figure 19.16: Probability of failure of network links.

19.5-5 In the ISI problem, show that the matched filter output y_k of (19.2) can be written as

$$y_k = \sum_n I_n x_{k-n} + v_k,$$

where v_k is a noise sequence. Determine the statistics of v_k if $z(t)$ is AWGN.

19.5-6 (Constrained transmission) Suppose a binary transmission system is constrained so that runs of more than two repetitions of the same symbol are not allowed. For example, the sequence 001010 is allowed, but 000101 is not, since there is a run of three zeros.

(a) Draw a state diagram that represents this constrained transmission.

(b) Draw the corresponding trellis. How many states are there?

19.5-7 (ISI) Suppose a channel has an impulse response

$$c(t) = \delta(t) + 0.75\delta(t-T) - 0.3\delta(t-2T).$$

(a) Determine $h(t)$, assuming that a flat pulse of duration T is used at the input of the channel.

(b) Determine the autocorrelation function x_n.

19.5-8 Show that (19.5) is an appropriate branch metric for the log-likelihood function shown in (19.4).

19.5-9 (CDMA) For the received signal of (19.6) and signal-transmission model of (19.7):

(a) Determine the log-likelihood function, assuming that the noise is AWGN with two-sided PSD $N_0/2$.

(b) Determine an appropriate branch metric for the Viterbi decoding algorithm.

19.5-10 Write a MATLAB function `hmmupdateym` that accepts multiple observation sequences `y{1}, y{2}, ..., y{N}` and computes an update to an HMM using Viterbi methods that use the ensemble of observations.

19.9 References

There is a tremendous literature on graph theory and algorithms on graphs for a variety of purposes; only the barest minimum is presented here. An excellent starting point from the algorithmic standpoint is [298]. An introductory text on graph theory and related combinatorial topics is [372]; [80] provides another good summary, while [143] provides a good description of forward and backward dynamic programming. An excellent introduction to the Viterbi algorithm in association with convolutional codes, and the important idea of the transfer function of a graph for systems analysis, is in [90]. A recent text dedicated to dynamic programming, particularly as it applies to controls, is [25]. A variety of interesting examples can be found in [254].

For the applications of the Viterbi algorithm, [68, chapter 12] provides a good summary of hidden Markov models and training using the Viterbi algorithm, as do [265, 266]. Maximum-likelihood sequence estimation and convolutional codes are standard topics in digital communications; see for example [261] or [198]. CDMA detection using the Viterbi algorithm is discussed, for example, in [353, 355, 311]. Dynamic (time) warping is discussed in [251].

Chapter 20

Linear Programming

20.1 Introduction to linear programming

Linear programming (LP) is a special case of the general constrained optimization problem we have already seen. In this case, the function to be minimized is linear, $\mathbf{c}^T\mathbf{x}$, where $\mathbf{x} \in \mathbb{R}^n$. The constraints are also linear, being of the form $A\mathbf{x} = \mathbf{b}$ or $A\mathbf{x} \leq \mathbf{b}$, where $\mathbf{b} \in \mathbb{R}^m$. Furthermore, there is an inequality constraint $\mathbf{x} \geq \mathbf{0}$. Generally A is not invertible (or else the problem becomes trivial). Thus we have the problem

$$\begin{aligned} \text{minimize } & \mathbf{c}^T\mathbf{x} \\ \text{subject to } & A\mathbf{x} = \mathbf{b}, \\ & \mathbf{x} \geq \mathbf{0}. \end{aligned} \qquad (20.1)$$

A problem of the form (20.1) is said to be in **standard form**. A solution that satisfies all of the constraints is said to be a **feasible solution**. The **value** of the linear program is $\mathbf{cx}$ when $\mathbf{x}$ is the optimal feasible solution.

The linear programming problem can also be expressed in terms of inequality constraints in the form

$$\begin{aligned} \text{minimize } & \mathbf{c}^T\mathbf{x} \\ \text{subject to } & A\mathbf{x} \leq \mathbf{b}, \\ & \mathbf{x} \geq 0. \end{aligned} \qquad (20.2)$$

We see in what follows that, despite the apparent difference between the form (20.1) with equality constraints and the form (20.2) with inequality constraints, the inequality constraints can be expressed as equality constraints.

Example 20.1.1 It may be easier to visualize the geometry of the problem with the constraints expressed as inequalities. Consider the problem

$$\begin{aligned} \text{maximize } & x_1 + x_2 \\ \text{subject to } & x_1 + 2x_2 \leq 10, \\ & 6x_1 + 5x_2 \leq 45, \\ & x_1 \geq 0 \quad x_2 \geq 0. \end{aligned}$$

The region determined by the constraints is shown in figure 20.1. Also shown is the direction of maximum increase of the objective $x_1 + x_2$ (the gradient vector for the function). If you think of the gradient pointing in the direction of a stream of water, then a marble in the stream will roll in the direction of the gradient until it runs into one of the walls, which represent the constraints. The marble will then slide along the constraint wall until it can go no further in the direction of the gradient. This is the optimum solution.

20.2 Putting a Problem into Standard Form

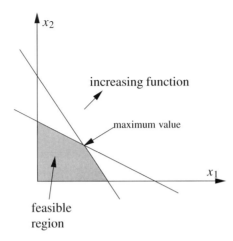

Figure 20.1: A linear programming problem.

From the geometry, it is clear that if one of the constraint lines lies orthogonal to the gradient of the objective function, there may be an infinite number of solutions: any solution along that constraint line is optimum. Otherwise, there will be a solution at the intersection of two or more of the constraints. □

From the geometry of the problem, the following cases may occur:

1. There may be no feasible solution, that is, no solution which satisfies all of the constraints. This occurs if the constraints are contradictory.
2. There may be no bounded optimum feasible solution. For example, the feasible region may extend infinitely in a direction of increase for the objective function.
3. There may be one solution.
4. There may be an infinite number of solutions.

In the general case, the feasible region will be bounded by hyperplanes. It can be shown that, provided that the feasible region where $\mathbf{x}$ satisfies the constraints

$$A\mathbf{x} = \mathbf{b},$$
$$\mathbf{x} \geq 0,$$

is bounded, the feasible region is convex.

20.2 Putting a problem into standard form

The algorithm presented in the following for solving linear programming problems requires that the optimization problem be placed into standard form. While the standard form may appear to be somewhat limited in applicability, by employing a few straightforward tricks a variety of problems can be placed into this form.

20.2.1 Inequality constraints and slack variables

By the introduction of variables called *slack* variables and *surplus* variables, inequality constraints can be converted to equality constraints, so that the two problems are effectively equivalent. For example, a constraint of the form

$$3x_1 + 5x_2 - 6x_3 \leq 7$$

with $x_1 \geq 0$, $x_2 \geq 0$, $x_3 \geq 0$, can be written as

$$3x_1 + 5x_2 - 6x_3 + y_1 = 7,$$

with the additional constraint that $y_1 \geq 0$. The new variable y_1 is known as a *slack* variable. A constraint of the form

$$5x_1 - 6x_2 - 4x_3 \geq 3$$

with $x_1 \geq 0$, $x_2 \geq 0$, $x_3 \geq 0$, can be written as

$$5x_1 - 6x_2 - 4x_3 - y_2 = 3,$$

with $y_2 \geq 0$. The new variable y_2 is known as a *surplus* variable.

A problem with inequality constraints on both sides, such as

$$-5 \leq 3x_1 + 5x_2 - 6x_3 \leq 7$$

can be treated by adding both slack and surplus variables:

$$3x_1 + 5x_2 - 6x_3 + y_1 = 7,$$
$$3x_1 + 5x_2 - 6x_3 - y_2 = -5,$$
$$y_1 \geq 0 \quad y_2 \geq 0.$$

A problem with **absolute value** constraints can be converted as follows:

$$|3x_1 + 4x_2 - 3x_3| \leq 2$$

is, of course, equivalent to

$$-2 \leq 3x_1 + 4x - 2 - 3x_3 \leq 2$$

which leads to the two constraints

$$3x_1 + 4x - 2 - 3x_2 + y_1 = 2,$$
$$3x_1 + 4x - 2 - 3x_2 - y_2 = -2,$$
$$y_1 \geq 0 \quad y_2 \geq 0.$$

20.2.2 Free variables

The problem (20.1) also imposes the constraint that $x_i \geq 0$ for $i = 1, 2, \ldots, n$. A variable x_i that is not so constrained is called a *free variable*. There are two ways of treating free variables. The first is by introducing two other variables that are not free; the second is by eliminating the free variables by Gaussian eliminations.

A problem with a free variable x_i can be put into standard form by introducing two new variables u_i and v_i and using

$$x_i = u_i - v_i, \qquad (20.3)$$

where $u_i \geq 0$ and $v_i \geq 0$.

20.2 Putting a Problem into Standard Form

Example 20.2.1 Place the following linear programming problem into standard form.

$$\text{maximize } 3x_1 - 2x_2 + 4x_3$$
$$\text{subject to } 7x_1 + 5x_2 - x_3 \leq 9,$$
$$2x_1 + 2x_2 + 3x_3 = 5,$$
$$5x_1 + 6x_2 - 8x_3 \geq 10,$$
$$x_1 \geq 0 \quad x_3 \geq 0,$$
$$x_2 \text{ free}.$$

By the introduction of a slack and a surplus variable, the first and last constraints become

$$7x_1 + 5x_2 - x_3 + y_1 = 9,$$
$$5x_1 + 6x_2 - 8x_3 - y_2 = 10.$$

Now, letting $x_2 = u_2 - v_2$, we can write the problem as

$$\text{minimize } -3x_1 + 2(u_2 - v_2) - 4x_3$$

$$\text{subject to } \begin{bmatrix} 7 & 5 & -5 & -1 & 1 & 0 \\ 2 & 2 & -2 & 3 & 0 & 0 \\ 5 & 6 & -6 & -8 & 0 & -1 \end{bmatrix} \begin{bmatrix} x_1 \\ u_2 \\ v_2 \\ x_3 \\ y_1 \\ y_2 \end{bmatrix} = \begin{bmatrix} 9 \\ 5 \\ 10 \end{bmatrix},$$

$$x_1 \geq 0, u_2 \geq 0, v_2 \geq 0, x_3 \geq 0, y_1 \geq 0, y_2 \geq 0. \quad \square$$

The method of dealing with free variables just described introduces two new variables into the problem for every free variable. If every variable is free, computation of the solution of the dual problem, as discussed in section 20.5, is advised. When there are a few free variables, another approach is advised. The variables may be eliminated—expressed in terms of other variables in the problem—then computed after the solution to the problem is known. This can be done using the row operations familiar from Gaussian elimination (section 5.1).

Example 20.2.2 The problem

$$\text{minimize } x_1 + 5x_2 + 3x_3$$
$$\text{subject to } x_1 + 2x_2 - x_3 = 3,$$
$$2x_1 + 3x_2 + 2x_3 = 8,$$
$$x_2 \geq 0 \quad x_3 \geq 0,$$
$$x_1 \text{ free},$$

is in standard form, except for the presence of the free variable x_1. This problem is in the form

$$\text{minimize } \mathbf{c}^T \mathbf{x},$$
$$\text{subject to } A\mathbf{x} = \mathbf{b},$$

where $\mathbf{c}^T = \begin{bmatrix} 1 & 5 & 2 \end{bmatrix}$ and

$$A = \begin{bmatrix} 1 & 2 & -1 \\ 12 & 3 & 2 \end{bmatrix} \quad \mathbf{b} = \begin{bmatrix} 3 \\ 8 \end{bmatrix}.$$

To eliminate the free variable, it is useful to represent the problem using an array of coefficients known as the **tableau**, which is the matrix formed by

$$\begin{bmatrix} A & \mathbf{b} \\ \mathbf{c} & 0 \end{bmatrix}.$$

For this problem, the tableau is

$$\begin{bmatrix} 1 & 2 & -1 & 3 \\ 2 & 3 & 2 & 8 \\ 1 & 5 & 3 & 0 \end{bmatrix}.$$

To eliminate the free variable x_1, we perform Gaussian elimination on the first column:

$$\boxed{\text{row 2}} \leftarrow \boxed{\text{row 2}} - (2)\boxed{\text{row 1}}$$

$$\boxed{\text{row 3}} \leftarrow \boxed{\text{row 3}} - \boxed{\text{row 1}}$$

to obtain the tableau

$$\begin{bmatrix} 1 & 2 & -1 & 3 \\ 0 & -1 & 4 & 2 \\ 0 & 3 & 4 & -3 \end{bmatrix}.$$

The last two rows of the resulting tableau correspond to an optimization problem that is equivalent to the original problem, except that the variable x_1 has been eliminated:

$$\text{minimize } 3x_2 + 4x_3 - 3$$
$$\text{subject to } -x_2 + 4x_3 = 2.$$

Once the solution (x_2, x_3) to this problem is obtained, x_1 can be obtained from the original first constraint $x_1 + 2x_2 - x_3 = 3$, or

$$x_1 = 3 - 2x_2 + x_3. \qquad \square$$

Multiple free variables can be eliminated using Gaussian elimination. The resulting smaller problem is solved, then the free variables are obtained by back-substitution from the original constraints.

20.2.3 Variable-bound constraints

The constraint in the normal form

$$\mathbf{x} \geq \mathbf{d}$$

can be converted readily to a constraint of the form

$$\mathbf{y} \geq \mathbf{0}$$

by letting $\mathbf{y} = \mathbf{x} - \mathbf{d}$ and making the appropriate substitutions. Thus, the problem

$$\text{minimize } \mathbf{c}^T \mathbf{x}$$
$$\text{subject to } A\mathbf{x} = \mathbf{b},$$
$$\mathbf{x} \geq \mathbf{d},$$

becomes

$$\text{minimize } \mathbf{c}^T \mathbf{y}$$
$$\text{subject to } A\mathbf{y} = \mathbf{b} - A\mathbf{d},$$
$$\mathbf{y} \geq \mathbf{0}.$$

The value of this is offset from the original value by $\mathbf{c}^T \mathbf{d}$.

A constraint of the form

$$\mathbf{x} \leq \mathbf{d}$$

is treated by letting $\mathbf{y} = \mathbf{d} - \mathbf{x}$, with the constraint that $\mathbf{y} \geq \mathbf{0}$, and making the appropriate substitutions.

20.2.4 Absolute value in the objective

The objective to

$$\text{minimize } |\mathbf{c}^T \mathbf{x}|$$

$$\text{subject to (other constraints)}$$

can be treated as two cases. In the first case, solve where $\mathbf{c}^T \mathbf{x}$ is positive by using $\mathbf{c}^T \mathbf{x}$ as the objective, adding in the extra constraint that $\mathbf{c}^T \mathbf{x} \geq 0$. This can be done, of course, using a surplus variable. That is

$$\text{minimize } \mathbf{c}^T \mathbf{x}$$
$$\text{subject to } \mathbf{c}^T \mathbf{x} - y = 0,$$
$$y \geq 0,$$
$$\text{(other constraints)}.$$

In the second case, where $\mathbf{c}^T \mathbf{x}$ is negative, we want to find a maximum of $\mathbf{c}^T \mathbf{x}$. We throw in the extra constraint. Thus we solve

$$\text{minimize } -\mathbf{c}^T \mathbf{x}^T$$
$$\text{subject to } \mathbf{c}^T \mathbf{x} + y = 0,$$
$$y \geq 0,$$
$$\text{(other constraints)}.$$

Both subcase problems must be solved, then the problem that leads to the minimum value of $|\mathbf{c}^T \mathbf{x}|$ is used.

20.3 Simple examples of linear programming

We present in this section some simple (and historic) examples of linear programming to demonstrate how such problems may arise. Examples of more interest in signal processing and systems theory in general are presented in section 20.6.3. The examples are presented in the form most natural to the statement of the problem, not in standard form.

Example 20.3.1 (Transportation problem) A producer produces a substance at m warehouses and desires to ship from these warehouses to each of n stores. At warehouse i there is s_i (supply) of the substance to be shipped, $i = 1, 2, \ldots, m$. Each store requires an amount d_j (demand), $j = 1, 2, \ldots, n$, of the substance to be delivered. In shipping from warehouse i to store j, there is a unit shipping cost c_{ij}. Let x_{ij} denote the amount of substance shipped from warehouse i to store j. We desire to determine x_{ij} in such a way as to minimize shipping cost and achieve the requirements of the stores, without exceeding the output capability of the warehouse. We have the constraints

$$\sum_{j=1}^{n} x_{ij} = s_i \quad i = 1, 2, \ldots, m \quad \text{(meet supply)},$$

$$\sum_{i=1}^{m} x_{ij} = d_j \quad j = 1, 2, \ldots, n \quad \text{(meet demand)}.$$

Also, since we cannot ship a negative amount, we also have $x_{ij} \geq 0$. The cost is

$$\sum_{i=1}^{m} \sum_{j=1}^{n} x_{ij} c_{ij}.$$

This is a linear programming problem in mn variables. □

Example 20.3.2 (Activity analysis) Suppose that a manufacturer has several different manufacturing resources, such as raw materials, labor, and various pieces of equipment. Each of these resources can be combined to produce any one of several commodities. The manufacturer knows the amounts of the various resources that are needed to produce each of the commodities, and he also knows how much profit he can make on each of the commodities. The problem is: how to apportion each of the resources among the various commodities in order to maximize profit?

Let m denote the number of resources and n the number of producible commodities. Let a_{ij} be the number of units of resource i required to produce one unit of the commodity j. The constraints on the resources are indicated by b_i. Let c_j denote the profit per unit of commodity j. Finally, let x_j denote the amount of the jth commodity produced. We want to

$$\text{maximize} \sum_{j=1}^{n} c_j x_j$$

subject to

$$\sum_{j=1}^{m} a_{ij} x_i \leq b_i, \quad i = 1, 2, \ldots, m,$$

and, since we cannot produce negative commodities,

$$x_j \geq 0, \quad j = 1, 2, \ldots, n.$$

□

20.4 Computation of the linear programming solution

Efficient algorithms exist for the solution of linear programming problems, the most famous of which is known as the *simplex method*. It relies upon several remarkable facts about linear programming problems, but ultimately proves to be essentially the same as Gaussian elimination on selected columns of the tableau.

20.4.1 Basic variables

In the set of equalities from the problem in standard form,

$$A\mathbf{x} = \mathbf{b} \qquad (20.4)$$

where A is $m \times n$, we assume that there are more variables than constraints, $m < n$, and also that A is full rank. If A is not full rank the problem is said to be *degenerate*. Degenerate problems arise either because the constraints are contradictory, or because there is redundancy in the constraints. In the first case there is no solution; in the second case the redundant constraints may be eliminated. We therefore assume in all cases that $\text{rank}(A) = m$.

For notational simplicity, we assume for the moment that the first m columns of A are linearly independent. Let us decompose A as

$$A = [\hat{A}\ \tilde{A}],$$

where $\hat{A}$ is $m \times m$ and is nonsingular. Then (20.4) can be written as

$$[\hat{A}\ \tilde{A}]\mathbf{x} = [\hat{A}\ \tilde{A}] \begin{bmatrix} \mathbf{x}_B \\ \tilde{\mathbf{x}} \end{bmatrix} = \mathbf{b},$$

which has the (not necessarily unique) solution

$$\mathbf{x} = \begin{bmatrix} \hat{A}^{-1}\mathbf{b} \\ \mathbf{0} \end{bmatrix} = \begin{bmatrix} \mathbf{x}_B \\ \tilde{\mathbf{x}} \end{bmatrix}. \qquad (20.5)$$

20.4 Computation of the Linear Programming Solution

A solution (not necessarily optimal) is obtained by choosing m independent columns of A, finding a solution involving those columns, and setting the other $n - m$ components of $\mathbf{x}$ to zero. A solution in which the $n - m$ components of $\mathbf{x}$ not associated with $\tilde{A}$ are set to zero is called a **basic solution** of (20.4). The components of $\mathbf{x}$ associated with the columns of $\tilde{A}$ are called the **basic variables.** Of course, in general, the basic variables do not need to be the first m variables; they may be any m variables provided that the corresponding columns of A are linearly independent (or more generally, so that $\mathbf{b}$ lies in the column space of these columns).

Is it also possible that a basic variable is zero. Such a basic variable is said to be *degenerate*, and the solution obtained is a degenerate basic solution.

Example 20.4.1 For the problem

$$\begin{bmatrix} 1 & 6 & 7 & 4 & 6 \\ 2 & 2 & 2 & 5 & 9 \\ 3 & 3 & 4 & 6 & 3 \end{bmatrix} \mathbf{x} = \begin{bmatrix} 3 \\ -3 \\ 7 \end{bmatrix},$$

let us take columns 1, 3, and 5 as the components in $\hat{A}$. These are linearly independent. Then a solution (with the components of $\mathbf{x}$ presented in the order of the original problem) is

$$\mathbf{x} = \begin{bmatrix} 2 \\ 0 \\ 1 \\ 0 \\ -1 \end{bmatrix}.$$

This is a basic solution, involving $n - m$ linearly independent columns of A.

If the right-hand side changes to $\begin{bmatrix} 15 \\ 6 \\ 11 \end{bmatrix}$, then the solution becomes

$$\mathbf{x} = \begin{bmatrix} 1 \\ 0 \\ 2 \\ 0 \\ 0 \end{bmatrix},$$

which is basic, but degenerate. □

The key fact that is used in linear programming is:

> The optimum solution of a linear programming problem is a basic feasible solution.

More formally, we have the following result.

Theorem 20.1 *For the linear programming problem (20.1), where A is $m \times n$ of rank m, if there is an optimal feasible solution, then there is an optimal basic feasible solution.*

Proof The proof of theorem 20.1 will set the stage for the simplex algorithm. Suppose that $x_1, x_2, \ldots, x_n$ is an optimal feasible solution. Then

$$x_1 \mathbf{a}_1 + x_2 \mathbf{a}_2 + \cdots + x_n \mathbf{a}_n = \mathbf{b},$$

where $\mathbf{a}_i$ is the ith column of A. Also, $x_i \geq 0$. Suppose that in the solution there are exactly p values of x_i that are nonzero, and assume (for notational convenience) that they are indexed so that they are the first p components. Then

$$x_1 \mathbf{a}_1 + x_2 \mathbf{a}_2 + \cdots + x_p \mathbf{a}_p = \mathbf{b}. \tag{20.6}$$

If $p < m$ and the vectors $\mathbf{a}_1, \mathbf{a}_2, \ldots, \mathbf{a}_p$ are linearly independent, then a basic solution can be found by finding an additional $m - p$ columns of A such that the resulting set of vectors $\mathbf{a}_1, \mathbf{a}_2, \ldots, \mathbf{a}_m$ is linearly independent (which can be done because A has rank m). Then a basic (although degenerate) solution can be found by setting p components of $\mathbf{x}$ as in (20.6), and the remaining components to zero.

If the vectors $\mathbf{a}_1, \mathbf{a}_2, \ldots, \mathbf{a}_p$ are not linearly independent, then (because A has rank m), it must be that $p > m$. We must find another solution with m nonzero components that has the same value. By the linear dependence of the vectors, there are values $y_1, y_2, \ldots, y_p$ such that

$$y_1 \mathbf{a}_1 + y_2 \mathbf{a}_2 + \cdots + y_p \mathbf{a}_p = 0, \tag{20.7}$$

where y_i is positive for some i. Multiplying (20.7) by ϵ and subtracting from (20.6), we find

$$(x_1 - \epsilon y_1)\mathbf{a}_1 + (x_2 - \epsilon y_2)\mathbf{a}_2 + \cdots + (x_P - \epsilon y_p)\mathbf{a}_p = \mathbf{b},$$

so the vector $\mathbf{x} - \epsilon \mathbf{y}$ still satisfies the constraint $A(\mathbf{x} - \epsilon \mathbf{y}) = \mathbf{b}$. In order to satisfy the constraint $\mathbf{x} - \epsilon \mathbf{y} > 0$, we note that since at least one value of y_i is positive, the ith component of $\mathbf{x} - \epsilon \mathbf{y}$ will decrease as ϵ increases. We find the value of ϵ to the first point where a component becomes zero:

$$\epsilon = \min_{y_i > 0} \frac{x_i}{y_i}.$$

By this means, the p components of $\mathbf{x}$ are reduced to (not more than) $p - 1$ components in $\mathbf{x} - \epsilon \mathbf{y}$. This process is repeated as necessary to obtain a solution that has m nonzero components.

In converting the optimal feasible solution to a basic solution, we must ensure that the optimality has not been compromised. The value of the solution $\mathbf{x} - \epsilon \mathbf{y}$ is

$$\mathbf{c}^T (\mathbf{x} - \epsilon \mathbf{y}).$$

On the basis of the optimality of $\mathbf{x}$, we conclude that

$$\mathbf{c}^T \mathbf{y} = 0. \tag{20.8}$$

Otherwise, for values of ϵ sufficiently small that $\mathbf{x} - \epsilon \mathbf{y}$ is still feasible, selection of an ϵ of appropriate sign $\epsilon \mathbf{c}^T \mathbf{y}$ would be positive, so that $\mathbf{c}^T (\mathbf{x} - \epsilon \mathbf{y})$ would be smaller than $\mathbf{c}^T \mathbf{x}$, violating the optimality. In light of (20.8), the basic solution has the same optimal value as the optimal feasible solution, hence the basic solution is optimal. □

The theorem means that the solution can be found simply by searching appropriately over only the basic solutions. Since there are n variables and m unknowns, a brute-force solution would require solving each of $_nC_m$ (the number of combinations of n things taken m at a time) systems of equations. Rather than using brute force over all possible feasible solutions, however, much more elegant computational algorithms exist, the most famous of which is the simplex algorithm.

20.4.2 Pivoting

Starting from an initial basic feasible solution, the simplex algorithm works by substituting one basic variable for another in such a way that $\mathbf{c}^T \mathbf{x}$ decreases. In the linear programming literature, the substitution step is known as *pivoting*, and it is essentially just a Gaussian elimination step. (It should not be confused with the pivoting step performed in the LU factorization.) Pivoting is used to move from one basic solution to another.

20.4 Computation of the Linear Programming Solution

We will introduce the concept by an example. We want to solve the following problem, which is already in a form such that a basic feasible solution can be readily identified.

minimize $2x_1 + 3x_2 - 4x_3 - 5x_4 + x_5 - 2x_6$

subject to
$$\begin{cases} x_1 - 2x_2 & + 2x_5 + 7x_6 = 3, \\ 3x_2 + x_4 - x_5 + 4x_6 = 5, \\ x_3 + -3x_5 + 2x_6 = 4, \end{cases} \quad (20.9)$$

$x_1 \geq 0, \; x_2 \geq 0, \; x_3 \geq 0, \; x_4 \geq 0, \; x_5 \geq 0, \; x_6 \geq 0.$

In this form, basic variables are those that appear uniquely in their respective columns. From (20.9), it is clear that x_1, x_3, and x_4 are basic variables, and that a basic solution is

$$x_1 = 3 \quad x_2 = 0 \quad x_3 = 4 \quad x_4 = 5 \quad x_5 = 0 \quad x_6 = 0.$$

It is convenient to represent these equations in an array known as a *tableau*. The tableau representing the constraints for this problem is

$$\begin{array}{cccccc} x_1 & x_2 & x_3 & x_4 & x_5 & x_6 \end{array}$$
$$\begin{bmatrix} 1 & -2 & 0 & 0 & 2 & 7 & 3 \\ 0 & 3 & 0 & 1 & \boxed{-1} & 4 & 5 \\ 0 & 0 & 1 & 0 & -3 & 2 & 4 \end{bmatrix}. \quad (20.10)$$

The columns of the tableau represent the variables $x_1, x_2, \ldots, x_6$, as indicated.

Suppose we now want to make x_5 basic instead of x_4 in the tableau (20.10). In the original tableau, the basic variable x_4 has its nonzero coefficient in the second row of the table. We make x_5 basic by performing Gaussian elimination row operations to modify the fifth column, so that all elements in it are zero except in the second row, in which a 1 is to appear (at the location of the circled element in (20.10)). The row operations are

$$\boxed{\text{row 1}} \leftarrow \boxed{\text{row 1}} + (2)\boxed{\text{row 2}}$$

$$\boxed{\text{row 3}} \leftarrow \boxed{\text{row 3}} + (-3)\boxed{\text{row 2}}$$

$$\boxed{\text{row 2}} \leftarrow (-1)\boxed{\text{row 2}}.$$

The result of the elimination step is the new tableau

$$\begin{array}{cccccc} x_1 & x_2 & x_3 & x_4 & x_5 & x_6 \end{array}$$
$$\begin{bmatrix} 1 & 4 & 0 & 2 & 0 & 15 & 13 \\ 0 & -3 & 0 & -1 & 1 & -4 & -5 \\ 0 & -9 & 1 & -3 & 0 & -10 & -11 \end{bmatrix}.$$

The basic variables are now x_1, x_3, and x_5, and the basic solution is

$$x_1 = 13 \quad x_2 = 0 \quad x_3 = -11 \quad x_4 = 0 \quad x_5 = -5 \quad x_6 = 0.$$

Let us now denote the elements of the tableau by y_{ij}, $i = 1, 2, \ldots, m$, $j = 1, 2, \ldots, n+1$, where m is the number of constraints and n is the number of variables. To replace the basic variable x_p with nonzero coefficient in the kth row of the table for a nonbasic variable x_q, we perform the following row operations:

$$\begin{cases} y_{ij} = y_{ij} - \frac{y_{iq}}{y_{kq}} y_{kj} & i \neq k, \; j = 1, 2, \ldots, n+1, \\ y_{kj} = \frac{y_{kj}}{y_{kq}}, & j = 1, 2, \ldots, n+1. \end{cases} \quad (20.11)$$

20.4.3 Selecting variables on which to pivot

Having the ability to select new basic variables by pivoting, the question now arises: How should the basic variables be selected to obtain the desired minimum of $\mathbf{c}^T\mathbf{x}$? For the sake of discussion, we assume for the moment that the variables are indexed so that the basic variables are the first m variables. The tableau can be written as

$$\begin{array}{cccccccc} \mathbf{a}_1 & \mathbf{a}_2 & \cdots & \mathbf{a}_m & \mathbf{a}_{m+1} & \cdots & \mathbf{a}_n & \mathbf{b} \end{array}$$
$$\begin{bmatrix} 1 & 0 & & 0 & y_{1,m+1} & & y_{1n} & y_{1,n+1} \\ 0 & 1 & & 0 & y_{2,m+1} & & y_{2n} & y_{2,n+1} \\ \vdots & & & & & & & \\ 0 & 0 & & 1 & y_{m,m+1} & & y_{mn} & y_{m,n+1} \end{bmatrix}. \qquad (20.12)$$

The elements $y_{i,j}$ are used to represent the elements of the tableau. The column headers indicate that the columns come from the A matrix. Thus $\mathbf{a}_1$ is the first column of A, and so forth. The identity matrix in the first m columns indicates that the first m variables are basic. The solution is

$$\mathbf{x}^T = \begin{bmatrix} \mathbf{x}_B \\ \mathbf{0} \end{bmatrix} = \begin{bmatrix} y_{1,n+1} \\ y_{2,n+1} \\ \vdots \\ y_{m,n+1} \end{bmatrix}.$$

The value of the objective function is

$$z = \sum_{i=1}^{n} c_i x_i. \qquad (20.13)$$

For the basic solution indicated in the tableau, the value is

$$z_b = \sum_{i=1}^{m} c_i x_i = \mathbf{c}_B^T \mathbf{x}_B = \sum_{i=1}^{m} c_i y_{i,n+1},$$

where $\mathbf{c}_B = [c_1, c_2, \ldots, c_m]^T$. For problem (20.9), the value of the basic solution shown is $z_b = -35$.

If we were to use *other* than a basic solution in (20.12), so that $x_{m+1}, \ldots, x_n$ are not all zero, we can express the first m variables in terms of these values as

$$\begin{aligned} x_1 &= y_{1,n+1} - \sum_{j=m+1}^{n} y_{1j} x_j, \\ x_2 &= y_{2,n+1} - \sum_{j=m+1}^{n} y_{2j} x_j, \\ &\vdots \\ x_m &= y_{m,n+1} - \sum_{j=m+1}^{n} y_{mj} x_j. \end{aligned} \qquad (20.14)$$

Now, we substitute (20.14) into (20.13). We obtain

$$z = z_b + x_{m+1}(c_{m+1} - w_{m+1}) + x_{m+2}(c_{m+2} - w_{m+2}) + \cdots + x_n(c_n - w_n), \qquad (20.15)$$

where

$$w_k = \sum_{i=1}^{n} y_{ik} c_i. \qquad (20.16)$$

20.4 Computation of the Linear Programming Solution

In (20.15), the basic variables have been eliminated. If we examine (20.15) with the intent of minimizing z, we see that we can reduce z below its value due to the current basic solution z_b by increasing any of the variables x_k for which $(c_k - w_k)$ is negative. (Recall that we have the constraint that $x_k \geq 0$.) Furthermore, if each $c_k - w_k$ is positive, then no further decrease in the value is possible, and the optimum solution has been reached. Let

$$r_k = c_k - w_k. \tag{20.17}$$

These values are known as the **reduced-cost coefficients**. Then the value can be decreased by increasing any variable for which $r_k < 0$. In practice, it is common to choose the smallest (most negative) r_k. Let q be the index of the variable chosen,

$$q = \arg\min r_k.$$

The next question is: How much can x_q be increased? It should be noted that x_q cannot be changed completely arbitrarily: as x_q is changed, the other variables also must change to still satisfy $A\mathbf{x} = \mathbf{b}$. Change in x_q can only go far enough to ensure that $\mathbf{x} \geq 0$. Let us consider how the basic variable x_1 can change. From (20.14),

$$x_1 = y_{1,n+1} - \sum_{i=m+1}^{n} y_{1j} x_j;$$

as x_q increases, the basic variable x_1 changes by $-y_{1q} x_q$. If $y_{1q} < 0$, then as x_q increases x_1 increases and the constraint on x_1 never becomes active. If this is true for each variable, then the feasible region is unbounded.

On the other hand, if $y_{1q} > 0$, then x_1 decreases as x_q increases. By the constraint, x_q can only increase until $x_1 = 0$. On this basis, the largest x_q can be is $y_{1,n+1}/y_{1q}$. Since similar constraints must hold for each basic variable, we must have

$$x_q \leq \min \frac{y_{i,n+1}}{y_{i,q}}, \quad i = 1, 2, \ldots, m.$$

Let

$$p = \arg\min_i \frac{y_{i,n+1}}{y_{iq}}$$

for those y_{iq} that are greater than 0, and set

$$x_q = \frac{y_{p,n+1}}{y_{pq}}.$$

Then, as x_q is introduced into the problem, the variable x_p is reduced to zero. By appropriate row operations, the operation takes place as a pivot. This is the key step of the simplex algorithm.

20.4.4 The effect of pivoting on the value of the problem

Computationally, the effect of pivoting variables on the value of the problem can be obtained as follows. Note that (20.13) can be written as

$$c_1 x_1 + c_2 x_2 + \cdots + c_n x_n - z = 0. \tag{20.18}$$

In this equation, z may be regarded as another basic variable. This variable remains basic through all the computations (pivoting never takes places on z). The variable z does not need to be explicitly added to the tableau. For the example problem (20.9), the tableau with the last row added to represent (20.18) is

$$\mathbf{c}^T: \begin{array}{c c c c c c c c} & x_1 & x_2 & x_3 & x_4 & x_5 & x_6 & \mathbf{b} \\ & \begin{bmatrix} 1 & -2 & 0 & 0 & 2 & 7 & 3 \\ 0 & 3 & 0 & 1 & -1 & 4 & 5 \\ 0 & 0 & 1 & 0 & -3 & 2 & 4 \\ 2 & 3 & -4 & -5 & 1 & -2 & 0 \end{bmatrix} \end{array}.$$

Since the coefficients of the basic variables x_1, x_2, and x_3 no longer appear alone in their columns in this augmented tableau, the variables no longer look like basic variables. However, standard row operations can be performed to make the elements in the last row corresponding to the basic elements be equal to zero. Before expressing this as row operations, however, let us examine its effect. Subtract from each side of (20.18) the quantity

$$z_b = \sum_{i=1}^{m} c_i x_i.$$

Recognizing from (20.16) that for $i = 1, 2, \ldots, m$ we have $c_i = w_i$, and using the fact that $r_k = c_k - w_k$, we find that (20.18) becomes

$$r_{m+1} x_{m+1} + r_{m+2} x_{m+2} + \cdots + r_n x_n - z = -z_b.$$

The coefficients in the last row of the tableau thus become the reduced-cost coefficients, and row operations carried out on the tableau can be applied to this row as to the other rows, as the basic variables are moved around.

For example (20.9), the row operations are

$$\boxed{\text{row 4}} \leftarrow \boxed{\text{row 4}} - (2)\boxed{\text{row 1}} - (-5)\boxed{\text{row 2}} - (-4)\boxed{\text{row 3}},$$

yielding the tableau

$$\mathbf{r}^T: \begin{array}{c c c c c c c c} & \mathbf{a}_1 & \mathbf{a}_2 & \mathbf{a}_3 & \mathbf{a}_4 & \mathbf{a}_5 & \mathbf{a}_6 & \mathbf{b} \\ & \begin{bmatrix} 1 & -2 & 0 & 0 & 2 & 7 & 3 \\ 0 & 3 & 0 & 1 & -1 & 4 & 5 \\ 0 & 0 & 1 & 0 & 3 & 2 & 4 \\ 0 & 22 & 0 & 0 & -20 & 12 & 35 \end{bmatrix} \end{array}. \qquad (20.19)$$

The value corresponding to the current basic variables is shown in the lower right corner.

20.4.5 Summary of the simplex algorithm

Once the tableau is in this form, with the basic variables identified and the reduced-cost coefficients on the last row obtained by row operations on $\mathbf{c}^T$, the simplex algorithm operates by the pivoting method described previously. Let y_{ij}, $i = 1, 2, \ldots, m+1$, $j = 1, 2, \ldots, n+1$ represent the current values of the tableau, including the last row.

1. If each $r_j \geq 0$, $j = 1, 2, \ldots, n$, then the current basic solution is optimal. Stop.
2. Otherwise, select a q such that $r_q < 0$. Commonly,

$$q = \arg \min_{i=1,2,\ldots,n} r_i.$$

3. If there are no $y_{iq} > 0$ for $i = 1, 2, \ldots, m$, the problem is unbounded. Otherwise,

$$p = \arg \min_{y_{iq} > 0} \frac{y_{i,n+1}}{y_{iq}}.$$

4. Pivot on the pqth element, updating rows $1, 2, \ldots, m+1$. Return to step 1.

20.4 Computation of the Linear Programming Solution

At each step of the algorithm, the current value of the solution appears in the lower right corner.

Example 20.4.2 We now complete the linear programming solution to the problem posed in (20.9). The initial basic solution is the tableau of (20.19),

$$\mathbf{r}^T: \begin{array}{c} \\ \\ \\ \end{array} \begin{array}{cccccc} \mathbf{a}_1 & \mathbf{a}_2 & \mathbf{a}_3 & \mathbf{a}_4 & \mathbf{a}_5 & \mathbf{a}_6 & \mathbf{b} \\ \begin{bmatrix} 1 & -2 & 0 & 0 & \boxed{2} & 7 & 3 \\ 0 & 3 & 0 & 1 & -1 & 4 & 5 \\ 0 & 0 & 1 & 0 & -3 & 2 & 4 \\ 0 & 22 & 0 & 0 & \underline{-20} & 12 & 35 \end{bmatrix} \end{array}.$$

The first column on which to pivot has the negative residual underlined. The row on which to pivot has the pivot circled. The result of pivoting on this element is the tableau

$$\begin{bmatrix} 0.5 & -1 & 0 & 0 & 1 & 3.5 & 1.5 \\ 0.5 & 2 & 0 & 1 & 0 & 7.5 & 6.5 \\ 1.5 & -3 & 1 & 0 & 0 & 12.5 & 8.5 \\ 10 & 2 & 0 & 0 & 0 & 82 & 65 \end{bmatrix}.$$

As all elements in the last row are nonnegative, the algorithm is complete. We identify the solution as

$$x_1 = 0 \quad x_2 = 0 \quad x_3 = 8.5 \quad x_4 = 6.5 \quad x_5 = 1.5 \quad x_6 = 0.$$

The value of the solution is

$$(8.5)(-4) + (6.5)(-5) + (1.5)(1) = -65,$$

the negative of the lower right corner of the tableau. □

20.4.6 Finding the initial basic feasible solution

We have seen that the optimal feasible solution is basic, and that pivoting can be used to move from one basic solution to another to decrease the objective function. There remains only one question: Given a general problem, how can an initial basic solution be obtained? For problems in which slack or surplus variables are introduced on each variable, the slack variables appear as basic variables. For example, the problem

$$\begin{aligned} \text{minimize } & 3x_1 + 4x_2 + 2x_3 \\ \text{subject to } & 3x_1 - 2x_2 \geq 2, \\ & 2x_1 + 4x_2 - 7x_3 \geq 1, \\ & x_1 \geq 0 \quad x_2 \geq 0 \quad x_3 \geq 0, \end{aligned}$$

can be written using slack variables as

$$\begin{aligned} \text{minimize } & 3x_1 + 4x_2 + 2x_3 \\ \text{subject to } & 3x_1 - 2x_2 - y_1 = 2, \\ & 2x_1 + 4x_2 - 7x_3 - y_2 = 1, \\ & x_1 \geq 0 \quad x_2 \geq 0 \quad x_3 \geq 0 \quad y_1 \geq 0 \quad y_2 \geq 0, \end{aligned}$$

with tableau

$$\begin{array}{cccccc} x_1 & x_2 & x_3 & y_1 & y_2 & \mathbf{b} \\ \begin{bmatrix} 3 & -2 & 0 & -1 & 0 & 2 \\ 2 & 4 & -7 & 0 & -1 & 1 \\ 3 & 4 & 2 & 0 & 0 & 0 \end{bmatrix} \end{array}.$$

The initial basic solution is apparent from the following:

$$x_1 = x_2 = x_3 = 0, \qquad y_1 = -2, y_2 = -1.$$

For problems without slack variables, a basic solution can be found by the introduction of *artificial variables*. The constraints

$$A\mathbf{x} = \mathbf{b},$$
$$\mathbf{x} \geq 0,$$

(where the problem is scaled so that $\mathbf{b} \geq 0$) are written using artificial variables as

$$\begin{aligned} &\text{minimize} \sum_{i=1}^{m} y_i \\ &\text{subject to } A\mathbf{x} + \mathbf{y} = \mathbf{b}, \\ &\qquad \mathbf{x} \geq 0 \qquad \mathbf{y} \geq 0. \end{aligned} \quad (20.20)$$

The vector $\mathbf{y} = [y_1, y_2, \ldots, y_m]^T$ is the vector of artificial variables. When a feasible solution to the constraints on $\mathbf{x}$ exists, the minimum value of $\sum_{i=1}^{m} y_i$ is zero. Problem (20.20) is again a linear programming problem, and in the tableau the $\mathbf{y}$ appear as basic variables. This provides a starting point. By applying pivoting to obtain the minimum solution, $\mathbf{y}$ becomes zero, and among the remaining variables a basic solution becomes apparent.

Example 20.4.3 Find a basic feasible solution to the set of constraints

$$-2x_1 + 2x_2 + 7x_3 = 3,$$
$$3x_1 - x_2 + 4x_3 = 5,$$

with

$$x_1 \geq 0 \qquad x_2 \geq 0 \qquad x_3 \geq 0.$$

We introduce the artificial variables y_1 and y_2 and the artificial objective function $y_1 + y_2$. The initial tableau with the artificial variables is

$$\begin{array}{c} \\ \\ \mathbf{c}^T: \end{array} \begin{array}{cccccc} x_1 & x_2 & x_3 & y_1 & y_2 & \mathbf{b} \\ \begin{bmatrix} -2 & 2 & 7 & 1 & 0 & 3 \\ 3 & -1 & 4 & 0 & 1 & 5 \\ 0 & 0 & 0 & 1 & 1 & 0 \end{bmatrix} \end{array}.$$

In order to arrive at basic variables, we apply row operations to the last row so that there are zeros under the basic variables, obtaining the tableau

$$\begin{array}{c} \\ \\ \mathbf{r}^T: \end{array} \begin{array}{cccccc} x_1 & x_2 & x_3 & y_1 & y_2 & \mathbf{b} \\ \begin{bmatrix} -2 & 2 & \boxed{7} & 1 & 0 & 3 \\ 3 & -1 & 4 & 0 & 1 & 5 \\ -1 & -1 & -11 & 0 & 0 & -8 \end{bmatrix} \end{array}.$$

20.4 Computation of the Linear Programming Solution

Pivoting is applied about the circled element to obtain the updated tableau

$$\begin{bmatrix} -2/7 & 2/7 & 1 & 1/7 & 0 & 3/7 \\ 29/7 & -15/7 & 0 & -4/7 & 1 & 23/7 \\ \underline{-29/7} & 15/7 & 0 & 11/7 & 0 & -23/7 \end{bmatrix}.$$

Now pivoting is applied about the underlined element to obtain the tableau

$$\begin{bmatrix} 0 & 0.1379 & 1 & 0.1034 & 0.0690 & 0.6552 \\ 1 & -0.5172 & 0 & -0.1379 & 0.2414 & 0.7931 \\ 0 & 0 & 0 & 1 & 1 & 0 \end{bmatrix}.$$

The artificial variables are now not basic variables, and we identify the initial basic solution as

$$x_1 = 0.7931 \qquad x_2 = 0 \qquad x_3 = 0.6552. \qquad \square$$

Once a basic solution is obtained using artificial variables, we can use it as the initial basic feasible solution. By this approach, there are two phases to the solution:

1. Phase I: introduce artificial variables, and solve the linear programming problem (20.20) to find an initial basic feasible solution.
2. Phase II: using this initial basic feasible solution, solve the original linear programming problem.

Example 20.4.4 We wish to minimize

$$2x_1 + 3x_2$$

subject to the same constraints as the previous example,

$$-2x_1 + 2x_2 + 7x_3 = 3,$$
$$3x_1 - x_2 + 4x_3 = 5,$$

$$x_1 \geq 0 \qquad x_2 \geq 0 \qquad x_3 \geq 0.$$

Having obtained an initial feasible basic solution from the last example, we can write an initial tableau for this problem as

$$\begin{array}{c} \\ \\ \mathbf{c}^T: \end{array} \begin{array}{cccc} x_1 & x_2 & x_3 & \mathbf{b} \\ \begin{bmatrix} 0 & 0.1379 & 1 & 0.6552 \\ 1 & -0.5172 & 0 & 0.7931 \\ 2 & 3 & 0 & 0 \end{bmatrix} \end{array}.$$

We now perform row operations to place zeros in the basic columns, and obtain the tableau

$$\begin{array}{c} \\ \\ \mathbf{r}^T: \end{array} \begin{array}{cccc} x_1 & x_2 & x_3 & \mathbf{b} \\ \begin{bmatrix} 0 & 0.1379 & 1.0000 & 0.6552 \\ 1 & -0.5172 & 0 & 0.7931 \\ 0 & 4.0345 & 0 & -1.5862 \end{bmatrix} \end{array}.$$

No additional pivoting is necessary in this case. The solution to the linear programming problem is

$$x_1 = 0.7931 \qquad x_2 = 0 \qquad x_3 = 0.6552. \qquad \square$$

20.4.7 MATLAB code for linear programming

The MATLAB code for linear programming shown in algorithm 20.1 accepts the matrices **A**, **b**, and **c** for a problem in standard form. It computes both phases of linear programming (without checking first to see if any variables are already basic). A vector indicating which variables are free (if any) may also be passed in. If there are free variables, they are eliminated using Gaussian elimination prior to obtaining the linear programming solution, then computed by backsubstitution after the linear programming problem has been solved. The functions that perform these operations for free variables are `reducefree` and `restorefree`, respectively. The main linear programming function is `simplex1`. Essentially, all this function does is set up the necessary tableau, and call `pivottableau`.

Algorithm 20.1 The simplex algorithm for linear programming
    File: `simplex1.m`

Algorithm 20.2 Tableau pivoting for the simplex algorithm
    File: `pivottableau.m`

Algorithm 20.3 Elimination and backsubstitution of free variables for linear programming
    File: `reducefree.m`
          `restorefree.m`

Example 20.4.5 Solve the linear programming problem

$$\text{minimize } 7x_1 + 4x_2 - 2x_3 + x_4 + 5x_5$$
$$\text{subject to } 3x_1 + 2x_2 - x_3 + x_4 + x_5 = 6,$$
$$2x_1 + x_2 - x_3 + x_4 + 2x_5 = 7,$$
$$x_1 + x_2 - x_3 + 2x_4 + x_5 = 4,$$
$$x_1 \geq 0, \quad x_2 \geq 0, \quad x_3 \text{ free}, \quad x_4 \geq 0, \quad x_5 \geq 0.$$

The data for MATLAB can be entered as

```
A = [3      2     -1     1     1
     2      1     -1     1     2
     1      1     -1     2     1];
b =[6;7;4];
c = [7;4;-2;1; 5];
freevar = [0 0 1 0 0];
```

20.4 Computation of the Linear Programming Solution

and the solution obtained by

```
[x,value] = simplex1(A,b,c,freevar);
```

The solution obtained is

$x_1 = 1 \quad x_2 = 0 \quad x_3 = -1 \quad x_4 = 0 \quad x_5 = 2 \quad \text{value} = -12.$ □

20.4.8 Matrix notation for the simplex algorithm

It is convenient for some theoretical explanations to express aspects of the simplex algorithm using matrix notation. In this notation, we assume without loss of generality that the basic vectors consist of the first m columns of A. We will write

$$A = [\hat{A} \; \tilde{A}]$$

and partition the vectors in the problem similarly:

$$\mathbf{x} = \begin{bmatrix} \hat{\mathbf{x}} \\ \tilde{\mathbf{x}} \end{bmatrix} \quad \mathbf{c} = \begin{bmatrix} \hat{\mathbf{c}} \\ \tilde{\mathbf{c}} \end{bmatrix},$$

where $\hat{\mathbf{x}}$ and $\hat{\mathbf{c}}$ correspond to the basic solution, and $\tilde{\mathbf{x}}$ and $\tilde{\mathbf{c}}$ correspond to variables not in the basic solution. The linear programming problem in standard form is

$$\text{minimize } \hat{\mathbf{c}}^T \hat{\mathbf{x}}^T + \tilde{\mathbf{c}}^T \tilde{\mathbf{x}}$$
$$\text{subject to } \hat{A}\hat{\mathbf{x}} + \tilde{A}\tilde{\mathbf{x}} = \mathbf{b},$$
$$\hat{\mathbf{x}} \geq \mathbf{0} \quad \tilde{\mathbf{x}} \geq \mathbf{0}.$$

The basic solution is found by setting $\tilde{\mathbf{x}}$ to $\mathbf{0}$ and solving

$$\hat{\mathbf{x}} = \hat{A}^{-1}\mathbf{b}.$$

However, other solutions can be found for arbitrary values of $\tilde{\mathbf{x}}$ by

$$\hat{\mathbf{x}} = \hat{A}^{-1}\mathbf{b} - \hat{A}^{-1}\tilde{A}\tilde{\mathbf{x}}.$$

The cost function is

$$z = \mathbf{c}^T \mathbf{x}$$
$$= \hat{\mathbf{c}}^T \hat{A}^{-1}\mathbf{b} + (\tilde{\mathbf{c}}^T - \mathbf{c}^T \hat{A}^{-1}\tilde{A})\tilde{\mathbf{x}}$$
$$= \hat{\mathbf{c}}^T \hat{\mathbf{x}} + (\tilde{\mathbf{c}}^T - \mathbf{c}^T \hat{A}^{-1}\tilde{A})\tilde{\mathbf{x}}.$$

Comparison of this with (20.17) shows that the reduced-cost coefficient vector is

$$\mathbf{r}^T = \tilde{\mathbf{c}}^T - \hat{\mathbf{c}}^T \hat{A}^{-1}\tilde{A}. \tag{20.21}$$

The operation of the simplex algorithm is such as to make $\mathbf{r}^T \geq 0$.

The simplex tableau (not necessarily expressed in terms of basic vectors) is

$$\begin{bmatrix} A & \mathbf{b} \\ \mathbf{c}^T & 0 \end{bmatrix} = \begin{bmatrix} \hat{A} & \tilde{A} & \mathbf{b} \\ \hat{\mathbf{c}}^T & \tilde{\mathbf{c}}^T & 0 \end{bmatrix}. \tag{20.22}$$

The result of the simplex algorithm is to write this tableau (by means of row operations) as

$$\begin{bmatrix} I & \hat{A}^{-1}\tilde{A} & \hat{A}^{-1}\mathbf{b} \\ \mathbf{0} & \tilde{\mathbf{c}}^T - \hat{\mathbf{c}}^T \hat{A}^{-1}\tilde{A} & -\hat{\mathbf{c}}^T \hat{A}^{-1}\mathbf{b} \end{bmatrix}.$$

20.5 Dual problems

Every linear programming problem has an associated problem called a **dual**, which is related by having the same value and involving the same parameters. By comparison with the dual, the original problem is called the *primal* problem. A linear program of the form (20.2) has the most symmetric dual:

$$\begin{array}{ll} \textit{Primal} & \textit{Dual} \\ \text{minimize } \mathbf{c}^T\mathbf{x} & \text{maximize } \mathbf{w}^T\mathbf{b} \\ \text{subject to } A\mathbf{x} \geq \mathbf{b} & \text{subject to } \mathbf{w}^T A \leq \mathbf{c}^T, \\ \mathbf{x} \geq 0 & \mathbf{w} \geq 0. \end{array} \qquad (20.23)$$

The dual of the standard form (20.1) can be obtained by expressing the equality constraint $A\mathbf{x} = \mathbf{b}$ as a pair of inequality constraints. The standard form can be written as

$$\begin{array}{l} \textit{Primal} \\ \text{minimize } \mathbf{c}^T\mathbf{x} \\ \text{subject to } A\mathbf{x} \geq \mathbf{b}, \\ \qquad -A\mathbf{x} \geq -\mathbf{b}, \\ \qquad \mathbf{x} \geq 0. \end{array}$$

This is of the form of the primal problem in (20.23), using the inequality coefficient matrix $\begin{bmatrix} A \\ -A \end{bmatrix}$. Let $\begin{bmatrix} \mathbf{u} \\ \mathbf{v} \end{bmatrix}$ be the dual variable. Then the dual problem can be expressed as

$$\begin{array}{l} \textit{Dual} \\ \text{maximize } \mathbf{u}^T\mathbf{b} - \mathbf{v}^T\mathbf{b} \\ \text{subject to } \mathbf{u}^T A - \mathbf{v}^T A \leq \mathbf{c}^T, \\ \qquad \mathbf{u} \geq 0 \quad \mathbf{v} \geq 0. \end{array}$$

Now let $\mathbf{w} = \mathbf{u} - \mathbf{v}$. Making the same observation that we made in association with (20.3), $\mathbf{w}$ is a free variable. The linear programming problem in standard form and its dual can now be expressed as

$$\begin{array}{ll} \textit{Primal} & \textit{Dual} \\ \text{minimize } \mathbf{c}^T\mathbf{x} & \text{maximize } \mathbf{w}^T\mathbf{b} \\ \text{subject to } A\mathbf{x} = \mathbf{b} & \text{subject to } \mathbf{w}^T A \leq \mathbf{c}^T, \\ \mathbf{x} \geq \mathbf{0} & \mathbf{w} \text{ free}. \end{array} \qquad (20.24)$$

In general, as linear inequalities in a primal problem are converted to equalities, the corresponding components of $\mathbf{w}$ in the dual problem become free variables. Conversely, if some of the components of $\mathbf{x}$ are free variables, then the corresponding inequalities $\mathbf{w}^T A \leq \mathbf{c}^T$ in the dual become equalities. One of the useful features about the dual problem in relation to the standard form is that the standard form requires satisfaction of the constraint $\mathbf{x} \geq \mathbf{0}$, whereas the dual problem does not. A problem involving linear inequalities, but lacking this positivity constraint, can often be solved most efficiently in terms of its dual. A little thought reveals that the dual to the dual is again the primal problem.

The basic result for duality is that both a problem and its dual have the same value. Consider the duality stated in (20.24). If $\mathbf{x}$ is feasible for the primal problem and $\mathbf{w}$ is feasible for the dual problem, then we have

$$\mathbf{w}^T\mathbf{b} = \mathbf{w}^T A\mathbf{x} \leq \mathbf{c}^T\mathbf{x}.$$

Thus, for any feasible solutions to the two problems, the quantity from the primal $\mathbf{c}^T\mathbf{x}$ is greater than or equal to the corresponding quantity from the dual problem $\mathbf{w}^T\mathbf{b}$. By these bounds, we can observe that if there is an $\mathbf{x}^*$ and a $\mathbf{w}^*$ such that $\mathbf{c}^T\mathbf{x}^* = (\mathbf{w}^*)^T\mathbf{b}$, then (since

20.5 Dual Problems

no higher values can be found) $\mathbf{x}^*$ and $\mathbf{w}^*$ must be the optimal solutions to their respective problems. Thus $\mathbf{x}$ works its way "down" while $\mathbf{w}$ works its way "up" until the two bounds meet at the middle. In fact, if a primal problem has a finite optimal solution, then so does its dual.

If the solution to the primal problem is known, then it is straightforward to determine the solution to the dual problem. Suppose that for the problem

$$\text{minimize } \mathbf{c}^T \mathbf{x}$$
$$\text{subject to } A\mathbf{x} = \mathbf{b},$$
$$\mathbf{x} \geq 0,$$

A is partitioned as $A = [\hat{A}, \tilde{A}]$ and the optimal solution is

$$\mathbf{x} = \begin{bmatrix} \hat{\mathbf{x}} \\ 0 \end{bmatrix},$$

where $\hat{\mathbf{x}} = \hat{A}^{-1} \mathbf{b}$. The value of this problem is $\hat{\mathbf{c}}^T \hat{\mathbf{x}}$. We will show that

$$\mathbf{w}^T = \hat{\mathbf{c}}^T \hat{A}^{-1} \quad (20.25)$$

is the solution to the dual problem

$$\text{maximize } \mathbf{w}^T \mathbf{b}$$
$$\text{subject to } \mathbf{w}^T A \leq \mathbf{c}^T.$$

Note first that, for this $\mathbf{w}$,

$$\mathbf{w}^T \mathbf{b} = \hat{\mathbf{c}}^T \hat{A}^{-1} \mathbf{b} = \hat{\mathbf{c}}^T \hat{\mathbf{x}},$$

so that the value of the dual problem is the same as the value of the primal problem. It remains to show that $\mathbf{w}$ is feasible. To this end, recall from (20.21) that

$$\mathbf{r}^T = \tilde{\mathbf{c}}^T - \hat{\mathbf{c}}^T \hat{A}^{-1} \tilde{A}.$$

By means of the simplex algorithm, the relative-cost vector is obtained such that $\mathbf{r} \geq \mathbf{0}$, so that $\tilde{\mathbf{c}}^T \geq \hat{\mathbf{c}}^T \hat{A}^{-1} \tilde{A}$. Then the constraint in the dual problem is

$$\mathbf{w}^T A = \mathbf{w}^T [\hat{A}, \tilde{A}] = [\hat{\mathbf{c}}^T, \mathbf{c}^T \hat{A}^{-1} \tilde{A}] \leq [\hat{\mathbf{c}}^T, \tilde{\mathbf{c}}^T] = \mathbf{c}^T.$$

So $\mathbf{w}$ is both feasible and optimal.

Computationally, the dual solution often can be obtained with few extra computations after the simplex algorithm has been applied to the primal problem. If an $m \times m$ identity matrix appears in the original A matrix then, after the simplex algorithm, those columns of the tableau in which the identity appeared will contain the matrix $\hat{A}^{-1}$. Let $\mathbf{c}_i^T$ denote the original elements in the last row of the tableau corresponding to the columns of the identity matrix. After the simplex algorithm, these elements are converted by row operations to $\mathbf{c}_f^T = \mathbf{c}_i^T - \mathbf{c}^T \hat{A}^{-1}$. Hence, by (20.25), the solution can be obtained as

$$\mathbf{w}^T = \mathbf{c}_i^T - \mathbf{c}_f^T.$$

Example 20.5.1 The problem

$$\text{minimize } x_1 - x_2 + 4x_3$$
$$\text{subject to } 5x_1 + x_3 = 2,$$
$$-3x_1 + x_2 = 6,$$
$$x_1 \geq 0, \quad x_2 \geq 0, \quad x_3 \geq 0,$$

has the tableau

$$\begin{bmatrix} 5 & 0 & 1 & 2 \\ -3 & 1 & 0 & 6 \\ 1 & \underline{-1} & 4 & 0 \end{bmatrix}.$$

The initial vector c_i corresponding to the columns of the identity is $c_i = [4, -1]^T$ (the underlined elements of the tableau). The tableau after pivoting is

$$\begin{bmatrix} 1 & 0 & 0.2 & 0.4 \\ 0 & 1 & 0.6 & 7.2 \\ 0 & \underline{0} & \underline{4.4} & 6.8 \end{bmatrix}.$$

Thus $c_f = [4.4, 0]^T$, and the dual solution is

$$\mathbf{w} = \mathbf{c}_i - \mathbf{c}_f = \begin{bmatrix} -0.4 \\ -1 \end{bmatrix}.$$

□

In order for this technique to find the solution to the dual problem, there must be an identity in the initial A matrix, which is the case when there are slack variables in the problem. When artificial variables are used to determine the initial basic feasible solution, the dual to the resulting primal problem has a different solution. However, it is possible to transform the solution of the modified dual solution into the solution of the original dual solution (see exercise 20.5-10).

20.6 Karmarker's algorithm for LP

In this section, we examine an alternative to the simplex algorithm for linear programming, which is known by the generic name of Karmarker's algorithm. This is a fairly recent algorithm, originally proposed in 1984, based upon extensions of gradient-descent techniques. (Karmarker's algorithm has been used in some signal-processing applications, such as [286], and is connected with the iterative reweighted least-squares technique [45]. Greater exposure to the signal processing community will lead to future applications.) A variety of reasons, both technical and pedagogical, can be given for examining alternate methods of computing solutions to linear programming problems.

Computational complexity. The simplex algorithm works by moving from vertex to vertex of the polytope of the constraints. Since the algorithm provides no *a priori* way of determining which vertices to examine, in the worst case it must examine a significant number of them. In some worst-case examples, a problem with n unknowns requires up to 2^n simplex iterations [182].

Karmarker's algorithm and its variants, on the other hand, have complexity that is polynomial in the number of variables in the problem.

(However, it should be pointed out that the exponential complexity of the simplex algorithm is only in the worst case, and best-case complexity is typically around $3n$ iterations [374, page 57]. Thus, Karmarker's algorithm is theoretically faster, but in many practical instances slower, then the simplex algorithm.)

Extra utilty from gradient descent. Consider the linear programing problem min $\mathbf{c}^T \mathbf{x}$, subject to some constraints. If a gradient-descent method is used, then movement is in the direction of $-\mathbf{c}$ as far as possible until the boundary of the constraint region. If the minimum has not been reached by the time the boundary is reached, conventional gradient techniques have nothing else to say about the problem! Karmarker's vital insight is that the domain of the problem and the function being minimized can be changed, providing more room

20.6 Karmarker's Algorithm for LP

for the gradient to descend, breathing new life into an old tool. He also showed that at each iteration, some improvement in the objective function is guaranteed, provided that an optimum point actually exists.

Our presentation also provides an introduction to new tools such as projective transformations and barrier functions.

Let the unknown vector be $\mathbf{x} \in \mathbb{R}^n$, and let A be $m \times n$. The Karmarker formulation of the linear programming problem is

$$\begin{aligned} \text{minimize } & \mathbf{c}^T \mathbf{x} \\ \text{subject to } & A\mathbf{x} = \mathbf{0}, \\ & \mathbf{1}^T \mathbf{x} = n, \\ & \mathbf{x} \geq \mathbf{0}, \end{aligned} \qquad (20.26)$$

where $\mathbf{1}$ is the vector of all ones of appropriate size. The matrix A is such that $A\mathbf{1} = \mathbf{0}$. Thus $\mathbf{x}^{[0]} = \mathbf{1}$ is an initial feasible point. We also assume that A has rank m. Finally, at the optimum value $\mathbf{x}^*$, $\mathbf{c}^T\mathbf{x} = 0$, so the optimal value of the problem is zero.

While this formulation of the LP problem seems fairly restrictive in comparison to the standard form introduced in section 20.1, we show that standard form LP problems can be mapped to the form of (20.26). We refer to an LP problem in the form of (20.26) as being in "Karmarker standard form."

The constraints $\mathbf{1}^T \mathbf{x} = n$ and $\mathbf{x} \geq \mathbf{0}$ mean that $\mathbf{x}$ lies in a *simplex*. We shall denote the *interior* of the simplex by

$$S = \{\mathbf{x} \in \mathbb{R}^n_{++} : \mathbf{1}^T \mathbf{x} = n\},$$

where $\mathbb{R}^n_{++} = \{\mathbf{x} \in \mathbb{R}^n : x_j > 0 \text{ for all } j\}$. (This region $\mathbb{R}^n_{++}$ is called the strictly positive octant.) In the Karmarker algorithm, we only search inside S, never quit getting to the solution that lies on the boundary. Methods such as this that search on the interior of the feasible region are termed *interior point methods*.

Let F denote the constraint set

$$F = \{\mathbf{x} \in \mathbb{R}^n_{++} : A\mathbf{x} = \mathbf{0}, \mathbf{1}^T \mathbf{x} = n\}.$$

Then the set of feasible solutions is

$$\Omega = S \cap F.$$

These sets are depicted in figure 20.2(a) for $n = 3$.

Karmarker's algorithm has the following outline, which is depicted in figure 20.2(b–c).

1. Starting from the initial point $\mathbf{x}^{[0]} = \mathbf{1}$, the point $\mathbf{x}^{[1]}$ is found by steepest descent (figure 20.2(b)).
2. By means of a projective transformation (not a projection) T, the point $\mathbf{x}^{[1]}$ is remapped to the center $\hat{\mathbf{x}}^{[1]}\hat{\mathbf{x}} = \mathbf{1}$ (figure 20.2(c)). This transformation maps $S \to S$, and creates a new constraint set $F \to \hat{F}$. (It is helpful to think of the domain S as a piece of rubber fixed at the edges, which stretches to move $\mathbf{x}^{[1]}$ to $\mathbf{1}$ under T.)
3. A new point $\hat{\mathbf{x}}^{[2]}$ is found by gradient descent in the new domain (figure 20.2(c)).
4. This new point $\hat{\mathbf{x}}^{[2]}$ is mapped back to the original domain to produce the updated point $\mathbf{x}^{[2]}$ (figure 20.2(d)).
5. The process repeats from step 2 starting with $\mathbf{x}^{[2]}$.

In order to make this sequence of steps work, the objective function must be modified: we must find a function that assists in the enforcement of the constraints and that decreases both in the transformed domain and also after it is transformed back.

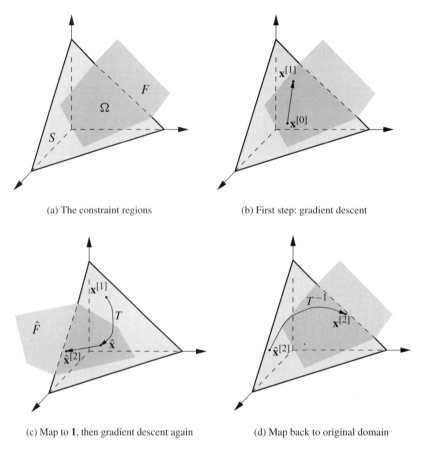

(a) The constraint regions
(b) First step: gradient descent
(c) Map to **1**, then gradient descent again
(d) Map back to original domain

Figure 20.2: Illustration of Karmarker's algorithm.

The objective function. Rather than dealing simply with minimizing $\mathbf{c}^T\mathbf{x}$, Karmarker suggested the use of the function

$$f(\mathbf{x}; \mathbf{c}) = n \log \mathbf{c}^T \mathbf{x} - \sum_{j=1}^{n} \log x_j = \sum_{j=1}^{n} \log(\mathbf{c}^T \mathbf{x}/x_j). \tag{20.27}$$

Observe that for any point on the boundary $x_j = 0$ where $\mathbf{c}^T\mathbf{x} \neq 0$, $f(\mathbf{x}; \mathbf{c})$ attains the value ∞. Because of this, any gradient descent operating on $f(\mathbf{x}; \mathbf{c})$ will not violate the constraint $\mathbf{x} \geq 0$. A function in which the constraint is enforced implicitly by the function values in this way is called a penalty function or barrier function.

Projective transformation. The transformation T employed in the algorithm has the property of picking up $\mathbf{x}^{[k]}$ and placing it in the center of the simplex region. For an arbitrary point $\mathbf{x} \in S$, the transformation is a *projective transformation*[1], defined as

$$\hat{\mathbf{x}} = T(\mathbf{x}) = \frac{nX_k^{-1}\mathbf{x}}{\mathbf{1}^T X_k^{-1}\mathbf{x}} \tag{20.28}$$

---

[1] The general form for a projective transformation is $T(\mathbf{x}) = \frac{C\mathbf{x}+\mathbf{d}}{\mathbf{f}^T\mathbf{x}+g}$ for some $(C, \mathbf{d}, \mathbf{f}, g)$, where $\begin{bmatrix} C & \mathbf{d} \\ \mathbf{f}^T & g \end{bmatrix}$ is nonsingular.

20.6 Karmarker's Algorithm for LP

where $X_k = \text{diag}(\mathbf{x}^{[k]})$. It is straightforward to show that T maps S (the simplex region) into itself, mapping $\mathbf{x}^{[k]}$ to $\mathbf{1}$. The set F maps under T to the set

$$\hat{F} = \{\mathbf{x} \in \mathbb{R}^n_{++} : \hat{A}\mathbf{x} = 0\},$$

where $\hat{A} = AX_k$. The inverse transformation is

$$\mathbf{x} = T^{-1}(\hat{\mathbf{x}}) = \frac{nX_k\hat{\mathbf{x}}}{\mathbf{1}^T X_k \hat{\mathbf{x}}}. \tag{20.29}$$

Furthermore, the function $f(\mathbf{x}; \mathbf{c})$ is essentially invariant under the transformation T, as follows: Let $\hat{\mathbf{c}} = X_k \mathbf{c}$, then

$$f(\hat{\mathbf{x}}; \hat{\mathbf{c}}) = \sum_{j=1}^n \log(\hat{\mathbf{c}}^T \hat{\mathbf{x}}/\hat{x}_j) = f(\mathbf{x}; \mathbf{c}) + \log \det X_k. \tag{20.30}$$

Thus, a decrease in the function $f(\hat{\mathbf{x}}, \hat{\mathbf{c}})$ leads to a corresponding decrease in $f(\mathbf{x}, \mathbf{c})$, up to the additive factor of $\log \det X_k$. We let $\hat{f}(\hat{\mathbf{x}}) = f(\hat{\mathbf{x}}; \hat{\mathbf{c}})$.

Gradient descent. Given a point $\hat{\mathbf{x}} = \mathbf{x}^{[k]} = T(\mathbf{x}^{[k]}) = \mathbf{1}$, we desire to determine a point $\hat{\mathbf{x}}^{[k+1]}$ to minimize $\hat{f}(\hat{\mathbf{x}})$ while satisfying the constraints $\hat{\mathbf{x}}^{[k+1]} \in S \cap \hat{F}$. Satisfaction of the constraint $\hat{\mathbf{x}}^{[k+1]} > 0$ is accomplished by virtue of the barrier implicit in $\hat{f}$. Minimization of $\hat{f}$ subject to the constraints $\hat{A}\hat{\mathbf{x}} = 0$ and $\mathbf{1}^T \hat{\mathbf{x}} = n$ is accomplished by computing the gradient of $\hat{f}$, then projecting the gradient on the space orthogonal to the nullspace of $\hat{A}$ and $\mathbf{1}$. We note that the gradient evaluated at $\hat{\mathbf{x}} = \mathbf{1}$ is

$$\frac{\partial}{\partial \hat{\mathbf{x}}} \hat{f}(\mathbf{1}; \hat{\mathbf{c}}) = \frac{n\hat{\mathbf{c}}}{\mathbf{1}^T \hat{\mathbf{c}}} - \mathbf{1}.$$

Let

$$P_{\hat{A}\perp} = I - \hat{A}^T(\hat{A}\hat{A}^T)^{-1}\hat{A} \qquad P_{\mathbf{1}\perp} = I - \mathbf{1}\mathbf{1}^T/n$$

denote the projectors onto the nullspace of $\hat{A}$ and $\mathbf{1}$, and let

$$P_{\hat{B}\perp} = P_{\hat{A}\perp} P_{\mathbf{1}\perp}.$$

Then

$$\tilde{\mathbf{d}}^{[k]} = -P_{\hat{B}\perp} \left(\frac{\partial}{\partial \hat{\mathbf{x}}} \hat{f}(\mathbf{1}; \hat{\mathbf{c}}) \right)$$

projects the gradient so that the updated point $\hat{\mathbf{x}}^{[k+1]}$ still satisfies the constraints, since $P_{\mathbf{1}\perp}\mathbf{e} = 0$, $\tilde{\mathbf{d}}$ is proportional to $-P_{\hat{B}\perp}\hat{\mathbf{c}}$. We absorb the constant of proportionality by normalization: let

$$\hat{\mathbf{d}}^{[k]} = -P_{\hat{B}\perp} X_k \mathbf{c} / \|P_{\hat{B}\perp} X_k \mathbf{c}\|.$$

A gradient descent update in the transformed domain can be computed by

$$\hat{\mathbf{x}}^{[k+1]} = \hat{\mathbf{x}} + \alpha^{[k]}\hat{\mathbf{d}}^{[k]} = \mathbf{1} + \alpha^{[k]}\hat{\mathbf{d}}^{[k]},$$

where $\alpha^{[k]}$ is the step size. In our implementation (as per Karmarker's suggestion), we choose $\alpha^{[k]} = 1/3$ for all k. More effectively, one might choose $\alpha^{[k]}$ by searching in the $\hat{\mathbf{d}}^{[k]}$ direction from $\mathbf{1}$ until a minimum point is reached (a line search), or take the largest $\alpha^{[k]}$ such that $\hat{\mathbf{x}}^{[k+1]} \geq 0$ and scale it back slightly to remain on the interior of the region.

From $\hat{\mathbf{x}}^{[k+1]}$, the solution in the original space is obtained by transforming back,

$$\mathbf{x}^{[k+1]} = T^{-1}(\hat{\mathbf{x}}^{[k+1]}).$$

Algorithm 20.4 illustrates an implementation, assuming that the problem is set up in the appropriate form.

Algorithm 20.4 Karmarker's algorithm for linear programming
File: `karmarker.m`
`karf.m`

20.6.1 Conversion to Karmarker standard form

In this section, we describe how to place a problem in the "Karmarker standard" form shown in (20.26). Assume that the problem is standard form in the variable $\bar{\mathbf{x}}$,

$$\text{minimize } \bar{\mathbf{c}}^T \bar{\mathbf{x}}$$
$$\text{subject to } A_0 \bar{\mathbf{x}} = \mathbf{b}_0,$$
$$\bar{\mathbf{x}} \geq \mathbf{0}.$$

As discussed in section 20.5, the equality constraints can be written as inequality constraint to produce a primal problem and its symmetric dual, of the form

$$\begin{array}{ll} \textit{Primal} & \textit{Dual} \\ \text{minimize } \bar{\mathbf{c}}^T \bar{\mathbf{x}} & \text{maximize } \bar{\mathbf{w}}^T \begin{bmatrix} \mathbf{b}_0 \\ -\mathbf{b}_0 \end{bmatrix} \\ \text{subject to } \begin{bmatrix} A_0 \\ -A_0 \end{bmatrix} \bar{\mathbf{x}} \geq \begin{bmatrix} \mathbf{b}_0 \\ -\mathbf{b}_0 \end{bmatrix} & \text{subject to } \begin{bmatrix} A_0^T & -A_0^T \end{bmatrix} \bar{\mathbf{w}} \leq \bar{\mathbf{c}}, \\ \mathbf{x} \geq 0 & \bar{\mathbf{w}} \geq 0. \end{array} \quad (20.31)$$

Let $\overline{A} = \begin{bmatrix} A_0 \\ -A_0 \end{bmatrix}$ and let $\overline{\mathbf{b}} = \begin{bmatrix} \mathbf{b} \\ -\mathbf{b} \end{bmatrix}$. The conversion to Karmarker standard form is accomplished in 4 steps:

1. Since the primal problem and the dual problem have the same value at the point of solution, we have $\bar{\mathbf{c}}_0^T \bar{\mathbf{x}} - \bar{\mathbf{b}}_0^T \bar{\mathbf{w}} = 0$ at a point of solution. By the introduction of slack variables $\bar{\mathbf{y}}$ and $\bar{\mathbf{v}}$, we can write the primal and dual problem combined as

$$\overline{A}\bar{\mathbf{x}} - \bar{\mathbf{y}} = \bar{\mathbf{b}},$$
$$\overline{A}^T \bar{\mathbf{w}} + \bar{\mathbf{v}} = \bar{\mathbf{c}},$$
$$\bar{\mathbf{c}}^T \bar{\mathbf{x}} - \bar{\mathbf{b}}^T \bar{\mathbf{w}} = 0,$$
$$\bar{\mathbf{x}} \geq 0 \quad \bar{\mathbf{w}} \geq 0 \quad \bar{\mathbf{v}} \geq 0 \quad \bar{\mathbf{y}} \geq 0. \quad (20.32)$$

2. The next step is to introduce an artificial variable v to create an initial interior point. Let $\mathbf{x}_0, \mathbf{y}_0, \mathbf{w}_0,$ and $\mathbf{v}_0$ be points satisfying (20.32). Consider the problem

$$\text{minimize } v$$
$$\text{subject to } \overline{A}\bar{\mathbf{x}} - \bar{\mathbf{y}} + (\bar{\mathbf{b}} - \overline{A}\mathbf{x}_0 + \mathbf{y}_0) v = \bar{\mathbf{b}},$$
$$\overline{A}^T \bar{\mathbf{w}} + \bar{\mathbf{v}} + \left(\bar{\mathbf{c}} - A_0^T \mathbf{w}_0 - \mathbf{v}_0\right) v = \bar{\mathbf{c}}, \quad (20.33)$$
$$\mathbf{c}_0^T \bar{\mathbf{x}} - \mathbf{b}_0^T \bar{\mathbf{w}} + \left(-\mathbf{c}_0^T \mathbf{x}_0 + \mathbf{b}_0^T \mathbf{w}_0\right) v = 0,$$
$$\bar{\mathbf{x}} \geq 0 \quad \bar{\mathbf{w}} \geq 0 \quad \bar{\mathbf{v}} \geq 0 \quad \bar{\mathbf{y}} \geq 0.$$

The points $\bar{\mathbf{x}} = \mathbf{x}_0, \bar{\mathbf{w}} = \mathbf{w}_0, \bar{\mathbf{v}} = \mathbf{v}_0,$ and $\bar{\mathbf{y}} = \mathbf{y}_0$, with $v = 1$, is a strictly interior feasible solution that can be used as a starting point. The minimum value of v is zero if and only if the problem in step 1 is feasible.

20.6 Karmarker's Algorithm for LP

3. We write problem (20.33) in matrix form as follows:

   minimize v

   subject to $\begin{bmatrix} \overline{A} & 0 & -I & 0 & (\overline{\mathbf{b}} - \overline{A}\mathbf{x}_0 + \mathbf{y}_0) \\ 0 & \overline{A}^T & 0 & I & (\overline{\mathbf{c}} - \overline{A}^T\mathbf{w}_0 - \mathbf{v}_0) \\ \overline{\mathbf{c}}^T & -\overline{\mathbf{b}}^T & 0 & 0 & (-\overline{\mathbf{c}}^T\mathbf{x}_0 + \overline{\mathbf{b}}^T\mathbf{w}_0) \end{bmatrix} \begin{bmatrix} \overline{\mathbf{x}} \\ \overline{\mathbf{w}} \\ \overline{\mathbf{y}} \\ \overline{\mathbf{v}} \\ v \end{bmatrix} = \begin{bmatrix} \overline{\mathbf{b}} \\ \overline{\mathbf{c}} \\ 0 \end{bmatrix}$. (20.34)

   This we write as

   $$\text{minimize } \tilde{\mathbf{c}}^T \tilde{\mathbf{x}}$$
   $$\text{subject to } \tilde{A}\tilde{\mathbf{x}} = \tilde{\mathbf{b}},$$
   $$\mathbf{x} \geq 0,$$

   where we identify $\tilde{\mathbf{x}}$ and $\tilde{A}$ as the stacked vector and matrix in (20.34), and $\tilde{\mathbf{c}}^T = [\mathbf{0}, 1]$. Assume that n is such that $\tilde{\mathbf{x}} \in \mathbb{R}^{n-1}$.

4. The next step is to project this problem so that the simplex enters as a constraint. Define a transformation $P: \mathbb{R}^{n-1}_{++} \to S$ by $\mathbf{x} = P(\tilde{\mathbf{x}})$ when

   $$x_i = n\tilde{x}_i/(\mathbf{1}^T \tilde{\mathbf{x}} + 1) \quad i = 1, 2, \ldots, n-1,$$
   $$x_n = n - \sum_{j=1}^{n-1} x_i.$$

   Then points $\tilde{\mathbf{x}} \geq 0$ map to the simplex $S = \{\mathbf{x} \in \mathbb{R}^n: \sum_{i=1}^n x_i = 1\}$. The inverse transformation $P^{-1}: S \to \mathbb{R}^{n-1}_{++}$ is

   $$\tilde{x}_i = \frac{x_i}{x_n} \quad i = 1, 2, \ldots, n-1. \quad (20.35)$$

   Write $\tilde{A}$ in terms of its columns as

   $$\tilde{A} = \begin{bmatrix} \tilde{\mathbf{a}}_1 & \tilde{\mathbf{a}}_2 & \cdots & \tilde{\mathbf{a}}_{n-1} \end{bmatrix}.$$

   Then the constraint $\tilde{A}\tilde{\mathbf{x}} = \tilde{\mathbf{b}}$ can be written as

   $$\frac{1}{x_n} \sum_{i=1}^{n-1} \tilde{\mathbf{a}}_i x_i = \tilde{\mathbf{b}}$$

   or, using (20.35),

   $$\sum_{i=1}^{n-1} \tilde{\mathbf{a}}_i x_i - \tilde{\mathbf{b}} x_n = 0.$$

   We define the matrix A by

   $$A = \begin{bmatrix} \tilde{A} & -\tilde{\mathbf{b}} \end{bmatrix}$$

   to create the constraint in the transformed coordinates $A\mathbf{x} = 0$.

   The objective $\tilde{\mathbf{c}}^T \tilde{\mathbf{x}}$ can be written as

   $$\tilde{\mathbf{c}}^T \tilde{\mathbf{x}} = \sum_{i=1}^{n-1} \tilde{c}_i \tilde{x}_i = \frac{1}{n} \sum_{i=1}^{n-1} \tilde{c}_i x_i.$$

   Since the objective is to be zero, the factor $1/x_n$ is unimportant, and we define

   $$\mathbf{c} = \begin{bmatrix} \tilde{\mathbf{c}} \\ 0 \end{bmatrix}.$$

   Thus $\mathbf{c}^T \mathbf{x} = 0$ implies $\tilde{\mathbf{c}}^T \tilde{\mathbf{x}} = 0$.

Algorithm 20.5 illustrates MATLAB code that implements the transformation from standard to Karmarker standard form.

Algorithm 20.5 Conversion of standard form to Karmarker standard form
File: `tokarmarker.m`

20.6.2 Convergence of the algorithm

In this section we will show that the method just outlined decreases f at each iteration. It can be shown (see exercise 20.6-17) that the dual to (20.26) can be written using the dual variables $\mathbf{y} \in \mathbb{R}^m$ and $z \in \mathbb{R}$ as

$$\text{maximize } nz$$
$$\text{subject to } A^T\mathbf{y} + \mathbf{1}z \leq \mathbf{c}, \tag{20.36}$$
$$\mathbf{y} \text{ free} \quad z \text{ free}.$$

For any value of $\mathbf{y}$, $z = \min_j (\mathbf{c} - A^T\mathbf{y})_j$ is feasible.

The function $f(x;c) = n\log \mathbf{c}^T\mathbf{x} - \sum_j \log x_j$. Our approach takes two steps. Starting from an initial value $\mathbf{x} = \mathbf{1}$, we first show that each step of the algorithm decreases $n \log \mathbf{c}^T\mathbf{x}$. Second, we show that the extra term $-\sum_j \log x_j$ does not increase f too much. We use the following notation:

$$B = \begin{bmatrix} A \\ \mathbf{1}^T \end{bmatrix} \qquad P_{B\perp} = I - B^T(BB^T)^{-1}B.$$

In the following theorem, we assume that the initial value is $\mathbf{x}^{[k]} = \mathbf{1}$, since for all other iterations, $\mathbf{c}$ and A can be replaced by $\hat{\mathbf{c}}$ and $\hat{A}$, with the statement applying to $\hat{\mathbf{x}}^{[k]} = \mathbf{1}$.

Theorem 20.2 *[339] Let $\mathbf{d}$ be the projected gradient, $\mathbf{d} = -P_{B\perp}\mathbf{c}$. Let $\mathbf{y}$ be the least-squares solution $\mathbf{y} = (AA^T)^{-1}A\mathbf{c}$, and let $z = \min_j(\mathbf{c} - A^T\mathbf{y})$. Starting from an initial value of $\mathbf{x} = \mathbf{1}$, then either $\mathbf{d} = 0$ (in which case $\mathbf{x} = \mathbf{1}$ is an optimal solution to (20.26)), or $\mathbf{x}_{\text{new}} = \mathbf{1} + \alpha\mathbf{d}/\|\mathbf{d}\|$ satisfies*

$$\mathbf{c}^T\mathbf{x}_{\text{new}} \leq \mathbf{c}^T\mathbf{x} - \frac{\alpha}{n}\mathbf{c}^T\mathbf{x} + \alpha z. \tag{20.37}$$

Proof If $\mathbf{d} = -P_{B\perp}\mathbf{c} = -P_{\mathbf{1}\perp}P_{A\perp}(\mathbf{c} - A^T\mathbf{y}) = -P_{\mathbf{1}\perp}(\mathbf{c} - A^T\mathbf{y})$ is zero, then $\mathbf{c} - A^T\mathbf{y}$ is a multiple of $\mathbf{1}$, say $\mathbf{c} - A^T\mathbf{y} = \beta\mathbf{1}$. Then

$$n\mathbf{z} = n\beta = (\mathbf{c} - A^T\mathbf{y})^T\mathbf{1} = \mathbf{c}^T\mathbf{1}.$$

Thus the value of the dual problem is equal to the value of the primal problem, and an optimal solution must exist.

Otherwise, we note that

$$\|\mathbf{d}\|^2 = \mathbf{c}^T P_{B\perp}\mathbf{c} = -\mathbf{c}\mathbf{d}.$$

Then

$$\mathbf{c}^T\mathbf{x}_{\text{new}} = \mathbf{c}^T\mathbf{1} + \alpha\mathbf{c}^T\mathbf{d}/\|\mathbf{d}\| = \mathbf{c}^T\mathbf{1} - \alpha\|\mathbf{d}\|.$$

If we can show that $\|\mathbf{d}\| \geq \mathbf{c}^T\mathbf{1} - z$, then the bound (20.37) will be established. We have

$$\mathbf{d} = -P(\mathbf{c} - A^T\mathbf{y}) = -(\mathbf{c} - A^T\mathbf{y} - \mathbf{1}\mathbf{1}^T(\mathbf{c} - A^T\mathbf{y})/n)$$
$$= -(\mathbf{c} - A^T\mathbf{y}) + (\mathbf{c}^T\mathbf{1})/n)\mathbf{1},$$

20.6 Karmarker's Algorithm for LP

where the last equality follows since $A\mathbf{1} = 0$. The quantity $\mathbf{c}^T \mathbf{1}$ is greater than or equal to the value of the primal problem, which in turn is greater than or equal to the value of the dual problem, so $\mathbf{c}^T \mathbf{1} \geq nz$, thus $\mathbf{c}^T \mathbf{1}/n \geq z$. By our choice of feasible solution, $z = (\mathbf{c} - A^T \mathbf{y})_i$ for some i. Then for that i,

$$d_i = (\mathbf{c}^T \mathbf{1})/n - z \geq 0.$$

So for that i, $d_i = |d_i|$. Now simply recognize that $\|\mathbf{d}\| \geq |d_i| = \mathbf{c}^T \mathbf{1}/n - z$. □

If $z < 0$ (that is, the optimal solution has not been obtained), then it is straightforward to show (see exercise 20.6-19) that

$$n \log \mathbf{c}^T \mathbf{x}_{\text{new}} \leq n \log \mathbf{c}^T \mathbf{x} - \alpha. \tag{20.38}$$

To proceed with the next step, we use the following inequality: If $|\epsilon| \leq \alpha < 1$, then

$$\epsilon - \frac{\epsilon^2}{2(1-\alpha)^2} \leq \ln(1+\epsilon) \leq \epsilon. \tag{20.39}$$

Lemma 20.1 *[172] If $\|\mathbf{x} - \mathbf{1}\| \leq \alpha < 1$ and $\mathbf{1}^T \mathbf{x} = n$ then*

$$0 \leq -\sum_j \ln x_j \leq \frac{\alpha^2}{2(1-\alpha)^2}. \tag{20.40}$$

Proof Using the inequality (20.39) with $\epsilon = x_j - 1$ we have

$$(x_j - 1) - \frac{(x_j - 1)^2}{2(1-\alpha)^2} \leq \ln(1 + (x_j - 1)) \leq (x_j - 1).$$

Summing both sides over, and using the fact that $\sum_j x_j = n$, we have

$$-\frac{1}{2(1-\alpha)^2} \sum_j (x_j - 1)^2 \leq \sum_j \ln x_j \leq 0. \tag{20.41}$$

Since $\|\mathbf{x} - \mathbf{1}\| \leq \alpha$, it follows that $\sum_j (x_j - 1)^2 \leq \alpha^2$. Using this in (20.41) we obtain

$$0 \leq -\sum_j \ln x_j \leq \frac{\alpha^2}{2(1-\alpha)^2}.$$

□

Now, using (20.38) and (20.40), we find that

$$f(\mathbf{x}_{\text{new}}; \mathbf{c}) = n\mathbf{c}^T \mathbf{x}_{\text{new}} - \sum_j \log x_j \leq \log \mathbf{c}^T \mathbf{x} - \alpha + \frac{\alpha^2}{2(1-\alpha)^2}.$$

The offset $-\alpha + \alpha^2/(2(1-\alpha^2))$ achieves a minimum at $\alpha = .3177$. Karmarker suggests a fixed value of $\alpha = 1/3$, giving the bound

$$f(\mathbf{x}_{\text{new}}; \mathbf{c}) \leq f(\mathbf{x}; \mathbf{c}) - \frac{1}{5}.$$

As f is reduced by at least a constant at every step, $\mathbf{c}^T \mathbf{x}$ goes to zero exponentially (leading to the reported covergence in polynomial time). If $\mathbf{c}^T \mathbf{x}$ does increase (as it might), the factor $\sum_j \log x_j$ must also increase. In so doing, the algorithm "gains altitude" for later steps, moving away from the boundaries $x_j = 0$.

Notwithstanding the proven convergence properties, the algorithm *must* remain on the interior, and the algorithm can at best get close. In practice, it may take *many* iterations to get close. In commercial software employing Karmarker's algorithm, once a sufficiently advanced point is identified, then another algorithm is used for final termination. This is not represented in the code demonstrated here (but see, for example, [374, chapter 7]).

Example 20.6.1 The following code

```
% testkarmarker.m: Test the Karmarker linear programming solution
c0 = [3 -2 4 0 0 0]';
A0 = [7 5 -1 1 0 0
 2 2 3 0 1 0
 -5 -6 8 0 0 1];
b0 = [9; 5; -10];
[x,value,w] = simplex1(A0,b0,c0)
[A,c] = tokarmarker(A0,b0,c0);
xk = karmarker(A,c)
```

demonstrates an example solution. Using the simplex algorithm, the optimum solution obtained is

$$\mathbf{x}^T = [0 \quad 1.8 \quad 0 \quad 0 \quad 1.4 \quad 0.8].$$

Using Karmarker's algorithm, the solution (the first six elements of xk) is

$$\mathbf{x}^T = [1.67378 \times 10^{-6} \quad 1.66904 \quad 2.69664 \times 10^{-6} \quad 2.42697 \times 10^{-5} \quad 1.29816 \quad 0.741796]$$

While not particularly close, the value $\mathbf{c}^T\mathbf{x} = 4.2 \times 10^{-6}$. □

20.6.3 Summary and extensions

Since the initial development of the algorithm, many extensions have been proposed. For example, using the **y** and z that arise in the proof of theorem 20.2, Todd and Burrell [339] have extended Karmarker's algorithm so that the solutions to both the primal and the dual are found simultaneously, providing an example of what is known as a primal/dual interior-point method. A variety of other extensions have also been developed; for an excellent survey, see [374].

Examples and applications of linear programming

Linear programming has been employed in a variety of signal-processing applications. We briefly introduce a few of these applications, making reference to the appropriate literature for more detailed investigations.

20.7 Linear-phase FIR filter design

Let the coefficients of a linear phase filter be g_n, $n = -(N-1)/2, \ldots, (N-1)/2$, with N odd and $g_n = g_{-n}$. (With suitable modifications, the results can be extended to filters with an even number of coefficients, and defining $h_n = g_{n-(N-1)/2}$ results in a causal filter with the same magnitude response.) The frequency response of the filter is

$$G(e^{j\omega}) = g_0 + 2 \sum_{n=1}^{(N-1)/2} g_n \cos \omega n.$$

We want to design a lowpass filter, with passband frequency F_p and stopband frequency F_s, as shown in figure 20.3. The ripple in the passband is $\pm \delta_1$ and the ripple in the stopband is $\pm \delta_2$. We will assume that δ_1 is specified (there are other design possibilities). Then in

20.7 Linear-Phase FIR Filter Design

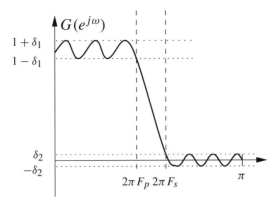

Figure 20.3: Filter design constraints.

the design of the filter, we desire to choose the filter coefficients and δ_2, subject to the constraints

$$0 \leq \omega \leq 2\pi F_p : \begin{cases} g_0 + 2\sum_{n=1}^{(N-1)/2} g_n \cos \omega n \leq 1 + \delta_1, \\ -g_0 - 2\sum_{n=1}^{(N-1)/2} g_n \cos \omega n \leq -1 + \delta_1, \end{cases}$$

$$2\pi F_s \leq \omega \leq \pi : \begin{cases} g_0 + 2\sum_{n=1}^{(N-1)/2} g_n \cos \omega n - \delta_2 \leq 0, \\ -g_0 - 2\sum_{n=1}^{(N-1)/2} g_n \cos \omega n - \delta_2 \leq 0, \end{cases}$$

in such a way as to maximize the negative stopband ripple, $-\delta_2$. This is clearly linear in the unknowns; it is a linear programming problem in the form of the dual to a problem in standard form. Matrices for a linear programming solution are obtained by suitable sampling in ω.

The MATLAB code in algorithm 20.6 accepts filter parameters and returns parameters for a causal filter. For this example, the stopband ripple and passband ripple are set equal, and are determined by the program. That is, $\delta_1 = \delta_2 = \delta$, and we wish to maximize $-\delta$.

Algorithm 20.6 Optimal filter design using linear programming
File: `lpfilt.m`

Figure 20.4 shows the frequency response and impulse response of a filter design with $F_s = 0.1$, $F_p = 0.2$, and $N = 45$ coefficients. The value of δ returned is $\delta = 0.0764$, which corresponds to -22 dB of attenuation in the stopband.

One advantage to filter design in the form of a linear programming problem is that a variety of additional constraints can be utilized. For example, the response at a given frequency can be completely nullified. Or a constraint can be added to limit the amount of overshoot in the impulse response. Filters can also be designed to minimize deviation from a given spectral prototype.

20.7.1 Least-absolute-error approximation

We have seen throughout this text several examples of approximation using an L_2 norm, leading to minimum mean-squared error or least-squares error designs (chapter 3 provides

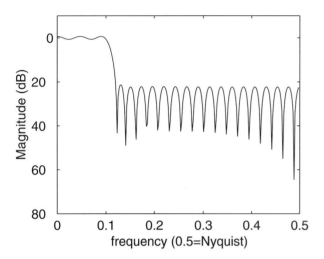
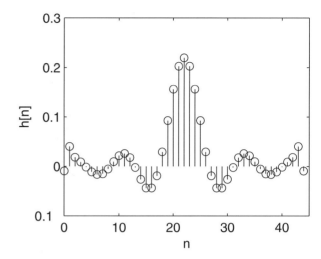

Figure 20.4: Frequency and impulse response of a filter designed using linear programming ($n = 45$ coefficients).

several examples). These approximation techniques are commonly employed because the quadratic terms arising in the norm are differentiable, and the method is defensible as a maximum-likelihood technique in Gaussian noise.

However, there are instances where an L_1 or L_∞ norm may be desirable. One reason for choosing another norm is that the L_2 optimization problems tend to provide "smoother" answers than may be desirable in some cases.

Example 20.7.1 We desire to find a solution to the underdetermined problem

$$A\mathbf{x} = \mathbf{b}, \tag{20.42}$$

subject to the constraint that $\|\mathbf{x}\|$ is minimized. If the norm is the 2-norm, $\|\mathbf{x}\|^2 = \mathbf{x}^T\mathbf{x}$, then the solution is given (see section 3.15) by

$$\hat{\mathbf{x}} = A^T(AA^T)^{-1}\mathbf{b}. \tag{20.43}$$

Let

$$A = \begin{bmatrix} 3 & 2 & 1 \\ 4 & 6 & 9 \end{bmatrix} \qquad \mathbf{b} = \begin{bmatrix} 2 \\ 6 \end{bmatrix}.$$

Suppose that the true solution is

$$\mathbf{x} = \begin{bmatrix} 0 \\ 1 \\ 0 \end{bmatrix},$$

which is "spiky," and that there are physical reasons for the spikiness of the data. If the true solution were not known, then the minimum squared norm solution from (20.43) would be

$$\hat{\mathbf{x}} = \begin{bmatrix} 0.35705 \\ 0.315653 \\ 0.297542 \end{bmatrix},$$

which is considerably smoother than the original vector.

Now we use something related to the L_1 norm: We desire to minimize $\|\mathbf{x}\| = |x_1 + x_2 + x_3|$, subject to the constraints (20.42). We also assume for convenience the constraint $\mathbf{x} \geq 0$. Using the techniques of section 20.2, this can be expressed as two linear programming problems:

$$\text{minimize } x_1 + x_2 + x_3$$
$$\text{subject to } A\mathbf{x} = \mathbf{b},$$
$$\mathbf{x} \geq 0,$$

and

$$\text{minimize } -(x_1 + x_2 + x_3)$$
$$\text{subject to } A\mathbf{x} = \mathbf{b},$$
$$\mathbf{x} \geq 0.$$

The linear programming solution to the second problem gives

$$\hat{\mathbf{x}} = \begin{bmatrix} 0 \\ 1 \\ 0 \end{bmatrix}.$$

This is, in fact, the true solution, and is observed to be more spiky than the L_2 solution. □

As this example illustrates, problems stated using the L_1 norm with linear constraints can often be expressed as linear programming problems. In addition to providing more "spiky" answers, optimization using the L_1 norm is natural when the noise in the problem is double exponential. For example, if the signal measured in noise is of the form

$$y[t] = \sum_{i=1}^{p} a_i x[t-i] + n[t],$$

where $n[t]$ has the distribution

$$n[t] = \frac{\alpha}{2} e^{-\alpha |n[t]|},$$

then the natural norm to use for maximum-likelihood estimation of the signal parameters $\{a_i\}$ is the L_1 norm. Such a problem can be expressed as a linear programming problem (see the references for illustrations and applications).

20.8 Linear optimal control

Given a discrete-time system with n-dimensional state $\mathbf{x}$ and scalar input

$$\mathbf{x}(k+1) = G\mathbf{x}(k) + \mathbf{h} u(k),$$

where G and $\mathbf{h}$ are known, we can write the state $\mathbf{x}$ as

$$\mathbf{x}(k) = G^k \mathbf{x}(0) + \sum_{i=0}^{k-1} G^i \mathbf{h} u(k-i-1),$$

where $\mathbf{x}(0)$ is the state at time 0. Suppose that the input is constrained so that $u_{\min} \leq u \leq u_{\max}$. Then, given various linear costs, we can determine an optimum input sequence $u_0, u_1, \ldots, u_k$, using linear programming.

For example, suppose that we desire the discrete-time system to reach as closely as possible some target state $\tilde{\mathbf{x}}$ at a fixed time K. Then we want to minimize the L_1 norm $\|\tilde{\mathbf{x}} - \mathbf{x}(k)\|_1$, that is,

$$\sum_{i=1}^{n} |\mathbf{x}_{D,i} - \mathbf{x}_i(K)|.$$

20.9 Exercises

20.2-1 Convert the following problems to standard form:

(a) maximize $2x_1 + 3x_2 + 4x_3$
subject to $2x_1 - 3x_3 \geq 0$,
$x_1 \geq 2, \quad x_2 \geq 4, \quad x_3 \geq 7.$

(b) minimize $x_1 + x_2 + 3x_3$
subject to $2 \leq 4x_1 - 2x_3 \leq 7$,
$6x_1 + 2x_3 - 5x_3 = 2$,
$-2 \leq x_1 \leq 3, \quad x_2 \geq 0, \quad x_3 \leq 0.$

(c) minimize $x_1 + x_2$
subject to $|x_1 - x_2| = 3$,
$x_1 \geq 1, \quad x_2 \leq 2.$

(d) minimize $|2x_1 + 3x_2|$
subject to $3x_1 - x2_3 \leq 7$,
$x_1 \geq 0, \quad x_2 \geq 0, \quad x_3 \leq 0.$

20.2-2 Place the following problem in standard form.

$$\text{minimize } |x_1| + |2x_2| + |3x_3|$$
$$\text{subject to } x_1 + x_2 \leq 2,$$
$$x_1 + 3x_3 = 2.$$

20.2-3 For the following problem with free variables,

$$\text{minimize } 2x_1 + 2x_2 - 3x_3$$
$$\text{subject to } x_1 - x_2 + 4x_3 \leq 2,$$
$$3x_2 + 4x_3 = 7,$$
$$x_1 \geq 0, \quad x_2 \geq 0, \quad (x_3 \text{ free}),$$

eliminate the free variable and find a tableau for the problem.

20.2-4 A materials scientist wishes to formulate a compound that is 20% element X and 80% element Y. Several compounds with different proportions of the elements available for different costs. The desired compound may be produced by combining some of the other compounds.

Compound	1	2	3	4
% X	15	20	30	70
% Y	85	80	70	30
Price/kg	$ $5	$4	$3	$2

Set up a linear programming problem to determine the amounts of the alloys needed to produce the least expensive compound. Solve the problem.

20.2-5 Consider the linear programming problem,

$$\text{maximize } x_1 + x_2$$
$$\text{subject to } x_1 + x_2 \leq 3,$$
$$-x_1 + x_2 \geq 2,$$
$$x_1 \geq 0, x_2 \geq 0.$$

(a) Solve this problem, using the simplex algorithm by hand.

(b) Draw the graphical representation of the problem and show that the solution makes sense.

(c) On the graph, indicate the points corresponding to the steps of the simplex algorithm.

20.9 Exercises

20.5-6 Show that the dual of a dual is the original problem.

20.5-7 Show that if a linear inequality in a primal problem is changed to an equality, the corresponding dual variable becomes free.

20.5-8 An inequality relationship that is useful in a variety of optimization studies [216] is the *Motzin transposition theorem*: Let A, B, and C be real constant matrices, with A being nonempty (B or C may be empty). Then either the system

$$\mathbf{y}^T A < 0 \qquad \mathbf{y}^T B \leq 0 \qquad \mathbf{y}^T C = 0$$

has a solution $\mathbf{y}$, or the system

$$A\mathbf{z} + B\mathbf{z}_2 + C\mathbf{z}_3 = 0 \qquad \mathbf{z}_1 \geq 0 \qquad \mathbf{z}_1 \neq 0 \qquad \mathbf{z}_2 \geq 0$$

has a solution, but never both.

20.5-9 Find the dual of

$$\text{maximize } \mathbf{c}^T \mathbf{x}$$
$$\text{subject to } A\mathbf{x} = \mathbf{b},$$
$$\mathbf{x} \geq \mathbf{d}.$$

20.5-10 Consider the primal linear programming problem

$$\text{minimize } \mathbf{c}^T \mathbf{x}$$
$$\text{subject to } A\mathbf{x} = \mathbf{b},$$
$$\mathbf{x} \geq \mathbf{0},$$

and its dual

$$\text{maximize } \mathbf{w}^T \mathbf{b}$$
$$\text{subject to } \mathbf{w}^T A \leq \mathbf{c}^T.$$

The primal problem is transformed to

$$\text{minimize } \mathbf{c}^T \mathbf{x}$$
$$\text{subject to } \tilde{A}\mathbf{x} = \tilde{\mathbf{b}},$$
$$\mathbf{x} \geq \mathbf{0},$$

where $\tilde{A} = X^{-1}A$ and $\tilde{\mathbf{b}} = X^{-1}\mathbf{b}$. The dual of the transformed problem is

$$\text{maximize } \mathbf{z}^T \tilde{\mathbf{b}}$$
$$\text{subject to } \mathbf{z}^T \tilde{A} \leq \mathbf{c}^T.$$

Suppose that a solution to the transformed problem $\mathbf{x}$ and its dual $\mathbf{z}$ are known. Show that $\mathbf{w} = X^{-T}\mathbf{z}$ is a solution to the original dual problem.

20.5-11 (Min/max optimzation and duality)

(a) Show that the primal minization problem of (3.4) can be expressed as the min/max problem

$$\min_{\substack{A\mathbf{x}=\mathbf{b} \\ \mathbf{x} \geq 0}} \mathbf{c}^T \mathbf{x} = \min_{\mathbf{x} \geq 0} \max_{\mathbf{w}} [\mathbf{c}^T \mathbf{x} - \mathbf{W}^T(A\mathbf{x} - \mathbf{b})]. \qquad (20.44)$$

What values of $\mathbf{x}$ lead to a finite maximum? (Note that the argument is simply the Lagrangian of the constrained problem, where $\mathbf{w}$ is the Lagrange multiplier; see section 18.8.)

(b) Now consider the interchanged problem, written as

$$\max_{\mathbf{w}} \min_{\mathbf{x} \geq 0} [(\mathbf{c}^T - \mathbf{w}^T A)\mathbf{x} + \mathbf{w}^T \mathbf{b}]. \qquad (20.45)$$

Show that this is equivalent to the maximization problem

$$\max_{\mathbf{c}^T - \mathbf{w}^T A} \mathbf{w}^T \mathbf{b}.$$

(For what value of **x** is the minimum finite?) Hence, assuming that the min/max equation of (20.44) is equal to the max/min equation of (20.45), we have shown that the value of the primal linear programming problem is the same as the value for the dual.

(c) The problem of (20.23) can be expressed similarly. Argue that the following is true:

$$\min_{\substack{A\mathbf{x} \geq \mathbf{b} \\ \mathbf{x} \geq 0}} \mathbf{c}^T \mathbf{x} = \min_{\mathbf{x} \geq 0} \max_{\mathbf{y} \geq 0} [\mathbf{c}^T - \mathbf{w}^T (A\mathbf{x} - \mathbf{b})]. \tag{20.46}$$

(d) Interchanging min and max in (20.46), argue that the following is true:

$$\max_{\mathbf{y} \geq 0} \min_{\mathbf{x} \geq 0} [(\mathbf{c}^T - \mathbf{w}^T A)\mathbf{x} + \mathbf{y}\mathbf{b}] = \max_{\substack{\mathbf{w}^T A \leq \mathbf{c} \\ \mathbf{w} \leq 0}} \mathbf{w}^T \mathbf{b}.$$

20.6-12 Write a MATLAB function `lpfilt2(Hdesired,Omegalist,n)` that, using linear programming, designs a linear-phase FIR filter that minimizes the amplitude error between a specified frequency response $H(e^{j\omega})$ and the frequency response

$$\hat{H}(e^{j\omega}) = \sum_{k=0}^{n} h_k e^{-j\omega k}.$$

That is, that chooses coefficients $h_0, h_1, \ldots, h_{n-1}$ and δ so that δ in

$$-\delta \leq |H(e^{j\omega}) - \hat{H}(e^{j\omega})| \leq \delta$$

is as small as possible. The parameters `Hdesired` and `Oomegalist` are, respectively, the desired magnitude and frequency value.

20.6-13 Show that:

(a) The transformation in (20.28) maps S into itself and maps the point $\mathbf{x}^{[k]}$ into **1**.

(b) The transformation (20.29) is an inverse of that in (20.28).

(c) Equation (20.30) is correct.

20.6-14 Show that the gradient-descent update rule can be written in terms of the original variables as

$$\hat{\mathbf{d}}^{[k]} = -P_{\hat{A}\perp} P_{\mathbf{1}\perp} X_k \mathbf{c},$$
$$\mathbf{d}^{[k]} = X_k \hat{\mathbf{d}}^{[k]},$$
$$\tilde{\mathbf{x}}^{[k+1]} = \mathbf{x}^{[k]} + \alpha \mathbf{d}^{[k]} / \|\hat{\mathbf{d}}^{[k]}\|,$$
$$\mathbf{x}^{[k+1]} = n\tilde{\mathbf{x}}^{[k+1]} / \mathbf{1}^T \tilde{\mathbf{x}}^{[k+1]}.$$

20.6-15 Let

$$\hat{B} = \begin{bmatrix} \hat{A} \\ \mathbf{1}^T \end{bmatrix}.$$

Show that $P_{\hat{B}\perp}$, the projector onto the nullspace of $\hat{B}$, can be written as

$$P_{\hat{B}\perp} = P_{\hat{A}\perp} P_{\mathbf{1}\perp}.$$

20.6-16 Let $\phi(\alpha) = f(\mathbf{x} + \alpha \mathbf{d}; \mathbf{c})$, where **d** is the gradient direction and **d** and **x** are not proportional. Show that $\phi(\alpha)$ has at most one stationary point where $\phi'(\alpha) = 0$, and if it has one it is a minimizer. (Hint: Take the derivative, and recognize the mean of the quantities $\delta_j = d_j/(x_j + \alpha d_j)$. Use the convexity of the function δ_j^2.)

20.6-17 Show that (20.36) is correct.

20.6-18 Show that (20.39) is correct.

20.6-19 Show that (20.38) is correct.

20.10 References

There is an enormous literature on linear programming. Our approach has been to introduce the salient concepts, while leaving out many of the fascinating details. A good starting point is [210]; [61] is a classic; and [100] has a variety of interesting examples.

FIR filter design using linear programming as presented in section 20.7 is discussed in [264]. Several other examples of filter design using linear programming appear in the literature; for example, design of IIR filters using linear programming is presented in [52]. Another discussion of FIR filter design using linear programming appears in [320], in which it is pointed out the linear programming techniques can be used to find filters of specified characteristic of minimum filter length. While the linear programming method is slower to design than the more common Remez exchange technique, it is straightforward under LP to incorporate other constraints. A discussion of filter design using linear programming for multirate filters is in [308]. For two-dimensional filter design, see, for example, [107]. Nonuniformly-spaced FIR filters appear in [180].

A stack filter minimizes a mean-absolute-error criterion (as opposed to the more conventional mean-squared error criterion). Design of stack filters can be accomplished using linear programming [57, 95].

The problem of optimization using a least-absolute-error (L_1) criterion instead of a least-squares (L_2) criterion has been examined in a variety of applications. For example, [295] has used L_1 optimization for robust sinusoidal frequency estimation. Power spectrum estimation using linear programming and an L_1 norm is discussed in [200], in which it is shown that the linear programming–based methods provide spectral resolution superior to conventional techniques based on solution of the normal equations. The study of least-absolute-error optimization, also called least-absolute deviations (LAD), has been examined in [378, 20, 359, 73]. In such applications, the solution is typically found using modifications to RLS algorithms known as iteratively reweighted least squares and residual steepest descent, which may converge faster than many algorithms based on steepest descent. A discussion of LAD methods, including regression, autoregression, and algorithms, appears in [37].

Karmarker's original algorithm appears in [172]. Our presentation follows the notation in [339]. A thorough treatment of related algorithms appears in [374]. Application of Karmarker's and simplex algorithms to approximation under L_1 and L_∞ norms is described in [286].

Application of linear programming to optimal control, as in section 20.8, is described in [254].

Appendix A

Basic Concepts and Definitions

> The words or the language as they are written or spoken, do not seem to play any role in my mechanism of thought. The psychical entities which seem to serve as elements in thought are certain signs and more or less clear images which can be "voluntarily" reproduced and combined.... The above-mentioned elements are, in my case, of visual and some muscular type. Conventional words or other signs have to be sought for laboriously only in a secondary stage.
> — *Albert Einstein*

Mathematics is a language with its own vocabulary, grammar, and rhetoric. Adeptness at mathematics requires a solid understanding of the basic vocabulary, and the use of this specialized vocabulary adds precision and conciseness to mathematical developments. In this appendix, we present a summary of concepts that should be generally familiar in topics related to sets and functions. This material provides a reference to topics used throughout the book and establishes several notational conventions.

A.1 Set theory and notation

Existential quantifiers

The notation $\exists$ means "there exists." The notation $\forall$ means "for all," or "for each," or "for every." The abbreviation s.t. means "such that."

Example A.1.1 The statement "$\exists\, x \in \mathbb{R}$ s.t. $x^3 > 23$" means: there is a real number x such that $x^3 > 23$.
   The statement "$\forall x \in \mathbb{R}, x^2 \geq 0$" means: for every real number x, x^2 is nonnegative. □

Notation for some common sets

- $\emptyset$ is the null set, the set containing no elements, which is a subset of every set.
- $\mathbb{Z}$ is the set of integers, $\ldots, -3, -2, -1, 0, 1, 2, 3, \ldots$. The set of nonnegative integers (including zero) is $\mathbb{Z}^+$. Sets of integers in a range may be denoted by $[a, b]$ which is the set $\{a, a+1, \ldots, b\}$. The notation $a{:}b$ may be used to indicate the same set. The notation $a{:}s{:}b$ indicates the set $\{a, a+s, a+2s, \ldots, d\}$, where $d \leq b$.
- $\mathbb{Q}$ is the set of rational numbers, that is, numbers that can be expressed as the ratio m/n, where $m \in \mathbb{Z}$ and $n \in \mathbb{Z}, n \neq 0$ (that is, they are both integers).

$\mathbb{R}$ is the set of real numbers. These are the numbers that we are most likely to be familiar with, such as $2, \pi, \sqrt{2}, -7.23$, and so forth. $\mathbb{R}^+$ is the set of nonnegative real numbers; $\mathbb{R}\setminus\{0\}$ is the set of real numbers excluding 0. Sets of real numbers formed by intervals are indicated by $[a, b]$, which is the set of numbers in the interval from a to b, including the endpoints, or $[a, b)$, which is the interval from a to b, which includes the endpoint a but excludes the endpoint b.

$\mathbb{C}$ is the set of complex numbers.

When dealing with n-dimensional space, or n-space, we may refer to $\mathbb{R}^n$. Points (elements) in n-space are denoted by an n-tuple $x = (x_1, x_2, \ldots, x_n)$. When regarded as vectors, however, they are represented as column vectors (see box 1.2 on page 6).

Basic concepts and notations of set theory

A **set** is a collection of objects. The notation $x \in A$ means x is an element of (or x is in) the set A. We write $x \notin A$ to indicate that x is not an element of A.

Intersection. $A \cap B$ denotes the intersection of A and B. An element $x \in A \cap B$ if and only if $x \in A$ and $x \in B$. The intersection of multiple sets $A_1, A_2, \ldots, A_n$ is denoted

$$\bigcap_{i=1}^{n} A_i.$$

Union. $A \cup B$ denotes the union of A and B. An element $x \in A \cup B$ if and only if $x \in A$ or $x \in B$ (or possibly both). The union of multiple sets $A_1, A_2, \ldots, A_n$ is denoted

$$\bigcup_{i=1}^{n} A_i.$$

Complement. The complement of the set A denoted, A^c, is such that $x \in A^c$ if and only if $x \notin A$. The complement is always with respect to some universal set X, where $A \subset X$.

Exclusion. The set $A\setminus B$ means those elements of A that are not common with B. $x \in A\setminus B$ if and only if $x \in A$ and $x \notin B$.

Subset. We say that $B \subset A$ if and only if every $x \in B$ is also in A. When B is possibly equal to A we write $B \subseteq A$. When B is a subset of A not equal to A then B is a **proper subset** of A.

If $A \subseteq B$ and $B \subseteq A$ then $A = B$. Two sets A and B are often shown to be equal by showing that each is a subset of the other.

Cartesian product. The set $A \times B$ is obtained by taking each element of A with each element of B. The elements may be displayed in a comma-separated list. For example, if

$$A = \{1, 2\} \quad \text{and} \quad B = \{z, y, x\}$$

then

$$A \times B = \{(1, z), (1, y), (1, x), (2, z), (2, y), (2, x)\}.$$

The same principle extends to multiple Cartesian products, such as $A \times B \times C$; if n sets are involved in the product, then the resulting set has elements described as n-tuples.

Cardinality of a set. The cardinality (or cardinal number) of a set A is the number of elements in the set, denoted $|A|$. (There is some notational ambiguity between the order of

A.1 Set Theory and Notation

a set and the determinant of a matrix. However, context will establish which is intended). For the set

$$A = \{2, 4, 6, 8\}$$

the cardinality is $|A| = 4$.

For the set $B = \{x \in \mathbb{R}: 0 < x \leq 1\} = (0, 1]$, what is the cardinality? There is clearly an infinite number of points. For the set $\mathbb{Z}$, there is also an infinite number of points. Is the cardinality of $(0, 1]$ the same as the cardinality of $\mathbb{Z}$? Interestingly, it can be shown that the answer is no: $|B| > |\mathbb{Z}|$. The set of integers is said to be *countable*; a set such as $(0, 1]$ is said to be noncountable.

The noncountability of the set $(0, 1]$ was established by the mathematician Georg Cantor using a subtle and powerful idea that has come to be known as Cantor's diagonal argument. To determine the cardinality of a set, we form a one-to-one mapping from the elements of the set to the positive integers. In other words, we simply count the elements in the set. To prove that the cardinality of the set $(0, 1]$ exceeds that of the integers, we try to establish a counting mapping which is ostensibly complete, then demonstrate that no matter what we do there are still elements in the set uncounted. Hence we conclude that no matter how we count, even up to an infinite number, there are still numbers uncounted, and hence the cardinality of elements in the set is too large to count.

To set up the mapping, we write a list of all the numbers in the interval $(0, 1]$ as decimals, and simply put an integer to indicate the row number. We might end up with a table like the following:

$$\begin{array}{ll} 1 & 0.3\underline{2}342212\cdots \\ 2 & 0.4\underline{3}235532\cdots \\ 3 & 0.32\underline{4}53232\cdots \\ 4 & 0.676\underline{5}4543\cdots \end{array}$$

(There are some minor technical arguments about the uniqueness of the decimal representation that we are overlooking.) If we presume that the list is a complete enumeration of all the numbers in the interval $(0, 1]$, then we simply count the number of rows in the table. We will now show that we cannot produce such a complete listing. We will create a new number in the interval that cannot be in the table. The first digit is formed by taking the number from the first row, and modifying its first digit (the one that is underlined). We will choose 7, since it is different from 3. The second digit of our new numbers is formed by modifying the second digit of the second row, and so forth. We can get a number $0.7216\cdots$. This number is different from *every* number in the table since it differs by at least one digit from every number. Hence it does not appear in the table, and the supposedly complete table is incomplete. This incompleteness of the table remains no matter how many new rows we add. We conclude (reluctantly) that there is no one-for-one mapping from the integers to the real numbers so there must be more real numbers in $(0, 1]$ than there are positive integers.

There are some counterintuitive notions associated with the cardinality of infinite sets. For example, the set of even integers $2\mathbb{Z}$ has the same cardinality as the set of integers. It can also be shown that the set $(0, 1]$ has the same cardinality as $\mathbb{R}$. The set of rational numbers $\mathbb{Q}$ is countable.

Convex sets. A set in $\mathbb{R}^n$ is said to be **convex** if all points of the line segment connecting any two points in the set remain in the set (see figure A.1). More precisely, if $S \subset \mathbb{R}^n$ is a convex set, then for any two points p and q in S, all points of the form

$$\lambda p + (1 - \lambda)q,$$

for $0 \leq \lambda \leq 1$, are in S. Observe that the point $\lambda p + (1 - \lambda)q$ is on the line adjoining p and q for $0 \leq \lambda \leq 1$.

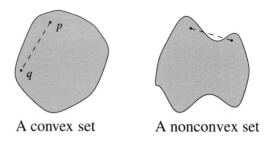

A convex set   A nonconvex set

Figure A.1: Illustration of convex and nonconvex sets.

Indicator functions. When representing sets, both mathematically and in computer implementations, it is occasionally useful to use an indicator function for a set. A binary indicator function $\chi_A(x)$ for a set A takes the value 1 for arguments that are in the set, and the value 0 for arguments that are not in the set:

$$\chi_A(x) = \begin{cases} 1 & \text{if } x \in A, \\ 0 & \text{if } x \notin A. \end{cases}$$

The basic set operations can be represented easily using indicator functions. Let A and B be sets with indicator functions $\chi_A(x)$ and $\chi_B(x)$, respectively. Then the indicator function for the set $C = A \cap B$ is

$$\chi_C(x) = \min(\chi_A(x), \chi_B(x)).$$

The indicator function for $D = A \cup B$ is

$$\chi_D(x) = \max(\chi_A(x), \chi_B(x)).$$

As a matter of notation, the operator $\vee$ is often used to indicate maximum,

$$\boxed{\max(x, y) = x \vee y}$$

and the operator $\wedge$ is often used to indicate minimum:

$$\boxed{\min(x, y) = x \wedge v}$$

So we can write $\chi_C(x) = \chi_A(x) \wedge \chi_B(x)$. The indicator for the set $E = A^c$ is

$$\chi_E(x) = 1 - \chi_A(x).$$

Example A.1.2 Let

$$S = \{x \in \mathbb{R} : x \geq 3\}$$

and

$$T = \{x \in \mathbb{R} : x \leq 4\}.$$

Then

a union: $S \cup T = \mathbb{R}$ (the entire real line),
an intersection: $S \cap T = [3, 4]$,
and a complement: $\overline{S \cap T} = (-\infty, 3) \cup (3, \infty)$.

The indicator functions for S, T, and $S \cap T$ are shown in figure A.2.   □

A.2 Mappings and Functions

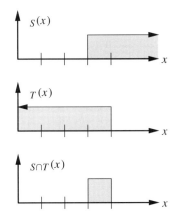

Figure A.2: Indicator functions for some simple sets.

A.2 Mappings and functions

Definition A.1 A mapping $M: S_1 \to S_2$ is a rule for associating elements of the set S_1 with elements of the set S_2.

A mapping M is said to be one-to-one if the images of distinct elements in S_1 are distinct in S_2. That is, if $x_1, x_2 \in S_1$ with $x_1 \neq x_2$, the mapping is one-to-one if $M(x_1) \neq M(x_2)$. □

Example A.2.1 Our first example is a familiar one. Let S_1 be the set of real signals with bounded energy,

$$S_1 = \{f(t): \int_{-\infty}^{\infty} f^2(t)\, dt < \infty\}.$$

There is a mapping $\mathcal{F}: S_1 \to S_2$ to the set of square integrable functions,

$$S_2 = \{F(\omega): \int_{\infty}^{\infty} |F(\omega)|^2 \, d\omega < \infty\},$$

defined by

$$\mathcal{F}(f(t)) = F(\omega) = \int_{-\infty}^{\infty} f(t) e^{-j\omega t}\, dt.$$

Strictly speaking, the Fourier transform is not one-to-one: there is a whole set of signals that have the same Fourier transform. For example, consider the two functions which are identical except at a single point. Both of these functions have the same Fourier transform, because the single point of difference which does not affect the value of the integral. (In fact, functions that are equal almost everywhere all have the same Fourier transform.) □

Definition A.2 A **function** f from a set A to a set B is a rule for assigning to each point $a \in A$ exactly one element $b \in B$. The **domain** of a function is the set of possible objects the function can be applied to (the set A). The **range** of a function is the set of possible values that may be mapped to. We write

$$f: A \to B$$

to explicitly indicate that f is a function from the domain A to (possibly a subset of) the range B. □

Definition A.3 A function f from A into B is **one-to-one** if each element of B has at most one element of A mapped into it. That is, if $f(x) = f(y)$, then $x = y$. Sometimes "one-to-one" is written **1-1.** A one-to-one function is also called an **injection**.

A function is **onto** if each element of B has at least one element of A that is mapped into it. That is, for every $b \in B$ there is an $a \in A$ such that $f(a) = b$. (Every element of B is covered.) An onto function is also called a **surjection**.

A function that is both one-to-one and onto is called a **bijection**. □

Example A.2.2 Consider $f\colon \mathbb{R} \to \mathbb{R}$ defined by $f(x) = x^2$. This function is *not* one-to-one, since $f(-3) = f(3)$.

The function $f(x)$ is also not onto, since there is no $x \in \mathbb{R}$ such that $f(x) = -2$. However, we could *redefine* the function with a different domain. Let $f\colon \mathbb{C} \to \mathbb{R}$ be defined by $f(x) = x^2$. This function is now onto (but still not one-to-one). We could also define the function with $f\colon \mathbb{R}^+ \to \mathbb{R}^+$. Using this restriction, the function is both one-to-one and onto. □

As this example shows, the domain and range should properly be part of the description of the function.

A **transformation** f is a mapping $f\colon \mathbb{R}^n \to \mathbb{R}^m$. For example, a mapping from $(u, v, w) \in \mathbb{R}^3$ to $(x, y) \in \mathbb{R}^2$ might be defined by

$$x = u^2 - 2vw,$$
$$y = v - u^2 w.$$

Definition A.4 Let $f\colon A \to B$ be a one-to-one function, and let $f(A)$ denote the range of f. Then the **inverse function** $f^{-1}\colon B \to A$ is defined as follows: if $y = f(x)$, then $x = f^{-1}(y)$. If a function is not one-to-one, then there is no unique inverse function. □

Example A.2.3 The function $f\colon \mathbb{R} \to \mathbb{R}$ defined by $f(x) = x^2$ has no inverse, since if $y = f(x) = 9$ there is no way to determine, given y, whether $x = 3$ or -3. However, we can talk about the *inverse mapping*. We could say that the inverse mapping $f^{-1}(9)$ is the set of values $\{3, -3\}$. □

A.3 Convex functions

Definition A.5 A function f is said to be **convex** over an open set D if for every $s, t \in D$,

$$f(\lambda s + (1 - \lambda)t) \leq \lambda f(s) + (1 - \lambda) f(t)$$

for all λ such that $0 \leq \lambda \leq 1$. If equality holds only for $\lambda = 0$ or $\lambda = 1$, then the function is **strictly convex**. □

Since $\lambda s + (1-\lambda)t$ is a point on the line segment connecting s and t, and $\lambda f(s) + (1-\lambda) f(t)$ is the chord connecting $f(s)$ and $f(t)$, we observe that a function is convex if the function lies below the chord. Figure A.3 illustrates a convex function. A function that is convex on

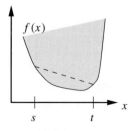

Figure A.3: Illustration of a convex function.

its whole domain is said to be simply convex. A function f is **concave** if $-f$ is convex. If $f(x)$ is a convex function, then $g(x) = af(x) + b$ is an affine transformation of f. A linear function of a convex function is also convex if $a > 0$. In performing optimization, the function f being optimized may be examined to determine if it is convex, because a minimum point in a function convex over a set D is guaranteed to be the global minimum over that set.

We pause now in our introduction of vocabulary for a statement of a geometrically obvious fact.

Theorem A.1 *If $f: \mathbb{R} \to \mathbb{R}$ has a second derivative that is nonnegative everywhere, then f is convex.*

Let $f(x)$ be a function that is convex over a set D, and let $x_i \in D, i = 1, 2, \ldots, m$, and let $p_i \geq 0$ be such that

$$\sum_{i=1}^{m} p_i = 1.$$

Then, by the convexity of $f(x)$ applied inductively,

$$f\left(\sum_{i=1}^{m} p_i x_i\right) \leq \sum_{i=1}^{m} p_i f(x_i). \tag{A.1}$$

If the x_i represent outcomes of a random variable X, occurring with probability p_i, then the sum on the left-hand side of (A.1) is recognized as $f(E[X])$, while the sum on the right-hand side is $E[f(X)]$. This gives us Jensen's inequality,

$$\boxed{f(E[X]) \leq E[f(X)]} \tag{A.2}$$

for a convex function f. Jensen's inequality also applies to random variables having continuous distributions (described by pdfs instead of pmfs).

A.4 O and o notation

The O and o notation are used to indicate "order of magnitude" for a function. Saying that f is $O(g)$ says that f is approximately the same "size" as g, in some limit. Saying that f is $o(g)$ means that f gets small faster than g, in some limit. More precisely, we have the following:

Definition A.6 Let f and g be real-valued functions. The function f is $O(g)$ as $x \to x_0$ if there is a constant C (independent of x) such that

$$|f(x)| < C|g(x)|$$

for all x in a neighborhood of x_0 or, to put it another way, if

$$\lim_{x \to x_0} \left|\frac{f(x)}{g(x)}\right| = C.$$

The function f is $o(g)$ as $x \to x_0$ if

$$\lim_{x \to x_0} \left|\frac{f(x)}{g(x)}\right| = 0.$$

$\square$

Example A.4.1

1. The function $f(x) = x^2 \ln x$ is $o(x)$ as $x \to 0^+$ (that is, approaching 0 from the right). To see this, take the limit using L'Hospital's rule twice:

$$\lim_{x \to 0^+} \frac{x^2 \ln x}{x} = \lim_{x \to 0^+} \frac{\ln x}{1/x} = 0.$$

2. The function $f(x) = x^2$ is $o(x)$ as $x \to 0$:

$$\lim_{x \to 0} \frac{x^2}{x} = 0.$$

3. Let $f(x) = x^2$ and $g(x) = -4x^2 + 3x$. Then f is $O(g)$ as $x \to 0$:

$$\lim_{x \to 0} \frac{f}{g} = -\frac{1}{4}.$$

□

Saying f is $O(1)$ means that f is bounded, and saying f is $o(1)$ as $x \to 0$ means that f is "infinitesimal."

The O notation is often used in describing the computational complexity of algorithms. For example, it is well known that the computational complexity of an n-point FFT algorithm is $O(n \log n)$. If the actual number of computations required by the computation is c, then this notation says

$$\frac{c}{n \log n} \approx C$$

for some constant C. What the exact constant is depends on many particulars of the algorithmic implementation. Thus $O(n \log n)$ is only an "order of magnitude," not an exact description.

A.5 Continuity

The basic concepts of continuity should be familiar from basic calculus. We present here a definition for functions defined on metric spaces.

Definition A.7 Let (X, d_1) and (Y, d_2) be metric spaces, and let $f \colon X \to Y$ be a function. Then f is continuous at a point $x_0 \in X$ if for every $\epsilon > 0$ there is a δ so that, for points $x \in X$ sufficiently close to x_0,

$$d_1(x_0, x) < \delta;$$

then the points in the range are close also,

$$d_2(f(x_0), f(x)) < \epsilon.$$

□

Intuitively, a function is continuous if there are no jumps.

The size δ of the neighborhood about x_0 may depend upon x_0. Consider the figure shown in figure A.4. The neighborhood around the point x_1 must be smaller than the neighborhood around the point x_0, since the function f is steeper near x_1. Let D be a domain of the function f. When the size of δ in the definition for continuity does *not* depend upon the x_0 for any $x_0 \in D$, then the function is said to be **uniformly continuous**.

Example A.5.1 The function $f(x) = 1/x$ is continuous over the domain $D = (0, 1)$. However, it is not uniformly continuous. Given an ϵ, no matter how small δ is, an x_0 near zero can be found such that

$$|f(x_0) - f(x_0 + \delta)| > \epsilon.$$

Note that the domain D is not compact, and the function does not achieve an extremum over D. □

A.5 Continuity

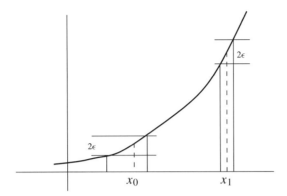

Figure A.4: Illustration of the definition of continuity.

Some useful facts (theorems) about continuous functions:

1. Continuity preserves convergence: if $x_n \to x$, and if f is continuous, then $f(x_n) \to f(x)$. Conversely, if $f(x_n) \to f(x)$ for any convergent sequence $\{x_n\}$ in the domain of f, then f is continuous.
2. A function $f: R \to D$ is continuous on an open set S if and only if the inverse image of every open set is open. That is, if for every open $T \subset D$, $f^{-1}(T) \subset S$ is open, then f is continuous on S.
3. If f and g are continuous functions, then so are $f(g)$ (f composed with g), $f + g$, fg, and f/g where $g \neq 0$, where these operations are defined.
4. A function continuous on a compact set S is uniformly continuous on that set.
5. If $f: D \to \mathbb{R}$ is a continuous function such that $f(p_0) > 0$ for some $p_0 \in D$, then there is a neighborhood N around p_0 such that $f(p) > 0$ for all $p \in N \cap D$. That is, if a continuous function is positive at a point, then it is locally positive around that point.
6. If a real-valued continuous function $f: [a, b] \to \mathbb{R}$ is one-to-one, then f is strictly monotonic on $[a, b]$.
7. (Intermediate value theorem) Let f be a real-valued continuous function defined on a connected set S, and let $f(x_1) = a$ and $f(x_2) = b$ for $x_1, x_2 \in S$ and $a < c < b$. Then there must be a point $y \in S$ such that $f(y) = c$. This theorem provides the theoretical basis for several algorithms that hunt for solutions to equations of the form $f(x) = 0$.
8. A continuous function $f: [a, b] \to \mathbb{R}$ can be uniformly approximated arbitrarily closely by a polynomial. By uniform approximation, we mean that for any $\epsilon > 0$, an approximating polynomial $p(x)$ can be found so that the maximum error between f and $p(x)$ is less than ϵ for any $x \in [a, b]$. This fact (known as the Weierstrass approximation theorem) points to the importance of polynomials in computational mathematics.
9. (Intermediate value theorem) If f is a continuous function on some interval $[a, b]$, and $f(a) > 0$ and $f(b) < 0$, then there is some $c \in (a, b)$ such that $f(c) = 0$.

One of the reasons for studying continuous functions, of all the mathematical functions that might exist, is a matter of practicality: continuity makes a lot of things possible.

Further, it may be argued, at least from an engineering point of view, that there are no truly discontinuous functions occurring in nature.

A.6 Differentiation

Differentiation arises in a variety of contexts, from linearization of functions to optimization. We first review the basic concepts of differentiation of real functions of a single real variable, then generalize to functions defined over several variables.

A.6.1 Differentiation with a single real variable

The basic definition of a derivative is assumed to be familiar for a continuous function $f \colon \mathbb{R} \to \mathbb{R}$:

$$\left. \frac{df}{dx} \right|_{x=x_0} = f'(x_0) = \lim_{x \to x_0} \frac{f(x) - f(x_0)}{x - x_0} = \lim_{\Delta x \to 0} \frac{f(x_0 + \Delta x) - f(x_0)}{\Delta x}. \quad (A.3)$$

If this limit exists, then f is said to be **differentiable** at x_0. A function that is continuous at a point x_0 may not be differentiable at x_0; the function $f(x) = |x|$ is not differentiable at $x = 0$, but it is continuous there.

Some basic facts about derivatives:

1. The derivative provides for a local linear representation of f. That is, near x_0, f can be approximated as

$$f(x_0 + \epsilon) = f(x_0) + f'(x_0)\epsilon + o(\epsilon). \quad (A.4)$$

2. If f has a local **extreme value** (maximum or minimum) at a point x_0 and f is differentiable at x_0, then $f'(x_0) = 0$.

   Proof Assume f has a local maximum. In (A.3), let $x \to x_0$ from the right, so $x - x_0 \geq 0$. Then $f'(x_0) \geq 0$. Now let $x \to x_0$ from the left, so $x - x_0 \leq 0$. Then $f'(x_0) \leq 0$. The only way both inequalities can be satisfied is if $f'(x_0) = 0$. The method is similar if f has a local minimum. □

3. (Rolle's theorem) If $f \colon \mathbb{R} \to \mathbb{R}$ is continuous on $[a, b]$ and $f'(x)$ exists for $x \in (a, b)$, then if $f(a) = f(b)$ there is a point $x_0 \in (a, b)$ such that $f'(x_0) = 0$.

4. (Mean value theorem) If f is continuous on $[a, b]$ and $f'(x)$ exists on (a, b), then there is a point $x_0 \in (a, b)$ such that

$$f(b) - f(a) = (b - a) f'(x_0). \quad (A.5)$$

   More generally, if $g'(x)$ also exists on (a, b), then there is a point $x_0 \in (a, b)$ such that

$$(f(b) - f(a))g'(x_0) = (g(b) - g(a))f'(x_0). \quad (A.6)$$

   Proof We will create a new function to which Rolle's theorem may be applied. Let $F(x) = f(x) - Kg(x)$, with K to be determined. To apply Rolle's theorem, we need $F(a) = F(b)$, or

$$f(a) - Kg(a) = f(b) - Kg(b).$$

   If $g(b) \neq g(a)$, we can solve for K as

$$K = \frac{f(b) - f(a)}{g(b) - g(a)}.$$

A.6 Differentiation

Now, by Rolle's theorem, there is a point x_0 such that $F'(x_0) = 0$, which means, that
$$f'(x_0) = Kg'(x_0).$$
Substituting the value of K gives (A.6). If $g(b) = g(a)$ then there is an x_0 such that $g'(x_0) = 0$; in this case both sides of (A.6) are zero.

The first form of the mean value theorem (A.5) follows from the second by letting $g(x) = x$. $\square$

The second derivative of f is denoted as $f''(x)$. The third derivative is $f^{(3)}(x)$. The nth derivative is $f^{(n)}(x)$, with the convention that $f^{(0)}(x) = f(x)$.

A.6.2 Partial derivatives and gradients on $\mathbb{R}^m$

We now move from functions on $\mathbb{R}$ to functions on $\mathbb{R}^m$. Let $f \colon \mathbb{R}^m \to \mathbb{R}$ be designated as $f(x_1, x_2, \ldots, x_m)$. The derivative of f with respect to x_i (regarding the others as fixed) is denoted as
$$\frac{\partial f}{\partial x_i}.$$
This partial derivative is the change in f as the ith argument is infinitesimally changed. For example, for x_1,
$$\frac{\partial f}{\partial x_1} = \lim_{h \to 0} \frac{f(x_1 + h, x_2, \ldots, x_m) - f(x_1, x_2, \ldots, x_m)}{h}.$$
Another notation for the partial derivative is to indicate the derivative with respect to the ith argument as f_i. Yet another notation is to indicate the derivative with respect to the ith argument as $D_i f$; then
$$\frac{\partial f}{\partial x_i} = f_i = D_i f.$$
The notation f_i may be confusing when f has several components, so it is not used in this text.

A function over $\mathbb{R}^m$ is said to be differentiable at a point $\mathbf{x}_0$ if all its partial derivatives exist at $\mathbf{x}_0$.

Definition A.8 Let f be continuous and defined on a domain D. Then f is said to be in **class** C^k in D, sometimes denoted as $f \in C^k(D)$, if all partial derivatives of f of order up to and including k exist and are continuous. Where the domain D of the function is significant, we write $C^k[D]$. $\square$

Example A.6.1

1. The set of all functions continuous over $[0, 2\pi]$ is denoted as $C[0, 2\pi]$.
2. A function $f \in C^0(D)$ is continuous, but does not have a derivative. The function $f(x) = |x|$ is a C^0 function.
3. A function $f \in C^1(D)$ is both continuous and differentiable.
4. The function $f(x) = \sin(x)$ is in $C^\infty(\mathbb{R})$; that is, it is infinitely differentiable over the whole real line. $\square$

Definition A.9 The **gradient** or **total derivative** of a function $f \colon \mathbb{R}^m \to \mathbb{R}$ at a point $\mathbf{x}_0 = (x_1, x_2, \ldots, x_m)$ is denoted as
$$\nabla f(\mathbf{x}_0) \quad \text{or} \quad \frac{\partial f(\mathbf{x}_0)}{\partial \mathbf{x}}.$$

The gradient is obtained by stacking the partial derivatives:

$$\nabla f(\mathbf{x}) = \begin{bmatrix} \frac{\partial f}{\partial x_1} \\ \frac{\partial f}{\partial x_2} \\ \vdots \\ \frac{\partial f}{\partial x_m} \end{bmatrix}. \tag{A.7}$$

Where the variables with respect to which f is differentiated are to be emphasized, they are indicated with a subscript, as in

$$\nabla_x f.$$

□

In other sources (see, for example, [132]), a distinction is made between ∇f and $\frac{\partial f}{\partial \mathbf{x}}$. In this text, they are always taken to have the same meaning. We will use both notations interchangeably.

In this book, **gradients are always column vectors**, as are vectors in general. (This may be inconsistent with other definitions, but it maintains an internal consistency.)

Example A.6.2 Let

$$f(x, y, z) = x^2 - 2xy + z.$$

Then

$$\frac{\partial f}{\partial x} = D_1 f = 2x - 2y,$$

$$\frac{\partial f}{\partial y} = D_2 f = -2x,$$

$$\frac{\partial f}{\partial z} = D_3 f = 1,$$

and

$$\nabla f = \begin{bmatrix} \frac{\partial f}{\partial x} \\ \frac{\partial f}{\partial y} \\ \frac{\partial f}{\partial z} \end{bmatrix} = \begin{bmatrix} 2x - 2y \\ -2x \\ 1 \end{bmatrix}.$$

□

We now state without proof a useful generalization of the mean value theorem.

Theorem A.2 (*Mean value theorem for multiple variables*) *Let f be differentiable on an open convex set $D \subset \mathbb{R}^m$, and let $\mathbf{x}, \mathbf{y} \in D$. Then there is a point $\mathbf{x}^* = [x_1^*, x_2^*, \ldots, x_m^*]^T$ that is on the line between $\mathbf{x}$ and $\mathbf{y}$ such that*

$$f(\mathbf{x}) - f(\mathbf{y}) = \frac{\partial f(\mathbf{x}^*)}{\partial x_1}(x_1 - y_1) + \frac{\partial f(\mathbf{x}^*)}{\partial x_2}(x_2 - y_2) + \cdots + \frac{\partial f(\mathbf{x}^*)}{\partial x_m}(x_m - y_m). \tag{A.8}$$

The expression (A.8) can be expressed more concisely using an inner product. An inner product of two vectors $\mathbf{a}$ and $\mathbf{b}$ may be defined as

$$\mathbf{a} \cdot \mathbf{b} = \mathbf{a}^T \mathbf{b} = \sum_{i=1}^{m} a_i b_i.$$

A.6 Differentiation

(Other concepts related to inner products are presented in section 2.4.) Using inner-product notation, the mean value theorem becomes

$$f(\mathbf{x}) - f(\mathbf{y}) = (\nabla f(\mathbf{x}^*))^T (\mathbf{x} - \mathbf{y}). \tag{A.9}$$

A.6.3 Linear approximation using the gradient

We observed in equation (A.4) that the derivative can be used to determine a linear approximation to a real-valued function defined on $\mathbb{R}$. In a similar way, the gradient can be used to define a linear approximation to a real-valued function defined on $\mathbb{R}^m$.

Theorem A.3 *Let $f \in C^1$ be in an open set $D \subset \mathbb{R}^m$. Let $\mathbf{x}_0 \in D$. Then at a point $\mathbf{x} = \mathbf{x}_0 + \Delta \mathbf{x}$ near the point $\mathbf{x}_0$, we can write*

$$f(\mathbf{x}_0 + \Delta\mathbf{x}) = f(\mathbf{x}_0) + (\nabla f(\mathbf{x}))^T \Delta\mathbf{x} + R, \tag{A.10}$$

where R is a remainder term such that $R = o(\|\mathbf{x} - \mathbf{x}_0\|)$.

Proof Comparison of (A.10) with (A.9) reveals that

$$R = (\nabla f(\mathbf{x}^*) - \nabla f(\mathbf{x}))^T \Delta\mathbf{x} = \sum_{i=1}^{m} \left(\frac{\partial f(\mathbf{x}^*)}{\partial x_i} - \frac{\partial f(\mathbf{x})}{\partial x_i} \right) \Delta x_i.$$

Since ∇f is continuous (because $f \in C^1$), there is a neighborhood $\mathcal{N}$ about $\mathbf{x}_0$ such that each

$$\frac{\partial f(\mathbf{x}^*)}{\partial x_i} - \frac{\partial f(\mathbf{x})}{\partial x_i}$$

is less than any positive ϵ. Hence

$$\|R\| \leq m\epsilon \|\Delta \mathbf{x}_i\|. \qquad \square$$

We now examine the geometry of the linear approximation described by theorem A.3. Recall from analytical geometry that a multidimensional plane is defined by a point on the plane and a vector orthogonal to the plane. We will use this notion to show that the gradient vector of a function is orthogonal to the tangent surface of that function at a point.

Let $f: \mathbb{R}^m \to \mathbb{R}$, and consider the "graph" of this function defined in $\mathbb{R}^{m+1}$ as the point $(\mathbf{x}, f(\mathbf{x}))$. At a point $\mathbf{x}_0$, let P be the plane in $\mathbb{R}^{m+1}$ tangent to $f(\mathbf{x})$, and let $\mathbf{n} \in \mathbb{R}^{m+1}$ be a vector normal (orthogonal) to the plane P at $\mathbf{x}_0$. Notationally, the plane is defined as

$$P = \{\mathbf{x} \in \mathbb{R}^{m+1} : (\mathbf{x} - \mathbf{x}_0)^T \mathbf{n} = 0\}.$$

We now show that the linear approximation (A.10) corresponds to a plane tangent at $f(\mathbf{x}_0)$. Let

$$\mathbf{n} = \begin{bmatrix} \nabla f(\mathbf{x}_0) \\ -1 \end{bmatrix}$$

and consider the plane which has $\mathbf{n}$ as the normal and the point

$$\begin{bmatrix} \mathbf{x}_0 \\ f(\mathbf{x}_0) \end{bmatrix}$$

as the point of intersection. Then the points $(\mathbf{x}, z)$ lying on the plane satisfy the equation

$$[\mathbf{x} - \mathbf{x}_0]^T \nabla f(\mathbf{x}_0) - z + f(\mathbf{x}_0) = 0,$$

or

$$z = f(\mathbf{x}_0) + [\mathbf{x} - \mathbf{x}_0]^T \nabla f(\mathbf{x}_0).$$

Comparison with (A.10) indicates that z is the linear approximation of $f(\mathbf{x})$.

Application of these principles to a variety of vector and matrix derivatives of use in practice is presented in appendix E.

A.6.4 Taylor series

The Taylor series provides a polynomial representation of a function in terms of the function and its derivatives at a point. Application of the Taylor series often arises when nonlinear functions are employed and we desire to obtain a linear approximation. In its simplest form, only the first two terms of the Taylor are employed, giving essentially the mean value theorem in which functions are represented as an offset and a linear term. Often the third term (the quadratic term) is also employed to provide an indication of the error in using the linear approximation.

For a function $f\colon \mathbb{R} \to \mathbb{R}$, the Taylor series should be familiar. Let $f\colon \mathbb{R} \to \mathbb{R}$ be such that $f \in C^{n+1}$ in an open neighborhood about a point x_0. Then

$$f(x) = f(x_0) + (x - x_0)f'(x_0) + \frac{1}{2}f''(x_0)(x - x_0)^2 \\ + \cdots + \frac{1}{n!}f^{(n)}(x_0)(x - x_0)^n + R_n(x) \qquad (A.11)$$

where

$$R_n(x) = f^{(n+1)}(x^*)\frac{(x - x_0)^{n+1}}{(n + 1)!} = o(|x - x_0|^n)$$

and $x^* \in [x, x_0]$.

Extension of Taylor's theorem to $f\colon \mathbb{R}^m \to \mathbb{R}$ is straightforward. Let $\Delta \mathbf{x} = \mathbf{x} - \mathbf{x}_0 = (\Delta x_1, \Delta x_2, \ldots, \Delta x_m)$. We define the operator $\mathcal{D}$ by

$$\mathcal{D} = \Delta x_1 \frac{\partial}{\partial x_1} + \Delta x_2 \frac{\partial}{\partial x_2} + \cdots + \Delta x_m \frac{\partial}{\partial x_m}.$$

Then

$$\mathcal{D} f(\mathbf{x}) = (\nabla f(\mathbf{x}))^T \Delta \mathbf{x} \\ = \frac{\partial f(\mathbf{x})}{\partial x_1}\Delta x_1 + \frac{\partial f(\mathbf{x})}{\partial x_2}\Delta x_2 + \cdots + \frac{\partial f(\mathbf{x})}{\partial x_m}\Delta x_m.$$

Using $\mathcal{D}$, the Taylor expansion formula is

$$f(\mathbf{x}_0 + \Delta \mathbf{x})$$
$$= f(\mathbf{x}_0) + \mathcal{D} f(\mathbf{x}_0) + \frac{1}{2!}\mathcal{D}^2 f(\mathbf{x}_0) + \cdots + \frac{1}{n!}\mathcal{D}^n f(\mathbf{x}_0) + \frac{1}{(n+1)!}\mathcal{D}^{n+1} f(\mathbf{x}^*)$$
$$= f(\mathbf{x}_0) + \mathcal{D} f(\mathbf{x}_0) + \frac{1}{2!}\mathcal{D}^2 f(\mathbf{x}_0) + \cdots + \frac{1}{n!}\mathcal{D}^n f(\mathbf{x}_0) + o(\|\Delta \mathbf{x}\|^n), \qquad (A.12)$$

where $\mathbf{x}^*$ is a point on the line segment adjoining $\mathbf{x}_0$ and $\mathbf{x}$.

Example A.6.3 Let $f(x, y) = x + \sin(xy)$. Then

$$\mathcal{D} = \Delta x \frac{\partial}{\partial x} + \Delta y \frac{\partial}{\partial y}$$

and

$$\mathcal{D}^2 = (\Delta x)^2 \frac{\partial}{\partial x^2} + 2\Delta x \Delta y \frac{\partial^2}{\partial x \partial y} + (\Delta y)^2 \frac{\partial}{\partial y^2}.$$

Then for $\mathbf{x}_0 = (x_0, y_0)$,

$$\mathcal{D}f(\mathbf{x}_0) = \Delta x(1 + y_0 \cos(x_0 y_0)) + \Delta y x_0 \cos(x_0 y_0)$$
$$\mathcal{D}^2 f(\mathbf{x}_0) = -(\Delta x)^2 y_0^2 \sin(x_0 y_0) + 2\Delta x \Delta y(\cos(x_0 y_0) - x_0 y_0 \sin(x_0 y_0)) - (\Delta y)^2 x_0^2 \sin(x_0 y_0)$$

and the first three terms of the Taylor series are

$$f(\mathbf{x}_0 + \Delta \mathbf{x}) = x_0 + \sin(x_0 y_0) + \Delta x(1 + y_0 \cos(x_0 y_0)) + \Delta y(x_0 \cos(x_0 y_0))$$
$$+ \Delta x \Delta y(\cos(x_0 y_0) - x_0 y_0 \sin(x_0 y_0)) - \frac{1}{2}(\Delta x)^2 y_0^2 \sin(x_0 y_0) - \frac{1}{2}(\Delta y)^2 x_0^2 \sin(x_0 y_0).$$

$\square$

A.7 Basic constrained optimization

Constrained optimization is employed at several points in book. We present here the basic concepts, leaving the justification for the method to chapter 18.

Let $f(t)$ be a function to be minimized or maximized, subject to a constraint $g(t) = 0$. The constraint is introduced by forming the function

$$J = f(t) + \lambda g(t),$$

where λ is a Lagrange multiplier. Then, sufficient conditions for an optimal value are

$$\frac{\partial J}{\partial t} = 0,$$

$$\frac{\partial J}{\partial \lambda} = 0.$$

The last equation always returns the original constraint.

When there are several variables, a partial derivative is introduced for each. When there are several constraints, each constraint is introduced using its own Lagrange multiplier. For example, if $g_1(t) = 0, g_2(t) = 0, \ldots, g_p(t) = 0$ are p constraints, then we form

$$J = f(t) + \lambda_1 g_1(t) = \lambda_2 g_2(t) + \cdots + \lambda_p g_p(t),$$

then find sufficient conditions for constrained optimality by

$$\frac{\partial J}{\partial t} = 0 \quad \frac{\partial J}{\partial \lambda_1} = 0 \quad \frac{\partial J}{\partial \lambda_2} = 0 \quad \cdots \quad \frac{\partial J}{\partial \lambda_p} = 0.$$

Example A.7.1 We will demonstrate the optimization with a single constraint and two variables. A fence is to be built along a river so as to enclose the maximum possible area, as shown in figure A.5. The total length of fencing available is R meters. The constraint is

$$x + 2y = R.$$

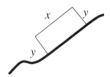

Figure A.5: A constrained optimization problem.

The function to be maximized is the area enclosed,

$$f(x, y) = xy.$$

We form the function
$$J = xy + \lambda(x + 2y - R)$$
and compute the derivatives
$$\frac{\partial J}{\partial x} = y + \lambda = 0 \qquad \frac{\partial J}{\partial y} = x + 2\lambda = 0 \qquad \frac{\partial J}{\partial \lambda} = x + 2y - R = 0.$$

From the first equation we find $\lambda = -y$, and from the second we find $\lambda = -x/2$. Equating these, we find
$$y = x/2.$$
Substituting this into the constraint we find that $x = R/2$ meters. $\square$

A.8 The Hölder and Minkowski inequalities

In this section we examine some useful inequalities, both to establish that the l_p and L_p norms are, in fact, norms, and to provide an introduction to the study of inequalities in general. The proofs introduce several additional inequalities that are useful in their own right. In addition, several other useful inequalities are examined in the exercises.

That the l_p and L_p norms satisfy the properties of norms given in section 2.1 must be established. All of the properties are straightforward to establish except the property that $\|\mathbf{x} + \mathbf{y}\| \leq \|\mathbf{x}\| + \|\mathbf{y}\|$. For the p-norms, this is established by means of the Minkowski inequality, which in turn follows from Hölder's inequality.

Lemma A.1 *(Hölder's inequality) For $x, y > 0$ such that $x + y = 1$, and (possibly complex) sequences $\{a_i, i = 1, 2, \ldots, n\}$ and $\{b_i, i = 1, 2, \ldots, n\}$,*

$$\left| \sum_{i=1}^{n} a_i b_i \right| \leq \left(\sum_{i=1}^{n} |a_i|^{1/x} \right)^x \left(\sum_{i=1}^{n} |b_i|^{1/y} \right)^y. \tag{A.13}$$

We note that when $x = y = 1/2$ we obtain the Cauchy–Schwarz inequality. Equation (A.13) can be written more succinctly (and more generally) as

$$|\langle \mathbf{a}, \mathbf{b} \rangle| \leq \|\mathbf{a}\|_{1/x} \|\mathbf{b}\|_{1/y}.$$

Proof If either $a_1 = a_2 = \cdots = a_n = 0$ or $b_1 = b_2 = \cdots = b_n = 0$, then the inequality (A.27) is trivial, so we exclude this case. Let us consider first the case that the numbers a_i and b_i, $i = 1, 2, \ldots, n$ are nonnegative real numbers. We begin with the inequality

$$y^\alpha \leq 1 + \alpha(y - 1), \tag{A.14}$$

for $y > 0$ and $0 < \alpha < 1$ (see exercise A.7-27). Let $y = A/B$ for $A, B > 0$. Then $A^\alpha B^{1-\alpha} \leq B + \alpha(A - B)$. Now let $x = \alpha$ and $y = 1 - \alpha$ to obtain

$$A^x B^y \leq xA + yB. \tag{A.15}$$

Equality holds in (A.15) if and only if $A = B$. Now let

$$A_i = \frac{a_i^{1/x}}{\sum_{i=1}^{n} a_i^{1/x}} \qquad B_i = \frac{b_i^{1/y}}{\sum_{i=1}^{n} b_i^{1/y}}. \tag{A.16}$$

Then, using (A.15), we get

$$\sum_{i=1}^{n} A_i^x B_i^x \leq x \sum_{i=1}^{n} A_i + y \sum_{i=1}^{n} B_i = x + y = 1. \tag{A.17}$$

A.9 Exercises

Substituting from (A.16) into (A.17), we obtain

$$\sum_{i=1}^{n} a_i b_i \leq \left(\sum_{i=1}^{n} a_i^{1/x}\right)^x \left(\sum_{i=1}^{n} b_i^{1/y}\right)^y. \quad (A.18)$$

In the case that the a_i and b_i are complex, we have

$$\left|\sum_{i=1}^{n} a_i b_i\right| \leq \sum_{i=1}^{n} |a_i b_i| = \sum_{i=1}^{n} |a_i| |b_i|,$$

from which, using (A.18), equation (A.13) follows.

From the equality condition for (A.15), equality holds in (A.13) if and only if $A_i = B_i$ for $i = 1, 2, \ldots, n$, which is equivalent to

$$\frac{a_1^y}{b_1^x} = \frac{a_2^y}{b_2^x} = \cdots = \frac{a_n^y}{b_n^x}. \qquad \square$$

Now the triangle inequality for p-norms is established by the Minkowski inequality:

Lemma A.2 *(Minkowski inequality)* *For positive numbers A_i and B_i, $i = 1, 2, \ldots, n$,*

$$\left(\sum_{i=1}^{n} (A_i + B_i)^p\right)^{1/p} \leq \left(\sum_{i=1}^{n} A_i^p\right)^{1/p} + \left(\sum_{i=1}^{n} B_i^p\right)^{1/p}, \quad (A.19)$$

with equality if and only if

$$\frac{a_1}{b_1} = \frac{a_2}{b_2} = \cdots = \frac{a_n}{b_n}.$$

The proof is discussed in exercise A.7-28.

The Hölder and Minkowski inequalities also have an integral form. Let $p, q > 0$ such that $\frac{1}{p} + \frac{1}{q} = 1$. For the Hölder inequality:

$$\left|\int_a^b f(t)g(t)\,dt\right| \leq \left(\int_a^b |f(t)|^p\,dt\right)^{1/p} \left(\int_a^b |g(t)|^q\,dt\right)^{1/q}. \quad (A.20)$$

Equality holds if at least one of f or g is identically zero, or if fg does not change sign on $[a, b]$ and there are positive constants α and β such that

$$\alpha |f|^p = \beta |g|^q \text{ on } [a, b].$$

For the Minkowski inequality:

$$\left(\int_a^b |f(t) + g(t)|^p\,dt\right)^{1/p} \leq \left(\int_a^b |f(t)|^p\,dt\right)^{1/p} + \left(\int_a^b |g(t)|^p\,dt\right)^{1/p}. \quad (A.21)$$

Proofs of these statements are examined in the exercises.

A.9 Exercises

A.1-1 Show that the following sets are convex:
   (a) The set of Toeplitz matrices.
   (b) The set of monic polynomials. (A polynomial is monic if the coefficient of the highest-order term is 1.)
   (c) The set of symmetric matrices.

A.1-2 The use of an indicator function can be generalized for the description of *fuzzy sets*. In this case, the indicator function is not simply binary valued, but takes on values in a continuum from 0 to 1. Figure A.6 provides the indicator function for the fuzzy set of $\tilde{T}$ = "set of real numbers near 10." Note that there is a degree of arbitrariness about the description; fuzzy sets can be used to represent human taste and variation to a much greater degree than can nonfuzzy (crisp) sets.

(a) Draw the indicator function for the fuzzy set of real numbers "near 3." Call this set $\tilde{A}$.

(b) Draw the indicator function for the fuzzy set of real numbers "near 4." Call this set $\tilde{B}$.

(c) Draw the indicator functions for $\tilde{A} \cap \tilde{B}$ and $\tilde{A} \cup \tilde{B}$.

(d) A fuzzy set defined over the real numbers is called (loosely) a fuzzy number. Devise a reasonable rule for the addition of fuzzy numbers. Show the result of the rule for $\tilde{A} + \tilde{B}$.

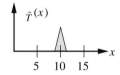

Figure A.6: The indicator function for a fuzzy number "near 10".

A.1-3 The set of even integers can be represented as $2\mathbb{Z}$. Show that $|2\mathbb{Z}| = |\mathbb{Z}|$ (that is, there are as many even integers as there are integers). Similarly show that there are as many odd integers as there are integers.

A.1-4 Show that $|(0, 1]| = |\mathbb{R}|$.

A.1-5 Show that the intersection of convex sets is convex.

A.1-6 If S and T are convex sets both in $\mathbb{R}^n$, show that the *set sum*

$$S + T = \{x : x = s + t, s \in S, t \in T\}$$

is convex. Figure A.7 illustrates the set sum.

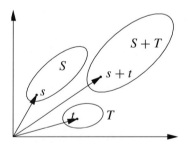

Figure A.7: The set sum.

A.1-7 If S and T are convex sets, show that $S \cap T$ is convex.

A.1-8 Show that the polytope in n dimensions defined by

$$P_n = \left\{ \mathbf{x} \in \mathbb{R}^n : x_i \geq 0, \sum_{i=1}^{n} x_i = 1 \right\}$$

is convex.

A.9 Exercises

A.1-9 For the polytope P_n of the previous example, let $(a_1, a_2, \ldots, a_n) \in P_n$. Show (by induction) that

$$n^2 \leq \sum_{i=1}^n \frac{1}{a_i}.$$

A.3-10 If $f(x)$ is convex, show that $af(x) + b$ is also convex for $a > 0$.

A.3-11 (Order notation)
  (a) Show that $f = \sin(x)$ is $O(x)$ as $x \to 0$.
  (b) Show that $f = x \sin(1 + 1/x)$ is $O(x)$ as $x \to 0$.
  (c) Show that $f = e^{-1/x}$ is $o(x^\alpha)$ as $x \to 0$ for all values of α.

A.5-12 Using $f(x, y) = x^2$, show that a continuous function does not necessarily map an open set to an open set.

A.5-13 Using $f(x) = x^2/(1 + x^2)$, show that a continuous function does not necessarily map a closed set to a closed set.

A.5-14 Find an example of a continuous function that maps an open set to a closed set.

A.5-15 Show that property 1 for continuity is true.

A.5-16 Show that property 2 for continuity is true.

A.5-17 Let (X, d_2) be the metric space of functions defined on $\mathbb{R}$ with the Euclidean metric, $d_2(f, g) = \int (f(t) - g(t))^2 \, dt$. Define the mapping $\Phi_\phi: X \to \mathbb{R}$ by

$$\Phi(x) = \int_{-\infty}^{\infty} x(t)\phi(t) \, dt.$$

Show that if ϕ is square integrable,

$$\int_{-\infty}^{\infty} \phi^2(t) \, dt < \infty,$$

then Φ_ϕ is a continuous mapping. Hint: Use the Cauchy–Schwarz inequality

$$\left| \int a(t)b(t) \right| \leq \left(\int a^2(t) \, dt \right)^{1/2} \left(\int b^2(t) \, dt \right)^{1/2}.$$

A.5-18 Prove Rolle's theorem.

A.5-19 Show that if f is differentiable on an interval (a, b) then the zeros of f are separated by the zeros of f'. Hint: Use Rolle's theorem.

A.5-20 Show that if f and g are both continuous on $[a, b]$ and both differentiable on (a, b), and if $f'(a) = g(a)$ and $f(b) = g(b)$, then there is a point $x_0 \in (a, b)$ such that $f'(x_0) = g'(x_0)$. Hint: Use Rolle's theorem.

A.5-21 Using Rolle's theorem, show that if f is a function differentiable on (a, b) with $f(a) = f'(a) = f''(a) = f^{(3)}(a) = 0$ and $f(b) = 0$, then there is a point $c \in (a, b)$ such that $f^{(4)}(c) = 0$.

A.5-22 Using the mean value theorem,
  (a) Show that if $f'(x) = 0$ on an interval, then f is constant on that interval.
  (b) Show that if $f'(x)$ never changes sign on an interval, then f is monotonic on that interval.

A.6-23 Determine a linear approximation to each of the following functions:
  (a) $f(x, y, z) = x^{y+z}$ at $(1, 1, 1)$.
  (b) $f(x, y) = \cos(x, y)$ at $(1, 2)$.
  (c) $f(x, y) = e^{x^2 - y^2}$ at $(1, 2)$.

A.6-24 Write a MATLAB function x = gradesc(f, df, n, x0, mu) that iterates the gradient descent algorithm on a function f with derivative df in n dimensions starting from x0. (The function and its derivative are MATLAB function-functions.)

A.6-25 Determine a complex vector **c** that minimizes
$$(\mathbf{b} - A\mathbf{c})^H (\mathbf{b} - A\mathbf{c})$$
using gradients.

A.6-26 Determine the Taylor series for the following functions up to the quadratic term. A symbolic manipulation program may be helpful.
  (a) $f(x, y, z) = x^{y+z}$ at $(1, 1, 1)$.
  (b) $f(x, y) = \cos(x, y)$ at $(1, 2)$.
  (c) $f(x, y) = e^{x^2 - y^2}$ at $(1, 2)$.

A.7-27 In this exercise, we develop a proof of (A.14), which introduces some other useful inequalities along the way.
  (a) For $x > 0$ and for n an integer, show that
  $$(1 + x)^n > 1 + nx.$$
  Hint: Use the binomial theorem (1.83).
  (b) The *geometric mean* of a set of numbers $z_1, z_2, \ldots, z_n$ is defined by
  $$G_m = \left(\prod_{i=1}^{n} z_i \right)^{1/m}.$$
  The *arithmetic mean* of this set of numbers is
  $$A_m = \frac{1}{m} \sum_{i=1}^{m} z_i.$$
  The following result holds:

  > The geometric mean of m positive real numbers is less than or equal to their arithmetic mean, with equality if and only if all of the numbers are equal.

  We now show that for $x \geq -1$ and $0 < \alpha < 1$,
  $$(1 + x)^\alpha \leq 1 + \alpha x.$$

A.9 Exercises

Assume that $\alpha = m/n$, then we can write

$$(1+x)^\alpha = \sqrt[n]{\underbrace{(1+x)(1+x)\cdots(1+x)}_{m \text{ factors}} \underbrace{1 \cdot 1 \cdots 1}_{n-m \text{ factors}}}.$$

Using the inequality relating the geometric and arithmetic means, we conclude that

$$(1+x)^{m/n} \leq \frac{m(1+x)+n-m}{n} = 1 + \frac{m}{n}x.$$

(c) Employ a continuity argument to extend this result for all α such that $0 < \alpha < 1$.

(d) Finally, to establish (A.14), let $x = y - 1$.

A.7-28 (Proof of Minkowski's inequality)

(a) Apply (A.13) using $a_k = A_k$ and $b_k = (A_k + B_k)^{p/q}$, with $x = 1/p$ and $y = 1 - 1/p = 1/q$.

(b) Similarly, apply (A.13) where now $a_k = B_k$ and $b_k = (A_k + B_k)^{p/q}$.

(c) Add the two equations just obtained, using $p = 1 + p/q$, to obtain

$$\sum_{i=1}^{n}(A_i + B_i)^p \leq \left[\left(\sum_{i=1}^{n} A_i^p\right)^{1/p} + \left(\sum_{i=1}^{n} B_i^p\right)^{1/p} \right] \left[\sum_{i=1}^{n}(A_i + B_i)^p\right]^{1/q}.$$

(d) Conclude, using $1/q = 1 - 1/p$, that

$$\left[\sum_{i=1}^{n}(A_i + B_i)^p\right]^{1/p} \leq \left(\sum_{i=1}^{p} A_i^p\right)^{1/p} + \left(\sum_{i=1}^{n} B_i^p\right)^{1/p}.$$

A.7-29 Prove (A.20). Hint: In (A.15), let

$$A = \frac{|f(t)|^p}{\int_a^b |f(t)|^p \, dt} \quad \text{and} \quad B = \frac{|g(t)|^p}{\int_a^b |f(t)|^p \, dt}$$

and integrate from a to b.

A.7-30 Prove (A.21).

A.7-31 [325] Show that if $p, q > 1$ with $p^{-1} + q^{-1} = 1$ then for all scalars α and β

$$|\alpha \beta| \leq \frac{|\alpha|^p}{p} + \frac{|\beta|^q}{q}.$$

Hint: Let $\phi(\tau) = \tau^p/p + \tau^{-q}/q$, and show that $\phi(\tau)$ satisfies $\phi(\tau) \geq 1$ for all positive τ. Then let $\tau = |\alpha|^{1/q}|\beta|^{-1/p}$.

A.7-32 Prove the "information theory inequality,"

$$\log x \leq 1 + x.$$

A.7-33 In the proof of lemma A.1, the inequality $A^x B^y \leq xA + yB$ for $x + y = 1$ with $A, B, x, y > 0$. Another inequality (not as tight) for these conditions is

$$A^x B^y \leq xA + yB + 1.$$

Prove this inequality. Hint: Use the information theory inequality of the preceding exercise.

A.7-34 Let

$$S = \{\mathbf{x} \in \mathbb{R}^n : \|\mathbf{x}\| \leq 1\}$$

be a unit sphere defined by the norm $\|\cdot\|$. Show that S is convex.

A.7-35 [56] Show for $0 \leq x, y \leq 1$ and $n > 0$ that

$$(1 - xy)^n \leq 1 - x + e^{-yn}.$$

Hint: Let $f(y) = e^{-y} - 1 + y$. Show that $f(y) > 0$ for $y > 0$. Show that this gives the result for $x = 1$. Then show that $g_y(x) = (1 - xy)^n$ is a convex function of x and hence $g_y(x) \leq (1 - x)g_y(0) + xg_y(1)$.

A.10 References

A comprehensive introduction to set theory is provided in [126]. The fuzzy sets alluded to in exercise A.1-2 were introduced in [383]. A recent treatment giving a variety of engineering applications is [279].

The analysis presented in this appendix is available from a variety of sources, including [42, 92]. Some of the material has also been taken from [209]. The material on derivatives with respect to complex vectors is from [132].

Our discussion of the Hölder and Minkowski inequalities is drawn from [176].

Appendix B

Completing the Square

Completing the square is a simple algebraic technique that arises frequently enough in both scalar and vector problems that it is worth illustrating.

B.1 The scalar case

The quadratic expression

$$J(x) = ax^2 + bx + c \tag{B.1}$$

can be written as

$$a\left(x^2 + \frac{b}{a}x\right) + c.$$

In completing the square, we write this as a perfect square with a constant offset. Taking the coefficient of x and dividing by 2, it is straightforward to verify that

$$\boxed{a\left(x^2 + \frac{b}{a}x\right) + c = a\left(x + \frac{b}{2a}\right)^2 - \frac{b^2}{4a} + c} \tag{B.2}$$

By means of completing the square, we can obtain both the minimizing value of x and the minimum value of $J(x)$ in (B.1). Examination of (B.2) reveals that the minimum must occur when

$$x = -\frac{b}{2a},$$

a result also readily obtained via calculus. In this case, we also get the minimum value as well, since if $x = -\frac{b}{2a}$ then

$$J_{\min} = c - \frac{b^2}{4a}.$$

Example B.1.1 We demonstrate the use of completing the square for the estimation of a Gaussian random variable observed in Gaussian noise. Suppose $X \sim \mathcal{N}(\mu_x, \sigma_x^2)$ and $N \sim \mathcal{N}(0, \sigma_n^2)$, where X is regarded as a signal and N is regarded as noise. We make an observation of the signal in the noise,

$$Y = X + N.$$

Given a measurement of $Y = y$, we desire to find $f(x \mid y)$. By the Bayes theorem,

$$f(x \mid y) = \frac{f(y \mid x) f(x)}{\int_{-\infty}^{\infty} f(y \mid x) f(x) \, dx}.$$

The density $f(y\,|\,x)$ can be obtained by observing that for a given value of $X = x$,

$$Y = x + N$$

is simply a shift of the random variable N, and hence is Gaussian with variance σ_n^2 and mean x. That is,

$$f(y\,|\,x) = f_n(y - x),$$

where f_n is the density of N. We can therefore write

$$f(x\,|\,y) = \frac{f_n(x - y)f(x)}{\int_{-\infty}^{\infty} f(y\,|\,x)f(x)\,dx}.$$

The constant in the denominator is simply a normalizing value to make the density $f(x\,|\,y)$ integrate to 1; we call it C and pay little attention to it. We can write

$$f(x\,|\,y) = \frac{1}{C}\frac{1}{\sigma_n\sqrt{2\pi}}e^{-(y-x)^2/2\sigma_n^2}\frac{1}{\sigma_x\sqrt{2\pi}}e^{-(x-\mu_x)^2/2\sigma_x^2}.$$

Let us focus our attention on the exponent, which we denote by E,

$$E = -\frac{1}{2\sigma_n^2}(y - x)^2 - \frac{1}{2\sigma_x^2}(x - \mu_x)^2.$$

This can be written as

$$E = x^2\left(-\frac{1}{2\sigma_n^2} - \frac{1}{s\sigma_x^2}\right) + x\left(\frac{y}{\sigma_n^2} + \frac{\mu_x}{\sigma_x^2}\right) + C_1,$$

where C_1 does not depend upon x. By completing the square, we have

$$E = -\frac{\sigma_x^2 + \sigma_n^2}{2\sigma_n^2\sigma_x^2}\left(x - \frac{y\sigma_x^2 + \mu_x\sigma_n^2}{\sigma_x^2 + \sigma_n^2}\right) + C_2,$$

where C_2 does not depend upon x. The density can thus be written

$$f(x\,|\,y) = \left(\frac{1}{2\pi C\sigma_n\sigma_x}e^{C_2}\right)\exp\left[-\frac{1}{2}\left(x - \frac{y\sigma_x^2 + \mu_x\sigma_n^2}{\sigma_x^2 + \sigma_n^2}\right)1/\left(\sigma_n^2\sigma_x^2/\left(\sigma_x^2 + \sigma_n^2\right)\right)\right].$$

This has the form of a Gaussian density, so the constants in front of the exponential must be such that this integrates to 1. The mean of this Gaussian density is

$$\mu_{x|y} = \frac{\sigma_x^2}{\sigma_x^2 + \sigma_n^2}y + \frac{\sigma_n^2}{\sigma_x^2 + \sigma_n^2}\mu_x$$

and the variance is

$$\sigma_{x|y}^2 = \frac{\sigma_n^2\sigma_x^2}{\sigma_x^2 + \sigma_n^2}.$$

Let is consider an interpretation of this result. If $\sigma_n^2 \gg \sigma_x^2$, then an observation of Y does not tell us much about X because the interfering noise N is too strong. The information we have about X given Y is thus about the same as the information we have about X alone. This observation is validated in the analysis: if $\sigma_n^2 \gg \sigma_x^2$, then

$$\mu_{x|y} \approx \mu_x$$

and

$$\sigma_{x|y}^2 \approx \sigma_x^2.$$

On the other hand, if the noise variance is small, so that $\sigma_n^2 \ll \sigma_x^2$, then an observation of Y is almost the same as an observation of X itself. In this case we have

$$\mu_{x|y} \approx y \quad \text{and} \quad \sigma_{x|y}^2 \approx \sigma_n^2. \qquad \square$$

B.2 The matrix case

The equation
$$\mathbf{x}^T A \mathbf{x} + \mathbf{x}^T \mathbf{y} + c$$
where A is symmetric and invertible can be written as
$$(\mathbf{x} + \mathbf{z})^T A (\mathbf{x} + \mathbf{z}) + d, \tag{B.3}$$
where
$$\mathbf{z} = \frac{1}{2} A^{-1} \mathbf{y}$$
and
$$d = c - \frac{1}{2} \mathbf{y}^T \mathbf{z}.$$

B.3 Exercises

B.1-1 The characteristic function of a random variable X is the (conjugate of the) Fourier transform of its density,
$$\Phi_X(\omega) = \int f_X(x) e^{j\omega x} \, dx.$$

(a) Show that a Gaussian density with
$$f_X(f) = \frac{1}{\sqrt{2\pi}\sigma} = e^{-(x-\mu)^2/2\sigma^2}$$
has the characteristic function
$$\Phi_X(\omega) = \exp\left[j\mu\omega - \frac{1}{2}\omega^2\sigma^2\right].$$

(b) To follow up the characteristic function idea, show that the nth moment of X can be obtained from its characteristic function by
$$E[X^n] = \frac{1}{j^n} \left. \frac{d^n \Phi_X(\omega)}{d\omega^n} \right|_{\omega=0}.$$

B.1-2 Show that the conditional density in (1.52) is correct.

B.2-3 Using (B.3), determine $f(\mathbf{x} \mid \mathbf{y})$ when
$$Y = X + N$$
and X and N are Gaussian-distributed random vectors with
$$X \sim \mathcal{N}(\boldsymbol{\mu}, R_x) \quad \text{and} \quad N \sim \mathcal{N}(\mathbf{0}, R_n).$$

Appendix C

Basic Matrix Concepts

C.1 Notational conventions

In this text, matrices are denoted by capital letters such as A. An $m \times n$ matrix consists of m rows and n columns. The elements are usually considered to be from $\mathbb{C}$ or $\mathbb{R}$. Where the underlying field is specified, we may use the notation $A \in M_{m,n}(F)$ to indicate that A is a $m \times n$ matrix with elements from F. Where the field is understood, we may simply write $A \in M_{m,n}$.

We can write a matrix A in the form

$$A = \begin{bmatrix} a_{11} & a_{12} & \cdots & a_{1n} \\ a_{21} & a_{22} & \cdots & a_{2n} \\ \vdots & \vdots & \ddots & \vdots \\ a_{m1} & a_{m2} & \cdots & a_{mn} \end{bmatrix} = [A]_{ij}.$$

In this display, the starting index is 1. However, in many instances it turns out to be convenient to allow other starting indices. Sometimes the convenience is a matter of adjusting to the expediencies of a programming language: by default, C arrays start indices at 0, while Fortran and MATLAB indices start at 1. When coding matrix operations, care should be taken to start and end at the correct index.

The ith element of the vector $\mathbf{x}$ is denoted by x_i or by $\mathbf{x}_i$, unless there is some notational confusion, in which case the notation $[\mathbf{x}]_i$ is employed. For example, the ith element of the vector $\mathbf{x}_j$ might be denoted $[\mathbf{x}_j]_i$. The (i, j)th element of the matrix A is denoted as a_{ij} or A_{ij}.

An $m \times n$ matrix can be considered as a side-by-side stack of n column vectors, each with m elements, in which case we could write

$$A = [\mathbf{a}_1 \quad \mathbf{a}_2 \quad \cdots \quad \mathbf{a}_n], \tag{C.1}$$

where each $\mathbf{a}_i \in \mathbb{C}^m$ is a vector. The notation employed to indicate the side-by-side stacking is simply juxtaposition, with the vectors separated by a space. A matrix can also be considered as a stack of rows, one on top of the other. This would be written as

$$A = \begin{bmatrix} \mathbf{b}_1^T \\ \mathbf{b}_2^T \\ \vdots \\ \mathbf{b}_m^T \end{bmatrix}, \tag{C.2}$$

where each $\mathbf{b}_i \in \mathbb{C}^n$.

C.1 Notational conventions

It is occasionally necessary to extract portions of a matrix to form submatrices. The notation for that employed here follows [114] and MATLAB. If A is a matrix, then $A(1:i, 1:j)$ is used to indicate the submatrix consisting of the first i rows and the first j columns. The colon : is a separator in a range unless it appears alone in a dimension, in which case it indicates the maximum possible extent in the index for that dimension.

Example C.1.1 For the matrix
$$A = \begin{bmatrix} 1 & 2 & 3 \\ 4 & 5 & 6 \\ 7 & 8 & 9 \end{bmatrix},$$
some representative submatrices are
$$A(1:2, 1:2) = \begin{bmatrix} 1 & 2 \\ 4 & 5 \end{bmatrix} \quad A(2, :) = [4 \ 5 \ 6] \quad A(:, 2) = \begin{bmatrix} 2 \\ 5 \\ 8 \end{bmatrix}.$$
□

When the rows and columns are not contiguous, the rows and columns can be presented in a list.

Example C.1.2 For the matrix
$$A = \begin{bmatrix} 3 & 5 & 7 & 9 & 11 \\ 2 & 4 & 6 & 8 & 10 \\ 3 & 1 & 4 & 1 & 5 \\ 2 & 7 & 1 & 8 & 2 \end{bmatrix},$$
the following are submatrices:
$$C_1 = \begin{bmatrix} 3 & 5 & 7 \\ 2 & 4 & 6 \end{bmatrix} = A(1:2, 1:3),$$

$$C_2 = \begin{bmatrix} 2 & 4 & 8 & 10 \\ 2 & 7 & 8 & 2 \end{bmatrix} = A(\{2, 4\}, \{1, 2, 4, 5\}),$$

$$C_3 = \begin{bmatrix} 3 & 7 \\ 2 & 6 \\ 3 & 4 \end{bmatrix} = A(1:3, \{1, 3\}).$$
□

The jth column of A may be represented as $A_{:,j}$, and the ith row of A may be represented as $A_{i,:}$.

A useful notation is the **unit vector** or **element vector** $\mathbf{e}_i$, which is a vector of length n (usually determined by context) that is all 0s except for a 1 in the ith location. Thus

$$\mathbf{e}_1 = \begin{bmatrix} 1 \\ 0 \\ 0 \\ \vdots \\ 0 \end{bmatrix}, \quad \mathbf{e}_2 = \begin{bmatrix} 0 \\ 1 \\ 0 \\ \vdots \\ 0 \end{bmatrix}, \quad \ldots, \quad \mathbf{e}_n = \begin{bmatrix} 0 \\ 0 \\ 0 \\ \vdots \\ 1 \end{bmatrix}.$$

Using the unit vectors, the jth column of A can be written as
$$A_{:,j} = A\mathbf{e}_j \tag{C.3}$$
and the ith row of A is
$$A_{i,:} = \mathbf{e}_i^T A. \tag{C.4}$$

A **unit element matrix** is a matrix $E_{r,s}$ of size $m \times n$ (usually determined by context) that is all 0s except for a 1 at location (r, s). For example, $E_{3,2} \in M_{4,4}$ is

$$E_{3,2} = \begin{bmatrix} 0 & 0 & 0 & 0 \\ 0 & 0 & 0 & 0 \\ 0 & 1 & 0 & 0 \\ 0 & 0 & 0 & 0 \end{bmatrix}.$$

Clearly $E_{ij} = \mathbf{e}_i \mathbf{e}_j^T$.

A **diagonal matrix** is a matrix of the form

$$D = \begin{bmatrix} d_{11} & & & \\ & d_{22} & & \\ & & \ddots & \\ & & & d_{nn} \end{bmatrix},$$

where 0s appear in all locations not on the diagonal. This is occasionally written in abbreviated form as

$$D = \text{diag}(d_{11}, d_{22}, \ldots, d_{nn}).$$

If the matrix is not square, then a diagonal matrix has rectangular section of 0s at the bottom if the number of rows exceeds the number of columns or a rectangular section of 0s at the right if the number of columns exceeds the number of rows.

Throughout the text it will be common to deal with vectors and matrices of 0s, for example $[0\ 0\ 0\ 0]^T$. Where there is little room for confusion, a vector or matrix of 0s will be indicated simply by the symbol 0, where the shape is determined by the context of the symbol. Where there may be confusion between a vector or matrix of 0s and a scalar 0, the vector or matrix is denoted with a bold font, as **0**.

Basic properties of matrix addition and multiplication are assumed to be understood. It is helpful to remember the shape of some matrix/vector products:

- A matrix times a column vector is a column vector:

$$\begin{bmatrix} \quad \\ \quad \end{bmatrix} \begin{bmatrix} \ \\ \ \end{bmatrix} = \begin{bmatrix} \ \\ \ \end{bmatrix}.$$

- A row vector times a matrix is a row vector:

$$[\quad] \begin{bmatrix} \quad \\ \quad \end{bmatrix} = [\quad].$$

- A row vector times a column vector is a scalar:

$$[\quad] \begin{bmatrix} \ \\ \ \end{bmatrix} = [\ ].$$

- A product such as $\mathbf{x}^T A \mathbf{y}$ yields a scalar:

$$[\quad] \begin{bmatrix} \quad \\ \quad \end{bmatrix} \begin{bmatrix} \ \\ \ \end{bmatrix} = [\ ].$$

C.2 Matrix identity and inverse

There is an **identity element for matrix multiplication**, usually denoted by I or, where the size must be explicitly shown for clarity, I_m. The identity is such that for an $m \times n$ matrix A,

$$IA = A,$$
$$AI = A,$$

C.3 Transpose and trace

where in the first case the identity is $m \times m$ and in the second case it is $n \times n$: without indication, the identity is usually taken to size itself as appropriate to be conformable. The $n \times n$ identity matrix is

$$I = \begin{bmatrix} 1 & 0 & 0 & \cdots & 0 \\ 0 & 1 & 0 & \cdots & 0 \\ 0 & 0 & 1 & \cdots & 0 \\ \vdots & & & & \\ 0 & 0 & 0 & \cdots & 1 \end{bmatrix} = \mathrm{diag}(1, 1, \ldots, 1).$$

The inverse of a matrix A (when it exists) is a matrix, denoted A^{-1}, such that

$$AA^{-1} = I \quad \text{and} \quad A^{-1}A = I.$$

If the inverse exists, the matrix must be square. (However, squareness is not sufficient for the existence of an inverse.) Unfortunately, for many matrices an inverse does not exist.

Example C.2.1 The matrix

$$A = \begin{bmatrix} 1 & 2 \\ 2 & 4 \end{bmatrix}$$

does not have an inverse. One way to show this is to presume an inverse $A^{-1} = \begin{bmatrix} a & b \\ c & d \end{bmatrix}$ and show that a consistent set of elements of the matrix cannot be found. Multiplying AA^{-1} and equating to the identity yields (among others) the two equations

$$a + 2c = 1,$$
$$2a + 4c = 0,$$

which are inconsistent: they cannot both be true. □

The matrix inverse is studied further in chapter 4.

To summarize the properties of addition and multiplication:

1. Matrix addition forms a commutative group: addition is associative and commutative, an additive identity exists, and each matrix has an additive inverse.
2. Matrix multiplication is associative.
3. There is a multiplicative identity, denoted as I.
4. There is not a multiplicative inverse for each matrix.
5. Matrix multiplication is not commutative.

Matrix multiplication is not a group operation, hence the arithmetic of matrices does not form a field.

C.3 Transpose and trace

Definition C.1 The **transpose** of an $m \times n$ matrix A is an $n \times m$ matrix A^T, where the (i, j)th element of A^T is A_{ji}: the rows of A become the columns of A^T. □

Example C.3.1 Suppose

$$A = \begin{bmatrix} 1 & 2 & 3 \\ 4 & 5 & 6 \\ 7 & 8 & 9 \end{bmatrix}.$$

Then
$$A^T = \begin{bmatrix} 1 & 4 & 7 \\ 2 & 5 & 8 \\ 3 & 6 & 9 \end{bmatrix}.$$

Definition C.2 A (real) matrix such that $A = A^T$ is said to be **symmetric**.

For complex matrices, it is often of interest to both transpose a matrix and conjugate its elements.

Definition C.3 The **Hermitian transpose** of an $m \times n$ matrix A is an $n \times m$ matrix A^H where the (i, j)th element of A^H is $\overline{A}_{ji}$. (The overline denotes complex conjugate.)

Example C.3.2 If
$$A = \begin{bmatrix} 1+2j & 3+4j & 2-4j \\ 1 & j & 4-2j \end{bmatrix}$$
then
$$A^H = \begin{bmatrix} 1-2j & 1 \\ 3-4j & -j \\ 2+4j & 4+2j \end{bmatrix}.$$

Definition C.4 A matrix such that $A = A^H$ is said to be **Hermitian** or Hermition symmetric.

The transpose of the product is the product of the transposes in reverse order:
$$(AB)^T = B^T A^T \tag{C.5}$$

$$(AB)^H = B^H A^H \tag{C.6}$$

Another useful fact is that the transpose of the inverse is the inverse of the transpose:
$$(A^{-1})^T = (A^T)^{-1} \tag{C.7}$$

$$(A^{-1})^H = (A^H)^{-1} \tag{C.8}$$

The inverse transpose is often written as A^{-H}.

The **trace** is an easily computed function of a square matrix:
$$\mathrm{tr}(A) = \sum_{i=1}^{n} a_{ii}$$

That is, the trace is the sum of the elements along the main diagonal. When AB and BA both exist,
$$\mathrm{tr}(AB) = \mathrm{tr}(BA) \tag{C.9}$$

Since we can write $a_{ii} = \mathbf{e}_i^T A \mathbf{e}_i$, where $\mathbf{e}_i$ is a unit vector, we can write using (C.4) and (C.3) as
$$\mathrm{tr}(A) = \sum_i \mathbf{e}_i^T A_{:,i} = \sum_i A_{i,:}^T \mathbf{e}_i^T.$$

The trace is sometimes used to define an inner product between matrices as $\langle A, B \rangle = \operatorname{tr}(A^T B)$.

C.4 Block (partitioned) matrices

In many applications it is convenient to partition a matrix into blocks. For example the matrix

$$A = \begin{bmatrix} 1 & 2 & 3 \\ 6 & 2 & 1 \\ 2 & 2 & 4 \end{bmatrix}$$

can be written as

$$A = \begin{bmatrix} A_{11} & A_{12} \\ A_{21} & A_{22} \end{bmatrix}$$

where the A_{ij} are the blocks

$$A_{11} = \begin{bmatrix} 1 & 2 \\ 6 & 2 \end{bmatrix} \quad A_{12} = \begin{bmatrix} 3 \\ 1 \end{bmatrix} \quad A_{21} = [2 \ \ 2] \quad A_{22} = [4].$$

Operations on block matrices go as if the blocks were scalars. For example,

$$\begin{bmatrix} A_{11} & A_{12} \\ A_{21} & A_{22} \end{bmatrix} + \begin{bmatrix} B_{11} & B_{12} \\ B_{21} & B_{22} \end{bmatrix} = \begin{bmatrix} A_{11} + B_{11} & A_{12} + B_{12} \\ A_{21} + B_{21} & A_{22} + B_{22} \end{bmatrix},$$

$$\begin{bmatrix} A_{11} & A_{12} \\ A_{21} & A_{22} \end{bmatrix} \begin{bmatrix} B_{11} & B_{12} \\ B_{21} & B_{22} \end{bmatrix} = \begin{bmatrix} A_{11}B_{11} + A_{12}B_{21} & A_{11}B_{21} + A_{12}B_{22} \\ A_{21}B_{11} + A_{22}B_{21} & A_{21}B_{21} + A_{22}B_{22} \end{bmatrix},$$

provided that all of the block products are conformable.

C.5 Determinants

A determinant of a square matrix is a scalar quantity that can provide some useful information about a matrix; for instance, about its invertability. However, since it provides only a scalar quantity, it cannot summarize all the useful information about a matrix.

C.5.1 Basic properties of determinants

To introduce the determinant, consider the simple matrix

$$A = \begin{bmatrix} a & b \\ c & d \end{bmatrix}.$$

It can be verified by direct multiplication that

$$\boxed{A^{-1} = \frac{1}{ad - bc} \begin{bmatrix} d & -b \\ -c & a \end{bmatrix}}$$

The quantity in the denominator $ad - bc$ is the determinant of this matrix. The determinant is denoted either by enclosing the matrix in straight lines, or by using "det" as an operator:

$$\det(A) = |A| = \begin{vmatrix} a & b \\ c & d \end{vmatrix}.$$

Permuting the order of the rows we obtain a new matrix

$$B = \begin{bmatrix} c & d \\ a & b \end{bmatrix}$$

with inverse

$$B^{-1} = \frac{1}{bc - ad} \begin{bmatrix} b & -d \\ -a & c \end{bmatrix}.$$

Observe that the determinant in this case is $bc - ad$, which is the negative of the determinant for A. Similarly if the columns of the matrix were permuted the determinant would change sign. This little demonstration reveals one of the properties of determinants:

1. The determinant changes sign when two rows are exchanged.
2. More generally, for an $n \times n$ matrix, if the rows or columns of a matrix are permuted by a permutation of order m, then the sign of the determinant changes by $(-1)^m$.

It is also straightforward to show for the 2×2 matrix, and is true in general, that

1. The determinant depends linearly on the first row:

$$\begin{vmatrix} a+e & b+f \\ c & d \end{vmatrix} = \begin{vmatrix} a & b \\ c & d \end{vmatrix} + \begin{vmatrix} e & f \\ c & d \end{vmatrix},$$

$$\begin{vmatrix} xa & xb \\ c & d \end{vmatrix} = x \begin{vmatrix} a & b \\ c & d \end{vmatrix}.$$

2. The determinant of the identity matrix is 1.

These three properties are axiomatically sufficient to establish exactly what the determinant is. Other important properties may be derived from the first three.

1. If two rows are the same, the determinant is 0.
2. The determinant is unchanged by elementary row operations. That is, if a multiple of one row is subtracted from another row, the determinant remains the same:

$$\begin{vmatrix} a - xc & b - xd \\ c & d \end{vmatrix} = \begin{vmatrix} a & b \\ c & d \end{vmatrix}.$$

3. If one of the rows is 0, then the determinant is 0.
4. If A is triangular (either upper or lower), then

$$\det(A) = \prod_{i=1}^{n} a_i i.$$

5. The determinant of the product of *square* matrices AB is the product of the determinants:

$$\boxed{\det(AB) = \det(A) \det(B)}$$

From this we determine that $\det(A^{-1}) = 1/\det(A)$.

6. If A is singular, then $\det(A) = 0$. If A is invertible, then $\det(A) \neq 0$. This provides a straightforward test for the invertability of a matrix.
7. The transpose A has the same determinant as A^T:

$$\boxed{\det(A) = \det(A^T)}$$

8. Based upon the previous result and property 5, it follows immediately that for an orthogonal matrix Q,

$$\det(Q) = 1.$$

C.5.2 Formulas for the determinant

The basic properties can be used to build up formulas for the determinant. This will be demonstrated for 2 × 2 matrices, and can be generalized. Employing the linearity property row-by-row we obtain

$$\begin{vmatrix} a & b \\ c & d \end{vmatrix} = \begin{vmatrix} a & 0 \\ c & 0 \end{vmatrix} + \begin{vmatrix} a & 0 \\ 0 & d \end{vmatrix} + \begin{vmatrix} 0 & b \\ c & 0 \end{vmatrix} + \begin{vmatrix} 0 & b \\ 0 & d \end{vmatrix} \quad \text{(by linearity)}$$

$$= \begin{vmatrix} a & 0 \\ 0 & d \end{vmatrix} + \begin{vmatrix} 0 & b \\ c & 0 \end{vmatrix} \quad \text{(since det(columns of 0) = 0)}$$

$$= ad \begin{vmatrix} 1 & 0 \\ 0 & 1 \end{vmatrix} - bc \begin{vmatrix} 0 & 1 \\ 1 & 0 \end{vmatrix} \quad \text{(by linearity of rows)}.$$

In the more general case, an $n \times n$ matrix can be expanded using linearity on each row into n^n determinants. All but $n!$ of these will have repeated columns and hence have determinant 0. From the $n!$ determinants that remain, the constants can be factored out, leaving only permutation matrices. A further illustration for the 3×3 determinant is

$$\det(A) = \begin{vmatrix} a_{11} & a_{12} & a_{13} \\ a_{21} & a_{22} & a_{23} \\ a_{31} & a_{32} & a_{33} \end{vmatrix}$$

$$= a_{11}a_{22}a_{33} \begin{vmatrix} 1 & & \\ & 1 & \\ & & 1 \end{vmatrix} + a_{12}a_{23}a_{31} \begin{vmatrix} & 1 & \\ & & 1 \\ 1 & & \end{vmatrix} + a_{13}a_{21}a_{32} \begin{vmatrix} & & 1 \\ 1 & & \\ & 1 & \end{vmatrix}$$

$$+ a_{11}a_{23}a_{32} \begin{vmatrix} 1 & & \\ & & 1 \\ & 1 & \end{vmatrix} + a_{12}a_{21}a_{33} \begin{vmatrix} & 1 & \\ 1 & & \\ & & 1 \end{vmatrix} + a_{13}a_{22}a_{31} \begin{vmatrix} & & 1 \\ & 1 & \\ 1 & & \end{vmatrix}.$$

The first permutation matrix comes from the columns (1,2,3); no rows are interchanged. The second matrix comes from the columns (2,3,1); four row interchanges are required, and so forth. The determinant of a permutation matrix is ± 1, with the sign depending on the parity of the number of permutations. From this, the 3×3 determinant can be determined as

$$\boxed{\det(A) = a_{11}a_{22}a_{33} + a_{12}a_{23}a_{31} + a_{13}a_{21}a_{32} - a_{11}a_{23}a_{32} - a_{12}a_{21}a_{33} - a_{13}a_{22}a_{31}}$$

For a general $n \times n$ matrix, the determinant can be written as

$$\boxed{\det(A) = \sum_\sigma \left(a_{1\sigma_1} \; a_{2\gamma_2} \; \cdots \; a_{2\gamma_n} \right)(-1)^{\tau(\sigma)}} \quad \text{(C.10)}$$

where the sum is taken over the $n!$ permutations of n integers, represented as $\sigma = (\sigma_1, \sigma_2, \ldots, \sigma_n)$, and $\tau(\sigma)$ is the number of transpositions in the permutation σ.

A more computationally oriented formula for the determinant is **expansion by cofactors**. This begins by expanding the determinant using linearity, as

$$\begin{vmatrix} a_{11} & a_{12} & a_{13} \\ a_{21} & a_{22} & a_{23} \\ a_{31} & a_{32} & a_{33} \end{vmatrix} = \begin{vmatrix} a_{11} & & \\ & a_{22} & a_{23} \\ & a_{32} & a_{33} \end{vmatrix} + \begin{vmatrix} & a_{12} & \\ a_{21} & & a_{23} \\ a_{31} & & a_{33} \end{vmatrix} + \begin{vmatrix} & & a_{13} \\ a_{21} & a_{22} & \\ a_{31} & a_{32} & \end{vmatrix}. \quad \text{(C.11)}$$

By paying attention to which terms vanish now in (C.11), it can be shown that

$$\begin{vmatrix} a_{11} & a_{12} & a_{13} \\ a_{21} & a_{22} & a_{23} \\ a_{31} & a_{32} & a_{33} \end{vmatrix} = a_{11} \begin{vmatrix} a_{22} & a_{23} \\ a_{32} & a_{33} \end{vmatrix} - a_{12} \begin{vmatrix} a_{21} & a_{23} \\ a_{31} & a_{33} \end{vmatrix} + a_{13} \begin{vmatrix} a_{21} & a_{22} \\ a_{31} & a_{32} \end{vmatrix}$$

$$\triangleq a_{11} \det(M_{11}) - a_{12} \det(M_{12}) + a_{13} \det(M_{13}). \qquad (C.12)$$

The matrix M_{ij} is the submatrix formed by deleting the ith row and the jth column of A. The (i, j)th **minor** is the determinant of M_{ij}. (More generally, the determinant of any square submatrix of A is said to be a minor of A. A **principal minor** is any minor whose diagonal elements are also the diagonal element of A.)

The **cofactor** A_{ij} is formed by taking the determinant of the minor M_{ij} with the sign as in (C.12),

$$A_{ij} = (-1)^{i+j} \det(M_{ij}).$$

Example C.5.1 Let

$$A = \begin{bmatrix} a_{11} & a_{12} \\ a_{21} & a_{22} \end{bmatrix}.$$

Then the minors and cofactors are

$$M_{11} = a_{22} \quad M_{12} = a_{21} \quad M_{21} = a_{21} \quad M_{22} = a_{11},$$
$$A_{11} = a_{22} \quad A_{12} = -a_{21} \quad A_{21} = -a_{21} \quad A_{22} = a_{11},$$

respectively. □

Using the cofactor notation, the determinant may be written as

$$\det(A) = a_{11}A_{11} + a_{12}A_{12} + a_{13}A_{13}.$$

More generally, the determinant can be computed using the cofactors from any row:

$$\boxed{\det(A) = \sum_{j=1}^{n} a_{ij} A_{ij}} \qquad (C.13)$$

for some row i. It is common to choose the row with the most zeros, to reduce the number of cofactors that have to be computed.

Example C.5.2 Find the determinant of the matrix

$$A = \begin{bmatrix} 4 & 3 & 3 & 1 \\ 0 & 0 & 2 & 4 \\ 5 & 2 & -1 & -2 \\ 4 & 6 & 9 & 2 \end{bmatrix}.$$

The second row has the most zeros, and the determinant can be expanded as

$$\det(A) = 2A_{23} + 4A_{24} = -2 \det(M_{23}) + 4 \det(M_{24})$$

$$= -2 \begin{vmatrix} 4 & 3 & 1 \\ 5 & 2 & -2 \\ 4 & 6 & 2 \end{vmatrix} + 4 \begin{vmatrix} 4 & 3 & 3 \\ 5 & 2 & -1 \\ 4 & 6 & 9 \end{vmatrix}.$$

□

C.5.3 Determinants and matrix inverses

The **adjugate** of a matrix A (not to be confused with the adjoint) is the transpose of the matrix formed of the cofactors,
$$\text{adj}(A) = [A_{ij}]^T.$$

Example C.5.3 For the matrix of example C.5.1, the adjugate is
$$\text{adj}(A) = \begin{bmatrix} a_{22} & -a_{21} \\ -a_{12} & a_{11} \end{bmatrix}.$$

Observe that
$$A\,\text{adj}(A) = \begin{bmatrix} a_{11}a_{22} - a_{12}a_{21} & 0 \\ 0 & a_{11}a_{22} - a_{12}a_{21} \end{bmatrix} = \det(A)I.\qquad\square$$

What holds for the 2×2 case holds in general: The adjugate is designed precisely so that it contains those portions of A that are the algebraic complements of the determinant of A, so that $A\,\text{adj}(A)$ gives the determinant on the diagonal.

Based on this observation, if A is invertible then
$$A^{-1} = \frac{\text{adj}(A)}{\det(A)}.$$

From this, we observe that
$$\det(A)I = A\,\text{adj}(A). \tag{C.14}$$

C.6 Exercises

C.1-1 Show that (C.5) is true.

C.1-2 To exercise understanding of unit vectors and their notation, show that the following are true.
  (a) $A_{i,:} = \sum_j a_{ij} \mathbf{e}_j$.
  (b) $A_{:,j} = \sum_i a_{ij} \mathbf{e}_i$.
  (c) $A_{i,:}^T = \sum_j a_{ij} \mathbf{e}_j^T$.
  (d) $A = \sum_j A_{:,j} \mathbf{e}_j^T$.
  (e) $A = \sum_i \mathbf{e}_i A_{i,:}^T$.
  (f) $E_{ij} A = \mathbf{e}_i A_{j,:}^T$.
  (g) $A E_{ij} = A_{:,j} \mathbf{e}_j^T$.
  (h) $E_{ij} A E_{rs} = a_{jr} E_{is}$.
  (i) $\delta_{ij} = \mathbf{e}_j^T \mathbf{e}_i$.

C.1-3 Show that $\text{tr}(E_{rs} X) = X_{sr} = \text{tr}(X E_{rs})$.

C.1-4 Show that
$$\text{tr}\left(E_{rs}^T X\right) = X_{rs}. \tag{C.15}$$

C.1-5 Show that
  (a) $\text{tr}(E_{rs}^T A) = a_{rs}$ and hence that
$$\begin{bmatrix} \text{tr}\left(E_{11}^T A\right) & \text{tr}\left(E_{12}^T A\right) & \cdots & \text{tr}\left(E_{1m}^T A\right) \\ \text{tr}\left(E_{21}^T A\right) & \text{tr}\left(E_{22}^T A\right) & \cdots & \text{tr}\left(E_{2m}^T A\right) \\ \vdots & & & \\ \text{tr}\left(E_{m1}^T A\right) & \text{tr}\left(E_{m2}^T A\right) & \cdots & \text{tr}\left(E_{mm}^T A\right) \end{bmatrix} = A.$$

(b) Show that $\text{tr}(E_{rs}A) = a_{sr}$ and hence that

$$\begin{bmatrix} \text{tr}(E_{11}A) & \text{tr}(E_{12}A) & \cdots & \text{tr}(E_{1m}A) \\ \text{tr}(E_{21}A) & \text{tr}(E_{22}A) & \cdots & \text{tr}(E_{2m}A) \\ \vdots & & & \\ \text{tr}(E_{m1}A) & \text{tr}(E_{m2}A) & \cdots & \text{tr}(E_{mm}A) \end{bmatrix} = A^T.$$

C.3-6 Show that (C.7) is true.

C.3-7 Show that (C.9) is true.

C.7 References

This material is summarized from [333] and [114].

Appendix D

Random Processes

In this appendix, we review the basic concepts and definitions of random processes. Notation for both continuous-time and discrete-time random processes is presented, insofar as possible, in parallel, with the continuous-time notation appearing on the *left* of a display, and the discrete-time notation appearing on the *right*. This should help to strengthen the awareness of the connections between discrete-time and continuous-time signals.

D.1 Definitions of means and correlations

A **random process** $x(t)$ or $x[t]$ (for continuous time) or x_t (for discrete time) is a family of functions, real or complex, scalar or vector, defined on a probability space. At specified times, such as t_1, t_2, and so forth, the samples $x(t_1), x(t_2)$, and so forth, are random variables (or random vectors).

The **mean** of a random process is $Ex(t)$, and may be a function of the time index. In these definitions, we will denote the mean by

$$\mu_x(t) = Ex(t) \qquad \mu_x[t] = Ex[t].$$

The **autocorrelation** of a random process is defined by

$$r_{xx}(t_1, t_2) = Ex(t_1)\overline{x}(t_2) \qquad r_{xx}[t_1, t_2] = Ex[t_1]\overline{x}[t_2],$$

where the subscript on r_{xx} indicates the random process whose autocorrelation is computed. This is also written as

$$r_x(t_1, t_2) \qquad r_x[t_1, t_2],$$

using a single subscript. For a vector random process $\mathbf{x}_t$,

$$R_{xx}(t_1, t_2) = E\mathbf{X}(t_1)\overline{\mathbf{X}}(t_2)^T \qquad R_{xx}[t_1, t_2] = E\mathbf{X}[t_1]\overline{\mathbf{X}}(t_2)^T,$$

which is a matrix.

The **autocovariance** is the autocorrelation of a centralized (mean-removed) random process,

$$c_{xx}(t_1, t_2) = E(x(t_1) - \mu_x(t_1))(\overline{x}(t_2) - \overline{\mu}_x(t_2))$$
$$= Ex(t_1)\overline{x}(t_2) - \mu_x(t_1)\overline{\mu}_x(t_2),$$

$$c_{xx}[t_1, t_2] = E(x[t_1] - \mu_x[t_1])(\overline{x}[t_2] - \overline{\mu}_x[t_2])$$
$$= Ex[t_1]\overline{x}[t_2] - \mu_x[t_1]\overline{\mu}_x[t_2].$$

The cross-correlation between two random processes x_t and y_t is

$$r_{xy}(t_1, t_2) = Ex(t_1)\overline{y}(t_2) \qquad r_{xy}[t_1, t_2] = Ex[t_1]\overline{y}[t_2].$$

For vector random processes we define the cross-correlation as

$$R_{xy}(t_1, t_2) = E\mathbf{x}(t_1)\overline{\mathbf{y}}^T(t_2) \qquad R_{xy}[t_1, t_2] = E\mathbf{x}[t_1]\overline{\mathbf{y}}^T[t_2]$$

and define the cross-covariance as

$$c_{xy}(t_1, t_2) = E(x(t_1) - \mu_x(t_1))(\overline{y}(t_2) - \overline{\mu}_y(t_2))$$
$$= Ex(t_1)\overline{y}(t_2) - \mu_x(t_1)\overline{\mu}_y(t_2),$$

$$c_{xy}[t_1, t_2] = E(x[t_1] - \mu_x[t_1])(\overline{y}[t_2] - \overline{\mu}_y[t_2])$$
$$= Ex[t_1]\overline{y}[t_2] - \mu_x[t_1]\overline{\mu}_y[t_2].$$

D.2 Stationarity

A stochastic process is **strict-sense stationary** if *all* of its statistical properties are invariant with respect to shifts of the time origin. For example, the joint distribution of the random vector $[x_{t_1}, x_{t_2}, \ldots, x_{t_m}]^T$ is the same as the joint distribution of the random vector $[x(t_1 + \tau), x(t_2 + \tau), \ldots, x(t_m + \tau)]$, for any dimensionality m and any shift τ. In particular, the mean and the correlation and covariance functions must all be invariant to shift; that is, the mean must be constant (with similar notation for discrete time).

A stochastic process is **wide-sense stationary** if its mean is constant, and the autocorrelation depends only on the time difference $\tau = t_1 - t_2$. By an abuse of notation, we write $r_x(t_1, t_2) = r_x(\tau)$, so that

$$r_{xx}(\tau) = Ex(t+\tau)\overline{x}(t) \qquad r_{xx}[\tau] = Ex[t+\tau]\overline{x}[t].$$

By a simple change of variable, equivalent forms are

$$r_xx(\tau) = Ex(t)\overline{x}(t-\tau) \qquad r_x(\tau) = E(t+\tau/2)\overline{x}(t+\tau/2),$$

with similar changes for discrete time. The auto-covariance is similarly defined for wide-sense stationary processes.

The autocorrelation function has the symmetry

$$r_{xx}(\tau) = \overline{r}_{xx}(-\tau) \qquad r_{xx}[\tau] = \overline{r}_{xx}[-\tau]. \qquad (D.1)$$

For real random processes, the autocorrelation function is *even*:

$$r_{xx}(\tau) = r_{xx}(-\tau) \qquad r_{xx}[\tau] = r_{xx}[-\tau].$$

Two processes $x(t)$ and $y(t)$ are **jointly stationary** if each is stationary and their cross-correlation $r_{xy}(t_1, t_2)$ depends only on $\tau = t_1 - t_2$. We then write $r_{xy}(\tau) = r_{xy}(t_1, t_2)$, or

$$r_{xy}(\tau) = Ex(t+\tau)\overline{y}_t = Ex(t)\overline{y}(t-\tau),$$

with similar notation for discrete-time.

In the general case, a strict-sense stationary process is wide-sense stationary, but a wide-sense stationary process is not necessarily strict-sense stationary. However, a Gaussian random process (in which each sample x_t is distributed as a Gaussian random variable) that is wide-sense stationary is also strict-sense stationary, since a Gaussian is characterized by only two moments.

D.3 Power spectral-density functions

For both continuous-time and discrete-time random processes, the **power spectral-density** of a wide-sense stationary random process is the Fourier transform of the autocorrelation function. For continuous-time random processes,

$$S_{xx}(\omega) = \int_{-\infty}^{\infty} r_{xx}(\tau) e^{-j\omega\tau} \, d\tau.$$

This is also denoted as $S_x(\omega)$. Because of symmetry (D.1), the PSD is a *real* function of ω. Furthermore, since real power cannot be negative, the PSD must satisfy $S_x(\omega) \geq 0$ for all ω. It is straightforward to show that the average power of a random process is

$$E|x_t|^2 = r_x(0) = \frac{1}{2\pi} \int_{-\infty}^{\infty} S(\omega) \, d\omega.$$

The **cross-spectral density** is

$$S_{xy}(\omega) = \int_{-\infty}^{\infty} r_{xy}(\tau) e^{-j\omega\tau} \, d\tau.$$

We shall also have occasion to use the Laplace transform of the autocorrelation function. We thus define (by an abuse of notation)

$$S_{xx}(s) = \int_{-\infty}^{\infty} r_{xx}(\tau) e^{-s\tau} \, d\tau,$$

$$S_{xy}(s) = \int_{-\infty}^{\infty} r_{xy}(\tau) e^{-s\tau} \, d\tau.$$

This is a bilateral Laplace transform.

For a real random process, since the autocovariance is real and even, its bilateral Laplace transform will be even; that is,

$$S_{xx}(s) = S_{xx}(-s).$$

Furthermore, when $s = j\omega$, the PSD will have the property

$$S_{xx}(-\omega) = S_{xx}^*(\omega).$$

For discrete-time functions, we have the PSD

$$S_{xx}(\omega) = \sum_k r_{xx}[k] e^{-j\omega k}.$$

This is also frequently written $S_{xx}(e^{j\omega})$. Similarly,

$$S_{xy}(\omega) = S_{xy}(e^{j\omega}) = \sum_k r_{xy}[k] e^{-j\omega k}.$$

Similarly, we define

$$S_{xx}(z) = \sum_k r_{xx}[k] z^{-k},$$

$$S_{xy}(z) = \sum_k r_{xy}[k] z^{-k}.$$

For a real random process,

$$S_{xx}(1/z) = S_{xx}(z)$$

and

$$S_{xx}(-\omega) = S_{xx}^*(\omega).$$

D.4 Linear systems with stochastic inputs

D.4.1 Continuous-time signals and systems

Let $x(t)$ be the input to a linear system having impulse response $h(t)$, and let $y(t)$ be its output. Then

$$y(t) = x(t) * h(t) = \int_{-\infty}^{\infty} x(t-\alpha) h(\alpha) \, d\alpha.$$

If $x(t)$ is a random process, then $y(t)$ is also, with mean

$$Ey(t) = \int_{-\infty}^{\infty} Ex(t-\alpha) h(\alpha) \, d\alpha = \mu_x(t) * h(t).$$

The cross-correlation is

$$r_{xy}(t_1, t_2) = \int E[x(t_1)\overline{x}(t_2 - \alpha)] \overline{h}(\alpha) \, d\alpha$$

$$= \int \int r_{xx}(t_1, t_2 - \alpha) \overline{h}(\alpha) \, d\alpha \tag{D.2}$$

$$= r_{xx}(t_1, t_2) * \overline{h}(t_2). \tag{D.3}$$

The output autocorrelation is

$$r_{yy}(t_1, t_2) = \int \int E[x(t_1 - \alpha)\overline{x}(t_2 - \beta)] h(\alpha) \overline{h}(\beta) \, d\alpha d\beta$$

$$= \int \int r_{xx}(t_1 - \alpha, t_2 - \beta) h(\alpha) \overline{h}(\beta) \, d\alpha d\beta$$

$$= h(t_1) * r_{xx}(t_1, t_2) \overline{h}(t_2).$$

For wide-sense stationary random processes, we have

$$\mu_y = \mu_x \int h(\alpha) \, d\alpha.$$

For the cross-correlation function, in the integral (D.2) we note that $r_{xx}(t_1, t_2 - \alpha) = r_{xx}(t_1 - t_2 + \alpha)$ for a wide-sense stationary process, so that for $\tau = t_1 - t_2$,

$$r_{xy}(\tau) = \int r_{xx}(\tau + \alpha) \overline{h}(\alpha) \, d\alpha$$

$$= \int r_{xx}(\tau - \alpha) \overline{h}(-\alpha) = r_{xx}(\tau) * \overline{h}(-\tau).$$

Similarly, we find that

$$r_{yy}(\tau) = r_{xy}(\tau) * h(\tau) = r_{xx}(\tau) * h(\tau) * \overline{h}(-\tau).$$

If $h(t)$ has Fourier transform $H(\omega)$ and Laplace transform $H(s)$ (using the familiar abuse of notation), then

$$\overline{h}(-t) \leftrightarrow \overline{H}(\omega) \qquad \text{(Fourier)},$$

$$\overline{h}(-t) \leftrightarrow \overline{H}(-\overline{s}) \qquad \text{(Laplace)}.$$

Then (in the Fourier transform context)

$$S_{xy}(\omega) = S_{xx}(\omega) \overline{H}(\omega) \qquad S_{yy}(\omega) = S_{xy}(\omega) H(\omega) \qquad S_{yy}(\omega) = S_{xx}(\omega) |H(\omega)|^2$$

and (in the Laplace transform context)

$$S_{xy}(s) = S_{xx}(s)\overline{H}(-\bar{s}) \qquad S_{yy}(s) = S_{xy}(s)H(s) \qquad S_{yy}(s) = S_{xx}H(s)\overline{H}(-\bar{s}).$$

If $h(t)$ is a *real* function, then $\overline{h}(-t) = h(-t)$ and

$$h(-t) \leftrightarrow H(-\omega) \qquad \text{(Fourier)},$$
$$h(-t) \leftrightarrow H(-s) \qquad \text{(Laplace)}.$$

D.4.2 Discrete-time signals and systems

Let $x[t]$ be the input and $y[t]$ be the output of a linear system with impulse response $h[t]$. Then

$$Ey[t] = \mu_x[t] * h[t];$$

for a stationary process, $Ey[t] = \mu_x \sum_t h[t]$.

The cross-correlation can be written (reasoning as before)

$$r_{xy}[t_1, t_2] = r_{xx}[t_1, t_2] * \overline{h}[t_2];$$

for a stationary process,

$$r_{xy}[\tau] = r_{xx}[\tau] * \overline{h}[-\tau].$$

The output autocorrelation is

$$r_{yy}[t_1, t_2] = r_{xx}[t_1, t_2] * h[t_1] * \overline{h}[t_2],$$

and for a stationary random process,

$$r_{yy}[\tau] = r_{xx}[\tau] * h(\tau) * \overline{h}[-\tau].$$

If $h[t]$ has discrete-time Fourier transform $H(e^{j\omega}) = H(\omega)$ (by the familiar abuse of notation) and Z-transform $H(z)$, then

$$\overline{h}[-t] \leftrightarrow \overline{H}(\omega) \qquad \text{(Fourier)},$$
$$\overline{h}[-t] \leftrightarrow \overline{H}(1/\bar{z}) \qquad \text{(Z)}.$$

Then (in the Fourier transform context)

$$S_{xy}(\omega) = S_{xx}(\omega)\overline{H}(\omega) \qquad S_{yy}(\omega) = S_{xy}(\omega)H(\omega) \qquad S_{yy}(\omega) = S_{xx}(\omega)|H(\omega)|^2$$

and (in the Z-transform context)

$$S_{xy}(z) = S_{xx}(z)\overline{H}(1/\bar{z}) \qquad S_{yy}(z) = S_{xy}(z)H(z) \qquad S_{yy}(z) = S_{xx}(z)H(z)\overline{H}(1/\bar{z}).$$

If $h[t]$ is a *real* function, then $\overline{h}[-t] = h[-t]$ and

$$h[-t] \leftrightarrow H(-\omega) = H(e^{-j\omega}) \qquad \text{(Fourier)},$$
$$h[-t] \leftrightarrow H(1/z) \qquad \text{(Z)}.$$

D.5 References

Our summary of concepts associated with random processes has been drawn largely from the summary provided in [248].

Appendix E

Derivatives and Gradients

E.1 Derivatives of vectors and scalars with respect to a real vector

The derivative of a scalar function y with respect to a real vector $\mathbf{x} \in \mathbb{R}^n$ is the *gradient*, and is defined as

$$\frac{\partial y}{\partial \mathbf{x}} = \begin{bmatrix} \frac{\partial y}{\partial x_1} \\ \frac{\partial y}{\partial x_2} \\ \vdots \\ \frac{\partial y}{\partial x_n} \end{bmatrix}.$$

This is also written as $\nabla_x y$.

The derivative of the real vector $\mathbf{y} \in \mathbb{R}^m$ with respect to a real vector $\mathbf{x} \in \mathbb{R}^n$ is defined as

$$\frac{\partial \mathbf{y}}{\partial \mathbf{x}} = \begin{bmatrix} \frac{\partial y_1}{\partial x_1} & \frac{\partial y_2}{\partial x_1} & \cdots & \frac{\partial y_m}{\partial x_1} \\ \frac{\partial y_1}{\partial x_2} & \frac{\partial y_2}{\partial x_2} & \cdots & \frac{\partial y_m}{\partial x_2} \\ \vdots & & & \\ \frac{\partial y_1}{\partial x_n} & \frac{\partial y_2}{\partial x_n} & \cdots & \frac{\partial y_m}{\partial x_n} \end{bmatrix} = \begin{bmatrix} \frac{\partial y_j}{\partial x_i} \end{bmatrix}.$$

Example E.1.1 Let $\mathbf{y} \in \mathbb{R}^3$ and $\mathbf{x} \in \mathbb{R}^2$, and define the following functions:

$$y_1 = x_1^3 - 2x_2^2,$$
$$y_2 = x_1 x_2,$$
$$y_3 = x_2 + 4.$$

Then

$$\frac{\partial \mathbf{y}}{\partial \mathbf{x}} = \begin{bmatrix} \frac{\partial y_1}{\partial x_1} & \frac{\partial y_2}{\partial x_1} & \frac{\partial y_3}{\partial x_1} \\ \frac{\partial y_1}{\partial x_2} & \frac{\partial y_2}{\partial x_2} & \frac{\partial y_3}{\partial x_2} \end{bmatrix} = \begin{bmatrix} 3x_1^2 & x_2 & 0 \\ -2x_2 & x_1 & 1 \end{bmatrix}. \tag{E.1}$$

□

A **chain rule** for vector derivatives may be derived using the foregoing definitions. If $\mathbf{z} \in \mathbb{R}^r$,

$$\mathbf{z} = \mathbf{y}(\mathbf{x}),$$

E.1 Derivatives of Vectors and Scalars with Respect to a Real Vector

then

$$\frac{\partial z_i}{\partial x_j} = \sum_{k=1}^{m} \frac{\partial z_i}{\partial y_k} \frac{\partial y_k}{\partial x_j}.$$

In matrix form,

$$\frac{\partial \mathbf{z}}{\partial \mathbf{x}} = \begin{bmatrix} \frac{\partial z_1}{\partial x_1} & \frac{\partial z_2}{\partial x_1} & \cdots & \frac{\partial z_r}{\partial x_1} \\ \frac{\partial z_1}{\partial x_2} & \frac{\partial z_2}{\partial x_2} & \cdots & \frac{\partial z_r}{\partial x_2} \\ \vdots & & & \\ \frac{\partial z_1}{\partial x_n} & \frac{\partial z_2}{\partial x_n} & \cdots & \frac{\partial z_r}{\partial x_n} \end{bmatrix} = \begin{bmatrix} \sum_{k=1}^{m} \frac{\partial z_1}{\partial y_k} \frac{\partial y_k}{\partial x_1} & \sum_{k=1}^{m} \frac{\partial z_2}{\partial y_k} \frac{\partial y_k}{\partial x_1} & \cdots & \sum_{k=1}^{m} \frac{\partial z_r}{\partial y_k} \frac{\partial y_k}{\partial x_1} \\ \sum_{k=1}^{m} \frac{\partial z_1}{\partial y_k} \frac{\partial y_k}{\partial x_2} & \sum_{k=1}^{m} \frac{\partial z_2}{\partial y_k} \frac{\partial y_k}{\partial x_2} & \cdots & \sum_{k=1}^{m} \frac{\partial z_r}{\partial y_k} \frac{\partial y_k}{\partial x_2} \\ \vdots & & & \\ \sum_{k=1}^{m} \frac{\partial z_1}{\partial y_k} \frac{\partial y_k}{\partial x_n} & \sum_{k=1}^{m} \frac{\partial z_2}{\partial y_k} \frac{\partial y_k}{\partial x_n} & \cdots & \sum_{k=1}^{m} \frac{\partial z_r}{\partial y_k} \frac{\partial y_k}{\partial x_n} \end{bmatrix}$$

$$= \begin{bmatrix} \frac{\partial y_1}{\partial x_1} & \frac{\partial y_2}{\partial x_1} & \cdots & \frac{\partial y_m}{\partial x_1} \\ \frac{\partial y_1}{\partial x_2} & \frac{\partial y_2}{\partial x_2} & \cdots & \frac{\partial y_m}{\partial x_2} \\ \vdots & & & \\ \frac{\partial y_1}{\partial x_n} & \frac{\partial y_2}{\partial x_n} & \cdots & \frac{\partial y_m}{\partial x_n} \end{bmatrix} \begin{bmatrix} \frac{\partial z_1}{\partial y_1} & \frac{\partial z_2}{\partial y_1} & \cdots & \frac{\partial z_r}{\partial y_1} \\ \frac{\partial z_1}{\partial y_2} & \frac{\partial z_2}{\partial y_2} & \cdots & \frac{\partial z_r}{\partial y_2} \\ \vdots & & & \\ \frac{\partial z_1}{\partial y_m} & \frac{\partial z_2}{\partial y_m} & \cdots & \frac{\partial z_r}{\partial y_m} \end{bmatrix}.$$

Thus

$$\frac{\partial \mathbf{z}}{\partial \mathbf{x}} = \frac{\partial \mathbf{y}}{\partial \mathbf{x}} \frac{\partial \mathbf{z}}{\partial \mathbf{y}}. \tag{E.2}$$

Example E.1.2 Suppose that functions are defined as in example E.1.1 and, additionally, that

$$z_1 = y_1 y_2,$$
$$z_2 = y_1 - y_2 + y_3.$$

Then

$$\frac{\partial \mathbf{z}}{\partial \mathbf{y}} = \begin{bmatrix} y_2 & 1 \\ y_1 & -1 \\ 0 & 1 \end{bmatrix},$$

so that, from (E.1) and (E.2),

$$\frac{\partial \mathbf{z}}{\partial \mathbf{x}} = \begin{bmatrix} 3x_1^2 & x_2 & 0 \\ -2x_2 & x_1 & 1 \end{bmatrix} \begin{bmatrix} y_2 & 1 \\ y_1 & -1 \\ 0 & 1 \end{bmatrix} = \begin{bmatrix} x_2 y_1 + 3x_1^2 y_2 & 3x_1^2 - x_2 \\ x_1 y_1 - 2x_2 y_2 & 1 - x_1 - 2x_2 \end{bmatrix}.$$

Substituting from (E.1.1), we obtain

$$\frac{\partial \mathbf{z}}{\partial \mathbf{x}} = \begin{bmatrix} 3x_1^3 x_2 + x_2 \left(x_1^3 - 2x_2^2\right) & 3x_1^2 - x_2 \\ -2x_1 x_2^2 + x_1 (x_1^3 - 2x_2^2) & 1 - x_1 - 2x_2 \end{bmatrix}. \qquad \square$$

E.1.1 Some important gradients

Linear and quadratic functions of multiple variables arise frequently in practice. We derive here formulas for the gradients of these functions. A more complete listing of gradient formulas is provided in appendix E. The formulas are for the gradients with respect to $\mathbf{x}$ of the linear functions

$$\mathbf{x}^H \mathbf{d} \quad \text{and} \quad \mathbf{d}^H \mathbf{x}$$

and

$$\mathbf{x}^H R \mathbf{x},$$

where R is a Hermitian symmetric matrix, $R^H = R$, with separate consideration for the case that $\mathbf{x}$ is real and the case that $\mathbf{x}$ is complex.

Gradient of a linear function of a real vector

The gradient of $\mathbf{x}^T \mathbf{d}$ is computed simply by

$$\nabla_{\mathbf{x}} \mathbf{x}^T \mathbf{d} = \begin{bmatrix} \frac{\partial}{\partial x_1} \\ \frac{\partial}{\partial x_2} \\ \vdots \\ \frac{\partial}{\partial x_m} \end{bmatrix} \sum_{i=1}^m x_i d_i = \begin{bmatrix} d_1 \\ d_2 \\ \vdots \\ d_m \end{bmatrix}$$

so that

$$\nabla_{\mathbf{x}} \mathbf{x}^T \mathbf{d} = \mathbf{d}. \tag{E.3}$$

Similarly,

$$\nabla_{\mathbf{x}} \mathbf{d}^T \mathbf{x} = \mathbf{d}. \tag{E.4}$$

Gradient of a quadratic function of a real vector

We want to compute

$$\nabla_{\mathbf{x}} \mathbf{x}^T R \mathbf{x} = \begin{bmatrix} \frac{\partial}{\partial x_1} \\ \frac{\partial}{\partial x_2} \\ \vdots \\ \frac{\partial}{\partial x_m} \end{bmatrix} \sum_{i=1}^m \sum_{j=1}^m x_i x_j R_{ij}$$

where R is symmetric. Let us consider the first component as typical:

$$\frac{\partial}{\partial x_1} \sum_{i=1}^m \sum_{j=1}^m x_i x_j R_{ij}.$$

When $i \neq 1$ and $j \neq 1$, there is no contribution to the gradient. When $i = 1$ and $j \neq 1$, we get

$$\frac{\partial}{\partial x_1} \sum_{j \neq 1}^m x_i x_j R_{ij} \bigg|_{i=1} = \sum_{j \neq 1} x_j R_{1j}. \tag{E.5}$$

Similarly when $i \neq 1$ and $j = 1$ we get

$$\frac{\partial}{\partial x_1} \sum_{i \neq 1}^m x_i x_j R_{ij} \bigg|_{j=1} = \sum_{i \neq 1} x_i R_{i1} = \sum_{j \neq 1} x_j R_{1j}, \tag{E.6}$$

where the last equality follows from the symmetry of R. When $i = j = 1$ we get

$$\frac{\partial}{\partial x_1} x_i x_j R_{ij} \bigg|_{i=1, j=1} = \frac{\partial}{\partial x_1} x_1^2 R_{11} = 2 x_1 R_{11}. \tag{E.7}$$

Adding (E.5), (E.6), and (E.7), we get

$$\frac{\partial}{\partial x_1} \sum_{i=1}^m \sum_{j=1}^m x_i x_j R_{ij} = 2 \sum_{i=1}^m R_{1j} x_j$$

E.2 Derivatives of Real-Valued Functions of Real Matrices

which is the first element in the product $2R\mathbf{x}$. Now, stacking the partial derivatives, we obtain

$$\nabla_{\mathbf{x}} \mathbf{x}^T R \mathbf{x} = 2R\mathbf{x}. \qquad (E.8)$$

Some commonly-used derivatives which may be developed using these rules are shown in table E.1.

Table E.1: Derivative of scalar and vector functions with respect to a real vector

1.	$\dfrac{\partial A\mathbf{x}}{\partial \mathbf{x}} = A'.$	
2.	$\dfrac{\partial \mathbf{x}^T A}{\partial \mathbf{x}} = A.$	
3.	$\dfrac{\partial \mathbf{x}^T \mathbf{x}}{\partial \mathbf{x}} = 2\mathbf{x}.$	
4.	$\dfrac{\partial \mathbf{x}^T A\mathbf{x}}{\partial \mathbf{x}} = A\mathbf{x} + A^T \mathbf{x}.$	
5.	$\dfrac{\partial \mathbf{x}^T A\mathbf{x}}{\partial \mathbf{x}} = 2A\mathbf{x}$	(if A is symmetric).
6.	$\dfrac{\partial \mathbf{z}}{\partial \mathbf{x}} = \dfrac{\partial \mathbf{y}}{\partial \mathbf{x}} \dfrac{\partial \mathbf{z}}{\partial \mathbf{y}}$	($\mathbf{z} = \mathbf{y}(\mathbf{x})$).

E.2 Derivatives of real-valued functions of real matrices

Let X be an $m \times n$ matrix and let $y = f(X)$ be a scalar function of X. The derivative of y with respect to X is defined as

$$\frac{\partial y}{\partial X} = \begin{bmatrix} \frac{\partial y}{\partial x_{11}} & \frac{\partial y}{\partial x_{12}} & \cdots & \frac{\partial y}{\partial x_{1n}} \\ \frac{\partial y}{\partial x_{21}} & \frac{\partial y}{\partial x_{22}} & \cdots & \frac{\partial y}{\partial x_{2n}} \\ \vdots & & & \\ \frac{\partial y}{\partial x_{m1}} & \frac{\partial y}{\partial x_{m2}} & \cdots & \frac{\partial y}{\partial x_{mn}} \end{bmatrix}. \qquad (E.9)$$

An application that arises when X is square is the computation of the derivative of the determinant with respect to the matrix (that is, if $y = \det(X)$, compute $\partial y / \partial X$). We consider a more general case of $y = \det(Y)$, where Y is an $n \times n$ matrix whose elements are functions of a $p \times q$ matrix X,

$$y_{ij} = f_{ij}(X).$$

We compute

$$\frac{\partial \det(Y)}{\partial x_{ij}},$$

from which $\partial \det(Y)/\partial X$ can be obtained by stacking according to (E.9). The key is to recall the determinant expansion by cofactors from (C.13),

$$\det(Y) = \sum_{m=1}^{n} y_{im} Y_{im},$$

where Y_{im} is the cofactor of the element y_{im} in $\det(Y)$. Note that

$$\frac{\partial \det(Y)}{\partial y_{kl}} = Y_{kl},$$

since the cofactor Y_{kl} is independent of y_{kl}. By the chain rule,

$$\frac{\partial \det(Y)}{\partial x_{ij}} = \sum_{k=1}^{n}\sum_{l=1}^{n} \frac{\partial \det(Y)}{\partial y_{kl}} \frac{\partial y_{kl}}{\partial x_{ij}} = \sum_{k=1}^{n}\sum_{l=1}^{n} Y_{kl} \frac{\partial y_{kl}}{\partial x_{ij}}.$$

Now form the matrix A from the cofactors of Y, and the matrix B from the partials as

$$a_{kl} = Y_{kl} \qquad b_{kl} = \frac{\partial y_{kl}}{\partial x_{ij}}$$

so

$$B = \frac{\partial Y}{\partial x_{ij}}.$$

Then

$$\frac{\partial \det(Y)}{\partial x_{ij}} = \sum_{k=1}^{n}\sum_{l=1}^{n} a_{ij}b_{ij} = \mathrm{tr}(AB^T) = \mathrm{tr}(B^T A) \tag{E.10}$$

(see exercise E.7-7).

Example E.2.1 As an example of the use of (E.10), we compute

$$\frac{\partial \det(X)}{\partial X}$$

for a 2×2 matrix X when X is a general matrix, and when X is symmetric. In this case,

$$Y = X = \begin{bmatrix} x_{11} & x_{12} \\ x_{21} & x_{22} \end{bmatrix}.$$

Let

$$A = \begin{bmatrix} X_{11} & X_{12} \\ X_{21} & X_{22} \end{bmatrix}$$

be the matrix of cofactors of X, and observe that

$$\frac{\partial X}{\partial x_{ij}} = E_{ij},$$

where E_{ij} is an elementary matrix (see C.1). Then

$$\frac{\partial \det(X)}{\partial x_{ij}} = \mathrm{tr}\left(E_{ij}^T A\right),$$

which may be written using (9.18) as

$$\frac{\partial \det(X)}{\partial x_{ij}} = (\mathrm{vec}(E_{ij}))^T \mathrm{vec}(A). \tag{E.11}$$

For the 2×2 example, stacking according to (E.9) leads to

$$\frac{\partial \det(X)}{\partial X} = \begin{bmatrix} X_{11} & X_{12} \\ X_{21} & X_{22} \end{bmatrix} = [X_{ij}],$$

where $[X_{ij}]$ is the matrix formed by the cofactors of X.

Using the matrix inverse formula (C.14), we obtain

$$\frac{\partial \det(X)}{\partial X} = \det(X) X^{-T}.$$

This formula also applies to a general $n \times n$ matrix X.

When X is symmetric, a little more care is needed. In this case,

$$X = \begin{bmatrix} x_{11} & x_{12} \\ x_{12} & x_{22} \end{bmatrix}$$

and

$$\frac{\partial X}{\partial x_{12}} = \frac{\partial X}{\partial x_{21}} = X_{12} + X_{21} = 2X_{12},$$

where X_{12} is the cofactor of X. Stacking the results as before we obtain

$$\frac{\partial \det(X)}{\partial X} = \begin{bmatrix} X_{11} & 2X_{12} \\ 2X_{21} & X_{22} \end{bmatrix} = 2\begin{bmatrix} X_{11} & X_{12} \\ X_{21} & X_{22} \end{bmatrix} - \begin{bmatrix} X_{11} & 0 \\ 0 & X_{22} \end{bmatrix}.$$

Notationally, this can be written for an $n \times n$ matrix as

$$\frac{\partial \det(X)}{\partial X} = 2[X_{ij}] - \text{diag}\{X_{ii}\} \qquad (X \text{ symmetric}),$$

where $[X_{ij}]$ denotes the matrix formed by the cofactors of X and $\text{diag}(X_{ii})$ denotes the diagonal matrix of indicated cofactors. Then since X is symmetric, $[X_{ij}] = X^{-1}\det(X)$, and we can write

$$\frac{\partial \det(X)}{\partial X} = \det(X)(2X^{-1} - \text{diag}(X^{-1})). \qquad \square$$

Example E.2.2 Computation of the derivative of $\log \det |X|$ is now straightforward:

$$\frac{\partial \log \det(X)}{\partial X} = \frac{1}{\det(X)} \frac{\partial \det(X)}{\partial X}$$

$$= \begin{cases} X^{-T} & \text{(general } X), \\ 2X^{-1} - \text{diag}(X^{-1}) & \text{(symmetric } X). \end{cases} \qquad \square$$

Example E.2.3 As a final example of this type, we have

$$\frac{\partial (\det(X))^r}{\partial X} = r(\det(X))^{r-1}\frac{\partial \det(X)}{\partial X}. \qquad \square$$

E.3 Derivatives of matrices with respect to scalars, and vice versa

In many problems, it is useful to be able to determine derivatives of the trace of a function of a matrix. This is discussed in section E.7; as shown there, computation of some of these derivatives can be facilitated by computing

$$\frac{\partial Y}{\partial x_{rs}} \quad \text{or} \quad \frac{\partial y_{ij}}{\partial X}$$

when Y is a function of X. Accordingly, the remainder of this section examines derivatives of this type and the relationship between them. The derivatives are defined, respectively, as

$$\frac{\partial Y}{\partial x_{rs}} = \begin{bmatrix} \frac{\partial y_{11}}{\partial x_{rs}} & \frac{\partial y_{12}}{\partial x_{rs}} & \cdots & \frac{\partial y_{1q}}{\partial x_{rs}} \\ \frac{\partial y_{21}}{\partial x_{rs}} & \frac{\partial y_{22}}{\partial x_{rs}} & \cdots & \frac{\partial y_{2q}}{\partial x_{rs}} \\ \vdots & & & \\ \frac{\partial y_{l1}}{\partial x_{rs}} & \frac{\partial y_{l2}}{\partial x_{rs}} & \cdots & \frac{\partial y_{lq}}{\partial x_{rs}} \end{bmatrix}, \qquad (E.12)$$

and

$$\frac{\partial y_{ij}}{\partial X} = \begin{bmatrix} \frac{\partial y_{ij}}{\partial x_{11}} & \frac{\partial y_{ij}}{\partial x_{12}} & \cdots & \frac{\partial y_{ij}}{\partial x_{1n}} \\ \frac{\partial y_{ij}}{\partial x_{21}} & \frac{\partial y_{ij}}{\partial x_{22}} & \cdots & \frac{\partial y_{ij}}{\partial x_{2n}} \\ \vdots & & & \\ \frac{\partial y_{ij}}{\partial x_{m1}} & \frac{\partial y_{ij}}{\partial x_{m2}} & \cdots & \frac{\partial y_{ij}}{\partial x_{mn}} \end{bmatrix}. \qquad (E.13)$$

Example E.3.1 Find $\frac{\partial y_{ij}}{\partial X}$ when $Y = AXB$ for a $p \times m$ matrix A and an $n \times q$ matrix B. The (i, j)th element of Y is

$$y_{ij} = \sum_{l=1}^{n} \sum_{k=1}^{m} a_{ik} x_{kl} b_{lj},$$

so that

$$\frac{\partial y_{ij}}{\partial x_{rs}} = a_{ir} b_{sj}. \qquad (E.14)$$

Stacking according to (E.13), we obtain

$$\frac{\partial y_{ij}}{\partial X} = \begin{bmatrix} a_{i1}b_{1j} & a_{i1}b_{2j} & \cdots & a_{i1}b_{nj} \\ a_{i2}b_{1j} & a_{i2}b_{2j} & \cdots & a_{i2}b_{nj} \\ \vdots & & & \\ a_{im}b_{1j} & a_{im}b_{2j} & \cdots & a_{im}b_{nj} \end{bmatrix} = \begin{bmatrix} a_{i1} \\ a_{i2} \\ \vdots \\ a_{im} \end{bmatrix} \begin{bmatrix} b_{1j} & b_{2j} & \cdots & b_{nj} \end{bmatrix}.$$

This can be written using elementary matrices as

$$\frac{\partial y_{ij}}{\partial X} = A^T E_{ij} B^T. \qquad (E.15)$$

□

Example E.3.2 Find $\frac{\partial Y}{\partial x_{rs}}$ when $Y = AXB$, as before. Stacking (E.14) as per (E.12) leads to

$$\frac{\partial Y}{\partial x_{rs}} = \begin{bmatrix} a_{1r}b_{s1} & a_{1r}b_{s2} & \cdots & a_{1r}b_{sq} \\ a_{2r}b_{s1} & a_{2r}b_{s2} & \cdots & a_{2r}b_{sq} \\ \vdots & & & \\ a_{pr}b_{s1} & a_{pr}b_{s2} & \cdots & a_{pr}b_{sq} \end{bmatrix} = \begin{bmatrix} a_{1r} \\ a_{2r} \\ \vdots \\ a_{pr} \end{bmatrix} \begin{bmatrix} b_{s1} & b_{s2} & \cdots & b_{sq} \end{bmatrix},$$

which can be written as

$$\frac{\partial Y}{\partial x_{rs}} = A E_{rs} B. \qquad (E.16)$$

□

Example E.3.3 When $Y = AX^T B$, we can proceed as before to determine that

$$\frac{\partial AX^T B}{\partial x_{rs}} = A E_{rs}^T B \qquad (E.17)$$

and

$$\frac{\partial (AX^T B)_{ij}}{\partial X} = B E_{ij}^T A. \qquad (E.18)$$

□

E.4 The transformation principle

Observe that there is a symmetry between (E.15) and (E.16), and a symmetry between (E.17) and (E.18). The rule for transforming between the partial of an element with respect to a matrix and the partial of a matrix with respect to an element, when the derivative involves linear operations, is called the *transformation principle* [117, 65]. It can be stated as follows.

1. When $\frac{\partial y_{ij}}{\partial X}$ is of the form

$$\frac{\partial y_{ij}}{\partial X} = CE_{ij}D$$

    the derivative $\frac{\partial Y}{\partial x_{rs}}$ may be obtained by

$$\frac{\partial Y}{\partial x_{rs}} = C^T E_{rs} D^T.$$

    Note that in this rule there is *no transpose* on the elementary matrix.

2. When $\frac{\partial y_{ij}}{\partial X}$ is of the form

$$\frac{\partial y_{ij}}{\partial X} = CE_{ij}^T D$$

    the derivative $\frac{\partial Y}{\partial x_{rs}}$ may be obtained by

$$\frac{\partial Y}{\partial x_{rs}} = DE_{rs}^T C.$$

    Note that in this rule there *is* a transpose on the elementary matrix.

These rules apply even when C and/or D are functions of X.

E.5 Derivatives of products of matrices

Another useful rule determines derivatives of products of matrices. Let $Y = UV$, where $U = U(X)$ is $m \times n$ and $V = V(X)$ is $n \times p$, and consider

$$\frac{\partial Y}{\partial x_{rs}} \quad \text{and} \quad \frac{\partial y_{ij}}{\partial X}.$$

Since $y_{ij} = \sum_{k=1}^{n} u_{ik} v_{kj}$, it follows that

$$\frac{\partial y_{ij}}{\partial x_{rs}} = \sum_{k=1}^{n} \frac{\partial u_{ik}}{\partial x_{rs}} v_{kj} + \sum_{k=1}^{n} u_{ik} \frac{\partial v_{kj}}{\partial x_{rs}}.$$

Stacking the partials leads to

$$\frac{\partial Y}{\partial x_{rs}} = \frac{\partial U}{\partial x_{rs}} V + U \frac{\partial V}{\partial x_{rs}}, \tag{E.19}$$

which is similar to the familiar product rule.

Example E.5.1 Let $Y = AX^{-1}B$ and find

$$\frac{\partial Y}{\partial x_{rs}} \quad \text{and} \quad \frac{\partial y_{ij}}{\partial X}.$$

First, note by (E.16) that

$$\frac{\partial Y^{-1}}{\partial x_{rs}} = \frac{\partial (B^{-1} X A^{-1})}{\partial x_{rs}} = B^{-1} E_{rs} A^{-1}. \tag{E.20}$$

Now differentiate $YY^{-1} = I$ using (E.19):

$$\frac{\partial Y}{\partial x_{rs}} Y^{-1} + Y \frac{\partial Y^{-1}}{\partial x_{rs}} = 0,$$

so that, solving and substituting from (E.20),

$$\frac{\partial Y}{\partial x_{rs}} = -Y \frac{\partial Y^{-1}}{\partial x_{rs}} Y = -YB^{-1} E_{rs} A^{-1}.$$

Finally, using $Y = AX^{-1}B$, we obtain

$$\frac{\partial Y}{\partial x_{rs}} = -AX^{-1} B E_{rs} X^{-1} B.$$

To find the other derivative, use the first rule of the transformation principle with $C = AX^{-1}$, $D = X^{-1}B$. Then,

$$\frac{\partial y_{ij}}{\partial X} = -C^T E_{ij} D^T = -X^{-T} A^T E_{ij} B^T X^{-T}. \qquad \square$$

Example E.5.2 Find $\frac{\partial Y}{\partial x_{rs}}$ and $\frac{\partial y_{ij}}{\partial X}$ when $Y = X^T AX$. Using (E.19), we can write

$$\frac{\partial Y}{\partial x_{rs}} = \frac{\partial X^T}{\partial x_{rs}} AX + X^T \frac{\partial AX}{\partial x_{rs}}.$$

By (E.17) and (E.16), we can write

$$\frac{\partial Y}{\partial x_{rs}} = E_{rs}^T AX + X^T A E_{rs}. \qquad (E.21)$$

To find $\frac{\partial y_{ij}}{\partial X}$, use the transformation principle. For the first term $E_{rs}^T AX$, let $C = I$ and $D = AX$ in the second transformation principle,

$$E_{rs}^T AX \leftrightarrow AX E_{rs}^T.$$

For the second term in (E.21), use $C^T = X^T A$ and $D^T = I$ in the first transformation principle, so that

$$X^T A E_{rs} \leftrightarrow A^T X E_{rs}.$$

Combining these, we obtain

$$\frac{\partial y_{ij}}{\partial X} = AX E_{ij}^T + A^T X E_{ij}. \qquad \square$$

E.6 Derivatives of powers of a matrix

A rule for powers of a matrix $Y = X^n$ may be obtained by repeated application of the rule for products (E.19). This leads to

$$\frac{\partial X^n}{\partial x_{rs}} = \sum_{k=0}^{n-1} X^k E_{rs} X^{n-k-1}, \qquad (E.22)$$

E.6 Derivatives of Powers of a Matrix

and application of the transformation rule leads to

$$\frac{\partial (X^n)_{ij}}{\partial X} = \sum_{k=1}^{n-1} (X^T)^k E_{ij} (X^T)^{n-k-1}. \tag{E.23}$$

It is also straightforward to show that

$$\frac{\partial X^{-n}}{\partial x_{rs}} = -X^{-n} \left(\sum_{k=0}^{n-1} X^k E_{rs} X^{n-k-1} \right) X^{-n}. \tag{E.24}$$

These results, and others which may be similarly derived, are summarized in table E.2.

Table E.2: Derivatives of matrices with respect to an element, or elements with respect to a matrix

1. $\dfrac{\partial X}{\partial x_{rs}} = E_{rs}.$

2. $\dfrac{\partial X^T}{\partial x_{rs}} = E_{rs}^T.$

3. $\dfrac{\partial (AXB)}{\partial x_{rs}} = A E_{rs} B.$

4. $\dfrac{\partial (AXB)_{ij}}{\partial X} = A^T E_{ij} B^T.$

5. $\dfrac{\partial (AX^T B)}{\partial x_{rs}} = A E_{rs}^T B.$

6. $\dfrac{\partial (AX^T B)_{ij}}{\partial X} = A^T E_{ij}^T B^T.$

7. $\dfrac{\partial (AX^{-1} B)}{\partial x_{rs}} = -AX^{-1} E_{rs} X^{-1} B.$

8. $\dfrac{\partial (AX^{-1} B)_{ij}}{\partial X} = -X^{-T} A^T E_{ij} B^T X^{-T}.$

9. $\dfrac{\partial UV}{\partial x_{rs}} = \dfrac{\partial U}{\partial x_{rs}} V + U \dfrac{\partial V}{\partial x_{rs}}.$

10. $\dfrac{\partial X^T A X}{\partial x_{rs}} = E_{rs}^T A X + X^T A E_{rs}.$

11. $\dfrac{\partial (X^T A X)_{ij}}{\partial X} = A X E_{ij}^T + A^T X E_{ij}.$

12. $\dfrac{\partial X^n}{\partial x_{rs}} = \sum_{k=0}^{n-1} X^k E_{rs} X^{n-k-1}.$

13. $\dfrac{\partial (X^n)_{ij}}{\partial X} = \sum_{k=1}^{n-1} (X^T)^k E_{ij} (X^T)^{n-k-1}.$

14. $\dfrac{\partial X^{-n}}{\partial x_{rs}} = -X^{-n} \left(\sum_{k=0}^{n-1} X^k E_{rs} X^{n-k-1} \right) X^{-n}.$

E.7 Derivatives involving the trace

Given that $Y = F(X)$ where Y is an $m \times m$ matrix and X is an $m \times m$ matrix, we want to find

$$\frac{\partial \operatorname{tr}(Y)}{\partial X} = \begin{bmatrix} \frac{\partial \operatorname{tr}(Y)}{\partial x_{11}} & \frac{\partial \operatorname{tr}(Y)}{\partial x_{12}} & \cdots & \frac{\partial \operatorname{tr}(Y)}{\partial x_{1m}} \\ \frac{\partial \operatorname{tr}(Y)}{\partial x_{21}} & \frac{\partial \operatorname{tr}(Y)}{\partial x_{22}} & \cdots & \frac{\partial \operatorname{tr}(Y)}{\partial x_{2m}} \\ \vdots & & & \\ \frac{\partial \operatorname{tr}(Y)}{\partial x_{m1}} & \frac{\partial \operatorname{tr}(Y)}{\partial x_{m2}} & \cdots & \frac{\partial \operatorname{tr}(Y)}{\partial x_{mm}} \end{bmatrix}. \quad (E.25)$$

When $Y = AX$ and X is a general matrix, the derivative with respect to an element can be found by expansion of the operations

$$\frac{\partial \operatorname{tr}(Y)}{\partial x_{rs}} = \frac{\partial}{\partial x_{rs}} \sum_i \sum_k a_{ik} x_{ki} = A_{sr},$$

so that

$$\frac{\partial \operatorname{tr}(AX)}{\partial X} = A^T.$$

When X is symmetric, somewhat more care is required.

$$\frac{\partial \operatorname{tr}(AX)}{\partial x_{rs}} = \frac{\partial}{\partial x_{rs}} \sum_i \sum_j a_{ij} x_{ji} = \sum_{i<j} a_{ij} x_{ji} + \sum_{i=j} a_{ii} x_{ii} + \sum_{i>j} a_{ij} x_{ij}$$

$$= \begin{cases} a_{sr} + a_{rs} & r \neq s, \\ a_{ss} & r = s. \end{cases}$$

Stacking these, we obtain

$$\frac{\partial \operatorname{tr} AX}{\partial X} = A^T + A - \operatorname{diag}(A).$$

It may be shown similarly that

$$\frac{\partial \operatorname{tr}(XX^T)}{\partial X} = 2X.$$

A rule that helps to derive a variety of other formulas is to observe (see exercise E.7-11) that

$$\frac{\partial \operatorname{tr}(Y)}{\partial x_{rs}} = \operatorname{tr} \frac{\partial Y}{\partial x_{rs}}. \quad (E.26)$$

Example E.7.1 The derivative when $Y = X^n$ is treated using (E.26) and (E.22):

$$\frac{\partial \operatorname{tr}(X^n)}{\partial x_{rs}} = \operatorname{tr} \frac{\partial X^n}{\partial x_{rs}}$$

$$= \operatorname{tr} \sum_{k=0}^{n-1} X^k E_{rs} X^{n-k-1}$$

$$= \operatorname{tr} E_{rs} X^{n-1} + \operatorname{tr} X E_{rs} X^{n-2} + \cdots + \operatorname{tr} X^{n-1} E_{rs}$$

$$= (n-1) \operatorname{tr} E_{rs} X^{n-1}.$$

Stacking these as in (E.25), we obtain

$$\frac{\partial \operatorname{tr}(X^n)}{\partial X} = (n-1) \begin{bmatrix} \operatorname{tr} E_{11} X^{n-1} & \operatorname{tr} E_{12} X^{n-1} & \cdots & \operatorname{tr} E_{1m} X^{n-1} \\ \operatorname{tr} E_{21} X^{n-1} & \operatorname{tr} E_{22} X^{n-1} & \cdots & \operatorname{tr} E_{2m} X^{n-1} \\ \vdots & & & \\ \operatorname{tr} E_{m1} X^{n-1} & \operatorname{tr} E_{m2} X^{n-1} & \cdots & \operatorname{tr} E_{mm} X^{n-1} \end{bmatrix}$$

E.7 Derivatives Involving the Trace

$$= (n-1) \begin{bmatrix} (X^{n-1})_{11} & (X^{n-1})_{21} & \cdots & (X^{n-1})_{m1} \\ (X^{n-1})_{12} & (X^{n-1})_{22} & \cdots & (X^{n-1})_{m2} \\ \vdots & & & \\ (X^{n-1})_{1m} & (X^{n-1})_{2m} & \cdots & (X^{n-1})_{mm} \end{bmatrix} = (n-1)(X^{n-1})^T.$$

□

Example E.7.2 When $Y = X^T A X$, then by (E.17) and (E.26),

$$\frac{\partial \operatorname{tr}(Y)}{\partial x_{rs}} = \operatorname{tr} E_{rs}^T A X + \operatorname{tr} X^T A E_{rs}.$$

Using (C.15) leads to

$$\frac{\partial \operatorname{tr}(X^T A X)}{\partial X} = A X + A^T X.$$

□

These results are summarized in table E.3.

Table E.3: Derivatives of scalar functions with respect to a real matrix

1. $\dfrac{\partial \det(X)}{\partial X} = \begin{cases} \det(X) X^{-T} & \text{(general } X\text{)}, \\ \det(X)(2X^{-1} - \operatorname{diag}(X^{-1})) & \text{(symmetric } X\text{)}. \end{cases}$

2. $\dfrac{\partial \log \det(X)}{\partial X} = \dfrac{1}{\det(X)} \dfrac{\partial \det(X)}{\partial X}.$

3. $\dfrac{\partial (\det(X))^r}{\partial X} = r \det(X)^r \dfrac{\partial \det(X)}{\partial X}.$

4. $\dfrac{\partial \operatorname{tr}(X)}{\partial X} = I.$

5. $\operatorname{tr} \dfrac{\partial Y}{\partial x_{rs}} = \dfrac{\partial \operatorname{tr}(Y)}{\partial x_{rs}}.$

6. $\dfrac{\partial \operatorname{tr}(AX)}{\partial X} = \begin{cases} A^T & \text{(general } X\text{)}, \\ A^T + A - 2\operatorname{diag}(A) & \text{(symmetric } X\text{)}. \end{cases}$

7. $\dfrac{\partial \operatorname{tr}(A^T X)}{\partial X} = A.$

8. $\dfrac{\partial \operatorname{tr}(AXB)}{\partial X} = A^T B^T.$

9. $\dfrac{\partial \operatorname{tr}(X^T A X B)}{\partial X} = A X B + A^T X B^T.$

10. $\dfrac{\partial \operatorname{tr}(X X^T)}{\partial X} = 2X.$

11. $\dfrac{\partial \operatorname{tr}(X^2)}{\partial X} = 2X^T.$

12. $\dfrac{\partial \operatorname{tr}(X^n)}{\partial X} = n(X^{n-1})^T.$

13. $\dfrac{\partial \operatorname{tr}(AX^{-1}B)}{\partial X} = -(X^{-1} B A X^{-1})^T.$

14. $\dfrac{\partial \operatorname{tr}(e^X)}{\partial X} = e^{X^T}.$

E.8 Modifications for derivates of complex vectors and matrices

Some minor modifications to the derivative formulas are necessary when dealing with complex vectors and matrices. In most circumstances in which complex vectors are used, a real cost functional such as $\mathbf{z}^H R \mathbf{z}$ appears. We will define a new derivative operator that is appropriate for optimization problems of the sort commonly encountered in signal processing. Let $z = x + jy$ be a complex number with real part

$$x = \frac{1}{2}(z + \bar{z})$$

and imaginary part

$$y = \frac{1}{2j}(z - \bar{z}).$$

We *define* the derivative with respect to z by the following operators [132, page 891] as

$$\frac{\partial}{\partial z} = \frac{1}{2}\left(\frac{\partial}{\partial x} - j\frac{\partial}{\partial y}\right) \quad \text{and} \quad \frac{\partial}{\partial \bar{z}} = \frac{1}{2}\left(\frac{\partial}{\partial x} + j\frac{\partial}{\partial y}\right). \quad (E.27)$$

By this definition,

$$\frac{\partial z}{\partial z} = 1$$

and

$$\frac{\partial z}{\partial \bar{z}} = \frac{\partial \bar{z}}{\partial z} = 0.$$

Properties of the operators in (E.27) are explored in exercise E.8-12.

For a complex vector $\mathbf{z} = [z_1, z_2, \ldots, z_n]^T$ with $z_k = x_k + jy_k$, we define the differentiation operators

$$\frac{\partial}{\partial \mathbf{z}} = \begin{bmatrix} \frac{\partial}{\partial x_1} - j\frac{\partial}{\partial y_1} \\ \frac{\partial}{\partial x_2} - j\frac{\partial}{\partial y_2} \\ \vdots \\ \frac{\partial}{\partial x_n} - j\frac{\partial}{\partial y_n} \end{bmatrix} \quad \partial \bar{\mathbf{z}} = \begin{bmatrix} \frac{\partial}{\partial x_1} + j\frac{\partial}{\partial y_1} \\ \frac{\partial}{\partial x_2} + j\frac{\partial}{\partial y_2} \\ \vdots \\ \frac{\partial}{\partial x_n} + j\frac{\partial}{\partial y_n} \end{bmatrix}. \quad (E.28)$$

By these definitions,

$$\frac{\partial \mathbf{z}}{\partial \mathbf{z}} = I$$

and

$$\frac{\partial \mathbf{z}}{\partial \bar{\mathbf{z}}} = \frac{\partial \bar{\mathbf{z}}}{\partial \mathbf{z}} = \mathbf{0}.$$

Application of these definitions can lead to results that are somewhat different than might be expected based on results from calculus on real vectors.

Gradient of a linear function of a complex vector

Let $\mathbf{a}$ and $\mathbf{z}$ be complex vectors of length n. Then

$$\frac{\partial}{\partial \bar{\mathbf{z}}} \mathbf{a}^H \mathbf{z} = \frac{\partial}{\partial \bar{\mathbf{z}}} \sum_{i=1}^{n} \bar{a}_i z_i = 0$$

E.8 Modifications for Derivates of Complex Vectors and Matrices

and
$$\frac{\partial}{\partial \mathbf{z}} \mathbf{a}^H \mathbf{z} = \frac{\partial}{\partial \mathbf{z}} \sum_{i=1}^{n} \overline{a}_i z_i = \overline{\mathbf{a}}$$

and
$$\frac{\partial}{\partial \overline{\mathbf{z}}} \mathbf{z}^H \mathbf{a} = \frac{\partial}{\partial \overline{\mathbf{z}}} \sum_{i=1}^{n} \overline{z}_i a_i = \mathbf{a}.$$

Another useful result is
$$\frac{\partial}{\partial \overline{\mathbf{z}}} \text{Re}(\mathbf{z}^H \mathbf{a}) = \frac{\partial}{\partial \overline{\mathbf{z}}} \frac{1}{2} (\mathbf{z}^H \mathbf{a} + \mathbf{a}^H \mathbf{z}) = \frac{1}{2} \mathbf{a}.$$

Gradient of a quadratic function of a complex vector

It may be verified that when R is a Hermitian symmetric matrix,
$$\frac{\partial \mathbf{z}^H R \mathbf{z}}{\partial \overline{\mathbf{z}}} = R \mathbf{z}$$
and
$$\frac{\partial \mathbf{z}^H R \mathbf{z}}{\partial \mathbf{z}} = R^T \overline{\mathbf{z}}.$$

Contrast these with row 4 from table E.1.

Example E.8.1 In the least-squares solution of
$$A \mathbf{z} = \mathbf{b},$$
we minimize the norm of the error $\mathbf{e} = \mathbf{b} - A \mathbf{z}$. Taking the norm as
$$\|\mathbf{e}\|^2 = (\mathbf{b} - A \mathbf{z})^H (\mathbf{b} - A \mathbf{z}) = \mathbf{b}^H \mathbf{b} - \mathbf{b}^H A \mathbf{z} - \mathbf{z}^H A^H \mathbf{b} + \mathbf{z}^H A^H A \mathbf{z},$$
we can minimize by taking the derivative with respect to the unknown vector $\mathbf{z}$. Using $\frac{\partial}{\partial \mathbf{z}}$ and equating to zero leads to
$$\frac{\partial}{\partial \mathbf{z}} \|\mathbf{e}\|^2 = A^T \overline{\mathbf{b}} + (A^H A)^T \overline{\mathbf{z}} = 0. \tag{E.29}$$
In this equation, observe that $\mathbf{z}$ appears conjugated. Solving for $\mathbf{z}$, we obtain $\mathbf{z} = (A^H A)^{-1} A^H \mathbf{b}$ as expected.
By the use of $\frac{\partial}{\partial \overline{\mathbf{z}}}$, we obtain the equation
$$\frac{\partial}{\partial \overline{\mathbf{z}}} \|\mathbf{e}\|^2 = A^H \mathbf{b} + A^H A \mathbf{z}.$$
In this equation $\mathbf{z}$ is not conjugated. The solution is the same as before. $\square$

As the last example illustrates, the operator $\partial / \partial \overline{\mathbf{z}}$ yields unconjugated results that are easier to deal with. This operator, known as the *conjugate derivative*, is therefore adopted as the derivative with respect to a complex vector.

For derivatives of matrices with respect to complex elements, the rules change somewhat. For example, modifying row 10 of table E.2, we obtain
$$\frac{\partial X^H A X}{\partial x_{rs}} = \frac{\partial X^H A}{\partial x_{rs}} + X^H A \frac{\partial X}{\partial x_{rs}} = X^H A E_{rs}$$
and, similarly,
$$\frac{\partial X^H A X}{\partial \overline{x}_{rs}} = E_{rs}^T A X. \tag{E.30}$$

For derivatives involving traces, the following equations illustrate some useful facts:

$$\frac{\partial \operatorname{tr}(X^H X)}{\partial X} = \overline{X} \qquad \frac{\partial \operatorname{tr}(X^H X)}{\partial \overline{X}} = X \qquad \frac{\partial \operatorname{tr}(X^H X)}{\partial X^H} = X^T \qquad \text{(E.31)}$$

$$\frac{\partial \operatorname{tr}(X^H A X B)}{\partial X} = A^T \overline{X} B^T \qquad \frac{\partial \operatorname{tr}(X^H A X B)}{\partial \overline{X}} = A X B \qquad \text{(E.32)}$$

$$\frac{\partial \operatorname{tr}(A X B)}{\partial X} = A^T B^T \qquad \frac{\partial \operatorname{tr}(A \overline{X} B)}{\partial \overline{X}} = A^T B^T \qquad \frac{\partial \operatorname{tr}(A X B)}{\partial \overline{X}} = 0 \qquad \text{(E.33)}$$

$$\frac{\partial \operatorname{tr}(A X^H B)}{\partial \overline{X}} = BA \qquad \text{(E.34)}$$

$$\frac{\partial \operatorname{tr}(A X X^H B)}{\partial X} = A^T B^T \overline{X} \qquad \frac{\partial \operatorname{tr}(A X X^H B)}{\partial \overline{X}} = B A X \qquad \text{(E.35)}$$

E.9 Exercises

E.1-1 Verify the derivatives in table E.1.

E.5-2 [117] Verify for $Y = AXB$, with $Y \in M_{lq}$ and $X \in M_{mn}$, that:

(a) The derivative of an element of Y with respect to an element of X is

$$\frac{\partial y_{ij}}{\partial x_{rs}} = a_{ir} b_{sj}.$$

(b) The derivative of an element of Y with respect to X, defined in (E.13), may be written as

$$\frac{\partial y_{ij}}{\partial X} = A^T E_{ij} B^T,$$

where $E_{ij} \in M_{lq}$ is an elementary matrix.

(c) The derivative of Y with respect to an element of X, defined in (E.12), may be written as

$$\frac{\partial Y}{\partial x_{rs}} = A E_{rs} B,$$

where $E_{rs} \in M_{mn}$ is an elementary matrix.

(d) i.
$$\frac{\partial X^T A X}{\partial x_{rs}} = E_{rs}^T A X + X^T A E_{rs}.$$

ii.
$$\frac{\partial (X^T A X)_{ij}}{\partial X} = A X E_{ij}^T + A^T X E_{ij}.$$

E.5-3 Show that

(a) $\dfrac{\partial X A X}{\partial x_{rs}} = E_{rs} A X + X A E_{rs}.$

(b) $\dfrac{\partial (X A X)_{ij}}{\partial X} = E_{ij} X^T A^T + A^T X^T E_{ij}.$

(c) $\dfrac{\partial X^T A X^T}{\partial x_{rs}} = E_{rs}^T A X^T + X^T A E_{rs}^T.$

(d) $\dfrac{\partial (X^T A X^T)_{ij}}{\partial X} = A X^T E_{ij}^T + E_{ij}^T X^T A.$

E.9 Exercises

E.6-4 Verify (E.22).

E.6-5 Verify (E.23).

E.6-6 Verify (E.24).

E.7-7 Show for a matrix $A = [a_{ij}]$ and $B = [b_{ij}]$ that
$$\sum_i \sum_j a_{ij} b_{ij} = \operatorname{tr}(AB^T).$$

E.7-8 If $y = \operatorname{tr}(X)$, show that
$$\frac{\partial y}{\partial X} = I$$
(the derivative of a scalar with respect to a matrix).

E.7-9 Show that
$$\frac{\partial \operatorname{tr}(AX)}{\partial X} = A^T.$$

E.7-10 Show that

(a) $\dfrac{\partial \operatorname{tr}(AXB)}{\partial X} = A^T B^T.$

(b) $\dfrac{\partial \operatorname{tr}(X^2)}{\partial X} = 2X^T.$

(c) $\dfrac{\partial \operatorname{tr}(X^n)}{\partial X} = n(X^{n-1})^T.$

(d) $\dfrac{\partial \operatorname{tr}(AX^{-1}B)}{\partial X} = -(X^{-1}BAX^{-1})^T.$

(e) $\dfrac{\partial \operatorname{tr}(e^X)}{\partial X} = e^{X^T}.$

E.7-11 Show that
$$\operatorname{tr}\left(\frac{\partial Y}{\partial x_{rs}}\right) = \frac{\partial \operatorname{tr}(Y)}{\partial x_{rs}}.$$

E.8-12 Show by direct application of the operators defined in (E.27) that the following properties are true for the derivative operator.

(a) $\dfrac{\partial z}{\partial \bar{z}} = 0.$ (b) $\dfrac{\partial z}{\partial z} = 1.$

(c) $\dfrac{\partial |z|^2}{\partial z} = \bar{z}.$ (d) $\dfrac{\partial |z|^2}{\partial \bar{z}} = z.$

(e) $\dfrac{\partial z^2}{\partial z} = 2z.$ (f) $\dfrac{\partial z^2}{\partial \bar{z}} = 0.$

(g) $\dfrac{\partial z^n}{\partial z} = nz^{n-1}.$ (h) $\dfrac{\partial z^n}{\partial \bar{z}} = 0.$

(i) $\dfrac{\partial}{\partial z} fg = \dfrac{\partial f}{\partial z} g + f \dfrac{\partial g}{\partial z}.$ (j) $\dfrac{\partial}{\partial z}\left(\dfrac{f}{g}\right) = \dfrac{g \dfrac{\partial f}{\partial z} - f \dfrac{\partial g}{\partial z}}{g^2}.$

(k) $\dfrac{\partial}{\partial z} az = a.$ (l) $\dfrac{\partial}{\partial z} e^z = e^z.$

(m) $\dfrac{\partial}{\partial z} \log z = \dfrac{1}{z}.$

E.8-13 Verify the following by application of the operators defined in (E.28):

(a) $\dfrac{\partial}{\partial \mathbf{z}} \mathbf{z}^H \mathbf{a} = 0$.

(b) $\dfrac{\partial}{\partial \bar{\mathbf{z}}} \mathbf{z}^H \mathbf{a} = \mathbf{a}$.

(c) $\dfrac{\partial}{\partial \mathbf{z}} \mathbf{a}^H \mathbf{z} = \bar{\mathbf{a}}$.

(d) $\dfrac{\partial}{\partial \bar{\mathbf{z}}} \mathbf{a}^H \mathbf{z} = 0$.

(e) $\dfrac{\partial}{\partial \mathbf{z}} \mathbf{z}^T R \mathbf{z} = R\mathbf{z} + R^T \mathbf{z}$.

(e) $\dfrac{\partial}{\partial \bar{\mathbf{z}}} \mathbf{z}^T R \mathbf{z} = 0$.

(f) $\dfrac{\partial}{\partial \mathbf{z}} \mathbf{z}^H R \mathbf{z} = R^T \bar{\mathbf{z}}$.

(g) $\dfrac{\partial}{\partial \bar{\mathbf{z}}} \mathbf{z}^H R \mathbf{z} = R\mathbf{z}$.

E.8-14 Verify (E.30).

E.8-15 Verify (E.31) and (E.32).

E.10 References

The information in this section has been drawn from [117, 132].

Appendix F

Conditional Expectations of Multinomial and Poisson r.v.s

F.1 Multinomial distributions

Let X_1, X_2, X_3 have a multinomial distribution with class probabilities (p_1, p_2, p_3), so that

$$P(X_1 = x_1, X_2 = x_2, X_3 = x_3) = \frac{(x_1 + x_2 + x_3)!}{x_1! x_2! x_3!} p_1^{x_1} p_2^{x_2} p_3^{x_3}.$$

This multinomial in three outcomes can be combined to form a multinomial in two outcomes, and in general a multinomial distribution with m outcomes can be similarly reduced to a multinomial with $m - 1$ outcomes.

Let $Y = X_1 + X_2$. The (binomial) probability $P(Y, X_3)$ can be determined as follows:

$$P(X_1 + X_2 = y, X_3 = x_3) = \sum_{i=0}^{y} P(X_1 = i, X_2 = y - i, X_3 = x_3)$$

$$= \frac{(y + x_3)!}{y! x_3!} p_3^{x_3} \sum_{i=0}^{y} \frac{y!}{i!(y-i)!} p_1^i p_2^{y-i}$$

$$= \frac{(y + x_3)!}{y! x_3!} (p_1 + p_2)^y p_3^{x_3}, \tag{F.1}$$

where the last step follows from the binomial theorem. So $(X_1 + X_2, X_3)$ is binomial with class probabilities $(p_1 + p_2, p_3)$. This generalizes by induction to other multinomials.

To compute the conditional expectation $E[X_1 \mid Y = y]$, it is first necessary to determine the conditional probability, $P(X_1 = x_1 \mid Y = y) = P(X_1 = x_1 \mid X_1 + X_2 = y)$. The conditional probability can be written as

$$P(X_1 = x_1 \mid Y = y) = \frac{P(X_1 = x_1, Y = y)}{P(Y = y)}$$

$$= \frac{P(X_1 = x_1, X_2 = y - x_1)}{P(Y = y)},$$

where the numerator probability is trinomially distributed out of $n = x_1 + x_2 + x_3$ trials, and the denominator probability binomially distributed out of n trials. Then,

$$P(X_1 = x_1 \mid Y = y) = \frac{y!}{x_1!(y - x_1)!} p_1^{x_1} p_2^{y - x_1} \frac{1}{(p_1 + p_2)^y}.$$

The conditional expectation is then

$$E[X_1 \mid X_1 + X_2 = y] = \sum_{x_1=0}^{y} x_1 \frac{y!}{x_1!(y-x_1)!} p_1^{x_1} p_2^{x_2} \frac{1}{(p_1+p_2)^y} \quad (F.2)$$

$$= y \frac{p_1}{p_1 + p_2}.$$

Similarly, it can be shown that

$$E[X_2 \mid X_1 + X_2 = y] = y \frac{p_2}{p_1 + p_2}.$$

F.2 Poisson random variables

Computations are similar for Poisson random variables. If X_1 and X_2 are independent Poisson random variables with means λ_1 and λ_2, respectively, and $Y = X_1 + X_2$, then

$$P[Y = y] = \sum_{i=0}^{y} P(X_1 = i, X_2 = y - i)$$

$$= \sum_{i=0}^{y} e^{-\lambda_1} \frac{\lambda_1^i}{i!} e^{-\lambda_2} \frac{\lambda_2^{y-i}}{(y-i)!}$$

$$= \frac{e^{-\lambda_1 - \lambda_2}}{y!} \sum_{i=0}^{y} \frac{y!}{i!(y-i)!} \lambda_1^i \lambda_2^{y-i}$$

$$= e^{-(\lambda_1 + \lambda_2)} \frac{(\lambda_1 + \lambda_2)^y}{y!}; \quad (F.3)$$

so Y is Poisson with mean $\lambda_1 + \lambda_2$. The conditional expectation $E[X_1 \mid Y]$ requires the conditional probability

$$P(X_1 = x_1 \mid Y = y) = \frac{P(X_1 = x_1, Y = y)}{P(Y = y)} = \frac{P(X_1 = x_1, X_2 = y - x_1)}{P(Y = y)}$$

$$= \frac{\lambda_1^{x_1} \lambda_2^{y-x_1}}{(\lambda_1 + \lambda_2)^y} \binom{y}{x_1}. \quad (F.4)$$

The conditional expectation can then be computed in a fashion similar to that for (F.2), to obtain

$$E[X_1 \mid Y] = y \frac{\lambda_1}{\lambda_1 + \lambda_2}. \quad (F.5)$$

F.3 Exercises

F.1-1 Show that F.1 is correct.

F.1-2 Show that the conditional expectation in (F.2) is correct.

F.2-3 Show that (F.3), (F.4), and (F.5) are correct.

Bibliography

[1] Aarts, E. H. L. *Simulated Annealing and Boltzmann Machines: A Stochastic Approach to Combinatorial Optimization and Neural Computing.* Wiley, New York, 1989.

[2] Abramowitz, Milton, and Irena A. Stegun. *Handbook of Mathematical Functions.* Dover, New York, 1964.

[3] Akaike, H. Block Toeplitz Matrix Inversion. *SIAM J. Appl. Math.*, 24:234–41, 1979.

[4] Anderson, Brian D. O., and J. B. Moore. *Linear Optimal Control.* Prentice-Hall, Englewood Cliffs, NJ, 1979.

[5] Ansari, A., and R. Viswanathan. Application of EM Algorithm to the Detection of Direct Sequence Signal in Pulsed Noise Jamming. *IEEE Trans. Comm.*, 41(8): 1151–54, 1993.

[6] Antoniadis, N., and A. O. Hero. Time-Delay Estimation for Filtered Poisson Processes Using an EM-type Algorithm. *IEEE Trans. Signal Processing*, 42(8): 2112–23, 1994.

[7] Axelrod, Robert. *The Evolution of Cooperation.* Basic Books, New York, 1984.

[8] Banavar, R. N., and J. L. Speyer. A Linear Quadratic Game Theory Approach to Estimation and Smoothing. *Proc. IEEE ACC*, 2818–22, 1991.

[9] Banham, Mark R., and Aggelos K. Katsaggelos. Digital Image Restoration. *IEEE Signal Processing Magazine*, 14(2): 24–41, March, 1997.

[10] Bansal, Rakesh K., and P. Papantoni-Kazakos. An Algorithm for Detecting a Change in a Stochastic Process. *IEEE Trans. Info. Theory*, 32(2): 227–35, March, 1986.

[11] Barkat, Mourad. *Signal Detection and Estimation.* Artech House, Boston, 1991.

[12] Barnsley, Michael. *Fractals Everywhere.* Academic Press, Boston, 1988.

[13] Barrett, R., et al. *Templates for the Solution of Linear Systems: Building Blocks for Iterative Methods.* SIAM, Philadelphia, 1994.

[14] Bartels, R. H., and G. W. Stewart. Solution of the Equation $AX + XB = C$. *Comm. ACM*, 15:820–26, 1972.

[15] Basar, T., and P. Bernhard. H_∞-Optimal Control and Related Minimax Design Problems: A Dynamic Game Approach. Birkhauser, Boston, 1991.

[16] Basor, E. L., and K. E. Morrison. The Fisher-Hartwig Conjecture and Toeplitz Eigenvalues. *Lin. Alg. Appl.*, 202:129–42, 1994.

[17] Basseville, Michele. Edge Detection Using Sequential Methods for Change in Level—Part 2: Sequential Detection of Change in Mean. *IEEE Trans. Acoust., Speech, Signal Processing*, 29(1): 32–50, 1981.

[18] Basseville, Michele, and Albert Benveniste. Sequential Detection of Abrupt Changes in Spectral Characteristics of Digital Signals. *IEEE Trans. Info. Theory*, 29(5): 709–24, 1983.

[19] Basseville, Michele, Bernard Espiaau, and Jacky Gasnier. Edge Detection Using Sequential Methods for Change in Level—Part 1: A Sequential Edge Detection Algorithm. *IEEE Trans. Acoust., Speech, Signal Processing*, 29(1): 24–31, 1981.

[20] Bednar, J. B., R. Yarlagadda, and T. L. Watt. l_1 Deconvolution and Its Application to Seismic Processing. *IEEE Trans. Acoust., Speech, Signal Processing*, 34:1655–58, 1986.

[21] Bell, A. J., and T. J. Sejnowski. An Information-Maximization Approach to Blind Separation and Blind Deconvolution. *Neural Computation*, 7:1129–59, 1995.
[22] Bellman, R. E., and S. E. Dreyfus. *Applied Dynamic Programming*. Princeton University Press, Princeton, NJ, 1962.
[23] Benedetto, Sergio, Ezio Bilieri, and Valentino Castellani. *Digital Transmission Theory*. Prentice-Hall, Englewood Cliffs, NJ, 1987.
[24] Berger, James. *Statistical Decision Theory and Baysian Analysis*. 2d ed. Springer-Verlag, New York, 1985.
[25] Bertsekas, Dimitri P. *Dynamic Programming: Deterministic and Stochastic Models*. Prentice-Hall, Englewood Cliffs, NJ, 1987.
[26] Bezdek, James C. *Pattern Recognition with Fuzzy Objective Function Algorithms*. Plenum Press, New York, 1981.
[27] Bezdek, James C., et al. Detection and Characterization of Cluster Substructure I. Linear Structure: Fuzzy c-Lines. *SIAM J. Appl. Math.*, 40(2): 339–57, April, 1980.
[28] Bezdek, James C., et al. Detection and Characterization of Cluster Substructure II. Fuzzy c-Varieties and Convex Combinations Thereof. *SIAM J. Appl. Math.*, 40(2): 358–72, April, 1980.
[29] Bickel, Peter J., and Kjell A. Doksum. *Mathematical Statistics*. Prentice-Hall, Englewood Cliffs, NJ, 1977.
[30] Billingsley, Patrick. *Probability and Measure*. Wiley, New York, 1986.
[31] Blahut, R. E. Computation of Channel Capacity and Rate-Distortion Functions. *IEEE Trans. Info. Theory*, 18(4): 460–73, July, 1972.
[32] Blahut, R. E. *Theory and Practice of Error Control Codes*. Addison-Wesley, Reading, MA, 1983.
[33] Blahut, R. E. *Fast Algorithms for Digital Signal Processing*. Addison-Wesley, Reading, MA, 1985.
[34] Blahut, R. E. *Principles and Practice of Information Theory*. Addison-Wesley, Reading, MA, 1987.
[35] Blahut, R. E. *Digital Transmission of Information*. Addison-Wesley, Reading, MA, 1990.
[36] Blahut, R. E. *Algebraic Methods for Signal Processing and Communications Coding*. Springer-Verlag, New York, 1992.
[37] Bloomfield, P., and W. L. Steiger. *Least Absolute Deviations*. Birkhauser, Boston, 1983.
[38] Box, G. E. P., and G. M. Jenkins. *Time Series Analysis: Forecasting and Control*. Holden-Day, San Francisco, 1978.
[39] Boyles, R. A. On the Convergence of the EM Algorithm. *J. Roy. Sta. B.*, 45(1): 47–50, 1983.
[40] Brunk, H. D. *Mathematical Statistics*. 2d ed. Blaisdell, Waltham, MA, 1965.
[41] Bryson, Arthur E., and Yu-Chi Ho. *Applied Optimal Control: Optimization, Estimation, and Control*. Hemisphere Publishing Corporation, Washington, D.C., 1975.
[42] Buck, R. C. *Advanced Calculus*. McGraw-Hill, New York, 1978.
[43] Burrus, C. S., J. A. Barreto, and I. W. Selesnick. Iterative Reweighted Least-Squares Design of FIR Filters. *IEEE Trans. Signal Processing*. 42(11): 2926–36, November, 1994.
[44] Burrus, C. S., Ramesh A. Gopinath, and Haitao Guo. *Introduction to Wavelets and Wavelet Transforms: A Primer*. Prentice-Hall, Upper Saddle River, NJ, 1998.
[45] Byrd, R. H., and D. A. Pyne. Convergence of the Iteratively Reweighted Least Squares Algorithm for Robust Regression. Technical Report 313, Dept. of Mathematical Science. Johns Hopkins University, June, 1979.
[46] Byrne, W. Alternating Minimization and Boltzman Machine Learning. *IEEE Trans. Neural Networks*, 3(4): 612–20, 1992.
[47] Cadzow, J. A. Blind Deconvolution via Cumulant Extrema. *IEEE Signal Processing Magazine*, 13(3): 24–42, May, 1996.
[48] Cadzow, J. A. Signal Enhancement—A Composite Property Mapping Algorithm. *IEEE Trans. Acoust., Speech, Signal Processing*, 36(1): 49–62, January, 1988.
[49] Cardoso, J.-F. Blind Separation of Sources. *IEEE Signal Processing Magazine*, 15(5): 48–49, September, 1998.
[50] Cardoso, J.-F. Blind Signal Separation: Statistical Principles. *Proc. IEEE* (forthcoming).

Bibliography

[51] Chan, T. F. An Improved Algorithm for Computing the Singular Value Decomposition. *ACM Trans. Math. Soft.*, 8:72–83, 1982.

[52] Chen, Xiangkun, and Thomas Parks. Design of IIR Filters in the Complex Domain. *IEEE Trans. Acoust., Speech, Signal Processing*, 38(6): 910–20, June, 1990.

[53] Chui, Charles K. *An Introduction to Wavelets*. Academic Press, Boston, 1992.

[54] Comon, P. Independent Component Analysis: A New Concept? *Signal Processing*, 36:287–314, 1994.

[55] Courant, R., and D. Hilbert. *Methods of Mathematical Physics*. Interscience Publishers, New York, 1953.

[56] Cover, Thomas M., and Joy A. Thomas. *Elements of Information Theory*. Wiley, New York, 1991.

[57] Coyle, E. J., and J.-H. Lin. Stack Filters and the Mean Absolute Error Criterion. *IEEE Trans. Acoust., Speech, Signal Processing*, 36:1244–54, 1988.

[58] Cramér, H. *Mathematical Methods of Statistics*. Princeton University Press, Princeton, NJ, 1946.

[59] Csiszar, I., and G. Tusnday. Information Geometry and Alternating Minimization Procedures. *Statistics and Decisions, Supplement Issue 1*, 205–337, 1984.

[60] Cybenko, G. Approximations by Superpositions of a Sigmoidal Function. *Math. of Control, Signals, and Systems*, 2:303–14, 1989.

[61] Dantzig, G. B. *Linear Programming and Extensions*. Princeton University Press, Princeton, NJ, 1963.

[62] Daubechies, Ingrid. Orthonormal Bases of Compactly Supported Wavelets. *Com. Pure Appl. Math.*, XLI:909–96, 1988.

[63] Daubechies, Ingrid. *Ten Lectures on Wavelets*. SIAM, Philadelphia, 1992.

[64] Davis, Philip J. *Circulant Matrices*. Wiley, New York, 1979.

[65] de Prony, Baron R. Essai Expérimental et Analytique: Sur les Lois de la Dilatabilité de Fluides Élastiques et sur Celles de la Force Expansive de la Vapeur de l'eau et de la Vapeur de l'alcool, à Différentes Températures. *Journal del l'École Polytechnique (Paris)*, 1(2): 24–76, 1795.

[66] de Souza, C. E., U. Shaked, and M. Fu. Robust H_∞ Filtering for Continuous Time Varying Uncertain Systems with Deterministic Signal. *IEEE Trans. Acoust., Speech, Signal Processing*, 43:709–19, 1995.

[67] DeGroot, M. *Optimal Statistical Decisions*. McGraw-Hill, New York, 1970.

[68] Deller, John R., John G. Proakis, and John H. L. Hansen. *Discrete–Time Processing of Speech Signals*. Macmillan, New York, 1993.

[69] Demmel, J., and W. Kahan. Jacobi's Method is More Accurate Than QR. Technical Report 468. Department of Computer Science, Courant Institute of Mathematical Sciences, New York University, New York, 1989.

[70] Demmel, J., and W. Kahan. Accurate Singular Values of Bidiagonal Matrices. *SIAM J. Sci. Stat. Comp.*, 11:873–912, 1990.

[71] Demmel, James Weldon. The Smallest Perturbation of a Submatrix Which Lowers the Rank and Constrained Total Least Squares. *SIAM J. Numer. Anal.*, 24(1): 199–206, February, 1987.

[72] Dempster, A. P., N. M. Laird, and D. B. Rubin. Maximum Likelihood from Incomplete Data via the EM Algorithm. *J. Royal Statistical Soc., Ser. B*, 39(1): 1–38, 1977.

[73] Denoel, E., and J. P. Solvay. Linear Prediction of Speech with a Least Absolute Error Criterion. *IEEE Trans. Acoust., Speech, Signal Processing*, 33:1397–403, 1985.

[74] Devaney, Robert L. *An Introduction to Chaotic Dynamical Systems*. Addison-Wesley, Redwood City, CA, 1989.

[75] Dologlou, I., S. Van Huffel, and D. Van Ormont. Improved Signal Enhancement Procedures Applied to Exponential Electron Modeling. *IEEE Trans. Signal Processing*, vol. 45, Mar. '97, 799–803.

[76] Duda, Richard O., and Peter E. Hart. *Pattern Classification and Scene Analysis*. Wiley, New York, 1973.

[77] Durbin, J. The Fitting of Time Series Models. *Rev. Inst. Int. Stat.*, 28:233–43, 1960.

[78] Elliot, Douglas F., and K. Ramamohan Rao. *Fast Transforms: Algorithms, Analysis, Applications*. Academic Press, New York, 1980.

[79] Erdélyi, A. *Higher Transcendental Functions*, vol. 2. McGraw-Hill, New York, 1953.
[80] Even, Shimon. *Graph Algorithms*. Computer Science Press, Potomac, MD, 1979.
[81] Eyuboğlu, M. V., and S. U. H. Qureshi. Reduced-State Sequence Estimation with Set Partitioning and Decision Feedback. *IEEE Trans. Comm.*, 36(1): 13–20, January, 1988.
[82] Eyuboğlu, M. V., and S. U. H. Qureshi. Reduced-State Sequence Estimation for Coded Modulation on Intersymbol Interference Channels. *IEEE J. on Selected Areas in Comm.*, 7(6): 989–95, August, 1989.
[83] Feder, M. Parameter Estimation and Extraction of Helicopter Signals Observed with a Wide-Band Interference. *IEEE Trans. Signal Processing*, 41(1): 232–44, 1993.
[84] Feder, M., A. V. Oppenheim, and E. Weinstein. Maximum Likelihood Noise Cancellation Using the EM Algorithm. *IEEE Trans. Acoust., Speech, Signal Processing*, 37(2): 204–16, 1989.
[85] Ferguson, T. S. *Mathematical Statistics*. Academic Press, New York, 1967.
[86] Fernando, K. V., and B. N. Parlett. Accurate Singular Values and Differential QD Algorithms. *Numerische Mathematik*, 67:191–229, 1994.
[87] Fessler, J. A., and A. O. Hero. Space-Alternating Generalized EM Algorithm. *IEEE Trans. Signal Processing*, 42(10): 2664–77, October, 1994.
[88] Fletcher, R. *Practical Methods of Optimization*. Wiley, New York, 1980.
[89] Fletcher, R., J. A. Grant, and M. D. Hebden. The Calculation of Linear Best L_p Approximations. *Comput. J.*, 14:276–79, 1971.
[90] Forney, G. David, Jr. The Viterbi Algorithm. *Proc. IEEE*, vol. 61, 268–78, March, 1973.
[91] Forsythe, George E., and Gene H. Golub. On the Stationary Values of a Second-Degree Polynomial on the Unit Sphere. *J. SIAM*, 13(4): 1050–68, December, 1965.
[92] Franks, L. E. *Signal Theory*. Prentice-Hall, Englewood Cliffs, NJ, 1969.
[93] Friedland, Bernard. *Control System Design: An Introduction to State-Space Design*. McGraw-Hill, New York, 1986.
[94] Fritz, John. Extremum Problems with Inequalities as Side Conditions. In *Studies and Essays, Courant Anniversay Volume*, K. O. Friedrichs, O. E. Neugebauer, and J. J. Stoker. ed. Wiley Interscience, New York, 1948.
[95] Gabbouj, Moncef, and Edward J. Coyle. On the LP Which Finds a MMAE Stack Filter. *IEEE Trans. Signal Processing*, 39(11): 2419–24, November, 1991.
[96] Gallager, R. G. *Information Theory and Reliable Communication*. Wiley, New York, 1968.
[97] Gander, Walter. Least Squares with a Quadratic Constraint. *Numerische Mathematik*, 36:291–307, 1981.
[98] Gantmacher, F. R. *The Theory of Matrices*, vol. 1. Chelsea Publishing Company, New York, 1959.
[99] Gardner, W. A. Learning Characteristics of Stochastic-Gradient-Descent Algorithms: A General Study, Analysis, and Critique. *Signal Processing*, 6:113–33, 1984.
[100] Gass, Saul I. *Linear Programming: Methods and Applications*. 4th ed. McGraw-Hill, New York, 1975.
[101] Gath, I., and B. Geva. Unsupervised Optimal Fuzzy Clustering. *IEEE Trans. Patt. Anal. Machine Intell.*, 11(7): 773–81, July, 1989.
[102] Gelb, Arthur. *Applied Optimal Estimation*. MIT Press, Cambridge, MA, 1986.
[103] Georghiades, C. N., and D. L. Snyder. The EM Algorithm for Symbol Unsynchronized Sequence Detection. *IEEE Trans. Comm.*, 39(1): 54–61, 1991.
[104] Gevers, Michel R., and Thomas Kailath. An Innovations Approach to Least-Squares Estimation—Part VI: Discrete-Time Innovations Representations and Recursive Estimation. *IEEE Trans. Automatic Control*, AC-18(6): 588–600, December, 1973.
[105] Gilbert, P. F. C. Iterative Methods for the Reconstruction of Three-Dimensional Objects from Projections. *J. Theor. Biol.*, 36:105–17, 1972.
[106] Gill, P. E., W. Murray, and M. H. Wright. *Practical Optimization*. Academic Press, London, 1981.
[107] Gislason, Eyjolfur, et al. Three Different Criteria for the Design of Two-Dimensional Zero Phase FIR Digital Filters. *IEEE Trans. Signal Processing*, 41(10): 3020–74, October, 1993.

[108] Goldberg, David. *Genetic Algorithms in Search, Optimization, and Machine Learning.* Addison-Wesley, Reading, MA, 1989.
[109] Golomb, S. W. *Shift Register Sequences.* Holden-Day, San Francisco, 1967.
[110] Golub, G. H., A. Hoffman, and G. W. Stewart. A Generalization of the Eckart-Young-Mirsky Matrix Approximation Theorem. *Lin. Alg. Appl.*, 88/89:317–27, 1987.
[111] Golub, G. H., and W. Kahan. Calculating the Singular Values and Pseudo-Inverse of a Matrix. *SIAM J. Numer. Anal.*, 2:205–24, 1965.
[112] Golub, G. H., and C. F. Van Loan. An Analysis of the Total Least Squares Problem. *SIAM J. Numer. Anal.*, 17(6): 883–93, December, 1980.
[113] Golub, G. H. Some Modified Matrix Eigenvalue Problems. *SIAM Rev.*, 15(2): 318–34, April, 1973.
[114] Golub, G. H., and Charles F. Van Loan. *Matrix Computations.* 3d ed. Johns Hopkins University Press, Baltimore, 1996.
[115] Goodwin, G. C., and R. L. Payne. *Dynamic Systems Identification.* Academic Press, New York, 1977.
[116] Gordon, Richard. A Tutorial on ART (Algebraic Reconstruction Techniques). *IEEE Trans. Nuclear Science*, 21:78–93, June, 1974.
[117] Graham, Alexander. *Kronecker Products and Matrix Calculus with Applications.* Halsted Press, New York, 1981.
[118] Gray, R. M. On the Asymptotic Eigenvalue Distribution of Toeplitz Matrices. *IEEE Trans. Info. Theory*, 18:725–30, November, 1972.
[119] Grenander, Ulf, and Gabor Szegö. *Toeplitz Forms and Their Applications.* Chelsea Publishing Company, New York, 1984.
[120] Greville, T. N. E. On Smoothing a Finite Table: A Matrix Approach. *J. Soc. Ind. Appl. Math.*, 5:137–54, 1957.
[121] Greville, T. N. E. On Stability of Linear Smoothing Formulas. *J. SIAM Numer. Anal.*, 3:157–70, 1966.
[122] Grimble, M. J., and A. Elsayed. Solutions of the H_∞ Optimal Linear Filtering Problem for Discrete-Time Systems. *IEEE Trans. Acoust., Speech, Signal Processing*, 38:1092–104, 1990.
[123] Gu, M. J., and S. C. Eisenstat. A Divide-and-Conquer Algorithm for the Bidiagonal SVD. Research Report YALEU/DCS RR-933, UC Berkeley, April, 1994.
[124] Gunderson, R. W., and R. Canfield. Piecewise Multilinear Prediction from FCV Disjoint Principal Component Models. *Int. J. General Systems*, 16:373–83, 1990.
[125] Haddad, W. M., D. S. Berstein, and D. Mustafa. Mixed-Norm H_2/H_∞ Regulation and Estimation: The Discrete-Time Case. *Syst. Contr. Lett.*, 16:235–47, 1991.
[126] Halmos, Paul R. *Naive Set Theory.* Van Nostrand, Princeton, NJ, 1960.
[127] Harmuth, Henning F. *Transmission of Information by Orthogonal Functions.* Springer-Verlag, New York, 1970.
[128] Harwit, M., and N. J. A. Sloane. *Hadamard Transform Optics.* Academic Press, New York, 1979.
[129] Hassibi, B., and T. Kailath. H_∞ Adaptive Filtering. *Proc. IEEE ICASSP*, 949–52, 1995.
[130] Haykin, S. *Blind Deconvolution.* Prentice-Hall, Englewood Cliffs, NJ, 1994.
[131] Haykin, S. *Neural Networks: A Comprehensive Foundation.* Macmillan, New York, 1994.
[132] Haykin, S. *Adaptive Filter Theory.* 3d ed. Prentice-Hall, Upper Saddle River, NJ, 1996.
[133] Hegland, Markus. An Implementation of Multiple and Multivariate Fourier Transforms on Vector Processors. *SIAM J. Sci. Comp.*, 16(2): 271–88, March, 1995.
[134] Hero, A. Advances in Detection and Estimation Algorithms for Signal Processing. *IEEE Signal Processing Magazine*, 15(5): 24–26, September, 1998.
[135] Hero, A. O., and J. A. Fessler. Convergence in Norm for Alternating Expectation-Maximization (EM) Type Algorithms. *Statistica Sinica*, 5(1): 41–54, 1995.
[136] Hestenes, M. R., and E. Stiefel. Methods of Conjugate Gradients for Solving Linear Systems. *J. Res. Nat. Bur. Stand.*, 49:409–36, 1952.
[137] Hildebrand, F. B. *Introduction to Numerical Analysis.* McGraw-Hill, New York, 1956.
[138] Hinkley, D. V. Inference About the Change-Point from Cumulative Sum Tests. *Biometrika*, 58(3): 509–23, 1971.

[139] Ho, B. L., and R. E. Kalman. Effective Construction of Linear State-Variable Models from Input/Output Data. *Proc. Third Allerton Conference*, 449–59, October, 1965.
[140] Hofstadter, Douglas. "The Prisoner's Dilemma, Computer Tournaments, and the Evolution of Cooperation." Chapter 29 in *Metamagical Themas*. Bantam, Toronto, 1985.
[141] Hogg, Robert V., and Allen T. Craig. *Introduction to Mathematical Statistics*. Macmillan, New York, 1978.
[142] Horn, Roger A., and Charles A. Johnson. *Matrix Analysis*. Cambridge University Press, Cambridge, 1985.
[143] Horowitz, E., and S. Sahni. *Fundamentals of Computer Algorithms*. Computer Science Press, Potomac, MD, 1978.
[144] Householder, A. S. Unitary Triangularization of a Nonsymmetric Matrix. *J. ACM*, 5:339–42, 1958.
[145] Howson, C., and P. Urbach. *Scientific Reasoning: The Bayesian Approach*. Open Court, La Salle, Illinois, 1989.
[146] Hu, Yu Hen. CORDIC-Based VLSI Architectures for Digital Signal Processing. *IEEE Signal Processing Magazine*, 9(3): 16–35, July, 1992.
[147] Hunt, B. R. A Matrix Theory Proof of the Discrete Convolution Theorem. *IEEE Trans. Audio Electroacoustics*, 19:285–88, December, 1971.
[148] Immink, K. A. S. Runlength-Limited Sequences. *Proc. IEEE*, 78: 1745–59, 1990.
[149] Immink, K. A. S. *Coding Techniques for Digital Recorders*. Prentice-Hall, Englewood Cliffs, NJ, 1991.
[150] Ipsen, Ilse C. F. Computing an Eigenvector with Inverse Iteration. *SIAM Rev.*, 39(2): 254–91, June, 1997.
[151] Isaksson, A. J. Identification of ARX Models Subject to Missing Data. *IEEE Trans. Automatic Control*, 38(5): 813–19, 1993.
[152] Jain, Anil K. *Fundamentals of Digital Image Processing*. Prentice-Hall, Englewood Cliffs, NJ, 1989.
[153] Jain, Anil K., and Richard C. Dubes. *Algorithms for Clustering Data*. Prentice-Hall, Englewood Cliffs, NJ, 1988.
[154] Jazwinski, A. H. *Stochastic Processes and Filtering Theory*. Academic Press, New York, 1970.
[155] Jiang, C. J. The Use of Mixture Models to Detect Effects of Major Genes on Quantitative Characteristics in a Plant-Breeding Experiment. *Genetics*, 136(1): 383–94, 1994.
[156] Jolliffe, I. T. *Principal Component Analysis*. Springer-Verlag, New York, 1986.
[157] Kyung Sub Joo, Tamal Bose, and Gus Fang Xu. Image Restoration Using a Conjugate Gradient-Based Adaptive Filtering Algorithm. *Circuits, Systems and Signal Processing*, 16(2), 1977.
[158] Jordan, M. I., and R. A. Jacobs. Hierarchical Mixtures of Experts and the EM Algorithm. *Neural Comp.*, 6(2): 181–214, 1994.
[159] Kaczmarz, S. Angenäherte Auflösung von Systemen linearer gleichungen. *Bull. Acad. Polon. Sci. Lett. A.*, 35:355–57, 1937.
[160] Kahan, W. Numerical Linear Algebra. *Candadian Math. Bull.*, 9:757–801, 1966.
[161] Kahaner, David, Cleve Moler, and Stephen Nash. *Numerical Methods and Software*. Prentice-Hall, Englewood Cliffs, NJ, 1989.
[162] Kahng, S. W. Best L_p Approximation. *Math. Comput.*, 26(118): 505–08, 1972.
[163] Kailath, T. The Innovations Approach to Detection and Estimation Theory. *Proc. IEEE*, 58(5): 680–95, May, 1970.
[164] Kailath, T. *Linear Systems*. Prentice-Hall, Englewood Cliffs, NJ, 1980.
[165] Kailath, T. *Lectures on Wiener and Kalman Filtering*. CISM Courses and Lectures 140. Springer-Verlag, Berlin, 1981.
[166] Kailath, T., and H. V. Poor. Detection of Stochastic Processes. *IEEE Trans. Information Theory*, 44(6): 2230–59, October, 1998.
[167] Kailath, T., A. Vieira, and M. Morf. Inverses of Toeplitz Operators, Innovations, and Orthogonal Polynomials. *SIAM Rev.*, 20(1): 106–19, March, 1978.
[168] Kaleh, G. H. The Baum-Welch Algorithm for the Detection of Time-Unsynchronized Rectangular PAM Signals. *IEEE Trans. Comm.*, 42(2–4): 127–33, 1994.

[169] Kalman, R. E. A New Approach to Linear Filtering and Prediction Problems. *J. Basic Engineering*, 35–45, March, 1960.
[170] Kalman, R. E., and R. S. Bucy. New Results in Linear Filtering and Prediction Theory. *J. Basic Engineering*, 95–108, March, 1961.
[171] Karlin, Samuel. *Mathematical Methods and Theory in Games, Programming and Economics*. Dover, New York, 1992.
[172] Karmarker, N. A New Polynomial-Time Algorithm for Linear Programming. *Combinatorica*, 4: 373–95, 1984.
[173] Katsaggelos, A. K., and K. T. Lay. Maximum Likelihood Blur Identification and Image Restoration Using the EM Algorithm. *IEEE Trans. Signal Processing*, 39(3): 729–33, 1991.
[174] Kay, Steven M. *Modern Spectral Estimation*. Prentice-Hall, Englewood Cliffs, NJ, 1988.
[175] Kay, Steven M. *Fundamentals of Statistical Signal Processing: Detection Theory*. Prentice-Hall, Upper Saddle River, NJ, 1998.
[176] Kazarinoff, N. D. *Analytic Inequalities*. Holt, Rinehart and Winston, New York, 1961.
[177] Keener, James P. *Principles of Applied Mathematics: Transformation and Approximation*. Addison-Wesley, Reading, MA, 1988.
[178] Kelly, J. L. A New Interpretation of Information Rate. *Bell System Tech. J.*, 35:917–26, 1956.
[179] Kesler, S. B. *Modern Spectrum Analysis II*. IEEE Press, New York, 1986.
[180] Kim, Joon Tae, Woo Jin Oh, and Yong Hoon Lee. Design of Nonuniformly Spaced Linear-Phase FIR Filters Using Mixed Integer Linear Programming. *IEEE Trans. Signal Processing*, 44(1): 123–26, January, 1996.
[181] Kincaid, David, and Ward Cheney. *Numerical Analysis*. Brooks/Cole, Pacific Grove, CA, 1996.
[182] Klee, V., and G. J. Minty. How Good is the Simplex Algorithm? In *Inequalities*, ed. O. Shisha, 159–75. Academic Press, New York, 1972.
[183] Klema, V. C., and A. J. Laub. The Singular Value Decomposition: Its Computation and Some Applications. *IEEE Trans. Automatic Control*, 25(2): 164–76, April, 1980.
[184] Kosko, Bart. *Neural Networks and Fuzzy Systems*. Prentice-Hall, Englewood Cliffs, NJ, 1992.
[185] Krishnapuram, Raghu, and James M. Keller. A Possibilistic Approach to Clustering. *IEEE Trans. Fuzzy Systems*, 1(2): 98–110, May, 1993.
[186] Kumaresan, R., L. L. Scharf, and A. K. Shaw. An Algorithm for Pole-Zero Modeling and Spectral Analysis. *IEEE Trans. Acoust., Speech, Signal Processing*, 34(3): 637–40, 1986.
[187] Kumaresan, R., and Donald W. Tufts. Singular Value Decomposition and Spectral Analysis. *Proc. 20th IEEE Conf. on Decision and Control*, 1–11, December, 1981.
[188] Kumaresan, R., and Donald W. Tufts. Estimating the Parameters of Exponentially Damped Sinusoids and Pole-Zero Modeling in Noise. *IEEE Trans. Acoust., Speech, Signal Processing*, 30(6): 833–40, December, 1982.
[189] Kumaresan, R., Donald W. Tufts, and Louis L. Scharf. A Prony Method for Noisy Data: Choosing the Signal Components and Selecting the Order in Exponential Signals. *Proc. IEEE*, 72(2): 230–33, February, 1984.
[190] Kundur, D., and D. Hatzinakos. Blind Image Deconvolution. *IEEE Signal Processing Magazine*, 13(3): 43–64, May, 1996.
[191] Kung, S. A New Identification and Model Reduction Algorithm via Singular Value Decompositions. In *Twelfth Asilomar Conference on Circuits, Systems, and Computers*, ed. Chi Chia Hsieh, 705–14, 1978.
[192] Kung, S. Y. *Digital Neural Networks*. Prentice-Hall, Englewood Cliffs, NJ, 1993.
[193] Kung, S. Y., and David W. Lin. Optimal Hankel-Norm Model Reductions: Multivariable Systems. *IEEE Trans. Automatic Control*, 26(4): 832–52, August, 1981.
[194] Kung, S. Y., and David W. Lin. A State–Space Formulation for Optimal Hankel–Norm Approximations. *IEEE Trans. Automatic Control*, 26(4): 942–46, August, 1981.
[195] Lagendijk, R. L., J. Biemond, and D. E. Boekee. Identification and Restoration of Noisy Blurred Images Using the Expectation-Maximization Algorithm. *IEEE Trans. Acoust., Speech, Signal Processing*, 38(7): 1180–91, 1990.

[196] Lancaster, P. Explicit Solutions of Linear Matrix Equations. *SIAM Rev.*, 12:544–66, 1970.
[197] Lawson, Charles L., and Richard J. Hanson. *Solving Least Squares Problems*. Prentice-Hall, Englewood Cliffs, NJ, 1974.
[198] Lee, Edward A., and David G. Messerschmitt. *Digital Communication*. 2d ed. Kluwer Academic, Boston, 1994.
[199] Levinson, N. The Weiner RMS Error Criterion in Filter Design and Prediction. *J. Math. Phys.*, 25:261–78, 1947.
[200] Levy, Shlomo, et al. A Linear Programming Approach to the Estimation of the Power Spectra of Harmonic Processes, *IEEE Trans. Signal Processing*, 30(4): 675–79, August, 1982.
[201] Lin, Shu, and Daniel J. Costello, Jr. *Error Control Coding: Fundamentals and Applications*. Prentice-Hall, Englewood Cliffs, NJ, 1983.
[202] Lind, Douglas, and Brian Marcus. *Symbolic Dynamics and Coding*. Cambridge University Press, Cambridge, 1995.
[203] Linde, Y., A. Buzo, and R. M. Gray, An Algorithm for Vector Quantizer Design. *IEEE Trans. Comm.*, 28(1): 84–95, January, 1980.
[204] Lippmann, Richard P. An Introduction to Computing with Neural Nets. *IEEE Signal Processing Magazine*, 4–22, April, 1987.
[205] Little, R. J. A., and D. B. Rubin. On Jointly Estimating Parameters and Missing Data by Maximizing the Complete-Data Likelihood. *Am. Statistn.*, 37(3): 218–200, 1983.
[206] Liu, C. H., D. B. Rubin, and Y. N. Wu. The ECMA Algorithm: A Simple Extension of EM and ECM with Fast Monotone Convergence. *Biometrika*, 81: 633–48, 1994.
[207] Lovitt, W. V. *Linear Integral Equations*. Dover, New York, 1950.
[208] Luce, R. D., and H. Raiffa. *Games and Decisions*. Wiley, New York, 1957.
[209] Luenberger, David G. *Optimization by Vector Space Methods*. Wiley, New York, 1969.
[210] Luenberger, David G. *Linear and Nonlinear Programming*. Addison-Wesley, Reading, MA, 1984.
[211] Lupas, Ruxandra, and Sergio Verdu. Near-Far Resistance of Multiuser Detectors in Asynchronous Channels. *IEEE Trans. Comm.*, 38: 496–508, April, 1990.
[212] Macchi, O. *Adaptive Processing: The LMS Approach with Applications in Transmission*. Wiley, New York, 1995.
[213] Macchi, O., and E. Eweda. Convergence Analysis of Self-Adaptive Equalizers. *IEEE Trans. Information Theory*, 30:161–76, 1984. (Special issue on linear adaptive filtering.)
[214] Macwilliams, F. J., and N. J. A. Sloane. *The Theory of Error Correcting Codes*. North-Holland Publishing Company, Amsterdam, 1977.
[215] Makhoul, J. Linear Prediction: A Tutorial Review. *Proceedings of the IEEE*, 63:561–80, April, 1975.
[216] Mangasarian, O. L., and S. Fromovitz. The Fritz John Necessary Optimality Conditions in the Presence of Equality and Inequality Constraints. *J. Math. Analysis Appl.*, 17:37–47, 1967.
[217] Mangasarian, O. L., and J. Ponstein. Duality in Nonlinear Programming. *Quarterly of Applied Math.*, 20:300–02, 1962.
[218] Mangasarian, O. L., and J. Ponstein. Minmax and Duality in Nonlinear Programming. *J. Math. Analysis Appl.*, 11:504–18, 1965.
[219] Mardia, K. V., J. T. Kent, and J. M. Bibby. *Multivariate Analysis*. Academic Press, New York, 1979.
[220] Marple, S. L. *Digital Spectral Analysis*. Prentice-Hall, Upper Saddle River, NJ, 1987.
[221] Massey, James L. Shift-Register Synthesis and BCH Decoding. *IEEE Trans. Info. Theory*, 15(1): 122–27, 1969.
[222] Mathews, V. J., and Z. Xie. A Stochastic Gradient Adaptive Filter with Gradient Adaptive Step Size. *IEEE Trans. Signal Processing*, 41:2075–87, 1993.
[223] Mathias, R. Accurate Eigen System Computation by Jacobi Methods. *SIMAX*, 16:977–1003, 1995.
[224] McLachlan, G., and T. Krishnan. *The EM Algorithm and Extensions*. Wiley, New York, 1997.
[225] Meng, X. L., and D. Van Dyk. The EM Algorithm—An Old Folk-Song Sung to a Fast New Tune. *J. Royal Statistical Society, Ser. B.*, 59(3): 511–67, 1997.

[226] Miller, M. I., and D. R. Fuhrmann. Maximum Likelihood Narrow-Band Direction Finding and the EM Algorithm. *IEEE Trans. Acoust., Speech, Signal Processing*, 38(9): 1560–77, 1990.
[227] Minc, Henrik. *Nonnegative Matrices*. Wiley-Interscience, New York, 1988.
[228] Minsky, M. L., and S. A. Papert. *Perceptrons*. (Expanded edition.) MIT Press, Cambridge, MA, 1988.
[229] Mohan, Seshadri, and John B. Anderson. Computationally Optimal Metric-First Code Tree Search Algorithms. *IEEE Trans. Comm.*, 32(6):710–17, June, 1984.
[230] Moler, C. B., and C. F. Van Loan. Nineteen Dubious Ways to Compute the Exponential of a Matrix. *SIAM Rev.*, 20:801–36, 1978.
[231] Moon, T. K. The EM Algorithm in Signal Processing. *IEEE Signal Processing Magazine*, 13(6): 47–60, November, 1996.
[232] Moon, T. K., et al. Epistemic Decision Theory Applied to Multiple–Target Tracking. *IEEE Trans. Systems, Man, and Cybernetics*, 24(2): 313–18, February, 1994.
[233] Moonen, Marc, and Joos Vandewalle. QSVD Approach to On- and Off-Line State-Space Identification. *Int. J. Control*, 51(5): 1133–46, 1990.
[234] Moore, B. C. Singular Value Analysis of Linear Systems. *Proc. IEEE Conf. on Decision and Control*, 66–73, 1978.
[235] Morrell, Darryll R., and Wynn C. Stirling. Set-Valued Filtering and Smoothing. *IEEE Trans. Systems, Man, and Cybernetics*, 21(1): 184–93, Jan/Feb 1991.
[236] Morrison, Donald F. *Multivariate Statistical Methods*. McGraw-Hill, New York, 1976.
[237] Nagpal, K. M., and P. P. Khargonekar. Filtering and Smoothing in an H_∞ Setting. *IEEE Trans. Automatic Control*, 36:152–66, 1991.
[238] Naylor, Arch W., and George R. Sell. *Linear Operator Theory in Engineering and Science*. Springer-Verlag, New York, 1982.
[239] Nelder, J. A., and R. Mead. A Simplex Method for Function Minimization. *Comp. J.*, 7:308–13, 1965.
[240] Nguyen, T. Q. Design of Arbitrary Digital Filters Using the Eigenfilter Method. *IEEE Trans. Signal Processing*, 41(3): 1128–39, 1993.
[241] Nguyen, T. Q. Eigenfilter Approach for the Design of Allpass Filter Approximations with a Given a Phase Response. *IEEE Trans. Signal Processing*, 42(9): 2257–63, 1994.
[242] Nussbaumer, Henri J. *Fast Fourier Transform and Convolution Algorithms*. Springer-Verlag, Berlin, 1980.
[243] Oppenheim, A. V., et al. Single-Sensor Active Noise Cancellation Based on the EM Algorithm. *Proc. ICASSP*, I-277–I-280, 1992.
[244] Oppenheim, Alan V., and Ronald W. Schafer. *Discrete-Time Signal Processing*. Prentice-Hall, Englewood Cliffs, NJ, 1989.
[245] Ortega, J. M. *Matrix Theory: A Second Course*. Plenum Press, New York, 1988.
[246] Ortega, J. M., and W. C. Rheinboldt. *Iterative Solution of Nonlinear Equations in Several Variables*. Academic Press, New York, 1979.
[247] Papoulis, Athanasios. A New Algorithm in Spectral Analysis and Band-Limited Extrapolation. *IEEE Trans. Circuits and Systems*, 22(9): 735–42, September, 1975.
[248] Papoulis, Athanasious. *Signal Analysis*. McGraw-Hill, New York, 1977.
[249] Parks, T. W., and J. H. McClellan. Chebyshev-Approximation for Nonrecursive Digital Filters with Linear Phase. *IEEE Trans. Circuit Theory*, 18:687–96, 1972.
[250] Parks, T. W., and J. H. McClellan. A Computer Program for Designing Optimum FIR Linear Phase Digital Filters. *IEEE Trans. Audio and Electroacoustics*, 21:506–26, 1973.
[251] Parsons, Thomas. *Voice and Speech Processing*. McGraw-Hill, New York, 1987.
[252] Pei, S. C., and J. J. Shyu. Design of FIR Hilbert Transformers and Differentiators by Eigenfilters. *IEEE Trans. Circuits and Systems*, 135(11): 1457, 1988.
[253] Pei, S. C., and Min-Hung Yeh. An Introduction to Discrete Finite Frames. *IEEE Signal Processing Magazine*, 14(6): 84–96, November, 1997.
[254] Pierre, D. A. *Optimization Theory with Applications*. Dover, New York, 1986.
[255] Pisarenko, V. F. The Retrieval of Harmonics from a Covariance Function. *Geophys. J. Roy. Astron. Soc.*, 33:347–66, 1973.

[256] Polya, George. *How to Solve It*. 2d ed. Princeton University Press, Princeton, NJ, 1971.
[257] Poor, H. V. *An Introduction to Signal Detection and Estimation*. Springer-Verlag, New York, 1988.
[258] Poor, H. V. On Parameter Estimation in DS/SSMA Formats. *Proceedings of the International Conference on Advances in Communications and Control Systems*, 1988.
[259] Powers, David L. *Boundary Value Problems*. 2d ed. Academic Press, New York, 1979.
[260] Press, William H., et al. *Numerical Recipes in C*. Cambridge University Press, Cambridge, 1988.
[261] Proakis, J. G. *Digital Communications*. 3d ed. McGraw-Hill, New York, 1995.
[262] Proakis, J. G., and Dimitris G. Manolakis. *Digital Signal Processing: Principles, Algorithms and Applications*. 3d ed. Prentice-Hall, Upper Saddle River, NJ, 1996.
[263] Proakis, J. G., et al. *Advanced Digital Signal Processing*. Macmillan, New York, 1992.
[264] Rabiner, L. R. The Design of Finite Impulse Response Digital Filters Using Linear Programming Techniques. *Bell Sys. Tech. J.*, 51(6): 1177–98, July–August, 1972.
[265] Rabiner, L. R., and Biing-Hwang Juang. *Fundamentals of Speech Recognition*. Prentice-Hall, Englewood Cliffs, NJ, 1993.
[266] Rabiner, L. R. A Tutorial on Hidden Markov Models and Selected Applications in Speech Recognition. *Proc. IEEE*, 77(2): 257–86, February, 1989.
[267] Rader, C. M. VLSI Systolic Arrays for Adaptive Nulling. *IEEE Signal Processing Magazine*, 13(4): 29–49, July, 1996.
[268] Rader, C. M., and A. O. Steinhardt. Hyperbolic Householder Transformations. *IEEE Trans. Acoust., Speech, Signal Processing*, 34(6): 1589–602, December, 1986.
[269] Rahman, M. A., and Kai-Bor Yu. Total Least Squares Approach for Frequency Estimation Using Linear Prediction. *IEEE Trans. Acoust., Speech, Signal Processing*, 35(10): 1440–54, October, 1987.
[270] Ralston, A., and Philip Rabinowitz. *A First Course in Numerical Analysis*. McGraw-Hill, 1978.
[271] Rauch, H. E., F. Tung, and C. T. Striebel. Maximum Likelihood Estimates of Linear Dynamic Systems. *AIAA Journal*, 3(8): 1445–50, August, 1965.
[272] Reddi, S. S. Eigenvector Properties of Toeplitz Matrices and Their Application to Spectral Analysis of Time Series. *Signal Processing*, 45–56, 1984.
[273] Redner, R. A., and H. F. Walker. Mixture Densities, Maximum-Likelihood Estimation and the EM Algorithm. *SIAM Rev.*, 26(2): 195–237, 1984.
[274] Rissanen, J. Recursive Identification of Linear Systems. *SIAM J. Control*, 9(3): 420–30, 1971.
[275] Rissanen, J. Algorithms for Triangular Decomposition of Block Hankel and Toeplitz Matrices with Application to Factoring Positive Matrix Polynomials. *Mathematics of Computation*, 27(121): 147–56, January, 1973.
[276] Rissanen, J. Solution of Linear Equations with Hankel and Toeplitz Matrices. *Numer. Math.*, 22: 361–66, 1974.
[277] Rissanen, J., and T. Kailath. Partial Realization of Random Systems. *Automatica*, 8: 389–96, 1972.
[278] Robbins, Herbert, and Sutton Monro. A Stochastic Approximation Method. *Ann. Math. Stat.*, 22: 400–07, 1951.
[279] Ross, T. *Fuzzy Logic with Engineering Applications*. McGraw-Hill, New York, 1995.
[280] Roy, R., and T. Kailath. Total Least Squares ESPRIT. *Proc. XXIst Asilomar Conf. on Circ., Syst., and Comp.*, 297–301, Pacific Grove, CA, November, 1978.
[281] Roy, R., and T. Kailath. ESPRIT—Estimation of Signal Parameters via Rotational Invariance Techniques. *IEEE Trans. Acoust., Speech, Signal Processing*, 37(7): 984–95, 1989.
[282] Roy, R., A. Paulraj, and T. Kailath. ESPRIT—A Subspace Rotation Approach to Estimation of Parameters of Cisoids in Noise. *IEEE Trans. Acoust., Speech, Signal Processing*, 34(5): 1340–42, October, 1986.
[283] Royden, H. L. *Real Analysis*. 3d ed. Macmillan, New York, 1988.
[284] Rugh, Wilson J. *Linear System Theory*. 2d ed. Prentice-Hall, Upper Saddle River, NJ, 1996.
[285] Rumelhart, D. E., and J. J. McClelland. *Parallel Distributed Processing*. MIT Press, Cambridge, MA, 1986.

[286] Ruzinsky, Steven A., and Elwood T. Olsen. L_1 and L_∞ Minimization via a Variant of Karmarker's Algorithm. *IEEE Trans. Signal Processing*, 37(2): 245–53, February, 1989.

[287] Santina, M. S., A. R. Stubberud, and G. H. Hostetter. *Digital Control System Design*. Saunders College Publishing, Fort Worth, TX, 1994.

[288] Sarwate, Dilip V., and Michael B. Pursley. Crosscorrelation Properties of Pseudorandom and Related Sequences. *Proc. IEEE*, 68(5): 593–619, May, 1980.

[289] Sayed, A. H., and T. Kailath. A State-Space Approach to Adaptive RLS Filtering. *IEEE Signal Processing Magazine*, 11: 18–60, 1994.

[290] Scharf, L. L., and D. W. Tufts. Rank Reduction for Modeling Stationary Signals. *IEEE Trans. Acoust., Speech, Signal Processing*, 35: 350–55, March, 1987.

[291] Scharf, L. L. *Statistical Signal Processing: Detection, Estimation, and Time Series Analysis*. Addison-Wesley, Reading, MA, 1991.

[292] Schmee, J., and G. J. Hahn. Simple Method for Regression Analysis with Censored Data. *Technometrics*, 21(4): 417–32, 1979.

[293] Schmidt, R. O. Multiple Emitter Location and Signal Parameter Estimation. *Proc. RADC Spectral Estimation Workshop*, 243–58, Rome, NY, 1979.

[294] Schroeder, J., and R. Yarlagadda. Linear Predictive Spectral Estimation Via the L1 Norm. *Signal Processing*, 17:19–29, 1989.

[295] Schroeder, Jim, and John Hershey. Suboptimal Robust Sinusoidal Frequency Estimation. *Proc. ICASSP*, 2555–58, 1990.

[296] Schroeder, Manfred. *Number Theory in Science and Communication*. Springer-Verlag, New York, 1986.

[297] Schutz, B. *Geometrical Methods of Mathematical Physics*. Cambridge University Press, Cambridge, 1980.

[298] Sedgewick, Robert. *Algorithms in C*. Addison-Wesley, Reading, MA, 1990.

[299] Segal, M., and E. Weinstein. Parameter Estimation of Continuous Dynamical Linear Systems Given Discrete Time Observations. *Proc. IEEE*, 75(5): 727–29, 1987.

[300] Segal, M., and E. Weinstein. The Cascade EM Algorithm. *Proc. IEEE*, 76(10): 1388–90, 1988.

[301] Sezan, M. I., and H. Stark. Image Restoration by the Method of Convex Projections. Part 2: Applications and Numerical Results. *IEEE J. Medical Imaging*, 1(2): 95–101, October, 1982.

[302] Shaked, U., and Y. Theodor. H_∞-Optimal Estimation: A Tutorial. *Proc 31st IEEE CDC*, 2278–86, 1992.

[303] Shannon, Claude. A Mathematical Theory of Communication. *Bell Sys. Tech. J.*, 27: 623–56, 1948.

[304] Shannon, Claude. *Collected Papers*. IEEE Press, New York, 1993.

[305] Shapiro, J. M. Embedded Coding Using Zerotrees of Wavelet Coefficients. *IEEE Trans. Acoust., Speech, Signal Processing*, 41(12): 3445–62, December, 1993.

[306] Shen, X., and L. Deng. Discrete H_∞ Filter Design with Application to Speech Enhancement. *Proceedings of the IEEE ICASSP*, 1504–07, 1995.

[307] Shen, X., and Li Deng. Game Theory Approach to Discrete H_∞ Filter Design. *IEEE Trans. Signal Processing*, 45:1092–95, 1997.

[308] Shenoy, Ram G., Daniel Burnside, and Thomas W. Parks. Linear Periodic Systems and Multirate Filter Design. *IEEE Trans. Signal Processing*, 42(9): 2242–55, September, 1994.

[309] Shepp, L. A., and Y. Vardi. Maximum Likelihood Reconstruction for Emission Tomography. *IEEE Journal on Medical Imaging*, 1(2): 113–22, October, 1982.

[310] Shim, Y. S., and Z. H. Cho. SVD Pseudoinverse Image Reconstruction. *IEEE Trans. Acoust., Speech, Signal Processing*, 29(4): 904–09, August, 1981.

[311] Short, Robert T. *Multiple-User Receiver Structures*. Ph.D. diss., University of Utah, 1989.

[312] Simon, Marvin K., and Dariush Divsalar. Some New Twists to Problems Involving the Gaussian Probability Integral. *IEEE Trans. Comm.*, 46(2): 200–10, February, 1998.

[313] Simon, Marvin K., Sami M. Hinedi, and William C. Lindsey. *Digital Communication Techniques: Signal Design and Detection*. Prentice-Hall, Englewood Cliffs, NJ, 1995.

[314] Singer, Y. Dynamical Encoding of Cursive Handwriting. *Biol. Cybern.*, 71(3): 227–37, 1994.

[315] Snyder, D. L., and D. G. Politte. Image Reconstruction from List-Mode Data in an Emission Tomography System Having Time-of-Flight Measurements. *IEEE Trans. Nuclear Science*, 30(3): 1843–49, 1983.

[316] Solodovnikov, V. V. *Introduction to Statistical Dynamics of Automatic Control Systems*. Dover, New York, 1960.

[317] Spjøtvoll, Emil. A Note on a Theorem of Forsythe and Golub. *SIAM J. Appl. Math.*, 23(2): 307–11, November, 1972.

[318] Stanley, B., S. Bialkowski, and D. Marshall. Analysis of First-Order Rate Constant Spectra with Regularized Least Squares and Expectation Maximization. *An. Chem.* vol. 65, Feb. 1, '93, 259–67.

[319] Steiglitz, K., and L. E. McBride. A Technique for the Identification of Linear Systems. *IEEE Trans. Automatic Control*, 10:461–64, October, 1965.

[320] Steiglitz, Kenneth, Thomas W. Parks, and James F. Kaiser. METEOR: A Constraint-Based FIR Filter Design Program. *IEEE Trans. Signal Processing*, 40(8): 1901–09, August, 1992.

[321] Steinhardt, Allan O. Householder Transforms in Signal Processing. *IEEE ASSP Magazine*, 5(3): 4–12, July, 1988.

[322] Stenger, Frank. Numerical Methods Based on Whittaker Cardinal, or Sinc Functions. *SIAM Rev.*, 2:165–224, April, 1981.

[323] Stenger, Frank. *Numerical Methods Based on Sinc and Analytic Functions*. Springer-Verlag, New York, 1993.

[324] Stevens, J. C., and K. K. Ahuja. Recent Advances in Active Noise Control. *AIAA Journal*, 29(7): 1058–67, 1991.

[325] Stewart, G. W. *Introduction to Matrix Computations*. Academic Press, New York, 1973.

[326] Stoer, J., and R. Bulirsch. *Introduction to Numerical Analysis*. Springer-Verlag, New York, 1993.

[327] Stoica, P., and R. L. Moses. *Introduction to Spectral Analysis*. Prentice-Hall, Upper Saddle River, NJ, 1997.

[328] Stoica, P., et al. Optimal High-Order Yule-Walker Estimation of Sinusoidal Frequencies. *IEEE Trans. Signal Processing*, 39(6): 1360–68, 1991.

[329] Stoica, P., and A. Nehorai. MUSIC, Maximum Likelihood, and Cramér-Rao Bound. *IEEE Trans. Signal Processing*, 37(5): 720–41, 1989.

[330] Stoica, P., and A. Nehorai. Performance Comparison of Subspace Rotation and MUSIC Methods for Direction Estimation. *IEEE Trans. Signal Processing*, 39(2): 446–53, 1991.

[331] Stoica, P., and T. Söderström. Statistical Analysis of MUSIC and Subspace Rotation Estimates of Sinusoidal Frequencies. *IEEE Trans. Signal Processing*, 39(8): 1836–47, 1991.

[332] Strang, Gilbert. *Introduction to Applied Mathematics*. Wellesley Cambridge, Wellesley, MA, 1986.

[333] Strang, Gilbert. *Linear Algebra and Its Applications*. 3d ed. Harcourt Brace Jovanovich, Fort Worth, TX, 1988.

[334] Strang, Gilbert. Wavelets and Dilation Equations: A Brief Introduction. *SIAM Rev.*, 31(4): 614–27, 1989.

[335] Streit, R. L., and T. E. Luginbuh. ML Training of Probabilistic Neural Networks. *IEEE Trans. Neural Networks*, 5(5): 764–83, 1994.

[336] Sunder, S., W. S. LU, and A. Antoniou. Design of Digital Differentiators Satisfying Prescribed Specifications Using Approximation Techniques. *IEEE Proc. G.*, 138(3): 315–20, 1991.

[337] Szegö, G. *Orthogonal polynomials*. 3d ed. American Mathematical Society, Providence, RI, 1967.

[338] Tang, K. S., K. F. Man, and S. Kwong. Genetic Algorithms and Their Applications. *IEEE Signal Processing Magazine*, 13(6): 22–37, November, 1996.

[339] Todd, Michael J., and Bruce P. Burrell. An Extension of Karmarker's Algorithm for Linear Programming Using Dual Variables. *Algorithmica*, 1:409–24, 1986.

[340] Tufts, D., C. Kot, and R. J. Vacaro. The Analysis of Threshold Behavior of SVD-Based Algorithms. *Proc. XXIst Asilomar Conf. on Circ., Syst., and Comp.*, 550–54, Pacific Grove, CA, November, 1978.

[341] Vaidyanathan, P. P. Multirate Digital Filters, Filter Banks, Polyphase Networks, and Applications: A Tutorial. *Proc. IEEE*, 78(1): 56–93, January, 1990.

[342] Vaidyanathan, P. P. *Multirate Systems and Filters Banks*. Prentice-Hall, Englewood Cliffs, NJ, 1993.

[343] Vaidyanathan, P. P., and T. Q. Nguyen. Eigenfilters: A New Approach to Least-Squares FIR Filter Design and Applications Including Nyquist Filters. *IEEE Trans. Circuits and Systems*, 34:11–23, January, 1987.

[344] van Huffel, S., and J. Vandewalle. The Use of Total Linear Least Squares Technique for Identification and Parameter Estimation. *Proc. IFAC/IFORS Symp. on Identification and Parameter Estimation*, 1167–72, July, 1985.

[345] van Huffel, S., and J. Vandewalle. The Partial Total Least Squares Algorithm. *J. Computational Appl. Math.*, 21:333–41, 1988.

[346] van Huffel, S., and J. Vandewalle. *The Total Least Squares Problem: Computational Aspects and Analysis*. Kluwer Academic, Boston, 1991.

[347] van Huffel, S., J. Vandewalle, and A. Haegemans. An Efficient and Reliable Algorithm for Computing the Singular Subspace of a Matrix, Associated with Its Smallest Singular Values. *J. Comput. Appl. Math.*, 21:313–20, 1987.

[348] van Huffel, S., and J. Vandewalle. Subset Selection Using the TLS Approach in Collinearity Problems with Errors in the Variables. *Lin. Alg. Appl.*, 88/89:695–714, April, 1987.

[349] van Trees, H. L. *Detection, Estimation, and Modulation Theory, Part I*. Wiley, New York, 1968.

[350] Varga, R. *Matrix Iterative Analysis*. Prentice-Hall, Englewood Cliffs, NJ, 1962.

[351] Vaseghi, S. V., and P. J. W. Rayner. Detection and Suppression of Impulsive Noise in Speech Communication Systems. *IEEE. Proc.-I*, 137(1): 38–46, 1990.

[352] Velleman, Daniel J. *How to Prove It*. Cambridge University Press, New York, 1994.

[353] Verdu, Sergio. *Optimum Multi-User Signal Detection*. Ph.D. diss., University of Illinois, 1984.

[354] Verdu, Sergio. Optimum Multiuser Asymptotic Efficiency. *IEEE Trans. Comm.*, 34(9): 890–96, September, 1986.

[355] Verdu, Sergio. Minimum Probability of Error for Asynchronous Gaussian Multiple-Access Channels. *IEEE Trans. Info. Theory*, 32(1): 85–96, January, 1986.

[356] Verhaegen, Michel, and Paul Van Dooren. Numerical Aspects of Different Kalman Filter Implementations. *IEEE Trans. Automatic Control*, 31(10): 907–17, October, 1986.

[357] von Neumann, J., and O. Morganstern. *Theory of Games and Economic Behavior*. 3d ed. Princeton University Press, Princeton, NJ, 1944.

[358] Walter, Gilbert G. *Wavelets and Other Orthogonal Systems with Applications*. CRC Press, Boca Raton, FL, 1994.

[359] Ward, R. K. An On-Line Adaptation for Discrete L_1 Linear Estimation. *IEEE Trans. Automatic Control*, 29:67–71, 1984.

[360] Weinstein, E., et al. Iterative and Sequential Algorithms for Multisensor Signal Enhancement. *IEEE Trans. Signal Processing*, 42(4): 846–59, 1994.

[361] Wicker, Stephen B. *Error Control Systems for Digital Communications and Storage*. Prentice-Hall, Englewood Cliffs, NJ, 1995.

[362] Wickerhauser, M. V. *Adapted Wavelet Analysis from Theory to Software*. A. K. Peters, Wellesley, MA, 1994.

[363] Widom, H. Toeplitz Matrices. In *Studies in Real and Complex Analysis*, I. I. Hirschmann, ed. MAA Studies in Mathematics. Prentice-Hall, Englewood, Cliffs, NJ, 1965.

[364] Widrow, B. Adaptive Noise Cancelling: Principles and Applications. *Proc. IEEE*, 63:1692–716, 1975.

[365] Widrow, B. Stationary and Nonstationary Learning Characteristics of the LMS Adaptive Filter. *Proc. IEEE*, 64:1151–62, 1976.

[366] Widrow, B., and S. D. Stearns. *Adaptive Signal Processing*. Prentice-Hall, Englewood Cliffs, NJ, 1985.

[367] Widrow, B., and E. Walach. On the Statistical Efficiency of the LMS Algorithm with Nonstationary Inputs. *IEEE Trans. Info. Theory*, 30:211–21, 1984.

[368] Widrow, B., and Samuel D. Stearns. *Adaptive Signal Processing*. Prentice-Hall, Englewood Cliffs, NJ, 1985.
[369] Wilkinson, J. H. *Rounding Errors in Algebraic Processes*. Dover, New York, 1963.
[370] Wilkinson, J. H. *The Algebraic Eigenvalue Problem*. Clarendon Press, Oxford, England, 1965.
[371] Willsky, Alan S., and Harold L. Jones. A Generalized Likelihood Approach to the Detection and Estimation of Jumps in Linear Systems. *IEEE Trans. Automatic Control*, 108–12, February, 1976.
[372] Wilson, Robin J. *Introduction to Graph Theory*. Longman Scientific, Essex, England, 1985.
[373] Wozencraft, John M., and Irwin M. Jacobs. *Principles of Communication Engineering*. Wiley, New York, 1965.
[374] Wright, Steven J. *Primal-Dual Interior Point Methods*. SIAM, Philadelphia, PA, 1997.
[375] Wu, C. F. J. On the Convergence Properties of the EM Algorithm. *Ann. Statist.*, 11(1): 95–103, 1983.
[376] Xuanli, Lisa Xie, and Gerarado Beni. A Validity Measure for Fuzzy Clustering. *IEEE Trans. Patt. Anal. Machine Intell.*, 13(8): 841–47, August, 1991.
[377] Yaesh, I., and U. Shaked. A Transfer Function Approach to the Problem of Discrete-Time Systems: H_∞ Optimal Linear Control and Filtering. *IEEE Trans. Automatic Control*, 16: 1264–71, 1991.
[378] Yarlagadda, R., J. B. Bednar, and T. L. Watt. Fast Algorithms for L_p Deconvolution. *IEEE Trans. Acoust., Speech, Signal Processing*, 33:174–82, 1985.
[379] Youla, D. C., and H. Webb. Image Restoration by the Method of Convex Projections: Part 1—Theory. *IEEE Journal on Medical Imaging*, 1(2): 81–94, October, 1982.
[380] Youla, D. C. Generalized Image Restoration by the Method of Alternating Orthogonal Projections. *IEEE Trans. Circuits and Systems*, 25(9): 694–702, September, 1978.
[381] Young, D. *Iterative Solution of Large Linear Systems*. Academic Press, New York, 1971.
[382] Zabin, S. M., and H. V. Poor. Efficient Estimation of Class-A Noise Parameters via the EM Algorithm. *IEEE Trans. Info. Theory*, 37(1): 60–72, 1991.
[383] Zadeh, Lofti. Fuzzy Sets. *Information and Control*, 8:338–53, 1965.
[384] Zangwill, W. I. *Nonlinear Programming: A Unified Approach*. Prentice-Hall, Englewood Cliffs, NJ, 1969.
[385] Zeiger, Paul H., and A. Julia McEwen. Approximate Linear Realizations of Given Dimension via Ho's Algorithm. *Proc. IEEE*, 153, 1974.
[386] Zhou, K. *Essentials of Robust Control*. Prentice-Hall, Upper Saddle River, NJ, 1998.
[387] Ziemer, Rodger E., and Roger L. Peterson. *Digital Communications and Spread Spectrum Systems*. Macmillan, New York, 1985.
[388] Ziskind, I., and D. Hertz. Maximum Likelihood Localization of Narrow-Band Autoregressive Sources via the EM Algorithm. *IEEE Trans. Signal Processing*, 41(8): 2719–24, 1993.
[389] Ziskind, I., and M. Wax. Maximum Likelihood Localization of Multiple Sources by Alternating Projection. *IEEE Trans. Acoust., Speech, Signal Processing*, 36(10): 1553–60, 1988.
[390] Zoltowski, M. D. Signal Processing Applications of the Method of Total Least Squares. *Proc. XXIst Asilomar Conf. on Signals, Systems and Computers*, 290–96, November, 1987.

Index

$AX - XB = C$, 432
$AXB = C$, 429
δ, 10
O and o notation, 861
χ^2 random variables, 478
Γ function, 478
∇, 865
$\otimes$, see Kronecker product
$\perp$, see orthogonal
$\mathbb{Q}$, 857
( ) (continuous time), 7
$\binom{n}{k}$, 65
$\oplus$, 73
$^{-H}$, 239
$^{-T}$, 239
: notation, 881
$\leftrightarrow$, 9, 188
[ ], 7
^, 23
|, 66
$|\cdot|$, 856
0, 9
1, 9
$\forall$, 855
β random variable, 575
$\mathbb{C}$, 856
$C[a, b]$, 75, 80, 86
$\partial \mathbf{x}$, 865
D_N, 198
$\exists$, 855
Γ random variable, 576
^{H}, see transpose
I, 9
i, 7
I_0, 510
j, 7
$\mathcal{L}$, 20
$L_\infty[a, b]$, 76
$L_p[a, b]$, 76, 82

$\mathbb{Q}$, 855
$\mathbb{R}$, 856
$\mathbb{R}^n$, 86, 856
^{T}, see transpose
$\mathbb{Z}$, 855
$\mathcal{Z}$, 9

A

absolute error loss (Bayesian), 571
absolute value (objective for LP), 823
  constraints (for LP), 820
active noise cancellation, 729
adaptive filter, 28
  LMS, 643
  RLS, 259
adjacency matrix (of a graph), 788
adjoint, 237
adjugate, 307, 889
affine transformation, 629, 861
algebraic complement, 111, 114
algebraic reconstruction techniques, 706
all-pole, see autoregressive
all-zero, see moving average
almost everywhere, 82
alternating projections, 673
annihilating polynomial, 344
antipodal signal constellation, 477
approximation, 130
AR, see autoregressive
arg max, 326
arg min, 326
arithmetic mean, 874

arithmetic–geometric mean inequality, 357, 874
ARMA, 8, 382
  parameter estimation, 561
art1.m, 707
asymptotic equivalence (of matrices), 414
attractive fixed point, 625–26
attractor, 625. *See also* attractive fixed point
autocorrelation, 13, 67, 891
  function, 25
"autocorrelation" method, 151
autocovariance, 891
autoregressive, 8
  moving average, 8

B

Bèzout theorem, 343
back substitution, 277
backpropagation algorithm, 649
backward predictor, *see* linear predictor
Banach space, 106
bandlimited signal, 109, 670
  reconstruction, 671, 675
$^-$, *see* complex conjugate
barrier function, 840
basic variables (for LP), 824
Bass–Gura formula, 347
Bayes
  envelope function, 520
  estimate, 267
  estimation theory, 568
  risk, 569
    function (r), 489
bayes3.m, 532

Bellman's principal of optimality, 790
Bernoulli random variable, 447
Bessel function I_0, 510
Bessel's inequality, 188
`bidiag.m`, 392
bidiagonal, 390
bijection, 860
bilinear transformation, 285
binary entropy function, 661
binary phase-shift keying, 213
binary symmetric channel, 47
binomial coefficient, 65
binomial random variable, 455, 457
binomial theorem, 65, 874
bisection method, 642
blind source separation, 660
`bliter1.m`, 675
block matrix, 885
   inverse, 264
BLUE, 141
boundary, 77
bounded, 233
branch, 787
`bss1.m`, 665

C

Cantor argument, 857
Cantor, Georg, 857
cardinal number, 856
cardinality, 857
Cartesian product, 112, 856
Cauchy sequence, 80
   bounded, 124
   convergence, 123
Cauchy–Schwartz inequality, 34, 100, 136, 213, 551, 553, 870, 873
Cayley transformation, 285, 301
Cayley–Hamilton theorem, 344, 366
central limit theorem, 556
chain rule, 755, 896
change detection, 540
channel capacity, 257
channel coding theorem, 42
chaos, 629
characteristic equation, 306
characteristic function, 478, 879
   Gaussian, 879
   moments using, 879
   χ^2, 478
characteristic polynomial, 306, 342
Chebyshev inequality, 548
Chebyshev polynomial, 104, 120
Cholesky factorization, 275, 283
`cholesky.m`, 285
Christoffel–Darboux formula, 224
circulant, 418
circulant matrix, 410, 412
   eigenvalues, 411, 413
class C^k, 124, 865
closed set, 77
closure, 247
cluster points, 78
cluster validity, 715
clustering, 695
cocktail party problem, 660
code division multiple access, 804
codimension, 180
cofactor, 888
colinear, 102
column space, 241. *See also* range
commutative diagram, 374–75
companion matrix, 17, 412
   properties, 367
complete, 81, 124, 187
complete data, 721
complete set, 121
complete sufficient statistic, 451
complete the square, 36, 101, 267, 877
complex conjugate, 7
complex constraints, 773
complex vectors, derivative with respect to, 908
`compmap2.m`, 682
composite mapping algorithm, 676
composition of mappings, 670
compressed likelihood function, 563
concave function, 861
`cond`, 256
condition number, 253–54
conditional expectation, 444
conditional probability, 36
   Gaussian, 36, 266
`conjgrad1.m`, 711
`conjgrad2.m`, 713
conjugate direction methods, 708
conjugate family, 575
conjugate gradient method, 710
consistent (estimator), 553
constrained least squares, 765
constrained optimization, 236, 321, 389, 544, 869
   definitions, 757
   waterfilling solution, 783
continuous (definition), 862
continuous time to discrete time conversion, 606
contraction mapping, 629
   theorem, 630
contrapositive, 43
contravariant vector, 760
control, 28
controllability test matrix, 347
controllable, 345
controller canonical form, 14
convergence
   definition, 79
   preserved by continuous functions, 863
convergent rate, 631
converse, 43
convex function, 860, 876
convex set, 44, 857
   projection on (POCS), 689
convolution, 10, 32, 108
convolutional coding, 806
CORDIC rotations, 297
correlation, 13
countable, 857
"covariance" method, 151
covariant vector, 760
Cramér–Rao lower bound, 550
cross-correlation, 133

D

data cleaning, using composite mapping, 687
data compression, 23, 44, 207
Daubechies, *see* wavelets
decimation, 67, 226
decision feedback equalization, 814
decision region, 211

Index

derivative
  functions of matrices, 899
  involving trace, 906
  matrices with respect to scalars, 901
  products of matrices, 903
  scalars with respect to matrices, 901
  with respect to complex vector, 908. *See also* gradient
descent function, 678
detection of change, 540–41
detection probability, 462
determinant, 248, 277, 885, 887
  cofactor expansion, 887
  product of eigenvalues, 356
diagonal matrix, 28, 34, 60, 882
diagonalization of a matrix, 309
diagonally dominant, 283, 327, 666, 702
difference equation, 7, 10
  eigenvectors and, 305
differential equation, 230
digital communications, 208
Dijkstra's algorithm, 798
dijkstra.m, 798
Diophantine equations, 757
direct product, 422
direct sum, 112
Dirichlet kernel, 224
dirtorefl.m, 407
discrete Fourier transform, 189
  using Kronecker product, 426
discrete noiseless channel, 348
discrete-time Fourier transform, 25
disjoint, 111
divides, *see* |
domain, 859
dot product, *see* inner product
dual approximation, 179
dual problems (for LP), 836
duality, 773
durbin.m, 403
dyadic number, 358
dynamic programming, 789
dynamic warping, 811

E

ECME, 725
edge (of a graph), 787
efficiency (of an estimator), 552
eigenfilter, 330
  desired spectral response, 332
  random signals, 330
eigenvalue, 236, 306
  disparity (spread), 320, 255, 641
  triangular matrix, 356
eigenvector, 306
eigfil.m, 334
eigfilcon.m, 336
eigmakePQ.m, 334
eigqrshiftstep.m, 354
elementary matrix, 279
EM algorithm, 37
em1.m, 720
emission tomography, 725
empiric distribution estimation, 543
energy signal, 25
energy spectral density, 25
entropy, 660
equality constraints, 758
equalizer
  least-squares, 153
  LMS (adaptive), 645
  minimum mean-square, 159
  RLS (adaptive), 261
ergodic, 162
error surface, 137
ESPRIT, 341
esprit.m, 342
essential supremum, 84
estimation, 36
et1.m, 729
Euclidean norm, 237
even function, 13
expectation conditional maximization either, 725
expectation–maximization algorithm, 717
exponential family, 453
  and EM algorithm, 722
exponential random variable, 458
extended Kalman filter, 608

F

factorization theorem, 448
false alarm, 462
Fano algorithm, 814
fblp.m, 156
feasible direction, 752
feasible solution, 818
Fejèr–Riesz theorem, 221
field, 48
  finite, 48
filter design, 21
  eigenfilters, 330
  using linear programming, 846
finite impulse response, 10
Fisher information matrix, 550
fixed point, 624
fixed-point problems, 358
fordyn.m, 791
forward predictor, *see* linear predictor
forward substitution, 277
four fundamental subspaces, 242, 304
Fourier series, 189
  generalized, 188
Fourier transform, 63
frame, 129
Fredholm alternative theorem, 243
frequency estimation, 382
frequency-shift keying, 213
Frobenius norm, 237
function (definition), 859
  space, 74
fundamental theorem of algebra, 278
fuzzy
  k-varieties, 715
  clustering, 700
  set, 872

G

gamble.m, 782
game (definition), 438
gamma function, 478
gamma random variable, 458
Gauss–Seidel iteration, 703
Gaussian
  elimination, 278
  mean factoring theorem, 667

Gaussian (*cont.*)
  quadrature, 224
  random number, 283
  random variable, 31
    attributes, 32
    conditional density, 34, 36, 266
  white noise, 513
`gaussseid.m`, 703
generalized eigenvalues, 340
generalized maximum-likelihood detection, 563
genetic algorithm, 643
geometric mean, 874
geometric random variable, 458
geometric series, 20
geometric sum, 66
Gershgorin circle theorem, 324
$GF(2)$, 50, 110
Givens rotations, 293
global minimum, 751, 861
Gödel, Kurt, 107
Golub–Kahan step, 391
`golubkahanstep.m`, 392
gradient, 137, 865, 896,
  chain rule, 896
  linear function, 898
  quadratic function, 898
  scalar with respect to vector, 896
  vector with respect to vector, 896. *See also* derivative
Gram–Schmidt process, 118, 209, 287, 370, 596
Grammian, 133, 215
`gramschmidt1.m`, 120
graph, 787
  multistage, 789
  (of a map), 678
  trellis, 792
greatest lower bound, 74
group, 85

H

Haar function, 198
Hadamard matrix, 425
Hadamard's inequality, 357
Hamel basis, 90, 111, 233
Hamming distance, 73
Hamming window, 682

Hankel matrix, 366, 379, 398, 564
  given rank, 687
Hermition, 418, 884
Hermitian symmetric, *see* symmetric
Hermitian transpose, 884
Hessenburg matrix, 300
Hessian matrix, 137, 636
hidden Markov model, 39
  EM training algorithm, 732
  Viterbi training algorithm, 808
Hilbert matrix, 143
Hilbert space, 106
Hilbert, David, 107
HMM, *see* hidden Markov model
`hmmabn.m`, 739
`hmmApiupn.m`, 739
`hmmdiscf.m`, 739
`hmmdiscfupn.m`, 739
`hmmf.m`, 739
`hmmfupdaten.m`, 739
`hmmgausf.m`, 739
`hmmgausfupn.m`, 739
`hmmgendat.m`, 739
`hmmgendisc.m`, 739
`hmmgengaus.m`, 739
`hmminitvit.m`, 809
`hmmlpyseqn.m`, 739
`hmmlpyseqv.m`, 810
`hmmnorm.m`, 809
`hmmnotes.m`, 738
`hmmtest2vb.m`, 811
`hmmupdaten.m`, 739
`hmmupdatev.m`, 810
`hmmupfv.m`, 810
Hoffman–Wielandt theorem, 361
Hölder's inequality, 870
Householder transformations, 287
`houseleft.m`, 291
`houseright.m`, 291
hypothesis testing, 441
  composite, 462, 483
  simple, 462

I

idempotent, 113
identity, 882
if and only if, 43
iff, *see* if and only if

`ifs3a.m`, 628
ill conditioned, 143, 254, 320, 375
implication, 43
implicit QR shift, 354
impulse response, 10
imputed data, 720
incidence matrix (of a graph), 788
incoherent detection, 507, 510
independent component analysis, 660
indicator function, 451, 858
induced norm, 135
inequalities,
  arithmetic–geometric mean, 357, 874
  Bessel's, 188
  Cauchy-Schwarz, 100
  Haramard's, 357
  Hölder's, 870
  information, 875
  Jensen's, 724, 861
  Minkowski, 871
  Motzkin transposition, 851
  triangle, 44, 72
inequality constraints, 777
inertia (of a matrix), 316
inf, 74
infinite dimensional space, 74
infinite impulse response, 10
information inequality, 668, 875
information theory, 660
`initcluster.m`, 699
`initpoisson.m`, 729
`initvit1.m`, 796
injection, *see* one-to-one
inner product, 97, 866
inner sum, 110
innovations, 167, 596
integral
  Lebesgue, 84
  Riemann, 83
integral equation, 230
interference cancellation, 30
interior, 76
interior point methods, 839
intermediate value theorem, 863
interpolating polynomial, 146
intersymbol interference channel, 800
invariance, 841

Index

invariant subspace, 316
invariant tests, 504
inverse mapping, 860
inverse system identification, 29
invertible, 248
`invwavetrans.m`, 207
`invwavetransper.m`, 207
irrational, 45
irrelevance, 517
`irwls.m`, 185
isometric, 269, 285
isomorphism, 112
iterative algorithm, 78
    LMS filter, 643
    neural network, 648
    RLS filter, 259

J

Jacobi iteration, 631, 702
`jacobi.m`, 703
Jacobian, 609, 663
Jensen's inequality, 724, 861
Jordan form, 61, 311

K

k-means clustering, 698
`kalex1.m`, 595
Kalman filter, 37, 591
    Bayesian derivation, 592
    continuous-time to discrete-time, 606
    extended, 608, 611
    innovations derivation, 595
    minimum-variance derivation, 619
    nonlinear systems, 607
    smoother, 613
Kalman gain, 260
`kalman1.m`, 594
`karf.m`, 842
Karhunen–Loève, 327
`karmarker.m`, 842
kernel, 109
`kissarma.m`, 565
Kronecker delta, 102
Kronecker product, 422
    and DFT, 426
    eigenvalues and, 424
    properties, 423
Kronecker sum, 424
    eigenvalues and, 424
Kuhn–Tucker conditions, 777

L

L_1-norm estimation, 228
Lagrange multiplier, 236, 321, 544, 758, 763, 869
    complex, 773
Lagrangian, 763
Laplace transform, 20
    two-sided, 165
Laplacian random variable, 535
lattice filter, 404
least-absolute error approximation, 847
least favorable prior, 524
least squares, 106, 133, 239, 248, 251
least upper bound, 74
    condition, 257
    filtering, 150
    normal equation approach, 284
    QR solution, 286
    VLSI-appropriate algorithms, 296
left nullspace, 242, 372
Legendre polynomial, 103, 120
Leverrier formulas, 365
Levinson algorithm, 408
`levinson.m`, 409
LGB algorithm, 698
`lgb.m`, 699
likelihood equation, 545
likelihood function, 503, 515, 543
likelihood ratio, 467
lim inf, 80
limit, 79
limit point, 79
lim sup, 79
line search, 642, 841
linear combination, 87
linear feedback shift register, 48
linear independence, 215
linear least squares, 133
linear observer, 366
linear optimal control (using LP), 849
linear prediction, 29
linear predictor, 21, 22, 154
    backward, 154, 219
    forward, 154, 218
    forward–backward, 155

linear programming, 818
linear structure matrix, 685
linear system, 7
linear variety, 179, 681
linearization, 608
linearly independent, 88, 134–35
LMS adaptive filter, 643
    normalized, 667
    variable step, 669
`lms.m`, 645
`lmsinit.m`, 645
local minimum, 751
log-likelihood function, 545
log-likelihood ratio, 467
logistic function, 627
`logistic.m`, 628
low-rank approximation, 328, 362
`lpfilt.m`, 847
`lsfilt.m`, 153
LU factorization, 275–276
lumped system, 163

M

M-ary detection, 499
machine epsilon, 255
Mahalonobis distance, 105
`makehankel.m`, 380
`makehouse.m`, 291
manifold, 671
mapping, 859
Markov
    model, 37
    parameters, 19, 21, 367, 378
    process, 619
Massey's algorithm, 50
matrix exponential, 20. *See also* state transition matrix
matrix factorizations
    Cholesky, 275, 283
    LU, 275, 276
    QR, 275
    SVD, 275
    identity, 882
    inverse, 247
    2×2, 18
    inversion lemma, 258
    multiplication
        matrix properties preserved, 418
    norms, *see* norm

matrix (*cont.*)
  notation, 9
  pencil, 340
  polynomial, 342
  properties preserved under multiplication, 418
  square root, 283
max (distribution of random variable), 452
`maxeig.m`, 351
maximal-length sequence, 50, 67
maximum-likelihood estimation, 41, 64, 503, 542
  iterative, 717
maximum-likelihood sequence estimation, 800
maximum *a posteriori* detection, 501
maximum *a posteriori* estimate, 573
mean value theorem, 626, 864, 866
measure, 82
measure zero, 82
median (as least-absolute error estimate), 572
metric, 72
  L_p, 74
  L_∞, 75
  l_1, 73
  l_2, 73
  l_p, 73
  See also norm
metric space, 74
`mineig.m`, 351
minimal polynomial, 342, 344
minimal sufficient statistic, 451
minimal system, 11
minimax decision, 524
minimax principle (for eigenvalues), 323
minimum error, 141
minimum mean squares, 106, 133, 156
  Bayesian estimate, 571
  filtering, 157
minimum variance, 141
Minkowski inequality, 871
minmax theorem, 776
minor, 888

ML estimation (continuous time), 566
ML estimator
  asymptotic efficiency, 556
  asymptotic normality, 555
  consistency, 554
modal analysis, 26. *See also* modal matrix
modal matrix, 396
mode, 11
  of a distribution, 572
modified covariance method, 156
modified Gram–Schmidt, 128, 287
moments of a random variable, 879
momentum (in neural networks), 654
monic, 871
monotone likelihood ratio, 484
Motzkin transposition, 851
moving average, 8
multinomial random variable, 458
MUSIC, 339
`musicfun.m`, 340
mutual information, 660, 662

N

Nash equilibrium, 456
necessary, 43
negative semidefinite, 134
neighborhood, 76
Nelder–Mead algorithm, *see* simplex algorithm
Neumann expansion, 20, 235
neural network, 648
`neweig.m`, 355
`newlu.m`, 282
`newsvd.m`, 392
Newton identities, 365
Newton's method, 632
Neyman–Pearson lemma, 463
`nn1.m`, 655
`nnrandw.m`, 655
`nntrain1.m`, 655
noise subspace, 336
nominal trajectory, 608
nonexpansive operator, 691
nonsingular, 248, 418
norm, 94
  l_2, 236, 371
  Euclidean, 237

  Frobenius, 237, 371
  spectral, *see* norm, l_2
  subordinate, 232
  See also metric
normal, *see* orthogonal,
normal equation, 766
normal equations, 133, 137, 139
normal matrix, 359, 411, 418
normal random variable, *see* Gaussian random variable
normalized vector, 95
nuisance parameters, 507
null hypothesis, 441
nullity, 242
nullspace, 109, 242, 372
numerical integration, *see* Gaussian quadrature
Nyquist filter, 363

O

observability test matrix, 366
observer canonical form, 60
on–off keying, 212
one-to-one, 860
open set, 77
operator, 108
orbit, 624
order (of a system), 378
order of convergence, 631
Ornstein–Uhlenbeck process, 174
orthogonal, 102, 135
  matrix, 269, 285
  polynomial, 104, 120, 190
  Procrustes problem, 389
  projection, 114
  signal constellation, 477
  subspace, 107
orthogonality principle, 135
orthonormal, 102
  basis, 121
over-relaxed operator, 692

P

parameter estimation, 41
Parseval's equality, 188
Parseval's theorem, 214
partial fraction expansion, 11
partial total least squares, 386
partitioned matrix, *see* block matrix

Index

pattern recognition, 41, 105, 697
  NN and ML compared, 659
penalty function, 840
periodic point, 625
permutation matrix, 399
`permutedata.m`, 665
perpendicular, *see* orthogonal
persymmetric matrix, 401
phase estimation (ML), 566
phase-shift keying, 211, 537
Pisarenko harmonic decomposition, 338
`pisarenko.m`, 339
pivoting, 280
  (for LP), 826
pivots, 277, 279
`pivottableau.m`, 834
point-to-set map, 678
Poisson random number generator (Matlab code), 728
Poisson random variable, 455, 457–58, 727
`poisson.m`, 729
pole, 10
polynomial approximation
  continuous polynomials, 143
  discrete polynomials, 145
polytope, 872
positive definite, 103, 134, 248, 359, 418
  comparing matrices, 548
positive semidefinite, 141, 359
  condition for minimality, 753
positive-semidefinite matrix, 142
positive sequence determination, 681
positive-semidefinite, 134
posterior distribution (for Bayes), 494
power method, 350
power of a test (detection), 462
power spectral density, 25, 893
prewindowing method, 152
principal component, 329
principle of optimality, 790
prior estimate, 36
probability density function, 31

probability mass function, 31
projection, 113, 132, 672
projection matrix, 115, 116, 139, 317
projection on convex sets, 689
projection theorem, 116
projective transformation, 839–840
Prony's method, 398
proof
  by "computation", 43
  by contradiction, 45, 54
  by induction, 46, 55, 402, 404
pseudoinverse, 139, 183, 248, 251, 373
pseudometric, 122
pseudonoise sequence, 50
`ptls1.m`, 387
`ptls2.m`, 388
pulse-position modulation, 213
Pythagorean theorem (statisticians), 102, 142

Q

Q function, 472
QR factorization, 252, 275
`qrgivens.m`, 295
`qrhouse.m`, 292
`qrmakeq.m`, 293
`qrmakeqgiv.m`, 296
`qrqtb.m`, 292
`qrqtbgiv.m`, 296
`qrtheta.m`, 295
quadratic form, 318
quadrature–amplitude modulation, 213, 537

R

random process, 891
random processes
  through linear systems, 894
random variable
  Bernoulli, 447
  beta, 575
  binomial, 455, 457
  exponential, 458
  gamma, 458, 576
  Gaussian, 31
  geometric, 458
  Laplacian, 535
  multinomial, 458
  Poisson, 455, 457–58
  Rayleigh, 446, 458

  transformations, 445
  χ^2, 478
range, 109, 241, 372, 859
rank, 244, 249
  deficient, 249
  one, 249
Rayleigh quotient, 322, 331, 334
Rayleigh random variable, 446, 458
`rcond`, 256
realization, 378
Receiver operating characteristic, 468
`reducefree.m`, 834
Reed–Muller codes, 425
reflection, 288
reflection coefficient, 403
regression, 147
repeated poles, 11
resolution of identity, 317
resolvent identities, 363
resolvent of a matrix, 363
`restorefree.m`, 834
`refltodir.m`, 407
Riccati equation, 600
Riesz representation theorem, 232
risk function (R), 487
RLS adaptive filter, 259
  comparison with Kalman filter, 603
`rlsinit.m`, 261
Rolle's theorem, 864, 873
Rosenbrock function, 636
rotation matrix, 294, 303
rotation of subspaces, 389
row operation, 278
row space, 242, 278, 372
runlength-constrained coding, 367

S

saddle point optimality, 775
SAGE, 725
sampling, 231
scaling, 195
  function, 195, 358
Schur complement, 216, 264
score function, 548
secant method, 642
second companion form, 60
secular equation, 766
separable operator, 430

sequence space, 74
sequential estimation, 580
set sum, 872
Sherman–Morrison formula, 258, 260
shifting, 195
shortest path algorithm, *see* dynamic programming or Viterbi algorithm
signal plus noise, 22
signal processing
 definition, 4
signal subspace, 336
 identification, 565
signature of a matrix, 316
similar matrix, 17, 309
simplex, 642, 839
 algorithm, 642
simplex1.m, 834
simulated annealing, 643
sinc function, 193
singular value decomposition, *see* SVD
singular values, 369
size of a test (false alarm), 462
skew Hermitian, 359
skew-symmetric, 301, 359
slack variables, 819
smoothing
 Rauch–Tung–Streibel, 613
sor.m, 706
source separation, 660
space alternating expectation maximization, 725
spectral
 decomposition, 317
 factorization, 166, 221
 norm, *see* norm, l_2
 radius, 236, 701
 theorem, 313
spectrum (of an operator), 306
spectrum analysis, 25
spectrum estimation, 154
speech processing, 23, 41
 pattern recognition, 697
spread-spectrum multi-user communication, 740, 804
square root
 of a matrix, 134, 275
 of a transfer function, 221
square-root Kalman filter, 284, 604

squared-error loss (Bayesian), 570
stability, 11
stack algorithm, 814
standard form (for LP), 818
state-transition matrix, 21, 348
state-space form, 15
stationary
 strict-sense, 892
 wide-sense, 892
statistic (definition), 444
stochastic approximation, 663, 644
stochastic gradient descent, 644
stochastic matrix, 39
 nearest, 684
stochastic processes, 12
 review, 162
strict-sense stationary, 892
strong law of large numbers, 555
subordinate norm, 232
subsequence, 79
successive approximation, 625
successive over-relaxation, 705
sufficient, 43
sufficient statistic, 446
sup, 74, 76
supervised training, 649
support, 78, 110
surjection, *see* onto
SVD, 252, 275, 369
Sylvester's equation, 432
Sylvester's law of inertia, 316
symmetric, 14, 238, 884
sysidsvd.m, 380
system function, *see* transfer function
system identification, 29, 264, 378, 384
Szegö's theorem, 416

T
t statistic, 506, 507
tableau, 821
tangent plane, 760, 867
Taylor series, 20, 868
 linearization, 608
testet.m, 729
testirwls.m, 185

testnn10.m, 656
time-varying system, 17, 21
tls.m, 384
Toeplitz matrix, 14, 51, 152, 159, 400, 418, 871
 eigenvalues, 413
tohankel.m, 687
tohanktoep.m, 688
tokarmarker.m, 844
tomography, 706, 725
tostoch.m, 686
total least squares, 381
total set, *see* complete set
trace, 106, 884
 sum of eigenvalues, 356
training
 data, 696
 phase, 41
 set, 162
transfer function, 10
transformation, 860
 linear, 108
 of random variables, 445
transitive closure, 799
transpose, 883
 ^{H} (Hermitian), 9
 ^{T}, 9
trellis, 792
triangle inequality, 325
triangular matrix, 276, 418
 eigenvalues, 356
tridiag.m, 352
tridiagonalization, 352
truncation
 in frequency, 109
 in time, 109
two-scale equation, 196
Type I error, 442
Type II error, 442

U
unbiased, 64, 140, 452
uncorrelated, 34
uniformly most powerful test, 483
uniqueness, 118, 244
unit element matrix E_{rs}, 882
unit vector, 95
 $\mathbf{e}_i$, 881
unit-step function, 11
unitary, 418
 determinant, 301
 matrix, 269, 285, 369, 374

Index

V

Vandermonde matrix, 146, 337, 397, 409, 412, 418
variable-step LMS, 669
vec (operator), 428
vector notation, 9
vector quantization, 44, 220, 695
vector space, 85
vertex (of a graph), 787
vitbestcost.m, 810
Viterbi algorithm, 37, 791
viterbi1.m, 796
vitflush.m, 796
Voronoi region, 44, 698

W

warp.m, 813
Warshall's algorithm, 798
warshall.m, 799
water filling solution, 783
wavecoeff.m, 199
wavelet transform, 199
wavelets, 194
wavetest.m, 204
wavetesto.m, 205
wavetrans.m, 207
weak duality, 785
Weierstrass theorem, 191, 416
weighted inner product, 103
weighted least squares, 140, 148

wftest.m, 161
wide-sense stationary, 33, 892
Wiener filter, 157
Wiener–Hopf, 219
  equations, 158
    continuous time, 169
Wilkinson shift, 354

Y

Yule–Walker equations, 13, 51, 401

Z

zero, 10
zerorow.m, 392